# SELECTED PAPERS
*of*
# C. C. LIN

CHIA-CHIAO LIN

# SELECTED PAPERS
*of*
# C. C. LIN

Volume 1

## Fluid Mechanics and Applied Mathematics

*Editors*

David J. Benney, Frank H. Shu and Chi Yuan

World Scientific
*Singapore • New Jersey • Hong Kong*

*Published by*

**World Scientific Publishing Co. Pte. Ltd.**
P.O. Box 128, Farrer Road, Singapore 9128

*U.S.A. office*: World Scientific Publishing Co., Inc.
687 Hartwell Street, Teaneck NJ 07666, USA

Library of Congress Cataloging-in-Publication data is available.

**SELECTED PAPERS OF C. C. LIN**



ISBN 9971-50-318-2

Printed in Singapore by Kim Hup Lee Printing Co. Pte. Ltd.

*A gift to C. C. Lin in celebration of his seventieth birthday from Friends, Colleagues, and Students*

# TABLE OF CONTENTS

## Volume I: Fluid Mechanics and Applied Mathematics

# PREFACE

These two volumes of the ***Selected Papers of C. C. Lin*** are a gift to him from his many friends, colleagues, and former students on the occasion of his retirement from active service as Institute Professor at MIT. It represents the most significant tribute that we could make to his long and productive scientific career. Of course, without his advice and cooperation, a publishing enterprise of this kind could not have been undertaken. In particular, he kindly agreed to write the sections entitled "Recollections and Comments", which portray the development of his thoughts on various subjects and place events in historical perspective. The publication of these comments and the original articles in two bound volumes should prove an invaluable resource for future researchers in science and in the history of science.

The selected papers are grouped according to subject. In Volume I, the papers belonging to Fluid Mechanics and Applied Mathematics are divided into five topics: stability theory, turbulence, general fluid dynamics and aerodynamics, mathematical physics, and essays on applied mathematics. Volume II contains Lin's work in Astrophysics. Within each major topic, the selected papers are arranged in chronological order. A complete list of C. C. Lin's publications at the time that this work went to press is placed at the end of Volume II.

We are grateful to the publishers and journal editors who granted us permission to reprint the articles in these two volumes. We would especially like to acknowledge the help and cooperation of Dr. K. K. Phua, the publisher of World Scientific Publishing Company, who has been exceedingly generous in his support of this project. We also thank Ms. P. Y. Chin, the managing editor of the present volumes for her careful work and suggestions.

David J. Benney, *Associate Editor*
Frank H. Shu, *Associate Editor*
Chi Yuan, *Chief Editor*

# RECOLLECTIONS AND COMMENTS

# I

A major part of this volume (Sections A and B) consists of papers on hydrodynamic stability and turbulence. These papers were published during the period 1945–1963. The resolution of the well-known mathematical issues raised in the linear theory were published somewhat later ([56] and [83]).

It is clear from these papers that, during the early period, there was a wide gap between the "linear" studies of flows with respect to infinitesimal disturbances and the "nonlinear" studies of fully developed turbulence, although some steps were already taken to close this gap. Since then, extensive studies have been carried out on the process of transition from linear disturbances to fully developed turbulence. Many of those concern flows with instabilities of somewhat different types. The papers here may therefore be regarded as giving an early history to such major developments.

With the aid of large scale computation, some researchers are still attempting to find out a complete description of the most basic experimental phenomena; e.g., what is the nature of the transition to turbulence of the flow in a pipe near the threshold conditions at Reynolds numbers as low as $10^3$. Practical applications are of course not the primary objective of such work. Rather, turbulence is regarded as a prime example of a nonlinear dynamical process, whose studies have become very attractive since the advent of large scale computational approach to fluid dynamics. Naturally, besides the computational approach, one must also understand the physical mechanisms on the basis of concepts developed from analytical studies. There is still a great deal of fascination in this old problem.

Several of the papers selected in Sections C and D formed the basis of further developments. For example, the method used for the systematic classification of linear and nonlinear water waves ([59]) can be generalized, and has been, to the cases of bottoms of variable height. The variational principle for fluid dynamics in the Eulerian formulation (which was developed in [67] to demonstrate the difficulty of proving the quantization of vortices through macroscopic experiments) has been generalized to cover hydromagnetics. The resolution of the well-known mathematical issues in the problem of hydrodynamical stability of parallel flows is an interesting extension of Langer's work on the WKB method in quantum mechanics. It is interesting to note that the mathematical controversy surrounding this problem actually arose even before quantum mechanics was developed. (See J. Mehra and H. Rechenberg, *The Historical Development of Quantum Theory*, pp. 49–63, Springer Verlag.) The history of the subsequent controversy was also sketched in this book as well as in my monograph *Theory of Hydrodynamic Stability* (Cambridge University Press).

The controversy became very intense, especially after Chaim Pekeris showed quite conclusively that instability was absent in the case of flow through a circular pipe. A numerical resolution of the problem was decided upon, and John Von

Neumann was actually devising a numerical procedure himself and writing the code. Eventually, the calculations of L. H. Thomas and his collaborators succeeded in verifying our numerical results. Heisenberg did not have numerical answers; he only gave a qualitative sketch of what the answers might be. His lecture at the 1950 International Congress of Mathematics, where I met him for the first time, probably had a significant favorable impact on my early career.

The complete analytical resolution of the mathematical issues, however, came much later. At one point, there were a number of asymptotic studies which indicate, but do not prove, that the flow was entirely stable. Among the earlier work, there were the papers of Sydney Goldstein and Rudolph Langer. The complication of the mathematical problem arises from a triple limiting process. (cf. [63]) Wolfgang Wasow was looking for uniformly valid solutions. I spent one summer at Madison, Wisconsin in 1958 when Rudolph Langer again became interested in this problem. He was trying to resolve the difficulty by using a reference differential equation of the third order. Eventually, I decided that the problem cannot be resolved unless we use an equation of the fourth order as a basis, and succeeded in getting a solution. Since our work, an alternative mathematical theory has been developed by W. D. Lakin and W. H. Reid and B. S. Ng (see monograph by P. G. Drazin and W. H. Reid, *Hydrodynamic Stability*, p. 285, Cambridge University Press).

C. C. Lin
*Massachusetts Institute of Technology*
*May 1987*

# SELECTED PAPERS
*of*
# C. C. LIN

Volume 1

## Fluid Mechanics
## and
## Applied Mathematics

# ON THE STABILITY OF TWO-DIMENSIONAL PARALLEL FLOWS

## PART I.—GENERAL THEORY*

BY

C. C. LIN

*Guggenheim Laboratory, California Institute of Technology*

**1. Introduction.** The study of the stability of laminar motion and its transition to turbulence dates back to the time of Helmholtz and Reynolds [46], and had already attracted great attention at the end of the last century.** Since that time, the subject has not only become a major problem for workers in hydrodynamics, but has also attracted the attention of people like Lord Rayleigh [43–45], Lord Kelvin [20–21], Lorentz [29], Sommerfeld [58], and Heisenberg [14], whose chief interest is not limited to the study of mechanics. Although numerous contributions have since been made, the subject has remained one of considerable dispute, as can be seen from the two general lectures given by Taylor [70] and by Synge [63] as late as 1938. Still more recently, there appeared the work of Görtler [8, 9] and of Thomas [71].

Most of the work on the stability of laminar motions has the following final aims.

1) The first aim is to determine whether a given flow (or a given class of flows) is ultimately unstable for sufficiently large Reynolds numbers. For this purpose, it is desirable to obtain some simple general criterion which will give a rapid classification of velocity profiles according to their stability.

2) The second purpose is to determine the minimum critical Reynolds number at which instability begins. It is often easier to find sufficient conditions for stability than to find the condition for passage from stability to instability.

3) Finally, we want to understand the physical mechanism underlying the phenomena by giving theoretical interpretations and experimental confirmations of the results obtained from mathematical analysis.

Although numerous attempts have been made in these directions, especially for the apparently simplest cases of parallel flows in two dimensions, our knowledge is still very meagre. The classical case of plane Pouiseuilla motion has remained an unsettled problem,† and no satisfactory general results have been reached regarding the stability of a real fluid. The best-known general criterion is that of Rayleigh (1880) and Tollmien [74], classifying profiles according to the occurrence of a flex†† with respect to the stability of a fluid at infinite Reynolds numbers. However, the significance of their results has been too much exaggerated and often misunderstood, and no physical interpretation has ever been given. The present work offers such an interpretation, but also shows that the results can only give some indication regarding

* Received March 3, 1945. An abstract of this paper has already appeared under the same title [27].

** In 1888, the problem was proposed by Rayleigh and Stokes as the subject for the Adams Prize Essay. Cf. p. 321 of Ref. [21], and also the footnote on p. 267 of Ref. [44].

† Cf. Synge's lecture [63].

†† Following Professors Frank Morley and H. Bateman, we shall use the word "flex" for "point of inflection."

the instability of a real (viscous) fluid. This will be discussed in more detail below.

The chief aim of the present work is to try to answer the three questions mentioned above for two-dimensional parallel flows. This work is divided into three parts. Part I (the present paper) deals with the general mathematical theory, with particular emphasis on attempting to clarify the mathematical difficulties involved in the solution of the equation of stability. Part II deals with the stability problem in an inviscid fluid (infinite Reynolds numbers). Part III deals with the problem in a real fluid. The following results have been obtained.

1) It is shown that all velocity distributions of the symmetrical type and of the boundary-layer type are unstable for sufficiently large (but finite) values of the Reynolds number (Part III). The plane Poiseuille motion is included as a special case.

2) A simple approximate method is obtained by which one can calculate the minimum Reynolds number marking the beginning of instability with very little numerical labor (Part III).

3) The tendency toward instability of a profile with a point of inflection is interpreted by considering the distribution of vorticity (Part II). The effect of viscosity is considered as diffusing the disturbance from the "critical layer" inside the fluid and from the solid boundary. A very simple quantity is thereby derived which serves as a measure of the effect of viscosity (Part III). This can also be easily connected with the general mathematical theory.

As numerical examples, we have worked out the curve of neutral stability for the Poiseuille case and the Blasius case. Comparisons with existing results are discussed (Part III). The relation between instability and transition to turbulence is also discussed in Part III of this work.

Since some of the present results differ markedly from customary beliefs, it is necessary to trace the history of the existing lines of thought in order to give proper recognition to earlier ideas and results used in the present work, and to analyze all the results in disagreement with present conclusions. This requires the repetition of some known results when they fall into the present line of treatment. The review of literature is not intended to be exhaustive; only the necessary references are cited. A more complete bibliography up to 1932 has been given by Bateman [2].

**2. Historical survey of existing theories.** There seem to be two schools of thought in regard to the cause of transition from steady to turbulent conditions. One school contends that transition is due to a definite instability of the flow, i.e., to a condition in which infinitesimal disturbances grow exponentially. The second school regards the motion in most cases as definitely stable for infinitesimal disturbances but liable to be made turbulent by suitable disturbances of finite magnitude or by a large enough pressure gradient. Both schools, however, generally agree that the fluid can be considered as incompressible and that its motion is governed by the Navier-Stokes equations of motion. Since the agreement between theory and experiment has not been very satisfactory, it has also been proposed that the cause of transition must be traced back to the effect of compressibility or to the possible failure of the Navier-Stokes equations. The present work tends to confirm the simplest point of view that the motion in most cases is definitely unstable for infinitesimal disturbances governed by the Navier-Stokes equations for an incompressible fluid.

The theory of finite disturbances dates back to Reynolds [46] and Kelvin [21].

It was developed by Schiller, Taylor and others.* Mathematical investigations of such finite disturbances are mainly based on considerations of energy or of the square of the vorticity of the disturbance, because the solution of the non-linear equations satisfied by the disturbance is extremely difficult. At the end of Part III we shall briefly discuss this line of thought together with the results of the present paper. For more details, the reader is referred to the lecture of Taylor [70] and the papers of Synge [62] and Thomas [71].

For small disturbances, positive definite integrals of the energy and vorticity of the disturbance have been extensively used. These considerations have been discussed by Orr [37], Lorentz [29], von Kármán [18], Synge [63, 64] and others. For excellent accounts of this phase of the theory, the reader is referred to the works of Noether [35], von Kármán [18], Prandtl [42], and Synge [64]. Additional references are cited at the end of this paper. As is now well-known, this method can only give sufficient conditions for stability. Also, since all disturbances are usually allowed, including those which do not satisfy the hydrodynamic equations of motion, a larger viscous decay is required to insure stability than when these disturbances are excluded. Consequently, the limit of stability is always found to be much lower than that indicated by experiment. However, from these considerations, Synge [63] has arrived at a very convenient form of a sufficient condition for the stability of two-dimensional parallel flows with respect to two-dimensional disturbances. This will be found very useful for the discussions in Part III.

To get more concrete results, we have to solve the linearized equations satisfied by the disturbance. The most successful case appeared to be Taylor's treatment of Couette flow [67] between concentric cylinders. His work was verified by the experiments carried out by himself [67, 69] and by others [28]. A rigorous mathematical investigation in this connection was made by Faxon [4]. In fact, it is now known that his analysis is a typical case of the stability of a fluid motion where the centrifugal force plays a dominant part. Such cases were first considered by Lord Rayleigh [45], who gave a condition for the stability of an inviscid fluid. Mathematical proof of a sufficient condition of stability of Couette flow was recently given by Synge [65]. Extension of Taylor's work to the boundary layer over a curved wall was carried out by Görtler [8, 9], who used numerical methods successfully.

While the investigation of curved flows was uneventful, the investigation of axially symmetrical flows was not extensive. The Poiseuille flow in a circular pipe was studied by Sexl [55] with a conclusion of stability. Prandtl [42] gave some discussions of the possible cause of instability in his article in the book "Aerodynamic Theory," edited by Durand.

The most extensive discussion of hydrodynamic stability seems to be the treatment of parallel flows by attempting to solve the eigen-value problem associated with the linearized equations governing the disturbance. This line of development can be easily traced in the work of Helmholtz, Lord Rayleigh [43, 44], Orr [37], Sommerfeld [58], von Mises [31, 32], Hopf [16], Prandtl [41], Tietjens [72], Heisenberg [14], Tollmien [73–75], and Schlichting [52–54]. Other contributions are those of Noether [36], Solberg [57], Southwell [59], Squire [60], Goldstein [6], Pekeris [39, 40], Synge [61–65] and Langer [25].

---

* See Taylor's lecture [70] for references to the works on finite disturbances.

For convenience, the theory deals with two-dimensional wavy disturbances propagated along the direction of the main flow. Squire [60] has shown that three-dimensional wavy disturbances are more stable than two-dimensional ones. However, Prandtl still mentions the possibility of greater instability of three-dimensional disturbances in his article [42] appearing after Squire's paper.

The first study of two-dimensional hydrodynamic stability seems to have been made by Helmholtz. He proved the instability of wavy disturbances over the surface of discontinuity of two parallel streams of different velocities. Later, Rayleigh [43] realized that Helmholtz's approximation was not good enough to bring out the main features of a flow with continuous velocity distributions. He therefore made an improved approximation consisting of several linear profiles joined up continuously. The vorticity distribution then has constant values in several layers, but has a discontinuity in passing from one layer to another. Investigations with continuous vorticity distributions were also made. Rayleigh's work was mainly concerned with an inviscid fluid. Two main results were obtained. The first is that instability (in an inviscid fluid) can only occur with velocity distributions having a point of inflection. It is usually believed that Lord Rayleigh has also proved that damped disturbances can also only occur with such profiles. The possibility of a disturbance for a profile without a flex then becomes a paradox [5]. Actually, Rayleigh's proof does not lead to such a conclusion. This point will be more fully discussed in §5 and Part II. Rayleigh's second result is obtained from the analysis of broken linear profiles; it substantiates the first result by demonstrating definite instability of broken linear velocity distributions of the type shown in Fig. 1(a), and only stability in the other cases. Rayleigh [43] supported his result by obtaining the condition determining stability in the approximate form

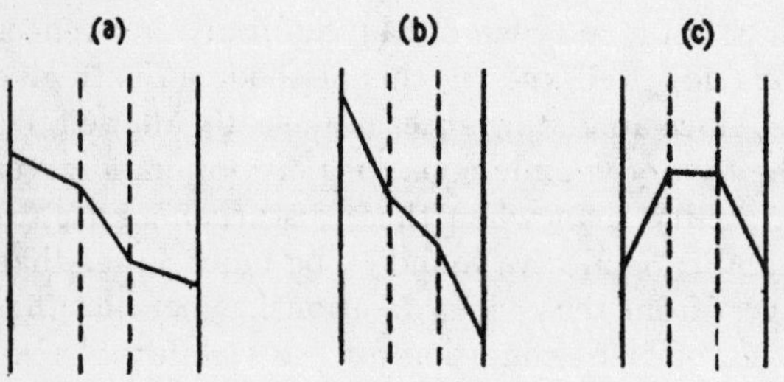


FIG. 1. Broken profiles investigated by Lord Rayleigh. Case (a) may be unstable; the other two cases are stable.

$$\int_{y_1}^{y_2} (w - c)^{-2} dy = 0, \tag{2.1}$$

where $w(y)$ is the velocity distribution, $y_1$ and $y_2$ are the coordinates of the solid boundaries, and $c$ is a constant the real part of which represents the wave velocity and the imaginary part of which gives damping or amplification.

Meanwhile, the exact analysis of linear velocity distributions including the effect of viscosity was given by von Mises [31, 32], and Hopf [16] and was also studied by Rayleigh [44]. The results indicate only stability. Prandtl and Tietjens [72] applied Rayleigh's method of approximation to the stability of the boundary layer, taking account of the effect of viscosity. In such an approximation, the inner friction layer mentioned above (§1) for continuous vorticity distributions is left out. The result of Tietjens did not give a minimum critical Reynolds number.

It was Heisenberg [14] who first successfully studied the stability of a variable continuous vorticity distribution. As a particular example, he demonstrated that the plane Poiseuille flow was unstable for sufficiently large Reynolds numbers. Also, using the same equation (2.1) with which Rayleigh supported his approximation with linear profiles, Heisenberg pointed out the fallacy in Rayleigh's method. The essential point is that the corners in the velocity profile introduce extraneous roots of the above equation for $c$. Consequently, the results of this type of analysis depend upon the manner in which the velocity distribution is approximated.

Heisenberg's numerical computation was, however, incomplete and very rough, and his theory was not generally accepted. Better known are the results of Tollmien and Schlichting. They studied the cases of Blasius [73] and plane Couette flow [48], using Heisenberg's theory essentially. The former case was pursued very much in detail. For the latter case, Schlichting followed the idea of Prandtl, asserting that the instability may be attributed to the initial unsteady distribution prior to the formation of the linear profile. Indeed, the same kind of idea was also suggested by Prandtl to account for the instability of Poiseuille flow by ascribing it to the entrance section where the profile is not yet parabolic [41]. This problem will be discussed in some detail later (§14, Part III).

For an inviscid fluid, Tollmien has also proved the instability of boundary-layer and symmetrical profiles with a point of inflection [74] For a viscous fluid, the present investigation shows that instability depends upon the general type of these profiles rather than on the appearance of the point of inflection. The inner friction layer plays a dominant role in determining the instability. Attempts to interpret this point physically are given by Prandtl [42] and in the present paper.

**3. General formulation of the problem.** We shall now formulate the problem of the stability of two-dimensional parallel flows mathematically. In the first place, we note that if the steady motion is strictly two-dimensional and parallel, the velocity distribution must be either linear or parabolic (if body forces are absent). We then have one of the following: 1) the plane Couette flows; 2) the plane Poiseuille flow; 3) a combination of these two flows. The problem would then be very restricted.

However, there are a large number of cases where the flow is *essentially* parallel to one direction. These are the cases where the boundary-layer consideration is permissible. The following are important special cases belonging to this class: 4) inlet flow between parallel walls, flow in a slightly convergent or divergent channel; 5) flow along a flat plate; 6) wake behind a cylindrical body, jet from a narrow slit. Whether these flows can be properly considered as belonging to the same class as the above three is a question of some controversy. Taylor has criticized Tollmien's work with the boundary layer on this ground [70]. In the Appendix to Part III of this work, we shall try to demonstrate that this treatment is generally permissible, but that the interpretation of the results must be taken up with care. A discussion of Tollmien's work will also be found there.

In considering the stability of the main flow, we superpose upon it a hydrodynamically possible small disturbance, and consider its behavior. The disturbance is small in the sense that the inertia forces corresponding to the disturbance alone are negligible and that its behavior is unaltered when its amplitude is (say) doubled or halved. It is then simplest to consider separate harmonic components with respect to time, which may be damped, neutral, or self-excited. By considering disturbances which are also spacially periodic both in the direction of flow and in the direction perpendicular to the plane of symmetry of the main motion, Squire [60] was able to show that two-dimensional disturbances are less stable than three-dimensional disturbances. Hence, important features of the stability problem can be obtained by considering two-dimensional disturbances alone. This is an essential difference between the stability of a parallel flow and of a curved flow. In the latter case, three-dimensional disturbances are of utmost importance.

The consideration of periodic disturbances alone is again a question of some controversy. Justification has been attempted and objection has been raised. We shall see later that at least the difficulties raised are chiefly caused by a misinterpretation of the mathematical results.

Admitting that we can consider two-dimensional disturbances alone, we have a much simplified physical picture at hand. If the effect of viscosity is negligible, we have the well-known fact of conservation of vorticity for two-dimensional motions. Actually, the stability problem is found to depend both on the inertia forces and on the viscous forces. However, the effect of viscosity is also well-known to be one of diffusion of vorticity. Thus, important results can be expected from considerations of vorticity transfer.

Let us now proceed with the mathematical formulation of the problem. We shall give a complete derivation of the stability equations so that we can see how to settle the disputes about the approximations in considering velocity distributions of the boundary-layer type.

Admitting Squire's work as a proper indication that only two-dimensional disturbances need be considered, we may conveniently consider the equation of vorticity

$$\Delta\psi_t + \psi_y \Delta\psi_x - \psi_x \Delta\psi_y = \nu\Delta\Delta\psi, \tag{3.1}$$

with the velocity components

$$u = \psi_y = \frac{\partial\psi}{\partial y}, \qquad v = -\psi_x = -\frac{\partial\psi}{\partial x}, \tag{3.2}$$

and the vorticity

$$\zeta = \frac{\partial v}{\partial x} - \frac{\partial u}{\partial y} = -(\psi_{xx} + \psi_{yy}) = -\Delta\psi. \tag{3.3}$$

As usual, $\nu$ is the kinematical viscosity. We may add that Squire's original proof was intended for flow *bounded* between two parallel walls. There is no difficulty in seeing that the proof holds also for a fluid *extending to infinity*,* because the boundary conditions for the disturbance are essentially the same.

Let us put

$$\psi = \Psi(x, y) + \psi'(x, y, t), \tag{3.4}$$

where $\Psi(x, y)$ represents the steady main flow and $\psi'(x, y, t)$ represents the disturbance. Main flows which vary but slowly with time can also be treated this way, but we shall restrict ourselves to steady flows in order to fix our ideas.

If we substitute (3.4) into (3.1) the terms corresponding to the main flow cancel out. If we then drop the terms quadratic in $\psi'(x, y, t)$ and its derivatives, we have the equation

$$\Delta\psi_t' + \Psi_y \Delta\psi_x' - \Psi_x \Delta\psi_y' + \psi_y' \Delta\Psi_x - \psi_x' \Delta\Psi_y = \nu\Delta\Delta\psi'. \tag{3.5}$$

* Cf. Ref. [15].

We shall now assume the flow to be essentially parallel to the $x$-axis. Using the boundary-layer approximation, we should drop the $x$-derivative of any quantity connected with the main flow compared with its $y$-derivative. But for the disturbance we would expect $\psi_x'$ and $\psi_y'$ to be of the same order of magnitude. This will be verified *a posteriori* in the specific examples. Further discussions will be found in the Appendix to Part III. With these considerations, (3.5) reduces to

$$\Delta\psi_t' + \Psi_y \Delta\psi_x' - \psi_x' \frac{\partial^3 \Psi}{\partial y^3} = \nu \Delta\Delta\psi'. \tag{3.6}$$

Now we shall make an approximation of the same order by taking for $w = \Psi_y$ and $\partial^2 w/\partial y^2 = \partial^3 \Psi/\partial y^3$ their *local* values at a given value $x_0$ of $x$. Then we may write

$$\Delta\psi_t' + w(y)\Delta\psi_x' - w''(y)\psi_x' = \nu\Delta\Delta\psi'. \tag{3.7}$$

For the boundary conditions, we shall also consider the *local* boundaries. The problem is then essentially simplified, and can be treated similarly to plane Couette and Poiseuille flows. We consider a main flow between two parallel planes $y = y_1$ and $y = y_2$ with a more or less arbitrary distribution of velocity $w(y)$. Then the disturbance $\psi'(x, y, t)$ must be found as a solution of (3.7) satisfying the conditions $u' = v' = 0$ over the boundaries.

The usual way of dealing with the solution of (3.7) subject to given boundary conditions is to consider periodic disturbances. We shall refer all velocities to a characteristic velocity $U$ and all lengths to a characteristic length $l$, and define the Reynolds number $R = Ul/\nu$. The two-dimensional periodic disturbance of a field of flow in which the main flow is $w(y)$ may be represented by the stream function $\psi' = \phi(y)e^{i\alpha(x-ct)}$, and the linearized differential equation for $\phi(y)$ is

$$(w - c)(\phi'' - \alpha^2\phi) - w''\phi = -\frac{i}{\alpha R}(\phi^{\text{iv}} - 2\alpha^2\phi'' + \alpha^4\phi), \tag{3.8}$$

as can be easily obtained from (3.7). We shall take $\alpha$ always real and positive, while $c$ may be complex; thus,

$$c = c_r + ic_i. \tag{3.9}$$

To fix our ideas about the boundary conditions, let us consider a flow between the planes $y = y_1$ and $y = y_2$. The equation (3.8) is then to be solved under the boundary conditions

$$\phi(y_1) = 0, \qquad \phi(y_2) = 0, \qquad \phi'(y_1) = 0, \qquad \phi'(y_2) = 0. \tag{3.10}$$

Let us now forget about the physical problem and consider the differential equation (3.8) as a linear differential equation of the fourth order in the *complex* $y$-plane. To be sure, the function $w(y)$ is defined only for real values of $y$ between $y_1$ and $y_2$. We can of course, consider it as defined for other values of $y$ by analytical continuation. We shall assume that the function thus defined is holomorphic in every finite region with which we shall be concerned. The equation (3.8) then has every point in the region under consideration as a regular point, and its coefficients are also *entire* functions of the parameters $c$, $\alpha$, and $\alpha R$ (regarded as complex variables). By a well-known theorem in the theory of differential equations, there exists a fundamental system of four solutions of (3.8) which are analytic functions of the variable $y$ and of the parameters $c$, $\alpha$, and $\alpha R$, being in fact *entire* functions of the parameters. The consequences of these simple general analytical considerations appear to have escaped serious attention from earlier investigators. In §§4, 5 of this paper, we shall find this type of consideration very important in settling the controversies about the question of convergence of the series used in the actual solution of equations (3.8) and (3.14).

Let us denote the above-mentioned system of solutions of (3.8) by $\phi_1(y)$, $\phi_2(y)$, $\phi_3(y)$, $\phi_4(y)$, the dependence upon the parameters $c$, $\alpha$, $\alpha R$ being understood. The conditions (3.10) then give rise to the secular equation

$$F(c, \alpha, \alpha R) = \begin{vmatrix} \phi_1(y_1) & \phi_2(y_1) & \phi_3(y_1) & \phi_4(y_1) \\ \phi_1(y_2) & \phi_2(y_2) & \phi_3(y_2) & \phi_4(y_2) \\ \phi_1'(y_1) & \phi_2'(y_1) & \phi_3'(y_1) & \phi_4'(y_1) \\ \phi_1'(y_2) & \phi_2'(y_2) & \phi_3'(y_2) & \phi_4'(y_2) \end{vmatrix} = 0. \tag{3.11}$$

Since the function $F(c, \alpha, \alpha R)$ is an entire function of the variables $c$, $\alpha$, $\alpha R$, we may solve for $c$, obtaining

$$c = c(\alpha, R). \tag{3.12}$$

There may be several branches of the solution, or there may be none as in the case when $F(c, \alpha, \alpha R)$ is (say) $\exp(\alpha Rc)$. In general, we would expect the solution to be unique, or we may consider only one branch of the solution.

Since $\alpha$ and $R$ are later taken to be real and positive, it is convenient to separate (3.12) into its real and imaginary parts. Thus.

$$c_r = c_r(\alpha, R), \qquad c_i = c_i(\alpha, R). \tag{3.13}$$

It is customary to plot curves of constant $c_i$ or $\alpha c_i$ in the $\alpha R$-plane. The curve $c_i = 0$ gives the limit of stability.

We are particularly interested in the case when the Reynolds number is very large. The study of this case is complicated by the fact that the functions $\phi_1$, $\phi_2$, $\phi_3$, $\phi_4$ involved have essential singularities at the infinite point of the $\alpha R$-plane. From the differential equation (3.8) itself, we see that when $\alpha R \to \infty$, we have the equation

$$(w - c)(\phi'' - \alpha^2\phi) - w''\phi = 0, \tag{3.14}$$

which is only of the second order. Thus, two solutions of (3.8) are lost. From detailed mathematical investigations, we shall find later that two linearly independent solutions of (3.8), say $\phi_1$ and $\phi_2$, will satisfy (3.14) in the limit of infinite $\alpha R$, except along certain straight lines through the point $w = c$. The other two linearly independent solutions $\phi_3$ and $\phi_4$ are highly oscillating for large $\alpha R$ and would therefore disappear in the limit of infinite $\alpha R$. Furthermore, we shall see that $\phi_3$ and $\phi_4$ can be so chosen that if $\phi_3(y_1) \gg \phi_4(y_1)$, then $\phi_3(y_2) \ll \phi_4(y_2)$, with corresponding relations for their derivatives. It then appears plausible that the limiting form of (3.11) for infinite $\alpha R$ is

$$\begin{vmatrix} \phi_1(y_1) & \phi_2(y_1) \\ \phi_1(y_2) & \phi_2(y_2) \end{vmatrix} = 0, \tag{3.15}$$

with $\phi_1(y)$, $\phi_2(y)$ satisfying (3.14).

The condition (3.15) states that we are looking for a solution $\phi(y)$ of (3.14) satisfy-

$$\phi(y_1) = 0, \qquad \phi(y_2) = 0, \tag{3.16}$$

with the other two conditions of (3.10) relaxed. Physically, this means that we allow a slipping along the walls $y = y_1$ and $y = y_2$. For very large Reynolds numbers, only a very thin layer of fluid will stick to the solid, and we have naturally an apparent slipping. These points will be taken up again more carefully (§6) after a thorough mathematical investigation of the solutions.

**4. Solution of the equation of Orr and Sommerfeld by methods of successive approximation.** The stability equation of Orr and Sommerfeld

$$(w - c)(\phi'' - \alpha^2\phi) - w''\phi = -\frac{i}{\alpha R}(\phi^{\text{iv}} - 2\alpha^2\phi'' + \alpha^4\phi) \tag{4.1}$$

has a fundamental system of four solutions, which are analytic functions of $y$ (wherever $w(y)$ is analytic) and which are entire functions of $\alpha$, $c$, and $\alpha R$. In order to obtain useful solutions, it is usual to expand the solutions as power series of a suitable small parameter, say, $(\alpha R)^{-1}$. However, since $(\alpha R)^{-1}$ occurs with the highest derivative in (4.1), the study of such an expansion becomes very complicated. It will be done later.

a) *Solution by convergent series.* An alternative method* is to choose a small parameter $\epsilon$ related to $(\alpha R)^{-1}$ and first make a change of variable ($y_0$ being an arbitrary point so far)

$$y - y_0 = \epsilon\eta, \qquad \phi(y) = \chi(\eta), \tag{4.2}$$

so that (4.1) becomes

$$(w - c)(\chi'' - \alpha^2\epsilon^2\chi) - \epsilon^2 w''\chi = -\frac{i}{\alpha R\epsilon^2}(\chi^{\text{iv}} - 2\alpha^2\epsilon^2\chi'' + \alpha^4\epsilon^4\chi), \tag{4.3}$$

where

$$\left.\begin{aligned} w - c &= (w_0 - c) + w_0'(\epsilon\eta) + \frac{w_0''}{2!}(\epsilon\eta)^2 + \cdots, \\ w'' &= w_0'' + w_0'''(\epsilon\eta) + \frac{w_0^{\text{iv}}}{2!}(\epsilon\eta)^2 + \cdots. \end{aligned}\right\} \tag{4.4}$$

The solution is then obtained in the form

$$\phi(y) = \chi(\eta) = \chi^{(0)}(\eta) + \epsilon\chi^{(1)}(\eta) + \epsilon^2\chi^{(2)}(\eta) + \cdots, \tag{4.5}$$

and the differential equations for the approximations of successive orders can be obtained by substituting (4.4) and (4.5) into (4.3) and equating all the coefficients of the various powers of $\epsilon$ to zero.

If we take $y_0$ to be the point where $w = c$, the proper choice of the parameter $\epsilon$ is

$$\epsilon = (\alpha R)^{-1/3}. \tag{4.6}$$

The differential equations for the functions $\chi^{(0)}(\eta)$, $\chi^{(1)}(\eta)$, $\chi^{(2)}(\eta)$, $\cdots$ are as follows:

$$\left.\begin{aligned} &\epsilon^0; \quad w_0'\eta\chi^{(0)''} + i\chi^{(0)\text{iv}} = 0, \\ &\epsilon^n; \quad w_0'\eta\chi^{(n)''} + i\chi^{(n)\text{iv}} = L_{n-1}(\chi), \quad (n \geqq 1), \end{aligned}\right\} \tag{4.7}$$

* This method was first used by Heisenberg, loc. cit. [14], p. 588.

where $L_{n-1}(\chi)$ is a linear combination of $\chi^{(0)}(\eta)$, $\chi^{(1)}(\eta)$, $\cdots$, $\chi^{(n-1)}(\eta)$ and their derivatives. In particular,

$$L_0(\chi) = w_0''(\chi^{(0)} - \tfrac{1}{2}\eta^2\chi^{(0)''}). \tag{4.8}$$

We note that the homogeneous part is the same for all the differential equations in (4.7). Hence, if we can solve for the first approximation, the rest can all be obtained by *quadratures*. Indeed, the first equation of (4.7) is Stokes' equation* for $\chi^{(0)''}$, and its solution can be readily expressed in terms of Bessel functions of the order 1/3. Thus, for the first equation of (4.7) we have the four particular integrals**

$$\left.\begin{aligned} \chi_1^{(0)} &= \eta, & \chi_3^{(0)} &= \int_{+\infty}^{\eta} d\eta \int_{+\infty}^{\eta} d\eta\, \eta^{1/2} H_{1/3}^{(1)}\left[\tfrac{2}{3}(i\alpha_0\eta)^{3/2}\right], \\ \chi_2^{(0)} &= 1, & \chi_4^{(0)} &= \int_{-\infty}^{\eta} d\eta \int_{-\infty}^{\eta} d\eta\, \eta^{1/2} H_{1/3}^{(2)}\left[\tfrac{2}{3}(i\alpha_0\eta)^{3/2}\right], \end{aligned}\right\} \tag{4.9}$$

where

$$\alpha_0 = (w_0')^{1/3}. \tag{4.10}$$

The higher approximations are given by

$$\chi_i^{(n)} = \frac{\pi}{6}\int^{\eta} d\eta \int^{\eta} d\eta \left\{\chi_4^{(0)''}\int^{\eta} d\eta\chi_3^{(0)''}L_{n-1}(\chi) - \chi_3^{(0)''}\int^{\eta} d\eta\chi_4^{(0)''}L_{n-1}(\chi)\right\}, \tag{4.11}$$

$$(i = 1, 2, 3, 4).$$

These are the explicit formulae for finding the approximations of various orders. In actual calculations, only the initial approximation (4.9) is required. Furthermore, *the series* (4.5) *is convergent provided* $\epsilon$ *is restricted so that the series* (4.4) *are convergent.* For then the differential equation (4.3) for $\chi(\eta)$, when normalized, has analytic functions of the parameter $\epsilon$ as its coefficients. Hence, a fundamental system of its solutions consists of four analytic functions of $\epsilon$.

It should be mentioned that if $y_0$ is not taken at the particular point for which $w=c$, the proper parameter to be taken is $(\alpha R)^{-1/2}$ instead of $(\alpha R)^{-1/3}$. In this case, all the approximations can be expressed in terms of elementary transcendental functions. However, it is not found particularly advantageous to do so, because the study of "crossing substitution" (§5) would not be easy. Also, the method is then too much different from those used by earlier investigators to allow an easy comparison of the results.

b) *Solution by asymptotic series.* Although the previous method is theoretically

---

* Cf. the exact treatment of (4.1) by Hopf [16] and Rayleigh [44] for the case $w''=0$.

** Note that $\chi_3^{(0)}$ and $\chi_4^{(0)}$ and also $\chi_i^{(n)}$ have *no* branch point at $\eta=0$. The order of the solutions $\{\phi_1, \phi_2, \phi_3, \phi_4\}$ agrees with Tollmien's notation. They are $\{\phi_3, \phi_4, \phi_1, \phi_2\}$ in Heisenberg's notation. Heisenberg gave the solutions $\phi_3$ and $\phi_4$ in terms of Hankel functions in the form

$$\phi_j = (w - c)\int^1 \frac{1}{\eta} H_{2/3}^{(j)}\left[\frac{2}{3}(i\alpha_0\eta)^{3/2}\right] d\eta, \qquad (j = 1, 2)$$

(p. 289, and Eq. (19a) p. 591). It can be easily verified that these are the same as $\chi_{3,4}^{(0)}$ up to a constant factor $w_0'\epsilon$ and to the proper order of approximation. Note that throughout Heisenberg's paper, $i$ is to be replaced by $-i$ in order to conform to our notation. This can be seen from a comparison of our Eq. (4.1) with his Eq. (7a). The difference arises from a difference of notation in the stream function $\psi'(x, y, t)$.

complete, it is usually more convenient to use asymptotic series for numerical purposes, particularly in dealing with boundary-value problems. Heisenberg has given two asymptotic methods, each of which gives only *two* particular solutions of (4.1). These methods will now be described and investigated mathematically in more detail, because Heisenberg's work has received some criticism in this connection.*

The first of these methods is to develop $\phi(y)$ in powers of $(\alpha R)^{-1}$. We put

$$\phi(y) = \phi^{(0)}(y) + (\alpha R)^{-1}\phi^{(1)}(y) + (\alpha R)^{-2}\phi^{(2)}(y) + \cdots, \tag{4.12}$$

and substitute in (4.1). Comparing corresponding powers of $(\alpha R)^{-1}$, we have the following differential equations

$$\left.\begin{aligned} &(w-c)(\phi^{(0)\prime\prime} - \alpha^2\phi^{(0)}) - w''\phi^{(0)} = 0,\\ &(w-c)(\phi^{(k)\prime\prime} - \alpha^2\phi^{(k)}) - w''\phi^{(k)} = -i[\phi^{(k-1)\mathrm{iv}} - 2\alpha^2\phi^{(k-1)\prime\prime} + \alpha^4\phi^{(k-1)}],\\ &\qquad\qquad (k \geqq 1). \end{aligned}\right\} \tag{4.13}$$

The initial approximation satisfies the inviscid equation and can be solved by developing $\phi^{(0)}$ in powers of $\alpha^2$. Indeed, two particular integrals of (4.13) are

$$\left.\begin{aligned} \phi_1^{(0)} &= (w-c)[h_0(y) + \alpha^2 h_2(y) + \alpha^4 h_4(y) + \cdots],\\ \phi_2^{(0)} &= (w-c)[k_1(y) + \alpha^2 k_3(y) + \alpha^4 k_5(y) + \cdots], \end{aligned}\right\} \tag{4.14}$$

where

$$\left.\begin{aligned} &h_0(y) = 1, \qquad h_{2n+2}(y) = \int_{y_1}^{y} dy(w-c)^{-2}\int_{y_1}^{y} dy(w-c)^2 h_{2n}(y), \quad (n \geqq 0)\\ &k_1(y) = \int_{y_1}^{y} dy(w-c)^{-2}, \quad k_{2n+3}(y) = \int_{y_1}^{y} dy(w-c)^2\int_{y_1}^{y} dy(w-c)^{-2}k_{2n+1}(y), \quad (n \geqq 0). \end{aligned}\right\} \tag{4.15}$$

The point $y_1$ might have been any fixed point instead of one of the end points; but it is found convenient to take it this way

Having found two particular integrals for $\phi^{(0)}$, we can obtain the higher approximations by *quadratures*. In actual calculations, this is not necessary

Because of the general nature of the Eq. (4.1), $\phi(y)$ is an entire function of $\alpha R$. Hence, the infinite point of the $\alpha R$-plane is a singular point, unless $\phi(y)$ is independent of $\alpha R$. Consequently, the series (4.12) is asymptotic, unless $\phi(y)$ is a polynomial in $(\alpha R)^{-1}$. We note also that (4.13) is of the second order, so that only *two* solutions are obtained by this method. The solutions of (4.13) are entire functions of $\alpha^2$ and hence the series (4.14) are uniformly convergent for any finite region of the complex $\alpha^2$-plane, for a fixed value of $y$, except when $y$ is the singular point $y_0$ of the differential equation (4.13).†

* Tollmien, loc. cit., 1929, p. 43.

† This can also be seen from the series itself. So long as it is possible to run a path of finite length from $y_1$ to $y$ on which $w-c\neq 0$, the general terms $\alpha^{2n}h_{2n}$ and $\alpha^{2n+1}k_{2n+1}$ of the two series are bounded by $A(\alpha M)^{2n}/(2n)!$ and $B(\alpha M)^{2n+1}/(2n+1)!$, respectively, ($A$, $B$, $M$ being (suitably) fixed constants), and

In fact, the differential equation (4.13) has a logarithmic singularity at the point $y_0$. This point is, however, an ordinary point in the exact equation (4.1), and the singularity is introduced purely by the method of asymptotic integration. However, the appearance of this singularity gives a serious ambiguity in the determination of the correct path leading from $y_1$ to $y$ in order that (4.14) may give valid approximations to integrals of (4.1) *all along the path*.* The proper way to settle this question is to compare the solutions (4.14) with the *asymptotic expansions of the regular solutions* obtained by the previous method. This will be done later after we have described the second asymptotic method of Heisenberg for the other *two* particular integrals; for the same kind of problem also arises there.

To obtain *two* other integrals of (4.1) in asymptotic forms. let us make the transformation

$$\phi = \exp\left\{\int g dy\right\}. \tag{4.16}$$

Then, we obtain the non-linear differential equation

$$(w-c)\{(g^2+g')-\alpha^2\} - w'' = -\frac{i}{\alpha R}\{g^4+6g^2g'+3g'^2+4gg''+g'''-2\alpha^2(g^2+g')+\alpha^4\} \tag{4.17}$$

for the function $g(y)$. We try to solve this by putting

$$g(y) = (\alpha R)^{1/2}g_0(y) + g_1(y) + (\alpha R)^{-1/2}g_2(y) + \cdots. \tag{4.18}$$

Then, we obtain the set of equations

$$(w-c)g_0^2 = -ig_0^4, \qquad (w-c)(g_0'+2g_0g_1) = -i(4g_0^3g_1+6g_0^2g_0'),$$

$$(w-c)(g_1'+g_1^2+2g_0g_2-\alpha^2) - w'' = -i(4g_0^3g_2+6g_1^2+2g_0g_1g_0'+3g_0'^2-2\alpha^2g_0^2),$$

$$\cdots\cdots\cdots\cdots\cdots\cdots\cdots\cdots\cdots$$

Hence we can obtain the successive approximations without integration. Thus,

$$g_0 = \pm\sqrt{i(w-c)}, \qquad g_1 = -\frac{5}{2}\frac{g_0'}{g_0}, \cdots. \tag{4.19}$$

For definiteness, we define

---

hence the series converge like the cosine and the sine series, respectively. Heisenberg did not *prove* the convergence of these series, but stated that their convergence can be hoped to be sufficiently rapid for $\alpha^2$ of the order of unity (loc. cit., 1924, pp. 584, 587). This was made a point of criticism by Tollmien (loc. cit., 1929, p. 43).

* Considerable dispute has arisen in this connection. Note that it is impossible to dispense with this difficulty by remarking that the two different determinations will differ only by a constant multiple of a particular integral. If we draw two paths from $y_1$ to $y$ and obtain such a difference in the solution, it is evident that the asymptotic solution cannot be valid on both paths, because the *exact* equation (4.1) has *no* singular point at $y=y_0$ and hence its solution must be single-valued. Although a mistake here would not cause serious difficulties so far as the numerical evaluation of the eigen-value problem is concerned, it does lead to misunderstanding and confusion elsewhere. Even after Heisenberg and Tollmien have analyzed this problem in some detail, they still take the very misleading step of taking the complex conjugate of the inviscid equation (Heisenberg, loc. cit., 1924, p. 596; Tollmien, loc. cit., 1935, p. 88). This point will be discussed more fully later.

$$\arg i = \frac{\pi}{2}, \qquad \arg(w - c) > 0 \quad \text{for} \quad w - c > 0. \tag{4.20}$$

For negative values of $w-c$, we cannot decide, without further investigation, whether $\arg(w-c)=+\pi$ or $\arg(w-c)=-\pi$. The point $y_0$, where $w=c$, appeared in the previous asymptotic solution as a logarithmic branch point; here it is an algebraic branch point. The determination of the correct path should follow the same criterion as the other two integrals, that (4.18) gives two asymptotic solutions of the exact equation (4.1) *all along the path*. This path might be expected to be the same as that in the previous case. All these will be discussed in the next section.

After such a question is settled, substitution of (4.19) into (4.16) and (4.17) gives the two asymptotic solutions

$$\left.\begin{aligned} \phi_3(y) &= (w - c)^{-5/4} \exp\left\{-\int_{y_0}^{y} \sqrt{i\alpha R(w - c)}\, dy\right\}, \\ \phi_4(y) &= (w - c)^{-5/4} \exp\left\{+\int_{y_0}^{y} \sqrt{i\alpha R(w - c)}\, dy\right\}, \end{aligned}\right\} \tag{4.21}$$

where factors of the order $\exp(\alpha R)^{-1/2}=1+O\{(\alpha R)^{-1/2}\}$ are taken as unity.

**5. Analytical properties of the solutions.** Having thus obtained four asymptotic solutions of the equation (4.1), we must try to correlate them with the four solutions (4.9) and (4.11), and above all to study the correct determination of path around the artificial singularity introduced by the asymptotic methods. For this purpose, we consider the asymptotic expansions of the four regular solutions obtained by the first method and transfer back to the independent variable $y$.

Let us recall that the asymptotic expansions of the Hankel functions $H_{1/3}^{(1)}(\xi)$, $H_{1/3}^{(2)}(\xi)$ are given by [76],

$$\left.\begin{aligned} H_{1/3}^{(1)}(\xi) &\sim \left(\frac{2}{\pi\xi}\right)^{1/2} \exp\left\{i\left(\xi - \frac{5\pi}{12}\right)\right\}\left\{1 + \sum_{r=1}^{\infty} \frac{(-)^r(1/3, r)}{(2i\xi)^r}\right\}, \\ &\qquad\qquad (-\pi < \arg \xi < 2\pi), \\ H_{1/3}^{(2)}(\xi) &\sim \left(\frac{2}{\pi\xi}\right)^{1/2} \exp\left\{-i\left(\quad - \frac{\pi}{12}\right)\right\}\left\{1 + \sum_{r=1}^{\infty} \frac{(1/3, r)}{(2i\xi)^r}\right\}, \\ &\qquad\qquad (-2\pi < \arg \xi < \pi). \end{aligned}\right\} \tag{5.1}$$

If we put $3\xi = 2(i\alpha_0\eta)^{3/2}$, then (5.1) becomes

$$\left.\begin{aligned} H_{1/3}^{(1)}[\tfrac{2}{3}(i\alpha_0\eta)^{3/2}] &\sim \left(\frac{3}{\pi}\right)^{1/2} (i\alpha_0\eta)^{-3/4} \exp\left\{\tfrac{2}{3}(\alpha_0\eta)^{3/2} e^{5\pi i/4} - \frac{5\pi}{12}\right\}\{1 + O(\eta^{-3/2})\}, \\ &\qquad \left(-\frac{7\pi}{6} < \arg(\alpha_0\eta) < \frac{5\pi}{6}\right), \\ H_{1/3}^{(2)}[\tfrac{2}{3}(i\alpha_0\eta)^{3/2}] &\sim \left(\frac{3}{\pi}\right)^{1/2} (i\alpha_0\eta)^{-3/4} \exp\left\{\tfrac{2}{3}(\alpha_0\eta)^{3/2} e^{\pi i/4} + \frac{5\pi}{12}\right\}\{1 + O(\eta^{-3/2})\}, \\ &\qquad \left(-\frac{11\pi}{6} < \arg(\alpha_0\eta) < \frac{\pi}{6}\right) \end{aligned}\right\} \tag{5.2}$$

With the help of these formulae and using the legitimate process of integrating the asymptotic expansions term by term, we obtain

$$\left.\begin{aligned}
\chi_1^{(0)} + \epsilon\chi_1^{(1)} &\sim \eta + \frac{\epsilon w_0''}{2w_0'}\eta^2 = \frac{1}{w_0'\epsilon}\left(w_0' y + \frac{w_0''}{2}y^2\right), \\
\chi_2^{(0)} + \epsilon\chi_2^{(1)} &\sim 1 + \epsilon\frac{w_0''}{w_0'}\eta\log\eta \sim 1 + \frac{w_0''}{w_0'}y\log y, \\
\chi_3^{(0)} &\sim \text{const. } \eta^{-5/4}\exp\left\{\tfrac{2}{3}(\alpha_0\eta)^{3/2}e^{5\pi i/4}\right\} \\
&= \text{const. } (y-y_0)^{-5/4}\exp\left\{-\int_{y_1}^{y}\sqrt{i\alpha R w_0'(y-y_0)}\,dy\right\} \\
\chi_4^{(0)} &\sim \text{const. } \eta^{-5/4}\exp\left\{\tfrac{2}{3}(\alpha_0\eta)^{3/2}e^{\pi i/4}\right\} \\
&= \text{const. } (y-y_0)^{-5/4}\exp\left\{\int_{y_0}^{y}\sqrt{i\alpha R w_0'(y-y_0)}\,dy\right\}
\end{aligned}\right\} \qquad (5.3)$$

These formulae can be easily seen to agree with the four asymptotic solutions (4.13) and (4.21) to the proper order of approximation, if we replace $y_1$ by $y_0$ in $\phi_1^{(0)}$ (which is permissible).

In evaluating the asymptotic expressions (5.3), the argument $\alpha_0\eta$ must satisfy both requirements specified in (5.2), i.e.,

$$-7\pi/6 < \arg(\alpha_0\eta) < \pi/6. \qquad (5.4)$$

*In this range, the asymptotic solutions* (4.13) *and* (4.21) *hold*. Having thus established the range of validity of these solutions, we no longer need to make further comparisons of the two methods of solution.

At least three plans are now possible for further numerical work. First, we may use the four solutions obtained in the approximate form (4.9). Secondly, we may use the four asymptotic solutions (4.14) and (4.21). Thirdly, we may approximate $\{\phi_1, \phi_2, \phi_3, \phi_4\}$ by the four functions $\{\phi_1^{(0)}, \phi_2^{(0)}, \chi_3^{(0)}, \chi_4^{(0)}\}$ given by (4.14) and (4.9). The first method is very similar to the method used by Hopf [16] and Tietjens [72] for linear velocity distributions, where the *exact* solutions are given by functions of the general nature of those in (4.9). For curved velocity distributions, the functions $\chi_1^{(0)}$, $\chi_2^{(0)}$ do not give $\phi_1$ and $\phi_2$ with sufficient accuracy, and this plan is not good. The second plan was used by Heisenberg in his investigation of the stability of the Poiseuile flow; but he also realized that it served only part of his purpose, and he stated that the third plan should be used.* Tollmien substantially adopted the third plan for his investigation of the stability of the boundary layer, although he did not point out the connection of his method with Heisenberg's work. Instead of the expressions (4.14) for $\phi_1$ and $\phi_2$, he used solutions in the forms of power series in $y$. These solutions are easily manageable only for linear and parabolic velocity distributions. Accordingly, he tried to approximate the Blasius profile with such profiles. Since such approximations are not good enough in the neighborhood of the point $y = y_0$, where $w = c$, his discussion becomes very complicated. In the present work, we base our cal-

* Loc. cit., p. 404.

culations upon the use of (4.14). It will be seen that our method can be applied to any profile with good accuracy. A comparison with Tollmien's method will be discussed in the Appendix to Part III.

It may be added that the adoption of the third plan leaves an error of the order of $(\alpha R)^{-1}$ in $\phi_1$ and $\phi_2$, and an error of the order of $(\alpha R)^{-1/3}$ in $\phi_3$ and $\phi_4$. These errors can be reduced by including the higher approximations. In practice, this is hardly necessary. A detailed discussion of numerical accuracy will be found in the Appendix to Part III.

Having thus established the region of validity of the asymptotic solutions, we shall try to settle a few questions of considerable dispute, namely, (a) the "crossing substitution," (b) the inner friction layers, and (c) the complex conjugate of the inviscid solution.

a) *The crossing substitution.* From previous discussions, it is evident that if we pass from $y>\mathrm{Re}(y_0)$ to $y<\mathrm{Re}(y_0)$ along a path *below* the point $y_0$, we are always in a region of the $y$-plane where the above asymptotic solutions hold, and no further investigation is necessary. In fact, if $c_i>0$ (and is small) and $\mathrm{Re}(w_0')>0$, the point $y_0$ is above the real axis, and the asymptotic solutions are *valid along the real axis of* $y$. In the case of real $c$, the point $y_0$ is on the real axis, and there is *one point on the real axis* where the asymptotic solutions fail to be valid. In the case $c_i<0$, and $\mathrm{Re}(w_0')>0$, the point $y_0$ is below the real axis, and the lines $\arg\{\alpha_0(y-y_0)\}=-7\pi/6,\ \pi/6$ intersect the real axis in two points $y_f'$, $y_f''$ with* $y_1<y_f'<y_f''<y_2$. Thus, the asymptotic expressions (4.14) and (4.21) represent one solution for $y_1\leqq y<y_f'$ and $y_f''<y\leqq y_2$, but not the same solution for $y_f'<y<y_f''$. It is necessary to obtain a suitable "crossing substitution" in order to obtain the correct solutions for $\pi/6<\arg\{\alpha_0(y-y_0)\}<5\pi/6$ (i.e., in *crossing* the lines $\arg\{\alpha_0(y-y_0)\}=-7\pi/6,\ \pi/6$). For this purpose, we must obtain the asymptotic expansion of the Hankel functions $H_{1/3}^{(j)}[2(i\alpha_0\eta^{3/2})/3]$, $(j=1,\ 2)$, proper to that region. The analytical expression for $H_{1/3}^{(2)}$ would then be quite different from that given in (5.2). Thus, in crossing the *two points* $y_f'$ *and* $y_f''$ *of the real axis*, the asymptotic solutions fail to be analytic. However, it is to be noted that the failure of the asymptotic solutions along the real axis does not exclude their use in the investigation of the boundary value problems to be considered below, so long as we are concerned only with the eigen-value problem. It is only necessary that these solutions be valid in a *connected* region containing the end-points $y_1$ and $y_2$. The calculation of the amplitude distribution of the disturbance (the eigen-function) in the neighborhood of the inner friction layers, however, is to be made with the regular solution, or we can calculate the eigen-function for $y_f'<y<y_f''$ by using a proper

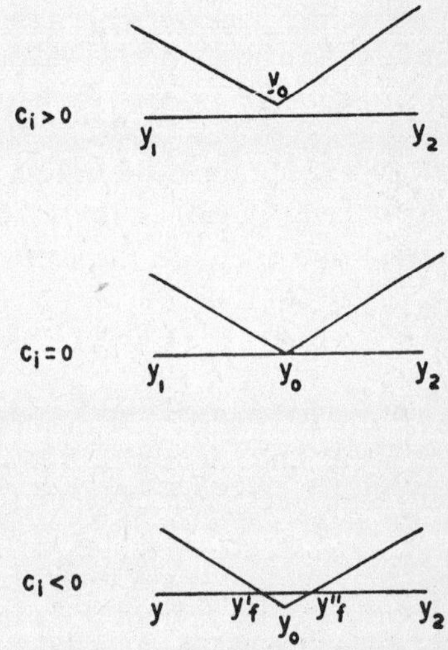


FIG. 2. Diagram showing the relative position of the real axis and the region of validity of the asymptotic solutions in each of the three cases.

* The whole theory must be modified for extremely highly damped solutions for which $y_f'<y_1<y_2<y_f''$.

"crossing substitution." Since we are chiefly concerned with the eigen-value problem, we shall not go into further details.

In order to make the situation still clearer, let us see what would happen if we try to obtain our solutions for $y_i' < y < y_i''$ by going along a path above the point $y_0$. For simplicity, let us take the case of real $c$ with $w_0' > 0$, and consider the asymptotic expressions $\phi_3^{(0)}$ and $\phi_4^{(0)}$ given by (5.3). We have ($A$, $B$ being arbitrary constants)

$$\phi_3^{(0)} \sim A\eta^{-5/4} \exp\{\tfrac{2}{3}(\alpha_0\eta)^{3/2} e^{5\pi i/4}\}, \qquad (\eta > 0),$$
$$\phi_3^{(0)} \sim A|\eta|^{-5/4} \exp\{\tfrac{2}{3}(\alpha_0|\eta|)^{3/2} e^{-\pi i/4} + 5\pi i/4\}, \qquad (\eta < 0);$$

$$\phi_4^{(0)} \sim B\eta^{-5/4} \exp\{\tfrac{2}{3}(\alpha_0\eta)^{3/2} e^{\pi i/4}\}, \qquad (\eta > 0),$$
$$\phi_4^{(0)} \sim B|\eta|^{-5/4} \exp\{\tfrac{2}{3}(\alpha_0|\eta|)^{3/2} e^{-5\pi i/4} + 5\pi i/4\}, \qquad (\eta < 0).$$

These are obtained by taking a path below the point $y_0$. If we had taken the other path, then arg $(\eta) = \pi$ for $\eta < 0$, and we would have the functions $\tilde{\phi}_3^{(0)}$ and $\tilde{\phi}_4^{(0)}$, which agree with $\phi_3$ and $\phi_4$ for $\eta > 0$, but are defined by

$$\tilde{\phi}_3^{(0)} \sim A|\eta|^{-5/4} \exp\{\tfrac{2}{3}(\alpha_0|\eta|)^{3/2} e^{3\pi i/4} - 5\pi i/4\},$$
$$\tilde{\phi}_4^{(0)} \sim B|\eta|^{-5/4} \exp\{\tfrac{2}{3}(\alpha_0|\eta|)^{3/2} e^{7\pi i/4} - 5\pi i/4\},$$

for $\eta < 0$. Thus, if $A$ and $B$ are taken to be the same, we have

$$\tilde{\phi}_3^{(0)} = -i\phi_4^{(0)}, \quad \tilde{\phi}_4^{(0)} = -i\phi_3^{(0)}, \quad \text{for} \quad \eta < 0.$$

Hence if we took $\tilde{\phi}_3^{(0)}$ and $\tilde{\phi}_4^{(0)}$ as the proper determinations, we would have to make the following "crossing substitution" corresponding to a passage from $\eta > 0$ to $\eta < 0$: $\tilde{\phi}_3^{(0)} \to i\tilde{\phi}_4^{(0)}$; $\tilde{\phi}_4^{(0)} \to i\tilde{\phi}_3^{(0)}$. If we note that $\tilde{\phi}_3^{(0)} \ll \tilde{\phi}_4^{(0)}$ both for $w-c>0$ and for $w-c<0$, we would also have the following equivalent change: $\tilde{\phi}_3^{(0)} \to \tilde{\phi}^{(0)} + i\tilde{\phi}_4^{(0)}$; $\tilde{\phi}_3^{(0)} - i\tilde{\phi}_4^{(0)} \to \tilde{\phi}_3^{(0)}$. These may be compared with Eq. (16), p. 589 of Heisenberg's paper. In making the comparison, we note his definition of the angle of $w-c$ (p. 585), and the difference of notation in the fundamental equation of stability.

The first study of "the crossing substitution" seems to be due to Stokes in connection with the asymptotic expansions of Bessel functions. It may therefore be properly designated as Stokes' phenomenon [76]. We should also compare our results with the work of Jeffreys [17], the *W-K-B* method [23] in quantum mechanics,* and the mathematical investigations of Langer [24] and others.** In those cases, a differential equation of the form $\epsilon^3\phi'' + q(y)\phi = 0$ is considered. If this equation is treated by the method of §4 by writing $\phi = \chi^{(0)}(\eta) + \epsilon\chi^{(1)}(\eta) + \cdots$, $y - y_0 = \epsilon\eta$, and $q(y) = q_0'(\epsilon\eta) + \frac{1}{2}q_0''(\epsilon\eta)^2 + \cdots$, the equation for $\chi^{(0)}(\eta)$ is $\chi^{(0)''} + q_0'\eta\chi^{(0)} = 0$ as compared with (4.7), $\chi^{(0)\text{iv}} - i\eta w_0'\chi^{(0)''} = 0$. It is evident that our $\eta$ corresponds to $i\eta$ in their case. Kramers has shown that the cuts in their asymptotic expansions are the lines arg $(\eta) = \pm\pi/3$. Thus, in our case, the cuts should be arg $(\eta) = \pi/6$, $5\pi/6$. This agrees with our previous discussions. An important difference is the following. In

* I am indebted to Professor P. S. Epstein for calling my attention to this comparison.

** For example, S. Goldstein, Proc. Lond. Math. Soc. (2) **28,** 81–90 (1928); C. C. Hurd, Tôhoku Math. Journ. **45,** 58–68 (1939), and the papers of W. J. Trjitzinsky and others quoted there.

their case, the two boundary points on the real axis are separated into two regions of the complex plane by the cuts, so that a *crossing substitution is absolutely necessary.* In our case, the two boundary points on the real axis belong to the same region, and a *crossing substitution is superfluous*, so far as the eigen-value problem is concerned.

b) *The inner friction layers*. There is also a very significant physical interpretation associated with the "crossing substitution" of the asymptotic solutions. The initial approximations $\phi_1^{(0)}$ and $\phi_2^{(0)}$ satisfy the inviscid equation. Hence, if $c_i>0$, these solutions hold throughout the interval $(y_1, y_2)$ of the real axis, and the effect of viscosity is entirely negligible inside the fluid for sufficiently large Reynolds numbers. If $c_i \leqq 0$, the inviscid solution can never hold all along the real axis, and hence the effect of viscosity inside the fluid is not negligible, however large the Reynolds number may be. The singularity of the asymptotic solutions means a very rapid change of velocity within a small distance so that the effect of viscosity is no longer negligible there. Physically, such a point on the real axis corresponds to a layer of fluid where the viscous forces play an important role.

Referring to the foregoing discussions, we see that *there are two inner friction layers for the damped oscillations, one for the neutral oscillations, and none for the self-excited oscillations.*

In the neutral case, the first term of (4.1) disappears at the critical layer $w=c$. The equation then represents a balancing of the vorticity transferred by the disturbance and that diffused away by the effect of viscosity. It is therefore understandable that the effect of viscosity must be predominant there. In the other two cases, $w-c$ never vanishes in the fluid, there is the vorticity carried by the main flow (relative to an observer moving with the phase velocity $c_r$) and there is always the change of vorticity due to amplification or damping. In the case of amplified oscillations, these two effects can be in equilibrium with the transfer of vorticity due to the disturbance, and the effect of viscosity is completely negligible at very large Reynolds numbers. In the case of damped oscillations, these effects presumably never balance each other, thus resulting in the formation of two critical layers, where the effect of viscosity is not negligible.

c) *The complex conjugate of the solution* $\phi(y)$. It is often argued,* that if $\phi(y)$ is a solution of the inviscid equation with an eigen-value $c$, then $\bar{\phi}(y)$ is another solution with the eigen-value $\bar{c}$, satisfying the same real boundary conditions on the real axis. Thus, to each damped solution, there is always a corresponding amplified solution, and vice versa. This argument is in direct contradiction to the foregoing discussions, because an amplified solution and a damped solution have entirely different characteristics with respect to inner friction layers. It appears, therefore, that $\bar{\phi}(y)$ should still represent a solution of the same nature as $\phi(y)$.

This is indeed the case, and can be seen more clearly from an examination of the complete equation (4.1). If we take the complex conjugate of that equation, and write $y$ for $\bar{y}$ (which is essentially done in the usual argument), we have

$$\{w(y) - c\}\{\bar{\phi}'' - \alpha^2\bar{\phi}\} - w''(y)\bar{\phi} = \frac{i}{\alpha R}\{\bar{\phi}^{iv} - 2\alpha^2\bar{\phi}'' + \alpha^4\bar{\phi}\}. \tag{5.5}$$

* Heisenberg, loc. cit. p. 596; Tollmien, loc. cit., 1935, p. 88. The failure of such an argument would indicate that Heisenberg's classification of velocity profiles on p. 597 of his paper is untenable.

The complete stream function $\bar{\Psi}'(x, y, t)$ satisfying this equation is

$$\bar{\Psi}'(x, y, t) = \bar{\phi}(y)e^{-i\alpha(x-\bar{c}t)}.$$

Thus, if $\text{Im}(c) < 0$, then $\text{Im}(\bar{c}) > 0$, and we still have a damped solution. This should also hold for the inviscid equation, since it is regarded as a *limiting* case of the viscous equation. From the inviscid equation itself, there is no way of telling whether $\text{Im}(\bar{c}) > 0$ corresponds to damping or to amplification. In fact, the asymptotic solutions of Eq. (5.5) (including the limiting inviscid solutions) hold for

$$-\pi/6 < \arg\left\{\bar{\alpha}_0(y - y_0)\right\} < 7\pi/6, \qquad w(y_0) = \bar{c}. \tag{5.6}$$

Thus, we have a solution $\bar{\phi}(y)$, valid in a region which is quite improper for an asymptotic solution of (4.1). [Compare (5.4) and (5.6).] Hence, it is not legitimate to conclude that a solution of a different nature can be obtained by taking the complex conjugate. The influence of these discussions upon the usual conclusions regarding stability in an inviscid fluid will be discussed fully in the next part of the paper.

**6. The boundary value problems.** Having fully investigated the solutions of the equation of disturbance, we shall now turn to a study of the boundary-value problems which have been taken up briefly at the end of §3. In general, the boundary conditions are essentially that the velocities of the disturbance should vanish on the solid boundaries, and also at infinity if the field of flow extends to infinity. However, it is often convenient to use equivalent boundary conditions for certain types of velocity distributions.

In order not to be lost in too much generality, we shall limit ourselves to three classes of velocity distributions (as specified below and shown in Fig. 3), and select our fundamental interval $(y_1, y_2)$ so that $w'(y) \geqq 0$ for $y_1 \leqq y \leqq y_2$. We shall define our characteristic length so that $y_2 - y_1 = 1$, and let $\phi_1(y;\, c,\, \alpha,\, \alpha R)$, $\phi_2(y;\, c,\, \alpha,\, \alpha R)$, $\phi_3(y;\, c,\, \alpha,\, \alpha R)$, $\phi_4(y;\, c, \alpha, \alpha R)$ be a fundamental system of solutions (4.1) arranged in the order discussed above.

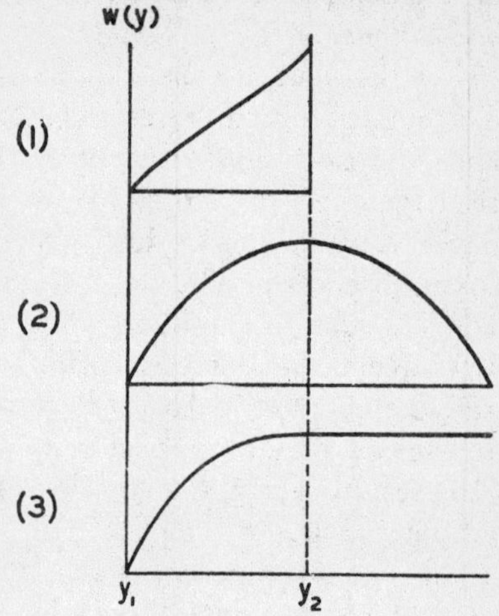


FIG. 3. The three types of velocity distributions.

CASE (1). *Flow between solid walls in relative motion.* In this case, the boundary conditions are given by

$$\phi(y_1) = \phi'(y_1) = \phi(y_2) = \phi'(y_2) = 0, \tag{6.1}$$

because the velocity of the disturbance should vanish on both the solid boundaries. The determinantal equation corresponding to these conditions is

$$F_1(\alpha, c, \alpha R) = \begin{vmatrix} \phi_{11} & \phi_{21} & \phi_{31} & \phi_{41} \\ \phi_{12} & \phi_{22} & \phi_{32} & \phi_{42} \\ \phi_{11}' & \phi_{21}' & \phi_{31}' & \phi_{41}' \\ \phi_{12}' & \phi_{22}' & \phi_{32}' & \phi_{42}' \end{vmatrix} = 0, \tag{6.2}$$

where $\phi_{11}$, $\phi_{11}'$, etc., stand for $\phi_1(y_1)$, $\phi_1'(y_1)$, etc. In this and all later discussions, a subscript 1 or 2 attached to a function of $y$ shall denote the value of that function at $y=y_1$ or $y=y_2$ respectively.

CASE (2). *Symmetrical flow between solid walls at rest.* In this case, it is easily seen from (4.1) that the disturbance can be separated into two independent parts, one symmetrical with respect to the line $y=y_2$ and the other antisymmetrical. (a) If $\phi(y)$ is a symmetrical function (antisymmetrical disturbance), then the conditions

$$\phi(y_1) = \phi'(y_1) = \phi'(y_2) = \phi'''(y_2) = 0 \tag{6.3}$$

hold, and we have the determinantal equation

$$F_2(\alpha, c, \alpha R) = \begin{vmatrix} \phi_{11} & \phi_{21} & \phi_{31} & \phi_{41} \\ \phi'_{11} & \phi'_{21} & \phi'_{31} & \phi'_{41} \\ \phi'_{12} & \phi'_{22} & \phi'_{32} & \phi'_{42} \\ \phi'''_{12} & \phi'''_{22} & \phi'''_{32} & \phi'''_{42} \end{vmatrix} = 0. \tag{6.4}$$

(b) If $\phi(y)$ is an odd function of $y-y_2$ (symmetrical disturbance), then the boundary conditions are

$$\phi(y_1) = \phi'(y_1) = \phi(y_2) = \phi''(y_2) = 0, \tag{6.5}$$

and we have the relation

$$F_3(\alpha, c, \alpha R) = \begin{vmatrix} \phi_{11} & \phi_{21} & \phi_{31} & \phi_{41} \\ \phi_{12} & \phi_{22} & \phi_{32} & \phi_{42} \\ \phi'_{11} & \phi'_{21} & \phi'_{31} & \phi'_{41} \\ \phi''_{12} & \phi''_{22} & \phi''_{32} & \phi''_{42} \end{vmatrix} = 0. \tag{6.6}$$

CASE (3). *Flow of the boundary-layer type.* In this case, the point $y_2$ is taken to correspond to the "edge" of the boundary layer, beyond which the velocity is substantially constant. The boundary conditions to be satisfied at $y_1$ are the usual ones;

$$\phi(y_1) = \phi'(y_1) = 0. \tag{6.7}$$

The boundary conditions for $y$ becoming infinite are to be replaced as follows.* Since the particular integral $\phi_4$ becomes infinite as $y$ becomes infinite, our boundary condition requires that $\phi$ is a linear combination of $\phi_1$, $\phi_2$, $\phi_3$ alone. Thus,

$$\phi \doteq C_1\phi_1 + C_2\phi_2 + C_3\phi_3, \tag{6.8}$$

where $C_1$, $C_2$, $C_3$ are constants of integration. Also, the integral $\phi_3$ makes practically no contribution for $y \geqq y_2$ so that we expect $\phi(y)$ to satisfy the inviscid equation for $y \geqq y_2$. Here $w''=0$, and hence two particular integrals are $e^{\pm\alpha y}$. The condition that $\phi \to 0$ as $y \to \infty$ excludes the integral $e^{\alpha y}$. Hence, $\phi$ must be proportional to $e^{-\alpha y}$ for $y > y_2$. This may be expressed as follows:

$$\phi' + \alpha\phi = 0 \quad \text{for} \quad y \geqq y_2. \tag{6.9}$$

Hence, we have the determinantal equation

* Cf. Tietjens [72] and Tollmien [73].

$$F_4(\alpha, c, \alpha R) = \begin{vmatrix} \phi_{11} & \phi_{21} & \phi_{31} \\ \phi_{12}' + \alpha\phi_{12} & \phi_{22}' + \phi\alpha_{22} & 0 \\ \phi_{11}' & \phi_{21}' & \phi_{31}' \end{vmatrix} = 0. \tag{6.10}$$

We note that the point $y_2$ can be replaced by any value of $y > y_2$. This is equivalent to the fact that the thickness of the boundary layer cannot be definitely defined. The larger this thickness is taken, the more accurate the results should be .*

The functions $F_1$, $F_2$, $F_3$, and $F_4$ are entire functions of the parameters $\alpha$, $c$, and $R$.

*Reduction of the equations for large values of $\alpha R$.* The equations (6.2), (6.4), and (6.6) can be substantially simplified for large values of $\alpha R$. Referring to (4.21), we see that $\phi_3(y) = A(y)e^{-Y}$, $\phi_4(y) = B(y)e^{Y}$, where $A(y)$ and $B(y)$ are of the order of unity, and $Y$ is defined by the relation

$$Y = \int_{y_1}^{y} \sqrt{i\alpha R(w - c)}\, dy.$$

Hence, we have the following relations, giving the order of magnitude of certain quantities:

$$\left.\begin{aligned}
&\frac{\phi_{31}'}{\phi_{31}} = -\sqrt{i\alpha R(w_1 - c)} + \frac{A_1'}{A_1}, \qquad \frac{\phi_{32}}{\phi_{31}} = \frac{A_2}{A_1} e^{-P}, \\
&\frac{\phi_{32}'}{\phi_{31}} = \left\{-\sqrt{i\alpha R(w_2 - c)}\, \frac{A_2}{A_1} + \frac{A_2'}{A_1}\right\} e^{-P}, \\
&\frac{\phi_{32}''}{\phi_{31}} = \left\{i\alpha R(w_2 - c)\, \frac{A_2}{A_1} + O(\sqrt{\alpha R})\right\} e^{-P}, \\
&\frac{\phi_{32}'''}{\phi_{31}} = \left\{-\,[i\alpha R(w_2 - c)]^{3/2} \frac{A_2}{A_1} + O(\alpha R)\right\} e^{-P};
\end{aligned}\right\} \tag{6.11}$$

$$\left.\begin{aligned}
&\frac{\phi_{41}'}{\phi_{41}} = \sqrt{i\alpha R(w_1 - c)} + \frac{B_1'}{B_1}, \qquad \frac{\phi_{42}}{\phi_{41}} = \frac{B_2}{B_1} e^{P}, \\
&\frac{\phi_{42}'}{\phi_{41}} = \left\{\sqrt{i\alpha R(w_2 - c)}\, \frac{B_2}{B_1} + \frac{B_2'}{B_1}\right\} e^{P}, \\
&\frac{\phi_{42}''}{\phi_{41}} = \left\{i\alpha R(w_2 - c)\, \frac{B_2}{B_1} + O(\sqrt{\alpha R})\right\} e^{P}, \\
&\frac{\phi_{42}'''}{\phi_{41}} = \left\{[i\alpha R(w_2 - c)]^{3/2} \frac{B_2}{B_1} + O(\alpha R)\right\} e^{P},
\end{aligned}\right\} \tag{6.12}$$

where

$$P = \int_{y_1}^{y_2} \sqrt{i\alpha R(w - c)}\, dy.$$

It then appears that the sign of the real part of $P$ is of consequence. It can be veri-

* In later calculation of the Blasius case, we shall take a boundary layer about 1.19 times as thick as that used by Tollmien.

fied that it is always positive when $c_i>0$. For then the path of integration can be taken along the real axis of $y$, and we have $-\pi<\arg(w-c)<0$; consequently, $-\pi/4<\arg(P)<\pi/4$. With reference to (4.21), (6.11) and (6.12), we see that the condition $P=n\pi i$ ($n$ = an integer), expresses the fact that $\phi_{31}'\phi_{42}'=\phi_{41}'\phi_{32}'$ when terms of the order $(\alpha R)^{1/2}$ are neglected. This is the corrected form of the first solution of Heisenberg as expressed by the condition (27) on p. 596 of his paper. Heisenberg also stated that such a condition can only be satisfied for damped solutions. In fact, from the condition just obtained for $c_i>0$, we see that $\mathrm{Re}(P)$ can be negative only for highly damped solutions, for which the whole discussion must be modified. (Cf. footnote on p. 29, §5.)

Neglecting quantities of the order $e^{-P}$ and $(\alpha R)^{-1}$ against quantities of the order of unity, we have the following simplifications of Eqs. (6.2), (6.4), and (6.6) for Cases (1) and (2).

CASE (1). *Flow between solid walls in relative motion.* We have

$$\frac{f_1(\alpha, c)}{f_3(\alpha, c)}=\frac{\phi_{31}}{\phi_{31}'}+\frac{\phi_{42}}{\phi_{42}'}\frac{f_2(\alpha, c)}{f_4(\alpha, c)}\,. \tag{6.13}$$

CASE (2a). *Antisymmetrical disturbance in a symmetrical flow between solid walls.* We have

$$f_2(\alpha, c)/f_4(\alpha, c) = \phi_{31}/\phi_{31}'. \tag{6.14}$$

CASE (2b). *Symmetrical disturbance in a symmetrical flow between solid walls.* We have

$$f_1(\alpha, c)/f_3(\alpha, c) = \phi_{31}/\phi_{31}'. \tag{6.15}$$

In these equations, the functions $f_1(\alpha, c)$, $f_2(\alpha, c)$, $f_3(\alpha, c)$ and $f_4(\alpha, c)$ are defined as follows:

$$\begin{aligned} f_1(\alpha, c) &= \begin{vmatrix} \phi_{11} & \phi_{12} \\ \phi_{21} & \phi_{22} \end{vmatrix}, & f_2(\alpha, c) &= \begin{vmatrix} \phi_{11} & \phi_{12}' \\ \phi_{21} & \phi_{22}' \end{vmatrix}, \\ f_3(\alpha, c) &= \begin{vmatrix} \phi_{11}' & \phi_{12} \\ \phi_{21}' & \phi_{22} \end{vmatrix}, & f_4(\alpha, c) &= \begin{vmatrix} \phi_{11}' & \phi_{12}' \\ \phi_{21}' & \phi_{22}' \end{vmatrix}. \end{aligned} \tag{6.16}$$

These functions depend only on $\alpha$ and $c$, because we may take the inviscid solutions for $\phi_1$ and $\phi_2$, which are accurate up to the order of $(\alpha R)^{-1}$. It may be reiterated that in computing $\phi_{12}$, $\phi_{22}$, $\phi_{12}'$, $\phi_{22}'$ in (6.16), we must take the path leading from $y_1$ to $y_2$ in the lower half plane.

CASE (3). *Flow in the boundary layer along a flat plate.* In this case we can reduce (6.10) to

$$\frac{f_2+\alpha f_1}{f_4+\alpha f_3}=\frac{\phi_{31}}{\phi_{31}'}, \tag{6.17}$$

if we also replace $\phi_1$ and $\phi_2$ by their inviscid approximations. The equations (6.13), (6.14), (6.15), and (6.17) are the final equations based on which the stability investigations are to be made

*The inviscid case.* In the limit $\alpha R\to\infty$. The above equations reduce to

$$f_1(\alpha, c) = 0 \quad \text{for Case (1) and Case } (2b), \tag{6.18}$$

$$f_2(\alpha, c) = 0 \quad \text{for Case (2a)}, \tag{6.19}$$

$$f_2 + \alpha f_1 = 0 \quad \text{for Case (3)}. \tag{6.20}$$

Mathematically, these are equivalent to the solution of the inviscid equation

$$(w - c)(\phi'' - \alpha^2\phi) - w''\phi = 0 \tag{6.21}$$

subjected to one of the following three sets of boundary conditions

$$\phi(y_1) = \phi(y_2) = 0, \quad \phi(y_1) = \phi'(y_2) = 0, \quad \phi(y_1) = \phi'(y_2) + \alpha\phi(y_2) = 0. \tag{6.22}$$

We have thus arrived at the conclusion that *some asymptotic behavior of the stability conditions can be obtained by neglecting the effect of viscosity* (provided proper care is given to the inner friction layer). This was tacitly assumed in the work of Rayleigh and others, while Heisenberg pointed out [loc. cit., p. 583] that a proof was necessary in accordance with some remarks of Oseen (38); he also virtually gave the proof.

In the next part of the paper, we shall therefore consider the simpler problem of an inviscid fluid. After a thorough investigation of that problem, we shall investigate the effect of viscosity by considering Eqs. (6.13), (6.14), (6.15), and (6.17) in greater detail. These results may be compared with the earlier ones of Heisenberg and Tollmien.

Numerical calculations of the stability limit based upon these equations will also be carried out for certain important special cases. For all these purposes, the evaluation of the six functions $f_1(\alpha, c), f_2(\alpha, c), f_3(\alpha, c), f_4(\alpha, c), \phi_{31}/\phi'_{31}, \phi_{42}/\phi'_{42}$ is necessary. We shall discuss this briefly here.

1) *Evaluation of* $f_1(\alpha, c), f_2(\alpha, c), f_3(\alpha, c), f_4(\alpha, c)$. These quantities are related to the inviscid solutions given by (4.14) with the path of integration subjected to the condition (5.4). Hence, we have

$$\phi_{11} = -c, \quad \phi_{21} = 0, \quad \phi'_{11} = w'_1, \quad \phi'_{21} = -\frac{1}{c}, \tag{6.23}$$

$$\left.\begin{aligned} \phi_{12} &= (1 - c)\sum_{n=0}^{\infty}\alpha^{2n}H_{2n}(c), \qquad \phi_{22} = (1 - c)\sum_{n=0}^{\infty}\alpha^{2n}K_{2n+1}(c), \\ \phi'_{12} &= (1 - c)^{-1}\sum_{n=0}^{\infty}\alpha^{2n}H_{2n-1}(c) + (1 - c)^{-1}w'_2\phi_{12}, \\ \phi'_{22} &= (1 - c)^{-1}\sum_{n=0}^{\infty}\alpha^{2n}K_{2n}(c) + (1 - c)^{-1}w'_2\phi_{22}, \end{aligned}\right\} \tag{6.24}$$

where

$$\left.\begin{aligned} H_{2n}(c) &= h_{2n}(y_2), \qquad H_{2n-1}(c) = (1 - c)^{-2}h'_{2n}(y_2), \\ K_{2n+1}(c) &= k_{2n+1}(y_2), \qquad K_{2n}(c) = (1 - c)^2 k'_{2n+1}(y_2) \end{aligned}\right\} \tag{6.25}$$

are functions of $c$ alone. In the above evaluations, we have put $w_1 = 0$, in accordance with the actual conditions in all the cases considered. We have also chosen the characteristic velocity so that $w_2=1$. Referring to (6.23), we have

$$\left.\begin{aligned} f_1(\alpha, c) &= - c\phi_{22}, & f_2(\alpha, c) &= - c\phi_{22}', \\ f_3(\alpha, c) &= w_1'\phi_{22} + \frac{1}{c}\phi_{12}, & f_4(\alpha, c) &= w_1'\phi_{22}' + \frac{1}{c}\phi_{12}'. \end{aligned}\right\} \tag{6.26}$$

The actual evaluation therefore depends upon the calculation of the integrals (6.25).

2) *Evaluation of* $\phi_{31}/\phi_{31}'$, $\phi_{42}/\phi_{42}'$. These quantities are related to the highly oscillating integrals $\phi_3$ and $\phi_4$. For values of $\alpha R$ so large that both $(\alpha R)^{1/3}c$, $(\alpha R)^{1/3}(1-c) \gg 1$, the approximate values of $\phi_{31}/\phi_{31}'$ and $\phi_{42}/\phi_{42}'$ are given by using (4.21). Thus,

$$\frac{\phi_{31}}{\phi_{31}'} = -\frac{e^{\pi i/4}}{\sqrt{\alpha R c}}, \qquad \frac{\phi_{42}}{\phi_{42}'} = \frac{e^{-\pi i/4}}{\sqrt{\alpha R(1-c)}}. \tag{6.27}$$

The condition $(\alpha R)^{1/3}(1-c)\gg 1$ is generally satisfied, because $c$ is usually small and $\alpha R$ is usually very large. This is especially true for Eq. (6.13), for it will be seen from considerations of an inviscid fluid that this case is relatively stable. For the condition $(\alpha R)^{1/3}c\gg 1$, the situation is different, because $c$ is usually small. It is then more convenient to approximate $\phi_3$ by $\chi_3^{(0)}(\eta)$ given by (4.9) (or with higher approximations, if so desired). We then have

$$\phi_{31}/\phi_{31}' = (y_1 - y_0)F(z), \tag{6.28}$$

where

$$F(z) = \frac{\int_{+\infty}^{-z} d\zeta \int_{+\infty}^{\zeta} \zeta^{1/2} d\zeta H_{1/3}^{(1)}\left[\tfrac{2}{3}(i\zeta)^{3/2}\right]}{-z\int_{+\infty}^{-z} d\zeta \zeta^{1/2} H_{1/3}^{(1)}\left[\tfrac{2}{3}(i\zeta)^{3/2}\right]}, \qquad z = -\alpha_0(\eta_1 - \eta_0). \tag{6.29}$$

This function has been calculated numerically by Tietjens for real values of $z$. We have made slightly more extensive and more accurate calculations. The present result differs slightly from that of Tietjens and is here tabulated in Table I and plotted in Fig. 4 together with the related function $\mathcal{J}(z)$ defined by

$$\mathcal{J}(z) = [1 - F(z)]^{-1}. \tag{6.30}$$

The asymptotic form of $F(z)$ is

$$F(z) \sim z^{-3/2}e^{\pi i/4}, \qquad (z \gg 1). \tag{6.31}$$

This agrees with (6.27) if $\alpha_0^3(y_1-y_0) = w_0'(y_1-y_0)$ is replaced by $-c$.

Acknowledgement. The author wishes to express his sincere gratitude to Professor Theodore von Kármán for suggesting this problem and for his invaluable help throughout this work; to Professor Clark B. Millikan for helpful suggestions and for the use of some of his unpublished notes, from which much inspiration and assistance have been derived; to Professors P. S. Epstein, H. Bateman, Dr. H. W. Liepmann and others for many valuable discussions and suggestions. The author is indebted to Mr. L. Lees for helpful criticism of the final manuscript.

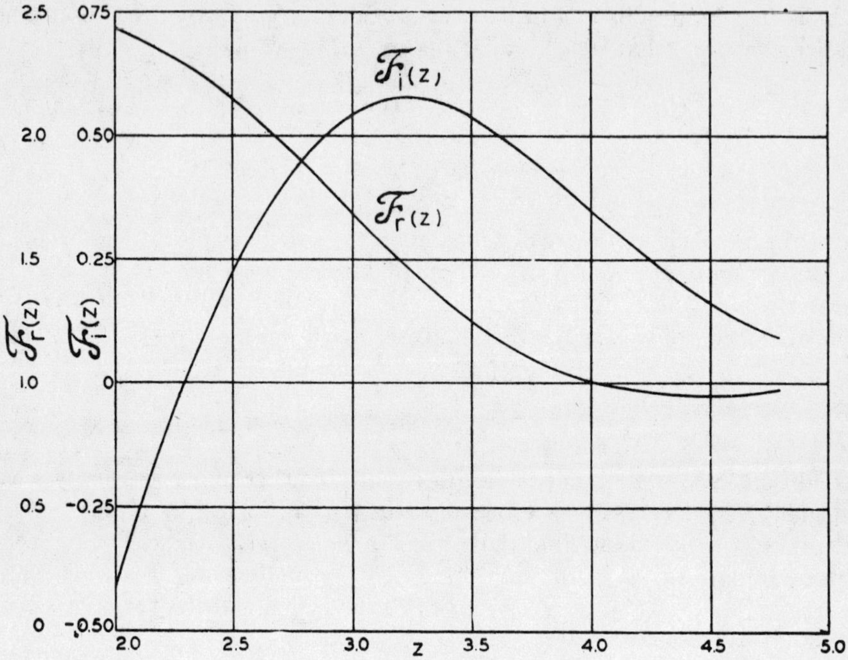


FIG. 4. The function $\mathcal{F}(z)$ shown in its real and imaginary parts (cf. Table I).

TABLE I.—The functions $F(z)$ and $\mathcal{F}(z)$.

| $z$ | $F_r$ | $F_i$ | $\mathcal{F}_r$ | $\mathcal{F}_i$ |
|---|---|---|---|---|
| 1.0 | 0.89161 | −0.35025 | 0.80630 | −2.60557 |
| 1.2 | 0.78969 | −0.27310 | 1.77012 | −2.29854 |
| 1.4 | 0.71970 | −0.21213 | 2.26836 | −1.71669 |
| 1.6 | 0.66931 | −0.16009 | 2.44985 | −1.18600 |
| 1.8 | 0.63143 | −0.11274 | 2.48104 | −0.75892 |
| 2.0 | 0.60144 | −0.06741 | 2.43927 | −0.41253 |
| 2.2 | 0.57599 | −0.02226 | 2.35196 | −0.12348 |
| 2.4 | 0.55230 | −0.02395 | 2.22724 | 0.11916 |
| 2.6 | 0.52773 | −0.07203 | 2.06929 | 0.31558 |
| 2.8 | 0.49952 | 0.12220 | 1.88566 | 0.46043 |
| 3.0 | 0.46456 | 0.17391 | 1.68938 | 0.54872 |
| 3.2 | 0.41947 | 0.22520 | 1.49726 | 0.58082 |
| 3.4 | 0.36110 | 0.27193 | 1.32516 | 0.56401 |
| 3.6 | 0.28802 | 0.30705 | 1.18429 | 0.51074 |
| 3.8 | 0.20352 | 0.32130 | 1.07982 | 0.43560 |
| 4.0 | 0.11800 | 0.30721 | 1.01118 | 0.35220 |
| 4.2 | 0.04698 | 0.26559 | 0.97361 | 0.27133 |
| 4.4 | 0.00240 | 0.20811 | 0.96056 | 0.20038 |
| 4.6 | 0.02160 | 0.14475 | 0.95989 | 0.13601 |
| 4.8 | 0.01477 | 0.09875 | 0.97659 | 0.09503 |

(*To be continued*)

## References

1. A. B. Bassett, Proc. Roy. Soc. London (A) **52,** 273–276 (1893).
2. H. Bateman, Bull. Nat. Research Council, Washington, no. **84,** 372–384 (1932).
3. O. Blumenthal, Sitz. Akad. Wiss., München, 563–595 (1913).
4. H. Faxen, Arkiv Mat. Ast. Fysik **21A,** no. 26, 11 p. (1928). (Curved.)
5. K. O. Friedrichs, *Fluid dynamics* (mimeographed lecture notes, Brown University, 1942), ch. 4, pp. 200–209
6. S. Goldstein, Proc. Camb. Phil. Soc. **32,** 40–54 (1936).
7. S. Goldstein, *Modern developments in fluid dynamics*, The Clarendon Press, Oxford, 1938, pp. 135–136, 157, 194–200, 564.
8. H. Görtler, Zeit. angew. Math. Mech. **20,** 138–147 (1940). (Curved.)
9. H. Görtler, Nachr. Ges. Wiss. Göttingen **2,** 1–26 (1940). (Curved.)
10. G. Hamel, Nachr. Ges. Wiss. Göttingen, 261–270 (1911).
11. W. H. Harrison, Proc. Camb. Phil. Soc. **20,** 455–459 (1920–1921).
12. O. Haupt, Sitz. Akad. Wiss., München, 289 (1912).
13. T. H. Havelock, Proc. Roy. Soc. London (A) **98,** 428–437 (1921).
14. W. Heisenberg, Ann. d. Phys. **74,** 577–627 (1924).
15. S. H. Hollingdale, Phil. Mag. (7) **29,** 209–257 (1940).
16. L. Hopf, Ann. d. Phys. **4,** 1–60 (1914).
17. H. Jeffreys, Proc. London Math. Soc. (2) **23,** 428–436 (1924).
18. Th. von Kármán, *Proc. 1st Int. Congress Appl. Mech., Delft*, 1924, p. 97; Abh. Aerodynam. Inst. Tech. Hochschule, Aachen, **4,** (1925).
19. Th. von Kármán, *Proc. 4th Int. Congress Appl. Mech., Cambridge, England*, 1934, Cambridge University Press, Cambridge, 1935, pp. 54–60, 77.
20. Lord Kelvin, *Mathematical and physical papers*, Cambridge University Press, vol. 4, pp. 186–187 (1880).
21. Lord Kelvin, *Mathematical and physical papers*, Cambridge University Press, vol. 4, pp. 321–330, 330–337 (1887).
22. L. V. King, Phil. Mag. (6) **31,** 332–338 (1916).
23. H. A. Kramers, Zeit. f. Phys. **39,** 828–836 (1926).
24. R. E. Langer, Trans. Amer. Math. Soc. **33,** 23–64 (1931); **34,** 447–480 (1932).
25. R. E. Langer, Bull. Amer. Math. Soc. **46,** 257–263 (1940).
26. L. Lichtenstein, *Hydromechanik*, Julius Springer, Berlin, 1929, pp. 409–414.
27. C. C. Lin, Proc. Nat. Acad. Sci. U.S.A. **30,** 316–324 (1944).
28. J. W. Lewis, Proc. Roy. Soc. London (A) **117,** 388–407 (1928). (Curved.)
29. L. A. Lorentz, Abh. theoret. Phys. (Leipzig) **1,** 43–71 (1907)
30. W. T. MacCreadie, Proc. Nat. Acad. Sci. U.S.A. **17,** 381–388 (1931).
31. R. von Mises, Deut. Math. Ver. **21,** 241–248 (1912).
32. R. von Mises, *Festschrift Heinrich Weber*, Teubner, Leipzig and Berlin, 1912, pp. 252–282.
33. F. Noether, Sitz. Akad. Wiss., München, 309–329 (1913).
34. F. Noether, Deut. Math. Ver. **23,** 138–144 (1914).
35. F. Noether, Zeit. angew. Math. Mech. **1,** 125–138, 218–219 (1921).
36. F. Noether, Zeit. angew. Math. Mech. **6,** 232–243, 339–340, 428, 497–498 (1926).
37. W. M. F. Orr, Proc. Roy. Irish Acad. **27,** 9–26, 69–138 (1906–1907).
38. C. W. Oseen, Arkiv. Mat. Ast. Fysik **7,** no. 15, 20 p. (1911).
39. C. L. Pekeris, Proc. Cambridge Phil. Soc. **32,** 55–66 (1936).
40. C. L. Pekeris, Jour. Aero. Sci. **5,** 237–240 (1938).
41. L. Prandtl, Zeit. angew. Math. Mech. **1,** 431–436 (1921); **11,** 407 (1931); Phys. Zeit. **23,** 19–23 (1922); *Vorträge aus dem Gebeite der Aerodynamik und Verwandter Gebeite, Aachen*, 1929, Julius Springer, Berlin, 1930, pp. 1–7.
42. L. Prandtl, *Aerodynamic theory*, edited by W. F. Durand, Julius Springer, Berlin, 1935, vol. 3, pp. 178–190.
43. Lord Rayleigh, *Scientific papers*, Cambridge University Press, Cambridge, vol. 1, pp. 474–487 (1880); vol. 3, pp. 17–23 (1887); vol. 3, pp. 575–584 (1892); vol. 4, pp. 203–209 (1895); vol. 6, p. 917 (1913).
44. Lord Rayleigh, *Scientific papers*, Cambridge University Press, Cambridge, vol. 6, p. 266 (1914); vol. 6, p. 341 (1915).

45. Lord Rayleigh, *Scientific papers*, Cambridge University Press, Cambridge, vol. 6, p. 447 (1916). (Curved.)

46. O. Reynolds, *Scientific papers*, Cambridge University Press, Cambridge, vol. 2, pp. 535–577 (1895).

47. P. Savic, Phil. Mag. (7) **32,** 245–252 (1941).

48. P. Savic and J. W. Murphy, Phil. Mag. (7) **34,** 139–144 (1943).

49. H. Schlichting, Ann. d. Phys. (5) **14,** 905–936 (1932).

50. H. Schlichting, Nachr. Ges. Wiss. Göttingen, 160–198 (1932). (Rotating cylinders.)

51. H. Schlichting, Zeit. angew. Math. Mech. **13,** 171–174 (1933).

52. H. Schlichting, Nachr. Ges. Wiss. Göttingen, 181–208 (1933).

53. H. Schlichting, Naturwissenschaften **22,** 376–381 (1934).

54. H. Schlichting, Nachr. Ges. Wiss. Göttingen **1,** 47–78 (1935).

55. Th. Sexl, Ann. Phys. **83,** 835–848 (1927); **84,** 807–822 (1927). (Circular tube.)

56. F. R. Sharpe, Trans. Amer. Math. Soc. **6,** 496–503 (1905).

57. H. Solberg, *Proc. 1st Int. Congress Appl. Mech., Delft*, 1924, pp. 387–394.

58. A. Sommerfeld, *Proc. 4th Int. Congress Math., Rome*, 1908, pp. 116–124.

59. R. V. Southwell and L. Chitty, Trans. Roy. Soc. London (A) **229,** 205–253 (1930); Aeronautical Research Committee R. and M. no. **1200,** (1930).

60. H. B. Squire, Proc. Roy. Soc. London (A) **142,** 621–628 (1933).

61. J. L. Synge, Trans. Roy. Soc. Canada **27,** 1–18 (1933).

62. J. L. Synge, J. Math. Physics **15,** 205–210 (1936).

63. J. L. Synge, *Semicentennial publications of the Amer. Math. Soc.*, 1938, vol. 2, pp. 227–269.

64. J. L. Synge, *Proc. 5th Int. Congress Appl. Mech., Cambridge, Mass., U.S.A.*, 1938, Wiley and Sons, New York, 1939, pp. 326–332.

65. J. L. Synge, Proc. Roy. Soc. London (A) **167,** 250–256 (1938). (Curved.)

66. K. Tamaki and W. J. Harrison, Trans. Cambridge Phil. Soc. **22,** 425–438 (1920). (Curved.)

67. G. I. Taylor, Phil. Trans. Roy. Soc. London (A) **223,** 289–343 (1923); also, *Proc. 1st Int. Congress Appl. Mech. Delft*, 1924, p. 89. (Curved.)

68. G. I. Taylor, Proc. Roy. Soc. London (A) **156,** 307–317 (1936).

69. G. I. Taylor, Proc. Roy. Soc. London (A) **157,** 546–564 (1936).

70. G. I. Taylor, *Proc. 5th Int. Congress Appl. Mech., Cambridge, Mass., U.S.A.*, 1938, Wiley and Sons, New York, 1939, pp. 304–310.

71. T. Y. Thomas, Amer. J. Math. **64,** 754–767 (1942); Proc. Nat. Acad. Sci. U.S.A. **29,** 243–246 (1943).

72. O. Teitjens, Zeit. angew. Math. Mech. **5,** 200–217 (1925).

73. W. Tollmien, Nachr. Ges. Wiss. Göttingen, 21–44 (1929); *Proc. 3rd Int. Congress Appl. Mech., Stockholm*, 1930, vol. 1, pp. 105–108; *Vorträge aus dem Gebeite der Aerodynamik und verwandte Gebeite, Aachen*, 1929, Julius Springer, Berlin, 1930, pp. 18–21.

74. W. Tollmien, Nachr. Ges. Wiss. Göttingen (Neue Folge) **1,** 70–114 (1935).

75. W. Tollmien, Zeit. angew. Math. Mech. **15,** 96–100 (1935).

76. G. N. Watson, *Theory of Bessel functions*, Cambridge University Press, Cambridge, ed. 1, 1922, or ed. 2, 1944, p. 198; for Stokes' phenomenon, see p. 201.

# ON THE STABILITY OF TWO-DIMENSIONAL PARALLEL FLOWS

## PART II.—STABILITY IN AN INVISCID FLUID*

BY

C. C. LIN

*Guggenheim Laboratory, California Institute of Technology*

**7. General considerations. The criteria of Rayleigh and Tollmien.** At the end of Part I, we have shown that the study of the stability problem in an inviscid fluid gives valuable information provided it is kept in mind that we are actually dealing with the limiting case where the Reynolds number becomes indefinitely large. The study of the stability of two-dimensional parallel flows in an inviscid fluid is usually regarded as being quite complete, through the work of Rayleigh and Tollmien. Their results show that instability depends very much upon the occurrence of a point of inflection in the velocity profile. However, it seems that physical interpretations of such general results are not well known. Such an interpretation will be given in §§9, 10 of this part. There are also several points in the mathematical theory which require further development and clarification. These will be brought out for further consideration in §§7, 8.

We now proceed to make a critical survey of some aspects of the stability problem in an inviscid fluid. First, let us summarize the conclusions obtained by Rayleigh and Tollmien. These can be conveniently described as the necessary and the sufficient conditions for the existence of a disturbance, self-excited, neutral, or damped.

1) *Necessary conditions for the existence of a disturbance.*

*a*) If the flow possesses a self-excited or neutral mode of disturbance with finite wave length, the velocity profile has a flex at some point $y=y_s$, where $y_1<y_s<y_2$. Furthermore, in the case of a *neutral* disturbance, the phase velocity must be $c=w(y_s)$.

*b*) If the flow possesses a *damped* mode of disturbance, no immediate conclusion can be drawn.

2) *Sufficient conditions for the existence of a disturbance.* So far, the sufficient conditions are known only for symmetrical and for boundary-layer velocity distributions. The results may be stated as follows.

*a*) There is always the neutral disturbance given by $c=0$, $\alpha=0$, $\phi(y)=w(y)$.

*b*) If $w''(y_s)=0$, for $y_1<y_s<y_2$, there is a neutral disturbance with $c=w(y_s)$; furthermore, if $w'''(y_s)\neq 0$, self-excited disturbances also exist.

*Discussion.* The condition $w'''(y_s)\neq 0$ involved in (2) (b) will be shown to be actually unnecessary, by an improved method of proof to be discussed in the next section. The statement in (1) (*b*) regarding damped disturbances differs from the original conclusion of Rayleigh and Tollmien. Indeed, in the work of Lord Rayleigh, the solution is taken to be valid all along the real axis. Hence, in accordance with the discussion of §5, Part I, such considerations do not include damped disturbances. However, Rayleigh and Tollmien did not distinguish between an amplified disturbance and a damped disturbance, because they regarded them as complex conjugates. As pointed out in §5, this is not permissible. In fact, if we accept the original conclusions

* Received May 18, 1945. Part I of this paper appeared in this QUARTERLY, 3, 117–142 (1945).

of Rayleigh and Tollmien, a profile without a flex could not execute any kind of disturbance. This can hardly be reconciled with our intuition regarding the state of affairs in a real fluid at infinitely large Reynolds numbers.[1] According to the present interpretation, only damped solutions can exist. Such a conclusion is also borne out by the investigations for a viscous fluid.[2] It is to be noted that the neutral and the self-excited disturbances, existing under the condition $w''(y_s)=0$, are free from the effect of viscosity inside the fluid, because the neutral solution is also regular at $y=y_s$ where $w=c$. Hence, we may conclude that *disturbances essentially free from the effect of viscosity inside a fluid can exist only for velocity distributions with a flex.*

The results of Rayleigh and of Tollmien discussed above tend to give the impression that the occurrence of a flex in the profile is the decisive factor in the determination of instability not only in the case of an inviscid fluid, but also in the case of a viscous fluid.[3] However, the investigation in Part III will show that this is by no means the case. When instability first occurs, as one increases the Reynolds number, viscous forces still play a dominant role, and the main characteristics of the behavior of the fluid with respect to a disturbance do not depend upon the occurrence of a flex in the velocity curve. Indeed, it is physically improbable that a slight change of the pressure gradient in the case of a boundary layer—which may cause a change from a velocity curve without a flex to one with a flex—should cause a radical change in the essential characteristics of stability. As we shall see later, the instability of a boundary layer depends more on the outside free stream than on the occurrence of a point of inflection. It might be argued that the free stream is analogous to a point of inflection in that a vanishing curvature is involved; but even if this is admitted, we must still note that the essential features in this case are not obtained from an investigation neglecting the effect of viscosity. Indeed, from inviscid investigations, it is concluded that a boundary layer with zero or favorable pressure gradient is stable, except for the very trivial type of disturbance with infinite wave-length and zero phase velocity. The present investigation shows that *all* boundary-layer profiles can be unstable, and exhibits results in agreement with the physical suggestion just discussed.

It thus seems that *any conclusion obtained from inviscid investigations must not be taken over directly to the case of the real fluid, where the stability phenomenon is largely controlled by the effect of viscosity and not decided primarily by the occurrence of a flex in the velocity curve.*

Indeed, even when we are mainly interested in the behavior in the limiting case of infinite Reynolds numbers, the existence of a flex is not as significant as it may appear to be at first sight. The existence of neutral or amplified disturbances has so far been proved only for *symmetrical* and *boundary-layer* types of velocity profiles. This may not be true for other types of velocity profiles, e.g., when the walls are in relative motion. The following example will bring out this point. Let us consider the velocity distribution $w(y)=A+B\sin y$, $y_1<y<y_2$, which has a flex at $y=0$ if $y_1<0<y_2$. According to the above necessary conditions, the only possible neutral

---

[1] This is the objection of Friedrichs, loc. cit. (Ref. [5]) p. 209. (The references are listed at the end of Part I.) It must also be noted that the non-linear terms are not negligible in the case of an *ideal* fluid. We shall consistently restrict the magnitude of our disturbances so that the effect of viscosity is always more important than the effect of non-linearity.

[2] See figures in Ref. [27].

[3] See Taylor's discussion on p. 308 of Ref. [70].

disturbance is the one with $c=A$. Then the equation of disturbance (6.21) reduces to $\phi''+(1-\alpha^2)\phi=0$. It has the solution

$$\phi(y) = C \sin \left\{\sqrt{1-\alpha^2}\,(y-y_1)\right\},$$

which vanishes at $y=y_1$. If $\phi(y_2)$ is also required to vanish, we must have

$$\sqrt{1-\alpha^2}\,(y_2-y_1) = n\pi, \qquad (n = \text{integer}),$$

and hence

$$\alpha^2 = 1 - [n\pi/(y_2-y_1)]^2.$$

Thus, if $y_2-y_1<\pi$, there is no possible neutral disturbance; if $y_2-y_1=\pi$, there is the one with $\alpha=0$; if $\pi<y_2-y_1<2\pi$, there is one with $\alpha\neq 0$; in general, if $m\pi<y_2-y_1<(m+1)\pi$, there are $m$ neutral disturbances with $\alpha\neq 0$. In the last case, there are also $m$ points of inflection in the velocity profile.

It can thus be seen that the general shape of the velocity profile plays a very important role even in the limit of infinite Reynolds numbers. Indeed, it will become clear from Part III that the eigen solution with eigen values $\alpha=c=0$ is not as trivial as it might appear at first sight, for it actually represents a limiting case with $R\to\infty$. This solution exists for symmetrical and boundary-layer profiles, but its existence is not immediately evident for other types of profiles.

In spite of all these points against the *decisive* nature of the flex, it must be admitted that its occurrence certainly makes the motion comparatively unstable. This can be expected from the original results of Rayleigh and Tollmien, and can be seen more clearly from the interpretation of the mechanism of inertia forces to be given in §§9, 10. However, these results must not be taken to indicate any decisive nature of a flex. The essential features of instability can only be obtained through consideration of the effect of viscosity.

We shall now conclude this section by making some critical discussions of Heisenberg's classification of velocity profiles and the use of broken linear profiles for the study of stability problems.

*Heisenberg's classification of velocity distributions.* Heisenberg attempted the case of flow between solid walls in relative motion with the condition that $\mathrm{Re}(w-c)$ vanishes only once for $y_1<y<y_2$ (loc. cit., p. 592). Regarding $\alpha^2$ as small, he approximated the condition (6.18) by $K_1(c)=0$ [cf. (6.26), (6.24)]. He then classified the profile into four classes: (*i*) those for which $K_1(c)=0$ has a complex root; (*ii*) those for which $K_1(c)=0$ has a real root; (*iii*) those for which the real part of $K_1(c)$ vanishes for a certain real value of $c$; (*iv*) those for which none of the above three cases is true. Heisenberg concluded that the first class is unstable, the second generally unstable, the rest stable.

In discussing the validity of these conclusions, the following point must be borne in mind. If we can show that a certain type of disturbance exists for $\alpha^2\doteq 0$ and $\alpha R\to\infty$, it may also be expected to exist for sufficiently large values of $\alpha R$ and sufficiently small values of $\alpha^2$. However, the non-existence of a certain type of disturbance for $\alpha^2=0$ and $\alpha R\to\infty$ does not exclude the possibility of its existence for finite values of $\alpha^2$ and $\alpha R$. It appears therefore that we can only expect to conclude the ***instability*** of a velocity distribution by discussing the roots of $K_1(c)=0$. Thus, apart from some flaws in Heisenberg's mathematical deductions, only the first two classes can have any decisive significance.

If $K_1(c)$ has a root with a *positive* imaginary part, the motion is unstable. If $K_1(c)$ has a *real* root, Heisenberg shows that the motion would be unstable when the effect of viscosity is considered. This will be studied more fully in a generalized form in §11. However, if $K_1(c)$ has a root with a *negative* imaginary part, we cannot conclude the instability of the flow by taking the complex conjugate of $K_1(c)=0$ (as Heinsenberg did). For if

$$\int_C dy(w-c)^{-2} = 0,$$

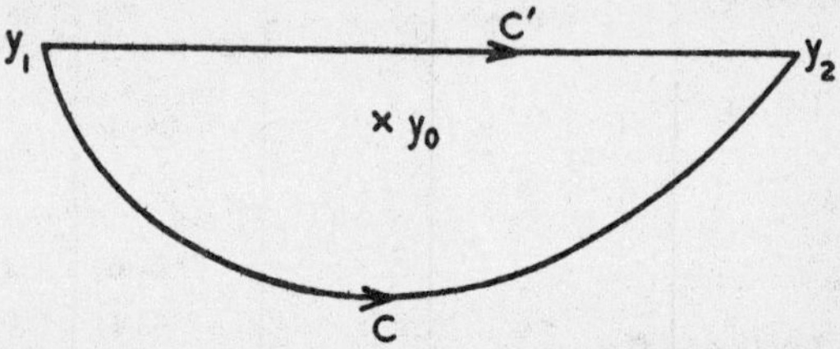

FIG. 5. Path around the critical point in the case $c_i<0$.

then (cf. Fig. 5)

$$\int_{C'} dy(w-c)^{-2} = -2\pi i R_0,$$

where $R_0$ is the residue of $(w-c)^{-2}$ at $y_0$. In fact $R_0 = -w_0''/w_0'^3$. Now $K_1(\bar{c})$ is the complex conjugate of $\int_{C'} dy(w-c)^{-2}$. Hence,

$$K_1(\bar{c}) = 2\pi i \bar{R}_0 = -2\pi i \bar{w}_0''/\bar{w}_0'^3,$$

which does not vanish unless $w_0''=0$. Hence, the equation $K_1(c)=0$ tells us nothing about the existence of the root $\bar{c}$ or any other root with a positive imaginary part.

Thus, Heisenberg's attempt appears to be not as successful as Tollmien's later work [75], which at least brings out the characteristic properties of symmetrical and boundary-layer distributions. A complete classification of velocity distributions, however, is not yet existent.

*Approximation using broken linear profiles.* Some investigations of Lord Rayleigh were carried out by approximating the velocity profile with straight-line segments. With this approximation, the solutions of (6.21) can be expressed in terms of elementary functions. Lord Rayleigh also tried to verify his conclusions by considering the roots of $K_1(c)=0$, using the same approximation for the velocity. However, the results of his investigations are doubtful, because the number of roots obtained for $K_1(c)$ is equal to the number of corners chosen in the approximation. This was demonstrated by Heisenberg to be inherent in the method of approximation. The general idea is as follows. As discussed above, the stability condition (6.18) may be approximated by $K_1(c)=0$ in certain cases. Although Rayleigh's approximation may be made very close so far as the velocity distribution is concerned, the approximation to $(w-c)^{-2}$ is always bad in the neighborhood of the corners. Consequently, the integral $K_1(c)$ is not properly approximated. In fact, a *continuous* broken profile $w(y)$ does not allow itself to be continued analytically to the complex $y$-plane without introducing *discontinuities* (cuts). It thus appears that all results deduced from the consideration of broken profiles must be regarded with reserve. The same criticism applies to Tietjen's work with the viscous fluid. His analysis failed to give a minimum Reynolds number below which all small disturbances are damped out.

**8. Rigorous proof and extension of Tollmien's result for the existence of unstable modes of oscillation.** In this section, we want to give a rigorous proof of the existence of amplified solutions of (6.21) satisfying the second and the third boundary conditions of (6.22) when the velocity profile $w(y)$ has a flex at $y=y_s$, i.e.,

$$w''(y_s) = 0. \tag{8.1}$$

The idea of the proof is essentially the same as that used by Tollmien, but the method is improved. It has the further advantage of enabling us to extend the results to cover cases where $w'''(y_s)=0$,—a condition which had to be excluded by Tollmien.

According to previous results, the neutral disturbance must have a phase velocity $c$ equal to

$$c_s = w_s = w(y_s) \tag{8.2}$$

Let the corresponding value of $\alpha$ be denoted by $\alpha_s$. The essential idea of the proof is (1) to show that there exist eigen-values of $c$ and $\alpha>0$ in the neighborhood of the values of $c_s$ and $\alpha_s$ such that the imaginary part of $c$ does not vanish, and then (2) to show that the imaginary part is actually positive. The first statement can be expected and can be readily established, if we can show that the left-hand sides of (6.18)–(6.20) are analytic functions $f(\alpha, c)$ of the two variables $\alpha$ and $c$ in the neighborhoods of $\alpha_s$ and $c_s$. For if this is true, we can always solve $f(\alpha, c)=0$ for $c$ as an analytic function of $\alpha$, (there may be more than one branch), by the implicit function theorem. Hence, there is at least one value of $c$ corresponding to every *real* value of $\alpha$ in the neighborhood of $\alpha=\alpha_s$. Furthermore, by (8.2), this value of $c$, being unequal to $c_s$, cannot be real, and the first part of our result is established.

To prove the analyticity of $f(\alpha, c)$ seems to be a trivial problem. Nevertheless, we shall find below that it is impossible to establish it in the neighborhood of $(\alpha, c)=(0, 0)$. The chief problem in the proof is to overcome the difficulty caused by the singular point of the differential equation (6.21).

If $w-c\neq 0$, we can write (6.21) in the form

$$\phi'' - \alpha^2\phi - \frac{w''}{w-c}\phi = 0. \tag{8.3}$$

Let us now consider a simply-connected region $R$ of the $y$-plane which encloses the

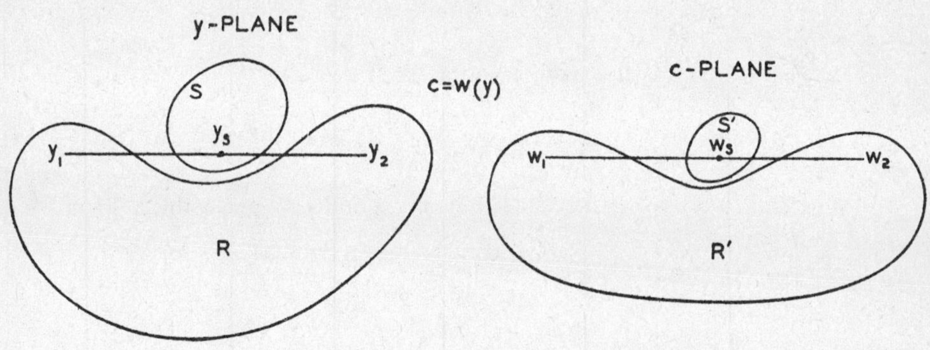


FIG. 6. The region of analyticity of the inviscid solutions.

points $y=y_1$ and $y=y_2$, but excludes the point $y_s$, the passage from $y_1$ to $y_2$ being taken in the *lower* half of the $y$-plane. We consider also a neighborhood $S$ of $y_s$, mutually exclusive with the region $R$. Let us regard the relation

$$c = w(y) \tag{8.4}$$

as mapping the regions $R$ and $S$ into two regions $R'$ and $S'$ of the $c$-plane (Fig. 6). If the mapping is one-to-one, (as can be expected if $w'(y)\neq 0$ for $y_1<y<y_2$), these

regions will also be mutually exclusive. Then, if we restrict $y$ to $R$ and $c$ to $S'$, the coefficients of (8.3) are analytic functions of the independent variable $y$ and the parameters $\alpha$ and $c$. Hence, a fundamental system of solutions of (8.3), which we denote by $\phi_1(y; \alpha, c)$ and $\phi_2(y; \alpha, c)$, are *analytic* functions of the *three* variables $y$, $\alpha$, and $c$. We understand that $y$ is restricted to the region $R$, $c$ is restricted to the region $S'$, while $\alpha$ may be in any finite region enclosing $\alpha_s$. Thus, (for example),

$$f_2(\alpha, c) \equiv \begin{vmatrix} \phi_1(y_1; \alpha, c) & \phi_2(y_1; \alpha, c) \\ \phi_1'(y_2; \alpha, c) & \phi_2'(y_2; \alpha, c) \end{vmatrix} \tag{8.5}$$

is an analytic function of the variables $\alpha$ and $c$, as we want to prove.

We note that in the neighborhood of $(\alpha, c) = (0, 0)$, the above reasoning fails. The region $R$ (which has to enclose the point $y = y_1$) and the region $S$ (which has to enclose the point where $w = c = 0$) cannot be taken to be mutually exclusive. In fact, $f(\alpha, c)$ presumably has a singular point at the point $\alpha = 0$ (a logarithmic branch point). We shall discuss this case a little more closely at the end of this section.

Let us proceed to show that there actually exist values of $c = c(\alpha^2)$ with a *positive* imaginary part corresponding to positive real values of $\alpha$. This is necessary because the usual argument of taking complex conjugates has been shown to be invalid. For this purpose, we consider the power series

$$c = c_s + \left(\frac{dc}{d\lambda}\right)_s (\lambda - \lambda_s) + \frac{1}{2!}\left(\frac{d^2c}{d\lambda^2}\right)_s (\lambda - \lambda_s)^2 + \cdots, \tag{8.6}$$

where $\lambda = \alpha^2$.[4] Since $\lambda$ is restricted to real values, the important point to be shown is that *the first of the derivatives in* (8.6) *for which the imaginary part does not vanish is of odd order*. Then, by taking values of $\lambda$ slightly greater or smaller than $\lambda_s$, we can always make $c_i > 0$. For these values of $c$ and $\alpha^2$, we can continue our solution $\phi(y)$ analytically so that it is given along the real axis between $y_1$ and $y_2$, thus obtaining an inviscid solution.

Let us now consider (8.3), writing $\lambda$ for $\alpha^2$. We have

$$L(\phi) \equiv \phi'' - \lambda\phi - \frac{w''}{w - c}\phi = 0. \tag{8.7}$$

Let $\phi$ be an eigen function with $\lambda$, $c$ as the corresponding eigen-values. Then

$$L(\phi_\lambda) \equiv \phi_\lambda'' - \alpha^2\phi_\lambda - \frac{w''}{w - c}\phi_\lambda = \left\{1 + \frac{w''}{(w - c)^2}\frac{dc}{d\lambda}\right\}\phi, \tag{8.8}$$

where

$$\phi_\lambda = \frac{\partial\phi}{\partial\lambda} + \frac{\partial\phi}{\partial c}\frac{dc}{d\lambda}.$$

We distinguish two cases: (1) the point $y = y_s$ is a simple root of $w''(y) = 0$; (2) the point $y = y_s$ is a multiple root of $w''(y) = 0$.

In the first case, $w'''(y_s) \neq 0$. In the limit $\lambda \to \lambda_s$, $c \to c_s$, Eqs. (8.7) and (8.8) become[5]

---

[4] Since $d\lambda/d\alpha \neq 0$ at $\alpha = \alpha_s$, the correspondence between $\alpha$ and $\lambda$ is one-to-one in the neighborhood of $\lambda_s = \alpha_s^2$.

[5] A subscript $s$ denotes that the parameters $\lambda$ and $c$ are put equal to $\lambda_s$ and $c_s$ respectively. A subscript $\lambda$ denotes differentiation with respect to $\lambda$.

$$L_s(\phi_s) \equiv \phi_s'' - \lambda_s\phi_s - \frac{w''}{w - c_s}\phi_s = 0, \tag{8.9}$$

$$L_s(\phi_{\lambda s}) \equiv \phi_{\lambda s}'' - \lambda_s\phi_{\lambda s} - \frac{w''}{w - c_s}\phi_{\lambda s} = \left\{1 + \frac{w''}{(w - c_s)^2}\left(\frac{dc}{d\lambda}\right)_s\right\}\phi_s. \tag{8.10}$$

From these, we deduce that

$$\phi_s L_s(\phi_{\lambda s}) - \phi_{\lambda s}L_s(\phi_s) = \frac{d}{dy}\left\{\phi_s\phi_{\lambda s}' - \phi_{\lambda s}\phi_s'\right\} = \left\{1 + \frac{w''}{(w - c_s)^2}\left(\frac{dc}{d\lambda}\right)_s\right\}\phi_s^2.$$

Now, $\phi_\lambda$ satisfies the same boundary conditions as $\phi$ does, because those conditions are satisfied by $\phi$ for each pair of values of $\lambda$ and $c$, and $\phi$ is an analytic function of them. Hence, integrating $\phi_s L_s(\phi_{\lambda s}) - \phi_{\lambda s}L_s(\phi_s)$ between the limits $(y_1, y_2)$, we have

$$\left(\frac{dc}{d\lambda}\right)_s \int_{y_1}^{y_2} \frac{w''}{(w - c_s)^2}\phi_s^2 dy + \int_{y_1}^{y_2} \phi_s^2 dy = 0,$$

or

$$\left(\frac{dc}{d\lambda}\right) = -\int_{y_1}^{y_2} \phi_s^2 dy \Big/ \int_{y_1}^{y_2} w''\phi_s^2(w - c_s)^{-2} dy. \tag{8.11}$$

The denominator of the above expression is equal to

$$\int_{y_1-y_s}^{y_2-y_s} (w_s'''y + \tfrac{1}{2}w^{iv}y^2 + \cdots)(w_s'y + \tfrac{1}{2}w_s'''y^2 + \cdots)^{-2}(\phi_{ss}^2 + 2\phi_{ss}\phi_{ss}'y + \cdots)dy$$

$$= \frac{w_s'''}{w_s'^2}\phi_{ss}^2 \int_{y_1-y_s}^{y_2-y_s} \left\{\frac{1}{y} + A_0 + A_1 y + \cdots\right\}dy,$$

where $\phi_{ss}$ is the value of $\phi_s$ at $y = y_s$, and $A_0, A_1, \cdots$ are real. Hence, the imaginary part of the above expression is $\pi\phi_{ss}w_s'''/w_s'^2$. Since $\phi_s(y)$ is real and $\phi_{ss}$ does not vanish,[6] we have arrived at the required result. The above argument is a rigorous formulation of Tollmien's work.

In case $w''(y)$ has a multiple root at $y = y_s$, the proof of Tollmien does not hold, but the above method can still be carried through. The restriction must be made, however, that the point $y_s$ is a point of inflection where $w''(y)$ actually changes its sign. Then, $y_s$ is a root of $w''(y)$ of odd multiplicity, and the first of the derivatives $w^{iv}(y_s), w^{v}(y_s), \cdots$ which does not vanish is of *odd* order. Such a point always exists when the curvature of the velocity curve has different signs at $y_1$ and $y_2$. If we differentiate (8.7) $n$ times with respect to $\lambda$, we have the following equation for each value of $n$:

$$L(\phi_{\lambda^{(n)}}) = n\phi_{\lambda^{(n-1)}} + \sum_{r=1}^{n} C_r^n \phi_{\lambda^{(n-r)}} \frac{d^r}{d\lambda^r}\left(\frac{w''}{w - c}\right). \tag{8.12}$$

Let $w''(y)$ have the root $y_s$ up to the multiplicity $2m+1$, $m > 0$. Then, (8.10) is regular in a neighborhood of $y = y_s$, and the value of $(dc/d\lambda)_s$ as given by (8.11) is *real*. Let us consider the boundary value problem of the differential equation (8.10), requiring $\phi_{\lambda s}$ to satisfy the same boundary conditions as $\phi_s$. The solution can be ob-

[6] Tollmien [74], p. 92.

tained from $\phi_\lambda(y)$ by making $\lambda \to \lambda_s$, and is moreover real along the real axis, by a direct consideration of (8.10).

Continuing the same argument with equations of the type (8.12) with $n=2, 3, \cdots, 2m$ and $c \to c_s$, $\lambda \to \lambda_s$, we find that

$$(\phi_{\lambda^{(2)}})_s,\ (\phi_{\lambda^{(3)}})_s,\ \cdots,\ (\phi_{\lambda^{(2m)}})_s;\ \left(\frac{d^2c}{d\lambda^2}\right)_s,\ \left(\frac{d^3c}{d\lambda^3}\right)_s,\ \cdots,\ \left(\frac{d^{2m}c}{d\lambda^{2m}}\right)_s$$

are all real. Finally, for $n=2m+1$, we obtain a relation of the type

$$(2m+1)!\left(\frac{dc}{d\lambda}\right)_s^{2m+1}\int_{y_1}^{y_2} w''\phi_s^2(w-c)^{-(2m+2)}dy + \left(\frac{d^{2m+1}c}{d\lambda^{2m+1}}\right)_s\int_{y_1}^{y_2}\phi_s^2 dy = \text{real}. \quad (8.13)$$

Just as in the case of the equation preceding (8.11), it can be easily seen that the above integral $\int_{y_1}^{y_2} w''\phi_s^2(w-c)^{-(2m+2)}dy$ has the imaginary part $w_s^{(2m+3)}\phi_{ss}^2/(2m+1)!(w_s')^{(2m+2)}$ while the other term on the left-hand side of (8.13) is real. Thus, $(d^{2m+1}c/d\lambda^{2m+1})_s$ has a non-vanishing imaginary part. This is the result desired.

This completes the proof of the existence of amplified solutions near the neutral solution $c=c_s$, $\alpha=\alpha_s$ when the velocity curve has a point of inflection.

The proof of the existence of amplified solutions near the neutral solution $c=0$, $\alpha=0$ cannot be so easily formulated into a rigorous form. From the solutions (4.14), it is very easy to obtain the solution $\phi_I$ which approaches the eigen solution $\phi=w(y)$ as $c\to 0$, $\alpha^2\to 0$, with $\alpha^2=O(c)$. The solution is

$$\phi_I = -cw_1'(w-c)\int_{y_1}^{y}(w-c)^{-2}dy \times\left\{1+\alpha^2\int_{y_1}^{y}dy(w-c)^2\int_{y_1}^{y}dy(w-c)^{-2}+\cdots\right\}. \quad (8.14)$$

As can be easily verified from (6.21) the condition that $\phi_I$ be an eigen function is

$$cw_1' + \alpha^2\int_{y_1}^{y_2}(w-c)\phi_I dy = 0. \quad (8.15)$$

From this, it follows that

$$\left(\frac{dc}{d\lambda}\right)_0 = \frac{1}{w_1'}\int_{y_1}^{y_2} w^2 dy, \quad (8.16)$$

and that the imaginary part of $(d^2c/d\lambda^2)_0$ is $2\pi(dc/d\lambda)_0 w_1'/w_1'^2$, which is positive if there is one flex in the velocity profile ($w_1'>0$). However, the real part of $(d^2c/d\lambda^2)_0$ becomes logarithmically infinite, and hence the argument is not rigorous. Also, it does not seem easy to make suitable modifications and extensions in case $w''(y_1)=0$. It should be remarked that Tollmien's proof is not essentially different from the argument just given.

Similar considerations can be applied to boundary-layer profiles, and similar results can be obtained confirming and extending Tollmien's original results.

**9. Physical interpretation of instability in an inviscid fluid.** The fact that the instability of a two-dimensional parallel laminar flow is so closely connected with the occurrence of a point of inflection in the velocity profile demands a physical interpretation. Since Eq. (6.21) is essentially the vorticity equation, we would expect $w''=0$ to indicate a maximum or minimum of the vorticity $-w'$ of the main flow. This is actually where the explanation is to be found.

Since we have neglected the effect of viscosity, a fluid element maintains its vorticity throughout the motion. From this point of view, a two-dimensional parallel flow may be regarded as the motion of a large number of vortex filaments under the action of each other. Filaments of equal vorticity are arranged in the same layer, and the whole flow is built up of a collection of such layers.

The following physical interpretation is based upon the fact that a fluid element is accelerated in such a field if it is associated with an excess or a defect of vorticity. These considerations were originally developed by von Kármán[7] for the interpretation of the failure of the simple vorticity-transfer theory of fully developed turbulence as applied to the case of parallel Couette flow. The idea is developed in greater mathematical detail here in this section and the next. It will be noticed that the consideration is essentially two-dimensional, and hence is even more suitable here than for fully developed turbulence, where the fluctuations are essentially three-dimensional. An alternative interpretation of the results of Rayleigh and Tollmien, but still based upon vorticity considerations, will also be given to demonstrate the role of the viscous forces.

Let us imagine a disturbance of the flow such that an element $E_1$ of fluid of the layer $L_1$ is interchanged with an element $E_2$ of a neighboring layer $L_2$. For definiteness, let us suppose that the layer $L_2$ has a higher vorticity than the layer $L_1$ in the undisturbed state. Since $E_1$ preserves its vorticity, it will appear to have a defect of vorticity when it is in $L_2$. Similarly, $E_2$ appears to have an excess of vorticity.

Let us fix our attention on one element, say $E_2$. It will be shown in §10 that a fluid element with an excess of vorticity is accelerated in the direction of the positive $y$-axis with an acceleration $\Gamma^{-1}\iint\{v'(x, y)\}^2\zeta_0'\,dxdy$, where $\zeta_0'(y)$ is the gradient of vorticity of the main flow, $v'(x, y)$ is the component of the disturbing velocity perpendicular to the direction of flow, and $\Gamma$ is the total strength of the vortex filaments corresponding to the disturbance. Examining the signs of the various quantities in the acceleration formula, we can easily see that $E_2$ is accelerated toward a region of higher vorticity if the gradient of vorticity does not change sign anywhere in the fluid. Thus, $E_2$ is accelerated toward $L_2$. A similar consideration holds for the element $E_1$. Hence, in either case, *the fluid element is returned to the layer where it belonged* (by the acceleration due to its interaction with other vortex filaments). *The motion is therefore stable when the gradient of the vorticity does not vanish.*

When there is an extremum of vorticity, an interchange of fluid elements on opposite sides of the extremum does not give rise to an excess or a defect of vorticity. Furthermore, the gradient of vorticity vanishes there, and has opposite signs on opposite sides of that layer. It can easily be seen from the above acceleration formula that the restoring tendency mentioned above is largely impaired in such a case. Thus, exchanged fluid elements are not as strongly forced back by the action discussed above. Such an exchange constitutes a disturbance because there is an exchange of momentum. Thus, a disturbance may tend to persist and perhaps to augment. The motion is not necessarily stable.

[7] Cf. discussions of the vorticity transfer theory of turbulence in his general lecture at the Fourth International Congress for Applied Mechanics [19]. Some developments in that direction were continued by C. B. Millikan (unpublished).

The above discussion is based on very general considerations and does not depend on the consideration of a periodic wavy disturbance as used in the mathematical analysis. We shall now support the above argument by considering a neutral wavy disturbance, with the understanding that if such a disturbance can persist (except for the exceptional case of infinite wave-length and zero phase velocity), the motion is presumably unstable. From these considerations, the importance of viscosity in the inner friction layer will also be brought out.

Let us consider an observer moving with the phase velocity of a neutral wavy disturbance. He will observe a stationary pattern of the flow (see Fig. 7).[8] Closed stream lines are inevitable unless the disturbance has no $v$-component of velocity in the critical layer $w=c$, for the flows on opposite sides of the critical layer are in opposite directions relative to the observer. It appears unlikely that the $v$-component of the disturbance should be zero throughout that layer. Indeed, it has been shown to be impossible mathematically.[9] Thus, whenever a neutral disturbance persists, it involves a steady exchange of fluid elements on opposite sides of the critical layer.

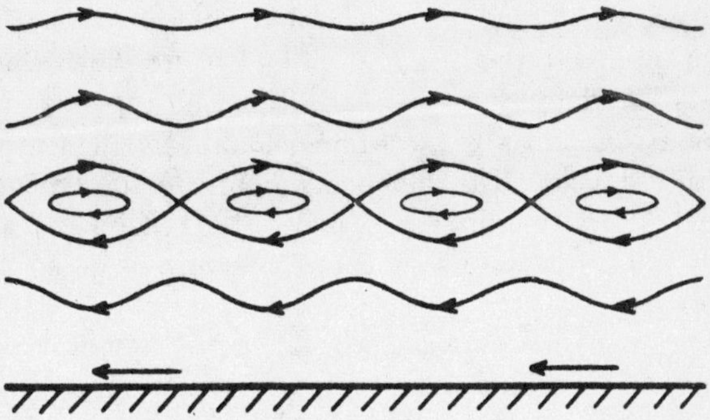

FIG. 7. Stream lines of a neutral disturbance as observed by an observer moving with the wave velocity.

If the effect of viscosity is to be negligible, fluid elements on the same stream line must have the same vorticity. If the gradient of vorticity of the main flow is zero or small near the critical point, it is easy to compensate this small difference of vorticity by the vorticity of the superposed flow, while the "scale" of disturbance [as measured in order of magnitude by $u'/(\partial u'/\partial y)$] remain the same as that of the main flow. It is thus *not impossible* to find a neutral disturbance *for which the effect of viscosity is negligible.* The motion may be unstable.

On the other hand, if the gradient of vorticity of the main flow is finite, the superposed small disturbance must also give a finite gradient of vorticity. This means that the "scale" of the disturbance must be very small in the critical layer. The diffusion of vorticity by the effect of viscosity is then inevitable. It is thus *impossible* to find a neutral disturbance *for which the effect of viscosity is negligible.* The motion is inertially stable.

**10. Acceleration of vortices in a non-uniform field of vorticity.** In the foregoing physical interpretation of inertial instability, we have considered the acceleration of an element of fluid in a two-dimensional parallel flow when this element of fluid does not have the same vorticity as the surrounding layer. We are now going to derive the explicit formula for the acceleration. The derivation shall be made in two different ways: (1) by kinematical considerations (using vorticity theorems); (2) by considera-

---

[8] This figure is due to Lord Kelvin (loc. cit.). He pointed out that the facts discussed here are "surprising," but did not attempt to explain their connection with the mechanism of hydrodynamic stability.

[9] This follows at once from Rayleigh's original results, if we apply it to the region between this layer and the solid wall (cf. Tollmien, loc. cit., 1935).

tions of the pressure gradient. In either method, we shall consider a perfect fluid in accordance with the stability problem under consideration.

1) *First derivation, by kinematical considerations*. For definiteness, let us consider a two-dimensional flow between two solid walls, which we shall take to be $y = \pm b$. Let the velocity components of the main flow be

$$U = w(y), \qquad V = 0, \tag{10.1}$$

and those of the secondary flow be

$$u' = u'(x, y), \qquad v' = v'(x, y). \tag{10.2}$$

The distribution of vorticity of the main flow is

$$\zeta_0 = \zeta_0(y) = -w'(y), \tag{10.3}$$

and that of the secondary flow is

$$\zeta' = \frac{\partial v'}{\partial x} - \frac{\partial u'}{\partial y}. \tag{10.4}$$

The latter distribution shall approximate a vortex "at" the point $(\xi_0, \eta_0)$. Thus, if the signs of $\zeta'$ and $\zeta_0$ are the same (or opposite), we have essentially a small element of fluid having an excess (or a defect) of vorticity near the point $(\xi_0, \eta_0)$.

The stream function for the secondary flow is

$$\psi'(x, y) = -\frac{1}{2\pi}\iint \zeta'(\xi, \eta)G(x, y; \xi, \eta)d\xi d\eta, \tag{10.5}$$

with

$$\left.\begin{aligned} u'(x, y) &= -\frac{\partial\psi'}{\partial y} = \frac{1}{2\pi}\iint \zeta'(\xi, \eta)\frac{\partial}{\partial y}G(x, y; \xi, \eta)d\xi d\eta, \\ v'(x, y) &= \frac{\partial\psi'}{\partial x} = -\frac{1}{2\pi}\iint \zeta'(\xi, \eta)\frac{\partial}{\partial x}G(x, y; \xi, \eta)d\xi d\eta. \end{aligned}\right\} \tag{10.6}$$

In these expressions, the integrals are extended over the whole region between the planes. The function $G(x, y; \xi, \eta)$ is the Green's function of the first kind for the region under consideration. It is defined by the following conditions:

$$\left.\begin{aligned} &\frac{\partial^2 G}{\partial x^2} + \frac{\partial^2 G}{\partial y^2} = 0 \text{ except at } (\xi, \eta), \\ &G(x, y; \xi, \eta) \sim -\log\{(x-\xi)^2 + (y-\eta)^2\}^{1/2} \text{ near } (\xi, \eta), \\ &G(x, y; \xi, \eta) = 0 \text{ over the solid boundaries.} \end{aligned}\right\} \tag{10.7}$$

As is well-known, it has the reciprocity property

$$G(x, y; \xi, \eta) = G(\xi, \eta; x, y). \tag{10.8}$$

For the case of a channel, it is given by the real part of

$$f(z) = -\left\{\log \operatorname{sh}\frac{\pi}{4b}(z - z_0) - \log \operatorname{ch}\frac{\pi}{4b}(z - \bar{z}_0)\right\},$$

$$(z = x + iy,\ z_0 = \xi + i\eta,\ \bar{z}_0 = \xi - i\eta). \tag{10.9}$$

Let us now consider the behavior of a particular element of fluid at $(\xi, \eta)$ having an excess (or a defect) of vorticity corresponding to the secondary flow (10.5) and (10.6). It causes a distortion of the main vorticity distribution as indicated in Fig. 8. After a very small interval of time $\delta t$ the vorticity at the point $(x, y)$ is changed by the amount

$$\delta\zeta(x, y) = -v'(x, y)\delta t\zeta_0'(y), \tag{10.10}$$

because it is replaced by a fluid element from below, which retains its original vorticity. This change produces an effect at the "vortex," i.e., at the element of fluid under

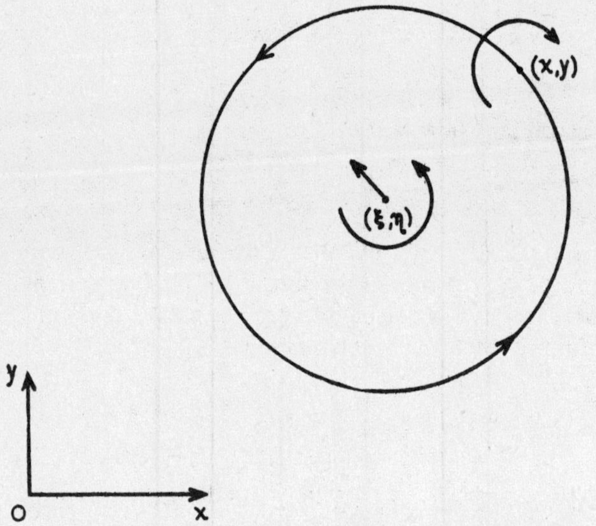


FIG. 8. Acceleration of vorticies in a non-uniform field of vorticity ($\zeta_0'(y)>0$, $\Gamma>0$).

consideration at $(\xi, \eta)$. It can be easily seen that the effect is a small velocity with components

$$\left.\begin{aligned} \delta u(\xi, \eta) &= \frac{1}{2\pi}\iint \frac{\partial}{\partial\eta} G(x, y; \xi, \eta)\delta\zeta(x, y)dxdy, \\ \delta v(\xi, \eta) &= -\frac{1}{2\pi}\iint \frac{\partial}{\partial\xi} G(x, y; \xi, \eta)\delta\zeta(x, y)dxdy, \end{aligned}\right\} \tag{10.11}$$

the integrals being extended over the whole region between the planes. Dividing these quantities by $\delta t$ and passing to the limit $\delta t \to 0$, we have the following components of acceleration at the point $(\xi, \eta)$:

$$\left.\begin{aligned} a_x(\xi, \eta) &= -\frac{1}{2\pi}\iint \frac{\partial}{\partial\eta} G(x, y; \xi, \eta)v'(x, y)\zeta_0'(y)dxdy, \\ a_y(\xi, \eta) &= \frac{1}{2\pi}\iint \frac{\partial}{\partial\xi} G(x, y; \xi, \eta)v'(x, y)\zeta_0'(y)dxdy. \end{aligned}\right\} \tag{10.12}$$

Let us first consider the $y$-component of this acceleration. From the special form in which $x$ and $\xi$ enter into the Green's function [cf. (10.9)], we can also write

$$a_y(\xi, \eta) = -\frac{1}{2\pi}\iint v'(x, y)\zeta_0'(y)\frac{\partial}{\partial x}G(x, y; \xi, \eta)dxdy. \tag{10.13}$$

If we multiply this equation by $\zeta'(\xi, \eta)$ and integrate over the whole region, we have the final formula

$$\iint a_y(\xi, \eta)\zeta'(\xi, \eta)d\xi d\eta = \iint \{v'(x, y)\}^2\zeta_0'(y)dxdy, \tag{10.14}$$

upon using (10.6). Before discussing its signifiance, let us first notice that

$$\iint a_x(\xi, \eta)\zeta'(\xi, \eta)d\xi d\eta = 0 \tag{10.15}$$

if $\zeta'(\xi, \eta)$ is an even function of $\xi-\xi_0$, i.e., if the vorticity distribution $\zeta'(\xi, \eta)$ has a symmetry about the line $\xi=\xi_0$. For then $v'(x, y)$ is an odd function of $x-\xi_0$ and $a_x(\xi, \eta)$ is an odd function of $\xi-\xi_0$, both being the consequence of the fact that $G(x, y; \xi, \eta)$ is an even function of $x-\xi$. Hence, we have the conclusion.

If we recall that the vorticity $\zeta'(\xi, \eta)$ is spread over a small region, we may take $\Gamma=\iint\zeta'(\xi, \eta)d\xi d\eta$ as the strength of the "superposed vortex." If we divide the left-hand side of (9.14) and (9.15) by $\Gamma$, we may consider the results as giving the components of the "average acceleration." The $x$-component of acceleration vanishes; the sign of the $y$-component depends upon the sign of the superposed vortex and the sign of $\zeta_0'(y)$. This component of acceleration is the one used in the above physical considerations.

It should be mentioned that in considering the stability of a motion we deal with a vortex pair. Although this makes it difficult to obtain a compact formula for the average accelerations of the individual vortices, a kinematical consideration such as that given above (cf. Fig. 8) shows that the general tendency is not changed. Furthermore, the two vortices are soon separated, because they are situated in layers of different mean velocity.

Another point should be mentioned. If we notice the tendency for the main vorticity to be swung around the secondary vortex, there is an acceleration of every element of fluid toward the vortex. Whatever this acceleration may be, it is expected to be of minor importance, because the effect is spread out over the whole field. This point will be brought out clearly in the following derivation of (10.14), where we shall study the whole phenomenon from the point of view of pressure forces. The acceleration will be identified with the negative of the pressure gradient divided by the density of the fluid, because the effect of viscosity has been neglected. Thus, if we can calculate the pressure disturbance corresponding to a given velocity disturbance, the left-hand side of (10.14) can be calculated.

2) *Second derivation, by consideration of pressure forces correlated with vorticity fluctuations.* To calculate the pressure distribution from a given velocity distribution, we use for the pressure a differential equation of Poisson's type obtained by taking the divergence of the equations of motion. Thus, if the equations of motion are[10]

[10] The usual notation is used: $x_i$ $(i=1, 2, 3)$ are the coordinates, $u_i$ are the components of velocity, $p$ is the pressure, and $\rho$ is the density of the fluid. Summation over a repeated index is understood. For a discussion of this type, see Lichtenstein's book [26].

$$\frac{\partial u_i}{\partial t} + u_j \frac{\partial u_i}{\partial x_j} = -\frac{1}{\rho}\frac{\partial p}{\partial x_i} + \nu \Delta u_i, \qquad (i = 1, 2, 3), \tag{10.16}$$

and the equation of continuity is

$$\partial u_i / \partial x_i = 0, \tag{10.17}$$

we have

$$\Delta(p/\rho) = -\sigma, \tag{10.18}$$

where

$$\sigma = \frac{\partial u_i}{\partial x_j}\frac{\partial u_j}{\partial x_i} = e_{ij}e_{ij} - \omega_{ij}\omega_{ij} = 2\left\{\frac{\partial(u_1, u_2)}{\partial(x_1, x_2)} + \frac{\partial(u_2, u_3)}{\partial(x_2, x_3)} + \frac{\partial(u_3, u_1)}{\partial(x_3, x_1)}\right\}, \tag{10.19}$$

and

$$e_{ij} = \frac{1}{2}\left(\frac{\partial u_i}{\partial x_j} + \frac{\partial u_j}{\partial x_i}\right), \qquad \omega_{ij} = \frac{1}{2}\left(\frac{\partial u_i}{\partial x_j} - \frac{\partial u_j}{\partial x_i}\right) \tag{10.20}$$

are the components of deformation and of rotation. If we integrate (10.18) under proper boundary conditions, $\sigma$ being known at the *initial* instant, we obtain the *initial* distribution of pressure. The initial acceleration field is then obtained from (10.16) as the negative gradient of the pressure, if we neglect the effect of viscosity.

For a perfect fluid, the only boundary condition at a solid wall is

$$u_i n_i = 0, \tag{10.21}$$

where $n_i$ is the outward normal of the boundary surface. If we multiply (10.16) by $n_i$, neglecting the effect of viscosity, we have

$$-\frac{1}{\rho}\frac{\partial p}{\partial n} = V_0 \frac{\partial u_i}{\partial s} n_i, \tag{10.22}$$

where $V_0$ is the velocity along a stream line on the boundary, and $ds$ is an element of its arc. If we write

$$u_i = V_0 l_i, \qquad \frac{\partial u_i}{\partial s} = V_0 \frac{\partial l_i}{\partial s} + l_i \frac{\partial V_0}{\partial s},$$

where $l_i$ are the direction cosines of the velocity over the boundary surface, we have

$$-\frac{1}{\rho}\frac{\partial p}{\partial n} = \frac{V_0^2}{R}, \tag{10.23}$$

where $R$ is the radius of curvature of the stream line, $R^{-1} = n_i \partial l_i / \partial s$. This relation expresses the balance of pressure and centrifugal force. With a given distribution of velocity, the right-hand side is known. We have thus a potential problem of the second kind for the pressure.

*Two-dimensional flow between parallel solid walls.* Returning to the problem at hand, we have the very simple boundary condition

$$\partial p / \partial y = 0 \quad \text{at} \quad y = \pm b. \tag{10.24}$$

Since the main motion is a two-dimensional parallel motion, we have

$$u_1 = w(y) + u'(x, y), \qquad u_2 = v'(x, y), \qquad u_3 = 0, \tag{10.25}$$

where $w(y)$ represents the main flow, and $u'$ and $v'$ give a secondary flow approximating a vortex. Equation (10.18) becomes

$$\frac{1}{\rho}\left(\frac{\partial^2 p}{\partial x^2} + \frac{\partial^2 p}{\partial y^2}\right) = -\sigma, \qquad \sigma = 2\left\{w'(y)\frac{\partial v'}{\partial x} - \frac{\partial(u', v')}{\partial(x, y)}\right\}. \tag{10.26}$$

We note that we can write $\sigma = \sigma_1 + \sigma_2$, where

$$\sigma_1 = -2\frac{\partial(u', v')}{\partial(x, y)}, \qquad \sigma_2 = -2\zeta_0\frac{\partial v'}{\partial x}. \tag{10.27}$$

$\sigma_1$ depends upon the structure of the secondary vortex itself, and $\sigma_2$ depends upon its interaction with the main flow. We shall also separate the pressure into two parts and require them to satisfy (10.24) separately. Thus,

$$\left.\begin{aligned} &p = p_1 + p_2, \qquad \sigma = \sigma_1 + \sigma_2, \\ &\frac{1}{\rho}\left(\frac{\partial^2 p_1}{\partial x^2} + \frac{\partial^2 p_1}{\partial y^2}\right) = -\sigma_1, \qquad \frac{\partial p_1}{\partial y} = 0 \quad \text{at} \quad y = \pm b, \\ &\frac{1}{\rho}\left(\frac{\partial^2 p_2}{\partial x^2} + \frac{\partial^2 p_2}{\partial y^2}\right) = -\sigma_2, \qquad \frac{\partial p_2}{\partial y} = 0 \quad \text{at} \quad y = \pm b. \end{aligned}\right\} \tag{10.28}$$

We can reduce our problem to that of the *first* kind by looking for the acceleration $a_y(x, y)$ in the $y$-direction, $a_y = -(1/\rho)\,\partial p/\partial y$. If we differentiate (10.28) with respect to $y$, we have

$$\left.\begin{aligned} &a_y = \alpha_1 + \alpha_2, \\ &\frac{\partial^2\alpha_1}{\partial x^2} + \frac{\partial^2\alpha_1}{\partial y^2} = \frac{\partial\sigma_1}{\partial y}, \qquad \frac{\partial^2\alpha_2}{\partial x^2} + \frac{\partial^2\alpha_2}{\partial y^2} = \frac{\partial\sigma_2}{\partial y}, \end{aligned}\right\} \tag{10.29}$$

with $\alpha_1 = 0$, $\alpha_2 = 0$ at $y = \pm b$. The $x$-component of acceleration is zero from symmetry considerations. The solutions of (10.29) are

$$\left.\begin{aligned} \alpha_1(x, y) &= -\frac{1}{2\pi}\iint G(x, y: \xi, \eta)\left(\frac{\partial\sigma_1}{\partial y}\right)_{\xi,\eta} d\xi d\eta, \\ \alpha_2(x, y) &= \frac{1}{2\pi}\iint G(x, y; \xi, \eta)\left(\frac{\partial\sigma_2}{\partial y}\right)_{\xi,\eta} d\xi d\eta, \end{aligned}\right\} \tag{10.30}$$

where the integrals are extended over the whole region between the planes. These formulae give the distribution of acceleration. Actually, it is more convenient to deal with the integrated quantities

$$I_1 = \iint \alpha_1(x, y)\zeta_0(x, y)dxdy, \tag{10.31}$$

$$I_2 = \iint \alpha_2(x, y)\zeta_0(x, y)dxdy, \tag{10.32}$$

$$J_1 = \iint \alpha_1(x, y)\zeta'(x, y)dxdy, \tag{10.33}$$

$$J_2 = \iint \alpha_2(x, y)\zeta'(x, y)dxdy. \tag{10.34}$$

The first two integrals correspond to the accelerations of the main flow by the secondary flow itself and by the interaction; the latter two quantities correspond to the accelerations of the secondary flow by itself and by the interaction. It can be verified (as will be done presently) that

$$I_1 = -\iint v'^2 \frac{d\zeta_0}{dy} dxdy, \quad (10.35) \qquad I_2 = 0, \quad (10.36)$$

$$J_1 = 0, \quad (10.37) \qquad J_2 = \iint v'^2 \frac{d\zeta_0}{dy} dxdy. \quad (10.38)$$

We note that (10.38) is essentially a reproduction of the formula (10.14), the significance of which has been discussed above. The integral $I_1$ is equal to the negative of $J_2$. This is the above-mentioned acceleration distributed among the fluid elements throughout the field. It is therefore relatively unimportant. Thus, all the statements made in the last section have been verified, if we can verify (10.35)–(10.38).

*Verification of* (10.35)–(10.38). To verify these equations, let us first examine the behavior of the quantities $u'$, $v'$, $\partial p/\partial x$, $\partial p/\partial y$ for large values of $x$. From the expression (10.9) for the Green's function, we see that if $\zeta'$ vanishes sufficiently rapidly as $x$ becomes infinite, we have

$$u' = O(|x|^{-3}), \qquad v' = O(|x|^{-2}), \tag{10.39}$$

for large values of $x$. From the equations of motion, we then find that

$$\frac{\partial p}{\partial x} = O(u') = O(|x|^{-3}), \qquad \frac{\partial p}{\partial y} = O(v') = O(|x|^{-2}). \tag{10.40}$$

This will assure the convergence of the integrals involved and the validity of the steps taken in the following transformations.

In the first method of derivation, we have been mainly concerned with $J_2$. We shall therefore consider it first. Referring to (10.30) and (10.5), we see that

$$J_2 = \iint \psi'(\xi, \eta)\left(\frac{\partial \sigma_2}{\partial y}\right)_{\xi,\eta} d\xi d\eta.$$

If we now introduce the value of $\sigma_2$ as given by (10.27) and replace $(\xi, \eta)$ by $(x, y)$, we have

$$J_2 = -2\iint \psi'(x, y)\frac{\partial^2}{\partial x \partial y}(v'\zeta_0)dxdy.$$

On integrating by parts with respect to $x$, we obtain

$$J_2 = 2\iint v' \frac{\partial}{\partial y}(v'\zeta_0)dxdy = \iint \left\{\frac{\partial}{\partial y}(v'^2\zeta_0) + v'^2\frac{d\zeta_0}{dy}\right\} dxdy.$$

The result (10.38) or (10.14) is thereby verified.

Similar calculations can be carried out for the integrals $I_1$, $I_2$, and $J_1$. Thus,

$$I_1 = \iint \alpha_1 \zeta_0 dxdy = \iint w(y) \frac{\partial \alpha_1}{\partial y} dxdy$$
$$= \iint w\left(\sigma_1 + \frac{1}{\rho}\frac{\partial^2 p_1}{\partial x^2}\right) dxdy,$$

by (10.28). If we note that

$$w\sigma_1 = -2w\frac{\partial(u', v')}{\partial(x, y)} = 2\frac{\partial}{\partial x}\left(wv'\frac{\partial u'}{\partial y}\right) - 2\frac{\partial}{\partial y}\left(wv'\frac{\partial u'}{\partial x}\right) - \frac{dw}{dy}\frac{\partial^2 v'}{\partial y^2},$$

the above integral is easily transformed into the form (10.35). Following an exactly analogous process, we have

$$I_2 = \iint w\left(\sigma_2 + \frac{1}{\rho}\frac{\partial^2 p_2}{\partial x^2}\right) dxdy = 0,$$

when we make use of (10.27). The integral $J_1$ has also the significance that it is the effect of the solid boundaries upon a general flow $\psi'(x, y)$ consistent with (10.40), because it is independent of $w(y)$. Using (10.30), (10.5), and (10.27), we have

$$J_1 = \iint \alpha_1 \zeta' dxdy = \iint \psi'(\xi, \eta)\left(\frac{\partial \sigma_1}{\partial y}\right)_{\xi,\eta} d\xi d\eta$$
$$= -2\iint u'\frac{\partial(u', v')}{\partial(x, y)} dxdy.$$

If we note that

$$u'\frac{\partial(u', v')}{\partial(x, y)} = \frac{\partial}{\partial y}\left(u'v'\frac{\partial u'}{\partial x}\right) - \frac{\partial}{\partial x}\left(u'v'\frac{\partial u'}{\partial y}\right),$$

we see that $J_1=0$. The results (10.35)–(10.38) are thereby verified. We have thus completed the investigations indicated at the beginning of this section.

NOTE ADDED IN PROOF. In a very early work, [Phil. Trans. Roy. Soc. London (A) **215,** 23–26 (1915)] G. I. Taylor gave a physical interpretation of Rayleigh's results on the stability of the laminar motion of an inviscid fluid, based on momentum considerations. He also indicated clearly that a motion, stable according to Rayleigh's criterion, may be unstable through the effect of viscosity.

*(To be continued)*

# ON THE STABILITY OF TWO-DIMENSIONAL PARALLEL FLOWS

## PART III.—STABILITY IN A VISCOUS FLUID*

BY

C. C. LIN**

*Guggenheim Laboratory, California Institute of Technology*

**11. General considerations.** The investigations in Part II* of the stability characteristics of a parallel flow in an inviscid fluid led to very useful information. In the first place, they enable us to visualize the effect of pressure forces very clearly. In the second place, the results can be used as a guide for studying the stability problem in a real fluid, since instability is expected to occur only for sufficiently large Reynolds numbers. Thus, if we know the general characteristics of "inviscid stability" for a given velocity distribution, some stability characteristics in a real fluid can be obtained by considering a *modification* of these results by the effect of viscosity. Such a consideration was first made by Heisenberg, [14]† who demonstrated that the effect of viscosity is generally *destabilizing* at very large Reynolds numbers. There are, however, a few points to be supplemented in his discussion. We shall therefore study this problem in some detail in §12.

To get a good understanding of the stability problem, we want to know the following points for any given *class* of velocity distribution. First of all, we want to know whether this class of flows is stable for *all* Reynolds numbers. Secondly, if it is stable for certain Reynolds numbers with respect to disturbances of certain wavelengths, but unstable under other conditions, we want to know the general nature of the curve $c_i(\alpha, R)=0$ which separates regions of stability and instability in the $\alpha$, $R$ plane. Thirdly, such a curve will be expected to show a minimum in $R$, below which all small disturbances are damped out. It is therefore desirable to be able to calculate this minimum critical Reynolds number rapidly.

In the next section, we shall solve these problems for two classes of flows; namely, (a) velocity distributions of the symmetrical type, and (b) velocity distributions of the boundary-layer type. Indeed, it will be shown that *in these cases the flow is always unstable for sufficiently large Reynolds numbers*, whether the velocity curve has a point of inflection or not. The curve of neutral stability $c_i(\alpha, R)=0$ will be shown to belong to either of the types shown in Fig. 9. When the velocity curve has no point of inflection, the two asymptotic branches of the curve have the common asymptote $\alpha=0$

* Received July 18, 1945. Parts I and II of this paper appeared in this Quarterly 3, 117–142, and 218–234 (1945).

** Now at Brown University.

† The figures in brackets refer to titles in the Bibliography at the end of Part I.

(Fig. 9a). When there is a point of inflection, one branch has the asymptote $\alpha=0$, while the other has the asymptote $\alpha=\alpha_s>0$ (Fig. 9b). In either case, it will be shown that the region inside the loop is the region of instability. It should be observed how these results fit in with those of Rayleigh and Tollmien for the inviscid case, and with the destabilizing effect of viscosity noted by Heisenberg for large Reynolds numbers. Simple formulae will be derived to express the asymptotic branches of the curves in Fig. 9. In fact, it is by means of these asymptotic behaviours and a criterion of stability of Synge [63], that the results mentioned above are established. Simple rules will also be given, by which the minimum critical Reynolds number can be very easily obtained from quantities involving very simple integral and differential expressions of the velocity distribution $w(y)$. Very little numerical labor is involved for the calculation in any particular case.

Heisenberg also discussed the general shape of the curve $c_i(\alpha, R)=0$. However, his argument does not appear to be very decisive, and some of his results are not well stated. He did not obtain the asymptotic forms of the $\alpha(R)$ curve as given below. He also tried to estimate the order of magnitude of the critical Reynolds number, but did not try to make an approximate calculation in terms of simple differential and integral expressions of $w(y)$ (loc. cit., p. 600).

In order to obtain definite numerical results, which may be subjected to experimental verification, we shall apply our theory to the stability problem of special velocity distributions. In §13, we shall give the results of calculation of the neutral curves in (a) the Blasius case and (b) the Poiseuille case. The method of calculation and its numerical accuracy will be given in the Appendix. Frequent reference to the equations in it will therefore be made in the following sections.

**12. Heisenberg's criterion and the general characteristics of the curve of neutral stability.** We shall now proceed to study the general stability characteristics in a viscous fluid as indicated above. In the first place, we shall develop Heisenberg's criterion in a slightly improved form. We shall then restrict ourselves to velocity distributions of the symmetrical type and of the boundary-layer type. For these cases, we shall prove the results summarized in the next section

a) *Heisenberg's criterion.* Along the neutral curve

$$c_i(\alpha, R) = 0$$

(if it exists), all the parameters $\alpha$, $c$, and $R$ are functions of one of them, say, $R$. Let us restrict ourselves to cases where $\alpha$ and $c$ do not approach zero along the neutral curve as $R$ becomes infinite. Then, the approximations (6.27) are valid for sufficiently large values of $R$. By using (6.26), (6.24) and (6.27), we can transform (6.13) into the following:

$$\sum_{n=0}^{\infty}\alpha^{2n}K_{2n+1}(c) = \frac{e^{-\pi i/4}}{\sqrt{(\alpha R(1-c)^5}}\left\{\sum_{n=0}^{\infty}\alpha^{2n}K_{2n}(c) + w_2'(1-c)\sum_{n=0}^{\infty}\alpha^{2n}K_{2n+1}(c)\right\}$$
$$+\frac{e^{\pi i/4}}{\sqrt{(\alpha Rc)^5}}\left\{\sum_{n=0}^{\infty}\alpha^{2n}H_{2n}(c) + w_1 c\sum_{n=0}^{\infty}\alpha^{2n}K_{2n+1}(c)\right\}.$$

If there is an inviscid neutral disturbance with $c=c_s$, $\alpha=\alpha_s$, we have

$$\sum_{n=0}^{\infty}\alpha_s^{2n}K_{2n+1}(c_s) = 0.$$

Hence, when $\alpha=\alpha_s$, and $c=c_s+\Delta c$, differing but slightly from $c_s$, we have

$$\Delta c = \frac{e^{-\pi i/4}}{\sqrt{\alpha_s R(1-c_s)^5}} \frac{\sum_{n=0}^{\infty} \alpha_s^{2n} K_{2n}(c_s)}{\sum_{n=0}^{\infty} \alpha_s^{2n} K'_{2n+1}(c_s)} + \frac{e^{\pi i/4}}{\sqrt{\alpha_s R c_s^5}} \frac{\sum_{n=0}^{\infty} \alpha_s^{2n} H_{2n}(c_s)}{\sum_{n=0}^{\infty} \alpha_s^{2n} K'_{2n+1}(c_s)}. \tag{12.1}$$

Similar considerations of (6.14), (6.15), and (6.17) give respectively

$$\Delta c = \frac{e^{\pi i/4}}{\sqrt{\alpha_s R c_s^5}} \frac{\sum_{n=0}^{\infty} \alpha_s^{2n} H_{2n+1}(c_s)}{\sum_{n=0}^{\infty} \alpha_s^{2n} K'_{2n+2}(c_s)}, \tag{12.2}$$

$$\Delta c = \frac{e^{\pi i/4}}{\sqrt{\alpha_s R c_s^5}} \frac{\sum_{n=0}^{\infty} \alpha_s^{2n} H_{2n}(c_s)}{\sum_{n=0}^{\infty} \alpha_s^{2n} K'_{2n+1}(c_s)}, \tag{12.3}$$

$$\Delta c = \frac{e^{\pi i/4}}{\sqrt{\alpha_s R c_s^5}} \frac{\sum_{n=1}^{\infty} \alpha_s^{2n} H_{2n-1}(c_s) + (1-c_s)^2 \sum_{n=0}^{\infty} \alpha_s^{2n+1} H_{2n}(c_s)}{\sum_{n=1}^{\infty} \alpha_s^{2n} K'_{2n}(c_s) + (1-c_s)^2 \sum_{n=0}^{\infty} \alpha_s^{2n+1} K'_{2n+1}(c_s)}. \tag{12.4}$$

In general, it is not very easy to determine whether $\Delta c_i > 0$ or $<0$. However, when $c_s$ and $\alpha_s$ are both small, but not zero, all the above equations will reduce to

$$\Delta c = e^{\pi i/4} / \sqrt{\alpha_s R c_s^5}\, K_1'(c_s), \tag{12.5}$$

after we make use of the reductions corresponding to (1) and (2) of the Appendix. Now, when $c$ is small,

$$K_1(c) = \int_{y_1}^{y_2} dy(w-c)^{-2} = -\frac{1}{w_1' c} - \frac{w_0''}{w_1'^3}(\log c + i\pi) + O(1);$$

hence, it can be easily verified that $K_1'(c)$ is approximately real and positive for small real values of $c$. Hence, (12.5) shows that $\Delta c_i > 0$ in every case. The disturbance with wave length $2\pi/\alpha_s$, neutrally stable in the inviscid case, is unstable when viscous forces are considered. This result was first obtained by Heisenberg, and may be formulated as follows:

HEISENBERG'S CRITERION. *If a velocity profile has an "inviscid" neutral disturbance with non-vanishing wave number and phase velocity, the disturbance with the same wave number is unstable in the real fluid when the Reynolds number is sufficiently large.*

In Heisenberg's original discussion, only the first type of motion is considered. The last equation on p. 597 of his paper is essentially our equation (12.1) with all the terms in $\alpha^2$ dropped. Evidently, the above arguments hold only for $\alpha_s, c_s \neq 0$. It will be seen later from Fig. 9 that one neutral disturbance with $\alpha=0$ and $c=0$ for infinite $R$, is actually stable for finite values of $R$.

b) *Asymptotic behavior of the neutral $\alpha(R)$ curve.* We shall now study the general asymptotic behavior of the neutral curve, assuming that it exists. The answer to the existence problem will be evident during the course of the investigation. For large values of $\alpha R$, we would generally expect $z$ of (6.28) to be much greater than 1, but it is also possible for $z$ to approach a finite value or zero. We shall therefore discuss both possibilities.

For large values of $z$, we have approximately,

$$\mathfrak{F}_r = 1, \qquad \mathfrak{F}_i = 1/\sqrt{2z^3} = w_1'/\sqrt{2\alpha Rc^3}, \tag{12.6}$$

where $\mathfrak{F}_i$ is small. If we refer to (5) and (7) of Appendix, we see that the imaginary part $v$ of the right-hand side member of those equations depends on that of $w_1' cK_1(c)$ and those of the integrals $H_1$, $H_2$, $M$'s and $N$'s. If $\alpha$ and $c$ are small, which will be verified *a posteriori*, we have only the contribution from the first term; thus,*

$$v = -\pi w_1 \frac{ww''}{w'^3} \quad \text{for} \quad w = c. \tag{12.7}$$

By using (12.6) and (12.7), we see that the equations (8) in the Appendix can be approximated by

$$u = 1, \tag{12.8}$$

and

$$v = -\frac{\pi}{w_1'} cw_0'' = w_1'/\sqrt{2\alpha Rc^3}. \tag{12.9}$$

These are the equations for determining a relation $\alpha(R)$, if we eliminate $c$ between them.

In the case where $\alpha Rc^3$ approaches a finite limit as $\alpha R \to \infty$, $c$ must approach zero. Hence, $v$ must approach zero, and from (8) of Appendix, $\mathfrak{F}_i$ must also approach zero. From the curve for $\mathfrak{F}_i(z)$, we see that this happens for $z = 2.294$, for which $\mathfrak{F}_r = 2.292$. Then, using (9) of Appendix, (12.6), and (12.7), we have

$$\alpha R = w_1'^2 z^3/c^3, \qquad z = 2.294, \tag{12.10}$$

and

$$u = \mathfrak{F}_r = 2.292. \tag{12.11}$$

The two types of relations (12.8), (12.9) and (12.10), (12.11) evidently correspond to two different branches of the $\alpha(R)$ curve (cf. Fig. 9). These conditions can be satisfied in cases (2a) and (3) of section 6 (cf. (11.5), (11.7) of Appendix), but it appears to be difficult in case (2b) (cf. (11.6) of Appendix).

CASE (2a). *Symmetrical velocity distribution with symmetrical $\phi(y)$.* We consider the cases where both $\alpha$ and $c$ are small. By using (12.5) and noting that $u$ takes on a finite value in either (12.8) or (12.11), we see that we must have

$$u = \frac{w_1' c}{H_{10}\alpha^2}, \quad \text{where} \quad H_{10} = H_1(0) = \int_{y_1}^{y_2} w^2 dy: \tag{12.12}$$

* In fact, the other terms never give considerable contributions to the imaginary part even for only moderately small values of $\alpha$ and $c$. This point will be discussed in the Appendix. The approximation (12.7) will be used for all later calculations.

i.e., $c$ must approach zero as fast as $\alpha^2$. The asymptotic behavior of the $\alpha(R)$ curves as given by (12.8)–(12.11) are as follows:

$$R = (w_1'^{11}/2\pi^2 H_{10}^5 w_0''^2)\alpha^{-11}, \qquad c = (H_{10}/w_1')\alpha^2, \quad \text{(first branch)}, \tag{12.13}$$

$$R = w_1'^5(z/\mathfrak{J}_r H_{10})^3\alpha^{-7}, \qquad c = (H_{10}\mathfrak{J}_r/w_1')\alpha^2, \quad \text{(second branch)}, \tag{12.14}$$

where $\mathfrak{J}_r = 2.292$, $z = 2.294$ approximately.

CASE (3). *Boundary-layer profile.* Here, the equation corresponding to (12.12) is (cf. (7) of Appendix)

$$u = w_1 c/\alpha, \tag{12.15}$$

i.e., $c$ must approach zero as fast as $\alpha$. Note that in the previous case, the relation (12.12) depends both on $w_1'$ and on $\int_{y_1}^{y_2} w^2 dy$. Here, it depends only on the initial slope of the velocity curve $w_1'$. The two branches of the $\alpha(R)$ curve for large values of $R$ are

$$R = (w_1'^{11}/2\pi^2 w_0''^2)\alpha^{-6}, \qquad c = \alpha/w_1', \qquad \text{(first branch)}, \tag{12.16}$$

and

$$R = w_1'(z/\mathfrak{J}_r)^3\alpha^{-4}, \qquad c = \alpha\mathfrak{J}_r/w_1' \qquad \text{(second branch)}, \tag{12.17}$$

where $\mathfrak{J}_r = 2.292$, $z = 2.294$ approximately.

*Effect of varying curvature in the curve of velocity distribution.* In either case, the second branch of our asymptotic curve depends very little upon the shape of the velocity profile, while the first branch depends very much upon it through the term $w_0''$. This fact will enable us to correlate our present results with the inviscid investigations of Rayleigh and Tollmien, as discussed in Part II.

In all the cases considered, we have $w'' < 0$ for $y < y_2$ but sufficiently near to it. If $w''(y)$ never vanishes for $y_1 < y < y_2$, we can replace $w_0''$ by $w_1''$ in the expressions (12.13) and (12.16). In general,

$$w_0'' = w_1'' + \frac{w_1''}{w_1} c + \left(\frac{w^{\text{iv}}}{2w_1^2} - \frac{w_1'' w_1'''}{2w_1^3}\right) c^2 + \cdots .$$

Now, for a flow which is essentially parallel, the boundary condition $\Delta\Delta\psi = 0$, which holds on the solid wall for all two-dimensional laminar flows, can be easily verified to be equivalent to $w_1''' = 0$. Hence, we have

$$w_0'' = w_1'' + \frac{w^{\text{iv}}}{2w_1'^2} c^2 + \cdots .$$

Thus, if $w_1'' = 0$, but $w''$ does not vanish for $y_1 < y < y_2$, we have

$$R = \left\{2w_1'^{19}/\pi^2 H_{10}^9 (w^{\text{iv}})^2\right\}\alpha^{-19}, \quad \text{for case (2a)}, \tag{12.18}$$

and

$$R = \left\{2w_1'^{19}/\pi^2 (w^{\text{iv}})^2\right\}\alpha^{-10}, \qquad \text{for case (3)}. \tag{12.19}$$

In case the velocity profile shows a point of inflection,

$$w_0'' = 0 \quad \text{for} \quad c = c_s.$$

Then, we have approximately

$$w_0'' = (w_1^{iv}/2w_1'^2)(c^2 - c_s^2). \qquad (12.20)$$

It can now be seen from (12.13) and (12.16) that $R$ becomes infinite as $c$ approaches $c_s$. Let the corresponding value of $\alpha$ be denoted by $\alpha_s$. Then instead of (12.18) and (12.19), the following relations hold:

$$R = \{2w_1'^{19}/\pi^2 H_{10}^9 (w_1^{iv})^2\}\alpha^{-11}(\alpha^4 - \alpha_s^4)^{-2}, \quad \alpha_s^2 = w_1' c_s / H_{10}, \quad \text{for case (2a)}, \qquad (12.21)$$

$$R = \{2w_1'^{19}/\pi^2 (w_1^{iv})^2\}\alpha^{-6}(\alpha^2 - \alpha_s^2)^{-2}, \quad \alpha_s = w_1' c_s, \quad \text{for case (3)}. \qquad (12.22)$$

Thus, for either a symmetrical or a boundary-layer distribution with a flex, we have

$$R \sim (\alpha - \alpha_s)^{-2} \qquad (12.23)$$

as $R \to \infty$, $\alpha \to \alpha_s$, $c \to c_s$. In all these approximations, we *assume $\alpha_s$ and $c_s$ to be so small* that the previous approximations still hold, but the qualitative nature of our conclusions cannot be changed for moderate values of $\alpha_s$ and $c_s$.

The general characteristics obtained from our foregoing discussions are summarized in Table II, and are indicated by the asymptotic branches of the solid lines in Fig. 9. Let us proceed to discuss their significance.

i) It may be expected that the region between the two asymptotic branches of the curves represents a region of instability. Thus, ***every symmetrical or boundary-layer profile is unstable for sufficiently large values of the Reynolds number***. This point will be substantiated below.

ii) In the cases where $w_1'' > 0$, the two branches of curves approach the two different asymptotes $\alpha = 0$ and $\alpha = \alpha_s$, leaving a *finite instable region* for infinite Reynolds number. In the other cases, the two branches approach the same asymptote $\alpha = 0$, leaving only the possibility of a neutral disturbance of infinite wave-length at infinite Reynolds number. These results agree with Heisenberg's criterion and the results obtained from the considerations of an inviscid fluid in Part II of this work. It thus appears that the inviscid disturbance with $\alpha = 0$, $c = 0$ is actually not as trivial as it may first appear to be, for it is actually the limiting case of neutral disturbances in a real fluid.

c) *Existence of self-excited disturbances.** To establish the actual existence of self-excited disturbances, we try to show that $c_i > 0$ at least in the neighborhood of the neutral curve. Indeed, we may regard $c$ as a function of the two independent parameters $\alpha$ and $R' = \alpha R$, and show that $(\partial c_i/\partial R')_\alpha < 0$ for the first branch of the curve. This is analogous to Heisenberg's criterion, and demonstrates the same general conclusion that the effect of viscosity is destabilizing at large Reynolds numbers. To fix our ideas, we shall carry out the analysis for the case of symmetrical profiles. The other case can be carried out in a similar manner.

We begin with the equation

$$\frac{f_2(\alpha, c)}{f_4(\alpha, c)} = \frac{\phi_{31}}{\phi_{31}'} \qquad (6.14)$$

where $f_2(\alpha, c)$ and $f_4(\alpha, c)$ are given by (6.26) and $\phi_{31}/\phi_{31}'$ is given by (6.28). By using those relations, we can transform (6.14) into

* This section was inserted late in 1944 after discussions with Prof. C. L. Pekeris. He mentioned the possibility that the neutral curve might be a curve of minimum damping with stable regions on both sides of it. See also Schlichting's calculations [52].

$$\mathcal{J}(z) = \phi_{22}' \Big/ \left(\phi_{22}' + \frac{1}{w_1' c}\phi_{12}'\right);$$

or, by further using (6.30) and (6.24),

$$\mathcal{J}(z) - 1 = w_1' c \sum_{n=1}^{\infty} \alpha^{2n} K_{2n}(c) \Big/ \sum_{n=1}^{\infty} \alpha^{2n} H_{2n-1}(c).$$

We now regard $\alpha$ as fixed and consider the variation of $c$ with $R'$ or $z$, which is a known function of $c$ and $R'$. Taking logarithmic derivatives on both sides, we have

$$\frac{\mathcal{J}'(z)}{\mathcal{J}(z) - 1} = \left\{\frac{1}{c} + \frac{\sum_{n=1}^{\infty} \alpha^{2n} K_{2n}'(c)}{\sum_{n=1}^{\infty} \alpha^{2n} K_{2n}(c)} - \frac{\sum_{n=1}^{\infty} \alpha^{2n} H_{2n-1}'(c)}{\sum_{n=1}^{\infty} \alpha^{2n} H_{2n-1}(c)}\right\} \frac{dc}{dz}. \tag{12.24}$$

So far, $\alpha$, $c$, and $z$ are arbitrary. On the neutral curve, $c$ and $z$ are real, and we may use the relation (12.13) if we are on the first branch of the neutral curve with large values of $R'$. Indeed, for large values of $z$, (12.6) gives

$$\frac{\mathcal{J}'(z)}{\mathcal{J}(z) - 1} = -\frac{3}{2z}$$

and

$$-\frac{3dz}{2z} = -\frac{1}{2}\frac{dR'}{R'} - \frac{3}{2}\frac{dc}{c}.$$

By using these relations, it can be easily verified that (12.24) leads to

$$\frac{dR'}{2R'} - \frac{3dc}{2c} = dc\left\{\frac{1}{c} + \frac{\sum_{n=1}^{\infty} \alpha^{2n} K_{2n}'(c)}{\sum_{n=1}^{\infty} \alpha^{2n} K_{2n}(c)} - \frac{\sum_{n=1}^{\infty} \alpha^{2n} H_{2n-1}'(c)}{\sum_{n=1}^{\infty} \alpha^{2n} H_{2n-1}(c)}\right\}.$$

Remembering that $c = O(\alpha^2)$ and noting that

$$K_1(c) = -\frac{1}{w_1 c} - \frac{w_0''}{w_0'^3}(\log c + i\pi) + O(1),$$

where $O(1)$ is real, we can reduce the above equation to

$$\frac{dR'}{2R'} + \frac{3dc}{2c} = dc\left\{1 + w_1' c \frac{w_0''}{w_0'^3}(\log c + i\pi)\right\}^{-1} \frac{d}{dc}\left\{\frac{w_1' c w_0''}{w_0^3}(\log c + i\pi)\right\},$$

or

$$\frac{1}{2}\frac{dR'}{R'} = -\frac{3dc}{2c}\left\{1 + c\frac{d}{dc}\left[\frac{w_1' c w_0''}{w_0'^3}(\log c + i\pi)\right]\right\}. \tag{12.25}$$

For small values of $c$, the expression in the brackets has a positive real part and a negative imaginary part. Hence, $(\partial c/\partial R')_\alpha$ has a negative imaginary part. This com pletes the proof. Indeed, if $w_1''$ does not vanish, we have

$$\left.\begin{aligned}\left(\frac{\partial c_r}{\partial R'}\right)_\alpha &= -\frac{c}{3R'} \sim R'^{-6/5},\\ \left(\frac{\partial c_i}{\partial R'}\right)_\alpha &= \frac{2\pi w_1''}{9w_1'}\frac{c^2}{R'} \sim R'^{-7/5}.\end{aligned}\right\} \tag{12.26}$$

In the above derivation, it should be noted that all approximations are made by neglecting small terms of higher orders in *comparison* with some terms which have been retained. Thus, the conclusion of stability or instability would not be altered by those terms of higher orders.

d) *The minimum critical Reynolds number and the minimum critical wave-length.* Having demonstrated the instability of the symmetrical and the boundary-layer profiles, we want to answer the following questions. First, does there exist a minimum critical Reynolds number, below which the flow is stable for disturbances of all wave-lengths? If so, can we get an approximate estimate of its magnitude? Secondly, does there exist a minimum wave-length of the disturbance (maximum $\alpha$) below which the flow is stable at all Reynolds numbers? If so, can we get an approximate estimate of its magnitude? We shall see that in trying to answer these questions, we can also roughly depict the complete $\alpha(R)$ curve, which separates stability from instability.

The existence of these minimum values can be most conveniently inferred from a condition of stability derived by Synge* from energy considerations. His condition reads

$$(qR)^2 < (2\alpha^2+1)(4\alpha^4+1)/\alpha^2, \qquad q = \max|w'|. \tag{12.27}$$

This condition insures the existence of a minimum critical Reynolds number. It permits $\alpha$ to become infinite only for $R\to\infty$. But we know from our previous considerations that $\alpha\to\alpha_s$ or 0 as $R\to\infty$. Hence, we would expect that there exists a maximum value of $\alpha$, above which there is always stability. The neutral curve must therefore take the general shape shown in Fig. 9. The asymptotic behaviors of the solid curves are drawn in qualitative accordance with (12.6), (12.7), and (12.23); the other parts of the solid curves are arbitrarily drawn to indicate the general shape of the curve. It is evident that the region outside the curve is the region of stability, and the enclosed region is the region of instability. Similar conclusions have been reached by Heisenberg† but his arguments and results appear to be somewhat obscure.

Having established the existence of the minimum critical value of $R$ and the maximum critical value of $\alpha$, we proceed to make an estimate of their magnitude. We shall see that our theory permits us to give a quite good approximation to the minimum value of $\alpha R$ (cf. (12.30)). Since this roughly corresponds to the minimum value of $R$ and also to the maximum value of $\alpha$ (as will be clear from the individual examples given below), we can get a rough estimate of these values by making a rough estimate of $\alpha$ corresponding to the minimum value of $\alpha R$.

Using the second equation of (8) of Appendix and the approximation (12.7) for $v$, we have approximately

$$\mathfrak{J}_i(z) = v(c) = -\pi w_1' \frac{ww''}{w'^2}. \tag{12.28}$$

* Synge, [63], eq. (11.23), p. 258. His $\lambda$ is our $\alpha$. The condition is originally stated for plane Couette and plane Poiseuille motion; but it is easily seen that it holds for a general velocity distribution with $q=\max|w'|$.

† Heisenberg, loc. cit., p. 601

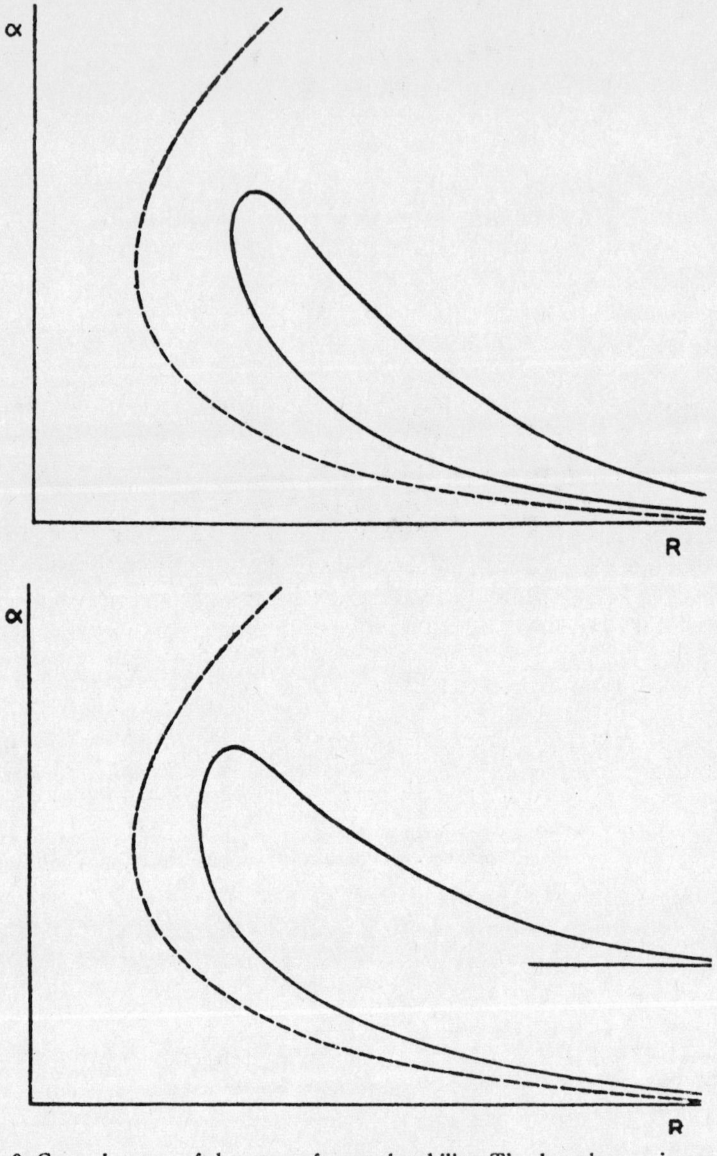


FIG. 9. General nature of the curve of neutral stability. The dotted curve is curve of stability given by Synge.

If we recall that $z$ is proportional to $c(\alpha R)^{1/3}$, this equation determines $(\alpha R)^{1/3}$ as a function of $c$. It can then be easily verified that the minimum value of $(\alpha R)^{1/3}$ occurs when

$$z\mathfrak{J}_i'(z) = cv'(c). \tag{12.29}$$

If the point where this holds is denoted by $z = z_0$, we have approximately from (11.28)*

$$\alpha R = w_1'^2 \left(\frac{z_0}{c}\right)^3. \tag{12.30}$$

* Cf. Heisenberg, loc. cit., eq. (29a), p. 602. He put $z_0 \sim 1$.

The point $z_0$ is roughly the value where $\mathcal{J}_i(z)$ takes its maximum value, because (12.29) is approximately $\mathcal{J}_i'(z)=0$, when $c$ is small. The corresponding value of $\alpha$ can be approximately obtained by taking

$$u = \mathcal{J}_r(z_0), \tag{12.31}$$

in accordance with the first equation of (8) of Appendix, where $u$ is given by the real part of (5) or (7) of Appendix, as the case may be.

e) *Approximate rules.* We now proceed to make some rough approximations in order to obtain simple rules, which are convenient for estimating the minimum critical Reynolds number. The condition $\mathcal{J}_i'(z_0)=0$ gives $z_0=3.21$, where $\mathcal{J}_r(z_0)=1.49$ and $\mathcal{J}_i(z_0)=0.58$. The corresponding value of $c$ can be obtained from the second equation of (8) of Appendix. Putting $u=1.5$ in accordance with (12.31) and neglecting second order terms of $\lambda$, we have

$$v(1-2\lambda) = \mathcal{J}_i(z_0) = 0.58,$$

i.e.,

$$-\pi w_1'(1-2\lambda)(ww''/w'^3) = 0.58, \tag{12.32}$$

where $\lambda$ is defined by (4) of the Appendix. To find the value of $c$ from this equation, it is convenient to plot its left-hand side together with $w(y)$ against $y$, and read the value off the latter curve where the former curve gives the value 0.58. The value of $c$ so obtained turns out to be very close to its maximum value along the neutral curve, and is approximately the value where $R$ is a minimum.

The values of $\alpha$ and $R$ must be obtained from more rough approximations. With a consultation of the values of the integrals $H(c)$, $K(c)$, $M(c)$ and $N(c)$ given in the Appendix, we may derive the following reasonable estimates of $\alpha$:

$$\alpha^2 = w_1'c/H_{10}, \qquad H_{10} = \int_{y_1}^{y_2} w^2 dy, \quad \text{for symmetrical profiles,} \tag{12.33}$$

$$\alpha = w_1'c, \qquad \text{for boundary layer profiles.} \tag{12.34}$$

These values turn out to be somewhat lower than the accurate values. With an approximate allowance for these inaccuracies and taking round numbers, we get the following approximate rules for the minimum critical Reynolds number:

$$R = \frac{30w_1'}{c^3}\sqrt{\frac{H_{10}w_1}{c}}, \quad \text{for symmetrical profiles,} \tag{12.35}$$

$$R = \frac{25w_1'}{c^4}, \qquad \text{for boundary layer profiles.} \tag{12.36}$$

The calculations have been carried out for the Blasius case and the plane Poiseuille case. In the first case, the thickness of the boundary layer is taken so that the initial slope is $w_1'=2$. It is found that

$$\left.\begin{aligned} R &= 5906 \quad \text{for Poiseuille case,} \\ R\delta_1 &= 502 \quad \text{for Blasius case.} \end{aligned}\right\} \tag{12.37}$$

The quantity $\delta_1$ is the displacement thickness

$$\delta_1 = \int_0^\infty (1-w)dy = 0.28673,$$

where $y$ is measured from the solid wall. These values for the minimum critical Reynolds number agree fairly well with those obtained below from more elaborate numerical calculations.

When these estimates of the minimum values of $R$ and the corresponding values of $\alpha$ (eqs. (12.32)–(12.36)) are combined with the asymptotic behavior of the $\alpha(R)$ curves (eqs. (12.16)–(12.17)), the curve of neutral stability in any case can be sketched with fair accuracy with very little labor.

The maximum value of $\alpha$ on the neutral curve cannot be very well estimated. It is usually somewhat higher than the values of $\alpha$ given by (12.33) and (12.34).

**13. Stability characteristics of special velocity distributions.** We shall now apply our theory to some special cases in order to obtain numerical results comparable with experiments. We take (a) the Blasius case as a typical boundary-layer profile, and (b) the plane Poiseuille motion as a typical symmetrical profile. In any case, the resultant curve of stability limit should have the general shape discussed in the last two sections. Only the results will be given here; the method of calculation and its accuracy will be discussed in the Appendix.

a) *Stability of plane Poiseuille flow.* The velocity distribution of plane Poiseuille motion is given by

$$w(y) = 2y - y^2, \quad \text{with} \quad w_1' = 2, \quad H_1(0) = 8/15. \tag{13.1}$$

TABLE II. Behavior of $R(\alpha)$ Curve for Large Values of $R$ for Velocity Distributions with $w''<0$ for the Main Part of the Profile.

| | First branch | | | Second branch |
|---|---|---|---|---|
| | $w_1''<0$ | $w_1''=0$ | $w_1''>0$ | |
| Symmetrical profile | $\alpha^{-11}$ | $\alpha^{-19}$ | $(\alpha-\alpha_s)^{-2}$ | $\alpha^{-7}$ |
| Boundary-layer profile | $\alpha^{-6}$ | $\alpha^{-10}$ | $(\alpha-\alpha_s)^{-2}$ | $\alpha^{-4}$ |

The two branches of the $\alpha(R)$ curve are given by (cf. (12.13), (12.14))

$$\left.\begin{aligned} R^{1/3} &= 8.44(\alpha^2)^{-11/6}, \qquad c = 4\alpha^2/15; \\ R^{1/3} &= 5.96(\alpha^2)^{-7/6}, \qquad c = 0.611\alpha^2. \end{aligned}\right\} \tag{13.2}$$

The numerical results are shown in Table III and Fig. 10. The significance of the column $s$ in the table will be explained in the next section. From the figure, we see that the minimum critical Reynolds number occurs at $R^{1/3}=17.45$, or $R=5314$, agreeing very well with our previous estimation.*

*Earlier results.* The stability of plane Poiseuille flow has been attempted by many authors. Comparatively recent papers are those of Heisenberg, [14], Noether [36], Goldstein [6], Pekeris [39, 40], Synge [64], and Langer [25]. The papers of Goldstein and Synge and one of the papers of Pekeris [39] give definite indication of stability at sufficiently low Reynolds numbers. Heisenberg's paper is in general agreement

* The values given here are somewhat different from those published before [27], because the computation of Tietjen's function has been revised.

with the present investigations. He gave only a very rough calculation, whose result is reproduced in the figure. It seems that his curve is

$$R^{1/3} = 13.4(\alpha^2)^{-11/6}.$$

This is different from our present result (13.2) by a numerical factor. It may be noted

TABLE III. Stability of Plane Poiseuille Flow.

| $c$ | $z$ | $\alpha$ | $R$ | $s$ | $\alpha^2$ | $R^{1/3}$ |
|---|---|---|---|---|---|---|
| 0 | 2.294 | 0 | $\infty$ | .7214 | 0 | $\infty$ |
| 0.05 | 2.363 | 0.3056 | $13.64\times10^5$ | .6901 | 0.0934 | 110.91 |
| 0.10 | 2.448 | 0.4603 | $1.243\times10^5$ | .6544 | 0.2119 | 49.90 |
| 0.15 | 2.540 | 0.6024 | 31048 | .6192 | 0.3629 | 31.43 |
| 0.20 | 2.668 | 0.7506 | 12024 | .5752 | 0.5634 | 22.91 |
| 0.25 | 2.868 | 0.9263 | 6108 | .5161 | 0.8580 | 18.28 |
| 0.266 | 3.012 | 1.0101 | 5369 | .4795 | 1.0203 | 17.51 |
| 0.270 | 3.080 | 1.0414 | 5314 | .4637 | 1.0845 | 17.45 |
| 0.272 | 3.21 | 1.0836 | 5659 | .4358 | 1.1741 | 17.82 |
| 0.270 | 3.240 | 1.0888 | 5920 | .4298 | 1.1854 | 18.09 |
| 0.266 | 3.320 | 1.1007 | 6602 | .4144 | 1.2115 | 18.76 |
| 0.25 | 3.495 | 1.1033 | 9287 | .3836 | 1.2173 | 21.02 |
| 0.20 | 3.857 | 1.0254 | 26597 | .3309 | 1.0514 | 29.85 |
| 0.15 | 4.152 | 0.8824 | 92529 | .2963 | 0.7787 | 45.23 |
| 0.10 | 4.458 | 0.6990 | $4.9435\times10^5$ | .2663 | 0.4886 | 79.07 |

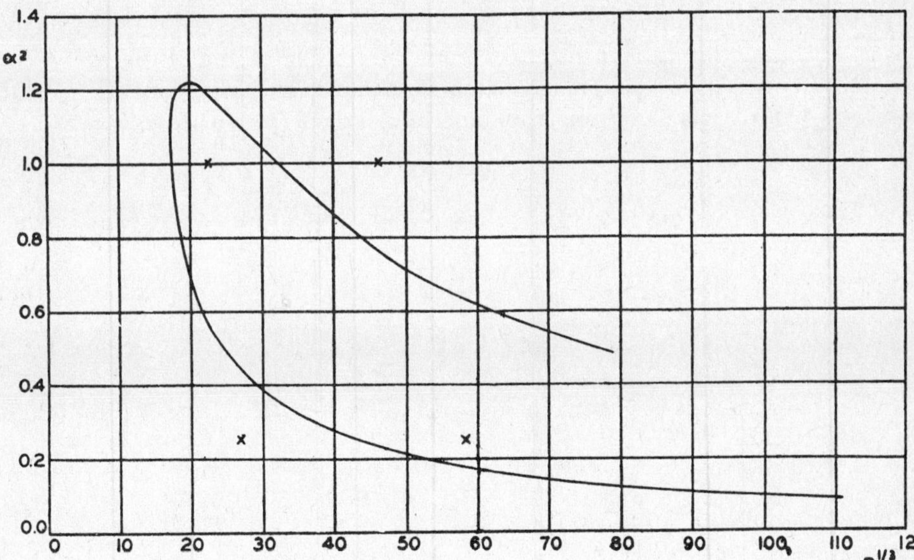


FIG. 10. Curve of neutral stability for the plane Poiseuille case.

that for the values of $\alpha$ for which his curve is drawn, the approximation used in deriving (13.2) is no longer legitimate. Noether's work is based upon a very good mathematical approach, which seems to promise further developments. However, in applying his method to particular examples, he neglected the terms in $\alpha^2$ in the inviscid solutions. As is evident from previous discussions, this is bound to lead to the wrong conclusion that the plane Poiseuille flow is stable (as Noether actually did). The mathematical analysis in Langer's work shows that the region of the $c$-plane for which $c_i > 0$ must go to zero as $\alpha R$ becomes infinite. This is in agreement with present results. Langer, however, concluded from his analysis that the motion is probably stable in general. This would be a natural deduction if the effect of viscosity were only *stabilizing*. The instability of the plane Poiseuille flow must therefore be attributed to the destabilizing effect of viscosity. This is a very significant fact and will be discussed in greater detail in §14.

Pekeris' second paper [40] is a numerical treatment of (4.1), replacing a derivative by a ratio of two finite differences. Unfortunately, his method is not suitable for the purpose, because he has virtually *neglected the inner friction layer*. In his approximation, he divided the half-width of the channel into (at most) four equal parts corresponding to $w=0$, 7/16, 3/4, 15/16, 1. From the present work, we know that the inner friction layer occurs definitely for $c<6/16$. We know also from our previous investigations that the function $\phi$ varies very rapidly in the neighborhood of the inner friction layer. Hence, it is not legitimate to replace $\phi'$ by $\Delta\phi/\Delta y$ for the interval (0, 1/4), $y$ being measured from the solid wall here. Also, most of the combinations of values $(\alpha, R)$ he selected do not correspond to a strong instability. These values are marked with crosses in Fig. 10.

b) *Stability of Blasius flow*. For this case, we choose the boundary-layer thickness to be defined by

$$\tilde{y} = \tilde{\delta} = 6\tilde{x}/\sqrt{R_x}, \qquad R_x = \tilde{u}_1 x/\nu, \tag{13.3}$$

where $\tilde{x}$, $\tilde{y}$ are the *dimensional* distances from the leading edge and the wall respectively, and $\tilde{u}_1$, $\nu$ are the *dimensional* free stream velocity and the kinematic viscosity respectively.* With this definition, the dimensional displacement thickness is

$$\tilde{\delta}_1 = 0.28673\tilde{\delta}. \tag{13.4}$$

Such a choice has the convenience that the initial part of the velocity curve can be very accurately represented by

$$w(y) = 2y - 3y^4, \tag{13.5}$$

$y$ being measured from the wall. Also, since the edge of the boundary layer is farther from the solid wall than that set by Tollmien and Schlichting, greater accuracy can be expected. To make it easy to compare with other results, all final values are presented in terms of

$$\alpha_1 = \alpha\delta_1, \qquad R_1 = R\delta_1. \tag{13.6}$$

The two asymptotic branches of the $\alpha(R)$ curve are given by the following formulae (cf. (12.16) and (12.17)):

$$R_1 = 2.21(10)^{-5}\alpha_1^{-10}, \qquad c = 1.74\alpha_1, \tag{13.7}$$

$$R_1 = 0.0622\alpha_1^{-4}, \qquad c = 4.00\alpha_1. \tag{13.8}$$

* Cf. Goldstein [7], vol. I, p. 135.

These formulae may be compared with those given by Tollmien.* The complete numerical result is shown in Table IV and Fig. 11. The minimum critical Reynolds number occurs at $R_1=420$, agreeing fairly well with our previous estimation and the earlier results of Tollmien and Schlichting.

TABLE IV. Stability of Blasius Flow.

| $c$ | $z$ | $\alpha$ | $R$ | $s$ | $\alpha_1$ | $R_1$ |
|---|---|---|---|---|---|---|
| 0 | 2.294 | 0 | $\infty$ | .7214 | 0 | $\infty$ |
| 0.05 | 2.294 | 0.0473 | $81.45\times10^5$ | .7214 | 0.0136 | $23.353\times10^5$ |
| 0.10 | 2.296 | 0.1040 | $4.655\times10^5$ | .7205 | 0.0298 | $1.335\times10^5$ |
| 0.15 | 2.311 | 0.1730 | 84555 | .7135 | 0.0496 | 24244 |
| 0.20 | 2.341 | 0.2588 | 24783 | .6998 | 0.0742 | 7106 |
| 0.25 | 2.396 | 0.3693 | 9536 | .6759 | 0.1059 | 2734 |
| 0.30 | 2.481 | 0.5156 | 4388 | .6414 | 0.1478 | 1258 |
| 0.35 | 2.624 | 0.7149 | 2358 | .5897 | 0.2050 | 676 |
| 0.40 | 2.942 | 1.0778 | 1477 | .4967 | 0.3090 | 423 |
| 0.411 | 3.21 | 1.2968 | 1470 | .3459 | 0.3718 | 421 |
| 0.40 | 3.540 | 1.4264 | 1944 | .3763 | 0.4090 | 557 |
| 0.35 | 4.219 | 1.2992 | 5392 | .2893 | 0.3725 | 1546 |
| 0.30 | 4.382 | 1.0055 | 12399 | .2733 | 0.2883 | 3555 |
| 0.25 | 4.685 | 0.7578 | 34739 | .2472 | 0.2173 | 9961 |

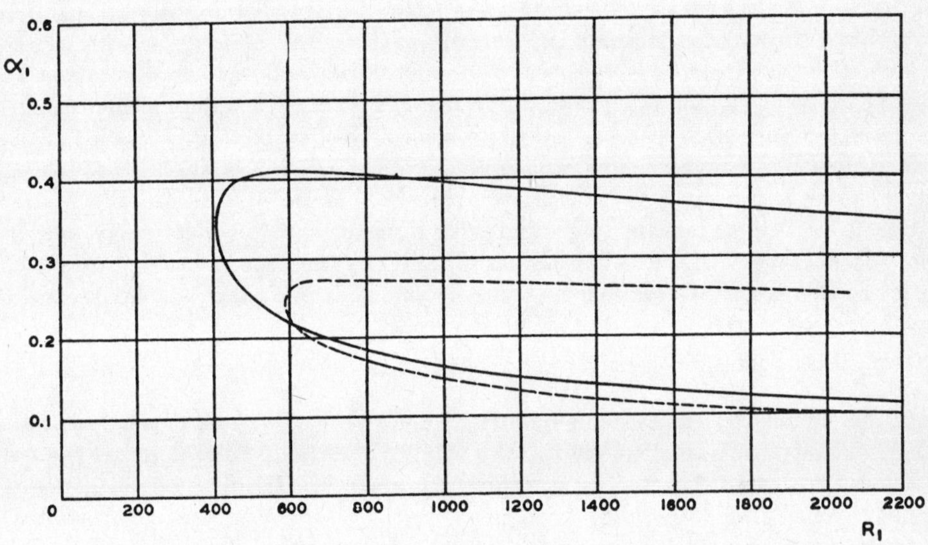


FIG. 11. Curve of neutral stability for the Blasius case: ——— present calculation, – – – Schlichting's calculation.

*Earlier results*. The stability of the boundary layer has been calculated by Tollmien and later by Schlichting, approximating the velocity distribution by linear and parabolic distributions. For the evaluation of the imaginary part corresponding to the inviscid solutions, they used the exact profile. The calculation of Schlichting is shown dotted in the figure. Tollmien's curve agrees fairly well with the present calculations,

* Loc. cit. [73], first paper, p. 42.

except for a somewhat lower peak. Schlichting also calculated the amplification of the unstable disturbances [52], and the amplitude distribution and energy balance of the neutral disturbances [54]. Since the neutral curve in his calculation is inexact, it might be desirable to repeat some of his work if experimental results were available. For those calculations, the present scheme promises less numerical labor than Schlichting's original work.

**14. Physical significance of the results. Prospect of further developments.** Let us now summarize all the results which have been obtained and discuss their physical significance. In the first place, we may conclude that all the inertia forces controlling the stability of two-dimensional parallel flows can be considered in terms of the distribution of vorticity. If the gradient of vorticity of the main flow does not vanish inside the fluid, then self-excited disturbances cannot exist except through the effect of viscosity.

In fact, the effect of viscosity is in general *destabilizing* for very large Reynolds numbers. Thus, if a wavy disturbance of *finite* wave-length can exist neutrally for an inviscid fluid, it will be *amplified through the effect of viscosity.* Indeed,* if the Reynolds number of a flow is continually decreased, a disturbance of finite wave-length, which is damped at very large Reynolds numbers, becomes amplified, unless the wave-length is so small as to cause excessive dissipation at any Reynolds number. For still smaller Reynolds numbers, the damping effect becomes predominant, and we have again a decay of the disturbance. However, for the particular disturbance of infinite wave-length (essentially a steady deviation), the effect of viscosity may be said to be always of the nature of a damping.

The effect of viscosity is essentially one of diffusion of vorticity. It can be seen more clearly from the following considerations. Let us imagine a disturbance originating from the inner friction layer where the phase velocity of the disturance is equal to the velocity of the main flow. During one period $2\pi l/\alpha c U$ of the disturbance, the viscous forces will propagate it side-wise through a distance of the order $\sqrt{2\pi \nu l/\alpha c U} = l\sqrt{2\pi/\alpha R c}$. It is significant to compare this distance with the distance between the inner friction layer and the solid boundary. For if they are nearly equal, it means that the effect of viscosity is dominant at least from the solid surface to that layer. This ratio is approximately

$$s = \sqrt{2\pi/z^3} \tag{14.1}$$

where $z$ is defined by (6.28). This quantity may be regarded as a measure of the effect of viscosity. Its value is included in Tables III and IV. We notice that the value of $s$ decreases from 0.7 to 0.5 as we follow the lower branch of the neutral curve of stability from infinite Reynolds number to the minimum critical Reynolds number. Then, as $s$ decreases to zero, we are following the other branch of the neutral curve to infinite Reynolds numbers. Thus (see Figs. 9, 10, 11), the lower branch is essentially controlled by the effect of viscosity. The effect here is stabilizing, since an increase of Reynolds number gives instability. On the other branch, the effect of viscosity on diffusion of vorticity is predominant in comparison with the effect of dissipation. Here, an increase of Reynolds number gives stability; i.e., the effect of viscosity is destabilizing. This destabilizing mechanism is essentially to shift the phase difference between the $u$ and $v$ components of the disturbance. It has been ex-

* See Fig. 9.

plained in some detail in Prandtl's article [42] from the point of view of energy balance.

If we consider disturbances from the wall and from the inner friction layer, we may regard the region in between to be wholly governed by the effect of viscosity, if these disturbances meet after a period. Thus, it is not without significance that the minimum critical Reynolds number occurs for $s=\frac{1}{2}$ approximately, which may be regarded as marking the passage from stabilizing effect to destabilizing effect of the viscous forces.

These discussions hold both for symmetrical velocity distributions and for boundary-layer distributions. In both cases, it has been demonstrated that instability is essentially caused by the effect of viscosity. These velocity distributions are unstable whether a point of inflection occurs in the velocity profile or not. Thus, although the gradient of vorticity plays a part in controlling the stability of the flow, it is by no means the dominant factor, particularly at low Reynolds numbers. There is thus no reason to associate a point of inflection in the profile directly with instability. This removes Taylor's objection of instability theories based on von Doenhoff's experiments.* Even if the point of inflection in the velocity profile occurs in the leading part of the plate in his experiments, the flow there is definitely stable.

There is another objection raised by Taylor against Tollmien's work on the stability of the boundary layer. He questions whether the change of boundary-layer thickness should not have a drastic influence. This point can best be settled experimentally. So far as mathematical considerations are concerned, it seems justifiable to consider a boundary layer as a parallel flow; the fractional variation of thickness is very small over a distance of one wave-length of the disturbance, and the error incurred is only a few per cent. A fuller discussion of all the errors involved in the theory will be given in the Appendix.

Another point should be settled by experimental investigations. Since the general impression had been that the plane Poiseuille flow was stable, Prandtl suggested that instability occurred at the entrance flow where the velocity distribution is not yet parabolic. The present work certainly concludes that such entrance flows are unstable, if they can be considered as approximately parallel. It is hard to decide theoretically whether a well-developed turbulence has already been reached before the parabolic profile is established. This presumably depends upon the conditions of disturbance at the inlet. The question can be best settled experimentally.

Of the six types of problems mentioned in section 3, Part I, the three types (1), (2), and (5) seem to be quite settled. The present work on the boundary layer checks Tollmien's result approximately, with a minimum critical Reynolds number $R_1=R\delta_1=420$. The minimum critical Reynolds number for plane Poiseuille flow is found to be 10600 based upon the width of the channel. These values are at least not in disagreement with the existing experimental results. It would be very interesting if experiments could be carried out to check the theoretical results so far obtained.

Since plane Couette motion is concluded to be stable while plane Poiseuille motion is concluded to be unstable, it seems interesting to investigate a combination of them to find out when does the instability begin as one changes both the pressure gradient and the relative motion of the plates.

* Taylor, loc. cit. [70], p. 308.

The stability of two-dimensional jets and wakes has never been investigated with the effect of viscosity included. It seems that a study of the stability of the two-dimensional wake might give us valuable information regarding the von Kármán vortex street,—particularly regarding the minimum Reynolds number of its occurrence and the width of the street as compared with the size of the body.*

*Transition to turbulence.* The success of Taylor's theory of transition [68, 70] to turbulence in the boundary-layer as caused by external turbulence seems to throw the instability theories at a disadvantage. However, it seems that Taylor's work should be regarded as only one phase of the problem, i.e., concerning cases where the external turbulence plays the dominant role. In fact, it is not impossible to construct an instability theory, taking account of the free turbulence outside the boundary-layer if this is the main cause of transition. The boundary condition $\phi'+\alpha\phi=0$ at the edge of the boundary layer signifies that the disturbance there has equal magnitudes in directions parallel and perpendicular to the wall. This can be easily reconciled with the nearly isotropic turbulence in the free stream. Of course, the theory can only be pushed to the point where ***non-linear*** effects ***begin*** to appear. Otherwise, we have to deal with a very difficult mathematical problem. It is possible that the beginning of non-linear effect is not far from the actual point of transition. Then the instability theory should give useful results regarding transition, which might be expected to check with experiment.

## Appendix

In the following paragraphs, we shall describe the methods by which the numerical calculations are carried out. We shall then give a discussion of the numerical accuracy involved in the calculations. Special emphasis will be placed on the case of Blasius flow.

a) *Transformation of equations.* The basic equations for the determination of the stability characteristics are given at the end of Part I. To carry out the numerical calculation in any particular case, we have to evaluate the functions (6.24) which occur in the equations (6.14), (6.15) and (6.17) through the relations (6.26). It is found convenient to transform (6.24) into

$$\left.\begin{aligned}
\phi_{12} &= (1-c)(1-\alpha^2H_2)^{-1}\left(1-\sum_{n=2}^{\infty}\alpha^{2n}M_{2n}\right),\\
\phi_{22} &= K_1\phi_{12}-(1-c)\sum_{n=0}^{\infty}\alpha^{2n}N_{2n+1},\\
\phi_{12}' &= (1-c)(1-\alpha^2H_2)^{-1}\left(\alpha^2H_1-\sum_{n=2}^{\infty}\alpha^{2n}M_{2n-1}\right)+(1-c)^{-1}w_2'\phi_{12},\\
\phi_{22}' &= K_1\phi_{12}'+(1-c)^{-1}\left(1-\sum_{n=1}^{\infty}\alpha^{2n}N_{2n}\right)+(1-c)^{-1}w_2'\phi_{22},
\end{aligned}\right\}\quad(1)$$

where the functions $M_n(c)$ and $N_n(c)$ are defined by

$$\left.\begin{aligned}
M_n &= H_n - H_2H_{n-2}, \qquad n \geqq 3,\\
N_n &= K_n - K_1H_{n-1}, \qquad n \geqq 2.
\end{aligned}\right\}\quad(2)$$

* This problem has been attempted by Heisenberg; see Goldstein's book [7].

The principal advantages of such transformations is to bring out the dominant terms in the functions $\phi_{12}$, $\phi_{12}'$, $\phi_{22}$ and $\phi_{22}'$. For the terms in $M$'s and $N$'s are usually very small (particularly for small values of $c$, with which we are usually concerned), while all the terms in the series of (6.24) are of considerable importance. This point will be fully discussed below.

The calculation of (6.13) (Case 1) is still quite complicated; it is found necessary to take (6.15) (Case 2b) as a first approximation. Since we are not going to make any actual calculation for this case, we shall not go into further details with (6.13). All the other equations (6.14), (6.15), and (6.17) (Cases 2a, 2b, 3) can be transformed into the form

$$\mathcal{J}(z) = \frac{(1+\lambda)(u+iv)}{1+\lambda(u+iv)}, \tag{3}$$

with $\mathcal{J}(z)$ defined by (6.31) and $\lambda=\lambda(c)$ defined by

$$w_1'(y_1 - y_0) = -c(1+\lambda). \tag{4}$$

Thus, $\lambda$ is usually very small. The quantities $u$ and $v$ are real functions of $\alpha$ and $c$, different for different cases. For Case 2a we have

$$\begin{aligned} u+iv &= 1+w_1'c\phi_{22}'/\phi_{12}' \\ &= w_1'c\left(K_1+\frac{1}{w_1'c}\right) \\ &\quad +\frac{w_1'c}{\alpha^2}(1-\alpha^2H_2)(1-\alpha^2H_2-\alpha^4N_4-\cdots)(H_1-\alpha^2M_3-\alpha^4M_5-\cdots)^{-1}, \end{aligned} \tag{5}$$

where the second form of the right-hand side is derived by using (1). Similarly, for Cases 2b and 3, we have, respectively,

$$\begin{aligned} u+iv &= 1+w_1'c\phi_{22}/\phi_{12} \\ &= w_1'c\left(K_1+\frac{1}{w_1'c}\right) \\ &\quad +w_1'c\alpha^2(1-\alpha^2H_2)(1-\alpha^4M_4-\alpha^6M_6-\cdots)^{-1}(N_3+\alpha^2N_5+\cdots) \end{aligned} \tag{6}$$

and

$$\begin{aligned} u+iv &= 1+w_1'c(\phi_{22}'+\alpha\phi_{22})/(\phi_{12}'+\alpha\phi_{12}) \\ &= w_1'c\left(K_1+\frac{1}{w_1'c}\right) \\ &\quad +\frac{w_1'c}{\alpha^2}\,\frac{(1-\alpha^2H_2)\{(1-\alpha^2H_2-\alpha^4N_4-\cdots)-(1-c)^2(\alpha^3N_3+\alpha^5N_5+\cdots)\}}{(1-c)^2(1-\alpha^4M_4-\alpha^6M_6-\cdots)+\alpha(H_1-\alpha^2M_3-\cdots)}. \end{aligned} \tag{7}$$

Equation (3) contains the two real equations

$$\left.\begin{aligned} \mathcal{J}_r(z) &= (1+\lambda)\{u(1+\lambda u)-\lambda v^2\}\{(1+\lambda u)^2+(\lambda v)^2\}^{-1}, \\ \mathcal{J}_i(z) &= (1+\lambda)v\{(1+\lambda u)^2+(\lambda v)^2\}^{-1}, \end{aligned}\right\} \tag{8}$$

for $z$, $\alpha$, $c$. Thus, for each value of $z$, we can determine corresponding values of $\alpha$ and $c$. Finally, the Reynolds number is given by (cf. (6.28))

$$\alpha R = \frac{1}{w_0'(1+\lambda)^3}\left(\frac{w_1' z}{c}\right)^3. \tag{9}$$

The actual procedure of calculation will be described presently.

b) *Procedure of calculation.* The calculations required in §13 are as follows: (a) to find the values of $\alpha$ and $z$ corresponding to each value of $c$ by using equations (8), with $u$ and $v$ defined by (7) and (9); and (b) to calculate $R$ from (9). To do this, we may take the following procedure. We first plot $\mathcal{J}_i$ against $\mathcal{J}_r$; then plot the corresponding right-hand side members of (8) in a similar manner in the same diagram. Noting that the latter are functions of $\alpha$ and $c$ only, we may plot by drawing curves of constant $\alpha$ (or constant $c$). The intersections of this set of curves with the $(\mathcal{J}_r, \mathcal{J}_i)$ curve give the desired results.

This procedure is however, very laborious. A simpler method is as follows: As will be seen below, the imaginary parts of $H$'s, $M$'s and $N$'s appearing in (5) and (7) are very small compared with that of $K_1(c)$, we can therefore use the approximation

$$v = v(c) = -\pi w_1'(ww''/w'^3) \quad \text{for} \quad w = c. \tag{10}$$

The following steps are then taken:

i) *Calculation of* $\alpha R$. In this step, the auxiliary functions

$$\lambda(c),\ w_0'(c),\ v(c)$$

are required. These are tabulated in Tables V and VI for the cases considered. Having calculated these functions, we can determine $z$ and $u$ for each value of $c$; $\alpha R$ is then

TABLE V. Auxiliary Functions for Calculating the Stability of the Plane Poiseuille Flow.

| $c$ | $w_0'$ | $\lambda$ | $v$ | $w_1' cRl$ | $H_1$ | $H_2$ | $M_3$ | $N_3$ |
|---|---|---|---|---|---|---|---|---|
| 0 | 2 | 0 | 0 | 0 | 0.53333 | 0.21817 | 0.06038 | 0.19340 |
| 0.05 | 1.94936 | 0.01282 | 0.08482 | 0.06499 | 0.46917 | 0.20696 | 0.05047 | 0.20401 |
| 0.10 | 1.89737 | 0.02633 | 0.18397 | 0.10187 | 0.41000 | 0.19516 | 0.04183 | 0.21636 |
| 0.15 | 1.84391 | 0.04061 | 0.30066 | 0.13015 | 0.35583 | 0.18276 | 0.03441 | 0.22777 |
| 0.20 | 1.78885 | 0.05573 | 0.43905 | 0.15351 | 0.30667 | 0.16982 | 0.02813 | 0.23918 |
| 0.25 | 1.73205 | 0.07180 | 0.60460 | 0.17356 | 0.26250 | 0.15626 | 0.02293 | 0.25000 |
| 0.30 | 1.67332 | 0.08893 | 0.80463 | 0.19121 | 0.22333 | 0.14209 | 0.01872 | 0.26191 |
| 0.35 | 1.61245 | 0.10728 | 1.04910 | 0.20699 | 0.18917 | 0.12729 | 0.01540 | 0.27000 |
| 0.40 | 1.54919 | 0.12701 | 1.35193 | 0.22130 | 0.16000 | 0.11182 | 0.01282 | 0.27062 |
| 0.45 | 1.48324 | 0.14836 | 1.73296 | 0.23438 | 0.13583 | 0.09563 | 0.01087 | 0.26232 |
| 0.50 | 1.41421 | 0.17157 | 2.22144 | 0.24645 | 0.11667 | 0.07875 | 0.00937 | 0.23366 |

determined from (9). In actual practice, a procedure of successive approximations is used. By taking

$$u^{(0)} = \mathcal{J}_r(z^{(0)}), \qquad z = z^{(0)}, \qquad \mathcal{J}_r^{(0)} = \mathcal{J}_r(z_r^{(0)}), \qquad \mathcal{J}_i^{(0)} = \mathcal{J}_i(z^{(0)}) = v$$

as the initial approximation for $u$, $z$, $\mathcal{J}_r$, $\mathcal{J}_i$, we can obtain the successive approxima tions by the formulae

$$\begin{cases} \mathcal{J}_i^{(n+1)} = v\{1+\lambda\}\{(1+\lambda u^{(n)})^2 + (\lambda v)^2\}^{-1} \\ u^{(n+1)} = \mathcal{J}_r^{(n+1)}\{(1+\lambda u^{(n)})^2 + (\lambda v)^2\}\{(1+\lambda)(1+\lambda u^{(n)})\}^{-1} + \lambda v(1+\lambda u^{(n)})^{-1} \end{cases}$$

TABLE VI. Auxiliary Functions for Calculating the Stability of Blasius Flow.

| $c$ | $v$ | $w_1' cRl$ | $H_1$ | $H_2$ | $M_3$ | $N_3$ |
|---|---|---|---|---|---|---|
| 0 | 0 | 0 | 0.60260 | 0.23513 | 0.07966 | 0.17615 |
| 0.05 | 0.00088 | 0.02194 | 0.53378 | 0.22392 | 0.06975 | 0.18676 |
| 0.10 | 0.00709 | 0.06121 | 0.46995 | 0.21212 | 0.06111 | 0.19911 |
| 0.15 | 0.02406 | 0.12108 | 0.41112 | 0.19972 | 0.05369 | 0.21052 |
| 0.20 | 0.05773 | 0.20469 | 0.35730 | 0.18678 | 0.04741 | 0.22193 |
| 0.25 | 0.11498 | 0.31568 | 0.30848 | 0.17322 | 0.04221 | 0.23275 |
| 0.30 | 0.20438 | 0.45835 | 0.26465 | 0.15905 | 0.03800 | 0.24466 |
| 0.35 | 0.33718 | 0.63751 | 0.22583 | 0.14425 | 0.03468 | 0.25275 |
| 0.40 | 0.52839 | 0.85766 | 0.19200 | 0.12878 | 0.03210 | 0.25337 |
| 0.45 | 0.79820 | 1.12133 | 0.16318 | 0.11259 | 0.03015 | 0.24507 |
| 0.50 | 1.17350 | 1.42576 | 0.13935 | 0.09571 | 0.02865 | 0.21641 |

In each approximation, $\mathcal{J}_r^{(k)}$ and $z^{(k)}$ are determined graphically from $\mathcal{J}_i^{(k)}$.

ii) *Calculation of* $\alpha$. For this purpose, the additional auxiliary functions

$$w_1' cRl = w_1' cRl\left\{K_1(c) + \frac{1}{w_1' c}\right\},\ H_1(c),\ H_2(c),\ M_3(c),\ N_3(c),\ \cdots$$

are required. These are tabulated in Tables V and VI for the cases considered. The methods of evaluating these functions and their accuracy will be discussed below. For sufficient accuracy in the final results, only the real parts of $H_2$, $M_3$, $N_3$ are required, besides $w_1' cRl$ and $H_1$. Having calculated these functions, we can determine the value of $\alpha$ from the real part of the equations (5) or (7). A similar method of suc cessive approximations may be used by writing those equations in the forms

$$\alpha^2 = \frac{w_1' c}{H_1(u - w_1' cRl)}(1-\alpha^2 H_2)\frac{1-\alpha^2 H_2 - \alpha^4 N_4 - \cdots}{1-\alpha^2 P_2 - \alpha^4 P_4 - \cdots}, \qquad P_{2n} = M_{2n+1}/H_1,$$

$$\alpha = \frac{w_1' c}{u - w_1' cRl}(1-\alpha^2 H_2)\frac{(1-\alpha^2 H_2 - \alpha^4 N_4 - \cdots) - (1-c)^2(\alpha^3 N_3 + \alpha^5 N_5 + \cdots)}{(1-c)^2(1-\alpha^4 M_4 - \alpha^6 M_6 - \cdots) + \alpha(H_1 - \alpha^2 M_3 - \alpha^4 M_5 \cdots)}.$$

An approximate value of $\alpha$ is put into the right-hand side to obtain an approximation of the higher order on the left-hand side. For the initial approximation, take $\alpha = 0$.

c) *Numerical accuracy of the calculations.* The numerical accuracy of our calculation as based upon the final equations given in section 6 are limited by several factors:

i) by neglecting quantities of the orders $e^{-P}$ and $(\alpha R)^{-1}$ in the reduction of the determinantal equations of the boundary-value problems,

ii) by using the inviscid solutions for $\phi_1$ and $\phi_2$ (error of the order $(\alpha R^{-1})$),

iii) by the approximations of the rapidly varying solutions $\phi_3$ and $\phi_4$ as discussed at the end of §6,

iv) by the boundary-layer approximation used in setting up the equation oı stability (except in the cases of plane Couette and Poiseuille flows).

Finally, certain numerical approximations have to be used in the actual evaluation of the quantities $u$ and $v$ in equations (11.11). We shall now discuss these factors one by one.

The inaccuracy due to (i) and (ii) is negligible in all the cases considered, because $\alpha R$ is always sufficiently large. In connection with (iii), the situation is more complicated. The first approximation of the asymptotic solution should give an error of the order of $(\alpha R)^{-1/2}$; while the first approximation using Hankel functions should give an error of the order $(\alpha R)^{-1/3}$. It might therefore be thought that the asymptotic method should always give a better approximation. However, this is not the case. For the order of accuracy of the first method is based upon a fixed value of $y$, while that of the second is based upon a fixed value of $\eta$. Thus, if $\alpha R$ may be allowed to become very large while $y-y_0$ remains to be of the order of unity, the first method is definitely better. This is the case with the quantities $\phi_{42}$ and $\phi'_{42}$. With $\phi_{31}$ and $\phi'_{31}$, the situation is different. Here, $y_1-y_0$ is always small. Except for one branch of the neutral curve for profiles with a flex, $y_1-y_0$ goes to zero as $\alpha R$ becomes infinite. Because of the smallness of $y_1-y_0$, the asymptotic solution (which fails to be accurate in the neighborhood of $y_0$) never gives a good approximation. This is why the other method has to be used in most of the calculations, and we are limited to an accuracy of $(\alpha R)^{-1/3}$. We note that the curvature of the velocity distribution does not come into this approximation. Thus, for better accuracy, a second approximation should be used, the error being then reduced to the order of $(\alpha R)^{-2/3}$. However, since the error in the method used is only a few per cent, and an improvement in accuracy would not alter the general conclusions, it does not seem worth while to improve the accuracy in the light of general interest. Indeed, the inaccuracy due to the other causes (to be discussed) is also of the same order of magnitude. Another support to the method used is that it does agree with the asymptotic method when $z$ is large; there is only a negligible difference (cf. eq. (6.31)).

The effect of the change of the thickness of the boundary layer might be taken to be more serious than a mere numerical inaccuracy. Taylor regarded this as invalidating the existing instability theory of the boundary layer. This question can best be settled experimentally. For the present, we only want to discuss its effect upon our boundary value problem. An approximate estimate of this effect may be obtained by considering the change of the thickness of the boundary layer for one wavelength of the disturbance. This can be easily verified* to be $\pi(1.72)^2/\alpha_1 R_1$. For the lowest value of $\alpha_1 R_1$ involved in the calculations of §13, this is about 6 per cent. Thus, the error is not large. Hence, in the physical interpretation of the results, we need only consider a change of Reynolds number as we pass down stream. One interesting point is the following: as the Reynolds number keeps on increasing, all disturbances finally become stable, if the linear theory holds throughout. Thus, the transition to turbulence depends upon the occurrence of the non-linear effect and hence must depend upon the amount of initial disturbance.

d) *Calculation of* $\phi_{12}^{(0)}$, $\phi_{22}^{(0)}$, *etc.* We shall now discuss the method by which these quantities are evaluated for the calculation of $u$ and $v$ in the equations (5) and (7). A discussion of the accuracy of the present method and of Tollmien's method of evaluating these quantities will also be made.

The original question is to evaluate the integrals $H_m(c)$ and $K_m(c)$ as occurring in (6.24). Various methods are possible for carrying out the calculation, including straightforward numerical integration. The method to be described is an attempt at a simple one. With the transformations (2), we hope to bring out the dominant terms of the series (6.24), and the calculations of $u$ and $v$ according to (7) and (9) are based upon the use of the transformed series (1). We make the following approximations.

* Cf. Goldstein [7], last column of table of p. 157.

i) The imaginary part $v$ is chiefly given by that of the first term, namely, $w_1 c(K_1+1/w_1' c)$; this implies that the imaginary part due to $H_2(c)$, $M_3(c)$, $N_3(c)$, etc. is negligible.

ii) The real part receives also little contribution from those of $H_2(c)$, $M_3(c)$, $N_3(c)$, etc., and hence these need be calculated only approximately.

iii) The series are cut short; terms like $N_4$, $M_4$, $\cdots$ are entirely neglected.

Let us proceed to justify these approximations.

The justification of (ii) and (iii) is based upon the following two facts.

a) The quantities in the series involved decrease roughly like $1/m!$, $m$ being the number of integrations involved in defining a certain term.

b) For $\alpha<1$, the terms also decrease as $\alpha^m$. Thus, the accuracy is not very good for $\alpha>1$, namely for low Reynolds numbers.

But from a consultation of Tables V and VI, and the manner in which the integrals $H_2(c)$, $M_3(c)$, $H_3(c)$, etc., enter (5) and (7), we see that an error of ten per cent in these integrals will cause a negligible error in the final results.

The justification of (i) needs more explanation. For definiteness, let us take $N_3(c)$ as an example. Now,

$$N_3(c) = \int_{y_1}^{y_2} dy(w-c)^{-2}\int_{y_1}^{y} dy(w-c)^2\int_{y_1}^{y} dy(w-c)^{-2}.$$

This can be expressed as the sum of the following three integrals:

$$N_{31}(c) = K_1(c)\cdot\int_{y_1}^{y_0} dy(w-c)^2\int_{y}^{y_2} dy(w-c)^{-2},$$

$$N_{32}(c) = \int_{y_1}^{y_2} dy(w-c)^{-2}\int_{y_0}^{y} dy(w-c)^2\int_{y}^{y_2} dy(w-c)^{-2},$$

$$N_{33}(c) = \int_{y_0}^{y_2} dy(w-c)^{-2}\int_{y_0}^{y} dy(w-c)^2\int_{y}^{y_2} dy(w-c)^{-2}.$$

The third integral is real, because $y>y_0$. A further transformation of the last integration in $N_{31}$ and $N_{32}$ like

$$\int_{y}^{y_2} dy(w-c)^{-2} = K_1(c) - \int_{y_1}^{y} dy(w-c)^{-2}$$

gives

$$N_{31} = \{K_1(c)\}^2\int_{y_1}^{y_0} dy(w-c)^2 - K_1(c)\int_{y_1}^{y_0} dy(w-c)^2\int_{y_1}^{y} dy(w-c)^{-2}\cdot$$

$$N_{32} = K_1(c)\int_{y_1}^{y_0} dy(w-c)^{-2}\int_{y_0}^{y} dy(w-c)^2$$

$$-\int_{y_1}^{y_0} dy(w-c)^{-2}\int_{y_0}^{y} dy(w-c)^2\int_{y_1}^{y} dy(w-c)^{-2}.$$

Now, the last integral is real because $y<y_0$. Further, it can be easily verified that

$$\int_{y_1}^{y_0} dy(w-c)^2 \int_{y_1}^{y} dy(w-c)^{-2} = \int_{y_1}^{y_0} dy(w-c)^{-2} \int_{y_0}^{y} dy(w-c)^2.$$

Hence, the only term which can contribute to the imaginary part of $N_3(c)$ is $\{K_1(c)\}^2 \int_{y_1}^{y_0} dy(w-c)^2$. Now, $c$ is usually small so that we may put

$$\int_{y_1}^{y_0} dy(w-c)^2 = \frac{w_1'^2}{3}(y_1-y_0)^3 = \frac{1}{3}\frac{(w_1'c)^3}{w_1'^4}.$$

Hence,

$$\{K_1(c)\}^2 \int_{y_1}^{y_0} dy(w-c)^2 = \{w_1' cK_1(c)\}^2 \frac{1}{3}\frac{w_1' c}{w_1'^4}.$$

Now we have approximately

$$w_1' cK_1(c) = 1 - vi.$$

Substituting into the above expression, we obtain the imaginary part of $N_3(c)$ as $-2w_1' cv/3w_1'^4$. This will give a contribution of approximately $-(2/3)\{\alpha c(1-c)/w_1'\}^2 v$ to the imaginary part of $v$. This is negligible, because the factor preceding $v$ is at most of the order of 0.02 in our calculations. Hence, it is justifiable to neglect the contribution of $N_3$ to $v$. With the other terms, the approximation is even better; thus, the imaginary part of $H_2(c)$ is of the order of $c^3$ times that of $K_1(c)$, and that of $M_3(c)$ is of the order of $c^6$ times that of $K_1(c)$.

Having thus justified the approximations described above, the task is to evaluate $K_1(c)$, $H_1(c)$, $H_2(c)$, $M_3(c)$, and $N_3(c)$ with proper degree of accuracy. For parabolic distribution, these integrals can be evaluated exactly; the approximation (ii) is not necessary. Thus,

$$H_1(c) = A, \tag{11}$$

$$K_1(c) = -\frac{1}{2a^2(1-a^2)} + \frac{A}{2a^3}\left\{\log\frac{1+a}{1-a} + i\pi\right\}, \tag{12}$$

$$H_2(c) = \frac{1}{30}\frac{4a^2-3}{a^2} + \frac{A}{2a^3}\log(1+a) - \frac{2a^2}{15}\log a^2 - \frac{B}{4a^3}\{\log(1-a^2) - i\pi\}, \tag{13}$$

$$\begin{aligned} M_3(c) = {} & \frac{B^2}{2a^2}K_1 + \frac{8a^2A}{15}\left\{\frac{1}{a+1} + \log\frac{a+1}{a}\right\} \\ & + \frac{1}{225}\left\{-\frac{54}{7} + \frac{108}{5}a^2 - \frac{38}{3}a^4 - 64a^6\right\}, \end{aligned} \tag{14}$$

$$\begin{aligned} Rl\, N_3(c) = {} & \frac{1}{24a^2}(1+3a^2) - \frac{(1-a^2)^2}{16a^5}\log\frac{1+a}{1-a} \\ & + \frac{1}{16a^6}\left\{\int_0^1 dx\left\{(a^2-x^2)\log\left|\frac{a+x}{a-x}\right|\right\}^2 - B\pi^2\right\}, \end{aligned} \tag{15}$$

$$ImN_3(c) = \frac{\pi}{16a^6}\left\{2B\log\frac{1+a}{1-a} + \frac{32a}{15}\log\frac{1+a}{2a}\right.$$

$$\left. - \frac{2a}{15}(1-a)(2+3a+18a^2+20a^3)\right\}, \tag{16}$$

where

$$a^2 = 1 - c, \qquad A = a^4 - \frac{2a^2}{3} + \frac{1}{5}, \qquad B = A - \frac{8a^5}{15}. \tag{17}$$

These are the equations on which Table V is based, where only the real parts are given. For any other profile, the rule is as follows:

i) Evaluate $K_1(c)$ with as much accuracy as possible. Usually, it is broken up into two parts. Thus,

$$K_1(c) = K_{11}(c) + K_{12}(c); \qquad K_{11}(c) = \int_{y_1}^{y_j} dy(w-c)^{-2},$$

$$K_{12}(c) = \int_{y_j}^{y_2} dy(w-c)^{-2}, \tag{18}$$

where $y_1 < y_j < y_2$. The value of $y_j$ is chosen so that $K_{11}(c)$ can be calculated with sufficient accuracy by developing $w$ as a power series of $(y-y_0)$, while $K_{12}(c)$ can be evaluated by developing the integrand as a power series of $c/w$.

ii) Evaluate by numerical integration the quantities

$$H_1(0) = \int_{y_1}^{y_2} dyw^2, \tag{19}$$

$$H_1'(0) = -2\int_{y_1}^{y_2} wdy \tag{20}$$

$$H_2(0) = \int_{y_1}^{y_2} w^{-2}dy\int_{y_1}^{y} w^2dy, \tag{21}$$

$$M_3(0) = \int_{y_1}^{y_2} w^2dy\int_{y}^{y_2} w^{-2}dy\int_{y_1}^{y} w^2dy, \tag{22}$$

$$N_3(0) = \int_{y_1}^{y_2} w^{-2}dy\int_{y_1}^{y} w^2dy\int_{y}^{y} w^{-2}dy. \tag{23}$$

iii) The integral $H_1(c)$ is then given by

$$H_1(c) = H_1(0) + H_1'(0)c + c^2. \tag{24}$$

iv) The real part of the integrals $H_2(c)$, $M_3(c)$, $N_3(c)$ are obtained by comparison with the corresponding quantities for parabolic distribution (Table V). For example,

$$M_3(c) - M_3(0) = \left(\frac{w'}{2}\right)^2 \times \text{corresponding quantity for parabolic distribution.} \tag{25}$$

The idea of the last step is essentially to approximate the given profile with a parabolic one.

For the Blasius profile, (with $w_1' = 2$, $y_j - y_1 = 0.4$). We obtain

$$\left.\begin{aligned} K_{11}(c) + \frac{1}{w_1'c} &= -0.5615 - 0.3937c - 1.543c^2 - 1.803c^3 - 1.368c^4 - 5.022c^5 \\ &\quad + \cdots + \frac{9}{8}\left(c^2 + \frac{21}{8}c^5 + \cdots\right)\left(\log\frac{0.8-c}{c} + i\pi\right), \\ K_{12}(c) &= 0.7080 + 1.3546c + 2.588c^2 + 3.860c^3 + 5.446c^4 + 7.455c^5 \\ &\quad + \cdots, \\ K_1(c) + \frac{1}{w_1'c} &= 0.1465 + 1.2467c + 1.045c^2 + 2.039c^3 + 4.078c^4 + 2.423c^5 \\ &\quad + \cdots + \frac{9}{8}\left(c^2 + \frac{21}{8}c^5 + \cdots\right)\left(\log\frac{0.8-c}{c} + i\pi\right). \end{aligned}\right\} \quad (26)$$

In evaluating these integrals, we take

$$\left.\begin{aligned} w &= 2(y-y_1) - 3(y-y_1)^4, && 0 \leqq y - y_1 \leqq 0.4, \\ w &= 1 - \{0.9 - (y-y_1)\}^2, && 0.4 < y - y_1 \leqq 0.9, \\ w &= 1 && 0.9 \leqq y - y_1 \leqq 1. \end{aligned}\right\} \quad (27)$$

For the integrals $H_1(0)$ and $H_1'(0)$, we make use of the known values of the displacement thickness and the momentum thickness $\delta_2$.

$$\left.\begin{aligned} \delta_1 &= \frac{1}{6}(1.7208) = 0.28673, \\ \delta_2 &= \frac{1}{12}(1.32824) = 0.11067. \end{aligned}\right\} \quad (28)$$

Thus,

$$H_1(0) = 1 - \delta_1 - \delta_2 = 0.6026, \qquad H_1'(0) = -2(1-\delta_1) = -1.4265. \quad (29)$$

The values of $H_2(0)$, $M_3(0)$, $N_3(0)$, as evaluated by numerical integration, are given in the first row of Table VI. The rest of the table is constructed by following the procedure described above.

We see that the method of approximation developed above is purely a numerical one, and the calculation can be done without excessive labor. In any case, even if the above method does not give satisfactory results, suitable approximations can always be devised for the evaluation of the necessary integrals. This is the advantage of using Heisenberg's form of the inviscid solutions. In the method used by Tollmien, it is necessary that the *profile* may be approximated by linear and parabolic parts; otherwise, the numerical labor is excessive. It is not clear at once what is the effect of such an approximation on the solutions $\phi(y)$. A more serious criticism of Tollmien's method is the joining of the inviscid solutions at the point of junction of the two approximate profiles. Mathematically speaking, such a junction presents an essential singularity in the coefficients of the differential equation (3.8) or (3.14). Numerically speaking, serious difficulty would be expected when $c$ is equal or even only very near to the velocity of junction, because the inviscid solution fails at the critical layer where $w=c$. Tollmien did not publish how he overcame this difficulty.*

* Tollmien [73], footnote, p. 37.

NATIONAL ADVISORY COMMITTEE FOR AERONAUTICS

TECHNICAL NOTE NO. 1115

# INVESTIGATION OF THE STABILITY OF THE LAMINAR BOUNDARY LAYER IN A COMPRESSIBLE FLUID

By Lester Lees and Chia Chiao Lin

## SUMMARY

In the present report the stability of two-dimensional laminar flows of a gas is investigated by the method of small perturbations. The chief emphasis is placed on the case of the laminar boundary layer.

Part I of the present report deals with the general mathematical theory. The general equations governing one normal mode of the small velocity and temperature disturbances are derived and studied in great detail. It is found that for Reynolds numbers of the order of those encountered in most aerodynamic problems, the temperature disturbances have only a negligible effect on those particular velocity solutions which depend primarily on the viscosity coefficient ("viscous solutions"). Indeed, the latter are actually of the same form in the compressible fluid as in the incompressible fluid, at least to the first approximation. Because of this fact, the mathematical analysis is greatly simplified. The final equation determining the characteristic values of the stability problem depends on the "inviscid solutions" and the function of Tietjens in a manner very similar to the case of the incompressible fluid. The second viscosity coefficient and the coefficient of heat conductivity do not enter the problem; only the ordinary coefficient of viscosity near the solid surface is involved.

Part II deals with the limiting case of infinite Reynolds numbers. The study of energy relations is very much emphasized. It is shown that the disturbance will gain energy from the main flow if the gradient of the product of mean density and mean vorticity near the solid surface has a sign opposite to that near the outer edge of the boundary layer.

A general stability criterion has been obtained in terms of the gradient of the product of density and vorticity, analogous to the Rayleigh-Tollmien criterion for the case of an incompressible fluid. If this gradient vanishes for some value of the velocity ratio of the main flow exceeding $1-1/M$ (where M is the free stream Mach number), then neutral and self-excited "subsonic" disturbances exist in the inviscid fluid. (The subsonic disturbances die out rapidly with distance from the solid surface.) The conditions for the existence of other types of disturbance have not yet been established to this extent of exactness. A formula has been worked out to give the amplitude ratio of incoming and reflected sound waves.

It is found in the present investigation that when the solid boundary is heated, the boundary layer flow is destabilized through the change in the distribution of the product of density and vorticity, but stabilized through the increase of kinematic viscosity near the solid boundary. When the solid boundary is cooled, the situation is just the reverse. The actual extent to which these two effects counteract each other can only be settled by actual computation or some approximate estimates of the minimum critical Reynolds number. This question will be investigated in a subsequent report.

Part III deals with the stability of laminar flows in a perfect gas with the effect of viscosity included. The method for the numerical computation of the stability limit is outlined; detailed numerical calculations will be carried out in a subsequent report.

## INTRODUCTION

In a recent paper (reference 1), one of the present authors has clarified the theory of the stability of two-dimensional parallel flows in a homogeneous viscous incompressible fluid. The experimental investigations of H. L. Dryden, G. B. Schubauer, H. K. Skramstad (reference 2) and H. W. Liepmann (reference 3) agree with the calculations made by Tollmien (reference 4), Schlichting (reference 5) and those given in the paper quoted (reference 1). Because of the increasing importance of phenomena of gas flow at high speeds, it seems natural that the investigation should be extended to cover the case of a gas, taking into account the effects of compressibility and heat transfer.

The interest in this problem is further enhanced by the fact that disturbances of finite amplitude in high-speed flows are known to have the tendency of building themselves up into shock waves. It is therefore possible that instability of high-speed laminar flows will lead to shock waves instead of turbulence.[1] Although an instability theory involving only small disturbances would not be able to settle this point, it at least paves the way to such an investigation.

[1] This possibility was first pointed out to the authors by Doctor H. W. Liepmann.

Reprinted from JOURNAL OF THE AERONAUTICAL SCIENCES

JULY, 1950 VOLUME 17, No. 7

# On the Stability of the Boundary Layer Over a Cone

R. H. Battin and C. C. Lin
*Department of Mathematics, Massachusetts Institute of Technology, Cambridge, Mass.*
April 19, 1950

RECENTLY, THERE HAS BEEN considerable interest in the stability of the boundary layer over a cone in supersonic flow. Since Hantzsche and Wendt[1] have shown that a suitable transformation of the equations for the steady flow in the boundary layer over a cone leads to those for a flat plate, it is generally recognized that the work of Lees and Lin for the two-dimensional case[2] should be applicable to the cone problem as well. However, a detailed analysis has so far been lacking.

The equations for the small disturbances in the boundary layer over a body of revolution have now been examined, and it has been found that the equations are indeed identical with those for a two-dimensional boundary layer, within the limitation of usual approximations. Thus, the local stability characteristics of the boundary layer over a body of revolution can be examined by existing theories. It is only to be remembered that comparison must be made for the same Reynolds Number based on the thickness of the boundary layer. Indeed, the other dimensionless quantities must also be defined with the boundary-layer thickness as the reference length.

In applying the theory to the *downstream* development of a disturbance of a given *time* frequency, the case of the cone is obviously different from the case of a flat plate, because the boundary-layer thickness varies with downstream distance in different manners. Since comparison must be made at corresponding Reynolds Numbers based on the thickness of the boundary layer, it must involve different distances downstream. In fact, if $x_1$ and $x_2$ are two stations along the flat plate and if $r_1$ and $r_2$ are two stations along the cone having corresponding Reynolds Numbers, the result of Hantzsche and Wendt shows that

$$r_2 - r_1 = 3(x_2 - x_1)$$

Based on this result and the present analysis, it can be shown that, if the amplification of a disturbance between two stations $x_1$ and $x_2$ over a flat plate is $A$, then the corresponding amplification (between $r_1$ and $r_2$), in the case of a cone, is $A^3$. In other words, a plot of log $A$ versus $x$ for the flat plate and the cone differ from each other only by scale factors of 3 for both coordinate axes.

## REFERENCES

1 Hantzsche, W., and Wendt, H., *Die laminare Grenzschicht bei einen mit Uberschellgeschwindigkeit angestroemten nichtangestellen Kreiskegel*, Jahrbuch der Deutschen Luftfahrtforschung, p. 176, 1941.

2 Lees, L., and Lin, C. C., *Investigation of the Stability of the Laminar Boundary Layer in a Compressible Fluid*, N.A.C.A. T.N. No. 1115, 1946; Lees, L., N.A.C.A. T.N. No. 1360, July, 1947, reissued as N.A.C.A. T.R. No. 876, January, 1950.

NATIONAL ADVISORY COMMITTEE FOR AERONAUTICS

TECHNICAL NOTE NO. 2887

# ON THE STABILITY OF THE LAMINAR MIXING REGION BETWEEN TWO PARALLEL STREAMS IN A GAS

By C. C. Lin

## SUMMARY

The stability of the mixing of two streams was studied both for the interest in the problem and for clarifying certain points in the basic stability theory. It is shown that, when the relative speed of the two parallel streams exceeds the sum of their velocities of sound, subsonic oscillations cannot occur and the mixing region may be expected to be stable with respect to small disturbances. It is further shown that, when viscosity and heat conductivity are neglected, if the flow can execute a small neutral subsonic oscillation of finite wave length, it can also execute self-excited oscillations of longer wave lengths and damped oscillations of shorter wave lengths.

Rigorous developments of the mathematical theory of asymptotic solutions confirm previous methods of solution of the stability equations in a compressible fluid. This theory also shows that, at high Reynolds numbers, the damped oscillations in a strictly parallel main flow have a structure similar to that of the vorticity field in fully developed turbulent flow.

Sample calculations are also included exhibiting various quantitative properties of these small oscillations.

## INTRODUCTION

The mixing of two parallel streams of gas occurs in a number of cases. An interesting example is furnished by the slip stream in a three-shock configuration. It has long been suggested that such laminar mixing zones could, at sufficiently high speeds, be stable with respect to small disturbances although they are known to be very unstable at low speeds. The purpose of the present investigation is to find out some of these stability characteristics. Apart from the development of the general theory, there are included the calculations of the neutral and unstable oscillations, the extent of the amplification, and other related properties.

Reprinted from JOURNAL OF THE AERONAUTICAL SCIENCES

JULY, 1955 VOLUME 22, No. 7

# On the Stability of the Laminar Boundary Layer in a Compressible Fluid*

D. W. DUNN† AND C. C. LIN‡

*Massachusetts Institute of Technology*

ABSTRACT

A more complete theory is developed for the stability of the laminar boundary layer in a compressible fluid, removing some limitations and modifying certain general conclusions in the earlier treatment of Lees and Lin. The main changes are the inclusion of three-dimensional disturbances and the demonstration that, in many cases, the stability characteristics depend on temperature fluctuations (contrary to previous conclusions). The former change is made possible by a more careful analysis and simplification of the basic equations of the disturbances; the latter modification is found through improved solutions of these equations.

At high supersonic free-stream velocities, the relation for the characteristic values is found to be dependent on temperature fluctuations and the boundary conditions imposed on them, except for three-dimensional disturbances traveling very obliquely to the direction of the free stream. For subsonic and slightly supersonic velocities, where the wave speed of the disturbances is usually not large, the earlier conclusion of Lees and Lin is still valid—i.e., temperature fluctuations are unimportant for the stability problem.

For the boundary-layer flow over a flat surface, three-dimensional disturbances are found to be of little importance at low subsonic free-stream Mach Numbers, since they usually have a higher minimum critical Reynolds Number than two-dimensional ones, except possibly under conditions of extreme surface cooling. As the Mach Number increases, three-dimensional disturbances become significant under conditions that are less and less extreme, until finally, at a Mach Number between one and two, they begin to play the leading role in many cases of practical interest.

At supersonic free-stream Mach Numbers the boundary layer can never be *completely* stabilized by cooling with respect to *all* three-dimensional disturbances. There is always present a class of unstable three-dimensional disturbances no matter how low the temperature of the solid surface becomes. However, in many cases, surface cooling is still a very effective means of stabilizing the boundary layer.

Numerical results illustrating these conclusions are presented for the case of the boundary layer on a flat plate.

Received February 10, 1954. Revised and received November 18, 1954.

* Part of the material in this paper is contained in a thesis submitted by D. W. Dunn in partial fulfillment of the requirements for the degree of Doctor of Philosophy at the Massachusetts Institute of Technology. Some of the conclusions given in a preliminary version of the paper[13] have been somewhat revised and extended. The numerical results remain the same.

The work was begun 1951–1952 while D. W. Dunn was holding a scholarship from the Research Council of Ontario. It was later completed with the support of the Office of Naval Research, under Contract N5 ori-07872. Some of the numerical calculations were carried out by W. V. Caldwell, also with the support of the Office of Naval Research, under Contract N5 ori-60.

† Formerly, Department of Mathematics, Massachusetts Institute of Technology. Now at the Department of Aeronautics, School of Engineering, The Johns Hopkins University.

‡ Professor of Mathematics, Massachusetts Institute of Technology.

## (1) INTRODUCTION

IN RECENT YEARS there has been considerable interest in the theory of boundary-layer stability for compressible fluids. Most of the recent work on the subject has been based on the original theoretical investigations of Lees and Lin.[1, 2] As has been pointed out on several occasions, there are certain inadequacies in this basic theory which limit its significance and which in some cases make unreliable the results of numerical calculations based on it. The purpose of the present investigation is to clarify some of these problems and, in general, to gain further insight into the stability theory for the laminar boundary layer in a compressible fluid.

One of the chief limitations of the theory up to now has been consideration of only two-dimensional disturbances by analogy with the procedure for an incompressible fluid. Squire[3] has shown by means of a transformation of the equations of the problem that for an incompressible fluid every three-dimensional disturbance is equivalent to a two-dimensional one at a lower Reynolds Number. Therefore, for an incompressible fluid the minimum critical Reynolds Number for two dimensional disturbances is the minimum critical Reynolds Number for all disturbances, so that the restriction to two-dimensional disturbances is justified. If an attempt is made to prove a similar result for a compressible fluid, it is found that the equations obtained by making the usual approximations (see reference 1) are too complicated to permit a transformation of the type used by Squire. In the present work an order-of-magnitude analysis of the terms in the complete equations for a three-dimensional disturbance is carried out. A consistent approximation scheme based on this now yields simplified equations of the same order of accuracy as the usual ones, but with fewer terms, so that a transformation to equations of the same form as for a two-dimensional disturbance is possible. However, these transformed two-dimensional equations are not the equations of a proper two-dimensional disturbance, so that no direct conclusion regarding the importance of three-dimensional disturbances can be made. For a compressible fluid the properties of three-dimensional disturbances have to be found by means of a more elaborate investigation than for an incompressible fluid. The transformed two-dimensional equations lead to a considerable simplification mathematically, however.

It is found in the present investigation that for subsonic free-stream Mach Numbers three-dimensional disturbances are not of much importance, but for supersonic Mach Numbers they may play a leading role in certain circumstances. In particular, for supersonic sonic Mach Numbers there is always present a class of unstable three-dimensional disturbances no matter how low the temperature of the solid surface is, so that complete stabilization of the boundary layer in such cases, as predicted in reference 2, is impossible in the strict sense.

It is known that at high Mach Numbers the usual form of the boundary-layer theory becomes inaccurate,[4, 5] so there is some question as to the proper form of the stability theory in such cases. An investigation of this problem[6] has shown that, as long as the proper velocity and temperature profiles for the mean flow are used, the simplified stability equations derived in this paper are still valid as a first approximation at high Mach Numbers. However, in such cases the accuracy of a first approximation may not be great, so that higher approximations and thus more complicated equations may have to be considered.†

Various limitations to some of the approximations made in the theoretical developments of reference 1 have been brought out. In fact, as pointed out in reference 7, the original theory of reference 1 has sometimes been used in situations in which it was not intended to be valid. This theory was formulated with the restriction to small values of the wave speed $c$, so that, strictly speaking, its range of reliability is limited to values of the Mach Number not too much larger than unity. In particular, the viscous solutions of reference 1 are accurate first approximations only in the immediate neighborhood of the point where $w = c$. However, at high Mach Numbers, $c$ is often large in the stability calculations, so that, in order to satisfy conditions at the solid boundary, the viscous solutions have to be accurate far away from this point. For an incompressible fluid, Tollmien[8] obtained viscous solutions that are uniformly accurate all across the boundary layer. This work has been extended to the case of a compressible fluid, and a new mathematical formulation of the stability problem in terms of the new viscous solutions has been obtained. In the present paper the modified formulas for the calculation of the characteristic values are given. The detailed mathematical discussion will be presented elsewhere.

The conclusion reached in reference 1—that the relation for the characteristic values is independent of the temperature fluctuations and the boundary conditions imposed on them—is now found to be restricted to subsonic and slightly supersonic Mach Numbers in general. At high Mach Numbers it is true only for three-dimensional disturbances whose direction of propagation is at a large angle with the free-stream direction. In all other cases the form of the relation for the characteristic values depends on the boundary condition for the temperature fluctuations at the solid boundary. However, for very small values of the wave number, the relation again takes on a simple form much like that of the earlier theory. This situation, which corresponds to very low frequency fluctuations, occurs in the limiting case of complete stabilization of the boundary layer by cooling, for example. Thus, it is still possible to obtain the critical temperature ratios for complete stabilization by fairly simple numerical calculations. In all cases at high Mach Numbers when the wave number is not small, the relation for the characteristic values is more complicated.

† As an example of the increase in complication, in a second approximation terms must be retained whose coefficients involve the $v$-component of the basic flow velocity, so that to this order of approximation the flow cannot be considered to be parallel. This conclusion was reached independently in reference 6 and in an investigation by Cheng.[12]

## (2) General Formulation of Problem

The basic differential equations for the motion of a compressible fluid, in Cartesian tensor notation, are as follows:

Equation of motion

$$(\partial u_i/\partial t^*) + [u_j(\partial u_i/\partial x_j)] = (1/\rho^*)\,(\partial \tau_{ij}/\partial x_j) \quad (2.1)$$

Equation of continuity

$$(\partial \rho^*/\partial t^*) + (\partial/\partial x_j)\,(\rho^* u_j) = 0 \quad (2.2)$$

Equation of energy

$$\rho^* C_v\{(\partial T^*/\partial t^*) + [u_j(\partial T^*/\partial x_j)]\} = (\partial/\partial x_j)\,[k^*(\partial T^*/\partial x_j)] + \tau_{ij}e_{ij} \quad (2.3)$$

Equation of state

$$p^* = R^*\rho^* T^* \quad (2.4)$$

In these equations $e_{ij}$ is the rate-of-strain tensor and $\tau_{ij}$ is the stress tensor, defined in the following way

$$e_{ij} = (1/2)\,[(\partial u_i/\partial x_j) + (\partial u_j/\partial x_i)] \quad (2.5)$$

$$\tau_{ij} = 2\mu^* e_{ij} + [(2/3)\,(\lambda^* - \mu^*)e_{kk} - p^*]\delta_{ij} \quad (2.6)$$

All quantities are dimensional, as defined in the List of Symbols (p. 477). The coefficients of viscosity $\mu^*$ and $\lambda^*$, the coefficient of heat conductivity $k^*$, the specific heats $C_p$ and $C_v$, and the Prandtl Number $\sigma$ are considered to be functions of the temperature alone. The gravitational force terms are omitted, since they can be shown to be negligible for the large values of the Froude Number to be expected in most practical problems (see reference 1).

We consider a time-independent basic flow and a small time-dependent disturbance. Thus, any quantity $Q(x_i, t^*)$ may be written as the sum of a part $\bar{Q}(x_i)$ and a part $Q'(x_i, t^*)$:

$$Q(x_i, t^*) = \bar{Q}(x_i) + Q'(x_i, t^*) \quad (2.7)$$

We are interested in determining whether the disturbances represented by $Q'$ are stable or unstable—that is, whether they decrease or increase with time. Both the quantities, $Q$ in the disturbed motion and $\bar{Q}$ in the steady motion, satisfy Eqs. (2.1) to (2.4). Thus, if expressions of the form of Eq. (2.7) for all the variables are inserted into these equations and quadratic terms in the small disturbance quantities are neglected, the following system of linearized equations for the fluctuations $Q'$ is obtained:

$$\bar{u}_j(\partial\bar{u}_i/\partial x_j)\rho^{*\prime} + \bar{\rho}^*\{(\partial u_i'/\partial t^*) + [(\partial\bar{u}_i/\partial x_j)u_j'] + [\bar{u}_j(\partial u_i'/\partial x_j)]\} = \partial\tau_{ij}'/\partial x_j \quad (2.8)$$

$$(\partial\rho^{*\prime}/\partial t^*) + (\partial/\partial x_j)(\bar{u}_j\rho^{*\prime} + u_j'\bar{\rho}^*) = 0 \quad (2.9)$$

$$\bar{u}_j\frac{\partial\bar{T}^*}{\partial x_j}\rho^{*\prime} + \bar{\rho}^*\left(\frac{\partial T^{*\prime}}{\partial t^*} + \frac{\partial\bar{T}^*}{\partial x_j}u_j' + \bar{u}_j\frac{\partial T^{*\prime}}{\partial x_j}\right) = \frac{1}{\bar{C}_v}\frac{\partial}{\partial x_j}\left[\frac{\bar{C}_p\bar{\mu}^*}{\bar{\sigma}}\frac{\partial T^{*\prime}}{\partial x_j} + \left(\frac{C_p\mu^*}{\sigma}\right)'\frac{\partial\bar{T}^*}{\partial x_j}\right] + \frac{1}{\bar{C}_v}(\bar{\tau}_{ij}e_{ij}' + \tau_{ij}'\bar{e}_{ij}) + \left(\frac{1}{C_v}\right)'\frac{\partial}{\partial x_j}\left(\frac{\bar{C}_p\bar{\mu}^*}{\bar{\sigma}}\frac{\partial\bar{T}^*}{\partial x_j}\right) + \left(\frac{1}{C_v}\right)'\bar{\tau}_{ij}\bar{e}_{ij} \quad (2.10)$$

$$p^{*\prime}/\bar{p}^* = (\rho^{*\prime}/\bar{\rho}^*) + (T^{*\prime}/\bar{T}^*) \quad (2.11)$$

The quantities $e_{ij}'$, $\tau_{ij}'$, $(C_p\mu^*/\sigma)'$, $(1/C_v)'$, $\lambda^{*\prime}$, and $\mu^{*\prime}$ are given by

$$e_{ij}' = (1/2)\,[(\partial u_i'/\partial x_j) + (\partial u_j'/\partial x_i)] \quad (2.12)$$

$$\tau_{ij}' = 2(\bar{\mu}^*e_{ij}' + \mu^{*\prime}\bar{e}_{ij}) + [(2/3)\,(\bar{\lambda}^* - \bar{\mu}^*)e_{kk}' + (2/3)\,(\lambda^{*\prime} - \mu^{*\prime})\bar{e}_{kk} - p^{*\prime}]\delta_{ij} \quad (2.13)$$

and

$$\left.\begin{aligned} (C_p\mu^*/\sigma)' &= \overline{(d/dT^*)\,(C_p\mu^*/\sigma)}T^{*\prime} \\ (1/C_v)' &= \overline{(d/dT^*)\,(1/C_v)}T^{*\prime} \\ \lambda^{*\prime} &= \overline{(d\lambda^*/dT^*)}T^{*\prime} \\ \mu^{*\prime} &= \overline{(d\mu^*/dT^*)}T^{*\prime} \end{aligned}\right\} \quad (2.14)$$

Equations (2.8) to (2.11) form a system of linear partial differential equations for the fluctuations. The coefficients are known functions given by the steady flow.

## (3) Stability of Parallel and Nearly Parallel Flows

When the fluid is incompressible, the basic steady flow can be strictly parallel, as in the case of two-dimensional flow through a channel. The formation of the stability problem is then simple. For compressible fluids, however, there is no such simple basic flow. Flows satisfying the boundary-layer approximation are the closest to parallel flows. It will be shown that for such flows the stability problem is essentially the same as that for a parallel flow which agrees locally with the basic flow of the boundary-layer type.

For proving the theorem in Section (4) and for application to boundary layers in three-dimensional flow, the development will be made in a somewhat more general manner than usual. We consider a semi-infinite flat plate in an oblique incoming stream. The free stream has components of velocity $U_1$ and $U_3$ parallel to the plate, and the flow inside the boundary layer is parallel to the free stream at each location $(x_1, x_3)$. Indeed, the boundary layer can be described in terms of strips parallel to the direction of the resultant free stream. Each strip behaves like a flat plate with the leading edge normal to the resulting stream. The location of the leading edge coincides with that of the particular strip in question. Thus, if $s$ is the distance of a point from the leading edge in the direction of the free stream, the boundary-layer thickness $\delta$ at that point is given by

$$\delta/s = 0(R_s^{-1/2}) = 0(R^{-1}) \quad (3.1)$$

where

$$R_s = (s\sqrt{U_1^2 + U_3^2})/\bar{\nu}_1^* \quad (3.2)$$

$$R = R_\delta = (\delta\sqrt{U_1^2 + U_3^2})/\bar{\nu}_1^* \quad (3.3)$$

and $\bar{\nu}_1^*$ is the kinematic viscosity in the free stream. As is well known, the quantities $\bar{Q}$ associated with the basic flow satisfy the following order-of-magnitude relations (if the free-stream Mach Number is not too large):

$$(\partial\bar{Q}/\partial x_2):\ [(\partial\bar{Q}/\partial x_1) \text{ or } (\partial\bar{Q}/\partial x_3)] = 0(R_s^{1/2}) = 0(R) \quad (3.4)$$

$$(\bar{u}_1 \text{ or } \bar{u}_3):\ \bar{u}_2 = 0(R_s^{1/2}) = 0(R) \quad (3.5a)$$

$$\bar{T}^* = 0(\bar{T}_1^*), \quad \bar{\rho}^* = 0(\bar{\rho}_1^*), \quad \bar{\mu}^* = 0(\bar{\mu}_1^*) \quad (3.5b)$$

We shall now simplify the equations for the small fluctuations by means of analogous considerations regarding the fluctuations and their derivatives. From physical considerations and also from the mathematical nature of the boundary-value problem associated with the system of disturbance equations [Eqs. (2.8) to (2.11)], it may be expected that the fluctuations have the typical boundary-layer type behavior in a region adjacent to the solid boundary. That is, near to the boundary, viscous effects and therefore the highest order derivatives in the direction normal to the solid surface are important, and changes in the normal direction are much more rapid than in any direction parallel to the surface.

It should first be recognized that the fluctuations have a general length scale of the order of the boundary-layer thickness, so that

$$(1/Q')\,(\partial Q'/\partial x_1) = (1/\bar{Q})\,(\partial\bar{Q}/\partial x_1)\,0(R) \quad (3.6)$$

In the viscous region next to the surface, the characteristic length in the $x_2$ direction (normal direction) is much smaller and is obtained by considerations analogous to the usual boundary-layer arguments for the basic flow. It is clear from Eq. (2.8) that the equation of motion for $u_1'$ contains a term of the form $\bar{\rho}^*\bar{u}_1\partial u_1'/\partial x_1$ from the inertial terms and a term of the form $\bar{\mu}^*\partial^2 u_1'/\partial x_2^2$ from the viscous terms. Thus, if these terms are to be of the same order, $x_2$-derivatives must be larger than $x_1$-derivatives. Now,

$$\bar{\rho}^*\bar{u}_1(\partial u_1'/\partial x_1) = 0[(\bar{\rho}_1^* U_1 u_1')/\delta] \tag{3.7}$$

$$\bar{\mu}^*(\partial^2 u_1'/\partial x_2^2) = 0[\bar{\mu}_1^*(u_1'/\delta^2)R^{2n}] \tag{3.8}$$

where $R^n$ is a factor indicating the difference of order of magnitude between $x_1$ and $x_2$ derivatives:

$$(\partial Q'/\partial x_2): \; [(\partial Q'/\partial x_1) \text{ or } (\partial Q'/\partial x_3)] = 0(R^n) \tag{3.9}$$

A comparison of Eqs. (3.7) and (3.8) gives $n = 1/2$. The relation (3.9) is now comparable to Eq. (3.4) with $R_s$ replaced by $R_\delta$. The relation analogous to Eq. (3.5a) is

$$(u_1' \text{ or } u_3'): \; u_2' = 0(R^{1/2}) \tag{3.10}$$

An examination of the equations also gives the following relations [which are used in Section (4)]:

$$T^{*\prime}/\bar{T}_1^* = 0(u_1'/U_1), \quad \rho^{*\prime}/\bar{\rho}_1^* = 0(u_1'/U_1) \tag{3.11}$$

*Therefore we have a boundary-layer phenomenon associated with the fluctuations within the boundary layer for the mean flow.* In a thin layer next to the solid surface with thickness of the order of $\delta R^{-1/2}$ the fluctuations have the viscous behavior characterized by rapid variation in the $x_2$-direction, while outside this layer they are usually independent of viscosity since the viscous terms in the equations are usually unimportant there. In general, the order-of-magnitude relations (3.9) and (3.10) are valid only in the viscous layer at the solid surface, while outside it the velocity fluctuations are all of the same order of magnitude and the characteristic length in all directions is the same—that is, of the order of the boundary-layer thickness $\delta$.

In all of these order-of-magnitude estimates $0(R^{1/2})$ may denote a quantity not actually of the order of magnitude of $R^{1/2}$ in the strict sense, but smaller. Such a situation occurs in certain layers of the fluid where it is shown by a more complete analysis (see reference 6) that $0(R^{1/2})$ is replaced by $0(R^{1/3})$. In fact, when the disturbance is analyzed into periodic components [see Section (5)] and the ordinary differential equations for the amplitudes of these components examined, it is found that such a layer ("critical layer") always occurs at the $x_2$-location in the boundary layer where the wave speed in the $x_1$-direction of the particular component considered is equal to the component of the velocity of the basic flow in that direction. There is the possibility that in certain cases this critical layer (where viscosity is also always important) may occur outside the viscous layer at the solid surface described above. In such cases, relation (3.9) with $0(R^{1/2})$ replaced by $0(R^{1/3})$ will also be valid in the critical layer, so that variations through this layer in the $x_2$-direction will also be extremely rapid, just as in the layer at the wall. The order-of-magnitude analysis of the disturbance equations carries through in just the same way, so that it is not necessary for present purposes to consider the more complicated properties of the solutions of these equations.

## (4) Simplified Equations for Small Disturbances in Flows of the Boundary-Layer Type

By making use of the boundary-layer approximations outlined above [corresponding to relations (3.4), (3.5), (3.9), (3.10), and (3.11)] it can be shown that the equations for the small disturbances can be simplified to the following:

$$\bar{\rho}^*\left(\frac{\partial u_1'}{\partial t^*} + \bar{u}_1\frac{\partial u_1'}{\partial x_1} + \bar{u}_3\frac{\partial u_1'}{\partial x_3} + u_2'\frac{\partial \bar{u}_1}{\partial x_2}\right) = -\frac{\partial p^{*\prime}}{\partial x_1} + \bar{\mu}^*\frac{\partial^2 u_1'}{\partial x_2^2} \tag{4.1}$$

$$\bar{\rho}^*\left(\frac{\partial u_2'}{\partial t^*} + \bar{u}_1\frac{\partial u_2'}{\partial x_1} + \bar{u}_3\frac{\partial u_2'}{\partial x_3}\right) = -\frac{\partial p^{*\prime}}{\partial x_2} + \bar{\mu}^*\frac{\partial^2 u_2'}{\partial x_2^2} \tag{4.2}$$

$$\bar{\rho}^*\left(\frac{\partial u_3'}{\partial t^*} + \bar{u}_1\frac{\partial u_3'}{\partial x_1} + \bar{u}_3\frac{\partial u_3'}{\partial x_3} + u_2'\frac{\partial \bar{u}_3}{\partial x_2}\right) = -\frac{\partial p^{*\prime}}{\partial x_3} + \bar{\mu}^*\frac{\partial^2 u_3'}{\partial x_2^2} \tag{4.3}$$

$$\frac{\partial \rho^{*\prime}}{\partial t^*} + \bar{u}_1\frac{\partial \rho^{*\prime}}{\partial x_1} + \bar{u}_3\frac{\partial \rho^{*\prime}}{\partial x_3} + u_2'\frac{\partial \bar{\rho}^*}{\partial x_2} = -\bar{\rho}^*\Delta' \tag{4.4}$$

$$\bar{\rho}^*C_v\left(\frac{\partial T^{*\prime}}{\partial t^*} + \bar{u}_1\frac{\partial T^{*\prime}}{\partial x_1} + \bar{u}_3\frac{\partial T^{*\prime}}{\partial x_3} + u_2'\frac{\partial \bar{T}^*}{\partial x_2}\right) = -\bar{p}^*\Delta' + \bar{k}\frac{\partial^2 T^{*\prime}}{\partial x_2^2} \tag{4.5}$$

$$p^{*\prime}/\bar{p}^* = (\rho^{*\prime}/\bar{\rho}^*) + (T^{*\prime}/\bar{T}^*) \tag{4.6}$$

where

$$\Delta' = (\partial u_1'/\partial x_1) + (\partial u_2'/\partial x_2) + (\partial u_3'/\partial x_3) \tag{4.7}$$

Equation (4.5) can also be written in the form

$$\bar{\rho}^*\left(\frac{\partial T^{*\prime}}{\partial t^*} + \bar{u}_1\frac{\partial T^{*\prime}}{\partial x_1} + \bar{u}_3\frac{\partial T^{*\prime}}{\partial x_3} + u_2'\frac{\partial \bar{T}^*}{\partial x_2}\right) = \frac{1}{C_p}\left(\frac{\partial p^{*\prime}}{\partial t^*} + \bar{u}_1\frac{\partial p^{*\prime}}{\partial x_1} + \bar{u}_3\frac{\partial p^{*\prime}}{\partial x_3}\right) + \frac{\bar{\mu}^*}{\sigma}\frac{\partial^2 T^{*\prime}}{\partial x_2^2} \tag{4.8}$$

where $\sigma = C_p\bar{\mu}^*/\bar{k}^*$. These equations are correct except for terms $0(R^{-1/2})$.

A few typical steps in the process of simplification are discussed below:

*Example 1.*—In the inertial terms of the equations of motion we have terms of the kind $\bar{u}_1(\partial u_1'/\partial x_1) + \bar{u}_2(\partial u_1'/\partial x_2)$. Only the first term is included in Eq. (4.1). This is justified by the fact that

$$\bar{u}_2(\partial u_1'/\partial x_2): \; \bar{u}_1(\partial u_1'/\partial x_1) = (\bar{u}_2/\bar{u}_1)\,[(\partial u_1'/\partial x_2)/(\partial u_1'/\partial x_1)] \tag{4.9}$$

which is $0(R^{-1})$ outside the viscous layer and $0(R^{-1/2})$ inside the viscous layer, according to Eqs. (3.4), (3.5), (3.9), and (3.10).

*Example 2.*—None of the terms arising from the dissipation function $(\tau_{ij}e_{ij} + p^*\partial u_i/\partial x_i)$ are included in Eq. (4.5). This is justified as follows. The typical terms of the largest order which could contribute to Eq. (4.5) are

$\mu^{*\prime}(\partial\bar{u}_1/\partial x_2)^2 + 2\bar{\mu}^*(\partial\bar{u}_1/\partial x_2)(\partial u_1'/\partial x_2) + \ldots$

Their orders of magnitude can be estimated by comparing them with $\bar{\mu}^*(\partial\bar{u}_1/\partial x_2)^2$, which is the largest term in the dissipation function for the mean flow. Thus,

$$\mu^{*\prime}(\partial\bar{u}_1/\partial x_2)^2 = (\mu^{*\prime}/\bar{\mu}^*)\bar{\mu}^*(\partial\bar{u}_1/\partial x_2)^2 \quad (4.10)$$

$$2\bar{\mu}^*(\partial\bar{u}_1/\partial x_2)(\partial u_1'/\partial x_2) = 2[(\partial u_1'/\partial x_2)/(\partial\bar{u}_1/\partial x_2)]\bar{\mu}^*(\partial\bar{u}_1/\partial x_2)^2 \quad (4.11)$$

On the other hand, the term $\bar{\rho}^*\bar{u}_1 C_v \partial T^{*\prime}/\partial x_1$ retained on the left-hand side of Eq. (4.5) can be estimated as follows:

$$\bar{\rho}^*\bar{u}_1 C_v(\partial T^{*\prime}/\partial x_1) = \bar{\rho}^*\bar{u}_1 C_v[(\partial T^{*\prime}/\partial x_1)/(\partial\bar{T}^*/\partial x_1)](\partial\bar{T}^*/\partial x_1) \quad (4.12)$$

Now, from the energy equation for the basic flow, $\bar{\rho}^*\bar{u}_1 C_v(\partial\bar{T}^*/\partial x_1)$ is of the same order as the dissipation term $\bar{\mu}^*(\partial\bar{u}_1/\partial x_2)^2$. Consequently, the various quantities given by Eqs. (4.10)–(4.12) are in the ratios $1:1:R$ outside the viscous layer and in the ratios $1:R^{1/2}:R$ within the viscous layer. We are therefore justified in neglecting the first two quantities, which are the largest contribution of the dissipation function to Eq. (4.5).

In addition to the approximations just described, we can also neglect the $x_1$ and $x_3$ dependence of the quantities in the basic flow, which determine the coefficients in Eqs. (4.1) to (4.6). This approximation, which has been repeatedly discussed by various writers on the stability problem, depends on the boundary-layer nature of the mean flow and can be shown to be equivalent to neglecting terms of the order of $R^{-1}$ (see references 6, 8, and 10).

The system of Eqs. (4.1) to (4.6) is essentially of the form expected from a strictly parallel flow. One important characteristic of the system is the absence of dissipative terms in the energy equation. This makes it possible, as we shall see in Section (5), to reduce the problem for a three-dimensional disturbance to a two-dimensional problem. In previous investigations such a reduction was not possible because of certain terms retained in the equations for the disturbances that prevent a proper transformation. Such terms are now shown to be negligible.

The discussion up to now has been restricted to the case of moderate Mach Numbers, for which relations (3.1), (3.4), and (3.5) are valid. For very high supersonic Mach Numbers, the order-of-magnitude relations in the boundary layer corresponding to relations (3.1), (3.4), and (3.5) are different (see reference 5), and, consequently, those for the disturbances corresponding to Eqs. (3.9), (3.10), and (3.11) are different. An order-of-magnitude analysis (see reference 6) similar to that described for the case of low Mach Numbers shows that the disturbance equations are again, in the first approximation, of just the same form as Eqs. (4.1) to (4.6). The proper velocity and temperature distributions given by a modified boundary-layer theory (see reference 4) would have to be used, of course, at very high Mach Numbers. The error estimate for the neglected terms in the disturbance equations is now different, however, being expressed not in terms of $R$ but in terms of a Reynolds Number

$$R_0 = (\bar{\nu}_1^*/\bar{\nu}_0^*)R \quad (4.13)$$

based on fluid properties at some representative temperature $\bar{T}_0^*$ *inside* the boundary layer. Since $\bar{T}_0^*$ is roughly proportional to $M_1^2$ at high Mach Numbers, $R_0$ is much less than $R$ in such cases. Consequently, the errors committed in the simplified equations [which are $0(R_0^{-1/2})$] might not be very small at high Mach Numbers.

The present analysis is carried out for any boundary layer over a flat surface, whether there is a pressure gradient or not. Thus, the results are valid, not only for the zero pressure-gradient flow over a flat plate, but also for the flows over wedge-shaped profiles. Although a pressure gradient is associated with the latter flows at subsonic Mach Numbers, terms due to pressure gradient do not appear explicitly in the disturbance equations to the accepted order of approximation.†

Although the simplified equations [Eqs. (4.1) to (4.6)] are obtained by rather crude order-of-magnitude arguments, an *a posteriori* check verifies that the approximations made are valid in the stability investigation of the boundary layer over a solid surface.‡

## (5) Periodic Solutions of the Equations for Small Disturbances

The Eqs. (4.1) to (4.7) contain $x_1$, $x_3$, and $t^*$ only through the partial derivatives, so that solutions of the form

$$Q' = q(x_2)\exp[i(\alpha_1 x_1 + \alpha_3 x_3 - \alpha_1 c^* t^*)] \quad (5.1)$$

are possible. It may be expected that a disturbance of a fairly general nature can be analyzed into normal modes of this type, with $\alpha_1$ and $\alpha_3$ real and $c^*$ in general complex. A normal mode is damped if $c_i^* < 0$, neutral if $c_i^* = 0$, and self-excited if $c_i^* > 0$.

Evidently solutions of the form (5.1) may be regarded as waves propagating in a direction in the $(x_1, x_3)$ plane specified by the numbers $(\alpha_1, \alpha_3)$ (with changing amplitude in time if $c^*$ is complex).

Consider in particular the case $\alpha_3 = 0$. Then, the terms involving the $x_3$-derivatives drop out of the Eqs. (4.1) to (4.7), and the equations for $u_1'$, $u_2'$, $p^{*\prime}$, $\rho^{*\prime}$, $T^{*\prime}$ are exactly the same as if the component $\bar{u}_3$ of the basic

† A more careful examination is necessary when there is surface curvature. Preliminary results indicate that for most three-dimensional disturbances curvature effects are negligible when the curvature is small, but that for those disturbances with propagation direction nearly normal to the free-stream direction [see Section (5)] additional terms have to be retained in Eq. (4.2).

‡ Free boundary layers (such as wakes and jets) are excluded from all of the above discussion. The effect of viscosity on the stability of such flows is quite different. Further investigation of these cases is being carried out.

flow were zero. Thus, for waves propagating in the $x_1$-direction only the component $\bar{u}_1$ of the basic flow (besides the distributions of the scalar quantities $\bar{p}^*$ and $\bar{T}^*$) has any influence. Now, a rotation of the coordinate system in the $(x_1, x_3)$ plane will not change the system of Eqs. (4.1) to (4.7). Hence, we may conclude that for waves propagating in any direction only the component of the basic flow in that direction has to be considered, in addition to the distributions of the scalar quantities. *Every periodic three-dimensional disturbance may be treated in terms of a corresponding two-dimensional problem.* One need only take the proper component of velocity in the basic flow to carry out the simplification.

If we reduce the above equations to dimensionless forms, it is clear that the waves propagating in the direction of the resultant free-stream velocity are associated with the Reynolds Number $R$ and the Mach Number $M_1$ of the basic flow. For a wave propagating at an angle $\Theta$ relative to this direction, these numbers are reduced to

$$\tilde{R} = R \cos \Theta, \quad \tilde{M}_1 = M_1 \cos \Theta \tag{5.2}$$

A direct transformation verifying this statement will be given in the next section.

The advantage of the present treatment lies in its generality and simplicity. Whether the incoming stream is oblique or normal to the leading edge of the flat plate makes no difference in the discussion. When specialized to the incompressible case, the present results cover both the original work of Squire[3] and the case of an oblique incoming flow discussed by Kuethe.[9] For two-dimensional parallel flows, Squire obtained a set of transformation formulas leading to a conclusion equivalent to Eq. (5.2). Supplementary discussions are required in Squire's method. Kuethe, dealing with the case of the oblique incoming stream by Squire's method, noted the variation of "effective" Reynolds Number with different directions. He did not, however, arrive at the conclusion reached here that the maximum "effective" Reynolds Number is reached by waves propagating in the direction of the free stream. Consequently, these waves first give rise to instability as the Reynolds Number of the flow is increased.

## (6) Equations for the Amplitude Functions

Consider now the stability of a two-dimensional boundary layer over a flat plate coinciding with the $(x_1, x_3)$ plane and introduce the dimensionless variables given in the List of Symbols (p. 477). The free-stream velocity is now $U_1$, since $U_3 = 0$. The free-stream reference quantities $\delta$, $U_1$, $\bar{\rho}_1^*$, $\bar{T}_1^*$, $\bar{p}_1^*$, $\bar{\mu}_1^*$ are evaluated at a fixed point $x_1 = x_1^0$—that is, at a definite distance from the leading edge of the plate—in the part of the boundary layer that is of interest.

The dimensionless velocity and temperature distributions, $w = \bar{u}_1/U_1$ and $T = \bar{T}^*/\bar{T}_1^*$, depend on a number of factors—in particular, the boundary conditions at the wall and the free-stream Mach Number $M_1$.

We consider a small disturbance of the boundary layer in the manner discussed previously, with solutions of the form

$$Q'(x, y, z, t) = q(y) \exp [i(\alpha x + \beta z - \alpha c t)] \tag{6.1}$$

for the disturbance quantities, as indicated in the List of Symbols. When relations of the type [Eq. (6.1)] are inserted into Eqs. (4.1) to (4.7), the following equations for the amplitude functions $f$, $\phi$, $h$, $\pi$, $r$, $\theta$ are obtained:

$$\rho[i\alpha(w - c)f + w'\alpha\phi] = -[(i\alpha)/(\gamma M_1^2)]\pi + (\mu/R)f'' \tag{6.2}$$

$$\rho[i\alpha^2(w - c)\phi] = -[1/(\gamma M_1^2)]\pi' + (\mu/R)\alpha\phi'' \tag{6.3}$$

$$\rho[i\alpha(w - c)h] = -[(i\beta)/(\gamma M_1^2)]\pi + (\mu/R)h'' \tag{6.4}$$

$$\alpha(w - c)r + \rho'\alpha\phi + \rho(i\alpha f + i\beta h + \alpha\phi') = 0 \tag{6.5}$$

$$\rho[i\alpha(w - c)\theta + T'\alpha\phi] = -(\gamma - 1)(i\alpha f + i\beta h + \alpha\phi') + [(\gamma\mu)/(\sigma R)]\theta'' \tag{6.6}$$

$$\frac{\pi}{p} = \frac{r}{\rho} + \frac{\theta}{T} \tag{6.7}$$

where a prime now refers to differentiation with respect to $y$.

The combination of terms $\alpha f + \beta h$ suggests its replacement by a new variable $\tilde{\alpha}\tilde{f}$. Indeed, the following transformations convert Eqs. (6.2) to (6.7) to those of an equivalent two-dimensional problem:

$$\left.\begin{aligned} \alpha f + \beta h &= \tilde{\alpha}\tilde{f} \\ \alpha\phi &= \tilde{\alpha}\tilde{\phi} \\ \alpha\pi &= \tilde{\alpha}\tilde{\pi} \\ \alpha r &= \tilde{\alpha}\tilde{r} \\ \alpha\theta &= \tilde{\alpha}\tilde{\theta} \\ c &= \tilde{c} \\ \sqrt{\alpha^2 + \beta^2} &= \tilde{\alpha} \\ \alpha R &= \tilde{\alpha}\tilde{R} \\ \alpha M_1 &= \tilde{\alpha}\tilde{M}_1 \end{aligned}\right\} \tag{6.8}$$

The two-dimensional equations are:

$$\rho[i\tilde{\alpha}(w - c)\tilde{f} + w'\tilde{\alpha}\tilde{\phi}] = -[(i\tilde{\alpha})/(\gamma\tilde{M}_1^2)]\tilde{\pi} + (\mu/\tilde{R})\tilde{f}'' \tag{6.9}$$

$$\rho[i\tilde{\alpha}^2(w - c)\tilde{\phi}] = -[1/(\gamma\tilde{M}_1^2)]\tilde{\pi}' + (\mu/\tilde{R})\tilde{\alpha}\tilde{\phi}'' \tag{6.10}$$

$$i\tilde{\alpha}(w - c)\tilde{r} + \rho'\tilde{\alpha}\tilde{\phi} + \rho(i\tilde{\alpha}\tilde{f} + \tilde{\alpha}\tilde{\phi}') = 0 \tag{6.11}$$

$$\rho[i\tilde{\alpha}(w - c)\tilde{\theta} + T'\tilde{\alpha}\tilde{\phi}] = -(\gamma - 1)(i\tilde{\alpha}\tilde{f} + \tilde{\alpha}\tilde{\phi}') + [(\gamma\mu)/(\sigma\tilde{R})]\tilde{\theta}'' \tag{6.12}$$

$$\tilde{\pi}/p = (\tilde{r}/\rho) + (\tilde{\theta}/T) \tag{6.13}$$

Evidently the transformations [Eqs. (6.8)] are equivalent to Eq. (5.2) with $\cos \Theta = \alpha/\sqrt{\alpha^2 + \beta^2}$, so that $\Theta$ does in fact give the direction of propagation of the waves, Eq. (6.1). The dimensionless wave speed

$c$ is not changed, which indicates that the phase velocity in the $x$-direction is actually sec $\Theta$ times the phase velocity in the direction of the wave normal.

To complete the analysis, it is necessary to verify the equivalence of the boundary conditions. In the stability problem the boundary conditions are homogeneous, so that there is no difficulty in this respect. The homogeneous boundary conditions of the three-dimensional problem clearly transform to the proper homogeneous boundary conditions of the equivalent two-dimensional problem. For forced oscillations, however, it is obvious that the equivalence would have to be more carefully examined.

We have now shown that the boundary value problem associated with the system of Eqs. (6.2) to (6.7) for a three-dimensional disturbance is equivalent to that associated with the system of two-dimensional Eqs. (6.9) to (6.13). Thus the stability problem for three-dimensional disturbances can always be formulated as a two-dimensional problem which is mathematically the same as that for two-dimensional disturbances. It should be realized, however, that the equivalence is purely mathematical since Eqs. (6.9) to (6.13) *do not represent a proper two-dimensional disturbance.* There is a decisive difference between the present results and those of Squire for an incompressible fluid. In the latter case, because the Mach Number does not enter, it is possible to transform the equations for a three-dimensional disturbance to equations that *actually do represent a real two-dimensional disturbance at a lower Reynolds Number.* It is then possible to conclude that three-dimensional disturbances can be ignored when the minimum critical Reynolds Number is being investigated. For a compressible fluid this is not true because of the necessity of transforming the free-stream Mach Number. It is not even true that every three-dimensional disturbance is equivalent to a two-dimensional one with a modified Mach Number as well as Reynolds Number, because with a reduced Mach Number the velocity and temperature distributions would also have to be modified. In the equivalence discussed above these distributions must remain the same.

To apply the results of a two-dimensional analysis to the three-dimensional problem we may use the transformations (6.8), with the quantities $\tilde{q}$ associated with the two-dimensional problem. It is convenient, therefore, to think of three-dimensional disturbances as waves propagating in various directions. For each value of $\tilde{M}_1$—that is, for each value of the direction angle of the waves—the stability problem is investigated by means of Eqs. (6.9) to (6.13) with the usual methods of the two-dimensional theory. Thus, for each value of $\tilde{M}_1$ there is a neutral curve in the $(\tilde{\alpha}, \tilde{R})$ plane which transforms to one in the $(\alpha, R)$ plane by means of the transformation (6.8). The family of all such neutral curves, for all possible values of $\tilde{M}_1$ in the range $0 \leq \tilde{M}_1 \leq M_1$, gives the stability characteristics of three-dimensional disturbances.

## (7) Formulation of the Characteristic-Value Problem

Having established a two-dimensional formulation of the problem for three-dimensional disturbances, we may make use of the theory of Lees and Lin[1] for further investigations. However, some modifications are necessary when the wave speed of the disturbance is not small compared with the free-stream speed. For supersonic free-stream velocities this is usually the most important case. Detailed mathematical discussions of the necessary modifications will be given in a separate paper. Here we shall merely give a summary of the theoretical arguments and indicate how the final formulas differ from the existing ones. For a brief outline of the mathematical theory see the Appendix.

To determine the analytical nature of the system of Eqs. (6.9) to (6.13), it is convenient to introduce the variables

$$\left.\begin{aligned} Z_1 &= \tilde{f} \\ Z_2 &= \tilde{f}' \\ Z_3 &= \tilde{\phi} \\ Z_4 &= \tilde{\pi}/\tilde{M}_1^2 \\ Z_5 &= \tilde{\theta} \\ Z_6 &= \tilde{\theta}' \end{aligned}\right\} \tag{7.1}$$

Then the equations transform to a system of *six* equations of the first order. The latter, when reduced to normal form, have coefficients analytic in the independent variable $y$ and the parameters $\tilde{M}_1^2$, $\tilde{\alpha}\tilde{R}$, $\tilde{\alpha}^2$, and $c$.

Six homogeneous boundary conditions must be satisfied for a natural oscillation. These are given by

$$\left.\begin{aligned} &\tilde{f}, \tilde{\phi}\ \tilde{\theta} \text{ bounded as } y \to \infty \\ &\tilde{f}(0) = \tilde{\phi}(0) = 0 \\ &a\tilde{\theta}'(0) + b\tilde{\theta}(0) = 0 \end{aligned}\right\} \tag{7.2}$$

where $a$ and $b$ are constants depending upon $\tilde{\alpha}$, $c$, $\Theta$ and the physical properties of the gas and the adjacent solid wall (see Appendix).

The condition of boundedness for $y \to \infty$ can be put into an analytical form with the aid of the differential equations [Eqs. (6.9) to (6.13)]. For large $y$, viscosity effects become negligible, so that the terms of order $1/R$ can be neglected, and the equations reduce to

$$(d/dy)\left\{[(w-c)\tilde{\phi}' - w'\tilde{\phi}]/[T - \tilde{M}_1^2(w-c)^2]\right\} - \left\{[\tilde{\alpha}^2(w-c)]/T\right\}\tilde{\phi} = 0 \tag{7.3}$$

For $y \to \infty$ this is easily seen to become

$$\left.\begin{aligned} &\tilde{\phi}'' - \beta^2\tilde{\phi} = 0 \\ &\beta^2 = \tilde{\alpha}^2[1 - \tilde{M}_1^2(1-c)^2] \end{aligned}\right\} \tag{7.4}$$

Equation (7.4) has two solutions $\exp(\pm\beta y)$, where (for real $c$) $\beta$ denotes the positive square root

$$\beta = \tilde{\alpha}\sqrt{1 - \tilde{M}_1^2(1-c)^2} \tag{7.5}$$

Thus, for real $c$ the condition of boundedness at infinity is automatically satisfied if

$$c < 1 - (1/\tilde{M}_1) \tag{7.6}$$

In this case, the characteristic values are continuous, and the solutions represent disturbances whose phase velocity in the $(x, z)$ plane is supersonic relative to the component of the free-stream velocity in the direction of propagation of the disturbances in this plane. These supersonic disturbances correspond to sound waves propagating into or out of the boundary layer (see reference 1).

For subsonic disturbances with

$$c > 1 - (1/\tilde{M}_1) \tag{7.7}$$

the solution of Eq. (7.3) which is bounded at infinity is $\exp(-\beta y)$. Therefore the boundary condition for $y \to \infty$ is

$$\tilde{\phi}'/\tilde{\phi} \to -\beta \text{ as } y \to \infty \tag{7.8}$$

In this case the characteristic values are discrete and are determined from a secular equation which can be reduced to the form

$$E(\tilde{\alpha}, c, \tilde{M}_1^2) = F(z)\,\{1 + \Delta\} \tag{7.9}$$

In this equation, $F(z)$ is the function of Tietjens defined by

$$F(z) = 1 + \left\{\int_\infty^{-z} \zeta^{3/2} H_{1/3}^{(1)}\left[\frac{2}{3}\,(i\zeta)^{3/2}\right] d\zeta \div z\int_\infty^{-z} \zeta^{1/2} H_{1/3}^{(1)}\left[\frac{2}{3}\,(i\zeta)^{3/2}\right] d\zeta\right\} \tag{7.10}$$

where

$$z = (\tilde{\alpha}\tilde{R})^{1/3}\left(\frac{3}{2}\int_0^{y_c}\sqrt{\frac{c - w}{\nu}}\,dy\right)^{2/3} \tag{7.11}$$

and the function $E(\tilde{\alpha}, c, \tilde{M}_1^2)$ is given by

$$\left(\frac{c - w}{\nu}\right)_w^{-1/2}\left[\frac{3}{2}\int_{y_c}^{0}\sqrt{\frac{c - w}{\nu}}\,dy\right] E(\tilde{\alpha}, c, \tilde{M}_1^2) = \begin{vmatrix} \tilde{\phi}_{1w} & \tilde{\phi}_{11}' + \beta\tilde{\phi}_{11} \\ \tilde{\phi}_{2w} & \tilde{\phi}_{21}' + \beta\tilde{\phi}_{21} \end{vmatrix} \div \begin{vmatrix} [(T_w\tilde{\phi}_{1w}' + \tilde{M}_1^2 w_w' c\tilde{\phi}_{1w})/(T_w - \tilde{M}_1^2c^2)] & \tilde{\phi}_{11}' + \beta\tilde{\phi}_{11} \\ [(T_w\tilde{\phi}_{2w}' + \tilde{M}_1^2 w_w' c\tilde{\phi}_{2w})/(T_w - \tilde{M}_1^2c^2)] & \tilde{\phi}_{21}' + \beta\tilde{\phi}_{21} \end{vmatrix} \tag{7.12}$$

The quantities $\tilde{\phi}_{1w}$, $\tilde{\phi}_{2w}$ are the wall values and $\tilde{\phi}_{11}$, $\tilde{\phi}_{21}$ the free-stream values of two linearly independent solutions of Eq. (7.3).

The function $\Delta = \Delta(\tilde{\alpha}, c, \tilde{M}_1^2, a, b, z, \sigma)$, whose analytical definition is given in the Appendix, depends on both the inviscid and viscous solutions of the disturbance equations and on the thermal boundary condition in Eq. (7.2).

The principal difference between the present formulation and that given by Lees and Lin[1] is in the presence of the function $\Delta$. The appearance of this new term in the secular equation is due essentially to the fact that we are now placing no restriction on the value of $c$. In fact, for finite values of $T_w$,† $\Delta$ can always be written as

$$\Delta = \{[(\gamma - 1)\tilde{M}_1^2c^2]/T_w\}\,\Delta_1 \tag{7.13}$$

where $\Delta_1$ is another function depending on the same quantities as $\Delta$ which is 0(1) at most. Thus, it is clear that $\Delta$ is always negligible when $c$ is small. On the other hand, at high Mach Numbers, when $c$ is usually large, $\Delta$ is in general 0(1) and cannot be neglected, since the value of $\Delta_1$ depends on the thermal boundary condition and is therefore small only in very special cases.

In reference 1 it was concluded that the characteristic value problem is independent of the temperature fluctuations and the boundary conditions imposed on them. The present considerations show that this conclusion is of general validity only when $c$ is small, since it is only then that $\Delta$ can be neglected without any restriction being placed on the value of $\Delta_1$. For practical purposes, this limits the range of validity of the conclusion to subsonic and slightly supersonic Mach Numbers, when $c$ is not large in most cases of interest. For example, the error corresponding to the neglect of $\Delta$ for an insulated wall at $M_1 = 1.6$ can be estimated from Eq. (7.13) to be about 10 per cent (on taking $c \approx 1 - 1/M_1$). This is still not very large, considering the usual order of accuracy of the asymptotic theory. However, for somewhat higher Mach Numbers the error increases rapidly.

The preceding conclusion is true in an extremely restricted sense at high Mach Numbers—namely, for those disturbances whose direction of propagation is at a large angle with the free-stream direction so that $\tilde{M}_1$ is small. For these, it is clear that $\Delta$ is again negligible without any restriction being placed on the value of $\Delta_1$.

In the cases just discussed the secular equation reduces to

$$E(\tilde{\alpha}, c, \tilde{M}_1^2) = F(z) \tag{7.14}$$

which has a form much like that in the theory of reference 1. In all other cases, the form of this relation depends on the thermal boundary condition in Eq. (7.2).

A detailed analysis of the thermal boundary condition leads to the important conclusion that *when the wave number $\tilde{\alpha}$ is very small the relation for the characteristic values again reduces to Eq. (7.14), since $\Delta_1$ then becomes negligible.* Such a situation corresponds to very low frequency fluctuations, since the time-frequency is given by $\tilde{\alpha}c \cos \Theta$. An important example is the limiting case of complete stabilization of the boundary layer by cooling, for which all values of $\tilde{\alpha}$ on the neutral curve must approach zero in the limit [see Section (10)]. Thus, the critical temperature ratios for complete stabilization can still be obtained by fairly simple calculations with the use of Eq. (7.14).

† The exceptional case $T_w = 0$ is discussed in the Appendix.

In all cases at high Mach Numbers when the wave number is not small, the relation for the characteristic value is more complicated than Eq. (7.14). We shall not consider these more complicated situations any further here and shall limit all subsequent calculations in this paper to situations in which Eq. (7.14) is a valid approximation. The detailed presentation of the extended theory will be given in future communications.

An examination of Eq. (7.14) shows that it reduces to the corresponding relation given by Lees and Lin for small $c$ but that it is somewhat different for large $c$ because of the changes in the definition Eq. (7.11) for $z$ and the first factor in Eq. (7.12). The results of the previous theory are reproduced by assuming $c$ to be small and keeping only the first term in each of the power series expansions of $(w - c)$ and $\nu$ about $y_c$. In the supersonic case, when $c$ is not small, such a procedure does not represent a good approximation for our purposes.† For a discussion of the differences between results obtained by the present and older theories, see Section (10).

## (8) Determination of the Curves of Neutral Stability for Oblique Waves

Consider oblique waves with direction angle arc cos $\tilde{M}_1/M_1$, where $\tilde{M}_1$ is some value in the range $0 \leq \tilde{M}_1 \leq M_1$. The condition (7.14) for the characteristic values $(\tilde{\alpha}, \tilde{R}, c)$ can be transformed to

$$\Phi(z) = (1 + \lambda)\,\{(u + iv)/[1 + \lambda(u + iv)]\} \quad (8.1)$$

with $\Phi(z) \equiv 1/[1 - F(z)]$, where $z$, $\lambda$, and $u + iv$ are given by

$$\tilde{\alpha}\tilde{R} = z^3 \left(\frac{3}{2}\int_0^{y_c} \sqrt{\frac{c - w}{\nu}}\, dy\right)^{-2} \quad (8.2a)$$

$$\lambda(c) = \frac{3}{2}\frac{w_w' \nu_w^{1/2}}{c^{3/2}} \int_0^{y_c} \sqrt{\frac{c - w}{\nu}}\, dy - 1 \quad (8.2b)$$

$$u + iv = 1 + (w_w'c/T_w)\,\{[\phi_{21}' + \tilde{\alpha}\sqrt{1 - \tilde{M}_1^2(1 - c)^2}\phi_{21}]/[\phi_{11}' + \tilde{\alpha}\sqrt{1 - \tilde{M}_1^2(1 - c)^2}\,\phi_{11}]\} \quad (8.2c)$$

and $\phi_{11}$, $\phi_{21}$, $\phi_{11}'$, $\phi_{21}'$ are as defined by Eqs. (8.6a)–(8.6d).

For some purposes it is more convenient to solve Eq. (8.1) for $u + iv$, so that

$$u + iv = \Psi_r(z, c) + \Psi_i(z, c) \quad (8.3)$$

where

$$\Psi_r(z, c) = [(1 + \lambda)\Phi_r - \lambda(\Phi_r^2 + \Phi_i^2)] \div \{[1 + \lambda(1 - \Phi_r)]^2 + \lambda^2\Phi_i^2\} \quad (8.4a)$$

$$\Psi_i(z, c) = [(1 + \lambda)\Phi_i]/\{[1 + \lambda(1 - \Phi_r)]^2 + \lambda^2\Phi_i^2\} \quad (8.4b)$$

Before discussing the determination of the neutral curve we shall write down the required formulas and make certain transformations that are useful in the calculations. The quantities $\phi_{11}$ and $\phi_{21}$ are the free-stream values of two linearly independent solutions of the inviscid equation

$$(d/dy)\,\{[(w - c)\phi' - w'\phi]/[T - \tilde{M}_1^2(w - c)^2]\} - \tilde{\alpha}^2[(w - c)/T]\phi = 0 \quad (8.5)$$

These two solutions may be obtained as power series in $\tilde{\alpha}^2$, as described in reference 1, so that

$$\phi_{11} = (1 - c)\sum_{n=0}^{\infty} \tilde{\alpha}^{2n}\tilde{H}_{2n} \quad (8.6a)$$

$$\phi_{21} = (1 - c)\sum_{n=0}^{\infty} \tilde{\alpha}^{2n}\tilde{K}_{2n+1} \quad (8.6b)$$

$$\phi_{11}' = (1 - c)\left[\frac{1 - \tilde{M}_1^2(1 - c)^2}{(1 - c)^2}\right]\sum_{n=1}^{\infty} \tilde{\alpha}^{2n}\tilde{H}_{2n-1} \quad (8.6c)$$

$$\phi_{21}' = (1 - c)\left[\frac{1 - \tilde{M}_1^2(1 - c)^2}{(1 - c)^2}\right]\sum_{n=0}^{\infty} \tilde{\alpha}^{2n}\tilde{K}_{2n} \quad (8.6d)$$

† The approximation is worse in $\nu$ than in $(w - c)$.

where

$$\left.\begin{aligned}\tilde{H}_0 = 1, \quad \tilde{H}_1 &= \int_0^1 \frac{(w - c)^2}{T}\, dy, \quad \tilde{H}_2 = \int_0^1 \frac{T - \tilde{M}_1^2(w - c)^2}{(w - c)^2}\, dy \int_0^y \frac{(w - c)^2}{T}\, dy,\\ \tilde{H}_3 &= \int_0^1 \frac{(w - c)^2}{T}\, dy \int_0^y \frac{T - \tilde{M}_1^2(w - c)^2}{(w - c)^2}\, dy \int_0^y \frac{(w - c)^2}{T}\, dy, \ldots\end{aligned}\right\} \quad (8.7a)$$

$$\left.\begin{aligned}\tilde{K}_0 = 1, \quad \tilde{K}_1 &= \int_0^1 \frac{T - \tilde{M}_1^2(w - c)^2}{(w - c)^2}\, dy, \quad \tilde{K}_2 = \int_0^1 \frac{(w - c)^2}{T}\, dy \int_0^y \frac{T - \tilde{M}_1^2(w - c)^2}{(w - c)^2}\, dy,\\ \tilde{K}_3 &= \int_0^1 \frac{T - \tilde{M}_1^2(w - c)^2}{(w - c)^2}\, dy \int_0^y \frac{(w - c)^2}{T}\, dy \int_0^y \frac{T - \tilde{M}_1^2(w - c)^2}{(w - c)^2}\, dy, \ldots\end{aligned}\right\} \quad (8.7b)$$

For values of $c$ that are not too large, the integrals (8.7a) and (8.7b) can be put into forms that are more suitable for calculation purposes. For example, in $\tilde{K}_2$ we use the relation

$$\int_0^y \frac{T - \tilde{M}_1^2(w-c)^2}{(w-c)^2} dy = \int_0^1 \frac{T - \tilde{M}_1^2(w-c)^2}{(w-c)^2} dy - \int_y^1 \frac{T - \tilde{M}_1^2(w-c)^2}{(w-c)^2} dy \quad (8.8)$$

to transform $\tilde{K}_2$ into

$$\tilde{K}_2 = \tilde{H}_1\tilde{K}_1 - \tilde{N}_2 \quad (8.9)$$

where

$$\tilde{N}_2 = \int_0^1 \frac{(w-c)^2}{T} dy \int_y^1 \frac{T - \tilde{M}_1^2(w-c)^2}{(w-c)^2} dy$$

Thus $\tilde{N}_2$ has an integrand

$$\frac{(w-c)^2}{T} \int_y^1 \frac{T - \tilde{M}_1^2(w-c)^2}{(w-c)^2} dy \quad (8.10)$$

which is not numerically large for any value of $y$, so that $\tilde{N}_2$ may be expected to be numerically smaller than $\tilde{K}_2$. Similar transformations are made for the other integrals as follows:

$$\left.\begin{aligned} \tilde{K}_n &= \tilde{K}_1\tilde{H}_{n-1} - \tilde{N}_n, \quad n \geq 3 \\ \tilde{N}_2 &= \tilde{H}_2 \\ \tilde{H}_n &= \tilde{H}_2\tilde{H}_{n-2} - \tilde{M}_n, \quad n \geq 3 \end{aligned}\right\} \quad (8.11)$$

In general, $\tilde{M}_n$, and $\tilde{N}_n$ will be numerically smaller than the corresponding $\tilde{H}_n$ and $\tilde{K}_n$ for small $c$. The preceding transformations were first used in the stability theory for incompressible fluids,[10] where $c$ is usually fairly small, and then extended to the case of compressible fluids.[2] Obviously they are not useful for reasonably large values of $c$, in which case other transformations may be more appropriate. This situation occurs for large enough Mach Numbers—for $M_1$ greater than about two, perhaps. When $M_1$ becomes very large, in fact, $c$ is often very close to unity.

The relation (8.2c) may now be written as

$$u + iv = \frac{w_w'c}{T_w}\left\{\tilde{K}_1 + [T_w/(w_w'c)] + (1/\tilde{\alpha})G(\tilde{\alpha}, c)\right\} \quad (8.12)$$

where

$$G(\tilde{\alpha}, c) = (1 - \tilde{H}_2\tilde{\alpha}^2)\left\{\left[\frac{\sqrt{1 - \tilde{M}_1^2(1-c)^2}}{(1-c)^2} \times \left(1 - \sum_1^\infty \tilde{\alpha}^{2n}\tilde{N}_{2n}\right) - \sum_1^\infty \tilde{\alpha}^{2n+1}\tilde{N}_{2n+1}\right] \div \left[\left(1 - \sum_2^\infty \tilde{\alpha}^{2n}\tilde{M}_{2n}\right) + \frac{\sqrt{1 - \tilde{M}_1^2(1-c)^2}}{(1-c)^2} \times \left(\tilde{\alpha}\tilde{H}_1 - \sum_1^\infty \tilde{\alpha}^{2n+1}\tilde{M}_{2n+1}\right)\right]\right\} \quad (8.13)$$

If $G(\tilde{\alpha}, c)$ is expanded as a power series in $\tilde{\alpha}$, the real and imaginary parts of Eq. (8.12) are given by the following expressions in the case of real $c$:

$$u = (w_w'c/T_w)\left\{\tilde{K}_{1_r} + [T_w/(w_w'c)] + (k/\tilde{\alpha})\,[1 - k\tilde{H}_1\tilde{\alpha} + (k^2\tilde{H}_1^2 - 2\tilde{H}_{2_r})\tilde{\alpha}^2 - +]\right\} \quad (8.14)$$

$$v = (w_w'c/T_w)\,[\tilde{K}_{1_i} + (k/\tilde{\alpha})\,(-2\tilde{H}_{2_i}\tilde{\alpha}^2 + - \ldots)] \quad (8.15)$$

where

$$k = [1 - \tilde{M}_1^2(1-c)^2]^{1/2}/(1-c)^2$$

Such series expressions are expected to be good approximations when $k\tilde{\alpha}$ is small. This is true for all cases at subsonic and slightly supersonic Mach Numbers, but at high supersonic Mach Numbers it appears to hold only for cases of extreme cooling since otherwise $k$ may be very large.

In the present work we shall neglect all terms of order $\tilde{\alpha}^2$ in Eqs. (8.14) and (8.15) for the sake of simplicity, so that

$$u = (w_w'c/T_w)\left\{\tilde{K}_{1_r} + [T_w/(w_w'c)] + (k/\tilde{\alpha})\,(1 - k\tilde{H}_1\tilde{\alpha})\right\} \quad (8.16a)$$

$$v = v_0(c) \equiv (w_w'c/T_w)\tilde{K}_{1_i} \quad (8.16b)$$

In the calculations of reference 2 the same approximation [Eq. (8.16b)] for $v$ is used, but terms of order $\tilde{\alpha}^3$ are retained in the relation (8.14) for $u$. The results show, however, that often the error corresponding to the neglect of terms of order $\tilde{\alpha}^2$ in Eq. (8.16a) is no larger than that in Eq. (8.16b). In any event, the approximations (8.16a) and (8.16b) appear to be reasonably adequate for all subsonic and slightly supersonic Mach Numbers, with errors within the limits of error of the asymptotic theory. Moreover, they are very accurate at *any* Mach Number *when $\tilde{\alpha}$ is very small* (which occurs, for example, when cooling at the solid surface is sufficient). [See Section (10)]. For Mach Numbers greater than about two, however, this is true only for cases of extreme cooling. The present approximations, therefore, are probably not accurate at such high Mach Numbers in most cases of practical importance, and the method of numerical calculation may have to be revised.†

Now let us consider the procedure for finding the characteristic values ($\tilde{\alpha}$, $\tilde{R}$, $c$) on the neutral curve. The various steps can be illustrated most clearly by geometrical considerations, with the aid of the condition for the characteristic values in the form of Eq. (8.3) rather than Eq. (8.1).

For the values of $c$ that occur in the calculations, $\lambda(c)$ is always small, so that $\Psi_r(z, c)$ and $\Psi_i(z, c)$ are approximated fairly closely by $\Phi_r(z)$ and $\Phi_i(z)$. In fact, for $c = 0$, $\Psi_r(z, 0) = \Phi_r(z)$ and $\Psi_i(z, 0) = \Phi_i(z)$. Therefore, for each value of $c$, the graphs of $\Psi_r$ and $\Psi_i$ against $z$ have much the same shapes as those of $\Phi_r$ and $\Phi_i$, respectively (see Fig. 1). The maximum $\Psi_{im}(c)$ on the graph of $\Psi_i(z, c)$ against $z$ is generally close to the maximum value $\Phi_i = 0.580$ at $z = 3.22$ on the $\Phi_i$ curve.

† All existing calculations suffer from limitations of this kind. In some investigations, numerical calculations were carried out by using methods adequate only for small $c$ and $\alpha$, even though the resulting values of these quantities were not small. Such results must be treated with reserve.

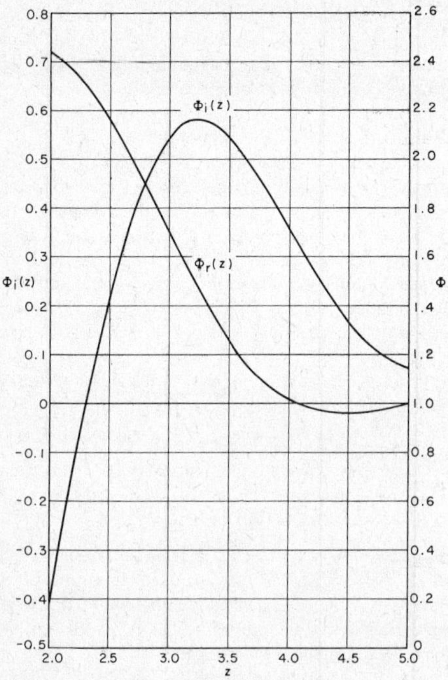


FIG. 1. The Functions $\Phi_r(z)$ and $\Phi_i(z)$.

For Mach Numbers that are not greater than about two, the graph of $v_0(c)$ against $c$ has one of the typical forms shown in Fig. 2a (which is plotted from the data given in reference 2). In general, the greater the cooling—that is, the lower the wall temperature—the higher the $v_0$ curve is in the $(c, v_0)$ plane. For larger Mach Numbers the $v_0$ curves may have more complicated shapes, as indicated in Fig. 2b. For the present, we shall consider only Mach Numbers for which the curves have the simple behavior shown in Fig. 2a.

The pairs of values $(\tilde{\alpha}, \tilde{R})$ are to be found for a series of real values of $c$. The minimum value of $c$ for subsonic Mach Numbers is $c = 0$, and for supersonic Mach Numbers it is either $c = 0$ or $c = 1 - 1/\tilde{M}_1$ if $\tilde{M}_1 > 1$.

For each value of $c$, relation (8.3) is solved for $z$ and $u$ with the aid of Eq. (8.16b). First, $z$ is obtained by solving $v_0(c) = \Psi_i(z, c)$, which is equivalent to finding the value of $z$ corresponding to the point of intersection of the curve $\Psi_i(z, c)$ against $z$ with a straight line parallel to the $z$-axis at a distance $v_0(c)$ from it. From the behavior of the graph of $\Psi_i(z, c)$ [which is similar to the graph of $\Phi_i(z)$ in Fig. 1], we see that if $v_0(c) < 0$ there is one intersection point (one value of $z$), and if $v_0(c) > 0$ there are generally two intersection points (two values of $z$) if $v_0(c)$ is not too large. For some value of $c$ large enough it is clear that the straight line will be just tangent to the corresponding graph of $\Psi_i(z, c)$ at its maximum point, and for larger values of $c$ there will be no intersection points at all. This maximum value of $c$, therefore, is given by the intersection of the graph of $\Psi_{im}(c)$ with that of $v_0(c)$ (see Figs. 2). Corresponding to it there is just one value of $z$. With the value of $z$ determined, $u = \Psi_r(z, c)$ is obtained from the graph of $\Psi_r(z, c)$ [which is similar to the graph of $\Phi_r(z)$ in Fig. 1]. Then $\tilde{\alpha}$ follows by solving Eq. (8.16a), since all the quantities in this equation besides $\tilde{\alpha}$ are now known functions of $c$, and finally $\tilde{R}$ is obtained from Eq. (8.2a). The corresponding values $(\alpha, R)$ are then determined by means of the transformations (6.8). The points $(\alpha, R)$ for all possible values of $c$ give the curve of neutral stability in the $(\alpha, R)$ plane.

Evidently $z(c)$ and $u(c)$ could be obtained graphically as just described by drawing graphs of $\Psi_r(z, c)$ and $\Psi_i(z, c)$ against $z$ for each value of $c$. However, for the purpose of the numerical calculations it is more convenient and more accurate to solve the two real relations obtained from Eq. (8.1) for $z$ and $u$ by an iteration procedure with the aid of the tabulated values of $\Phi_r(z)$ and $\Phi_i(z)$ (see Table 1), as described in reference 2. The maximum value of $c$ and the corresponding values of $z$ and $u$ can also be found accurately by an iteration method (see reference 13).†

Present calculations are carried out for the boundary layer on a flat plate at a Mach Number of $M_1 = 1.6$

† A detailed discussion of the procedure for the numerical calculations described in this and following sections and complete tables of various functions used in the calculations are given in reference 13.

TABLE 1
The Functions $\Phi_r(z)$, $\Phi_i(z)$, $\Phi_r'(z)$, $\Phi_i'(z)$

| z | $\Phi_r(z)$ | $\Phi_i(z)$ | $\Phi_r'(z)$ | $\Phi_i'(z)$ |
|---|---|---|---|---|
| 1.0 | 0.80630 | -2.60557 | | |
| 1.1 | | | | |
| 1.2 | 1.77012 | -2.29854 | 3.71 | 2.781 |
| 1.3 | | | 2.54 | |
| 1.4 | 2.26836 | -1.71669 | 1.505 | 2.937 |
| 1.5 | | | .850 | 2.633 |
| 1.6 | 2.44985 | -1.18600 | .4245 | 2.384 |
| 1.7 | | | .1390 | 2.127 |
| 1.8 | 2.48104 | - .75892 | - .06428 | 1.907 |
| 1.9 | | | - .2140 | 1.726 |
| 2.0 | 2.43927 | - .41253 | - .3337 | 1.573 |
| 2.1 | | | - .4377 | 1.443 |
| 2.2 | 2.35196 | - .12348 | - .5333 | 1.325 |
| 2.3 | | | - .6242 | 1.213 |
| 2.4 | 2.22724 | + .11916 | - .7108 | 1.1007 |
| 2.5 | | | - .7912 | .9833 |
| 2.6 | 2.06929 | .31558 | - .8625 | .8584 |
| 2.7 | | | - .9209 | .7254 |
| 2.8 | 1.88566 | .46043 | - .9627 | .5853 |
| 2.9 | | | - .9850 | .4414 |
| 3.0 | 1.68938 | .54872 | - .9853 | .2978 |
| 3.1 | | | - .9640 | .1590 |
| 3.2 | 1.49726 | .58082 | - .9223 | .0296 |
| 3.3 | | | - .8629 | - .0867 |
| 3.4 | 1.32516 | .56401 | - .7893 | - .1862 |
| 3.5 | | | - .7054 | - .2695 |
| 3.6 | 1.18429 | .51074 | - .6151 | - .3332 |
| 3.7 | | | - .5222 | - .3786 |
| 3.8 | 1.07982 | .43560 | - .4302 | - .4067 |
| 3.9 | | | - .3422 | - .4193 |
| 4.0 | 1.01118 | .35220 | - .2609 | - .4189 |
| 4.1 | | | - .1865 | - .4060 |
| 4.2 | .97361 | .27133 | - .1172 | - .3793 |
| 4.3 | | | - .0625 | - .3538 |
| 4.4 | .96056 | .20038 | - .01470 | - .3225 |
| 4.5 | | | - .0061 | |
| 4.6 | .95989 | .13601 | - .0034 | |
| 4.7 | | | + .0476 | |
| 4.8 | .97659 | .09503 | + .1794 | |
| 4.9 | | | | |
| 5.0 | .99582 | .07266 | | |

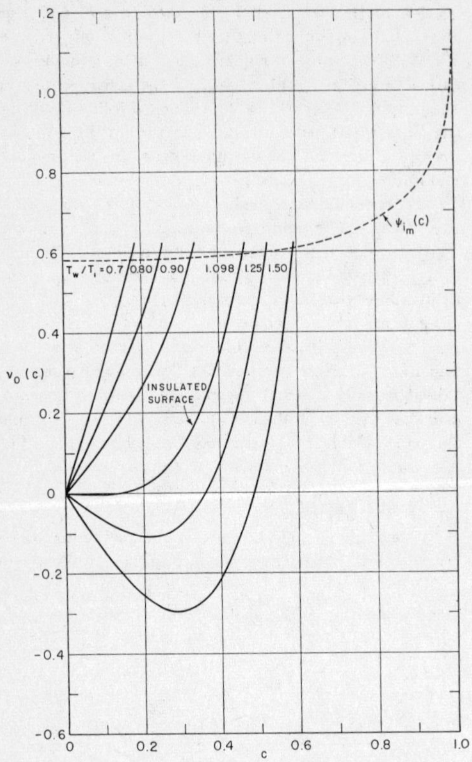


FIG. 2a. The function $v_0(c)$ versus $c$ for various ratios $T_w/T_1$ of wall to free-stream temperature at a free-stream Mach Number of $M_1 = 0.7$.

and ratio of wall to free-stream temperature of $T_w/T_1 = 1.073$. This temperature ratio is the critical ratio below which the boundary layer is completely stable with respect to two-dimensional disturbances [see Section (10)]. The Prandtl Number is given the value 0.75, and, for simplicity of calculation, the viscosity coefficient is taken to be directly proportional to the absolute temperature, with the constant of proportionality chosen so as to give the correct result at the wall. This procedure for the viscosity coefficient is known to be a good approximation in the case of the boundary layer itself when the Mach Number is not too large. Although it probably is not as accurate in the stability calculations, estimates show that it does not significantly alter the general trend of the results.

In Table 3, $\alpha$, $\alpha_x$, $\alpha_\theta$, $R$, $R_x$, $R_\theta$ are tabulated as functions of $c$ for several values of $\tilde{M}_1$ in the range $0 < \tilde{M}_1 < 1.6$, where $\alpha_x$ and $R_x$ are wave number and Reynolds Number based on a reference length $x_1$ (the distance from the leading edge of the plate) and $\alpha_\theta$ and $R_\theta$ are those based on the momentum thickness as reference length.

In Fig. 3 neutral curves in the $(\alpha_\theta, R_\theta)$ plane are shown for the three cases $\tilde{M}_1 = 1.4, 1.0, 0.6$ corresponding to direction angles of 29.0°, 51.3°, 68.0°, respectively.

The approximation is now made that the minimum critical Reynolds Number is very close to the value of $R$ corresponding to the maximum value of $c$. This was found to be true in previous investigations for an incompressible fluid[10] and for a compressible fluid.[2] The neutral curves in Fig. 3 show that in the present case it is also an excellent approximation, with only a slight error. The minimum critical Reynolds Number $R_{xc}$ approximated in this way is plotted against the direction angle of the oblique waves in Fig. 4 and tabulated in Table 4. As might have been expected, $R_{xc}$ becomes infinite at direction angles of 0° and 90°. The minimum value of $R_{xc}$ occurs at $\tilde{M}_1 = 1.030$ (corsponding to a direction angle of 49.9°) and is given by $R_{xc} = 1.64 \times 10^6$. This value $R_{xc} = 1.64 \times 10^6$, therefore, is the minimum critical Reynolds Number for all three-dimensional disturbances in this particular case.

## (9) The Dependence of the Minimum Critical Reynolds Number on the Direction of Propagation of the Waves

For those cases in which the approximations (8.16a) and (8.16b) are sufficiently accurate, it is possible to derive fairly simple analytical results for the directional dependence of the minimum critical Reynolds Number —that is, dependence on the parameter $\tilde{M}_1$. By solving Eq. (8.16a) for $\tilde{\alpha}$ we have

$$\tilde{\alpha} = \tilde{\alpha}_0/(1 + k\tilde{H}_1\tilde{\alpha}_0) \qquad (9.1)$$

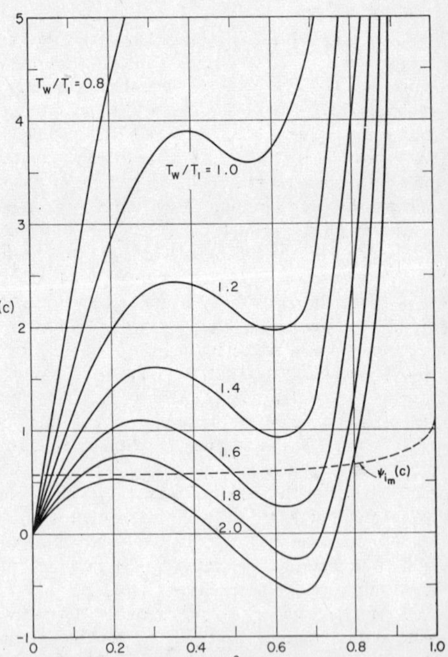


FIG. 2b. The function $v_0(c)$ versus $c$ for various ratios $T_w/T_1$ of wall to free-stream temperature at a free-stream Mach Number of $M_1 = 4$.

where

$$\tilde{\alpha}_0 = \left[\frac{w_w'c}{T_w}\frac{\sqrt{1-\tilde{M}_1^2(1-c)^2}}{(1-c)^2}\right] \div \left[u - \frac{w_w'c}{T_w}\left(\tilde{K}_{1r} + \frac{T_w}{w_w'c}\right)\right] = \frac{(w_w'c/T_w)k}{u-\tilde{L}} \quad (9.2)$$

Then the minimum value of $R$ on the neutral curve in the $(\alpha, R)$ plane is given, with the aid of Eqs. (6.8) and (8.2a), by

$$R_c = \frac{M_1}{\tilde{M}_1}\tilde{R}_c = \frac{M_1}{\tilde{M}_1} z^3 \left(\frac{3}{2}\int_0^{y_c}\sqrt{\frac{c-w}{\nu}}\,dy\right)^{-2} \times \frac{1 + k\tilde{H}_1\tilde{\alpha}_0}{\tilde{\alpha}_0} \quad (9.3)$$

where $c$ is approximated by the maximum permissible value of $c$ described previously. Once the particular boundary-layer flow is specified—i.e., the functions $w(y)$ and $T(y)$—this value of $c$ and therefore $z(c)$, $u(c)$, and all the other functions of $c$ in Eq. (9.3) are determined. For small $\lambda$, previous discussions show that $c$ is given approximately by $v_0(c) \approx 0.580$, and also that $z(c) \approx 3.22$, $u(c) \approx 1.48$ (see Fig. 1).

We now have

$$R_c = K\{[1/(\tilde{M}_1\tilde{\alpha}_0)] + (k\tilde{H}_1/\tilde{M}_1)\} \quad (9.4)$$

where $K$ depends on the given boundary-layer flow but is *independent of* $\tilde{M}_1$. In order to determine the dependence on $\tilde{M}_1$ of the other quantities entering into $R_c$, we note that

$$\tilde{H}_1 = \int_0^1 \frac{(w-c)^2}{T}\,dy = H_1 \quad (9.5)$$

is independent of $\tilde{M}_1$, and, with the use of Eqs. (8.7b) and (9.2), that

$$u - \tilde{L} = u - L - (w_w'c/T_w)(M_1^2 - \tilde{M}_1^2) \quad (9.6)$$

where

$$L = \frac{w_w'c}{T_w}\left[Re\int_0^1 \frac{T - M_1^2(w-c)^2}{(w-c)^2}\,dy + \frac{T_w}{w_w'c}\right] \quad (9.6a)$$

Then, with the aid of the definition for $k$, Eq. (9.4) can be written as

$$R_c = (A + B\tilde{M}_1^2)/[\tilde{M}_1\sqrt{1-\tilde{M}_1^2(1-c)^2}] \quad (9.7)$$

where

$$A = K\left((1-c)^2\{[T_w/(w_w'c)](u-L) - M_1^2\} + [H_1/(1-c)^2]\right) \quad (9.8a)$$

$$B = K[(1-c)^2 - H_1] \quad (9.8b)$$

The quantities $A$, $B$, and $c$ are independent of $\tilde{M}_1$ and are determined from the distributions $w(y)$ and $T(y)$ of velocity and temperature, respectively, in the boundary layer.

From Eq. (9.7) we find

$$dR_c/d\tilde{M}_1 = \{\tilde{M}_1^2[2A(1-c)^2 + B] - A\}/\{\tilde{M}_1^2 \times [1 - \tilde{M}_1^2(1-c)^2]^{3/2}\} \quad (9.9)$$

Therefore

$$dR_c/d\tilde{M}_1 = 0$$

when

$$\tilde{M}_1 = 1/\sqrt{2(1-c)^2 + (B/A)} \quad (9.10)$$

and

$$dR_c/d\tilde{M}_1 < 0 \quad (9.11)$$

when

TABLE 2(a)
The Function $v_0(c)$ for Various Ratios of Wall to Free-Stream Temperature at a Free-Stream Mach Number of 0.7

The function $v_0(c)$; $M_1 = 0.7$, $\rho\mu = 1$, $\sigma = 1$

| $T_w/T_1 = 0.70$ | | $T_w/T_1 = 0.80$ | | $T_w/T_1 = 0.90$ | | $T_w/T_1 = 1.098$ | | $T_w/T_1 = 1.25$ | |
|---|---|---|---|---|---|---|---|---|---|
| c | $v_0$ | c | $v_0$ | c | $v_0$ | c | $v_0$ | c | $v_0$ |
| .0262 | .0825 | .0237 | .0486 | .0433 | .0517 | .0353 | -.0009 | .0346 | -.0237 |
| .0521 | .1645 | .0472 | .0965 | .0863 | .1028 | .0705 | -.0024 | .0692 | -.0476 |
| .0777 | .2466 | .0705 | .1443 | .1291 | .1568 | .1058 | -.0025 | .1040 | -.0698 |
| .1030 | .3297 | .0937 | .1925 | .1714 | .2173 | .1410 | .0006 | .1389 | -.0886 |
| .1281 | .4146 | .1168 | .2417 | .2135 | .2885 | .1762 | .0090 | .1738 | -.1021 |
| .1529 | .5023 | .1397 | .2926 | .2551 | .3746 | .2114 | .0248 | .2088 | -.1085 |
| .1701 | .5661 | .1625 | .3457 | .2963 | .4805 | .2464 | .0501 | .2439 | -.1057 |
| .1726 | .5754 | .1851 | .4017 | .3166 | .5426 | .2813 | .0872 | .2789 | -.0917 |
| | | .2075 | .4614 | .3268 | .5762 | .3161 | .1389 | .3138 | -.0641 |
| | | .2298 | .5253 | | | .3505 | .2082 | .3485 | -.0203 |
| | | .2409 | .5592 | | | .3847 | .2985 | .3831 | .0427 |
| | | .2475 | .5801 | | | .4185 | .4137 | .4174 | .1286 |
| | | | | | | .4352 | .4821 | .4512 | .2414 |
| | | | | | | .4452 | .5270 | .4846 | .3859 |
| | | | | | | .4559 | .5790 | .5092 | .5184 |
| | | | | | | | | .5190 | .5779 |

TABLE 2(b)
The Function $v_0(c)$ for Various Ratios of Wall to Free-Stream Temperature at a Free-Stream Mach Number of 4.0

The function $v_0(c)$; $M_1 = 4$, $\rho\mu = \rho_w\mu_w$, $\sigma = 0.75$

| c \ $T_w/T_1$ | 0.8 | 1.0 | 1.2 | 1.4 | 1.6 | 1.8 | 2.0 |
|---|---|---|---|---|---|---|---|
| 0.1 | 2.275 | 1.572 | 1.144 | .858 | .654 | .502 | .385 |
| 0.2 | 4.367 | 2.856 | 1.976 | 1.409 | 1.019 | .735 | .521 |
| 0.3 | 5.830 | 3.632 | 2.388 | 1.605 | 1.083 | .710 | .433 |
| 0.4 | 6.571 | 3.888 | 2.406 | 1.498 | .899 | .480 | .177 |
| 0.5 | 6.207 | 3.645 | 2.164 | 1.192 | .563 | .133 | -.172 |
| 0.6 | 6.828 | 3.741 | 1.969 | .925 | .264 | -.179 | -.487 |
| 0.7 | 9.095 | 4.809 | 2.557 | 1.245 | .422 | -.135 | -.496 |
| 0.8 | 18.98 | 10.94 | 6.676 | 4.172 | 2.587 | 1.365 | .793 |
| 0.9 | 84.3 | 52.6 | 35.5 | 25.3 | 18.71 | 14.23 | 11.00 |
| 0.95 | 343 | 218 | 150.8 | 110.0 | 83.6 | 65.6 | 52.6 |

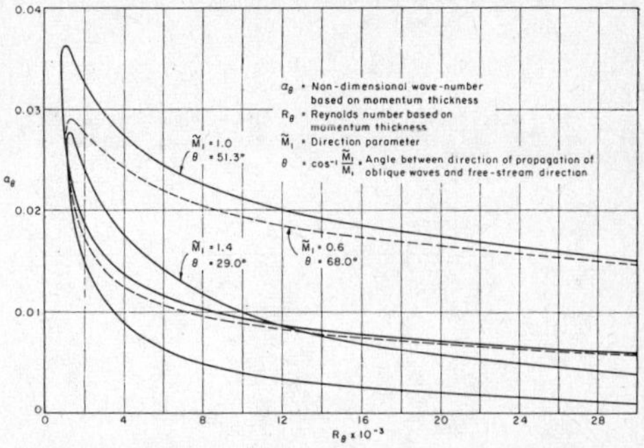

FIG. 3. Curves of neutral stability at a free-stream Mach Number of $M_1 = 1.6$ and a ratio of wall to free-stream temperature of $T_w/T_1 = 1.073$.

TABLE 3

Wave speeds, wave numbers, and Reynolds Numbers for Neutral Subsonic Disturbances at a Free-Stream Mach Number of 1.6 and Ratio of Wall to Free-Stream Temperature of 1.073

(a) $\widetilde{M}_1 = 0.6,\ M_1 = 1.6,\ T_w/T_1 = 1.073$

| $c$ | $\alpha$ | $R\times10^{-4}$ | $\alpha_\theta$ | $R_\theta\times10^{-3}$ | $\alpha_x$ | $R_x\times10^{-6}$ |
|---|---|---|---|---|---|---|
| 0 | 0 | ∞ | 0 | ∞ | ∞ | ∞ |
| .15 | .0515 | 40.7 | .00534 | 42.1 | 512 | 4240 |
| .175 | .0642 | 21.3 | .00665 | 22.1 | 334 | 1110 |
| .2 | .0779 | 12.2 | .00807 | 12.7 | 233 | 365 |
| .25 | .110 | 4.82 | .0114 | 5.00 | 129 | 56.7 |
| .275 | .129 | 3.23 | .0134 | 3.35 | 102 | 25.5 |
| .325 | .178 | 1.63 | .0184 | 1.69 | 70.8 | 6.52 |
| .3625 | .231 | 1.09 | .0239 | 1.13 | 61.7 | 2.92 |
| .375 | .268 | 1.05 | .0278 | 1.09 | 68.8 | 2.70 |
| .3625 | .278 | 1.40 | .0288 | 1.45 | 94.7 | 4.76 |
| .325 | .253 | 2.63 | .0262 | 2.73 | 162 | 16.9 |
| .275 | .208 | 6.23 | .0216 | 6.45 | 317 | 94.8 |
| .25 | .186 | 9.89 | .0193 | 10.3 | 450 | 239 |
| .2 | .144 | 28.2 | .0149 | 29.2 | 988 | 1938 |
| 0 | 0 | ∞ | 0 | ∞ | ∞ | ∞ |

Table 3 (Continued)

(b) $\widetilde{M}_1 = 1.0,\ M_1 = 1.6,\ T_w/T_1 = 1.073$

| $c$ | $\alpha$ | $R\times10^{-4}$ | $\alpha_\theta$ | $R_\theta\times10^{-3}$ | $\alpha_x$ | $R_x\times10^{-6}$ |
|---|---|---|---|---|---|---|
| 0 | 0 | ∞ | 0 | ∞ | ∞ | ∞ |
| .05 | .00869 | 4980 | .000900 | 5160 | 10600 | $606\times10^5$ |
| .1 | .0264 | 249 | .00273 | 258 | 1600 | 151000 |
| .15 | .0515 | 40.7 | .00534 | 42.2 | 513 | 4050 |
| .175 | .0671 | 20.4 | .00695 | 21.1 | 335 | 1017 |
| .2 | .0849 | 11.2 | .00880 | 11.6 | 233 | 307 |
| .25 | .128 | 4.14 | .01330 | 4.28 | 130 | 41.8 |
| .275 | .155 | 2.70 | .0161 | 2.79 | 103 | 17.8 |
| .3 | .186 | 1.84 | .0193 | 1.90 | 83.6 | 8.25 |
| .325 | .222 | 1.31 | .0230 | 1.35 | 70.9 | 4.17 |
| .35 | .267 | .969 | .0277 | 1.00 | 63.3 | 2.30 |
| .3625 | .295 | .856 | .0306 | .887 | 61.7 | 1.79 |
| .375 | .343 | .822 | .0355 | .852 | 68.8 | 1.65 |
| .3625 | .349 | 1.11 | .0362 | 1.15 | 95.0 | 3.02 |
| .35 | .337 | 1.40 | .0350 | 1.45 | 115 | 4.76 |
| .325 | .307 | 2.17 | .0318 | 2.25 | 163 | 11.5 |
| .3 | .274 | 3.38 | .0284 | 3.50 | 227 | 27.9 |
| .275 | .241 | 5.39 | .0250 | 5.58 | 318 | 71.0 |
| .25 | .209 | 8.81 | .0217 | 9.13 | 449 | 189 |
| .2 | .151 | 26.8 | .0156 | 27.8 | 990 | 1760 |
| .15 | .0993 | 109 | .0103 | 113 | 2640 | 28900 |
| .1 | .0553 | 748 | .00573 | 775 | 10120 | 137x10[illegible] |
| .05 | .0193 | 19180 | .00200 | 19900 | 90300 | $9\times10^8$ |
| 0 | 0 | ∞ | 0 | ∞ | ∞ | ∞ |

Table 3 (Continued)

(c) $\widetilde{M}_1 = 1.4,\ M_1 = 1.6,\ T_w/T_1 = 1.073$

| $c$ | $\alpha$ | $R\times10^{-4}$ | $\alpha_\theta$ | $R_\theta\times10^{-3}$ | $\alpha_x$ | $R_x\times10^{-6}$ |
|---|---|---|---|---|---|---|
| .2857 | 0 | ∞ | 0 | ∞ |  | ∞ |
| .2875 | .0219 | 17.18 | .00227 | 17.80 | 92.0 | 721 |
| .292 | .0421 | 8.63 | .00436 | 8.94 | 88.8 | 182 |
| .3 | .0662 | 5.17 | .00686 | 5.36 | 83.7 | 65.3 |
| .325 | .125 | 2.33 | .0129 | 2.41 | 70.9 | 13.2 |
| .35 | .183 | 1.41 | .0189 | 1.46 | 63.3 | 4.89 |
| .3625 | .216 | 1.17 | .0224 | 1.21 | 61.8 | 3.35 |
| .375 | .263 | 1.07 | .0272 | 1.11 | 68.8 | 2.81 |
| .3625 | .250 | 1.55 | .0259 | 1.61 | 94.9 | 5.87 |
| .35 | .225 | 2.09 | .0233 | 2.17 | 115 | 10.7 |
| .325 | .166 | 4.01 | .0172 | 4.15 | 163 | 39.3 |
| .3 | .0936 | 9.92 | .00969 | 10.28 | 227 | 241 |
| .292 | .0605 | 17.05 | .00627 | 17.67 | 252 | 711 |
| .2875 | .0318 | 34.5 | .00329 | 35.7 | 268 | 2904 |
| .2857 | 0 | ∞ | 0 | ∞ |  | ∞ |

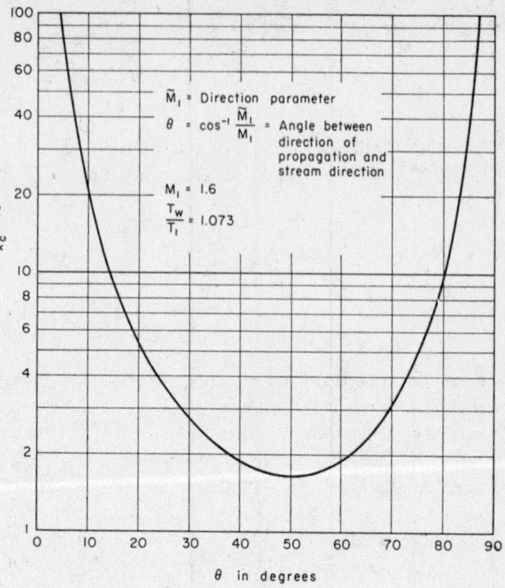


FIG. 4. Minimum critical Reynolds Number versus direction of propagation of oblique waves.

TABLE 4

Minimum Critical Reynolds Numbers and Direction Angles for Oblique Waves at a Free-Stream Mach Number of 1.6 and Ratio of Wall to Free-Stream Temperature of 1.073

$M_1 = 1.6,\ T_w/T_1 = 1.073$

Direction angle = $\cos^{-1}\widetilde{M}_1/M_1$

| $\widetilde{M}_1$ | $\cos^{-1}\widetilde{M}_1/M_1$ | $R_{x_c}\times10^{-6}$ |
|---|---|---|
| 0 | 90° | ∞ |
| .1 | 86.4 | 75.1 |
| .2 | 82.8 | 19.2 |
| .4 | 75.5 | 5.23 |
| .6 | 68.0 | 2.70 |
| .8 | 60.0 | 1.89 |
| 1.0 | 51.3 | 1.65 |
| 1.030 | 49.9 | 1.64 |
| 1.2 | 41.4 | 1.80 |
| 1.4 | 29.0 | 2.81 |
| 1.5 | 20.4 | 5.08 |
| 1.576 | 10 | 19.6 |
| 1.594 | 5 | 76.7 |
|  | 0 | ∞ |

$$\tilde{M}_1 < 1/\sqrt{2(1-c)^2 + (B/A)}$$

Since $0 \leq \tilde{M}_1 \leq M_1$ it follows that

$$M_1 \sqrt{2(1-c)^2 + (B/A)} < 1 \qquad (9.12)$$

is the condition that all three-dimensional disturbances have a *higher* minimum critical Reynolds Number than two-dimensional ones. Thus, if it can be shown that the parameter $C = M_1\sqrt{2(1-c)^2 + (B/A)}$ for a specified boundary-layer flow is less than unity, then three-dimensional disturbances are more stable and are not very important in the stability problem.

In Table 5 and Fig. 5 are given the results of calculations of $C$ as a function of $M_1$ for an insulated flat plate from $M_1 = 0$ to $M_1 = 2.4$ and of $C$ as a function of $T_w/T_1$ at a Mach Number of $M_1 = 0.7$ from $T_w/T_1 = 0.70$ to $T_w/T = 1.25$. For the insulated plate, it is clear that three-dimensional disturbances are more stable than two up to a Mach Number of about 1.8 when they become slightly more unstable. The fact that $C$ decreases and again becomes less than unity when $M_1$ increases beyond 1.8 may not be of much significance because of the large errors at such Mach Numbers due to the approximations used in the deductions.†

### (10) The Influence of Surface Cooling on Boundary-Layer Stability

The problem of boundary-layer stability at supersonic speeds is different from that at subsonic speeds in one important respect—there is always a class of oblique waves (those with $1 < \tilde{M}_1 \leq M_1$) for which the values of $c$ are bounded away from zero (i.e., $c \geq 1 - 1/\tilde{M}_1 > 0$). This leads to interesting consequences in the case of surface cooling.

Let us first consider strictly two-dimensional disturbances (for which $\tilde{M}_1 = M_1$), since the situation is then somewhat simpler. The minimum value of $c$ on the neutral curve is $1 - 1/M_1$, and the maximum value is given by $v_0(c) = \Psi_{im}(c)$—that is, it is the value of $c$ corresponding to the intersection point of the graph of $v_0(c)$ with that of $\Psi_{im}(c)$ (see Figs. 2). For sufficiently small $T_w/T_1$, $v_0(c) > 0$ for $c = 1 - 1/M_1$ and $v_0(c)$ is monotonically increasing for $c > 1 - 1/M_1$. Also, further decreases in $T_w/T_1$ raise the whole of the $v_0$ curve higher in the $(c, v_0)$ plane. This behavior of the $v_0$ curve is typical for Mach Numbers up to $M_1 = 7.48$.‡ For Mach Numbers greater than $M_1 = 7.48$, no amount of cooling is sufficient to raise the $v_0$ curve sufficiently to make $v_0(c)$ positive at $c = 1 - 1/M_1$.

We see, therefore, that in this range of Mach Numbers, for $T_w/T_1$ small enough, the range of values of $c$ on the neutral curve is bounded below by $1 - 1/M_1$ and above by an upper limit given by $v_0(c) = \Psi_{im}(c)$, which can be made as close to $1 - 1/M_1$ as desired by making $T_w/T_1$ smaller. Thus, there is a critical temperature ratio $(T_w/T_1)_c$ given by $v_0(1 - 1/M_1) = \Psi_{im}(1 - 1/M_1)$ for which the whole neutral curve disappears. This can be seen by solving Eq. (8.16a) for $\tilde{\alpha}$ and noting that $\tilde{\alpha} \to 0$ as $c \to 1 - 1/M_1$. Therefore, when $T_w/T_1 \to (T_w/T_1)_c$, all values of $c$ on the

Table 5

The Function $C = M_1\sqrt{2(1-c)^2 + B/A}$ for Various Mach Numbers and Rates of Heat Transfer

Table 5(a)

The function $C = M_1\sqrt{2(1-c)^2 + B/A}$ for zero heat transfer; $\sigma = 1$, $\rho\mu = 1$.

| $M_1$ | 0 | 0.5 | 0.9 | 1.3 | 1.7 | 2.472 |
|---|---|---|---|---|---|---|
| C | 0 | 0.455 | 0.759 | 0.941 | 1.006 | 0.885 |

Table 5(b)

The function $C = M_1\sqrt{2(1-c)^2 + B/A}$ for various temperature ratios at $M_1 = 0.7$; $\sigma = 1$, $\rho\mu = 1$

| $T_w/T_1$ | 0.70 | 0.80 | 0.90 | 1.25 |
|---|---|---|---|---|
| C | 0.871 | 0.797 | 0.722 | 0.544 |

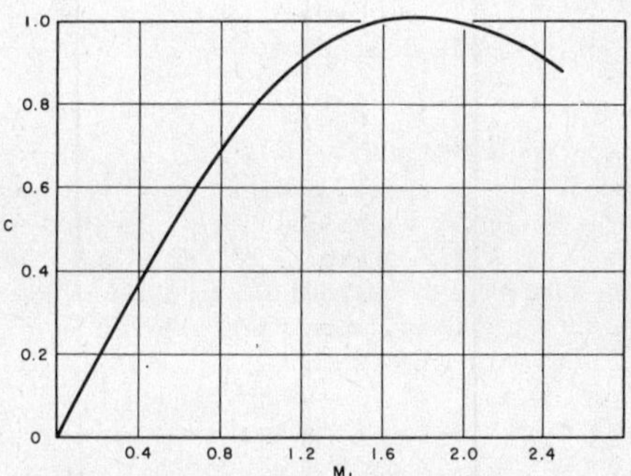


THE FUNCTION $C = M_1\sqrt{2(1-c)^2 + \frac{B}{A}}$ FOR ZERO HEAT TRANSFER AT VARIOUS MACH NUMBERS

FIGURE 5(a)

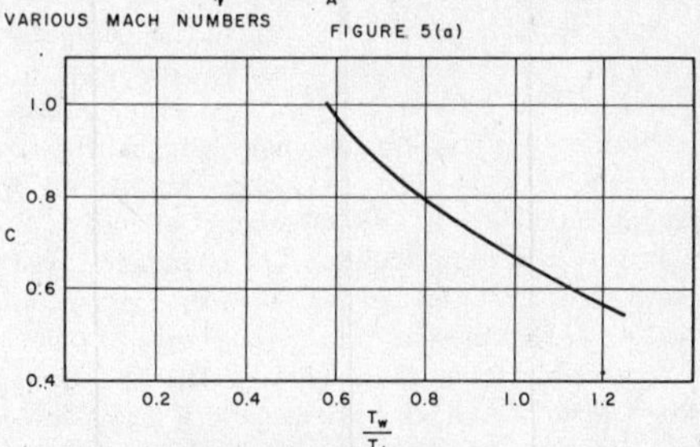


THE FUNCTION $C = M_1\sqrt{2(1-c)^2 + \frac{B}{A}}$ FOR VARIOUS TEMPERATURE RATIOS AT A FREE-STREAM MACH NUMBER OF $M_1 = 0.7$

FIGURE 5(b)

† It should be emphasized that, because of the use of the approximate relations (8.16a) and (8.16b), the results of this section are in most practical cases valid only for Mach Numbers not too much larger than unity. Their implications in regard to the influence of surface cooling at low Mach Numbers are discussed at the end of the next section.

‡ The actual value of this upper limit to the Mach Number depends on the value of the Prandtl Number and the particular viscosity law used. The value $M_1 = 7.48$ corresponds to a Prandtl Number of 0.75 and a linear viscosity-temperature relationship.

TABLE 6
Critical Temperature Ratios for Complete Stability with Respec to Two-Dimensional Disturbances for Free-Stream Mach Numbers from 1 to 8

$\sigma = 0.75, \quad \rho\mu = \rho_w\mu_w$

| $M_1$ | $T_w/T_1$ |
|---|---|
| 1 | 0 |
| 1.1 | .582 |
| 1.2 | .766 |
| 1.25 | .827 |
| 1.5 | 1.018 |
| 2 | 1.262 |
| 2.5 | 1.453 |
| 3 | 1.590 |
| 3.5 | 1.678 |
| 4 | 1.700 |
| 4.5 | 1.641 |
| 5. | 1.548 |
| 6 | 1.146 |
| 6.5 | .842 |
| 7 | .453 |
| 7.25 | .240 |
| 7.48 | 0 |

neutral curve approach $1 - 1/M_1$, so that all values of $\tilde{\alpha} \to 0$ and all values of $\tilde{R} \to \infty$.

It is to be noticed that in this limiting situation relations (8.16a) and (8.16b) become *exact* since the errors are $0(\tilde{\alpha}^2)$, so that the relation

$$v_0[1 - (1/M_1)] = \Psi_{im}[1 - (1/M_1)] \quad (10.1)$$

for the critical temperature ratio $(T_w/T_1)_c$ is exact.

When $T_w/T_1 < (T_w/T_1)_c$ evidently $v_0(c) > \Psi_{im}(c)$ for $c \geq 1 - 1/M_1$, so that no solution is possible, at least for small $\alpha$. The inequality $T_w/T_1 \leq (T_w/T_1)_c$ is usually interpreted as the condition for the complete stabilization of the boundary layer with respect to two-dimensional disturbances.† The preceding arguments are essentially the same as those used by Lees.[2]

Calculations have been carried out to determine the variation of $(T_w/T_1)_c$ with $M_1$ for the boundary layer on a flat plate with Prandtl Number equal to 0.75 and viscosity coefficient directly proportional to absolute temperature. The results are tabulated in Table 6, and the curve of critical temperature ratio against Mach Number for two-dimensional disturbances is shown in Fig. 6. Since the critical temperature ratio approaches zero at $M_1 = 7.48$, no amount of cooling is sufficient to completely stabilize the boundary layer with respect to two-dimensional disturbances when $M_1 \geq 7.48$. The present numerical results do not differ appreciably from those of Van Driest,[11] even though corrected viscous solutions [which enter into the determination of $\lambda(c)$ and thus of $\Psi_{im}(c)$] are used. The explanation is probably traceable to the fact that $v_0(c)$ is so extremely sensitive to variations in $T_w/T_1$ that small errors in $\Psi_{im}$ in Eq. (10.1) do not affect the solution $T_w/T_1$ much.‡

† This interpretation will be discussed later.

The conclusions depend, of course, strongly on the value of the Prandtl Number and the particular viscosity-temperature relationship used.[11] However, for the same reason as given above, it is not likely that the use of the present viscous solutions would significantly alter the results for these other cases obtained in reference 11.

Now consider oblique waves with direction angle arc cos $\tilde{M}_1/M_1$, where $1 < \tilde{M}_1 < M_1$. Just as in the case of strictly two-dimensional disturbances, there is again the possibility of completely stabilizing the boundary layer with respect to oblique waves propagating in this direction for small enough $T_w/T_1$. The arguments carry through much as before except for one complication due to the fact that when the Mach Number is large the curve $v_0(c)$ against $c$ may have an inflection point and thus a maximum and minimum in the interval $0 < c < 1$ (see Fig. 2b). Thus, even when $T_w/T_1$ is small enough for the relation

$$v_0[1 - (1/\tilde{M}_1)] = \Psi_{im}[1 - (1/\tilde{M}_1)] \quad (10.2)$$

to hold, the corresponding $v_0(c)$ may *not* be monotonically increasing for $c > 1 - 1/\tilde{M}_1$, when $\tilde{M}_1$ is small

‡ Van Driest's computations, in addition to the use of the inaccurate $\lambda(c)$ of reference 1, also involve what is essentially an approximation for $\Psi_{im}(c)$ in terms of $\lambda(c)$. It is interesting to observe that if the exact procedure of calculation is used, in combination with the inaccurate value of $\lambda(c)$ given by reference 1, several critical cooling ratios will be found for each value of $c = 1 - 1/M_1$. This has been pointed out by Bloom.[14] However, with the present function $\lambda(c)$ whose influence is small, such difficulties do not arise.

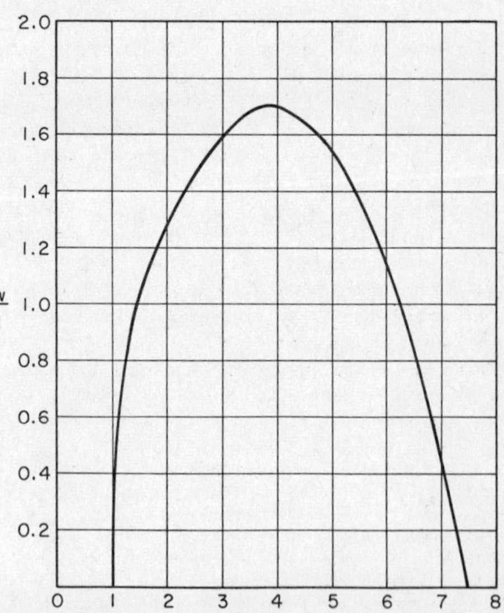


FIG. 6. Critical temperature ratios for complete stability with respect to two-dimensional disturbances.

enough.† There could still be values of $c > 1 - 1/\tilde{M}_1$ for which $v_0(c) < \Psi_{im}(c)$ (see Fig. 2b), so that a neutral curve might still exist. We shall make the tentative assumption‡ that the critical temperature ratio, below which the boundary layer is completely stable with respect to oblique waves with direction angle arc cos $\tilde{M}_1/M_1$, is the largest $T_w/T_1$ for which

$$v_0(c) \geq \Psi_{im}(c), \quad c \geq 1 - (1/\tilde{M}_1) \qquad (10.3)$$

This value of $T_w/T_1$ is evidently the one for which the corresponding $v_0(c)$ curve is just tangent near its minimum point to the $\Psi_{im}(c)$ curve (see Fig. 2b).

Obviously if $\tilde{M}_1 \leq 1$ the minimum value of $c$ on the neutral curve is just $c = 0$. Therefore, it is not possible by any amount of cooling to completely stabilize the boundary layer with respect to oblique waves with direction angles greater than arc cos $1/M_1$. The limiting directions are thus the normals to the Mach lines in a plane parallel to the solid surface.

For the boundary layer on a flat plate at a Mach Number $M_1 = 4$ with Prandtl Number equal to 0.75 and a linear viscosity temperature relation, the critical temperature ratio $T_w/T_1$ has been calculated as a function of direction angle arc cos $\tilde{M}_1/M_1$ and plotted as a full-line curve in Fig. 7. The calculations are carried out by first finding $T_w/T_1$ as a function of $\tilde{M}_1$ from Eq. (10.2). The resulting values of $T_w/T_1$ when plotted against arc cos $\tilde{M}_1/M_1$ give a curve with a maximum and a minimum for $\tilde{M}_1$ in the range $0 < \tilde{M}_1 < M_1$. The proper curve for the critical temperature ratio is then obtained by drawing a horizontal tangent to this curve at its minimum point (thus cutting off the dashed part around the maximum point in Fig. 7).** The numerical results are listed in Table 7.

Figure 7 shows that for a temperature ratio $T_w/T_1 = 1.700$ all two-dimensional disturbances are completely stabilized, and, for a somewhat smaller ratio $T_w/T_1 = 1.474$, the three-dimensional disturbances corresponding to oblique waves with direction angles from 0° to 74.3° are completely stabilized. The critical temperature ratio $T_w/T_1$ decreases to zero rapidly, however, when the direction angle increases from 74.3° to arc cos $1/M_1 = 75.5°$.

When the temperature ratio $T_w/T_1$ is as low as 1.474, therefore, all oblique waves are stabilized except those with direction angles greater than 74.3°. Estimates of the minimum critical Reynolds Number of the latter (all of which have extremely small values of the wave speed $c$) show that $R_c \sim 10^6$ or $R_{xc} \sim 10.^{12}$ Since such large values of the Reynolds Number are often far beyond the range of practical importance, it follows that if the wall temperature ratio can be reduced to as low a value as 1.474 the boundary layer will be highly stabilized from the practical point of view. However, since the boundary layer is sensitive to changes in $T_w/T_1$, even a small increase above $T_w/T_1 = 1.474$ might make the boundary layer relatively unstable.

For each value of $M_1$ up to $M_1 = 7.48$ it is probable that the nature of the oblique waves is about the same as at $M_1 = 4$; that is, for a sufficiently small temperature ratio $T_w/T_1$, it is possible to completely stabilize oblique waves with direction angles less than arc cos $1/M_1$ but not those with larger direction angles.

At the present time the state of knowledge regarding the general problem of boundary-layer stability at high Mach Numbers is mainly restricted to such results as the preceding regarding complete stabilization by cooling. It is only in such limiting cases, for which $\tilde{\alpha}$

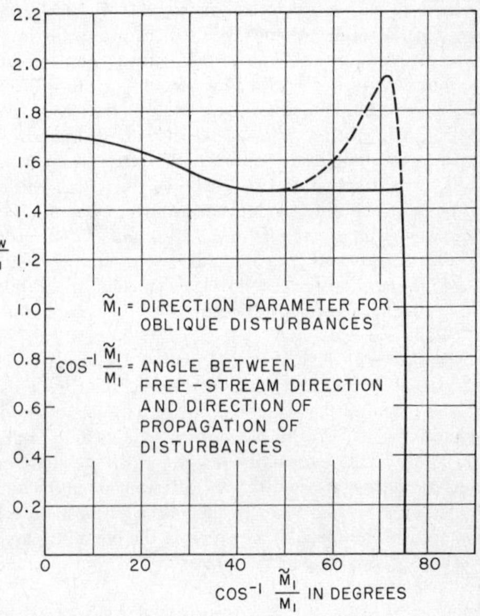


FIG. 7. Critical temperature ratios for complete stability with respect to three-dimensional disturbances at a free-stream Mach Number of $M_1 = 4$.

TABLE 7

Critical Temperature Ratios for Complete Stability with Respect to Three-Dimensional Disturbances at a Free-Stream Mach Number of 4

$M_1 = 4$, cos $\tilde{M}_1/M_1$ = direction angle, $\rho/\mu = \rho_w/\mu_w$, $\sigma = 0.75$

| $\tilde{M}_1$ | cos $\tilde{M}_1/M_1$ | $T_w/T_1$ | $\tilde{M}_1$ | cos $\tilde{M}_1/M_1$ | $T_w/T_1$ |
|---|---|---|---|---|---|
| 4. | 0° | 1.700 | 1.429 | 69.1° | 1.885 |
| 3.61 | 25.4° | 1.600 | 1.25 | 71.8 | 1.935 |
| 3.33 | 33.7 | 1.540 | 1.143 | 73.5 | 1.800 |
| 2.67 | 48.1 | 1.474 | 1.096 | 74.1 | 1.600 |
| 2.5 | 51.3 | 1.483 | 1.068 | 74.5 | 1.400 |
| 2.0 | 60.0 | 1.585 | 1.051 | 74.8 | 1.200 |
| 1.968 | 60.5 | 1.600 | 1.036 | 75.0 | 1.000 |
| 1.667 | 65.4 | 1.740 | 1.026 | 75.2 | 0.800 |
| 1.567 | 66.9 | 1.800 | 1.000 | 75.5 | 0 |

† This difficulty does not occur when $\tilde{M}_1$ is as large as $M_1$, at least for the boundary layer on a flat plate.

‡ This assumption seems the most reasonable one to make, but whether or not it is always justified will depend on results of calculation of neutral curves for large values of $\tilde{\alpha}$. See comments further on in this section.

** This is equivalent to the use of the condition mentioned in connection with relation (10.3). The exact significance of the dashed part of the curve is not yet entirely understood.

TABLE 8

The Functions $\lambda(c)$, $\psi_{im}(c)$ for $\rho\mu = \rho_w\mu_w$, $\sigma = 0.75$ and Zero Pressure Gradient

| c | $\lambda(c)$ | $\psi_{i_m}(c)$ | c | $\lambda(c)$ | $\psi_{i_m}(c)$ |
|---|---|---|---|---|---|
| 0 | 0 | .58114 | .6 | .0273 | .61307 |
| .1 | $.11520 \times 10^{-3}$ | .58216 | .7 | .0462 | .63625 |
| .2 | $.9237 \times 10^{-3}$ | .58251 | .8 | .0769 | .67622 |
| .3 | $3.136 \times 10^{-3}$ | .58529 | .9 | .1298 | .75297 |
| .4 | .0068 | .58891 | .95 | .1775 | .83165 |
| .5 | .0152 | .59868 | 1.0 | .320 | 1.15 |

approaches zero in the limit, that the present calculation procedures are satisfactory.

It should be emphasized that for high Mach Numbers the proper interpretation of the critical temperature ratios calculated by present methods is not completely certain. All that is known in any particular case is that for a temperature ratio *slightly* above the critical value a neutral curve exists at all points of which $\tilde{\alpha}$ is very small, and, when the temperature ratio is reduced to the critical value, this neutral curve recedes to infinite values of $\tilde{R}$ with all values of $\tilde{\alpha}$ on it approaching zero at the same time. It is possible that if we start with a temperature ratio *considerably* above the critical value and allow it to approach the latter the neutral curve might split into two branches, one of which disappears in the limit as described above, *while the other continues to exist at finite values of* $\tilde{\alpha}$ *and* $\tilde{R}$. In order to determine whether or not this is so, a method would have to be devised to allow the calculation at high Mach Numbers of neutral curves for values of $\tilde{\alpha}$ which are not very small.

For Mach Numbers not too much larger than unity, the situation is clearer, since neutral curves can be calculated by existing methods for finite values of $\tilde{\alpha}$ and thus for all temperature ratios. It appears that in these cases the possibility suggested above does *not* occur and the *entire* neutral curve disappears in the limit so that the critical temperature ratios actually do correspond to complete stabilization of the boundary layer with respect to the particular class of disturbances under consideration.

For low Mach Numbers (up to about $M_1 = 2$), the theoretical results of Section (9), which are then satisfactory approximations, also provide considerable useful information. The numerical calculations indicate that the parameter $C$ increases as $T_w/T_1$ decreases (see Fig. 5). Thus, the effect of cooling a flat plate at a given Mach Number is evidently to make three-dimensional disturbances relatively more unstable, so that with sufficient cooling some of them can actually become more unstable than two-dimensional disturbances. Since $C$ for an insulated plate increases from $C = 0$ at $M_1 = 0$ to a value close to unity when $M_1 \approx 2$, it appears that the amount of cooling required to make any of the three-dimensional disturbances more unstable than two-dimensional ones is extreme at low subsonic Mach Numbers but gradually becomes less and less extreme as $M_1$ increases, until, at a Mach Number of the order of two, little is required. In particular, at $M_1 = 0.7$, the wall temperature ratio must be reduced below $T_w/T_1 = 0.58$ (see Fig. 5).

It is known that at subsonic Mach Numbers cooling of the solid surface is effective in stabilizing the boundary layer if only two-dimensional disturbances are considered.[2] The preceding remarks indicate that probably this conclusion is still valid even if three-dimensional disturbances are considered, since the latter usually have a higher minimum critical Reynolds Number than two-dimensional disturbances (except under conditions of extreme cooling).

Even at the moderate supersonic Mach Numbers when three-dimensional disturbances begin to play the leading role, cooling still has an important stabilizing effect, although complete stabilization at supersonic Mach Numbers as predicted in reference 2 is impossible in the strict sense, according to the present theory. For example, estimates show that for an insulated flat plate at $M_1 = 1.6$, $R_{xc}$ is of the order of $10^5$, whereas the calculated results show that for a cooled plate at the same Mach Number $M_1 = 1.6$, with a temperature ratio $T_w/T_1 = 1.073$ which is just small enough for two-dimensional disturbances to be completely stabilized, $R_{xc}$ is of the order of $10^6$.

## (11) CONCLUSIONS

(a) An order of magnitude analysis of the complete linearized equations for a three-dimensional disturbance superimposed on a two-dimensional boundary layer shows that they can be reduced to much simpler equations in the first approximation. These simplified equations are still valid at high free-stream Mach Numbers if correct velocity and temperature profiles for the boundary layer in such cases are used. At very high Mach Numbers, however, the accuracy of a first approximation may not be great, so that higher order approximations (and therefore more complicated equations with more terms retained) may have to be considered.

(b) The present equations for a three-dimensional disturbance transform to equations that have the same form mathematically as those for a two-dimensional disturbance but which do not, however, correspond to a real two-dimensional disturbance. Although no direct conclusion regarding the importance of three-dimensional disturbances is now possible, the use of these transformed two-dimensional equations leads to a considerable mathematical simplification of the theory.

(c) Corrected viscous solutions have been obtained which lead to a new formulation of the stability theory. The conclusion reached by Lees and Lin—that the relation for the characteristic values is independent of the temperature fluctuations and the boundary conditions imposed on them—is now found to be restricted in general validity to subsonic and slightly supersonic Mach Numbers and is true only in a restricted sense—namely, for very obliquely traveling waves at high Mach Numbers. At high Mach Numbers the relation for the characteristic values depends in general on the thermal boundary conditions. It is found, however, that for very small values of the wave number the relation again has the same simple form as at low Mach Numbers. In all cases at high Mach Numbers when the wave number is not small, the relation is more complicated. The detailed mathematical development of this improved theory is to be presented elsewhere.

(d) Consideration of the present analytical and numerical results suggests that probably three-dimensional disturbances are of little importance at subsonic free-stream Mach Numbers, since they usually have a higher minimum critical Reynolds Number than two-dimensional ones, except possibly under conditions of extreme surface cooling. However, further numerical calculations for oblique waves in this range of Mach Numbers would be useful.

(e) As the Mach Number increases, three-dimensional disturbances enter the picture under conditions which become less and less extreme, until finally for free-stream Mach Number $M_1$ between one and two they begin to play the leading role in many problems of practical interest.

(f) At supersonic free-stream Mach Numbers, the boundary layer can never be completely stabilized with respect to *all* three-dimensional disturbances. Although oblique waves whose directions of propagation are at an angle less than arc cos $1/M_1$ with the free-stream direction can be completely stabilized by sufficient cooling when $M_1$ is not too large, those whose directions of propagation are at a greater angle cannot.

(g) The graph of critical temperature ratio (below which strictly two-dimensional disturbances are completely stabilized) versus Mach Number (Fig. 6) is not significantly different from that found by Van Driest,[11] in spite of the corrections to the viscous solutions. Although these corrections are very important for the accurate determination of the curve of neutral stability at supersonic Mach Numbers, they are not as important for the determination of the critical temperature ratios.

(h) For Mach Numbers up to about two, cooling of the solid surface is effective in stabilizing the boundary layer. For higher Mach Numbers, reliable calculations of the minimum critical Reynolds Number over wide ranges of the wall temperature ratio have not yet been made; but this general conclusion would probably remain unchanged for Mach Numbers up to six, say.

(i) Little more can be said about the problem of boundary-layer stability at Mach Numbers greater than two, since the present methods of numerical calculation for the curve of neutral stability (and thus the minimum critical Reynolds Number) do not appear to be adequate in such cases. When $c$ and $\tilde{\alpha}$ are both large, there are difficulties in working with the more complicated relation for the characteristic values and also in evaluating the inviscid solutions of the disturbance equations. These problems are being studied further.

(j) At very high Mach Numbers there is an even more serious difficulty than the preceding—namely, that the present asymptotic theory, which is essentially a first-order theory, may in itself be inaccurate [as pointed out in paragraph (a) above]. More information about this matter will be provided by numerical results (including estimates of certain parameters) for high Mach Numbers, which can be obtained as soon as the difficulties mentioned above in paragraph (i) are resolved.

(k) It is believed that the approximate method of calculation of the neutral curve described in Section (8) is adequate up to a Mach Number of the order of two, especially in cases involving cooling, when $c$ and $\tilde{\alpha}$ are both small. Additional neutral curves for the oblique waves in this range of Mach Numbers or perhaps further calculations of the type described in Sections (9) and (10) would be of considerable value. It might also be of interest to determine in a few typical cases the effect of using a more accurate approximation than Eq. (8.16a) by keeping terms of higher order in $\tilde{\alpha}$, as Lees has done.[2]

(l) For practical purposes, the transition Reynolds Number is much more important than the minimum critical Reynolds Number. Thus, information concerning the rate of amplification of the boundary-layer disturbances after they enter the unstable region would be extremely useful. Further investigations are being directed toward developing the amplification theory for three-dimensional disturbances, which may be somewhat different from that for strictly two-dimensional disturbances.

(m) Applications of any of the present theory to cases involving finite wall curvature should be made with caution. Although there are no restrictions regarding pressure gradient, the basic equations derived here are strictly correct only for flow over a flat surface. Although curvature effects are usually extremely small for moderate values of the direction angle of the disturbances (see references 6 and 15), they may become important for values of this angle approaching 90°. For example, whereas the present theory predicts an infinite minimum critical Reynolds Number for disturbances with direction angle of 90°, it might be expected from the work of Görtler[16] that for small but finite concave curvature such disturbances would be unstable at finite Reynolds Numbers.

## Appendix

### Derivation of the Relation for the Characteristic Values

The basic relation (7.9) is obtained as the practical form of the secular equation on which all subsequent deductions and calculations are based. A complete derivation involves a detailed study of the asymptotic solutions of the system of Eqs. (6.9) to (6.13) and will be presented elsewhere. Here we shall merely outline the major steps and point out the nature of the approximations involved.

For large values of $\tilde{\alpha}\tilde{R}$ and finite values of $(y - y_c)$, formal expansions give rise to solutions of the following three types:

(a) Two solutions for which each of the functions has an asymptotic form

$$Z_i(y) = Z_i^{(0)}(y) + [1/(\tilde{\alpha}\tilde{R})]Z_i^{(1)}(y) + \ldots \quad \text{(A1)}$$

(b) Two solutions for which each of the functions has an asymptotic form

$$\left.\begin{aligned} Z_i(y) &= \{\exp[\pm(\tilde{\alpha}\tilde{R})^{1/2}P]\}\,[\bar{Z}_i^{(0)} + (\tilde{\alpha}\tilde{R})^{-1/2}\bar{Z}_i^{(1)} + \ldots] \\ P(y) &= \int_{y_c}^{y}[(w - c)/\nu]^{1/2}\,dy \end{aligned}\right\} \quad \text{(A2)}$$

(c) Two solutions for which each of the functions has an asymptotic form

$$\left.\begin{aligned} Z_i(y) &= \{\exp[\pm(\tilde{\alpha}\tilde{R})^{1/2}Q]\}\,[\hat{Z}_i^{(0)} + (\tilde{\alpha}\tilde{R})^{-1/2}\hat{Z}_i^{(1)} + \ldots] \\ Q(y) &= \int_{y_c}^{y}[\sigma(w - c)/\nu]^{1/2}\,dy \end{aligned}\right\} \quad \text{(A3)}$$

In the above, $y_c$ is the point where $w = c$. The approximate solution $Z_i^{(0)}(y)$ is the inviscid approximation and has the exponential behavior $\exp[\pm\beta y]$ for large $y$, as discussed in Section (7).

Consider the case of real $c$. It is clear that, if the solution is to remain bounded at infinity, only the negative signs are allowed in *all* the exponentials. Thus, we have *one* permissible solution from each of the three types, (a), (b), and (c). Let us denote these solutions by

(a) $(\tilde{f}, \tilde{\phi}\ \tilde{\theta}) = (F, \Phi, \Theta)$
(b) $(\tilde{f}, \tilde{\phi}, \tilde{\theta}) = (f_3, \phi_3, \theta_3)$
(c) $(\tilde{f}, \tilde{\phi}, \tilde{\theta}) = (f_5, \phi_5, \theta_5)$

To satisfy the conditions at the wall, i.e.,

$$\tilde{f} = \tilde{\phi} = 0, \quad a\tilde{\theta}' + b\tilde{\theta} = 0 \quad \text{(A4)}$$

where $a$ and $b$ are given constants, we must have

$$a\begin{vmatrix} F(0) & f_3(0) & f_5(0) \\ \Phi(0) & \phi_3(0) & \phi_5(0) \\ \Theta'(0) & \theta_3'(0) & \theta_5'(0) \end{vmatrix} + b\begin{vmatrix} F(0) & f_3(0) & f_5(0) \\ \Phi(0) & \phi_3(0) & \phi_5(0) \\ \Theta(0) & \theta_3(0) & \theta_5(0) \end{vmatrix} = 0 \quad \text{(A5)}$$

Actual calculations reveal that the general orders of magnitudes of the functions are given by [see Eq. (A9)]

$$\left.\begin{aligned} (F, \Phi, \Theta) &\sim (1, 1, 1) \\ (f_3, \phi_3, \theta_3) &\sim (1, \epsilon, 0) \\ (f_5, \phi_5, \theta_5) &\sim (0, \epsilon, 1) \end{aligned}\right\} \quad \text{(A6)}$$

to the order of $\epsilon = (\tilde{\alpha}\tilde{R})^{-1/2}$. Hence, we obtain

$$\Phi(0)/F(0) = [\phi_3(0)/f_3(0)]\,(1 + \Delta) \quad \text{(A7)}$$

where

$$\Delta = \frac{a\Theta'(0) + b\Theta(0)}{F(0)}\ \frac{\theta_5(0)}{a\theta_5'(0) + b\theta_5(0)}\ \frac{\phi_5(0)\,f_3(0)}{\theta_5(0)\,\phi_3(0)} \quad \text{(A8)}$$

We shall consider $\Delta$ in more detail later and just observe at this point that it is always $0(1)$ at most.

From Eq. (A6), it follows that $\phi_3(0)/f_3(0) = 0(\epsilon)$, which shows that in the limit $\epsilon \to 0$ Eq. (A7) reduces to the inviscid formulation $\Phi(0) = 0$, as it should.

On the basis of the general nature of the solutions (A1), (A2), and (A3) and because of the orders of magnitude of the solutions as shown in Eq. (A6), we may refer to these solutions as (a) the inviscid solution, (b) the viscous solution of velocity variation, and (c) the viscous solution of thermal variation.

For practical purposes it is found to be desirable to use asymptotic forms of the viscous solutions which remain accurate even for $(y - y_c) \to 0$. For example, for the viscous solution of velocity variation, detailed studies give

$$\left.\begin{aligned} f_3 &= A(y)\int_{\infty}^{\zeta}\zeta^{1/2}H_{1/3}^{(1)}\left[\frac{2}{3}(i\zeta)^{3/2}\right]d\zeta \\ \phi_3 &= (\tilde{\alpha}\tilde{R})^{1/2}B(y)\int_{\infty}^{\zeta}d\zeta\int_{\infty}^{\zeta}\zeta^{1/2}H_{1/3}^{(1)}\left[\frac{2}{3}(i\zeta)^{3/2}\right]d\zeta \end{aligned}\right. \quad \text{(A9)}$$

where

$$B(y) = -i\left(\frac{w - c}{\nu}\right)^{-1/2}\left[\frac{3}{2}\int_{y_c}^{y}\left(\frac{w - c}{\nu}\right)^{1/2}dy\right]A(y) \quad \text{(A10)}$$

and

$$\zeta = (\tilde{\alpha}\tilde{R})^{1/3}\left[\frac{3}{2}\int_{y_c}^{y}\left(\frac{w - c}{\nu}\right)^{1/2}dy\right]^{2/3} = (\tilde{\alpha}\tilde{R})^{1/3}\left(\frac{3P}{2}\right)^{2/3} \quad \text{(A11)}$$

This solution agrees asymptotically with Eq. (A2) when $\zeta$ is large but is also valid for small $\zeta$. Furthermore, if each of the integrals (A10) and (A11) is evaluated in the neighbourhood of $y_c$ by retaining only the first term in the power series expansion of the integrand, the solution given in reference 1 is obtained. An improved version of the viscous solution of thermal variation may also be developed.*

By substituting Eq. (A9) into Eq. (A7), the relation (7.9) can be easily obtained. The only additional step is to express $\Phi(y)$ as a linear combination of two linearly independent inviscid solutions $\phi_1(y)$ and $\phi_2(y)$—i.e., two such solutions of Eq. (7.3)—and then to require that this linear combination shall be proportional to $\exp[-\beta y]$ for $y \geq 1$, where $y = 1$ corresponds to the edge of the boundary layer.

Now let us return to Eq. (A8) for $\Delta$ and investigate the conditions under which it may be neglected and those under which it must be retained. It may be shown that, for $T_w$ bounded away from zero,† Eq. (A8) becomes

$$\Delta = [(\gamma - 1)\tilde{M}_1^2c/T_w]\,(\{a[(T_w'c - T_ww_w')/T_w]\} + bc)\,\{[\theta_5(0)]/[a\theta_5'(0) + b\theta_5(0)]\}\chi(z, \sigma) \quad \text{(A12)}$$

where $\chi(z, \sigma)$ is a function depending on the viscous solutions which is always $0(1)$ ($z$ is defined by Eq.

* The improved solutions are obtained by extending a method of Tollmien in the incompressible case.[8] Their use does not alter any of the order-of-magnitude estimates in the present discussion.

† The exceptional case $T_w \to 0$ is to be considered later.

(7.11) and $\sigma$ is the Prandtl Number). It might be mentioned that the factor $c$ in the first term is due to $\phi_5(0)$ being proportional to $c$.

In reference 1 the conclusion was reached that the characteristic-value problem is independent of the temperature fluctuations and the boundary conditions imposed on them, to the order of accuracy of the theory. From Eqs. (A7) and (A12) it is clear that this will be true when $\Delta$ can be neglected without any restrictions being placed on the values of $a$ and $b$. Since the factors multiplying $(\gamma - 1)\tilde{M}_1^2 c/T_w$ in Eq. (A12) are all $0(1)$ at most, it follows that the conclusion is valid in two main cases:

$$c \ll 1 \tag{A13}$$

which was considered in Reference 1, and

$$\tilde{M}_1^2 \ll 1 \tag{A14}$$

which corresponds to oblique waves with direction angles close to 90°.

Case (A13) covers most practical situations at subsonic and slightly supersonic Mach Numbers, when $c$ is usually not large. Case (A14) extends the conclusion to very obliquely traveling disturbances at high Mach Numbers.

It might be remarked that since $\phi_5(0)$ is proportional to $c$ and since the theory of reference 1 was restricted to values of $c$ of order $\epsilon$, it is understandable that the preceding conclusion should be a natural consequence of the asymptotic theory [in which quantities $0(\epsilon)$ are neglected], as developed in reference 1.

In all situations at high Mach Numbers when Eq. (A14) does not hold, the exact form of Eq. (A7) depends on a knowledge of the thermal boundary conditions—i.e., a knowledge of the constants $a$ and $b$. If it is observed that $\theta_5'(0)/\theta_5(0) = 0(1/\epsilon)$, an examination of Eq. (A12) shows that three cases may be distinguished:

$$b/a = 0(1), \quad \Delta = 0(\epsilon) \tag{A15}$$

$$\begin{aligned} a/b &= 0(\epsilon) \\ \Delta &= [(\gamma - 1)\tilde{M}_1^2 c^2/T_w]\,\{[\theta_5(0)]/[(a/b)\theta_5'(0) + \theta_5(0)]\}\,\chi(z, \sigma) \end{aligned} \tag{A16}$$

$$\begin{aligned} a/b &= 0(\epsilon^2) \\ \Delta &= [(\gamma - 1)\tilde{M}_1^2 c^2/T_w]\,\chi(z, \sigma) \end{aligned} \tag{A17}$$

*In case (A15), $\Delta$ may again be neglected, so that Eq. (A7) again reduces to the same simple form as at low Mach Numbers.* In the other two cases, $\Delta$ cannot be neglected, and in Eq. (A16), moreover, a new parameter $a/b$ enters into the theory.

It may be noted that when $\Delta$ is not $0(\epsilon)$, it is always of the form

$$\Delta = [(\gamma - 1)\tilde{M}_1^2 c^2/T_w]\,\Delta_1 \tag{A18}$$

where $\Delta_1 = 0(1)$. Thus, the error in neglecting $\Delta$ on the basis of small $c$ is somewhat smaller than might have been expected from the discussion regarding condition (A13).

To determine the proper form of the thermal boundary condition (A4) and thus the values of the constants $a$ and $b$ in practical situations, an analysis of the heat conduction problem in the medium below the interface between the gas and solid wall is necessary. Since this is straightforward but fairly lengthy, we shall just give the final conclusions here for a few typical cases.

Consider a boundary-layer flow of a gas extending from the solid surface at $y = 0$ to $y = \infty$, and suppose that from $y = 0$ to $y = -h_1$ is a thin metal wall, from $y = -h_1$ to $y = -h$ a channel through which a cooling liquid is flowing, and finally from $y = -h$ to $y = -\infty$ an insulating material. Although the problem is considerably idealized, it illustrates some of the essential features in certain cases of interest. For this problem, it can be shown that the condition for the temperature fluctuations in the boundary layer is given by

$$\theta'(0) - \{\omega - [(1/k)\,(dk/dT)\,(\partial T/\partial y)]_w\}\theta(0) = 0 \tag{A19}$$

where

$$\omega = \omega_m \frac{k_m}{k_0}\left\{\left[\left(1 + S\frac{k_i\omega_i}{k_m\omega_m}\right) - \left(1 - S\frac{k_i\omega_i}{k_m\omega_m}\right)e^{-2\omega_m h}\right]\Big/\left[\left(1 + S\frac{k_i\omega_i}{k_m\omega_m}\right) + \left(1 - S\frac{k_i\omega_i}{k_m\omega_m}\right)e^{-2\omega_m h}\right]\right\} \tag{A20}$$

in which the various quantities appearing are defined as follows:

$$\frac{1}{S} = \frac{k_i\omega_i}{k_l\omega_l}\left\{\left[\left(1 + \frac{k_i\omega_i}{k_l\omega_l}\right)e^{-2\omega_l(h_1 - h)} + \left(1 - \frac{k_i\omega_i}{k_l\omega_l}\right)\right]\Big/\left[\left(1 + \frac{k_i\omega_i}{k_l\omega_l}\right)e^{-2\omega_l(h_1 - h)} - \left(1 - \frac{k_i\omega_i}{k_l\omega_l}\right)\right]\right\} \tag{A21}$$

$$\omega_m^2 = -i\alpha c\,(\rho_m c_m/k_m) \tag{A22}$$

with corresponding definitions for $\omega_l$ and $\omega_i$. The symbols $\rho_m$, $c_m$, and $k_m$ denote density, specific heat, and thermal conductivity, respectively. The subscripts 0, $m$, $l$, and $i$ refer to gas at $y = 0$, metal, liquid, and insulator, respectively. The quantity $\alpha c$ is the nondimensional time frequency of the fluctuations [see Eq. (6.1)].

Relation (A20) is appropriate for a *cooled wall*. It reduces to

$$\omega = \omega_m(k_m/k_0) \tag{A23}$$

for high frequencies, and to

$$\omega = \omega_i(k_i/k_0) \tag{A24}$$

for low frequencies. The case of low frequencies is of most interest here. In Section (10) it is shown that if the simple eigenvalue relation (7.14) obtained by neglecting $\Delta$ is used, then in the limiting case of complete stabilization $\tilde{\alpha} \to 0$ everywhere on the neutral curve. Since the frequency is given by $\alpha c = \tilde{\alpha} c \cos \Theta$, where $\Theta$ is the direction angle of the waves, it follows that $\omega \to 0$ in the limit and thus from Eq. (A15) that $\Delta$ is indeed negligible. Thus, we see that the use of the simple relation (7.14) in calculating the critical temperature ratios is justified.

The definition of $\omega$ for an *insulated wall* is easily obtained from Eq. (A20) by letting $h_1 = h$. For high frequencies this also reduces to Eq. (A23). If there is no metal layer so that the insulator extends right to the surface, $h = 0$, and $\omega$ is given by

$$\omega = \omega_i(k_i/k_0) \qquad \text{(A25)}$$

for all frequencies.

In all cases at high Mach Numbers when a neutral curve exists, $\tilde{\alpha} = 0(1)$, $c = 0(1)$, are reasonable estimates. Then, on taking $\cos\Theta = 0(1)$, we find that Eqs. (A23) and (A25) are accurate approximations and that $\omega$ is *extremely large*. This is true of Eq. (A23) even when $\tilde{\alpha} = 0(10^{-2})$, say. Thus, in most cases of interest at high Mach Numbers when cooling is not sufficient to give complete stabilization, $\omega$ is extremely large, and the relation for the characteristic values is that corresponding to Eq. (A17).

The exceptional case $T_w \to 0$ has not been considered so far. This occurs in the calculations of the critical temperature ratios for $\tilde{M}_1 \to 1$ and for $M_1 \to 7.48$ (see Figs. 6 and 7). We may justify the procedure of using the simple relation (7.14) for these calculations *a posteriori*—that is, we assume Eq. (7.14) to hold in these limiting cases and then check the consequences to see if they are consistent with the assumption. In the case $\tilde{M}_1 \to 1$, $c = 1 - 1/\tilde{M}_1 \to 0$, and an examination of Eqs. (10.2) and (8.16b) shows that $T_w \propto c$. Then Eq. (A12) shows that $\Delta$ is now $0(c)$, so that the procedure is justified. In the case $M_1 \to 7.48$, $c$ remains finite, and relations (10.1) and (8.16b) imply that $K_{1_i} = 0$—i.e., that $[d/dy(w'/T)]_c = 0$. This indicates that the solution obtained by the use of Eqs. (7.14) or (8.1) reduces to that obtained from an inviscid formulation. On the other hand, the more accurate relation (7.9), when put into the form (8.1), has $\Phi(z)$ replaced by $[1 - F(z)(1 + \Delta)]^{-1}$. At the same time, Eq. (A12) shows that $\Delta = 0(1/T_w)$. It can be shown that for $\Delta$ with such a behavior the more accurate relation also leads to the inviscid solution in the limit $T_w \to 0$. Thus, the results obtained by the use of Eq. (7.14) are again justified.

Summarizing our main conclusions, we have found that for very low frequencies the boundary condition for the temperature fluctuations is

$$\theta'(0) + [(1/k)(dk/dT)(\partial T/\partial y)]_w\,\theta(0) = 0 \qquad \text{(A26)}$$

and the corresponding secular equation is

$$E(\tilde{\alpha}, c, \tilde{M}_1^2) = F(z) \qquad \text{(A27)}$$

while for very high frequencies, the boundary condition is

$$\theta(0) = 0 \qquad \text{(A28)}$$

and the corresponding secular equation is

$$E(\tilde{\alpha}, c, \tilde{M}_1^2) = F(z)\{1 + [(\gamma - 1)\tilde{M}_1^2c^2/T_w] \times \chi(z, \sigma)\} \qquad \text{(A29)}$$

The boundary conditions (A26) and (A28) are actually just what would be expected physically, as the following considerations will show. In the stability theory, the temperature fluctuations are essentially natural oscillations originating in the boundary layer. Although these fluctuations can penetrate into the adjacent medium (gas, liquid, solid, etc.), they must go to zero at large enough distances. For very low frequencies, it is clear that such fluctuations will go far into the surrounding medium before dying out. Thus, for low enough frequencies, the change in amplitude will be negligible over distances in the solid of the order of the boundary-layer thickness, so that the fluctuating part of the heat flux will be negligible. Therefore, since the left-hand side of Eq. (A26) is proportional to the latter quantity expressed in terms of boundary-layer properties at the wall, relation (A26) will hold closely.

On the other hand, for very high frequencies the thermal inertia of the medium will prevent the fluctuations at any given location from reaching the maximum amplitude attained near the origin of the disturbance. Thus, the amplitude will die away to zero within a short distance inside the solid wall, for example.* Ultimately, for high enough frequencies, even the amplitude at the surface will become negligible, so that the condition $\theta(0) = 0$ will be a very accurate approximation.

## References

[1] Lees, L., and Lin, C. C., *Investigation of the Stability of the Laminar Boundary Layer in a Compressible Fluid*, NACA TN 1115, 1946.

[2] Lees, L., *The Stability of the Laminar Boundary Layer in a Compressible Fluid*, NACA Report No. 876, 1947 (formerly TN 1360).

[3] Squire, H. B., *On the Stability for Three-Dimensional Disturbances of Viscous Fluid Flow Between Parallel Walls*, Proc. Roy. Soc. (London), Vol. A142, pp. 621–628, 1933.

[4] Lees, L., and Probstein, R. F., *Hypersonic Viscous Flow over a Flat Plate*, Princeton University, Aeronautical Engineering Laboratory, Report No. 195, April, 1952.

[5] Shen, S. F., *On the Boundary-Layer Equations in Hypersonic Flow*, Journal of the Aeronautical Sciences, Vol. 19, No. 7, pp. 500–501, July, 1952.

[6] Dunn, D. W., Thesis for Ph.D., Massachusetts Institute of Technology, June, 1953.

[7] Lin, C. C., *On the Stability of the Boundary Layer with Respect to Disturbances of Large Wave Velocity*, Journal of the Aeronautical Sciences, Vol. 19, No. 2, pp. 138–139, February, 1952.

[8] Tollmien, W., *Asymptotische Integration der Störungsdifferentialgleichung ebener laminarer Strömungen bei hohen Reynoldschen Zahlen*, Zeitschrift für Ang. Math. und Mech., Vols. 25–27, pp. 33–50, 70–83, 1947.

[9] Kuethe, A. M., *Some Aspects of Boundary-Layer Transition and Flow Separation on Cylinders in Yaw*, Proceedings of the Mid-western Conference on Fluid Dynamics, 1st conference, May, 1950, pp. 44–55; J. W. Edwards, Ann Arbor, Mich., 1951.

* This "surface-layer effect" or "skin effect" is well known in other problems involving temperature fluctuations.

[10] Lin, C. C., *On the Stability of Two-Dimensional Parallel Flows*, Parts I, II, III; Quart. Appl. Math., Nos. 2, 3, 4; pp. 117–142, pp. 218–234, pp. 277–301, 1945–1946.

[11] Van Driest, E. R., *Calculation of the Stability of the Laminar Boundary Layer in a Compressible Fluid on a Flat Plate with Heat Transfer*, Journal of the Aeronautical Sciences, Vol. 19, No. 12, pp. 801–812, December, 1952.

[12] Cheng, Sin-I, *On the Stability of Laminar Boundary Layer Flow*, Quart. Appl. Math., Vol. 11, No. 3, pp. 346–350, October, 1953.

[13] Dunn, D. W., and Lin, C. C., *On the Stability of the Laminar Boundary Layer in a Compressible Fluid*, (Preliminary Report), Department of Mathematics, Massachusetts Institute of Technology, December, 1953.

[14] Bloom, M., *On the Calculation of Laminar Boundary-Layer Stability*, Journal of the Aeronautical Sciences, Vol. 21, No. 3, pp. 207–210, March, 1954.

[15] Görtler, H., *Über den Einflusz der Wandkrümmung auf die Entstehung der Turbulenz*, Zeit für Ang. Math. und Mech., Vol. 20, pp. 138–147, 1940.

[16] Görtler, H., *Über eine Dreidimensionale Instabilität Laminarer Grenzschichten an Konkaven Wänden*, Nachr. Ges. Wiss., Göttingen, Vol. 2, pp. 1–26, 1940 (Translated as NACA TM 1375, 1954).

List of Symbols

| Dimensional quantities | Nondimensional quantities | Reference quantities |
|---|---|---|
| Positional co-ordinates | | |
| (1) $x_1$ | $x$ | $\delta$ |
| (2) $x_2$ | $y$ | $\delta$ |
| (3) $x_3$ | $z$ | $\delta$ |
| Time | | |
| (4) $t^*$ | $t$ | $\delta/U_1$ |
| Velocity components | | |
| (5) $u_1 = \bar{u}_1 + u_1'$ | $w(y) + f(y) \exp [i(\alpha x + \beta z - \alpha ct)]$ | $U_1$ |
| (6) $u_2 = \bar{u}_2 + u_2'$ | $v(y) + \alpha\phi(y) \exp [i(\alpha x + \beta z - \alpha ct)]$ | $U_1$ |
| (7) $u_3 = u_3'$ | $h(y) \exp [i(\alpha x + \beta z - \alpha ct)]$ | $U_1$ |
| Density | | |
| (8) $\rho^* = \bar{\rho}^* + \rho'^*$ | $\rho(y) + r(y) \exp [i(\alpha x + \beta z - \alpha ct)]$ | $\bar{\rho}_1^*$ |
| Pressure | | |
| (9) $p^* = \bar{p}^* + p'^*$ | $p(y) + \pi(y) \exp [i(\alpha x + \beta z - \alpha ct)]$ | $\bar{p}_1^*$ |
| Temperature | | |
| (10) $T^* = \overline{T}^* + T'^*$ | $T(y) + \theta(y) \exp [i(\alpha x + \beta z - \alpha ct)]$ | $\overline{T}_1^*$ |
| Viscosity coefficients | | |
| (11) $\mu^* = \bar{\mu}^* + \mu'^*$ | $\mu(y) + \overline{(d\mu/dT)}\, \theta(y) \exp [i(\alpha x + \beta z - \alpha ct)]$ | $\bar{\mu}_1^*$ |
| (12) $\lambda^* = \bar{\lambda}^* + \lambda'^*$ | $\lambda(y) + \overline{(d\lambda/dT)}\, \theta(y) \exp [i(\alpha x + \beta z - \alpha ct)]$ | $\bar{\mu}_1^*$ |
| Thermal conductivity | | |
| (13) $k^* = \bar{k}^* + k'^*$ | $[\mu(y)/\sigma(y)] + \overline{(d/dT)\, (\mu/\sigma)}\theta(y) \exp [i(\alpha x + \beta z - \alpha ct)]$ | $C_p\bar{\mu}_1^*$ |
| Wave numbers of the disturbance | | |
| (14) $\alpha_1$ | $\begin{cases} \alpha \\ \alpha_\theta \\ \alpha_x \end{cases}$ | $\delta^{-1}$ <br> $\theta^{*-1}$ <br> $x_1^{-1}$ |
| (15) $\alpha_3$ | $\beta$ | $\delta^{-1}$ |
| Wave velocity of the disturbance | | |
| (16) $c^*$ | $c$ . | $U_1$ |
| Specific heat at constant volume | | |
| (17) $C_v$ | 1 | $C_v$ |
| Specific heat at constant pressure | | |
| (18) $C_p$ | $\gamma$ | $C_v$ |
| Gas constant per gram | | |
| (19) $R^*$ | $\gamma - 1$ | $C_v$ |
| Acceleration due to gravity | | |
| (20) $g$ | $1/F^2$ | $U_1^2/\delta$ |
| Momentum thickness of boundary layer | | |
| (21) $\theta^* = \int_0^\infty \rho w(1 - w)\, dx_2$ | | |
| Reynolds Number | | |
| (22) Based on boundary layer thickness $\delta$ | $R = \bar{\rho}_1^* U_1\delta/\bar{\mu}_1^*$ | |
| (23) Based on momentum thickness $\theta^*$ | $R_\theta = \bar{\rho}_1^* U_1\theta^*/\bar{\mu}_1^*$ | |
| (24) Based on distance $x_1$ from leading edge | $R_x = \bar{\rho}_1^* U_1 x_1/\bar{\mu}_1^*$ | |
| Mach Number | | |
| (25) | $M_1 = U_1/\sqrt{\gamma_1 R^* \overline{T}_1^*}$ | |
| Froude Number | | |
| (26) | $F = U_1/\sqrt{g\delta}$ | |
| Prandtl Number | | |
| (27) | $\sigma = C_p\mu^*/k^*$ | |

(a) A bar (———) denotes mean value, a prime denotes fluctuation. The subscript 1 refers to the value at the edge of the boundary layer at the particular $x$-location considered, and the subscript $w$ refers to wall value. The subscripts $r$ and $i$ denote the real and imaginary parts of a quantity, respectively.

(b) The nondimensional mean quantities $u_1(x_1, x_2)/U_1$, $u_2(x_1, x_2)/U_1$, $\bar{\rho}(x, y)$, $\bar{p}(x, y)$, . . . are represented as functions of $y$ alone—that is, as $w(y)$, $v(y)$, $\rho(y)$, $p(y)$, . . . , respectively. This corresponds to an approximation depending on the boundary-layer nature of the mean flow [see Section (4)].

(c) The quantities $C_p$, $C_v$, $\mu$, $\lambda$, $k$, $\sigma$, $\gamma$ are all considered to be functions of $T$ only. The fluctuation $Q'$ of any such quantity $Q$ is given by $Q' = \overline{dQ/dT}\, T'$.

(d) For moderate Mach Numbers and rates of heat transfer $C_p$, $C_v$, $\gamma$, $\sigma$ vary only slightly in the boundary layer. They are therefore regarded as constants equal to their free-stream values in the numerical calculations.

(e) The reference length $\delta$ is the boundary-layer thickness at the particular value of $x$ considered. Its definition is somewhat arbitrary and may be taken in any convenient manner. In the present numerical computations it is taken to be the value of $x_2$ at which $u_1/U_1 = 0.999$.

*Reprinted from*

THE PHYSICS OF FLUIDS VOLUME 3, NUMBER 4 JULY-AUGUST, 1960

# On the Secondary Motion Induced by Oscillations in a Shear Flow

D. J. BENNEY AND C. C. LIN
*Department of Mathematics,*
*Massachusetts Institute of Technology,*
*Cambridge, Massachusetts*
(Received May 20, 1960)

ONE of the most basic and challenging problems in fluid mechanics is to reach an understanding of the various physical mechanisms involved in the process of transition from laminar to turbulent flow. It is now well known that in many instances the initial process is one of instability with respect to infinitesimal disturbances. This initial disturbance could take on the form of a steady secondary flow, as in the case of convective instability, or the form of a small oscillation, as in the case of certain shear flows. It is quite likely that the processes governing the subsequent development of such disturbances are also different. The present letter deals with the case of shear flows.

We consider, as usual, a small disturbance of the basic flow by writing the velocity and pressure of the disturbed motion in the form

$$
\begin{aligned}
u_i(x_k, t) &= u_i^{(0)}(x_k, t) + a u_i^{(1)}(x_k, t) + a^2 u_i^{(2)}(x_k, t) + \cdots, \\
p(x_k, t) &= p^{(0)}(x_k, t) + a p^{(1)}(x_k, t) + a^2 p^{(2)}(x_k, t) + \cdots \qquad (i, k = 1, 2, 3),
\end{aligned} \tag{1}
$$

where $a$ is a measure of the amplitude of the disturbance, and all the variables are in dimensionless form. These expressions must satisfy the equation of continuity and the Navier-Stokes equations, from which we obtain a sequence of equations governing the motion of each order. The equations of order $a^0$ give the basic flow, those of order $a$ give the usual linear theory, and those of the order $a^2$ give the nonlinear effects under consideration.

In the case of shear flows, the linear terms give rise to stable and unstable normal modes in the form of waves. There is now ample experimental evidence to support the correctness of this theory. The nonlinear effect has been examined theoretically by Meksyn and Stuart,[1] and again by Stuart[2] using a different approach. A theory of secondary instability has been proposed by Görtler and Witting[3] to account for the nonlinear effects. Experimental observations, especially those by Schubauer, Klebanoff, and Tidstrom,[4] however, show remarkable characteristics in the behavior of the flow not accountable by the theories of these authors. In particular, the experimental observations indicate the existence of longitudinal vortical motions (with axis along the direction of the basic flow) which give a redistribution of momentum of the mean flow downstream in the plane normal to it. Such a vortical motion is very similar to that described in the Görtler-Witting theory, but the location of the vortex relative to the primary oscillation is found to be exactly *opposite in phase* to the theoretical prediction. This does not mean that the Görtler-Witting mechanism is not present, but it does show that there are other effects beyond those indicated by their theory. The Meksyn-Stuart theory and the later Stuart theory are based on two-dimensional primary oscillations, and therefore fail to provide for the three-dimensional vortical motions of the type just described.

In the present theory, we examine the general equations governing the terms of the second order for a first-order primary oscillation whose amplitude varies periodically in the cross-wind direction. Without making any extra physical assumptions, we found that the nonlinear effects naturally divide themselves into two categories: (1) the "two-dimensional" effects examined by Meksyn and Stuart, and (2) the intrinsically three-dimensional effects, which include the longitudinal vortex motion discussed in the foregoing.

Detailed calculations were made with a basic profile

$$u_1^0(x_k) = U(y) = \tanh y, \tag{2}$$

containing a primary oscillation of the form

$$
\begin{aligned}
u_1^{(1)}(x_k, t) &= \mathrm{Re}\{[\lambda \hat{u}_1(y) \cos \beta z + \mu \hat{U}_1(y)] e^{i\alpha(x-ct)}\}, \\
u_2^{(1)}(x_k, t) &= \mathrm{Re}\{[\lambda \hat{v}_1(y) \cos \beta z + \mu \hat{V}_1(y)] e^{i\alpha(x-ct)}\}, \\
u_3^{(1)}(x_k, t) &= \mathrm{Re}\{[\lambda w_1(y) \sin \beta z] e^{i\alpha(x-ct)}\},
\end{aligned} \tag{3}
$$

where $\alpha, \beta, \lambda, \mu$, and $c$ are constants. In the case $\mu/\lambda \gg 1$, the longitudinal vortex motion is as shown in Fig. 1.

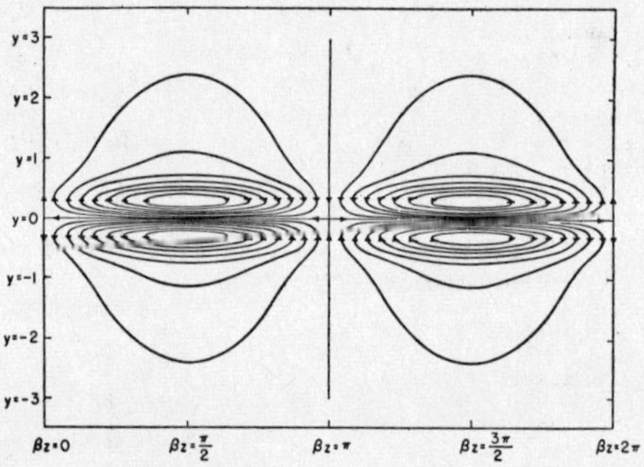


FIG. 1. Streamline pattern of secondary motion. $\mu/\lambda \gg 1$.

The combined effect of this longitudinal vortical motion and that due to the primary oscillations yields a net motion similar to that observed by Schubauer, Klebanoff, and Tidstrom.[4] The vortical motion is found to be strongest at points where the primary oscillation yields a *convex* streamline, in contrast to the predictions of the Görtler-Witting theory. Other agreements in the general features of the flow are also found between the theoretical calculations and experimental observations. These agreements are all the more remarkable, since there is a substantial difference in the basic profile between the theoretical and the experimental cases. It is believed that this indicates the generality of the features revealed by the present investigation, as indeed the nature of the theory would lead us to expect.

Another general feature of the theoretical predictions is the important role of the critical layer. It is found that the secondary motion is generally larger than the order $a^2$ by factors involving powers of $(\alpha R)^{1/3}$ where $R$ is the Reynolds number of the basic flow. This is the familiar parameter that occurs with inner friction layers. Thus the effect of these secondary terms becomes large through the "weakness" of the critical layer; but, once this large secondary motion occurs, its magnitude remains of the same order even outside the critical layer. This important role of the critical layer in the transition process has been conjectured in a previous discussion of other nonlinear effects.[5]

The authors are indebted to Dr. G. B. Schubauer and Mr. P. S. Klebanoff for their generous help in supplying unpublished experimental results and for several helpful and delightful discussions.

This work was partly supported by the Office of Naval Research.

[1] D. Meksyn and J. T. Stuart, Proc. Roy. Soc. (London) **A208**, 517 (1951).

[2] J. T. Stuart, J. Fluid Mech. **4**, 1 (1958).

[3] H. Görtler and H. Witting, Boundary Layer Research Symposium (Freiburg) 1957.

[4] G. B. Schubauer, Boundary Layer Research Symposium (Freiburg) 1957; P. S. Klebanoff and K. D. Tidstrom, Nat. Bureau of Standards, Rept. 5741 (1958).

[5] C. C. Lin, Boundary Layer Research Symposium (Freiburg) 1957.

*Reprinted without change of pagination from the*
Journal of Fluid Mechanics, *volume* 10, *part* 3, *pp*. 430–438, 1961

# Some mathematical problems in the theory of the stability of parallel flows

By C. C. LIN
Department of Mathematics, Massachusetts Institute of Technology

(Received 17 February 1960 and in revised form 6 January 1961)

By applying the method of initial values to the theory of stability of shear flows, Case has recently found certain results which are in apparent conflict with those obtained by the theory of normal modes. It is shown how these differences may be reconciled. Some new features in the theory of normal modes are also brought out. The relative merits of the two theories are compared.

## 1. Introduction

The usual theory of hydrodynamical stability is based on a study of the normal modes. There has been frequent controversy over this theory, because the behaviour of these normal modes are indeed very complicated at very large Reynolds numbers. For example, this writer (Lin 1955) has shown, from purely mathematical analysis, that '. . . there are certain damped solutions in a viscous fluid which, in the limit of vanishing viscosity, do not reduce to solutions of the inviscid equation throughout the whole region of the flow.' To those familiar with the theory of turbulence, such a behaviour is not at all surprising. Indeed, it is welcome as a concrete example of one of the fundamental characteristics of turbulent flow—i.e. that the flow field has an *intermittent spotty structure*. However, it is clear that the mathematical analysis leading to such a conclusion must be very complicated. In trying to clarify this situation as much as possible (and for other reasons), several attempts have been made to develop the theory of uniformly valid asymptotic solutions (see Langer 1940). Indeed, the old conclusions are found to be justified.

Recently, Case (1960, 1961) investigated the stability of parallel flows with respect to infinitesimal disturbances by considering the initial value problem. It is, of course, to be expected that there should be complete equivalence (cf. § 2) between this approach and the theory based on normal modes. Yet Case found that there were apparent inconsistencies, especially in connexion with the above-mentioned problem of the asymptotic limit as the Reynolds number becomes infinite. These apparent discrepancies disappear upon closer inspection. However, since they concern the fundamentals of the theories of stability and turbulence, and the problem is of general interest in the theory of singular perturbations of boundary-value problems of partial differential equations, it seems worth while to discuss the relationship between the two approaches in some detail so as to gain a deeper understanding of the issues. This is the purpose of the present note. The relative merits of the two approaches will be discussed in the concluding section of this paper.

It might be noted in passing, since the issue has been raised by Case (1961), that there is as yet no experimental evidence to suggest that the Navier–Stokes equations are inadequate to describe the phenomena of turbulent motion under ordinary conditions. For rarefied gases, the molecular structure should be considered once the Kolmogoroff scale of turbulence becomes comparable with the scales of molecular phenomena, e.g. the mean free path.

## 2. Some fundamental issues

We shall first formulate some of the fundamental questions that may be raised about the normal-mode theory of hydrodynamic stability, as a result of the conclusions reached by Case from the initial value theory. We consider the stability of a parallel flow†, for example, the pressure flow in a channel between parallel plates placed at $y = \pm 1$. The flow is in the $x$-direction with a velocity distribution $U(y) = 1-y^2$ (or some other parabolic function). The equation for small disturbances is

$$\frac{\partial \zeta'}{\partial t} + U\frac{\partial \zeta'}{\partial x} + v'\frac{d^2U}{\partial y^2} = \nu\Delta\zeta', \tag{2.1}$$

where $\zeta'$ is the disturbance vorticity, related to the disturbance stream function $\psi'(x,y,t)$ by

$$\zeta' = \frac{\partial^2\psi'}{\partial x^2} + \frac{\partial^2\psi'}{\partial y^2} = \Delta\psi', \tag{2.2}$$

and the velocity components are given by

$$u' = \frac{\partial\psi'}{\partial y}, \quad v' = -\frac{\partial\psi'}{\partial x}. \tag{2.3}$$

The constant $\nu$ is the kinematic viscosity coefficient, or the inverse of the Reynolds number $R$ in the present dimensionless formulation. The boundary conditions are

$$u' = v' = 0 \quad \text{at} \quad y = \pm 1. \tag{2.4}$$

The solution of (2.1) can be treated either in terms of the theory of normal modes or by the initial value approach. In the first approach, we superpose particular solutions of the form

$$\psi'(x,y,t) = \mathrm{Re}\{\phi(y)\, e^{i\alpha(x-ct)}\}, \tag{2.5}$$

where $\phi(y)$ satisfies the familiar Orr–Sommerfeld equation

$$\phi^{\mathrm{iv}} - 2\alpha^2\phi'' + \alpha^4\phi = i(\alpha/\nu)\,[(U-c)\,(\phi''-\alpha^2\phi) - U''\phi], \tag{2.6}$$

which is to be solved with the boundary conditions

$$\phi(\pm 1) = \phi'(\pm 1) = 0. \tag{2.7}$$

In the initial value approach used by Case, we consider the Laplace transform of $\psi'(x,y,t)$ with respect to $t$ and its Fourier transform with respect to $x$ ($\alpha$ real, $\mathrm{Re}(p) > 0$):

$$\overline{\psi}'(y;\alpha,p) = \iint \psi'(x,y,t)\, e^{-i\alpha x}\, dx\, e^{-pt}\, dt. \tag{2.8}$$

† Since Case did not formulate his theory in terms of the stream function, we develop below such a formulation in some detail to facilitate comparison with the normal-mode theory.

After solving for $\overline{\psi}'(y;\alpha,p)$ in terms of the initial conditions and the boundary conditions, we calculate $\psi'(x,y,t)$ by the inverse transform

$$\psi'(x,y,t) = \iint \overline{\psi}'(y,\alpha,p)\, e^{-i\alpha x} d\alpha\, e^{pt} dp, \tag{2.9}$$

where the integration with respect to $p$ is taken along the line $\mathrm{Re}(p) = p_0 > 0$ in the direction of increasing $\mathrm{Im}(p)$. The normal mode representation is obtained by evaluating the $p$-integral at the singularities of $\overline{\psi}'$ in the $p$-plane by the theory of residues.

Apparent differences arise when one considers the *inviscid problem*, which is obtained by *formally* putting $\nu = 0$ in the above formulation and dropping the boundary condition $u' = 0$ in (2.4). For this case, Case has shown, by the consideration of the initial value problem, that normal modes associated with certain *continuous* spectra† of eigenvalues must be included. The eigenfunctions found are continuous functions with discontinuous first derivatives.

Case has further shown that for a set of initial conditions *independent of the Reynolds' number*, the solution of the viscous problem, in the limit $\nu \to 0$, approaches the solution of the inviscid problem except for a boundary-layer correction near the walls.

It might be tempting to suggest, as was done at one time, that the above conclusions combined would contradict the existence of the type of solutions described by this writer, as quoted in §1. However, since the type of solutions considered, being of the type $\mathrm{Re}\{\phi(y,\nu)\, e^{i\alpha(x-ct)}\}$, is *dependent* on $\nu$ at *all* times, it is clear that they fall outside the scope of Case's theorem, because of his restriction on the initial conditions. There is therefore no contradiction. (In this connexion, it should be emphasized that the existence of the type of solution under discussion was established on *mathematical* grounds and *supplemented* by *physical* arguments—not the reverse.)

However, other questions may be raised about the theory of normal modes. For example, since the inviscid problem has normal modes associated with a continuous spectrum, and the solution of the viscous problem in general approaches the inviscid solutions as $\nu \to 0$, one may naturally pose the following question:

(1) are there continuous spectra of eigenvalues for the Orr–Sommerfeld equation?

We shall show in §4 that the answer is negative. This answer leads to the next question:

(2) what is the viscous solution that corresponds to an inviscid normal mode, especially the one associated with the continuous spectrum?

We may also pose this question in terms of an initial value problem. Suppose we start with a set of *initial conditions* corresponding to an inviscid normal mode $\psi_0'(x,y,t) = \mathrm{Re}\{\phi_0(y)\, e^{i\alpha(x-ct)}\}$, i.e. suppose we require

$$\psi'(x,y,0) = \phi_{0r}(y)\cos\alpha x - \phi_{0i}(y)\sin\alpha x,$$

† The importance of studying the continuous spectra was pointed out earlier by Friedrichs (1941), but he did not give any detailed analysis.

and we consider the solution of (2.1) both in the inviscid case and in the viscous case. In the inviscid case, we naturally get the solution $\psi_0'(x,y,t)$ back. In the viscous case, the solution should also be close to $\psi_0'(x,y,t)$, according to Case. If there exists a normal mode

$$\psi_N'(x,y,t) = \mathrm{Re}\{\phi_N(y,\nu)\,e^{i\alpha(x-ct)}\}$$

that approaches the normal mode $\psi_0'(x,y,t)$, the natural answer to the initial value problem is then $\psi'(x,y,t) = \psi_N'(x,y,t) + \text{small correction terms}$. But suppose such a normal mode does not exist,† what is the nature of the solution? The logical conjecture is also at hand; namely, the solutions should be represented as an infinite sum of normal modes (as any solution can be so represented), the sum being *not* dominated by a single mode. But is this logical answer also *plausible*?

In the following section, we shall show that this answer is indeed plausible by considering some simple examples where the solutions are explicitly known. These examples will also show the various ways in which the eigenfunctions of the viscous problem can behave in the limit of vanishing viscosity. In particular, we shall see that the limiting eigenfunctions may still form a discrete set while the inviscid problem has a continuous set of eigenfunctions. It is also possible for the inviscid problem to have eigenfunctions, while the limit of every viscous eigenfunction does not exist.

## 3. Examples of singular perturbation

In this section we shall examine the limiting behaviour of the solution of certain well-known parabolic equations as the 'diffusion coefficient' approaches zero. These properties are essentially well known; here they are presented in such a manner as to bring out the points of greatest interest to the issues at hand.

*Example I.* Consider the diffusion equation (which is applicable to the unsteady parallel flow between two parallel plates)

$$\frac{\partial u}{\partial t} = \nu \frac{\partial^2 u}{\partial y^2} \tag{3.1}$$

with the boundary conditions $u(0) = u(1) = 0$. The 'inviscid' problem is simply

$$\frac{\partial u_0}{\partial t} = 0 \tag{3.2}$$

with no boundary conditions required. The normal modes (solutions obtained by the method of separation of variables) in the viscous case are

$$u_n(y,t,\nu) = \sin n\pi y\, e^{-n^2\pi^2\nu t}, \tag{3.3}$$

where $n$ must take on the *discrete* set of values $1, 2, 3, \ldots$. For the inviscid equation (3.2), any continuous function $f(y)$ is a normal mode.

† This occurs not only in the case of continuous spectra of eigenvalues but also in the case when the inviscid problem has an unstable mode. The complex conjugate operation yields a damped inviscid solution

$$\tilde{\psi}'(x,y,t) = \mathrm{Re}\,\{\phi_0^*(y)\,e^{i(x-c^*t)}\}$$

which has no counterpart among the normal modes of this viscous problem.

In the limit $\nu \to 0$, the normal modes (3.3) become

$$u_n(y, t, 0) = \sin n\pi y \quad (n = 1, 2, 3, \ldots). \tag{3·4}$$

This is but one *discrete* set of normal modes of the *continuous* set $f(y)$ of the normal modes of (3.2). *Thus, the existence of a continuous set of eigenvalues and eigenfunctions in the inviscid case does not necessarily imply the existence of a corresponding set in the viscous case, even in the inviscid limit.*

Consider now the initial value problem, with

$$u(y, 0) = F(y). \tag{3.5}$$

Clearly, the inviscid solution is simply

$$u_0(y, t) = F(y). \tag{3.6}$$

The viscous solution is in general given by the series

$$u(y, t) = \sum_{n=1}^{\infty} A_n \sin n\pi y \, e^{-n^2\pi^2\nu t}, \tag{3.7}$$

where $A_n$ is given by
$$A_n = \frac{2}{\pi}\int_0^{\pi} F(y) \sin n\pi y \, dy. \tag{3.8}$$

Clearly, as $\nu \to 0$, the viscous solution (3.7) approaches the inviscid solution (3.6) for all finite $t$. This is the gist of the theorem proved by Case for the hydrodynamical problem. On the other hand, if the initial condition (3.5) should contain $\nu$, e.g. if the condition were $u(y, 0) = \sin(y/\nu)$, such a correspondence would not exist.

We note in passing that the limiting processes $\nu \to 0$ and $t \to \infty$ are not interchangeable for the solution (3.7), exactly as Case (1961) noted in his treatment o the hydrodynamical problem.

The solution (3.7) shows clearly that *all* the normal modes are in general excited in an initial value problem in the viscous case, even though the solution in the inviscid case is just a *single* mode. [The initial condition $u(y, 0) = \delta(y - y_c)$ yields the set of coefficients $A_n = \sin n\pi y_c$, which remain finite even as $n \to \infty$. In this case, a clear description of the solution for small $t$ is the spread of a singularity, and not conveniently given by the method of normal modes. Our understanding can also be improved by examining the limiting process $\nu \to 0$ in the Laplace integral representation of the solution (see Appendix).]

Exceptional cases occur when the initial conditions correspond to the *inviscid limit* of a single viscous normal mode, $\sin n\pi y$ (or a finite sum of such terms). In this case, only that particular mode is excited. This example thus gives plausibility to the answers given at the end of the last section.

We note that the *inviscid limit* of the viscous normal modes *happens* to form a complete set of *inviscid* normal modes. In the next example, we shall see that *all* the viscous normal modes do not even have inviscid limits.

*Example II.* Consider now the equation

$$\frac{\partial u}{\partial t} + k\frac{\partial u}{\partial y} = \nu\frac{\partial^2 u}{\partial y^2}. \tag{3.9}$$

The normal modes are

$$u_n(y) = w_n(y)\, e^{p_n t}, \tag{3.10}$$

with

$$w_n = e^{(k/2\nu)\, y} \sin n\pi y,$$

$$p_n = -\frac{k^2}{4\nu} - \nu n^2 \pi^2. \tag{3.11}$$

The inviscid problem has the normal modes

$$u = e^{\alpha y}\, e^{-k\alpha t}. \tag{3.12}$$

For the initial condition $e^{\alpha y}$, the viscous problem has a solution of the form

$$u(y,t,\nu) = \sum_1^\infty A_n(\nu)\, w_n(y,\nu)\, e^{p_n(\nu) t},$$

where the dependence of $A_n$, $w_n$ and $p_n$ on $\nu$ is specifically noted. It is obviously a complicated problem if one tries to calculate the limit of $u(y,t,\nu)$ as $\nu \to 0$ from the above series, since each *individual eigenfunction* $w_n(y,\nu)$ *does not have a limit as* $\nu \to 0$.

## 4. Continuous and discrete characteristic values

The examples examined in the last section serve to illustrate the type of behaviour that may occur in the asymptotic limit of the solution of a partial differential equation of the second order. Since the hydrodynamical problem involves an equation of the fourth order, it is not surprising that the behaviour of the solution could be even more complicated. (See §5 for the description of some important normal modes.) With this general background, let us return to the characteristic value problem (2.6), (2.7), and consider the question of a continuous spectrum of characteristic values.

Since the equation (2.6) is regular in the independent variable $y$ as well as in the parameters $R$, $\alpha$, $c$, there exists a fundamental system of four solutions which are analytic in $y$ and the parameters $R$, $\alpha$, $c$. Indeed they are entire functions. Furthermore since the boundary conditions (2.7) are to be satisfied for finite values of $y$, the secular equation for the determination of eigenvalues

$$F(R, \alpha, c) = 0 \tag{4.1}$$

is an entire function of the three parameters $R$, $\alpha$, $c$, regarded as complex variables. If now there were a continuous set of eigenvalues for $c$, with $R$, $\alpha$ given, the equation (4.1) must reduce to an identity, by the theory of analytic continuation. Thus, the eigenvalues of the problem defined by (2.6) and (2.7) must all be *discrete*.

The above is merely an elementary proof, for our special case, of a general theorem for regular operators. The conclusion is thus obvious once it is pointed out.

Continuous spectra are expected for the inviscid equation,

$$(U-c)(\phi_0'' - \alpha^2 \phi_0) - U'' \phi_0 = 0, \tag{4.2}$$

(at least for the case $U'' = 2 \neq 0$), with the boundary conditions

$$\phi_0(\pm 1) = 0, \tag{4.3}$$

because the operator is now singular. Let us try to see this fact more directly in terms of the solutions of equation (4.2). This equation is well known to have solutions of the forms

$$\phi_1(y) = (y-y_c)[1+\dots] \tag{4.4}$$

and

$$\phi_2(y) = 1+\dots+\frac{U_c''}{U_c'}\phi_1(y)\log(y-y_c), \tag{4.5}$$

where $1+\dots$ denotes a regular power series in $(y-y_c)$, and $U(y_c)=c$. In the case of real eigenvalues $c$, the *real* and *imaginary* parts of $\phi_2(y)$ are separately solutions of (4.2). The real part is the *continuous* function

$$\phi_{2r}(y) = 1+\dots+\frac{U_c''}{U_c'}\phi_1(y)\log|y-y_c|, \tag{4.6}$$

and the imaginary part is another *continuous* function

$$\left.\begin{aligned} \phi_{2i}(y) &= 0, & y &\geqslant y_c; \\ \phi_{2i}(y) &= -\pi\frac{U_c''}{U_c'}\phi_1(y) & y &\leqslant y_c. \end{aligned}\right\} \tag{4.7}$$

We have thus *three* linearly independent solutions (4.4), (4.6) and (4.7) by permitting the solutions to have discontinuous or singular derivatives. With three linearly independent solutions for (4.2) and two boundary conditions (4.3), it is obvious that the characteristic values have a continuous spectrum.

However, if one regards the inviscid solution as the limit of the viscous solution as $\nu\to 0$, only the linear combination $\phi_{2r}(y)+i\phi_{2i}(y)$ is a solution, not $\phi_{2r}(y)$ and $\phi_{2i}(y)$ separately. Continuous spectra are therefore not obtained. The point is that the inviscid limit *does* exist in a properly defined region,† and remains an *analytic* function of the complex variable $y$. (This region, however, does not necessarily include the complete physical region.) Thus, the solution for $y>y_c$, when continued to the range $y<y_c$ must represent the same analytic function. The functions $\phi_r(y)$ and $\phi_i(y)$ do not satisfy this requirement.

## 5. Concluding remarks

Let us now summarize our conclusions and compare the relative merits of the methods of normal modes and initial values.

(i) Mathematically speaking, it should first be borne in mind that the Laplace integral representation of the solution can be decomposed into an infinite series of normal modes by the method of residues, and that the two approaches are essentially equivalent (see (vi) below).

(ii) Secondly, a great deal of apparent contradictions can be avoided if the following facts are kept in mind. A normal mode in the inviscid theory may not be the limit of a normal mode in the viscous theory. Conversely, a normal mode in the viscous theory may not have an inviscid limit. These conclusions are based on mathematical investigations of the solutions of the Orr–Sommerfeld equation. They are illustrated by simple examples in § 3. The calculations in the Appendix

† This region is defined by two of the three lines $\mathrm{Re}\int_{y_c}^{y}\sqrt{\{i(U-c)\}}\,dy = 0$, chosen in such a manner as to include the two end points of our boundary-value problem. See Lin (1955, figure 8·1, p. 130).

illustrate fully the lack of direct correspondence between the inviscid eigenfunctions and the inviscid limit of the viscous ones.

(iii) If we are dealing with disturbances where the viscous effects are restricted to a boundary-layer region of the order of $(\alpha R)^{-\frac{1}{2}}$ near the solid surface, the initial value approach leads to the familiar boundary-layer theory; i.e. the solution is composed of an inviscid solution plus a boundary-layer correction. In this method of description, the normal modes belonging to the continuous spectrum (or spectra), which occur in the inviscid case, must be used. This description is simpler than in terms of the normal modes in the viscous theory (which are discrete). The type of simplification is similar to that occurring with the examples of § 3.

(iv) However, there are important types of disturbances not of the type described above, In particular, the unstable disturbances responsible for the initiation of turbulence in the channel or the boundary layer are of a different type. These disturbances extend over a layer of the order of $(\alpha R)^{-\frac{1}{3}}$, which is thicker than the layer in the previous case. For these disturbances, the viscous forces have a well-known destabilizing influence typical of the instability of shear flows.

(v) There are also exponentially damped normal modes which have the intermittent spotty structure characteristic of turbulence. There are also normal modes which have no inviscid limit anywhere in the physical region. When, to attempt a description of the transition process, the non-linear theory is considered these normal modes are expected to play an important role.

(vi) The initial value theory has not yet been developed to study the two types of disturbances just mentioned. It would be interesting to carry out these investigations to see whether different features can be revealed by the Laplace integral representation. In particular, it should be possible to bring out the destabilizing influence of the viscous forces mentioned in (iv). It would appear, from the fact that there are only a finite number of unstable modes, that the two methods would yield essentially the same results with regard to the unstable behaviour of the flow. On the other hand, it is likely that one would find only stable disturbances for the flow through a channel, if the viscous forces are restricted to boundary-layer regions which have a thickness of the order of $(\alpha R)^{-\frac{1}{2}}$ (cf. Lin & Rabenstein 1960).

This work was done while the author was at the Institute for Advanced Study, Princeton, N.J. The author is grateful to Professor J. R. Oppenheimer for the hospitality extended to him during his stay. He wishes also to thank Professors F. J. Dyson and C. N. Yang for their interest and discussions. The partial support of the office of Naval Research is acknowledged.

## Appendix A

To get a closer comparison with Case's investigation (1961) of the hydrodynamical problem, let us explicitly use the method of the Laplace transformation (for the equation (3.1)). The solution is then given by

$$u(y,t) = \frac{1}{2\pi i}\int_C \bar{u}(y,p)\, e^{pt}\, dp, \tag{A 1}$$

where $\bar{u}(y, p)$ satisfies the differential equation

$$\frac{d^2\bar{u}}{dy^2} - q^2\bar{u} = \frac{1}{k}u(y, 0), \quad q^2 = \frac{p}{\nu}, \tag{A 2}$$

and the contour $C$ is the line $\mathrm{Re}(p) = p_0 > 0$, taken in the direction of increasing $\mathrm{Im}(p)$. For the purposes in hand, we need only consider the simple case $u(y, 0) = \text{const}$. We then have

$$\bar{u} = \frac{1}{p \sinh q}\{\sinh qy + \sinh q(1-y) - \sinh q\}. \tag{A 3}$$

We make two observations.

(i) If we deform the contour $C$ so as to evaluate the integral (A 1) in terms of the residues at the poles of (A 3), $p = -\nu n^2\pi^2 (n = 1, 2, \ldots)$, we recover the series solution (3.7) in terms of the eigenfunctions. Notice that $\bar{u}$ does not have a singularity at $p = 0$.

(ii) In the limit $\nu \to 0$, the singularities of (A 3) become dense along the whole negative part of the real axis in the $p$-plane, but the limiting function is

$$\lim_{\nu \to 0} \bar{u} = -\frac{1}{p} \quad \text{for} \quad 0 < y < 1. \tag{A 4}$$

This limiting form (A 4) agrees with the inviscid transform

$$\bar{u}_0 = -\frac{1}{p}, \tag{A 5}$$

but its singularity lies at $p = 0$, and does not coincide with any one of the singularities of (A 3) for $\nu > 0$.

## REFERENCES

Case, K. M. 1960 Stability of inviscid plane Couette flow. *Phys. Fluids*, **3**, 143.

Case, K. M. 1961 Hydrodynamic stability and the inviscid limit. *J. Fluid Mech* **10**, 420.

Friedrichs, K. O. 1941 *Fluid Dynamics* (Brown University Lecture Notes), p. 209.

Langer, R. E. 1940 *Bull. Amer. Math. Soc.* **46**, 257.

Lin, C. C. 1955 *The Theory of Hydrodynamic Stability*. Cambridge University Press.

Lin, C. C. & Rabenstein, A. L. 1960 Asymptotic solutions of a class of differential equations of the fourth order. *Trans. Amer. Math. Soc.* **94**, 24.

Reprinted from
Proceedings of Symposia in Applied Mathematics, Vol. 13
Hydrodynamic Instability


# ON THE INSTABILITY OF SHEAR FLOWS

BY

C. C. LIN AND D. J. BENNEY

1. **Introduction.** One of the most basic and challenging problems in fluid mechanics is to reach an understanding of the various physical mechanisms involved in the process of transition from laminar to turbulent flow. It is now well known that in many instances the initial process is one of instability with respect to infinitesimal disturbances. This initial disturbance could take on the form of a steady secondary flow, as in the case of convective instability, or the form of a small oscillation, as in the case of certain shear flows. It is quite likely that the processes governing the subsequent development of such disturbances are also different. The present note deals with the case of shear flows.

We consider a small disturbance of the basic flow by writing the velocity and pressure of the disturbed motion in the form

$$\begin{aligned} u_i(x_k,t) &= u_i^{(0)}(x_k) + au_i^{(1)}(x_k,t) + a^2u_i^{(2)}(x_k,t) + \cdots, \\ p(x_k,t) &= p^{(0)}(x_k) + ap^{(1)}(x_k,t) + a^2p^{(2)}(x_k,t) + \cdots \end{aligned} \tag{1.1}$$

$(i,k = 1,2,3)$ where $a$ is a measure of the amplitude of the disturbances and all the variables are dimensionless. The above expressions must satisfy the equation of continuity and the Navier–Stokes equations, from which we obtain a sequence of equations governing the motion of each order. The equations of order $a^0$ give the basic flow, those of order $a$ give the usual linear theory, and those of the order $a^2$ give the nonlinear effects under consideration.

In the case of shear flow, the linear terms give rise to stable and unstable normal modes in the form of waves. There is now ample evidence to support the correctness of the theory. The mathematical basis of the theory involves, however, many subtle points, especially with respect to the relationship between the solutions at finite Reynolds numbers and the inviscid limit. The uniformly valid asymptotic solutions for large Reynolds numbers have been studied by Wasow [5], by Langer [6] and by Lin and Rabenstein [7]. Other issues have been investigated by Case [8] and by Lin [9]. The major results of these investigations will be discussed in §§2, 3.

The nonlinear effect has been examined theoretically by Meksyn and Stuart [1], and later by Stuart [2], using a somewhat different approach. A theory of secondary instability has also been proposed by Görtler and Witting [3]. Experimental observations, especially those by Schubauer, Klebanoff and Tidstrom [4], however, show remarkable characteristics in the behavior of the flow not accountable by the theories of these authors. In particular, the experimental observations indicate the existence of longitudinal vortical motions (with axis along the direction of the basic flow) which give a redistribution of the momentum of mean flow in the plane normal to it. Such a vortical motion is very similar to that calculated by the Görtler–Witting theory, but its location relative to the primary oscillation is found to be exactly opposite in phase to the theoretical prediction. The theories of Meksyn and Stuart and of Stuart are based on two-dimensional primary oscillations and therefore fail to provide for such motions.

In the present theory, we examine the general equations governing the terms of the second order for a first-order primary oscillation whose amplitude varies periodically in the cross-wind direction. We found that the nonlinear effects naturally divide themselves into two categories: (1) the "two-dimensional" effects examined by Meksyn and Stuart with a variation in intensity in the spanwise direction, and (2) the intrinsically three-dimensional effects which include the longitudinal vortical motions. Detailed calculations were made with a basic profile

$$u_1^{(0)}(x_k) = U(y) = \tanh y.$$

It is found that the combined effect of the nonlinear longitudinal vortical motion and that due to the primary oscillations yields a net motion similar to that observed by Schubauer, Klebanoff and Tidstrom. The vortices are found to be at points where the primary oscillation yields a convex streamline. Other general features of the flow are also similar between the theoretical and the experimental results. The agreement is all the more remarkable since there is a substantial difference in the basic profile between the theoretical and experimental cases. It is believed that this indicates the generality of the features revealed by these investigations.

Another general feature is the important role of the critical layer. It is found that the secondary motion is generally larger than the order of $a^2$ by factors involving powers of $(\alpha R)^{1/3}$ where $R$ is the Reynolds number of the basic flow. This is the familiar parameter that occurs with inner friction layers. Thus, the effect of these second-order terms becomes large through the weakness of the critical layer; but once this large secondary flow occurs, its magnitude remains of the same order even outside of the critical layer. The importance of the critical layer in the transition process has been conjectured in a previous discussion of other nonlinear effects. A brief description of these nonlinear effects found from the theoretical investigations and their comparison with experiments will be presented in §§4–6.

2. **The linear theory.** We shall now discuss the linear theory of hydrodynamic stability especially in connection with the behavior of the solution at large Reynolds numbers. We consider the stability of a parallel flow, for example, the pressure flow through a channel between parallel plates placed at $y = \pm 1$. The flow is in the $x$-direction with a velocity distribution $U(y) = 1 - y^2$ (or some other parabolic function). The equation for a two-dimensional small disturbance is

$$\frac{\partial \zeta'}{\partial t} + U\frac{\partial \zeta'}{\partial x} + v'\frac{d^2U}{dy^2} = \nu\Delta\zeta', \tag{2.1}$$

where $\zeta'$ is the disturbance vorticity, related to the disturbance stream function $\psi'(x,y,t)$ by

$$\zeta' = \frac{\partial^2 \psi'}{\partial x^2} + \frac{\partial^2 \psi'}{\partial y^2} = \Delta \psi', \tag{2.2}$$

and the velocity components are given by

$$u' = \frac{\partial \psi'}{\partial y}, \qquad v' = -\frac{\partial \psi'}{\partial x}. \tag{2.3}$$

The constant $\nu$ is the kinematic viscosity coefficient, or the inverse of the Reynolds number in the present dimensionless formulation. The boundary conditions are

$$u' = v' = 0 \quad \text{at} \quad y = \pm 1. \tag{2.4}$$

The solution of (2.1) can be treated either in terms of the theory of normal modes or as an initial value problem. In the first approach, we superpose particular solutions of the form

$$\psi'(x,y,t) = \mathrm{Re}\,\{\phi(y)e^{i\alpha(x-ct)}\}, \tag{2.5}$$

where $\phi(y)$ satisfies the familiar Orr–Sommerfeld equation

$$\phi^{iv} - 2\alpha^2\phi'' + \alpha^4\phi = i(\alpha/\nu)[(U-c)(\phi'' - \alpha^2\phi) - U''\phi], \tag{2.6}$$

which is to be solved with the boundary conditions

$$\phi(\pm 1) = \phi'(\pm 1) = 0. \tag{2.7}$$

In the initial value approach used by Case [**8**], we consider the Laplace transform of $\psi'(x,y,t)$ with respect to $t$ and its Fourier transform with respect to $x$ ($\alpha$ real, $\mathrm{Re}\,(p) > 0$):

$$\bar{\psi}'(y;\alpha,p) = \int_0^\infty \int_{-\infty}^\infty \psi'(x,y,t) \cdot e^{i\alpha x} dx \cdot e^{-pt} dt. \tag{2.8}$$

After solving for $\bar{\psi}'(y;\alpha,p)$ in terms of the initial conditions and the boundary conditions, we calculate $\psi'(x,y,t)$ by the inverse transform

$$\psi'(x,y,t) = \int\int \bar{\psi}'(y;\alpha,p) \cdot e^{-i\alpha x} d\alpha \cdot e^{pt} dp, \tag{2.9}$$

where the integration with respect to $p$ is taken along the line $\mathrm{Re}\,(p) = p_0 > 0$ in the direction of increasing $\mathrm{Im}\,(p)$. The normal mode representation is obtained by evaluating the $p$-integral at the singularities of $\bar{\psi}'$ in the $p$-plane by the theory of residues.

A natural approach to the solution of (2.1) for the case of small viscosity is to begin with the case $\nu = 0$. However, since the resultant equation contains spatial derivatives of lower orders, the perturbation procedure is singular, and one has to guard against all the associated pitfalls that may occur. For example, the complete equation (2.6), being regular, has only *discrete* eigenvalues; whereas the reduced equation

$$(U - c)(\phi'' - \alpha^2\phi) - U''\phi = 0, \tag{2.10}$$

in general has a singularity at $y = y_c$ where $U = c$, and therefore possesses *continuous* eigenvalues when the boundary conditions

$$\phi(\pm 1) = 0 \tag{2.11}$$

are imposed. The question may then be raised: in which sense does the inviscid limit represent an adequate approximation to the physical situation (which is after all always viscous)?

An obvious answer is the following. The solutions of the boundary value problem (2.1) fall into two classes: (1) those which approach the inviscid solution in the limit $\nu \to 0$; and (2) those which do not. *A priori*, either class may be empty; but closer examination reveals that there are important solutions in both classes. The members of the first class, which are expected to exhibit the boundary layer behavior, can be further divided into two sub-classes: (a) those with a boundary layer thickness of the order of $\nu^{1/2}$, and (b) those with some other thickness. The initial value method easily yields solution of class (a), whereas solutions of class (b), with a boundary layer thickness of the order of $\nu^{1/3}$, have been known through the method of normal modes. It is noteworthy that the unstable disturbances responsible for the initiation of turbulence in the channel or the boundary layer is of this latter type.

It is perhaps illuminating to examine a simple example to illustrate the variety of behavior of solutions when a process of singular perturbation is involved. Consider the equation

$$\frac{\partial u}{\partial t} + k\frac{\partial u}{\partial y} = \nu\frac{\partial^2 u}{\partial y^2} \qquad (k > 0), \tag{2.12}$$

to be solved in the region $0 \leqq y \leq 1$ with the boundary conditions $u(0) = u(1) = 0$. The inviscid problem is

$$\frac{\partial u_0}{\partial t} + k\frac{\partial u_0}{\partial y} = 0, \tag{2.13}$$

with no boundary conditions required. The normal modes (solutions obtained by the method of separation of variables) of the viscous problem are the *discrete* set of functions

$$U_n(y,t,\nu) = w_n(y)e^{p_n t}, \tag{2.14}$$

where

$$\begin{aligned} w_n &= e^{ky/2\nu}\sin n\pi y, \\ p_n &= -\frac{k^2}{4\nu} - \nu n^2\pi^2, \end{aligned} \tag{2.15}$$

and $n = 0, 1, 2, \cdots$. None of these eigenfunctions has a limit as $\nu \to 0$. On the other hand, the inviscid problem has the normal modes

$$u_0 = e^{\alpha y}e^{-k\alpha t}, \tag{2.16}$$

with continuous spectrum for the eigenvalue $\alpha$. It is clear that if we attempt to solve the complete equation (2.12), with the initial condition $e^{\alpha y}$, we should get a solution $u(y,t,\nu)$ which approaches the above inviscid solution (2.16) as $\nu \to 0$. On the other hand, the representation of this solution in terms of the normal modes $u_n(y,t,\nu)$ is of the form

$$u(y,t,\nu) = \sum_{1}^{\infty} A_n(\nu) w_n(y,\nu) e^{p_n(\nu)t}, \tag{2.17}$$

where we have emphasized the dependences of $A_n$, $w_n$ and $p_n$ on $\nu$. It is obviously a complicated problem if one tries to calculate the limit of $u(y,t,\nu)$ as $\nu \to 0$ from the above series.

The above example illustrates the following. *A normal mode in the inviscid theory may not be the limit of a normal mode in the viscous theory. Conversely, a normal mode in the viscous theory may not have an inviscid limit.* This exemplifies the classification of solutions into the two classes (1) and (2) above. The example is, however, too simple to illustrate the further subdivision of solutions of class (1).

We shall not go into further discussion of the relative merits of the method of normal modes versus the method of initial values—which is made elsewhere. Since we are going to hear from Professor Case on the method of initial values, we shall proceed to the method of normal modes, which depends on the solution of the eigenvalue problem (2.6). We are especially interested in the cases where $\nu$ is small.

3. **Asymptotic solutions of the Orr–Sommerfeld equation.** It is well known that the equation (2.6) has formal asymptotic solutions of the forms

$$\phi = \phi^{(0)} + \frac{1}{\alpha R}\phi^{(1)} + \cdots \tag{3.1}$$

and

$$\phi = \exp{(\alpha R)^{1/2} Q(y)} \left[ f^{(0)} + \frac{1}{(\alpha R)^{1/2}} f^{(1)} + \cdots \right], \tag{3.2}$$

$$Q(y) = \int_{y_c}^{y} (i(U-c))^{1/2} dy.$$

However, these solutions have singularities at the point $y_c$ where $U - c = 0$, whereas the original equation (2.6) is perfectly regular at that point. Search for a system of regular solutions with tractable asymptotic behavior has been made by Wasow [**5**], by Langer [**6**], and by Lin and Rabenstein [**7**]. The relationship between the methods used has been discussed in the last mentioned paper, the main ideas of which will now be summarized.

To put it in a more general perspective, we consider the linear differential equation of the fourth order

$$\mathscr{L}(\phi) = \frac{d^4\phi}{dx^4} + \lambda^2 \left\{ P(x,\lambda)\frac{d^2\phi}{dx^2} + Q(x,\lambda)\frac{d\phi}{dx} + R(x,\lambda)\phi \right\} = 0, \tag{3.3}$$

when the parameter $\lambda$ is large. In the above expression all the three functions $P(x,\lambda)$, $Q(x,\lambda)$ and $R(x,\lambda)$ are analytic functions in the complex variable $x$, and they depend on $\lambda$ in such a manner that asymptotic expansions of the following form hold:

$$F(x,\lambda) = \sum_{n=0}^{\infty} \lambda^{-n} F_n(x). \tag{3.4}$$

Especially we are interested in the behavior of the solution in the neighborhood of a *simple* zero of the function $P_0(x)$. Such a point is called a *turning point* (of the first order). It is of special interest because asymptotic solutions of (3.3) in the form

$$\phi = e^{\lambda S(x)} \sum_{n=0}^{\infty} \lambda^{-n} \phi_n(x) \tag{3.5}$$

exhibits singularities at such a point even though the point is a regular point of the differential equation (3.3). Other solution forms must therefore be found, if we wish to find a complete set of asymptotic solutions *uniformly* valid in a neighborhood containing the turning point.

We base our solution on the simplest of equations of the type (3.3); namely,

$$L(u) \equiv u^{iv} + \lambda^2(zu'' + \alpha u' + \beta u) = 0, \tag{3.6}$$

where $\alpha$ and $\beta$ may depend on $\lambda$ asymptotically in the following manner:

$$\alpha = \sum_{n=0}^{\infty} \lambda^{-n} \alpha^{(n)}, \tag{3.7a}$$

$$\beta = \sum_{n=0}^{\infty} \lambda^{-n} \beta^{(n)}. \tag{3.7b}$$

The solutions of this *basic reference equation* (3.6) can be studied by the method of Laplace transformation, as was carried out by Rabenstein **[10]**. Our aim is to show that we can obtain asymptotic solutions of the form

$$\phi(x) = c_0(x)u(z) + c_1(x)u'(z) + c_2(x)u''(z) + c_3(x)u'''(z), \tag{3.8}$$

where the variables $z$ and $x$ are connected by a suitable analytic function (e.g., (3.10) below), and the functions $c_i(x,\lambda)$ all have the asymptotic behavior (3.4). Actually, it turns out that

$$c_0, c_1 = O(1), \quad \text{and} \quad c_2, c_3 = O(\lambda^{-2}).$$

The method for establishing such a solution will be described below. For practical purposes, once we know the existence of the solutions of the form (3.8), the coefficients $c_i$ can be calculated by using the formal asymptotic solutions of $\phi$ and $u$ of the type (3.5).

The derivation of the solution (3.8) is made as follows. We first carry out *finite* transformations such that the equation (3.3) is reduced to a form as close to

(3.6) as possible (no approximation being made in this step). It can be shown that one can always achieve the *normal form*,

$$L_0(\chi) \equiv \chi^{iv} + \lambda^2(z\chi'' + \alpha_0\chi' + \beta_0\chi) \\ = \lambda(\tilde{a}\chi + \tilde{b}\chi' + \lambda^{-1}\tilde{c}\chi'' + \lambda^{-2}\tilde{d}\chi'''), \tag{3.9}$$

where $\alpha_0$ and $\beta_0$ are constants while $\tilde{a}$, $\tilde{b}$, $\tilde{c}$ and $\tilde{d}$ have the behavior (3.4). Such a transformation is carried out in two steps:

**First transformation.** We introduce into (1.1) the new independent variable,

$$z = \left[\frac{3}{2}\int_{x_0}^{x}[P_0(x)]^{1/2}dx\right]^{2/3}, \tag{3.10}$$

and the new dependent variable,

$$\psi(z,\lambda) = \phi(x,\lambda)[P_0(x)/z]^{3/4}. \tag{3.11}$$

This will have the effect of retaining the general form of the equation (1.1) but casting $P_0(x)$ into the independent variable $z$ itself.

**Second transformation.** Next we consider a transformation of the form,

$$\chi = A(z,\lambda)\psi + B(z,\lambda)\psi' + \lambda^{-1}C(z,\lambda)\psi'' + \lambda^{-2}D(z,\lambda)\psi''', \tag{3.12}$$

where $A$, $B$, $C$, $D$ has the behavior (3.4). By formal differentiation, we obtain the transformation in the vector form,

$$\boldsymbol{\chi} = G\boldsymbol{\psi}, \tag{3.13}$$

where

$$\boldsymbol{\chi} = (\chi,\chi',\lambda^{-1}\chi'',\lambda^{-2}\chi'''), \tag{3.14}$$

$$\boldsymbol{\psi} = (\psi,\psi',\lambda^{-1}\psi'',\lambda^{-2}\psi'''), \tag{3.15}$$

and $G(z,\lambda)$ is a nonsingular matrix with asymptotic dependence on $\lambda$ in the manner,

$$G(z,\lambda) = \sum_{n=0}^{\infty} \lambda^{-n}G^{(n)}(z). \tag{3.16}$$

It can be shown by explicit calculation that this is indeed possible with a *finite* number of terms, all regular in $z$. In the process of carrying out this transformation, it will become clear that the constant $\alpha_0$ in (3.9) is indeed completely determined by the given equation, and that the constant $\beta_0$ is also determined in the cases where $\alpha_0$ is an integer, positive, negative or zero.

**Construction of the asymptotic solutions.** The construction of the asymptotic solutions of the normalized equation (3.9) then follows usual methods. If we introduce the vector

$$\boldsymbol{u} = (u,u',\lambda^{-1}u'',\lambda^{-2}u''), \tag{3.17}$$

where $u$ is a solution of (3.6), then we may expect it possible, by proper determination of $\alpha$ and $\beta$ to obtain formal solutions of the structure,

$$\boldsymbol{\chi} = H\boldsymbol{u}, \tag{3.18}$$

and with $H$ expected to be of the form,[1]

$$H = I + \sum_{m=1}^{\infty} \lambda^{-m} h^{(m)}(z), \tag{3.19}$$

where $I$ is the identity matrix. Combining (3.13) and (3.18), we may write

$$X = G\Psi = HU, \tag{3.20}$$

or

$$\Psi = G^{-1}HU, \tag{3.21}$$

where $U$, $\Psi$, $X$ are fundamental matrix solutions of the system of equations corresponding to the fourth order equations for $u$, $\psi$ and $\chi$. For convenience of reference, we list the equations for $\boldsymbol{u}$ and $\boldsymbol{\chi}$ explicitly. The equation for $\boldsymbol{u}$ may be written

$$\frac{d\boldsymbol{u}}{dz} = M\boldsymbol{u}, \tag{3.22}$$

where

$$M = \begin{bmatrix} 0 & 1 & 0 & 0 \\ 0 & 0 & \lambda & 0 \\ 0 & 0 & 0 & \lambda \\ -\beta & -\alpha & -\lambda z & 0 \end{bmatrix}. \tag{3.23}$$

The equation for $\boldsymbol{\chi}$ may be written as

$$\frac{d\boldsymbol{\chi}}{dz} = (M_0 + \epsilon)\boldsymbol{\chi}, \tag{3.24}$$

where $M_0$ is obtained from $M$ by putting $\alpha = \alpha_0$, $\beta = \beta_0$, and

$$\epsilon = \lambda^{-1} \begin{bmatrix} 0 & 0 & 0 & 0 \\ 0 & 0 & 0 & 0 \\ 0 & 0 & 0 & 0 \\ \bar{a} & \bar{b} & \tilde{c} & \tilde{d} \end{bmatrix}, \tag{3.25}$$

in which

$$\bar{a} = \tilde{a} + (\beta - \beta_0), \qquad \bar{b} = \tilde{b} + (\alpha - \alpha_0). \tag{3.26}$$

If we terminate the series (3.19) with $h^{(n)}(z)$ as the last term, we obtain the $n$th approximation,

$$\boldsymbol{\chi}_n = H_n\boldsymbol{u}, \tag{3.27}$$

which satisfies the differential equation,

$$\frac{d\boldsymbol{\chi}_n}{dz} = (M_0 + \epsilon_n)\boldsymbol{\chi}_n, \tag{3.28}$$

[1] In certain special cases, fractional powers of $\lambda$ may occur; but the general nature of the theory remains unchanged.

with

$$\epsilon - \epsilon_n = \lambda^{-(n+1)}\Delta, \tag{3.29}$$

where $\Delta$ is a matrix of the same form as $\lambda\epsilon$. The equation (3.28) is the approximating *related equation* of (3.24). Heuristically, it would appear natural to think of (3.7) and (3.9) as *asymptotically equivalent*, and to expect that once the construction (3.27) is made, it should be easy to prove its asymptotic validity; namely, to show that there are actual solutions of (3.24) which is approximated by (3.27). Although we found it possible to do this, the proof was *not easy*. It depends on the examination of each individual solution in a particular fundamental set of four solutions. Possibly a more general theorem can be established which corresponds more closely to heuristic expectations.

4. **The classification of nonlinear effects.** As we have mentioned earlier the linear theory gives an adequate representation of the motion when the disturbances are very small and it is also capable of predicting the onset of instability. However, as these disturbances grow the nonlinear terms in the equations of motion must be included in the analysis. For finite amplitude oscillations it is well known that two new features appear: (1) the excitation of higher harmonics of the primary oscillation and (2) a modification of the original flow profile by the action of the Reynolds stresses in producing a redistribution of momentum. Meksyn and Stuart have calculated this modification for flow between two parallel walls when the primary oscillation is strictly two dimensional.

One point of interest is the relative importance of two and three dimensional disturbances. On the basis of linearized theory, Squire's result applies. Thus, to estimate the onset of instability one need only consider two dimensional oscillations. These two dimensional waves are indeed observed experimentally during the initial instability. However, one must not be tempted to read more into Squire's theorem than it actually implies. For example, at a Reynolds number beyond the critical Reynolds number there will be unstable three dimensional waves and there is no guarantee that the most highly amplified disturbance will still be two dimensional, even on the basis of linear theory. The details would depend on the particular profile. Indeed the simple observation that turbulence is an essentially three dimensional phenomenon suggests that this two dimensional supremacy cannot be expected to persist through the finite amplitude regime. Recent experimental evidence (to which we will refer later) points strongly to the desirability of a theoretical investigation of three dimensional disturbances.

We now examine finite amplitude oscillations in a given parallel flow, paying special attention to the three dimensionality. This is done by a straightforward perturbation from the linear theory as we shall describe below.

If $p$, $\mathbf{q}$, $\boldsymbol{\omega}$ denote the pressure, velocity and vorticity respectively and we set up the perturbation (suggested in the introduction) by writing

$$p = p^{(0)} + ap^{(1)} + a^2p^{(2)} + \cdots, \tag{4.1}$$

$$q = q^{(0)} + aq^{(1)} + a^2q^{(2)} + \cdots, \tag{4.2}$$

$$\omega = \omega^{(0)} + a\omega^{(1)} + a^2\omega^{(2)} + \cdots, \tag{4.3}$$

where $q^{(0)}$ is the original basic flow, $q^{(1)}$ the primary oscillation, etc. The symbol $a$ is used to denote a perturbation amplitude, and $p^{(0)}$ is the pressure distribution associated with the basic flow. We take $xyz$ as rectangular coordinates in a given parallel basic flow $q^{(0)} = (u_0^{(0)}(y),0,0)$, and consider waves of small amplitude propagating downstream having a possible $z$ variation of amplitude. We write

$$q^{(n)} = (u^{(n)}, v^{(n)}, w^{(n)}), \tag{4.4}$$

$$\omega^{(n)} = (\xi^{(n)}, \eta^{(n)}, \zeta^{(n)}) = \operatorname{curl} q^{(n)}. \tag{4.5}$$

Equating successive powers of $a$ to zero in the continuity equation and in the Navier–Stokes equations we have

$$\operatorname{div} q^{(n)} = 0, \tag{4.6}$$

$$\frac{\partial q^{(n)}}{\partial t} + \sum_{r=0}^{n} (q^{(r)} \cdot \nabla) q^{(n-r)} = -\nabla p^{(n)} + \frac{1}{R}\Delta q^{(n)}. \tag{4.7}$$

If $\alpha$ is the downstream wave number of the primary oscillation, we may write,

$$p^{(1)} = p_1^{(1)}(y,z,t)e^{i\alpha x} + p_1^{(1)*}(y,z,t)e^{-i\alpha x}, \tag{4.8}$$

$$q^{(1)} = q_1^{(1)}(y,z,t)e^{i\alpha x} + q_1^{(1)*}(y,z,t)e^{-i\alpha x}, \tag{4.9}$$

$$\omega^{(1)} = \omega_1^{(1)}(y,z,t)e^{i\alpha x} + \omega_1^{(1)*}(y,z,t)e^{-i\alpha x}, \tag{4.10}$$

where an asterisk denotes a complex conjugate. The equations governing the first order motion are

$$i\alpha u_1^{(1)} + \frac{\partial v_1^{(1)}}{\partial y} + \frac{\partial w_1^{(1)}}{\partial z} = 0,$$

$$\begin{aligned} \frac{\partial u_1^{(1)}}{\partial t} + i\alpha u_0^{(0)} u_1^{(1)} + v_1^{(1)} \frac{du_0^{(0)}}{dy} &= -i\alpha p_1^{(1)} + \frac{1}{R}\left(\frac{\partial^2}{dy^2} + \frac{\partial^2}{dz^2} - \alpha^2\right) u_1^{(1)}, \\ \frac{\partial v_1^{(1)}}{\partial t} + i\alpha u_0^{(0)} v_1^{(1)} &= \frac{-\partial p_1^{(1)}}{\partial y} + \frac{1}{R}\left(\frac{\partial^2}{\partial y^2} + \frac{\partial^2}{\partial z^2} - \alpha^2\right) v_1^{(1)}, \\ \frac{\partial w_1^{(1)}}{\partial t} + i\alpha u_0^{(0)} w_1^{(1)} &= \frac{-\partial p_1^{(1)}}{\partial z} + \frac{1}{R}\left(\frac{\partial^2}{\partial y^2} + \frac{\partial^2}{\partial z^2} - \alpha^2\right) w_1^{(1)}. \end{aligned} \tag{4.11}$$

The second order motion which we shall refer to as the secondary flow can be written as the sum of a mean motion and a second harmonic oscillation. Superscripts have been used to denote the order of the motion and subscripts for the harmonic content. Thus we have

$$p^{(2)} = p_0^{(2)}(y,z,t) + p_2^{(2)}(y,z,t)e^{2i\alpha x} + p_2^{(2)*}(y,z,t)e^{-2i\alpha x}, \tag{4.12}$$

$$q^{(2)} = q_0^{(2)}(y,z,t) + q_2^{(2)}(y,z,t)e^{2i\alpha x} + q_2^{(2)*}(y,z,t)e^{-2i\alpha x}, \tag{4.13}$$

$$\omega^{(2)} = \omega_0^{(2)}(y,z,t) + \omega_2^{(2)}(y,z,t)e^{2i\alpha x} + \omega_2^{(2)*}(y,z,t)e^{-2i\alpha x}. \tag{4.14}$$

The equations that determine the mean secondary flow are

$$\frac{\partial v_0^{(2)}}{\partial y} + \frac{\partial w_0^{(2)}}{\partial z} = 0,$$

$$\frac{\partial u_0^{(2)}}{\partial t} + v_0^{(2)} \frac{du_0^{(0)}}{dy} + \frac{\partial}{\partial y}(u_1^{(1)}v_1^{(1)*} + u_1^{(1)*}v_1^{(1)}) + \frac{\partial}{\partial z}(u_1^{(1)}w_1^{(1)*} + u_1^{(1)*}w_1^{(1)}) = \frac{1}{R}\left(\frac{\partial^2}{\partial y^2} + \frac{\partial^2}{\partial z^2}\right) u_0^{(2)},$$

(4.15)
$$\frac{\partial v_0^{(2)}}{\partial t} + 2\frac{\partial}{\partial y}(v_1^{(1)}v_1^{(1)*}) + \frac{\partial}{\partial z}(v_1^{(1)}w_1^{(1)*} + v_1^{(1)*}w_1^{(1)}) = \frac{-\partial p_0^{(2)}}{\partial y} + \frac{1}{R}\left(\frac{\partial^2}{\partial y^2} + \frac{\partial^2}{\partial z^2}\right) v_0^{(2)},$$

$$\frac{\partial w_0^{(2)}}{\partial t} + \frac{\partial}{\partial y}(v_1^{(1)}w_1^{(1)*} + v_1^{(1)*}w_1^{(1)}) + 2\frac{\partial}{\partial z}(w_1^{(1)}w_1^{(1)*}) = \frac{-\partial p_0^{(2)}}{\partial z} + \frac{1}{R}\left(\frac{\partial^2}{\partial y^2} + \frac{\partial^2}{\partial z^2}\right) w_0^{(2)}.$$

By eliminating the pressure term from the last two equations, we find that the $x$ component $\xi_0^{(2)}$ of the mean second order vorticity satisfies the equation

(4.16)
$$\frac{\partial \xi_0^{(2)}}{\partial t} + \left(\frac{\partial^2}{\partial y^2} - \frac{\partial^2}{\partial z^2}\right)(v_1^{(1)}w_1^{(1)*} + v_1^{(1)*}w_1^{(1)}) + 2\frac{\partial^2}{\partial y \partial z}(w_1^{(1)}w_1^{(1)*} - v_1^{(1)}v_1^{(1)*}) = \frac{1}{R}\left(\frac{\partial^2}{\partial y^2} + \frac{\partial^2}{\partial z^2}\right) \xi_0^{(2)}.$$

This quantity $\xi_0^{(2)}$ will play a key role in the subsequent analysis.

In a similar way the equations governing the second order second harmonic motion can be written down:

$$2i\alpha u_2^{(2)} + \frac{\partial v_2^{(2)}}{\partial y} + \frac{\partial w_2^{(2)}}{\partial z} = 0,$$

$$\frac{\partial u_2^{(2)}}{\partial t} + v_2^{(2)} \frac{du_0^{(0)}}{dy} + 2i\alpha u_0^{(0)}u_2^{(2)} + i\alpha u_1^{(1)^2} + v_1^{(1)} \frac{\partial u_1^{(1)}}{\partial y} + w_1^{(1)} \frac{\partial u_1^{(1)}}{\partial z} = -2i\alpha p_2^{(2)} + \frac{1}{R}\left(\frac{\partial^2}{\partial y^2} + \frac{\partial^2}{\partial z^2} - 4\alpha^2\right) u_2^{(2)},$$

(4.17)
$$\frac{\partial v_2^{(2)}}{\partial t} + 2i\alpha u_0^{(0)}v_2^{(2)} + i\alpha u_1^{(1)}v_1^{(1)} + v_1^{(1)} \frac{\partial v_1^{(1)}}{\partial y} + w_1^{(1)} \frac{\partial v_1^{(1)}}{\partial z} = -\frac{\partial p_2^{(2)}}{\partial y} + \frac{1}{R}\left(\frac{\partial^2}{\partial y^2} + \frac{\partial^2}{\partial z^2} - 4\alpha^2\right) v_2^{(2)},$$

$$\frac{\partial w_2^{(2)}}{\partial t} + 2i\alpha u_0^{(0)}w_2^{(2)} + i\alpha u_1^{(1)}w_1^{(1)} + v_1^{(1)} \frac{\partial w_1^{(1)}}{\partial y} + w_1^{(1)} \frac{\partial w_1^{(1)}}{\partial z} = -\frac{\partial p_2^{(2)}}{\partial z} + \frac{1}{R}\left(\frac{\partial^2}{\partial y^2} + \frac{\partial^2}{\partial z^2} - 4\alpha^2\right) w_2^{(2)}.$$

Clearly the higher order terms in the perturbation series will contain higher order harmonics. Indeed, $q^{(n)}$ will contain harmonics of orders $n, n-2, n-4$, etc.; the mean flow contributions occurring only for $n$ even. In the present investigation we neglect terms of $O(a^3)$ and so we are primarily interested in $q^{(2)}$, and in particular, the mean second order flow $q_0^{(2)}$.

In many situations instability is not apparent until the Reynolds number is large. As we have mentioned earlier in this paper the velocity field will then conform to the inviscid equation, as a first approximation, except in a region close to the critical layer. This inviscid limit is in fact justified for the case of amplified disturbances and for neutral disturbances as a limiting case. That is, if $c_i \geqq 0$, where $c_i$ is the imaginary part of the complex wave velocity $c$.

If we admit that the basic early mechanism of damping or amplification of flow disturbances is that described by the linear stability theory, then for the case of high Reynolds number flows we can find quite simple explicit formulas for the rates of growth of the second order mean velocity modifications. For this purpose, it is convenient to rewrite equations for the mean secondary flow by introducing the amplitude functions for the primary oscillation. These are denoted by a circumflex. Thus,

$$u_1^{(1)}(y,z,t) = \hat{u}_1^{(1)}(y,z)e^{-i\alpha ct}, \tag{4.18}$$

with similar expressions for $v_1^{(1)}$, $w_1^{(1)}$, $p_1^{(1)}$, $\xi_1^{(1)}$. By dropping the viscous terms, equations (4.15) can be rewritten as,

$$\frac{\partial v_0^{(2)}}{\partial y} + \frac{\partial w_0^{(2)}}{\partial z} = 0,$$

$$\frac{\partial w_0^{(2)}}{\partial y} - \frac{\partial v_0^{(2)}}{\partial z} = \xi_0^{(2)},$$

$$\frac{\partial u_0^{(2)}}{\partial t} + v_0^{(2)}\frac{du_0^{(0)}}{dy} - \hat{r}_0^{(2)}(y,z)e^{2\alpha c_i t} = 0, \tag{4.19}$$

$$\frac{\partial \xi_0^{(2)}}{\partial t} = \hat{\xi}_0^{(2)}(y,z)e^{2\alpha c_i t},$$

$$\hat{r}_0^{(2)}(y,z) = -\frac{\partial}{\partial y}(\hat{u}_1^{(1)}\hat{v}_1^{(1)*} + \hat{u}_1^{(1)*}\hat{v}_1^{(1)}) - \frac{\partial}{\partial z}(\hat{u}_1^{(1)}\hat{w}_1^{(1)*} + \hat{u}_1^{(1)*}\hat{w}_1^{(1)}),$$

$$\hat{\xi}_0^{(2)}(y,z) = \left(\frac{\partial^2}{\partial z^2} - \frac{\partial^2}{\partial y^2}\right)(\hat{v}_1^{(1)}\hat{w}_1^{(1)*} + \hat{v}_1^{(1)*}\hat{w}_1^{(1)}) + 2\frac{\partial^2}{\partial y \partial z}(\hat{v}_1^{(1)}\hat{v}_1^{(1)*} - \hat{w}_1^{(1)}\hat{w}_1^{(1)*}).$$

Integrating, we have

$$\xi_0^{(2)} = \hat{\xi}_0^{(2)}(y,z)\frac{(e^{2\alpha c_i t} - 1)}{2\alpha c_i} + \hat{C}(y,z), \tag{4.20}$$

where the form of the solution is so written that the limiting case $c_i \to 0$ can be readily discussed. The arbitrary function of integration gives the initial value of $\xi_0^{(2)}$, which we may take as zero for the purpose at hand. In a similar way we find from (4.19),

$$
\begin{aligned}
u_0^{(2)} &= \hat{s}_0^{(2)}(y,z)\,\frac{(e^{2\alpha c_i t} - 1 - 2\alpha c_i t)}{(2\alpha c_i)^2} + \hat{r}_0^{(2)}(y,z)\,\frac{(e^{2\alpha c_i t} - 1)}{2\alpha c_i},\\
v_0^{(2)} &= \hat{v}_0^{(2)}(y,z)\,\frac{(e^{2\alpha c_i t} - 1)}{2\alpha c_i},\\
w_0^{(2)} &= \hat{w}_0^{(2)}(y,z)\,\frac{(e^{2\alpha c_i t} - 1)}{2\alpha c_i},\\
\xi_0^{(2)} &= \hat{\xi}_0^{(2)}(y,z)\,\frac{(e^{2\alpha c_i t} - 1)}{2\alpha c_i},
\end{aligned}
\tag{4.21}
$$

where

$$
\begin{aligned}
&\frac{\partial \hat{v}_0^{(2)}}{\partial y} + \frac{\partial \hat{w}_0^{(2)}}{\partial z} = 0,\\
&\frac{\partial \hat{w}_0^{(2)}}{\partial y} - \frac{\partial \hat{v}_0^{(2)}}{\partial z} = \hat{\xi}_0^{(2)}(y,z),\\
&\hat{s}_0^{(2)}(y,z) = -\,\hat{v}_0^{(2)}(y,z)\,\frac{du_0^{(0)}}{dy},
\end{aligned}
\tag{4.22}
$$

and $\hat{r}_0^{(2)}(y,z)$, $\hat{\xi}_0^{(2)}(y,z)$, are given by the last two equations in (4.19).

We notice that the second order mean velocities grow exponentially for amplified waves, and as powers of $t$ for neutral waves. Thus even for neutral disturbances there will be a buildup of this secondary flow. Another interesting feature of the analysis is that it shows that a three dimensional primary oscillation induces a mean second order vorticity into the flow having a component $\xi_0^{(2)}$ in the downstream direction. It is this mechanism which is responsible for a spanwise momentum exchange and causes a crumbling of the original velocity profile.

5. **The mean secondary flow.** Thus far the analysis has been kept quite general and the equations are applicable to the instability at high Reynolds numbers of a perfectly general type of parallel flow. Knowledge of the functions $\hat{\xi}_0^{(2)}(y,z)$ and $\hat{r}_0^{(2)}(y,z)$ is sufficient to calculate the mean secondary flow. This in turn requires that the amplitude functions for the primary oscillation corresponding to the given basic flow be known. That is, the linear stability problem must first be solved and the eigenfunctions so found used to calculate the second order velocities.

To be more definite, we must specify the form of the primary oscillation. As we mentioned in part 3, this infinitesimal oscillation must be initially two dimensional; but three dimensional waves will appear at some later stage during the development. We assume this happens before the oscillation leaves the linear range. There is strong experimental support for the conjecture. Schubauer and Klebanoff report that a strong warping of the wave was invariably a prerequisite to transition. Therefore, we take the primary oscillation to be compounded of a pure two dimensional component together with a standing wave in the spanwise direction (wave number $\beta$). That is, we consider a wave propagating downstream having a sinusoidal amplitude variation across the wave.

At this stage it is convenient to revert to a less involved notation and write

$$
\begin{aligned}
u_0^{(0)}(y) &= u_0(y), \\
\hat{u}_1^{(1)}(y,z) &= \lambda \hat{u}_1(y) \cos \beta z + \mu \hat{U}_1(y), \\
\hat{v}_1^{(1)}(y,z) &= \lambda \hat{v}_1(y) \cos \beta z + \mu \hat{V}_1(y), \\
\hat{w}_1^{(1)}(y,z) &= \lambda \hat{w}_1(y) \sin \beta z, \\
\hat{p}_1^{(1)}(y,z) &= \lambda \hat{p}_1(y) \cos \beta z + \mu \hat{P}_1(y), \\
\hat{\xi}_1^{(1)}(y,z) &= \lambda \hat{\xi}_1(y) \sin \beta z,
\end{aligned} \tag{5.1}
$$

where the ratio $\mu/\lambda$ is a measure of the relative importance of two and three dimensional oscillations.

The two sets of equations governing the primary oscillation are obtained from (4.11). They are

$$
\begin{aligned}
i\alpha \hat{u}_1 + \frac{d\hat{v}_1}{dy} + \beta \hat{w}_1 &= 0, \\
i\alpha(u_0 - c)\hat{u}_1 + \frac{du_0}{dy}\hat{v}_1 &= -i\alpha \hat{p}_1 + \frac{1}{R}\left(\frac{d^2}{dy^2} - \alpha^2 - \beta^2\right)\hat{u}_1, \\
i\alpha(u_0 - c)\hat{v}_1 &= -\frac{d\hat{p}_1}{dy} + \frac{1}{R}\left(\frac{d^2}{dy^2} - \alpha^2 - \beta^2\right)\hat{v}_1, \\
i\alpha(u_0 - c)\hat{w}_1 &= \beta \hat{p}_1 + \frac{1}{R}\left(\frac{d^2}{dy^2} - \alpha^2 - \beta^2\right)\hat{w}_1.
\end{aligned} \tag{5.2}
$$

$$
\begin{aligned}
i\alpha \hat{U}_1 + \frac{d\hat{V}_1}{dy} &= 0, \\
i\alpha(u_0 - c)\hat{U}_1 + \frac{du_0}{dy}\hat{V}_1 &= -i\alpha \hat{P}_1 + \frac{1}{R}\left(\frac{d^2}{dy^2} - \alpha^2\right)\hat{U}_1, \\
i\alpha(u_0 - c)\hat{V}_1 &= -\frac{d\hat{P}_1}{dy} + \frac{1}{R}\left(\frac{d^2}{dy^2} - \alpha^2\right)\hat{V}_1.
\end{aligned} \tag{5.3}
$$

The downstream component of vorticity of the primary oscillation has the amplitude function $\hat{\xi}_1$ where

$$
\hat{\xi}_1 = \frac{d\hat{w}_1}{dy} + \beta \hat{v}_1. \tag{5.4}
$$

The mean secondary flow components must then be of the form

$$
\begin{aligned}
\hat{v}_0^{(2)}(y,z) &= \lambda^2 \hat{v}_a(y) \cos 2\beta z + \lambda\mu \hat{v}_b(y) \cos \beta z + \lambda^2 \hat{v}_c(y) + \mu^2 \hat{v}_d(y), \\
\hat{w}_0^{(2)}(y,z) &= \lambda^2 \hat{w}_a(y) \sin 2\beta z + \lambda\mu \hat{w}_b(y) \sin \beta z + \lambda^2 \hat{w}_c(y) + \mu^2 \hat{w}_d(y), \\
\hat{\xi}_0^{(2)}(y,z) &= \lambda^2 \hat{\xi}_a(y) \sin 2\beta z + \lambda\mu \hat{\xi}_b(y) \sin \beta z + \lambda^2 \hat{\xi}_c(y) + \mu^2 \hat{\xi}_d(y), \\
\hat{r}_0^{(2)}(y,z) &= \lambda^2 \hat{r}_a(y) \cos 2\beta z + \lambda\mu \hat{r}_b(y) \cos \beta z + \lambda^2 \hat{r}_c(y) + \mu^2 \hat{r}_d(y), \\
\hat{s}_0^{(2)}(y,z) &= \lambda^2 \hat{s}_a(y) \cos 2\beta z + \lambda\mu \hat{s}_b(y) \cos \beta z + \lambda^2 \hat{s}_c(y) + \mu^2 \hat{s}_d(y), \\
\hat{p}_0^{(2)}(y,z) &= \lambda^2 \hat{p}_a(y) \cos 2\beta z + \lambda\mu \hat{p}_b(y) \cos \beta z + \lambda^2 \hat{p}_c(y) + \mu^2 \hat{p}_d(y).
\end{aligned} \tag{5.5}
$$

Thus the secondary flow may be thought of as the superposition of four distinct flow patterns as indicated by the attached subscripts. These we shall henceforth refer to as the "$a$," "$b$," "$c$" and "$d$" flows. Their respective origins are clear. The "$a$" flow gives a three dimensional distortion of the mean flow due entirely to the original three dimensionality of the oscillation and having half the spanwise wavelength of the oscillation. The "$c$" flow is a purely two dimensional modification; but originates from the three dimensional part of the disturbance. The "$b$" flow is again a three dimensional distortion due to the interaction of the two and three dimensional components of the oscillation, and has a spanwise wavelength equal to that of the oscillation. The "$d$" flow is a purely two dimensional flow and is due entirely to the two dimensional component of the oscillation. It is this "$d$" flow that has been considered previously by Meksyn and Stuart.

We note in passing that a similar classification is possible for the second order harmonic $q_2^{(2)}$; but we shall not discuss it in this paper.

On referring to the general equations (4.15) for the mean secondary flow we can now obtain sets of equations for determining the "$a$," "$b$," "$c$" and "$d$" flows. These are listed for completeness.

The "$a$" flow is determined by

$$\hat{r}_a = -\frac{1}{2}\frac{d}{dy}(\hat{u}_1\hat{v}_1^* + \hat{u}_1^*\hat{v}_1) - \beta(\hat{u}_1\hat{w}_1^* + \hat{u}_1^*\hat{w}_1),$$

$$\hat{s}_a = -\hat{v}_a\frac{du_0}{dy},$$

$$\hat{\xi}_a = -\left(\frac{1}{2}\frac{d^2}{dy^2} + 2\beta^2\right)(\hat{v}_1\hat{w}_1^* + \hat{v}_1^*\hat{w}_1) - 2\beta\frac{d}{dy}(\hat{v}_1\hat{v}_1^* + \hat{w}_1\hat{w}_1^*), \tag{5.6}$$

$$\frac{d\hat{v}_a}{dy} + 2\beta\hat{w}_a = 0, \quad \frac{d\hat{w}_a}{dy} + 2\beta\hat{v}_a = \hat{\xi}_a.$$

The "$b$" flow is determined by

$$\hat{r}_b = -\frac{d}{dy}(\hat{u}_1\hat{V}_1^* + \hat{u}_1^*\hat{V}_1 + \hat{U}_1\hat{v}_1^* + \hat{U}_1^*\hat{v}_1) - \beta(\hat{U}_1\hat{w}_1^* + \hat{U}_1^*\hat{w}_1),$$

$$\hat{s}_b = -\hat{v}_b\frac{du_0}{dy},$$

$$\hat{\xi}_b = -\left(\frac{d^2}{dy^2} + \beta^2\right)(\hat{V}_1\hat{w}_1^* + \hat{V}_1^*\hat{w}_1) - 2\beta\frac{d}{dy}(\hat{V}_1\hat{v}_1^* + \hat{V}_1^*\hat{v}_1), \tag{5.7}$$

$$\frac{d\hat{v}_b}{dy} + \beta\hat{w}_b = 0, \quad \frac{d\hat{w}_b}{dy} + \beta\hat{v}_b = \hat{\xi}_b.$$

The "$c$" and "$d$" flows are independent of $z$ and so

$$\hat{\xi}_c = \hat{\xi}_d = 0.$$

Also

$$\frac{d\hat{v}_c}{dy} = \frac{d\hat{v}_d}{dy} = \frac{d\hat{w}_c}{dy} = \frac{d\hat{w}_d}{dy} = 0.$$

Hence for the case of a flow in the presence of a solid boundary or extending to infinity in $y$ direction, these four velocity components must vanish.

There the "$c$" flow is given by

$$\hat{r}_c = -\frac{1}{2}\frac{d}{dy}(\hat{u}_1\hat{v}_1^* + \hat{u}_1^*\hat{v}_1), \tag{5.8}$$

$$\hat{s}_c = \hat{\xi}_c = \hat{v}_c = \hat{w}_c = 0.$$

Similarly the "$d$" flow is given by

$$\hat{r}_d = -\frac{1}{2}\frac{d}{dy}(\hat{U}_1\hat{V}_1^* + \hat{U}_1^*\hat{V}_1), \tag{5.9}$$

$$\hat{s}_d = \hat{\xi}_d = \hat{v}_d = \hat{w}_d = 0.$$

Thus the "$c$" and "$d$" flows contribute only to the $x$ momentum through the gradient of the Reynolds stress and have no $y$ or $z$ components.

We see that the calculation of the mean secondary flow is a comparatively simple matter for a given flow, provided the amplitude functions for the primary oscillation are known. To illustrate the theory we have performed these calculations for the shear profile $u_0(y) = \tanh y$. Such a shear flow can be readily approximated experimentally by the mixing of two parallel streams. This profile has an inflection point at $y = 0$ and there is a simple known analytical expression for a neutral disturbance. The actual computations are lengthy and will not be reported in any detail here.

However, some interesting features of the motion become apparent in doing these calculations. The solutions involve a Lommel function

$$L(Y) = L_r(Y) + iL_i(Y),$$

where $L(Y)$ is the solution of the equation

$$\frac{d^2L}{dY^2} - iYL = -1, \tag{5.10}$$

which behaves like $1/iY$ for $Y$ large. $Y$ is the familiar variable $Y = (\alpha R)^{1/3}y$. The Lommel functions occur as a correction of the boundary layer type, in the neighborhood of the critical layer $y = 0$ (where the inviscid solutions exhibit singularities). Retaining only the highest powers of $(\alpha R)^{1/3}$ we find that the primary oscillation, putting $c = 0$, is given by

$$
(5.11)\quad
\begin{aligned}
u^{(1)} &= 2\lambda\left[\frac{-\beta^2(\alpha R)^{1/3}}{\alpha} L_r(Y)\cos\alpha x\right.\\
&\qquad \left. + \frac{1}{\alpha}(\beta^2 \operatorname{cosech} y - \operatorname{sech} y \tanh y) Y L_i(Y)\sin\alpha x\right]\cos\beta z\\
&\quad + 2\mu[\operatorname{sech} y \tanh y \sin\alpha x],\\
v^{(1)} &= 2\lambda[\operatorname{sech} y \cos\alpha x]\cos\beta z + 2\mu[\operatorname{sech} y\cos\alpha x],\\
w^{(1)} &= 2\lambda[-\beta \operatorname{cosech} y\; Y L_i(Y)\cos\alpha x - \beta(\alpha R)^{1/3} L_r(Y)\sin\alpha x]\sin\beta z,\\
\xi^{(1)} &= 2\lambda\left[-\beta \operatorname{cosech}^2 y \operatorname{sech} y\, Y^2 \frac{dL_i}{dY}(Y)\cos\alpha x\right.\\
&\qquad \left. - \beta(\alpha R)^{2/3}\frac{dL_r}{dY}(Y)\sin\alpha x\right]\sin\beta z.
\end{aligned}
$$

Note that $\xi^{(1)}$ is $O(\alpha R)^{2/3}$ at the outer edge of the critical layer. The graph of $\xi^{(1)}$, at $\beta z = \pi/2$, is shown for different values of the phase in Figure 1. In all calculations the values $(\alpha R)^{1/3} = 5$, and $\beta = 1/2$, have been taken to be typical.

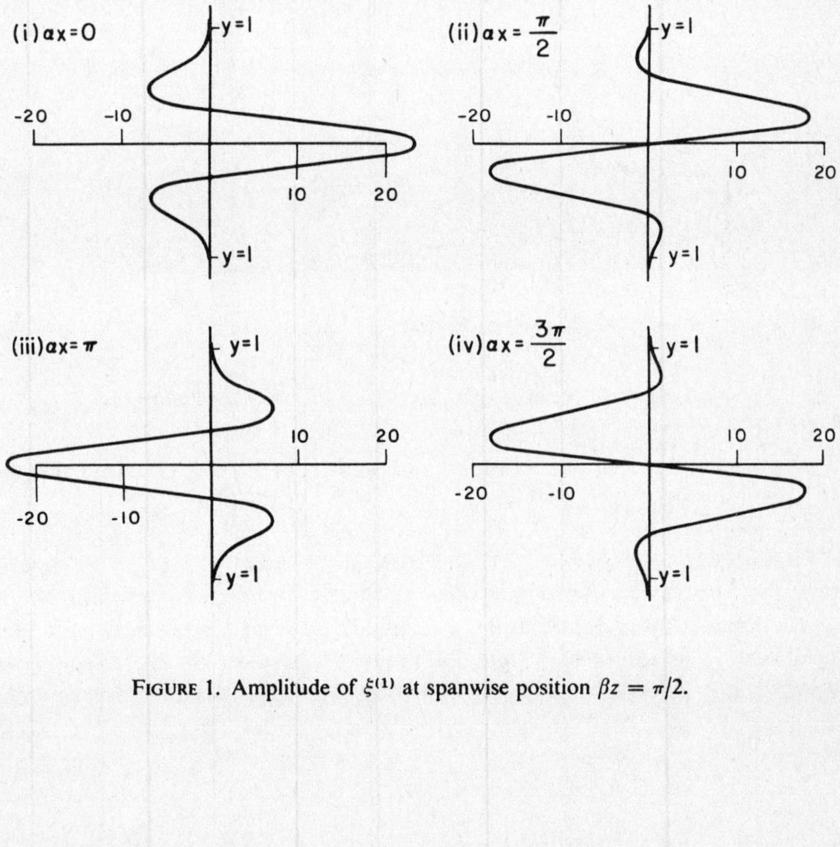


FIGURE 1. Amplitude of $\xi^{(1)}$ at spanwise position $\beta z = \pi/2$.

In a similar way the contributions to the downstream component of the mean secondary vorticity are found to be

$$(5.12)\quad \begin{aligned} \hat{\xi}_a &= -\frac{16\beta}{\sinh^3 2y}[(1-\beta^2)\cosh 2y - \beta^2]G(Y), \\ \hat{\xi}_b &= -\frac{2\beta}{\sinh^3 2y}[16\cosh 2y + \beta^2\cosh 4y - \beta^2]H(Y), \end{aligned}$$

where $G(Y)$ and $H(Y)$ depend on the Lommel function and have the property that they are both even functions of $Y$, have a zero of order four at $Y = 0$ (and so remove the apparent singularities in $\hat{\xi}_a$ and $\hat{\xi}_b$), and tend to unity as $Y \to \infty$.

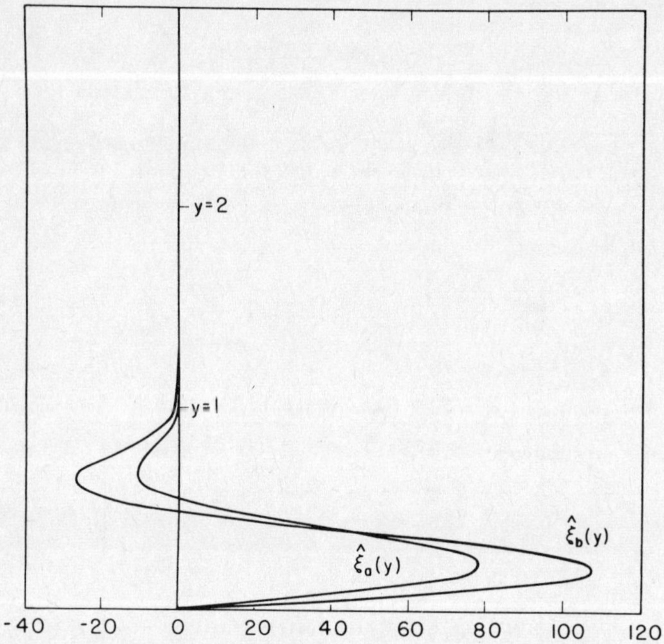


FIGURE 2. Mean secondary vorticity amplitude functions.

The vorticity distributions $\hat{\xi}_a$ and $\hat{\xi}_b$ are shown in Figure 2. Note that both these functions are $O(\alpha R)$ at the outer edge of the critical layer.

Another interesting point arises in solving for $\hat{v}_a$, $\hat{w}_a$ (and likewise for $\hat{v}_b$ and $\hat{w}_b$). It is found that one part of this motion is a potential motion of intensity $O(a^2\lambda^2(\alpha R)^{1/3})$ and this determines the direction of the circulation for the "$a$" flow at infinity. Thus although the viscous region is of thickness $O(\alpha R)^{-1/3}$, the critical layer is indeed responsible for inducing this two dimensional potential flow which is dominant far away from $y = 0$. A similar phenomenon occurs for

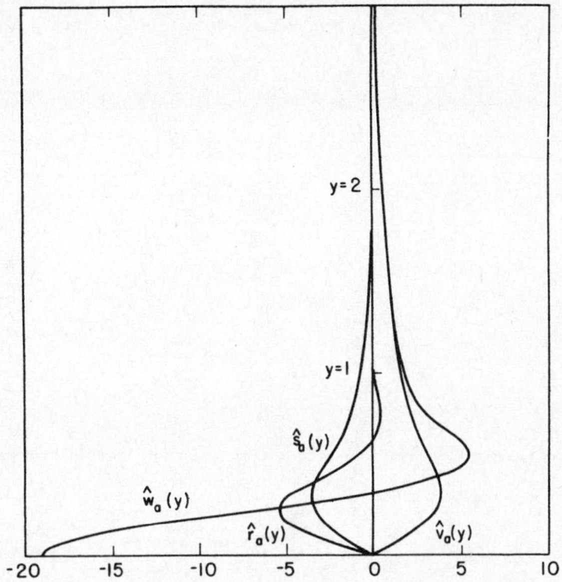


FIGURE 3. Amplitude functions for the "a" flow.

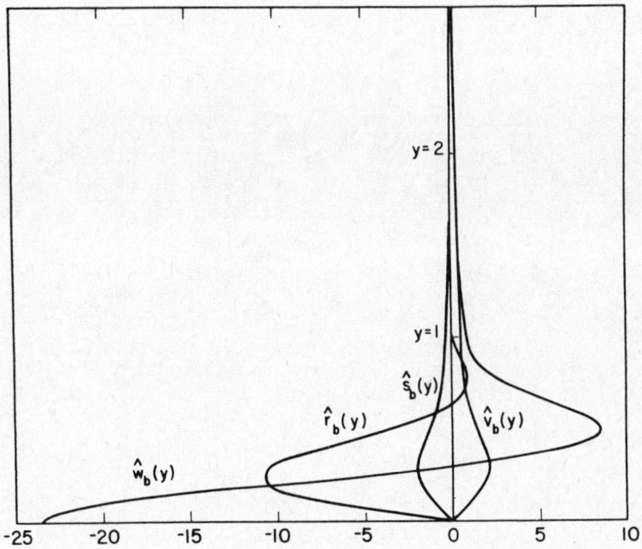


FIGURE 4. Amplitude functions for the "b" flow.

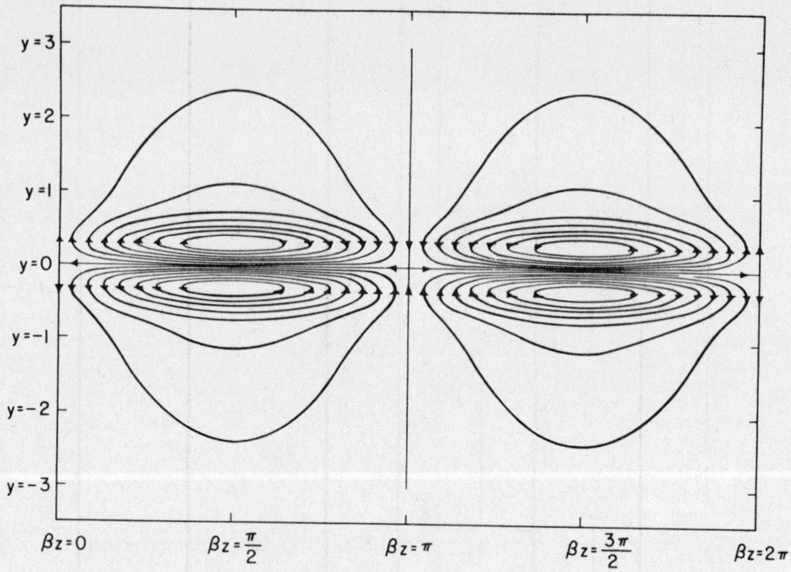


FIGURE 5. Streamline pattern for secondary motion $\frac{\mu}{\lambda} \gg 1$.

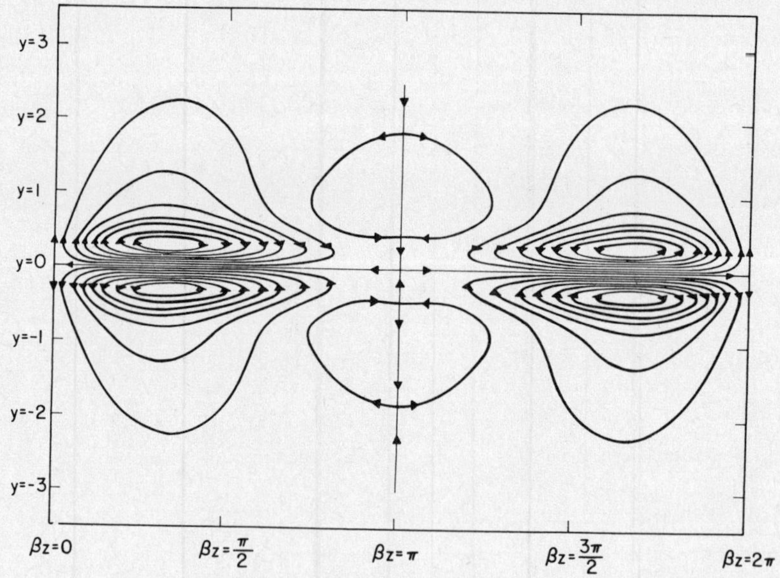


FIGURE 6. Streamline pattern for secondary motion $\frac{\mu}{\lambda} = 2$.

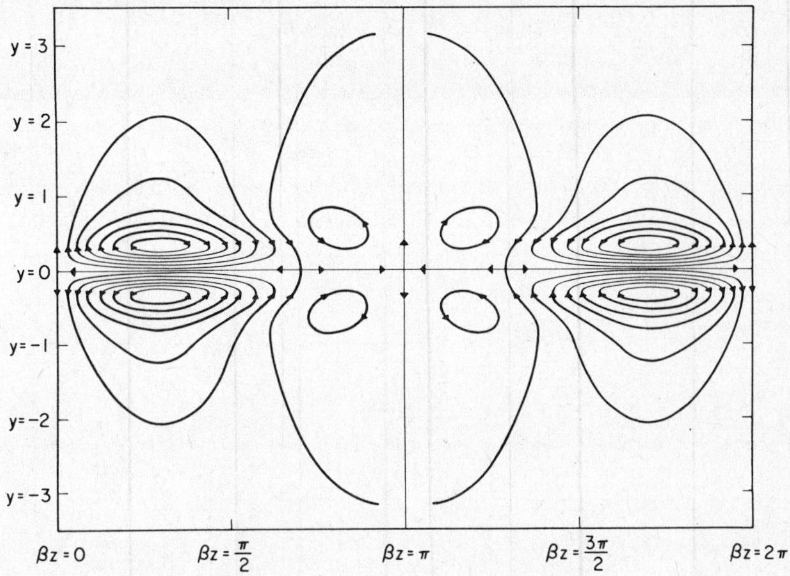


FIGURE 7. Streamline pattern for secondary motion $\frac{\mu}{\lambda} = 1$.

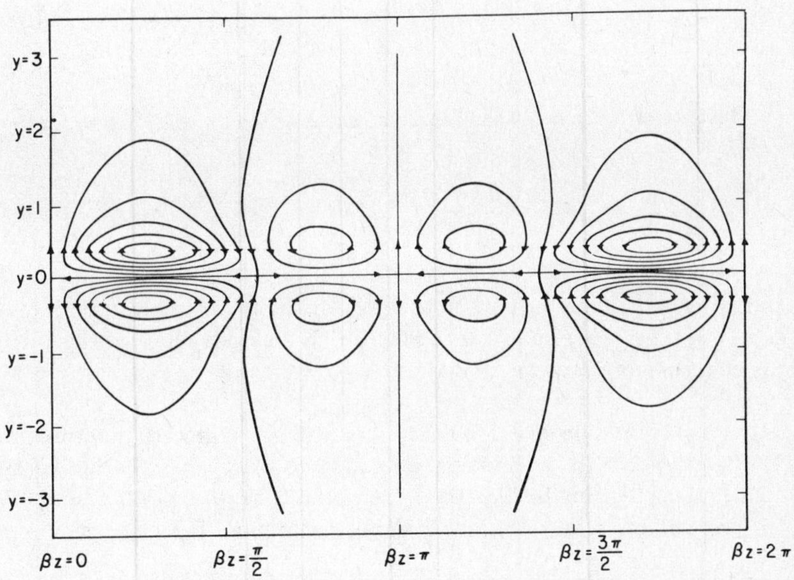


FIGURE 8. Streamline pattern for secondary motion $\frac{\mu}{\lambda} = \frac{1}{2}$.

the "$b$" flow. However, in this case, although the intensity appears to be $O(a^2\lambda\mu(\alpha R)^{1/3})$, it can be shown to be merely $O(a^2\lambda\mu(\alpha R)^{-1/3})$.

The various amplitude functions for the "$a$" and "$b$" flows are shown in Figures 3 and 4. In this example it so happens that $\hat{r}_c = \hat{r}_a$, and that $\hat{r}_d = 0$, so that to the order of the approximation used in the present example the "$d$" flow, due to the two dimensional component of the oscillation, does not affect the mean flow.

The longitudinal vortex structure of the secondary flow is illustrated by plotting the projections on the $yz$ plane of the streamlines for the mean secondary flow. These are shown in Figures 5, 6, 7, 8, 9 for various ratios of $\mu/\lambda$.

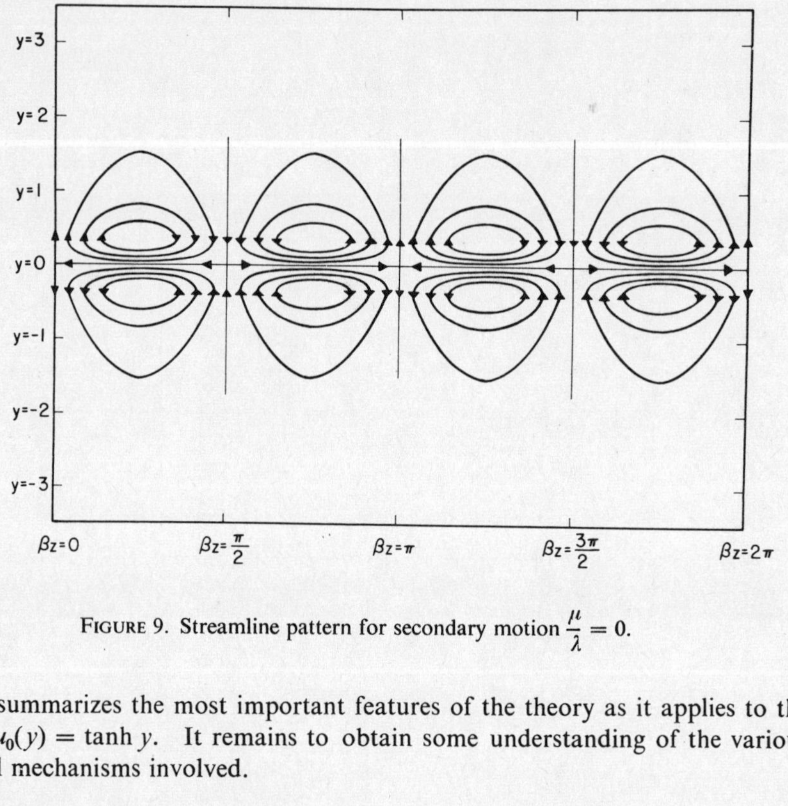


FIGURE 9. Streamline pattern for secondary motion $\frac{\mu}{\lambda} = 0$.

This summarizes the most important features of the theory as it applies to the profile $u_0(y) = \tanh y$. It remains to obtain some understanding of the various physical mechanisms involved.

6. **A description of the complete flow field.** We now give a physical interpretation to these results. Although we limit ourselves to the case of the profile $u_0(y) = \tanh y$, a similar mechanism will exist for oscillations in any parallel flow. We would again remind the reader that the effect of second and higher harmonics are neglected in this discussion.

During the onset of instability one must expect the oscillation to be mainly two dimensional. This corresponds to a situation in which $\mu/\lambda \gg 1$, and so the mean secondary flow will be essentially the "$b$" flow. This "$b$" flow induces a steady cellular vortex structure on the motion as indicated in Figure 5. Momentum is extracted from the viscous region at spanwise positions $\beta z = 0, \pm 2\pi, \pm 4\pi, \cdots$ where the amplitude of the $u^{(1)}$ oscillation is a maximum and fed back into the region at intermediate spanwise positions $\beta z = \pm \pi, \pm 3\pi, \cdots$. This large scale exchange process produces an alternate defect and excess of $x$ momentum at these points respectively. In turn it is responsible for a spanwise alternate thinning and bulging of the original velocity profile. From Figure 4 it is seen that this effect should be most pronounced at the outer edges of the critical layer, the profile gradually becoming more and more warped.

This picture is somewhat modified by the primary oscillation itself. The oscillation is periodic in $x$ and has the same spanwise period as that of the "$b$" flow. The downstream component of the vorticity for this oscillation is an even or odd function of $y$ according to whether $\alpha x$ is an even or odd multiple of $\pi/2$. Clearly from Figures 1 and 5 this vortex system will tend to reinforce and cancel that of the secondary "$b$" flow once each downstream wavelength. This vortex intensification will take place when $\alpha x = \pi/2, 5\pi/2, 9\pi/2, \cdots$, corresponding to the superposition of longitudinal vortices having their centers at spanwise positions $\beta z = \pi/2, 3\pi/2, \cdots$. At such positions the streamline is convex. The maximum cancellation occurring where the streamline is concave, contrary to the Görtler–Witting model. Existing experimental evidence lends support to the vortex intensification at the convex part of the streamline, as predicted by the present theory.

If the three dimensionality of the oscillation increases, then the "$a$" flow contribution to the secondary motion will become more prominent. The extreme case for an entirely standing wave oscillation is shown in Figure 9. The "$a$" flow also induces a cellular vortex structure on the motion, the spanwise period being $\pi/\beta$. The vorticity is again an odd function of $y$. However, this motion is unlikely to be observed without a considerable "$b$" flow contribution being superposed.

If we now trace the growth of the secondary motion supposing the three dimensionality of the oscillation becomes stronger, the secondary flow then goes through the stages sketched in Figures 5 through 9. The original vortices move closer towards the velocity defect regions and much weaker vortices begin to appear outside the critical layer near the excess positions. The centers of these latter vortices gradually move down closer to the critical layer as the motion becomes more three dimensional. It is to be recalled that the "$a$" flow has a dominant potential component $O(a^2\lambda^2(\alpha R)^{1/3})$. For the stronger vortices the same periodic intensification will be present at the downstream stations $\alpha x = \pi/2$, $5\pi/2, \cdots$; but in the case of the weaker newly formed vortices, which may be difficult to observe, the reinforcement will be less marked. However, any such detectable intensification should now take place at $\alpha x = 3\pi/2, 7\pi/2, \cdots$, due to the change in sign of the circulation. Should these two systems of vortices become of comparable strengths (before being obscured by other effects), they will show alternate partial reinforcement and cancellation over each half wavelength in the downstream direction. For example, the four vortices grouped around the defect region $z = 0$ will intensify when $\alpha x = \pi/2, 5\pi/2, \cdots$, while the weaker set of four clustered around $z = \pi/\beta$ will be most prominent when $\alpha x = 3\pi/2, 7\pi/2, \cdots$.

This longitudinal vortex structure and the associated mechanism of momentum transfer produces spanwise alternate excesses and defects in the $x$ momentum. The directions of the "$a$" and "$b$" vortices are such as to accentuate the spanwise points of velocity defect, which should therefore be quite sharply defined at $\beta z = 0, \pm 2\pi, \pm 4\pi, \cdots$. The "$c$" flow adds a two dimensional defect to the profile at all spanwise positions. Although the thinning of the profile should be quite pronounced, the bulging at the intermediate positions $\beta z = \pm \pi, \pm 3\pi, \cdots$, will be more evenly distributed about these points. In fact as the wave becomes more warped these regions of excess may show a small defect region at their centers. These conclusions can readily be seen by referring to the streamline patterns.

The great importance of nonlinear effects during transition makes it a problem clearly worthy of a detailed theoretical and experimental study. The complexity of the situation makes it all the more desirable that any theoretical approach should be a systematic one. It is believed that the present investigation is a step in this direction.

ACKNOWLEDGEMENT. During the completion of this work we have had some very pleasant informal discussions with Dr. G. B. Schubauer and Mr. P. S. Klebanoff of the National Bureau of Standards. They have shown us the results of some recent experiments on a Blasius profile. Despite the vast difference between the case of shear flow and boundary layer instability, the agreement of theory with experiment is most distinct. An application of the present theory to boundary layer transition is planned in the near future.

## REFERENCES

1. D. Meksyn and J. T. Stuart, Proc. Roy. Soc. London Ser. A. vol. 208 (1951) pp. 517–526.
2. J. T. Stuart, J. Fluid Mech. vol. 4 (1958) pp. 1–21.
3. H. Görtler and H. Witting, *Boundary layer research symposium* (Freiburg), 1957, pp. 110–126.
4. G. B. Schubauer, *Boundary layer research symposium* (Freiburg), 1957, pp. 85–109; P. S. Klebanoff and K. D. Tidstrom, Nat. Bur. Standards, Rep. 5741 (1958).
5. W. Wasow, Ann. of Math. vol. 58 (1953) pp. 222–252.
6. R. E. Langer, Trans. Amer. Math. Soc. vol. 84 (1957) pp. 144–191.
7. C. C. Lin and A. L. Rabenstein, Trans. Amer. Math. Soc. vol. 94 (1960) pp. 24–57.
8. K. M. Case, *Hydrodynamic stability and the inviscid limit*, J. Fluid Mech. vol. 10 (1961) pp. 420–429.
9. C. C. Lin, *Some mathematical problems in the theory of the stability of parallel flows*, J. Fluid Mech. vol. 10 (1961) pp. 430–438.
10. A. L. Rabenstein, Arch. Rational Mech. Anal. vol. 1 (1958) pp. 418–435.

MASSACHUSETTS INSTITUTE OF TECHNOLOGY,
CAMBRIDGE, MASSACHUSETTS

Reprinted from REVIEWS OF MODERN PHYSICS, Vol. 21, No. 3, pp. 516–519, July, 1949

# On the Concept of Similiarity in the Theory of Isotropic Turbulence

THEODORE VON KÁRMÁN*
*Columbia University, New York, New York*

AND

C. C. LIN**
*Massachusetts Institute of Technology, Cambridge, Massachusetts*

MUCH recent work has been done in the study of isotropic turbulence, particularly from the point of view of its spectrum. But the underlying concept is still the assumption of the similarity of the spectrum during the process of decay, which is equivalent to the idea of self-preservation of the correlation functions introduced by the senior author. It is however generally recognized that the correlation function does change its shape during the process of decay, and hence the concept of self-preservation or similarity must be interpreted with suitable restrictions. Under the limitation to low Reynolds numbers of turbulence, the original idea of Kármán-Howarth has been confirmed. Then the decay consists essentially of viscous dissipation of energy separately in each individual frequency interval. However, when turbulent diffusion of energy, i.e., transfer of energy between frequency intervals, occurs at a significant rate, the interpretation of the decay process and the spectral distribution is quite varied. This can be seen by a comparison of the recent publications of Heisenberg,[1] Batchelor,[2] Frenkiel,[3] and the present authors.[4,5] The purpose of the present paper is an attempt to clarify this situation.

Since some of these discussions are presented in terms of the correlation functions and others in terms of the spectrum, we shall begin by giving a systematic demonstration of the relation between these two theories. This can be easily done by a three-dimensional Fourier transform of the equation for the change of the double correlation tensor:

$$\partial/\partial t(u'^2 R_{ik}) - u'^3(\partial/\partial\xi_j)(T_{ijk}+T_{kji}) = 2\nu u'^2 R_{ik}, \quad (1)$$

where $u'^2$ is the mean square of the turbulent of velocity, $t$ is the time, $\nu$ is the kinematic viscosity coefficient, and $R_{ik}(\xi_l, t)$ and $T_{ijk}(\xi_l, t)$ are the double and the triple correlation tensor defined by Kármán and Howarth for two points $P$ and $P'$ separated by a space vector $\xi_l$. By contracting the resultant equation and multiplying it with $4\pi\kappa^2/3$, where $\kappa$ is the wave number, we obtain the following equation for the change of spectrum:

$$\partial F/\partial t + W = -2\nu\kappa^2 F, \quad (2)$$

where

$$\begin{aligned} F &= (4\pi\kappa^2/3)F_{nn}, \\ F_{ik}(\kappa_l) &= u'^2/(2\pi)^3 \int\int\int R_{ik}(\xi_l, t)e^{i(\kappa_n\xi_n)}d\tau(\xi), \\ W &= (4\pi\kappa^2/3)\cdot 2i\kappa_j W_{njn}, \\ W_{ijk}(\kappa_l) &= u'^3/(2\pi)^3 \int\int\int T_{ijk}(\xi_l, t)e^{i(\kappa_n\xi_n)}d\tau(\xi). \end{aligned} \quad (3)$$

Evaluating these integrals in terms of spherical coordinates in the $\xi$-space we obtain

$$\begin{aligned} F &= \tfrac{1}{3}\{\kappa^2 F_1''(\kappa) - \kappa F_1'(\kappa)\}, \\ F_1(\kappa) &= 2u'^2/\pi \int_0^\infty f(r, t)\cos\kappa r dr; \\ W &= \tfrac{2}{3}\{\kappa^2 H_1''(\kappa) - \kappa H_1'(\kappa)\}, \\ \kappa H_1(\kappa) &= 2u'^3/\pi \int_0^\infty h(r, t)\sin\kappa r dr. \end{aligned} \quad (4)$$

where $f(r, t)$ and $h(r, t)$ are the double and triple correlation functions satisfying the Kármán-Howarth equation:

$$\partial/\partial t(u'^2 f) + 2u'^3(\partial h/\partial r + 4h/r) = 2\nu u'^2(\partial^2 f/\partial r^2 + (4/r)(\partial f/\partial r)). \quad (5)$$

The relations (4) which connect (2) and (5) have been obtained previously (1947) by the junior author.[6] The

---

* Honorary professor, Columbia University, New York, New York.

** Associate professor, Massachusetts Institute of Technology, Cambridge, Massachusetts; consultant, Naval Ordnance Laboratory.

[1] W. Heisenberg, Proc. Roy. Soc. London, A **195**, 402 (1948).

[2] G. K. Batchelor, "Recent developments on turbulence research," general lecture at the Seventh International Congress for Applied Mechanics, September, 1948.

[3] F. N. Frenkiel, "Comparison between theoretical and experimental laws of decay of turbulence," presented at the Seventh International Congress for Applied Mechanics, September, 1948.

[4] Th. v. Kármán, Comptes Rendus **226**, 2108 (1948); Proc. Nat. Acad. Sci. **34**, 531 (1948); Sverdrup anniversary volume (1949).

[5] C. C. Lin, Proc. Nat. Acad. Sci. **34**, 540 (1948); "On the law of decay and the spectrum of isotropic turbulence," presented at the Seventh International Congress for Applied Mechanics, September, 1948.

[6] C. C. Lin, "Remarks on the spectrum of turbulence," presented at the First Symposium of Applied Mathematics, American Mathematical Society, August, 1947; to appear in the Proceedings of the symposium.

function $F$ is essentially identical with Heisenberg's spectral function, whereas $F_1$ is the spectrum function introduced by Taylor about a decade ago on the basis of a one-dimensional Fourier analysis of wind-tunnel turbulence. The tensor $F_{ik}$ in (3) was introduced and studied by Batchelor and Kampé de Férièt in 1948.

In this note, we shall restrict ourselves to the spectral theory. We propose to analyse the spectrum and its change during the process of decay. We distinguish two extreme cases: (a) the Reynolds number of turbulence is initially very large, and (b) very small initial Reynolds numbers. The latter case is very much simpler, and can be explained in a few words after the first case is investigated. We shall therefore now consider the case of very large Reynolds numbers.

There is no general principle known which would determine the most probable energy distribution over the spectrum. The problem we deal with is not a question of statistical equilibrium in the proper sense. We shall base our investigations on the concept that during the process of decay the spectrum shows a tendency to become similar. Similarity in this case means that the spectrum can be expressed in the form

$$F = U^2 l \varphi(\kappa l), \tag{6}$$

where $U$ is a typical velocity and $l$ a typical length. The problem is to connect these typical quantities with measurable ones, such as the kinematic viscosity $\nu$, Loitsiansky's invariant $J_0$, and the rate of energy dissipation $\epsilon$. Full similarity would mean that it is possible to express $U$ and $l$ by unique relations for all values of $\kappa$ and for the whole process of decay. Dealing with the experimental evidences and by dimensional considerations, one readily recognizes that this is not possible. Hence, one looks for a solution which best satisfies the similarity requirement.

All the authors agree in the following picture: the low frequency ranges contain the bulk of energy, while the viscous dissipation is negligible. They furnish energy by the action of inertial forces to the high frequency ranges, where it is converted into heat. Physically, this was seen by Taylor in the early stages of the development of the theory, but Kolmogoroff made these ideas more precise.

Kolmogoroff recognized that based on this physical concept the parameters which determine $U$ and $l$ for the high frequency range are the coefficient of kinematic viscosity $\nu$ and the rate of energy dissipation $\epsilon$. The rate of dissipation in any case is equal to $10\nu u'^2/\lambda^2$, where $\lambda$ is the microscale in the dissipation mechanism. It is essential in Kolmogoroff's concept that $u'$ and $\lambda$ do *not* appear explicitly in the similarity analysis. Thus,

$$U = (\nu\epsilon)^{\frac{1}{4}}, \quad l = (\nu^3/\epsilon)^{\frac{1}{4}}. \tag{7}$$

It is reasonable to assume that at the lower end of this $\kappa$-range, the spectrum becomes independent of $\nu$. Then necessarily

$$F \sim \epsilon^{\frac{2}{3}} \kappa^{-5/3}, \tag{8}$$

as it was concluded by several authors.

On the other hand, concerning the low frequencies, the junior author has first shown[6] that the lowest frequencies involve a fixed parameter, i.e., Loitsansky's invariant $J_0$, and indeed that the spectrum is of the form $J_0\kappa$.[4]

This situation clearly shows the infeasibility of one similarity range extending from $\kappa = 0$ to $\kappa = \infty$, and furthermore makes it necessary to consider at least three ranges: that of the lowest frequencies essentially determined by Loitsiansky's invariant, an intermediary range which not necessarily complies with the similarity characteristics of the high range but is significantly influenced by the rate of dissipation $\epsilon$ (though not directly by $\nu$), and the high frequency range in the sense of Kolmogoroff.

We therefore introduce three sets of characteristic quantities, namely, $V^*, L^*$; $V, L$; and $v, \eta$ for the lowest, the medium, and the high frequency ranges respectively. The problem is how to connect them with each other and with other physical quantities.

Equation (7) gives

$$v = (\nu\epsilon)^{\frac{1}{4}}, \quad \eta = (\nu^3/\epsilon)^{\frac{1}{4}}. \tag{7a}$$

For the lowest and the medium ranges, we postulate the existence of a parameter (in general variable with time) which is common to these ranges, and governs the complicated mechanism of energy transfer in these low frequency ranges as the viscosity governs transfer of energy into heat in the high range. The formal statement of this hypothesis is

$$V^* L^* = VL = D. \tag{9}$$

We may call this parameter $D$ the transfer coefficient or eddy diffusion coefficient of the turbulence mechanism. Obviously, it has a significance only when turbulent diffusion is active.

By introducing this hypothesis, we come to the following picture. In the lowest range, as it was pointed out, the invariant $J_0$ has a decisive influence. Hence, the characteristic parameters must be determined by $J_0$ and $D$. In the medium range, they must depend not only on $D$, but also on $\epsilon$, since this range supplies the energy to be dissipated in the high frequency range. Thus, we have

$$V = (D\epsilon)^{\frac{1}{4}}, \quad \text{and} \quad L = (D^3/\epsilon)^{\frac{1}{4}}, \tag{10}$$

for the medium range, analogous to (7a), and

$$V^* = (D^5/J_0)^{\frac{1}{3}} \quad \text{and} \quad L^* = (J_0/D^2)^{\frac{1}{3}}, \tag{11}$$

for the range of lowest frequencies.

The three ranges, with characteristic quantities (7a), (10), and (11) appear clearly separated when their scales are much different from each other. Thus, when

$$\eta/L = (D/\nu)^{-\frac{3}{4}} \ll 1, \tag{12}$$

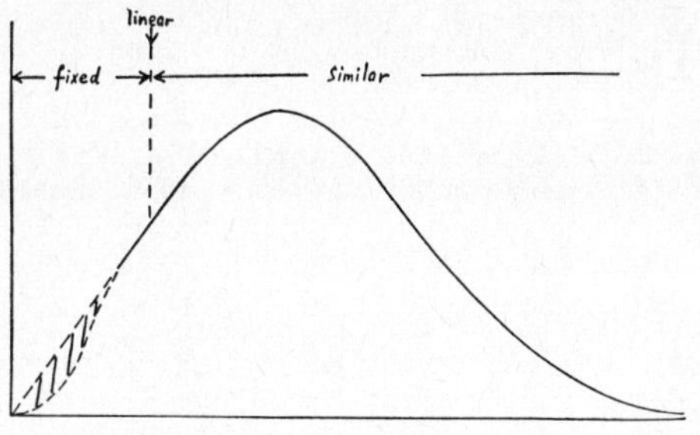


FIG. 1. Spectrum during early period.

the high and intermediate ranges appear clearly defined over significant parts of the whole spectrum. In between, there is a transition range depending only on the parameter $\epsilon$ common to both ranges. From dimensional arguments, we have for this transition range

$$F \sim \epsilon^{2/3}\kappa^{-5/3} \tag{13}$$

in accordance with Kolmogoroff's result for high Reynolds numbers. We shall see later that the condition (12) warrants the high value for the Reynolds number. Similarly, when

$$L/L^* = (T/T^*)^{1/4} \ll 1, \quad (T = L/V, \quad T^* = L^*/V^*), \tag{14}$$

we have a transition range between the low and medium frequencies. In this transition range, the spectrum depends only on $D$. Again, dimensional arguments show that there

$$F \sim D^2\kappa. \tag{15}$$

The physical significance of (14) will be explained below.

The exact behavior in the medium range will be determined by the fact which of the fixed parameters $\nu$ or $J_0$ has the predominating influence. We believe that we can arrive at a satisfactory description of the actual process by assuming that a change over takes place. We may divide the process into three stages: (I) the early stage, in which we shall see that $\eta : L =$ constant, (II) the intermediate stage, in which we shall see that $L : L^*$ constant, and (III) the final stage, in which the distinction of several scales is impossible. This last case is the well-understood case of complete similarity at extremely low Reynolds numbers.

(I) *The early stage.* The turbulence field is actually created by some mechanism, natural or artificial, which produce individual eddies. Apparently, these eddies converge toward a kind of statistical balance through exchange of energy. The first period after homogeneity and isotropy are established shall be designated the early stage of the decay process. Experimental evidence on the decay law in these early stages shows that the similarity prevailing at high frequencies extends to the medium range. This statement is identical with the conclusion reached by the junior author,[5] assuming that a perfect similarity of the correlation function exists with the exception of correlations of points between very large distances. Accordingly $V$ and $L$ are constant multiples of $v$ and $\eta$. Hence,

$$D = V^*L^* = VL \sim v\eta \tag{16}$$

and because $v\eta = \nu$, $D$ is a constant. On the other hand,

$$V^{*2}L^{*5} = J_0. \tag{17}$$

Thus, in the early stage, the spectral function for the lowest frequency range appears to be independent of time. This fixed range extends as far as the linear part of the spectrum described by (15).

The spectrum is thus as shown in Fig. 1 (after Batchelor). By integration, one arrives at

$$u'^2 = \int_0^\infty F(\kappa)d\kappa = \text{const.}V^2 - u_D^2, \tag{18}$$

where $u_D^2$ is represented by the shaded area. By computing $\epsilon = -du'^2/dt$ and using the relation (10), we see that $V^2 \sim t^{-1}$, and

$$\left.\begin{aligned} u'^2 &= \frac{D_0}{10}t^{-1} - u_D^2, \quad \lambda^2 = 10\nu t\{1 - 10u_D^2 t/D_0\}, \\ R_\lambda &= R_{\lambda 0}\{1 - 10u_D^2 t/D_0\}, \end{aligned}\right\} \tag{19}$$

where $R_{\lambda 0}$ is the initial Reynolds number

$$R_{\lambda 0} = \lim_{t\to 0}(u'\lambda/\nu), \tag{20}$$

and $D_0$ is a quantity proportional to $D$, defined by

$$D_0 = \lim_{t\to 0}(u'^2\lambda^2/\nu). \tag{21}$$

A formula of the type of Eq. (21) for the diffusion coefficient has been suggested previously (1937) by the senior author.[7] The law of decay (19) was given by the junior author.[5] The role of the relatively invariant low frequency components has also been discussed by Heisenberg and Batchelor.

By using the law (19), one can see from (14) that the linear range of the spectrum would exist essentially for small values of $t/T^*$. This sets a limit to the period of validity of the law of decay (19). One can easily see the same limitation from the law (19) itself.

It can now be seen that the condition (12) is essentially the requirement that the Reynolds number is large (see (21)). It should be noted however that the existence of the $\kappa^{-5/3}$-range is not essential in the above discussion of the decay process. Hence, the Reynolds number need not be large, in order that the law (19) holds. In fact, for small initial Reynolds numbers, the Reynolds number at the end of the early period may be so small that the final period already sets in. This explains the

[7] Th. v. Kármán, J. Ae. Sci. 4, 131 (1937).

agreement obtained by the junior author for almost the whole process of decay in comparing his theory with the experiments of Batchelor and Townsend.[8]

(II) *The intermediate stage.* For large Reynolds numbers, after the disappearance of the linear part of the spectrum, the scales $L$ and $L^*$ become of the same order of magnitude, and it may be expected that the bulk of turbulent energy of scale $L$ shares the behavior of the large eddies of scale $L^*$. The ratios $L/L^*$ and $V/V^*$ are expected to be constants. These conditions lead at once to the law of decay discussed by the senior author.[4] It is characterized by $d\lambda^2/d(\nu t)=7$.

During this period, the spectrum at low and medium frequencies depends only on the parameters $J_0$ and $\nu$, and must therefore be of the form

$$F=J_0\kappa^4\Phi(\kappa L), \tag{22}$$

where $\Phi(\kappa L)$ behaves as $(\kappa L)^{-17/3}$ for $\kappa L\gg 1$, and approaches unity when $\kappa L\to 0$. An interpolation formula for $\Phi$ has been suggested and checked by correlation measurements by the senior author.

The diffusion coefficient in this range is easily seen to be proportional to $u'^2\lambda^2/\nu$. Hence, the condition (12) is again that the Reynolds number should be large. When the Reynolds number becomes very small, the scales $\eta$ and $L$ are of the same order, so that there is only one scale for all frequencies. We then approach a complete similarity, and are at the beginning of the final period. With reference to (12), we see that it should happen when $R_\lambda$ is of the order of unity. According to the experiments of Batchelor and Townsend, the final period sets in at $R_\lambda\sim 5$.

The intermediate stage is very long, if the initial Reynolds number is very large. It begins with some value of $R_\lambda$ close to $R_{\lambda 0}$. During this period, $R_\lambda$ changes according to the power law $t^{-3/14}$. Although the supposed origin of time in this formula is unknown, it must be before the beginning of the early period, since the slope of the $\lambda^2$ *versus* $\nu t$ curve decreases. Thus, $R_\lambda$ become of the order of unity only when $t$ is of the order of $T^*R_{\lambda 0}{}^{14/3}$. One expects therefore to find an intermediate stage many times the early period for high initial Reynolds numbers.

The above predictions are based on some simple hypotheses and physical picture, and should be confirmed experimentally. Unfortunately, there does not seem to be any experimental data available for sufficiently high Reynolds numbers and over a sufficiently long period. Most of the decay measurements at high Reynolds numbers hardly extend beyond the early period, when the law of decay (19) is quite adequate. Also, it must be kept in mind that the above discussions hold only for an infinite field of turbulence. In an actual experiment, the scale of the apparatus might become comparable with the scale of turbulence. In such cases, the significance of Loitsiansky's invariant becomes uncertain.

[8] G. K. Batchelor and A. A. Townsend, Proc. Roy. Soc. London, A **194**, 527 (1948).

# On the Statistical Theory of Isotropic Turbulence[1]

By Th. von Kármán[2] and C. C. Lin[3]

*Columbia University, New York, New York, and the Massachusetts Institute of Technology, Cambridge, Massachusetts*

CONTENTS



Much recent work has been done in the study of isotropic turbulence, particularly from the point of view of its spectrum. But the underlying concept is still the assumption of the similarity of the spectrum during the process of decay, which is equivalent to the idea of self-preservation of the correlation functions introduced by the senior author. It is, however, generally recognized that the correlation function does change its shape during the process of decay, and hence the concept of self-preservation or similarity must be interpreted with suitable restrictions. Under the limitation to low Reynolds numbers of turbulence, the original idea of Kármán-Howarth has been confirmed. Then the decay consists essentially in viscous dissipation of energy separately in each individual frequency interval. However, when turbulent diffusion of energy, i.e., transfer of energy between frequency intervals, occurs at a significant rate, the interpretation of the decay process and the spectral distribution is quite varied. This can be seen by a comparison of the recent publications of Heisenberg (3), Batchelor (4), Frenkiel (5), and the present authors (6,7). The purpose of the present paper is an attempt to clarify this situation.

Since some of these discussions are presented in terms of the correlation functions and others in terms of the spectrum, we shall begin by

[1] Invited paper at joint meeting of Fluid Mechanics Division, American Physical Society and Institute of Aeronautical Sciences, January 26, 1949. An abbreviated version of this paper appeared in the *Revs. Mod. Phys.*, **21,** 516–519 (1949).

[2] Honorary professor, Columbia University.

[3] The work of Lin was partly sponsored by the Naval Ordnance Laboratory.

giving a simple systematic demonstration of the relation between the correlation theory and the spectral theory (Section I). This will serve as a background for the later developments. In Section II, we discuss the various considerations of similarity, their mutual relations, and a general comparison of their consequences with experiments. In Section III we propose a simple analysis of the spectrum, which gives a unified point of view regarding the types of similarity that should prevail at the various stages of the decay process. A detailed study of the early period of decay is given in Section IV.

It is concluded that the balance of energy among the fluctuations of various frequencies leads to the following conditions. At an early stage, all but the largest eddies participate in an equilibrium. There is a similarity of the spectrum if the lowest frequencies are excluded. In terms of correlation functions, this is the assumption proposed by the junior author (7). At an intermediate stage (to which the work of the senior author applies), the low-frequency components share the equilibrium of the largest eddies, and the high-frequency components form a separate regime of equilibrium. This separate high-frequency range is in general clearly defined for sufficiently high Reynolds number of turbulence, according to the concept of Kolmogoroff. Finally, when the Reynolds number becomes very low, the high-frequency components also share the equilibrium of the lowest frequencies and complete similarity is realized. There is then practically a separate viscous dissipation in each individual frequency range. When the initial Reynolds number is very high, the intermediate stage is very long. When the initial Reynolds number is very low, the intermediate stage practically disappears. Thus, to check the theory, experiments should be carried out at very high initial Reynolds number of turbulence, and the flow should be allowed to decay until the Reynolds number becomes very low. So far, no such experimental data seem to be available. Some suggestions for experimental measurements are made in Section V.

## I. Relation between Correlation and Spectral Theories

It has been shown by von Kármán and Howarth that the change of the double correlation tensor $R_{ik}$ is governed by the relation

$$\frac{\partial}{\partial t}(u'^2R_{ik}) - u'^3\frac{\partial}{\partial \xi_j}(T_{ijk} + T_{kji}) = 2\nu u'^2\nabla^2 R_{ik} \qquad (1.1)$$

where $u'^2$ is the mean square of the turbulent velocity, $t$ is the time, $\nu$ is the kinematic viscosity coefficient, and $R_{ik}(\xi_l,t)$ and $T_{ijk}(\xi_l,t)$ are the double and triple correlation tensors for two points $P$ and $P'$ separated by a spatial vector $\xi_l$:

$$\begin{aligned} \overline{u_i(P)u_k(P')} &= u'^2R_{ik} \\ \overline{u_i(P)u_j(P)u_k(P')} &= u'^3T_{ijk} \end{aligned} \tag{1.2}$$

The above equation has been derived for homogeneous and isotropic turbulence, but it is not difficult to extend it to homogeneous anisotropic turbulence.

Another approach to the theory of turbulence is to make a Fourier analysis of the turbulent fluctuations. One may introduce this through the correlation theory by the spectral tensor $\mathfrak{F}_{ik}(\kappa_l)$ defined by the pair of Fourier transform relations:

$$\begin{aligned} \mathfrak{F}_{ik}(\kappa_l) &= \frac{1}{(2\pi)^3}\iiint u'^2R_{ik}(\xi_m)e^{i(\kappa_l\xi_l)}d\tau(\xi_m) \\ u'^2R_{ik}(\xi_l) &= \iiint \mathfrak{F}_{ik}(\kappa_m)e^{-i(\kappa_l\xi_l)}d\tau(\kappa_m) \end{aligned} \tag{1.3}$$

It is clear from the second equation that $\frac{1}{3}\mathfrak{F}_{nn}$ represents the density of energy $u'^2$ in the $\kappa$-space at frequency $\kappa_i$. One may now obtain an equation for the change of the spectrum $\mathfrak{F}_{nn}$ by a Fourier transform of (1.1). To complete the calculation one has to introduce the Fourier transform $\mathfrak{W}_{ijk}$ of the triple correlation tensor $T_{ijk}$:

$$\begin{aligned} \mathfrak{W}_{ijk}(\kappa_l) &= \frac{1}{(2\pi)^3}\iiint u'^3T_{ijk}(\xi_m)e^{i(\kappa_l\xi_l)}d\tau(\xi_m) \\ u'^3T_{ijk}(\xi_l) &= \iiint \mathfrak{W}_{ijk}(\kappa_m)e^{-i(\kappa_l\xi_l)}d\tau(\kappa_m) \end{aligned} \tag{1.4}$$

A simple calculation then shows that (1.1) becomes

$$\frac{\partial\mathfrak{F}_{ik}}{\partial t} + \mathfrak{W}_{ik} = -2\nu\kappa^2\mathfrak{F}_{ik} \tag{1.5}$$

where

$$\mathfrak{W}_{ik} = i\kappa_j(\mathfrak{W}_{ijk} + \mathfrak{W}_{kji}) \tag{1.6}$$

Each of the equations (1.1) and (1.5) consists of six equations. These are in general dependent on the coordinate system chosen. Invariant relations are obtained by a contraction, and equation (1.5) then becomes an equation for the change of energy spectrum.

Further results on the spectral tensor have been obtained by J. Kampé de Fériet (8) in the case of homogeneous anisotropic turbulence; we shall however limit ourselves to the case of homogeneous isotropic turbulence. Here von Kármán and Howarth have shown that, by virtue of the incompressibility of the fluid, there is only one scalar correlation function $f(r,t)$ needed for characterizing $R_{ik}$ and another correlation function $h(r,t)$ for $T_{ijk}$. Hence, equation (1.1) is essentially given by a single equation, which has been shown to be

$$\frac{\partial}{\partial t}(u'^2 f) + 2u'^3\left(\frac{\partial h}{\partial r} + \frac{4h}{r}\right) = 2\nu u'^2\left(\frac{\partial^2 f}{\partial r^2} + \frac{4}{r}\frac{\partial f}{\partial r}\right) \tag{1.7}$$

It is obvious that this equation can be obtained by a contraction of (1.1). Similarly, in the spectral representation, one scalar function, such as $\mathfrak{F}_{kk}$, is able to characterize the whole spectral tensor $\mathfrak{F}_{ik}$; and the energy equation, obtained from a contraction, gives all the information contained in (1.5). This equation is more conveniently expressed in terms of the spectral function

$$\mathfrak{F} = \frac{4\pi\kappa^2}{3}\mathfrak{F}_{nn}, \tag{1.8}$$

since the second equation of (1.3) shows that

$$u'^2 = \int_0^\infty \mathfrak{F}(\kappa,t)d\kappa \tag{1.9}$$

If one further introduces the notation

$$\mathfrak{W} = \frac{4\pi\kappa^2}{3}\mathfrak{W}_{nn} \tag{1.10}$$

by analogy with (1.8), the final equation for energy transfer is then

$$\frac{\partial\mathfrak{F}}{\partial t} + \mathfrak{W} = -2\nu\kappa^2\mathfrak{F} \tag{1.11}$$

The relation between this equation and the Kármán-Howarth equation is not hard to find. In fact, if $\mathfrak{F}$ and $\mathfrak{W}$ are calculated with the help of the first formulas of (1.3) and (1.4), straightforward evaluation of the integrals by the use of spherical coordinates in the $\xi$-space gives

$$\begin{aligned}
\mathfrak{F} &= \frac{1}{3}\{\kappa^2\mathfrak{F}_1''(\kappa) - \kappa\mathfrak{F}_1'(\kappa)\},\\
\mathfrak{F}_1(\kappa) &= \frac{2u'^2}{\pi}\int_0^\infty f(r)\cos\kappa r dr, \quad f(r) = \int_0^\infty \mathfrak{F}_1(\kappa)\cos\kappa r d\kappa\\
\mathfrak{W} &= \frac{2\kappa^2}{3}\{\kappa^2 H_1''(\kappa) - \kappa H_1'(\kappa)\},\\
\kappa H_1(\kappa) &= \frac{2u'^3}{\pi}\int_0^\infty h(r)\sin\kappa r dr, \; h(r) = \int_0^\infty \kappa H_1(\kappa)\sin\kappa r d\kappa
\end{aligned} \tag{1.12}$$

These formulas for $\mathfrak{F}$ and $\mathfrak{W}$ have been obtained by Lin (9) by a somewhat different approach. The above discussion is essentially the one made by Batchelor (10). It is clear from this kind of discussion that the transfer of energy among the various frequencies is independent of the pressure fluctuations.

The spectral equation (1.11) merely gives a different aspect of the theory without removing the fundamental difficulty inherent in this kind of approach to the theory of turbulence. There are still two functions occurring in (1.11): $\mathfrak{F}(\kappa,t)$ and $\mathfrak{W}(\kappa,t)$. To resolve this difficulty, one must try to establish some relation between the transfer function $\mathfrak{W}(\kappa,t)$, the spectrum $\mathfrak{F}(\kappa,t)$, and the frequency $\kappa$. The senior author proposed the relation

$$\int_0^\kappa \mathfrak{W}(\kappa,t)d\kappa = C\left\{\int_\kappa^\infty \{\mathfrak{F}(\kappa')\}^\alpha \kappa'^\beta d\kappa'\right\}\left\{\int_0^\kappa \{\mathfrak{F}(\kappa')\}^{3/2-\alpha}\kappa'^{1/2-\beta}d\kappa'\right\} \quad C > 0 \tag{1.13}$$

where $\alpha$ and $\beta$ are constants. This includes Heisenberg's form

$$\int_0^\kappa \mathfrak{W}(\kappa,t)d\kappa = C\left\{\int_\kappa^\infty \{\mathfrak{F}(\kappa')\}^{1/2}\kappa'^{-3/2}d\kappa'\right\} \cdot \left\{\int_0^\kappa \mathfrak{F}(\kappa')\kappa'^2 d\kappa''\right\} \tag{1.14}$$

when $\alpha = \frac{1}{2}$, $\beta = -\frac{3}{2}$, and a modified Obukhoff's form

$$\int_0^\kappa \mathfrak{W}(\kappa,t)d\kappa = C\left\{\int_\kappa^\infty \mathfrak{F}(\kappa')d\kappa'\right\}\left\{\int_0^\kappa \{\kappa'\mathfrak{F}(\kappa')\}^{1/2}d\kappa'\right\} \tag{1.15}$$

when $\alpha = 1$, $\beta = 0$. It is essential from dimensional considerations that $\mathfrak{W}(\kappa,t)$ must be of degree three-halves in both $\mathfrak{F}$ and $\kappa$.

With these assumptions, the future behavior of the spectrum is completely known, once it is given at a certain instant. In the present paper, we shall however not pursue this line of attack any farther. We shall turn back to more general considerations, such as the similarity of turbulent fields.

A very important step in the study of the spectrum has been taken by Kolmogoroff. Whereas the previous similarity considerations postulated that the correlation functions and therefore also the spectral functions be expressed by $u'$ as characteristic velocity and a characteristic length for the scale, Kolmogoroff postulates that at least at sufficiently high Reynolds numbers of turbulence, the local turbulence shall depend only upon the kinematic viscosity $\nu$ of the fluid and the total rate of energy dissipation

$$\epsilon = -\frac{du'^2}{dt} = 2\nu \int_0^\infty \kappa^2 \mathfrak{F}(\kappa,t)d\kappa \tag{1.16}$$

In other words, the characteristic velocity and length of the turbulent state at any instant shall be fully determined by these two parameters. It then appears from purely dimensional arguments that in such a range, the characteristic velocity and scale are

$$v = (\nu\epsilon)^{1/4}, \qquad \eta = (\nu^3/\epsilon)^{1/4} \tag{1.17}$$

If one further assumes that when $\kappa\eta \ll 1$, the spectrum is independent of $\nu$, it then follows that

$$\mathfrak{F} = \text{constant } \epsilon^{2/3}\kappa^{-5/3} \tag{1.18}$$

which may be obtained from purely dimensional arguments and as a matter of fact was found by Obukhoff and developed independently by Onsager (11) and Weizsäcker (12). Kolmogoroff applied his postulate to the correlation function and concluded that

$$f(r,t) = 1 - \text{constant } u'^{-2}(\epsilon r)^{2/3} \tag{1.19}$$

in the limit of infinite Reynolds number. Equations (1.18) and (1.19) are essentially equivalent.

Another general result is due to Loitsiansky (13). He noted that the Kármán-Howarth equation (1.7) can be integrated to give the invariant relation

$$u'^2 \int_0^\infty f(r)r^4dr = \text{constant} \tag{1.20}$$

provided the integral is convergent and the triple correlation function vanishes so rapidly at infinity that $\lim_{r\to\infty} r^4h = 0$. It has been first shown by the junior author (9) with the help of equation (1.12) that this is equivalent to the conclusion that

$$\lim_{\kappa\to 0} \frac{\mathfrak{F}}{\kappa^4} = J_0, \quad \text{a constant} \tag{1.21}$$

This relation has also been found independently by Batchelor (10).

The results of Kolmogoroff and Loitsiansky will play an important role in the developments of the present paper; but we shall chiefly be guided by the idea of self-preserving correlation functions, or what is equivalent, the similarity of the spectrum.

## II. Consideration of Similarity

The energy distribution among the various frequencies is changing through the transfer mechanism and through the effect of viscous dissipation [equation (1.11)]. It is reasonable to expect that the exchange of energy among the various frequencies will cause the spectrum of turbulence to tend to be similar in the course of time. This is the equivalent to the idea of self-preserving correlation functions introduced by von Kármán.

Let us consider the equation for the change of spectrum

$$\frac{\partial \mathfrak{F}}{\partial t} + \mathfrak{W} = -2\nu\kappa^2\mathfrak{F} \tag{2.1}$$

and try to find a similarity solution. If $V$ is a characteristic velocity, and $l$ is a characteristic length, then, from dimensional arguments.[4]

$$\mathfrak{F} = V^2 l\psi(\xi), \qquad \mathfrak{W} = V^3 w(\xi), \quad \xi = \kappa l \tag{2.2}$$

Thus, (2.1) becomes

$$\frac{1}{V}\frac{dl}{dt}\{\xi\psi'(\xi) + \psi(\xi)\} + \frac{2l}{V^2}\frac{dV}{dt}\psi(\xi) + w(\xi) = -\frac{2\nu}{Vl}\xi^2\psi(\xi) \tag{2.3}$$

If the similarity solution is to be valid, one must have

$$\frac{1}{V}\frac{dl}{dt} = a_1, \quad \text{a constant} \tag{2.4}$$

$$\frac{2l}{V^2}\frac{dV}{dt} = a_2, \tag{2.5}$$

$$\frac{\nu}{Vl} = a_3 \tag{2.6}$$

and the equation (2.3) becomes

$$a_1\xi\psi'(\xi) + (a_1 + a_2)\psi(\xi) + 2a_3\xi^2\psi(\xi) + w(\xi) = 0 \tag{2.7}$$

Besides (2.4)–(2.6) it is evident that the quantities $u'^2$ and $\epsilon$, according to the definitions (1.9) and (1.16) have to satisfy the relations

$$u'^2 = V^2\int_0^\infty \psi(\xi)d\xi \tag{2.8}$$

$$-\frac{du'^2}{dt} = \epsilon = 2\nu\frac{V^2}{l^2}\int_0^\infty \xi^2\psi(\xi)d\xi \tag{2.9}$$

Finally if the convergence criteria for Loitsiansky's relation (1.21) are assumed to be valid, we have

$$V^2 l^5 \lim_{\xi\to 0}\left\{\frac{\psi(\xi)}{\xi^4}\right\} = J_0 \tag{2.10}$$

This system of equations presumes that the transfer term in (2.3) is considered generally of equal importance with the term expressing the viscous dissipation. It has been shown by Dryden (14) in the equivalent

[4] This type of discussion is used in a memorandum by Lin to the Mechanics Division of the Naval Ordnance Laboratory during the spring of 1948. Abstract of paper appeared in the *Bull. Am. Math. Soc.*, **55**, 49 (1949). (Submitted June 28, 1948.)

problem of self-preserving correlation functions that such a solution is connected with the statement that the square of the characteristic length is proportional to the time $t$ and the law of decay is expressed by $u'^2 \sim t^{-1}$. Heisenberg (3) indicated in a recent paper an equivalent solution for the spectral problem. It is easily seen that Dryden's and Heisenberg's solutions are at variance with equation (2.10). In other words, *full similarity is only possible when we reject Loitsiansky's theorem.* This fact has been noted by Heisenberg (3). In addition, experimental evidence clearly indicates that the law of decay and the behavior of the characteristic length during decay exclude the possibility to adopt full similarity as a *general valid assumption.*

Let us consider now two opposite specific cases: ($a$) that the transfer term is considered negligible for all frequencies; ($b$) that the influence of viscous dissipation is restricted to high frequencies, whereas for low frequencies the transfer term is the prevailing factor.

Case ($a$), $w(\xi) = 0$, leads to a solution of equation (2.1) which furnishes full similarity for all frequencies and also satisfies Loitsiansky's relation. One obtains with $\xi = \kappa l$ and $l = \sqrt{\nu t}$

$$\mathfrak{F} = \text{constant } V^2 l \xi^4 e^{-2\xi^2} \tag{2.11}$$

or

$$\mathfrak{F} = \text{constant } V^2 l^5 \kappa^4 e^{-2\kappa^2 \nu t}. \tag{2.12}$$

By the definition of $J_0$ in (1.21) we write

$$\mathfrak{F} = J_0 \kappa^4 e^{-2\nu\kappa^2 t} \tag{2.13}$$

The corresponding correlation function can be easily shown to be

$$f(r,t) = e^{-r^2/8\nu t} \tag{2.14}$$

by using the first set of formulas in (1.12). This correlation function was noted by von Kármán and Howarth, and discussed by Millionshchikov (15), Loitsiansky (13), and Batchelor and Townsend (16). The law of decay in this case is the five-fourths power law:

$$u'^2 \sim (t - t_0)^{-5/4}, \quad \lambda^2 = 4\nu(t - t_0) \tag{2.15}$$

This law of decay and the corresponding correlation function has been verified experimentally by Batchelor and Townsend (16) for the final stage of decay.

It is generally believed that this solution furnishes a fair approximation for the final stage of the decay in any case and it is believed that it is correct for almost the whole process of decay if the initial Reynolds number of the turbulent motion is very small.

Case ($b$) has also been treated in the theory of self-preserving correlations by von Kármán and Howarth and later by Kolmogoroff (17). The former authors came to the conclusion that any power law for the decay-time relation may prevail in the decay process. Kolmogoroff pointed out that if one assumes the validity of Loitsiansky's theorem, the relations

$$u'^2 = \text{constant } t^{-10/7} \quad \text{and} \quad \lambda^2 = 7\nu t \tag{2.16}$$

must apply.

The senior author dealt with the corresponding spectral problem in two communications (6) assuming the specific decay law (2.16). He obtained results for the spectral distribution which are in good accordance with measurements. Especially it appears that the spectral distribution has the form $\mathfrak{F} \sim \kappa^4$ for small values of $\kappa$ and $\mathfrak{F} \sim \text{constant } \kappa^{-5/3}$ for large values of $\kappa$ provided in the whole $\kappa$-range considered the viscous dissipation remains small. We shall therefore refer to this case as the case with nonviscous similarity.

The junior author (7) was led by his consideration of general similarity to a law of decay of the form

$$u'^2 = a(t - t_0)^{-1} + b \tag{2.17}$$

where $a$ is a positive and $b$ a negative constant.

Subsequent discussion between the authors resulted in an analysis which is believed to clarify the situation essentially and constitute the main substance of this paper. It requires the introduction of two elements characterizing the process of decay: the time history and the initial Reynolds number.

We consider now as *case* ($c$) the assumption that similarity extends over the whole frequency range, with the exception of the lowest. In order to satisfy Loitsiansky's relation we assume that $\mathfrak{F}(\kappa)$ behaves for small $\kappa$ as proportional to $\kappa^4$.

The assumption is essentially the following. The deviation from similarity shall occur only for such small values of $\kappa$ that whereas the contribution of the deviation is negligible for computation of $\epsilon$ [cf. equation (2.9)], it enters in the calculation of the energy [cf. (2.8)].

It appears that this assumption is equivalent to Kolmogoroff's similarity hypothesis in Lin's extended formulation. By using the first set of formulas in (1.12) one can easily verify that

$$u'^2(1 - f) = C_2 \frac{r^2}{2!} - C_4 \frac{r^4}{4!} + \cdots \tag{2.18}$$

where

$$C_{2n} = \frac{1}{(2n + 1)(2n + 3)} \int_0^\infty \mathfrak{F}(\kappa,t)\kappa^{2n} d\kappa \tag{2.19}$$

Thus, if the deviation from similarity has a negligible effect in $\epsilon$, i.e., $C_2$, then $C_{2n}$'s are all proportional to $V^2 l^{-2n}$. This is the assumption made in Lin's paper. Conversely, if the similarity of $u'^2(1-f)$ is assumed, the spectrum is similar to the extent that the moments $C_{2n}$ are all proportional to $V^2 l^{-2n}$.

The law of decay can be obtained from the general relations (2.4), (2.5), (2.6), and (2.9), which are valid for any similarity hypothesis. One obtains the positive and negative half-power laws for the change of the characteristic length $l$ and the characteristic velocity $V$, and the inverse square law for the rate of dissipation $\epsilon$. To be more specific, one may identify $l$ and $V$ with Kolmogoroff's characteristic quantities [cf. (2.6) and (2.9)]

$$\eta = (\nu^3/\epsilon)^{1/4} \quad \text{and} \quad v = (\nu\epsilon)^{1/4} \tag{2.20}$$

It can be easily seen by introducing these relations into (2.4)–(2.6), that the law of decay is of Lin's form (2.17).

For convenience of reference, the results are listed below with definite physical interpretation attached to the constants. The law of decay is given by

$$\epsilon = (D_0/10)t^{-2}, \qquad u'^2 = (D_0/10)t^{-1} - u_D^2, \qquad \lambda^2 = 10\nu t\left(1 - \frac{10u_D^2}{D_0}\,t\right) \tag{2.21}$$

where $u_D{}^2$ is the additive constant giving the departure from similarity, and $D_0$ is the initial diffusion coefficient

$$D_0 = \lim_{t\to 0} \frac{u'^2\lambda^2}{\nu} \tag{2.22}$$

defined according to a formula of the kind suggested by von Kármán (1) in 1937. The change of the characteristic velocity and scale, and of the Reynolds number of turbulence are given by

$$v^2 = (10)^{-1/2} R_{\lambda 0}\nu t^{-1}, \qquad \eta^2 = (10)^{1/2} R_{\lambda 0}^{-1}\nu t, \qquad R_\lambda = R_{\lambda 0}\left(1 - \frac{10u_D^2}{D_0}\,t\right) \tag{2.23}$$

where $R_{\lambda 0}$ is the initial Reynolds number of turbulence,

$$R_{\lambda 0} = \lim_{t\to 0} \frac{u'\lambda}{\nu} \tag{2.24}$$

It is evident from the equations (2.21) and (2.23) that the solutions obtained can only be applied to an early stage of the decay process, in which $10u_D^2 t/D_0$ remains small.

Most experiments on decay have been carried out for relatively small values of $R_{\lambda 0}$. They confirm the conclusions of Lin's theory at least for the early stage of decay. In particular, the linear decrease of $R_\lambda$ is indicated. Also the few experiments where $R_{\lambda 0}$ is not small, show that in the initial stage the relation $\lambda^2 = 10\nu t$ approximately holds. Therefore although no a priori reason can be found why the decay should not start according to the law specified by the nonviscous similarity theory, we propose that for the early stage the hypothesis of case (*c*) holds, whatever the initial Reynolds number may be.

In the next section a general discussion is given covering the whole process of decay.

## III. Proposed Theory

The linear decrease of $R_\lambda$ as given by Lin's theory cannot last very long, because the slope of the $(\lambda^2, \nu t)$ curve is given by $10[2R_\lambda/R_{\lambda 0} - 1]$ and therefore when $R_\lambda$ becomes equal to $17R_{\lambda 0}/20$, the law of decay is already similar to that governed by the nonviscous kind of similarity. [Compare (2.16) and (2.23).] One may reasonably suppose that this kind of similarity then actually prevails. The physical basis for this change of nature of similarity will become clear after an analysis of the change of spectrum—as we shall do later.

During this period of nonviscous similarity, the Reynolds number $R_\lambda$ keeps on decreasing and eventually becomes so small that the final period of decay is reached. If the initial Reynolds number $R_{\lambda 0}$ is very large, the time required for this change from $17R_{\lambda 0}/20$ to small Reynolds numbers is very long. One would estimate a very long nonviscous period in which $\lambda^2 = 7\nu t$. This has never been thoroughly investigated experimentally: most of the measurements with high Reynolds numbers of turbulence scarcely extend beyond the early period. On the other hand, those experiments which do cover both the early and the final periods are performed at such small Reynolds numbers, that a complete omission of the range of nonviscous similarity gives a rather good description of the decay process.

We shall therefore devote our later discussions to cases where the initial Reynolds number is large. We shall consider the spectrum and the law of decay during the various stages in the idealized case of an infinite field of turbulence. In comparing the theory with experimental investigations, one must bear in mind that the actual size of the apparatus may be a limitation to the behavior of the largest eddies.

First we note that Kolmogoroff's theory as discussed above, gives a definite meaning to the high-frequency components, at least for large Reynolds numbers. It is the range in which viscous effects are dominating and the major part of dissipation is occurring. On the other hand, it contains a negligible fraction of the energy. The spectrum in this range depends on $\epsilon$ and $\nu$.

The part of the spectrum below this range is related to the large eddies which are the carriers of turbulent diffusion. It should be possible to characterize the spectrum in this range by a turbulent diffusion coefficient $D$. It includes the spectrum for very small values of $\kappa$ which, as shown above, has the form $J_0\kappa^4$ and therefore is practically invariant in time. Then it includes those somewhat smaller eddies which significantly decrease in intensity, since they supply energy to the high-frequency range to be dissipated at the rate $\epsilon$. We shall therefore distinguish again two frequency ranges: (1) the largest eddies characterized by $J_0$ and $D$, (2) the medium frequency range characterized by $D$ and $\epsilon$. In other words, if $V^*$ and $L^*$ are the characteristic velocity and length of the largest eddies, and $V$ and $L$ are corresponding quantities for the medium eddies, we are assuming that

$$VL = V^*L^* \tag{3.1}$$

which quantity is then regarded as the turbulent diffusion coefficient $D$, characteristic for low-frequency phenomena.

We have thus divided the spectrum into three ranges: (1) the high-frequency range, depending on $\epsilon$ and $\nu$; (2) the low-frequency range, depending on $\epsilon$ and $D$; (3) the range of largest eddies, depending on $D$ and $J_0$. By dimensional arguments, one can obtain the characteristic quantities and other general properties of the spectrum in the various ranges. These are tabulated in Table I. The formulas for the spectrum

TABLE I

| Frequency range | (1) High | (2) Medium | (3) Low |
|---|---|---|---|
| Basic characteristic quantities | $\nu, \epsilon$ | $\epsilon, D$ | $D, J_0$ |
| Length | $\eta = (\nu^3/\epsilon)^{1/2}$ | $L = (D^3/\epsilon)^{1/4}$ | $L^* = (J_0/D^2)^{1/3}$ |
| Velocity | $v = (\nu\epsilon)^{1/4}$ | $V = (D/\epsilon)^{1/4}$ | $V^* = (D^5/J_0)^{1/3}$ |
| Time | $\tau = (\nu/\epsilon)^{1/2}$ | $T = (D/\epsilon)^{1/2}$ | $T^* = (J_0^2/D^7)^{1/3}$ |
| Form of spectrum | $v^2\eta\phi(\kappa\eta)$ | $V^2LF(\kappa L)$ | Fixed $V^{*2}L^*F_1(\kappa L^*)$ |
| lower end | $\mathfrak{F} = \text{const. } \epsilon^{2/3}\kappa^{-5/3}$ | $\mathfrak{F} = \text{const. } D^2\kappa$ | $\mathfrak{F} = J_0\kappa^4$ |

in the transition ranges can be obtained from purely dimensional arguments. They can exist only when the scales $\eta$, $L$, and $L^*$ are very much different from each other. To examine the conditions required, let us note that

$$\begin{aligned} \frac{\eta}{L} &= \left(\frac{D}{\nu}\right)^{-3/4} \\ \frac{L}{L^*} &= \left(\frac{T}{T^*}\right)^{1/2} \end{aligned} \tag{3.2}$$

We shall see that these conditions show that the $\kappa^{-5/3}$-range would exist if the Reynolds number $R_\lambda$ is large, and that the linear range will exist during the early part of the history of decay.

The above analysis of the spectrum enables us to give a very simple physical description of the process of decay. We divide the process into three stages: (1) the early stage, (2) the intermediate stage, and (3) the final stage, in which the Reynolds number is small. This last stage has been fully clarified before and will not be discussed here.

### 1. *The Early Stage*

At the beginning of the process of decay, the small eddies, range (1), first reached a state of equilibrium. To this the low-frequency components, range (2), may be expected to participate through the mechanism of energy transfer. The ratios $\eta/L$, $v/V$, $\tau/T$ must therefore be constants. This is the kind of similarity discussed as case (*c*) of Section II. The law of decay is characterized by (2.21) and (2.23). One can easily verify from (2.23) that the diffusion coefficient

$$D = VL \sim v\eta \tag{3.3}$$

is proportional to $D_0$. We shall therefore identify $D$ and $D_0$ since both are a measure of diffusion. This means that $V^*$ and $L^*$ are fixed, since on the one hand we have $V^{*2}L^{*5} = J_0$, and on the other hand $V^*L^* = D_0$. The large eddies are invariant up to the linear part of spectrum.

The physical interpretation of the additive constant $u_D^2$ in (2.23) is now clear. From the above discussions, the spectrum must be of the type shown in Fig. 1. The part of the spectrum above the linear range shrinks radially inversely proportional to the square root of time [cf. (2.23) and the formula for the spectrum in Table I]. The linear part and the lower frequencies are fixed. The additive constant is simply the shaded area, representing the deviation of the lowest frequencies from complete similarity. This can be seen by calculating $u'^2$ by integration over the spectrum:

$$u'^2 = \int^\infty \mathfrak{F}(\kappa)d\kappa = V^2 \int_0^\infty \phi(\xi)d\xi - u_D^2 \tag{3.4}$$

which is the law of decay (2.17) first given by the junior author.

The role of these relatively invariant large eddies has been independently discussed by Heisenberg (3), Batchelor (4), and the junior author (7). In particular, Fig. 1 is implied in Heisenberg's discussion and has been shown explicitly by Batchelor.

One can easily verify, with $D = D_0$, that

$$\frac{\eta}{L} = \frac{v}{V} = \left(\frac{D}{\nu}\right)^{-3/4} = R_{\lambda 0}^{-3/2} \tag{3.5}$$

Comparing (3.2) with (3.5) this means that the $\kappa^{-5/3}$ range would exist when $R_{\lambda 0}$ is large, in agreement with the idea of Kolmogoroff. The second relation in (3.2), which is the condition for the existence of the

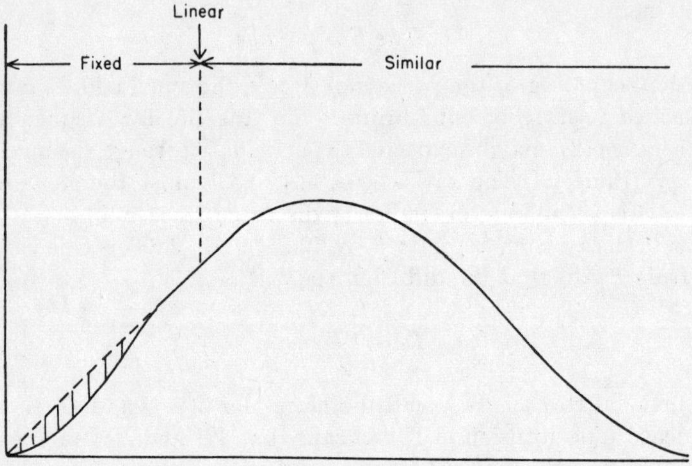


FIG. 1. Spectrum during early stage of decay process.

linear part of the spectrum, can also be easily worked out explicitly. One obtains

$$\frac{L}{L^*} = \left(\frac{\lambda^2}{\sqrt{10}\,\nu T^*}\right)^{1/2} \tag{3.6}$$

Hence, by noting the fact that $d\lambda^2/d(\nu t) \leq 10$, one can see that $L/L^* \ll 1$ essentially for the early period when $t/T^*$ is small. That is why the above kind of similarity, which depends on the existence of the linear range of spectrum, can exist only during the early period. This is equivalent to the condition reached in Section II that $10u_D^2 t/D_0$ should remain small, as will be demonstrated more fully with an example in Section IV.

It may be noted that, in the above discussions, the existence of the $\kappa^{-5/3}$ range is not essential, and hence the Reynolds number need not be large. In fact, for small initial Reynolds numbers, the Reynolds number at the end of the early period may be so small that the final period already sets in. This explains the agreement obtained by the junior author in comparing his theory with the experiments made by Batchelor and Townsend over the whole process of decay (7).

### 2. *The Intermediate Stage*

For large Reynolds numbers, after the disappearance of the linear part of the spectrum, the scales $L$ and $L_0$ become of the same order of magnitude, and it may be expected that the bulk of turbulent energy of scale $L$ shares the behavior of the large eddies of scale $L^*$. The ratios $L/L^*$, $V/V^*$, and $T/T^*$ are expected to be constants. These conditions lead at once to the law of decay discussed by the senior author. It is characterized by $d\lambda^2/d(\nu t) = 7$.

During this period, the spectrum at low and medium frequencies depends only on the parameters $J_0$ and $\epsilon$, and must therefore be of the form

$$J_0\kappa^4\Phi(\kappa L)$$

where $\Phi(\kappa L)$ behaves as $(\kappa L)^{-17/3}$ for $\kappa L \gg 1$, and approaches unity when $\kappa L \rightarrow 0$. An interpolation formula for $\Phi$ has been suggested and checked by correlation measurements by the senior author (6).

The diffusion coefficient in this range is easily seen to be proportional to $u'^2\lambda^2/\nu$. Hence, the condition (3.2) is again that the Reynolds number should be large. When the Reynolds number becomes very small, the scales $\eta$ and $L$ are of the same order, so that there is only one scale for all frequencies. We then approach a complete similarity and are at the beginning of the final period. With reference to (3.2), we see that it should happen when $R_\lambda$ is of the order of unity. According to the experiments of Batchelor and Townsend (10), the final period sets in at $R_\lambda \sim 5$.

The intermediate stage is very long, if the initial Reynolds number is very large. It begins with some value of $R_\lambda$ close to $R_{\lambda 0}$. During this period, $R_\lambda$ changes according to the power law $t^{-3/14}$. Although the supposed origin of time in this formula is unknown, it must be before the beginning of the early period, since the slope of the $\lambda^2$ versus $\nu t$ curve decreases. Thus, $R_\lambda$ becomes of the order of unity only when $t$ is of the order of $T^*R_{\lambda 0}{}^{14/3}$. One expects therefore to find an intermediate stage many times the early period for very high initial Reynolds numbers. In such a case, the final period could probably never be observed in practice. Such predictions can only be checked by experiments at very high Reynolds numbers performed over an extended period of decay. For although the initial period is relatively short, it extends over a very long distance in the wind tunnel when the wind speed is high.

## IV. Detailed Study of the Early Period

It is instructive to work out in some detail the above general discussions during the early period, in the case of high initial Reynolds numbers. In particular, we shall demonstrate explicitly that $u_D^2$ is proportional to $V^{*2}$. We shall also demonstrate the precise meaning of the statement that at high Reynolds numbers of turbulence, the major part of energy lies in the low-frequency range and the major part of dissipation lies in the high-frequency range.

The spectrum in the early period is given in Table I, with all the three ranges present. The intermediate range with $\mathfrak{F} \sim \kappa^{-5/3}$ obtains when $x = \kappa\eta \ll 1$ and $X = \kappa L \gg 1$. Let us choose a frequency $\kappa_2$ somewhere in this range. Then

$$X_2 = \kappa_2 L \gg 1, \qquad x_2 = \kappa_2 \eta \ll 1 \tag{4.1}$$

Since the ratio $x_2/X_2$ is of the order of $R_{\lambda 0}^{-3/2}$ [see (3.5)], one may so choose $\kappa_2$ that $X_2$ is of the order of $R_\lambda^{3/4}$, say, and $x_2$ of the order of $R_{\lambda 0}^{-3/4}$. The numbers $x_2$ and $X_2$ are parameters depending only on $R_{\lambda 0}$ but invariant with time. Similarly, one may choose a frequency $\kappa_1$ in the linear range of the spectrum satisfying the conditions

$$X_1 = \kappa_1 L \ll 1, \qquad X^* = \kappa_1 L^* \gg 1 \tag{4.2}$$

The magnitude of a quantity here is, however, determined by the time history rather than the Reynolds number.

Our present interest centers around the relative contribution to energy and dissipation in the high-frequency range as compared with those in all the other parts of the spectrum. Thus, the exact nature of the spectrum around the linear range is not too important; and for convenience, we shall approximate the spectrum by

$$\begin{aligned} \mathfrak{F}(\kappa,t) &= J_0\kappa^4 && \text{for } \kappa < \kappa_1 \\ &= V^2 L F(\kappa L) && \text{for } \kappa_1 < \kappa < \kappa_2 \\ &= v^2 \eta \phi(\kappa\eta) && \text{for } \kappa_2 < \kappa \end{aligned} \tag{4.3}$$

where $\kappa_1$ may be taken to be the intersection of $\mathfrak{F} = J_0\kappa^4$ with the linear law constant $D_0^2\kappa$. The function $F$ and $\phi$ outside the range specified above will be defined by their limiting behavior. Thus, $F(X)$ is proportional to $X$ for small $X$, and proportional to $X^{-5/3}$ for large $X$; and $\phi(x)$ is proportional to $x^{-5/3}$ for small $x$. The energy and its rate of dissipation are respectively given by

$$u'^2 = V^2 \int_0^{X_2} F(X)dX + v^2 \int_{x_2}^{\infty} \phi(x)dx - \frac{3}{10} J_0 \kappa_1^5 \qquad (4.4)$$

$$\frac{\epsilon}{2\nu} = \frac{V^2}{L^2} \int_0^{X_2} F(X)X^2dX + \frac{v^2}{\eta^2} \int_{x_2}^{\infty} \phi(x)x^2dx - \frac{3}{28} J_0 \kappa_1^7$$

for the spectrum (4.3).

We want to show that, as expected, in the expression for the energy $u'^2$, the part

$$u_l'^2 = V^2 \int_0^{X_2} F(X)dX \qquad (4.5)$$

related to the low-frequency components, is much larger than the part

$$u_h'^2 = v^2 \int_{x_2}^{\infty} \phi(x)dx \qquad (4.6)$$

related to the high-frequency components. On the other hand, in the expression for the dissipation $\epsilon/2\nu$, the part

$$\frac{\epsilon_l}{2\nu} = \frac{V^2}{L^2} \int_0^{X_2} F(X)X^2dX \qquad (4.7)$$

related to the low-frequency components, is much smaller than the part

$$\frac{\epsilon_h}{2\nu} = \frac{v^2}{\eta^2} \int_{x_2}^{\infty} \phi(x)x^2dx \qquad (4.8)$$

related to the high-frequency components. In fact, we shall see that

$$\begin{aligned} u_l'^2 &= u_h'^2 O(X_2^{2/3}) \\ \epsilon_l &= \epsilon_h O(x_2^{4/3}) \end{aligned} \qquad (4.9)$$

It is easily seen from direct calculation that

$$u_l'^2 = V^2 \left[ \int_0^{\infty} F(X)dX + O(X_2^{-2/3}) \right] \qquad (4.10)$$

and

$$u_h'^2 = v^2 \left\{ Cx_2^{-2/3} + \int_{x_2}^{\infty} [\phi(x) - \tfrac{2}{3}Cx^{-5/3}]dx \right\}$$

where we have stipulated that $\phi(x) \sim \frac{2}{3}Cx^{-5/3}$ for small $x$. The second term may consequently be approximated by the constant value

$$\int_0^{\infty} [\phi(x) - \tfrac{2}{3}Cx^{-5/3}]dx$$

for small values of $x_2$, and is therefore much smaller compared with $Cx_2^{-2/3}$. Now from Table I, we see that

$$\frac{v}{V} = \left(\frac{\eta}{L}\right)^{1/3} = \left(\frac{x}{X}\right)^{1/3} \qquad (4.11)$$

Hence, the above expression for $u_h'^2$ may be written

$$u_h'^2 = CV^2X_2^{-2/3}[1 + O(x_2^{2/3})] \tag{4.12}$$

Equations (4.10) and (4.12) give the first relation of (4.9) desired. Indeed, by making use of (4.10) and (4.12), the energy expression (4.4) may be written as

$$u'^2 = (1 - \delta_1)V^2 \int_0^\infty F(X)dX - \tfrac{3}{10}J_0\kappa_1^5, \quad \delta_1 = O(X_2^{-2/3}) \tag{4.13}$$

Similar arguments lead to the second relation in (4.9) and indeed give

$$\frac{\epsilon}{2\nu} = (1 - \delta_2)\frac{v^2}{\eta^2}\int_0^\infty \phi(x)x^2dx - \frac{3}{28}J_0x_1^7, \quad \delta_2 = O(x_2^{2/3}) \tag{4.14}$$

Since $V^2 \sim t^{-1}$ and $v^2/\eta^2 \sim t^{-2}$, the additive constants in (4.13) and (4.14) are not important. For small but finite values of $t$, one should retain the additive constant in $u'^2$ but reject it in $\epsilon/2\nu$, because it is only part of the low frequency contribution, which is wholly negligible in the case of dissipation. Thus (4.12) and (4.14) agree with (2.21) if

$$(1 - \delta_1)\cdot\int_0^\infty F(X)dX = 1, \qquad (1 - \delta_2)\cdot\int_0^\infty \phi(x)x^2dx = \tfrac{1}{2}$$

and

$$u_D^2 = \tfrac{3}{10}J_0\kappa_1^5$$

Since $J_0\kappa_1^4$ = constant $D_0^2\kappa_1$, it can be easily seen that (cf. Table I)

$$u_D^2 = \text{constant } V^{*2} \tag{4.15}$$

From this, it follows that the estimation of the initial period made in Sections II and III are identical.

## V. Conclusion

The proposed simple physical picture described in Section III explains all experimental data so far available. However, experimental data are yet lacking regarding a transition from the law $\lambda^2 = 10\nu t$ to the law $\lambda^2 = 7\nu t$ during the early period. The experiments must be performed at very high Reynolds numbers over an extended period of decay.

More careful experimental examination of the variation of $R_\lambda$ is also desired to see whether the law asserting its linear decrease in the early period is confirmed in all cases. According to the discussion of the junior author (7), the quantity $G = f^{iv}(0)\lambda^4$ should also decrease linearly in time, but the ratio $G/R_\lambda$ should be constant in any given experiment. It varies with the initial Reynolds number according to the relation

$$\frac{G}{R_\lambda} = \frac{S}{2} + \frac{30}{7R_{\lambda 0}}$$

where $S = k'''(0)\lambda^3$ should be an absolute constant. The above formula has been verified by Batchelor and Townsend over a very limited period during which the variation of $R_\lambda$ cannot be definitely established because of the experimental scatter.

The best experimental check would be a direct examination of the similarity hypotheses by measurements of the spectrum and the correlation functions. In particular, it has been asserted (Heisenberg) that the appearance of a linear part of the spectrum is associated with the early period but not related to the Reynolds number of turbulence, while the appearance of the five-thirds power range depends essentially on the instantaneous value of Reynolds number.

It must be kept in mind that all the above discussions hold only for an infinite field of turbulence. If the scale becomes comparable with that of the experimental apparatus, e.g., the cross-sectional dimension of the wind tunnel, one would be determining the influence of the wind tunnel instead of the natural properties of the turbulent field.

### *References*


1. von Kármán, Th., *J. Aeronaut. Sci.*, **4,** 131–138 (1937).
2. von Kármán, Th., and Howarth, L., *Proc. Roy. Soc.* (*London*), **A164,** 192–215 (1938).
3. Heisenberg, W., *Proc. Roy. Soc.* (*London*), **A195,** 402–406 (1948).
4. Batchelor, G. K., "Recent developments on turbulence research," General lecture at the Seventh International Congress for Applied Mechanics, 1948.
5. Frenkiel, F. N., "Comparison between some theoretical and experimental results of the decay of turbulence," presented at the Seventh International Congress for Applied Mechanics, 1948.
6. von Kármán, Th., *Compt. rend.*, **226,** 2108–2110 (1948); *Proc. Nat. Acad. Sci.*, **34,** 531–539 (1948).
7. Lin, C. C., *Proc. Nat. Acad. Sci.*, **34,** 540–543 (1948); also "On the law of decay and the spectrum of isotropic turbulence," presented at the Seventh International Congress for Applied Mechanics, 1948.
8. Kampé de Fériet, J., *Compt. rend.*, **227,** 760–761 (1948).
9. Lin, C. C., "Remarks on the spectrum of turbulence," presented at the First Symposium of Applied Mathematics, AMS, 1947.
10. Batchelor, G. K., *Proc. Roy. Soc.* (*London*), **A195,** 513–532 (1949).
11. Onsager, L., *Phys. Rev.*, **68,** 286 (1945) (abstract only).
12. von Weizsäcker, C. F., *Z. Physik*, **124,** 614–627 (1948).
13. Loitsiansky, L. G., C.A.H.I. (Moscow), *Rept.* **440,** 1939; also N.A.C.A. *Tech.* Memo. **1079.**
14. Dryden, H. L., *Quart. Applied Math.*, **1,** 7–42 (1943).
15. Millionshchikov, M. D., *Compt. rend. acad. sci. U.R.S.S.*, **22,** 231–237 (1939).
16. Batchelor, G. K., and Townsend, A. A., *Proc. Roy. Soc.* (*London*), **194,** 527–543 (1948).
17. Kolmogoroff, A. N., *Compt. rend. acad. sci. U.R.S.S.*, **31,** 538–540 (1941).

*Reprinted from*
QUARTERLY OF APPLIED MATHEMATICS
Vol. X, No. 4, January, 1953

QUARTERLY OF APPLIED MATHEMATICS

Vol. X January, 1953 No. 4

# ON TAYLOR'S HYPOTHESIS AND THE ACCELERATION TERMS IN THE NAVIER-STOKES EQUATIONS*

BY
C. C. LIN**
*Massachusetts Institute of Technology*

**1. Introduction.** For turbulence produced behind the grid in a wind tunnel, Taylor[1] introduced the assumption that the spatial pattern of turbulent motion is carried past a fixed point by the mean wind speed without any essential change. The assumption has been applied to connect the time spectrum with space correlation, and Taylor found indeed good agreement between theory and experiment.

The success of Taylor's hypothesis in the case of wind tunnel turbulence has led people to use it also for shear flow. This is obviously a tentative step until justification can be found. In wind-tunnel turbulence, the justification of this hypothesis is essentially based upon the following two conditions:

(a) the main flow is uniform,

(b) the level of turbulence is low.

One purpose of this paper is to demonstrate this justification explicitly as a basis for a similar discussion for shear flow, in which case neither of these conditions are as well satisfied. It seems plausible that an eddy of considerable size would change considerably by the shearing motion while it is carried past a given point by the main flow. In §6, estimates are made for the limitation, due to this cause, of the eddy size in the boundary layer, to which Taylor's hypothesis can be applied; and it appears that the major part of the shear contributing fluctuations are of such low time frequencies that Taylor's hypothesis cannot be immediately justified for its use in the determination of their space scale.

In preparation for the analysis of Taylor's hypothesis, it is necessary to give an estimate of the orders of magnitude of the various acceleration terms in the equations of motion. Such an estimate also gives us some insight into the nature of turbulence. The analysis is carried out in §§2–5. One main point is to estimate the acceleration due to pressure gradient. This is first done in §2 in a general manner without using Heisenberg's specific hypothesis[2] of the independence of the various harmonic components of velocity fluctuations. In §§3–4, general analyses are made of the correlation of the various acceleration terms. All the correlations are expressed explicitly in terms of double and triple

Received February 27, 1952.

*This work has been sponsored by the Office of Naval Research and was carried out under project number NR-044-003, entitled "Numerical Analysis and Theoretical Mechanics" at the U. S. Naval Ordnance Laboratory.

**Consultant, U. S. Naval Ordnance Laboratory.

velocity correlations and certain two-point quadruple correlations. In §5, Heisenberg's hypothesis is used in the equivalent form given by Batchelor[3] for correlation functions, and his results for pressure correlations are reproduced by a slightly different derivation. In addition, we have calculated the magnitudes of the inertial acceleration and the instantaneous local acceleration. At large Reynolds numbers, they are the dominant terms in the equations of motion, the acceleration due to pressure forces being negligible in comparison. This suggests that, in a sense, the fluid particles are moving with very little interaction among them. Speculations based on this notion should be made cautiously.

**2. Pressure fluctuations.** The fluctuations of pressure are related to those of velocity through the equations of motion. By making use of the hypothesis of independence of the harmonic components of the velocity fluctuations, Heisenberg obtained certain relations, in the case of isotropic turbulence, connecting the spectrum of the pressure fluctuations with the spectrum for kinetic energy. An equivalent development using correlation functions was later given by Batchelor. In this section, we shall develop two relations regarding pressure fluctuations without making specific assumptions.

The fluctuations of velocity $u_i$ and pressure $p$ satisfy the Navier-Stokes equations

$$\frac{\partial u_i}{\partial t} + u_j \frac{\partial u_i}{\partial x_j} = -\frac{1}{\rho}\frac{\partial p}{\partial x_i} + \nu \, \Delta u_i \, , \tag{2.1}$$

and the equation of continuity

$$\frac{\partial u_k}{\partial x_k} = 0, \tag{2.2}$$

where $\rho$ is the density of the fluid, $\nu$ is the kinematic viscosity coefficient, and $x_i$'s are the coordinates of a point $P$. By multiplying (2.1) with the pressure fluctuation $p'$ at another point $p'$, and taking averages, we obtain

$$\left\langle p'u_j \frac{\partial u_i}{\partial x_j}\right\rangle = -\frac{1}{\rho}\left\langle p' \frac{\partial p}{\partial x_i}\right\rangle; \tag{2.3}$$

where we have made use of the relations

$$\left\langle p' \frac{\partial u_i}{\partial t}\right\rangle = 0, \qquad \text{and} \qquad \langle p' \, \Delta u_i \rangle = 0 \tag{2.4}$$

since they are solenoidal tensors of the first rank[4]. From (2.3), we see that

$$\left\langle -\frac{1}{\rho}\frac{\partial p'}{\partial x_i'} u_j \frac{\partial u_i}{\partial x_j}\right\rangle = \left\langle \left(-\frac{1}{\rho}\frac{\partial p'}{\partial x_i'}\right)\left(-\frac{1}{\rho}\frac{\partial p}{\partial x_i}\right)\right\rangle \tag{2.5}$$

(either with or without summation with respect to the index $i$). We shall introduce the notation

$$\alpha_i = u_j \frac{\partial u_i}{\partial x_j}, \qquad \pi_i = -\frac{1}{\rho}\frac{\partial p}{\partial x_i}.$$

Then, (2.5) can be written as

$$\langle \pi_i \alpha_i' \rangle = \langle \pi_i \pi_i' \rangle, \tag{2.6}$$

which yields

$$\langle \pi_i \alpha_i \rangle = \langle \pi_i \pi_i \rangle. \tag{2.7}$$

Thus, the correlation of inertial acceleration and the acceleration due to pressure gradient is equal to the mean square of the latter. By Schwarz inequality,

$$\langle \pi_i \alpha_i \rangle^2 \leq \langle \pi_i^2 \rangle \cdot \langle \alpha_i^2 \rangle$$

we have then

$$\langle | \pi_i |^2 \rangle \leq \langle | \alpha_i |^2 \rangle, \tag{2.8}$$

giving an upper bound for the pressure gradient in terms of velocity correlations. It may be expected that

$$\langle | \alpha_i |^2 \rangle \sim \langle u^2 \rangle^2 / \lambda^2$$

so that $\langle | \pi_i |^2 \rangle$ cannot be larger than the order of $\langle (u^2) \rangle^2 / \lambda^2$. Thus, all the terms in (1.1) are limited to this order of magnitude for reasonable large Reynolds numbers, since, in that case, $\langle (\nu \Delta u_i)^2 \rangle$ is definitely of a smaller order of magnitude.

For very large Reynolds numbers, one may get the order of magnitude of $\langle | \pi_i |^2 \rangle$ by extending Kormogoroff's concept to include pressure fluctuations. If we assume that $1/\rho^2 \langle (p - p')^2 \rangle$ depends, for small values of $r$, only on the velocity scale $v = (\nu\epsilon)^{1/4}$ and the length scale $\eta = (\nu^3/\epsilon)^{1/4}$, where $\epsilon$ is the rate of energy dissipation, then

$$\frac{1}{\rho^2} \langle (p - p')^2 \rangle = v^4 F(r/\eta), \tag{2.9}$$

where $F$ is a universal function of its argument. From this, it follows that

$$\frac{1}{\rho^2} \left\langle \left( \frac{\partial p}{\partial x} \right)^2 \right\rangle = \frac{v^4}{\eta^2} F''(0)$$

or

$$\frac{1}{\rho^2} \left\langle \left( \frac{\partial p}{\partial x} \right)^2 \right\rangle = \frac{\langle u^2 \rangle^2}{\lambda^2} \cdot \frac{\text{const.}}{R_\lambda}, \qquad R_\lambda = \frac{\sqrt{\langle u^2 \rangle}\, \lambda}{\nu}. \tag{2.10}$$

This is in conformity with the result obtained above. Equation (2.10) has been previously obtained by Heisenberg with a specific value for the constant, but the derivation depends on specific assumptions. Batchelor used an equivalent set of assumptions, but he also pointed out that these assumptions may be questionable for small distances. Since the evaluation of the magnitude of the pressure gradient definitely depends on pressure correlations at small distances, it seems desirable that an independent argument be given.

In the following, we shall examine all the acceleration terms in the equations of motion.

**3. Some remarks on isotropic tensors of rank two.** We shall first develop a few general relations for isotropic tensors of the second rank, which will be useful for following developments. Robertson[4] has shown that such a tensor can always be represented by

$$R_{ik} = R_1(r) \frac{\xi_i \xi_k}{r^2} + R_2(r) \delta_{ik}, \tag{3.1}$$

where $(\xi_1, \xi_2, \xi_3)$ are the coordinates, and

$$r^2 = \xi_1^2 + \xi_2^2 + \xi_3^2. \tag{3.2}$$

Thus, there are two scalar functions required for the definition of such a tensor. He has also shown that, for a tensor of vanishing divergence (a solenoidal tensor), there is only one defining scalar function. These results are generalizations of those for velocity correlations given by von Kármán[5]. The general form of such a tensor is

$$Q_{ik} = -\frac{1}{2r}\frac{\partial F}{\partial r}\xi_i\xi_k + \left(F + \frac{r}{2}\frac{\partial F}{\partial r}\right)\delta_{ik}\,. \tag{3.3}$$

There is another kind of tensor defined by only one scalar function. This is constructed by starting with a single scalar function $P(r)$ and perform the gradient operation twice; thus,

$$P_{ik} = \frac{\partial^2 P(r)}{\partial\xi_i\,\partial\xi_k}. \tag{3.4}$$

Such a tensor shall be called a *gradient* tensor. By carrying out the calculations indicated in (3.4), we find

$$P_{ik} = \frac{1}{r}\frac{\partial G}{\partial r}\xi_i\xi_k + G\delta_{ik}\,, \qquad G = P'(r)/r. \tag{3.5}$$

By comparing (3.3) and (3.5), it is clear that a solenoidal tensor is a gradient tensor if both the conditions

$$\frac{\partial G}{\partial r} = -\frac{1}{2}\frac{\partial F}{\partial r}, \qquad G = F + \frac{r}{2}\frac{\partial F}{\partial r}$$

are satisfied. Then

$$F + r\frac{\partial F}{\partial r} = \text{const.},$$

and hence

$$F = C_1 + \frac{C_2}{r}. \tag{3.6}$$

If we are restricted to tensors finite everywhere, the only tensors which are both a solenoidal tensor and a gradient tensor are constant multiples of the Kronecker delta $\delta_{ik}$.

From the number of scalar functions involved, it seems reasonable to expect that *a general correlation tensor of rank two can be expressed as the sum of one gradient part and one solenoidal part.* We shall now show that this can be done in a *unique* manner, if all the tensors vanish for $r \to \infty$.

First, if there were two such decompositions, then

$$R_{ik} = P_{ik}^{(1)} + Q_{ik}^{(1)} = P_{ik}^{(2)} + Q_{ik}^{(2)},$$

and

$$\Delta_{ik} = P_{ik}^{(1)} - P_{ik}^{(2)} = Q_{ik}^{(2)} - Q_{ik}^{(1)}$$

is both a gradient tensor and a solenoidal tensor, since linear combinations of such tensors obviously retain the original properties (*cf.*(3.3) and (3.5)). Thus, $\Delta_{ik}$ can only be a multiple of $\delta_{ik}$. However, by letting $r \to \infty$, it is clear that the constant multiplier must be zero. Hence, the uniqueness of our decomposition is proved.

The decomposition is effected by setting

$$\left.\begin{aligned} R_1 &= r\frac{\partial}{\partial r}\left(G - \frac{F}{2}\right), \\ R_2 &= G + F + \frac{r}{2}\frac{\partial F}{\partial r}, \end{aligned}\right\} \tag{3.7}$$

which yields

$$\left.\begin{aligned} F &= -\frac{2}{3}\left\{\frac{1}{r^3}\int_r^\infty r^2(R_1 + 3R_2)\,dr - \int_r^\infty \frac{R_1}{r}\,dr\right\}, \\ G &= -\frac{1}{3}\left\{\frac{1}{r^3}\int_r^\infty r^2(R_1 + 3R_2)\,dr + 2\int_r^\infty \frac{R_1}{r}\,dr\right\}. \end{aligned}\right\} \tag{3.8}$$

For convenience of reference, some of the properties of gradient and solenoidal tensors are summarized in the following table.

| | Gradient tensor $P_{ik}$ | Solenoidal tensor $Q_{ik}$ |
|---|---|---|
| General form | $P_{ik} = \frac{1}{r}\frac{\partial G}{\partial r}\xi_i\xi_k + G\delta_{ik}$<br>$(P' = rG)$ | $Q_{ik} = -\frac{1}{2r}\frac{\partial F}{\partial r}\xi_i\xi_k$<br>$+ \left(F + \frac{r}{2}\frac{\partial F}{\partial r}\right)\delta_{ik}$ |
| Defining scalar<br>($R_{11}$ for $\xi_i = (r, 0, 0)$) | $P''(r) = G + r\frac{\partial G}{\partial r}$ | $F(r)$ |
| Laplacian<br>$\Delta R_{ik}$ | $P \to \Delta P$ | $F \to \frac{\partial^2 F}{\partial r^2} + \frac{4}{r}\frac{\partial F}{\partial r}$ |
| Trace | $\Delta P = 3G + r\frac{\partial G}{\partial r}$ | $3F + r\frac{\partial F}{\partial r}$ |

If $P_{ik}$ and $Q_{ik}$ are the components of the general tensor $R_{ik}$, then $F$ and $G$ are given by (3.8) and the traces of $P_{ik}$ and $Q_{ik}$ are given by

$$\left.\begin{aligned} \Delta P &= (R_1 + R_2) - 2\int_r^\infty \frac{R_1}{r}\,dr, \\ \frac{1}{r^2}\frac{\partial}{\partial r}(r^3 F) &= +2R_2 + 2\int_r^\infty \frac{R_1}{r}\,dr. \end{aligned}\right\} \tag{3.9}$$

**4. Correlation of the acceleration terms in the equations of motion.** The equations of motion

$$\frac{\partial u_i}{\partial t} + u_i \frac{\partial u_i}{\partial x_j} = -\frac{1}{\rho}\frac{\partial p}{\partial x_i} + \nu\,\Delta u_i \tag{4.1}$$

contain the four types of terms

$$a_i = \frac{\partial u_i}{\partial t}, \tag{4.2}$$

$$\alpha_i = u_j \frac{\partial u_i}{\partial x_j}, \tag{4.3}$$

$$\pi_i = -\frac{1}{\rho}\frac{\partial p}{\partial x_i}, \tag{4.4}$$

$$\mu_i = \nu\,\Delta u_i\,. \tag{4.5}$$

Among them, sixteen two-point correlations can be formed. By symmetry, only ten of them are distinct. Two of them, namely

$$\langle \pi_i a_k' \rangle \qquad \text{and} \qquad \langle \pi_i \mu_k' \rangle,$$

are identically zero, since they can be expressed in terms of the solenoidal tensors $\langle p a_k' \rangle$ and $\langle p u_k' \rangle$ which are identically zero themselves. We are, therefore, left with eight independent tensors.

By forming correlations of (4.1) at $P(x_i)$ with each of the four vectors (4.2) – (4.5) at the other point $P'(x_i)$, we obtain four relations among the eight tensors, thus reducing the number of independent tensors to four. For systematic analysis, we may choose these to be $\langle a_i a_k' \rangle$, $\langle \alpha_i \alpha_k' \rangle$, $\langle \pi_i \pi_k' \rangle$ and $\langle \mu_i \mu_k' \rangle$. The other tensors are then expressible in terms of these as follows:

$$\langle a_i \alpha_k' \rangle = \langle \alpha_i a_k' \rangle = \tfrac{1}{2}\{-\langle a_i a_k' \rangle + \langle \mu_i \mu_k' \rangle - V_{ik}\}, \tag{4.6}$$

$$\langle a_i \pi_k' \rangle = \langle \pi_i a_k' \rangle = 0, \tag{4.7}$$

$$\langle a_i \mu_k' \rangle = \langle \mu_i a_k' \rangle = \tfrac{1}{2}\{\langle a_i a_k' \rangle + \langle \mu_i \mu_k' \rangle - V_{ik}\}, \tag{4.8}$$

$$\langle \alpha_i \pi_k' \rangle = \langle \pi_i \alpha_k' \rangle = \langle \pi_i \pi_k' \rangle, \tag{4.9}$$

$$\langle \alpha_i \mu_k' \rangle = \langle \mu_i \alpha_k' \rangle = \tfrac{1}{2}\{-\langle a_i a_k' \rangle + \langle \mu_i \mu_k' \rangle + V_{ik}\}, \tag{4.10}$$

$$\langle \pi_i \mu_k' \rangle = \langle \mu_i \pi_k' \rangle = 0, \tag{4.11}$$

where $V_{ik}$ is given by

$$\langle \alpha_i \alpha_k' \rangle = \langle \pi_i \pi_k' \rangle + V_{ik}\,. \tag{4.12}$$

We notice that, from (4.6), $V_{ik}$ is a tensor of vanishing divergence. On the other hand $\langle \pi_i \pi_k' \rangle$ is obviously a gradient tensor. Hence (4.12) is the decomposition of $\langle \alpha_i \alpha_k' \rangle$ into its component parts, and is, therefore, equivalent to *two* relations. These are given by (3.8) of the last section, with the correspondence given by the following relations ($\xi_i = x_i' - x_i$):

$$\left.\begin{aligned} P(r) &= \langle p p' \rangle, \\ P_{ik} &= \rho^2 \langle \pi_i \pi_k' \rangle, \\ Q_{ik} &= \rho^2 V_{ik}\,, \\ R_{ik} &= \rho^2 \langle \alpha_i \alpha_k' \rangle. \end{aligned}\right\} \tag{4.13}$$

Thus, $\langle \pi_i \pi_k' \rangle$ and $V_{ik}$ are both expressible in terms of $\langle \alpha_i \alpha_k' \rangle$. The independent tensors are then $\langle a_i a_k' \rangle$, $\langle \alpha_i \alpha_k' \rangle$ and $\langle \mu_i \mu_k' \rangle$.

*Expression in terms of velocity correlations*

We shall now show that all the above correlations can be expressed in terms of velocity correlations. First, let us note that,

$$\langle \mu_i \mu_k' \rangle = \nu^2 \Delta\Delta \langle u_i u_k' \rangle, \tag{4.14}$$

$$\langle \alpha_i \mu_k' \rangle = -\nu\Delta \frac{\partial}{\partial \xi_i} \langle u_i u_j u_k' \rangle, \tag{4.15}$$

$$\langle \alpha_i \alpha_k' \rangle = -\frac{\partial^2}{\partial \xi_j \, \partial \xi_l} \langle u_i u_j u_k' u_l' \rangle. \tag{4.16}$$

As mentioned above, (4.12) enables us to express both $\langle \pi_i \pi_k' \rangle$ and $V_{ik}$ in terms of $\langle \alpha_i \alpha_k' \rangle$. Furthermore, (4.10) and (4.12) give

$$\langle a_i a_k' \rangle = \langle \mu_i \mu_k' \rangle - 2\langle \alpha_i \mu_k' \rangle + \langle \alpha_i \alpha_k' \rangle - \langle \pi_i \pi_k' \rangle \tag{4.17}$$

allowing $\langle a_i a_k' \rangle$ to be expressible in terms of correlations of velocity components at a given instant. The other equations in (4.6) – (4.10) then give all the other correlations. Thus, there are altogether *four* scalar functions involved in expressing all the correlations of accelerations; two of them are the usual double and triple velocity correlations introduced by (4.14) and (4.15), and the other two relate to quadruple velocity correlations, *cf.* (4.16).

*Magnitude of various acceleration terms.*

By contraction and putting $\xi_j = 0$, into (4.9), (4.14), (4.16) and (4.17), we may obtain the various quantities entering into the consideration of the magnitude of the acceleration terms in (3.1). We obtain

$$\langle | \pi_k |^2 \rangle = \langle \alpha_k \pi_k \rangle \leq \langle | \alpha_k |^2 \rangle, \tag{4.18}$$

$$\langle | \mu_k |^2 \rangle = 35\nu^2 \langle u^2 \rangle f_0^{iv}, \tag{4.19}$$

$$\langle \alpha_k \mu_k \rangle = +35\nu \langle u^2 \rangle^{3/2} h_0''', \tag{4.20}$$

$$\langle | \alpha_k |^2 \rangle = -\left[ \frac{\partial^2}{\partial \xi_j \, \partial \xi_l} \langle u_i u_j u_i' u_l' \rangle \right]_{\xi=0}, \tag{4.21}$$

$$\langle | a_k |^2 \rangle = 35\nu^2 \langle u^2 \rangle^2 f_0^{iv} - 70\nu \langle u^2 \rangle^{3/2} h_0''' + \langle | \alpha_k |^2 \rangle - \langle | \pi_k |^2 \rangle, \tag{4.22}$$

where $f(r)$ and $h(r)$ are the double and triple correlation functions introduced by von Kármán and Howarth, and the subscript zero denotes evaluation at $r = 0$. The second relation in (4.18) is obtained by Schwarz's inequality, as noted in §1.

It is quite clear from (4.22) that for large $R_\lambda = u'\lambda/\nu$, the mean square value $\langle | a_i |^2 \rangle$ of the magnitude of $a_k$ is at most of the order of $\langle | \alpha_k |^2 \rangle$, which may be estimated to be of the order of $\langle u^2 \rangle^2 / \lambda^2$. To make a closer analysis, one has to introduce suitable approximations for $\langle | \alpha_k |^2 \rangle$ and $\langle | \pi_k |^2 \rangle$. This will be carried out in the next section. It is convenient to put the above relations in dimensionless forms. For example, (4.22) may be written as

$$\frac{\langle | a_k |^2 \rangle}{\langle u^2 \rangle^2 / \lambda^2} = \frac{35}{R_\lambda^2} (G - R_\lambda S) + \frac{\langle | \alpha_k |^2 \rangle}{\langle u^2 \rangle^2 / \lambda^2} - \frac{\langle | \pi_k |^2 \rangle}{\langle u^2 \rangle^2 / \lambda^2}, \tag{4.23}$$

where

$$G = f_0^{iv}\lambda^4, \qquad S = 2h_0'''\lambda^3, \tag{4.24}$$

and (4.19) may be written as

$$\frac{\langle | \mu_k |^2\rangle}{\langle u^2\rangle^2/\lambda^2} = \frac{35G}{R_\lambda^2}. \tag{4.25}$$

**5. Hypothesis for reducing the quadruple correlations.** The above relations are all exact. To get more useful results, we shall make use of the assumption[3]

$$\langle u_i u_j u_k' u_l'\rangle = \langle u_i u_k'\rangle\langle u_j u_l'\rangle + \langle u_i u_l'\rangle\cdot\langle u_j u_k'\rangle + \langle u_i u_j\rangle\langle u_k u_l'\rangle. \tag{5.1}$$

Then $\langle \alpha_i \alpha_k'\rangle$ can be expressed in terms of a single double correlation function. The total number of scalars involved would reduce to two, say $f(r)$ and $h(r)$. In fact, calculations* show that

$$\langle u^2\rangle^{-2}\langle \alpha_i \alpha_k'\rangle = (4sff''' + 14ff'' - 16f'^2)(\xi_i\xi_k - s\delta_{ik}) - (4sff'' + 14sf'^2 + 10ff')\delta_{ik} \tag{5.2}$$

where

$$s = r^2, \qquad f' = \frac{df}{ds}, \qquad \text{etc.} \tag{5.3}$$

Then

$$\langle | \alpha_k |^2\rangle = 15\langle u^2\rangle^2/\lambda^2. \tag{5.4}$$

By (3.9), we may obtain $\Delta P$ from (5.1). Thus,

$$(\rho\langle u^2\rangle)^{-2}\,\Delta P(r) = 24\int_\infty^s f'^2(s)\,ds + 16sf'(s), \tag{5.5}$$

which gives

$$\langle | \nabla p |^2\rangle = 24(\rho\langle u^2\rangle)^2\int_0^\infty f'^2(s)\,ds = 12(\rho u^2)^2\int_0^\infty \left(\frac{df}{dy}\right)^2\frac{dy}{y}. \tag{5.6}$$

This formula for the pressure gradient was first obtained by Batchelor.

The magnitude of the other acceleration terms can then be easily obtained. We list below all these relations in dimensionless form:

$$\frac{\left\langle \left| u_j \dfrac{\partial u_i}{\partial x_j}\right|^2\right\rangle}{\langle u^2\rangle^2/\lambda^2} = 15, \tag{5.7}$$

$$\frac{\langle | \nabla p |^2\rangle}{\langle \rho u^2\rangle^2/\lambda^2} = 3\left(\frac{\lambda}{\lambda_p}\right)^2 = 12\lambda^2\int_0^\infty\left(\frac{df}{dy}\right)^2\frac{dy}{y}, \tag{5.8}$$

$$\frac{\left\langle \left| \dfrac{\partial u_i}{\partial t}\right|^2\right\rangle}{\langle u^2\rangle^2/\lambda^2} = \frac{35}{R_\lambda^2}(G - R_\lambda S) + 15 - 12\lambda^2\int_0^\infty\left(\frac{df}{dy}\right)^2\frac{dy}{y}. \tag{5.9}$$

$$\langle | A_i |^2\rangle/\{\langle u^2\rangle^2/\lambda^2\} = 12\lambda^2\int_0^\infty\left(\frac{df}{dy}\right)^2\frac{dy}{y} + \frac{35}{R_\lambda^2}(G - R_\lambda S) \tag{5.10}$$

---

*It has been found convenient to introduce $s = r^2$ as the variable in the process of calculation.

where $A_i$ is the total acceleration

$$A_i = \frac{\partial u_i}{\partial t} + u_j \frac{\partial u_i}{\partial x_j}. \tag{5.11}$$

The symbol $\lambda_p$ is defined by (5.8) and has been estimated by Batchelor for various Reynolds numbers. In particular, for very large Reynolds numbers, he found

$$\lambda_p/\lambda = 0.11R_\lambda^{1/2}. \tag{5.12}$$

This is in agreement with the general discussion given in §1, and shows that the pressure gradient has a magnitude much smaller than the inertial acceleration.

**6. Taylor's hypothesis.** The above results will now be used to examine Taylor's hypothesis. Consider first the case of a hot wire used to measure wind tunnel turbulence behind a grid. To the extent that the field may be regarded as homogeneous and isotropic, we may consider an observer moving with the mean wind speed $U$ and reduce the problem to that of a registering instrument carried at a velocity $-U$ through a field of homogeneous isotropic turbulence at rest. Then, the instrument registers $u(x + Ut, t)$, and

$$\frac{du}{dt} = \frac{\partial u}{\partial t} + U \frac{\partial u}{\partial x}$$

is the time rate of change as observed by the instrument. Taylor's hypothesis that the instrument registers the space variation is essentially based on the argument that

$$\left\langle \left(\frac{\partial u}{\partial t}\right)^2 \right\rangle \ll U^2 \left\langle \left(\frac{\partial u}{\partial x}\right)^2 \right\rangle, \tag{6.2}$$

It is seen from the above analysis that this is indeed the case provided

$$\frac{\langle u^2 \rangle}{U^2} \ll 1.$$

To be more precise, (5.9) gives for large Reynolds numbers,

$$\left\langle \left(\frac{\partial u}{\partial t}\right)^2 \right\rangle : U^2 \left\langle \left(\frac{\partial u}{\partial x}\right)^2 \right\rangle = 5 \frac{\langle u^2 \rangle}{U^2} \tag{6.3}$$

This may then serve as a guide for deciding on the accuracy of Taylor's hypothesis. Note that the number 5 is obtained by using the specific hypothesis in §5, but it should be correct in its order of magnitude.

Similar remarks apply to derivatives of higher orders.

Another way of looking at Taylor's hypothesis is to say that the main terms in the equations of motion

$$\frac{\partial u_i}{\partial t} + U \frac{\partial u_i}{\partial x} + u_j \frac{\partial u_i}{\partial x_j} = -\frac{1}{\rho} \frac{\partial p}{\partial x_i} + \nu \, \Delta u_i \tag{6.4}$$

are the first two, the others being much smaller in comparison. This is seen from the rough argument that the pressure terms are of the same order or smaller than the convective terms $u_j \partial u_i / \partial x_j$ as justified by the previous analysis. Indeed, the pressure terms are often much smaller than indicated by such a crude argument.

For the case of shear flow, it is difficult to make an exact analysis, and one may first proceed with the kind of rough arguments just given, with its plausibility supported by its correctness in the case of uniform flow. However, the pressure fluctuations are not only related to the quadratic terms in the fluctuations but also to the convection of momentum of the mean motion due to turbulent fluctuations. Thus, if $U(y)$ is the basic parallel flow, in the equations of motion

$$\left.\begin{aligned} \frac{\partial u}{\partial t} + U\frac{\partial u}{\partial x} + v\frac{dU}{dy} + u_j\frac{\partial u}{\partial x_j} &= -\frac{1}{\rho}\frac{\partial p}{\partial x} + \nu\Delta u, \\ \frac{\partial v}{\partial t} + U\frac{\partial v}{\partial x} \qquad\quad + u_j\frac{\partial v}{\partial x_j} &= -\frac{1}{\rho}\frac{\partial p}{\partial y} + \nu\Delta v, \\ \frac{\partial w}{\partial t} + U\frac{\partial w}{\partial x} \qquad\quad + u_j\frac{\partial w}{\partial x_j} &= -\frac{1}{\rho}\frac{\partial p}{\partial z} + \nu\Delta w, \end{aligned}\right\} \tag{6.5}$$

there is the term $v\, dU/dy$. If Taylor's hypothesis were true. Then $u = f(x - Ut)$, $v = g(x - Ut)$, $w = h(x - Ut)$, and we have approximately, for large $R_\lambda$ and small turbulence levels,

$$\left.\begin{aligned} \frac{\partial p}{\partial x} &= -\rho v\frac{dU}{dy}, \\ \frac{\partial p}{\partial y} &= 0, \\ \frac{\partial p}{\partial z} &= 0. \end{aligned}\right\} \tag{6.6}$$

Hence,

$$p = -\rho\frac{dU}{dy}G(x - Ut), \qquad \text{where} \qquad G'(\xi) = g(\xi),$$

and the second equation of (6.6) gives

$$\frac{\partial p}{\partial y} = -\rho\frac{d^2U}{dy^2}G(\xi) + \rho\left(\frac{dU}{dy}\right)^2 g(\xi)\frac{x - \xi}{U} = 0$$

this is generally true only when

$$\frac{dU}{dy} = 0, \qquad \frac{d^2U}{dy^2} = 0,$$

i.e., when the main flow is uniform. There is, therefore, no general justification of extending Taylor's hypothesis to the case of shear flow.

One may still attempt to justify it on the basis that

$$\left|U\frac{\partial u}{\partial x}\right| \gg \left|v\frac{dU}{dy}\right|.$$

Then it may be expected to hold only for components of wave length $k$ such that

$$kU \gg \frac{dU}{dy} \qquad \text{or} \qquad ky \gg \frac{y}{U}\frac{dU}{dy}.$$

The latter form is useful when the experimental data are presented in scales of log $y$, For boundary layers, experimental values at $y/\delta = 0.2$ (Townsend[6]) give roughly

$$ky \gg 0.9 \qquad \text{or} \qquad k\delta \gg 4.5.$$

In terms of the wave-length $\Lambda$ of the disturbance, this means that

$$\frac{\Lambda}{\delta} \ll 0.7.$$

Thus, the wave-length must be much less than the boundary layer thickness in order that Taylor's hypothesis may apply.

For smaller values of $y/\delta$, the restriction is even more severe. For larger values $y/\delta$, the condition

$$ky \gg 1$$

may be taken as a rough approximation, until one comes to the region where intermittency becomes appreciable ($\xi = 0.02$ in Townsend's experiment). Before this point is reached, and with $y \geq 0.2\ \delta$, $y(dU/dy)$ and $U$ both increase slightly with $y$, making the resultant change in the quantity $(y/U)\ (dU/dy)$ fairly small.

Thus, at $y/\delta = 0.6$, we must have

$$k\delta \gg 1.5,$$

or

$$\frac{\Lambda}{\delta} \ll 4.$$

This means that for harmonic components with wave-lengths of the order of magnitude of the boundary layer thickness, Taylor's assumption is barely applicable at $y/\delta = 0.6$.

Klebanoff and Diehl[7] presented their results in terms of $nL_x/U$, where $n$ is the time frequency in cycles, and $L_x = 0.4\ \delta$. The above condition becomes then

$$\frac{nL_x}{U} \gg 0.1.$$

With reference to their Figure 25, it is seen that this condition is satisfied from the $n^{-5/3}$ range upwards. Shear-contributing fluctuations, however, do not satisfy this requirement.

It should be noted, however, that the wave-length $U/n$ that may be assigned to the $n^{-5/3}$ range is of the order of magnitude of the thickness of the boundary layer ($\delta/2$ to $\delta$). This is rather large for the application of Kolmogoroff's theory. The possibility is not to be excluded that the disturbances are actually propagating downstream at a lower speed analogous to oscillations in a laminar boundary layer.

Referring to (6.3), it is seen that for the observed level of turbulence in the boundary layer, the non-linear terms does not cause much inaccuracy in the application of Taylor's hypothesis. This is contrary to the conjecture that might be based on the fact that the quadratic terms are important in the phenomena of turbulent motion.

## References


1. G. I. Taylor, Proc. Roy. Soc. **A 164**, 476-490 (1938).
2. W. Heisenberg, Z. f. Phys. 124, 628 (1948).
3. G. K. Batchelor, Proc. Camb. Phil. Soc. **47**, 359-374 (1951).
4. H. P. Robertson, Proc. Camb. Phil. Soc. **36**, 209-223 (1940).
5. Th. von Kármán, J. Aero. Sci. **4**, 131-138 (1937); Th. von Kármán and L. Howarth, Proc. Roy. Soc. Lond. **A 164**, 192-215 (1938).
6. A. A. Townsend, Proc. Camb. Phil. Soc. **47**, 375-395, (1951).
7. P. S. Klebanoff and Z. W. Diehl, NACA Tech. Note 2475 (October, 1951).

Reprinted from the Proceedings of the NATIONAL ACADEMY OF SCIENCES
Vol. 46, No. 4, pp. 566–570. April, 1960.

# *ON A THEORY OF DISPERSION BY CONTINUOUS MOVEMENTS*

BY C. C. LIN*

INSTITUTE FOR ADVANCED STUDY

*Communicated by S. Chandrasekhar, February 10, 1960*

1. *Introduction.*—One of the central relations in the theory of turbulent diffusion is Richardson's law of dispersion.[1] It has often been derived from Kolmogoroff's theory[2] with the help of dimensional reasoning.[3, 4] But reservations on the range of applicability of this derivation have also been expressed.[3, 5] In the present note, we show how Richardson's law can be derived by applying the general approach of Taylor's theory of diffusion by continuous movements[6] when we consider dispersion from a point source. This derivation may be valid even when not all the requirements of Kolmogoroff's theory are fulfilled.

The analysis gives rise to a Lagrangian parameter $B$ which is of the same dimension as the rate of energy dissipating $\epsilon$. However, unlike previous analyses involving $\epsilon$, the parameter $B$ is now defined by a different physical process, and it is not necessarily identifiable with $\epsilon$ (numerical factors being allowed). The parameter $B$ also enters the dispersion relations with definite numerical coefficients. In this respect, the gain in our theoretical understanding is similar to that attained by Taylor's precise definition of the turbulent diffusion coefficient.

The results obtained in this paper are comparable with those of Obukhov,[7] who, however, started with the assumption that "the evolution of the state of the selected particle in time forms a Markov process and can be described by the Fokker-Planck equation in the space $(x, v)$." Such a restrictive assumption is not made in this paper. But it is clear that the two theories are related, and that our parameter $B$ is essentially the diffusion coefficient in the velocity space in the Fokker-Planck equation, if the latter is indeed applicable. The exact relationship between the present results and Obukhov's theory will be discussed in a more detailed communication by referring to the general theory of Brownian motion (see Uhlenbeck and Ornstein, and Chandrasekhar[8]).

Obukhov also assumed that his parameter $B$ is proportional to $\epsilon$ with a constant coefficient of the order of unity. Such an assumption is certainly plausible from the point of view of Kolmogoroff's theory. However, it is also possible that the ratio $B/\epsilon$ is a function of the Reynolds number of turbulence $R_\lambda$ and some other dimensionless parameter. An attempt to find this ratio might be a crucial step in connecting the Eulerian and the Lagrangian descriptions of turbulence. It appears that experiments can be set up for various values of $R_\lambda$ to test (with $B$ determined from equation (15)) whether there is a universal relationship of the form $B/\epsilon = f(R_\lambda)$.

The theory has been extended to the case of homogeneous anisotropic turbulence so that the horizontal diffusion coefficient may even be many times different from the vertical. This brings out more clearly the difference between the present approach and Kolmogoroff's theory, in which the concept of local isotropy plays an important role. The details of this anisotropic theory will be presented in later communications.

2. *One-Particle Analysis.*—Consider first the motion of a single particle released in a stationary homogeneous field of turbulent motion. Instead of considering the statistical distribution of the particles in space for an ensemble of such experiments, as Taylor did in his original theory of diffusion by continuous movements, we apply that approach to the determination of the velocity correlation of the same particle at an interval of time $\tau$ apart. Indeed, we know that the velocity $\mathbf{u}(t)$ may be obtained by an integration of the acceleration $\mathbf{a}(t)$ so that

$$\mathbf{u}(t + \tau) - \mathbf{u}(t) = \int_t^{t+\tau} \mathbf{a}(t')dt'$$

Thus,

$$\frac{1}{2}\frac{d}{d\tau}\overline{|\mathbf{u}(t+\tau) - \mathbf{u}(t)|^2} = \int_t^{t+\tau} \overline{\mathbf{a}(t+\tau)\cdot\mathbf{a}(t')}\, dt'$$

where the overhead bar denotes the ensemble average. If we now introduce the correlation coefficient $R_0(\tau)$ for the acceleration

$$\overline{\mathbf{a}(t+\tau)\cdot\mathbf{a}(t)} = \overline{|\mathbf{a}|^2}R_0(\tau), \qquad (1)$$

we obtain

$$\frac{1}{2}\frac{d}{d\tau}\overline{|\mathbf{u}(t+\tau)-\mathbf{u}(t)|^2} = \overline{|\mathbf{a}|^2}\int_0^{\tau} R_0(s)\,ds. \tag{2}$$

Suppose that $R_0(\tau)$ is small for $\tau \geq \tau_1$, then for $\tau > \tau_1$, we may approximate the above relation by

$$\frac{1}{2}\frac{d}{d\tau}\overline{|\mathbf{u}(t+\tau)-\mathbf{u}(t)|^2} = B_0,\ B_0 = \overline{|\mathbf{a}|^2}\int_0^{\tau_1} R_0(s)\,ds. \tag{3}$$

Thus,

$$\overline{|\mathbf{u}(t+\tau)-\mathbf{u}(t)|^2} = 2B_0\tau, \text{ for } \tau \gg \tau_1. \tag{4}$$

Clearly this implies that $\tau$, while large compared with $\tau_1$, must be appropriately small in a certain sense. This is exactly the kind of requirement in Kolmogoroff's theory.[9] Indeed, equation (4) has often been written down by dimensional arguments based on Kolmogoroff's theory, with $B$ replaced by $\epsilon$, and with an indefinite numerical coefficient. However, it appears that not all the requirements in Kolmogoroff's theory, such as an extremely large Reynolds number of turbulence, must be fulfilled to make the present derivation valid.

3. *Two-Particle Analysis.*—Consider now two particles simultaneously released in a turbulent wind. They will meander together with the wind according to the large-scale motions, but will also drift apart due to the small-scale motions. It is the latter we are interested in.

Let the initial coordinates of the particles be $\mathbf{x}_1^{(0)}$ and $\mathbf{x}_2^{(0)}$. After a time interval $t$, let their coordinates be $\mathbf{x}_1$ and $\mathbf{x}_2$ and their velocities $\mathbf{u}_1$ and $\mathbf{u}_2$. Then their separation $\mathbf{x} = \mathbf{x}_2 - \mathbf{x}_1$ and their relative velocity $\mathbf{v} = \mathbf{u}_2 - \mathbf{u}_1$ bear the relationship

$$\mathbf{v} = \frac{d\mathbf{x}}{dt}. \tag{5}$$

If, now, the particles are released at such small distances apart that their initial separation $\mathbf{x}^0$ and their initial relative velocity $\mathbf{v}^{(0)}$ are negligibly small compared with their values at the time under consideration, we may omit them in our present analysis.[10]

The relative velocity and separation of the two particles at the present time are then caused by the differences in the forces experienced by the two particles during the history of their flight.

Consider now the results of the statistical analysis involving an ensemble of experiments, each involving a pair of particles. We have a situation very similar to Taylor's one-particle analysis, except that

$$\mathbf{v}(0) = 0. \tag{7}$$

For a short time interval $t$ over which the particle velocity has not changed very much, Taylor obtained

$$\overline{\mathbf{x}^2} = \overline{\mathbf{u}^2}t^2,$$

while no dispersion will be predicted in the present situation. The particles simply meander together with the large-scale motions.

For longer time intervals, $\mathbf{v}^2$ will be built up according to the law (cf. equation (4))

$$\overline{\mathbf{v}^2} = 2Bt, \quad B = \overline{\mathbf{a}_r{}^2} \int_0^{\tau_1} R(\tau)d\tau, \tag{8}$$

for $t \gg \tau_1$, where $t$ is the time interval after the release of the particles, and $\mathbf{a}_r$ is the relative acceleration. To calculate the mean square dispersion $\overline{\mathbf{x}^2}$, we note that for $\mathbf{v}(0) = 0$, $\mathbf{x}(0) = 0$, we have

$$\mathbf{x}(t) = \int_0^t (t - t')\mathbf{a}_r(t')dt' \tag{9}$$

and hence

$$\overline{\mathbf{a}_r(t) \cdot \mathbf{x}(t)} = \overline{|\mathbf{a}_r|^2} \int_0^t \tau R(\tau)d\tau, \tag{10}$$

which may be approximated by

$$\overline{\mathbf{a}_r(t) \cdot \mathbf{x}(t)} = B\tau_2 \tag{11}$$

for $t > \tau_1$, where

$$\tau_2 = \int_0^{\tau_1} \tau R(\tau)\, d\tau / \int_0^{\tau_1} R(\tau)d\tau. \tag{12}$$

By combining (8) and (11), we have, for $t \gg \tau_1$,

$$\frac{d^2}{dt^2}\left(\frac{1}{2}\overline{\mathbf{x}^2}\right) = B(2t + \tau_2) \cong 2\,Bt,$$

since $\tau_2$ is expected to be of the order of $\tau_1$. Thus, the dispersion coefficient is

$$D = \frac{d}{dt}\left(\frac{\overline{\mathbf{x}^2}}{2}\right) = Bt^2 \tag{13}$$

and

$$\overline{\mathbf{x}^2} = \frac{2}{3} Bt^3. \tag{14}$$

Eliminating $t$ between these two equations, we obtain

$$D = \left(\frac{3}{2}\right)^{2/3} \cdot B^{1/3} (\overline{\mathbf{x}^2})^{2/3} \tag{15}$$

which is Richardson's law. Notice that for the strict validity of this law, one must be comparing various situations with the same value of $B$.

I wish to thank Professor W. Feller for his interest in this problem and for his advice on the general theory of diffusion and Markov processes.

* On leave of absence from Massachusetts Institute of Technology, Cambridge, Massachusetts. The author is grateful to Professor J. R. Oppenheimer for the hospitality extended to him at the Institute for Advanced Study.

[1] Richardson, L. F., *Proc. Roy. Soc.* (*London*), **A110**, 709 (1926).

[2] Kolmogoroff, A. N., *C. R. Acad. Sci. U.R.S.S.*, **30**, 301 (1941). Throughout this note, we shall refer to Kolmogoroff's theory in the "nonviscous form."

[3] Batchelor, G. K., and A. A. Townsend, in *Surveys in Mechanics* (Cambridge University Press, 1956), p. 384.

[4] Taylor, Sir Geoffrey in *Advances in Geophysics* (New York: Academic Press, Inc., 1959), vol. 6, p. 101.

[5] Batchelor, G. K., *ibid.*, p. 449.

[6] Taylor, Sir Geoffrey, *Proc. London Math. Soc.*, **20**, 196 (1921).

[7] Obukhov, A. M., in *Advances in Geophysics* (New York: Academic Press, Inc., 1959), vol. 6, p. 113.

[8] Uhlenbeck, G. E., and L. S. Ornstein, *Phys. Rev.*, **36**, 823 (1930); Chandrasekhar, S., *Rev. Mod. Phys.*, **15**, 1 (1943).

[9] It is also the requirement in Taylor's theory of diffusion by continuous movements. Generally speaking, $\tau$ must not be so large that $\int_{\tau_1}^{\tau} R(\tau)\, d\tau$ is no longer negligible compared with $\int_0^{\tau_1} R(\tau)\, d\tau$.

[10] It is not difficult to include $v^{(0)}$ and $r^{(0)}$ in our analysis. The details have been carried out and will be presented in a later communication where other refinements and extensions will be included.

Reprinted from the Proceedings of the NATIONAL ACADEMY OF SCIENCES
Vol. 46, No. 8, pp. 1147–1150. August, 1960.

# ON A THEORY OF DISPERSION BY CONTINUOUS MOVEMENTS, II. STATIONARY ANISOTROPIC PROCESSES

BY C. C. LIN*

THE INSTITUTE FOR ADVANCED STUDY

*Communicated by S. Chandrasekhar, April 6, 1960*

1. *Introduction.*—In a previous note,[1] we have shown how Richardson's law of turbulent dispersion can be derived from purely Lagrangian concepts by a slight extension of the concepts used in the theory of Brownian motion, without reference to the energy spectrum in an Eulerian analysis. The process of dispersion has been assumed to be isotropic in spacial orientation. In this note, we shall remove the assumption of isotropy so that the theory may be used as a basic framework for application to atmospheric turbulence in which the horizontal coefficient of dispersion may be many times larger than that in the vertical. This brings out even more clearly the contrast between the present treatment and those based on Kolmogoroff's theory in which the concept of local isotropy plays an essential role. The results obtained here also furnish more relations for observational verification.

2. *The Basic Correlations.*—We consider the relative motion of two particles released at the same point in a field of turbulent motion which is homogeneous but not isotropic. We then consider the statistical average over an ensemble of such pairs. Let $x_i(t)$ denote the relative displacement of a typical pair at time $t$, and $v_i(t) = \dot{x}_i(t)$, $a_i(t) = \dot{v}_i(t)$, where a dot denotes time differentiation. As before, we assume that the process is stationary in the sense that the forces acting on the particles may be described by a stationary random process. Thus, we assume

$$\overline{a_i(t')a_j(t)} = \alpha_{ij}(\tau) = \alpha_{ij}(-\tau), \tag{2.1}$$

where $\tau = t' - t$; it follows that $\alpha_{ij} = \alpha_{ji}$. We attempt to describe the statistical behavior of a pair of such particles in terms of the correlations

$$V_{ij}(t) = \overline{v_i(t)v_j(t)}, \quad X_{ij}(t) = \overline{x_i(t)x_j(t)}. \tag{2.2}$$

For the pairs of particles under consideration, we have

$$x_i(0) = 0, v_i(0) = 0. \tag{2.3}$$

Thus, we do not have a relation of the form

$$\overline{v_i(t')v_j(t)} = v_{ij}(t' - t)$$

since these pairs of particles form a restricted ensemble with the property $V_{ij}(0) = 0$. If we had considered a statistical average over *all* pairs of particles, we should admit the above relationship. On the other hand, we assume that (2.1) holds even for this restricted ensemble. This is an approximation asserting that the pressure forces acting on a fluid particle are essentially independent of its velocity.

In the next two sections we shall show that

$$V_{ij}(t) = \overline{v_i(t)v_j(t)} = 2\int_0^t (t - \tau)A_{ij}(\tau)d\tau, \tag{2.4}$$

and

$$\frac{d^2}{dt^2}X_{ij}(t) = \frac{d^2}{dt^2}\overline{x_i(t)x_j(t)} = 4\int_0^t A_{ij}(\tau)d\tau - 2\int_0^t \tau A_{ij}(\tau)\,d\tau, \tag{2.5}$$

where

$$A_{ij}(\tau) = \frac{1}{2}\left[\alpha_{ij}(\tau) + \alpha_{ji}(\tau)\right] \tag{2.6}$$

is symmetric in the indices and is, of course, an even function of $\tau$. In Section 4, we shall give a generalized form of Richardson's law by applying customary approximations to (2.4) and (2.5).

3. *The Correlation of Relative Velocity.*—With the restriction $v_i(0) = 0$, we have

$$v_i(t) = \int_0^t a_i(t')\,dt' \tag{3.1}$$

and consequently

$$\overline{v_i(t)a_j(t)} = \int_0^t \overline{a_i(t')a_j(t)}dt' = \int_0^t \alpha_{ij}(\tau)d\tau.$$

By adding two such equations with interchanged indices, we have

$$\frac{d}{dt}\overline{v_i(t)v_j(t)} = 2\int_0^t A_{ij}(\tau)\,d\tau \tag{3.2}$$

Since $\overline{v_i(0)v_j(0)} = 0$, we have

$$V_{ij}(t) = \overline{v_i(t)v_j(t)} = 2\int_0^t (t - \tau)A_{ij}(\tau)d\tau. \tag{3.3}$$

Clearly, the above derivation holds for any vector variable $v_i(t)$ satisfying the general statistical properties specified, provided the initial value of this vector is zero. The principal point is that the time derivative $v_i(t)$ of the vector variable considered has a definite statistical correlation in time, independent of the values of $v_i(t)$. Essentially, it is the same calculation made by Taylor[2] in his theory of diffusion by continuous movements. In that theory, it is the displacement vector $y_i(t)$ that takes the place of $v_i(t)$ here.

4. *The Correlation of Relative Displacement.*—Under the conditions (2.3), the relative displacement between a pair of particles is given by

$$x_i(t) = \int_0^t (t - t')a_i(t')dt'. \tag{4.1}$$

From this, we derive that

$$\overline{x_i(t)a_j(t)} = \int_0^t (t - t')\overline{a_i(t')a_j(t)}dt' = \int_0^t \tau\alpha_{ij}(\tau)d\tau. \tag{4.2}$$

It then follows that

$$\overline{x_i(t)a_j(t)} + \overline{x_j(t)a_i(t)} = 2\int_0^t \tau A_{ij}(\tau)d\tau. \tag{4.3}$$

Notice that when (4.3) and (3.3) are combined, we have

$$\overline{x_ia_j} + \overline{x_ja_i} + \overline{v_iv_j} = 2t\int_0^t A_{ij}(\tau)d\tau. \tag{4.4}$$

However, we are more interested in $\overline{x_i(t)x_j(t)}$ whose second derivative involves the quantity occurring in (4.3). Indeed, by straightforward differentiation and the use of (4.3) and (3.3), we get immediately

$$\frac{d^2}{dt^2}\overline{x_i(t)x_j(t)} = 4\int_0^t tA_{ij}(\tau)d\tau - 2\int_0^t \tau A_{ij}(\tau)d\tau. \tag{4.5}$$

This equation can be integrated exactly to give $\overline{x_i(t)x_j(t)}$, but it is physically more meaningful to treat it approximately in the usual manner, introducing a finite correlation time beyond which $A_{ij}(\tau)$ becomes negligible in a proper sense.

5. *Generalized Richardson's Law of Dispersion.*—Let $\tau_1$ be a time interval such that $A_{ij}(\tau)$ is "negligibly small" for $\tau \geq \tau_1$ (with exact meaning to be specified). In particular, we shall assume it a good approximation to assert

$$\int_0^t A_{ij}(\tau)d\tau = \int_0^{\tau_1} A_{ij}(\tau)d\tau, \text{ for all } t > \tau_1. \tag{5.1}$$

Clearly, we have to impose an upper limit on $t$. For, as $t \to \infty$, we expect $\overline{v_iv_j}$ to approach the statistical average for the unrestricted ensemble, which has a definite value independent of $t$. Thus, by (3.2), we have

$$\lim_{t\to\infty}\int_0^t A_{ij}(\tau)d\tau = 0. \tag{5.2}$$

However, if we restrict $t$ to be only many times $\tau_1$ but still sufficiently short for the initial conditions (2.3) to matter, we are free from this restriction. Specifically, we shall restrict $\tau_1 < t < \tau_2$ such that

$$\int_{\tau_6}^t A_{ij}(\tau)d\tau \ll \int_0^{\tau_1} A_{ij}(\tau)d\tau, \text{ for } \tau_1 < t < \tau_2, \tag{5.3}$$

and choose $\tau_2$ as large as permissible and $\tau_1$ as small as permissible within the desired accuracy of our analysis. Clearly the method is one of asymptotic approximation.[3]

With this kind of approximation in mind, including in particular an approximation similar to (5.1) for the second integral on the right-hand side of (4.5), we have, from (3.2) and (4.5), respectively,

$$\frac{d}{dt}\overline{v_iv_j} = 2B_{ij},\ B_{ij} = \int_0^{\tau_1} A_{ij}(\tau)d\tau \tag{5.4}$$

and

$$\frac{d^2}{dt^2}\overline{x_ix_j} = 4B_{ij}t. \tag{5.5}$$

From these relations, we derive immediately

$$V_{ij} = \overline{v_i v_j} = 2B_{ij}t, \tag{5.6}$$

$$D_{ij} = \tfrac{1}{2}\frac{dX_{ij}}{dt} = B_{ij}t^2 \tag{5.7}$$

and

$$X_{ij} = \overline{x_i x_j} = \tfrac{2}{3}B_{ij}t^3 = \tfrac{2}{3}D_{ij}t, \tag{5.8}$$

where $D_{ij}$ may be called the "dispersion tensor." If we now contract the formula (5.8) for $X_{ij}$, we obtain

$$X = \tfrac{2}{3}Bt^3, \qquad \text{or } t = \left(\frac{3X}{2B}\right)^{1/3}.$$

Thus, solving for $D_{ij}$ from (5.8) we get

$$D_{ij} = X_{ij}\left(\frac{4X}{9B}\right)^{-1/3}. \tag{5.9}$$

If we now remember that $B$ is a characteristic of the field of turbulence and that $X_{ij}$ varies with the square of the linear scale, we see that $D_{ij}$ varies as the $^4/_3$ power of the linear scale, which is Richardson's law. Notice that (5.9) predicts the *same* ratio between all the components of $X_{ij}$ and those of $\dot{X}_{ij} = 2D_{ij}$. Thus, the results obtained here can be subjected to a more detailed observational test.

* On leave of absence from Massachusetts Institute of Technology, Cambridge, Massachusetts. The author is grateful to Professor J. R. Oppenheimer for the hospitality extended to him at the Institute for Advanced Study.

[1] Lin, C. C., "On a theory of dispersion by continuous movements," these PROCEEDINGS, **46,** 566 (1960).

[2] Taylor, Sir Geoffrey, *Proc. Lond. Math. Soc.*, **20,** 196 (1921).

[3] There are methods of refining the formulation such that the complications discussed above do not occur; e.g., the introduction of a small resistance force proportional to the velocity. This will be discussed in the full paper.

# Turbulent Flow, Theoretical Aspects.

By

C. C. LIN and W. H. REID

With 27 Figures

## I. General concepts.

**1. Statistical description of turbulent motion.** A body of fluid is a mechanical system with an infinite number of degrees of freedom. It may therefore be expected to execute a rather random motion comparable to that of the molecules in a gas. If one regards such a chaotic motion as analyzed into harmonic components of various scales, one recognizes that frictional forces tend to dissipate the small-scale oscillations and keep the motion more or less regular. Thus, when viscous forces are sufficiently strong, i.e., at sufficiently low Reynolds numbers, the motion will become laminar. On the other hand, at sufficiently high Reynolds numbers the motion will tend to become random and fluctuating, even when the external conditions are steady. Instability of laminar motion at sufficiently high Reynolds numbers has been discussed in the article by H. SCHLICHTING in Vol. VIII/1.

It is clear that unsteady motion of a rather random nature cannot be described in detail, and one has to be satisfied with the description of their statistical properties, such as the mean velocity distribution. Indeed, in many cases of unsteady flow at high Reynolds numbers, such statistical properties are found to be experimentally reproducible, and it is often desirable to limit the term turbulence only to such random fluctuating motions of a fluid having statistical properties which are definite and experimentally reproducible.

Although the first systematic application of statistical concepts was made by REYNOLDS in connection with shear flow, current literature in the statistical theory of turbulence is mainly limited to the treatment of homogeneous turbulence[1] without any essential mean motion. Superficially, one might think that there is little to be known about such fluid motions. Actually, the very absence of a mean flow allows one to go more deeply into the inherent nature of the turbulent motion itself. Many basic concepts have been developed in the study of homogeneous turbulence, and these concepts now gradually find their way into the study of turbulent shear flow.

In the present article, emphasis will be placed on the development of the basic concepts and relations that are essential for the study of the dynamics of turbulent motion. To get an idea of the type of concepts needed in such studies, we shall consider, in the next section, the theory of shear flow, first following the classical approach of REYNOLDS. This introduces us to a discussion of the various statistical concepts and methods, including a brief description of HOPF's work. The main body of the article that follows will be devoted to the treatment of homogeneous turbulence, using statistical correlations and their Fourier transforms. Toward

[1] This concept will be explained more precisely in the following pages.

the end of this article, there will be given a discussion of the theory of turbulent diffusion.

There are other interesting and important phenomena associated with turbulent motion, such as the turbulent motion in a compressible fluid, particularly the noise generated by turbulence, and the interaction of turbulent motion with an electromagnetic field. Each topic deserves a complete treatment of its own. Owing to the limitation of space, it has been necessary to omit such discussions entirely and to list only a few important references. Throughout the present treatment, the fluid will be regarded as incompressible unless the contrary is explicitly stated.

**2. The mean flow and the Reynolds stresses.** It is generally assumed that the motion can be separated into a mean flow whose components are $U_1, U_2, U_3$ and a superposed turbulent flow whose components are $u_1, u_2, u_3$ the mean values of which are zero. In taking average values, the following principles will be adopted. If $A$ and $B$ are dependent variables which are being averaged, and $S$ is any one of the space variables $x_1, x_2, x_3$ or the time $t$, then $\overline{\partial A/\partial S} = \partial \bar{A}/\partial S$, and $\overline{\bar{A}\,B} = \bar{A}\,\bar{B}$ where the bar denotes a mean value. There are various ways of defining this average.

When the mean flow is not varying, that is, when the average value defined by

$$\bar{A}(x_i, t) = \lim_{\tau\to\infty} \frac{1}{2\tau} \int\limits_{t-\tau}^{t+\tau} A(x_i, t')\, dt'$$

is independent of the time $t$, the time average is the natural mean value to use. Difficulties arise with such a definition when the flow is variable, and other types of averages have to be introduced. For instance, in the problem of turbulent flow near an infinite plate moving with variable velocity, the mean values could be taken over planes parallel to the plate. In more general cases, neither the time nor the space mean values can be conveniently defined to possess all the desired properties. We then consider the statistical average over a large (infinite) number of identical systems—the ensemble average.

The equation of continuity of an incompressible fluid, when averaged, becomes[1]

$$\frac{\partial U_j}{\partial x_j} = 0. \tag{2.1}$$

The Navier-Stokes equations of motion are

$$\varrho \frac{\partial v_i}{\partial t} = \frac{\partial}{\partial x_j}(\sigma_{ij} - \varrho\, v_i\, v_j), \quad v_i = U_i + u_i \tag{2.2}$$

where $\sigma_{ij}$ is the stress tensor due to pressure and viscous forces. If the mean value is taken, Eq. (2.2) becomes

$$\varrho \frac{\partial U_i}{\partial t} = \frac{\partial}{\partial x_j}(\bar{\sigma}_{ij} - \varrho\, U_i\, U_j - \varrho\, \overline{u_i u_j}). \tag{2.3}$$

This equation has the same form as (2.2), if $v_i$ is replaced by $U_i$, and the stress $\sigma_{ij}$ is replaced by $\bar{\sigma}_{ij} - \varrho\, \overline{u_i u_j}$. Thus, the equations of mean flow are the same as the ordinary equations of motion except that there are the additional virtual stresses

$$\tau_{ij} = -\varrho\, \overline{u_i u_j}, \tag{2.4}$$

[1] The sum convention (summation over subscripts which occur twice) has been used throughout.

which represent the mean rate of transfer of momentum across a surface due to the velocity fluctuations. These virtual stresses were first introduced by REYNOLDS[1], and are known by his name.

In the case of a turbulent flow with steady mean motion, the time average may be taken at every point, and the above physical interpretation of the Reynolds stresses is clear. The time average is also the one most readily observed experimentally. In the case of variable mean motion, such as the case of the infinite plate mentioned above, where mean values are defined as averages over parallel planes, the interpretation of $-\varrho\overline{u_i u_j}$ as a local stress is not so direct. Especially in the case of general variable motion, where the averaging process is the arithmetical mean, taken over a large (infinite) number of unrelated identical systems (ensemble average), the physical interpretation of an average quantity as an apparent stress requires even more careful examination, the average momentum transfer being not directly associated with any *one* particular system. Again, in the usual measurements of $\tau_{ij}$, i.e., in the case of steady mean flow, the time average is taken. Thus, if a general theory is developed on the basis of statistical averages, an ergodic hypothesis must be introduced to identify these two in that case. In the present treatment, the statistical average will be adopted, and the validity of such a hypothesis will be implied. Further investigations of such basic problems are beyond the scope of the present treatment.

**3. Conventional approaches to the statistical theory of turbulence.** It is clear that the Reynolds equations (2.3) do not enable us to make predictions even concerning the mean flow alone. In order to obtain a closed system of equations, one needs certain relations between the mean flow and the Reynolds stresses. There is no *a priori* reason that such relations exist, but attempts to find them, even allowing for crude approximations, seem to be definitely a logical first step in treating the problem of turbulent flow. Indeed, the classical theories of BOUSSINESQ, PRANDTL, TAYLOR and v. KÁRMÁN all belong to this category. For a detailed account of such theories, reference may be made to GOLDSTEIN'S "Modern Developments in Fluid Dynamics".

A more systematic approach to the problem would be the development of equations governing the behavior of the Reynolds stresses. This can be accomplished as follows. We first subtract (2.3) from (2.2) to obtain the equations for the turbulent fluctuations:

$$\frac{\partial u_i}{\partial t} + U_j \frac{\partial u_i}{\partial x_j} + u_j \frac{\partial u_i}{\partial x_j} + u_j \frac{\partial U_i}{\partial x_j} = -\frac{1}{\varrho}\frac{\partial p}{\partial x_i} - \frac{1}{\varrho}\frac{\partial \tau_{ij}}{\partial x_j} + \nu \nabla^2 u_i. \tag{3.1}$$

If we multiply (3.1) by $u_j$ and add to the product a similar equation with $i$ and $j$ interchanged, the resultant equation has a leading term $\frac{\partial}{\partial t}(u_i u_j)$. The statistical average of such an equation would then be the equations governing the behavior of the Reynolds stresses. However, the non-linear terms in (3.1) would clearly introduce *triple velocity correlations*, and *velocity-pressure correlations*, which are again new unknowns. It is theoretically possible to eliminate velocity-pressure correlations by expressing the pressure fluctuations in terms of the velocity fluctuations. We take the divergence of (3.1) to obtain a Poisson equation for the pressure:

$$\nabla^2 p = -2\varrho \frac{\partial U_m}{\partial x_n}\frac{\partial u_n}{\partial x_m} + \varrho \frac{\partial^2}{\partial x_m \partial x_n}(\overline{u_m u_n} - u_m u_n) \tag{3.2}$$

which can then be solved with a boundary condition on the normal derivative of $p$, again derived from (3.1). Clearly the process becomes very complicated,

[1] O. REYNOLDS: Phil. Trans. Roy. Soc. Lond., Ser. A **186**, 123 (1895).

and it also introduces correlations of velocities at *two* points. Even after all this is accomplished, one is still faced with the difficulty of eliminating the triple velocity correlations. Again, one may use (3.1) to derive the equations governing them. The non-linear terms will now give rise to correlations involving *four* velocity components. The process can then be continued indefinitely to correlations of higher and higher orders. In order to stop this process, one may try to introduce suitable hypotheses, for example, a hypothesis to connect the correlations of second and fourth orders. Such a step was taken by CHOU[1]. When applied to the pressure flow through a channel, the theory gave reasonable predictions of the distribution of the mean velocity and turbulence level. However, in view of the inherent complications of the problem, other approximations were introduced for carrying through the calculations, and the theory is not free from uncertainties. A modified form of CHOU's theory was given by ROTTA[2]. Although certain improvements have been made, the intrinsic difficulties have not been overcome.

From the above discussions, it is clear that in dealing with problems of turbulent motion, one has to consider correlations of velocities at two or more points. Indeed, such velocity correlations would then give us an idea of the *structure* of the turbulent field. If we now turn our attention to this inherent structure, we would naturally be led to consider the simplest field of turbulent motion, namely, *homogeneous isotropic* turbulence in the absence of mean motion. Homogeneity means that the *statistical* properties of the field are invariant with respect to a translation, isotropy requires the further restriction to invariance with respect to a rotation (and reflection). Thus, in the isotropic case, the mean square value of the three components of velocity are the same,

$$\overline{u_1^2} = \overline{u_2^2} = \overline{u_3^2} = \overline{u^2}, \tag{3.3}$$

and is each equal to $\overline{u_i^2}/3$. A similar observation applies to the vorticity components ($\boldsymbol{\omega} = \operatorname{curl} \boldsymbol{u}$):

$$\overline{\omega_1^2} = \overline{\omega_2^2} = \overline{\omega_3^2} = \overline{\omega^2} = \tfrac{1}{3}\,\overline{\omega_i^2}. \tag{3.4}$$

(We shall often find it convenient also to use the notation $u$ and $\omega$ defined by $u^2 = \overline{u^2}$, $\omega^2 = \overline{\omega^2}$.)

In this article, we shall restrict our attention to the case of homogeneous turbulence, and indeed usually the isotropic case[3]. In this restricted case, it is possible to develop approximate theories in detail by introducing assumptions to connect correlation functions of third or fourth orders with double correlations. Such theories will be described briefly in Sect. 19 and treated in detail in Chaps. IV and V.

It is clear that the type of turbulent flow just discussed must necessarily decay in the course of time. Another type of homogeneous turbulence is that supported by a uniform shear. Such a field is statistically stationary, and would also lead us to the original problems considered by REYNOLDS. But the presence of the mean shear makes the theory more complicated. The development of such theories has often followed the ideas developed in the theory of decaying homogeneous turbulence. Analogous to the theories that will be presented in Chap. IV,

---

[1] P. Y. CHOU: Quart. Appl. Math. **3**, 198 (1945).

[2] J. ROTTA: Z. Physik **129**, 547 (1951); **131**, 51 (1951).

[3] Fortunately, the turbulent motion behind a grid in a wind tunnel satisfies this restriction approximately.

there exist the nearly identical theories of REIS[1] and of BURGERS and MITCHNER[2]. Analogous to the theory that will be presented in Chap. V, there is the work of CRAYA[3], which will be briefly reproduced in Chap. VI.

An entirely different approach to the problem of turbulent shear flow and turbulent convection has been advanced by MALKUS[4].

**4. Probability distributions and statistical averages.** In the discussion of statistical averages, one basic concept is the frequency of occurrence, or the distribution function. For example, in the classical kinetic theory of gases, one considers a distribution function $f(\boldsymbol{u}, \boldsymbol{x}, t)$ such that $f(\boldsymbol{u}, \boldsymbol{x}, t)\, d\boldsymbol{u}\, d\boldsymbol{x}$ gives the fraction of molecules at time $t$, having velocities in the range $(\boldsymbol{u}, \boldsymbol{u}+d\boldsymbol{u})$ and lying in the element of volume $(\boldsymbol{x}, \boldsymbol{x}+d\boldsymbol{x})$. The kinetic theory of gases may then be based on the law governing the change of this function $f(\boldsymbol{u}, \boldsymbol{x}, t)$.

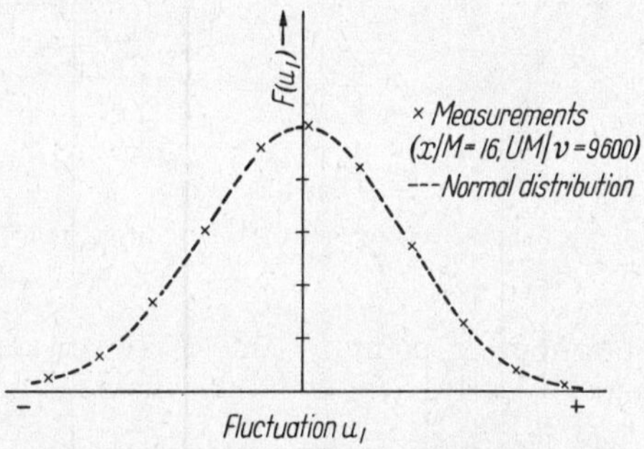


Fig. 1. Probability density function of the velocity component $u_1$ in the direction of the stream for the turbulence generated by a square mesh grid in a wind tunnel (from TOWNSEND).

In the case of turbulent motion, a similar function $F(\boldsymbol{u}, \boldsymbol{x}, t)$ can be introduced giving, for each point $P(\boldsymbol{x})$ and each instant $t$, the probability that the turbulent velocity shall lie in the range $(\boldsymbol{u}, \boldsymbol{u}+d\boldsymbol{u})$. If this function is known, then the Reynolds stress is given by formulas of the type

$$-\varrho\,\overline{u_i u_j} = -\varrho \int_{-\infty}^{\infty} F(\boldsymbol{u}, \boldsymbol{x}, t)\, u_i u_j\, d\boldsymbol{u} \quad (4.1)$$

for each point $P(\boldsymbol{x})$ at each instant $t$.

To analyze the structure of turbulence, one needs to know also the joint probability distribution for quantities observed at several points. For example, if we are interested in the correlation of velocities at two points $P$ and $P'$, then we must know a distribution function of the form $F(\boldsymbol{u}, \boldsymbol{x}; \boldsymbol{u}'\, \boldsymbol{x}'; t)$. The correlation of the velocities at these points is then given by

$$\overline{u_i(\boldsymbol{x})\, u_j(\boldsymbol{x}')} = \int\!\!\int F(\boldsymbol{u}, \boldsymbol{x}; \boldsymbol{u}', \boldsymbol{x}'; t)\, u_i u_j'\, d\boldsymbol{u}\, d\boldsymbol{u}'. \quad (4.2)$$

Further generalization of joint probability distributions will involve quantities observed at more than two points, and quantities other than velocity fluctuations.

Physically speaking, it is often convenient to use the notation $F(\boldsymbol{u})$ for the probability density function of the velocity vector $\boldsymbol{u}(\boldsymbol{x}, t)$ at the point $\boldsymbol{x}$ and $t$ without writing out the dependence of the function on those coordinates. The dependence of the statistical average quantity, such as the Reynolds stress (2.4), on $\boldsymbol{x}$ and $t$ is, however, obscured. Similarly, one finds it often convenient to write $F(\boldsymbol{u}, \boldsymbol{u}')$ for the distribution function occurring in (4.2). In the following discussion, we shall use the notation $F(\boldsymbol{u})$ etc. unless we want to exhibit the dependence of time and space coordinates explicitly.

Some experimental evidence is available regarding the distribution function $F(\boldsymbol{u})$. The Gaussian distribution has been found to be a fairly good approximation for each component of the velocity fluctuation in the turbulent motion behind a grid (Fig. 1). The closeness of the fitting can be measured in terms of the skew-

[1] F. B. REIS: Ph.D. Diss., Mass. Inst. Tech., 1951.

[2] J. M. BURGERS and M. MITCHNER: Proc. Kon. Ned. Akad. Wet., Ser. B **56**, 228, 343 (1953).

[3] A. CRAYA: C. R. Acad. Sci., Paris **244**, 560, 847, 1448, 1609 (1957).

[4] W. V. R. MALKUS: J. Fluid Mech. **1**, 521 (1956).

ness factor

$$s = \overline{u_1^3}/(\overline{u_1^2})^{\frac{3}{2}} \tag{4.3}$$

and the flatness factor

$$q = \overline{u_1^4}/(\overline{u_1^2})^2 \tag{4.4}$$

of the distribution function $F(u_1)$. These factors should have the values zero and three respectively for a Gaussian distribution. Experimental measurements of turbulent fluctuations behind a grid conform to such values.

The symmetry of the distribution is, however, found not to hold for the velocity derivative $-\partial u_1/\partial x_1$. Thus, its skewness factor

$$S = -\overline{\left(\frac{\partial u_1}{\partial x_1}\right)^3} \Big/ \left[\overline{\left(\frac{\partial u_1}{\partial x_1}\right)^2}\right]^{\frac{3}{2}} \tag{4.5}$$

does not vanish. This factor suggests the more general quantity

$$S(r) = -\frac{\overline{(u_1' - u_1)^3}}{[\overline{(u_1' - u_1)^2}]^{\frac{3}{2}}} \tag{4.6}$$

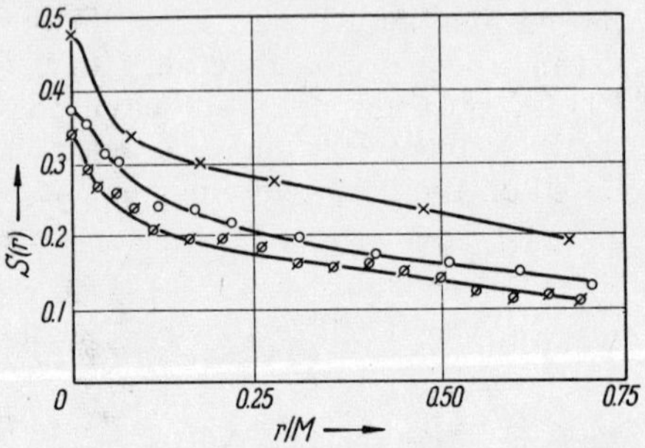


Fig. 2. Skewness factor for turbulent fluctuations behind a grid (from STEWART).

for the relative velocity $u_1' - u_1$ of two neighboring points $P$ and $P'$ separated by a distance $r$ in the direction of the $x_1$-axis. Clearly, $S(0)$ is the skewness factor for the velocity derivative in Eq. (4.5).

Fig. 2 shows the skewness factor for the relative velocity as determined by STEWART[1]. Clearly, there is a deviation of the distribution function for the relative velocity from the Gaussian function. On the other hand, the flatness factor (Fig. 3) for the distribution function of the relative velocity $u_1' - u_1$ is found to be close to the Gaussian value of 3 except for small distances. This suggests that an application of the joint Gaussian distribution for the calculation of fourth-order correlations may be sufficiently accurate for certain purposes, but that such an approximation should not be applied to correlations of nearby points.

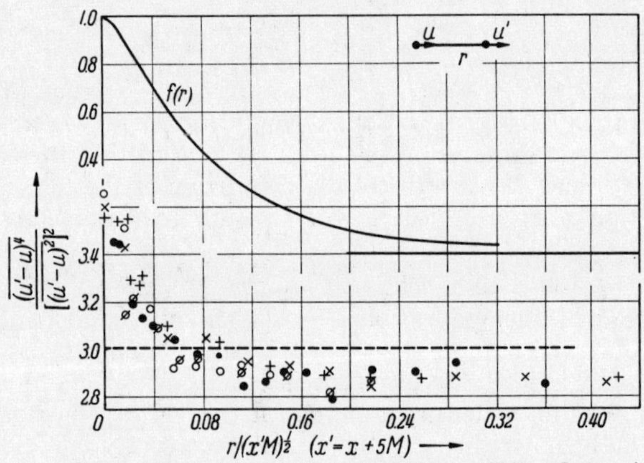


Fig. 3. Flatness factor for turbulent fluctuations behind a grid (from STEWART).

**5. HOPF's theory of turbulence.** To describe the structure of the velocity field fully, one would need to know the joint probability distribution of the velocity at $n$ points, where $n$ may take any integral value. Alternatively, one can employ the characteristic function which is just the Fourier transform of the distribution function, a simple example of which is

$$\varphi(\boldsymbol{\alpha}) = \int e^{i\boldsymbol{\alpha}\cdot\boldsymbol{u}} F(\boldsymbol{u})\, d\boldsymbol{u} = \overline{\exp(i\boldsymbol{\alpha}\cdot\boldsymbol{u})}. \tag{5.1}$$

[1] R.W. STEWART: Proc. Cambridge Phil. Soc. **47**, 146 (1951).

If $\varphi(\boldsymbol{\alpha})$ is now expanded as a power series in the components of $\boldsymbol{\alpha}$, then

$$\varphi(\boldsymbol{\alpha}) = \sum_{m=0}^{\infty} \frac{i^m}{m!} \overline{(\boldsymbol{\alpha}\cdot\boldsymbol{u})^m} \tag{5.2}$$

and the coefficients in this series are just the complete set of the moments of $\boldsymbol{u}$ at a single point. If it were possible to do so it would therefore be advantageous to work directly with $F(\boldsymbol{u})$ or $\varphi(\boldsymbol{\alpha})$ rather than with the moments themselves. Unfortunately, however, it does not appear possible to derive a dynamical equation for either $F(\boldsymbol{u})$ or $\varphi(\boldsymbol{\alpha})$.

It is easy to generalize the above ideas to $n$ points. We define

$$\left.\begin{aligned} \varphi(\boldsymbol{\alpha}_1, \ldots, \boldsymbol{\alpha}_n) &= \int \exp\left\{i\sum_{i=1}^{n} (\boldsymbol{\alpha}_i\cdot\boldsymbol{u}_i)\right\} F(\boldsymbol{u}_1, \ldots, \boldsymbol{u}_n)\, d\boldsymbol{u}_1 \ldots d\boldsymbol{u}_n \\ &= \overline{\exp\left\{i\sum_{i=1}^{n} \boldsymbol{\alpha}_i\cdot\boldsymbol{u}_i\right\}}\,. \end{aligned}\right\} \tag{5.3}$$

In other words, we simply replace $\boldsymbol{\alpha}$ by the $3n$-dimensional vector $(\boldsymbol{\alpha}_1, \ldots, \boldsymbol{\alpha}_n)$ and $\boldsymbol{u}$ by the $3n$-dimensional vector $(\boldsymbol{u}_1, \ldots, \boldsymbol{u}_n)$. The power series expansion of the exponential function will now yield all the correlations at $n$ points. By a generalization of the above ideas to include *all* the points, HOPF [*14*], [*15*] was able to obtain a closed statistical representation for the velocity field in terms of a single characteristic functional, the functional derivatives of which yield the mean values of $\boldsymbol{u}$ at any set of $n$ points for arbitrary integral values of $n$. Furthermore, HOPF was able to derive from the Navier-Stokes equation the dynamical equation governing the characteristic functional. Unfortunately, it has not been possible as yet to obtain solutions of this equation which correspond to the problem of homogeneous turbulence. The theory is somewhat difficult and even a brief outline of it would lead us too far afield. But it is to be hoped that further progress in this direction will not be too long delayed.

## II. Mathematical representation.

**6. Kinematics and dynamics of homogeneous turbulence.** Consider a homogeneous turbulent motion of an incompressible fluid, in which the mean velocity is zero. The condition of statistical homogeneity requires that the mean value of any function of the velocities and their derivatives be invariant for arbitrary translations. Thus, if $\boldsymbol{u}(\boldsymbol{x})$ denotes the velocity at the point $\boldsymbol{x}$, then the double velocity correlation tensor

$$R_{ij}(\boldsymbol{r}) = \overline{u_i(\boldsymbol{x})\, u_j(\boldsymbol{x}+\boldsymbol{r})} \tag{6.1}$$

is a function only of the single variable indicated[1]. From its definition, this tensor satisfies the symmetry condition

$$R_{ij}(\boldsymbol{r}) = R_{ji}(-\boldsymbol{r}), \tag{6.2}$$

and from the continuity equation,

$$\frac{\partial u_i}{\partial x_i} = 0, \tag{6.3}$$

it follows that the tensor (6.1) must satisfy the solenoidal conditions

$$\frac{\partial}{\partial r_i} R_{ij}(\boldsymbol{r}) = \frac{\partial}{\partial r_j} R_{ij}(\boldsymbol{r}) = 0. \tag{6.4}$$

[1] And, of course, of the time. But, since physical quantities under a mean value sign are taken at the same instant, except in Sect. 39, explicit reference to the time in this context can generally be omitted.

By virtue of the symmetry condition (6.2), however, only one of these solenoidal conditions is independent.

The (three-dimensional) Fourier transform of $R_{ij}(\boldsymbol{r})$ can now be defined by the relation

$$\Phi_{ij}(\boldsymbol{\varkappa}) = (2\pi)^{-3} \int R_{ij}(\boldsymbol{r})\, e^{-i\boldsymbol{\varkappa}\cdot\boldsymbol{r}}\, d\boldsymbol{r}, \tag{6.5}$$

where $i=\sqrt{-1}$, $d\boldsymbol{r}=dr_1\,dr_2\,dr_3$ and the integration is over all $\boldsymbol{r}$-space. By virtue of the reality of the correlation tensor $R_{ij}(\boldsymbol{r})$, the spectrum tensor $\Phi_{ij}(\boldsymbol{\varkappa})$ is complex in general, with the property

$$\Phi_{ij}^*(\boldsymbol{\varkappa}) = \Phi_{ij}(-\boldsymbol{\varkappa}), \tag{6.6}$$

where the asterisk denotes a complex conjugate. The transformed version of the symmetry condition (6.2) follows directly from the definition (6.5); thus

$$\Phi_{ij}(\boldsymbol{\varkappa}) = \Phi_{ji}(-\boldsymbol{\varkappa}), \tag{6.7}$$

which, in conjunction with Eq. (6.6), shows that $\Phi_{ij}(\boldsymbol{\varkappa})$ possesses Hermitian symmetry. The solenoidal conditions (6.4), resulting from the equation of continuity, transform into conditions of orthogonality in Fourier space; thus,

$$\varkappa_i\,\Phi_{ij}(\boldsymbol{\varkappa}) = \varkappa_j\,\Phi_{ij}(\boldsymbol{\varkappa}) = 0. \tag{6.8}$$

As in the physical representation, the Hermitian symmetry of $\Phi_{ij}(\boldsymbol{\varkappa})$ shows that only one of the orthogonality conditions (6.8) is independent.

The spectrum tensor $\Phi_{ij}(\boldsymbol{\varkappa})$ is subject to the further requirement[1] that the real Hermitian form

$$X_i X_j^*\,\Phi_{ij}(\boldsymbol{\varkappa}) \tag{6.9}$$

be non-negative for an arbitrary choice of the complex numbers $\boldsymbol{X}$. This requirement, together with the continuity conditions (6.8), has been used by KAMPÉ DE FÉRIET[2] to derive the functional form of $\Phi_{ij}(\boldsymbol{\varkappa})$ in homogeneous turbulence, a result that can be written in the form

$$\Phi_{ij}(\boldsymbol{\varkappa}) = \left(\delta_{ij} - \frac{\varkappa_i\varkappa_j}{\varkappa^2}\right)\Phi(\boldsymbol{\varkappa}) + \Phi_i(\boldsymbol{\varkappa})\,\Phi_j^*(\boldsymbol{\varkappa}), \tag{6.10}$$

where $\Phi_i(\boldsymbol{\varkappa})$ is some vector orthogonal to $\boldsymbol{\varkappa}$, i.e. $\varkappa_i\Phi_i(\boldsymbol{\varkappa})=0$.

The dynamical equation for the rate of change with time of the double velocity correlation tensor $R_{ij}(\boldsymbol{r})$ can be obtained by the methods developed first by v. KÁRMÁN and HOWARTH [*17*]; the dynamical equation for the spectral tensor $\Phi_{ij}(\boldsymbol{\varkappa})$ then follows by Fourier transformation. Thus, if the rate of change with time of $R_{ij}(\boldsymbol{r})$ is written in the form

$$\frac{\partial}{\partial t} R_{ij}(\boldsymbol{r}) = O_2\left\{\overline{\frac{\partial u_i(\boldsymbol{x})}{\partial t}\,u_j(\boldsymbol{x}+\boldsymbol{r})}\right\}, \tag{6.11}$$

where the operator $O_2$ acting on a second order tensor $C_{ij}(\boldsymbol{r})$ is such that

$$O_2\{C_{ij}(\boldsymbol{r})\} = C_{ij}(\boldsymbol{r}) + C_{ji}(-\boldsymbol{r}), \tag{6.12}$$

then by substituting for $\partial u_i(\boldsymbol{x})/\partial t$ from the Navier-Stokes equation,

$$\frac{\partial u_i}{\partial t} + u_j\frac{\partial u_i}{\partial x_j} = -\frac{1}{\varrho}\frac{\partial p}{\partial x_i} + \nu\,\nabla^2 u_i, \tag{6.13}$$

where $p(\boldsymbol{x})$ denotes the deviation of the pressure at the point $\boldsymbol{x}$ from its uniform mean value and $\varrho$ and $\nu$ denote the uniform density and kinematic viscosity of

[1] H. CRAMÉR: Ann. Math. **41**, 215 (1940).

[2] J. KAMPÉ DE FÉRIET: Proc. 7th Int. Cong. Appl. Mech. introd. vol. 1948. — C. R. Acad. Sci., Paris **227**, 760 (1948).

the fluid respectively, we obtain

$$\frac{\partial}{\partial t} R_{ij}(\boldsymbol{r}) = S_{ij}(\boldsymbol{r}) + 2\nu \nabla^2 R_{ij}(\boldsymbol{r}), \tag{6.14}$$

where

$$S_{ij}(\boldsymbol{r}) = O_2 \left\{ \frac{\partial}{\partial r_k} \overline{u_i(\boldsymbol{x})\, u_j(\boldsymbol{x}+\boldsymbol{r})\, u_k(\boldsymbol{x})} + \frac{1}{\varrho} \frac{\partial}{\partial r_i} \overline{p(\boldsymbol{x})\, u_j(\boldsymbol{x}+\boldsymbol{r})} \right\}. \tag{6.15}$$

It may be noted that the divergence of the Navier-Stokes equation (6.13) yields the result

$$\frac{1}{\varrho} \nabla^2 p = - \frac{\partial^2 u_i u_j}{\partial x_i \partial x_j}, \tag{6.16}$$

so that, in the boundary-free problem of homogeneous turbulence, the pressure is determined uniquely by the velocity field according to the equation

$$\frac{1}{\varrho} p(\boldsymbol{x}) = \frac{1}{4\pi} \int \frac{\partial^2 u_i(\boldsymbol{x}')\, u_j(\boldsymbol{x}')}{\partial x_i' \partial x_j'} \cdot \frac{d\boldsymbol{x}'}{|\boldsymbol{x}-\boldsymbol{x}'|}. \tag{6.17}$$

This relationship is too cumbersome to permit a simple elimination of the pressure correlation occurring in Eq. (6.15). In the corresponding spectral representation, however, the Fourier transform of this pressure correlation can be eliminated both simply and explicitly. Nevertheless, when only the single equation (6.14) is employed in the dynamics, very little is gained from such considerations.

Turning now to a consideration of the dynamical equation for the spectrum tensor $\Phi_{ij}(\boldsymbol{\varkappa})$, we observe that the transformed version of Eq. (6.14) can be written in the form

$$\frac{\partial}{\partial t} \Phi_{ij}(\boldsymbol{\varkappa}) = \Sigma_{ij}(\boldsymbol{\varkappa}) - 2\nu \varkappa^2 \Phi_{ij}(\boldsymbol{\varkappa}), \tag{6.18}$$

where $\Sigma_{ij}(\boldsymbol{\varkappa})$ is the Fourier transform of $S_{ij}(\boldsymbol{r})$ in the same sense that $\Phi_{ij}(\boldsymbol{\varkappa})$ is the Fourier transform of $R_{ij}(\boldsymbol{r})$. Furthermore, by virtue of Eq. (6.18), $\Sigma_{ij}(\boldsymbol{\varkappa})$ satisfies exactly the same symmetry and orthogonality conditions as $\Phi_{ij}(\boldsymbol{\varkappa})$.

So long as we confine our attention to mean values at only two points in the field, Eqs. (6.14) and (6.18) include all the information that can be obtained from the Navier-Stokes equation and their indeterminate character is evident. In fact, in the general case of homogeneous turbulence there are relatively few dynamical results that can be reported. Among these, recent work on the large-scale structure of homogeneous turbulence is described in Sect. 13 and the general dynamical equations which result from the assumption of a normal joint-probability distribution are given in Chap. V. But it is only in the case of isotropic turbulence that any real progress with the dynamical problem has been achieved and in the following two sections we present those relations that are basic to any discussion of the dynamics.

**7. Kinematics and dynamics of isotropic turbulence.** In isotropic turbulence, the mean value of any function of the velocities and their derivatives are invariant not only for arbitrary translations but also for arbitrary rotations and reflections of the configuration defining the mean value. Under these conditions, the double velocity correlation tensor $R_{ij}(\boldsymbol{r})$, say, has a particularly simple form; it has, in fact, only one independent component.

The representation of $R_{ij}(\boldsymbol{r})$ in terms of a single defining scalar in isotropic turbulence was given first by v. KÁRMÁN and HOWARTH [*17*] who used an essentially geometrical method of analysis. The present discussion of this problem, using the methods of invariant theory, follows more closely the method of analysis given by ROBERTSON[1].

[1] H.P. ROBERTSON: Proc. Cambridge Phil. Soc. **36**, 209 (1940).

Consider two arbitrary unit vectors $\boldsymbol{a}$ and $\boldsymbol{b}$. Then the correlation between the velocity components $u_i a_i$ at the point $\boldsymbol{x}$ and $u'_j b_j$ at the point $\boldsymbol{x}'$ is a bilinear form

$$R(\boldsymbol{r};\boldsymbol{a},\boldsymbol{b}) = \overline{u_i a_i u'_j b_j} = R_{ij}(\boldsymbol{r})\, a_i b_j. \tag{7.1}$$

We must determine, therefore, the most general function $R(\boldsymbol{r};\boldsymbol{a},\boldsymbol{b})$ which is a scalar invariant under arbitrary rotations and reflections of the configuration defined by the vectors $(\boldsymbol{r},\boldsymbol{a},\boldsymbol{b})$ and which, by Eq. (7.1), is bilinear in $\boldsymbol{a}$ and $\boldsymbol{b}$. The scalar products available for its construction are $(\boldsymbol{r}\cdot\boldsymbol{r})$, $(\boldsymbol{r}\cdot\boldsymbol{a})$, $(\boldsymbol{r}\cdot\boldsymbol{b})$ and $(\boldsymbol{a}\cdot\boldsymbol{b})$; the determinant formed from the vectors $(\boldsymbol{r},\boldsymbol{a},\boldsymbol{b})$ must be excluded since its sign changes under reflection; likewise, the square of this determinant must be rejected since it is not a homogeneous bilinear function of the vectors $\boldsymbol{a}$ and $\boldsymbol{b}$. Hence, the bilinear isotropic form $R(\boldsymbol{r};\boldsymbol{a},\boldsymbol{b})$ must be of the form

$$R(\boldsymbol{r};\boldsymbol{a},\boldsymbol{b}) = R_1(\boldsymbol{r}\cdot\boldsymbol{a})\,(\boldsymbol{r}\cdot\boldsymbol{b}) + R_2(\boldsymbol{a}\cdot\boldsymbol{b}), \tag{7.2}$$

where $R_1$ and $R_2$ are arbitrary scalar functions of $(\boldsymbol{r}\cdot\boldsymbol{r})=r^2$. By comparing this result with Eq. (7.1) we see that the general form of $R_{ij}(\boldsymbol{r})$ in isotropic turbulence is

$$R_{ij}(\boldsymbol{r}) = R_1 r_i r_j + R_2\,\delta_{ij}, \tag{7.3}$$

where $R_1$ and $R_2$, which we may now consider as even functions of $r$, are called the "defining scalars" of $R_{ij}(\boldsymbol{r})$.

In differentiating expressions such as (7.3) with respect to $r_j$ (say), it is convenient to introduce the differential operator

$$D_r = \frac{1}{r}\frac{\partial}{\partial r} \tag{7.4}$$

so that

$$\frac{\partial}{\partial r_j} = r_j D_r. \tag{7.5}$$

Thus, differentiating Eq. (7.3) with respect to $r_j$ gives

$$\frac{\partial}{\partial r_j} R_{ij}(\boldsymbol{r}) = [(r^2 D_r + 4)\,R_1 + D_r R_2]\, r_i. \tag{7.6}$$

Now the differential operator

$$r^2 D_r + n = r\frac{\partial}{\partial r} + n,$$

which occurs in the expression (7.6) for the divergence of $R_{ij}(\boldsymbol{r})$, satisfies the identity

$$D_r(r^2 D_r + n) = (r^2 D_r + n + 2)\, D_r. \tag{7.7}$$

The condition that $R_{ij}(\boldsymbol{r})$ be solenoidal is [cf. Eq. (7.6)]

$$(r^2 D_r + 4)\,R_1 + D_r R_2 = 0;$$

and, from the identity (7.7), it then follows that $R_1$ and $R_2$ can be expressed in terms of a single defining scalar in the form

$$\left.\begin{aligned} R_1 &= -\,D_r \tfrac{1}{2}\overline{u^2} f(r) \\ \text{and}\qquad R_2 &= +\,(r^2 D_r + 2)\,\tfrac{1}{2}\overline{u^2} f(r), \end{aligned}\right\} \tag{7.8}$$

where $\overline{u^2} f(r)$ is the usual longitudinal velocity correlation (see Fig. 4) introduced first by v. Kármán and Howarth [*17*] and defined by the equations

$$\overline{u^2} f(r) = R_{11}(\boldsymbol{r}), \qquad \boldsymbol{r} = (r, 0, 0). \tag{7.9}$$

Thus, Eq. (7.3) becomes

$$R_{ij}(\boldsymbol{r}) = -\tfrac{1}{2} D_r \overline{u^2} f(r)\, r_i r_j + \tfrac{1}{2}(r^2 D_r + 2)\overline{u^2} f(r)\, \delta_{ij}, \tag{7.10}$$

so that in incompressible isotropic turbulence $R_{ij}(\boldsymbol{r})$ is completely characterized by the single defining scalar $\overline{u^2} f(r)$.

The transverse velocity correlation (see Fig. 4),

$$\overline{u^2}\, g(r) = R_{22}(\boldsymbol{r}), \qquad \boldsymbol{r} = (r, 0, 0), \tag{7.11}$$

also appears occasionally in the analysis. In isotropic turbulence it is related to $f(r)$ by the relation

$$g(r) = f(r) + \frac{r}{2}\frac{d f(r)}{d r} \tag{7.12}$$

which follows from the continuity condition. All existing experimental evidence seems to suggest that $f(r)$ is everywhere positive, though no proof of this fact has yet been given[1]; on the other hand, provided that $r^2 f(r) \to 0$ as $r \to \infty$, Eq. (7.12) can be integrated to give

$$\int_0^\infty r\, g(r)\, dr = 0 \tag{7.13}$$

which shows that $g(r)$ must become negative.

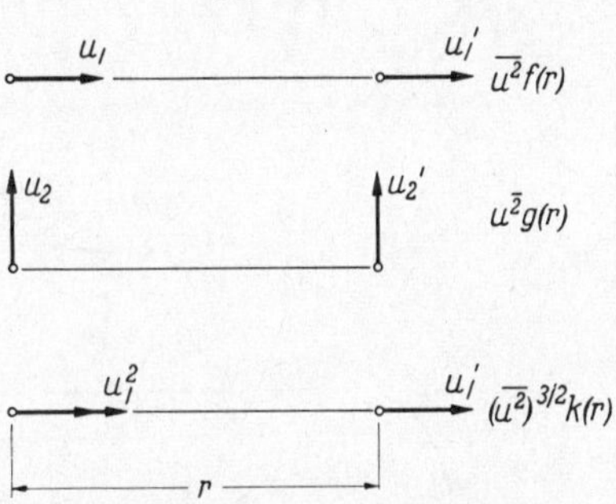

Fig. 4. The two-point correlation functions in isotropic turbulence.

The other mean values occurring in the definition of $S_{ij}(\boldsymbol{r})$, Eq. (6.15), can also be treated in a similar fashion. Since all solenoidal isotropic vectors vanish identically [17], we see that

$$\frac{1}{\varrho}\overline{p(\boldsymbol{x})\, u_j(\boldsymbol{x}+\boldsymbol{r})} \equiv 0, \tag{7.14}$$

and hence that mean values involving the pressure do not appear in the dynamical equation for $\overline{u^2} f(r)$. The isotropic form of the triple correlation $\overline{u_i(\boldsymbol{x})\, u_j(\boldsymbol{x}+\boldsymbol{r})\, u_k(\boldsymbol{x})}$ can be obtained by the methods outlined above for $R_{ij}(\boldsymbol{r})$, and it is not difficult to show that this tensor, which is symmetric in the indices $i$ and $k$ and solenoidal with respect to $r_j$, can also be represented in terms of a single defining scalar in the form

$$\left.\begin{aligned} &\overline{u_i(\boldsymbol{x})\, u_j(\boldsymbol{x}+\boldsymbol{r})\, u_k(\boldsymbol{x})} \\ &\quad = (\overline{u^2})^{\frac{3}{2}}\left[\frac{k - r k'}{2r^3} r_i r_j r_k + \frac{2k + r k'}{4r}(r_i \delta_{kj} + r_k \delta_{ij}) - \frac{k}{2r} r_j \delta_{ik}\right] \end{aligned}\right\} \tag{7.15}$$

where $(\overline{u^2})^{\frac{3}{2}} k(r)$ is the usual longitudinal triple correlation (see Fig. 4) defined by the equations

$$(\overline{u^2})^{\frac{3}{2}} k(r) = \overline{u_1^2(\boldsymbol{x})\, u_1(\boldsymbol{x}+\boldsymbol{r})}, \qquad \boldsymbol{r} = (r, 0, 0), \tag{7.16}$$

and a prime denotes differentiation with respect to $r$. By substituting this result into Eq. (6.15), we find that $S_{ij}(\boldsymbol{r})$ is given by

$$S_{ij}(\boldsymbol{r}) = -\tfrac{1}{2} D_r K(r)\, r_i r_j + \tfrac{1}{2}(r^2 D_r + 2)\, K(r)\, \delta_{ij}, \tag{7.17}$$

where

$$K(r) = \left(\frac{\partial}{\partial r} + \frac{4}{r}\right)(\overline{u^2})^{\frac{3}{2}} k(r). \tag{7.18}$$

Thus, comparing Eq. (7.17) with Eq. (7.10), we see that $K(r)$ is the defining scalar of $S_{ij}(\boldsymbol{r})$ in exactly the same sense as $\overline{u^2} f(r)$ is the defining scalar of $R_{ij}(\boldsymbol{r})$. The general shape of the correlation functions is shown in Fig. 5.

[1] See, however, Sect. 33.

The representation of the spectrum tensor $\Phi_{ij}(\boldsymbol{\varkappa})$ in terms of a single defining scalar is very much simpler than in the case of $R_{ij}(\boldsymbol{r})$. Since the Fourier transform (6.5) involves only the scalar product $(\boldsymbol{\varkappa}\cdot\boldsymbol{r})$, $\Phi_{ij}(\boldsymbol{\varkappa})$ must possess the same invariance properties as $R_{ij}(\boldsymbol{r})$; and, in particular, if $R_{ij}(\boldsymbol{r})$ is isotropic then $\Phi_{ij}(\boldsymbol{\varkappa})$ must be isotropic also, and it must therefore be of the general form [cf. Eq. (7.3)]

$$\Phi_{ij}(\boldsymbol{\varkappa}) = \Phi_1 \varkappa_i \varkappa_j + \Phi_2 \delta_{ij}, \tag{7.19}$$

where the defining scalars $\Phi_1$, and $\Phi_2$ are arbitrary functions of $(\boldsymbol{\varkappa}\cdot\boldsymbol{\varkappa}) = \varkappa^2$. The orthogonality conditions (6.8) then lead to a relationship between $\Phi_1$ and $\Phi_2$ of the form

$$\varkappa^2 \Phi_1 + \Phi_2 = 0$$

and, accordingly, $\Phi_{ij}(\boldsymbol{\varkappa})$ can be written in terms of a single defining scalar in the form

$$\Phi_{ij}(\boldsymbol{\varkappa}) = \left(\delta_{ij} - \frac{\varkappa_i \varkappa_j}{\varkappa^2}\right) \frac{E(\varkappa)}{4\pi\varkappa^2}, \tag{7.20}$$

where the defining scalar has been chosen to correspond with the usual isotropic energy spectrum function $E(\varkappa)$, i.e. $E(\varkappa)$ gives the spectral distribution of the total energy,

$$\tfrac{1}{2}\overline{\boldsymbol{u}^2} = \int_0^\infty E(\varkappa)\, d\varkappa. \tag{7.21}$$

Since $\Sigma_{ij}(\boldsymbol{\varkappa})$ satisfies exactly the same symmetry and orthogonality properties as $\Phi_{ij}(\boldsymbol{\varkappa})$, it too can be represented in terms of a single defining scalar in the form

$$\Sigma_{ij}(\boldsymbol{\varkappa}) = \left(\delta_{ij} - \frac{\varkappa_i \varkappa_j}{\varkappa^2}\right) \frac{T(\varkappa)}{4\pi\varkappa^2}, \tag{7.22}$$

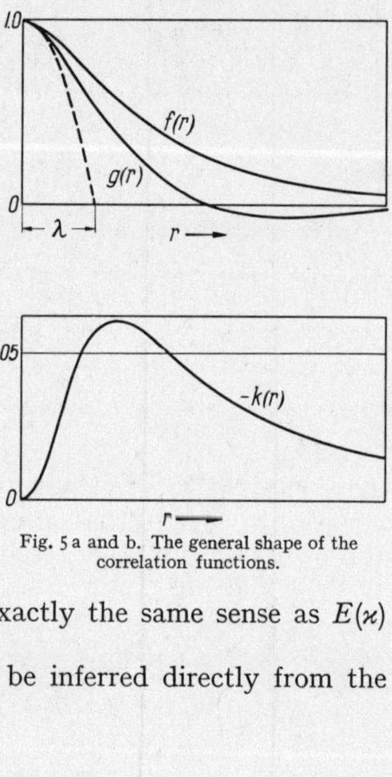


a

b

Fig. 5 a and b. The general shape of the correlation functions.

so that $T(\varkappa)$ is the defining scalar of $\Sigma_{ij}(\boldsymbol{\varkappa})$ in exactly the same sense as $E(\varkappa)$ is the defining scalar of $\Phi_{ij}(\boldsymbol{\varkappa})$.

The connection between $\overline{u^2} f(r)$ and $E(\varkappa)$ can be inferred directly from the Fourier transform (6.5) with the results

$$\overline{u^2} f(r) = 2 \int_0^\infty \frac{E(\varkappa)}{\varkappa^3 r^3} (\sin \varkappa r - \varkappa r \cos \varkappa r)\, d\varkappa \tag{7.23}$$

and

$$E(\varkappa) = \frac{1}{\pi} \int_0^\infty \overline{u^2} f(r)\, \varkappa r (\sin \varkappa r - \varkappa r \cos \varkappa r)\, dr. \tag{7.24}$$

Clearly, $K(r)$ and $T(\varkappa)$ are related in exactly the same way.

When the turbulence is isotropic, the tensor equation (6.14) must reduce to a single scalar equation, one form of which can be obtained directly simply by contracting the indices in Eq. (6.14). Alternatively, we may note that if $\overline{u^2} f(r)$ is the defining scalar of $R_{ij}(\boldsymbol{r})$, as in Eq. (7.10), then the defining scalar of $\nabla^2 R_{ij}(\boldsymbol{r})$ is $\left(\frac{\partial^2}{\partial r^2} + \frac{4}{r}\frac{\partial}{\partial r}\right)\overline{u^2} f(r)$ and, accordingly, we have

$$\frac{\partial}{\partial t}\overline{u^2} f(r) = \left(\frac{\partial}{\partial r} + \frac{4}{r}\right)(\overline{u^2})^{\frac{3}{2}} k(r) + 2\nu \left(\frac{\partial^2}{\partial r^2} + \frac{4}{r}\frac{\partial}{\partial r}\right)\overline{u^2} f(r), \tag{7.25}$$

which is the well-known equation for the rate of change of the double correlation function first given by V. KÁRMÁN and HOWARTH [17].

The scalar form of Eq. (6.18) follows directly from Eqs. (7.20) and (7.22) in the form

$$\frac{\partial E(\varkappa)}{\partial t} = T(\varkappa) - 2\nu\varkappa^2 E(\varkappa). \tag{7.26}$$

The same result can be obtained[1] by transforming Eq. (7.25) directly by means of Eq. (7.24).

It should perhaps be emphasized once again that, so long as we confine our attention to mean values at two points only, Eq. (7.25) or Eq. (7.26) contains all of the information that can be obtained from the Navier-Stokes equation and again the indeterminate nature of the problem is evident. But while the basic equations of the problem are indeterminate they do provide a framework into which many of the experimental results can be fitted. Thus, for example, STEWART[2] has measured each of the terms in the Kármán-Howarth equation (7.25) and found them to be self-consistent within the limits of experimental error.

**8. The viscous dissipation of energy.** In homogeneous turbulence, the rate of viscous dissipation is given by

$$\varepsilon = \nu\,\overline{\left(\frac{\partial u_i}{\partial x_j}\right)^2}. \tag{8.1}$$

The condition of isotropy leads to simple relations between the velocity derivatives $\partial u_i/\partial x_j$ of the form

$$\left.\begin{aligned}\overline{\left(\frac{\partial u_1}{\partial x_1}\right)^2} &= \overline{\left(\frac{\partial u_2}{\partial x_2}\right)^2} = \overline{\left(\frac{\partial u_3}{\partial x_3}\right)^2} = \frac{1}{2}\overline{\left(\frac{\partial u_1}{\partial x_2}\right)^2} = \frac{1}{2}\overline{\left(\frac{\partial u_1}{\partial x_3}\right)^2}\\ &= \frac{1}{2}\overline{\left(\frac{\partial u_2}{\partial x_1}\right)^2} = \frac{1}{2}\overline{\left(\frac{\partial u_2}{\partial x_3}\right)^2} = \frac{1}{2}\overline{\left(\frac{\partial u_3}{\partial x_1}\right)^2} = \frac{1}{2}\overline{\left(\frac{\partial u_3}{\partial x_2}\right)^2},\end{aligned}\right\} \tag{8.2}$$

so that Eq. (8.1) becomes simply

$$\varepsilon = 15\nu\,\overline{\left(\frac{\partial u_1}{\partial x_1}\right)^2}. \tag{8.3}$$

The relations (8.2), apart from their theoretical interest, have two very important practical uses: they provide a simple criterion for determining whether or not the structure of the dissipating eddies is isotropic; or, conversely, assuming isotropy of the small scale components of the motion, these relations permit the determination of the viscous dissipation.

Consider now the behavior of $f(r)$ and $k(r)$ for small values of $r$. By using the condition of homogeneity (that $\partial\overline{u_i^2}/\partial x_j = 0$), it can be shown that $f(r)$ is an even function of $r$ with a power series expansion of the form

$$f(r) = 1 + f_0''\frac{r^2}{2!} + \cdots + f_0^{(2n)}\frac{r^{2n}}{(2n)!} + \cdots. \tag{8.4}$$

If we now introduce a length $\lambda$ defined by the relation

$$-f_0'' = \frac{1}{\lambda^2} = \frac{1}{\overline{u^2}}\overline{\left(\frac{\partial u_1}{\partial x_1}\right)^2}, \tag{8.5}$$

then, provided that the total dissipation is finite, $f(r)$ will have a finite curvature near the origin and

$$f(r) \approx 1 - \frac{r^2}{2\lambda^2}. \tag{8.6}$$

[1] C. C. LIN: Proc. 1st Symp. in Appl. Math., Amer. Math. Soc. 1947, p. 81.
[2] R. W. STEWART: Proc. Cambridge Phil. Soc. **47**, 146 (1951).

The corresponding behavior of $g(r)$ is $g(r) \approx 1 - r^2/\lambda^2$ for small values of $r$ (see Fig. 5a). Furthermore, by using the fact that $\overline{u_1^3} = 0$ by isotropy, one can also show that $k(r)$ is an odd function of $r$ and that it is of order $r^3$ for small values of $r$; hence, we may write

$$k(r) = k_0''' \frac{r^3}{3!} + \cdots + k_0^{(2n+1)} \frac{r^{2n+1}}{(2n+1)!} + \cdots, \tag{8.7}$$

where $k_0'''$ is related to the skewness factor of $-\partial u_1/\partial x_1$ by the relation

$$k_0''' = -S/\lambda^3. \tag{8.8}$$

The coefficients in the power series (8.4) and (8.7) are of course related to the integral moments of $E(\varkappa)$ and $T(\varkappa)$ in the manner

$$\overline{u^2}\, f_0^{(2n)} = \frac{(-1)^n\, 2}{(2n+1)(2n+3)} \int_0^\infty \varkappa^{2n}\, E(\varkappa)\, d\varkappa \tag{8.9}$$

and

$$(\overline{u^2})^{\frac{3}{2}}\, k_0^{(2n+1)} = \frac{(-1)^n\, 2}{(2n+3)(2n+5)} \int_0^\infty \varkappa^{2n}\, T(\varkappa)\, d\varkappa. \tag{8.10}$$

Note, in particular, that due to $k(r)$ being of order $r^3$ for small values of $r$ we have the integral condition

$$\int_0^\infty T(\varkappa)\, d\varkappa = 0. \tag{8.11}$$

Thus, the reason for calling $T(\varkappa)$ the "transfer function" is clear: it is because $T(\varkappa)$ does not affect the rate of decay of the energy directly [cf. Eq. (8.12) below] but causes only a redistribution of energy between the different components of the motion.

For $r=0$, the Kármán-Howarth equation (7.25) gives simply the decay equation

$$\frac{d\overline{u^2}}{dt} = -\frac{10\,\nu\,\overline{u^2}}{\lambda^2}, \tag{8.12}$$

whence $\lambda$ is called the "dissipation length parameter". While the length $\lambda$ is associated with the smaller eddies, as may be seen from Eq. (8.6), it is not the length associated with the eddies responsible for the viscous dissipation and, for sufficiently large Reynolds numbers, $\lambda$ corresponds to eddies which in fact make negligible contributions to both the energy and the dissipation (cf. Chap. III, Sects. 16 and 17). Its principal value lies in the ease with which it can be measured experimentally; but, theoretically, as TOWNSEND [*74*] has emphasized, it is more useful as a measure of the "reaction time" of the turbulence as a whole, $\lambda^2/10\nu$. If the kinetic energy decreases according to a simple power law of the form $\overline{u^2} \propto t^{-n}$, then

$$\lambda^2 = 10\nu\, t/n. \tag{8.13}$$

**9. The role of vorticity.** The vorticity, though less fundamental from a theoretical standpoint, has played a central role in all physical thinking on the subject and its relationship to the velocity field must therefore be described in some detail. The vorticity vector $\boldsymbol{\omega}(\boldsymbol{x})$ is of course defined in the usual way

$$\boldsymbol{\omega}(\boldsymbol{x}) = \operatorname{curl} \boldsymbol{u}(\boldsymbol{x}). \tag{9.1}$$

29*

The correlation between the vorticity at two different points in space is then related to $R_{ij}(r)$ in the manner

$$\overline{\omega_i(\boldsymbol{x})\,\omega_j(\boldsymbol{x}+\boldsymbol{r})} = \left(\frac{\partial^2}{\partial r_i\,\partial r_j} - \delta_{ij}\nabla^2\right) R_{kk}(\boldsymbol{r}) + \nabla^2 R_{ji}(\boldsymbol{r}); \tag{9.2}$$

a contraction of the indices then gives

$$\overline{\omega_i(\boldsymbol{x})\,\omega_i(\boldsymbol{x}+\boldsymbol{r})} = -\nabla^2 R_{ii}(\boldsymbol{r}); \tag{9.3}$$

and letting $\boldsymbol{r}=0$ we have for one component of the mean square vorticity

$$\overline{\omega^2} = -5\overline{u^2}\,f_0''. \tag{9.4}$$

Now, if we let $\Omega_{ij}(\boldsymbol{\varkappa})$ be the Fourier transform of $\overline{\omega_i(\boldsymbol{x})\,\omega_j(\boldsymbol{x}+\boldsymbol{r})}$ in the sense of Eq. (6.5), then the relationship between $\Omega_{ij}(\boldsymbol{\varkappa})$ and $\Phi_{ij}(\boldsymbol{\varkappa})$ follows directly from Eq. (9.2) in the form

$$\Omega_{ij}(\boldsymbol{\varkappa}) = (\delta_{ij}\varkappa^2 - \varkappa_i\varkappa_j)\,\Phi_{kk}(\boldsymbol{\varkappa}) - \varkappa^2\,\Phi_{ji}(\boldsymbol{\varkappa}) \tag{9.5}$$

and by again contracting the indices, we obtain an expression for the density in wave-number space of the contributions to the mean-square total vorticity,

$$\Omega_{ii}(\boldsymbol{\varkappa}) = \varkappa^2\,\Phi_{ii}(\boldsymbol{\varkappa}). \tag{9.6}$$

Thus the vorticity spectrum is identical with the spectrum of viscous dissipation and, from Eqs. (9.4) and (8.9), we have

$$\tfrac{1}{2}\,\overline{\omega^2} = \int_0^\infty \varkappa^2\,E(\varkappa)\,d\varkappa. \tag{9.7}$$

From Eqs. (9.4) and (8.5) we may note a further interpretation for $\lambda$, viz., it is just the dispersion of energy in wave number space,

$$\frac{1}{\lambda^2} = \frac{1}{5}\cdot\frac{\int_0^\infty \varkappa^2\,E(\varkappa)\,d\varkappa}{\int_0^\infty E(\varkappa)\,d\varkappa}. \tag{9.8}$$

The coefficient of $r^2$ in the expansion of the Kármán-Howarth equation is

$$\frac{d}{dt}\,\overline{u^2}\,f_0'' = \frac{7}{3}\,(\overline{u^2})^{\frac{3}{2}}\,k_0''' + \frac{14}{3}\,\nu\,\overline{u^2}\,f_0^{\mathrm{iv}}. \tag{9.9}$$

But, by Eq. (9.4), this is simply the equation for the rate of change of the mean square vorticity; it can therefore be written in the equivalent form

$$\frac{d\,\overline{\omega^2}}{dt} = \frac{7}{3\sqrt{15}}\left(S - \frac{2G}{R_\lambda}\right)(\overline{\omega^2})^{\frac{3}{2}}, \tag{9.10}$$

where $S$ is the skewness factor of $-\partial u_1/\partial x_1$ defined by Eq. (4.5), $R_\lambda$ is the Reynolds number of the turbulence $(\overline{u^2})^{\frac{1}{2}}\lambda/\nu$ and

$$G = \overline{u^2}\,\frac{\overline{(\partial^2 u_1/\partial x_1^2)^2}}{[\overline{(\partial u_1/\partial x_1)^2}]^2} = \lambda^4\,f_0^{\mathrm{iv}}. \tag{9.11}$$

The meaning of Eq. (9.10) will perhaps be made somewhat clearer if we consider its derivation directly from HELMHOLTZ' equation for the rate of change of vorticity, obtained by taking the curl of the Navier-Stokes equation,

$$\frac{\partial\omega_i}{\partial t} + u_j\,\frac{\partial\omega_i}{\partial x_j} - \omega_j\,\frac{\partial u_i}{\partial x_j} = \nu\,\nabla^2\omega_i. \tag{9.12}$$

If this equation is multiplied by $\omega_i$ and mean values are taken, then an equation for the rate of change of mean-square vorticity is obtained that can be written in the form

$$\frac{\partial \overline{\omega^2}}{\partial t} = 2\overline{\omega_i \omega_j \frac{\partial u_i}{\partial x_j}} + 2\nu \overline{\omega_i \nabla^2 \omega_i}. \tag{9.13}$$

The second term on the right-hand side of this equation can be written

$$2\nu \overline{\omega_i \nabla^2 \omega_i} = \nu \left[\frac{\partial^2 \overline{\omega_i^2}}{\partial x_j^2} - 2\overline{\left(\frac{\partial \omega_i}{\partial x_j}\right)^2}\right] \tag{9.14}$$

and, since $\partial^2 \overline{\omega_i^2}/\partial x_j^2$ vanishes by homogeneity, it is essentially negative and describes the viscous dissipation of vorticity. The first term on the right-hand side of Eq. (9.13) can be written

$$2\overline{\omega_i \omega_j \frac{\partial u_i}{\partial x_j}} = 2\overline{\omega^2 \left(\frac{\partial u}{\partial s}\right)_s}, \tag{9.15}$$

where $\omega$ is the magnitude of the local vorticity vector and $(\partial u/\partial s)_s$ represents the gradient of the component of the velocity along the local vortex line. Thus, $(\partial u/\partial s)_s$ gives the instantaneous rate of extension of the local vortex line. Now, it is a fundamental property of turbulent flows that neighboring points tend to move apart (cf. Chap. VII) so that, on the average, vortex lines tend to be extended rather than contracted, and thus the term (9.15) represents a growth of vorticity due to the random stretching of vortex lines, the magnitude of which is limited ultimately only by the presence of viscous dissipation.

**10. One-dimensional energy spectra.** The spectrum functions $\Phi_{ij}(\boldsymbol{\varkappa})$ and $E(\varkappa)$ considered thus far are three-dimensional, in keeping with the (necessarily) three-dimensional character of the turbulence itself. But, while these are the appropriate quantities to employ in theoretical investigations, unfortunately they are not quantities that can be measured directly in the laboratory. What is actually done experimentally is to Fourier analyze the velocity fluctuations that are swept past a fixed point, and the time spectrum obtained in this way can be related to a theoretical one-dimensional spatial spectrum by the use of TAYLOR's hypothesis[1] which states that the spatial pattern of the motion is swept past a fixed point by the mean velocity without any essential change. The wave number $\varkappa$ corresponding to a frequency $n$ is therefore given by

$$\varkappa = 2\pi n/U, \tag{10.1}$$

where $U$ is the mean velocity.

Thus, a Fourier analysis of the velocity fluctuations along a line in the $x_1$ direction, the direction of mean flow, leads to the one-dimensional spectrum functions

$$\varphi_{ij}(\varkappa_1) = \frac{1}{2\pi}\int_{-\infty}^{\infty} R_{ij}(r_1, 0, 0)\, e^{-i\varkappa_1 r_1}\, dr_1 = \int_{-\infty}^{\infty} \Phi_{ij}(\varkappa)\, d\varkappa_2\, d\varkappa_3. \tag{10.2}$$

When the turbulence is isotropic, $\varphi_{ij}(\varkappa_1)$, like $\Phi_{ij}(\boldsymbol{\varkappa})$, is completely characterized by a single component $\varphi_{11}(\varkappa_1) = \varphi(\varkappa_1)$ say, where

$$\varphi(\varkappa) = \frac{1}{\pi}\int_0^{\infty} \overline{u^2} f(r) \cos \varkappa r\, dr = \frac{1}{2}\int_{\varkappa}^{\infty} \left(1 - \frac{\varkappa^2}{\varkappa'^2}\right) \frac{E(\varkappa')}{\varkappa'}\, d\varkappa', \tag{10.3}$$

---

[1] G.I. TAYLOR: Proc. Roy. Soc. Lond., Ser. A **164**, 476 (1938). See also C.C. LIN: Quart. Appl. Math. **10**, 295 (1953).

to which there correspond the inverse relations

$$\overline{u^2} f(r) = 2\int_0^\infty \varphi(\varkappa) \cos \varkappa r \, d\varkappa \tag{10.4}$$

and

$$E(\varkappa) = \varkappa^3 \frac{d}{d\varkappa}\left[\frac{1}{\varkappa}\,\frac{d\varphi(\varkappa)}{d\varkappa}\right]. \tag{10.5}$$

The function $\varphi(\varkappa)$ is called the longitudinal one-dimensional energy spectrum. Closely related to it is the lateral one-dimensional energy spectrum defined by

$$\left.\begin{aligned}\varphi_{22}(\varkappa) &= \frac{1}{\pi}\int_0^\infty \overline{u_2}\, g(r) \cos \varkappa r \, dr \\ &= \frac{1}{4}\int_\varkappa^\infty \left(1 + \frac{\varkappa^2}{\varkappa'^2}\right) \frac{E(\varkappa')}{\varkappa'}\, d\varkappa'.\end{aligned}\right\} \tag{10.6}$$

If $\varphi(\varkappa)$ and $\varphi_{22}(\varkappa)$ are both measured, then, because of the relationship between them,

$$\varphi_{22}(\varkappa) = \frac{1}{2}\left(1 - \varkappa\frac{d}{d\varkappa}\right)\varphi(\varkappa), \tag{10.7}$$

only a single differentiation of the experimental data is required to determine $E(\varkappa)$. By comparing direct measurements of the spectrum with those obtained by transforming the correlation function, Fig. 6, Stewart and Townsend [26] were able to show that Taylor's hypothesis is valid for all but the large-scale motions associated with small wave-numbers. The agreement between the two methods over a considerable range of wave-numbers is quite impressive and the deviations for small wave-numbers is very likely due to the fact that the large-scale motions may change significantly during the time required for them to pass the fixed point at which the measurements are being made.

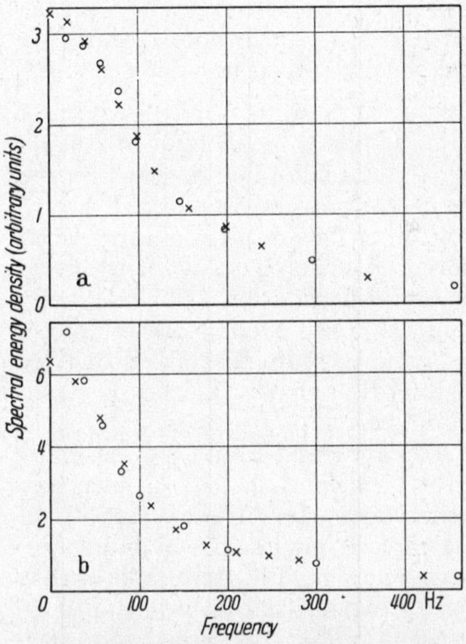


Fig. 6a and b. One-dimensional spectra: ○, direct measurement; ×, transformed double correlation.

| | $M$ (cm) | $x/M$ | $U$ (cm/sec) | $R_M$ |
|---|---|---|---|---|
| (a) | 1.27 | 60 | 620 | 5250 |
| (b) | 5.08 | 30 | 620 | 21000 |

From Stewart and Townsend [26].

**11. Loitsiansky's integral.** The behavior of the spectrum tensor $\Phi_{ij}(\boldsymbol{\varkappa})$ and the energy spectrum function $E(\varkappa)$ at small wave numbers has been the subject of considerable theoretical research. This has been largely due to the *apparent* statistical invariance of the largest eddies. The nature of this apparent invariance can most easily be derived directly from the Kármán-Howarth equation

$$\frac{\partial}{\partial t}\overline{u^2} f(r) = \left(\frac{\partial}{\partial r} + \frac{4}{r}\right)(\overline{u^2})^{\frac{3}{2}} k(r) + 2\nu\left(\frac{\partial^2}{\partial r^2} + \frac{4}{r}\frac{\partial}{\partial r}\right)\overline{u^2} f(r); \tag{11.1}$$

for, if this equation is multiplied by $r^4$ and integrated over all values of $r$, then one obtains the exact result

$$\frac{d}{dt}\overline{u^2}\int_0^\infty r^4 f(r)\, dr = (\overline{u^2})^{\frac{3}{2}} \lim_{r\to\infty} r^4 k(r) + 2\nu\,\overline{u^2} \lim_{r\to\infty} r^4 f'(r). \tag{11.2}$$

If it is assumed that the integral on the left-hand side exists and that $k(r)$ is of smaller order than $r^{-4}$ as $r \to \infty$, then the right-hand side of this equation vanishes and the integral

$$L = \overline{u^2} \int_0^\infty r^4 f(r)\, dr \tag{11.3}$$

is an invariant of the motion. In its original form this result is due to LOITSIANSKY[1]; the corresponding spectral results were obtained by LIN[2] for isotropic turbulence and by BATCHELOR[3] for homogeneous turbulence.

PROUDMAN and REID [*22*], however, have shown that the assumptions which lead to the invariance of the integral (11.3) are not satisfied when the joint-probability distribution of the velocity field is approximately normal (cf. Chap. V), and they have presented plausible arguments for believing this result to be true in general. The whole question has been extensively studied more recently by BATCHELOR and PROUDMAN [*4*] and their results confirm the earlier suggestions[4]. This work will be described in the following three sections.

**12. The long-range effect of pressure forces.** In the work of PROUDMAN and REID it was shown that even if $\Phi_{ij}(\boldsymbol{\varkappa})$ possessed an analytic expansion at some initial instant of time, the dynamical action of the pressure would give rise to a non-analytic form at subsequent instants of time. This result was established rigorously only for the special case of a normal joint-probability distribution of the velocity field. But, since such a probability distribution is kinematically admissible, it is clear that some important point had been overlooked heretofore. This point involves the long-range influence of the pressure that arises directly through dynamical action and leads to non-analytic behavior of the spectrum tensors and the related power-law forms for the correlation tensors.

In their analysis of the general problem BATCHELOR and PROUDMAN assumed that there exists some initial instant, $t_0$ say, at which time the turbulence is homogeneous and possesses convergent integral moments of all orders of all velocity correlations[5]. At first sight, this assumption might appear unduly restrictive. Actually, it is probably a conservative one; for, if it can be shown that power-law forms for the correlations develop for exponential initial conditions, then *a fortiori* they may be expected to develop for power-law initial conditions. In any event, the assumption mentioned is a mathematically convenient one and permits a systematic and fairly complete description to be made of the effect of pressure forces on the large-scale structure of the turbulence.

The precise nature of the difficulty, heretofore overlooked, can easily be demonstrated by considering the asymptotic behavior of the pressure-velocity correlation occurring in the dynamical equation (6.14) for $R_{ij}(\boldsymbol{r})$. By taking the divergence of that equation we obtain

$$\nabla^2 \frac{1}{\varrho} \overline{p\, u_j'} = - \frac{\partial^2 \overline{u_i u_j' u_k}}{\partial r_i \partial r_k}, \tag{12.1}$$

[1] L. G. LOITSIANSKY: Rep. Cent. Aero. Hydrodyn. Inst. (Moscow), No. 440 (1939). Translated as N.A.C.A. Tech. Memo., No. 1079 (1945).

[2] C.C. LIN: Proc. 1st Symp. in Appl. Math., Amer. Math. Soc. 1947, p. 81.

[3] G.K. BATCHELOR: Proc. Roy. Soc. Lond., Ser. A **195**, 513 (1949).

[4] See also G.K. BATCHELOR: Proc. 7th Symp. in Appl. Math., Amer. Math. Soc. 1955, p. 67.

[5] Or, more precisely, of all velocity *cumulants*, an example of which is the general fourth-order cumulant defined by the expression

$$\overline{u_i u_j' u_k'' u_l'''} - \overline{u_i u_j'} \cdot \overline{u_k'' u_l'''} - \overline{u_i u_k''} \cdot \overline{u_j' u_l'''} - \overline{u_i u_l'''} \cdot \overline{u_j' u_k''},$$

where the single, double and triple primes refer to the spatially distinct points $\boldsymbol{x}'$, $\boldsymbol{x}''$, and $\boldsymbol{x}'''$ respectively. The cumulants, unlike the correlations, have the property that they approach zero when the distance between any two of the points becomes large. In the absence of a mean velocity, the second- and third-order cumulants are identical with the correlations.

the solution of which is

$$\frac{1}{\varrho}\overline{p u_j'} = \frac{1}{4\pi}\int \frac{\partial^2 \overline{u_i'' u_j' u_k''}}{\partial x_i'' \partial x_k''} \cdot \frac{d\boldsymbol{x}''}{|\boldsymbol{x}'' - \boldsymbol{x}|}. \tag{12.2}$$

Thus, $\overline{p u_j'}$ is a quantity whose behavior for large values of $r$ need not be exponential, even at the initial instant $t_0$. By hypothesis, all integral moments of $\overline{u_i'' u_j' u_k''}$ exist at time $t_0$ and, accordingly, it is permissible to expand the right-hand side of Eq. (12.2) in inverse powers of $r$; the first non-zero term in this expansion is found to be

$$\frac{1}{\varrho}\overline{p u_j'} \sim - T_{ijkl}\frac{\partial^3}{\partial r_i \partial r_k \partial r_l}\left(\frac{1}{r}\right), \tag{12.3}$$

where

$$T_{ijkl} = \frac{1}{4\pi}\int \overline{u_i u_j' u_k} \, r_l \, d\boldsymbol{r}. \tag{12.4}$$

Thus, even if the velocity correlations are assumed initially to be exponentially small for large values of $r$, the mean values involving the pressure, depending as they do on the whole velocity distribution, will be algebraic in general for large values of $r$. This is what is meant by the long-range effect of pressure forces.

Furthermore, by substituting the result (12.3) into the dynamical equation for $R_{ij}(\boldsymbol{r})$ we obtain

$$\left[\frac{\partial R_{ij}(\boldsymbol{r})}{\partial t}\right]_{t=t_0} = - O_2\left\{T_{\alpha jkl}\frac{\partial^4}{\partial r_i \partial r_\alpha \partial r_k \partial r_l}\left(\frac{1}{r}\right)\right\} \tag{12.5}$$

so that, even if the behavior of $R_{ij}(\boldsymbol{r})$ for large values of $r$ is initially exponentially small, at subsequent instants it will in general be of order $r^{-5}$. Analogous results can also be established for the vorticity correlations $\overline{\omega_i u_j'}$ and $\overline{\omega_i \omega_j'}$ which play an important role in the subsequent analysis. It is found, for example, that the first time derivative of $\overline{\omega_i u_j'}$ initially is of order $r^{-6}$; in the case of $\overline{\omega_i \omega_j'}$, however, a power-law behavior does not appear until we consider the third time derivative, a result that can be partially traced to the less direct influence of the pressure on the vorticity, and it is found to be of order $r^{-8}$. By considering time derivatives of higher order, it can also be shown that no terms of larger order than those indicated will ever appear.

Thus, having indicated the types of power-law behavior to be expected in homogeneous turbulence as a result of the long-range effect of pressure forces we turn now to consider some of the consequences of these results insofar as they relate to the kinematics and dynamics of the large-scale motion.

**13. The large-scale structure of the turbulence.** The asymptotic behavior of the correlation tensors for large values of $r$ and the related behavior of the spectrum tensors for small values of $\varkappa$, with which the large-scale structure of the turbulence is associated, have been studied by BATCHELOR and PROUDMAN [*4*] taking into account the long-range effects of the pressure. The problem is one of considerable intricacy and no attempt can be made here to describe their analysis in any great detail; all that can be done is to give a few of their major results.

The fact that $R_{ij}(\boldsymbol{r})$ is of order $r^{-5}$ for large values of $r$ insures that its integral and first integral moment, taken over all space, converge absolutely, and this, in turn, insures the continuity of $\Phi_{ij}(\boldsymbol{\varkappa})$ and $\partial\Phi_{ij}(\boldsymbol{\varkappa})/\partial\varkappa_k$ for all values of $\varkappa$, including $\varkappa = 0$. But the second derivative $\partial^2\Phi_{ij}(\boldsymbol{\varkappa})/\partial\varkappa_k\partial\varkappa_l$ exists as a continuous function of $\boldsymbol{\varkappa}$ only for $\varkappa > 0$, and is in fact singular for $\varkappa = 0$. It is this singularity, heretofore overlooked, that constitutes the chief difficulty of the problem.

The results now to be described can be expressed in terms of the generalized Loitsiansky integral[1]

$$L_{ijkl} = \frac{1}{4\pi} \int R_{ij}(\boldsymbol{r})\, r_k r_l \, d\boldsymbol{r} \tag{13.1}$$

and the first integral moment of the two-point triple correlation, $T_{ijkl}$, defined by Eq. (12.4), above. It is also convenient to define the projection operator

$$\Delta_i^\alpha(\boldsymbol{\varkappa}) = \delta_{i\alpha} - \varkappa_i \varkappa_\alpha / \varkappa^2 \tag{13.2}$$

and the related operator in physical space

$$D_i^\alpha(\boldsymbol{r}) = \delta_{i\alpha} \nabla^2 - \partial^2/\partial r_i \, \partial r_\alpha . \tag{13.3}$$

If one regards $-k^2 \Delta_i^\alpha(\boldsymbol{\varkappa})$ and $i\boldsymbol{\varkappa}$ as Fourier transforms of $D_i^\alpha(\boldsymbol{r})$ and $\partial/\partial \boldsymbol{r}$ respectively, then the solenoidal representation of physical tensors in terms of the operator (13.3) becomes formally identical (apart from factors of $i$) with the orthogonal representation of spectral tensors in terms of the operator (13.2).

The behavior of $\Phi_{ij}(\boldsymbol{\varkappa})$ for small values of $\varkappa$ can now be written in the form

$$\Phi_{ij}(\boldsymbol{\varkappa}) \to \frac{1}{4\pi^2} C_{\alpha\beta kl} \Delta_i^\alpha(\boldsymbol{\varkappa})\, \Delta_j^\beta(\boldsymbol{\varkappa})\, \varkappa_k \varkappa_l \quad \text{as} \quad \varkappa \to 0, \tag{13.4}$$

where the tensor coefficient $C_{\alpha\beta kl}$ is related to the fourth integral moment of the double vorticity correlation. This tensor is symmetric in the indices $(\alpha, \beta)$ and $(k, l)$ and $C_{\alpha\beta\beta l}$ vanishes. It is also possible to relate it to $L_{ijkl}$ but the relationship is not simple. Eq. (13.4) shows explicitly that $\Phi_{ij}(\boldsymbol{\varkappa})$ is of order $\varkappa^2$ for small values of $\varkappa$ but contains a singularity at $\varkappa = 0$ that depends on the *direction* but not the magnitude of $\boldsymbol{\varkappa}$. The form of this singularity is in fact identical with the earlier results [22] obtained from dynamical considerations. In that work it was possible to trace the presence of the term $\varkappa_i \varkappa_\alpha / \varkappa^2$ in the projection operator $\Delta_i^\alpha(\boldsymbol{\varkappa})$ to the influence of the pressure.

The corresponding behavior of $R_{ij}(\boldsymbol{r})$ for large values of $r$ can be obtained by direct but rather lengthy methods; alternatively, it can be obtained more easily by simply taking the Fourier transform of Eq. (13.4) with the result

$$R_{ij}(\boldsymbol{r}) \sim \frac{1}{4} C_{\alpha\beta kl} D_i^\alpha(\boldsymbol{r})\, D_j^\beta(\boldsymbol{r}) \frac{\partial^2 r}{\partial r_k \, \partial r_l} \quad \text{as} \quad r \to \infty. \tag{13.5}$$

It now only remains necessary to quote the corresponding results for the tensors $\overline{u_i u_j' u_k}$ and $\overline{p u_j'}$ which appear in the dynamical equation for $R_{ij}(\boldsymbol{r})$. From a consideration of the dynamical equation for the first derivative of $\overline{u_i u_j' u_k}$, it can easily be shown that this tensor is in general of order $r^{-4}$ for large values of $r$, but that the integral

$$T_{ijkl} = \frac{1}{4\pi} \int \overline{u_i u_j' u_k}\, r_l \, d\boldsymbol{r},$$

which has already appeared in Eq. (12.3) for the *initial* behavior of $\overline{p u_j'}$, converges by arguments similar to those used to prove the convergence of the second moment of $R_{ij}(\boldsymbol{r})$. The asymptotic behavior of $\overline{u_i u_j' u_k}$ can then be written in the form

$$\overline{u_i u_j' u_k} \sim -\frac{5}{3} T_{iklm} \frac{\partial^3}{\partial r_j \, \partial r_l \, \partial r_m} \left(\frac{1}{r}\right) \quad \text{as} \quad r \to \infty, \tag{13.6}$$

[1] The conditional convergence of this integral follows from the fact that the leading term in the expansion of $R_{ij}(\boldsymbol{r})$ can be written as a third derivative of a tensor of order $r^{-2}$ and, consequently, this term contributes nothing to the total volume integral (13.1).

where the further condition $T_{ikll}=0$ is required by the fact that $\overline{u_i u_j' u_k}$ is solenoidal in the index $j$. The corresponding behavior of its Fourier transform,

$$\Upsilon_{ijk}(\boldsymbol{\varkappa}) = i\,(2\pi)^{-3}\int \overline{u_i\, u_j'\, u_k}\, e^{-i\boldsymbol{\varkappa}\cdot\boldsymbol{r}}\, d\boldsymbol{r}, \tag{13.7}$$

for small values of $\varkappa$ is then given by

$$\Upsilon_{ijk}(\boldsymbol{\varkappa}) \to \frac{1}{6\pi^2}\,(T_{ikl\beta} + 4\,T_{ik\beta l})\,\Delta_j^{\beta}(\boldsymbol{\varkappa})\,\varkappa_l \quad \text{as} \quad \varkappa \to 0. \tag{13.8}$$

Now it is a consequence of the incompressibility condition that $\Pi_j(\boldsymbol{\varkappa})$, the Fourier transform of $\overline{p u_j'}$, is related to $\Upsilon_{ijk}(\boldsymbol{\varkappa})$ by the relation

$$\Pi_j(\boldsymbol{\varkappa}) = -\frac{\varkappa_i \varkappa_k}{\varkappa^2}\,\Upsilon_{ijk}(\boldsymbol{\varkappa}), \tag{13.9}$$

and its behavior for small values of $\varkappa$ follows directly therefore from Eq. (13.8). From this result, the asymptotic behavior of $\overline{p u_j'}$ then follows by a simple Fourier transformation in the form

$$\frac{1}{\varrho}\,\overline{p\, u_j'} \sim \frac{1}{6}\,(T_{ikl\beta} + 4\,T_{ik\beta l})\,D_j^{\beta}(\boldsymbol{r})\,\frac{\partial^3 r}{\partial r_i\,\partial r_k\,\partial r_l} \quad \text{as} \quad r \to \infty. \tag{13.10}$$

We have thus given the general asymptotic behavior of all of the terms appearing in the dynamical equation for $R_{ij}(\boldsymbol{r})$ and the related behavior of their Fourier transforms for small values of $\varkappa$. These results have all been expressed in terms of the tensors $C_{ijkl}$, which is related to the second integral moment of $\overline{u_i u_j'}$, and $T_{ijkl}$, which is just the first integral moment of $\overline{u_i u_j' u_k}$. And we now turn to consider briefly the dynamical behavior of these tensors which describe the large-scale structure of the turbulence.

The dynamical behavior of the tensors $C_{ijkl}$ and $T_{ijkl}$ can be obtained of course by considering either the equation for $R_{ij}(\boldsymbol{r})$ or the one for $\Phi_{ij}(\boldsymbol{\varkappa})$. In the latter case, however, the analytical details are somewhat simpler. Since $\Sigma_{ij}(\boldsymbol{\varkappa})$ is given by

$$\Sigma_{ij}(\boldsymbol{\varkappa}) = O_2\,\{\varkappa_k\,\Delta_i^{\alpha}(\boldsymbol{\varkappa})\,\Upsilon_{\alpha jk}(\boldsymbol{\varkappa})\}, \tag{13.11}$$

its behavior for small values of $\varkappa$ follows directly from Eq. (13.8), and the form of Eq. (6.18) as $\varkappa \to 0$ is therefore given by

$$O_2\left\{\left[\frac{dC_{\alpha\beta kl}}{dt} - \frac{1}{3}\,(T_{\alpha kl\beta} + 4\,T_{\alpha k\beta l})\right]\Delta_i^{\alpha}(\boldsymbol{\varkappa})\,\Delta_j^{\beta}(\boldsymbol{\varkappa})\,\varkappa_k\,\varkappa_l\right\} = 0. \tag{13.12}$$

Thus, the large-scale structure of the turbulence, whether characterized by $C_{ijk}$ itself or, more correctly, by the whole leading term in the expansion of $\Phi_{ij}(\boldsymbol{\varkappa})$ is not invariant throughout the decay process.

When the turbulence is assumed to be isotropic, many of the analytica complications encountered in the more general treatment disappear. Thus, for example, the tensor coefficients introduced in the preceding analysis reduce to the simple forms

$$C_{lmnp} = \tfrac{1}{3}\,L\,[\delta_{lm}\,\delta_{np} - \tfrac{1}{4}\,(\delta_{ln}\,\delta_{mp} + \delta_{lp}\,\delta_{mn})], \tag{13.13}$$

$$L_{ijkl} = \tfrac{1}{3}\,L\,[-\,\delta_{ij}\,\delta_{kl} + \tfrac{1}{2}\,(\delta_{ik}\,\delta_{jl} + \delta_{il}\,\delta_{jk})] \tag{13.14}$$

and

$$T_{ijkl} = \tfrac{1}{30}\,T\,[\delta_{ij}\,\delta_{kl} - \tfrac{3}{2}\,(\delta_{ik}\,\delta_{jl} + \delta_{il}\,\delta_{jk})], \tag{13.15}$$

where the defining scalars $L$ and $T$ are defined by

$$L = \overline{u^2}\int_0^{\infty} r^4 f(r)\,dr \quad \text{and} \quad T = (\overline{u^2})^{\frac{3}{2}} \lim_{r\to\infty} r^4\,k(r), \tag{13.16}$$

there being no reason now to suppose that $T$ vanishes.

When the tensor $C_{ijkl}$ is of the form (13.13), the leading term in the expansion of $R_{ij}(\mathbf{r})$, Eq. (13.5), vanishes identically. In fact, by considering the initial behavior of $\partial^n R_{ij}(\mathbf{r})/\partial t^n$ ($n=1, 2, \ldots$), BATCHELOR and PROUDMAN were unable to find any algebraic term when the condition of isotropy was imposed. Nor were they able to show that such a term could not occur. Thus the behavior of $R_{ij}(\mathbf{r})$, and of its defining scalar $\overline{u^2} f(r)$, for large values of $r$ in isotropic turbulence still remains an open question and the possibility of their being exponentially small cannot be excluded. The fact that $R_{ij}(\mathbf{r})$ is now not larger than order $r^{-6}$ ensures that $L_{ijkl}$ (and $L$) are absolutely convergent and that $\Phi_{ij}(\boldsymbol{\varkappa})$ is no longer singular at $\varkappa=0$ but has the analytic form

$$\Phi_{ij}(\boldsymbol{\varkappa}) \to \frac{L}{12\pi^2} (\delta_{ij} \varkappa^2 - \varkappa_i \varkappa_j) \quad \text{as} \quad \varkappa \to 0. \tag{13.17}$$

Even when the turbulence is isotropic, the triple correlation $\overline{u_i u_j' u_k}$ remains of order $r^{-4}$ for large values of $r$ and its asymptotic behavior is given by

$$\overline{u_i u_j' u_k} \sim -\frac{1}{6} T \frac{\partial^3}{\partial r_i \partial r_j \partial r_k} \left(\frac{1}{r}\right). \tag{13.18}$$

The corresponding behavior of its Fourier transform for small $\varkappa$ then follows from Eq. (13.8) in the form

$$\Upsilon_{ijk}(\boldsymbol{\varkappa}) \to \frac{1}{24\pi^2} T \Delta_j^\beta(\boldsymbol{\varkappa}) (\varkappa_i \delta_{\beta k} + \varkappa_k \delta_{\beta i}) \quad \text{as} \quad \varkappa \to 0 \tag{13.19}$$

which has the required orthogonality property with respect to $\varkappa_j$ and is symmetric in the indices $i$ and $k$. The pressure correlation $\overline{p u_j'}$ and its Fourier transform $\Pi_j(\boldsymbol{\varkappa})$, it may be recalled, vanish identically in isotropic turbulence.

The dynamical behavior of the defining scalars $L$ and $T$ can, of course, be determined directly from Eq. (13.12) but it is somewhat simpler to go back to Eq. (6.18). For this purpose we may note that $\Sigma_{ij}(\boldsymbol{\varkappa})$ is given by

$$\Sigma_{ij}(\boldsymbol{\varkappa}) \to \frac{1}{12\pi^2} T (\delta_{ij} \varkappa^2 - \varkappa_i \varkappa_j) \quad \text{as} \quad \varkappa \to 0 \tag{13.20}$$

and hence, like $\Phi_{ij}(\boldsymbol{\varkappa})$, it is analytic for small values of $\varkappa$. By substituting this result, and the corresponding one for $\Phi_{ij}(\boldsymbol{\varkappa})$, Eq. (13.17), into Eq. (6.18) we obtain simply

$$\frac{dL}{dt} = T, \tag{13.21}$$

which is just Eq. (11.2) with the viscous term neglected,

$$\frac{d}{dt} \overline{u^2} \int_0^\infty r^4 f(r)\, dr = (\overline{u^2})^{\frac{3}{2}} \lim_{r \to \infty} r^4 k(r). \tag{13.22}$$

Now, in Sect. 34 it will be shown that for an approximately normal joint-probability distribution of the velocity field we have

$$\frac{d}{dt} \left[(\overline{u^2})^{\frac{3}{2}} \lim_{r\to\infty} r^4 k(r)\right] > 0 \quad \text{and} \quad \frac{d^2}{dt^2} \left[\overline{u^2} \int_0^\infty r^4 f(r)\, dr\right] > 0. \tag{13.23}$$

Existing measurements of $k(r)$ and $f(r)$ so far as they go suggest that these functions are everywhere negative and positive respectively and it appears therefore that the behavior of $T$ and $L$, as functions of the time, is dominated by the decay of $\overline{u^2}$. The parameter $L$ would therefore be a decreasing function of $t$ during the period of decay for which inertia forces are not negligible but would tend asymptotically to a constant value as the final period of decay is approached. During this so-called "final period" of decay, inertia forces are negligible, $L$ is constant, and it is possible therefore under these conditions to obtain a few simple theoretical predictions, which will be discussed in the following section.

**14. The final period of decay.** At a sufficiently advanced stage in the decay process, the inertia terms in the Navier-Stokes equation can be neglected relative to the viscous terms and we are then dealing with what is called the "final period" of decay (see also Sect. 18). Under these conditions the velocity satisfies the linear equation

$$\frac{\partial \boldsymbol{u}}{\partial t} = \nu \nabla^2 \boldsymbol{u} \tag{14.1}$$

and the equations satisfied by $R_{ij}(\boldsymbol{r})$ and $\Phi_{ij}(\boldsymbol{\varkappa})$ then become

$$\frac{\partial R_{ij}(\boldsymbol{r}, t)}{\partial t} = 2\nu \nabla^2 R_{ij}(\boldsymbol{r}, t) \tag{14.2}$$

and

$$\frac{\partial \Phi_{ij}(\boldsymbol{\varkappa}, t)}{\partial t} = -2\nu \varkappa^2 \Phi_{ij}(\boldsymbol{\varkappa}, t). \tag{14.3}$$

This latter equation possesses a solution of the form

$$\Phi_{ij}(\boldsymbol{\varkappa}, t) = \Phi_{ij}(\boldsymbol{\varkappa}, t_0)\, \mathrm{e}^{-2\nu\varkappa^2(t-t_0)} \tag{14.4}$$

where $t_0$ is a time for which Eq. (14.1) is valid. Now, as $t-t_0 \to \infty$ the exponential factor in Eq. (14.4) will be small except for small values of $\varkappa$; accordingly, we can replace $\Phi_{ij}(\boldsymbol{\varkappa}, t_0)$ by the leading term in its power series expansion and the solution then becomes, for $t-t_0 \to \infty$,

$$\Phi_{ij}(\boldsymbol{\varkappa}, t) = \frac{1}{4\pi^2} C_{\alpha\beta kl}\, \Delta_i^\alpha(\boldsymbol{\varkappa})\, \Delta_j^\beta(\boldsymbol{\varkappa})\, \varkappa_k \varkappa_l\, \mathrm{e}^{-2\nu\varkappa^2(t-t_0)}, \tag{14.5}$$

where $C_{\alpha\beta kl}$ is now constant. The corresponding behavior of $R_{ij}(\boldsymbol{r}, t)$ then follows by direct transformation in the form

$$R_{ij}(\boldsymbol{r}, t) = \frac{C_{\alpha\beta kl}}{4\sqrt{\pi}\,[8\nu(t-t_0)]^{\frac{5}{2}}} D_i^\alpha(\boldsymbol{y})\, D_j^\beta(\boldsymbol{y}) \times \frac{\partial^2}{\partial y_k\, \partial y_l} \left[\pi^{\frac{1}{2}}\left(y + \frac{1}{y}\right) \operatorname{erf} y + \mathrm{e}^{-y^2}\right], \tag{14.6}$$

where

$$\boldsymbol{y} = \boldsymbol{r}/[8\nu(t-t_0)]^{\frac{1}{2}} \quad \text{and} \quad \operatorname{erf} y = \frac{2}{\sqrt{\pi}} \int_0^y \mathrm{e}^{-t^2}\, dt.$$

From Eq. (14.6) we see that the energy tensor, $\overline{u_i u_j} = R_{ij}(0, t)$, decreases as $[\nu(t-t_0)]^{-\frac{5}{2}}$, the so-called final period decay law. Furthermore, for large values of $r$, the function of $y$ within the square brackets is equal to $\pi^{\frac{1}{2}} y$ asymptotically and we thus recover the correct behavior for $R_{ij}(\boldsymbol{r})$ as $r \to \infty$.

Although $C_{ijkl}$ is constant in the final period, its value cannot be related to the conditions prevailing at the initial instant of creation of the turbulence. The observed anisotropy in the final period[1], which was explained so well by the older theory since it could be related directly to the initial anisotropy of $C_{ijkl}$, therefore, requires further discussion. The argument advanced most strongly by BATCHELOR and PROUDMAN [4] is that the measurements were made at such low Reynolds numbers that the period during which inertia forces were appreciable was very short and, under such conditions, they argue that "the turbulence in

[1] G.K. BATCHELOR and R.W. STEWART: Quart. J. Mech. Appl. Math. 3, 1 (1950).

the final period of decay will not be very different from that which would be predicted with the assumption of an analytic spectrum". Such arguments are not wholly convincing, however, and the explanation of the anisotropy observed in the final period of decay must be left open for the present.

When the turbulence is isotropic, the foregoing results can be greatly simplified. Thus, the isotropic energy spectrum function and the longitudinal correlation coefficient are given by

$$E(\varkappa, t) = \frac{L}{3\pi} \varkappa^4 \, e^{-2\nu\varkappa^2(t-t_0)} \quad \text{and} \quad f(r) = e^{-r^2/8\nu(t-t_0)}, \tag{14.7}$$

and the decay law is

$$\overline{u^2} \sim \frac{L}{48(2\pi)^{\frac{1}{2}}[\nu(t-t_0)]^{\frac{5}{2}}}. \tag{14.8}$$

## III. General nature of homogeneous turbulence.

**15. Energy balance in decaying homogeneous turbulence.** We have just seen that the study of the behavior of the large-scale structure of turbulence is complicated by the long-range effect of the pressure forces. In this chapter we shall see more clearly why we cannot expect to get important precise results concerning the behavior of large eddies in homogeneous turbulent motion in the absence of a basic shear flow. Indeed, the key to the understanding of the problem of turbulent motion is in attempting to find general laws governing the behavior of the small eddies. This culminates in KOLMOGOROFF's theory of turbulence (Sect. 16), which we shall present in the next section. The physical basis of the theory becomes clear from the following consideration of the process of energy balance.

Consider an infinite field of homogeneous turbulence with no mean motion. Since there is no external supply of energy, the intensity of turbulence must decay in the course of time. The process of decay is generally described by Eq. (7.26), where $T(\varkappa)\,d\varkappa$ gives the net rate of transfer of energy into the wave-number range $(\varkappa, \varkappa+d\varkappa)$ while $2\nu\varkappa^2 E(\varkappa)\,d\varkappa$ gives the rate of dissipation of energy in this range into heat by molecular viscosity. Clearly, the factor $\varkappa^2$ indicates that the viscous forces are more important at high wave-numbers or small scales. However, there is no *a priori* reason to decide which range of wave-numbers will contribute to the major part of the energy dissipation, since $E(\varkappa)$ may become very small at large values of $\varkappa$.

From an analysis of his experimental data, TAYLOR[1] found that for moderately large values of the Reynolds number of the turbulence,

$$R_\lambda = u\,\lambda/\nu, \tag{15.1}$$

*the dissipation of energy is practically all associated with the high wave-number components that contribute a negligible amount of energy.* Such experimental results allow us to form a simple concept of energy balance in decaying turbulence. We may divide the spectrum into the energy-containing part and the dissipative part. During the decay process, the large energy-containing eddies lose their energy to the smaller dissipative eddies through the action of the transfer term $T(\varkappa)$. The mechanical energy of turbulent motion at the small scales is then dissipated into heat by the action of viscosity.

The transfer of energy from large eddies to small eddies may be visualized in terms of the mechanism of vortex stretching discussed in Sect. 9. Due to the diffusive nature of turbulent motion (Chap. VII), vortex lines tend to be extended

[1] G. I. TAYLOR: Proc. Roy. Soc. Lond., Ser. A **164**, 15 (1938).

and the mean-square vorticity tends to increase. If this stretching mechanism is sufficiently strong, it may even over-compensate the effect of viscosity which tends to reduce the mean-square vorticity. A net increase of the mean-square vorticity

$$\overline{\omega^2} = 2 \int_0^\infty \varkappa^2 E(\varkappa)\, d\varkappa \tag{15.2}$$

may be associated with an increase of energy at high wave numbers or small scales.

Taylor[1] first suggested the above mechanism as fundamental to turbulent motion. He measured two of the terms in Eq. (9.9) to substantiate his argument. To demonstrate the importance of the Reynolds number of the turbulence, Taylor and Green[2] made detailed calculations with a model of homogeneous turbulence. They considered the subsequent motion corresponding to the initial velocity distributions

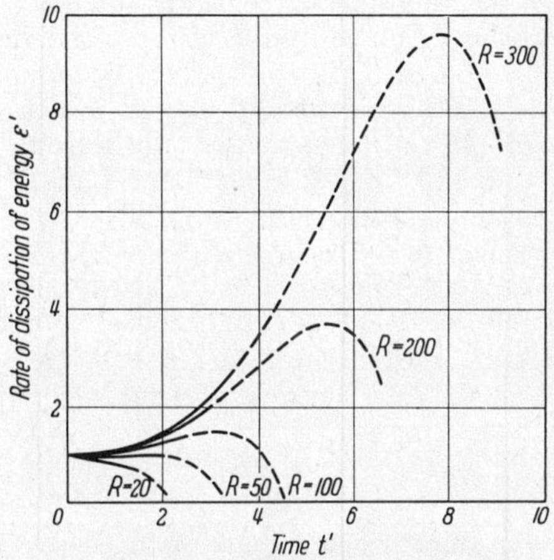


Fig. 7. The rate of dissipation of energy as a function of time in the Taylor-Green model of homogeneous turbulence.

$$\left.\begin{aligned} u &= A \cos a x \sin b y \sin c z, \\ v &= B \sin a x \cos b y \sin c z, \\ w &= C \sin a x \sin b y \cos c z, \end{aligned}\right\} \tag{15.3}$$

where the constants $A$, $B$, $C$ and $a$, $b$, $c$ satisfy the relation

$$A a + B b + C c = 0. \tag{15.4}$$

A power series solution in time was used[3]. At each instant, the pressure distribution is determined from the Poisson equation

$$\nabla^2 p = -\varrho \frac{\partial u_i}{\partial x_j} \frac{\partial u_j}{\partial x_i}. \tag{15.5}$$

Detailed calculations were made for the special case where $a=b=c$, $A=-B$ and $C=0$. The results are shown in Fig. 7 where the rate of energy dissipation (in the dimensionless form $\varepsilon' = 4\varepsilon/3\mu A a^2$) is plotted as a function of time (in the dimensionless form $t' = A a t$), for various values of the Reynolds number $R = A/a\nu$. Only qualitative accuracy is claimed for the dotted parts of the curves. It is seen that for very low Reynolds numbers, the stretching mechanism is not strong enough so that the mean-square vorticity (and hence also the rate of energy dissipation) steadily decreases under the action of viscosity. On the other hand, if the motion starts out at a fairly high Reynolds number, the mean-square vorticity first increases to several times its original value due to the stretching mechanism. The kinetic energy of the motion, however, decreases steadily. Eventually, the kinetic energy becomes very low and the stretching process is so weakened that the vorticity of the motion also steadily decreases.

**16. Kolmogoroff's theory.** The discussions of the last section lead to the following general picture of turbulent motion. Due to the inertial forces, the large eddies tend to break down into smaller eddies, and these in turn break down

[1] Loc. cit.

[2] G. I. Taylor and A. E. Green: Proc. Roy. Soc. Lond., Ser. A **158**, 449 (1937).

[3] S. Goldstein [Phil. Mag. (7) **30**, 85 (1940)] solved the same problem by using a power series expansion in the Reynolds number, and obtained results in essential agreement with Taylor and Green. Recently, G. R. Stoodley applied the same type of method to study energy partition in magnetohydrodynamic turbulence.

into still smaller eddies, and so on. The concepts of classical statistical mechanics would then predict the ultimate concentration of the energy into smaller and smaller scales (ultraviolet catastrophe). However, at the same time, the viscous forces are more effective in dissipating the energy at the very small scales and tend to bring about an equilibrium distribution of energy. In the long process of reaching the smallest eddies, the turbulent motion tends to adjust itself to some definite state independent of the source of energy at the large scales. The further down the scale, the less the dependence.

In line with these ideas, KOLMOGOROFF[1] postulates that, at very large Reynolds numbers of the turbulence, the *local* properties of the turbulent motion should have a universal character described by the following concepts. (1) The turbulent motion is locally isotropic whether the large scale motions are isotropic or not. (2) The motion at the very small scales is chiefly governed by the viscous forces and the rate at which energy is handed down to them from the large eddies. (3) At the large-scale end of these small-scale eddies, the effect of viscous forces is negligible, and the statistical properties of the turbulent motion depend on a single parameter, i.e., the rate of energy dissipation. Indeed, these eddies mark the transition between the energy-containing eddies and the dissipative eddies. Nearly *all* the dissipation of energy occurs at smaller scales whereas nearly *all* the energy is contained in eddies of larger scales.

To formulate these concepts mathematically, KOLMOGOROFF introduced correlation functions of the type $\overline{(u_1'-u_1)^n}$, which are the statistical moments of the *relative* velocity at two points $P$ and $P'$. Thus, the *local* nature of the turbulent field is stressed. We consider the quantity,

$$\overline{(u_1'-u_1)^2} = u^2[1-f(r)], \tag{16.1}$$

which has physical dimensions, instead of the dimensionless correlation coefficient, because $u^2$ has no direct bearing on the *local* structure.

The second step in the formulation of the theory is to introduce the assumption that, for small values of $r$, these moments depend only on the rate of energy dissipation $\varepsilon$ and the kinematic viscosity $\nu$. One can then construct the length scale

$$\eta = (\nu^3/\varepsilon)^{\frac{1}{4}}, \tag{16.2}$$

and the velocity scale

$$v = (\nu\,\varepsilon)^{\frac{1}{4}}. \tag{16.3}$$

From dimensional arguments, one then obtains

$$\overline{(u_1'-u_1)^2} = (\nu\,\varepsilon)^{\frac{1}{2}}\beta_{dd}(r/\eta), \tag{16.4}$$

$$\overline{(u_1'-u_1)^3} = (\nu\,\varepsilon)^{\frac{3}{4}}\beta_{ddd}(r/\eta), \tag{16.5}$$

where $\beta_{dd}$ and $\beta_{ddd}$ are *universal* functions for restricted values of $r$.

For very high Reynolds numbers, KOLMOGOROFF visualizes that, for $r/\eta$ very large, there is a range of $r$ for which viscous forces do not play a direct role. Eq. (16.4) then implies that

$$\overline{(u_1'-u_1)^2} \sim (\varepsilon\,r)^{\frac{2}{3}}. \tag{16.6}$$

A definite form is thereby predicted for the correlation function. It should be noted in passing that this form of the correlation function in no way contradicts the parabolic form discussed earlier, which holds when $r/\lambda$ is small. The relation (16.6) holds only for *larger* values of the relative separation of the two points (cf. Sect. 8).

[1] A. N. KOLMOGOROFF: C. R. Acad. Sci. URSS. **30**, 301 (1941).

The concept of KOLMOGOROFF can also be introduced into the spectral formulation of the theory. Thus, at very high Reynolds numbers, the spectrum $E(\varkappa)$ at *high* frequencies should be expressible in the form

$$E(\varkappa) = v^2 \eta \, \varphi(\varkappa \eta), \tag{16.7}$$

where $\varphi(x)$ has a universal form. For small values of $x$, we can again determine the form of the spectral function by assuming $E(\varkappa)$ to be independent of $\nu$ and then using dimensional arguments. Indeed, we obtain

$$E(\varkappa) \sim \varepsilon^{\frac{2}{3}} \varkappa^{-\frac{5}{3}}. \tag{16.8}$$

This form was first given by OBUKHOFF[1]. Clearly (16.6) and (16.8) are related by the usual Fourier transform relations. The range of the wave number (relative distance) for which Eq. (16.8) (and Eq. (16.6)) holds shall be referred to as the *inertial subrange* of KOLMOGOROFF'S *universal equilibrium range.*

**17. Energy-containing eddies and dissipative eddies.** The theories for predicting the form of the spectral function in the universal equilibrium range will be discussed in Sect. 18 and presented in detail in Chap. IV. Before going into these discussions, we shall first show that the spectral form (16.8) for an intermediate range is indeed compatible with the virtually complete separation of eddies into the energy-containing components and the dissipative components, and examine the behavior of these eddies during the process of decay.

For the sake of convenience, let us write the spectral function in the energy-containing range in the form

$$E(\varkappa) = V^2 L \, \Phi(\varkappa L), \tag{17.1}$$

where the scales $V$ and $L$ are suitably chosen so that $X = \varkappa L$ and $\Phi(X)$ are of the order of unity. Later in this section, we shall discuss the dynamical basis of (17.1) during the process of decay.

The spectral formula (17.1) should agree with (16.7) when $X = \varkappa L$ is large. Indeed, if we have, in (16.7) and (17.1) respectively,

$$\left.\begin{aligned} \varphi(x) &\approx c\, x^{-\frac{5}{3}} \quad \text{for } x \text{ small}, \\ \Phi(X) &\approx C X^{-\frac{5}{3}} \quad \text{for } X \text{ large}, \end{aligned}\right\} \tag{17.2}$$

where $c$ and $C$ are constants of the order of unity, we may identify (16.7) and (17.1) in the overlapping range where (16.8) holds. We then obtain

$$c\, v^2 \eta^{-\frac{2}{3}} = C\, V^2 L^{-\frac{2}{3}},$$

which may be rewritten as

$$\frac{v}{V} = \left(\frac{\eta}{L}\right)^{\frac{1}{3}} = \left(\frac{x}{X}\right)^{\frac{1}{3}}, \tag{17.3}$$

provided we so choose $V$ and $L$ that $C = c$.

We notice in passing that since

$$\frac{v}{u} \approx \frac{\eta}{\lambda} \approx R_{\lambda}^{-\frac{1}{2}}, \tag{17.4}$$

we have

$$\eta : \lambda : L \approx 1 : R_{\lambda}^{\frac{1}{2}} : R_{\lambda}^{\frac{3}{2}}. \tag{17.5}$$

Let us choose some wave number in the overlapping range and denote it by $\varkappa_*$, whose precise meaning we shall specify later. The total energy of the turbulent

[1] A.M. OBUKHOFF: C. R. Acad. Sci. URSS. **32**, 19 (1941).

motion is proportional to [cf. (7.21)]

$$\tfrac{3}{2}\,u^2 = \int_0^\infty E(\varkappa)\,d\varkappa = V^2 \int_0^{X_*} \Phi(X)\,dX + v^2 \int_{x_*}^\infty \varphi(x)\,dx, \tag{17.6}$$

whereas the rate of energy dissipation is proportional to

$$\frac{\varepsilon}{2\nu} = \int_0^\infty \varkappa^2 E(\varkappa)\,d\varkappa = \frac{V^2}{L^2}\int_0^{X_*} X^2\,\Phi(X)\,dX + \frac{v^2}{\eta^2}\int_{x_*}^\infty x^2\,\varphi(x)\,dx. \tag{17.7}$$

Here, $X_* = \varkappa_* L$ is large while $x_* = \varkappa_* \eta$ is small. We want to show that the dominant term in (17.6) is the first term on the right, whereas in (17.7) it is the second term.

Consider first (17.6). Since the integral

$$J = \int_0^\infty \Phi(X)\,dX$$

converges, we need only discuss the integral involving $\varphi(x)$. This may be written as

$$K = \int_{x_*}^\infty \{\varphi(x) - c\,x^{-\frac{5}{3}}\}\,dx + \int_{x_*}^\infty c\,x^{-\frac{5}{3}}\,dx.$$

The first part of $K$ may be replaced by an integral over the range $(0, \infty)$, while the second part has the value $\frac{3c}{2}\,x_*^{-\frac{2}{3}}$. Thus, (17.6) yields

$$\frac{3u^2}{2V^2} = \int_0^\infty \Phi(X)\,dX + \left(\frac{v}{V}\right)^2 O\left(x_*^{-\frac{2}{3}}\right). \tag{17.8}$$

If we handle (17.7) in a similar manner by writing

$$\Phi(X) = C X^{-\frac{5}{3}} + [\Phi(X) - C X^{-\frac{5}{3}}], \tag{17.9}$$

we obtain

$$\frac{\varepsilon}{2\nu\, v^2/\eta^2} = \int_0^\infty x^2\,\varphi(x)\,dx + \left(\frac{v}{V}\right)^4 O\left(X_*^{\frac{4}{3}}\right). \tag{17.10}$$

Thus, our desired results are obtained if

$$\left(\frac{v}{V}\right)^2 x_*^{-\frac{2}{3}} \ll 1 \quad\text{and}\quad \left(\frac{v}{V}\right)^4 X_*^{\frac{4}{3}} \ll 1,$$

or

$$\varkappa_*\,\eta \gg \frac{\eta}{L} \quad\text{and}\quad \varkappa_*\,L \ll \frac{L}{\eta}. \tag{17.11}$$

Both conditions can be satisfied within the restriction of the choice of $\varkappa_*$ specified above. Thus, the complete separation of the two ranges of eddies is demonstrated. It might be worth noting that the first moment of $E(\varkappa)$, $\int_0^\infty \varkappa E(\varkappa)\,d\varkappa$, will receive contributions of equal orders from the two ranges. On the other hand, an integral length scale $L_0$ defined by

$$u^2 L_0 = \int_0^\infty \varkappa^{-1} E(\varkappa)\,d\varkappa, \tag{17.12}$$

is indeed wholly determined by the energy-containing eddies.

Turning now to the consideration of the dynamical process, we shall also find that the two ranges of eddies behave quite differently. The spectrum of the energy-containing eddies undergoes a rather rapid change, whereas the spectrum of the dissipative eddies stays in quasi-equilibrium and changes only to the extent that the total rate of dissipation is altered.

*α) Decay of energy-containing eddies.* Let us now examine the dynamical process, assuming that the spectral form $\Phi(X)$ in (17.1) is preserved, but that the velocity and length scales change in the course of time. This can be true only if the transfer function $T(\varkappa, t)$ has a similar behavior:

$$T(\varkappa, t) = V^3 \Psi(\varkappa L). \tag{17.13}$$

The equation for energy transfer then gives us

$$\frac{1}{V}\frac{dL}{dt}[X\Phi'(X) + \Phi(X)] + \frac{2L}{V^2}\frac{dV}{dt}\Phi(X) - \Psi(X) = -\frac{2\nu}{VL}X^2\Phi(X). \tag{17.14}$$

If the self-preserving solution is to be valid, we must then have

$$\frac{1}{V}\frac{dL}{dt} : \frac{2L}{V^2}\frac{dV}{dt} : 1 = \text{constant ratios}. \tag{17.15}$$

No conclusion need be drawn from the right-hand side term in (17.14) because its magnitude is negligibly small. Eq. (17.15) shows that $V^2$ and $L$ satisfy power laws in time:

$$\left.\begin{aligned} V^2 &\sim (t-t_0)^{-n}, \\ L &\sim (t-t_0)^{1-n/2}. \end{aligned}\right\} \tag{17.16}$$

Furthermore, by virtue of (17.6), we may identify $V$ with the turbulence level $u$ within a constant factor. Then (17.16) leads to [cf. (8.12), (8.13)]

$$\lambda^2 = \frac{10}{n}\nu(t-t_0). \tag{17.17}$$

A power law of decay of the form (17.16) was first obtained by v. Kármán [*17*] from considering self-preserving correlation functions. Efforts to determine the exact value for the power $n$ will lead to more involved considerations, which we shall examine in Sect. 18. At the moment, let us notice that the time scale of decay $V^2\Big/\frac{dV^2}{dt}$ and the general time scale $\frac{L}{V}$ for the turbulent motion are both given by $(t-t_0)$. Thus, there is barely enough time for the spectral distribution to adjust itself to a quasi-equilibrium condition during the whole history of decay. There is therefore reasonable doubt whether we could expect an accurate fulfillment of the hypothesis of self-preservation of the spectrum in the energy-containing range, especially for the large eddies.

The formulas (17.15) can also be written in the following form to exhibit their physical meaning:

$$\left.\begin{aligned} \frac{dL}{dt} &\sim u, \\ \varrho\,\varepsilon = -\frac{3\varrho}{2}\frac{du^2}{dt} &\sim \varrho\,u^2\frac{u}{L}. \end{aligned}\right\} \tag{17.18}$$

The first relation asserts that the rate of increase of scale is given by the intensity of turbulence; the second relation shows that the decay process may be considered as the action of the Reynolds stress on the mean rate of strain of the large eddies.

*β) Quasi-equilibrium of the dissipative eddies.* If we now attempt a similar analysis for the range of dissipative eddies, we find that the term $\frac{\partial}{\partial t}E(\varkappa, t)$ is negligible compared with the dissipative term $-2\nu\varkappa^2 E$ as long as $(\varkappa\lambda)^2 \gg 1$. Indeed, if we write

$$E(\varkappa) = v^2\eta\,\varphi(\varkappa\eta), \qquad T(\varkappa) = v^3\,\psi(\varkappa\eta), \tag{17.19}$$

we get, as a very good approximation,

$$\psi(x) = -\,2x^2\,\varphi(x). \tag{17.20}$$

On the other hand, if we do not adopt the Kolmogoroff scales to begin with, but go through the analysis presented above, we shall be able to derive his formulas. Indeed, with unspecified scales $v'$ and $\eta'$ in place of $v$ and $\eta$, we can show that

$$v'^3\,\psi(x') = -\,2\nu\,(v'^2/\eta'^2)\,x'^2\,\varphi(x') \tag{17.21}$$

analogous to (17.20), and

$$\varepsilon = -\,2\nu(\,v'^2/\eta'^2)\int\limits_0^\infty x'^2\,\varphi(x')\,dx'. \tag{17.22}$$

Eqs. (17.21) and (17.22) show that the Kolmogoroff scales must be adopted, and that consequently the spectral formula (16.8) must hold.

**18. General consideration of self-preserving solutions—comparison with experiments.** In the last section, we considered self-preserving solutions only for such large values of $R_\lambda$ that there is an intermediate range to mark the separation of the energy-containing eddies and the dissipative eddies. In Sect. 14, we considered a solution for such small Reynolds numbers that the transfer term $T(\varkappa, t)$ is negligible. That solution can also be obtained by considering self-preserving solutions (see [*18*]). In most experimental situations, however, neither requirement is fulfilled. Indeed, there do not exist any accurate measurements which are suitable for comparison with the solutions of Sect. 17; and there is only a limited amount of data for comparison with the solution of Sect. 14. One must therefore extend the consideration of self-preservation to such Reynolds numbers for which the transfer term is not negligible while at the same time there does not exist an inertial range of the spectrum to mark the separation between energy-containing eddies and dissipative eddies. In such a case, if we should attempt a solution of the type discussed in Sect. 17, i.e., if we assume

$$E(\varkappa) = V^2 L\,\Phi(\varkappa L), \qquad T(\varkappa) = V^3\,\Psi(\varkappa L), \tag{18.1}$$

it should be regarded as valid for both the energy-containing eddies *and* the dissipative eddies. Thus, in addition to the relations (17.15), we should add the relation

$$\frac{VL}{\nu} = \text{const}. \tag{18.2}$$

This shows that the exponent $n$ in (17.16) must be equal to unity, and we have

$$\left.\begin{aligned} V^2 \sim (t-t_0)^{-1}, \qquad L \sim (t-t_0)^{\frac{1}{2}} \\ \lambda^2 = 10\nu(t-t_0). \end{aligned}\right\} \tag{18.3}$$

From (18.1), we have

$$\tfrac{3}{2}\,u^2 = \int\limits_0^\infty E(\varkappa)\,d\varkappa = V^2\int\limits_0^\infty \Phi(X)\,dX. \tag{18.4}$$

One could then identify $V^2$ with $u^2$ (up to a constant factor). However, the arguments in Sect. 17 suggest that this may not be a good approximation, because

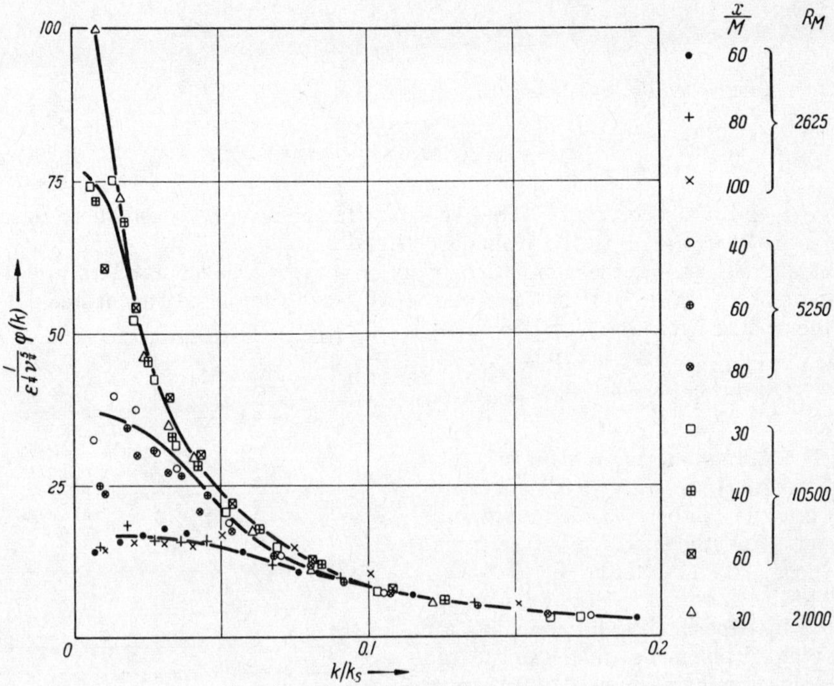


Fig. 8. The velocity spectrum in isotropic turbulence.

the large-scale eddies may not have enough time to adjust themselves to a quasi-equilibrium. This difficulty can be avoided by calculating the higher moments

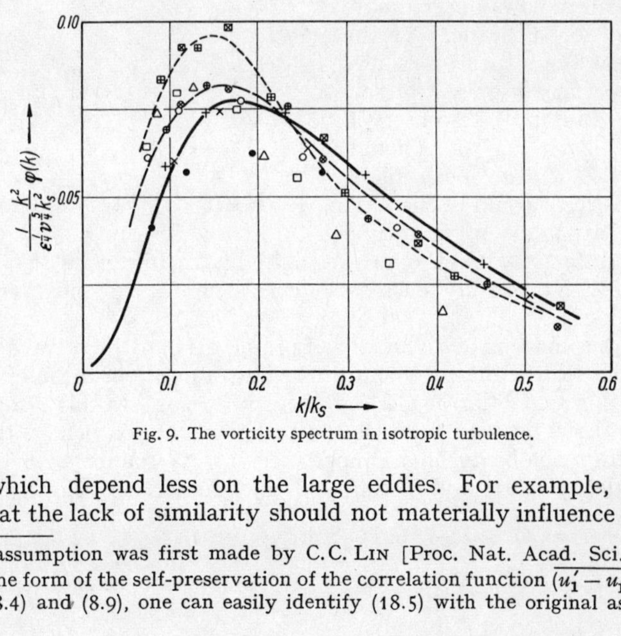


Fig. 9. The vorticity spectrum in isotropic turbulence.

of $E(\varkappa)$ which depend less on the large eddies. For example, it may be assumed[1] that the lack of similarity should not materially influence the calculation

[1] This assumption was first made by C. C. LIN [Proc. Nat. Acad. Sci., Wash. **34**, 540 (1948)] in the form of the self-preservation of the correlation function $\overline{(u_1' - u_1)^2} = u^2 [1 - f(r)]$. By using (8.4) and (8.9), one can easily identify (18.5) with the original assumption.

of the rate of energy dissipation

$$\varepsilon = 2\nu \int_0^\infty \varkappa^2 E(\varkappa)\, d\varkappa = 2\nu \frac{V^2}{L^2} \int_0^\infty X^2 \Phi(X)\, dX. \tag{18.5}$$

It then follows that

$$-\frac{du^2}{dt} = a(t-t_0)^{-2}, \quad \text{and} \quad u^2 = a(t-t_0)^{-1} + b, \tag{18.6}$$

where $a$ and $b$ are constants. The relations (18.2) and (18.5) also show that $V$ and $L$ are *proportional* to the Kolmogoroff scales $v$ and $\eta$.

One could assume that similarity exists in an even weaker form, as was suggested by GOLDSTEIN[1]. If the similarity of the spectrum is only accurate enough for the calculation of the fourth moment, then the law of decay becomes

$$u^2 = a(t-t_0)^{-1} + b + c(t-t_0), \tag{18.7}$$

which contains one more arbitrary constant than (18.6). The basis for these assumptions will be discussed more fully in Sect. 27 for the specific physical transfer theory of HEISENBERG.

The ideas presented above are supported by experimental observations for decaying turbulence at moderately large Reynolds numbers. Figs. 8 and 9 show the spectra observed by STEWART and TOWNSEND [26] for $u$ and $\partial u/\partial t$ (cf. Sect. 10). It is interesting to note that the latter spectrum shows approximate self-preservation for all ranges of the frequency whereas the $u$-spectrum shows considerable departure from self-preservation at low wave numbers. The observed law of decay is shown in Fig. 10. Although the variation of $\lambda^2$ and $u^{-2}$ both follow the linear law, as they would in the case of complete similarity, the origin of time must be taken differently for the two straight lines. It can be seen that the Reynolds number $R_\lambda$ steadily decreases [corresponding to the case $b<0$ in (18.6)].

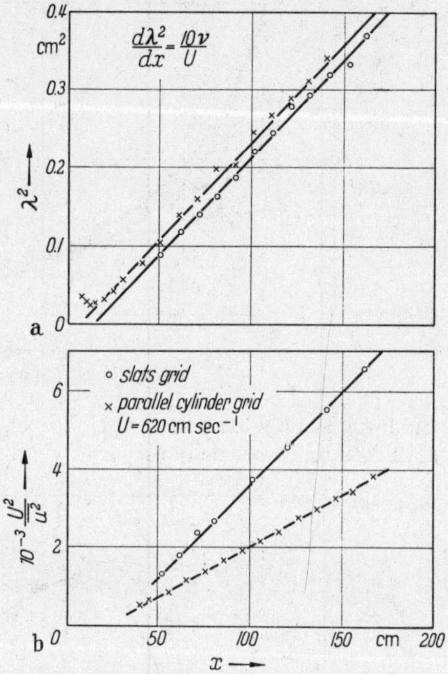

Fig. 10 a and b. The law of decay of isotropic turbulence.

To examine the concept of a partial self-preservation at high wave numbers, GOLDSTEIN[2] proposed the measurement of the law of decay of turbulence behind a grid when another grid of larger mesh is placed upstream. In this case, the large eddies from the first grid tend to cause the turbulent motion behind the second grid to depart from similarity in the range of low wave numbers. Such experiments were made by TSUJI and HAMA[3] and their results are shown in Fig. 11.

[1] S. GOLDSTEIN: Proc. Cambridge Phil. Soc. **47**, 554 (1951).
[2] Loc. cit.
[3] H. TSUJI and F. R. HAMA: J. Aeronaut. Sci. **20**, 848 (1953). — H. TSUJI: J. Phys. Soc. Japan **2**, 1096 (1956).

The results show strong departure from the simple law of decay $u^2 \sim t^{-1}$, but tend to agree with the concepts and conclusions discussed in the present section.

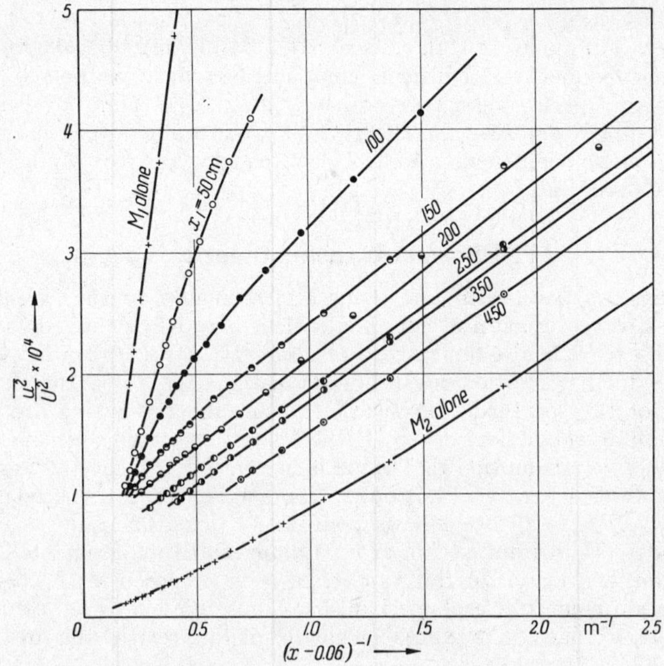


Fig. 11. The decay of turbulence behind two grids $M_1$ and $M_2$ with mesh widths 5 cm and 1 cm respectively ($M_2$ is downstream from $M_1$). $x$ is the distance behind the second grid $M_2$, $x_1$ is the distance between the grids, and $U$ is the mean velocity. From TSUJI and HAMA.

**19. Dynamical theories of homogeneous turbulence.** The general considerations of the dynamical process discussed so far do not depend on the explicit forms of the spectral function $E(\varkappa, t)$ and the transfer function $T(\varkappa, t)$; only the hypothesis is made that the spectrum should have some self-preserving character during the process of decay. In order to obtain a closed system of equations to describe the complete dynamical process, the simplest approach would be to assume, on the basis of physical arguments, a relationship connecting $E(\varkappa, t)$ and $T(\varkappa, t)$. Such theories have been proposed by OBUKHOFF[1], HEISENBERG [*13*], v. KÁRMÁN[2] and KOVÁSZNAY[3]. The hypothesis of HEISENBERG turned out to be the most successful, and will be discussed in detail in Chap. IV, where the other hypotheses will also be briefly discussed.

In terms of correlation functions, these physical transfer theories amount to the introduction of a relationship between the double and triple correlations in the Kármán-Howarth equation. A more elaborate systematic approach would be the development of equations involving correlations of higher orders and the derivation of a closed system either by neglecting the higher correlations or by adopting reasonable approximations connecting them to correlations of lower orders. A natural assumption of this kind is the quasi-Gaussian hypothesis first

[1] A.M. OBUKHOFF: C. R. Acad. Sci. URSS. **30**, 301 (1941).

[2] TH. v. KÁRMÁN: C. R. Acad. Sci., Paris **226**, 2108 (1948). — Proc. Nat. Acad. Sci. **34**, 530 (1948).

[3] L. S. G. KOVÁSZNAY: J. Aeronaut. Sci. **15**, 745 (1948).

introduced by MILLIONSHCHIKOV[1] connecting correlations of the fourth order with double correlations. This theory has been developed in detail by PROUDMAN and REID [*22*] and independently by TATSUMI [*28*]. It will be presented in Chaps. V and VI.

In the spectral formulation of the theory, HEISENBERG introduced a hypothesis of statistical independence of harmonic components, which can be shown to be equivalent to the quasi-Gaussian hypothesis. From this point of view, it is natural to attempt to improve the theory by introducing a "weak interaction" among the harmonic components. A theory of this type has recently been developed by KRAICHNAN [*39*].

## IV. Physical transfer theories.

**20. The physical mechanism underlying HEISENBERG'S transfer theory.** The incompleteness of the dynamical equation satisfied by $E(\varkappa, t)$ has already been remarked and, except in the final period of decay when $T(\varkappa, t)$ can be neglected entirely, some further relation must be postulated. We come therefore to a consideration of the so-called "physical" transfer theories which are capable of yielding definite predictions for both the energy spectrum and the transfer function. The basic assumption that is made in deriving a theory of this type is that $T(\varkappa, t)$ is determined by $E(\varkappa, t)$ only. The specific form of the relationship between $T(\varkappa, t)$ and $E(\varkappa, t)$ then follows from the particular mechanism of energy transfer considered and from certain general dimensional requirements[2].

Several transfer theories of this type have been proposed in the past, each based upon a different physical mechanism. Thus, for example, there is the *shear transfer theory* due to OBUKHOFF[3] which leads to a relationship of the form

$$T(\varkappa) = -2K_O \frac{d}{d\varkappa} \int_{\varkappa}^{\infty} E(\varkappa')\, d\varkappa' \left[ \int_0^{\varkappa} \varkappa''^2 E(\varkappa'')\, d\varkappa'' \right]^{\frac{1}{2}} \tag{20.1}$$

and the *local transfer theory* due to KOVÁSZNAY[4] which leads to the particularly simple relationship

$$T(\varkappa) = -2K_K \frac{d}{d\varkappa} \varkappa^{\frac{5}{2}} E^{\frac{3}{2}}(\varkappa). \tag{20.2}$$

On the basis of existing evidence, it seems clear that the eddy viscosity transfer theory due to HEISENBERG [*13*] is to be preferred to either of the other two. HEISENBERG argued that the mechanism governing the transfer of energy is similar to the one governing the viscous dissipation, provided the molecular viscosity is replaced by a suitable "eddy" viscosity depending only on the smaller eddies. While a mechanism of this type may be expected to give reasonable results for the transfer of energy at wave-numbers small compared with the wave-numbers associated with the eddy viscosity, it is likely to be in error when these two wave-number regions are of the same order of magnitude. Now the energy lying between wave-numbers zero and $\varkappa$ can be lost in two ways: by direct viscous

---

[1] M. MILLIONSHCHIKOV: C. R. Acad. Sci. URSS. **22**, 231, 235 (1939); **32**, 615, 619 (1941).

[2] Transfer theories of this same general type can also be derived in physical space by postulating a relationship between $(\overline{u^2})^{\frac{3}{2}} k(r, t)$ and $\overline{u^2} f(r, t)$. Attempts along this line have been made by K. HASSELMANN („Über die Trägheitskräfte der isotropen Turbulenz", Diss. Institut für Schiffbau der Universität Hamburg, May 1955).

[3] A.M. OBUKHOFF: C. R. Acad. Sci. URSS. **32**, 19 (1941). — Izv. Akad. Nauk SSSR., Ser. Geogr. i Geofiz. **5**, 453 (1941). Min. of Supply, England, Translation no. P21109T.

[4] L. S. G. KOVÁSZNAY: J. Aeronaut. Sci. **15**, 745 (1948).

dissipation which takes place linearly at each value of $\varkappa$ and by transfer to the smaller eddies lying between wave-numbers $\varkappa$ and infinity. The rate at which this energy is lost is governed therefore by the equation

$$\frac{\partial}{\partial t}\int_0^\varkappa E(\varkappa')\,d\varkappa' = \int_0^\varkappa T(\varkappa')\,d\varkappa' - 2\nu\int_0^\varkappa \varkappa'^2 E(\varkappa')\,d\varkappa', \tag{20.3}$$

where the terms on the right-hand side of this equation correspond to the energy transfer and viscous dissipation respectively. The existence of an eddy viscosity $\nu_t(\varkappa)$ say, as postulated by HEISENBERG, requires that the energy transfer be of the form

$$\int_0^\varkappa T(\varkappa')\,d\varkappa' = -2\nu_t(\varkappa)\int_0^\varkappa \varkappa'^2 E(\varkappa')\,d\varkappa'. \tag{20.4}$$

Since $\nu_t(\varkappa)$ must depend only on the smaller eddies, it is natural to require that it be of the form of an integral over all wave-numbers larger than $\varkappa$. The further assumption that it depend only on $\varkappa$ and $E(\varkappa)$ then leads, from purely dimensional requirements, to the relation

$$\nu_t(\varkappa) = K_H\int_\varkappa^\infty \sqrt{\frac{E(\varkappa')}{\varkappa'^3}}\,d\varkappa', \tag{20.5}$$

where $K_H$, like $K_0$ and $K_K$, is by hypothesis a numerical constant of order unity. We thus obtain a single equation for $E(\varkappa)$ alone of the form

$$\frac{\partial}{\partial t}\int_0^\varkappa E(\varkappa')\,d\varkappa' = -2\left[\nu + K_H\int_\varkappa^\infty \sqrt{\frac{E(\varkappa')}{\varkappa'^3}}\,d\varkappa'\right]\int_0^\varkappa \varkappa''^2 E(\varkappa'')\,d\varkappa''. \tag{20.6}$$

There have been other more or less formal generalizations of the foregoing transfer functions[1], but while they are of an admissible (i.e. dimensionally correct) form, they contribute little further insight into the mechanism of energy transfer.

If $E(\varkappa)$ were specified at some initial instant in time, $t=0$ say, then Eq. (20.6) could be solved, at least in principle, to give the subsequent behavior of $E(\varkappa)$ and from it the corresponding behavior of $T(\varkappa)$. This has not been done as yet[2]; instead, appealing to the belief that, given sufficient time, the turbulence will tend towards a universal state that is at least partially independent of the initial conditions, most of the solutions of Eq. (20.6) that have been obtained thus far involve a further similarity assumption of some type. These similarity solutions will be considered in some detail in the following sections.

**21. The equilibrium spectrum.** It is quite a simple matter to show that when $\varkappa$ lies in the equilibrium range, the solution of Eq. (20.6) leads to a spectrum in complete agreement with the predictions of KOLMOGOROFF's universal equilibrium theory. For values of $\varkappa$ so large that

$$\int_0^\varkappa E(\varkappa')\,d\varkappa' \gg \int_\varkappa^\infty E(\varkappa')\,d\varkappa',$$

[1] T. v. KÁRMÁN: Proc. Nat. Acad. Sci., Wash. **34**, 530 (1948). — S. GOLDSTEIN: Proc. Cambridge Phil. Soc. **47**, 554 (1951).

[2] Except for the work by W. TOLLMIEN: Proc. 8th Inter. Cong. Theor. Appl. Mech. Istanbul **2**, 23 (1952).

the left hand side of Eq. (20.6) approximates to the total rate of dissipation $\varepsilon$ and the solution of the resulting equation has been given by BASS[1] and CHANDRASEKHAR [9] in the form

$$E(\varkappa) \sim \left(\frac{8\varepsilon}{9K_H}\right)^{\frac{2}{3}} \varkappa^{-\frac{5}{3}}\left(1 + \frac{8\nu^3}{3K_H^2\varepsilon}\varkappa^4\right)^{-\frac{4}{3}}, \tag{21.1}$$

and contains (apart from the numerical factor $K_H$) only the parameters $\varepsilon$ and $\nu$ as required.

When $\varkappa \ll (3K_H^2\,\varepsilon/8\nu^3)^{\frac{1}{4}} = (3K_H^2/8)^{\frac{1}{4}}/\eta$, where $\eta = (\nu^3/\varepsilon)^{\frac{1}{4}}$ is the Kolmogoroff length scale, the spectrum (25.1) reduces to the form appropriate to the inertial subrange,

$$E(\varkappa) \sim \left(\frac{8}{9K_H}\right)^{\frac{2}{3}} \varepsilon^{\frac{2}{3}}\varkappa^{-\frac{5}{3}}, \tag{21.2}$$

i.e. it depends only on $\varepsilon$. On the other hand, when $\varkappa \gg (3K_H^2\,\varepsilon/8\nu^3)^{\frac{1}{4}}$, the spectrum reduces to

$$E(\varkappa) \sim \left(\frac{K_H\,\varepsilon}{2\nu^2}\right)^2 \varkappa^{-7}, \tag{21.3}$$

a form which is peculiar to HEISENBERG's transfer theory.

For large-wave numbers in the viscous region the energy spectrum may be expected to fall off much more rapidly than the $\varkappa^{-7}$ law given by Eq. (21.3). A crude approximation to the spectrum in this region can be obtained, therefore, by introducing a cut-off wave-number $\varkappa_c$ above which $E(\varkappa)$ is zero. On this hypothesis the equilibrium spectrum is given by

$$E(\varkappa) = \begin{Bmatrix} \left(\frac{8\varepsilon}{9K_H}\right)^{\frac{2}{3}}\varkappa^{-\frac{5}{3}} & \text{for} \quad \varkappa < \varkappa_c \\ 0 & \text{for} \quad \varkappa > \varkappa_c, \end{Bmatrix} \tag{21.4}$$

where

$$\varkappa_c = (3K_H^2\,\varepsilon/8\nu^3)^{\frac{1}{4}}. \tag{21.5}$$

Thus the cut-off wave-number is proportional to the inverse of the Kolmogoroff length and it is only here that the viscosity appears in the solution.

Thus, the predictions of KOLMOGOROFF'S universal equilibrium theory are all recovered within the framework of HEISENBERG'S transfer theory. These results all refer, of course, to large values of $\varkappa$ for which the equilibrium assumption is valid. The further approximation leading to Eq. (21.2) is one of large Reynolds number, and in the limit of infinite Reynolds number its validity extends to infinitely large values of $\varkappa$. For a finite value of the Reynolds number, however, the asymptotic behavior of $E(\varkappa)$ as $\varkappa$ approaches infinity is given by Eq. (21.3). The algebraic nature of this result seems to be at variance with existing experimental results[2] for which a more rapid decrease is observed and suggests the use of the cut-off spectrum (21.4). Thus, while HEISENBERG'S theory does not agree with experimental results respecting *detailed* characteristics of the turbulence associated primarily with large wave-numbers, it does provide, as will be shown in Sect. 24, a reasonably accurate description of many *overall* properties of the turbulence associated with the energy-containing range of wave-numbers.

**22. The similarity spectra in the initial period of decay.** During the initial period of decay of the turbulence (cf. Sect. 18) the decay law can be approximated by a simple power law of the form $\overline{u^2} \propto t^{-1}$. For a decay law of this form, the dissi-

[1] J. BASS: C. R. Acad. Sci., Paris **228**, 228 (1949).
[2] A. A. TOWNSEND: Proc. Roy. Soc. Lond., Ser. A **208**, 534 (1951).

pation length parameter $\lambda$ is given by $\lambda^2 = 10\nu t$ and hence the Reynolds number of the turbulence, $R_\lambda$, is constant. Now, if the Reynolds number is at all large, then the dominant mechanism in the decay process is the inertial transfer of energy. Under these conditions then it seems reasonable to assume that the motion is independent of the initial conditions and, accordingly, we consider the similarity solutions of Eq. (20.6) that are compatible with the given law of decay.

The appropriate similarity transformation can be written in the form

$$E(\varkappa, t) = K_H^{-2} \alpha^{\frac{3}{2}} \nu^{\frac{3}{2}} t^{-\frac{1}{2}} F(x) \tag{22.1}$$

and

$$T(\varkappa, t) = K_H^{-2} \alpha^{\frac{3}{2}} \nu^{\frac{3}{2}} t^{-\frac{3}{2}} U(x), \tag{22.2}$$

where

$$x = (\alpha \nu \varkappa^2 t)^{\frac{1}{2}}$$

and $K_H$ and $\alpha$ are inserted simply for convenience[1]. The transformation (22.1) is essentially the same as the one given first by HEISENBERG[2] in his discussion of the decay of turbulence behind a grid in a wind tunnel. When this transformation is substituted into Eq. (20.6) one obtains an ordinary integro-differential equation for $F(x)$,

$$\int_0^x F(x')\,dx' - \frac{1}{2} x F(x) = 2\left[\frac{1}{\alpha} + \int_x^\infty \sqrt{\frac{F(x')}{x'^3}}\,dx'\right] \int_0^x x''^2 F(x'')\,dx'', \tag{22.3}$$

the solutions of which form a one-parameter family, the parameter being $\alpha$ which now replaces the Reynolds number. This equation must now be integrated subject to the boundary conditions

$$F'(0) = 4 \quad \text{and} \quad (K_H R_\lambda)^2 = \tfrac{40}{3} I_2, \quad \text{where} \quad I_2 = \int_0^\infty x^2 F(x)\,dx. \tag{22.4}$$

Since $K_H$ always occurs in the combination $K_H R_\lambda$, this product is therefore the fundamental parameter of the problem.

In the method of solution used by CHANDRASEKHAR [9] and PROUDMAN [21], Eq. (22.3) is first transformed into an ordinary differential equation and it is this equation that is integrated numerically; the results of the integration are then transformed back to give $F(x)$ itself. The details of this work are fairly intricate, but since the final results for $F(x)$ itself can be understood without reference to them, they can safely be omitted here.

For small values of $x$, it can be shown from Eq. (22.3) that $F(x)$ possesses a series expansion of the form

$$F(x) = 4x + 16x^3 \log x + (3a - \tfrac{16}{3})\,x^3 + O(x^5 \log^2 x), \tag{22.5}$$

where $a$ is a constant of integration that is related to the Reynolds number. CHANDRASEKHAR [4] has shown that physically meaningful solutions exist only for $a \leqq a_*$: the solutions for $a < a_*$ correspond to finite values of the Reynolds number while the solution for $a = a_*$ corresponds to an infinite Reynolds number. BLANCH and FERGUSON [8] have shown that $a_* = 1.823945$ with an uncertainty of $\pm 1$ unit in the last decimal place.

[1] The parameter $\alpha$ is introduced for the purpose of adjusting the general scale of $F(x)$. More precisely, the numerical solutions of Eq. (22.3) given by CHANDRASEKHAR [9] and PROUDMAN [21] are scaled for convenience so that $F'(0) = 4$ for all Reynolds numbers and this condition serves to determine $\alpha$ as a function of the Reynolds number.

[2] W. HEISENBERG: Proc. Roy. Soc. Lond., Ser. A **195**, 402 (1948).

For large values of $x$, the asymptotic solutions of Eq. (22.3) are

$$\left.\begin{aligned} F(x) &= \frac{\alpha^4 I_0^2}{4x^7}\left[1 + \frac{4\alpha}{x^2} + \frac{\alpha^2(32-\alpha I_0)}{2x^4} + \frac{9\alpha^3(16-\alpha I_0)}{2x^6} + O\left(\frac{1}{x^8}\right)\right] \\ &\qquad (x\to\infty,\ a<a_*) \end{aligned}\right\} \tag{22.6}$$

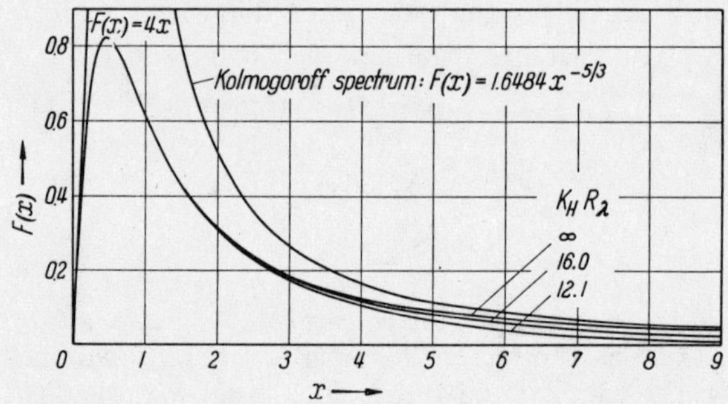


Fig. 12. The energy spectrum according to HEISENBERG's transfer theory with initial period similarity.

and

$$F(x) = \frac{4 I_0^{\frac{2}{3}}}{3^{\frac{4}{3}} x^{\frac{5}{3}}}\left[1 - \frac{8}{21}\cdot\frac{3^{\frac{2}{3}}}{I_0^{\frac{1}{3}}}\cdot\frac{1}{x^{\frac{2}{3}}} + \cdots\right] \qquad (x\to\infty,\ a=a_*), \tag{22.7}$$

the leading terms of which are identical with Eqs. (21.3) and (21.2) respectively when the latter are expressed in terms of the similarly variables. The integral

$$I_0 = \int_0^\infty F(x)\,dx, \tag{22.8}$$

which gives a measure of the total energy, is of course a function of $K_H R_\lambda$.

Numerical solutions of Eq. (22.3) having the behavior (22.5) for small values of $x$ and which meet either (22.6) or (22.7) for large values of $x$ have been obtained by CHANDRASEKHAR [9] for $K_H R_\lambda = 1.29$, 1.52, 1.95, 3.00, 5.41, 16.0 and $\infty$; a solution for $K_H R_\lambda = 12.1$ has also been given by PROUDMAN [21]; three of these solutions are shown in Fig. 12. Some of the solutions obtained more recently by BLANCH[1] using a UNIVAC computer are shown in Fig. 13. The parameters $I_0$ and $I_2$ must be determined by direct integration and the value of $\alpha$ then follows from the relation

$$\alpha I_0 = 2 I_2, \tag{22.9}$$

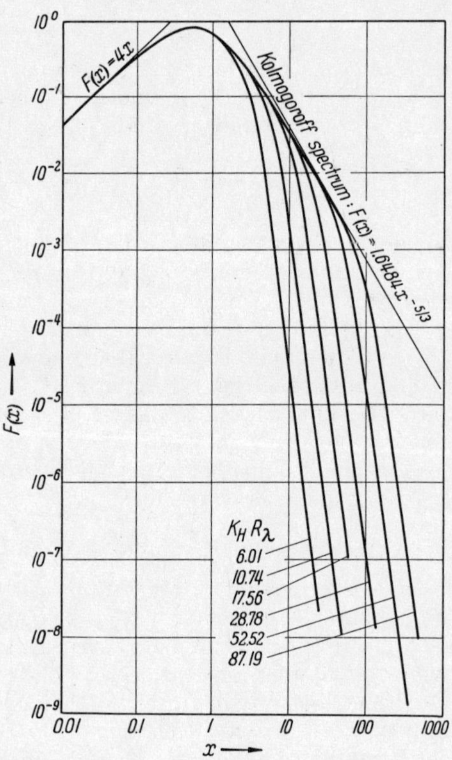


Fig. 13. The energy spectrum according to HEISENBERG's transfer theory with initial period similarity. From unpublished calculations by G. BLANCH (1958).

[1] G. BLANCH: Unpublished 1958. We are indebted to Dr. BLANCH for providing us with these results.

which is itself a direct consequence of applying the initial period similarity law to the energy equation (8.12).

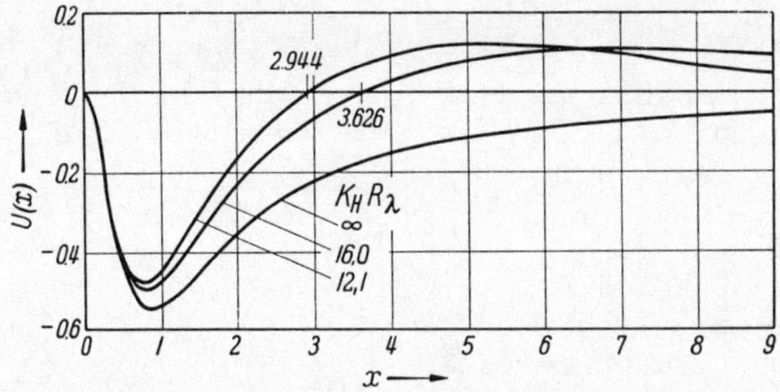


Fig. 14. The transfer function according to HEISENBERG's transfer theory with initial period similarity.

From the foregoing results, the corresponding behavior of the transfer function can be obtained by direct quadrature. Thus, from Eq. (22.2) and Eqs. (20.4) and (20.5) we have

$$U(x) = -2\frac{d}{dx}\left[\int_x^\infty \sqrt{\frac{F(x')}{x'^3}}\,dx' \int_0^x x''^2 F(x'')\,dx''\right]. \tag{22.10}$$

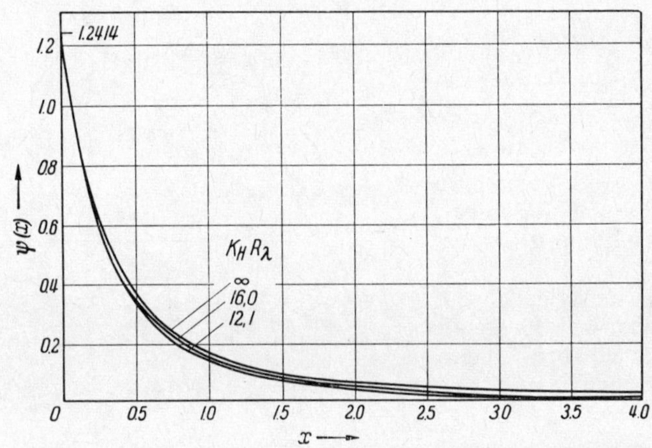


Fig. 15. The one-dimensional energy spectrum according to HEISENBERG's transfer theory with initial period similarity.

This function is shown in Fig. 14 for three values of the Reynolds number. It will be noticed that for $R_\lambda < \infty$, $U(x)$ satisfies the integral requirement

$$\int_0^\infty U(x)\,dx = 0,$$

but that it is violated for $R_\lambda = \infty$; this is the first, but not the only, singular feature of the solution for an infinite Reynolds number.

In addition to the three-dimensional energy spectrum $E(\varkappa)$, considerable interest attaches to the longitudinal one-dimensional energy spectrum $\varphi(\varkappa)$; for it is this latter quantity that can be measured most easily in the laboratory. Before the numerical calculation of $\varphi(\varkappa)$ can be undertaken, it is necessary to make the same type of similarity transformation with respect to $\varphi(\varkappa)$ as was made in Eq. (22.1) for $E(\varkappa)$; thus, since $\varphi(\varkappa)$ is of the same dimensions as $E(\varkappa)$, we may write

$$\varphi(\varkappa, t) = K_H^{-2}\,\alpha^{\frac{3}{2}}\,\nu^{\frac{3}{2}}\,t^{-\frac{1}{2}}\,\psi(x). \tag{22.11}$$

Using this relationship and the corresponding one for $E(\varkappa)$, Eq. (10.3) becomes

$$\psi(x) = \frac{1}{2}\int_x^\infty \left(1 - \frac{x^2}{x'^2}\right)\frac{F(x')}{x'}\,dx'. \tag{22.12}$$

This function is shown in Fig. 15 for three values of the Reynolds number.

A tabulation of the functions $F(x)$, $U(x)$ and $\psi(x)$ for $K_H R_\lambda = 12.1$, 16.0 and $\infty$ is given in the following section.

Any discussion of HEISENBERG's theory would be incomplete without some reference to the corresponding results for the correlation functions in physical space and their relation to experimental results. The requirement of initial period similarity, as applied to the correlation functions, is

$$\left.\begin{aligned} f(r,t) &= f(y) \\ k(r,t) &= k(y), \end{aligned}\right\} \tag{22.13}$$

where

$$y = r/(\alpha\nu t)^{\frac{1}{2}}. \tag{22.14}$$

The correlation functions $f(y)$ and $k(y)$ have been computed by PROUDMAN [21] for two values of the Reynolds number, and these results are shown in Fig. 16.

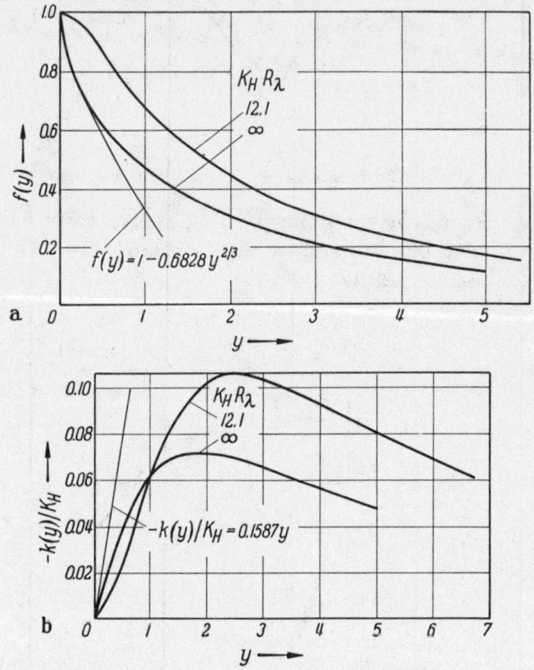


Fig. 16 a and b. The longitudinal double velocity correlations (a) and the longitudinal triple velocity correlations (b) according to HEISENBERG's transfer theory with initial period similarity. From PROUDMAN [21].

Since the behavior of the correlation functions for small values of $y$ are closely related to the asymptotic behavior of the spectra for large values of $x$, we find that, for $R_\lambda = \infty$,

$$f(y) = 1 - \frac{81}{55}\cdot\frac{(\frac{1}{3})!}{(3I_0)^{\frac{1}{3}}}\,y^{\frac{2}{3}} + O(y^{\frac{4}{3}}) \tag{22.15}$$

and

$$-\frac{k(y)}{K_H} = \frac{1}{5}\left(\frac{3}{2I_0}\right)^{\frac{1}{2}} y + O(y^{\frac{5}{3}}); \tag{22.16}$$

for a finite value of the Reynolds number, however, we have

$$f(r) = 1 - \frac{1}{2}\,\frac{r^2}{\lambda^2} + O(r^4) \tag{22.17}$$

and

$$-k(r) = \frac{S}{6}\cdot\frac{r^3}{\lambda^3} + O(r^5). \tag{22.18}$$

The $y^{\frac{2}{3}}$ behavior of $f(y)$ in Eq. (22.15) and the linear behavior of $k(y)$ in Eq. (22.16) are simply the consequences in physical space of KOLOMOGOROFF's universal equilibrium theory[1]. For values of $y$ approaching zero, however, they violate the requirements of continuity and thus provide a further example of the singular behavior of the solution for an infinite Reynolds number. On the other hand, for a finite value of the Reynolds number, the parabolic behavior of the double cor-

[1] A. N. KOLMOGOROFF: C. R. Acad. Sci. URSS 32, 16 (1941).

relation is essentially independent of the form of the transfer function and is a consequence of continuity; the corresponding behavior of the triple correlation likewise satisfies continuity but, in this case, the constant of proportionality involves the skewness factor which does depend on the form of the transfer function.

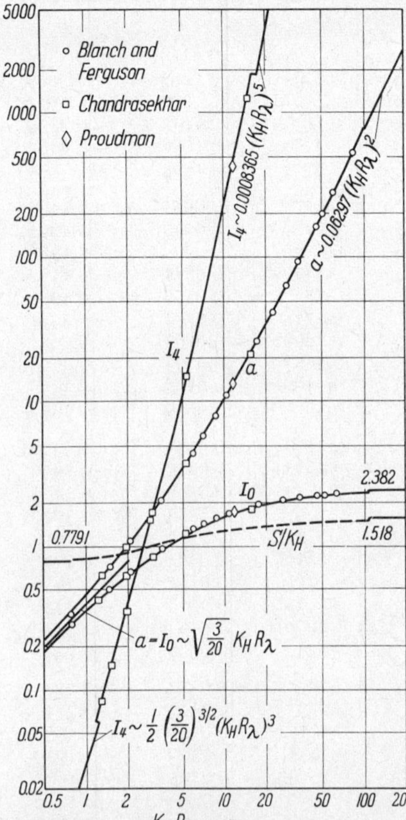


Fig. 17. The parameters associated with HEISENBERG's transfer theory.

The asymptotic behavior of the correlation functions for large values of $y$ is related to the behavior of the spectrum functions for small values of $x$; and it is not difficult to show that

$$f(y) = \frac{12}{I_0} \cdot \frac{1}{y^2} + O\left(\frac{\log y}{y^4}\right) \tag{22.19}$$

and

$$-\frac{k(y)}{K_H} = \frac{12}{I_0}\left(\frac{3}{2I_0}\right)^{\frac{1}{2}} \frac{\log y}{y^3} + O\left(\frac{1}{y^3}\right), \tag{22.20}$$

and these relations are valid for all values of the Reynolds number.

On the basis of the discussion given in this section, we may reasonably conclude that, under conditions of initial period similarity, HEISENBERG's theory is now well understood from a mathematical point of view (see also the following section), and we present, therefore, in Sect. 24 a brief comparison between these theoretical results and the experimental data.

**23. Numerical results for HEISENBERG'S transfer theory.** In the comparison of the theory with the results of experiment, it is necessary to know the variation of the various parameters occurring in the theory as a function of the Reynolds number. The values of $\alpha$, $I_0$, $I_2$ and $I_4$ have been computed from each of the solutions tabulated by CHANDRASEKHAR [9] and PROUDMAN [21] and these results are given in Table 1 and Fig. 17. While great care has been taken in the computation of these values, their accuracy is still not sufficient to permit an equally accurate determination of the skewness factor and, accordingly, the curve for $S/K_H$ has been shown dashed.

The energy spectrum function $F(x)$, tabulated in Table 2 for three values of the Reynolds number, was determined by interpolation from the tables of CHAN-

Table 1. *The parameters associated with Heisenberg's transfer theory.* Based on the solutions given by CHANDRASEKHAR [9] and PROUDMAN [21].

| $a$ | $K_H R_\lambda$ | $\alpha$ | $I_0$ | $I_2$ | $I_4$ |
|---|---|---|---|---|---|
| 0 | 1.29 | 0.591 | 0.423 | 0.125 | 0.0815 |
| 0.5 | 1.52 | 0.711 | 0.490 | 0.174 | 0.147 |
| 1.0 | 1.95 | 0.945 | 0.605 | 0.286 | 0.339 |
| 1.5 | 3.00 | 1.628 | 0.828 | 0.674 | 1.511 |
| 1.75 | 5.41 | 3.693 | 1.191 | 2.20 | 14.25 |
| 1.8 | 12.1 | 13.05 | 1.681 | 10.97 | 415.5 |
| 1.81 | 16.0 | 20.31 | 1.783 | 18.30 | 1229 |
| $a_*$ | $\infty$ | $\infty$ | 2.382 | $\infty$ | $\infty$ |

Table 2. *The three-dimensional energy spectrum, one-dimensional energy spectrum, and transfer function according to Heisenberg's theory with initial period similarity.*

| $x$ | $F(x)$ | $\psi(x)$ | $U(x)$ | $x$ | $F(x)$ | $\psi(x)$ | $U(x)$ |
|---|---|---|---|---|---|---|---|
| | | | $K_H R_\lambda = \infty$ | | | | |
| 0.00 | 0.0000 | 1.2414 | −0.0000 | 1.20 | 0.5450 | 0.1338 | −0.5104 |
| 0.01 | 0.0399 | 1.2023 | −0.0001 | 1.25 | 0.5256 | 0.1267 | −0.5009 |
| 0.02 | 0.0795 | 1.1650 | −0.0004 | 1.30 | 0.5070 | 0.1202 | −0.4910 |
| 0.03 | 0.1185 | 1.1293 | −0.0013 | 1.35 | 0.4892 | 0.1142 | −0.4809 |
| 0.04 | 0.1567 | 1.0951 | −0.0027 | 1.40 | 0.4721 | 0.1086 | −0.4705 |
| 0.05 | 0.1941 | 1.0624 | −0.0048 | 1.45 | 0.4558 | 0.1035 | −0.4601 |
| 0.06 | 0.2305 | 1.0311 | −0.0077 | 1.50 | 0.4402 | 0.0987 | −0.4495 |
| 0.07 | 0.2658 | 1.0011 | −0.0113 | 1.55 | 0.4253 | 0.0943 | −0.4391 |
| 0.08 | 0.2999 | 0.9723 | −0.0157 | 1.60 | 0.4110 | 0.0902 | −0.4286 |
| 0.09 | 0.3329 | 0.9446 | −0.0208 | 1.65 | 0.3973 | 0.0864 | −0.4182 |
| 0.10 | 0.3646 | 0.9180 | −0.0266 | 1.70 | 0.3843 | 0.0828 | −0.4081 |
| 0.11 | 0.3951 | 0.8925 | −0.0332 | 1.75 | 0.3718 | 0.0794 | −0.3981 |
| 0.12 | 0.4243 | 0.8680 | −0.0405 | 1.80 | 0.3600 | 0.0763 | −0.3884 |
| 0.13 | 0.4523 | 0.8444 | −0.0484 | 1.85 | 0.3487 | 0.0733 | −0.3790 |
| 0.14 | 0.4790 | 0.8217 | −0.0568 | 1.90 | 0.3379 | 0.0705 | −0.3698 |
| 0.15 | 0.5044 | 0.7999 | −0.0659 | 1.95 | 0.3276 | 0.0679 | −0.3608 |
| 0.16 | 0.5285 | 0.7788 | −0.0754 | 2.0 | 0.3178 | 0.0655 | −0.3522 |
| 0.17 | 0.5514 | 0.7586 | −0.0854 | 2.1 | 0.2993 | 0.0610 | −0.3353 |
| 0.18 | 0.5730 | 0.7391 | −0.0957 | 2.2 | 0.2824 | 0.0569 | −0.3196 |
| 0.19 | 0.5935 | 0.7203 | −0.1066 | 2.3 | 0.2669 | 0.0533 | −0.3047 |
| 0.20 | 0.6129 | 0.7021 | −0.1175 | 2.4 | 0.2527 | 0.0500 | −0.2908 |
| 0.21 | 0.6314 | 0.6847 | −0.1288 | 2.5 | 0.2396 | 0.0471 | −0.2778 |
| 0.22 | 0.6490 | 0.6678 | −0.1404 | 2.6 | 0.2275 | 0.0444 | −0.2656 |
| 0.23 | 0.6656 | 0.6515 | −0.1522 | 2.7 | 0.2163 | 0.0419 | −0.2539 |
| 0.24 | 0.6810 | 0.6358 | −0.1641 | 2.8 | 0.2059 | 0.0397 | −0.2430 |
| 0.25 | 0.6955 | 0.6206 | −0.1761 | 2.9 | 0.1963 | 0.0376 | −0.2328 |
| 0.26 | 0.7090 | 0.6059 | −0.1882 | 3.0 | 0.1874 | 0.0357 | −0.2233 |
| 0.27 | 0.7216 | 0.5917 | −0.2003 | 3.1 | 0.1791 | 0.0340 | −0.2143 |
| 0.28 | 0.7333 | 0.5780 | −0.2123 | 3.2 | 0.1713 | 0.0324 | −0.2057 |
| 0.29 | 0.7441 | 0.5648 | −0.2243 | 3.3 | 0.1641 | 0.0309 | −0.1979 |
| 0.30 | 0.7540 | 0.5520 | −0.2363 | 3.4 | 0.1573 | 0.0295 | −0.1903 |
| 0.31 | 0.7629 | 0.5396 | −0.2480 | 3.5 | 0.1509 | 0.0282 | −0.1830 |
| 0.32 | 0.7710 | 0.5276 | −0.2596 | 3.6 | 0.1450 | 0.0270 | −0.1764 |
| 0.33 | 0.7785 | 0.5159 | −0.2711 | 3.7 | 0.1395 | 0.0259 | −0.1703 |
| 0.34 | 0.7853 | 0.5047 | −0.2824 | 3.8 | 0.1342 | 0.0248 | −0.1642 |
| 0.35 | 0.7915 | 0.4938 | −0.2936 | 3.9 | 0.1293 | 0.0239 | −0.1587 |
| 0.36 | 0.7969 | 0.4833 | −0.3045 | 4.0 | 0.1246 | 0.0229 | −0.1532 |
| 0.37 | 0.8017 | 0.4730 | −0.3152 | 4.1 | 0.1202 | 0.0221 | −0.1481 |
| 0.38 | 0.8060 | 0.4631 | −0.3257 | 4.2 | 0.1160 | 0.0213 | −0.1431 |
| 0.39 | 0.8099 | 0.4535 | −0.3361 | 4.3 | 0.1121 | 0.0205 | −0.1387 |
| 0.40 | 0.8133 | 0.4442 | −0.3462 | 4.4 | 0.1084 | 0.0198 | −0.1344 |
| 0.45 | 0.8227 | 0.4016 | −0.3923 | 4.5 | 0.1049 | 0.0191 | −0.1304 |
| 0.50 | 0.8225 | 0.3648 | −0.4322 | 4.6 | 0.1015 | 0.0184 | −0.1262 |
| 0.55 | 0.8150 | 0.3327 | −0.4650 | 4.7 | 0.0983 | 0.0178 | −0.1224 |
| 0.60 | 0.8022 | 0.3048 | −0.4914 | 4.8 | 0.0953 | 0.0173 | −0.1189 |
| 0.65 | 0.7855 | 0.2804 | −0.5118 | 4.9 | 0.0925 | 0.0167 | −0.1159 |
| 0.70 | 0.7660 | 0.2585 | −0.5269 | 5.0 | 0.0897 | 0.0162 | −0.1123 |
| 0.75 | 0.7447 | 0.2392 | −0.5374 | 5.1 | 0.0871 | 0.0157 | −0.1092 |
| 0.80 | 0.7222 | 0.2221 | −0.5438 | 5.2 | 0.0846 | 0.0152 | −0.1062 |
| 0.85 | 0.6990 | 0.2067 | −0.5466 | 5.3 | 0.0822 | 0.0148 | −0.1034 |
| 0.90 | 0.6757 | 0.1930 | −0.5467 | 5.4 | 0.0799 | 0.0143 | −0.1007 |
| 0.95 | 0.6527 | 0.1806 | −0.5445 | 5.5 | 0.0777 | 0.0139 | −0.0979 |
| 1.00 | 0.6300 | 0.1693 | −0.5404 | 5.6 | 0.0756 | 0.0135 | −0.0956 |
| 1.05 | 0.6078 | 0.1592 | −0.5346 | 5.7 | 0.0737 | 0.0132 | −0.0933 |
| 1.10 | 0.5861 | 0.1499 | −0.5274 | 5.8 | 0.0718 | 0.0128 | −0.0910 |
| 1.15 | 0.5652 | 0.1415 | −0.5193 | 5.9 | 0.0700 | 0.0125 | −0.0889 |

Table 2. (Continuation).

| $x$ | $F(x)$ | $\psi(x)$ | $U(x)$ | $x$ | $F(x)$ | $\psi(x)$ | $U(x)$ |
|---|---|---|---|---|---|---|---|
| | | | $K_H R_\lambda = \infty$ | | | | |
| 6.0 | 0.0682 | 0.0121 | −0.0865 | 6.3 | 0.0633 | 0.0112 | −0.0807 |
| 6.1 | 0.0665 | 0.0118 | −0.0845 | 6.4 | 0.0618 | 0.0110 | −0.0786 |
| 6.2 | 0.0649 | 0.0115 | −0.0829 | 6.5 | 0.0603 | 0.0107 | −0.0765 |
| | | | $K_H R_\lambda = 16.0$ | | | | |
| 0.00 | 0.0000 | 1.2236 | −0.0000 | 0.65 | 0.7849 | 0.2629 | −0.4797 |
| 0.01 | 0.0399 | 1.1846 | −0.0001 | 0.70 | 0.7655 | 0.2413 | −0.4907 |
| 0.02 | 0.0795 | 1.1473 | −0.0004 | 0.75 | 0.7441 | 0.2221 | −0.4969 |
| 0.03 | 0.1185 | 1.1116 | −0.0013 | 0.80 | 0.7214 | 0.2050 | −0.4990 |
| 0.04 | 0.1567 | 1.0775 | −0.0027 | 0.85 | 0.6983 | 0.1897 | −0.4979 |
| 0.05 | 0.1941 | 1.0448 | −0.0048 | 0.90 | 0.6752 | 0.1760 | −0.4942 |
| 0.06 | 0.2305 | 1.0135 | −0.0076 | 0.95 | 0.6522 | 0.1637 | −0.4880 |
| 0.07 | 0.2658 | 0.9834 | −0.0111 | | | | |
| 0.08 | 0.2999 | 0.9546 | −0.0154 | 1.00 | 0.6291 | 0.1525 | −0.4796 |
| 0.09 | 0.3329 | 0.9269 | −0.0205 | 1.05 | 0.6066 | 0.1424 | −0.4697 |
| 0.10 | 0.3646 | 0.9004 | −0.0262 | 1.10 | 0.5849 | 0.1333 | −0.4589 |
| 0.11 | 0.3951 | 0.8748 | −0.0327 | 1.15 | 0.5642 | 0.1249 | −0.4474 |
| 0.12 | 0.4243 | 0.8504 | −0.0398 | 1.20 | 0.5440 | 0.1172 | −0.4350 |
| 0.13 | 0.4523 | 0.8267 | −0.0476 | 1.25 | 0.5244 | 0.1102 | −0.4219 |
| 0.14 | 0.4790 | 0.8040 | −0.0559 | 1.30 | 0.5057 | 0.1038 | −0.4086 |
| 0.15 | 0.5044 | 0.7822 | −0.0647 | 1.35 | 0.4877 | 0.0979 | −0.3950 |
| 0.16 | 0.5285 | 0.7612 | −0.0741 | 1.40 | 0.4704 | 0.0924 | −0.3812 |
| 0.17 | 0.5514 | 0.7409 | −0.0838 | 1.45 | 0.4540 | 0.0873 | −0.3676 |
| 0.18 | 0.5730 | 0.7214 | −0.0939 | 1.5 | 0.4383 | 0.0827 | −0.3540 |
| 0.19 | 0.5935 | 0.7026 | −0.1044 | 1.6 | 0.4090 | 0.0743 | −0.3275 |
| 0.20 | 0.6129 | 0.6845 | −0.1151 | 1.7 | 0.3823 | 0.0670 | −0.3017 |
| 0.21 | 0.6315 | 0.6670 | −0.1262 | 1.8 | 0.3578 | 0.0607 | −0.2756 |
| 0.22 | 0.6491 | 0.6501 | −0.1374 | 1.9 | 0.3354 | 0.0551 | −0.2517 |
| 0.23 | 0.6657 | 0.6339 | −0.1489 | 2.0 | 0.3151 | 0.0502 | −0.2293 |
| 0.24 | 0.6811 | 0.6181 | −0.1604 | 2.1 | 0.2965 | 0.0458 | −0.2081 |
| 0.25 | 0.6954 | 0.6030 | −0.1719 | 2.2 | 0.2792 | 0.0419 | −0.1879 |
| 0.26 | 0.7087 | 0.5883 | −0.1835 | 2.3 | 0.2634 | 0.0385 | −0.1687 |
| 0.27 | 0.7212 | 0.5741 | −0.1951 | 2.4 | 0.2490 | 0.0353 | −0.1510 |
| 0.28 | 0.7330 | 0.5604 | −0.2067 | 2.5 | 0.2357 | 0.0325 | −0.1343 |
| 0.29 | 0.7439 | 0.5472 | −0.2183 | 2.6 | 0.2233 | 0.0300 | −0.1184 |
| 0.30 | 0.7539 | 0.5344 | −0.2297 | 2.7 | 0.2119 | 0.0277 | −0.1033 |
| 0.31 | 0.7628 | 0.5220 | −0.2409 | 2.8 | 0.2014 | 0.0256 | −0.0895 |
| 0.32 | 0.7710 | 0.5100 | −0.2520 | 2.9 | 0.1916 | 0.0238 | −0.0764 |
| 0.33 | 0.7785 | 0.4984 | −0.2630 | 3.0 | 0.1824 | 0.0220 | −0.0638 |
| 0.34 | 0.7853 | 0.4872 | −0.2737 | 3.1 | 0.1738 | 0.0205 | −0.0520 |
| 0.35 | 0.7915 | 0.4763 | −0.2843 | 3.2 | 0.1658 | 0.0190 | −0.0409 |
| 0.36 | 0.7970 | 0.4657 | −0.2946 | 3.3 | 0.1583 | 0.0177 | −0.0304 |
| 0.37 | 0.8018 | 0.4555 | −0.3047 | 3.4 | 0.1513 | 0.0165 | −0.0206 |
| 0.38 | 0.8061 | 0.4456 | −0.3146 | 3.5 | 0.1447 | 0.0153 | −0.0112 |
| 0.39 | 0.8099 | 0.4360 | −0.3242 | 3.6 | 0.1385 | 0.0143 | −0.0024 |
| 0.40 | 0.8131 | 0.4267 | −0.3336 | 3.7 | 0.1326 | 0.0133 | +0.0063 |
| 0.41 | 0.8160 | 0.4177 | −0.3428 | 3.8 | 0.1271 | 0.0125 | 0.0137 |
| 0.42 | 0.8182 | 0.4089 | −0.3516 | 3.9 | 0.1219 | 0.0116 | 0.0212 |
| 0.43 | 0.8201 | 0.4004 | −0.3602 | 4.0 | 0.1169 | 0.01088 | 0.0283 |
| 0.44 | 0.8216 | 0.3922 | −0.3685 | 4.5 | 0.0958 | 0.00779 | 0.0575 |
| 0.45 | 0.8227 | 0.3841 | −0.3766 | 5.0 | 0.0791 | 0.00562 | 0.0793 |
| 0.46 | 0.8233 | 0.3764 | −0.3843 | 5.5 | 0.0656 | 0.00409 | 0.0946 |
| 0.47 | 0.8238 | 0.3688 | −0.3919 | 6.0 | 0.0546 | 0.00298 | 0.1045 |
| 0.48 | 0.8235 | 0.3614 | −0.3989 | 6.5 | 0.0454 | 0.00218 | 0.1100 |
| 0.49 | 0.8230 | 0.3543 | −0.4057 | 7.0 | 0.0377 | 0.00160 | 0.1117 |
| 0.50 | 0.8222 | 0.3474 | −0.4123 | 7.5 | 0.0311 | 0.00117 | 0.1103 |
| 0.55 | 0.8146 | 0.3154 | −0.4411 | 8.0 | 0.0255 | 0.00086 | 0.1063 |
| 0.60 | 0.8019 | 0.2875 | −0.4635 | 8.5 | 0.0209 | 0.00064 | 0.1004 |

Table 2. (Continuation).

| $x$ | $F(x)$ | $\psi(x)$ | $U(x)$ | $x$ | $F(x)$ | $\psi(x)$ | $U(x)$ |
|---|---|---|---|---|---|---|---|
| | | | $K_H R_\lambda = 16.0$ | | | | |
| 9.0 | 0.0170 | 0.00046 | 0.0933 | 12.0 | 0.0042 | 0.00008 | 0.0454 |
| 9.5 | 0.0136 | 0.00034 | 0.0852 | 12.5 | 0.0033 | 0.00006 | 0.0392 |
| 10.0 | 0.0109 | 0.00026 | 0.0766 | 13.0 | 0.0026 | 0.00005 | 0.0338 |
| 10.5 | 0.0087 | 0.00019 | 0.0683 | 13.5 | 0.0020 | 0.00004 | 0.0287 |
| 11.0 | 0.0070 | 0.00014 | 0.0606 | 14.0 | 0.0016 | 0.00003 | 0.0248 |
| 11.5 | 0.0055 | 0.00010 | 0.0529 | 14.5 | 0.0012 | 0.00002 | 0.0208 |
| | | | $K_H R_\lambda = 12.1$ | | | | |
| 0.00 | 0.0000 | 1.2187 | −0.0000 | 4.4 | 0.0850 | 0.00451 | 0.1092 |
| 0.05 | 0.1940 | 1.0412 | −0.0053 | 4.5 | 0.0809 | 0.00413 | 0.1125 |
| 0.10 | 0.3669 | 0.9011 | −0.0269 | 4.6 | 0.0770 | 0.00378 | 0.1153 |
| 0.15 | 0.5091 | 0.7767 | −0.0651 | 4.7 | 0.0731 | 0.00346 | 0.1174 |
| 0.20 | 0.6201 | 0.6773 | −0.1158 | 4.8 | 0.0695 | 0.00316 | 0.1192 |
| 0.25 | 0.7012 | 0.5949 | −0.1750 | 4.9 | 0.0661 | 0.00290 | 0.1207 |
| 0.30 | 0.7590 | 0.5258 | −0.2329 | 5.0 | 0.0627 | 0.00265 | 0.1217 |
| 0.35 | 0.7977 | 0.4674 | −0.2794 | 5.1 | 0.0596 | 0.00242 | 0.1225 |
| 0.40 | 0.8172 | 0.4176 | −0.3331 | 5.2 | 0.0566 | 0.00222 | 0.1231 |
| 0.45 | 0.8273 | 0.3748 | −0.3723 | 5.3 | 0.0535 | 0.00202 | 0.1234 |
| 0.5 | 0.8285 | 0.3379 | −0.4070 | 5.4 | 0.0508 | 0.00184 | 0.1231 |
| 0.6 | 0.8071 | 0.2775 | −0.4533 | 5.5 | 0.0482 | 0.00168 | 0.1222 |
| 0.7 | 0.7695 | 0.2311 | −0.4760 | 5.6 | 0.0457 | 0.00154 | 0.1197 |
| 0.8 | 0.7265 | 0.1947 | −0.4797 | 5.7 | 0.0434 | 0.00140 | 0.1181 |
| 0.9 | 0.6790 | 0.1657 | −0.4765 | 5.8 | 0.0411 | 0.00128 | 0.1174 |
| 1.0 | 0.6325 | 0.1422 | −0.4509 | 5.9 | 0.0389 | 0.00116 | 0.1160 |
| 1.1 | 0.5879 | 0.1231 | −0.4252 | 6.0 | 0.0369 | 0.00106 | 0.1147 |
| 1.2 | 0.5455 | 0.1702 | −0.3977 | 6.1 | 0.0349 | 0.00096 | 0.1133 |
| 1.3 | 0.5063 | 0.0940 | −0.3673 | 6.2 | 0.0331 | 0.00088 | 0.1116 |
| 1.4 | 0.4690 | 0.0828 | −0.3362 | 6.3 | 0.0313 | 0.00079 | 0.1100 |
| 1.5 | 0.4355 | 0.0732 | −0.2998 | 6.4 | 0.0295 | 0.00072 | 0.1082 |
| 1.6 | 0.4057 | 0.0650 | −0.2747 | 6.5 | 0.0279 | 0.00065 | 0.1062 |
| 1.7 | 0.3786 | 0.0580 | −0.2483 | 6.6 | 0.0263 | 0.00059 | 0.1039 |
| 1.8 | 0.3547 | 0.0518 | −0.2194 | 6.7 | 0.0249 | 0.00054 | 0.1016 |
| 1.9 | 0.3327 | 0.0465 | −0.1939 | 6.8 | 0.0236 | 0.00048 | 0.0997 |
| 2.0 | 0.3122 | 0.0418 | −0.1654 | 6.9 | 0.0223 | 0.00043 | 0.0975 |
| 2.1 | 0.2928 | 0.0376 | −0.1409 | 7.0 | 0.0209 | 0.00038 | 0.0950 |
| 2.2 | 0.2745 | 0.0340 | −0.1232 | 7.1 | 0.0196 | 0.00034 | 0.0923 |
| 2.3 | 0.2578 | 0.0307 | −0.1033 | 7.2 | 0.0186 | 0.00030 | 0.0900 |
| 2.4 | 0.2430 | 0.0278 | −0.0840 | 7.3 | 0.0175 | 0.00027 | 0.0875 |
| 2.5 | 0.2293 | 0.0252 | −0.0666 | 7.4 | 0.0163 | 0.00024 | 0.0848 |
| 2.6 | 0.2165 | 0.0229 | −0.0496 | 7.5 | 0.0154 | 0.00021 | 0.0824 |
| 2.7 | 0.2045 | 0.0208 | −0.0338 | 7.6 | 0.0147 | 0.00018 | 0.0802 |
| 2.8 | 0.1935 | 0.0190 | −0.0193 | 7.7 | 0.0138 | 0.00016 | 0.0778 |
| 2.9 | 0.1833 | 0.0173 | −0.0057 | 7.8 | 0.0128 | 0.00014 | 0.0752 |
| 3.0 | 0.1737 | 0.01574 | +0.0067 | 7.9 | 0.0119 | 0.00012 | 0.0725 |
| 3.1 | 0.1646 | 0.01436 | 0.0189 | 8.0 | 0.0111 | 0.00010 | 0.0699 |
| 3.2 | 0.1561 | 0.01310 | 0.0298 | 8.1 | 0.0104 | 0.00008 | 0.0675 |
| 3.3 | 0.1482 | 0.01198 | 0.0400 | 8.2 | 0.0097 | 0.00007 | 0.0652 |
| 3.4 | 0.1408 | 0.01094 | 0.0495 | 8.3 | 0.0091 | 0.00006 | 0.0629 |
| 3.5 | 0.1337 | 0.01000 | 0.0579 | 8.4 | 0.0086 | 0.00005 | 0.0610 |
| 3.6 | 0.1270 | 0.00915 | 0.0661 | 8.5 | 0.0081 | 0.00004 | 0.0588 |
| 3.7 | 0.1208 | 0.00837 | 0.0741 | 8.6 | 0.0076 | 0.00003 | 0.0564 |
| 3.8 | 0.1149 | 0.00766 | 0.0797 | 8.7 | 0.0071 | 0.00002 | 0.0540 |
| 3.9 | 0.1093 | 0.00702 | 0.0856 | 8.8 | 0.0066 | 0.00001 | 0.0519 |
| 4.0 | 0.1038 | 0.00642 | 0.0920 | 8.9 | 0.0061 | | 0.0500 |
| 4.1 | 0.0988 | 0.00588 | 0.0972 | 9.0 | 0.0058 | | 0.0479 |
| 4.2 | 0.0940 | 0.00538 | 0.1016 | 9.1 | 0.0054 | | 0.0462 |
| 4.3 | 0.0894 | 0.00492 | 0.1054 | 9.2 | 0.0051 | | 0.0445 |

DRASEKHAR [9] and PROUDMAN [21]. The tables of the one-dimensional spectrum $\psi(x)$ and of the transfer function $U(x)$ were then obtained by numerical integration[1]. For $K_H R_\lambda = \infty$ and 16.0 we have tabulated the same number of significant figures as CHANDRASEKHAR [9]; but, for $K_H R_\lambda = 12.1$, the last figure is uncertain[2].

Some of the more recent (and more accurate) results of BLANCH and FERGUSON [8] are also shown in Fig. 17.

**24. The comparison with experiment.** In comparing the results of HEISENBERG'S theory with experiment, our main purpose is to determine to what extent the arbitrary "constant" $K_H$ that appears in the theory is in fact a constant. From the calculations of the skewness factor given in Sect. 26, it would appear that $K_H$ is essentially independent of the Reynolds number and of the decay law.

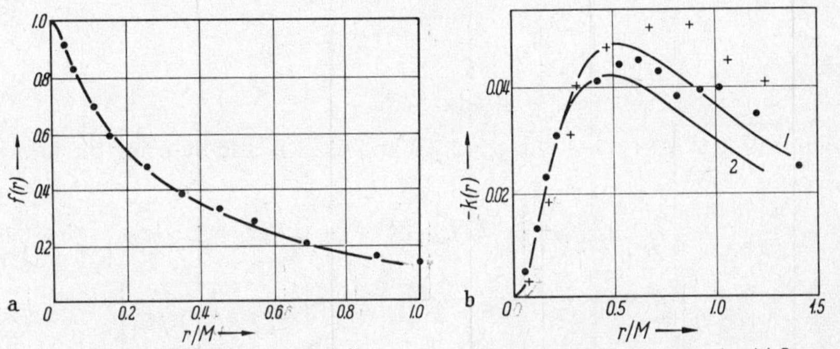


• experimental, $R_\lambda = 56.2$; —theoretical, $K_H = 0.45$.

• experimental, $R_\lambda = 56.2$; + experimental, $R_\lambda = 20.0$; 1 theoretical, $K_H = 0.45$, $R_\lambda = 26.9$; 2 theoretical, $K_H = 0.40$, $R_\lambda = 30.2$.

Fig. 18a and b. Comparison of the correlation functions with experiment. From PROUDMAN [21].

A more detailed comparison between theory and experiment has been given by PROUDMAN [21] using correlation functions. He found that when $K_H$ lies within the range

$$K_H = 0.45 \pm 0.05, \tag{24.1}$$

the agreement between theory and experiment is surprisingly good with respect to both shape and scale, provided $x/M$ is of the order of 30 (see Fig. 18). This agreement is not altogether convincing, however, since correlation functions are not very sensitive to theoretical assumptions.

An even more sensitive test could be made by using the longitudinal one-dimensional energy spectrum for which relatively extensive experimental data are now available [26]. Such a comparison would make possible a test of the hypothesis that $K_H$ is independent of $\varkappa$. Unfortunately, however, the absolute values of the experimental spectra are too uncertain to make possible a conclusive comparison.

**25. The approach to the final period of decay.** While solutions of Eq. (22.3) can be obtained for small values of the Reynolds numbers[3] they can at most be of mathematical interest; for the existence of an initial period of decay depends

[1] Except for $U(x)$ for $K_H R_\lambda = 12.1$ which was obtained by interpolation from an unpublished table kindly provided by Dr. PROUDMAN.

[2] We are indebted to Miss JOANNE ARRANT, Mrs. NOEL WILLIS, and Mr. JACK HOLT of the Ballistic Research Laboratories, Aberdeen Proving Ground, Maryland for their assistance with these calculations.

[3] W.H. REID: Quart. Appl. Math. **14**, 201 (1956).

very basically on the Reynolds number being large. During the final period of decay, however, the situation is just reversed[1]. Here the decay law is of the form $\overline{u^2} \propto t^{-\frac{5}{2}}$ which implies that $\lambda^2 = 4\nu t$ and hence $R_\lambda \propto t^{-\frac{3}{4}}$. Under these conditions then, all that we are really interested in is the asymptotic behavior of $E(\varkappa)$ and $T(\varkappa)$ as $t \to \infty$; i.e. for small values of the Reynolds number[2].

At the onset of the final period of decay, as defined by the decay law $\overline{u^2} \propto t^{-\frac{5}{2}}$, the Reynolds number of the turbulence, though small, is not negligible, being of the order of 5, and this suggests that a transfer of energy exists that is compatible with the observed final period decay law. Under these conditions a similarity solution can be obtained in the form[3]

$$E(\varkappa, t) = \nu^{\frac{3}{2}} t^{-\frac{1}{2}} R_\lambda^{\frac{2}{3}} F(x, R_\lambda) \tag{25.1}$$

and

$$T(\varkappa, t) = \nu^{\frac{3}{2}} t^{-\frac{3}{2}} R_\lambda^{\frac{2}{3}} U(x, R_\lambda) \tag{25.2}$$

where

$$x = (\nu \varkappa^2 t)^{\frac{1}{2}}.$$

By substituting these results in Eqs. (7.26) and (22.10), and by using the fact that $R_\lambda \propto t^{-\frac{3}{4}}$, we obtain

$$\left(x \frac{\partial}{\partial x} - \frac{3}{2} R_\lambda \frac{\partial}{\partial R_\lambda}\right) F(x) + 4(x^2 - 1) F(x) = 2U(x) \tag{25.3}$$

and

$$U(x) = -2K_H R_\lambda \frac{d}{dx}\left[\int_x^\infty \sqrt{\frac{F(x')}{x'^3}}\, dx' \int_0^x x''^2 F(x'')\, dx''\right]. \tag{25.4}$$

For small values of the Reynolds number, Eq. (25.4) shows that $U(x)$ is of smaller order than $F(x)$. Accordingly, we can assume a solution in the form of a power series in the Reynolds number of the form

$$\left.\begin{aligned} F(x, R_\lambda) &= \sum_{n=0}^{\infty} F_n(x)\,(K_H R_\lambda)^n \\ \text{and}\quad U(x, R_\lambda) &= \sum_{n=1}^{\infty} U_n(x)\,(K_H R_\lambda)^n. \end{aligned}\right\} \tag{25.5}$$

Since $R_\lambda \propto t^{-\frac{3}{4}}$, this power series in the Reynolds number is equivalent to an asymptotic expansion in the time, and it is this fact which makes a solution of the form (25.5) meaningful in the final period of decay; for, with an increasing Reynolds number, we are simply moving backward in time.

By substituting the series (25.5) into Eqs. (25.3) and (25.4) and by equating like powers of $R_\lambda$, one is led to a system of equations from which the functions $F_n(x)$ and $U_n(x)$ can be obtained by quadrature. From the relation

$$\overline{u^2} = \tfrac{2}{3} \int_0^\infty E(\varkappa)\, d\varkappa,$$

the boundary conditions on the functions $F_n(x)$ follow in the form

$$\int_0^\infty F_n(x)\, dx = \begin{cases} \frac{3}{8} \text{ for } n = 0 \\ 0 \text{ for } n \geqq 1, \end{cases} \tag{25.6}$$

[1] W.H. Reid: Proc. Nat. Acad. Sci., Wash. **42**, 559 (1956).

[2] A solution for an infinite Reynolds number in the final period of decay has been given by J.C. Rotta: J. Aeronaut. Sci. **20**, 769 (1953).

[3] Note that the similarity transformation and the definition of $x$ employed here differ slightly from those used in Sect. 22.

which show that the general shape of $F(x)$ is determined by $F_0(x)$ and that the functions $F_n(x)$, for $n \geqq 1$, cause only a detailed modification of that shape as the Reynolds number is increased and leave the decay law unaffected.

The functions $F_0(x)$ and $U_1(x)$ evaluated in this way are given by

$$F_0(x) = 4\,(2/\pi)^{\frac{1}{2}}\, x^4\, e^{-2x^2} \tag{25.7}$$

and

$$U_1(x) = -\,16\,(2/\pi)^{\frac{3}{4}} \frac{d}{dx}\left[\int_x^\infty x'^{\frac{1}{2}} e^{-x'^2}\, dx' \int_0^x x''^6 e^{-2x''^2}\, dx''\right], \tag{25.8}$$

which are shown graphically in Fig. 19. The spectrum (25.7) is simply the usual final period spectrum rewritten in terms of the present similarity variables. The transfer function (25.8) is typical of the small Reynolds number conditions prevailing in the final period of decay and has a shape that is distinctly different from the large Reynolds number transfer functions shown in Fig. 14 with which it may be compared.

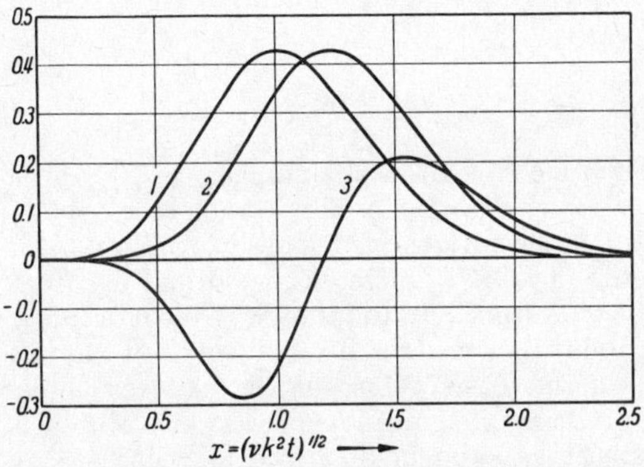


Fig. 19. The distribution of energy transfer in the final period of decay according to HEISENBERG's transfer theory. Curve 1, the energy spectrum, $F_0(x)$; curve 2, the energy-dissipation spectrum, $0.8054\,x^2 F_0(x)$; curve 3, the transfer function, $2\,U_1(x)$.

**26. The skewness factor.** The skewness factor $S$ of $-\partial u_1/\partial x_1$, defined by Eq. (4.5), provides an important measure of the non-linear effects responsible for the transfer of energy between diffterent wave numbers. By using ihe solutions given in the foregoing sections it is possible to say something about the variation of $S$ or, more precisely, of the ratio $S/K_H$ during the decay process.

By definition, the skewness factor $S$ of $-\partial u_1/\partial x_1$ is

$$S = -\,\lambda^3\, k'''(0), \tag{26.1}$$

which can also be written entirely in terms of the spectrum functions in the form

$$S = \frac{3}{14}\sqrt{30}\,\frac{\int_0^\infty \varkappa^2\, T(\varkappa)\, d\varkappa}{\left[\int_0^\infty \varkappa^2\, E(\varkappa)\, d\varkappa\right]^{\frac{3}{2}}}; \tag{26.2}$$

alternatively, this can be written entirely in terms of $E(\varkappa)$,

$$S = \frac{3}{7}\sqrt{30}\,\nu\,\frac{\int_0^\infty \varkappa^4\, E(\varkappa)\, d\varkappa}{\left[\int_0^\infty \varkappa^2\, E(\varkappa)\, d\varkappa\right]^{\frac{3}{2}}} + \frac{3}{14}\sqrt{30}\,\frac{\frac{d}{dt}\int_0^\infty \varkappa^2\, E(\varkappa)\, d\varkappa}{\left[\int_0^\infty \varkappa^2\, E(\varkappa)\, d\varkappa\right]^{\frac{3}{2}}}, \tag{26.3}$$

which is a convenient form for some purposes.

Consider first the behavior of $S$ during the initial period of decay. For this purpose we introduce the initial period similarity (22.1) into Eq. (26.3) and obtain

$$\frac{S}{K_H} = \frac{3}{7}\sqrt{30}\left(\frac{I_4}{\alpha\, I_2^{\frac{3}{2}}} - \frac{1}{I_2^{\frac{1}{2}}}\right), \tag{26.4}$$

where

$$I_n = \int_0^\infty x^n F(x)\, dx.$$

Now both of the integrals occurring in Eq. (26.4) are weighted heavily in favor of the small scale components. Thus, if the Reynolds number is sufficiently large, we can use the equilibrium spectrum without a cut-off, Eq. (21.1), the similarity form of which is

$$F(x) = \frac{4}{3^{\frac{4}{3}}} I_0^{\frac{2}{3}} x^{-\frac{5}{3}} \left(1 + \frac{8}{3\alpha^3 I_0} x^4\right)^{-\frac{4}{3}}, \tag{26.5}$$

to evaluate these integrals. Furthermore, when $R_\lambda \to \infty$, the second term in the parentheses in Eq. (26.4) becomes negligible compared with the first. The first term can then be evaluated to give

$$\frac{S}{K_H} \to \frac{3}{7} \sqrt{30}\, \frac{2^{\frac{1}{3}} \pi^2}{\Gamma^3(\frac{1}{3})} = 1.52 \qquad (R_\lambda \to \infty), \tag{26.6}$$

a result due to LEE[1]. For definite values of the Reynolds number, $S/K_H$ can be obtained by direct numerical integration (see Sect. 23). For completeness, however, it may be remarked that as $R_\lambda \to 0$ (in the initial period of decay), $S/K_H \to 0.78$. Thus, in the initial period of decay, the ratio $S/K_H$ increases monotonically from 0.78 to 1.52 as the Reynolds number varies from zero to infinity. Now it is found experimentally that $S$ varies from about 0.48 for small values of $R_\lambda$ to about 0.30 for large values of $R_\lambda$ so that, on this basis, $K_H$ would lie between about 0.62 and 0.20 which are of the same order of magnitude as the values obtained from the more detailed comparisons with experiment described in Sect. 24.

It is also possible to obtain an estimate of $S/K_H$ in the final period of decay from the solution given in Sect. 25. What is perhaps most surprising about that solution is that $S/K_H$ does not tend to zero as $t \to \infty$ as one might have expected but approaches a non-zero constant value. This can easily be seen by introducing the similarity transformations (25.1) and (25.2) into Eq. (26.2) to give

$$S = \frac{3}{14} \sqrt{30}\, \frac{1}{R_\lambda}\, \frac{\int_0^\infty x^2 U(x)\, dx}{\left[\int_0^\infty x^2 F(x)\, dx\right]^{\frac{3}{2}}} \tag{26.7}$$

which shows that, as $R_\lambda \to 0$,

$$\frac{S}{K_H} = \frac{128}{35} \int_0^\infty x^2 U_1(x)\, dx + O(K_H R_\lambda). \tag{26.8}$$

A numerical evaluation of the integral occurring in this equation then shows that, as $R_\lambda \to 0$,

$$S/K_H \to 0.58. \tag{26.9}$$

These results suggest, then, that the value of $S/K_H$ varies by less than a factor of 3 throughout the entire life-history of the turbulence, whatever may be the initial value of the Reynolds number.

**27. The stability of the similarity spectra in the initial period of decay.** It has been argued in Sect. 18 that while the smaller eddies may adjust themselves into

[1] T.D. LEE: Phys. Rev. **77**, 842 (1950). When $R_\lambda$ *equals* infinity, the behaviour of $S/K_H$ is singular with the value 0.60 [W.H. REID: Quart. Appl. Math. **14**, 201 (1956)].

a quasi-equilibrium state during the decay process the larger eddies cannot do so. One may question therefore the validity of applying the similarity transformation (22.1) to all wave-numbers; this is particularly so in view of the resulting linear behavior of the spectra near $x=0$ in direct violation of the requirements of continuity.

Some light has been cast on this problem by LIN[1] who considered the stability of the similarity spectrum to a small perturbation. By considering the rate at which such a perturbation dies out compared with the overall rate of viscous dissipation, one can obtain a measure of the ability of the spectrum to adjust itself to the quasi-equilibrium state during the process of decay.

Thus, consider a small perturbation $\varphi(\varkappa, t)$ superimposed on $E(\varkappa, t)$. The linearized equation for $\varphi(\varkappa, t)$ then follows from Eq. (20.6) in the form

$$\left.\begin{aligned}\frac{\partial}{\partial t}\int\limits_0^{\varkappa}\varphi(\varkappa)\,d\varkappa = -2\left[\nu + K_H\int\limits_{\varkappa}^{\infty}\sqrt{\frac{E(\varkappa)}{\varkappa^3}}\,d\varkappa\right]\int\limits_0^{\varkappa}\varkappa^2\,\varphi(\varkappa)\,d\varkappa - {}\\ - K_H\int\limits_{\varkappa}^{\infty}\frac{\varphi(\varkappa)}{\sqrt{\varkappa^3\,E(\varkappa)}}\,d\varkappa\int\limits_0^{\varkappa}\varkappa^2\,E(\varkappa)\,d\varkappa.\end{aligned}\right\} \tag{27.1}$$

A detailed study of this equation has not been made as yet, but it is sufficient for the present purpose to consider only the initial rate of disappearance of a perturbation that occurs only in a narrow band near a wave number $\varkappa_*$ say. Thus, if $e$ denotes the total energy of the perturbation, then by putting $\varkappa=\varkappa_*+0$ at the initial instant, we obtain for the rate of disappearance of the perturbation

$$\frac{1}{e}\,\frac{de}{dt} = 2\varkappa_*^2\left[\nu + K_H\int\limits_{\varkappa_*}^{\infty}\sqrt{\frac{E(\varkappa)}{\varkappa^3}}\,d\varkappa\right]. \tag{27.2}$$

The ratio of this quantity to the overall rate of decay,

$$\frac{1}{\overline{u^2}}\,\frac{d\overline{u^2}}{dt} = -\frac{10\nu}{\lambda^2},$$

is then simply

$$\gamma(x_*) = 2x_*^2\left[\frac{1}{\alpha} + \int\limits_{x_*}^{\infty}\sqrt{\frac{F(x)}{x^3}}\,dx\right]. \tag{27.3}$$

The quantity $\gamma(x_*)$ is in the nature of a "reaction time"; for, if $\gamma(x_*)$ is large then evidently the perturbation will die out much faster than $\overline{u^2}$, and those components of the motion for which this occurs will have sufficient time to achieve some degree of equilibrium. On the other hand, those components of the motion for which $\gamma(x_*)$ is small can never reach an equilibrium condition, simply because the time scale of decay of the perturbation is long compared with the time scale of decay of the total energy.

In Fig. 20, the reaction time $\gamma(x_*)$ is plotted against

$$P(x_*) = \frac{\int_0^{x_*} F(x)\,dx}{\int_0^{\infty} F(x)\,dx} \tag{27.4}$$

[1] C.C. LIN: Proc. 4th Symp. Appl. Math., Amer. Math. Soc. 1951, p. 19.

the fraction of the total energy below $x_*$. This figure should be considered in conjunction with Fig. 21, where the fraction of the total energy dissipation,

$$\delta(x_*) = \frac{\int_0^{x_*} x^2 F(x)\, dx}{\int_0^{\infty} x^2 F(x)\, dx}, \tag{27.5}$$

is plotted against $P(x_*)$. From these figures it is evident that over a considerable part of the spectrum $\gamma$ is of the order of one or less and, as argued above, this part of the spectrum would be unable to attain full similarity. This region, for which the dissipation is small, includes about one-third of the total energy.

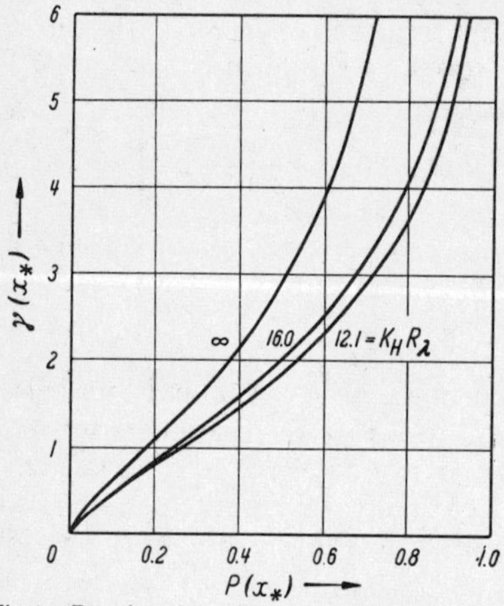


Fig. 20. Reaction time $\gamma(x_*)$ against fraction of energy $P(x_*)$.

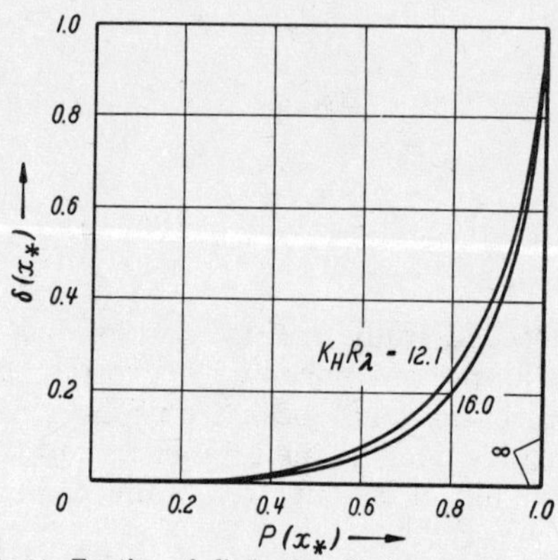


Fig. 21. Fraction of dissipation $\delta(x_*)$ against fraction of energy $P(x_*)$.

**28. An evaluation of the theory.** In the foregoing sections of this chapter we have attempted to give a reasonably complete discussion of HEISENBERG'S physical transfer theory and to describe some of the deductions that have been made from it. Our understanding of the theory is now sufficiently complete that an overall assessment of its worth can reasonably be made.

On the whole, the purely mathematical aspect of the theory may be considered essentially complete. For example, the quantity $K_H$ shows no systematic variation with wave-number, Reynolds number or decay time and is, in fact, a constant of order unity as originally postulated.

From a physical point of view, however, there are certain aspects of the theory which are not entirely satisfactory. The algebraic behavior of the spectra for large wave-numbers, for example, or full similarity in the initial period of decay are at variance with experimental results. Nevertheless, for many purposes the theory is perfectly adequate and, indeed, the only real question that still remains is: *why* is the theory so good? More fundamentally, of course, we should prefer a theory that could be derived purely deductively from the Navier-Stokes and continuity equations and, in the following chapter, we shall discuss some further efforts that have been made in this direction.

## V. The hypothesis of a normal joint-probability distribution for the velocity field.

**29. Introduction.** As indicated in Sect. 1, it is possible, at least in principle, to construct an infinite set of dynamical equations for the correlation functions (or, more precisely, for the *cumulant* correlation functions) of all orders from which the exact dynamical behavior of the turbulence could be determined. No definite set of these equations is complete, but the coupling between the equations is such that by introducing an appropriate approximation we can obtain a finite, closed system of equations. Thus, for example, if we neglect all cumulants of the $n$-th order and higher, then we obtain $n-2$ equations for the cumulants of order 2, $3, \ldots, n-1$.

It is not possible, however, to choose $n$ as low as three; for this corresponds to neglecting the triple-velocity correlations which are known not to be zero. For $n=4$, however, which is the case with which we shall be concerned here, the triple correlations need not be zero, but the vanishing of the fourth-order cumulants then requires that fourth-order mean values be related to second-order mean values as they would be for a normal joint-probability distribution. For example,

$$\overline{u_i u_j' u_k'' u_l'''} = \overline{u_i u_j'} \cdot \overline{u_k'' u_l'''} + \overline{u_i u_k''} \cdot \overline{u_j' u_l'''} + \overline{u_i u_l'''} \cdot \overline{u_j' u_k''}, \tag{29.1}$$

an approximation introduced into the theory of turbulence first by MILLIONSHTCHIKOV[1]. Furthermore, there exists a limited amount of experimental information[2] respecting the validity of Eq. (29.1) which, so far as it goes, suggests that it should be reliable over the energy-containing range of wave-numbers. The experiments do indicate a significant departure from the relation (29.1) for small spatial separations and this suggests that a theory based on the relation (29.1) may not agree with the results of KOLMOGOROFF'S theory for large wave-numbers[3] (see Sect. 35).

By using the relation (29.1), however, one can derive a complete system of partial differential equations with the same number of equations as unknowns: the equation for the rate of change of the double correlation involves the (two-point) triple correlation; the equation for the rate of change of the (three-point) triple correlation involves the (three-point) quadrupole correlation; and the quadruple correlation can then be reduced to double correlations by using the relation (29.1). In this manner a closed system of equations is obtained, some of the details of which are given in the following sections.

**30. The kinematics and dynamics of three-point mean values.** In addition to the two-point correlation $R_{ij}(\boldsymbol{r})$ and $P_i(\boldsymbol{r})=\overline{p(\boldsymbol{x})\, u_i(\boldsymbol{x}+\boldsymbol{r})}/\varrho$ introduced in Sect. 6, we must now consider the three-point correlations

$$R_{ijk}(\boldsymbol{r}, \boldsymbol{r}') = \overline{u_i(\boldsymbol{x})\, u_j(\boldsymbol{x}+\boldsymbol{r})\, u_k(\boldsymbol{x}+\boldsymbol{r}')} \tag{30.1}$$

and

$$P_{ij}(\boldsymbol{r}, \boldsymbol{r}') = \frac{1}{\varrho} \overline{p(\boldsymbol{x})\, u_i(\boldsymbol{x}+\boldsymbol{r})\, u_j(\boldsymbol{x}+\boldsymbol{r}')}, \tag{30.2}$$

which, by homogeneity, are functions only of the variables indicated. Each of these tensors satisfies certain symmetry conditions, obtained by permuting the products under the mean-value signs; as a result of the incompressibility condition, they are both solenoidal in every index. But, for the present purposes, it is not necessary to write out these expressions explicitly.

Now, as has already been noted, the fundamental approximation to be made in the dynamical equations is one that is not equally good for all eddy sizes. By employing their Fourier transforms rather than the correlation tensors themselves, however, it becomes possible to keep track of those regions of the spectrum in which the theory is likely to be applicable. Furthermore, the introduction of

---

[1] M. MILLIONSHTCHIKOV: C. R. Acad. Sci. URSS. **32**, 615, 619 (1941).

[2] M. S. UBEROI: J. Aeronaut. Sci. **20**, 197 (1953).

[3] E. HOPF: Proc. 7th Symp. in Appl. Math., Amer. Math. Soc. 1955, p. 41.

Fourier transforms leads to considerable mathematical simplification, inasmuch as spatial differential operators transform into algebraic operators.

Accordingly, the six-dimensional Fourier transforms of $R_{ijk}(\boldsymbol{r},\boldsymbol{r}')$ and $P_{ij}(\boldsymbol{r},\boldsymbol{r}')$ can now be defined by the equations

$$\Phi_{ijk}(\boldsymbol{\varkappa},\boldsymbol{\varkappa}') = i(2\pi)^{-6}\int R_{ijk}(\boldsymbol{r},\boldsymbol{r}')\,\mathrm{e}^{-i(\boldsymbol{\varkappa}\cdot\boldsymbol{r}+\boldsymbol{\varkappa}'\cdot\boldsymbol{r}')}\,d\boldsymbol{r}\,d\boldsymbol{r}' \tag{30.3}$$

and

$$\Pi_{ij}(\boldsymbol{\varkappa},\boldsymbol{\varkappa}') = (2\pi)^{-6}\int P_{ij}(\boldsymbol{r},\boldsymbol{r}')\,\mathrm{e}^{-i(\boldsymbol{\varkappa}\cdot\boldsymbol{r}+\boldsymbol{\varkappa}'\cdot\boldsymbol{r}')}\,d\boldsymbol{r}\,d\boldsymbol{r}'. \tag{30.4}$$

In the first of these equations the multiplicative factor $i=\sqrt{-1}$ has been introduced simply for convenience, and in both equations the integrations are over all the spaces of the $\boldsymbol{r}$'s. The tensors occurring in the integrands of Eqs. (30.3) and (30.4) are of course real; the spectrum tensors are therefore complex in general. The spectrum tensors also satisfy certain simple symmetry conditions corresponding to, but not identical with, the symmetry conditions satisfied by $R_{ijk}(\boldsymbol{r},\boldsymbol{r}')$ and $P_{ij}(\boldsymbol{r},\boldsymbol{r}')$ and, as a result of the incompressibility conditions, they are both orthogonal in every index. But, again, it is not necessary for the present purposes to write these out explicitly.

Consider now the dynamical equation satisfied by $R_{ij}(\boldsymbol{r})$, Eq. (6.14); it can also be written in the form

$$\frac{\partial}{\partial t}R_{ij}(\boldsymbol{r}) = O_2\left\{\frac{\partial}{\partial r_k}R_{ijk}(\boldsymbol{r},0) + \frac{\partial}{\partial r_i}P_j(\boldsymbol{r}) + \nu\frac{\partial^2}{\partial r_k^2}R_{ij}(\boldsymbol{r})\right\}. \tag{30.5}$$

Since the operator $O_2$ is invariant with respect to a Fourier transformation, the dynamical equation satisfied by $\Phi_{ij}(\boldsymbol{\varkappa})$ follows in the form

$$\frac{\partial}{\partial t}\Phi_{ij}(\boldsymbol{\varkappa}) = O_2\{\varkappa_k\int\Phi_{ijk}(\boldsymbol{\varkappa},\boldsymbol{\varkappa}')\,d\boldsymbol{\varkappa}' + \varkappa_i\Pi_j(\boldsymbol{\varkappa}) - \nu\varkappa^2\Phi_{ij}(\boldsymbol{\varkappa})\}. \tag{30.6}$$

The pressure transform $\Pi_j(\boldsymbol{\varkappa})$ can now be eliminated explicitly by using the incompressibility conditions (6.8). Thus, multiplying through by $\varkappa_i$ gives

$$\varkappa^2\Pi_j(\boldsymbol{\varkappa}) = -\varkappa_i\varkappa_k\int\Phi_{ijk}(\boldsymbol{\varkappa},\boldsymbol{\varkappa}')\,d\boldsymbol{\varkappa}', \tag{30.7}$$

so that Eq. (30.6) becomes

$$\frac{\partial}{\partial t}\Phi_{ij}(\boldsymbol{\varkappa}) = O_2\{\varkappa_k\Delta_i^\alpha(\boldsymbol{\varkappa})\int\Phi_{\alpha jk}(\boldsymbol{\varkappa},\boldsymbol{\varkappa}')\,d\boldsymbol{\varkappa}' - \nu\varkappa^2\Phi_{ij}(\boldsymbol{\varkappa})\}, \tag{30.8}$$

where

$$\Delta_i^\alpha(\boldsymbol{\varkappa}) = \delta_{i\alpha} - \varkappa_i\varkappa_\alpha/\varkappa^2. \tag{30.9}$$

Eq. (30.8) is the first of the two dynamical equations for the second- and third-order spectrum tensors. The appearance of the "orthogonal operator" (30.9) in that equation is particularly noteworthy; for this operator has precisely the same type of singularity at $\boldsymbol{\varkappa}=0$ as $\Phi_{ij}(\boldsymbol{\varkappa})$ (cf. Sect. 13).

The dynamical equation for $R_{ijk}(\boldsymbol{r},\boldsymbol{r}')$ can now be derived by the method of v. KÁRMÁN and HOWARTH [*17*]; the equation for $\Phi_{ijk}(\boldsymbol{\varkappa},\boldsymbol{\varkappa}')$ then follows by Fourier transformation. Thus, if we let

$$\frac{\partial}{\partial t}R_{ijk}(\boldsymbol{r},\boldsymbol{r}') = O_3\left\{\overline{\frac{\partial u_i(\boldsymbol{x})}{\partial t}u_j(\boldsymbol{x}+\boldsymbol{r})\,u_k(\boldsymbol{x}+\boldsymbol{r}')}\right\} \tag{30.10}$$

where the operator $O_3$ acting on a third-order tensor is such that

$$O_3\{C_{ijk}(\boldsymbol{r},\boldsymbol{r}')\} = C_{ijk}(\boldsymbol{r},\boldsymbol{r}') + C_{jki}(\boldsymbol{r}'-\boldsymbol{r},-\boldsymbol{r}) + C_{kij}(-\boldsymbol{r}',\boldsymbol{r}-\boldsymbol{r}'), \tag{30.11}$$

then by using the Navier-Stokes equation we obtain

$$\left.\begin{aligned}\frac{\partial}{\partial t} R_{ijk}(\boldsymbol{r}, \boldsymbol{r}') = O_3\Big\{\Big(\frac{\partial}{\partial r_l} + \frac{\partial}{\partial r_l'}\Big) \overline{u_i(\boldsymbol{x})\, u_j(\boldsymbol{x}+\boldsymbol{r})\, u_k(\boldsymbol{x}+\boldsymbol{r}')\, u_l(\boldsymbol{x})} + \\ + \Big(\frac{\partial}{\partial r_i} + \frac{\partial}{\partial r_i'}\Big) P_{jk}(\boldsymbol{r}, \boldsymbol{r}') + \nu\Big(\frac{\partial}{\partial r_l} + \frac{\partial}{\partial r_l'}\Big)^2 R_{ijk}(\boldsymbol{r}, \boldsymbol{r}')\Big\}.\end{aligned}\right\} \tag{30.12}$$

The appearance of fourth-order velocity-product mean values on the right-hand side of this equation indicates clearly the well-known incompleteness of any finite set of dynamical equations for mean values. As has already been indicated, the present theory is based on the hypothesis that the fourth-order mean values occurring in Eq. (30.12) are related to second-order mean values in the manner appropriate to a normal joint-probability distribution. Thus, we write

$$\left.\begin{aligned}\overline{u_i(\boldsymbol{x})\, u_j(\boldsymbol{x}+\boldsymbol{r})\, u_k(\boldsymbol{x}+\boldsymbol{r}')\, u_l(\boldsymbol{x})} = \overline{u_i(\boldsymbol{x})\, u_j(\boldsymbol{x}+\boldsymbol{r})} \cdot \overline{u_k(\boldsymbol{x}+\boldsymbol{r}')\, u_l(\boldsymbol{x})} + \\ + \overline{u_i(\boldsymbol{x})\, u_k(\boldsymbol{x}+\boldsymbol{r}')} \cdot \overline{u_j(\boldsymbol{x}+\boldsymbol{r})\, u_l(\boldsymbol{x})} + \\ + \overline{u_i(\boldsymbol{x})\, u_l(\boldsymbol{x})} \cdot \overline{u_j(\boldsymbol{x}+\boldsymbol{r})\, u_k(\boldsymbol{x}+\boldsymbol{r}')}\end{aligned}\right\} \tag{30.13}$$

and Eq. (30.12) then becomes

$$\left.\begin{aligned}\frac{\partial}{\partial t} R_{ijk}(\boldsymbol{r}, \boldsymbol{r}') = O_3\Big\{\Big(\frac{\partial}{\partial r_l} + \frac{\partial}{\partial r_l'}\Big) [R_{ij}(\boldsymbol{r})\, R_{lk}(\boldsymbol{r}') + R_{ik}(\boldsymbol{r}')\, R_{lj}(\boldsymbol{r})] + \\ + \Big(\frac{\partial}{\partial r_i} + \frac{\partial}{\partial r_i'}\Big) P_{jk}(\boldsymbol{r}, \boldsymbol{r}') + \nu\Big(\frac{\partial}{\partial r_l} + \frac{\partial}{\partial r_l'}\Big)^2 R_{ijk}(\boldsymbol{r}, \boldsymbol{r}')\Big\}.\end{aligned}\right\} \tag{30.14}$$

The Fourier transform of Eq. (30.14) can be obtained quite simply by employing the following device. If $\Gamma_{ijk}(\boldsymbol{\varkappa}, \boldsymbol{\varkappa}')$ is the Fourier transform of $C_{ijk}(\boldsymbol{r}, \boldsymbol{r}')$, in the sense of Eq. (30.3), then the Fourier transform of $O_3\{C_{ijk}(\boldsymbol{r}, \boldsymbol{r}')\}$ is

$$\Omega_3\{\Gamma_{ijk}(\boldsymbol{\varkappa}, \boldsymbol{\varkappa}')\} = \Gamma_{ijk}(\boldsymbol{\varkappa}, \boldsymbol{\varkappa}') + \Gamma_{jki}(\boldsymbol{\varkappa}', \boldsymbol{\varkappa}'') + \Gamma_{kij}(\boldsymbol{\varkappa}'', \boldsymbol{\varkappa}), \tag{30.15}$$

where the vector $\boldsymbol{\varkappa}''$ is defined by

$$\boldsymbol{\varkappa} + \boldsymbol{\varkappa}' + \boldsymbol{\varkappa}'' = 0. \tag{30.16}$$

The transformed version of Eq. (30.14) then becomes

$$\left.\begin{aligned}\frac{\partial}{\partial t} \Phi_{ijk}(\boldsymbol{\varkappa}, \boldsymbol{\varkappa}') = \Omega_3\{\varkappa_l''[\Phi_{ij}(\boldsymbol{\varkappa})\, \Phi_{lk}(\boldsymbol{\varkappa}') + \Phi_{ik}(\boldsymbol{\varkappa}')\, \Phi_{lj}(\boldsymbol{\varkappa})] + \\ + \varkappa_i'' \Pi_{jk}(\boldsymbol{\varkappa}, \boldsymbol{\varkappa}') - \nu \varkappa''^2\, \Phi_{ijk}(\boldsymbol{\varkappa}, \boldsymbol{\varkappa}')\}.\end{aligned}\right\} \tag{30.17}$$

By using the incompressibility condition in the form $\varkappa_i'' \Phi_{ijk}(\boldsymbol{\varkappa}, \boldsymbol{\varkappa}') = 0$, an explicit expression for the pressure transform $\Pi_{jk}(\boldsymbol{\varkappa}, \boldsymbol{\varkappa}')$ can be obtained from Eq. (30.17) in the form

$$\varkappa''^2 \Pi_{jk}(\boldsymbol{\varkappa}, \boldsymbol{\varkappa}') = -\varkappa_i'' \varkappa_l'' [\Phi_{ij}(\boldsymbol{\varkappa})\, \Phi_{lk}(\boldsymbol{\varkappa}') + \Phi_{ik}(\boldsymbol{\varkappa}')\, \Phi_{lj}(\boldsymbol{\varkappa})] \tag{30.18}$$

and thus Eq. (30.17) becomes

$$\left.\begin{aligned}\frac{\partial}{\partial t} \Phi_{ijk}(\boldsymbol{\varkappa}, \boldsymbol{\varkappa}') = \Omega_3\{\varkappa_l'' \Delta_i^\alpha(\boldsymbol{\varkappa}'')\, [\Phi_{\alpha j}(\boldsymbol{\varkappa})\, \Phi_{lk}(\boldsymbol{\varkappa}') + \Phi_{\alpha k}(\boldsymbol{\varkappa}')\, \Phi_{lj}(\boldsymbol{\varkappa})] - \\ - \nu \varkappa''^2\, \Phi_{ijk}(\boldsymbol{\varkappa}, \boldsymbol{\varkappa}')\}.\end{aligned}\right\} \tag{30.19}$$

Eqs. (30.8) and (30.19) thus constitute a complete set of equations for the determination of $\Phi_{ij}(\boldsymbol{\varkappa})$ and $\Phi_{ijk}(\boldsymbol{\varkappa}, \boldsymbol{\varkappa}')$, and for given initial conditions they could be solved, at least in principle, to give the behavior of these functions at later times. The complexity of the equations is such that little further progress has been made in the general case of homogeneous turbulence; it is only in the case of isotropic turbulence that any reasonably detailed results have been obtained and these will be discussed in the following sections.

**31. The dynamical equations in isotropic turbulence.** The scalar form of Eq. (30.8) in isotropic turbulence is the familiar equation

$$\frac{\partial E(\varkappa)}{\partial t} = T(\varkappa) - 2\nu\varkappa^2 E(\varkappa), \tag{31.1}$$

where the transfer function is now given by

$$T(\varkappa) = 4\pi\varkappa^2\varkappa_k \int \Phi_{iik}(\boldsymbol{\varkappa}, \boldsymbol{\varkappa}')\, d\boldsymbol{\varkappa}'. \tag{31.2}$$

The form of this last equation is such that we can also write

$$T(\varkappa) = \int_0^\infty \int_{-1}^{+1} \Phi(\varkappa, \varkappa', \mu)\, d\mu\, d\varkappa' \quad \text{say}, \tag{31.3}$$

where $\mu$ is defined by

$$(\boldsymbol{\varkappa}\cdot\boldsymbol{\varkappa}') = \varkappa\varkappa'\mu. \tag{31.4}$$

For some purposes it is more convenient to use the variable $\varkappa'' = |\boldsymbol{\varkappa}''|$ instead of $\mu$, the relationship between these two variables being simply

$$\varkappa''^2 = \varkappa^2 + 2\varkappa\varkappa'\mu + \varkappa'^2. \tag{31.5}$$

The scalar function $\Phi$ can be regarded, therefore, as a function of $(\varkappa, \varkappa', \mu)$ or $(\varkappa, \varkappa', \varkappa'')$ interchangeably. By comparing Eqs. (31.2) and (31.3) we see that

$$\Phi(\varkappa, \varkappa', \varkappa'') = 8\pi^2\varkappa^2\varkappa'^2\varkappa_k\,\Phi_{iik}(\boldsymbol{\varkappa}, \boldsymbol{\varkappa}') \tag{31.6}$$

and by performing the indicated contractions on Eq. (30.19) we obtain the dynamical equation for $\Phi(\varkappa, \varkappa', \varkappa'')$ in the form

$$\left.\begin{aligned}
&\frac{\partial}{\partial t}\Phi(\varkappa, \varkappa', \varkappa'') \\
&\quad = \frac{Q}{16\varkappa''^2}\left[\frac{E(\varkappa)}{\varkappa^2}\frac{E(\varkappa')}{\varkappa'^2}(\varkappa^4 + \varkappa'^4 - 2\varkappa^2\varkappa'^2 - \varkappa'^2\varkappa''^2 + 3\varkappa''^2\varkappa^2) - \right. \\
&\qquad - \frac{E(\varkappa')}{\varkappa'^2}\frac{E(\varkappa'')}{\varkappa''^2}(\varkappa'^4 + \varkappa''^4 - \varkappa^2\varkappa'^2 - 2\varkappa'^2\varkappa''^2 + 3\varkappa''^2\varkappa^2) + \\
&\qquad \left. + \frac{E(\varkappa'')}{\varkappa''^2}\frac{E(\varkappa)}{\varkappa^2}(\varkappa''^4 - \varkappa^4 + \varkappa^2\varkappa'^2 - \varkappa'^2\varkappa''^2)\right] - \\
&\qquad - \nu(\varkappa^2 + \varkappa'^2 + \varkappa''^2)\,\Phi(\varkappa, \varkappa', \varkappa''),
\end{aligned}\right\} \tag{31.7}$$

where $Q$ is the symmetric quartic

$$Q = \varkappa^4 + \varkappa'^4 + \varkappa''^4 - 2\varkappa^2\varkappa'^2 - 2\varkappa'^2\varkappa''^2 - 2\varkappa''^2\varkappa^2. \tag{31.8}$$

Thus, Eqs. (31.1) and (31.7) together with the relation (31.3) between $T(\varkappa)$ and $\Phi(\varkappa, \varkappa', \varkappa'')$ provide a complete system of equations for the determination of the two scalar functions $E(\varkappa)$ and $\Phi(\varkappa, \varkappa', \varkappa'')$. It should be emphasized, however, that while $E(\varkappa)$ determines completely the tensor $\Phi_{ij}(\boldsymbol{\varkappa})$ in isotropic turbulence, the single three-point scalar function $\Phi(\varkappa, \varkappa', \varkappa'')$ is insufficient to determine $\Phi_{ijk}(\boldsymbol{\varkappa}, \boldsymbol{\varkappa}')$ completely; for, as PROUDMAN and REID [*22*] have shown, $\Phi_{ijk}(\boldsymbol{\varkappa}, \boldsymbol{\varkappa}')$ requires two independent defining scalars in general. It was also shown by these authors, however, that one of these scalars satisfies a purely viscous type equation, a result due presumably to the assumption of a normal joint-probability distribution. The other defining scalar is affected of course by the inertial transfer of energy and is equivalent to the function $\Phi(\varkappa, \varkappa', \varkappa'')$ introduced here. Thus, $\Phi(\varkappa, \varkappa', \varkappa'')$ is sufficient to determine $\Phi_{ijk}(\boldsymbol{\varkappa}, \boldsymbol{\varkappa}')$ to within a purely viscous component. Furthermore, so long as we restrict our attention primarily to $E(\varkappa)$ there is no essential loss of generality in using the single equation (31.7) as was first done by TATSUMI [28].

Eq. (31.7) is still of a very complicated form and it has not been possible as yet to solve it, even partially, in its full form. Nor is it even clear at present that we would want a complete solution for $\Phi(\varkappa, \varkappa', \varkappa'')$; for this is a quantity for which there are no experimental results. Rather we should like to be able to determine something less, namely, $T(\varkappa)$; for this is a quantity whose physical significance is reasonably well understood. Unfortunately, the form of the viscous term in Eq. (31.7) precludes the possibility of simply integrating this equation to obtain an equation for $T(\varkappa)$ alone. This is not altogether surprising; for it was to avoid a difficulty of just this type that mean values at three points were introduced originally. Such an equation can be derived of course if we set the viscosity equal to zero, in which case we can derive some simple and interesting results.

The inviscid equations then with which we shall deal are, first, from Eq. (31.1),

$$\frac{\partial E(\varkappa)}{\partial t} = T(\varkappa) \tag{31.9}$$

and, second, the equation for the rate of change of $T(\varkappa)$, obtained by integrating the inviscid form of Eq. (31.7),

$$\left.\begin{aligned}\frac{\partial T(\varkappa)}{\partial t} = &-\frac{E(\varkappa)}{12\varkappa^2}\int_0^\infty (13\varkappa^4 - 8\varkappa^2\varkappa'^2 + 3\varkappa'^4)\,E(\varkappa')\,d\varkappa' - \\ &-\frac{E(\varkappa)}{8\varkappa^3}\int_0^\infty \frac{(\varkappa^2-\varkappa'^2)^3}{\varkappa'}\log\left(\frac{\varkappa+\varkappa'}{|\varkappa-\varkappa'|}\right)E(\varkappa')\,d\varkappa' - \\ &-\frac{1}{8\varkappa}\int_0^\infty \frac{E(\varkappa')}{\varkappa'}\int_{|\varkappa-\varkappa'|}^{\varkappa+\varkappa'} \overline{Q}\,\frac{E(\varkappa'')}{\varkappa''}\,d\varkappa''\,d\varkappa',\end{aligned}\right\} \tag{31.10}$$

where

$$\overline{Q} = 3\varkappa'^2\varkappa''^2 - (3\varkappa^4 - \varkappa^2\varkappa'^2 + 4\varkappa'^4) + (\varkappa^6 - \varkappa^4\varkappa'^2 - \varkappa^2\varkappa'^4 + \varkappa'^6)/\varkappa''^2, \tag{31.11}$$

and, in the following sections, we shall give certain simple solutions of these equations.

**32. The growth of vorticity.** When the Reynolds number of the turbulence is large, the mechanism which dominates the decay process is the inertial transfer of energy. According to the cascade theory of ONSAGER and KOLMOGOROFF this transfer of energy takes place predominately from large to small eddies. Now, unlike HEISENBERG's transfer theory, for example, which explicitly incorporates this aspect of the transfer mechanism, the present theory makes no such direct assumption respecting the transfer mechanism, and there is no *a priori* reason, therefore, to suppose that the present theory leads to a physically acceptable transfer mechanism. But that it does lead to a transfer of energy in the right direction can be demonstrated quite simply.

For this purpose we consider the second moment of $E(\varkappa)$, which is proportional to the mean-square vorticity. From Eq. (31.9), the inviscid rate of change of this quantity is

$$\frac{d}{dt}\int_0^\infty \varkappa^2 E(\varkappa)\,d\varkappa = \int_0^\infty \varkappa^2 T(\varkappa)\,d\varkappa. \tag{32.1}$$

This equation is nothing more than the Fourier transform of Eq. (9.9) and the physical mechanism that it describes is simply the stretching of vortex lines by the local strain field. For a transfer of energy from large to small eddies, the terms in (32.1) must be positive, corresponding to a stretching of the vortex lines.

An equation for the rate of change of the second moment of $T(\varkappa)$ can now be obtained from Eq. (31.10). By multiplying that equation by $\varkappa^2$ and integrating over all values of $\varkappa$, we obtain the very simple equation

$$\frac{d}{dt}\int_0^\infty \varkappa^2\, T(\varkappa)\, d\varkappa = \frac{2}{3}\left[\int_0^\infty \varkappa^2\, E(\varkappa)\, d\varkappa\right]^2. \tag{32.2}$$

In terms of $\overline{\omega^2}$, the mean-square value of one component of the vorticity, Eqs. (32.1) and (32.2) can be combined to give

$$\frac{d^2\overline{\omega^2}}{dt^2} = (\overline{\omega^2})^2, \tag{32.3}$$

which is remarkable for its simplicity. The solution of this equation is the Weierstrassian elliptic function

$$\overline{\omega^2}/\overline{\omega_0^2} = 2^{\frac{2}{3}}\,\wp\,(x;0,1), \tag{32.4}$$

where, with a suitable choice of the origin of time,

$$x = 2^{\frac{1}{3}}\,(\overline{\omega_0^2}/6)^{\frac{1}{2}}\, t. \tag{32.5}$$

In the physical problem, only one real period of the doubly periodic elliptic function is relevant, i.e. $0 \leqq x \leqq 2x_0$, where $x_0 = 1.53$. The solution then has the form shown in Fig. 22.

In solving Eq. (32.3) two initial conditions are required. If, for example, $\overline{\omega^2}$ and $d\overline{\omega^2}/dt$ are specified initially, then the solution is a transcendental function of these constants of integration. But, by a suitable re-scaling and re-interpretation, all such solutions can be reduced essentially to the form shown in Fig. 22. Points lying on the left-hand side of the figure are then found to correspond to initial conditions for which $d\overline{\omega^2}/dt < 0$, while points lying on the right-hand side of the figure correspond to $d\overline{\omega^2}/dt > 0$ initially. Since the slope of the curve is proportional to

$$\int_0^\infty \varkappa^2\, T(\varkappa)\, d\varkappa,$$

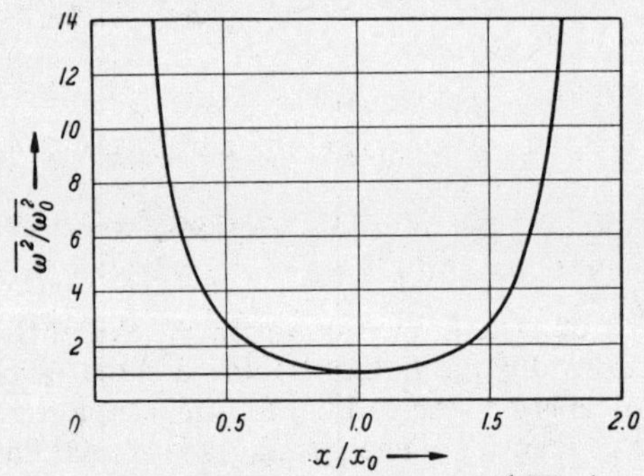


Fig. 22. The production of vorticity. $x = 2^{\frac{1}{3}}(\overline{\omega_0^2}/6)^{\frac{1}{2}}\, t$. $x_0 \approx 1.53$. From PROUDMAN and REID [22].

the former situation corresponds to a transfer of energy in the "wrong direction", i.e. from small to large eddies. The solution (32.4) shows, however, that even when this situation is prescribed initially the subsequent effect of the inertia forces ultimately leads to a transfer of energy in the "correct direction", i.e. from the large to small eddies.

Now, if $t_0$ is the time corresponding to $x = 2x_0$, then as $t \to t_0$ the vorticity becomes infinite according to the universal law

$$\overline{\omega^2} = \frac{6}{(t_0 - t)^2} + O\{(t_0 - t)^4\} \tag{32.6}$$

which is entirely independent of the initial conditions. The time required to reach this asymptotic state does depend on the initial conditions, however, since

$$t_0 = \frac{2\,x_0}{2^{\frac{1}{3}}(\overline{\omega_0^2}/6)^{\frac{1}{2}}} = \frac{5.9}{(\overline{\omega_0^2})^{\frac{1}{2}}}. \tag{32.7}$$

These results then show that in an inviscid fluid it is not possible to maintain the vorticity at a finite level for an infinite time.

The foregoing results for the vorticity were derived on the assumption of zero viscosity; they are, therefore, in the nature of a singular solution. For a small but finite viscosity one would expect an initial growth of the vorticity until a balance is achieved between stretching and dissipation followed by a decay of the vorticity. In this process, the controlling parameter is the initial value of the Reynolds number.

It is worth noting, however, that when the vorticity has reached the asymptotic state described by Eq. (32.6), we have $\overline{\omega^2} \gg \overline{\omega_0^2}$ and the first integral of Eq. (32.3) then becomes

$$\frac{d\overline{\omega^2}}{dt} = \sqrt{\frac{2}{3}}\,(\overline{\omega^2})^{\frac{3}{2}}. \tag{32.8}$$

On comparing this with the exact formula, Eq. (9.10) with $\nu = 0$,

$$\frac{d\overline{\omega^2}}{dt} = \frac{7}{3\sqrt{5}}\,S\,(\overline{\omega^2})^{\frac{3}{2}}, \tag{32.9}$$

we obtain for the skewness factor the result

$$S = \tfrac{1}{7}\sqrt{30} = 0.78. \tag{32.10}$$

As mentioned above, this is the singular value for $\nu = 0$. But it is interesting, since there are no adjustable parameters in the theory, that this result differs from experimental values by a factor of only about two.

**33. The distribution of energy transfer.** The results of the preceding section were concerned only with integral properties of energy transfer that were at least partially independent of initial conditions and they cannot, therefore, shed any light on the *distribution* of energy transfer. A complete description of the inertial transfer of energy is still beyond the scope of theoretical prediction, even within the framework of the present theory, but it is possible to discuss the initial growth of energy transfer for certain simple initial conditions and to obtain thereby some understanding of the dynamical processes underlying the inertial transfer of energy.

For this purpose it is assumed that initially $E(\varkappa)$ has a given form and that $T(\varkappa)$ is identically zero; the initial rate of change of $T(\varkappa)$ is then calculated from Eq. (31.10). These initial conditions correspond, in the discussion of the previous section, to setting $\overline{\omega^2} = \overline{\omega_0^2}$ and $d\overline{\omega^2}/dt = 0$, at the initial instant $t = 0$, which then corresponds to the minimum point in Fig. 22, and the growth of vorticity then proceeds along the right-hand half of that curve.

Perhaps the simplest initial energy spectrum is the one considered by TATSUMI [28][1],

$$E(\varkappa) = E_0\,\delta(x-1), \quad x = \varkappa/\varkappa_0, \tag{33.1}$$

[1] It may be noted that the double correlation corresponding to this energy distribution is

$$f(r) = 3\,\frac{\sin y - y\cos y}{y^3},$$

where $y = \varkappa_0 r$, and it becomes negative for sufficiently large values of $y$. Notice further that this initial correlation function does not satisfy the assumptions made by BATCHELOR and PROUDMAN [4] in their discussion of the large-scale structure of turbulence (see Sect. 13).

where $\varkappa_0$ and $E_0$ are parameters independent of $\varkappa$ and $\delta(x)$ denotes DIRAC's delta function defined by the conditions

$$\int_{-\infty}^{\infty} \delta(x)\, dx = 1 \quad \text{and} \quad \delta(x) = 0 \quad \text{for } x \neq 0. \tag{33.2}$$

This spectrum corresponds to a random isotropic distribution of waves, all of wave number $\varkappa_0$. It is particularly well suited, therefore, to show the manner in which the non-linear terms tend to spread the energy over the spectrum. Unlike the physical transfer theories which cannot excite modes outside those excited initially, the present theory will ultimately excite all modes. For the initial spectrum (33.1), the initial rate of change of $T(\varkappa)$ follows simply from Eq. (31.10) in the form

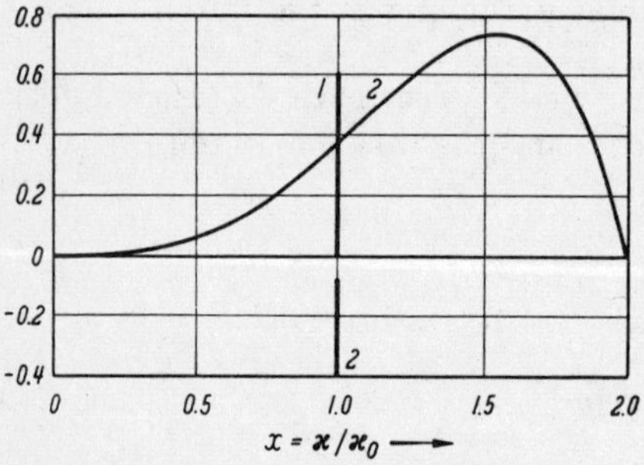


Fig. 23. A simple example of the rate of growth of energy transfer. Curve 1, the initial energy spectrum, $E(\varkappa)/E_0 = \delta(x-1)$; curve 2, $[\partial T(\varkappa)/\partial t]_{t=0}/\varkappa_0^3 E_0^2 = -\frac{2}{3}\delta(x-1) + \frac{1}{2}x^3(1-\frac{1}{4}x^2)$, $(0 \leqq x \leqq 2)$. From TATSUMI [28].

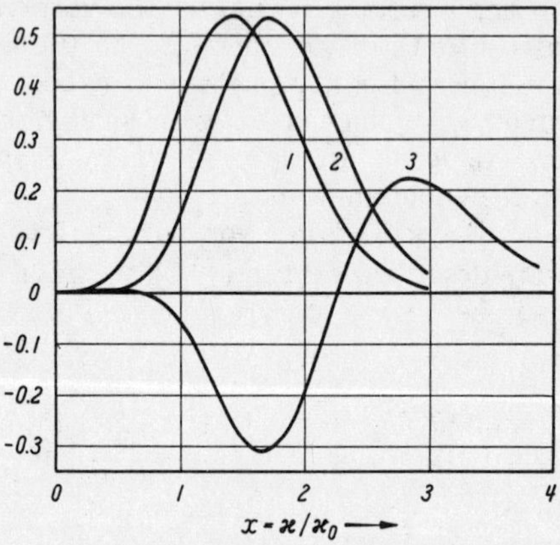


Fig. 24. A further example of the rate of growth of energy transfer. Curve 1, the initial energy spectrum, $E(\varkappa)/E_0$; curve 2, the initial energy-dissipation spectrum, $0.4\varkappa^2 E(\varkappa)/\varkappa_0^2 E_0$; curve 3, $[\partial T(\varkappa)/\partial t]_{t=0}/\varkappa_0^3 E_0^2$. From PROUDMAN and REID [22].

$$[\partial T(\varkappa)/\partial t]_{t=0}/\varkappa_0^3 E_0^2 = -\tfrac{2}{3}\delta(x-1) + \tfrac{1}{2}x^3\left(1 - \tfrac{1}{4}x^2\right), \quad (0 \leqq x \leqq 2), \tag{33.3}$$

which is shown graphically in Fig. 23. Since, in the absence of viscosity, $\partial T(\varkappa)/\partial t = \partial^2 E(\varkappa)/\partial t^2$, energy that was initially concentrated at $\varkappa = \varkappa_0$ is first spread over the interval $0 \leqq \varkappa \leqq 2\varkappa_0$ and, by evaluating higher-order time derivatives, TATSUMI succeeded in showing that (again in the absence of viscosity) the energy distribution represented by $\partial^3 T(\varkappa)/\partial t^3 = \partial^4 E(\varkappa)/\partial t^4$ would be spread over an interval $0 \leqq \varkappa \leqq 3\varkappa_0$ and so on.

Another suitable initial energy spectrum is

$$E(\varkappa) = E_0\, x^4\, e^{-x^2}, \qquad x = \varkappa/\varkappa_0. \tag{33.4}$$

This spectrum is the one associated with the final period of decay. It is, however, a kinematically admissible spectrum and, provided $\varkappa_0$ and $E_0$ are adjusted so that the initial Reynolds number is large, it would be expected to lead to a distribution of energy transfer not unlike that observed experimentally. For this initial spectrum, Eq. (31.10) gives

$$\left.\begin{aligned} \frac{1}{\varkappa_0^3 E_0^2}\left[\frac{\partial T(\varkappa)}{\partial t}\right]_{t=0} = \sqrt{\pi}\Big[\frac{1}{4}x^2(-12 + 5x^2 - 2x^4)\, e^{-x^2} + \\ + \frac{3}{4}x(4 + x^2)\, e^{-2x^2}\int_0^x e^{\xi^2}\, d\xi + \frac{1}{64\sqrt{2}}\, x^4(7 + x^2)\, e^{-x^2/2}\Big], \end{aligned}\right\} \tag{33.5}$$

which is shown graphically in Fig. 24. This distribution of energy transfer is typical of the low Reynolds number conditions associated with the energy distribution (33.4) and may be compared with the related low Reynolds number transfer function as given by HEISENBERG's theory in Fig. 19.

In order to compare this result with experiment it is convenient first to calculate the triple correlation corresponding to the transfer function (33.5). From an equation analogous to (7.23) this result is obtained in the form

$$\left.\begin{aligned}\frac{1}{\sqrt{\varkappa_0^3 E_0}}\left[\frac{\partial k(r)}{\partial t}\right]_{t=0} = \frac{\pi^{\frac{1}{4}}}{8y^4}\Big[21\sqrt{\frac{\pi}{2}}\,\mathrm{erf}\Big(\frac{y}{\sqrt{2}}\Big) + \\ + 12\sqrt{2\pi}\,(52-y^2)\,e^{-y^2/8}\,\mathrm{erf}\Big(\frac{y}{2\sqrt{2}}\Big) - \\ -4y(156+10y^2+y^4)e^{-y^2/4} - y(21+7y^2-y^4)\,e^{-y^2/2}\Big],\end{aligned}\right\} \tag{33.6}$$

where $y=\varkappa_0 r$. This function is shown graphically in Fig. 25.

Now, during the initial stages of the decay process, we may write

$$k(r) = t\left[\frac{\partial k(r)}{\partial t}\right]_{t=0} + O(t^3). \tag{33.7}$$

According to Fig. 25, the theoretical maximum of $-[\partial k(r)/\partial t]_{t=0}/\sqrt{\varkappa_0^3 E_0}$ is about 0.10 whereas the measured value[1] of $-k(r)$ during the initial period of decay is about 0.05. The time required for $k(r)$ to grow from its initial value of zero to the measured value is, therefore,

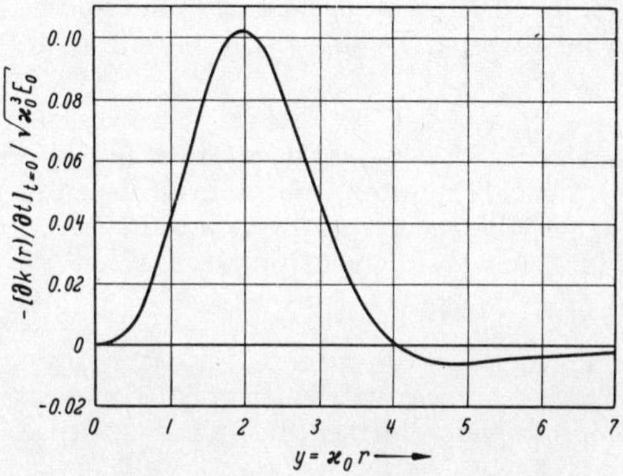


Fig. 25. The triple correlation function corresponding to Fig. 24. From PROUDMAN and REID [22].

$$t \sim \frac{1}{2\sqrt{\varkappa_0^3 E_0}} \tag{33.8}$$

which is also of the order of the characteristic period of the energy containing eddies. During this interval of time the proportion of the total energy lost by viscous dissipation is

$$\frac{\overline{u^2}(0)-\overline{u^2}(t)}{\overline{u^2}(0)} = \frac{2.4}{R_\lambda}, \tag{33.9}$$

where $R_\lambda$ is here the initial value of the Reynolds number. Thus, in order for a transfer mechanism of the initial period type to develop, $R_\lambda$ would need to be greater than about 20; otherwise the decay would be dominated by viscous action of the final period type[2].

The numerical integration of Eqs. (31.1) and (31.7) has recently been undertaken by OGURA[3]. The initial energy spectrum was assumed to be of the final period type [cf. Eq. (33.4)] and $\Phi(\varkappa,\varkappa',\varkappa'')$ was assumed to be identically zero initially. In terms of these initial conditions, the initial value of the Reynolds number is given by

$$R_\lambda(0) = \left(\frac{\sqrt{\pi}}{2}\right)^{\frac{1}{2}}\frac{(E_0/\varkappa_0)^{\frac{1}{2}}}{\nu} \tag{33.10}$$

[1] R. W. STEWART: Proc. Cambridge Phil. Soc. **47**, 146 (1951).

[2] G. K. BATCHELOR and A. A. TOWNSEND: Proc. Roy. Soc. Lond., Ser. A **194**, 527 (1948).

[3] OGURA, Y.: J. Geophys. Res. **67**, 3143 (1962). We are also indebted to Dr. OGURA for making available to us some additional results prior to publication.

and, in OGURA's calculations, was assigned the values $R_\lambda(0)=7.22$, 14.43, 28.86, and $\infty$. In discussing his results it is also convenient to introduce the time-scale $t_0=(E_0\varkappa_0^3)^{-\frac{1}{2}}$ and the non-dimensional time $\tau=t/t_0$. Note that on this time-scale and for these initial conditions, the vorticity would become infinite in an inviscid fluid after a time $\tau=2.83$ [cf. Eq. (32.5)] and this may be compared with the time $\tau=0.5$ [cf. Eq. (33.8)] required for the growth of the triple correlation from an initial value of zero.

The most striking feature of these calculations is that, for all but the lowest Reynolds number, the energy distribution develops negative values after a finite time[1]. For $R_\lambda(0)=\infty$, this first occurs at $\tau\approx 2.0$; for times larger than this (but less than 2.83) it becomes difficult to control the accuracy of the numerical work due to the transfer of energy to large wave-numbers and it is not possible to speculate about the asymptotic form of the energy distribution as $\tau$ approaches the limiting value of 2.83. For $R_\lambda(0)=28.86$, the mean-square vorticity decreases slightly at first, then rises to a maximum value of $1.29\,\overline{\omega_0^2}$ at $\tau\approx 2.3$ and thereafter decreases until the energy distribution takes on negative values at $\tau\approx 2.8$. For $R_\lambda(0)=14.43$, however, the mean-square vorticity is monotonically decreasing and negative values of the energy density do not appear until $\tau\approx 4.3$, at which time $\overline{\omega^2}$ has decreased to about $0.2\,\overline{\omega_0^2}$. At the lowest value of the initial Reynolds number, 7.22, there is no evidence of the energy distribution tending to become negative even after a time $\tau\approx 3.7$. From these calculations it would appear that, for Reynolds number less than about 10, a purely viscous calculation would be entirely adequate.

Thus, from the work of OGURA and O'BRIEN and FRANCIS it is now conclusively clear that the transfer theory discussed in this chapter does lead to negative energy densities at sufficiently large Reynolds numbers and after sufficiently large times[2]. For some purposes, however, the theory may still provide useful results over times which are not too long and at Reynolds numbers which are not too large.

**34. Energy transfer at small wave numbers.** The discussion of the large-scale structure of the turbulence presented in Sect. 13 was largely kinematical in nature. Though less general, the present theory provides further dynamical evidence for the results reached there.

Thus, in the case of homogeneous turbulence, if $\Phi_{ij}(\boldsymbol{\varkappa})$ is assumed to be initially analytic, then the initial inviscid rate of change of $\Sigma_{ij}(\boldsymbol{\varkappa})$ is

$$\left.\begin{aligned}\frac{\partial}{\partial t}\Sigma_{ij}(\boldsymbol{\varkappa}) = 2\varkappa_l\varkappa_m \Delta_i^\alpha(\boldsymbol{\varkappa})\,\Delta_j^\beta(\boldsymbol{\varkappa})\int[\Phi_{\alpha l}(\boldsymbol{\varkappa}')\,\Phi_{m\beta}(\boldsymbol{\varkappa}') +{}\\ + \Phi_{\alpha\beta}(\boldsymbol{\varkappa}')\,\Phi_{lm}(\boldsymbol{\varkappa}')]\,d\boldsymbol{\varkappa}' + O(\varkappa^3),\end{aligned}\right\}\tag{34.1}$$

so that at subsequent instants $\Phi_{ij}(\boldsymbol{\varkappa})$ possesses a singularity of the type found before.

In isotropic turbulence, the initial inviscid rate of change of $T(\varkappa)$ can be obtained either directly from Eq. (34.1) or by expanding Eq. (31.10) in powers

[1] This phenomenon had, in fact, been observed somewhat earlier by E. E. O'BRIEN and G. C. FRANCIS [J. Fluid Mech. **13**, 369 (1962)] in their work on the convection of a passive scalar field in isotropic turbulence (see the discussion of this problem given in the first half of Sect. 38) and by OGURA himself [Phys. Fluids **5**, 395 (1962)] in his work on isotropic turbulence in two-dimensions.

[2] See also KRAICHNAN, R. H.: Proc. 13th Symp. in Appl. Math., Amer. Math. Soc. 1962, p. 199.

of $\varkappa$ with the result

$$\frac{\partial T(\varkappa)}{\partial t} = \frac{14}{15}\varkappa^4 \int_0^\infty \frac{E^2(\varkappa')}{\varkappa'^2}\,d\varkappa' + O(\varkappa^6). \tag{34.2}$$

Consequently, the scalars $L$ and $T$ defined by Eq. (13.16) have the behavior

$$\frac{d^2 L}{dt^2} = \frac{dT}{dt} = \frac{14}{5}\pi \int_0^\infty \frac{E^2(\varkappa)}{\varkappa^2}\,d\varkappa \tag{34.3}$$

and are therefore non-zero for a normal joint-probability distribution of the velocity field.

**35. The energy spectrum in the inertial range.** The important question of whether or not the present theory is in agreement with KOLMOGOROFF's universal equilibrium theory was raised in Sect. 29. For wave numbers in the inertial range, TATSUMI [*49*] has shown that $E(\varkappa) \propto \varkappa^{-2}$ and KRAICHNAN [*39*] has given the same result in the more explicit form

$$E(\varkappa) = \text{const}\,[\varepsilon\,(\overline{u^2})^{\frac{1}{2}}\varkappa_0]^{\frac{1}{2}}\varkappa^{-2}, \tag{35.1}$$

where $\varkappa_0$ is a wave number characteristic of the energy-containing eddies.

The fact that this result is at variance with the Kolmogoroff spectrum is perhaps not altogether surprising. For it is only in the energy-containing range of wave numbers that the approximation of a normal probability distribution has received experimental support, and a discrepancy in the inertial range of the type noted here should not be considered prejudicial to the theory in the energy-containing range. It may also be noticed that the calculation of the skewness factor given in Sect. 32 *is* in agreement with KOLMOGOROFF's theory. Further clarification of the relationship between these two approaches is highly desirable.

## VI. Further applications of the quasi-Gaussian hypothesis.

**36. Pressure fluctuations.** The hypothesis of a normal joint-probability distribution of the velocity components at two points has also been used by HEISENBERG [*13*], OBUKHOFF[1], and BATCHELOR[2] to discuss the statistical properties of the pressure field. For this purpose consider the pressure correlation $\overline{p p'}$ which, by Eq. (6.16), is related to the velocity field in the manner

$$\frac{1}{\varrho^2}\,\nabla^4\,\overline{p p'} = \frac{\partial^4\,\overline{u_i u_j u_k' u_l'}}{\partial r_i\,\partial r_j\,\partial r_k\,\partial r_l}. \tag{36.1}$$

If the turbulence is isotropic then

$$\frac{1}{\varrho^2}\,\overline{p p'} = P(r) \quad \text{and} \quad \frac{\partial^4\,\overline{u_i u_j u_k' u_l'}}{\partial r_i\,\partial r_j\,\partial r_k\,\partial r_l} = Q(r) \tag{36.2}$$

are even functions of $r$ and Eq. (36.1) becomes

$$\left(\frac{d^4}{dr^4} + \frac{4}{r}\,\frac{d^3}{dr^3}\right) P(r) = Q(r). \tag{36.3}$$

---

[1] A. M. OBUKHOFF: Dokl. Akad. Nauk SSSR. **66**, 17 (1949); translation issued by Ministry of Supply, England, as TPA 3/TIB Trans. no. T 3757.

[2] G. K. BATCHELOR: Proc. Cambridge Phil. Soc. **47**, 359 (1951).

For a normal joint-probability distribution of the velocity components at two points, we have

$$\overline{u_i u_j u'_k u'_l} = \overline{u_i u_j} \cdot \overline{u'_k u'_l} + \overline{u_i u'_k} \cdot \overline{u_j u'_l} + \overline{u_i u'_l} \cdot \overline{u_j u'_k} \tag{36.4}$$

and $Q(r)$ can therefore be expressed entirely in terms of $f(r)$ in the form

$$Q(r) = (\overline{u^2})^2 \frac{4}{r^2} \frac{d}{dr}\left[\frac{1}{r} \frac{d}{dr}(r^3 f'^2)\right]. \tag{36.5}$$

Eq. (36.3) can now be integrated to give

$$P(r) = 2(\overline{u^2})^2 \int_r^\infty \left(1 - \frac{r^2}{r'^2}\right) r' f'^2(r')\, dr'. \tag{36.6}$$

In particular, we have

$$\frac{1}{\varrho^2} \overline{p^2} = 2(\overline{u^2})^2 \int_0^\infty r f'^2\, dr \tag{36.7}$$

and

$$\frac{1}{\varrho^2} \overline{(\mathrm{grad}\, p)^2} = 12(\overline{u^2})^2 \int_0^\infty \frac{1}{r} f'^2\, dr. \tag{36.8}$$

By using the values of $f(r)$ given by HEISENBERG's transfer theory for an infinite Reynolds number, BATCHELOR has evaluated the integral in Eq. (36.6) with the results shown in Fig. 26. He further found that

$$\frac{1}{\varrho^2} \overline{p^2} = 0.34(\overline{u^2})^2 \tag{36.9}$$

and

$$\frac{1}{\varrho^2} \overline{(\mathrm{grad}\, p)^2} = 3.9(\varepsilon^3/\nu)^{\frac{1}{2}}. \tag{36.10}$$

If, instead of considering the correlation between the pressure itself, we consider the correlation between pressure differences, then we have

$$\frac{1}{\varrho^2} \overline{(p - p')^2} = 4(\overline{u^2})^2 \left[\int_0^r r' f'^2(r')\, dr' + r^2 \int_r^\infty \frac{1}{r'} f'^2(r')\, dr'\right]. \tag{36.11}$$

For very high Reynolds numbers, the correlation function $f(r)$ has a universal form for small values of $r$ given by

$$2\overline{u^2}\,[1 - f(r)] = C(\varepsilon\nu)^{\frac{1}{2}} (r/\eta)^{\frac{2}{3}}, \tag{36.12}$$

where $C$ is an absolute constant and $\eta$ is the Kolmogoroff length, and the integrals which appear on the right-hand side of Eq. (36.11) are dominated by the form of $f(r)$ in this range. Thus, on substituting for $f(r)$ from Eq. (36.12) we obtain the simple relation

$$\frac{1}{\varrho^2} \overline{(p - p')^2} = [\overline{(u - u')^2}]^2, \tag{36.13}$$

a result due originally to OBUKHOFF.

The pressure velocity correlation [cf. Eq. (30.2)]

$$P_{ij}(\boldsymbol{r}, \boldsymbol{r}) = \frac{1}{\varrho} \overline{p\, u_i' u_j'} \tag{36.14}$$

can also be treated in a similar manner. In isotropic turbulence, $P_{ij}(\boldsymbol{r}, \boldsymbol{r})$ must be of the form

$$P_{ij}(\boldsymbol{r}, \boldsymbol{r}) = P_1(r)\, r_i r_j + P_2(r)\, \delta_{ij} \tag{36.15}$$

and, by a similar use of Eq. (36.4), LIMBER[1] has shown that $P_1(r)$ and $P_2(r)$ can be expressed in the form

$$P_1(r) = (\overline{u^2})^2 \left[\frac{6}{5} \int_r^\infty \frac{1}{r'} f'^2(r')\, dr' + \frac{7}{10 r^5} \int_0^r r'^4 f'^2(r')\, dr'\right] \tag{36.16}$$

and

$$P_2(r) = -\tfrac{1}{3} r^2 P_1(r) - \tfrac{4}{3} (\overline{u^2})^2 \int_r^\infty r' f'^2(r')\, dr'. \tag{36.17}$$

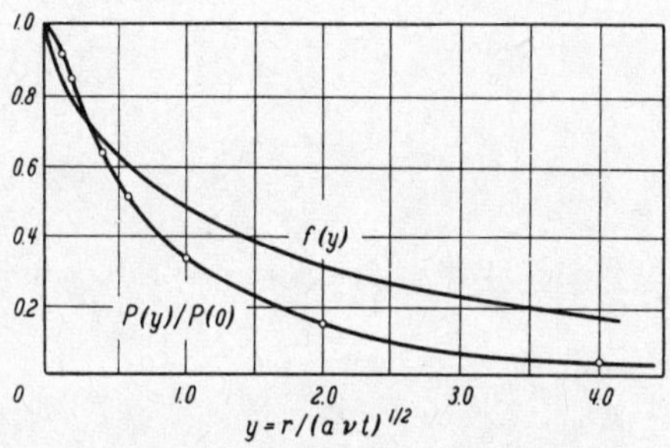


Fig. 26. Pressure correlation at $R_\lambda = \infty$. From G. K. BATCHELOR: Proc. Cambridge Phil. Soc. **47**, 359 (1951).

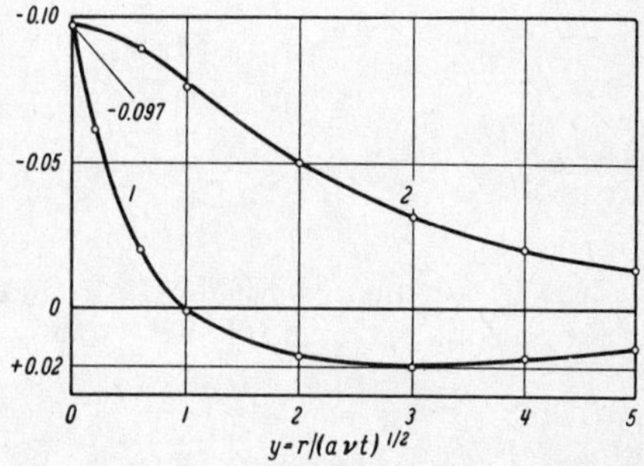


Fig. 27. Pressure-velocity correlations. Curve 1, $\overline{p u_\parallel'^2}/(\overline{p^2})^{\frac{1}{2}}\overline{u^2}$; curve 2, $\overline{p u_\perp'^2}/(\overline{p^2})^{\frac{1}{2}}\overline{u^2}$. After D.N. LIMBER: Proc. Nat. Acad. Sci., Wash. **37**, 230 (1951).

By using PROUDMAN's tables [*21*] of $f(r)$ for an infinite Reynolds number, LIMBER has computed the correlations

$$r^2 P_1(r) + P_2(r) = \frac{1}{\varrho} \overline{p\, u_\parallel'^2} \tag{36.18}$$

and

$$P_2(r) = \frac{1}{\varrho} \overline{p\, u_\perp'^2}, \tag{36.19}$$

and these results are shown in Fig. 27.

We may also note the results analogous to Eq. (36.13) for the correlation $\overline{(p - p')\, u_i' u_j'}$. For the contracted correlation we have

$$\overline{(p - p')\, u_i'^2} = 4 (\overline{u^2})^2 \int_0^r r' f'^2(r')\, dr' \tag{36.20}$$

and by using Eq. (36.12) we obtain

$$\overline{(p - p')\, u_i'^2} = \tfrac{1}{3} [\overline{(u - u')^2}]^2. \tag{36.21}$$

[1] D.N. LIMBER: Proc. Nat. Acad. Sci., Wash. **37**, 230 (1951).

In a similar way, we obtain

$$\overline{(p-p')\,u_{\|}'^2} = \tfrac{10}{39}\,[\overline{(u-u')^2}]^2 \tag{36.22}$$

and

$$\overline{(p-p')\,u_{\perp}'^2} = \tfrac{1}{26}\,[\overline{(u-u')^2}]^2. \tag{36.23}$$

**37. The acceleration terms in the equations of motion.** The hypothesis of a normal joint-probability distribution of the velocity components at two points has been used by LIN[1] in a systematic study of the sixteen two-point correlations that can be formed from the four terms in the equations of motion. Here we shall indicate only the results for the mean-square values of these four terms and of the related total acceleration $A_i = Du_i/Dt$.

LIN's results can be written in the form

$$\overline{\left(\frac{\partial u_i}{\partial t}\right)^2} \Big/ \frac{(\overline{u^2})^2}{\lambda^2} = 15 - 12\,\frac{\lambda^2}{\lambda_p^2} + \frac{35}{R_\lambda}\left(\frac{G}{R_\lambda} - S\right), \tag{37.1}$$

$$\overline{\left(u_j\,\frac{\partial u_i}{\partial x_j}\right)^2} \Big/ \frac{(\overline{u^2})^2}{\lambda^2} = 15, \tag{37.2}$$

$$\overline{\left(\frac{1}{\varrho}\,\frac{\partial p}{\partial x_i}\right)^2} \Big/ \frac{(\overline{u^2})^2}{\lambda^2} = 3\,\frac{\lambda^2}{\lambda_p^2}, \tag{37.3}$$

and

$$\overline{(\nu\,\nabla^2 u_i)^2} \Big/ \frac{(\overline{u^2})^2}{\lambda^2} = \frac{35\,G}{R_\lambda^2}, \tag{37.4}$$

where $\lambda_p$ is defined[2] by Eq. (37.3). On the hypothesis of a normal joint-probability distribution, we have [cf. Eq. (36.8)]

$$\frac{1}{\lambda_p^2} = 4\int_0^\infty \frac{1}{r}\,f'^2\,dr. \tag{37.5}$$

For large Reynolds numbers and for sufficiently small values of $r$, the mean-square pressure difference must be of the form

$$\frac{1}{\varrho^2}\,\overline{(p-p')^2} = v^4\,F(r/\eta), \tag{37.6}$$

where $v$ and $\eta$ are the Kolmogoroff velocity and length scales and $F$ is a universal function of its argument. From this statement it follows that

$$\overline{\left(\frac{1}{\varrho}\,\frac{\partial p}{\partial x_i}\right)^2} = 3\,\frac{v^4}{\eta^2}\,F''(0) \tag{37.7}$$

and hence that

$$\overline{\left(\frac{1}{\varrho}\,\frac{\partial p}{\partial x_i}\right)^2} \Big/ \frac{(\overline{u^2})^2}{\lambda^2} = \frac{\text{const}}{R_\lambda}. \tag{37.8}$$

A similar argument can be used to show that $G/R_\lambda \to$ const. as $R_\lambda$ becomes large. For this purpose we note that

$$G = \lambda^4 f_0^{\text{iv}} = \frac{\lambda^4}{\overline{u^2}}\,\frac{2}{35}\int_0^\infty \varkappa^4\,E(\varkappa)\,d\varkappa, \tag{37.9}$$

---

[1] C. C. LIN: Quart. Appl. Math. **10**, 295 (1953).

[2] G. K. BATCHELOR: Proc. Cambridge Phil. Soc. **47**, 359 (1951).

and that the fourth moment of $E(\varkappa)$ appearing in this equation is a quantity determined by the equilibrium range of wave numbers. Thus, writing

$$E(\varkappa, t) = v^2 \eta \, E_e(\eta \varkappa), \tag{37.10}$$

we see that

$$G = \text{const} \, \frac{\lambda^4/\overline{u^2}}{\eta^4/v^2} \tag{37.11}$$

and hence that [cf. Eq. (17.4)]

$$G = \text{const} \cdot R_\lambda. \tag{37.12}$$

Thus we have that, for large values of the Reynolds number, the mean-square viscous term is given by

$$\overline{(\nu \nabla^2 u_i)^2} \Big/ \frac{(\overline{u^2})^2}{\lambda^2} = \frac{\text{const}}{R_\lambda} \tag{37.13}$$

and is of the same order of magnitude as the mean-square pressure gradient.

To determine the value of the constant appearing in Eq. (37.8) requires a knowledge of the universal function $F(r/\eta)$. No information is available at present on the behavior of this function. An approximate calculation has been made by BATCHELOR, however, using Eq. (37.5). He found that

$$\overline{\left(\frac{1}{\varrho} \frac{\partial p}{\partial x_i}\right)^2} \Big/ \frac{(\overline{u^2})^2}{\lambda^2} = \frac{248}{R_\lambda}. \tag{37.14}$$

An estimate of the constant appearing in Eq. (37.13) can easily be obtained. For large values of the Reynolds number, $G = \frac{1}{2} S R_\lambda$, and if we choose a typical value for $S$ of 0.3 say, then we have

$$\overline{(\nu \nabla^2 u_i)^2} \Big/ \frac{(\overline{u^2})^2}{\lambda^2} = \frac{5.25}{R_\lambda}. \tag{37.15}$$

Thus, the mean-square pressure gradient is larger than the mean-square viscous term by a factor of about 50.

We may also note that the mean-square total acceleration is given by

$$\overline{A_i^2} \Big/ \frac{(\overline{u^2})^2}{\lambda^2} = 3 \, \frac{\lambda^2}{\lambda_p^2} + \frac{35 G}{R_\lambda^2} = \frac{\text{const}}{R_\lambda}, \tag{37.16}$$

being just the sum of the mean-square pressure and viscous terms since the correlation between $p$ and $u_i$ vanishes identically.

We notice that the acceleration $A_i$ of the particle has a magnitude smaller than that of the local rate of change of the velocity $\partial u_i/\partial t$ by a factor of $R_\lambda^{\frac{1}{2}}$. Thus, in a sense, the particles are moving almost freely and subjected only to "small" random forces (cf. Sect. 44). Furthermore, we notice that the "resistive" part of the force is even smaller; for we have, by definition,

$$\overline{A_i u_i} = -\varepsilon; \tag{37.17}$$

and hence, by (37.16), the correlation coefficient $\overline{A_i u_i}/(\overline{A_i^2} \cdot \overline{u_i^2})^{\frac{1}{2}}$ is of the order of $R_\lambda^{-\frac{1}{2}}$. Indeed, if we write

$$A_i = \alpha_i - \beta u_i \tag{37.18}$$

where $\alpha_i$ is uncorrelated with the velocity while $\beta u_i$ represents the resistive part, then

$$\frac{\beta^2 \overline{u_i^2}}{\overline{A_i^2}} = \frac{(\beta \overline{u_i^2})^2}{\overline{A_i^2} \cdot \overline{u_i^2}} = O(R_\lambda^{-1}). \tag{37.19}$$

Notice that the resistive force comes from the viscous term alone, for

$$-\frac{1}{\varrho} \overline{\frac{\partial p}{\partial x_i} u_i} = 0.$$

We may summarize by writing the Navier-Stokes equation in the form

$$\left.\begin{aligned} \frac{D u_i}{D t} &= \frac{\partial u_i}{\partial t} + u_j \frac{\partial u_i}{\partial x_j} = \alpha_i - \beta u_i, \\ \text{where} \quad \alpha_i &= -\frac{1}{\varrho} \frac{\partial p}{\partial x_i} + \nu \nabla^2 u_i + \beta u_i. \end{aligned}\right\} \tag{37.20}$$

Then the various terms in the dynamical equations stand in the following ratios

$$\left|\frac{\partial u_i}{\partial t}\right| : \left|u_j \frac{\partial u_i}{\partial x_j}\right| : |\alpha_i| : |\beta u_i| = 1 : 1 : R_\lambda^{-\frac{1}{2}} : R_\lambda^{-1}. \tag{37.21}$$

This result will be useful in the studies of turbulent dispersion (see especially Sect. 44).

**38. The convection of scalar and vector fields.** The hypothesis of a normal joint-probability distribution can also be used to study the convective effect of the turbulence on local properties of the fluid[1]. These properties include scalar quantities, such as temperature, for which the total amount in a material volume element remains constant. In the absence of molecular diffusion, a scalar field satisfies the equation

$$\frac{\partial \vartheta}{\partial t} + u_i \frac{\partial \vartheta}{\partial x_i} = 0. \tag{38.1}$$

By assuming a strong form of the normality hypothesis, namely, that *all* fourth-order mean values involving $\vartheta$ and $\boldsymbol{u}$ are related to second-order mean values in the same way as for a normal probability distribution, one obtains a closed system of two equations for the scalar correlations $\overline{\vartheta \vartheta'}$ and $\overline{u_i \vartheta' \vartheta''}$ in which the velocity correlation $\overline{u_i u_j'}$ also appears.

The dynamical equations obtained in this manner admit a simple exact integral for the mean-square gradient of the scalar field that is analogous to the result obtained previously for the vorticity [cf. Eq. (32.3)]. Thus, if we let $\boldsymbol{G} = \nabla \vartheta$, this result can be written in the form

$$\frac{d^2 \overline{\boldsymbol{G}^2}}{d t^2} = \frac{2}{3} \overline{\omega^2}\, \overline{\boldsymbol{G}^2}. \tag{38.2}$$

Two solutions of this equation are of interest. First, consider decaying turbulence for which the characteristic time of the dissipation eddies is small compared with that of the energy-containing eddies. In this case we may consider $\overline{\omega^2}$ to be constant and the solution of Eq. (38.2) for large values of $t$ is then

$$\overline{\boldsymbol{G}^2} = \text{const} \times \exp\left[(2\overline{\omega^2}/3)^{\frac{1}{2}} t\right]. \tag{38.3}$$

[1] W. H. Reid: Proc. Cambridge Phil. Soc. **51**, 350 (1955).

Second, consider the case in which there is a simultaneous growth of $\overline{\omega^2}$ and $\overline{G^2}$. The mean-square vorticity is then given by Eq. (32.4) and the corresponding solution for $\overline{G^2}$ is

$$\overline{G^2} = \overline{G_0^2}\,\wp'(x), \tag{38.4}$$

where $\overline{G_0^2}$ is an arbitrary constant of integration and $x=2^{\frac{1}{3}}(\overline{\omega_0^2}/18)^{\frac{1}{2}}t$. When $\overline{\omega^2}$ has reached the asymptotic state given by Eq. (32.6), we have

$$\overline{G^2} = \frac{\text{const}}{(t_0-t)^3}. \tag{38.5}$$

From these results we can obtain two estimates for the joint skewness factor

$$S_\vartheta = -\overline{\left(\frac{\partial\vartheta}{\partial x_1}\right)^2 \frac{\partial u_1}{\partial x_1}} \Big/ \overline{\left(\frac{\partial\vartheta}{\partial x_1}\right)^2} \left[\overline{\left(\frac{\partial u_1}{\partial x_1}\right)^2}\right]^{\frac{1}{2}}. \tag{38.6}$$

For the case of steady turbulence, the solution (38.3) gives $S_\vartheta = \frac{1}{5}\sqrt{10} = 0.63$; for the case of simultaneous growth, the asymptotic solution (38.5) gives $S_\vartheta = \frac{1}{10}\sqrt{30} = 0.55$. Unfortunately, the existing experimental measurements[1] of $S_\vartheta$ are not sufficiently accurate for a quantitative comparison with these theoretical results to be made.

A similar type of analysis can be carried through for vector fields for which the total flux across a material surface remains constant. Again, in the absence of molecular diffusion, a vector field of this type satisfies the equation

$$\frac{\partial F_i}{\partial t} = \frac{\partial}{\partial x_j}(F_i u_j - u_i F_j) = 0. \tag{38.7}$$

These results may therefore be interpreted as applying, for example, to a weak magnetic field in a fluid with infinite conductivity for which the magnetic field has no effect on the velocity field itself. The analysis in this case is somewhat more complicated than in the case of scalar fields; for, one must consider not only the correlations $\overline{F_i F_j'}$ and $\overline{u_i F_j' F_k''}$ but also the correlation $\overline{u_i F_j'}$ which appears when the normal distribution hypothesis is invoked. If one treats $\boldsymbol{F}$ as an axial vector, then one obtains two exact integrals of the governing equations:

$$\frac{d^2\overline{\boldsymbol{F}^2}}{dt^2} = \frac{2}{3}\overline{\boldsymbol{F}^2}\,\overline{\omega^2} - \frac{1}{3}(\overline{\boldsymbol{F}\cdot\omega})^2 \tag{38.8}$$

and

$$\frac{d^2\overline{\boldsymbol{F}\cdot\omega}}{dt^2} = \frac{1}{3}\overline{\boldsymbol{F}\cdot\omega}\,\overline{\omega^2}. \tag{38.9}$$

If $\boldsymbol{F}$ is identified with the velocity itself, then we again recover Eq. (32.3).

The solution of these equations can again be discussed in the same way as in the case of a scalar field. Thus, if $\overline{\omega^2}$ is considered to be constant, the relevant solutions of Eqs. (38.8) and (38.9) are

$$\overline{\boldsymbol{F}^2} = \text{const}\times\exp\left[(2\,\overline{\omega^2}/3)^{\frac{1}{2}}t\right] \tag{38.10}$$

and

$$\overline{\boldsymbol{F}\cdot\omega} = \text{const}\times\exp\left[-(\overline{\omega^2}/3)^{\frac{1}{2}}t\right]. \tag{38.11}$$

[1] R.R. MILLS Jr., A.L. KISTLER, V. O'BRIEN and S. CORRSIN: N.A.C.A. Tech. Note, No. 4288 (1958).

Similarly, when we consider the simultaneous growth of $\overline{\omega^2}$ and $\overline{F^2}$, we have in the asymptotic state the results

$$\overline{F^2} = \frac{\text{const}}{(t_0 - t)^3} \quad \text{and} \quad \overline{F \cdot \omega} = O([t_0 - t]^4). \tag{38.12}$$

In both cases we see that $\overline{F \cdot \omega}$ tends to zero. This result tends to support the assumption sometimes made in the study of hydromagnetic turbulence that $\overline{F \cdot \omega}$ vanishes identically.

The results obtained by considering scalar and vector fields are also relevant to the study of the deformation of material lines and surfaces[1]. The change in length $\delta L$ of a material line element is proportional to $F$ and the change in area $\delta A$ of a material surface element is proportional to $G$. If we consider then an initially spherical material volume element subjected to a uniform strain, and if $(\xi, \eta, \zeta)$ are the parameters describing the asymptotic shape of the ellipsoid, then we have

$$\frac{1}{\overline{\delta L^2}} \frac{\partial \overline{\delta L^2}}{\partial t} \to 2\xi \tag{38.13}$$

and

$$\frac{1}{\overline{\delta A^2}} \frac{\partial \overline{\delta A^2}}{\partial t} \to 2(\xi + \eta), \tag{38.14}$$

where $\eta$ is the larger of the two strain rates in the plane orthogonal to the direction of $\xi$. The strain rates $(\xi, \eta, \zeta)$, which must of course satisfy the condition for conservation of volume $\xi + \eta + \zeta = 0$, can therefore be calculated from the foregoing results. For the case in which $\overline{\omega^2}$ is constant we have

$$\xi = -\zeta = (\overline{\omega^2}/6)^{\frac{1}{2}} \quad \text{and} \quad \eta = 0 \tag{38.15}$$

and when there is build-up of vorticity we have

$$\xi = -\zeta = (\overline{\omega^2}/8)^{\frac{1}{2}} \quad \text{and} \quad \eta = 0. \tag{38.16}$$

These results are surprisingly similar and in both cases the strain rates are in the ratio (1, 0, −1). For further discussion of these results and their application to turbulent diffusion see BATCHELOR and TOWNSEND [*59*].

**39. Stationary isotropic turbulence.** The quasi-Gaussian approximation has also been used by CHANDRASEKHAR [*34*] to develop a theory of *stationary* isotropic turbulence. In this theory, one considers correlations between the velocity components at two different points in space ($\boldsymbol{x}$ and $\boldsymbol{x}'$, say) and at two different times ($t$ and $t'$, say). These correlations are then assumed to be homogeneous in space and stationary in time, so that they can depend only on the variables $\boldsymbol{r} = \boldsymbol{x}' - \boldsymbol{x}$ and $\tau = t' - t$.

The assumption of homogeneity effectively excludes any source of energy input, while the assumption of stationarity implies an absence of decay. The two assumptions would, therefore, be incompatible if applied to the energy-containing region of the spectrum. If, however, attention is restricted to those small-scale aspects of the motion which make negligible contributions to the total energy, then the two assumptions of homogeneity and stationarity can be imposed without inconsistency.

---

[1] G. K. BATCHELOR: Proc. Roy. Soc. Lond., Ser. A **213**, 349 (1952).

By assuming that the joint-probability of the velocity field in space-time is approximately normal, CHANDRASEKHAR [34] has derived a single dynamical equation for the longitudinal velocity correlation function,

$$\overline{u^2}\, f(r, \tau) = \overline{u_1(\boldsymbol{x}, t)\, u_1(\boldsymbol{x}+\boldsymbol{r}, t+\tau)}, \qquad \boldsymbol{r} = (r, 0, 0), \tag{39.1}$$

which can be written in the form

$$\frac{\partial}{\partial r}\left(\frac{\partial^2}{\partial \tau^2} - \nu^2 D_5^2\right) \overline{u^2}\, f = \overline{u^2}\, f \frac{\partial}{\partial r} D_5\, \overline{u^2}\, f, \tag{39.2}$$

where

$$D_5 = \frac{\partial^2}{\partial r^2} + \frac{4}{r}\frac{\partial}{\partial r}. \tag{39.3}$$

Since $\overline{u^2} f$ is determined by the energy-containing range of wave numbers, it is not a suitable function to employ in the present theory; for it is just this range of wave numbers that must be excluded. CHANDRASEKHAR [35] has argued that a more meaningful quantity to employ is $\chi(r, \tau) = -\partial \overline{u^2} f(r, \tau)/\partial r$; alternatively, we can consider the correlation between velocity differences after the original manner of KOLMOGOROFF. Thus, if we let

$$\psi(r, \tau) = \overline{u^2}\,[1 - f(r, \tau)], \tag{39.4}$$

then $\psi$ does not depend on the large eddies, and it is the equation satisfied by $\psi$,

$$\frac{\partial}{\partial r}\left[\frac{\partial}{\partial r}\left(\frac{\partial^2}{\partial \tau^2} - \nu^2 D_5^2\right)\psi \Big/ \frac{\partial}{\partial r} D_5\, \psi\right] = -\frac{\partial \psi}{\partial r}, \tag{39.5}$$

that is the fundamental dynamical equation of the theory. Since $\chi = \partial\psi/\partial r$, we also have

$$\frac{\partial}{\partial r}\left[\left(\frac{\partial^2}{\partial \tau^2} - \nu^2 \frac{\partial}{\partial r} D_5\, \mathscr{D}\right)\chi \Big/ \frac{\partial}{\partial r}\mathscr{D}\chi\right] = -\chi, \tag{39.6}$$

where

$$\mathscr{D} = \frac{\partial}{\partial r} + \frac{4}{r}. \tag{39.7}$$

Since $\chi$ is a quantity that does not depend on the energy-containing range of wave numbers, its general form can be determined from KOLMOGOROFF's principles. Thus, with the usual restrictions on $r$ and $\tau$, we see that $\chi$ must be of the form

$$\chi(r, \tau) = (\varepsilon^3/\nu)^{\frac{1}{4}}\, \mathrm{X}(x, y), \tag{39.8}$$

where

$$x = (\varepsilon/\nu^3)^{\frac{1}{4}}\, r \quad \text{and} \quad y = (\varepsilon/\nu)^{\frac{1}{2}}\, \tau, \tag{39.9}$$

and $\mathrm{X}(x, y)$ is a universal function of its arguments.

In the limit of zero viscosity, the right-hand side of Eq. (39.8) must become independent of $\nu$. This requirement implies that $\chi$ must be of the form

$$\chi(r, \tau) = \varepsilon^{\frac{2}{3}}\, r^{-\frac{1}{3}}\, \sigma(z), \tag{39.10}$$

where

$$z = y/x^{\frac{2}{3}} = \varepsilon^{\frac{1}{3}}\, t/r^{\frac{2}{3}}. \tag{39.11}$$

If the viscous term in Eq. (39.6) is neglected altogether, then $\sigma(z)$ satisfies the equation

$$\left(1 - z\frac{d}{dz}\right)\frac{\sigma''}{44\sigma + 10z\sigma' - 4z^2\sigma''} = \frac{1}{6}\sigma. \tag{39.12}$$

This equation has been integrated numerically by CHANDRASEKHAR [*35*] subject to the boundary conditions

$$\sigma(0) = \sigma_0 > 0, \quad \sigma'(0) = 0, \quad \text{and} \quad \sigma(\infty) = 0; \tag{39.13}$$

the existence and uniqueness of the solution was proved later by BACKUS [*32*]. The value of $\sigma_0$ is not determined in this non-viscous approximation, but it has been argued [*36*] that its value can be determined from a full viscous solution.

The spectral form of the theory may also be briefly noted. If $\Phi_{ij}(\boldsymbol{\varkappa}, \tau)$ is the Fourier transform of $\overline{u_i(\boldsymbol{x}, t)\, u_j(\boldsymbol{x}+\boldsymbol{r}, t+\tau)}$ in the sense of Eq. (6.5) and $E(\varkappa, \tau)$ is its defining scalar in the sense of Eq. (7.20), then the dynamical equation satisfied by $E(\varkappa, \tau)$ is

$$\left.\begin{aligned}\left(\frac{\partial^2}{\partial\tau^2} - \nu^2\varkappa^4\right)E(\varkappa) = \frac{1}{32\varkappa}\int\limits_0^\infty \frac{E(\varkappa')}{\varkappa'^3}\,d\varkappa' \int\limits_{|\varkappa-\varkappa'|}^{\varkappa+\varkappa'} [2\varkappa''^2\varkappa^2 + \\ + (\varkappa''^2 - \varkappa'^2)(\varkappa''^2 + \varkappa^2 - \varkappa'^2)]\, Q\,\frac{E(\varkappa'')}{\varkappa''^3}\,d\varkappa'',\end{aligned}\right\} \tag{39.14}$$

where $Q$ is the symmetric quartic

$$Q = \varkappa^4 + \varkappa'^4 + \varkappa''^4 - 2\varkappa^2\varkappa'^2 - 2\varkappa'^2\varkappa''^2 - 2\varkappa''^2\varkappa^2. \tag{39.15}$$

An erroneous form of this result was first given by CHAMBERLAIN and ROBERTS [*33*]. The form given here is in agreement with the later result of MÜNCH and WHEELON [*41*]. Some numerical results have also been obtained by WENTZEL [*42*].

There is one important point that should be specially noted in connection with the foregoing theory. As has already been pointed out, a theory which incorporates both homogeneity and stationarity cannot be applicable to the energy-containing range of wave numbers. Yet it is only in that range of wave numbers that the quasi-Gaussian hypothesis has received experimental support. This is a point that still requires further examination.

The problem of stationary isotropic turbulence at very large Reynolds numbers has also been considered by KRAICHNAN [*39*] from a somewhat different point of view. His work differs from that described above by an allowance for weak dependence between the Fourier coefficients. One of the important consequences of this theory is that the inertial spectrum is of the form

$$E(\varkappa) = \text{const} \times [\varepsilon(\overline{u^2})^{\frac{1}{2}}]^{\frac{1}{2}}\varkappa^{-\frac{3}{2}}. \tag{39.16}$$

The dependence of this result on $\overline{u^2}$ implies a coupling between the energy-containing and inertial ranges of wave-numbers. The theory also predicts that

$$S \to \text{const} \times R_\lambda^{-\frac{1}{2}} \quad \text{as } R_\lambda \to \infty. \tag{39.17}$$

Both of these results are at variance with the predictions of KOLMOGOROFF'S theory. The reasons for these differences, however, are not as yet entirely clear.

**40. Homogeneous shear flow.** Throughout this article so far we have considered only homogeneous turbulence. Thus we have excluded all flows in which there is a variation in mean velocity or in which solid boundaries are present. In attempting to extend the theory to shear flows both of these effects must be considered.

There would appear to be no conceptual difficulty in deriving a heirarchy of moment equations for shear flows of which the Reynolds equation (2.3) would be the first. There would, of course, still be the usual difficulty of obtaining closure of the system. In the lowest approximation, one would make a direct hypothesis about the behavior of the Reynolds stresses themselves and obtain thereby theories of the familiar "mixing-length" type. Such theories often provide satisfactory descriptions of mean velocity distributions but are not so satisfactory in predictions of the turbulent energy distributions.

In the next approximation one would also include the dynamical equations for the (two-point) double correlations and then make some supplementary assumption about the behavior of the (two-point) triple correlations appearing in those equations. Theories of this type are in the spirit of the physical transfer theories discussed in Chap. IV. They would be expected to provide a better description of the turbulent velocity distributions than the mixing-length theories but would still be unsatisfactory with respect to the detailed transfer of energy from one wave-number to another.

One might even consider going one stage further by including not only the equations for the (two-point) double correlations but also the equations for the (three-point) triple correlations. Closure of this system of equations would again be obtained by an application of the quasi-Gaussian approximation, after the manner described in Chap. V, though there is now perhaps less justification for employing this approximation. Unfortunately, the analytical difficulties that must be overcome in developing such a theory are so great that very little progress has been made as yet.

The one case in which some progress has been made is when the mean velocity is of the form

$$U_i = \lambda_{ij}\, x_j, \tag{40.1}$$

where $\lambda_{ij}$ is independent of position. By a suitable choice of the $\lambda$'s, a number of cases can be considered. For example, the case of a *uniform rotation* can be represented by $\lambda_{12} = -\lambda_{21} = \text{const}$ with all other $\lambda_{ij} = 0$; a *uniform shear* is obtained by letting $\lambda_{12} = \text{const}$, with all other $\lambda_{ij} = 0$; and for a *uniform irrotational distortion* we let $\lambda_{ij} = 0$ if $i \neq j$, with $\lambda_{ii} = 0$ as required by continuity.

For a mean velocity field of the form (40.1), and in the absence of rigid boundaries, the turbulence will still be homogeneous though, of course, not isotropic. The dynamical equations for this problem have been derived by CRAYA[1] by considering the two- and three-point correlations together with the quasi-Gaussian hypothesis. The spectral forms of these equations are given by

$$\left.\begin{aligned} \frac{\partial \Phi_{ij}}{\partial t} &= O_2\left\{\varkappa_k\, \Delta_i^{\alpha}(\boldsymbol{\varkappa}) \int \Phi_{\alpha jk}\, d\boldsymbol{\varkappa}' - \nu\, \varkappa^2\, \Phi_{ij}\right\} + \\ &\quad + \lambda_{\alpha\beta}\left[\left(2\,\frac{\varkappa_i \varkappa_\alpha}{\varkappa^2} - \delta_{i\alpha}\right)\Phi_{\beta j} + \left(2\,\frac{\varkappa_j \varkappa_\alpha}{\varkappa^2} - \delta_{j\alpha}\right)\Phi_{i\beta} + \varkappa_\alpha\,\frac{\partial}{\partial \varkappa_\beta}\,\Phi_{ij}\right] \end{aligned}\right\} \tag{40.2}$$

---

[1] A. CRAYA: C. R. Acad. Sci., Paris **244**, 560, 847, 1448, 1609 (1957).

and

$$\left.\begin{aligned}\frac{\partial \Phi_{ijk}}{\partial t} = {} & \Omega_3 \{\varkappa_l'' \Delta_i^\alpha(\boldsymbol{\varkappa}'') [\Phi_{\alpha j}(\boldsymbol{\varkappa}) \Phi_{lk}(\boldsymbol{\varkappa}') + \Phi_{\alpha k}(\boldsymbol{\varkappa}') \Phi_{lj}(\boldsymbol{\varkappa})] - \nu \varkappa''^2 \Phi_{ijk}\} + \\ & + \lambda_{\alpha\beta} \Big[\Big(2 \frac{\varkappa_i'' \varkappa_\alpha''}{\varkappa''^2} - \delta_{i\alpha}\Big) \Phi_{\beta jk} + \Big(2 \frac{\varkappa_j \varkappa_\alpha}{\varkappa^2} - \delta_{j\alpha}\Big) \Phi_{i\beta k} + \\ & + \Big(2 \frac{\varkappa_k' \varkappa_\alpha'}{\varkappa'^2} - \delta_{k\alpha}\Big) \Phi_{ij\beta} + \Big(\varkappa_\alpha \frac{\partial}{\partial \varkappa_\beta} + \varkappa_\alpha' \frac{\partial}{\partial \varkappa_\beta'}\Big) \Phi_{ijk}\Big].\end{aligned}\right\} \tag{40.3}$$

The form of these equations is still too complicated to admit of any simple general results. The reduction of these tensor equations to an equivalent set of scalar equations has also been discussed by CRAYA with special reference to the cases of uniform shear and plane deformation ($\lambda_{11} = -\lambda_{22} = \text{const}$, with all other $\lambda_{ij} = 0$).

A linearized form of Eq. (40.2) has been used by PEARSON [*51*] in his discussion of the effect of a uniform distortion on weak homogeneous turbulence. He has shown that for a uniform shear the total turbulent energy ultimately decays with time, a result that is consistent with the known stability of plane Couette flow. He has also obtained, on this linearized approximation, the general solution for an irrotational distortion. In studying the effect of a distortion on homogeneous turbulence, one would of course like to be able to include inertial effects. But to do so would require at least a partial consideration of Eq. (40.3), and this is a very formidable task indeed.

## VII. Turbulent diffusion.

**41. General remarks.** One of the basic characteristics of turbulent motion is its diffusive nature. When smoke emerges from a chimney in a turbulent wind, not only is the smoke carried downstream, but it also meanders and spreads sidewise. However, a fully developed theory of turbulent diffusion does not yet exist. One of the essential difficulties is the following. While the theory of turbulence is usually developed in the Eulerian description, the diffusion process must be treated from a Lagrangian point of view, since it deals with the transport of material particles.

In 1921, TAYLOR [*65*] introduced his concept of diffusion by continuous mouvements, and was able to define a diffusion coefficient $D$ such that the mean square displacement $\overline{Z^2}$ is given by

$$\overline{Z^2} = 2DT, \tag{41.1}$$

where $T$ is the time elapsed. But it is also clear from his discussion (which we shall present in the next section) that a diffusion equation of the form

$$\frac{\partial n}{\partial t} = \nabla \cdot (D \nabla n) \tag{41.2}$$

is not satisfactory[1] for the description of the concentration $n(\boldsymbol{x}, t)$.

In 1926, RICHARDSON[2] first proposed the idea that turbulent diffusion should be treated in terms of the distance between neighboring particles. Thus, instead

[1] For empirical purposes, a formula of the form (41.2) has frequently been used, sometimes even with $D = D(\boldsymbol{x}, t)$. For a discussion of this point, see the remarks of TAYLOR and SUTTON in the 1958 Oxford conference: "Atmospheric Diffusion and Air Pollution," Academic Press (1959), esp. pp. 105, 111.

[2] L.F. RICHARDSON: Proc. Roy. Soc. Lond., Ser. A **110**, 709 (1926).

of a direct consideration of the spread of smoke in space, one regards the spread as composed of two parts: (1) a meandering of the whole cloud with the large scale wind, and (2) a dispersion process of the particles of the cloud relative to each other, leading to a growth of the size of this cloud. He was able to show that a dispersion coefficient, defined by the rate of increase of the mean square scale

$$\mathcal{D} = \frac{1}{2} \frac{d\overline{l^2}}{dt}, \tag{41.3}$$

is not a constant, but is proportional to the $\frac{4}{3}$ power of the scale, i.e.,

$$\mathcal{D} \propto (\overline{l^2})^{\frac{2}{3}}. \tag{41.4}$$

This empirical relationship has been the subject of many investigations[1]. In particular, it has been suggested that (41.4) can be derived on the basis of dimensional arguments if one applies KOLMOGOROFF'S theory. Indeed, one argues that the only other parameter entering the problem should be the rate of energy dissipation $\varepsilon$. It then follows from a simple calculation that

$$\mathcal{D} \propto \varepsilon^{\frac{1}{3}} (\overline{l^2})^{\frac{2}{3}}. \tag{41.5}$$

However such a dimensional argument has its drawbacks. The quantities $\mathcal{D}$, and $\overline{l^2}$ under consideration are Lagrangian quantities, whereas $\varepsilon$ has a more precise meaning in the Eulerian formulation. It is thus not immediately clear why it should enter a process of dispersion. It could be argued, of course, that $\varepsilon$ in fact governs *all* the statistical properties of the turbulent motion, but this is true only in a certain range of the spectrum, and it is not clear why, or under what circumstances, the diffusion process should depend solely upon the motions at such scales (see [*59*]).

In the present chapter, we shall treat the problem of turbulent diffusion by using TAYLOR'S concept of diffusion by continuous movements. We shall indeed extend this concept to the discussion of turbulent dispersion. Such a treatment brings out the very close analogy between turbulent dispersion and the process of Brownian motion. The principal reason for this analogy is the fact that the pressure forces in turbulent motion are very small in a certain sense and have a very short-ranged correlation, as we shall explain in Sect. 44. The direct application of the theory of Brownian motion to the process of turbulent dispersion has been recently proposed by OBUKHOFF[2]. The relationship between the present point of view and his views will be discussed in Sect. 43.

**42. Diffusion by continuous movements.** Consider an idealized case of an infinite body of fluid in homogeneous isotropic turbulent motion which is not decaying[3]. Let there be a sheet of smoke particles initially concentrated in the plane $z=0$. In the course of time, the sheet will become wavy and also appear to have a finite thickness. At the moment, we shall not try to separate these two aspects, but consider only the overall spread of the smoke from the initial plane. For each individual particle, the displacement $Z$ at time $T$ is given by

$$Z(T) = \int_0^T W(t)\, dt, \tag{42.1}$$

[1] A more recent experimental verification of (41.4) was made by L. F. RICHARDSON and H. STOMMEL: J. Meteorology 5, 238 (1948).

[2] A.M. OBUKHOFF: Advances in Geophysics (6): Atmospheric Diffusion and Air Pollution. p. 113 New York: Academic Press 1959.

[3] The energy could be imagined as being supplied at such large scales that it does not influence the following discussions.

where $W(t)$ is the velocity of the particle at time $t$. Let us now consider the average quantities associated with the ensemble of all the particles. Then $\bar{Z}=0$, and

$$\frac{1}{2}\frac{d}{dT}\overline{Z^2}=\overline{\frac{dZ}{dT}Z}=\int_0^T\overline{W(T)\,W(t)}\,dt. \tag{42.2}$$

We now assume that the velocity correlation $\overline{W(T)\,W(t)}$ is a function of $\tau=T-t$ alone, and write

$$\overline{W(T)\,W(t)}=W_*^2\,R(\tau),\qquad \tau=T-t, \tag{42.3}$$

where $W_*$ is the root-mean-square value of the turbulent velocity. Then we have

$$\frac{1}{2}\frac{d}{dT}\overline{Z^2}=W_*^2\int_0^T R(\tau)\,d\tau, \tag{42.4}$$

and

$$\frac{1}{2}\overline{Z^2}=W_*^2\int_0^T(T-\tau)\,R(\tau)\,d\tau. \tag{42.5}$$

In general, we may expect $R(\tau)$ to decrease with increasing $\tau$. Suppose that, for all times $\tau>T_1$, $R(\tau)$ is practically zero. Then

$$\int_0^T R(\tau)\,d\tau=\int_0^{T_1}R(\tau)\,d\tau\quad\text{for } T>T_1. \tag{42.6}$$

Eq. (42.5) then gives

$$\overline{Z^2}=2D(T-T_0), \tag{42.7}$$

where

$$D=W_*^2\int_0^{T_1}R(\tau)\,d\tau, \tag{42.8}$$

and

$$T_0=\int_0^{T_1}\tau\,R(\tau)\,d\tau\Big/\int_0^{T_1}R(\tau)\,d\tau. \tag{42.9}$$

Clearly, one may expect $T_0\ll T$ when $T\gg T_1$. In such a case, we have

$$\overline{Z^2}=2DT, \tag{42.10}$$

analogous to a process of molecular diffusion with a coefficient $D$.

The fact that (42.10) holds only after the elapse of a sufficiently long time has its counterpart in the process of molecular diffusion. A molecule has to go through several collisions before its velocity is uncorrelated with its initial velocity. Thus, $T_1$ corresponds to several times the *mean free time* in molecular processes. The latter is however usually many orders of magnitude smaller than the time of observation involved, and hence (42.10) holds for molecular diffusion without any essential qualification. In the case of turbulent diffusion, one may even be interested in the behavior after a very small lapse of time, so that $R(\tau)\approx 1$ for $0\leqq\tau\leqq T$. Eq. (42.4) then yields

$$\overline{Z^2}=W_*^2\,T^2,\quad\text{or}\quad\sqrt{\overline{Z^2}}=W_*\,T. \tag{42.11}$$

Indeed, since the velocity of the particle has not changed very much over the time interval under consideration, we have

$$Z = W T . \tag{42.12}$$

Thus, not only is the root-mean-square deviation $Z_*$ proportional to $W_*$, but the frequency distribution of $Z$ is the same as the frequency distribution of $W$.

For short periods of time, the frequency distribution of the particles is nearly Gaussian because the velocity fluctuations follow such a law. On the other hand, after a very long period, no matter what the frequency distribution of $W$ may be, the distribution of the particles will follow the Gaussian distribution $T^{-1} \exp(-Z^2/4DT)$, just as in the process of molecular diffusion.

**43. Dispersion by continuous movements.** The above method of analyzing turbulent diffusion does not give a good description of the process when the time involved is neither very short nor very long. If the turbulent motion has very large scale components, it would indeed take a very long time $T_1$ for the proper definition of the diffusion coefficient $D$. One must, therefore, attempt a more refined analysis which would be adequate for times shorter than $T_1$.

As explained in Sect. 41, one way of accomplishing this is to pay special attention to the relative motion of two particles which are caused by small-scale turbulent motion. The effect of the components of turbulent motion at scales much larger than their distance apart is essentially to carry the particles around together. To be more precise, let us consider the motion of the sheet of smoke, discussed in the previous section, for a small interval of time during which (42.12) applies. The shape of the smoke sheet is essentially given by the initial velocity. Any area of the sheet having a diameter much less than TAYLOR's microscale $\lambda$ will essentially be moving together. Particles at distances of several macroscales apart move up or down independently of each other. Thus, we have large-scale waviness together with small-scale ripples. If we attempt to treat the small-scale ripples statistically, the best way would be to analyze the *relative* motion in the $z$-direction of particles which are close to each other and share the same large-scale motions; e.g., two particles which were originally at distances of (say) $d \approx \lambda/10$ apart. These particles have essentially the same initial *velocity*, but they may have different initial accelerations if $d \gg \eta$, the Kolmogoroff microscale; for the Eulerian correlation of pressure gradient has a scale of the order of $\eta$. Thus, the two particles will begin to have different velocities, and consequently they will drift apart. This will further accentuate the differences in their acceleration and their velocity, and the two particles will separate from each other at an accelerated rate.

These ideas can be easily put into mathematical form[1]. Let $Z_1(t)$ and $Z_2(t)$ be the $z$-position of two particles which were close to each other at $t=0$. Then

$$Z_1(0) = Z_2(0) = 0, \qquad W_2(0) - W_1(0) = 0.$$

Let their *relative* $z$-displacement at time $t$ be

$$z(t) = Z_2(t) - Z_1(t) \tag{43.1}$$

and the $z$-component of their relative velocity be

$$w(t) = W_2(t) - W_1(t) . \tag{43.2}$$

[1] The ideas presented here are essentially contained in the papers by C.C. LIN: Proc. Nat. Acad. Sci., Wash. **46**, 566, 1147 (1960).

Then, the initial conditions can be stated as

$$z(0) = 0, \quad w(0) = 0. \tag{43.3}$$

Let their *relative* acceleration be

$$a(t) = \dot{W}_2(t) - \dot{W}_1(t), \tag{43.4}$$

then $a(0) \neq 0$, and

$$\frac{dz}{dt} = w, \quad \frac{dw}{dt} = a. \tag{43.5}$$

From (43.3) and (43.5) it follows that

$$w(t) = \int_0^t a(t')\,dt' \tag{43.6}$$

$$z(t) = \int_0^t (t - t')\,a(t')\,dt'. \tag{43.7}$$

Multiplying each of these equations by $a(t)$ and averaging over an ensemble of pairs of such particles, we have

$$\frac{1}{2}\frac{d}{dt}\overline{w^2(t)} = \int_0^t \overline{a(t)\,a(t')}\,dt', \tag{43.8}$$

$$\frac{1}{2}\frac{d^2}{dt^2}\overline{z^2(t)} = \overline{w^2(t)} + \int_0^t (t - t')\,\overline{a(t)\,a(t')}\,dt'. \tag{43.9}$$

If we now assume the acceleration to be statistically stationary, we may set

$$\overline{a(t)\,a(t')} = a_*^2\,\mathscr{R}(\tau), \quad \tau = t' - t, \tag{43.10}$$

and obtain

$$\frac{1}{2}\frac{d}{dt}\overline{w^2(t)} = a_*^2\int_0^t \mathscr{R}(\tau)\,d\tau, \tag{43.11}$$

and

$$\frac{1}{2}\frac{d^2}{dt^2}\overline{z^2(t)} = \overline{w^2(t)} + a_*^2\int_0^t \tau\,\mathscr{R}(\tau)\,d\tau. \tag{43.12}$$

From these, it follows that

$$\overline{w^2(t)} = 2a_*^2\int_0^t (t - \tau)\,\mathscr{R}(\tau)\,d\tau, \tag{43.13}$$

and

$$\overline{z^2(t)} = \frac{2a_*^2}{3}\int_0^t (t-\tau)^3\,\mathscr{R}(\tau)\,d\tau + a_*^2\int_0^t (t-\tau)^2\,\tau\,\mathscr{R}(\tau)\,d\tau \tag{43.14}$$

or

$$\overline{z^2(t)} = \frac{2a_*^2}{3}\left\{t^3\int_0^t \mathscr{R}(\tau)\,d\tau - \frac{3}{2}t^2\int_0^t \tau\,\mathscr{R}(\tau)\,d\tau + \frac{1}{2}\int_0^t \tau^3\,\mathscr{R}(\tau)\,d\tau\right\}. \tag{43.14a}$$

If $\mathscr{R}(\tau)$ vanishes very rapidly for large values of $\tau$, we have approximately

$$\int_0^t \mathscr{R}(\tau)\,d\tau = \int_0^{T_2} \mathscr{R}(\tau)\,d\tau \quad \text{for } t > T_2. \tag{43.15}$$

Also for such large values of $t$, we may expect

$$t \gg \int_0^t \tau \mathcal{R}(\tau)\, d\tau \Big/ \int_0^t \mathcal{R}(\tau)\, d\tau, \qquad t^3 \gg \int_0^t \tau^3 \mathcal{R}(\tau)\, d\tau \Big/ \int_0^t \mathcal{R}(\tau)\, d\tau. \tag{43.16}$$

Under such circumstances, we have the following approximate relations

$$\overline{z^2(t)} = \tfrac{2}{3} B t^3, \tag{43.17}$$

$$\overline{w^2(t)} = 2Bt. \tag{43.18}$$

where

$$B = a_*^2 \int_0^{T_2} \mathcal{R}(\tau)\, d\tau. \tag{43.19}$$

If we now define a dispersion coefficient by

$$\mathcal{D} = \frac{1}{2} \frac{d\overline{z^2}}{dt} \tag{43.20}$$

we have

$$\mathcal{D} = B t^2 = (\tfrac{3}{2})^{\frac{2}{3}} B^{\frac{1}{3}} \{\overline{z^2(t)}\}^{\frac{2}{3}}. \tag{43.21}$$

This relation states that *the dispersion coefficient increases as the $\frac{4}{3}$ power of scale.* As mentioned above, such a relationship was first discovered in 1926 by L.F. RICHARDSON from a purely empirical analysis of atmospheric dispersion.

The applicability of the above analysis rests on two conditions. On the one hand, the time scale of the correlation of the forces must be *very short* compared with the time scale of our observations [cf. (43.15) and (43.16)]. On the other hand, there must be an *upper* limit to this latter time scale. These we shall discuss in detail in the next section. At the moment, let us make the following observations.

1. We may summarize the above results by stating that, under the restrictions of the time scale just discussed, the process of dispersion depends on the single parameter $B$. This is *analogous* to the introduction of the parameter $\varepsilon$ in the KOLMOGOROFF's theory of the Eulerian structure of the field. Indeed, apart from a constant factor, the result (43.18) follows from a dimensional argument just as the Eulerian relationship $\overline{(u-u')^2} \sim \varepsilon^{\frac{2}{3}} r^{\frac{2}{3}}$ is often derived. As a matter of fact, the parameters $B$ and $\varepsilon$ have the same dimensions. The tantalizing possibility then exists that they might be identifiable, apart from a *constant* factor. But in general, one can only surmize that perhaps $B/\varepsilon = f(R_\lambda)$.

2. The restriction to accelerations of short correlation reminds us of the classical theory of Brownian motion. Indeed, OBUKHOFF has suggested a theory of turbulent dispersion on the hypothesis that *the evolution of the state of a selected particle in time forms a Markoff process and can be described by the Fokker-Planck equation in the space* ($\boldsymbol{z}$, $\boldsymbol{w}$), i.e., the equation

$$\frac{\partial \varphi}{\partial t} + (\boldsymbol{w} \cdot \nabla)\, \varphi = B\, \Delta_{\boldsymbol{w}}\, \varphi \tag{43.22}$$

is satisfied by the function $\varphi(\boldsymbol{z}, \boldsymbol{w})$ which gives the probability density for finding a pair of particles at a distance $\boldsymbol{z}$ apart and moving at a relative velocity $\boldsymbol{w}$.

This is a more specific hypothesis[1] than that used above, but it is presumably a good approximation. We should note, however, that the validity of RICHARDSON'S law does not depend on it.

A further difference between OBUKHOFF'S concept and that presented here lies in the fact that OBUKHOFF depends on the Kolmogoroff theory and hence identifies $B$ with $\varepsilon$ up to a constant factor of the order of unity. On the other hand, as mentioned before, in the present point of view, it is not necessarily true that these two quantities are essentially the same. Still, future investigations might identify them. In that case, the above considerations could be developed into a dynamical basis for KOLMOGOROFF'S theory.

When one extends the above analysis to the three-dimensional case (Sect. 46), a further difference appears. The concept of local isotropy is not needed in the present theory. Indeed, one may begin to wonder whether one could possibly add some angular parameters to the Kolmogoroff theory.

3. It should be added that the relationship between the present analysis and the derivation of RICHARDSON'S law by using the Kolmogoroff spectrum is not obvious[2]. Among other points to be clarified, it should be shown that the ensemble of particles under consideration would distribute themselves in such a manner as to provide an unbiased sample in the Eulerian description.

**44. Physical basis of the theory: considerations for short periods of time.** As mentioned above, the validity of the theory requires that the time scale involved in the correlation of the forces be very short compared with the time scale of our observations. Indeed, the function describing the dependence of force as a function of time should resemble a $\delta$-function. This is indeed the basis for drawing the analogy with the phenomenon of Brownian motion. Let us now examine the validity of this assertion.

The properties of the various terms in the Navier-Stokes equations:

$$\frac{\partial u_i}{\partial t} + u_j \frac{\partial u_i}{\partial x_j} = -\frac{1}{\varrho} \frac{\partial p}{\partial x_i} + \nu \nabla^2 u_i \tag{44.1}$$

have been examined in Eq. (37.21), Sect. 37. In particular, it has been shown that the root mean square magnitudes of the four terms are in the ratios

$$1:1:O(R_\lambda^{-\frac{1}{2}}):O(R_\lambda^{-\frac{1}{2}}) \tag{44.2}$$

and that the viscous term is many times smaller than the pressure term. It has, however, the distinguishing feature of containing a resistive component whose magnitude is still smaller by a factor $O(R_\lambda^{-\frac{1}{2}})$. Thus, we may qualitatively describe the motion of a particle as nearly free, but being acted upon by a *small non-resistive force* (largely a pressure force) and an even *smaller resistive force*. If we would define a time scale $\tau_0$ by

$$\tau_0^2 = \overline{u_i^2} \Big/ \overline{\left(\frac{\partial u_i}{\partial t}\right)^2}, \tag{44.3}$$

[1] The implications of this hypothesis can be seen from the development of the theory of Brownian motion by UHLENBECK and ORNSTEIN: Phys. Rev. **36**, 823 (1930).

[2] R. H. KRAICHNAN pointed out (private communication) that the theory developed here does not depend on the exact $\frac{5}{3}$ power law for the Eulerian spectrum, but would hold for any power law with index between one and two.

then the total effect of the non-resistive force may be said to be small until it has accumulated over an interval of time at least[1] of the order of $\tau_0 \cdot R_\lambda^{\frac{1}{2}}$; and the total effect of the resistive force may be said to be small unless the time elapsed is at least of the order of $\tau_0 R_\lambda$.

It can be easily verified that $\tau_0$ is of the order of magnitude of the Kolmogoroff time scale[2]. On the other hand, the pressure forces cannot be expected to have a time scale longer than $\tau_0$. Thus, if we consider dispersion times of the order of $R_\lambda^{\frac{1}{2}} \tau_0$ or larger, we may visualize that the fluid particle has been subjected to successive small impulsive forces, like those experienced by a Brownian particle; each impulse lasts for a period of time of the order of $\tau_0$. The resistive forces are as yet ineffective until much longer times have elapsed. Their important role for such periods of time will be discussed in the next section.

The arguments presented above hold *a fortiori* for the relative motion of two particles since the effect of large scale motions tends to be eliminated.

**45. Physical basis of the theory: considerations for long periods of time.** We shall now consider the limitations of our theory when the time of observation is so long that the viscous forces acting on an individual particle become significant.

We first observe that, from (43.18), we would predict that $\overline{w^2(t)}$ must eventually become indefinitely large. This means that the velocity of each individual particle must eventually become infinite. Indeed, such a behavior is expected when we assume that the forces acting on the particles follow a stationary process. In reality, there is a dissipative mechanism which limits the magnitude of the velocity. The results of the above sections are then applicable to such time scales that are sufficiently long compared with $\tau_*$, and yet sufficiently short so that the cumulative effect of this dissipative mechanism is not yet appreciable.

To visualize this situation more explicitly, let us consider the usual model in the theory of Brownian motion[3]. We consider the motion of an ensemble of particles governed by the equations[4]

$$\frac{ds}{dt} = v, \qquad \frac{dv}{dt} + \beta v = \alpha(t), \tag{45.1}$$

where $\alpha(t)$ follows a stationary random process, and $\beta$ is a constant coefficient of resistance. The Eq. (37.20) for the motion of a fluid particle is exactly of this form. The solution of (45.1) for a particle having an initial velocity $v_0$ is given by

$$v(t) = v_0 \,\mathrm{e}^{-\beta t} + \mathrm{e}^{-\beta t} \int_0^t \mathrm{e}^{\beta\xi} \alpha(\xi)\, d\xi \tag{45.2}$$

$$s(t) = \frac{v_0}{\beta}(1 - \mathrm{e}^{-\beta t}) - \frac{1}{\beta} \mathrm{e}^{-\beta t} \int_0^t \mathrm{e}^{\beta\xi} \alpha(\xi)\, d\xi + \frac{1}{\beta} \int_0^t \alpha(\xi)\, d\xi. \tag{45.3}$$

---

[1] Since the pressure forces involved are fluctuating forces of short correlation, the time required for an effective modification of the velocity is presumably of the order of $\tau_0 R_\lambda$. This statemant does not contradict the discussion given in Sect. 12, since we are here dealing with pressure gradients.

[2] Indeed, $\tau_0^2 \approx 3\nu/\varepsilon$ [cf. Eq. (37.1)].

[3] Although, as we shall see below, the results of Sect. 43 do not depend on this specific model.

[4] The analysis is the same whether we are dealing with the motion of single particles or that of relative motion of pairs. See UHLENBECK and ORNSTEIN, quoted on p. 515, footnote 1.

If we now consider the mean square velocity of an ensemble of particles having the same initial velocity $v_0$, we have

$$\overline{v^2} = v_0^2\, e^{-2\beta t} + e^{-2\beta t} \int_0^t d\eta \int_0^t d\xi\, e^{\beta(\xi+\eta)}\, \overline{\alpha(\xi)\,\alpha(\eta)}. \tag{45.4}$$

By taking $\xi+\eta=\sigma$, $\xi-\eta=\tau$ as new variables, we get

$$\overline{v^2} = v_0^2\, e^{-2\beta t} + \frac{1}{\beta}\left[\int_0^t \mathcal{R}_0(\tau)\, d\tau\right](1 - e^{-2\beta t}) - \int_0^t 2\,\frac{\sinh\beta\tau}{\beta}\,\mathcal{R}_0(\tau)\, d\tau. \tag{45.5}$$

If $\mathcal{R}_0(\tau)$ vanishes very rapidly with increasing $\tau$, and $\beta$ is sufficiently small, the above equation simplifies to

$$\overline{v^2} = v_0^2\, e^{-2\beta t} + \frac{B_0}{\beta}(1 - e^{-2\beta t}) - 2 B_0 \tau_*, \tag{45.6}$$

where

$$B_0 = \int_0^\infty \mathcal{R}_0(\tau)\, d\tau, \qquad \tau_* = \int_0^\infty \tau\,\mathcal{R}_0(\tau)\, d\tau \Big/ \int_0^\infty \mathcal{R}_0(\tau)\, d\tau. \tag{45.7}$$

We have implicitly assumed that $\tau_* \ll 1/\beta$ for we have not asserted that $\beta t$ is large. If we neglect $\tau_*$ in (45.6) we have

$$\overline{v^2} = v_0^2\, e^{-2\beta t} + \frac{B_0}{\beta}(1 - e^{-2\beta t}). \tag{45.6a}$$

For small values of $\beta t$, i.e., for such times that

$$\tau_* \ll t \ll \frac{1}{\beta}, \tag{45.8}$$

Eq. (45.6) reduces to

$$\overline{v^2} = v_0^2(1 - 2\beta t) + 2 B_0 (t - \tau_*). \tag{45.9}$$

For the special case $v_0=0$, this reduces to (43.18), which is the basic equation for the simple theory discussed in Sect. 43. On the other hand, for $\beta t \gg 1$, (45.6) reduces to

$$\overline{v^2} = \frac{B_0}{\beta}(1 - 2\beta\tau_*) \tag{45.10}$$

which depends only on the cumulative effect of the random forces and the resistance coefficient $\beta$ ($\beta\tau_*$ being negligible). Thus, as long as we are concerned only with the behavior over a period during which the dissipative forces have not yet produced any significant influence, the theory of Sect. 43 applies. The paradox of infinite velocity raised above does not occur, because the relative velocity will eventually be limited to finite values by the dissipative forces.

Clearly, the nature of the force of resistance does not influence the results of Sect. 43; but the ultimate value of $\overline{v^2}$ does depend on it.

The paradox of infinite $\overline{v^2(t)}$ for a stationary random force has often been presented in a slightly different manner. If the velocity of the particle follows a stationary random process, and if (43.10) holds in addition, then we must have

$$\int_0^\infty \mathcal{R}(\tau)\, d\tau = 0,$$

for the integral of the acceleration correlation is the derivative of the velocity correlation. However, the discussions of Sect. 43 must be modified by the viscous forces for large $\tau$ and thus we expect the correlation coefficient $\mathscr{R}(\tau)$ to have a long tail for very large $\tau$, whose effect we are neglecting in Sect. 43.

The effect of the dissipative forces in limiting the mean square displacement is similar. Thus, the equation analogous to (45.6a) is

$$\overline{s^2} = \frac{2B_0}{\beta^2}t + \frac{v_0^2}{\beta^2}(1-e^{-\beta t})^2 + \frac{B_0}{\beta^3}(-3+4e^{-\beta t}-e^{-2\beta t}). \tag{45.11}$$

In the limit of $\beta t \to \infty$, we get

$$\overline{s^2} = \frac{2B_0}{\beta^2}\left(t-\frac{3}{2\beta}\right) \approx \frac{2B_0}{\beta^2}t, \tag{45.12}$$

in agreement with Eq. (42.10) in TAYLOR's theory. Also, for $\beta t \ll 1$, and $v_0 \neq 0$, we have

$$\overline{s^2} = v_0^2 t^2, \tag{45.13}$$

in agreement with (42.11). On the other hand, for $\beta t \ll 1$ and $v_0 = 0$, we have

$$\overline{s^2} = \tfrac{2}{3}B_0 t^3 \tag{45.14}$$

in agreement with (43.17).

We notice in passing that the relative velocity of the two particles as a function of their relative distance apart [as given by (45.6) and (45.11), with $v_0=0$, $\tau_*=0$] has a shape similar to the Eulerian correlation function expected from KOLMOGOROFF's theory in the limit of $R_\lambda \to \infty$. The significance, if any, of such an observation is as yet unclear.

**46. Dispersion in three dimensions**[1]. Having clarified the physical basis of the theory, we can easily extend our considerations to more general situations. In this section, we shall discuss the application of our concepts to three-dimensional dispersion in an anisotropic field of turbulence. In the next section, we shall discuss the application of the general theory of Brownian motion to turbulent dispersion.

We consider the relative motion of two particles released at the same point in a three-dimensional field of turbulent motion which is homogeneous but not isotropic. We then consider the statistical average over an ensemble of such pairs. Let $x_i(t)$ denote the relative displacement of a typical pair at time $t$, and let $v_i(t) = \dot{x}_i(t)$, $a_i(t) = \dot{v}_i(t)$, where a dot denotes time differentiation. As before, we assume that the process is stationary in the sense that the forces acting on the fluid particles (apart from a small dissipative component) may be described by a stationary random process. Thus, we assume

$$\overline{a_i(t')\,a_j(t)} = A_{ij}(\tau) = A_{ij}(-\tau), \tag{46.1}$$

where $\tau = t' - t$; it then follows that $A_{ij} = A_{ji}$. We attempt to describe the statistical behavior of a pair of such particles in terms of the statistical correlations

$$V_{ij}(t) = \overline{v_i(t)\,v_j(t)}, \quad X_{ij}(t) = \overline{x_i(t)\,x_j(t)}. \tag{46.2}$$

[1] C.C. LIN: Proc. Nat. Acad. Sci., Wash. **46**, 1147 (1960).

For the pairs of particles under consideration, we have

$$x_i(0) = 0, \qquad v_i(0) = 0. \tag{46.3}$$

Thus,

$$X_{ij}(0) = 0, \qquad V_{ij}(0) = 0. \tag{46.4}$$

By using methods similar to those used in Sect. 43, we can easily show that

$$V_{ij}(t) = 2\int_0^t (t-\tau)\, A_{ij}(\tau)\, d\tau, \tag{46.5}$$

and

$$\frac{d^2}{dt^2} X_{ij}(t) = 4\int_0^t A_{ij}(\tau)\, d\tau - 2\int_0^t \tau\, A_{ij}(\tau)\, d\tau. \tag{46.6}$$

Again, if we assume that $A_{ij}(t)$ is negligibly small for $t > T_1$, we may introduce the quantity

$$B_{ij} = \int_0^{T_1} A_{ij}(\tau)\, d\tau, \tag{46.7}$$

and obtain the following relations:

$$V_{ij}(t) = 2 B_{ij} t, \tag{46.8}$$

$$X_{ij}(t) = \tfrac{2}{3} B_{ij} t^3, \tag{46.9}$$

and

$$\mathcal{D}_{ij} = \frac{1}{2}\frac{dX_{ij}}{dt} = B_{ij} t^2, \tag{46.10}$$

where $\mathcal{D}_{ij}$ may be called the *dispersion tensor*. The relations (46.9) and (46.10) yield the *generalized Richardson's law*:

$$\mathcal{D}_{ij} = X_{ij}\left(\frac{4X}{9B}\right)^{-\frac{1}{3}}, \tag{46.11}$$

where $X = X_{kk}$, $B = B_{kk}$. Notice that (46.11) predicts the *same* ratio between all the components of $X_{ij}$ and $\mathcal{D}_{ij}$, thus presenting another result which may be subjected to a more detailed observational test.

The significance of the above extension is two-fold. First, for application to atmospheric turbulence, the large scale horizontal dispersion may be many times as strong as the vertical dispersion. Secondly, it is clear that the concept of local isotropy, as used in KOLMOGOROFF's theory is not involved in the above analysis. Thus, if the connection between the present theory and KOLMOGOROFF's theory could be established (which is an open question), it would suggest that the concept of local isotropy is not essential for the inertial range. Indeed, such a suggestion could be made even now in the absence of a demonstration of the interrelationship between the Eulerian and the Lagrangian descriptions.

**47. Application of the general theory of Brownian motion.** In the theory of Brownian motion, we consider the *probability density* function $\Phi(\boldsymbol{r}, \boldsymbol{u}, t)$ for finding a particle at the point $\boldsymbol{r}$ and having the velocity $\boldsymbol{u}$ at time $t$. This function satisfies the generalized[1] Fokker-Planck equation

$$\frac{\partial \Phi}{\partial t} + \boldsymbol{u}\cdot \operatorname{grad}_{\boldsymbol{r}} \Phi + \boldsymbol{K}\cdot \operatorname{grad}_{\boldsymbol{u}} \Phi = \beta \operatorname{div}_{\boldsymbol{u}}(\Phi \boldsymbol{u}) + q\, \nabla_{\boldsymbol{u}}^2 \Phi, \tag{47.1}$$

[1] S. CHANDRASEKHAR: Rev. Mod. Phys. **15**, 1 (1943), Eq. (249).

where $\boldsymbol{K}$ denotes the acceleration per unit mass caused by the external field of force, $\beta$ is the resistance coefficient, and $q$ is the parameter which characterizes the properties of the random forces.

If we are dealing with relative motions, we have a similar equation

$$\frac{\partial \varphi}{\partial t} + \boldsymbol{v} \cdot \text{grad}_{\boldsymbol{s}} \varphi + \boldsymbol{k} \cdot \text{grad}_{\boldsymbol{v}} \varphi = \beta \, \text{div}_{\boldsymbol{v}} (\varphi \boldsymbol{v}) + B \nabla_{\boldsymbol{v}}^2 \varphi \tag{47.2}$$

provided the external force $\boldsymbol{K}$ is *linear* in positional coordinates. In that case,

$$\boldsymbol{k} = \boldsymbol{K}(\boldsymbol{r}_2) - \boldsymbol{K}(\boldsymbol{r}_1) = c(\boldsymbol{r}_2 - \boldsymbol{r}_1) = c\, \boldsymbol{s}. \tag{47.3}$$

The problem is then that of the Brownian motion of the simple harmonic oscillator, which has been treated extensively in the literature. In the context of turbulent dispersion, this corresponds to the case of vertical dispersion in the presence of a uniform density gradient. Indeed, in this case,

$$c = g \frac{1}{\varrho} \frac{d\varrho}{dz} \tag{47.4}$$

where $\varrho(z)$ is the mean density distribution and $g$ is the acceleration due to gravity. Clearly, the same formalism holds whether the density stratisfication is stable ($c<0$) or unstable ($c>0$). In the latter case, the "restoring force" is negative, and we no longer have an oscillator.

The details of this analysis will be given elsewhere. It is of course to be expected that the dispersion will be less rapid in the stable case and more rapid in the unstable case. The whole system of stratified air will of course undergo large scale oscillations (and convective motions in the unstable case) which control the "meandering" of the diffusing substances as a whole.

## References.

### Decaying turbulence.

[1] AGOSTINI, L., and J. BASS: Les théories de la turbulence. Publications Scientifiques et Techniques du Ministère de l'Air, No. 237 (1950). Translated as N.A.C.A. Tech. Memo., No. 1377 (1955).

[2] BATCHELOR, G. K.: The theory of homogeneous turbulence. Cambridge: Cambridge University Press 1953.

[3] BATCHELOR, G. K.: The singularity in the spectrum of homogeneous turbulence. Proc. Symposia Appl. Math., Vol. VII, p. 67. New York: McGraw-Hill Book Company, Inc. 1957.

[4] BATCHELOR, G. K., and I. PROUDMAN: The large-scale structure of homogeneous turbulence. Phil. Trans. Roy. Soc. Lond., Ser. A **248**, 369 (1956).

[5] BETCHOV, R.: An inequality concerning the production of vorticity in isotropic turbulence. J. Fluid Mech. **1**, 497 (1956).

[6] BIRKHOFF, G.: Fourier synthesis of homogeneous turbulence. Comm. Pure Appl. Math. **7**, 19 (1954).

[7] BIRKHOFF, G., and J. KAMPÉ DE FÉRIET: Kinematics of homogeneous turbulence. J. Math. Mech. **7**, 663 (1958).

[8] BLANCH, G., and H. FERGUSON: Remarks on CHANDRASEKHAR'S results relating to HEISENBERG'S theory of turbulence. Phys. Fluids **2**, 79 (1959).

[9] CHANDRASEKHAR, S.: On HEISENBERG'S elementary theory of turbulence. Proc. Roy. Soc. Lond., Ser. A **200**, 20 (1949).

[10] CORRSIN, S.: Outline of some topics in homogeneous turbulent flow. J. Geophys. Res. **64**, 2134 (1959).

[11] Deissler, R. G.: Decay of homogeneous turbulence before the final period. Phys. Fluids **1**, 111 (1958).

[12] Deissler, R. G.: Theory of decaying homogeneous turbulence. Phys. Fluids **3**, 176 (1960).

[13] Heisenberg, W.: Zur statistischen Theorie der Turbulenz. Z. Physik **124**, 628 (1948). Translated as N.A.C.A. Tech. Memo., No. 1431 (1958).

[14] Hopf, E.: Statistical hydromechanics and functional calculus. J. Rational Mech. Anal. **1**, 87 (1952).

[15] Hopf, E.: On the application of functional calculus to the statistical theory of turbulence. Proc. Symposia Appl. Math., Vol. VII, p. 41. New York: McGraw-Hill Book Company, Inc. 1957.

[16] Hopf, E., and E. W. Titt: On certain special solutions of the $\Phi$-equation of statistical hydrodynamics. J. Rational Mech. Anal. **2**, 587 (1953).

[17] Kármán, Th. v., and L. Howarth: On the statistical theory of isotropic turbulence. Proc. Roy. Soc. Lond., Ser. A **164**, 192 (1938).

[18] Kármán, Th. v., and C. C. Lin: On the concept of similarity in the theory of isotropic turbulence. Adv. Appl. Mech. **2**, 1 (1951).

[19] Lin, C. C.: Statistical theories of turbulence. High Speed Aerodynamics and Jet Propulsion, Vol. **5**, pp. 196—253 Princeton: Princeton University Press 1959.

[20] Meecham, W. C.: Relation between time symmetry and reflection symmetry of turbulent fluids. Phys. Fluids **1**, 408 (1958).

[21] Proudman, I.: A comparison of Heisenberg's spectrum of turbulence with experiment. Proc. Cambridge Phil. Soc. **47**, 158 (1951).

[22] Proudman, I., and W. H. Reid: On the decay of a normally distributed and homogeneous turbulent velocity field. Phil. Trans. Roy. Soc. Lond., Ser. A **247**, 163 (1954).

[23] Reid, W. H.: One-dimensional equilibrium spectra in isotropic turbulence. Phys. Fluids **3**, 72 (1960).

[24] Reid, W. H., and D. L. Harris: Similarity spectra in isotropic turbulence. Phys. Fluids **2**, 139 (1959).

[25] Rosen, G.: Turbulence theory and functional integration. Phys. Fluids **3**, 519, 525 (1960).

[26] Stewart, R. W., and A. A. Townsend: Similarity and self-preservation in isotropic turbulence. Phil. Trans. Roy. Soc. Lond., Ser. A **243**, 359 (1951).

[27] Tanenbaum, B. S., and D. Mintzer: Energy transfer in a turbulent fluid. Phys. Fluids **3**, 529 (1960).

[28] Tatsumi, T.: A statistical theory of incompressible isotropic turbulence. Proc. Roy. Soc. Lond., Ser. A **239**, 16 (1957).

[29] Taylor, G. I.: Statistical theory of turbulence. Proc. Roy. Soc. Lond., Ser. A **151**, 421 (1935); **156**, 307 (1936).

[30] Taylor, G. I.: Some recent developments in the study of turbulence. Proc. 5th Int. Congr. Appl. Mech. 1938, p. 294.

[31] Weizsäcker, C. F. v.: Das Spektrum der Turbulenz bei großen Reynoldsschen Zahlen. Z. Physik **124**, 614 (1948).

## Stationary turbulence.

[32] Backus, G.: The existence and uniqueness of the velocity correlation derivative in Chandrasekhar's theory of turbulence. J. Math. Mech. **6**, 215 (1957).

[33] Chamberlain, J. W., and P. H. Roberts: Turbulence spectrum in Chandrasekhar's theory. Phys. Rev. [2], **99**, 1674 (1955).

[34] Chandrasekhar, S.: A theory of turbulence. Proc. Roy. Soc. Lond., Ser. A **229**, 1 (1955).

[35] Chandrasekhar, S.: Theory of turbulence. Phys. Rev. [2], **102**, 941 (1956).

[36] Chandrasekhar, S.: The theory of turbulence. J. Madras Univ., Sect. B **27**, 251 (1957).

[37] Kraichnan, R. H.: Relation of fourth-order to second-order moments in stationary isotropic turbulence. Phys. Rev. [2], **107**, 1485 (1957).

[38] Kraichnan, R. H.: Higher order interactions in homogeneous turbulence. Phys. Fluids **1**, 358 (1958).

[*39*] KRAICHNAN, R.H.: The structure of isotropic turbulence at very high Reynolds numbers. J. Fluid Mech. **5**, 497 (1959).

[*40*] KRAICHNAN, R.H.: Comments on "Space-time correlations in stationary isotropic turbulence". Phys. Fluids **2**, 334 (1959). Cf. [*41*].

[*41*] MÜNCH, G., and A.D. WHEELON: Space-time correlations in stationary isotropic turbulence. Phys. Fluids **1**, 462 (1958). Cf. [*40*].

[*42*] WENTZEL, D.G.: On the spectrum of turbulence. Phys. Fluids **1**, 213 (1958).

## Hydromagnetic turbulence.

[*43*] BATCHELOR, G.K.: On the spontaneous magnetic field in a conducting liquid in turbulent motion. Proc. Roy. Soc. Lond., Ser. A **201**, 405 (1950).

[*44*] CHANDRASEKHAR, S.: The invariant theory of isotropic turbulence in magneto-hydrodynamics. Proc. Roy. Soc. Lond., Ser. A **204**, 435; **207**, 301 (1951).

[*45*] CHANDRASEKHAR, S.: Hydromagnetic turbulence. Proc. Roy. Soc. Lond., Ser. A **233**, 322, 330 (1955).

[*46*] CHANDRASEKHAR, S.: The partition of energy in hydromagnetic turbulence. Ann. Physics **2**, 615 (1957).

[*47*] KRAICHNAN, R.H.: Irreversible statistical mechanics of incompressible hydromagnetic turbulence. Phys. Rev. [2], **109**, 1407 (1958). Errata: **111**, 1747.

[*48*] ROBERTS, P.H., and T. TATSUMI: The decay of magnetohydrodynamic turbulence. J. Math. Mech. **9**, 697 (1960).

[*49*] TATSUMI, T.: Energy spectra in magneto-fluid dynamic turbulence. Rev. Mod. Phys. **32**, 807 (1960).

## Distortion of homogeneous turbulence.

[*50*] BATCHELOR, G.K., and I. PROUDMAN: The effect of a rapid distortion of a fluid in turbulent motion. Quart. J. Mech. Appl. Math. **7**, 83 (1954).

[*51*] PEARSON, J.R.A.: The effect of uniform distortion on weak homogeneous turbulence. J. Fluid Mech. **5**, 274 (1959).

[*52*] RIBNER, H. S., and M. TUCKER: Spectrum of turbulence in a contracting stream. N.A.C.A. Tech. Note, No. 2606 (1952); reissued as N.A.C.A. Rep., No. 1113 (1953).

[*53*] TOWNSEND, A.A.: The uniform distortion of homogeneous turbulence. Quart. J. Mech. Appl. Math. **7**, 104 (1954).

[*54*] TUCKER, M.: Combined effect of damping screens and stream convergence on turbulence. N.A.C.A. Tech. Note, No. 2878 (1953).

[*55*] UBEROI, M.S.: Effect of wind-tunnel contraction on free-stream turbulence. J. Aeronaut. Sci. **23**, 754 (1956).

## Turbulent convection and diffusion.

[*56*] BATCHELOR, G.K.: Diffusion in free turbulent shear flows. J. Fluid Mech. **3**, 67 (1957).

[*57*] BATCHELOR, G.K.: Small-scale variation of convected quantities like temperature in turbulent fluid. Part 1. General discussion and the case of small conductivity. J. Fluid Mech. **5**, 113 (1959).

[*58*] BATCHELOR, G.K., I.D. HOWELLS and A.A. TOWNSEND: Small-scale variation of convected quantities like temperature in turbulent fluid. Part 2. The case of large conductivity. J. Fluid Mech. **5**, 134 (1959).

[*59*] BATCHELOR, G.K., and A.A. TOWNSEND: Turbulent diffusion. Surveys in Mechanics. The G.I. Taylor Anniversary Volume, p. 352. Cambridge: Cambridge University Press 1956.

[*60*] CORRSIN, S.: Statistical behavior of a reacting mixture in isotropic turbulence. Phys. Fluids **1**, 42 (1958).

[*61*] FRENKIEL, F.N., and P.A. SHEPPARD (editors): Atmospheric Diffusion and Air Pollution. Advances in Geophysics, Vol. 6. New York: Academic Press 1959.

[*62*] HOWELLS, I.D.: An approximate equation for the spectrum of a conserved scalar quantity in a turbulent fluid. J. Fluid Mech. **9**, 104 (1960).

[*63*] ROBERTS, P.H.: On the application of a statistical approximation to the theory of turbulent diffusion. J. Math. Mech. **6**, 781 (1957).

[64] SAFFMAN, P. G.: On the effect of the molecular diffusivity in turbulent diffusion. J. Fluid Mech. **8**, 273 (1960).

[65] TAYLOR, G. I.: Diffusion by continuous movements. Proc. Lond. Math. Soc. [2], **20**, 196 (1921).

## Noise generation.

[66] KULSRUD, R. M.: Effect of magnetic fields on generation of noise by isotropic turbulence. Astrophys. J. **121**, 461 (1955).

[67] LIGHTHILL, M. J.: On the energy scattered from the interaction of turbulence with sound or shock waves. Proc. Cambridge Phil. Soc. **49**, 531 (1953).

[68] LIGHTHILL, M. J.: On sound generated aerodynamically. Proc. Roy. Soc. Lond., Ser. A **211**, 564 (1952); **222**, 1 (1954).

[69] PHILLIPS, O. M.: On the aerodynamic surface sound from a plane turbulent boundary layer. Proc. Roy. Soc. Lond., Ser. A **234**, 327 (1956).

[70] PHILLIPS, O. M.: On the generation of sound by supersonic turbulent shear layers. J. Fluid Mech. **9**, 1 (1960).

[71] PROUDMAN, I.: The generation of noise by isotropic turbulence. Proc. Roy. Soc. Lond., Ser. A **214**, 119 (1952).

## Turbulent shear flow.

[72] HINZE, J. O.: Turbulence. New York: McGraw-Hill Book Company, Inc. 1959.

[73] SCHUBAUER, G. B., and C. M. TEHEN: Turbulent flow. High Speed Aerodynamics and Jet Propulsion, vol. **5**, pp. 75—195. Princeton: Princeton University Press 1959.

[74] TOWNSEND, A. A.: The structure of turbulent shear flow. Cambridge: Cambridge University Press 1956.

SIAM J. APPL. MATH.
Vol. 25, No. 3, November 1973

# NOTE ON STRETCHING OF LINE AND SURFACE ELEMENTS IN ISOTROPIC TURBULENCE*

C. C. LIN†

**Abstract.** Certain inequalities are rigorously established connecting the statistical stretching of line and surface elements with the statistical dilatation of volume in a compressible fluid in isotropic turbulent motion. The method of approach differs only slightly from those used by Cocke (1969) and by Orszag (1970) for the incompressible case, the only novel feature being the introduction of the general theory of isotropic tensors. In the incompressible case, the earlier results are reproduced in the form given by Orszag.

**1. Introduction.** Observations of the spread of smoke from a chimney suggests that turbulent motion is diffusive; that is, the separation between two fluid particles increases in a statistical sense. The important implications of this behavior on the stretching of vortex lines and vortex sheets and the consequent degradation of energy into smaller and smaller scales have been extensively discussed in the literature, starting with a paper by G. I. Taylor and A. E. Green (1937). A rigorous proof of such a stretching process due to *isotropic* turbulent motion in an *incompressible* fluid however remained wanting until recently when W. J. Cooke (1969) first established that

$$\langle \log (\delta x/\delta\xi) \rangle \geqq 0, \tag{1}$$

where $\delta x$ is the length at time $t$ of the infinitesimal material line element between two fluid particles, separated by distance $\delta\xi$ at $t = t_0$, and $\langle \cdot \rangle$ denotes the statistical average over an ensemble of turbulent flow fields. Orszag (1970) pointed out that Cocke's result implies that

$$\langle (\delta x)^2 \rangle \geqq (\delta\xi)^2, \tag{2}$$

and gave a proof of (2) "using a simple variant of Cocke's method." Both authors also provide a proof for the stretching of surface elements in a rather abbreviated form.

In this note, whose purpose is largely expository, we treat the fluid as *compressible.* In this general case, certain inequalities are established connecting the statistical changes of length and of area with the statistical change of volume. In the incompressible case, there is no change of volume; then it follows that the length of a material element and the area of a material surface always increase in the statistical sense.

By making use of the general theorems governing isotropic tensors, we arrive at the relations (9) and (20) which are basic to the theory. In this respect, the present approach is conceptually perhaps a little more direct and clearer than earlier proofs. This approach also allows us to generalize our studies to homogeneous turbulence with axial symmetry, although we did not carry out the details.

---

* Received by the editors March 8, 1973.

† Massachusetts Institute of Technology, Cambridge, Massachusetts 02139. This work was supported in part by the National Science Foundation.

**2. Stretching of line elements.** We shall first reproduce the result on line elements in Orszag's form by making use of a general theorem on isotropic tensors of order two. This serves as an introduction to the discussion of the stretching of surface elements, where the same reasoning will be followed except that the general theorem used will concern isotropic tensors of order four. In both cases, we begin with the Lagrangian description of fluid motion

$$\mathbf{x} = \mathbf{x}(t, \boldsymbol{\xi}) \tag{3}$$

which gives the position $\mathbf{x}$ at time $t$ of a fluid particle which was located at $\boldsymbol{\xi}$ at time $t_0$. Then the separation $\delta\mathbf{x}$ at time $t$ of two particles initially separated by $\delta\boldsymbol{\xi}$ is given by

$$\delta x_i = \frac{\partial x_i}{\partial \xi_\alpha}\,\delta\xi_\alpha, \tag{4}$$

and the square of the length $\delta x$ at time $t$ is given by

$$\delta x^2 = \delta x_i\,\delta x_i = W_{\alpha\beta}\,\delta\xi_\alpha\,\delta\xi_\beta, \tag{5}$$

where

$$W_{\alpha\beta}(t, \boldsymbol{\xi}) = \frac{\partial x_i}{\partial \xi_\alpha}\frac{\partial x_i}{\partial \xi_\beta}. \tag{6}$$

Note that $W_{\alpha\beta}$ is a real symmetric matrix whose determinant is the square of the Jacobian $J$ of the transformation (3). Thus,

$$|W_{\alpha\beta}| = J^2 = (\rho_0/\rho)^2, \tag{7}$$

where $\rho$ is the density of the fluid element at time $t$, and $\rho_0$ is the density at the initial instant $t_0$.

Consider now an ensemble of flow fields (3). For the *same* separation $\delta\boldsymbol{\xi}$, consider an ensemble average of (5); we then have

$$\langle \delta x^2 \rangle = \langle W_{\alpha\beta} \rangle\,\delta\xi_\alpha\,\delta\xi_\beta, \tag{8}$$

where the ensemble average is taken over the random part. Now, since the turbulent motion is statistically homogeneous, $\langle W_{\alpha\beta} \rangle$ cannot be dependent on the initial position vector $\boldsymbol{\xi}$, and is therefore primarily a function of time. Its exact form depends on the anisotropic properties of the turbulent field. In the isotropic case, we must have

$$\langle W_{\alpha\beta} \rangle = c\,\delta_{\alpha\beta}, \tag{9}$$

where $c$ depends only on time. Note that here the isotropy is taken in the Lagrangian description, quite different from that in the Eulerian description in the usual theory. From (8) and (9) it follows that

$$\langle \delta x^2 \rangle = c(t)\,\delta\xi^2, \tag{10}$$

and that

$$c(t) = \tfrac{1}{3}\langle W_{\beta\beta} \rangle. \tag{11}$$

Now, for each single realization, we can show that

$$\tfrac{1}{3}W_{\beta\beta} \geqq |W_{\alpha\beta}|^{1/3}, \quad \text{or} \quad \tfrac{1}{3}W_{\beta\beta} \geqq (\rho_0/\rho)^{2/3}. \tag{12}$$

The details of the calculations are given in Appendix A. Thus, if we introduce (7) into (12) and then take the ensemble average, we obtain

$$\frac{\langle \delta x^2 \rangle}{\delta \xi^2} = c(t) \geqq \left\langle \left(\frac{\rho_0}{\rho}\right)^{2/3} \right\rangle \tag{13}$$

if $c(t)$ is introduced from (10) and (11). Equation (13) is the basic result in the statistical theory of stretching of fluid line elements.

**3. Stretching of surface elements.** Let us now consider a material surface element originally defined by two infinitesimal displacement vectors $\delta\boldsymbol{\xi}$ and $\delta'\boldsymbol{\xi}$. At time $t$, this material surface element is defined by the two vectors $\delta\mathbf{x}$ and $\delta'\mathbf{x}$ such that

$$\delta x_i = \frac{\partial x_i}{\partial \xi_\alpha}\delta\xi_\alpha, \qquad \delta' x_k = \frac{\partial x_k}{\partial \xi_\beta}\delta'\xi_\beta. \tag{14}$$

The vector of this surface area is then

$$\begin{aligned} \delta S_i &= \varepsilon_{ijk}\,\delta x_j\,\delta' x_k \\ &= \left(\varepsilon_{ijk}\frac{\partial x_j}{\partial \xi_\alpha}\frac{\partial x_k}{\partial \xi_\beta}\right)\delta\xi_\alpha\,\delta'\xi_\beta. \end{aligned} \tag{15}$$

Notice that since the quantity in parentheses is antisymmetrical, the vector $\delta S_i$ bears a simple linear relationship to the initial vector of surface area

$$\delta\sigma_\gamma = \varepsilon_{\gamma\alpha\beta}\,\delta\xi_\alpha\,\delta'\xi_\beta, \tag{16}$$

as given in Appendix B. However, we shall use (15) directly. From (15), it can be easily verified (see Appendix B) that

$$\delta S^2 = \delta S_i\,\delta S_i = (W_{\alpha\mu}W_{\beta\nu} - W_{\alpha\nu}W_{\beta\mu})\,\delta\xi_\alpha\delta'\xi_\beta\,\delta\xi_\mu\,\delta'\xi_\nu, \tag{17}$$

and hence

$$\langle \delta S^2 \rangle = \langle W_{\alpha\mu}W_{\beta\nu} - W_{\alpha\nu}W_{\beta\mu} \rangle\,\delta\xi_\alpha\,\delta'\xi_\beta\,\delta\xi_\mu\,\delta'\xi_\nu, \tag{18}$$

where the ensemble average is taken, as before, for the random part of equation (17).

We note that

$$C_{\alpha\beta\mu\nu} \equiv \langle W_{\alpha\mu}W_{\beta\nu} - W_{\alpha\nu}W_{\beta\mu} \rangle \tag{19}$$

is an isotropic tensor of order four, with antisymmetry in either pair of the indices $(\alpha, \beta)$ and $(\mu, \nu)$. Thus, if we refer to the general theorem on isotropic tensors (see Jeffreys and Jeffreys (1966, p. 103)), $C_{\alpha\beta\mu\nu}$ must be of the form

$$C_{\alpha\beta\mu\nu} = C(t)(\delta_{\alpha\mu}\,\delta_{\beta\nu} - \delta_{\alpha\nu}\,\delta_{\beta\mu}) \tag{20}$$

since the two other possible terms have the wrong kind of symmetry and therefore must have zero coefficient. Thus, introducing (20) into (18), we obtain

$$\langle \delta S^2 \rangle = C(t)(\delta\sigma)^2, \tag{21}$$

where

$$(\delta\sigma)^2 = (\delta\xi_\mu)^2(\delta'\xi_\nu)^2 - (\delta\xi_\nu\,\delta'\xi_\nu)^2 \tag{22}$$

is the square of the initial area element, since

$$\begin{aligned} \delta\sigma^2 &= (\delta\boldsymbol{\xi})^2(\delta'\boldsymbol{\xi})^2\{1 - \cos^2(\delta\boldsymbol{\xi}, \delta'\boldsymbol{\xi})\} \\ &= (\delta\boldsymbol{\xi})^2(\delta'\boldsymbol{\xi})^2 \sin^2(\delta\boldsymbol{\xi}, \delta'\boldsymbol{\xi}). \end{aligned} \tag{23}$$

Equation (21) will give us an idea of the behavior of the surface element, if we have an idea of the bounds on $C(t)$, to which we now turn.

**Estimation of $C(t)$.** In equations (19) and (20), let us contract the tensors with respect to the indices $(\alpha, \mu)$, and again with respect to the indices $(\beta, \nu)$. We obtain

$$\langle W_{\alpha\alpha} W_{\beta\beta} - W_{\alpha\beta} W_{\beta\alpha} \rangle = 6C(t). \tag{24}$$

For each realization, let us calculate the contracted tensors in terms of their diagonal form. Then, we get (see Appendix A for the definition of $w_1, w_2, w_3$)

$$\begin{aligned} W_{\alpha\alpha} W_{\beta\beta} - W_{\alpha\beta} W_{\beta\alpha} &= (w_1 + w_2 + w_3)^2 - (w_1^2 + w_2^2 + w_3^2) \\ &= 2w_1 w_2 w_3 \left( \frac{1}{w_1} + \frac{1}{w_2} + \frac{1}{w_3} \right). \end{aligned} \tag{25}$$

Now, $w_1 w_2 w_3 = |W_{\alpha\beta}| = (\rho_0/\rho)^2$, by (7). On the other hand,

$$\frac{1}{3}\left( \frac{1}{w_1} + \frac{1}{w_2} + \frac{1}{w_3} \right) \geqq \left( \frac{1}{w_1} \cdot \frac{1}{w_2} \cdot \frac{1}{w_3} \right)^{1/3} = \left( \frac{\rho_0}{\rho} \right)^{-2/3}, \tag{26}$$

according to the inequality established in Appendix A. Thus, for each realization, we have

$$W_{\alpha\alpha} W_{\beta\beta} - W_{\alpha\beta} W_{\beta\alpha} \geqq 6(\rho_0/\rho)^{4/3}, \tag{27}$$

according to (25). The statistical average of (27) then shows, through the use of (21) and (24), that

$$\frac{\langle \delta S^2 \rangle}{\delta \sigma^2} = C(t) \geqq \left\langle \left( \frac{\rho_0}{\rho} \right)^{4/3} \right\rangle. \tag{28}$$

This is the analogue of (13) for line elements. Notice that the bounds in (13) and (28) are related to each other.

**4. Incompressible fluids.** In the incompressible case, we have $\rho = \rho_0$ under all circumstances, even when homogeneity of the fluid is not assumed. However, the latter is unlikely to be compatible with homogeneous and isotropic fields of turbulence. The inequalities (13) and (28) now become

$$\langle \delta x^2 \rangle \geqq (\delta \xi)^2, \tag{29}$$

$$\langle \delta S^2 \rangle \geqq (\delta \sigma)^2. \tag{30}$$

These inequalities state that, in the statistical sense, the line elements and surface elements are always increasing in extent. Notice that there is no requirement that $t > t_0$. The unique property of time instant $t_0$ is that all members of the ensemble have exactly the *same* specification of the location of the two particles at that instant.

A few remarks are in order here:

Although we have demonstrated the diffusive nature of turbulence in general, we have not shown any positive statistical correlation between a vortex line and the stretching process *along* it. Stretching in two directions *across* a vortex line would tend to decrease the vorticity in an incompressible fluid. Indeed, the author is unaware of any rigorous theorem which would support Taylor's original idea of vortex line stretching, although the physical process does seem to justify it.

We have not examined the problem for anisotropic turbulence. General considerations would suggest that the final conclusions ought not to be seriously modified. Neither have we pursued the matter of density change in a compressible fluid. One might pose this question: Is it possible, in the particular manner of taking statistical averages, that the quantities on the right-hand sides of equations (13) and (28) are always at least as large as unity? Such questions seem to deserve the attention of workers desiring a completely deductive understanding of the nature of turbulent diffusion. (For further heuristic discussions of turbulent diffusion, see Lin (1960) and Lin and Reid (1963, p. 509).)

**Appendix A. Proof of the inequality (12).** The inequality (12) is a well-known relationship for symmetric matrices with real positive roots. Since both sides of (12) are invariant under a rotation of axes, we may establish it for the coordinate system in which $W_{\alpha\beta}$ is diagonal. Let the three roots of $W_{\alpha\beta}$ be $w_1, w_2, w_3$. Then (12) becomes

$$\frac{1}{3}(w_1 + w_2 + w_3) \geqq (w_1 w_2 w_3)^{1/3}. \tag{A.1}$$

Since all the three quantities $(w_1, w_2, w_3)$ are positive, we may write them in the form $(e^{x_1}, e^{x_2}, e^{x_3})$, and (A.1) becomes

$$\frac{1}{3}(e^{x_1} + e^{x_2} + e^{x_3}) \geqq e^{(1/3)(x_1 + x_2 + x_3)} \tag{A.2}$$

or

$$M_a \equiv \frac{1}{3}\{\exp[\tfrac{2}{3}x_1 - \tfrac{1}{3}(x_2 + x_3)] + \text{c.p.}\} \geqq 1, \tag{A.3}$$

where (c.p.) denotes the sum of two more terms obtained from cyclic permutation of the indices in the term given explicitly.

Equation (A.3) must now be established. It is easy to see that

$$e^x \geqq 1 + x, \tag{A.4}$$

where the equality holds only when $x = 0$. Hence, (A.3) implies that

$$M_a \geqq \frac{1}{3}\{[1 + \tfrac{2}{3}x_1 - \tfrac{1}{3}(x_2 + x_3)] + \text{c.p.}\}. \tag{A.5}$$

The right-hand side is clearly identically equal to unity. Q.E.D.

The inequality (A.1) can be generalized to show that the arithmetical mean of any number of *positive* quantities is at least as large as the geometrical mean of the same quantities. Equality holds only when all the quantities are the same. This is the theorem referred to by Orszag as the arithmetical-mean–geometrical-mean inequality. The above proof was shown by him to the author.

**Appendix B.** Equation (17) can be verified from (15) by a direct calculation, if we recall the $\varepsilon$-$\delta$ identity

$$\varepsilon_{ijk}\varepsilon_{ipq} = \delta_{jp}\delta_{kq} - \delta_{iq}\delta_{kp}. \tag{B.1}$$

Indeed,

$$\delta S_i\, \delta S_i = C_{\alpha\beta\mu\nu}\, \delta\xi_\alpha\, \delta'\xi_\beta\, \delta\xi_\mu\, \delta'\xi_\nu, \tag{B.2}$$

where

(B.3)
$$C_{\alpha\beta\mu\nu} = (\delta_{jp}\delta_{kq} - \delta_{iq}\delta_{kp})\left(\frac{\partial x_j}{\partial \xi_\alpha}\frac{\partial x_k}{\partial \xi_\beta}\frac{\partial x_p}{\partial \xi_\mu}\frac{\partial x_q}{\partial \xi_\nu}\right) = W_{\alpha\mu}W_{\beta\nu} - W_{\alpha\nu}W_{\beta\mu},$$

where the $W$'s are defined by equation (6).

The connection between the surface elements defined by (15) and (16) is

(B.4)
$$\delta S_i = A_{i\alpha}\,\delta\sigma_\alpha,$$

where

(B.5)
$$A_{i\alpha} = \tfrac{1}{2}\varepsilon_{ijk}\varepsilon_{\alpha\mu\nu}\frac{\partial x_j}{\partial \xi_\mu}\frac{\partial x_k}{\partial \xi_\nu}.$$

To see this, note the antisymmetry of

(B.6)
$$B_{\alpha\beta} = \varepsilon_{ijk}\frac{\partial x_j}{\partial \xi_\alpha}\frac{\partial x_k}{\partial \xi_\beta}$$

in the indices $\alpha$ and $\beta$. Hence

(B.7)
$$\delta S_i = \tfrac{1}{2}B_{\alpha\beta}(\delta\xi_\alpha\,\delta'\xi_\beta - \delta\xi_\beta\,\delta'\xi_\alpha).$$

Now, from (16), by using (B.1), it can be easily verified that

(B.8)
$$\delta\sigma_\alpha\varepsilon_{\alpha\mu\nu} = \delta a_\mu\,\delta' a_\nu - \delta a_\mu\,\delta' a_\nu$$

and hence (B.7) leads to (B.5).

## REFERENCES

W. J. COCKE (1969), Phys. Fluids, 12, pp. 2488–2492.

H. JEFFREYS AND B. S. JEFFREYS (1966), *Methods of Mathematical Physics*, 3rd ed., Cambridge University Press, London.

C. C. LIN (1960), *On a theory of dispersion by continuous movements. I: Stationary anisotropic processes*, Proc. Nat. Acad. Sci. U.S.A., 46, pp. 1147–1150.

C. C. LIN AND W. H. REID (1963), *Turbulent flow, theoretical aspects*, Handbuch der Physik, vol. VIII, part 2, Springer, Berlin, pp 438–523.

S. A. ORSZAG (1970), Phys. Fluids, 13, pp. 2203–2204.

G. I. TAYLOR AND A. E. GREEN (1937), Proc. Roy. Soc., 158, pp. 499–521. Reprinted as Paper 38 in *Scientific Papers by Sir Geoffrey Taylor*, vol. 2, Cambridge University Press, London, 1960, pp. 409–426.

Reprinted from the Proceedings of the NATIONAL ACADEMY OF SCIENCES,
Vol. 27, No. 12, pp. 570–575. December, 1941

## *ON THE MOTION OF VORTICES IN TWO DIMENSIONS—I. EXISTENCE OF THE KIRCHHOFF-ROUTH FUNCTION*

BY C. C. LIN

DEPARTMENT OF APPLIED MATHEMATICS, UNIVERSITY OF TORONTO

Communicated October 20, 1941

1. *Introduction.*—Kirchhoff[1] was the first to establish the existence of "a stream function giving the motion of vortices" in an *unbounded* region. Later, in 1881, Routh[2] enunciated a theorem regarding the transformation of a function of that nature for the case of a *single* vortex moving in a *bounded* region. No proof was given and the existence of such a stream function was not established.[3] In 1921, Lagally[4] established the existence of this "*Routh's stream function*" and the more general "*Kirchhoff's path function*" for a *simply connected bounded* region. An independent proof for the case of a single vortex in such a region was also given by Masotti[5] in 1931, by using the Green function of the first kind. However, the most general case of the motion of a number of vortices in a *multiply connected*

region is not covered by their work. The present article establishes the existence of the "*Kirchhoff-Routh function*"[6] for the general case. The proof makes use of a generalized Green function and may therefore be regarded as a generalization of Masotti's work. Explicit formula of this function is also given. A more detailed treatment of this work will appear elsewhere.

2. *Statement of the Problem.*—Consider a number of isolated free vortices of strengths $\kappa_i$ ($i = 1, 2, ..., n$) at the points $P_i(x_i, y_i)$ ($i = 1, 2, ..., n$) in an incompressible fluid moving irrotationally in a region $R$. This region has a number of internal boundaries $C_k$ ($k = 1, 2, ..., m$) and it is (*a*) bounded externally by a closed curve $C_0$, or (*b*) unbounded from the outside, or (*c*) limited by curves $C_0$ extending to infinity.

If the ordinary stream function of fluid motion

$$\psi = \psi(x, y;\ x_1, y_1;\ ...;\ x_n, y_n) \tag{2.1}$$

is known (which is obviously independent of the time $t$ explicitly), the components of velocity of the $i$th vortex ($i = 1, 2, ..., n$) are given by

$$\frac{dx_i}{dt} = u_i = -\left(\frac{\partial\psi_{(i)}}{\partial y}\right)_{P_i}, \frac{dy_i}{dt} = v_i = \left(\frac{\partial\psi_{(i)}}{\partial x}\right)_{P_i} \tag{2.2}$$

where

$$\psi_{(i)} = \psi - \frac{\kappa_i}{2\pi}\log r_i,\ r_i = \sqrt{(x - x_i)^2 + (y - y_i)^2}, \tag{2.3}$$

and the suffixes $P_i$ denote that the results of differentiation are to be evaluated at the point $P_i$. For the case where the region $R$ has no solid boundaries and where steady streaming is absent, Kirchhoff[1] has shown that there exists a function

$$W = W(x_1, y_1;\ x_2, y_2;\ ...;\ x_n, y_n), \tag{2.4}$$

such that the motion of the $i$th vortex is given by

$$\kappa_i\frac{dx_i}{dt} = \kappa_i u_i = -\frac{\partial W}{\partial y_i},\ \kappa_i\frac{dy_i}{dt} = \kappa_i v_i = \frac{\partial W}{\partial x_i}. \tag{2.5}$$

The object of this article is to show that this result of Kirchhoff can be generalized to the motion of vortices in a region $R$ of the general type described above. In another article, we shall derive the law of transformation of the *Kirchhoff-Routh function* $W$ under a conformal transformation.

We shall first define a generalized Green function particularly suited to the study of vortex motion. The essence of this paper, like Masotti's work, lies in the application of the reciprocity property of a properly defined Green function. (Cf. equations (3.5) and (3.6)).

3. *The Green Function.*—Let us define a function $G(x, y;\ x_0, y_0)$ with respect to two points $P(x, y)$ and $P_0(x_0, y_0)$ in the region $R$ by the following three conditions.

(i) The function

$$g(x, y;\ x_0, y_0) = G(x, y;\ x_0, y_0) - \frac{1}{2\pi}\log r_0 \qquad \left(r_0 = \sqrt{(x - x_0)^2 + (y - y_0)^2}\right) \quad (3.1)$$

is harmonic with respect to $(x, y)$ throughout the region $R$ including the point $P_0$; thus,

$$\frac{\partial^2 g}{\partial x^2} + \frac{\partial^2 g}{\partial y^2} = 0. \quad (3.2)$$

(ii) If $\partial G/\partial n$ is the normal derivative of the function $G$ on a curve (with $x, y$ as variables), then

$$G = A_k \text{ on } C_k, \quad \int_{C_k'} \frac{\partial G}{\partial n} ds = 0 \quad (k = 1, 2, \ldots, m) \quad (3.3)$$

for each of the inner boundaries $C_k$; where $ds$ is an element of arc of the closed analytical curve $C_k'$, which encloses $C_k$ but neither the other boundary curves nor the point $P_0$. The positive normal to $C_k'$ is drawn toward $C_k$ and the positive direction of $C_k'$ is taken so that the main part of $R$ (including $P_0$) lies to its left. Similar positive directions are used for any other curve.

(iii*a*) If the region $R$ has a closed outer boundary $C_0$, then

$$G(x, y;\ x_0, y_0) = 0 \quad \text{over } C_0. \quad (3.4a)$$

(iii*b*) If the region $R$ extends to infinity in all directions, the function $G(x, y;\ x_0, y_0)$ behaves as follows:[7]

$$\left.\begin{aligned} &G(x, y;\ x_0, y_0) = \frac{1}{2\pi}\log r_0 + O\left(\frac{1}{r_0}\right), \\ &\frac{\partial G}{\partial s} = O\left(\frac{1}{r_0^2}\right), \frac{\partial G}{\partial n} = \frac{1}{2\pi r_0} + O\left(\frac{1}{r_0^2}\right), \end{aligned}\right\} \text{over a very large circle of radius } r_0. \quad (3.4b)$$

where $\partial G/\partial s$ is the tangential derivative along the circle.

(iii*c*) If the region $R$ has boundaries $C_0$ extending to infinity, the function $G(x, y;\ x_0, y_0)$ behaves as follows:[7]

$$\left.\begin{aligned} &G(x, y;\ x_0, y_0) = O \quad \text{over } C_0, \\ &G(x, y;\ x_0, y_0) = o(1) \quad \text{over a very large circle of radius } r_0. \end{aligned}\right\} \quad (3.4c)$$

Koebe[8] established the existence and uniqueness of this Green function $G$ by resolving it into a linear combination of two sets of basic functions: (*a*) the Green function of the first kind and (*b*) the set of harmonic functions each of which takes the value unity on a particular one of the boundary curves and the value zero on all the others. With the help of this resolution, the present writer finds it possible to prove the symmetry property of this Green function by using standard methods.[9]

We summarize the results in the following lemma.

LEMMA I. *The function* $G(x, y;\ x_0, y_0)$ *defined by the conditions* (3.2)-(3.4) *exists uniquely, and is a generalized Green function satisfying the reciprocity property*

$$G(x, y;\ x_0, y_0) = G(x_0, y_0;\ x, y). \tag{3.5}$$

The reciprocity property immediately leads to the following important result (cf. (3.1)):

$$\left.\begin{aligned} \frac{\partial}{\partial x_0} g(x_0, y_0;\ x_0, y_0) &= 2 \lim_{P \to P_0} \frac{\partial}{\partial x} g(x, y;\ x_0, y_0), \\ \frac{\partial}{\partial y_0} g(x_0, y_0;\ x_0, y_0) &= 2 \lim_{P \to P_0} \frac{\partial}{\partial y} g(x, y;\ x_0, y_0). \end{aligned}\right\} \tag{3.6}$$

4. *Kirchhoff's Equations.*—Let us now apply our function $G$ to the hydrodynamical problem of vortex motion with the stream function (2.1).

The motion given by the stream function

$$\Psi = \kappa_0 G(x, y;\ x_0, y_0) \tag{4.1}$$

may be called the motion *due* to a vortex $\kappa_0$ at the point $P_0(x_0, y_0)$. It is a possible potential motion in the region $R$ with the required singularity, and has no circulation around any one of the inner boundaries. Furthermore, if the region $R$ extends to infinity, the flow *across* an arc of the circle $r_0$ = constant (contributing to outward flux) approaches zero and the flow *along* it (contributing to circulation) is finite (or approaches zero), as $r_0$ becomes infinite.

If we subtract the stream functions *due* to all the vortices from the complete stream function (2.1), the remainder is a stream function giving a possible potential motion in the region $R$. Now, this motion is uniquely defined when the circulation around each of the curves $C_k$ and the fluid velocity at infinity are given. Since these conditions are the same for the complete stream function as for this part, this motion is actually independent of the (variable) positional coordinates $(x_i, y_i)$ of the vortices. It depends on the (constant) strengths $\kappa_i$, only if the conditions defining it happen to be so. We may therefore call it the motion "*due* to outside agencies."

We summarize our results in the following basic lemma.

LEMMA II. *If n vortices of strengths* $\kappa_i$ $(i = 1, 2, \ldots, n)$ *are present in an* *incompressible fluid at the points $P_i(x_i, y_i)$ $(i = 1, 2, ..., n)$ in a general region bounded by fixed boundaries, the stream function of fluid motion is given by*

$$\psi(x, y; x_1, y_1, ...; x_n, y_n) = \psi_0(x, y) + \sum_{i=1}^{n} \kappa_i G(x, y; x_i, y_i), \quad (4.2)$$

*where the properties of $G(x, y; x_i, y_i)$ are given in Lemma I, and $\psi_0(x, y)$ is the stream function of the motion due to outside agencies, independent of $P_i(x_i, y_i)$ (and of $\kappa_i$).*

With this result at hand, we can at once establish the existence of the Kirchhoff-Routh function.

THEOREM I. *For the motion of vortices of strengths $\kappa_i (i = 1, 2, ..., n)$ in a general region R bounded by fixed boundaries, there exists a Kirchhoff-Routh unction $W(x_1, y_1; x_2, y_2; ...; x_n, y_n)$ such that*

$$\left.\begin{aligned} \kappa_i \frac{dx_i}{dt} &= \kappa_i u_i = -\frac{\partial W}{\partial y_i}, \\ \kappa_i \frac{dy_i}{dt} &= \kappa_i v_i = \frac{\partial W}{\partial x_i}, \end{aligned}\right\} \quad (4.3)$$

*where $P_i(x_i, y_i)$ $(i = 1, 2, ..., n)$ are the instantaneous positions of the vortices. The function W is given by*

$$W = \sum_{i=1}^{n} \kappa_i \psi_0(x_i, y_i) + \sum_{\substack{i,j=1 \\ (i>j)}}^{n} \kappa_i \kappa_j G(x_i, y_i; x_j, y_j) + \frac{1}{2} \sum_{i=1}^{n} \kappa_i^2 g(x_i, y; x_i, y_i). \quad (4.4)$$

This can be at once verified by comparing the results for $u_i$ and $v_i$ obtained from (4.3), (4.4) and (3.6) with those obtained from (2.2), (2.3) and (4.2). We note that the system of *Kirchhoff's equations* (4.3) is a Hamiltonian system of differential equations in the set of variables $\sqrt{\kappa_i} x_i$ and $\sqrt{\kappa_i} y_i$, $(i = 1, 2, ..., n)$.

It is not difficult to show that when the vortices are fixed the *total* force reacting on all the solid boundaries is given by

$$X = \sum_{i=1}^{n} \rho(\partial W/\partial x_i), \; Y = \sum_{i=1}^{n} \rho(\partial W/\partial y_i),$$

$\rho$ being the density of the fluid. It can also be shown that $\rho W$ is a *measure* of the kinetic energy of fluid motion, so that equation (4.3) leads to the law of conservation of energy, $W$ = constant.

[1] Kirchhoff, G., *Vorlesungen über mathematische Physik, Mechanik*, p. 255 ff. See also Lamb, H., *Hydrodynamics*, 1932, p. 230.

[2] Routh, E. J., *Proc. Lond. Math. Soc.*, **12**, 83 (1881).

[3] Even in the standard books of recent years, the treatment is far from complete; cf. Lamb, H., *Hydrodynamics*, 1932, pp. 219–236; Ramsey, A. S., *A Treatise of Hydro-*

*mechanics, Pt. II, Hydrodynamics*, 1935, pp. 219–232; Milne-Thomson, L. M., *Theoretical Hydrodynamics*, 1938, pp. 323–348.

[4] Lagally, M., *Math. Zeits.*, **10**, 231–239 (1921).

[5] Masotti, A., *Atti Pontif. Accad. Sci. Nuovi Lincei*, **84**, 209–216, 235–245, 464–467, 468–473, 623–631 (1931). Also, *Seminario Matematico e Fisico di Milano*, **6**, 3–53 (1932).

[6] The function called by Lagally the Kirchhoff's path function shall be called in this paper the "Kirchhoff-Routh function." The study of the function called by him the Routh's stream function is not of much importance, because it is merely a special application of the other (cf. equation (6.1)).

[7] For the definition of $O(\ )$ and $o(\ )$, cf. Titchmarsh, E. C., *Theory of Functions*, 1932, p. 1, Oxford.

[8] Koebe, P., *Acta Math.*, **41**, 306–344 (1918). Note that our function in case (*b*) is the function with two singularities, one at $P_0$, the other at infinity.

[9] Cf. Kellogg, O. D., *Foundations of Potential Theory*, 1929, pp. 238–240, Berlin. Note that no assumption is made regarding the nature of the boundaries $C_0, C_1, \ldots, C_k$.

Reprinted from the Proceedings of the NATIONAL ACADEMY OF SCIENCES,
Vol. 27, No. 12, pp. 575–577. December, 1941

# ON THE MOTION OF VORTICES IN TWO DIMENSIONS—II SOME FURTHER INVESTIGATIONS ON THE KIRCHHOFF-ROUTH FUNCTION

BY C. C. LIN

DEPARTMENT OF APPLIED MATHEMATICS, UNIVERSITY OF TORONTO

Communicated October 20, 1941

5. *Conformal Transformation.*—We shall now investigate the behavior of the Kirchhoff-Routh function (whose existence we have established in the preceding article) under a conformal transformation of fluid motion.

THEOREM II (*Generalized Routh's theorem*).—*Under a conformal transformation*

$$\tilde{z} = f(z) \tag{5.1}$$

*which derives the motion in the $\tilde{z}$-plane from that in the z-plane, the Kirchhoff-Routh function for the new motion is given by*

$$\tilde{W} = W + \sum_{i=1}^{n} \frac{\kappa_i^2}{4\pi} \log \left|\frac{dz}{d\tilde{z}}\right|_{P_i} \tag{5.2}$$

*Proof.* If $F(z)$ is the complex stream function in the $z$-plane, we have (*cf.*(2.3))

$$-u_i + iv_i = \lim_{P \to P_i} \frac{d}{dz}\left\{F(z) - \frac{i\kappa_i}{2\pi}\log(z - z_i)\right\}. \tag{5.3}$$

We mark every quantity in the $\tilde{z}$-plane with a curl. The complex stream function for the new motion is then

$$\tilde{F}(\tilde{z}) = F(z), \tag{5.4}$$

by definition of conformal transformation. From this it follows that $\tilde{\kappa}_i = \kappa_i$, $(i = 1, 2, ..., n)$. It can be verified that, by (5.3) and (5.4),

$$-\tilde{u}_i + i\tilde{v}_i = (-u_i + iv_i)\frac{dz_i}{d\tilde{z}_i} + \frac{i\kappa_i}{4\pi}\frac{d}{d\tilde{z}_i}\log\left(\frac{dz_i}{d\tilde{z}_i}\right). \tag{5.5}$$

Multiplying (5.5) by $\delta z_i$ and taking imaginary parts, we have

$$\tilde{v}_i\delta x_i - \tilde{u}_i\delta\tilde{y}_i = v_i\delta x_i - u_i\delta y_i + \frac{\kappa_i}{4\pi}\delta\log\left|\frac{dz}{d\tilde{z}}\right|_{P_i}. \tag{5.6}$$

Multiplying (5.6) by $\kappa_i$ and summing for the index $i$, we have, by (4.3),

$$\delta\tilde{W} = \delta W + \delta\sum_{i=1}^{n}\frac{\kappa_i^2}{4\pi}\log\left|\frac{dz}{d\tilde{z}}\right|_{P_i}. \tag{5.7}$$

Equation (5.2) then follows at once (up to an additive constant).

6. *Discussion.*—(i) The above results hold when the solid boundaries are moving, for this affects the stream function $\psi_0$ alone. The Green function $G$, of course, depends upon the instantaneous configuration of the solid boundaries. Furthermore, the results hold also when the function $\psi_0$ has *fixed* singularities.

(ii) The theory gives explicitly the Kirchhoff-Routh function $W$ by the formula (4.4) when the ordinary stream function (4.2) is known, for the functions $\psi_0$, $G$ and $g$ can then be written down at once. In actual applications, we must be careful to see that $\psi$ is actually in the form (4.2), with $\psi_0$ and $G$ satisfying required conditions.

(iii) *Routh's stream function and Routh's stream.*—By (4.3), the motion of the $i$th vortex may be derived from a *Routh's theorem function*

$$\chi_{(i)} = \frac{W}{\kappa_i} \tag{6.1}$$

by the formulas

$$u_i = -\frac{\partial\chi_{(i)}}{\partial y_i},\ v_i = \frac{\partial\chi_{(i)}}{\partial x_i}, \tag{6.2}$$

just as the fluid velocity is derived from the ordinary stream function $\psi$. By (4.4) and (5.2), we see that

$$\chi_{(i)} = \psi_0(x_i, y_i) + \sum_{j\neq i}\kappa_j G(x_i, y_i;\ x_j, y_j) + \frac{\kappa_i}{2}g(x_i, y_i;\ x_i, y_i) + h_i, \tag{6.3}$$

and transforms according to the law

$$\chi_{(i)} = \chi_{(i)} + \frac{\kappa_i}{4\pi}\log\left|\frac{dz}{d\tilde{z}}\right|_{P_i} + k_i, \tag{6.4}$$

under the transformation (5.1). In these equations, $h_i$ and $k_i$ are independent of $x_i$ and $y_i$, and are therefore unimportant so far as the motion of the $i$th vortex is concerned. Equation (6.4) is *Routh's theorem* generalized to the case of a number of vortices.

*Routh's original special case.*—For a single vortex $\kappa_0$ at the point $P_0(x_0, y_0)$, the equations corresponding to (6.3) and (6.4) are

$$\chi = \frac{W}{\kappa_0} = \psi_0(x_0, y_0) + \frac{\kappa_0}{2} g(x_0, y_0;\ x_0, y_0) \qquad (6.5)$$

and

$$\tilde{\chi} = \chi + \frac{\kappa_0}{4\pi} \log \left| \frac{dz}{d\tilde{z}} \right|_{P_0}. \qquad (6.6)$$

Equation (6.6) is *Routh's theorem* in its original form. In this case, the path of the vortex is given by

$$\chi = \text{const.}, \qquad (6.7)$$

if the boundaries are fixed.

(iv) *Kirchhoff's original special case.*—If all the solid boundaries are absent, we have $g \equiv 0$. If, furthermore, there is no motion beyond that due to the vortices themselves, (4.4) reduces to the simple result

$$W = \frac{1}{2\pi} \sum_{\substack{i, j = 1 \\ (i > j)}}^{n} \kappa_i \kappa_j \log r_{ij}, \quad r_{ij} = \sqrt{(x_i - x_j)^2 + (y_i - y_j)^2}, \qquad (6.8$$

first derived directly by Kirchhoff.[1]

In conclusion, the author wishes to express his sincere thanks to Prof. J. L. Synge, for suggesting the problem treated in these two articles and for his help and encouragement throughout the work. The author is also indebted to Dr. A. Weinstein for his kind help.

[1] Kirchhoff, G., *Vorlesungen über Mathematische Physik, Mechanik*, p. 225 ff. See also Lamb, H., *Hydrodynamics*, 1932, p. 230.

Reprinted from QUARTERLY OF APPLIED MATHEMATICS
Vol. IV, No. 3, October, 1946 

# ON AN EXTENSION OF THE VON KÁRMÁN-TSIEN METHOD TO TWO-DIMENSIONAL SUBSONIC FLOWS WITH CIRCULATION AROUND CLOSED PROFILES*

BY

C. C. LIN

*Brown University*

**1. Introduction.** The method for treating compressible flows, as developed by Chaplygin,[1] von Kármán[2] and Tsien[3] leads to a successful solution of the flow pattern past solid bodies when the flow has no circulation. When the flow has a finite circulation, as in the case of airfoils, the profile shapes furnished by this theory are not closed. It is doubtless desirable to develop the theory so as to remove this difficulty.

Recently, Bers[4] succeeded in obtaining flows with circulation around closed profiles. As is usual in the case of a first success, the new method has a few disadvantages. In the first place, the mapping between the actual compressible flow and the associated incompressible flow is not regular at the stagnation points. Thus, if the profile in the associated incompressible flow is regular everywhere, angular points would appear in the profile in the compressible case; and vice versa. The application of the method is further complicated by the fact that the angle thus generated depends on the free-stream Mach number. For the engineer, the treatment has the additional inconvenience of involving the concepts of Riemannian geometry (which are avoided in the present treatment).

In the present article we shall describe a method which is free from the disadvantages mentioned above. The derivation is *very* simple, and no reference is made to Riemannian geometry. Yet the result includes all the previous ones as special cases. Indeed, the treatment seems to be now in the most natural and the most general form which is obtainable from the line of study of Chaplygin, von Kármán and Tsien. It also has great flexibility. Given one incompressible flow, there is still an analytic function at our disposal for constructing compressible flows. This freedom of choice enables us to avoid much unnecessary numerical labor in constructing flows of certain general types. A large number of compressible flows can be derived from a given incompressible flow by the present method without numerical integration.

Apart from giving a useful method for constructing compressible flow patterns, the present development has the following significance. First, the freedom of disposing of one analytic function leads to the solution of the direct problem,—namely to

---

* Received May 18, 1946.

[1] S. A. Chaplygin, *On gas jets.* Scientific Memoirs, Moscow Univ., Math. Phys. Sec. **21,** 1–121 (1902). (English translation published by Brown University, 1944. Also NACA TM No. 1063, 1944.)

[2] Th. von Kármán, *Compressibility effects in aerodynamics.* Jour. Aero. Sci. **8,** 337–356 (1941).

[3] Hsue-Shen Tsien, *Two-dimensional subsonic flow of compressible fluids.* Jour. Aero. Sci. **6,** 339–407 (1939).

[4] L. Bers, *On a method of constructing two-dimensional subsonic compressible flows around closed profiles,* NACA TN No. 969 (1945); *On the circulatory subsonic flow of a compressible fluid past a circular cylinder,* NACA TN No. 970 (1945). See also S. Bergman, *On two-dimensional flows of compressible fluids,* NACA TN No. 972 (1945).

calculate the compressible flow past a *given* profile. Indeed, *the solution of problems of compressible flow, either direct or inverse,*—namely, the construction of flows around profiles either given beforehand or not—*is now an a parallel footing with the incompressible case.* In either case, the direct problem requires a method of successive approximations.[5] Secondly, the application of the pressure coefficient formula of von Kármán and Tsien to flows with circulation is justified on the same basis as in the circulation-free case. Experimentally, the formula has been found to be successful even when there is circulation, although the theory has so far been incomplete. The original development of von Kármán and Tsien leaves the body not closed, while the profiles given by the method of Bers do not have the same analytic nature in the incompressible and the compressible cases. The present method removes these difficulties.

Further investigations of the significance of the present method are being carried out. The present article contains only the essential result and its proof. It is hoped that a complete discussion of further developments may be published very soon.

**2. Method of constructing two-dimensional subsonic flows with circulation around profiles.** Let $p$, $\rho$, $u$, $v$ be the pressure, density and components of velocity of a steady two-dimensional irrotational flow in the $x$, $y$ plane. Let $p$ be a function of the density $\rho$ only (given either by the isentropic relation or any other approximate relation). Then there exist the velocity potential $\phi$ and the stream function $\psi$ defined by the following differential relations:

$$d\phi = udx + vdy, \tag{2.1}$$

$$d\psi = -\rho vdx + \rho udy. \tag{2.2}$$

The velocity components $u$, $v$ and the density $\rho$ are further connected by Bernoulli's equation

$$\frac{q^2}{2} + \int \frac{dp}{\rho} = \text{const.}, \qquad (q^2 = u^2 + v^2). \tag{2.3}$$

It is convenient to refer the density of the gas to that at the stagnation point and to refer all the velocities to the velocity of sound at stagnation. The coordinates $x$, $y$ may be regarded as referred to the size of the body, and the pressure as referred to the product of stagnation density and the square of the velocity of sound at stagnation. Throughout this article, this process shall be implied, and all the quantities under discussion are dimensionless.

As is well-known, the problem simplifies greatly if the pressure-density relation is approximated by

$$p = A - \frac{B}{\rho}\cdot \tag{2.4}$$

This is the basis of the method of Chaplygin, von Kármán and Tsien. A discussion of its physical interpretation may be found in the papers of these authors. Equation (2.3) leads to

[5] For the incompressible case, see T. Theodorsen, *Theory of wing sections of arbitrary shapes*, NACA Rep. No. 411 (1931).

T. Theodorsen and I. E. Garrick, *General potential theory of arbitrary wing sections*, NACA Rep. No. 452 (1933).

$$c = \frac{1}{\rho} = \sqrt{1+q^2}, \tag{2.5}$$

where $c$ is the local velocity of sound. Indeed, $B$ must be equal to unity if the reference stagnation quantities are calculated from (2.3) by the use of (2.4)

Under the approximation (2.4), the following method may be used for constructing two-dimensional subsonic flows with circulation around closed profiles.

*Given an incompressible flow past a profile $P_0$ in the $\zeta$-plane ($\zeta=\xi+i\eta$) described by the complex stream function $F(\zeta)$ and the complex velocity $w_0(\zeta)$, choose a function $k(\zeta)$, regular in the region exterior to $P_0$ and including the point at infinity, having no root in $R_0$, and such that $R_0$*

$$\left|\tfrac{1}{2}w_0(\zeta)\right| < |k(\zeta)| < \infty \quad on \quad P_0, \tag{2.6}$$

*and that*

$$\oint k(\zeta)d\zeta - \frac{1}{4}\oint \frac{\overline{w_0^2(\zeta)}}{k(\zeta)}\, d\zeta = 0, \tag{2.7}$$

*where the integration is performed along any contour enclosing $P_0$. Then*

$$x + iy = \int k(\zeta)d\zeta - \frac{1}{4}\int \overline{\frac{w_0^2(\zeta)}{k(\zeta)}}\, d\zeta, \tag{2.8}$$

$$\frac{2q}{1+\sqrt{1+q^2}}\, e^{-i\theta} = \frac{w_0(\zeta)}{k(\zeta)}, \tag{2.9}$$

$$\phi + i\psi = F(\zeta) \tag{2.10}$$

*gives the parametric representation of a compressible flow past a profile $P$ in the $x$, $y$ plane with $\zeta$ as parameter, where $P$ has the same general analytic nature (e.g. same number of corners, etc.) as the original profile $P_0$. In these formulae, $\phi$, $\psi$ are the velocity potential and stream function defined by (2.1) and $q$, $\theta$ denote the magnitude and direction of the velocity of the compressible flow.*

**3. Proof.** The proof consists of two parts. First, it is necessary to show that, after the auxiliary variable $\zeta$ is eliminated from (2.8)–(2.10), we obtain proper functional relations between $\phi$, $\psi$, $q$, $\theta$ and $x$, $y$. Secondly, we must show that the profile $P$ is a closed curve and is mapped into $P_0$ by a regular mapping such that the regions $R$, exterior to $P$, and $R_0$, exterior to $P_0$, are mapped into each other in a one-to-one manner.

($a$) The first part of the proof is simple. It is well known[6] that under the approximation (2.4), the relations

$$dz = \frac{dF}{w^*} - \frac{1}{4}\,\overline{w^* dF}, \qquad F = F(w^*) \tag{3.1}$$

with $z$, $F$, $w^*$ defined by

$$\left.\begin{aligned} z &= x + iy, \\ F &= \phi + i\psi, \\ w^* &= \frac{2q}{1+\sqrt{1+q^2}}\, e^{-i\theta} \end{aligned}\right\} \tag{3.2}$$

[6] Cf., for example, Eqs. (23), (26) of Tsien's paper quoted in Footnote 3.

would give a solution of the *differential* equations of compressible flow. Instead of trying to establish a relation between $F$ and $w^*$ directly, we introduce an *auxiliary variable* $\zeta$ such that $F(\zeta)$ and $w^*(\zeta)$ are analytic functoins. It is well known that great simplification is obtained by taking $F(\zeta)$ to be the complex potential of an incompressible flow similar to the compressible flow we desire. However, the extent of arbitariness in the choice of $w^*(\zeta)$ has not been carefully examined. It is clear that any choice of $w^*(\zeta)$ will be sufficient so far as satisfying the *differential* equations is concerned.

In the present case we dispose of the arbitrary function by writing

$$w^*(\zeta) = w_0(\zeta)/k(\zeta). \tag{3.3}$$

The only requirements on $k(\zeta)$ are the general conditions of regularity and the relations (2.6) and (2.7), which will be discussed immediately.

(*b*) The second part of the proof is also very simple. In the first place, the profile $P$ is closed by virtue of (2.7).[6] The regularity of the mapping is established if the Jacobian of the transformation maintains the same sign and does not vanish or become infinite in the region $R_0$, including the boundary $P_0$ and the point at infinity. It can be easily verified that the Jacobian is

$$J = \frac{\partial(x, y)}{\partial(\xi, \eta)} = Im\{\bar{z}_\xi z_\eta\} = |k|^2\left\{1 - \frac{1}{16}\left|\frac{w_0}{k}\right|^4\right\}. \tag{3.4}$$

From this expression, it is clear that the requirement is satisfied when $k$ satisfies the restrictions specified in the last section.*

**4. Discussion.** (a) *The function* $k(\zeta)$. To make use of the freedom in choosing the function $k(\zeta)$ is the essential improvement made in the present paper. Von Kármán and Tsien gave an interpretation of $w^*(\zeta)$ by identifying it with the complex velocity $w_0(\zeta)$ in the associated incompressible flow. This means that they put

$$k(\zeta) \equiv 1. \tag{4.1}$$

They were therefore unable to meet the requirement (2.7) for closing the profile, for flows with circulation. Bers overcame this difficulty by virtually taking

$$k(\zeta) = \text{const.}\ \{w_0(\zeta)\}^{1-\sqrt{1-M_\infty^2}}, \tag{4.2}$$

where $M_\infty$ is the free-stream Mach number. However, at the stagnation points of the incompressible flow

$$|k(\zeta)| = 0, \tag{4.3}$$

and the mapping between the profiles $P$ and $P_0$ is not regular there. In applying his method to the calculation of compressible flow past a circle, Bers had to start with the rather complicated problem of finding the incompressible flow past two intersecting circular arcs with the stagnation points at the points of intersection and with the

* *Note added in proof*: Professor K. O. Friedricks pointed out to the author that, in the compressible as well as the incompressible case, a mapping with non-vanishing Jacobian does not always lead to a useful result: the region obtained may be simply-connected but multiply covered. This difficulty has not been experienced so far in some numerical examples which have been worked out.

angles properly adjusted in relation to the free-stream Mach number. Indeed, it seems that after reaching the relation

$$dz = \frac{dF}{w^*} - \frac{1}{4}\overline{w^* dF},$$

the most natural development of the Kármán-Tsien method is to leave $w^*(\zeta)$ quite free, as we have done here, instead of connecting it definitely with $w_0$, as was done by previous authors. The present method of leaving $k(\zeta)$ free seems to be the most general scheme.

If we deliberately want to introduce some singular points in $P$ by starting from a profile $P_0$ without a singular point, $|k(\zeta)|$ should be allowed to take the limiting values specified in (2.6); e.g., $k(\zeta)=0$ where $w_0(\zeta)=0$.

Although $k(\zeta)$ cannot be taken to be unity when the flow has a circulation, it should not depart from unity very much if the profiles $P$ and $P_0$ are not to differ much from each other. This is easily seen from a comparison of terms in (cf. (2.8))

$$dz = k d\zeta - \frac{1}{4}\overline{\frac{w_0^2}{k} d\zeta}. \tag{4.4}$$

They have the ratio (cf. (2.9) and (2.5))

$$\lambda = \left(\frac{q}{1+\sqrt{1+q^2}}\right)^2 = \frac{M^2}{(1+\sqrt{1-M^2})^2}, \tag{4.5}$$

where $M$ is the local Mach number $q/c$. This value is much smaller than unity, except for values of $M$ close to unity.[7] Hence, (4.4) is approximately the identity transformation if $k$ is very close to unity. This approximately preserves the shape of the profile during the transformation.

(b) *Conformal mapping of compressible flows.* If we make a conformal transformation of the $\zeta$-plane into the $\tilde{\zeta}$-plane by the analytic relation

$$\zeta = h(\tilde{\zeta}), \tag{4.6}$$

we are merely making a change of the auxiliary variable in (2.8)–(2.10). Indeed, we have

$$x + iy = \int \tilde{k} d\tilde{\zeta} - \frac{1}{4}\overline{\int \frac{\tilde{w}_0^2(\tilde{\zeta})}{\tilde{k}(\tilde{\zeta})} d\tilde{\zeta}}, \tag{4.7}$$

$$\frac{2q}{1+\sqrt{1+q^2}} e^{-i\theta} = \frac{\tilde{w}_0(\zeta)}{\tilde{k}(\tilde{\zeta})}, \tag{4.8}$$

$$\phi + i\psi = \tilde{F}(\tilde{\zeta}); \tag{4.9}$$

where

$$\tilde{F}(\tilde{\zeta}) = F(\zeta), \tag{4.10}$$

$$\tilde{w}_0(\tilde{\zeta}) = \tilde{F}'(\tilde{\zeta}) = F'(\zeta)\frac{d\zeta}{d\tilde{\zeta}}, \tag{4.11}$$

[7] Cf., Fig. 3 of Tsien's paper quoted in Footnote 3.

and

$$\tilde{k}(\tilde{\zeta}) = k(\zeta)\frac{d\zeta}{d\tilde{\zeta}}\cdot \tag{4.12}$$

We note that the equations (4.7)–(4.9) are of the same nature as (2.8)–(2.10). However, the profile $\tilde{P}_0$, into which the profiles $P$ and $P_0$ are mapped, may bear no resemblance at all to the original profiles. Indeed, there is no loss in generality in taking $P_0$ to be a circle. *The relations* (4.7)–(4.9) *thus serve to transform the incompressible flow past a circle into a compressible flow past a profile of a quite arbitrary shape.* Referring to (4.6) and (4.12) and to the fact that $k(\zeta)$ should be taken not far from unity, we see that $\tilde{k}(\tilde{\zeta})$ should be chosen so that it is not very much different from the derivative of the function mapping $P$ into a circle.

(c) *Formulation of the direct problem.* If we disregard the intermediate step of the $\zeta$-plane and drop the tilde, we have a mapping of the nature described in Section 2, but with $k(\zeta)$ so chosen that a circle will be mapped into some profile $P$. The function $k(\zeta)$ must satisfy the requirements established there, but it should not be very much different from the function $k_0(\zeta)$ which maps the circle into $P_0$ by the relation

$$z = \int k_0(\zeta)d\zeta.$$

Thus for each profile $P$, the determination of the compressible flow past it is equivalent to the determination of a proper $k(\zeta)$ mapping it into a circle by (2.8), where $w_0(\zeta)$ is the flow past the circle. There is no question about the existence of such a mapping function. It is clear that to each purely subsonic flow, where the approximation (2.4) is accurate enough and therefore $F = F(w^*)$, we may find a certain $k_1(\zeta)$ mapping the compressible flow into *some* incompressible flow. By considering successive conformal transformations, we can therefore *always* map the flow into a circle. The actual method of finding $k(\zeta)$ is all that remains to be done in the direct problem.

The theory of subsonic compressible flows (so long as the approximation (2.4) is valid) is now put on an equal footing with the incompressible flows. The inverse problem is complete, the direct problem of finding a mapping function for a given profile can only be solved (practically) by successive approximations, *even in the incompressible case.*[8] To develop a method of successive approximations for the direct problem in the compressible case seems to be a natural next step.*

**5. Application of von Mises' method of generating airfoils.** In the incompressible case, von Mises transforms a circle in the $\zeta$-plane into an airfoil of a quite general shape by the transformation

$$\frac{dz}{d\zeta} = \left(1 - \frac{\lambda_0}{\zeta}\right)\left(1 - \frac{\lambda_1}{\zeta}\right)\cdots\left(1 - \frac{\lambda_n}{\zeta}\right) \tag{5.1}$$

[8] Cf., e.g. the papers quoted in Footnote 5.

* *Note added in proof*: The essential difference of the two cases lies in the existence problem. While the existence of the incompressible flow follows from well-known results concerning the Laplace equation, very little seems to be known about the existence problem for compressible flows.

where $\lambda_1, \cdots, \lambda_r$ are points *inside* the circle, and $\lambda_0$ is a point on the boundary (which transforms into the trailing edge of the airfoil). A similar method can be used here. In (2.8)–(2.10), we put

$$k(\zeta) = \left(1 - \frac{\lambda_0}{\zeta}\right)\left(1 - \frac{\lambda_1}{\zeta}\right) \cdots \left(1 - \frac{\lambda_n}{\zeta}\right) \tag{5.2}$$

with the same general restrictions on the points $\lambda$. The condition that the points $\lambda$ are inside the circle is exactly the condition required of $k(\zeta)$ as stated in Section 2.

The condition (2.7) for the closure of the body has also its equivalence in the case considered by von Mises. There it is[9]

$$\oint k(\zeta) d\zeta = 0.$$

Here, it differs by the presence of another term. The condition that

$$| k(\zeta) | > | \tfrac{1}{2} w_0(\zeta) |$$

on the circle is the only additional restriction in the present case. As it is a mere inequality, there is no great difficulty in ensuring it to be satisfied.

The integration required in establishing the relation between $z$ and $\zeta$ can be readily performed, as it involves only rational functions. This ease of calculation, together with the flexibility of the choice of the points $\lambda$ in controlling the shape of the airfoil, are the advantages of the method of von Mises which are still preserved in the present application.

The author is greatly indebted to Dr. J. B. Diaz for very helpful discussions in the course of the investigation and to Professors W. Prager and K. O. Friedrichs for their interest and discussions.*

---

[9] Cf. W. F. Durand, *Aerodynamic theory*, vol. 2, J. Springer, Berlin, 1933, p. 78, Eq. (20.4).

* ***Note added in proof***: After the paper was completed and presented in a colloquium at Brown University, Professor E. Reissner informed the author that Professor A. Gelbart had recently presented a somewhat similar approach in a lecture at M.I.T.

Reprinted from QUARTERLY OF APPLIED MATHEMATICS
Vol. VI, No. 1, April, 1948 

# ON THE NATURE OF THE BOUNDARY LAYER NEAR THE LEADING EDGE OF A FLAT PLATE*

BY

G. F. CARRIER (*Brown University*) AND C. C. LIN (*Massachusetts Institute of Technology*)

**1. Introduction.** The customary treatment of boundary layer flow[1] furnishes what is essentially an asymptotic expansion of the stream function (or an equivalent function defining the velocities). This solution is valid, of course, only in a region far downstream of the leading edge of the obstacle. Since one knows, however, that to all practical purposes a potential flow exists sufficiently far from the obstacle, there remains only a closed region in the neighborhood of the origin in which the nature of the flow is not known. It is the purpose of the present paper to show the nature of the flow of an incompressible fluid in this region when the obstacle is a flat plate.

**2. General development.** The equations governing an incompressible, viscous, fluid flow are, in conventional notation[2],

$$u_1 \frac{\partial u_1}{\partial x_1} + v_1 \frac{\partial u_1}{\partial y_1} + \left(\frac{1}{\rho}\right) \frac{\partial p_1}{\partial x_1} = \nu \Delta u_1 , \tag{1}$$

$$u_1 \frac{\partial v_1}{\partial x_1} + v_1 \frac{\partial v_1}{\partial y_1} + \left(\frac{1}{\rho}\right) \frac{\partial p_1}{\partial y_1} = \nu \Delta v_1 , \tag{2}$$

$$\frac{\partial u_1}{\partial x_1} + \frac{\partial v_1}{\partial y_1} = 0. \tag{3}$$

These may be reduced to a single equation in one unknown function in any of several ways. For convenience, we first introduce the following changes of variable: $p = p_1/\rho u_0^2$, $x = x_1 u_0/\nu$, $y = y_1 u_0/\nu$, $u = u_1/u_0 = \partial\psi/\partial y$, $v = v_1/u_0 = -\partial\psi/\partial x$, $x + iy = z = re^{i\theta}$. Here, $u_0$ is the free stream velocity. With the introduction of these variables, Eqs. (1), (2), (3), are readily reduced to the form

$$\partial^4\psi/\partial z^2\,\partial \bar{z}^2 = \frac{1}{2i}\left[\partial\psi/\partial\bar{z}\;\partial^3\psi/\partial z^2\,\partial\bar{z} - \partial\psi/\partial z\;\partial^3\psi/\partial z\,\partial\bar{z}^2\right]. \tag{4}$$

Here $\bar{z}$ is the complex conjugate of $z$ and, we might note now, $u + iv = -2i\partial\psi/\partial\bar{z}$.

If we restrict our analysis to the flow past a flat plate such that as $x \to -\infty$, $\psi \to y$, we must impose the boundary conditions $\partial\psi/\partial\bar{z} \equiv 0$ when $\theta = 0$ and when $\theta = 2\pi$. Furthermore, we must anticipate that the line $\theta = 0$ will be a branch line of our solution since the higher derivatives of $\psi$ will be discontinuous along this line. We may construct such a function which is a solution of Eq. (4) in the following manner. We note that

$$\begin{aligned}\psi_0 &= A[\bar{z}\,z^{1/2} + z\,\bar{z}^{1/2} - z^{3/2} - \bar{z}^{3/2}] \\ &= Ar^{3/2}\left(\cos\frac{\theta}{2} - \cos\frac{3\theta}{2}\right)\end{aligned} \tag{5}$$

*Received Nov. 17, 1947.

[1]See, for instance, L. Prandtl, *The mechanics of viscous fluids*, in vol. 3 of W. F. Durand's *Aerodynamic theory*, p. 34, Springer, Berlin, 1934.

[2]The subscript unity is adopted so that we may use the variable without subscripts for the quantities to be introduced later.

is a biharmonic function for which $\theta = 0$ is a branch line, along which the velocity $\partial\psi/\partial\bar{z}$ vanishes. If we write Eq. (4) in the form

$$L(\psi) = L^*(\psi), \tag{6}$$

where $L$ is the biharmonic operator and $L^*(\psi)$ is the rest of Eq. (4), then we may define

$$\begin{aligned} &L(\psi_0) = 0, \\ &L(\psi_1) = L^*(\psi_0), \\ &L(\psi_n) = L^*(\psi_0 + \cdots + \psi_{n-1}) - L^*(\psi_0 + \cdots + \psi_{n-2}), \\ &\cdots . \end{aligned} \tag{7}$$

Over any region in which the series

$$\psi = \psi_0 + \psi_1 + \cdots \tag{8}$$

converges then, we may take $\psi$ (as defined above) as a solution to Eq. (4).[3] In this case, a particular integral for $\psi_1$ is

$$\psi_1^{(1)} = A^2 r^3[(\sin 2\theta)/40 - (\theta \cos \theta/64], \tag{9}$$

and a biharmonic function which when added to $\psi_1^{(1)}$ adjusts the conditions at the plate, is

$$\begin{aligned} \psi_1^{(2)} = \frac{A^2}{128} r^3[\ln r(3 \sin \theta - \sin 3\theta) + \theta(3 \cos \theta - \cos 3\theta) \\ - \frac{A^2 r^3}{20} (\alpha \sin \theta + \beta \sin 3\theta)], \end{aligned} \tag{10}$$

where $\alpha + 3\beta = 1$. As we shall see later, $\psi_1^{(2)} \ll \psi_0$ in the matching region so that the choice $\alpha = 1$ serves our purpose. The function $\psi_1$ becomes then

$$\begin{aligned} \psi_1 = \psi_1^{(1)} + \psi_1^{(2)} = [A^2 r^3/128][\theta(\cos \theta - \cos 3\theta) + \ln r(3 \sin \theta - \sin 3\theta)] \\ + [A^2 r^3/40][\sin 2\theta - 2 \sin \theta]. \end{aligned} \tag{11}$$

It is fairly evident that the higher approximations will contain terms in $r^{9/2}$, $r^{12}$, $r^{9/2} \ln r$, $\cdots$, and will also involve coefficients with higher powers of $A$. As we (again) shall see later, $A = .083$. Thus, it is evident that within some circle this development is convergent and represents *some* flow. It is not immediately evident that the expansion for $\psi$ converges for $y \ll 1$, $x > 1$. However, since our solution is of the Stokes type, it is to be anticipated that the leading term of the expansion describes the flow in a sufficiently thin neighborhood of the plate. Since we do not need to establish such a region of convergence, however, we shall not pursue the question further.[4]

**3. The matching procedure.** The solution obtained by Blasius for the velocity profile in the downstream region is usually presented in the form

[3]This implies that we base the expansion on the Stokes type slow-flow solutions.

[4]We note, however, that the choice of $\alpha$ for Eq. (10) and the corresponding "arbitrary" coefficients in $\psi_n$, where $n > 1$, would be governed largely by this convergence consideration.

$$u = f'(\eta), \qquad v = \frac{1}{2(x)^{1/2}} [\eta f'(\eta) - f(\eta)], \tag{12}$$

where, in our notation, $\eta = y/x^{1/2}$. In Eqs. (12), $f(\eta)$ can be given by a power series of the form[1]

$$f = \alpha\eta^2/2! - \alpha^2\eta^5/2\cdot 5! + \cdots, \tag{13}$$

where $\alpha = .332$.

Thus, $u = \alpha\eta - \alpha^2\eta^4/2\cdot 4! = \alpha r^{1/2} \sin\theta/(\cos\theta)^{1/2}$

$$-\alpha^2 r^2 \sin^4\theta/48\cos^2\theta + \cdots. \tag{14}$$

That is

$$u \simeq \alpha r^{1/2}\theta - \alpha^2 r^2\theta^4/48 + \cdots \tag{15}$$

in the neighborhood of $\theta = 0$.

The solution of the preceding section implies that the velocity component $u$ (i.e. Re $(-2i\partial\psi/\partial\bar{z})$ be given by

$$u = Ar^{1/2}(5\sin\theta/2 + \sin 3\theta/2) + \cdots$$

or

$$u \simeq 4Ar^{1/2}\theta + \gamma r^2\theta^4 + \delta r^2 \ln r\theta^4 + \cdots. \tag{16}$$

Two facts are evident. If these functions are to represent the same phenomenon in any overlapping regions where $r$ is not large, then $4A \equiv \alpha$. Furthermore, we note that these two functions can never be identically the same since the $r^2 \ln r$ term is lost in the asymptotic representation. The latter fact is quite in order since *neither* solution has the pretense of being a complete solution. Therefore, we may establish validity of the leading term of this representation (i.e. $\psi_0$) if we can find a region in which both $\psi_0$ and $f$ are valid approximations to the rigorous solution. That is, we must find a region in which $L(\psi) - L^*(\psi) \ll \partial^4\psi/\partial r^4$ (say). Our method of solution and its results indicate that $r \simeq 1$, $\theta \ll 1$ is such a region. But for $\theta$ sufficiently small and $r$ of order unity, $x^{1/2}f(\eta)$ is almost precisely the same function as $\psi_0$ (see Eq. (15)) and hence is as valid a solution. This, of course, can be worked out by computing $L[x^{1/2}f(\eta)] - L^*[x^{1/2}f(\eta)]$, but this is really not necessary. Hence, we have that in the neighborhood of $r = 0$, Eq. (5) demonstrates the nature of the flow field provided $A = .083$. Figure 1 shows the four regions into which the flow field may be divided and the solution applicable to each. In region IV, it is doubtful if any reasonably simple analytic expression could be found for $\psi$. However, it is suggested that the flow field could be computed if one combined some intelligent "first guessing" (interpolation essentially) with the relaxation procedure and applied these to the region IV.

**4. A modified Blasius solution.** The alternative procedure[5] for developing the foregoing solution leads to a slightly modified boundary layer solution which establishes more clearly the overlap of the slow-flow solution (which includes the leading edge) and the boundary layer solution.

---

[5]Actually, the two authors arrived at the solution independently by the two procedures described here, but felt that a joint presentation provided the most lucid account of the matter.

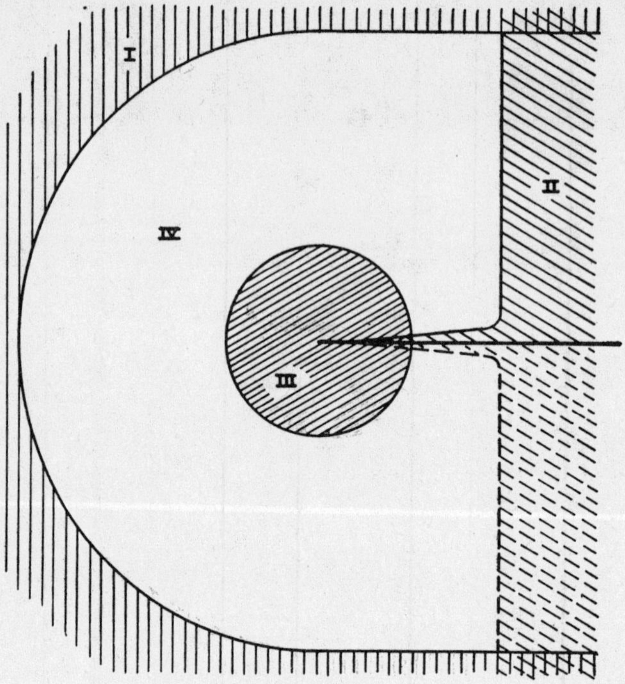


FIG. 1.

(I) Potential Solution (II) Blasius Solution (III) Slow Flow Solution

It seems that the most natural coordinate system to use for the flow past a flat plate is the parabolic coordinate system $(\xi, \eta)$ defined by

$$x + iy = (\xi + i\eta)^2 \tag{17}$$

or

$$\xi + i\eta = r^{1/2} \exp (i\theta/2)$$

By using this system of coordinates, it is possible to give a unified discussion of the boundary layer solution and the slow motion solution in the neighborhood of the plate, including the leading edge.

In the parabolic system of coordinates, the equation for the stream function $\psi$ becomes

$$L(\psi) = (\xi^2 + \eta^2)(\psi_\xi \Delta\psi_\eta - \psi_\eta \Delta\psi_\xi) - 2(\eta\psi_\xi - \xi\psi_\eta)\Delta\psi, \tag{18}$$

where

$$L(\psi) \equiv (\xi^2 + \eta^2)\Delta\Delta\psi - 4(\xi\Delta\psi_\xi + \eta\Delta\psi_\eta - \Delta\psi). \tag{19}$$

From a consideration of the diffusive effect of the viscous forces, it is obvious that the effect of viscosity is essentially limited to regions with $\eta \sim 1$, close to the solid boundary. For very small values of $\eta$, the motion is slow. Since $\psi = 0$ and $\partial\psi/\partial\eta = 0$ at $\eta = 0$ it is obvious that in this region a good approximation to (18) is

$$L(\psi) = 0. \tag{20}$$

A polynomial solution satisfying the boundary conditions $\psi = \psi_\eta = 0$ at $\eta = 0$ is

$$\psi = A\xi\eta^2 \tag{21}$$

which agrees with Eq. (5) of Sec. 2. This solution can also be obtained from Eq. (19) by writing $\psi = \eta^2(f(\xi) + 0(\eta))$ for small $\eta$.

If $\xi$ is large while $\eta \sim 1$, one may introduce a new variable

$$\xi' = \epsilon\xi \tag{22}$$

where $\epsilon$ is a small parameter.[6] In the limit $\epsilon \to 0$, the equation (3) becomes

$$\xi'^2 \frac{\partial^4\psi^{(0)}}{\partial\eta^4} = \xi'^2(\psi^{(0)}_{\xi'}\psi^{(0)}_{\eta^3} - \psi^{(0)}_{\eta}\psi^{(0)}_{\xi'\eta^2}) + 2\xi'\psi^{(0)}_{\eta}\psi^{(0)}_{\eta^2} \tag{23}$$

where

$$\psi^{(0)} = \epsilon\psi \tag{24}$$

This equation may be solved by putting

$$\psi^{(0)} = \xi' f(\eta); \tag{25}$$

then $f(\eta)$ satisfies

$$f^{iv} = ff''' + f'f'', \tag{26}$$

or

$$f''' = ff''; \tag{27}$$

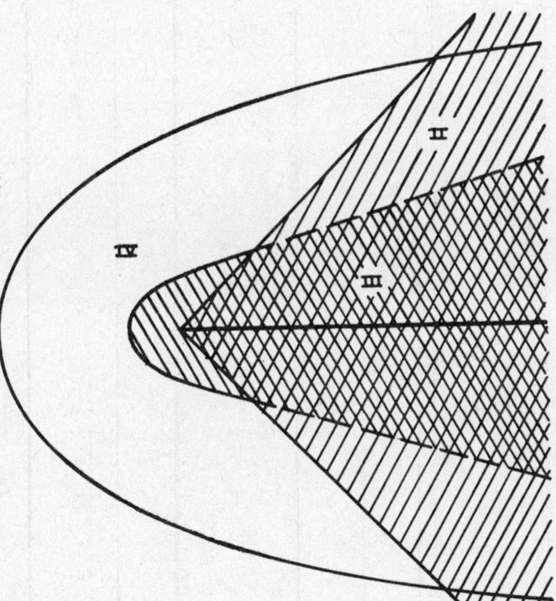


FIG. 2.

(I) Potential Solution (II) Modified Blasius Solution (III) Slow Flow Solution

[6]To give $\epsilon$ an explicit meaning, we consider the region $\eta \leq \epsilon\xi$.

this is the usual boundary layer equation except that $\eta$ is not quite the conventional parameter.[7] Higher approximations may be obtained by solving (18) in powers of $\epsilon$.

Thus, the regions of approximate validity of these solutions are as follows:

(*i*) potential solution: $\eta \gg 1$;
(*ii*) boundary layer equation: $\eta \sim 1$, $\eta \ll \xi$;
(*iii*) Stokes equation: $\eta \ll 1$, irrespective of $\xi$.

These regions are indicated in Fig. 2. It is obvious that there is an overlap of regions (*ii*) and (*iii*) where the solution (5) agrees with the boundary layer solution. Again, we have not established the convergence of the slow flow solution. However, the physical considerations which lead to the slow flow and the boundary layer solutions indicate that they are at least valid asymptotic approximations, if not convergent processes. This is further strengthened by the identical nature of the leading terms of the two types of solutions for $\eta \ll 1$.

[7]Here, $\eta = r^{1/2} \sin \theta/2$, conventionally $\eta = r^{1/2} \sin \theta/(\cos \theta)^{1/2}$.

Reprinted from JOURNAL OF MATHEMATICS AND PHYSICS
Vol. 27, No. 2, July, 1948

# ON THE FLOW BEHIND CURVED SHOCKS[1]

C. C. LIN AND S. I. RUBINOV

**1. Introduction.** Although flows involving curved and detached shocks occur very frequently in practice, and such phenomena are very important in drag studies, our knowledge is very limited. Experimentally, they have been observed for a long time, but the quantitative determination of the pressure distribution behind a curved shock has been carried out only recently by Professor R. Ladenburg [1] and his group at Princeton University by means of interferometric methods. Some limited theoretical work has been carried out by Crocco [2], Hentsche [3], and Thomas [4], some extensive numerical calculations[2] have been made by Maccoll [5], and finally there is the semi-empirical work of Laitone and Pardee [6]. However, a systematic analytical approach, with a view of getting some understanding of the general aspects of the problem, seems to be still lacking. In the present paper, we try to investigate the problem of the flow behind curved shocks from the general point of view. Yet concrete results are obtained which may be directly compared with experiments, and which are also useful for further theoretical investigations.

The investigation contained in this paper may be roughly divided into three parts. Part I (§§2–3) deals with general relations for flow behind curved shocks. These include a variational principle for iso-energetic flows (with vorticity and variable entropy), and a discussion of the differential relations of the first order across a shock. Part II (§§4–5) deals with shocks attached to the solid boundary, applying relations developed in Part I. In particular, relations developed by Crocco and others are reproduced as a specific case. The most interesting case, however, concerns the formation of a normal shock over the solid boundary.[3] It is shown here that a normal shock can be formed over a convex surface with continuous curvature only when the Mach number exceeds a certain supersonic value. For a concave surface, the Mach number must be below that value. The physical interpretation of this peculiar behaviour is also given. Part III deals with the flow behind a detached shock. A scheme is developed by which the calculation of the subsonic flow between the shock and the body may be carried out. Some preliminary results thus obtained agree very well with available experimental data. This work will be reported elsewhere by Mr. John Dugundji. In conclusion, it is pointed out that much further extension of this work should be carried out. These are now being investigated by some with one of the authors.

[1] Part of this work was carried out as a thesis of Rubinov for the master's degree (1947) when both authors were at Brown University. Further work done by Lin was partly sponsored by the Mechanics Division of the Naval Ordnance Laboratory.

[2] Prof. Emmons and Mr. Drebinger of Harvard University informed us that they have further extended Maccoll's work. The results are as yet unpublished.

[3] Cf. Ref. 7, 8.

**2. Fundamental equations for steady iso-energetic flows.** For convenience of reference, we collect below the fundamental equations for steady iso-energetic flows. For flow in two-dimensions or with axial symmetry, a stream function may be defined, and a variational principle will be developed in terms of this stream function.

For steady three-dimensional motions, we have six hydrodynamical variables; the pressure $p$, the density $\rho$, the entropy $S$ and the three components of velocity $u_i$ $(i = 1, 2, 3)$. All these are to be determined as functions of the three spacial coordinates $x_i$ $(i = 1, 2, 3)$ by the following six equations:

(i) the equations of motion (3),

$$u_j(\partial u_i/\partial x_j) = -\rho^{-1}(\partial p/\partial x_i), \tag{2.1}$$

(ii) the equation of continuity (1),

$$u_j(\partial \rho/\partial x_j) = -\rho(\partial u_j/\partial x_j) \tag{2.2}$$

(iii) the equation of state (1),

$$p = e^{S/c_v}\rho^{\gamma}, \tag{2.3}$$

where $c_v$ is the specific heat at constant volume and $\gamma$ is the ratio of specific heats; and finally

(iv) the equation of energy (1), which may be expressed in the form

$$u_j(\partial S/\partial x_j) = 0. \tag{2.4}$$

In addition, we often have the condition of iso-energetic flow

$$\gamma p/(\gamma - 1)\rho + q^2/2 = C, \tag{2.5}$$

where $q$ is the magnitude of the velocity. The fact that this is a constant along each stream line may be derived from Eqs. (2.1) – (2.4). For flows derived from isentropic irrotational flows by a curved shock, $C$ is actually a constant throughout the whole field. This is the case of isoenergetic flow under consideration. In general, (2.5) implies (2.4) when (2.1) and (2.2) hold. We note that (2.1) (2.2) and (2.5) form a system of five equations for $p$, $\rho$ and $u_i$ without reference to the entropy. One may therefore disregard the entropy at first and then supplement the results with the statement that the entropy, as given by (2.3), is a constant along each stream line, in accordance with (2.4).

For two-dimensional and axially symmetric flows, we have only five variables. We take the usual $\epsilon$-convention and let $(x, y)$ denote the distances along and perpendicular to the axis and $u$, $v$ the velocity components in these directions. The four of the fundamental equations not involving entropy are:

$$u(\partial u/\partial x) + v(\partial u/\partial y) = (-1/\rho)(\partial p/\partial x) \tag{2.6}$$

$$u(\partial v/\partial x) + v(\partial v/\partial y) = (-1/\rho)(\partial p/\partial y) \tag{2.7}$$

$$\partial(\rho u)/\partial x + \partial(\rho v)/\partial y = -\epsilon\rho v/y, \tag{2.8}$$

where

$\epsilon = 0$ for two-dimensional motion, and

$\epsilon = 1$ for axial symmetry.

(2.9) $$\gamma p/(\gamma - 1)\rho + q^2/2 = C.$$

The two equations involving entropy are the equation of state

(2.10) $$p = e^{S/c_v}\rho^\gamma$$

and an integrated form of the equation of energy,

(2.11) $$S = c_v f(\psi)$$

where $\psi$ is the stream function defined by using (2.8):

(2.12) $$\rho u y^\epsilon = \partial\psi/\partial y, \qquad \rho v y^\epsilon = -\,\partial\psi/\partial x$$

It is known (See, e.g., Vazsonyi [9], pp. 36, 37) that the vorticity is

(2.13) $$\omega = (p/R)y^\epsilon(dS/d\psi).$$

In intrinsic form with $\theta$ denoting the direction of the stream line, and $s$, $n$ denoting the arc lengths along and perpendicular to the stream line, the equations are

(2.14) $$q\partial q/\partial s = (-\,1/\rho)(\partial p/\partial s),$$

(2.15) $$\rho q^2(\partial\theta/\partial n) = -\,\partial p/\partial n,$$

(2.16) $$\frac{1 - M^2}{q}\frac{\partial q}{\partial s} + \frac{\partial\theta}{\partial n} = -\epsilon\frac{\sin\theta}{y},$$

(2.17) $$\omega = \frac{\partial q}{\partial n} - q\frac{\partial\theta}{\partial s} = \frac{p}{\rho q R}\frac{\partial S}{\partial n},$$

(2.18) $$\partial\psi/\partial n = \rho q y^\epsilon.$$

In addition, equations (2.10) and (2.11) hold.

For irrotational flows,

(2.19) $$\omega = 0,$$

and hence $S$ is constant everywhere. Equation (2.10) may be written as

(2.20) $$p = K\rho^\gamma$$

where $K$ is a constant.

*Variational principle.* While variational principles [10] for irrotational isentropic flows have been well-known, it does not seem to have been pointed out that iso-energetic flows are also subject to a similar variational relation. It can be easily verified that

(2.21) $$\delta I = \delta\iint\left\{\left(C\rho - \frac{e^{f(\psi)}}{\gamma - 1}\rho^\gamma\right)y^\epsilon + \frac{1}{2\rho y^\epsilon}\,(\psi_x^2 + \psi_y^2)\right\}dx\,dy = 0,$$

with a preassigned function $f$ and a fixed constant, $C$ leads to the desired relations, provided $\rho(x, y)$ and $\psi(x, y)$ are allowed to vary independently, and proper boundary conditions are satisfied. As usual, the condition $\delta I = 0$ leads to

$$\iint \left\{ \left( C - \frac{\gamma e^{f(\psi)}}{\gamma - 1} \rho^{\gamma-1} \right) y^{\epsilon} - \frac{1}{2\rho^2 y^{\epsilon}} (\psi_x^2 + \psi_y^2) \right\} \delta\rho \, dx \, dy$$

$$- \iint \left\{ \frac{\rho^{\gamma}}{\gamma - 1} e^{f(\psi)} f'(\psi) y^{\epsilon} + \frac{\partial}{\partial x}\left( \frac{1}{\rho y^{\epsilon}} \frac{\partial \psi}{\partial x} \right) + \frac{\partial}{\partial y}\left( \frac{1}{\rho y^{\epsilon}} \frac{\partial \psi}{\partial y} \right) \right\} \delta\psi \, dx \, dy$$

$$+ \int \frac{1}{y^{\epsilon}} \frac{1}{\rho} \frac{\partial \psi}{\partial n} \delta\psi \, dS = 0$$

Hence, if the boundary is a stream line with $\psi$ preassigned, $\delta I = 0$ leads to

$$\frac{1}{y^{2\epsilon}} \frac{1}{2\rho^2} (\psi_x^2 + \psi_y^2) + \frac{\gamma e^{f(\psi)} \rho^{\gamma-1}}{\gamma - 1} = C,$$

and

$$\frac{\partial}{\partial x}\left( \frac{1}{\rho y^{\epsilon}} \frac{\partial \psi}{\partial x} \right) + \frac{\partial}{\partial y}\left( \frac{1}{\rho y^{\epsilon}} \frac{\partial \psi}{\partial y} \right) = - \frac{\rho^{\gamma} e^{f(\psi)}}{\gamma - 1} f'(\psi) y^{\epsilon}.$$

With $p$, $u$, $v$ defined by (2.10) and (2.12), the first is the Bernoulli's equation (2.9), while the second equation is the vorticity equation (2.17). Thus, the fundamental equations are satisfied.

It may be noticed that the integrand in (2.21) (with $\epsilon = 0$) is again equal to the quantity $p + pq^2$ exactly as in the isentropic case.[4] Moreover, as formulated above, the integral can only be extended over a finite region. For an infinite region, a constant or a function independent of $\rho(x, y)$ and $\psi(x, y)$ has to be subtracted from the integrand to make the integral convergent.

**3. Conditions across a shock.** It is known that the pressure, density, and velocity behind a shock are obtainable from the corresponding quantities before the shock as long as the inclination of the shock to the incoming stream line is known. It will now be shown that if the curvature of the shock is known, the first spacial derivatives of the hydrodynamical quantities are also known. In general, if the flow in front of the shock is completely known, the spacial derivatives behind the shock are known to an order equal to that known for the derivatives of the angle of the shock.

One may regard this process as setting up the initial value problem after the shock. If the shock is given, the hydrodynamic quantities $u, v, p, \rho, s$ are known behind it. The fundamental system of equations (2.6) — (2.11) can then be solved in terms of these initial conditions.

The calculations below will largely be carried out for the two-dimensional case; the results will be applied to discuss several problems in §4. The extension to the case with axial symmetry is almost immediate. Along the axis of symmetry,

[4] See Bateman, loc. cit., Eqs. (1) and (16).

however, the situation is more complicated. Thus, the extension of (4.1) and (4.2) to the conditions near a conical tip requires further investigation.

The pressure $p$, the density $\rho$, and the magnitude $q$ and direction $\theta$ of the velocity in front and behind the shock satisfy the following relations (Fig. 1)

$$\rho_1 q_1 \sin \alpha = \rho_2 q_2 \sin (\alpha - \theta) \tag{3.1}$$

$$p_1 + \rho_1 q_1^2 \sin^2 \alpha = p_2 + \rho_2 q_2^2 \sin^2 (\alpha - \theta) \tag{3.2}$$

$$q_1 \cos \alpha = q_2 \cos (\alpha - \theta) \tag{3.3}$$

$$q_1 q_2 \sin \alpha \sin (\alpha - \theta) = c_*^2 - \frac{\gamma - 1}{\gamma + 1} q_1^2 \cos^2 \alpha \tag{3.4}$$

where

$$\theta = \theta_2 - \theta_1 , \qquad \alpha = \alpha_1 - \theta_1 , \qquad \tfrac{1}{2}[(\gamma + 1)/(\gamma - 1)]c_*^2 = C. \tag{3.5}$$

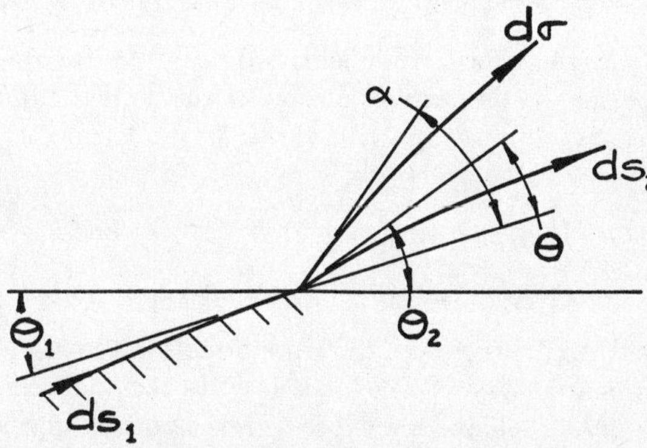


FIG. 1. CONDITIONS AT A CURVED SHOCK

In general, a quantity in front of the shock will carry the subscript 1; behind a shock, the subscript 2. This convention will be used throughout the paper. From (3.1) – (3.5) it can be shown that the jump in entropy is

$$S_2 - S_1 = c_v \log \left( \frac{2\gamma}{\gamma + 1} M_1^2 \sin^2 \alpha - \frac{\gamma - 1}{\gamma + 1} \right) \cdot \left( \frac{\gamma - 1}{\gamma + 1} \sin^2 \alpha + \frac{2}{(\gamma + 1) M_1^2} \right)^{\gamma}. \tag{3.6}$$

There are *ten* spacial derivatives, two for each of the *five* quantities $p$, $\rho$, $q$, $\theta$, and $S$. By (2.10), the calculation of the derivatives of $S$ may be expressed in terms of those for $p$ and $\rho$. More explicitly, the derivatives of $S$ are $\partial S/\partial s = 0$, $\partial S/\partial n = \rho q dS/d\psi$ with $dS/d\psi$ calculated from (3.6). Thus, we need to consider only the eight derivatives of $p$, $\rho$, $q$, $\theta$. Equations(3.1)–(3.4) will furnish the four derivatives along the shock. These furnish four relations for the tangential and normal derivatives defined with respect to a stream line. The other four are supplied by (2.9), and (2.14)–(2.16).

Furthermore, if we eliminate $\partial q/\partial s$ between (2.14) and (2.16) we have the relations

$$
(3.7) \qquad \begin{cases} \rho q^2(\partial\theta/\partial n) = (1 - M^2)(\partial p/\partial s), \\ \rho q^2(\partial\theta/\partial s) = -\ \partial p/\partial n, \end{cases}
$$

which are two relations among the four spatial partial derivatives of $p$ and $\theta$. If we now eliminate $\rho_2$ and $q_2$ from Eqs. (3.1) to (3.4), we have only to deal with the two quantities $p$ and $\theta$. After the derivatives of $p$ and $\theta$ are found, one can easily calculate those for $q$ from (1.14) and (1.17) with $dS/d\psi$ from (3.6) and those for $\rho$ from (2.9). Let us now carry out these calculations.

Elimination of $\rho_2$ and $q_2$ from (3.1) and (3.4) gives

$$
(3.8) \qquad \begin{cases} p_2 = p_1 + \rho_1\left\{q_1^2\left(1 - \dfrac{2}{\gamma + 1}\cos^2\alpha\right) - c_*^2\right\}. \\ p_2 = p_1 + \rho_1 q_1^2\{\sin^2\alpha - \sin\alpha\cos\alpha\tan(\alpha - \theta)\}, \end{cases}
$$

Differentiation along the shock will yield two relations for $\partial p_2/\partial\sigma$ and $\partial\theta_2/\partial\sigma$ in terms of $\partial\alpha_1/\partial\sigma$ and the derivatives along the shock of the quantities in front of it. One can easily express $\partial p_2/\partial\sigma$ and $\partial\theta_2/\partial\sigma$ in terms of $\partial p_2/\partial s$, $\partial p_2/\partial n$, $\partial\theta_2/\partial S$, $\partial\theta_2/\partial n$:

$$
(3.9) \qquad \begin{cases} \partial p_2/\partial\sigma = (\partial p_2/\partial n_2)\sin(\alpha - \theta) + (\partial p_2/\partial s_2)\cos(\alpha - \theta), \\ \partial\theta_2/\partial\sigma = (\partial\theta_2/\partial n_2)\sin(\alpha - \theta) + (\partial\theta_2/\partial s_2)\cos(\alpha - \theta). \end{cases}
$$

Thus, together with (3.7) we have four equations for $\partial p/\partial s_1 \cdots, \partial\theta/\partial n$. Actually we shall eliminate $\partial p/\partial n$ and $\partial\theta/\partial n$ and write the final relations in terms of two for $\partial p/\partial s$ and $\partial\theta/\partial s$. These are expressed in terms of spacial derivatives of the quantities on the left-hand side of the shock. Again, in front of the shock we may eliminate all other derivatives in favor of $\partial p/\partial s$ and $\partial\theta/\partial s$. The calculation is even simpler here because the motion is irrotational. It can be easily verified that

$$
(3.10) \qquad \begin{cases} \partial p_1/\partial\sigma = (\rho_1 q_1^2/r_1)\sin\alpha + (\partial p_1/\partial s_1)\cos\alpha, \\ \partial\rho_1/\partial\sigma = (\rho_1 M_1^2/r_1)\sin\alpha + (M_1^2/q_1^2)(\partial p_1/\partial s_1)\cos\alpha, \\ \partial q_1/\partial\sigma = -\ (q_1/\gamma_1)\sin\alpha - (1/\rho_1 q_1)(\partial p_1/\partial s_1)\cos\alpha, \\ \partial\theta_1/\partial\sigma = -\ (M_1^2 - 1)(\rho_1 q_1^2)^{-1}(\partial p_1/\partial s_1)\sin\alpha - (1/r_1)\cos\alpha. \end{cases}
$$

The final equations obtained are

$$
(3.11) \qquad \begin{aligned} &A_1(\rho_1 q_1^2)^{-1}(-\ \partial p_1/\partial s_1) + B_1(\partial\theta_1/\partial s_1) \\ &\qquad = A_2(\rho_1 q_1^2)^{-1}(-\ \partial p_2/\partial s_2) + B_2(\partial\theta_2/\partial s_2) + C(\partial\alpha_1/\partial\sigma), \end{aligned}
$$

$$
(3.12) \qquad \begin{aligned} &A_1'(\rho_1 q_1^2)^{-1}(-\ \partial p_1/\partial s_1) + B_1'(\partial\theta_1/\partial s_1) \\ &\qquad = A_2'(\rho_1 q_1^2)^{-1}(-\ \partial p_2/\partial_2 s) + B_2'(\partial\theta_2/\partial s_2) + C'(\partial\alpha_1/\partial\sigma), \end{aligned}
$$

where the coefficients are functions of $M_1$, $\alpha$, and $\theta$ given in Table I.

TABLE 1

| | $\alpha$, $\theta$ general | $\alpha = \pi/2, \theta = 0$ |
|---|---|---|
| $A_1$ | $\dfrac{2}{\gamma+1}\cos\alpha\left[(3M_1^2-4)\sin^2\alpha+\dfrac{1-\gamma}{2}\right]$ | 0 |
| $B_1$ | $\dfrac{2}{\gamma+1}\sin\alpha\left[(M_1^2-4)\sin^2\alpha+\dfrac{5-\gamma}{2}\right]$ | $\dfrac{2}{\gamma+1}\left[M_1^2-\dfrac{\gamma+3}{2}\right]$ |
| $A_2$ | $\cos(\alpha-\theta)$ | 0 |
| $B_2$ | $\sin 2\alpha/2\cos(\alpha-\theta)$ | $q_2/q_1$ |
| $C$ | $\dfrac{2}{\gamma+1}\sin 2\alpha$ | 0 |
| $A_1'$ | $M_1^2\cos^2\alpha\cos\theta-(M_1^2-1)\cos(2\alpha+\theta)$ | $M_1^2-1$ |
| $B_1'$ | $M_1^2\sin^2\alpha\sin\theta+\sin(2\alpha+\theta)$ | 0 |
| $A_2'$ | $1+(M_2^2-2)\sin 2(\alpha-\theta)$ | $M_2^2-1$ |
| $B_2'$ | $\sin 2\alpha$ | 0 |
| $C'$ | $\sin 2\theta/2\cos(\alpha-\theta)$ | $(q_2/q_1)-1$ |

The predominance of $p$ and $\theta$ may be regarded physically as follows. The isentropic flow in front can be described by one scalar ($p$) and one direction ($\theta$). The shock relations allow all the quantities to be expressed in terms of them, and the shock angle $\alpha_1$. Thus,

$$p_2 = p_2(p_1, \theta_1, \alpha_1), \qquad \theta_2 = \theta_2(p_1, \theta_1, \alpha_1)$$

The pressure gradient along the shock thus introduces the spatial derivatives which may be handled by (3.7) and (3.9). Notice that (3.7) follows directly from the intrinsic forms (2.14)–(2.16) of the equations of motion and continuity and holds either behind or in front of the shock.

**4. Shocks attached to the solid boundary.** In this section, we shall apply our results to a few special cases of interest. The first is the case of a uniform supersonic flow past a body with a high enough Mach number. The shock is then attached to the nose of the body. This case has been considered by Crocco [2] and Thomas [4]. Next we shall consider the shock attached to a body with continuous directions but possibly with discontinuities in curvature. In this case $\theta$ approaches zero. The weak family of the shock is thus expected to become a Mach line (characteristic), while the strong family is expected to become a normal shock. The latter case has been considered by Hantsche [3], Emmons [7] and Tsien [8]. The results obtained here are somewhat different from theirs. Here it is concluded that a shock can exist in the case of continuous convex curvature only for Mach number above a certain critical value. This might be understood to mean that the interaction between shock and boundary layer would be different for Mach numbers above or below this critical value. Detailed discussions, including physical interpretations, will be given below.

(a) Shock attached to the sharp nose of a body in uniform supersonic stream.

The simplest problem to which the above considerations may be applied is the shock formed by a uniform supersonic flow past a symmetrical body at a sufficiently high Mach number. In this case, all the quantities bearing the subscript 1 are constants, and the equations (3.11) and (3.12) become

$$\begin{aligned}&\frac{1}{r_2}\sin 2\alpha\left\{\frac{2}{\gamma+1}\sin 2\alpha - \frac{\sin 2\theta}{4\cos^2(\alpha-\theta)}\right\}\\ &\qquad = -\frac{1}{\rho_1 q_1^2}\frac{\partial p_2}{\partial s_2}\left\{\frac{2}{\gamma+1}\sin 2\alpha[1+\sin^2(\alpha-\theta)(M_2^2-2)] - \frac{\sin 2\theta}{2}\right\},\end{aligned} \tag{4.1}$$

$$\begin{aligned}&\frac{1}{r_2}\sin 2\alpha\left\{\cos(\alpha-\theta) - \frac{1+(M_2^2-2)\sin^2(\alpha-\theta)}{2\cos(\alpha-\theta)}\right\}\\ &\qquad = -\frac{\partial\alpha_1}{\partial\sigma}\left\{\frac{2}{\gamma+1}\sin 2\alpha[1+(M_2^2-2)\sin^2(\alpha-\theta)] - \frac{\sin 2\theta}{2}\right\}.\end{aligned} \tag{4.2}$$

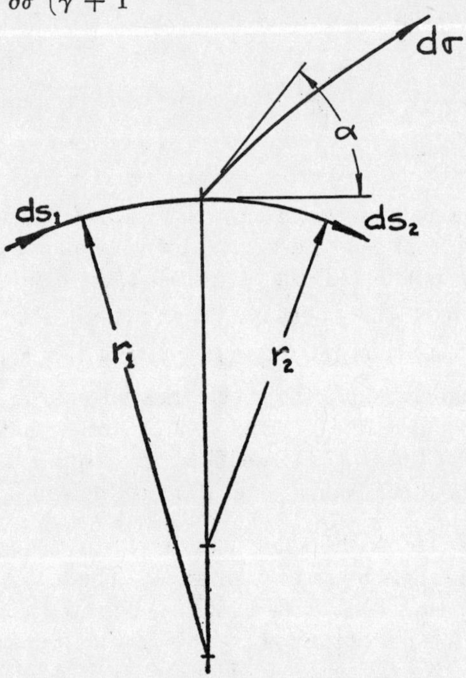


FIG. 2. CONDITIONS AT A CURVED MACH LINE

These equations were obtained by Crocco and Thomas. It is interesting to note that the curvature of the shock and the pressure gradient behind it are both proportional to the curvature of the body, with coefficients of proportionality depending on the free stream Mach number and the angle of the body.

(b) Shocks attached to boundary with continuous slope—weak family.

In the limit $\theta \to 0$, the discontinuity becomes one of higher order. The first three shock conditions (3.1)–(3.4) are satisfied by

$$p_1 = p_2, \qquad \rho_1 = \rho_2, \qquad q_1 = q_2, \tag{4.3}$$

while the last one leads to the conclusion that $\alpha$ becomes equal to the Mach angle $\sin^{-1}(1/M_1)$. The first order relations (3.11) and (3.12) lead to

$$\frac{\partial p_2}{\partial s_2} - \frac{\partial p_1}{\partial s_1} = \rho_1 q_1^2 \left(\frac{1}{r_1} - \frac{1}{r_2}\right) \tan \alpha \tag{4.4}$$

$$\frac{\partial \alpha_1}{\partial \sigma} = \frac{4 + (\gamma - 3)M_1^2}{2\sqrt{M_1^2 - 1}} \left(\frac{\sin \alpha}{r_1} + \frac{1}{\rho_1 q_1^2} \frac{\partial p_1}{\partial s_1} \cos \alpha\right). \tag{4.5}$$

The curvature of the Mach line is conditioned by two factors, the value of the pressure gradient along the Mach line and the multiplicative coefficient. Note that the latter quantity changes sign when $M_1^2 = 4/(3 - \gamma)$(for air, $\gamma = 1.4$ and $M_1 = \sqrt{2.5}$), and that the curvature may be positive or negative. Note also that the curvature of the Mach line is independent of the curvature $1/r_2$ of the stream line. If $r_1 = r_2$, there would not be any discontinuity in the first derivatives of physical quantities.

The relation (4.4) can be obtained more directly from the condition for characteristics. It is known that the discontinuity in derivatives of velocity must propagate along the characteristic, which is at an angle $\alpha = \sin^{-1}(1/M)$ with the direction of the stream line (Fig. 2). The pressure is continuous across it. Hence, if the jump is denoted by [ ], $[\partial p/\partial \sigma] = 0$. On the other hand,

$$\partial p/\partial \sigma = (\partial p/\partial s) \cos \alpha + (\partial p/\partial n) \sin \alpha,$$

and hence

$$0 = [\partial p/\partial s] \cos \alpha + [\partial p/\partial n] \sin \alpha$$

which then yields (4.4), if $\left[\frac{\partial p}{\partial n}\right]$ is evaluated by (2.15).

(c) Shocks attached to boundary with continuous direction—strong family.

The limit of the strong family is given by $\theta \to 0$, $\alpha \to \pi/2$. In this case, the coefficients $A_1$, $B_1$, $A_2$, $B_2$, C and $A_1'$, $B_1'$, $A_2'$, $B_2'$, $C'$ tend to the values shown in the last column of Table I. If all the other quantities in equations (3.11) and (3.12) remain finite, those equations become

$$\frac{2}{\gamma + 1}\left\{M_1^2 - \frac{\gamma + 3}{2}\right\}\left(-\frac{1}{r_1}\right) - \frac{1}{\gamma + 1}\left\{\frac{2}{M_1^2} + (\gamma - 1)\right\}\left(-\frac{1}{r_2}\right) = 0 \tag{4.6}$$

This is the relation discussed by previous authors.[5] However, if the curvature of the shock is allowed to become infinite, then it turns out that

$$\frac{2}{\gamma + 1}\left\{M_1^2 - \frac{\gamma + 3}{2}\right\}\left(-\frac{1}{r_1}\right) - \frac{1}{\gamma + 1}\left\{\frac{2}{M_1^2} + (\gamma - 1)\right\}\left(-\frac{1}{r_2}\right) \leq 0. \tag{4.7}$$

[5] See, e.g. Tsien, loc. cit., p. 72, Eq. (16).

The detailed calculations leading to this relation are given below. The implications of this relation are very interesting. They will be discussed in detail in the next section. Physical interpretations of the results will also be indicated there.

*Calculation of the limit form of* (3.11) *and* (3.12).

In carrying out the limiting process, we shall take the quantities in front of the shock to be finite, as the shock can only influence flow conditions downstream. Thus, the only quantities which may become infinite are

(i) the curvature of the shock, $\partial\alpha_1/\partial\sigma$

and

(ii) the pressure gradient behind the shock, $-(\rho_1 q_1^2)^{-1}(\partial p_2/\partial s_2)$.

The limit of (3.11) becomes

$$\frac{2}{\gamma+1}\left\{M_1^2 - \frac{\gamma+3}{2}\right\}\left(-\frac{1}{r_1}\right) = \lim A_2\left(-\frac{1}{\rho_1 q_1^2}\frac{\partial p_2}{\partial s_2}\right) + \frac{q_2}{q_1}\left(-\frac{1}{r_2}\right) + \lim C\left(\frac{\partial\alpha_1}{\partial\sigma}\right),$$

or

$$\frac{2}{\gamma+1}\left(M_1^2 - \frac{\gamma+3}{2}\right)\left(-\frac{1}{r_1}\right) - \frac{1}{\gamma+1}\left(\frac{2}{M_1^2}+\gamma-1\right)\left(-\frac{1}{r_2}\right) = \lim A_2\left(-\frac{1}{\rho_1 q_1^2}\frac{\partial p_2}{\partial s_2}\right) + \lim C\left(\frac{\partial\alpha_1}{\partial\sigma}\right). \tag{4.8}$$

The limit of (3.12) does not lead to a finite result. However, if we multiply the equation by $A_2$ before taking the limit, there is obtained

$$0 = A_2' \lim A_2\left(-\frac{1}{\rho q_1^2}\frac{\partial p_2}{\partial s_2}\right) + C' \lim A_2\left(\frac{\partial\alpha_1}{\partial\sigma}\right). \tag{4.9}$$

If we eliminate the term in pressure gradient in (4.8) by means of (4.9), we obtain

$$\frac{2}{\gamma+1}\left\{M_1^2 - \frac{\gamma+3}{2}\right\}\left(-\frac{1}{r_1}\right) - \frac{1}{\gamma+1}\left\{\frac{2}{M_1^2}+\gamma-1\right\}\left(-\frac{1}{r_2}\right) = \left\{1 - \frac{C'}{A_2'}\lim\frac{A_2}{C}\right\}\lim C\left(\frac{\partial\alpha_1}{\partial\sigma}\right). \tag{4.10}$$

Now, by (3.5),

$$\lim C\left(\frac{\partial\alpha_1}{\partial\sigma}\right) = \lim\frac{2}{\gamma+1}\sin 2\alpha\left(\frac{\partial\alpha}{\partial\sigma} - \frac{\partial\theta_1}{\partial\sigma}\right) = -\lim\frac{1}{\gamma+1}\frac{\partial}{\partial\sigma}\cos 2\alpha.$$

Hence,

$$\lim C(\partial\alpha_1/\partial\sigma) \le 0, \tag{4.11}$$

since $\cos 2\alpha$ is least for $\alpha = \pi/2$. By using the explicit expressions for $A_2$, $C$ and $A_2'$, $C'$ and the relation

$$\lim_{\substack{\alpha \to \pi/2 \\ \theta \to 0}} \frac{\sin 2\alpha}{2 \cos (\alpha - \theta)} = \frac{q_2}{q_1},$$

(See the values of $B_2$ in Table I), it can be easily verified that

$$1 - \frac{C'}{A_2'} \lim \frac{A_2}{C} = 1 + \left(\frac{q_1}{q_2} - 1\right) \frac{1}{1 - M_2^2} \frac{1}{\gamma + 1} > 0. \tag{4.12}$$

By putting (4.11) and (4.12) into (4.10), the relation (4.7) is verified. It should further be noted that a finite value for (4.10) may be obtained only when

$$\lim \partial(\cos 2\alpha)/\partial\sigma = \text{finite} \tag{4.13}$$

This point will be discussed further in the next section.

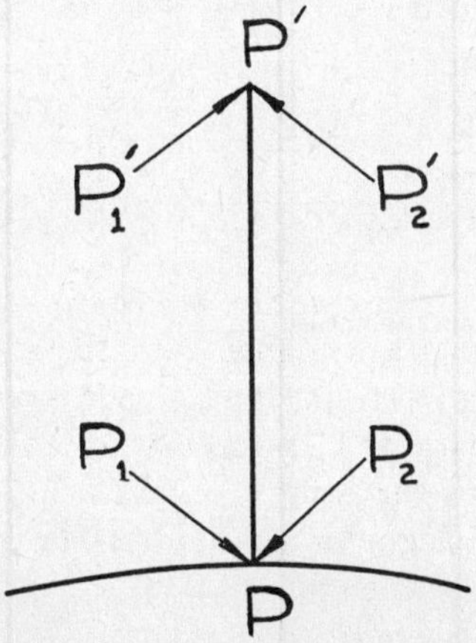


FIG. 3. PRESSURE RELATIONS AT A STRAIGHT NORMAL SHOCK

**5. Further studies of normal shocks over solid boundaries.** In this section, we shall discuss further the results obtained in §3(c). The implications and physical interpretations will be given. First, we shall consider the case of the finite curvature; next, the modification of the situation due to infinite curvature will be discussed. The shape of a shock with infinite curvature will then be discussed.

(i) *Shocks with finite curvature.*

In the case of finite curvature, the equality sign holds in (4.7). We then have

$$r_2 = r_1 \left(\frac{1}{M_1^2} + \frac{\gamma - 1}{2}\right) \Big/ \left(M_1^2 - \frac{\gamma + 3}{2}\right). \tag{5.1}$$

This is the result obtained by previous authors. It shows that for a given Mach number, the *ratio* of the curvatures[6] must have a certain value, in order that a

[6] Note that in the case of a straight boundary, regarded as a limiting case of curved

normal shock may be formed. We propose to give here the physical interpretation of this result, which does not seem to have been fully discussed.

For definiteness, let us first consider a straight shock intersecting the solid boundary at $P$, and let us compare the conditions at $P$ with that at a neighboring point $P'$ (Fig. 3). If the conditions at $P_1$ are known, the normal pressure gradient is also known. Thus, we have a knowledge of the flow conditions at $P_1'$. Then, from the shock conditions, the flow conditions at $P_2$ and $P_2'$ are also known. In particular, the pressure gradient along $P_2P_2'$ is known. As this must balance the centrifugal force, the curvature of the solid boundary cannot be arbitrary. This explains why $r_2$ is completely determined by $r_1$ in the case of a straight shock.

If the shock has a finite curvature, then the shock angle at $P'$ differs from $\pi/2$ by a quantity of the order of $PP'$, a small quantity of the first order. However, the change of pressure jump across the shock at $P'$ is second order in the change of the angle, as we are in the neighborhood of the normal shock. A reference to the shock polar shows this clearly. Thus a finite curvature of the shock does not modify the situation, as it is also clear from the way eq. (3.11) simplifies to (4.6).

Eq. (5.1) shows that with $r_1 = r_2$, $M_1$ must take on a critical value $M_c$ defined by

$$M_c^2 = \tfrac{1}{2}\{\sqrt{(\gamma + 1)^2 + 4} + (\gamma + 1)\}. \tag{5.2}$$

If $M_1 < M_c$, then $r_2 > r_1$. Physically, this means that the pressure gradient developed after the shock in accordance with the shock conditions is able to balance a centrifugal force corresponding to the smaller curvature $1/r_2$ ($r_1$, $r_2$ both positive[7]). The pressure at $P_2'$ is too low for the shock to be formed over a boundary with continuous curvature. If $M_1 > M_c$, the opposite is true; i.e., the pressure at $P_2'$ is too high for continuous curvature of boundary. The pressure is just suitable only when $M_1 = M_c$.

(ii) *Shocks with infinite curvature.*

If the shock has an infinite curvature, the difference of shock angle between $P$ and $P'$ is no longer of the same order as $PP'$. Therefore, the normal pressure gradient after the shock may be changed by finite values. This is borne out by the fact that the limiting values of $C(\partial\alpha_1/\partial\sigma)$ in (4.8) may not vanish, if infinite curvature is allowed. However, the adjustment appears to be one-sided as the inequality sign in (4.7) indicates. Indeed, in the case of continuous curvature ($r_1 = r_2$), it can be easily deduced from (4.7) that (cf. Fig. 4)

$$\begin{aligned} M_1^2 - \frac{1}{M_1^2} &> \gamma + 1 \qquad \text{if} \qquad r_1 = r_2 > 0, \\ M_1^2 - \frac{1}{M_1^2} &< \gamma + 1 \qquad \text{if} \qquad r_1 = r_2 < 0 \end{aligned} \tag{5.3}$$

boundary with $1/r_1$, $1/r_2$ approaching zero, the limit may take on any value. Hence, the shock may occur with any Mach number.

[7] This is taken to fix our ideas. The statement can be easily modified to suit other cases.

For the equality sign to hold, $M_1 = M_c$. Thus, to satisfy these requirements, we must have

$$\begin{cases} M_1 > M_c & \text{if} \quad r_1 = r_2 > 0 \\ M_1 < M_{c_1} & \text{if} \quad r_1 = r_2 < 0. \end{cases} \tag{5.4}$$

That is, *over a convex (concave) surface, a normal shock can be formed only when the Mach number of the flow is above (below) a certain critical Mach number.* This conclusion is interesting enough to require a further clarification of the physical situation.

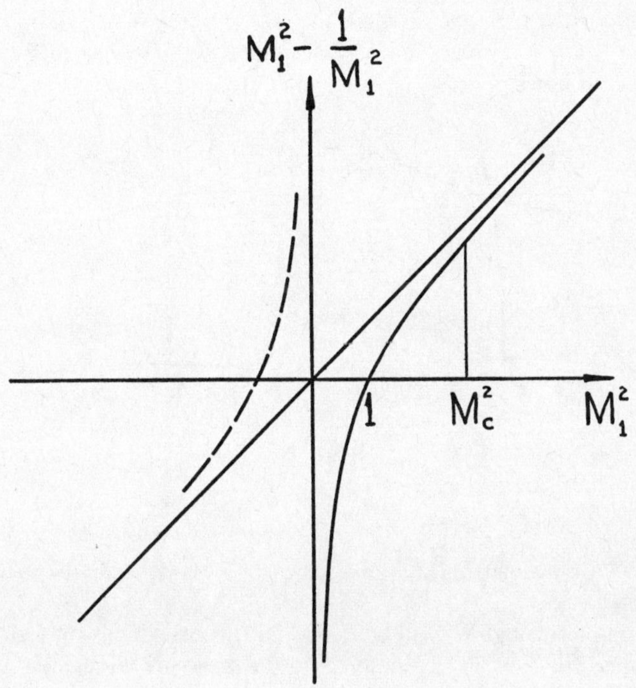


FIG. 4. DIAGRAM SHOWING FUNCTION DETERMINING THE CRITICAL MACH NUMBER

To fix our ideas, let us restrict ourselves to the case of convex surface ($r_1 = r_2 > 0$). It has been pointed out above that the pressure at $P_2'$ is too high for continuous curvature of the boundary if $M_1 > M_c$. Thus, if the shock at $P'$ is oblique, the pressure rise across the shock would be reduced, and the pressure gradient after the shock would balance the centrifugal acceleration at $P_2$ after the shock. If $M_1 < M_c$, the pressure at $P_2'$ is too low even with a normal shock. Thus, a finite shock angle at $P'$ would cause the pressure after it even more unbalanced.

However, this is not the whole story. We are considering the pressure gradient along the shock, and not the normal pressure gradient. Because of the infinite curvature of the shock, these two could be different. The geometry and the pressure distribution after the shock must be examined more carefully. This is also borne out by the fact that there are two types of terms in the limiting process of (4.8): one is related to the infinite shock curvature, the other is related to the infinite pressure gradient. The former has been discussed above, the latter has just been brought up.

Let us fix our attention on Fig. 5(a) first. We notice that the normal pressure gradient must be calculated on the basis of the pressure difference $p(P_2') - p(P'')$ and not for $p(P_2') - p(P_2)$. The difference between these two might be significant if the pressure gradient along the solid boundary is infinite; and this is indeed the case. It is interesting to note that this infinite pressure gradient influences the normal pressure gradient in the same way as the obliqueness of the shock. It is always to *reduce* the force acting *toward* the solid boundary. This may be seen as follows.

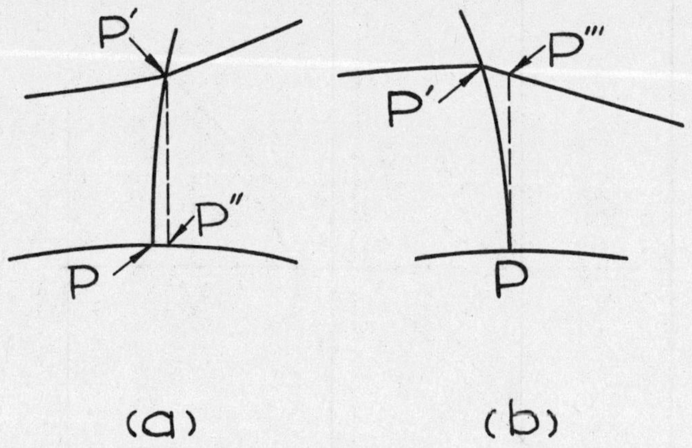


FIG. 5. Pressure relations at a curved normal shock with infinite curvature.

The stream line through $P'$ will be deflected upwards, causing a divergence of the flow. Since the flow is subsonic here, the pressure along the stream line will then increase downstream. Thus,

$$p(P'') > p(P_2)$$

Hence,

$$-\{p(P'') - p(P_2')\} - \{p(P_2') - p(P_2)\} = p(P_2) - p(P'') < 0$$

In the limiting process of taking the pressure gradient, although the distance $P_2P''$ goes to zero as a higher order of $P_2P_2'$, the infinite pressure gradient would cause the difference $p(P'') - p(P_2)$ to make a finite contribution. This explains the two types of terms considered in (4.8). However, because of the inequality sign in the above formula, it is clear that the normal pressure gradient can only be *smaller* than that along the shock. Thus, the infinite pressure gradient along the boundary acts in the same way as the obliqueness of the shock.

It might be thought that if the shock were curved in the other direction (Fig.

5(b)), the conclusion would be different. However, in this case, the flow is convergent after the shock; hence,

$$p(P''') < p(P_2')$$

Thus,

$$p(P''') - p(P_2) < p(P_2') - p(P_2),$$

and the normal pressure gradient is again smaller than the pressure gradient along the shock.

Having clarified the physical interpretation of the one-sided nature of the influence of infinite shock curvature, let us examine the geometry of the shock somewhat further. A curve with an infinite curvature does not show corners in a usual graph. Indeed, if one takes a coordinate system $(\xi, \eta)$ with origin at the intersection of shock and boundary and $\xi$-axis tangent to the boundary, it can be easily verified that the curve must be of the form

$$\xi = k\eta^{3/2} \qquad (k = \text{constant}) \tag{5.5}$$

to make

$$\alpha = \pi/2$$

and

$$d \cos 2\alpha / d\sigma = \text{constant}$$

at $(\xi, \eta) = (0, 0)$. Indeed, after a little calculation, (4.10) can be reduced to

$$\begin{aligned} \{2M_1^2 - (\gamma + 3)\} \frac{1}{r_1} + \left\{ \frac{2}{M_1^2} + (\gamma - 1) \right\} \frac{1}{r_2} \\ = \frac{9k^2}{2} \left\{ 1 + \frac{1}{(\gamma + 1)^2} \frac{2\gamma M_1^2 - (\gamma - 1)}{1 + (\gamma - 1)M_1^2/2} \right\}. \end{aligned} \tag{5.6}$$

In this reduction, the formulae

$$\frac{q_1}{q_2} = \frac{(\gamma + 1)M_1^2}{2 + (\gamma - 1)M_1^2}, \qquad 1 - M_2^2 = \frac{(\gamma + 1)(M_1^2 - 1)}{2\gamma M_1^2 - (\gamma - 1)}$$

for normal shocks are used.

**6. Flow behind detached shocks.** So far we have been dealing with shocks attached to solid bodies. When a sharp-nosed body is placed in a supersonic stream, this is the case, when the Mach number of the flow is very high. In fact, the flow behind the shock is then often supersonic, and can be worked out by stepwise numerical integration with the help of characteristics. Equations (3.1) and (3.2) would then be very useful in getting better accuracy. However, for axially symmetrical flows, the corresponding formulae have not yet been worked out.

If the Mach number of the flow is relatively low, or if the body is blunt, the shock detaches from the body, and there is always a subsonic region behind the shock. Maccoll and others have considered the cases of a supersonic flow past a body with straight boundaries followed by a sharp corner at the shoulder. In this way, he is able to define the subsonic region to a certain extent. A numerical procedure is then developed to study this region, treating the problem as a boundary value problem.

This method does not apply to bodies with blunt noses and smooth boundaries. Also in common with all numerical methods, it is difficult to obtain general conclusions. A number of specific cases have to be computed in order that general conclusions may be drawn.

On the other hand, experiments have been carried out for supersonic flow past spheres (cf. Fig. 6). Also, if a hot wire is ever used in a supersonic stream, such a fine hot wire is presumably in the form of a circular cylinder. It therefore seems desirable that methods be developed to investigate such flows.

The investigations below are aimed at such problems. The method of approach can be best understood by considering the limiting case of very high Mach numbers. It is known that the shock would be then very close to the body. Considering the local situation at the nose of the body, it is clear that the most important characteristic length is the radius of curvature of the body. Thus, it might be expected that the radius of curvature of the shock and the distance of the shock from the body are both proportional to the radius of curvature of the body, with coefficients depending on the Mach number of the uniform stream. Hence, if one would investigate the local nature of the flow in front of the nose of the body and that behind the "'nose" of the shock, and then try to fit the two flows together, these ratios can be found.

To improve on this first-order theory, one may develop the power series expansions in front of the nose of the body and behind the nose of the shock. It will be shown below that definite results are obtainable to any order of approximation desired; i.e., the power series may be terminated at any power, still the shape of the shock may be determined in terms of the shape of the body, and the flow in between would also be known, including the distance of the shock from the body.

The validity of the method depends on two conditions. First, the power series developed in the neighborhood of the nose of the body may be applied up to and including the neighborhood of the nose of the shock, and vice versa. Secondly, these power series may include part of the supersonic region. If this be the case, it is clear that any modification of the shape of the body downstream from the region of interest to us will not disturb the flow conditions in the subsonic region (cf. Fig. 6).

These conditions cannot be verified a priori. However, in the case of the sphere, the results obtained along these lines seem to be in reasonable agreement with experiments. In the case of bodies of extreme shape, such as a finite cylinder with one end placed against the stream, it is obvious that the method would break down. In the extreme case of very high Mach numbers, the argument put forth about a local characteristic length obviously does not hold.

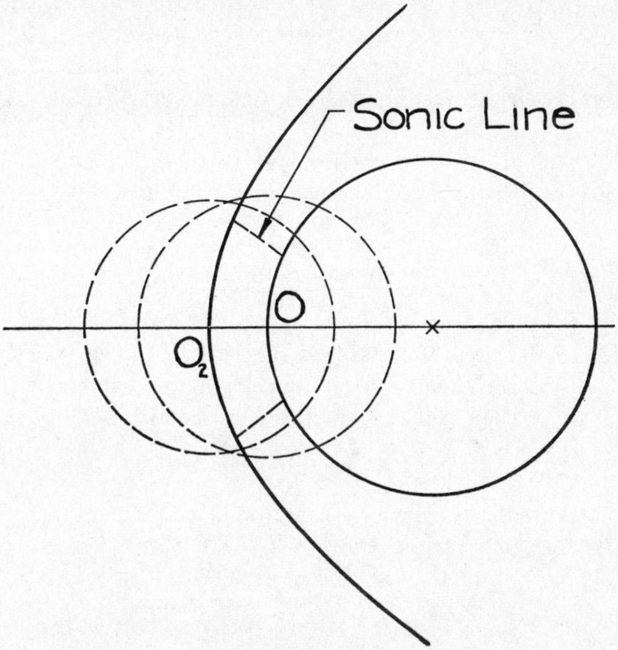


FIG. 6. Detached shock in front of a symmetrical body placed in a uniform stream.

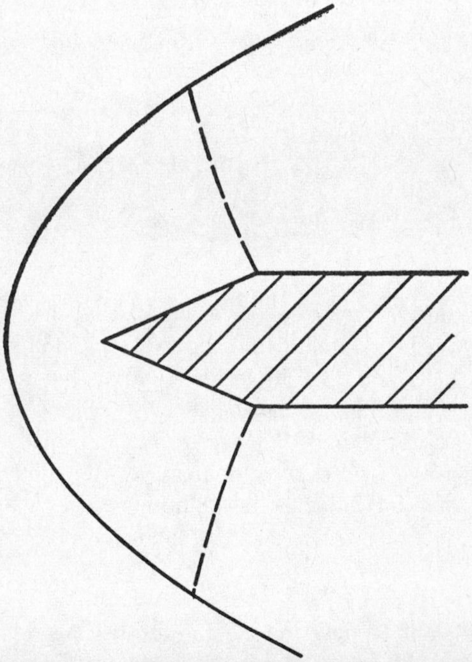

FIG. 7. Detached shock showing the sonic line through the sharp shoulder of the body.

In this paper, we shall indicate the scheme of calculation and state some results. The detailed treatment for the case of a sphere will be published in a separate paper by Mr. John Dugundji. Some of his results will be stated here.

We shall center our discussion on the determination of the shape of the shock and the distance of the shock in front of the body. It is obvious that the calculation of the whole field can be carried out along these lines.

Let the pressure distribution along the axis of symmetry behind the shock be represented by the power series

$$p = p_2 + p_2' x_2 + \tfrac{1}{2} p_2'' x_2^2 + \cdots \tag{6.1}$$

where $x_2$ is the distance measured from the nose of the shock, and $p_2$, $p_2'$, $p_2''$, $\cdots$ are the values of the pressure, the first derivative, the second derivative, etc. with respect to $x_2$ at the shock. From our calculations above, it is obvious that $p_2'$ should depend on the curvature of the shock. Indeed it is of the form

$$p_2'/p_0 = c_1(M_1)(\partial\alpha/\partial\sigma) \tag{6.2}$$

where $c_1(M_1)$ is a coefficient, being a function of the free-stream Mach number, and $(\partial\alpha/\partial\sigma)$ is the curvature of the shock. This can be seen from (3.11) and (3.12) by taking the limit $\alpha = \pi/2$, $\theta = 0$ with

$$-(\rho_1 q_1^2)^{-1}(\partial p_1/\partial s_1) = (\partial\theta_1/\partial s_1) = (\partial\theta_2/\partial s_2) = 0.$$

We have

$$(M_2^2 - 1)(-\rho_1 q_1^2)^{-1}(\partial p_2/\partial s_2) + (q_2/q_1 - 1)(\partial\alpha/\partial\sigma) = 0,$$

from (3.12), and an identity from (3.11). Further differentiation will then lead to relations of the following kind

$$\begin{cases} p_2''/p_0 = C_2(M_1)(\partial\alpha/\partial\sigma)^2 \\ p_2'''/p_0 = C_3(M_1)(\partial\alpha/\partial\sigma)^3 + C_4(M_1)(\partial^3\alpha/\partial\sigma^3) \\ p_2^{iv}/p_0 = C_5(M_1)(\partial\alpha/\partial\sigma)^4 + C_6(M_1)(\partial\alpha/\partial\sigma)(\partial^3\alpha/\partial\sigma^3) \\ \cdots. \end{cases} \tag{6.3}$$

We notice that the derivatives of even orders of $\alpha$ with respect to $\sigma$ all vanish identically, because of the symmetry of the problem. Thus, the order of derivatives introduced is really equal to the order of derivatives of $p_2$. Thus, because of the symmetry of the problem, *$n$ parameters are introduced in order that we may represent the pressure by a polynomial of degree $2n$.*

Before going on with the series development around the nose of the body, and the fitting together of the two series, let us notice a simplification if we stop the series (6.1) at terms in $x_2^4$. In this case, the conditions

$$p = p_0, \quad p' = 0$$

at the nose are sufficient to determine some interesting relations. Let the distance in between be $\delta$. Then, we obtain two relations of the type

$$p(\delta)/p_0 = F\left(\delta\frac{\partial\alpha}{\partial\sigma}, \delta^3\frac{\partial^3\alpha}{\partial\sigma^3}, M_1\right) = 1,$$

$$p'(\delta)/p_0 = F\left(\delta\frac{\partial\alpha}{\partial\sigma}, \delta^3\frac{\partial^3\alpha}{\partial\sigma^3}, M_1\right) = 0.$$

These then determine

$$\delta\frac{\partial\alpha}{\partial\sigma} = f(M_1), \qquad \delta^3\frac{\partial^3\alpha}{\partial\sigma^3} = g(M_1) \tag{6.4}$$

independent of the shape of the body. The second relation is presumably not accurate, but one may expect the first one to hold reasonably well.

The calculations of Mr. John Dugundji give

$$p_2/p_0 = 1 + \pi(M_1), p_2'/p_0 = \mu(M_1)(\partial\alpha/\partial\sigma), p_2''/p_0 = \nu(M_1)(\partial\alpha/\partial\sigma)^2, \tag{6.5}$$

with

$$\begin{aligned} \pi &= \{1 + (\gamma - 1)M_2^2/2\}^{-\gamma/(\gamma-1)} - 1 \\ \mu &= -4\gamma(\gamma + 1)^{-1}\{1 + (\gamma - 1)M_2^2/2\}^{-\gamma/(\gamma-1)} \end{aligned} \tag{6.6}$$

$$\nu = \frac{8\gamma}{(1 - M_2^2)^2}\frac{\{1 + (\gamma - 1)M_2^2/2\}^{-\gamma/(\gamma-1)}}{2\gamma M_1^2 - (\gamma - 1)}\left\{\frac{(M_1^2 - 1)^2[2(\gamma M_2^4 + 1) + (1 - M_2^2)^2]}{(M_2^2 - 1)[2 + (\gamma - 1)M_1^2]} + (M_1^2 - 1) + M_1^2(1 - M_2^2)\right\}.$$

The expansion in the neighborhood of the nose of the body is more complicated, because it depends not only on the shape of the body, but also on the entropy distribution, which in turn is determined by the curvature of the shock. However, it seems clear from detailed calculations (cf. following section), that the $(2n)$th derivative of pressure at the nose involves at most $\partial^n\alpha/\partial\sigma^n$. The other additional parameters are the $n$ derivatives of odd orders of the angle of the boundary with respect to its arc length, and the first derivative of the velocity along the axis. The higher derivatives may be eliminated by requiring the boundary to be a stream line. The detailed calculations have been carried through including third powers in the expression for pressure (see eqs. (7.27) and (7.28)). It is seen that up to this order of approximation, the change of entropy has not yet entered the problem in the two-dimensional case, but has already influenced the results in the axially symmetrical case.

By fitting the series developments around the nose of the shock and the nose of the body, one can determine all of the parameters entering the problem. Suppose one uses both series up to the degree $2n$. Then there are $n$ parameters from the series (6.1). Applying this series at the nose, one introduces an additional parameter $\delta$. There is only one additional unknown introduced by the series near the body. Thus, we have $(n + 2)$ unknowns all together. If we differentiate the power series (6.1) $(n + 1)$ times to obtain the value of $p(\delta)$, $p'(\delta)$, $\cdots$ $p^{(n+1)}(\delta)$ and equate them to the value at the nose, we can obtain all these parameters. The worst approximation in this calculation is presumably in $p^{(n+1)}(\delta)$. But even here, we have a polynomial of degree $n - 1$, and the accuracy can be improved by increasing $n$.

In practice, it is probably better to use derivatives of lower orders, and compare their values at both noses by differentiating both series. Thus, the accuracy can be improved.

From dimensional considerations, it is obvious that the results should be expressible in the form

$$\begin{cases} \delta \dfrac{\partial \alpha}{\partial \sigma} = f_1 \left\{ \delta \dfrac{\partial \alpha_0}{\partial \sigma_0}, \delta^3 \dfrac{\partial^3 \alpha_0}{\partial \sigma_0^3}, \cdots, \delta^{2n-1} \dfrac{\partial^{2n-1} \alpha_0}{\partial \sigma_0^{2n-1}}, M_1 \right\}, \\ \cdots \\ \delta \dfrac{\partial^{2n-1} \alpha}{\partial \sigma^{2n-1}} = f_n \left\{ \delta \dfrac{\partial \alpha_0}{\partial \sigma_0}, \delta^3 \dfrac{\partial^3 \alpha_0}{\partial \sigma_0^3}, \cdots \delta^{2n-1} \dfrac{\partial^{2n-1} \alpha_0}{\partial \sigma_0^{2n-1}}, M_1 \right\}. \end{cases}$$

where $\alpha_0$ and $\sigma_0$ are the angle and arc length along the body. These are obtainable by the calculation of $p(\delta), \cdots p^{(n-1)}(\delta), p^{(n)}(\delta)$. The actual determination of $\delta$(or $\delta(\partial\alpha_0/\partial\sigma_0)$) on the right hand side will require another relation.

The calculations by Dugundji include second order terms only. They agree with experimental results, indicating the applicability of the method. Further work seems to be desirable for the calculation of quantities of higher order.

**7. Iso-energetic flow in the neighborhood of the nose.** In this section, we outline the calculations of the flow conditions in the neighborhood of the nose, which is needed for the discussion of §6.

Although the flow is subsonic, the conditions of flow at the nose can be treated locally, at least to a certain extent. The hydrodynamic quantities, pressure, density, and velocity, must satisfy the equations of motion and the equation of continuity. Furthermore, if the flow is that behind a shock, the Bernoulli function is a constant (iso-energetic), even though the motion is not irrotational. These furnish four relations for the determination of the four hydrodynamic quantities.

If one refers the pressure and density to the stagnation quantities $p_0$ and $\rho_0$ and the velocity to $\sqrt{p_0/\rho_0}$, these equations are

$$u \frac{\partial \tau}{\partial x} + v \frac{\partial \tau}{\partial y} = \tau \left( \frac{\partial u}{\partial x} + \frac{\partial v}{\partial y} + \epsilon \frac{v}{y} \right), \tag{7.1}$$

$$u(\partial u/\partial x) + v(\partial u/\partial y) = -\tau(\partial p/\partial x), \tag{7.2}$$

$$u(\partial v/\partial x) + v(\partial v/\partial y) = -\tau(\partial p/\partial y), \tag{7.3}$$

$$p\tau + [(\gamma - 1)/2\gamma](u^2 + v^2) = 1 \tag{7.4}$$

where $\tau$ is the specific volume, $\epsilon = 0$ for two dimensional flows, and $\epsilon = 1$ for flows with axial symmetry. In the latter case, $y$ stands for the radial distance from the axis of symmetry. In the two-dimensional case, the flow shall also be assumed to be symmetrical. The shape of the nose may be specified as

$$y^2 = g(x) \tag{7.5}$$

in either case.

We shall take the nose to be the origin of our coordinate system, so that we shall be concerned with small values of $x$ and $y$. Taking into account the symmetry nature of the flow, we write

$$\begin{cases} p = 1 + p_1 x + (p_{11}x^2 + p_{22}y^2) + (p_{111}x^3 + p_{122}xy^2) + \cdots , \\ \tau = 1 + \tau_1 x + (\tau_{11}x^2 + \tau_{22}y^2) + (\tau_{111}x^3 + \tau_{122}xy^2) + \cdots , \\ u = u_1 x + (u_{11}x^2 + u_{22}y^2) + \cdots , \\ v = v_1 y + v_{12}xy + \cdots . \end{cases} \tag{7.6}$$

By putting these expressions into the equations (7.1) and (7.4), one obtains a series of relations among the coefficients. In particular, one obtains

$$p_1 = \tau_1 = 0 \tag{7.7}$$

from (7.2) and (7.4). The other relations, after making use of (7.7), are as follows.

From the equation of continuity (7.1), one obtains

$$\begin{cases} u_1 + (1 + \epsilon)v_2 = 0, \\ 2u_{11} + (1 + \epsilon)v_{12} = 0, \\ \cdots \end{cases} \tag{7.8}$$

after (7.7) has been used. Equation (7.2) leads to

$$\begin{cases} -2p_{11} = u_1^2 , \\ -p_{111} = u_1 u_{11} , \\ -p_{122} = (u_1 + 2v_2)u_{22} , \\ \cdots . \end{cases} \tag{7.9}$$

equations (7.3) leads to

$$\begin{cases} -2p_{22} = v_2^2 \\ -2p_{122} = (u_1 + 2v_2)v_{12}, \\ \cdots , \end{cases} \tag{7.10}$$

while equation (7.4) gives, besides $\tau_1 = 0$,

$$\begin{cases} p_{11} + \tau_{11} = -\dfrac{\gamma - 1}{2\gamma} u_1^2 , \\ p_{22} + \tau_{22} = -\dfrac{\gamma - 1}{2\gamma} v_2^2 , \\ p_{111} + \tau_{111} = -\dfrac{\gamma - 1}{2\gamma} u_1 u_{11} , \\ p_{122} + \tau_{122} = -\dfrac{\gamma - 1}{\gamma} (u_1 u_{22} + v_{12} v_2), \\ \cdots , \end{cases} \tag{7.11}$$

In addition, on the boundary (7.5), we must have

$$v/u = dy/dx \tag{7.12}$$

Hence, if one writes

$$y^2 = (2x/\kappa_0) + \cdots \tag{7.13}$$

where $\kappa_0$ is the curvature of the nose, there is obtained

$$\begin{cases} (u_1 - 2v_2)k_0/2 + u_{22} = 0, \\ \cdots \end{cases} \tag{7.14}$$

Higher approximations in (7.8)–(7.11) and (7.14) can be easily written down in a similar manner.

*The change of entropy.* If one now considers the change of entropy, by using the relation

$$p\tau^\gamma = e^{S/c_v} = e^{f(\psi)} \tag{7.15}$$

and expands $f$ as a power series

$$f = s_1 x + (s_{11}x^2 + s_{22}y^2) + (s_{111}x^3 + s_{122}xy^2) + \cdots \tag{7.16}$$

it can be easily shown that,

$$s_1 = s_{11} = s_{22} = s_{111} = 0,$$

after using (7.9)–(7.11). On the other hand, for $p_{122}$, $\tau_{122}$, and $s_{122}$, one obtains the relation

$$p_{122} + \gamma\tau_{122} = s_{122} \tag{7.17}$$

As the expressions for $p_{122}$ in (7.9) and (7.10) depend on $u_1 + 2v_2$, which vanishes in the axially symmetrical case but not in the two-dimensional case (cf. (7.8)), the exact relations depend on the particular case we are concerned with.

*Case* 1. Two dimensional case. Here

$$u_1 + 2v_2 \neq 0,$$

and hence (7.9) and (7.10) give

$$v_{12} = 2u_{22} \tag{7.18}$$

Putting this expression in (7.11), one obtains

$$p_{122} + \tau_{122} = -[(\gamma - 1)/\gamma](u_1 + 2v_2)u_{22}.$$

Making use of (7.9) again, this gives

$$p_{122} = -\gamma\tau_{122},$$

and hence (7.17) gives

$$s_{122} = 0.$$

Thus, in this case

$$S/c_v = f = s_{1111}x^4 + s_{1122}x^2y^2 + s_{2222}y^4 + \cdots . \tag{7.19}$$

*Case* 2. Axially symmetrical case. Here

$$u_1 + 2v_2 = 0,$$

and hence (7.9) and (7.10) lead to the identical result

$$p_{122} = 0. \tag{7.20}$$

Eqs. (7.11) and (7.17) then lead to

$$\tau_{122} = -[(\gamma - 1)/\gamma](u_1u_{22} + v_{12}v_2) \tag{7.21}$$

and

$$s_{122} = -(\gamma - 1)(u_1u_{22} + v_{12}v_2), \tag{7.22}$$

so that

$$S/c_v = s_{122}xy^2 + \cdots . \tag{7.23}$$

The difference between these two cases can be understood more clearly by the following consideration. In either case,

$$S/c_v = y^2\varphi(x,y)$$

by symmetry. Also,

$$S/c_v = f(\psi)$$

where $\psi$ is the stream function. Now, in the two-dimensional case,

$$\psi = u_1xy + \cdots .$$

Hence, in order that $S \sim y^2$ one must have

$$S \sim \psi^2 \sim x^2y^2 \tag{7.24}$$

In the axially symmetrical case, we have the Stokes stream function

$$\psi = \tfrac{1}{2}u_1xy^2 + \cdots .$$

Hence,

$$S \sim \psi \sim xy^2. \tag{7.25}$$

It is now also clear that

$$s_{122} = (df/d\psi)_0u_1/2. \tag{7.26}$$

The quantity $(df/d\psi)_0$ can be calculated at any point in the field of flow. In particular, it can be most conveniently calculated at the shock, where the change of entropy is produced.

Summarizing the above results, we find that, in the two-dimensional case,

$$
(7.27)\quad \begin{cases} p = 1 - \frac{1}{2}a^2(x^2 + y^2) - \dfrac{3a^2}{2}\kappa_0(x^3 + xy^2) + \cdots, \\ \tau = 1 + \dfrac{1}{2\gamma}a^2(x^2 + y^2) + \dfrac{3a^2}{2\gamma}\kappa_0(x^3 + \frac{1}{2}xy^2) + \cdots, \\ u = a\left\{x + \dfrac{3\kappa_0}{2}(x^2 - y^2) + \cdots\right\}, \\ v = -a\{y + 3\kappa_0 xy + \cdots\}, \end{cases}
$$

which depend on a single parameter $a$. In the axially symmetrical case,

$$
(7.28)\quad \begin{cases} p = 1 - \frac{1}{2}a^2(x^2 + \frac{1}{4}y^2) - abx^3 + \cdots \\ \tau = 1 + \dfrac{1}{2\gamma}a^2(x^2 + \frac{1}{4}y^2) + \dfrac{1}{\gamma}abx^3 + \dfrac{\gamma - 1}{\gamma}\left(a^2\kappa_0 - \dfrac{ab}{2}\right)xy^2 + \cdots \\ S/c_v = f = (\gamma - 1)(a_2\kappa_0 - \frac{1}{2}ab)xy^2 + \cdots \\ u = ax + bx^2 - a\kappa_0 y^2 + \cdots \\ v = -\frac{1}{2}ay - bxy + \cdots \end{cases}
$$

which depend on two parameters $a$ and $b$. In addition, the condition

$$
(7.29)\quad (df/d\psi)_0 = 2s_{122}/u_1 = 2(\gamma - 1)(a\kappa_0 - \tfrac{1}{2}b)
$$

must be satisfied. Hence, if $(df/d\psi)_0$ is calculated, there is again only a single parameter, since

$$
(7.30)\quad b = 2a\kappa_0 - [(df/d\psi)_0/(\gamma - 1)]
$$

It is interesting to note the fact that $(df/d\psi)_0$ vanishes in the two-dimensional case, but does not vanish in the axially symmetrical case. This causes a difference between the two cases in the relation of the sonic line with respect to a straight solid boundary. Over a straight boundary, $\partial\theta/\partial s = 0$ and hence (2.15) and (2.17) give

$$
(7.31)\quad \frac{\partial p}{\partial n} = 0, \qquad \frac{\partial q}{\partial n} = \omega = \frac{p}{R}\frac{dS}{d\psi}y^{\epsilon}.
$$

In the two-dimensional case, $(\partial S/\partial\psi)_0 = 0$, and we have

$$
(7.32)\quad \partial q/\partial n = 0,
$$

and hence (2.9) gives

$$
(7.33)\quad \partial\rho/\partial n = 0.
$$

Thus, the normal to the boundary is locally a line of constant $p$, $q$, $\rho$, and in particular the sonic line in perpendicular to the straight boundary in the two-dimensional case. In the axially symmetrical case, the perpendicular line is not a line of constant velocity.

## REFERENCES

1. LADENBURG, R., VAN VOORHIS, C. C., AND WINKLER, J., Interferometric Study of supersonic phenomena NAVORD Reports 69–46 (Part I) 93–46 (Part II), 7–47 (Part III).
2. CROCCO, L., L'Aerotechnia **17**, 519 (1937).
3. HENTZSCHE, Lilienthal Gessellschaft Bericht **156**, 174 (1942).
4. THOMAS, T. Y., J. Math. and Phys. **26**, 62–68 (1947).
5. MACCOLL, J. W., AND CODD, J., Theoretical investigation of the flow around various bodies in the sonic region of velocities, Theoretical Research Reports No. 17/45 (Sept. 1945) and No. 7/46 (Sept., 1946).
6. LAITONE, E. V., AND PARDEE, O. O'M., Location of detached shock wave in front of a body moving at supersonic speeds. NACA (Restricted) RM No. A7B10 (May 6, 1947).
7. EMMONS, H. W., The theoretical flow of a frictionless adiabatic perfect gas inside of a two-dimensional hypersonic nozzle. Appendix II, NACA Technical Note No. 1003 (1946).
8. TSIEN, H. S., J. Math. and Phys. **26**, 69–75 (1947).
9. VAZSONYI, A., Quarterly of Appl. Math. **3**, 29–37 (1945).
10. BATEMAN, H., Proc. Nat. Acad. Sci. **16**, 816–825 (1930).

MASSACHUSETTS INSTITUTE OF TECHNOLOGY
UNIVERSITY OF PENNSYLVANIA

(Received March 15, 1948)

Reprinted from Journal of Mathematics and Physics
Vol. XXVIII, No. 2, July, 1949

# ON THE SUBSONIC FLOW THROUGH CIRCULAR AND STRAIGHT LATTICES OF AIRFOILS

By C. C. Lin

**1. Introduction.** In the study of gas turbines and axial flow compressors, one often approximates the situation by considering a two-dimensional flow past an infinite lattice of identical airfoils (Fig. 1). In the centrifugal compressor, a vaned-diffusor is usually installed around the impellor to remove the circulation of the flow. There arises then the problem of flow through a group of identical airfoils arranged regularly around a circle (Fig. 2). The problem of incompressible flow is well-known in the first case, but does not seem to be extensively treated in the second case. In practice, the compressibility of the air can hardly be neglected, and the problem becomes very complicated in either case. The purpose of this paper is to try to provide methods for calculating flows of an incompressible and a compressible fluid through circular and straight lattices.

The difficulty in these cases can be appreciated by considering the much simpler problem of compressible flow past a single airfoil. The only exact analytical method known is the hodograph method, which is however rather lengthy. To bring out the essential aerodynamic features, one usually introduces suitable approximations. The most convenient of them are the Prandtl-Glauert approximation for nearly parallel flows and the von Kármán-Tsien approximation. The Janzen-Rayleigh approximation is suitable only for small Mach numbers. In the present cases, the Prandtl-Glauert approximation is also not suitable, because the flow involved is usually not nearly parallel. Even in the case of the straight lattice, there is often a finite deflection of the direction of flow.

It is thus obvious that the only feasible method is the hodograph method with the linear pressure-volume relation. The validity of this method does not depend on the nearly parallel nature of the flow, but on the fact that the compression involved is never too great. This method has been known for some time, but the complete formulation for flows with circulation has been solved only recently by a slight generalization of the original ideas.[1,2] It turns out that this generalization throws a new light on the problem, and the method can be now easily applied to flows through multiply connected regions, such as those we are concerned with here.

Realizing the complication involved, we are here to present the general inverse method for generating such flows. The direct method requires very complicated calculations, even in the case of an incompressible flow past a single thick airfoil. The present method is general in the sense that the final results involve an arbitrary function which may be used at our disposal to take care of the shape of the airfoils. It thus has great flexibility to be adopted to any design problem. On the other hand, no specific effort has been made in this paper to formulate a design procedure, although it is not expected to be difficult to do so for any set of design conditions.

It might be pertinent to add one word to the line of argument which led to the

FIG. 1. THE STRAIGHT LATTICE

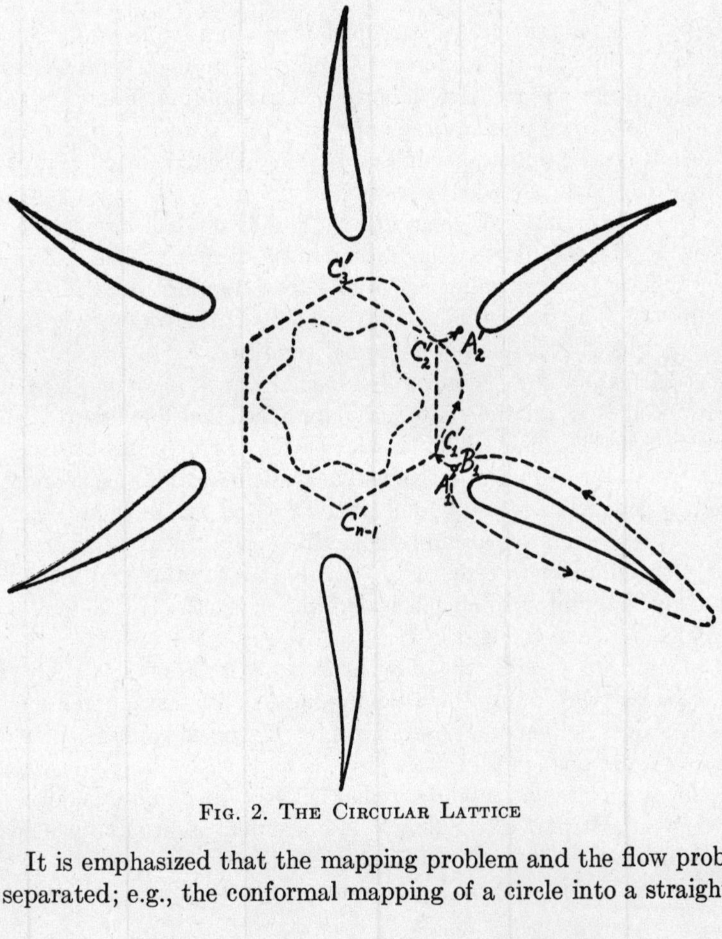


FIG. 2. THE CIRCULAR LATTICE

method. It is emphasized that the mapping problem and the flow problem are entirely separated; e.g., the conformal mapping of a circle into a straight lattice of identical airfoils is a purely geometrical problem. By examining the most general of such mapping functions, one can easily see the essential part of its nature. The generalization to the circular lattice of a finite number of airfoils is then very easy. By combining these ideas with the recent improvement of the Kármán-Tsien approximation, the method presented below is evolved.

It is almost needless to say that the method is limited to the subsonic range, —an inherent restriction in the linear pressure-volume approximation (when the flow has a stagnation point).

In the next section (§2), we shall first discuss the general characteristics of flows through lattices of airfoils, both circular and straight. It will be followed by a summary of our methods for their calculation (§3). The proof of these methods and the discussion of their consequences will be respectively treated in §4 and §5. In §6, alternative methods will be discussed.

**2. General nature of flows through lattices.** Before giving the details of our method, let us first say a few words about the general nature of the flows through lattices.

The incompressible flow past a straight lattice of an infinite number of airfoils is well-known. The flow is uniform at infinity to the left, and becomes again uniform at infinity to the right, with however a generally different magnitude and direction. This change in magnitude and direction is related to the circulation around the airfoils. While the circulation around a single airfoil cannot produce any change of the magnitude and direction of flow at infinity, the aggregate influence of an infinite number of them does. The amount of circulation is determined by the Kutta-Joukowski condition at the trailing edge. The subsonic compressible flow does not differ essentially from the incompressible case in its general aspects. The restriction to finite velocity at the trailing edge is however a necessity in this case, instead of an added condition.

The general character of the incompressible flow past a circular lattice may be described as follows. At the center of symmetry, the flow essentially has the nature of a complex source. The circulation may or may not be zero. A contour enclosing all the airfoils at large distances encloses the same source strength, but would in general have a different circulation, the difference being equal to the sum of the circulation around the individual airfoils. The flow at large distances is therefore again essentially that due to a vortex and a source. The circulation around individual airfoils is determined by the Kutta-Joukowski condition, as in the case of straight lattices.

In the compressible case, the flow near the center of symmetry must be changed, because there cannot be a point source or vortex in compressible flow. However, this does not cause any serious difficulty in the application to diffusors, because the central portion is occupied by the impeller itself. We start out with a certain flow with circulation and a radial velocity, at a finite distance from the center. The outside part of the flow does not differ essentially from that in the incompressible case. The circulation around the individual airfoils is again determined by the Kutta-Joukowski condition, and the flow at large distances is essentially that due to a vortex and a source. In diffusors, it is often required that the circulation outside be zero.

On the other hand, the continuation of the outside part of the flow inwards to the center of symmetry leads to the limiting line at some distance from it. The solution then ceases to exist for the inside part, and further continuation of the solution turns back to the outside part, and leads to the second branch of a multiple-valued solution.

**3. Methods for calculating flows through lattices of airfoils.** In the present section, we shall present the methods for calculating the flows discussed in the last section. We shall formulate below the method in full for the case of compressible flow through a circular cascade. The other three cases can all be obtained by simplifications of this case, as will be explained below.

Let $p$, $\rho$, $u$, $v$ be the pressure, density and components of velocity of a steady two-dimensional irrotational flow in the $x$, $y$ plane. Let $p$ be a function of the density $\rho$ only (given either by the isentropic relation or any other approximate relation). Then there exist the velocity potential $\phi$ and the stream function $\psi$ defined by the following differential relations:

$$d\phi = udx + vdy, \tag{3.1}$$

$$d\psi = -\rho vdx + \rho udy. \tag{3.2}$$

The velocity components $u$, $v$ and the density $\rho$ are further connected by Bernoulli's equation

$$\tfrac{1}{2}q^2 + \int dp/\rho = \text{const.}, \qquad (q^2 = u^2 + v^2). \tag{3.3}$$

It is convenient to refer the density of the gas to that at the stagnation point and to refer all the velocities to the velocity of sound at stagnation. The coordinates $x$, $y$ may be regarded as referred to the size of the body, and the pressure as referred to the product of stagnation density and the square of the velocity of sound at stagnation. Throughout this article, this process shall be implied, and all the quantities under discussion are dimensionless.

As is well-known, the problem simplifies greatly if the pressure-density relation is approximated by

$$p = A - B/\rho. \tag{3.4}$$

This is the basis of the method of Chaplygin, von Kármán and Tsien. A discussion of its physical interpretation may be found in the papers of these authors. Equation (3.3) leads to

$$c = 1/\rho = \sqrt{1 + q^2}, \tag{3.5}$$

where $c$ is the local velocity of sound. Indeed, $B$ must be equal to unity if the reference stagnation quantities are calculated from (3.3) by the use of (3.4).

Under the approximation (3.4), the following method may be used for constructing two-dimensional subsonic flows through lattices.

Let $F(\zeta)$ be the complex potential function of the incompressible flow due to two complex sources in the $\zeta$-plane ($\zeta = \xi + i\eta$) at $\zeta = a_1$ and $\zeta = a_2$ outside the circular boundary $|\zeta| = 1$ (Fig. 3):

$$F(\zeta) = A_1 \log(\zeta - a_1) + \bar{A}_1 \log(\zeta - 1/\bar{a}_1) + B_1 \log(\zeta - a_2) + \bar{B}_1 \log(\zeta - 1/\bar{a}_2), \qquad (3.5)$$

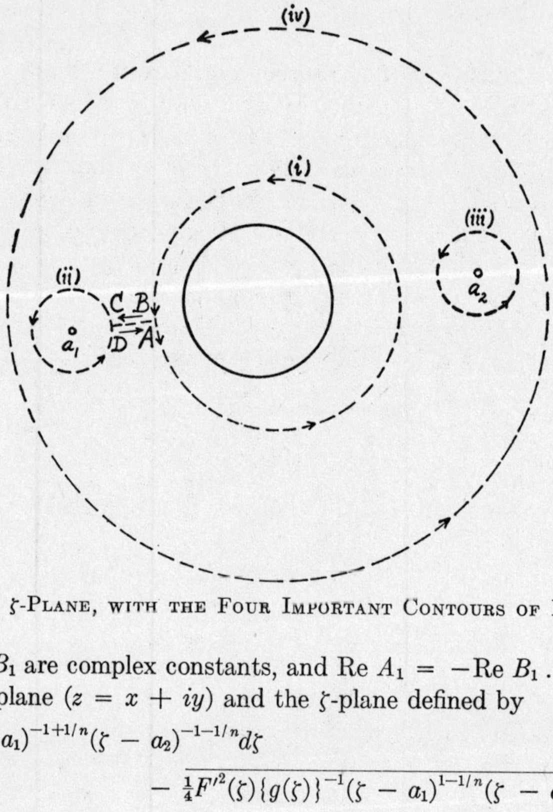


FIG. 3. THE $\zeta$-PLANE, WITH THE FOUR IMPORTANT CONTOURS OF INTEGRATION

where $A_1$ and $B_1$ are complex constants, and $\mathrm{Re}\, A_1 = -\mathrm{Re}\, B_1$. The mapping between the $z$-plane ($z = x + iy$) and the $\zeta$-plane defined by

$$dz = g(\zeta)(\zeta - a_1)^{-1+1/n}(\zeta - a_2)^{-1-1/n}d\zeta - \overline{\tfrac{1}{4}F'^2(\zeta)\{g(\zeta)\}^{-1}(\zeta - a_1)^{1-1/n}(\zeta - a_2)^{1+1/n}d\zeta} \qquad (3.6)$$

gives a compressible flow, with the linear pressure volume approximation, past a circular lattice of $n$ identical blades in the $z$-plane (Fig. 2) with velocity potential $\phi$ and stream function $\psi$ given by

$$\phi + i\psi = F(\zeta), \qquad (3.7)$$

provided that $g(\zeta)$ is chosen to satisfy the following general requirements:

(a) It is regular in the region $R$, defined by $|\zeta| \geqq 1$.

(b) It does not vanish in $R$ except possibly at one point on the circle, where $F'(\zeta) = 0$; the order of the zero does not exceed one.

(c) Along the circle $|\zeta| = 1$, $\oint dz = 0$.

(d) Except in the neighborhood of $\zeta = a_1$, the following inequality holds in $R$:

$$|F'(\zeta)\{g(\zeta)\}^{-1}(\zeta - a_1)^{1-1/n}(\zeta - a_2)^{1+1/n}| < 2.$$

Condition (d) sets a limit to the region of validity of the calculated flow field in the $z$-plane. The magnitude $q$ and direction $\theta$ of the velocity at any point is given by

$$\frac{2q}{1+\sqrt{1+q^2}}\, e^{-i\theta} = \frac{F'(\zeta)}{g(\zeta)}\,(\zeta - a_1)^{1-1/n}(\zeta - a_2)^{1+1/n} \tag{3.8}$$

*Simplification to other cases.*

In the limiting case $n \to \infty$, condition (d) is to be satisfied throughout $R$, and the above scheme defines the compressible flow past a *straight lattice* of an infinite number of blades (Fig. 1).

The incompressible flow through circular and straight lattices can be obtained from the same general scheme by simply omitting the second term in $dz$ (in Eq. (3.6)) and dropping the condition (d). Equation (3.8) also simplifies to

$$qe^{-i\theta} = \frac{F'(\zeta)}{g(\zeta)}\,(\zeta - a_1)^{1-1/n}(\zeta - a_2)^{1+1/n} \tag{3.9}$$

**4. Proof.** The proof of the above statements may be divided into three parts:

($\alpha$) to show that the desired geometry of the lattice is obtained,

($\beta$) to show that the differential equations for the velocity potential and the stream function are satisfied by the function (3.7) in the $z$-plane,

($\gamma$) to show that the required boundary conditions and singularities are fulfilled.

Again, we shall carry through the proof in detail only for the case of compressible flow through a circular lattice. For the other cases, we shall point out the modifications necessary.

($\alpha$) *Geometry of the mapping.*

To examine the nature of the mapping (3.6), let us first examine its Jacobian for the singularities. The Jacobian

$$J = \frac{\partial(x, y)}{\partial(\xi, \eta)} = \operatorname{Im}\left\{\frac{\partial\bar z}{\partial\xi}\cdot\frac{\partial z}{\partial\eta}\right\}$$

can be easily shown to be

$$J = \{\,|g|\cdot|\zeta - a_1|^{-1+1/n}\,|\zeta - a_2|^{-1-1/n}\}^2 \cdot\left\{1 - \frac{1}{16}\left|\frac{F'(\zeta)}{g(\zeta)}\,(\zeta - a_1)^{1-1/n}(\zeta - a_2)^{1+1/n}\right|^4\right\}. \tag{4.1}$$

Thus, if the condition (d) of §3 is satisfied, the Jacobian will always remain *non-negative* in $R$. It may vanish at a single point on the circle, which point is then mapped into a corner or a cusp. This is the point where $g(\zeta) = 0$. It is necessary that the condition (b) of §3 be satisfied in order that condition (d) may be satisfied in the neighborhood of this point. In specifying condition (d), it is necessary that a neighborhood of $\zeta = a_1$ be excluded, because the function $F'(\zeta)\{g(\zeta)\}^{-1}(\zeta - a_1)^{1-1/n}(\zeta - a_2)^{1+1/n}$ becomes infinite as $\zeta \to a_1$.

The Jacobian becomes infinite as $\zeta \to a_1$ or $\zeta \to a_2$. As the point $\zeta = a_1$ is to be excluded, we need only pay our attention to the infinity as $\zeta \to a_2$. However, notice that the mapping (3.6) does exhibit branch-point nature both around $\zeta = a_1$ and $\zeta = a_2$.

With these behaviour of the mapping function in mind, let us trace out the actual mapping of the circle $|\zeta| = 1$ into the circular lattice in the $z$-plane. To do this, we consider four closed paths in the $\zeta$-plane described respectively: (i) around the circle, (ii) around the point $\zeta = a_1$, (iii) around the point $\zeta = a_2$, and (iv) enclosing both singular points. (See Fig. 3.)

(i) Path around the circle. Because of condition (c) of §3, it is clear that, as $\zeta$ runs around such a path, the corresponding point $z$ will also describe a closed path. In particular, the circle is mapped into one of the airfoils desired. The boundary will have a sharp edge, if the function $g(\zeta)$ vanishes on the circle. It is not obvious *a priori* that the closed curve will be free from double points. However, some calculated example did not encounter this difficulty.

(ii) Path around $\zeta = a_1$. As $\zeta$ describes a closed path $CD$ in the $\zeta$-plane, its image in the $z$-plane does not return to its starting point, but rather moves from $C_1'$ to $C_2'$ along some path such as that indicated, because $\int dz$ along such a path is not zero. If we make the point $\zeta$ describe the same path again, it must be noted that the angle of $\zeta - a_1$ has been increased by $2\pi$, and hence the angle of $dz$ has been increased by $2\pi/n$ (cf. (3.6)). The first term of (3.6) gets this factor $e^{2\pi i/n}$ by virtue of the factor $(\zeta - a_1)^{-1+/n}$ while the second term gets it from $(\zeta - a_1)^{-1-1/n}$. Thus, if $C_3'$ is the image of $D$ after the second circling of the point $\zeta = a_1$, the vector $\overrightarrow{C_2'C_3'}$ is to be obtained from $\overrightarrow{C_1'C_2'}$ by a rotation through an angle of $2\pi/n$. In fact, the whole path $C_2'C_3'$ may be obtained from $C_1'C_2'$ this way. Thus, after the path $CD$ has been described $n$-times, we have a series of points $C_1', C_2', \cdots C_{n-1}', C_n'$, which are the vertices of a regular polygon of $n$ sides, if $C_n'$ coincides with $C_1'$. To show that this is the case, we introduce an auxiliary variable

$$\zeta_1 = (\zeta - a_1)^{1/n}$$

and consider the neighborhood of the point $\zeta_1 = 0$. Describing the path $CD$ $n$ times is equivalent to a closed path around the origin of the $\zeta_1$-plane. It can be easily seen that the first term in (3.6) becomes regular in $\zeta_1$, and that the second term has a pole with a Laurent development of the form $\zeta_1^{-2}(\alpha + \beta\zeta_1^n + \cdots)$. The residue around $\zeta_1 = 0$ is therefore zero if $n > 1$. Hence, a closed path around $\zeta_1 = 0$ leads to a closed path in the $z$-plane.

Thus, the mapping of a path $CD$ is obtained by first tracing the curve $C_1'C_2'$ and then rotating the whole figure through an angle of $2\pi/n$ about a certain point $n$ times. Although this center of rotation will not be reached by our mapping, it is fixed and the same for *all* points. To see this, it is only necessary to note that the path $CD$ is an arbitrary path and therefore may be made to cover any point.

The net effect of describing a path around $\zeta = a_1$ is therefore to rotate the whole picture through an angle of $2\pi/n$ in the same sense as the path is taken in the $\zeta$-plane. Thus $n$ identical airfoils are generated by the circle $|\zeta| = 1$ if $\zeta$ keeps on describing closed paths like $ABCDA$ (in the sense indicated in Fig. 3) with the points $A$, $B$ on the circle.

(iii) *Path around* $\zeta = a_2$. It is clear that when $\zeta$ describes a closed path around $\zeta = a_2$, the effect is similar. The only difference is that the rotation through an angle $2\pi/n$ in the $z$-plane is in a sense opposite to that of the motion of the point $\zeta$. The question arises whether the rotation takes place around the same center. This is equivalent to say whether the effect of describing a path around $\zeta = a_2$ exactly cancels that of describing a path around $\zeta = a_1$ in the same sense. As path (i) gives a nil effect, the question is really whether path (iv) also gives a nil effect. This is indeed the case, as shown below.

(iv) *Path enclosing the circle and the singularities.* To examine this path, we concentrate our attention to the behaviour of (3.6) at the point of infinity. In the first place, the coefficients are single-valued in this neighborhood, as can be easily seen. Let us examinate whether there is a finite residue.

Since $g(\zeta)$ does not vanish as $\zeta \to \infty$,

$$g(\zeta) = g(\infty) + 0(\zeta^{-1}). \tag{4.2}$$

The behaviour of $F'(\zeta)$ at infinity is given by

$$\begin{aligned} F'(\zeta) &= A_1(\zeta - a_1)^{-1} + \bar{A}_1(\zeta - 1/\bar{a}_1)^{-1} + B_1(\zeta - a_2)^{-1} + \bar{B}_1(\zeta - 1/\bar{a}_2)^{-1} \\ &= (A_1 + \bar{A}_1 + B_1 + \bar{B}_1)\zeta^{-1} + 0(\zeta^{-2}). \end{aligned}$$

Now, by the condition

$$\mathrm{Re}\, A_1 = -\,\mathrm{Re}\, B_1$$

we see that the coefficient of $\zeta^{-1}$ vanishes. Hence,

$$F'(\zeta) = C\zeta^{-2} + 0(\zeta^{-3}), \tag{4.3}$$

at infinity, where $C$ is finite or zero. Thus, the two terms in $dz$ of (3.6) have the following behaviour:

$$\left.\begin{aligned} g(\zeta)(\zeta - a_1)^{-1+1/n}(\zeta - a_2)^{-1-1/n} &\sim g(\infty)\zeta^{-2}, \\ F'^2(\zeta)\{g(\zeta)\}^{-1}(\zeta - a_1)^{1-1/n}(\zeta - a_2)^{1+1/n} &\sim C^2\{g(\infty)^{-1}\}\zeta^{-2} \end{aligned}\right\} \tag{4.4}$$

Thus, the residue is zero, and hence a closed path is described in the $z$-plane as we describe a closed path in the $\zeta$-plane at large distances. In fact, as $\zeta \to \infty$, $z$ does not become infinite. The point at infinity in the $\zeta$-plane is mapped into some point at finite distances from the airfoils.

The description of the mapping of such paths thus demonstrate that (3.6) is the desired mapping for the circular lattice of $n$ airfoils. Two remarks are to be added. (1) The point $\zeta = a_2$ maps into the point at infinity in the $z$-plane as one can easily see from (3.6). (2) Without the second term of (3.6), as it is indeed the case for incompressible flows, the point $\zeta = a_1$ maps into a finite point in the $z$-plane, which may be conveniently taken to be the origin. In the compressible case, one must exclude a certain region in the $\zeta$-plane where condition (d) is not satisfied. Thus, as one approaches the boundary of this region from $\zeta = a_2$, the point $z$ moves inwards from infinity to the image of this boundary in the $z$-plane. When the boundary is crossed, the path in the $z$-plane reverses its direction of motion, and we have the outer region covered again. This is the familiar phenomenon associated with a vanishing Jacobian, or specifically, the limiting line in compressible flow.

The geometrical part of the proof is now complete.

($\beta$) *Satisfaction of the differential equation.* Having clarified the geometry of the mapping, we want to show now that (3.7) and (3.8) give the velocity potential, the stream function and the magnitude and direction of velocity of the compressible flow in the $z$-plane. We shall first show that the differential equations for $\phi$ and $\psi$ are satisfied, leaving the discussion of the boundary conditions to the next section.

It is well known that under the approximation (3.4), the relations

$$dz = dF/w^* - \tfrac{1}{4}\,\overline{w^*\,dF} \qquad F = F(w^*) \tag{4.5}$$

with $z$, $F$, $w^*$ defined by

$$\left.\begin{aligned} z = x + iy, \qquad F &= \phi + i\psi \\ w^* = \frac{2q}{1 + \sqrt{1 + q^2}}\, e^{-i\theta} & \end{aligned}\right\} \tag{4.6}$$

would give a solution of the *differential* equations of compressible flow. Instead of trying to establish a relation between $F$ and $w^*$ directly, we introduce an *auxiliary* variable $\zeta$ such that $F(\zeta)$ and $w^*(\zeta)$ are analytic functions. This is done in (3.6) by taking $F(\zeta)$ to be given by (3.5) and

$$w^*(\zeta) = F'(\zeta)\{g(\zeta)\}^{-1}(\zeta - a_1)^{1-1/n}(\zeta - a_2)^{1+1/n}, \tag{4.7}$$

which is indeed (3.8). Thus, *differential* relations are satisfied, and $\phi$, $\psi$, $q$, $\theta$ as calculated indeed define a field of compressible flow.

($\gamma$) *Satisfaction of conditions in the large.*

The essential conditions to be satisfied in the large are the following:

(i) The flow field is invariant, when rotated through an angle of $2\pi/n$;

(ii) A contour such as $C_1'C_2' \cdots C_{n-1}'C_1'$ should give a certain amount of circulation and enclose a certain source strength;

(iii) A contour drawn to enclose all the airfoils has a similar property; and

(iv) Each airfoil should be a stream line, around which the circulation is equal to one-$n$th the difference of those in (ii) and (iii).

It is not difficult at all to verify that these conditions are satisfied. The first point follows directly from the geometry of the mapping. The flow field in the $\zeta$-plane is mapped onto $1/n$ of the $z$-plane corresponding to one determination of the $n$-th roots in (3.6). The other portions of the $z$-plane are obtained by rotating the first portion over angles of $2\pi/n$, $4\pi/n$, etc., and corresponds to the other determinations of the $n$-th roots in (3.6). The stream line patterns are identical, because $F(\zeta)$ is merely changed by additive constants when the angles of $\zeta - a_1$ and $\zeta - a_2$ change by $2\pi$.

To verify the second point, let us recall that the contour $C_1'C_2' \cdots C_{n-1}'C_1'$ corresponds to the contour $CD$ around $\zeta = a_1$ described $n$ times. It is therefore obvious that the source strength and the circulation are given by the real and imaginary parts of $n\bar{A}_1$.

A contour drawn to enclose all the airfoils corresponds to the contour (iii) of Fig. 3 described $n$ times. The source strength and the circulation are given by the real and imaginary parts of $-n\bar{B}_1$. The negative sign comes from the fact that we are mapping $\zeta = a_2$ to $z = \infty$, and therefore the positive sense of the contour has been reversed. Notice that

$$Re(n\bar{A}_1) = Re(-n\bar{B}_1),$$

by the original condition imposed on $A_1$ and $B_1$. The difference of circulation is

$$\mathrm{Im}(-n\bar{B}_1 - n\bar{A}_1) = n \,\mathrm{Im}(A_1 + B_1).$$

The airfoils are stream lines, because $\psi$ is constant along each of them. The circulation around each of them is the same as that around the circle in the $\zeta$-plane, which is $\mathrm{Im}(A_1 + B_1)$. This is exactly $1/n$ the circulation difference just discussed.

The proof is now complete for the compressible flow through a circular lattice.

*Simplification to Straight Lattice.** When $n \to \infty$, the mapping function (3.6.) becomes

$$dz = g(\zeta)(\zeta - a_1)^{-1}(\zeta - a_2)^{-1}d\zeta - \tfrac{1}{4}\overline{F'^2(\zeta)\{g(\zeta)\}^{-1}(\zeta - a_1)(\zeta - a_2)d\zeta}. \tag{4.8}$$

Thus, the branch point nature of the singularity at $\zeta = a_1$ and $\zeta = a_2$ disappears and a path around $\zeta = a_1$ and $\zeta = a_2$ changes $z$ by a constant value without changing the mapping function. Thus, a translation in the $z$ plane is produced, which may be regarded as the limit of the previous rotation through an angle $2\pi/n$, as $n$ goes to infinity. The lattice may be regarded as arranged on a circle of infinite radius.

Both the points $\zeta = a_1$ and $\zeta = a_2$ are mapped to infinity, the first to the left of the lattice, the second to the right, if the value of $g(\zeta)$ is properly chosen at $\zeta = a_1$. The velocity of flow at $z = \mp \infty$ are respectively given by (cf. (3.8)).

$$\frac{2q}{1 + \sqrt{1 + q^2}}\, e^{-i\theta} = \frac{A_1}{g(a_1)}\,(a_1 - a_2), \tag{4.9}$$

and

$$\frac{2q}{1 + \sqrt{1 + q^2}}\, e^{-i\theta} = \frac{B_1}{g(a_2)}\,(a_2 - a_1). \tag{4.10}$$

* In a yet unpublished paper, Prof. L. Bers showed that the integral equation method developed by him (Ref. 3) for the case of a single profile can be extended to the case of a straight cascade.

These formulae give the change both in the magnitude and in the direction of the velocity when the fluid passes through the lattice.

*Incompressible fluid.* If the fluid is incompressible, the second terms on the right-hand sides of (3.6) and (4.8) are to be dropped, leaving a conformal mapping between the $z$-plane and the $\zeta$-plane. The mapping covers the *whole* of the $z$-plane up to the center of symmetry. One can see this by noting that the Jacobian (4.1) will no longer vanish on account of the second factor (which is now identically unity).

In the case of a straight lattice, condition (c) of §3 leads to

$$g(a_1) = g(a_2). \tag{4.11}$$

The velocities of flow at infinity are respectively

$$qe^{-i\theta} = \frac{A_1}{g(a_1)}(a_1 - a_2) \qquad z = -\infty$$

$$qe^{-i\theta} = \frac{B_1}{g(a_2)}(a_2 - a_1) \qquad z = +\infty$$

The change of velocity is given by $-B_1/A_1$. A similar factor for change can be obtained from (4.9) and (4.10).

**5. Choice of the function** $g(\zeta)$**.** A few words about the choice of the function $g(\zeta)$ is in order. A simple function regular outside the unit circle is

$$g(\zeta) = \left(1 - \frac{\lambda_1}{\zeta}\right)\left(1 - \frac{\lambda_2}{\zeta}\right)\cdots\left(1 - \frac{\lambda_n}{\zeta}\right). \tag{5.1}$$

where $\lambda_1, \lambda_2, \cdots, \lambda_n$ are points *inside* the circle. The mapping is completely regular on the circle, and the body produced would be void of a sharp trailing edge, which is needed for an airfoil. By making one of the points, say $\lambda_1$, to lie on the circle, a cusp will be produced, provided

$$F'(\lambda_1) = 0$$

so that the Jacobian (4.1) does not become infinite through the second factor. Otherwise, the Jacobian changes its sign in the neighborhood of $\zeta = \lambda_1$ outside the circle, and the whole mapping procedure is made useless. If a finite angle is desired, one may choose, say,

$$g(\zeta) = \left(1 - \frac{\lambda_1}{\zeta}\right)^{1-\epsilon}\left(1 - \frac{\lambda_2}{\zeta}\right)^{1+\epsilon}\cdots\left(1 - \frac{\lambda_n}{\zeta}\right)$$

where $\epsilon > 0$. The angle would then be $\epsilon\pi$. The power $1 + \epsilon$ is necessary for some factor in order that the function $g(\zeta)$ be single-valued outside of $|\zeta| = 1$.

It is however difficult to know in advance whether the mapping will be multiply covered; i.e., whether the contour of the airfoil will be a curve with double points. It seems that experience should help in such a problem. So far the author has not encountered such difficulties.

**6. Discussion.** The previous treatment gives only the basis of the theory. It should not be difficult to work out the aerodynamic properties of the lattice in terms of the function $g(\zeta)$ analogous to the general airfoil theory. Thus, it might be expected that $g(a_1)$, $g'(a_1)$, $\cdots$ and $g(a_2)$, $g'(a_2)$, $\cdots$ would determine many of the aerodynamic characteristics. From these discussions, design procedures are to be evolved.

It is also noteworthy that the present formulation (as well as the compressible flow past a single airfoil) is actually independent of the hodograph plane, although the historical development has been that way. The essential step is a mapping

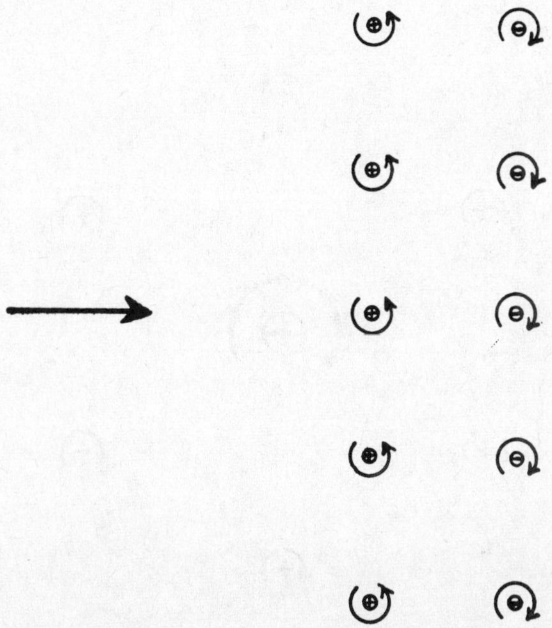

FIG. 4. GENERATION OF A STRAIGHT LATTICE BY SUPERPOSING A UNIFORM FLOW WITH A SYSTEM OF COMPLEX SOURCES.

of an incompressible flow into a compressible flow. More precisely, mappings of the kind (3.6), which depends explicitly on $F'(\zeta)$, make $F(\zeta)$ at once the complex velocity potential of an incompressible flow in the $\zeta$-plane and of a compressible flow in the $z$-plane with velocity given by (3.8). In general, one may consider

$$\left.\begin{aligned} dz &= k(\zeta)\, d\zeta - \tfrac{1}{4}[F'^2(\zeta)][k(\zeta)]^{-1}\, d\zeta, \qquad \oint dz = 0, \\ \phi + i\psi &= F(\zeta), \qquad \frac{2qe^{-i\theta}}{1 + \sqrt{1+q^2}} = \frac{F'(\zeta)}{k(\zeta)}, \end{aligned}\right\} \tag{6.1}$$

to be mapping an incompressible flow in the $\zeta$-plane into a compressible flow in the $z$-plane. If $k(\zeta)$ is not much different from unity, the two flows will be similar in nature. The reduction of (6.1) to conformal mapping of incompressible flows is very easy.

This immediately suggests other methods of obtaining flows of the types discussed above. One can easily construct a *particular* incompressible flow of the general nature desired. Two infinite rows of source-vortex combinations of equal and opposite source strength superposed with a uniform velocity will produce an incompressible flow past a straight lattice (Fig. 4). Similarly, the

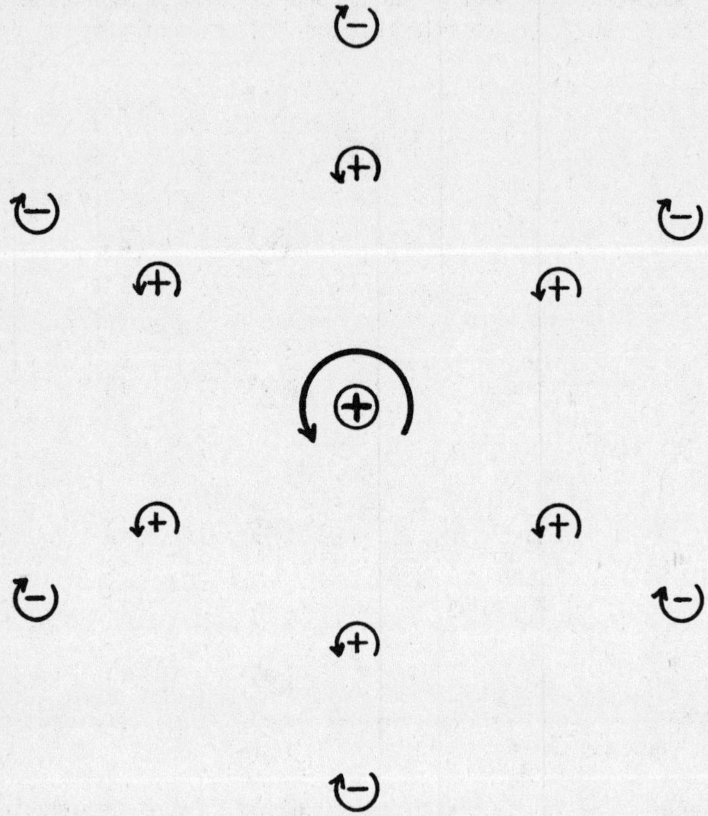

Fig. 5. Generation of a Circular Lattice by Superposing a System of Complex Sources.

circular arrangement of such pairs with a complex source at the center will produce the flow past a circular lattice (Fig. 5). The shape of the bodies thus obtained may be not very desirable. Nevertheless, one can make them the basis for producing desired flows by mappings of the kind (6.1) just described.

Doubtless, this basic incompressible flow can be mapped into the incompressible flow outside a unit circle with certain singularities in the field, such as that given by (3.6), with the second term on the right dropped. The direct mapping of the final compressible flow from the plane of the circle will then be of the kind (3.6). Only experience will tell whether these alternative methods just proposed will be more advantageous than the general theory for certain types of design problems. It is probably safe to say that at least the general theory of §3 will be easier to apply for the determination of the general flow characteristics.

**Acknowledgment.** This work was carried out at Brown University and sponsored by Pratt & Whitney Aircraft Division of United Aircraft Corporation. The author is indebted to these organizations and the Navy Department Bureau of Aeronautics for their permission to publish the material.

REFERENCES

1. Lin, C. C., Quarterly of Applied Mathematics 4: 291–297, 1946.
2. Gelbart, A., NACA Tech. Note No. 1170, 1947.
3. Bers, L., NACA Tech. Note No. 1006, 1946.

Massachusetts Institute of Technology.

(Received December 9, 1948)

# ON A PERTURBATION THEORY BASED ON THE METHOD OF CHARACTERISTICS

C. C. LIN

**1. Introduction*.** It is known that a perturbation theory in the physical plane for supersonic flow and wave propagation is not free from difficulties. A method for improvement has been given by Lighthill[1] and Whitham[2] for dealing with outgoing waves. On the other hand, the characteristic parameters are the natural independent variables to be used for hyperbolic differential equations and it may be expected that in a perturbation theory developed in terms of the characteristic form of the differential equations, such difficulties will not occur.

In the method of characteristics, the desired solution functions $u(x, y)$ and $v(x, y)$ are given in the parametric representation. We introduce a system of characteristic parameters $\alpha$ and $\beta$, and express $x$, $y$, as well as $u$, $v$, in terms of them. In a perturbation theory, the solution thus appears in the form

$$
\begin{aligned}
u &= u^{(0)}(\alpha, \beta) + \epsilon u^{(1)}(\alpha, \beta) + \cdots \\
v &= v^{(0)}(\alpha, \beta) + \epsilon v^{(1)}(\alpha, \beta) + \cdots \\
x &= x^{(0)}(\alpha, \beta) + \epsilon x^{(1)}(\alpha, \beta) + \cdots \\
y &= y^{(0)}(\alpha, \beta) + \epsilon y^{(1)}(\alpha, \beta) + \cdots
\end{aligned}
\tag{1.1}
$$

Now, such parametric representations have two obvious disadvantages, both of which are often emphasized in the hodograph method. In the first place, the boundary conditions are often specified in the physical plane, and becomes somewhat difficult to impose in the parametric form. Secondly, a single-valued solution in the physical plane could become multiple-valued in the parametric representation (See Cragg's[3] discussion of branch lines in the hodograph method). These difficulties might partly account for the fact that the method of characteristics has not been more extensively used for the development of a perturbation theory. However, if the flow is a perturbation from a nearly uniform state, our freedom in the choice of characteristic parameters makes it possible to avoid multiple-valued mappings of the nature of the branch line in the hodograph method. We observe that, in this case, the characteristic curves are nearly straight in the physical plane. Consequently, the mapping between the plane of characteristic parameters and the physical plane can be made an approximately affine transformation. The mapping between the two planes is expected to be one-to-one and can be easily visualized. In the case of shock formation, associated with the convergence of characteristics in the physical plane, the solution is still expected to be single-valued in the plane of characteristic parameters. Here, the mapping is not one-to-one; but the

* The author wishes to thank Professor K. O. Friedrichs for his suggestions in the final revision of this section.

multiple-valued solution in the physical plane is necessary for the correct representation of the physical phenomena.

A systematic development of a perturbation theory along the lines indicated above is not yet available in the literature, although the ideas are implied in the work of Donov[4] and of Meyer[5], and suggested explicitly by Friedrichs[6,7].

Donov treated the supersonic flow past a two-dimensional thin airfoil, and carried out the calculations to the fourth-order. However, the treatment depends much on the specific nature of the problem at hand, and is not formulated into a general scheme. Furthermore, the characteristic parameters chosen are related to the velocity variables, and hence, when other problems are treated, this approach may suffer from the inconvenience discussed on the preceding page. Meyer treated the problem of waves of finite amplitude in ducts. The method, like that of Lighthill and Whitham, is formulated in terms of successive corrections, rather than treating $u$, $v$ and $x$, $y$ as four functions of two parameters. Thus, his second approximation (to which his solution is carried out) actually corresponds to the first approximation in the form (1.1).

Friedrichs used a coordinate system with one physical distance and one characteristic parameter for the development of a perturbation theory for the flow past a two-dimensional thin airfoil. In this scheme, the solution is found to be very simple. In fact, the flow is approximated by a simple wave up to the second approximation. For higher approximations, he suggested the use of characteristic parameters related to certain linear distances in the physical plane[7].

In the present paper, we shall carry out a systematic development of this idea in the formulation (1.1) with emphasis on the proper choice of the parameters $\alpha$ and $\beta$. The method is applied to a few specific problems to show its advantage over the usual perturbation theory in the physical plane even in the *first* approximation. A convergence proof is given for the case of two-dimensional supersonic flow. The problems of propagation of plane, cylindrical and spherical waves are treated by Miss Phyllis A. Fox[8] by similar methods. This work will be reported separately.*

It should be emphasized here that Lighthill does not limit his approach to the hyperbolic system of differential equations. In fact, he gave examples where the coordinate perturbation leads to improvement of the solution of ordinary differential equations and partial differential equations of the elliptic type. In such cases, there is no "natural" coordinate system, and it appears that the method is sometimes not applicable in the usual form.† However, the Lighthill-Whitham idea is extremely important and valuable, since the method removes serious difficulties in a very elegant and simple manner whenever it can be applied.

* During a visit to M.I.T. on December 17, Dr. Keller showed me some of his old notes dealing with development of suitable methods of successive approximation similar to that in Fox's thesis.

† Cf. Fox's thesis [8]. This kind of conclusion is also reached by Mr. Tozer, working under Professor Lighthill.

Some further comments on coordinate perturbation will be found in the last section.

**2. Perturbation theory in characteristic coordinates.** Consider steady supersonic flow of a gas in two-dimensions. Let $\rho$ be the density, and $u$, $v$ be the components of velocity along the direction of coordinate axes $x$, $y$. We shall limit our discussion to the case of isentropic irrotational flow. Then the pressure $p$ and the speed of sound $c$ are functions of $\rho$. This is also true of the speed $q$, the Mach number $M = q/c$ and the Mach angle

$$A = \sin^{-1}(1/M). \tag{2.1}$$

In the following, we shall use the Mach angle $A$ as the principal variable.

The basic equations may be written in the form

$$\begin{aligned} v_x - u_y &= 0 \\ (c^2 - u^2)u_x - uv(u_y + v_x) + (c^2 - v^2)v_y &= 0. \end{aligned} \tag{2.2}$$

For a small deviation from a state of uniform flow, it is customary to consider a perturbation theory of the form

$$\begin{aligned} u &= U + \epsilon u^{(1)}(x, y) + \cdots \\ v &= \epsilon v^{(1)}(x, y) + \cdots \end{aligned} \tag{2.3}$$

Bernoulli's equation would then give a similar perturbation series for $c^2$. The first approximation of such a perturbation procedure gives solutions of the form

$$\begin{aligned} u^{(1)} &= f(y - \lambda x) + g(y + \lambda x), \\ v^{(1)} &= (1/\lambda)\{-f(y - \lambda x) + g(y + \lambda x)\} \end{aligned} \tag{2.4}$$

where $\lambda = 1/\sqrt{M_0^2 - 1}$, and $M_0$ is the Mach number in the undisturbed stream. The lines

$$y \pm \lambda x = \text{const.} \tag{2.5}$$

are the characteristics in the first approximation.

However, these solutions are inadequate for the representation of certain important cases of small deviations from a uniform flow; e.g., the Prandtl-Meyer expansion. In such a case, the characteristics are definitely not parallel. The reflection of a Prandtl-Meyer flow from a solid boundary is another case in point. These problems require a more exact treatment of the characteristics themselves.

In the $(x, y)$ plane, the characteristic directions $C_\pm$ are given by

$$dy/dx = \tan(\theta \pm A), \tag{2.6}$$

where $\theta$ is the direction of flow. These are only approximately satisfied by (2.5), which would be obtained by setting $\theta = 0$ and $A = A_0$ in (2.6). However, this small deviation in the direction would amount to a large deviation in the coordinates $x$, $y$ if the exact and approximate characteristics are respectively followed over a long distance. While one family of the characteristics in the case of Prandtl-Meyer flow converges to a point at a finite distance, the approximation (2.5) gives parallel characteristics. In the case of reflection mentioned above, *both* families of characteristics are inadequately approximated by (2.5).

Another way of seeing the inadequacy of the usual perturbation theory is to observe that the velocity and density gradients become infinite as the corner is approached.

The above difficulties would disappear if we use the proper set of coordinates; namely, the characteristic system. If $\alpha$ and $\beta$ are a set of characteristic parameters, the basic equations (2.2) may be written as[7]

$$\begin{cases} y_\alpha \cos(\theta + A) = x_\alpha \sin(\theta + A) \\ y_\beta \cos(\theta - A) = x_\beta \sin(\theta - A) \end{cases} \tag{2.7}$$

$$\begin{cases} v_\alpha \sin(\theta - A) = -u_\alpha \cos(\theta - A) \\ v_\beta \sin(\theta + A) = -u_\beta \cos(\theta + A) \end{cases} \tag{2.8}$$

We may now apply a perturbation theory in the form (1.1), with $u$, $v$, $x$, $y$ as functions of the independent variables $\alpha$, $\beta$; $\theta$ and $A$ being functions of $u$ and $v$. These will be developed in detail for specific examples in the next section. We shall first make the following general observations.

(i) *Relation with other methods.* It is well-known that the characteristics are fixed in the hodograph plane in the present case. This means that (2.8) can be integrated. In fact, we obtain

$$\theta + \psi(A) = F(\alpha), \qquad \theta - \psi(A) = G(\beta) \tag{2.9}$$

where

$$\psi(A) = +\int \frac{\sqrt{M^2 - 1}\, dq}{q} \tag{2.10}$$

In the case $p = K\rho^\gamma$, we have

$$\psi(A) = A - \tfrac{1}{2}\pi + (1/\mu)\tan^{-1}(\mu \cot A), \quad \mu^2 = (\gamma - 1)/(\gamma + 1). \tag{2.11}$$

The constant of integration is here chosen so that $\psi(A) = 0$ at $A = 0$, i.e., the sonic point.

With the solutions (2.9), only the pair of equations (2.7) need be integrated. The present method therefore becomes closely related to the hodograph method. However, it is more useful for the following two reasons. First, the basic importance of the characteristics is explicitly recognized, and hence less trouble is expected with discontinuities. Secondly, the characteristic parameters are still at our disposal, and we may choose them so that the mapping between the characteristic plane and the physical plane is simple. This would avoid unnecessary multiple-valued functions in the mapping process, as it often happens in the hodograph method. Indeed, such a choice may be made so that the usual perturbation theory can be used as a guide. We note that the initial approximation to (2.7) is given by

$$y_\alpha^{(1)} = x_\alpha^{(0)} \tan A_0 , \qquad y_\beta^{(0)} = -x_\beta^{(0)} \tan A_0 ; \tag{2.12}$$

and hence

$$y^{(0)} - \lambda x^{(0)} = k(\beta), \qquad y^{(0)} + \lambda x^{(0)} = h(\alpha), \qquad \lambda = \tan A_0 . \tag{2.13}$$

Now if we take $h(\alpha) = h\alpha$, $k(\beta) = k\beta$, the coordinates $\alpha$ and $\beta$ form an oblique Cartesian system obtained from $x^{(0)}$ and $y^{(0)}$ by a linear transformation. In fact, they are exactly the ones suggested by (2.5).

In other problems, such as flow with axial symmetry or spherical waves, the hodograph method is not always useful. The present theory, however, remains essentially the same.

(ii) *The simple wave.* The Prandtl-Meyer problem becomes trivial in the present method, but we shall give the solution here in a form convenient for later reference. Consider the special case $F(\alpha) = F_0$, a constant. Then

$$\theta = \tfrac{1}{2}[F_0 + G(\beta)], \qquad \psi(A) = \tfrac{1}{2}[F_0 - G(\beta)], \tag{2.14}$$

and the first equation of (2.7) gives a family of straight-line characteristics $C_+$. If the angle between the $x$-axis and these characteristics is denoted by $\phi$, then $\phi = \theta + A$. Since (2.14) gives $\theta = F_0 - \psi(A)$, we obtain $\phi = F_0 + A - \psi(A)$, or, by (2.11)

$$\begin{aligned} \tan A &= -\mu \cot \mu(\phi - \phi_*), \qquad \phi_* = \psi(A_0) + \pi/2, \\ \theta &= -[\psi(A) - \psi(A_0)], \end{aligned} \tag{2.15}$$

where the constants have been chosen so that $\theta = 0$ when $A = A_0$. Eq. (215) holds for general simple waves; the dependence of the positional coordinates on $\phi$ changes with the particular solution considered. For waves centered at the origin, it is convenient to introduce the polar coordinate system $(r, \phi)$ by

$$x = r \cos \phi, \qquad y = r \sin \phi. \tag{2.16}$$

The streamlines are given by*

$$r = R \cos^{-\mu^{-2}} \mu(\phi - \phi_*) \tag{2.17}$$

and the cross Mach lines are

$$r = R \cos^{-\frac{1}{2}\mu^{-2}} \mu(\phi - \phi_*) \sin^{-\frac{1}{2}} \mu(\phi - \phi_*). \tag{2.18}$$

**3. Application to certain problems in steady supersonic flow.** We shall now apply our method to a few specific problems which cannot be conveniently solved by the usual perturbation theory in the physical plane. The problems are as follows:

(1) interaction of two simple waves,
(2) reflection of a simple wave from a solid boundary,
(3) reflection of a simple wave from a free surface,
(4) the wave length of a supersonic jet.

* Courant-Friedrichs,[7] p. 273. The notation here is somewhat different.

Before we go into the details of any of these problems, let us first develop the general perturbation theory up to the first order.

According to (2.10), the direction and magnitude of the flow are given by

$$\theta = \tfrac{1}{2}[F(\alpha) + G(\beta)], \qquad \psi(A) = \tfrac{1}{2}[F(\alpha) - G(\beta)]. \tag{3.1}$$

If we want to apply these to (2.7), we have to evaluate $\tan(\theta \pm A)$. Now suppose $\theta$ and $A$ differ but slightly from $\theta_0$ and $A_0$, then we may develop $\tan(\theta \pm A)$ into power series in $\theta - \theta_0$ and $A - A_0$. To the first order, we have

$$\tan(\theta \pm A) = \pm\{\tan A_0 + \sec^2 A_0 (A - A_0)\} + \sec^2 A_0 (\theta - \theta_0). \tag{3.2}$$

We also have

$$\begin{aligned} \psi(A) - \psi(A_0) &= \psi'(A_0)(A - A_0), \\ \psi'(A_0) &= -(1 - \mu^2)/(\mu^2 + \tan^2 A_0). \end{aligned} \tag{3.3}$$

Thus, combining (3.1), (3.2) and (3.3), we obtain

$$\begin{aligned} \tan(\theta + A) &= \tan A_0 + \tfrac{1}{2}m[F(\alpha) - F(\alpha_0)] + \tfrac{1}{2}n[G(\beta) - G(\beta_0)], \\ \tan(\theta - A) &= -\tan A_0 + \tfrac{1}{2}n[F(\alpha) - F(\alpha_0)] + \tfrac{1}{2}m[G(\beta) - G(\beta_0)] \end{aligned} \tag{3.4}$$

where $\theta = \theta_0$ and $A = A_0$ for $\alpha = \alpha_0$ and $\beta = \beta_0$, and

$$m = \sec^2 A_0 [1 + 1/\psi'(A_0)], \qquad n = \sec^2 A_0 [1 - 1/\psi'(A_0)]. \tag{3.5}$$

Notice that $\psi'(A_0) < 0$ and therefore $n > 0$. However, $m$ changes sign as $A_0$ passes through $A_c$, which is given by

$$\tan^2 A_c = 1 - 2\mu^2, \tag{3.6}$$

or

$$M_c = \sqrt{2(1 - \mu^2)/(1 - 2\mu^2)} = \sqrt{4/(3 - \gamma)}. \tag{3.7}$$

For $\gamma = 1.4$, $M_c = 1.58$. For this critical Mach number, *the characteristics are straight lines.*

Let us now substitute (3.4) into (2.7). As indicated in the last section ((2.12) and (2.13)), the initial approximation should be taken as

$$y^{(0)} - \lambda x^{(0)} = k\beta, \qquad y^{(0)} + \lambda x^{(0)} = h\alpha \tag{3.8}$$

where $h$ and $k$ are constants. Now certain conditions will be prescribed for $x$ and $y$, since they are two of the dependent variables sought. Thus, (3.8) in effect imposes certain restrictions on the choice of $\alpha$ and $\beta$. In effect, it means that (at least to an initial approximation) $\alpha$ and $\beta$ must be chosen to be some linear distance along a straight line onto which the points on the initial curve are projected.

The particular choice will depend on the problems in question. It is often found convenient to impose the condition

$$y - \lambda x = k\beta, \qquad y + \lambda x = h\alpha. \tag{3.9}$$

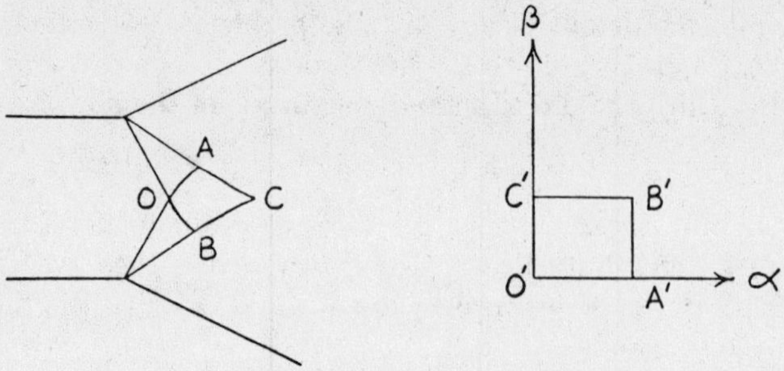


FIG. 1. Interaction of two centered simple waves

Then some of the initial conditions for $x^{(i)}$, $y^{(i)}$ become $x^{(i)} = 0$, $y^{(i)} = 0$ for $i \geqq 1$.

Let us now consider a few specific problems.

(1) *Interaction of two simple waves.* The region of iteration is bounded by four characteristics forming a rectangle in the $(\alpha, \beta)$ plane. Along $OA$ and $OC$, the conditions are specified by the simple wave relations; i.e., $x$, $y$, $A$ and $\theta$ are known as a function of a single parameter along $OA$ and $OC$ (cf. Eqs. (2.15) and (2.18)). We may take this parameter to be $x$ itself and identify it with $\alpha$ along $O'A'$. Then, we have, along $O'A'$,

$$x = \alpha, \qquad \theta = \epsilon f(\alpha), \tag{3.10}$$

while $y$ and $A$ can be calculated from the relations along a characteristic $C_+$. Similarly, along $OC$,

$$x = \beta, \qquad \theta = \epsilon g(\beta). \tag{3.11}$$

We have thus a characteristic initial value problem.

Since $\theta = 0$ at the point 0, $f(0) = g(0) = 0$, and Eq. (3.1) becomes

$$\begin{aligned} \theta &= \epsilon[f(\alpha) + g(\beta)], \\ \psi(A) - \psi(A_0) &= \epsilon[f(\alpha) - g(\beta)]. \end{aligned} \tag{3.12}$$

The initial approximation (3.8) for the coordinates gives

$$x^{(0)} = \alpha + \beta, \qquad y^{(0)} = (\alpha - \beta) \tan A_0 . \tag{3.13}$$

For the next approximation, (2.7) gives

$$\begin{aligned} y_\alpha^{(1)} - \lambda x_\alpha^{(1)} &= mf(\alpha) + ng(\beta), \\ y_\beta^{(1)} + \lambda x_\beta^{(1)} &= nf(\alpha) + mg(\beta). \end{aligned} \tag{3.14}$$

The initial conditions for $x^{(1)}$ are

$$x^{(1)} = 0 \text{ for } \alpha = 0 \text{ and } \beta = 0. \tag{3.15}$$

Now, from (3.14),

$$2\lambda x^{(1)} = n[\beta f(\alpha) - \alpha g(\beta)] + m\left[\int_0^\beta g(\beta)\, d\beta - \int_0^\alpha f(\alpha)\, d\alpha\right] + C_2(\alpha) - C_1(\beta),$$

where $C_1(\beta)$ and $C_2(\alpha)$ are additive "constants" of partial integration. By applying conditions (3.15) for $\alpha = 0$ and $\beta = 0$ successively, we obtain finally

$$(3.16)\quad \begin{cases} 2\lambda x^{(1)} = n[\beta f(\alpha) - \alpha g(\beta)], \\ 2y^{(1)} = n[\beta f(\alpha) + \alpha g(\beta)] + m\left[\int_0^\alpha f(\alpha)\,d\alpha + \int_0^\beta g(\beta)\,d\beta\right]. \end{cases}$$

The solution is therefore given in the parametric form as follows:

$$(3.17)\quad \begin{cases} \theta = \epsilon[f(\alpha) + g(\beta)], \\ \psi(A) - \psi(A_0) = \epsilon[f(\alpha) - g(\beta)], \\ x = (\alpha + \beta) + \epsilon(n/2\lambda)[\beta f(\alpha) - \alpha g(\beta)], \\ y = (\alpha - \beta)\lambda + \epsilon(n/2)[\beta f(\alpha) + \alpha g(\beta)] \\ \qquad\qquad + \epsilon(m/2)\left[\int_0^\alpha f(\alpha)\,d\alpha + \int_0^\beta g(\beta)\,d\beta\right], \end{cases}$$

where the functions $f$ and $g$ are given by the conditions

$$(3.18)\quad \begin{aligned} \theta &= \epsilon f(x) \text{ along the initial characteristic } C_+ \\ \theta &= \epsilon g(x) \text{ along the initial characteristic } C_- \end{aligned}$$

(2) *Reflection of a simple wave from a solid boundary.* This is easily obtained from the last case by taking $g(x) = -f(x)$. Then (3.17) gives $\theta = 0$ along $\alpha = \beta$, which is also the line $y = 0$, by symmetry. Thus, for the case of reflection,

$$(3.19)\quad \begin{cases} \theta = \epsilon[f(\alpha) - f(\beta)] \\ \psi(A) - \psi(A_0) = \epsilon[f(\alpha) + f(\beta)] \\ x = (\alpha + \beta) + \epsilon(n/2\lambda)[\beta f(\alpha) + \alpha f(\beta)] \\ y = (\alpha - \beta)\lambda + \epsilon(n/2)[\beta f(\alpha) - \alpha f(\beta)] \\ \qquad\qquad + \epsilon(m/2)\left[\int_0^\alpha f(\alpha)\,d\alpha - \int_0^\beta f(\beta)\,d\beta\right]. \end{cases}$$

In particular, along $\alpha = \beta$, $y = 0$, $\theta = 0$, and

$$(3.20)\quad \begin{aligned} \psi(A) - \psi(A_0) &= 2\epsilon f(\alpha) \\ x &= 2\alpha + \epsilon(n/\lambda)\alpha f(\alpha) \end{aligned}$$

The first formula shows that the pressure deviation is doubled by reflection, in agreement with the usual linear theory. The second formula shows that the width of the interaction region along the solid boundary is *longer* than that indicated by the usual linear theory if the incoming wave is an *expansion*, such as the Prandtl-Meyer flow. It is longer or shorter than that indicated by a geometrical extension of the Prandtl-Meyer fan to the wall depending on whether $M_0$ is larger or smaller than $M_c$. In fact, the following simple geometrical construction can be easily verified.

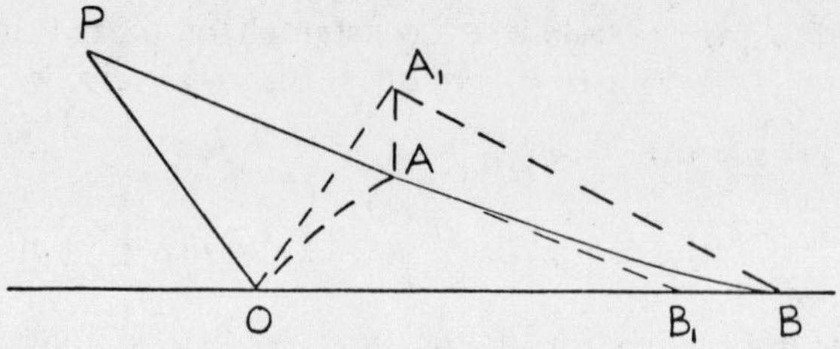


(Mo>Mc)

FIG. 2. Reflection of a Prandtl-Meyer expansion from a solid boundary

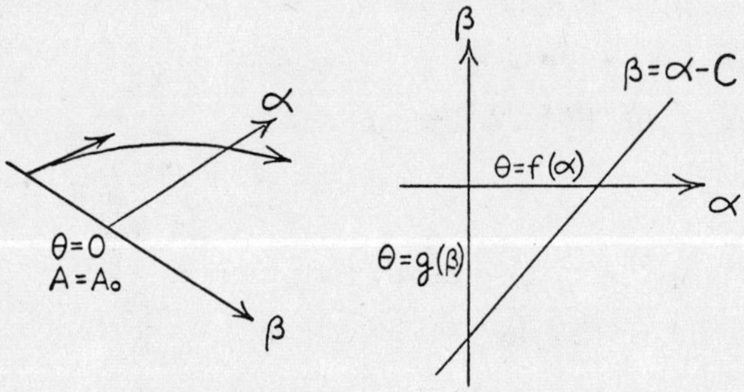


FIG. 3. Reflection from a state of rest

Suppose the Prandtl-Meyer flow is given by the triangle $OPB_1$ in the absence of the wall, with $OA$ as a cross characteristic. The characteristic $PA$ will intersect the wall at $B$ which is obtained by drawing the lines $OA_1$ and $A_1B$ parallel to the exact characteristics at $O$ and $A$ respectively; the point $A_1$ is on a vertical line through $A$. Fig. 2 shows the case $M_0 > M_c$, with the point $A_1$ above $A$. In the case $M_0 = M_c$, the characteristics are straight lines, and the point $B$ coincides with $B_1$.

(3) *Reflection from a region of uniform flow.* This is obtained from (I) by taking

$$g(\beta) = f(\beta + C) + D, \qquad f(C) + D = 0, \tag{3.21}$$

and consider the condition along a curve

$$\beta = \alpha - C + h(\alpha) \tag{3.22}$$

slightly different from the straight line $\beta + C = \alpha$. Then (3.16) give

$$\left\{\begin{aligned} \theta &= \epsilon[f(\alpha) + f(\beta + C) + D] \\ \psi(A) - \psi(A_0) &= \epsilon[f(\alpha) - f(\beta + C) - D] \\ x &= \alpha + \beta + \epsilon(n/2\lambda)[\beta f(\alpha) - \alpha f(\beta + C) - \alpha D] \\ y &= (\alpha - \beta)\lambda + \epsilon(n/2)[\beta f(\alpha) + \alpha f(\beta + C) + 2D] \\ &\quad + \epsilon(m/2)\left[\int_0^\alpha f(\alpha)\, d\alpha + \int_0^\beta f(\beta + C)\, d\beta + D\beta\right]. \end{aligned}\right. \tag{3.23}$$

Along $\beta = \alpha - C + h(\alpha)$,

$$(3.24)\quad \begin{cases} \theta = \epsilon[2f(\alpha) + D] \\ \psi(A) - \psi(A_0) = -\epsilon D \\ x = 2\alpha - C + \epsilon h(\alpha) + \epsilon(n/2\lambda)[-Cf(\alpha) - D\alpha] \\ y = C\lambda - \epsilon\lambda h(\alpha) + \epsilon(n/2)[(2\alpha - C)f(\alpha) + D\alpha] \\ \quad + \epsilon(m/2)\left[2\int_0^\alpha f(\alpha)\,d\alpha - \int_0^C f(\alpha)\,d\alpha + D(\alpha - C)\right] \end{cases}$$

The function $h(\alpha)$ must be determined so that

$$\frac{dy}{d\alpha}\Big/\frac{dx}{d\alpha} = \theta$$

In fact, this gives

$$\lambda h'(\alpha) = \tfrac{1}{2}n(2\alpha - C)f'(\alpha) + (m + n - 4)f(\alpha) + [\tfrac{1}{2}(m + n) - 4]D$$

We note that from (3.22) the two points on the axes $(C, 0)$ and $(0, -C)$ give

$$(3.25)\qquad \theta = \epsilon f(C) \text{ and } \theta = -\epsilon f(C)$$

respectively. Such a symmetry relation does not exist in general for other points on the free streamline.

(4) *Wave length of a supersonic jet.* As a final application, we consider a plane supersonic jet when a gas stream is discharged into a gas at rest at a somewhat lower pressure (Fig. 4). If all the characteristics involved were straight, it would be very easy to calculate the length of the jet. Although this is not exactly the case, it is sufficiently accurate for the purpose, if errors of the *second* order are tolerated. This may be seen as follows.

Consider the broken curve $POADEQ$. The portion $PO$ is straight and has a slope

$$\lambda_0 = \sqrt{M_0^2 - 1}.$$

The part $OA$ is curved, while the part $AD$ is straight with the slope

$$\lambda_0 + m\Delta\theta,$$

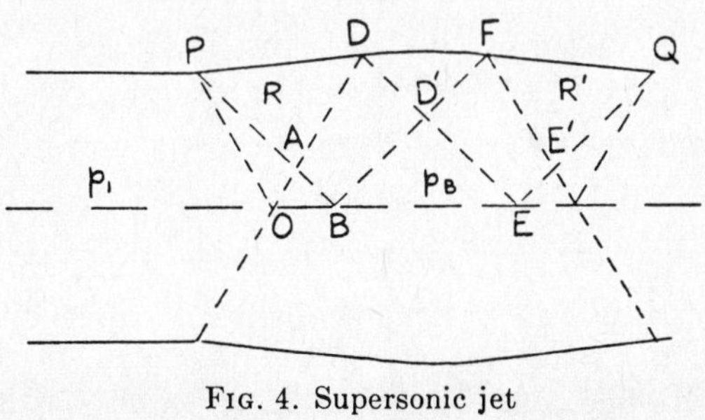


FIG. 4. Supersonic jet

where $\Delta\theta$ is the direction of flow in the region $R$ relative to the original stream. Now, the slope of the curve $OA$ differs from that of $AD$ by quantities of the first order. The length $OA$ itself is a small quantity of the first order compared with the length $PO$, say. Therefore, in trying to locate the point $A$ accurate up to the first order, the slope of $OA$ may be taken to be that of $AD$; i.e., we may regard $OD$ as a straight line having the slope $\lambda_0 + m\Delta\theta$ for the purposes of calculating the length of the jet. The same remark may be applied to the portions $DD'$ and $EE'$ as portions of $DE$ and $EQ$. We then have the following list of slopes:

| | |
|---|---|
| $PO$ | $\lambda_0$ |
| $OD$ | $\lambda_0 + m\Delta\theta$ |
| $DE$ | $\lambda_0 + 2\sec^2 A_0(\Delta A)$ |
| $EQ$ | $\lambda_0 - n\Delta\theta$ |

This list can be easily constructed by noting the following two results established previously. From (3.20), we have

$$P_B - P_A = P_A - P_1\,;$$

and from (3.25), we see that the direction of flow in the region $R'$ is obtained from that in the region $R$ by a reflection about a horizontal axis.

We can now easily calculate the distance $PQ$ from simple geometrical considerations. If the semiwidth of the jet is denoted by $W$, and the distance of the point $D$ above the horizontal line $PQ$ is denoted by $h$, (a quantity of the first order), then the length $PQ$ is given by $\Lambda$, where

$$\Lambda - 2\lambda_0 h - \tfrac{1}{2}W[4\lambda_0 + 2\sec^2 A_0(\Delta A) + (m-n)\Delta\theta] = 2[\lambda_0 + \sec^2 A_0(\Delta A)]W$$

Now $h$ is given to the desired degree of approximation by

$$h = (\overline{PD})\cdot\Delta\theta = \lambda_0 W(\Delta\theta).$$

Putting this into the last formula, we find that

$$\Lambda/W = 2[\lambda_0 + \sec^2 A_0(\Delta A)] + 2\lambda_0^2(\Delta\theta),$$

or

$$\Lambda/W = 2\sqrt{\bar{M}^2 - 1}(1 + \sqrt{\bar{M}^2 - 1}\,\Delta\theta), \tag{3.26}$$

where $\bar{M}$ is the Mach number of the flow in the Region $R$.

The formula (3.26) gives the length of the supersonic jet with an error of the second order. Previously, Prandtl gave the formula

$$\Lambda/W = 2\sqrt{M_0^2 - 1} \tag{3.27}$$

and Pack suggested the replacement of $M_0$ by $\bar{M}$. Our formula differs from that of Pack's by an additional correction factor.

Now, Pack used the usual linear theory, and had to be satisfied with an approximation to the jet consisting of sections of trapezoidal form, and he remarked that "this result does not represent physical reality." Since the difference between Prandtl's formula and Pack's formula is of the order $\epsilon$, it is essential that the improved formula be certainly established including terms of the order $\epsilon$. Unless the solution is accurate up to the order of $\epsilon$, there is always the uncertainty of the reference condition up to the same order. Thus, the choice between $M_0$ and $\bar{M}$ cannot be made conclusive without an accurate solution of the field. From the present standpoint, the usual linear theory is equivalent to the use of the solutions for $(A, \theta)$ accurate to the first approximation and the solutions for $(x, y)$ to the zeroth order. Thus, the accuracy of the calculation of a length can hardly be expected to be conclusive up to the first order. In the present treatment, such difficulties do not occur.

One of Prandtl's experiments gives the value $\Lambda/W = 3.8$. This corresponds to a Mach number $M_0 = 1.85$ and a reservoir pressure 7.74 times the external pressure. Taking $\gamma = 1.40$, we find that $\bar{M} = 1.99$ from this pressure ratio. Prandtl's formula gives the theoretical value 3.1 for $\Lambda/W$, Pack's modification gives the value 3.4 while our formula (3.26) gives the value 3.8, in agreement with experiment. (The value of $\Delta\theta$ can be easily obtained from a table for simple waves.)

Another more complicated formula was obtained by Hasimoto[10] by the hodograph method; it also gives the correct value 3.8 in this case.

**4. Solution by perturbation series.** The characteristic initial value problem formulated in §2 concerns a system of differential equations of the type

$$y_\alpha = P(\alpha, \beta, \epsilon)x_\alpha, \qquad y_\beta = Q(\alpha, \beta, \epsilon)x_\beta; \tag{4.1}$$

where $P(\alpha, \beta, 0)$ and $Q(\alpha, \beta, 0)$ are unequal constants. The initial conditions are

$$x = \alpha \quad \text{along } \beta = 0, \qquad x = \beta \quad \text{along } \alpha = 0. \tag{4.2}$$

The variable $y$ is then fully determined by the condition $y = 0$ at $\alpha = \beta = 0$. If $P(\alpha, \beta, \epsilon)$ and $Q(\alpha, \beta, \epsilon)$ are analytic functions of the parameter $\epsilon$, then according to existing theory,[11] one may expect the solutions $x(\alpha, \beta, \epsilon)$ and $y(\alpha, \beta, \epsilon)$ to be also analytic functions of $\epsilon$, expressible as power series in $\epsilon$, which are convergent for a finite region in the $(\alpha, \beta)$ plane and sufficiently small values of $\epsilon$. Actually, however, it can be shown that the convergence of such series holds for all positive values of $\alpha$ and $\beta$, provided the power series

$$\begin{aligned} P(\alpha, \beta, \epsilon) &= P^{(0)}(\alpha, \beta) + \epsilon P^{(1)}(\alpha, \beta) + \cdots, \\ Q(\alpha, \beta, \epsilon) &= Q^{(0)}(\alpha, \beta) + \epsilon Q^{(1)}(\alpha, \beta) + \cdots, \end{aligned} \tag{4.3}$$

and

$$\begin{aligned} P_\alpha(\alpha, \beta, \epsilon) &= P_\alpha^{(0)}(\alpha, \beta) + \epsilon P_\alpha^{(1)}(\alpha, \beta) + \cdots, \\ Q_\beta(\alpha, \beta, \epsilon) &= Q_\beta^{(0)}(\alpha, \beta) + \epsilon Q_\beta^{(1)}(\alpha, \beta) + \cdots, \end{aligned} \tag{4.4}$$

are convergent for all positive values of $\alpha$ and $\beta$. The usual proof can be extended to cover this case, but a simple direct proof is given below, since it also gives an indication of the order of magnitude of the successive terms in the solution series. Before proving this, let us first calculate the successive approximations in

$$x = x^{(0)}(\alpha, \beta) + \epsilon x^{(1)}(\alpha, \beta) + \cdots, \quad y = y^{(0)}(\alpha, \beta) + \epsilon y^{(1)}(\alpha, \beta) + \cdots \tag{4.5}$$

explicitly.

For the initial approximation, we have

$$y_\alpha^{(0)} = P^{(0)} x_\alpha^{(0)}, \qquad y_\beta^{(0)} = Q^{(0)} x_\beta^{(0)}, \tag{4.6}$$

and the conditions

$$x^{(0)} = \alpha \text{ for } \beta = 0, \qquad x^{(0)} = \beta \text{ for } \alpha = 0. \tag{4.7}$$

The solution is easily verified to be

$$x^{(0)} = \alpha + \beta, \qquad y^{(0)} = P^{(0)}\alpha + Q^{(0)}\beta \tag{4.8}$$

For the higher approximations, $n \geqq 1$, we have

$$y_\alpha^{(n)} - P^{(0)} x^{(n)} = \textstyle\sum_{k=1}^{n} P^{(k)}(\alpha, \beta) x_\beta^{(n-k)},$$

$$y_\beta^{(n)} - Q^{(0)} x_\beta^{(n)} = \textstyle\sum_{k=1}^{n} Q^{(k)}(\alpha, \beta) x_\beta^{(n-k)},$$

and the conditions

$$x^{(n)} = 0 \qquad \text{for } \alpha = 0 \text{ and } \beta = 0 \tag{4.10}$$

These equations may be integrated as follows:

$$\begin{cases} y^{(n)} - P^{(0)} x^{(n)} = \displaystyle\sum_{k=1}^{n} \int_0^\alpha P^{(k)} x_\alpha^{(n-k)}\, d\alpha + F^{(n)}(\beta), \\ y^{(n)} - Q^{(0)} x^{(n)} = \displaystyle\sum_{k=1}^{n} \int_0^\beta Q^{(k)} x_\beta^{(n-k)}\, d\beta + G^{(n)}(\alpha); \end{cases} \tag{4.11}$$

which gives

$$\begin{aligned} x^{(n)} = \frac{1}{Q^{(0)} - P^{(0)}} \sum_{k=1}^{n} \left\{ \int_0^\alpha P^{(k)} x_\alpha^{(n-k)}\, d\alpha - \int_0^\beta Q^{(k)} x_\beta^{(n-k)}\, d\beta \right\} \\ + \frac{1}{Q^{(0)} - P^{(0)}} \{F^{(n)}(\beta) - G^{(n)}(\alpha)\}. \end{aligned} \tag{4.12}$$

From the initial conditions (4.10), we see that $F^{(n)}(0) = G^{(n)}(0) = 0$. Putting $\alpha = 0$ and $\beta = 0$ in turn into (4.12), we then obtain

$$F^{(n)}(\beta) = \int_0^\beta Q^{(n)}(0, \beta)\, d\beta, \qquad G^{(n)}(\alpha) = \int_0^\alpha P^{(n)}(0, \alpha)\, d\alpha. \tag{4.13}$$

Equations (4.12) and (4.13) give the explicit solutions for $x^{(n)}(\alpha, \beta)$. From

(4.11), we see that the values of $y^{(n)}$ along the axes are given by (4.13). The solution $y^{(n)}$ is in general given by

$$y^{(n)} = \frac{1}{Q^{(0)} - P^{(0)}} \sum_{k=1}^{n} \left\{ Q^{(0)} \int_0^\alpha P^{(k)} x_\alpha^{(n-k)}\, d\alpha - P^{(0)} \int_0^\beta Q^{(k)} x_\beta^{(n-k)}\, d\beta \right\} \tag{4.14}$$
$$+ \frac{1}{Q^{(0)} - P^{(0)}} \{ Q^{(0)} F^{(n)}(\beta) - P^{(0)} G^{(n)}(\alpha) \}.$$

*Convergence proof.* The convergence proof shall now be carried out for the $x$-series. After this is done, the extension to the $y$-series is obvious.

We first carry out a partial integration for the integrals in (4.12); thus,

$$\int_0^\alpha P^{(k)} x_\alpha^{(n-k)}\, d\alpha = [P^{(k)} x^{(n-k)}]_0^\alpha - \int_0^\alpha P_\alpha^{(k)} x^{(n-k)}\, d\alpha.$$

Since $x^{(n-k)}(0, \beta) = 0$ except when $k = n$, we have

$$\int_0^\alpha P^{(k)} x_\alpha^{(n-k)}\, d\alpha = P^{(k)} x^{(n-k)} - \int_0^\alpha P_\alpha^{(k)} x^{(n-k)}\, d\alpha, \qquad \text{if } k \neq n$$
$$= P^{(n)}(\alpha, \beta) x^{(n-k)} - P^{(n)}(0, \beta)\beta - \int_0^\alpha P_\alpha^{(n)} x^{(0)}\, d\alpha, \quad \text{if } k = n.$$

Thus, for $n \geqq 1$,

$$x^{(n)} = -\frac{1}{Q^{(0)} - P^{(0)}} \sum_{k=1}^{n} \left\{ \int_0^\alpha P_\alpha^{(k)} x^{(n-k)}\, d\alpha - \int_0^\beta Q_\beta^{(k)} x^{(n-k)}\, d\beta \right\} \tag{4.15}$$
$$+ \frac{1}{Q^{(0)} - P^{(0)}} \{ [F^{(n)}(\beta) - \beta P^{(n)}(0, \beta)] - [G^{(n)}(\alpha) - \alpha Q^{(n)}\alpha, 0)] \};$$

from which we derive

$$| x^{(n)} | \leqq \frac{1}{| Q^{(0)} - P^{(0)} |} \sum_{k=1}^{n} \left\{ \int_0^\alpha | P_\alpha^{(k)} | | x^{(n-k)} | \, d\alpha + \int_0^\beta | Q_\beta^{(k)} | | x^{(n-k)} | \, d\beta \right\}$$
$$+ \frac{1}{| Q^{(0)} - P^{(0)} |} \left\{ \int_0^\beta | Q^{(n)}(0, \beta) | \, d\beta + \beta \, | P^{(n)}(0, \beta) | \right. \tag{4.16}$$
$$\left. + \int_0^\alpha | P^{(n)}(\alpha, 0) | \, d\alpha + \alpha \, | Q^{(n)}(\alpha, 0) | \right\}.$$

Consider now the sum

$$S_N = | x^{(0)} | + \textstyle\sum_{n=1}^{N} \epsilon^n \, | x^{(n)} |.$$

By using identities of the form

$$\textstyle\sum_{n=1}^{N} \sum_{k=1}^{n} \epsilon^n a^{(k)} \, | x^{(n-k)} | = \sum_{k=1}^{N} \sum_{n=k}^{N} \epsilon^n a^{(k)} \, | x^{(n-k)} | = \sum_{k=1}^{N} \epsilon^k a^{(k)} S_{N-k},$$

We can deduce from (4.8) and (4.13) that

$$S_N \leqq (\alpha+\beta) + \frac{1}{|Q^{(0)} - P^{(0)}|}\sum_{k=1}^{N} \cdot\left\{\int_0^\alpha \epsilon^k |P_\alpha^{(k)}| S_{N-k}\, d\alpha + \int_0^\beta \epsilon^k |Q_\beta^{(k)}| S_{N-k}\, d\beta\right\} + \frac{\epsilon}{|Q^{(0)} - P^{(0)}|}\left\{\int_0^\beta \left\{\hat{Q}_N(0,\beta)\, d\beta + \beta\hat{P}_N(0,\beta) + \int_0^\alpha \hat{P}_N(\alpha,0)\, d\alpha + \alpha\hat{Q}_N(\alpha,0)\right\}\right\}, \tag{4.17}$$

where

$$\begin{aligned} \epsilon\hat{P}_N(\alpha,\beta) &= \epsilon\,|P^{(1)}(\alpha,\beta)| + \cdots + \epsilon^N\,|P^{(N)}(\alpha,\beta)|,\\ \epsilon\hat{Q}_N(\alpha,\beta) &= \epsilon\,|Q^{(1)}(\alpha,\beta)| + \cdots + \epsilon^N\,|Q^{(N)}(\alpha,\beta)|. \end{aligned} \tag{4.18}$$

From the convergence of the series (4.3), we see that these sums increase to certain limits as $N \to \infty$. Thus, we can find a uniform upper bound to the sums $\hat{P}_N(\alpha, \beta)$ and $\hat{Q}_N(\alpha, \beta)$. This bound may be taken to be independent of $\alpha$, $\beta$ if we are dealing with a finite region in the $\alpha$, $\beta$ plane. Otherwise, we have to introduce the specific conditions that $\lim P_N$ and $\lim Q_N$ are uniformly bounded. These are usually satisfied in practice. Thus if the series (4.3) are bounded as follows

$$\bar{P}(\alpha,\beta,\epsilon) \leqq |P_0| + \epsilon J, \qquad \bar{Q}(\alpha,\beta,\epsilon) \leqq |Q_0| + \epsilon J, \tag{4.19}$$

we may write (4.14) as

$$S_N \leqq \left(1 + \frac{2\epsilon J}{|Q^{(0)} - P^{(0)}|}\right)(\alpha+\beta) + \frac{1}{|Q^{(0)} - P^{(0)}|}\sum_{k=1}^{N}\left\{\int_0^\alpha \epsilon^k\,|P_\alpha^{(k)}|\,S_{N-k}\,d\alpha + \int_0^\beta \epsilon^k\,|Q_\beta^{(k)}|\,S_{N-k}\,d\beta\right\}. \tag{4.20}$$

We now introduce similar restrictions for the series of the derivative $P_\alpha$ and $Q_\beta$:

$$\begin{aligned} [\epsilon\,|P_\alpha^{(1)}| + \epsilon^2\,|P_\alpha^{(2)}| + \cdots] &< \epsilon\tilde{J},\\ [\epsilon\,|Q_\beta^{(1)}| + \epsilon^2\,|Q_\beta^{(2)}| + \cdots] &< \epsilon\tilde{J}. \end{aligned} \tag{4.21}$$

Then, noting that $S_i$ increases monotonically with the index $i$, we may write (3.24) in the form

$$S_N \leqq \left(1 + \frac{2\epsilon J}{|Q^{(0)} - P^{(0)}|}\right)(a+\beta) + \frac{\epsilon\tilde{J}}{|Q^{(0)} - P^{(0)}|}\left(\int_0^\alpha S_N\,d\alpha + \int_0^\beta S_N\,d\beta\right). \tag{4.22}$$

We note that, roughly speaking, the right-hand side of (4.19) differs from $(\alpha + \beta)$ by quantities of the order of $\epsilon$ if we can find a uniform bound for $S_i$. This is the key to the convergence proof, which we carry out by induction.

We try a uniform bound of the form $Ke^{\alpha+\beta}$. This has the advantage of preserving its form upon integration. It is clear that

$$S_0 = \alpha + \beta < Ke^{\alpha+\beta} \tag{4.23}$$

for $K > 1$. Suppose now

$$S_j \leqq Ke^{\alpha+\beta} \text{ for } j = 0, 1, \cdots, N-1. \tag{4.24}$$

Then (4.22) gives

$$S_N \leqq \left(1 + \frac{2\,\epsilon\,(J + \bar{J})}{|\,Q^{(0)} - P^{(0)}\,|} K\right) e^{\alpha+\beta}$$

This is less than $Ke^{\alpha+\beta}$ if

$$1 + \frac{2\epsilon(J + \bar{J})}{|\,Q^{(0)} - P^{(0)}\,|} K < K,$$

or

$$K > \left\{1 - \frac{2\epsilon(J + \bar{J})}{|\,Q^{(0)} - P^{(0)}\,|}\right\}^{-1} > 1. \tag{4.25}$$

Thus, if

$$\epsilon < \tfrac{1}{2}\,|\,Q^{(0)} - P^{(0)}\,|/(J + \bar{J}), \tag{4.26}$$

We can indeed find a suitable $K$.

**5. Remarks on coordinate perturbation.** The Lighthill-Whitham approach, with correction made on *one* family of characteristics, leads essentially into the choice of coordinate system used in the theory of Friedrichs. In the present scheme, if the coordinates $(x, y)$ are calculated to the zeroth approximation, we obtain essentially the perturbation theory in the physical plane. If they are calculated to the first approximation, we obtain a solution corresponding to a generalization of the Lighthill-Whitham idea to the correction of both families of characteristics, after the perturbation theory has been developed to the first approximation in the physical plane. (The correction of both families of characteristics was not explicitly mentioned.) When higher approximations are desired, the connection between the Lighthill-Whitham method and a perturbation series is not easily seen, and it is convenient to start with the method of characteristics *ab initio*.

In the case of the Laplace equation, such as the thin airfoil problem considered by Lighthill,[12] the method of coordinate perturbation may be regarded as a perturbation theory in conformal mapping. If $F^{(0)}(z)$ is an analytic function whose imaginary part is the solution for a boundary problem associated with the boundary $C_0$, and $F(z)$ gives the solution for the corresponding problem with a slightly different boundary $C$, the usual perturbation theory would suggest writing

$$F(z) = F^{(0)}(z) + \epsilon F^{(1)}(z) + \cdots . \tag{5.1}$$

On the other hand, using a conformal mapping between the curves $C_0$ and $C$, we would try to represent the solution as

$$\begin{aligned} F(z) &= F^{(0)}(\zeta) + \cdots, \\ z &= \zeta + \epsilon z^{(1)}(\zeta) + \epsilon^2 z^{(2)}(\zeta) + \cdots . \end{aligned} \tag{5.2}$$

This is essentially the parametric type of solution mentioned in §1.

Such methods can be applied to other problems, e.g., the characteristic value problem for the vibration of a membrane or the Schrödinger equation. Consider, for example, the vibration of a membrane for a nearly circular boundary $C$. Let $C_0$ be the circular boundary in the $\zeta$-plane. The conformal mapping may be written as

$$z = \zeta + \epsilon h(\zeta). \tag{5.3}$$

The problem

$$u_{xx} + u_{yy} + \lambda u = 0, \qquad u = 0 \text{ on } C \tag{5.4}$$

is mapped into

$$u_{\xi\xi} + u_{\eta\eta} + \lambda(1 + \epsilon[h'(\zeta) + \bar{h}'(\zeta)] + \epsilon^2 \mid h'(\zeta) \mid^2)u = 0, \quad u = 0 \text{ on } C_0 . \tag{5.5}$$

Thus, we have the solution in the form

$$\begin{aligned} u &= u^{(0)}(\xi, \eta) + \epsilon u^{(1)}(\xi, \eta) + \epsilon^2 u^{(2)}(\xi, \eta) \\ \lambda &= \lambda^{(0)} + \epsilon\lambda^{(1)} + \cdots \end{aligned} \tag{5.6}$$

The problem now conforms to the usual perturbation theory developed in standard text books.

The method indicated above differs somewhat from the asymptotic process used by Feshbach,[13] which is based on the use of the Green's function for the region in the physical space and is applicable to three-dimensional problems. In fact, the present process is convergent if $\epsilon$ is small enough, and $h(\zeta)$ satisfies certain necessary regularity requirements on the boundary. It is certainly true if $h(\zeta)$ is analytic in a circle slightly larger than $C_0$.

*Note added in proof.* In all the references quoted in §1, the perturbation procedure is based on a small parameter occuring essentially in the *boundary conditions.* After this paper had been sent to the press, Mr. Roger D. Sullivan of the NACA called my attention to his paper *A comparison of two methods of linearized characteristics for a simple unsteady flow*, NACA Technical Note No. 2794 (1952). Although the title would seem to stress the linear approximation, higher approximations are given for a particular problem. Apart from the difference in the physical problems treated, the successive approximations in Mr. Sullivan's paper are based on a parameter in the *differential equations* rather than in the boundary conditions. Nor is the convergence proof given. But his paper, as well as the present one, is clearly based on the explicit use of the characteristic parameters along the lines suggested by Friedrichs.

## REFERENCES


1. Lighthill, M. J., "A technique for rendering approximate solutions to physical problems uniformly valid", Phil. Mag., ser. 7, vol. **50**, pp. 1179–1201 (1949).
2. Whitham, G. B., "The flow pattern of a supersonic projectile", Comm. Pure and Appl. Math., vol **5**, pp. 301–349 (1952), and the references quoted.
3. Craggs, J. M., "The breakdown of the hodograph transformation for irrotational compressible fluid flow in two dimensions", Proc. Camb. Phil. Soc. **44**, pp. 360–379 (1948).
4. Donov, A., A plane wing with sharp edges in a supersonic stream. Izvestia Akademia Nauk U.S.S.R. Serie Mathematique (1939).
5. Meyer, R. E., "On waves of finite amplitude in ducts", Quart. J. Mech. Appl. Math. **5**, pp. 257–269, 270–291, 1952.
6. Friedrichs, K. O., "Formation and decay of shock waves", Comm. Pure and Appl. Math., vol. **1**, pp. 211–246 (1948).
7. Courant, R. and Friedrichs, K. O., "Supersonic Flow and Wave Propagation", Interscience, New York, 1948, esp. p. 361.
8. Fox, P. A., "On the use of coordinate perturbations in the solution of physical problems", doctoral thesis, M.I.T., November 13, (1953).
9. Pack, D. C. "A note on Prandtl's formula for the wave length of a supersonic jet", Quart. Journ. Mech. and Applied Math., vol. **3**, pp. 173–181, (1950).
10. Hasimoto, Z., "Plane jet of gas emitted from a nozzle with supersonic velocity", Journ. Phys. Soc. Japan, vol. **8**, pp. 394–399, (1953).
11. See Courant, R. and Hilbert, D., "Methoden der Mathematischen Physik" (Interscience, 1937), Chapter V, §§4–7, esp. p. 322, and the references to Lewy and Friedrichs on p. 326.
12. Lighthill, M. J., "A new approach to thin airfoil theory", The Aeronautical Quarterly, vol. **3**, 193–210 (1951).
13. See Morse, P. and Feshbach, H., "Methods of Theoretical Physics" (McGraw-Hill, 1953), pp. 1052–1064; also Chapter IX for bibliography and other related discussions.


Massachusetts Institute of Technology

(Received Dec. 10, 1953)

Reprinted from JOURNAL OF THE AERONAUTICAL SCIENCES

MARCH, 1954 VOLUME 21, No. 3

# On a Perturbation Theory Based on the Method of Characteristics

C. C. Lin
*Professor of Mathematics, Massachusetts Institute of Technology, Cambridge, Mass.*
December 1, 1953

It is known that a perturbation theory in the physical plane for supersonic flow and wave propagation is not free from difficulties. A method for improvement has been proposed by Lighthill and Whitham[1] for dealing with outgoing waves. On the other hand, the characteristic parameters are the natural independent variables to use for hyperbolic differential equations, and it may be expected that, in a perturbation theory developed in terms of the characteristic form of the differential equations, such difficulties will not occur.

The method of characteristics is well known, but it does not seem to be extensively used in connection with the perturbation theory.[2] In fact, in this case, an additional convenience can be easily introduced. Since the characteristics are nearly straight in the physical plane, the mapping between the characteristic parameters and the physical coordinates can be made an approximately affine transformation, which is easy to visualize. The complication that sometimes arises with the hodograph method is avoided.

On the other hand, when the hodograph method is useful the characteristics are fixed in the hodograph plane. Thus, the present method would also embody the advantages of the hodograph method.

Briefly, the method works as follows. If $u(x, y)$ and $v(x, y)$ are the solutions sought, we introduce the characteristic parameters $\alpha$ and $\beta$ and seek the solution in the form

$$\left.\begin{aligned} u &= u^{(0)}(\alpha,\beta) + \epsilon u^{(1)}(\alpha,\beta) + \ldots \\ v &= v^{(0)}(\alpha,\beta) + \epsilon v^{(1)}(\alpha,\beta) + \ldots \\ x &= x^{(0)}(\alpha,\beta) + \epsilon x^{(1)}(\alpha,\beta) + \ldots \\ y &= y^{(0)}(\alpha,\beta) + \epsilon y^{(1)}(\alpha,\beta) + \ldots \end{aligned}\right\} \tag{1}$$

Thus, the solution is given by a parametric representation. The initial approximations to $x$ and $y$ are expected to be linear functions of $\alpha$ and $\beta$.

Several examples have been worked out by the present writer and Miss Phyllis Fox,[3] using the method just explained. Convergence proof is given in certain cases. The reflection of a Prandtl-Meyer expansion from the plane boundary will now be described to show how the method works. It is clear that the divergence of the characteristics is important here, and hence the perturbation theory in the physical plane does not work satisfactorily.

The problem is equivalent to that of the interaction of two expansion waves of equal strength, and we shall therefore consider the general case of interaction of two simple waves. The interaction region is bounded by four characteristics forming a rectangle in the $(\alpha, \beta)$ plane.

Along the two characteristics OA and OC, the solution is known; thus, we have a characteristic initial value problem. The choice of the scales of $\alpha$ and $\beta$ is still at our disposal. It is important, for convenience of application, that this choice be properly made to simplify the mapping between the $(x, y)$ plane and the $(\alpha, \beta)$ plane. This can be done simply by taking $x = \alpha$ along OA and $x = \beta$ along OC. The given physical conditions

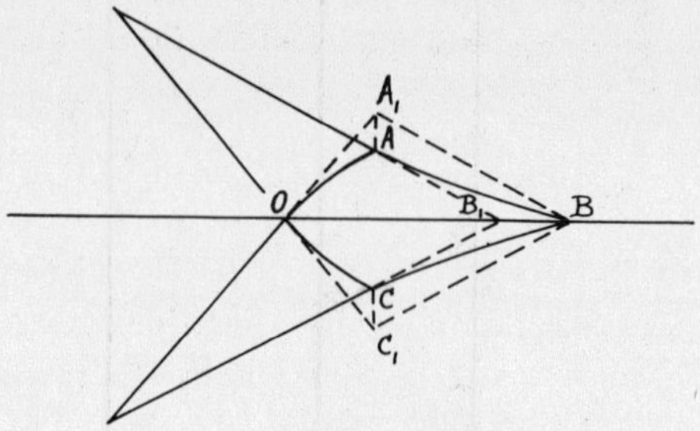


FIG. 1. Interaction of two simple waves

along OA and OC should be presented as functions of $x$. It has been found convenient to describe the state of the gas by using the Mach angle $A$ and the angle $\theta$ for the direction of flow. Thus, if $\theta = \epsilon f(x)$ along OA and $\theta = \epsilon g(x)$ along OC, the characteristic value problem is formulated as follows:

$$\left.\begin{aligned} \theta &= \epsilon f(\alpha), \quad x = \alpha \qquad \text{along OA} \\ \theta &= \epsilon g(\beta), \quad x = \beta \qquad \text{along OC} \end{aligned}\right\} \tag{2}$$

The equations for $\theta$ and $A$ can be integrated in closed form, and we have

$$\left.\begin{aligned} \theta &= \epsilon[f(\alpha) + g(\beta)] \\ \psi(A) - \psi(A_0) &= \epsilon[f(\alpha) - g(\beta)] \end{aligned}\right\} \tag{3}$$

where

$$\left.\begin{aligned} \psi(A) &= A - (\pi/2) + (1/\mu)\tan^{-1}(\mu \cot A) \\ \mu^2 &= (\gamma - 1)/(\gamma + 1) \end{aligned}\right\} \tag{4}$$

The equations for $x$ and $y$ then give

$$\left.\begin{aligned} x &= (\alpha + \beta) + \epsilon(n/2\lambda)\,[\beta f(\alpha) - \alpha g(\beta)] \\ y &= \lambda(\alpha - \beta) + \epsilon/(n/2)\,[\beta f(\alpha) + \alpha g(\beta)] + \\ &\qquad \epsilon\,(m/2)\left[\int_0^\alpha f(\alpha)\,d\alpha + \int_0^\beta g(\beta)\,d\beta\right] \end{aligned}\right\} \tag{5}$$

where

$$\left.\begin{aligned} m &= \sec^2 A_0 \{1 + [1/\psi'(A_0)]\} \\ n &= \sec^2 A_0 \{1 - [1/\psi'(A_0)]\} \end{aligned}\right\} \quad (\lambda = \tan A_0) \tag{6}$$

The reflection of a simple wave from a solid boundary can be obtained by taking $g(\beta) = -f(\beta)$. The boundary $y = 0$ becomes $\alpha = \beta$, where $\theta$ also vanishes. We have then

$$\left.\begin{aligned} \psi(A) - \psi(A_0) &= 2\epsilon f(\alpha) \\ x &= 2\alpha + \epsilon(n/\lambda)\alpha f(\alpha) \end{aligned}\right\} \tag{7}$$

for the pressure distribution along the solid boundary.

A detailed account of the method and the discussion of the results will be published later.

## References

[1] Whitham, G. B., Comm. on Pure and Appl. Math., Vol. 5, pp. 301–349 (1952), and the references quoted.

[2] Courant, R., and Friedrichs, K. O., *Supersonic Flow and Wave Propagation*, Interscience, 1948, esp. p. 361.

[3] Fox, Phyllis, Sc.D. Thesis, M.I.T., November 13, 1953.

*Offprint from "Archive for Rational Mechanics and Analysis", Volume 1, Number 5, 1958, P. 391—395*

*Springer-Verlag, Berlin · Göttingen · Heidelberg*

# *Note on a Class of Exact Solutions in Magneto-hydrodynamics*

C. C. LIN

## 1. Introduction

The purpose of this note is to describe a class of exact solutions of the magneto-hydrodynamic equations which are applicable to various flow configurations. Some examples of these solutions have been worked out in some detail by PETER GOTTLIEB; his solutions and some others will be reported in future communications.

The class of solutions considered here is of the "layer-type", including many classical exact solutions of the Navier-Stokes equations as special cases; *e.g.*, the flow near a stagnation point in the two-dimensional and the axially symmetrical cases, and the flow of a viscous fluid over a rotating disc. It represents a further generalization of a class of solutions discussed by the present writer for the steady or unsteady motion of an ordinary viscous fluid.

## 2. Basic equations

For a uniform electrically conducting incompressible fluid, the equations of magneto-hydrodynamics are as follows*:

$$\frac{\partial v_i}{\partial t} + v_j \frac{\partial v_i}{\partial x_j} - \frac{\mu}{4\pi\varrho} H_j \frac{\partial H_i}{\partial x_j} = -\frac{\partial \Pi}{\partial x_i} + \nu \Delta v_i, \qquad \frac{\partial v_k}{\partial x_k} = 0; \tag{2.1}$$

$$\frac{\partial H_i}{\partial t} + v_j \frac{\partial H_i}{\partial x_j} - H_j \frac{\partial v_i}{\partial x_j} = \eta \Delta H_i, \qquad \frac{\partial H_k}{\partial x_k} = 0. \tag{2.2}$$

In these equations, $x_i$ $(i = 0, 1, 2)$ are the space coordinates, $t$ is the time, $v_i$ are the components of velocity, $H_i$ are the components of the magnetic field. The properties of the fluid are characterized by the density $\varrho$, the kinematic viscosity $\nu$, the magnetic permeability $\mu$ and the electric conductivity $\sigma$, in terms of which we define the magnetic diffusivity

$$\eta = (4\pi\mu\sigma)^{-1}. \tag{2.3}$$

The quantity $\Pi$ is essentially the total pressure and is defined

$$\varrho \Pi = p + \mu H^2/8\pi, \tag{2.4}$$

where $p$ is the ordinary pressure of the fluid. The summation convention is used throughout this paper.

* *Cf.* COWLING (1957), p. 93

We seek a class of solutions in which the $x_0$-axis occupies a special position. We allow the velocity field, the magnetic field, and the pressure gradient to depend on the coordinate $x_0$ and the time $t$ in a fairly general manner, but only *linearly* on the two other coordinates $x_\alpha$ ($\alpha=1, 2$). To be more specific, we assume

$$\begin{aligned} v_0 &= u_0(x_0, t) \\ v_\alpha &= u_\alpha(x_0, t) + u_{\alpha\beta}(x_0, t)\, x_\beta, \end{aligned} \tag{2.5}$$

$$\begin{aligned} H_0 &= h_0(x_0, t) \\ H_\alpha &= h_\alpha(x_0, t) + h_{\alpha\beta}(x_0, t)\, x_\beta, \end{aligned} \tag{2.6}$$

and

$$\begin{aligned} \frac{\partial \Pi}{\partial x_0} &= \varpi_0(x_0\, t) \\ \frac{\partial \Pi}{\partial x_\alpha} &= \varpi_\alpha(x_0, t) + \varpi_{\alpha\beta}(x_0, t)\, x_\beta. \end{aligned} \tag{2.7}$$

Since we must have

$$\frac{\partial}{\partial x_\alpha}\left(\frac{\partial \Pi}{\partial x_0}\right) = \frac{\partial}{\partial x_0}\left(\frac{\partial \Pi}{\partial x_\alpha}\right), \tag{2.8}$$

the functions $\varpi_\alpha(x_0, t)$ and $\varpi_{\alpha\beta}(x_0, t)$ must actually be independent of $x_0$. The function $\Pi$ then takes on the form

$$\begin{aligned} \Pi &= \tfrac{1}{2}\,\varpi_{\alpha\beta}(t)\, x_\alpha\, x_\beta + \varpi_\beta(t)\, x_\beta + \varpi(x_0, t), \\ \varpi_0 &= \frac{\partial \varpi}{\partial x_0}. \end{aligned} \tag{2.9}$$

## 3. Equations in two independent variables

The particular choice of the form of the solutions is made so that the other terms in the basic equations (2.1) and (2.2) are also *linear* in the variables $x_\alpha$ ($\alpha=1, 2$). Consider, for example, $v_j \frac{\partial H_i}{\partial x_j}$. By using (2.5) and (2.6), it is easily seen that

$$v_i \frac{\partial H_i}{\partial x_i} = v_0 \frac{\partial h_0}{\partial x_0} \tag{3.1}$$

and

$$v_j \frac{\partial H_\alpha}{\partial x_j} = \left(v_0 \frac{\partial h_\alpha}{\partial x_0} + v_\gamma\, h_{\alpha\gamma}\right) + \left(v_0 \frac{\partial h_{\alpha\beta}}{\partial x_0} + v_{\gamma\beta}\, h_{\alpha\gamma}\right) x_\beta. \tag{3.2}$$

Thus, these basic equations can be satisfied by comparing coefficients of $x_1$ and $x_2$ and terms independent of them.

For convenience of reference, we list below the formulae useful for such reductions. Each of the vectors $v_i$ and $H_i$ is of the form

$$\{A_i\} = \{a_0,\, a_\alpha + a_{\alpha\beta}\, x_\beta\}, \tag{3.3}$$

where $a_0$, $a_\alpha$, $a_{\alpha\beta}$ are functions of $x_0$ and $t$. The vector $\partial A_i/\partial t$ is then of the same form. In fact

$$\left\{\frac{\partial A_i}{\partial t}\right\} = \left\{\frac{\partial a_0}{\partial t},\, \frac{\partial a_\alpha}{\partial t} + \frac{\partial a_{\alpha\beta}}{\partial t}\, x_\beta\right\}. \tag{3.4}$$

Also, the Laplacian of $A_i$ is of the form

(3.5) $$\{\Delta A_i\} = \left\{\frac{\partial^2 a_0}{\partial x_0^2},\ \frac{\partial^2 a_\alpha}{\partial x_0^2} + \frac{\partial^2 a_{\alpha\beta}}{\partial x_0^2} x_\beta\right\}.$$

Furthermore, if $B_i$ is another vector of the same form as $A_i$, then

(3.6) $$C_i = B_j \frac{\partial A_i}{\partial x_j}$$

is also of the same form; indeed,

(3.7) $$c_0 = \frac{\partial a_0}{\partial x_0}, \qquad c_\alpha = b_0 \frac{\partial a_\alpha}{\partial x_0} + a_{\alpha\gamma} b_\gamma, \qquad c_{\alpha\beta} = b_0 \frac{\partial a_{\alpha\beta}}{\partial x_0} + a_{\alpha\gamma} b_{\gamma\beta}.$$

With the help of these formulae, it is not difficult to write down the equations for the functions of $x_0$ and $t$ defined in (2.5) and (2.6).

The equation of continuity in (2.1) is satisfied if

(I) $$\frac{\partial u_0}{\partial x_0} + u_{\beta\beta} = 0.$$

Similarly, the condition for the absence of free magnetic charges is satisfied if

(II) $$\frac{\partial h_0}{\partial x_0} + h_{\beta\beta} = 0.$$

The equations of motion (2.1) lead to

(IIIa) $$\frac{\partial u_0}{\partial t} + u_0 \frac{\partial u_0}{\partial x_0} - \nu \frac{\partial^2 u_0}{\partial x_0^2} = -\varpi_0 + \frac{\mu}{4\pi\varrho}\left(h_0 \frac{\partial h_0}{\partial x_0}\right),$$

(IIIb) $$\frac{\partial u_\alpha}{\partial t} + u_0 \frac{\partial u_\alpha}{\partial x_0} + u_{\alpha\gamma} u_\gamma - \nu \frac{\partial^2 u_\alpha}{\partial x_0^2} = -\varpi_\alpha + \frac{\mu}{4\pi\varrho}\left(h_0 \frac{\partial h_\alpha}{\partial x_0} + h_{\alpha\gamma} h_\gamma\right),$$

(IIIc) $$\frac{\partial u_{\alpha\beta}}{\partial t} + u_0 \frac{\partial u_{\alpha\beta}}{\partial x_0} + u_{\alpha\gamma} u_{\gamma\beta} - \nu \frac{\partial^2 u_{\alpha\beta}}{\partial x_0^2} = -\varpi_{\alpha\beta} + \frac{\mu}{4\pi\varrho}\left(h_0 \frac{\partial h_{\alpha\beta}}{\partial x_0} + h_{\alpha\gamma} h_{\gamma\beta}\right).$$

Similarly, the equations (2.2) for the change of magnetic field lead to

(IVa) $$\frac{\partial h_0}{\partial t} + u_0 \frac{\partial h_0}{\partial x_0} - \eta \frac{\partial^2 h_0}{\partial x_0^2} = h_0 \frac{\partial u_0}{\partial x_0},$$

(IVb) $$\frac{\partial h_\alpha}{\partial t} + u_0 \frac{\partial h_\alpha}{\partial x_0} + h_{\alpha\gamma} u_\gamma - \eta \frac{\partial^2 h_\alpha}{\partial x_0^2} = h_0 \frac{\partial u_\alpha}{\partial x_0} + u_{\alpha\gamma} h_\gamma,$$

(IVc) $$\frac{\partial h_{\alpha\beta}}{\partial t} + u_0 \frac{\partial h_{\alpha\beta}}{\partial x_0} + h_{\alpha\gamma} u_{\gamma\beta} - \eta \frac{\partial^2 h_{\alpha\beta}}{\partial x_0^2} = h_0 \frac{\partial u_{\alpha\beta}}{\partial x_0} + u_{\alpha\gamma} h_{\gamma\beta}.$$

## 4. Consistency of the system

It is necessary to examine the above system of equations to see that there is indeed the right number of equations for the unknowns specified in (2.5), (2.6) and (2.7). Clearly, there are seven unknowns in each of (2.5) and (2.6) and only one unknown, *i.e.*, $\varpi_0(x_0, t)$, in (2.7), since $\varpi_\alpha$ and $\varpi_{\alpha\beta}$ are functions of $t$ alone and are therefore expected to be specified by boundary conditions at specified values of $x_0$ (*e.g.* at infinity). Thus, there are *fifteen* unknowns altogether.

27*

On the other hand, we have one equation for each of the types (I) and (II) and seven equations for each of the types (III) and (IV), making up a total of *sixteen* equations. Thus, one of them must be redundant if we are not going to have the difficulty of too many equations. Indeed, a closer examination shows that the equations (I), (II), (IVa) and (IVc) are not all independent.

If we contract (IVc), we obtain

$$\frac{\partial h_{\beta\beta}}{\partial t}+u_0\frac{\partial h_{\beta\beta}}{\partial x_0}+h_{\beta\gamma}u_{\gamma\beta}-\eta\frac{\partial^2 h_{\beta\beta}}{\partial x_0^2}=h_0\frac{\partial u_{\beta\beta}}{\partial x_0}+u_{\beta\gamma}h_{\gamma\beta},$$

or

$$\text{(IVc}'\text{)}\qquad \frac{\partial h_{\beta\beta}}{\partial t}+u_0\frac{\partial h_{\beta\beta}}{\partial x_0}-\eta\frac{\partial^2 h_{\beta\beta}}{\partial x_0^2}=h_0\frac{\partial u_{\beta\beta}}{\partial x_0}.$$

In this form, it is clear that, by using (I) and (II), $u_{\beta\beta}$ and $h_{\beta\beta}$ can be eliminated to give an equation in $h_0$ and $u_0$. The resultant equation must then be consistent with (IVa). This fact can be easily verified by noting that the $x_0$-derivative of (IVa), when added onto (IVc′), leads to an identity by virtue of (I) and (II).

## 5. Concluding Remarks

We have thus found a large class of exact solutions of the hydromagnetic equations, involving fifteen unspecified functions of the variables $x_0$ and $t$. If one restricts oneself to the steady case, the equations above reduce to ordinary differential equations which can be integrated (numerically, if necessary) under appropriate boundary conditions. Symmetry conditions also help to simplify the equations in various cases.

These simplifications are all present in the examples worked out by PETER GOTTLIEB. He considered the flow near a stagnation point both in the two-dimensional case and in the case of axial symmetry with a magnetic field parallel to a solid boundary. These examples serve to illustrate the behavior of the magnetic field as it is carried along a solid surface due to convection currents in the core of the earth. It is clear, from a consideration of the path of the fluid particles, that there can be a considerable stretching of the magnetic lines. In the ideal case of zero viscosity and zero magnetic diffusivity, the field may be expected to become infinite as the boundary is approached. It is to be expected that this tendency to increase will be counteracted by both the hydrodynamic and the magnetic diffusive effects. The resultant picture in the case of steady motion is worked out in a paper to follow. Other interesting problems may be developed within the present framework. For example, one might consider an infinite disc oscillating about its axis (taken to be the $x_0$-axis) in the presence of a magnetic field in that direction. Some hydromagnetic damping is certainly to be expected. The steady rotation of a disc would be another possible problem, extending that worked out by VON KÁRMÁN & COCHRAN years ago* in the purely hydrodynamic case. Indeed, this may be combined with the problems studied by GOTTLIEB to describe the combined effect of earth rotation and convection near a stagnation point in the problem of geomagnetism. Finally, one might consider the problem of a uniform magnetic field perpendicular to the $x_0$-axis

* See GOLDSTEIN (1938), p. 110.

and rotating about it (perhaps in an unsteady manner), and attempt to find out whether this would indeed provide an effective "magnetic wall" to hold an approaching fluid. It is clear from elementary considerations of Eq. (2.1) and (2.2) that a magnetic field can act in the manner of a static pressure to hold a fluid in balance, but the added effect of rotation may have the advantage of making the equilibrium more stable. These and other problems are subjects for future studies.

**Acknowledgement.** This work was partly sponsored by the Office of Naval Research and may be reproduced for any purpose of the United States Government.

## References

COWLING, T. G.: Magnetohydrodynamics. New York: Interscience 1957.
GOLDSTEIN, S.: Modern Developments in Fluid Dynamics. Oxford 1938.
GOTTLIEB, PETER: Some Exact Solutions in Hydromagnetics (to be published).

Massachusetts Institute of Technology,
Cambridge, Massachusetts

*(Received March 31, 1958)*

# ON THE THEORY OF SHALLOW WATER WAVES*

C. C. LIN and ALFRED CLARK, JR.**

## 1. Introduction

One of the most interesting problems in the theory of water waves is the existence of waves of permanent form when the amplitude of the waves is no longer infinitesimal. While existence proofs have been given by Levi-Civita[a], Struik[b], Friedrichs and Hyers[c], and Littman[d] for various cases, there still seems to be considerable difference of opinion of the theoretical treatment in the case of shallow water waves. In this case, it is well-known that in a certain approximation the theory is identical with that for compression waves in a gas for which the polytropic index is 2. From such a theory, it would follow that a progressive wave with decreasing wave height in the direction of propagation eventually breaks. How can this prediction be reconciled with the existence of permanent waves in shallow waters? This is the question that has been referred to as the "long wave paradox" by Ursell[e], and it has been answered in somewhat different ways by different authors[1].

In the present treatment, we shall present a systematic analysis of the problem of shallow water waves, which allows the improvement of the calculation by successive approximations. The method is essentially a development of that used by Lord Rayleigh[g]. Two parameters will be found to be important in this theory. The first parameter (denoted by $h$) is the depth of water, measured in terms of a characteristic horizontal dimension. This parameter has to be small for the general concept of "shallow" water to hold. The second parameter (denoted by $a$) is the amplitude of the wave measured in terms of the height of the water. It will be found that there are three principal classes of problems characterized as follows:

(a) the compression wave approximation (Airy theory), where $a=0(1)\gg h^2$

(b) non-linear theory of waves of small amplitudes (including the theory of permanent waves) where $a=0(h^2)$

* Work sponsored in part by the Office of Naval Research

** Department of Mathematics, Massachusetts Institute of Technology

(c) linear theory of waves in moderately shallow water where $a \ll h^2$

We shall give a set of differential equations which is applicable to all three cases under the conditions indicated. It is hoped that this will help to clarify our understanding of the problem of shallow water waves.

## 2. Formulation of the problem

Let the water be of uniform depth H in the undisturbed state. Consider a three-dimensional potential motion with a horizontal scale L. The water is shallow when

$$h=\frac{H}{L} \tag{2.1}$$

is small. This is the fundamental parameter of our problem.

We shall now formulate our problem in dimensionless form with the characteristic length L and the characteristic velocity

$$c_0=\sqrt{gH} \tag{2.2}$$

where $g$ is the gravitational constant. Then, the differential equations for the determination of the motion are

$$\triangle\varphi=0 \tag{2.3}$$

and

$$\frac{\partial\varphi}{\partial t}+\frac{1}{2}|\nabla\varphi|^2+p+\frac{z}{h}=C(t) \tag{2.4}$$

Here we have taken a Cartesian coordinate system $(x, y, z)$ with $z=0$ coinciding with the solid bottom. The boundary conditions are

$$\frac{\partial\varphi}{\partial z}=0 \text{ at } z=0 \tag{2.5}$$

and, at the free surface $z=Z\ (x, y, t)$,

$$w=\frac{\partial Z}{\partial t}+u\frac{\partial Z}{\partial x}+v\frac{\partial Z}{\partial y} \tag{2.6}$$

and

$$p=0 \text{ or } \frac{\partial\varphi}{\partial t}+\frac{1}{2}(u^2+v^2+w^2)+\frac{z}{h}=C(t), \tag{2.7}$$

where $(u, v, w)$ are the three components of velocity.

We note that Z is in general expected to be of the order of $h$, but there is indeed no such a limitation as long as it is small. This means that the total depth of water has to be small. Since $0\leq z\leq Z$, the harmonic function $\varphi\ (x, y, z, t)$, can be expanded in powers of $z$, and the resultant series would be expected to be rapidly convergent for the range of values of $z$ under consideration(2). This is then the basis for our approximate theory. Indeed, we have

$$\varphi(x, y, z, t)=\chi(x, y, t)+z^2\chi_2(x, y, t)+z^4\chi_4(x, y, t)+\dots. \tag{2.8}$$

where, by (2.3),

$$\frac{\partial^2\chi}{\partial x^2}+\frac{\partial^2\chi}{\partial y^2}=-2\chi_2, \tag{2.9}$$

$$\frac{\partial^2 \chi_4}{\partial x^2}+\frac{\partial^2 \chi_2}{\partial y^2}=-12\chi_4, \ldots;$$

and the odd powers in $z$ are absent by (2.5). We then proceed to use (2.8) for the satisfaction of the boundary conditions (2.6) and (2.7). Since the resultant expressions are power series in Z, we can terminate the series at appropriate steps, and obtain higher and higher approximations by including more and more terms in the series. Clearly, the order of differentiation increases with each approximation, and the process must therefore be *asymptotic*. In the next section, we shall see that the lowest approximation yields the compression wave analogy.

### 3. The gas-dynamics approximation

To carry out the calculation of the boundary conditions, let us note that

$$u=\frac{\partial \varphi}{\partial x}=\frac{\partial \chi}{\partial x}+z^2\frac{\partial \chi_2}{\partial x}+\ldots.$$

$$v=\frac{\partial \varphi}{\partial y}=\frac{\partial \chi}{\partial y}+z^2\frac{\partial \chi_2}{\partial y}+\ldots.$$

$$w=\frac{\partial \varphi}{\partial z}=2z\chi_2+4z^3\chi_4+\ldots.$$

and thus (2.6) may be written as

$$2Z\chi_2+4Z^3\chi_4+\ldots.=\frac{\partial Z}{\partial t}+[\frac{\partial \chi}{\partial x}+Z^2\frac{\partial \chi_2}{\partial x}+\ldots.]\frac{\partial Z}{\partial x}+[\frac{\partial \chi}{\partial y}+Z^2\frac{\partial \chi_2}{\partial y}+\ldots.]\frac{\partial Z}{\partial y} \tag{3.1}$$

and thus to a first approximation for small Z we have

$$\frac{\partial Z}{\partial t}+\frac{\partial \chi}{\partial x}\frac{\partial Z}{\partial x}+\frac{\partial \chi}{\partial y}\frac{\partial Z}{\partial y}=-Z(\frac{\partial^2 \chi}{\partial x^2}+\frac{\partial^2 \chi}{\partial y^2}) \tag{3.2}$$

This is exactly the equation of continuity for the two-dimensional unsteady potential motion of a gas with density Z (x, y, t) and velocity components $(\chi_x, \chi_y)$. Similarly, Eq. (2.8) gives

$$\frac{\partial \chi}{\partial t}+Z^2\frac{\partial \chi_2}{\partial t}+\frac{1}{2}[(\frac{\partial \chi}{\partial x})^2+(\frac{\partial \chi}{\partial y})^2]+2Z^2\chi_2^2+Z^2\left\{\frac{\partial \chi}{\partial x}\frac{\partial \chi_2}{\partial x}+\frac{\partial \chi}{\partial y}\frac{\partial \chi_2}{\partial y}\right\}+O(Z^4)+\frac{Z}{h}=C(t) \tag{3.3}$$

To a first approximation, we then have

$$\frac{\partial \chi}{\partial t}+\frac{1}{2}[(\frac{\partial \chi}{\partial x})^2+(\frac{\partial \chi}{\partial y})^2]+\varpi=C(t), \tag{3.4}$$

where

$$\varpi=Z/h=\int_0^Z dP/Z, \qquad \text{if } P=Z^2/2h. \tag{3.5}$$

The equation (3.4) is here again the equation for potential motion if the pressure P is proportional to the square of the density $Z(x, y, t)$.

Thus, *the lowest approximation leads to the compression wave analogy.*

## 4. Limitations of the compression wave analogy, the general system of equations

In the above calculations, the only approximation used is the regularity of the expansion (2.8) and the smallness of Z. Thus, the compression wave analogy holds even when the wave height $Z-h$ exceeds $h$ in order of magnitude, provided only that $Z \ll 1$. However, there is one limitation when the wave height

$$\zeta = Z - h \tag{4.1}$$

is small. To be sure, the relations (3.2) and (3.4) still hold as a good approximation. However, they are no longer sufficiently accurate for predicting the non-linear effect once

$$a = \zeta/h = (Z-h)/h \tag{4.2}$$

becomes of the order of $h^2$. Take (3.2) for example. Some of the non-linear terms are clearly of the order of $a^2$.

The terms

$$4Z^3\chi_4, \qquad Z^2\frac{\partial \chi_2}{\partial x}\frac{\partial Z}{\partial x}, \qquad Z^2\frac{\partial \chi_2}{\partial y}\frac{\partial Z}{\partial y}$$

neglected from (3.3), are respectively of the orders

$$h^3 a,\ h^2 a^2,\ h^2 a^2$$

Thus, the term $4Z^3\chi_4$ must be retained if

$$a = O(h^2) \text{ or smaller,} \tag{4.3}$$

We then have in place of (3.2)

$$\frac{\partial Z}{\partial t} + \frac{\partial}{\partial x}\Big[Z\frac{\partial \chi}{\partial x}\Big] + \frac{\partial}{\partial y}\Big[Z\frac{\partial \chi}{\partial y}\Big] = \underline{4Z^3\chi_4}. \tag{4.4}$$

This is an equation of continuity with a source term (underlined) proportional to the cube of the density and the Laplacian of the divergence of velocity (cf. (2.9)).

Similarly, the equation (3.3) leads to

$$\frac{\partial \chi}{\partial t} + \frac{1}{2}\Big[\Big(\frac{\partial \chi}{\partial x}\Big)^2 + \Big(\frac{\partial \chi}{\partial y}\Big)^2\Big] + \frac{Z}{h} + \underline{Z^2\frac{\partial \chi_2}{\partial t}} = C(t) \tag{4.5}$$

The underlined term is the extra term included. Equations (4.4) and (4.5) then form ***a basic approximate system which holds under all conditions*** as long as the concept of shallow water theory is valid. Before we examine them further let us rewrite them in terms of the amplitude parameter $a = \zeta/h$. Then (4.4) reads

$$\frac{\partial a}{\partial t} + \frac{\partial}{\partial x}\Big(a\frac{\partial \chi}{\partial x}\Big) + \frac{\partial}{\partial y}\Big(a\frac{\partial \chi}{\partial y}\Big) + \Big(\frac{\partial^2 \chi}{\partial x^2} + \frac{\partial^2 \chi}{\partial y^2}\Big) = 4h^2(1+a)^3\chi_4 \tag{4.6}$$

$$\frac{\partial \chi}{\partial t} + \frac{1}{2}\Big[\Big(\frac{\partial \chi}{\partial x}\Big)^2 + \Big(\frac{\partial \chi}{\partial y}\Big)^2\Big] + a + h^2(1+a)^2\frac{\partial \chi^2}{\partial t} = C_1(t), \tag{4.7}$$

where $C_1(t) = C(t) - 1$.

## 5. Non-linear theory of waves of small amplitudes

If we now specialize to the case of small amplitudes, the initial approximation to (4.6) and (4.7) gives (neglecting $C_1(t)$ for the moment)[(3)].

(5.1) $$\frac{\partial a}{\partial t}=-(\frac{\partial^2\chi}{\partial x^2}+\frac{\partial^2\chi}{\partial y^2})=-\nabla_1{}^2\chi,$$

and

(5.2) $$\frac{\partial\chi}{\partial t}=-a.$$

Thus, both $a$ and $\chi$ satisfy the linear wave equation, as is well-known for linear theory of shallow water waves.

In the non-linear theory, we have to retain terms of higher orders. But we may use the approximation (5.1) and (5.2) on terms which are multiplied by other small terms. Thus, in (4.7) we may put

(5.3) $$h^2(1+a)^2\frac{\partial\chi_2}{\partial t}\cong h^2\frac{\partial\chi_2}{\partial t}\cong h^2(+\frac{1}{2}\nabla_1{}^2a)$$

where the last step is carried out by using (5.2). Thus, (4.6) and (4.7) become

(5.4) $$\frac{\partial a}{\partial t}+\nabla_1\cdot(a\nabla_1\chi)+\nabla_1{}^2\chi=\frac{h^2}{6}\nabla_1{}^4\chi$$

(5.5) $$\frac{\partial\chi}{\partial t}+\frac{1}{2}|\nabla_1\chi|^2+a+\frac{1}{2}h^2\nabla_1{}^2a=0$$

It is easy to get an equation for $\chi$ by solving for $a$ from (5.5); but it is more interesting to get an equation for the amplitude function $a$.

In trying to accomplish this, let us differentiate (5.4) with respect to time. Then we get

$$\frac{\partial^2 a}{\partial t^2}+\nabla_1{}^2\dot\chi=-\nabla_1\cdot(\dot a\nabla_1\chi)-\nabla_1\cdot(a\nabla\dot\chi)+\frac{h^2}{6}\nabla_1{}^4\dot\chi$$

where a dot denotes differentiation with respect to time. If we replace $\dot\chi$ by using (5.5) and $\dot a$ by using (5.4) and neglect unimportant terms, we get

(5.6) $$\frac{\partial^2 a}{\partial t^2}-\nabla_1{}^2a=\frac{h^2}{3}\nabla_1{}^2\nabla_1{}^2a+\nabla_1{}^2(\frac{a^2}{2})+\nabla_1{}^2\left\{\frac{1}{2}(\nabla_1\chi)^2\right\}+\nabla\cdot(\nabla_1{}^2\chi\nabla_1\chi)$$

within the accuracy of our approximation. In the case of one spatial coordinate, we have

(5.7) $$\frac{\partial^2 a}{\partial t^2}-\frac{\partial^2 a}{\partial x^2}=\frac{h^2}{3}\frac{\partial^4 a}{\partial x^4}+\frac{\partial^2}{\partial x^2}[\frac{1}{2}a^2+(\frac{\partial\chi}{\partial x})^2].$$

This is to be contrasted with the equation of Boussinesq which reads (in our notation)

(5.8) $$\frac{\partial^2 a}{\partial t^2}-\frac{\partial^2 a}{\partial x^2}=\frac{h^2}{3}\frac{\partial^4 a}{\partial x^4}+\frac{\partial^2}{\partial x^2}(\frac{3}{2}a^2).$$

The two forms (5.7) and (5.8) can be reconciled in the case of ***progressive waves***, for then

(5.9) $$(\frac{\partial\chi}{\partial x})^2\cong(\frac{\partial\chi}{\partial t})^2\cong a^2$$

within the accuracy of our approximation.

Thus, in the general case, it seems impossible to get an equation for $a$ alone. It is easy, on the other hand to get an equation for $\chi$. By solving for $a$ from (5.5) [with $\nabla_1^2 a$ expressed back in terms of $\chi$ by (5.2)], we can eliminate $a$ completely to get

$$\frac{\partial^2\chi}{\partial t^2}-\nabla_1^2\chi=C_1'(t)+C_1(t)\nabla_1^2\chi-\frac{\partial\chi}{\partial t}\nabla_1^2\chi-\frac{\partial}{\partial t}[(\nabla_1\chi)^2]+\frac{h^2}{3}\nabla_1^2\nabla_1^2\chi. \tag{5.10}$$

In the linear theory, only the left-hand side of (5.10) is to be kept; and this relation has been used to get rid of the higher order time-derivatives on the right-hand side. A solution for (5.10) to yield permanent waves in the one-dimensional case is given in the appendix. Also given are the basic equations in the one-dimensional case where we retain terms of one order higher than those kept in the above equations.

## 6. Linear theory of waves in moderately shallow water

If we now make the linear approximation in (4.6) and (4.7) but retain the terms in $h^2$, we obtain

$$\frac{\partial a}{\partial t}+\left(\frac{\partial^2\chi}{\partial x^2}+\frac{\partial^2\chi}{\partial y^2}\right)-\frac{h^2}{6}\left(\frac{\partial^2}{\partial x^2}+\frac{\partial^2}{\partial y^2}\right)^2\chi=0 \tag{6.1}$$

$$\frac{\partial\chi}{\partial t}+a=\frac{1}{2}h^2\frac{\partial}{\partial t}\left(\frac{\partial^2\chi}{\partial x^2}+\frac{\partial^2\chi}{\partial y^2}\right) \tag{6.2}$$

Thus, we may eliminate $a$ to get

$$\frac{\partial^2}{\partial t^2}\left\{\chi-\frac{1}{2}h^2\left(\frac{\partial^2\chi}{\partial x^2}+\frac{\partial^2\chi}{\partial y^2}\right)\right\}=\left(\frac{\partial^2}{\partial x^2}+\frac{\partial^2}{\partial y^2}\right)\left\{\chi-\frac{1}{6}h^2\left(\frac{\partial^2\chi}{\partial x^2}+\frac{\partial^2\chi}{\partial y^2}\right)\right\} \tag{6.3}$$

The above relations clearly hold in the case when

$$a\ll h^2 \tag{6.4}$$

so that the non-linear terms are completely negligible even compared with those terms involving $h^2$. This is obviously the case of waves of infinitesimal amplitude where the depth is not so small that the terms of the order $h^2$ must be retained. The above equations can be more readily derived by using (2.6) and (2.7) in their linear form

$$\frac{\partial\phi}{\partial t}+\frac{h+\zeta}{h}=C(t) \qquad \text{at } z=h \tag{6.5}$$

and

$$w=\frac{\partial\zeta}{\partial t} \qquad \text{at } z=h \tag{6.6}$$

and substituting (2.8).

## Appendix[4]

A. *Permanent Wave Solutions.*

In the case of one spatial coordinate, (5. 10) becomes

$$\frac{\partial^2\chi}{\partial t^2}-\frac{\partial^2\chi}{\partial x^2}=C_1'(t)+C_1(t)\frac{\partial^2\chi}{\partial x^2}-\frac{\partial\chi}{\partial t}\frac{\partial^2\chi}{\partial x^2}-\frac{\partial}{\partial t}\left[\left(\frac{\partial\chi}{\partial x}\right)^2\right]+\frac{h^2}{3}\frac{\partial^4\chi}{\partial x^4}. \tag{A. 1}$$

We assume a solution of the form $\chi=\chi(x-ct)$, where $c=1+O(h^2)$, and take $C_1(t)=\delta$, where $\delta=0(h^2)$ is constant. Then (A. 1) becomes

(A. 2) $$\frac{h^2}{3}\beta'''+\beta'(3c\beta+1+\delta-c^2)=0,$$

where $\beta=\chi'$. We can readily integrate (A. 2) twice to obtain

(A. 3) $$\beta'^2=m^2\left[-\beta^3+\frac{\beta^2}{c}(c^2-1-\delta)+A_1\beta+A_2\right],$$

where $m^2=\frac{3c}{h^2}$, and $A_1=0(h^4)$ and $A_2=0(h^6)$ are constants of integration. For proper choices of $A_1$ and $A_2$, the cubic expression in $\beta$ has three real roots, $\beta_1\leq\beta_2\leq\beta_3$, and (A. 3) can be written as

(A. 4) $$\beta'^2=m^2(\beta_1-\beta)(\beta_2-\beta)(\beta_3-\beta).$$

The solution of (A. 4) is given by

(A. 5) $$\beta=\beta_2+(\beta_3-\beta_2)\,\mathrm{cn}^2\left(\frac{x-ct}{a},\,K\right),$$

where $$a=\frac{2}{m\sqrt{\beta_3-\beta_1}}\text{ and }K^2=\frac{\beta_3-\beta_2}{\beta_3-\beta_1}.$$

The equation for the free surface can be obtained from (4.7) and (A. 5); within the accuracy of the approximation, the latter equation is

$$\alpha=\delta+c\beta,\text{ or}$$

(A. 6) $$\alpha=\alpha_{min}+(\alpha_{max}-\alpha_{min})\,\mathrm{cn}^2\left(\frac{x-ct}{a},\,K\right),$$

where $\alpha_{min}=\beta_2+\delta$, $\alpha_{max}=\beta_3+\delta$. The horizontal and vertical (dimensionless) velocity components are given by

(A. 7) $$u=\beta=\beta_2+(\beta_3-\beta_2)\,\mathrm{cn}^2\left(\frac{x-ct}{a},\,K\right),$$

$$\text{and } w=-z\beta'=\frac{z}{a}(\beta_3-\beta_2)\,\mathrm{sn}\left(\frac{x-ct}{a},\,K\right)\mathrm{dn}\left(\frac{x-ct}{a},\,K\right).$$

The waves described by (A. 6) and (A. 7) are periodic with wavelength

$$\lambda=2a\int_0^{\pi/2}\frac{ds}{\sqrt{1-K^2\sin^2 S}},$$

and are the so-called cnoidal waves, first obtained by Korteweg and de Vries[h].

A limiting solution of particular interest is obtained by setting $\delta=\beta_1=\beta_2=0$. This is the case of the solitary wave[i]. In this case $K^2=1$, $\beta_3=\frac{c^2-1}{c}$, and the solution is given by

$$\beta=\frac{c^2-1}{c}\,\mathrm{sech}^2\left(\frac{x-ct}{a}\right),$$

(A, 8) $$\alpha=(c^2-1)\,\mathrm{sech}^2\left(\frac{x-ct}{a}\right),$$

$$u=\frac{c^2-1}{c}\,\mathrm{sech}^2\left(\frac{x-ct}{a}\right),$$

$$w=z\frac{2(c^2-1)}{a}\,\mathrm{sech}^2\left(\frac{x-ct}{a}\right)\tanh\left(\frac{x-ct}{a}\right),\qquad\text{where } a=\frac{2h}{\sqrt{3(c^2-1)}}.$$

B. *Higher Approximations.*

If it is desired to retain terms of higher order, then equations (4.6) and (4.7) must be replaced by

$$(\text{A. } 9)\quad \frac{\partial\alpha}{\partial t}+\frac{\partial}{\partial x}\left(\alpha\frac{\partial\chi}{\partial x}\right)+\frac{\partial^2\chi}{\partial x^2}=\frac{1}{6}h^2\frac{\partial^4\chi}{\partial x^4}+3h^2\alpha\frac{\partial^4\chi}{\partial x^4}+\frac{1}{2}h^2\frac{\partial^3\chi}{\partial x^3}\frac{\partial\alpha}{\partial x}-\frac{h^4}{5'}\frac{\partial^6\chi}{\partial x^6}$$

and

$$(\text{A. } 10)\quad \frac{\partial\chi}{\partial t}+\frac{1}{2}h^2\left(\frac{\partial\chi}{\partial x}\right)^2+\alpha-\frac{1}{2}h^2\frac{\partial^3\chi}{\partial x^2\partial t}=C_1(t)+h^2\alpha\frac{\partial^3\chi}{\partial x^2\partial t}+\frac{1}{2}h^2\frac{\partial\chi}{\partial x}\frac{\partial^3\chi}{\partial x^3}$$
$$-\frac{1}{2}h^2\left(\frac{\partial^2\chi}{\partial x^2}\right)^2-\frac{h^4}{4!}\frac{\partial^5\chi}{\partial x^4\partial t},$$

where we are considering the case of one dimension. The above equations lead to a single equation for $\chi$. If a solution of the form $\chi=\chi(x-\text{ct})$ is assumed, where $c=O(1)+O(h^2)$, then this single equation for $\chi$ reduces to

$$(\text{A. } 11)\quad \beta^{(5)}\frac{4h^4}{5'}+\beta'''\,h^2\left[\frac{1}{2}\beta-\frac{1}{2}\delta+\frac{1}{6}-\frac{1}{2}c^2\right]-\frac{1}{2}\beta'\beta''h^2$$
$$+\beta'\left[\frac{3}{2}\beta^2-3c\beta+c^2-1-\delta\right]=0,$$

where $\beta=\chi'$, and $C_1(t)=\delta=0\ (h^2)$, a constant. If only the largest terms are retained this reduces to (A. 2).

## References


(a) Levi-Civita, T. Math. Ann. *93* (1925), pp. 264–314.

(b) Struik, D. J. Math. Ann. *95* (1926), pp. 595–634.

(c) Friedrichs, K. O. and Hyers, D. H., Comm. Pure and Appl. Math. *7* (1954), pp. 517–550.

(d) Littman, W. Comm. Pure and Appl. Math. *10* (1957) 241–269.

(e) Ursell, F., Proc. Camb. Phil. Soc. *49* (1953).

(f) Stoker, J. J., "Water Waves" Interscience (1957).

(g) Rayleigh, Lord, Phil. Mag. (5), i. 257 (1876) [Papers, i. 251].

(h) Korteweg, D. J. and de Vries, G., Phil. Mag. (5) *39* (1895) p. 422.

(i) Lamb, H., "Hydrodynamics" Cambridge (1932), p. 423.


## NOTES

(1) See Stoker [*f*] p. 342 and the references to Birkhoff and Ursell [*e*]

(2) We exclude rapid variations along the surface of a scale much smaller than L.

(3) In (5.1) and the following equations, the subscript ( $)_1$ indicates the del-operator in twodimensions.

(4) In this appendix, $\alpha$ is used as the amplitude variable in place of $a$; and the symbol $a$ is used for a length scale.

# 淺水波之理論

林家翹 Alfred Clark, Jr.

淺水波之理論，雖已有數十年之歷史，然諸家意見，仍不一致。 本文用不旋轉水流理論，逐漸用近似法，引出淺水波基本方程式三種。

在兩向流動情形下，本文結論與 Ursell 1953 年之結論，大體相同，而有別於 Stoker 1957年書中意見。 惟在最有興趣一種情形下，本文仍與 Ursell 結果有異處。作者認為 Ursell 之方法有不適當處。

三向流動之淺水波理論尚未有人澈底討論過。 在此文中，其基本方程式，亦已引出；可用以解釋橫波現象(卽水波傳佈方向與外力推水方向垂直者)。

# ON THE DEPENDENCE OF INTERACTION ENERGY UPON ATOMIC ARRANGEMENTS IN SUPERLATTICES OF BINARY ALLOYS

BY CHIA-CHIAO LIN (林家翘)

Department of Physics, National South-west Associated

University, Kunming Yunnan.

### *Abstract*

In the statistical theory of superlattices in binary alloys, the dependence of the interaction energy upon atomic arrangements is taken into account by regarding the interaction energy in Bethe's theory as an average quantity depending on the degree of order and the composition of the alloy. Two simple assumptions concerning the functional relationship of the interaction energy with order and composition are made. The first is a linear function of order and composition. The second is a linear function of the average numbers of pairs of atoms. The result of applying these assumptions to superlattices of the type AB is that the critical temperature as a function of the composition is a maximum for equal sumber of A and B atoms only when a certain relation between the coefficients in the assumed function is satisfied. In the case of superlattices of type $AB_3$, the theory of Bragg and Williams is used for simplicity. It is shown that when the composition varies, the maximum of the critical temperature may occur at any desired composition by a suitables adjustment of the coefficients in the assumed. functions. There is thus a hope of removing the discrepancy between theory and experiment on this line. The anomalous specific heat at the critical temperature is also calculated for different compositions. In the case of the AB type of superlattices, Bethe's formula for the energy is no longer valid, and in order to calculate the specific heat, an approximate formula for the energy is obtained by analogy with the theory of Bragg and Williams. Finally, the problem of separation into more than one phase is briefly discussed.

§1. Introduction.

The statistical theory of superlattice formation in alloys has been developed by Bethe[1] and Peierls[2] assuming interaction between nearest neighbors. The effect of interaction between atoms other than nearest neighbors has been discussed by Chang[3] and by Wang.[4] In all these theories, it is tacitly assumed that the interaction energy is independent of the atomic arrangements in the lattice. Since an atom will be affected in different ways in the presence of different kinds of atoms, the assumption of constant interaction energy cannot be true. We must therefore expect the interaction energy to depend upon atomic arrangements. The exact nature of this dependence is however unknown in the present state of the knowledge of interatomic forces. And even if this were known, the partition function would be too complicated to be useful. We are thus led to make simple assumptions concerning this dependence of interaction energy upon the atomic arrangements and to show what sort of results we can obtain. In the following treatment, we confine ourselves to the case of interaction only between nearest neighbors. It is important to point out that the theory based on a constant interaction energy is symmetrical in the two kinds of atoms on account of the symmetry of the expression for the interaction energy involved in the partition function. This is also true for long range interactions.[5] Any asymmetry can therefore occur only in a theory based on variable interaction energy.

§2. Simple assumptions on the dependence of the interaction energy upon atomic arrangements.

To take into account the dependence of the interaction energy upon atomic arrangements, the most obvious way seems

1. H. A. Bethe, Proc. Roy. Soc. A *150*, 552-572, (1935).
2. R. Peierls, Proc. Roy. Soc. A *154*, 207-222, (1936).
3. T. S. Chang, Proc. Roy. Soc. A *161*, 546-563, (1937).
4. J. S. Wang, Proc. Roy. Soc. A *168*, 56-67 and 68-77, (1938).
5. Wang, loc. cit. 4 and W. Shockley, J. Chem. Phys. 6, 523-525, (1938).

to consider the interaction energy in Bethe's theory as a function of the degree of order by regarding it as an average interaction energy for all possible arrangements in a definite state of order. When the relative abundance of the two kinds of atoms in the alloy is not a simple stoichiometric ratio, the change of the composition of the alloy will evidently influence the atomic arrangements and hence the average interaction energy. Therefore, we must also regard the average interaction energy as a function of composition.

It seems that the simplest assumption concerning the dependence of the average interaction on the degree of order and the composition of the alloy is to take a linear relation. Let c be the fraction expressing the number of A atoms and 1-c that of B atoms. Denoting Bethe's interaction energy by $V_1$,

$$V_1=\tfrac{1}{2}(V_{AA}+V_{BB})-V_{AB},$$

and expressing the degree of order in terms of the quantity $r_a$-$c$, we have for the assumption of a linear relation[6]

$$V_1=V_0\{1+\alpha c+\beta(r_a-c)\}, \tag{1}$$

where $V_o$, $\alpha$ and $\beta$ are constants independent of $c$ and $r_a$.

It might appear more reasonable to assume that the dependence of $V^1$ on order and composition is through the average numbers of pairs of atoms. Then we may make the simple assumption that $V_1$ depends linearly on the average numbers $\overline{M}_{AA}$, $\overline{M}_{BB}$, $\overline{M}_{AB}$ of the pairs AA, BB and AB. It is more convenient to employ, instead of the large numbers $\overline{M}_{AA}$, etc., the fractions $m_{AA}$ defined by

$$m_{AA}=\overline{M}_{AA}/\tfrac{1}{2}zN,$$

where $\frac{1}{2}zN$ is the total number of pairs:

$$\tfrac{1}{2}zN=\overline{M}_{AA}+\overline{M}_{BB}+\overline{M}_{AB}\cdot$$

The number $N$ is the total number of atoms and $z$ is the number of nearest neighboring sites to any given lattice site. For the assumption of a linear relation, we have

6. The notations used in this paper will mainly follow those of C. E. Easthope, Proc., Camb. Phil. Soc., *33* 502-517, (1937).

$$V_1=\alpha_0+\alpha_1 m_{AA}+\alpha_2 m_{BB}+\alpha_{12} m_{AB}\cdot \qquad (2A)$$

Of the three quantities, $m_{AA}$, $m_{BB}$ and $m_{AB}$, only one is independent; they are connected by two relations:[7]

$$m_{AA}+\tfrac{1}{2}m_{AB}=c,$$

$$m_{BB}+\tfrac{1}{2}m_{AB}=1-c.$$

Hence, $V_1$ is reduced to

$$V_1=V_0(1+\alpha c+\beta m_{AA}) \qquad (2)$$

The constants $\alpha$ and $\beta$ in (1) and (2) will be small if the change of interaction energy with atomic arrangements is small.

We may conveniently use the average interaction energy in the sense that the values of $r_a$, $r_b$,.... are still given by Bethe's formulae. For the case r=½, we have,[6]

$$w_b=\frac{\varepsilon x}{1+\varepsilon x}r_a+\frac{\varepsilon}{\varepsilon+x}w_a, \qquad (3)$$

$$w_a=\frac{\eta x}{1+\eta x}r_b+\frac{\eta}{\eta+x}w_b, \qquad (4)$$

$$\frac{r_a r_b}{w_a w_b}=\left(\frac{1+\varepsilon x}{\varepsilon+x}\right)^z\left(\frac{1+\eta x}{\eta+x}\right)^z, \qquad (5)$$

Where $x=e^{-V_1/kT}$. In addition, the quantities $r_a$ and $w_b$ must satisfy the relation

$$r_a+w_b=2c. \qquad (6)$$

Eliminating $r_a$, $r_b$, $w_a$, $w_b$ and $x$ from (3), (4) and (6), we obtain a relation between $\eta$ and $\varepsilon$ from which we find

$$\eta=\frac{\varepsilon x-(2c-1)(1+\varepsilon x)}{x+(2c-1)(\varepsilon+x)}\cdot \qquad (7)$$

For the case $r=\frac{1}{4}$, the formulae for $r_a$,.... become too complicated. As our assumptions (1) and (2) on the average interaction energy are very rough, it is not worth while to go into the detailed calculations on Bethe's theory in this case. We shall therefore apply the simpler theory of Bragg and Williams.[8] This is done in §4.

7. These relations can be obtained by an argument similar to that which leads to Eq. (2) in §2 of Wang's paper, P. 58. loc. cit. 4.

8. W. L. Bragg, and E. J. Williams, Proc. Roy. Soc. A *145*, 699-730 (1934).

It should be mentioned that the expression for the total configurational energy in Bethe's theory does not apply under the above interpretation of the average interaction energy. A difficulty arises when we want to calculate the specific heat. This difficulty does not occur in the case $r=1/4$ when the theory of Bragg and Williams is applied. In the case $r=1/2$, this difficulty is overcome by analogy with the theory of Bragg and Williams, as is explained in the next section.

§3. Superlattices of type AB.

With the assumption of constant interaction energy, Easthope[6] has calculated the effect of the change of composition upon the critical temperature of the alloy and Chang[9] has investigated the effect upon the anomalous specific heat. We shall now see how their results should be modified if we assume the average interaction energy to depend upon composition and order of the alloy.

This dependence has been assumed to be given by (1) or (2). In order to apply (2), we must first express $m_{AA}$ in terms of the parameters in (3) and (4). If we assume that all the neighboring sites of an a-site are b-sites and vice versa, then $M_{AA}$ is equal to the product of $\frac{1}{2} r_a N$ by the average number of A-neighbors of an A atom in a-site, thus,

$$m_{AA} = r_a \frac{\varepsilon x}{1+\varepsilon x} \,. \tag{8}$$

The same argument with a-sites replaced by b-sites leads to

$$m_{AA} = w_b \frac{x}{\eta + x} \,. \tag{8'}$$

That these expressions are identical can be easily verified with the help of (3), (4), (6) and (7).

We can now proceed to calculate the critical temperature. Since there is no explicit occurrence of $T$ in equations (3)-(6), we may regard them as defining $\varepsilon$, $\eta$, $r_a$, .... in terms of $x$. Analogous to Easthope's argument,[6] we obtain the critical value

9. T. S. Chang, Proc. Camb. Phil. Soc. *34*, 224-237. (1938).

of x as that value for which the derivatives of the two sides of (5) with respect to $r_a$ ($x$ considered constant) are equal at $r_a = c$. It is now clear that the critical values of $x$ and $\varepsilon$ are not affected by the dependence of $V_1$ on composition and order. They are given by Easthope's equations (14) an d(13):

$$x_o^2 = \frac{(c-\frac{1}{z})(1-c-\frac{1}{z})}{c(1-c)}, \tag{9}$$

$$\varepsilon_o = \frac{c-\frac{1}{z}}{x_o(1-c)}, \quad \eta_o = \frac{1}{\varepsilon_o}. \tag{10}$$

From (9), the critical temperature is given by

$$T_o = \frac{2V_1}{k} \left| \log (1-\frac{1}{cz})(1-\frac{1}{(1-c)z}) \right|^{-1} \tag{11}$$

where $V_1$ is however no longer a constant but is given by (1) or (2); the critical values for $r_a$ and $m_{AA}$ must be substituted in the expressions for $V_1$, of course. Referring to quantities for $V_1 = V_o$ by a suffix zero, we may conventiently write

$$T_o = (T_o)_0 \, \chi(c), \tag{12}$$

where $\chi$ (c) stands for

$$1 + \alpha c,$$

or

$$1 + \alpha c + \beta c \frac{cz-1}{z-1},$$

according to whether (1) or (2) is assumed.

The composition at which the critical temperature is a maximum is given by

$$\frac{dT_o}{dc} = \frac{d(T_o)_0}{dc} \chi(c) + (T_o)_0 \chi'(c) = o.$$

Since $d(T_o)_o = 0$ at $c = \frac{1}{2}$, the maximum critical temperature will occur at $c = \frac{1}{2}$ only when $\chi'(c) = O$ at $c = \frac{1}{2}$. If (1) is assumed this requires $\alpha = O$, and the critical temperature is not influenced by the dependence of interaction energy upon atomic arrangements. If (2) is assumed, a little calculation will show that

$$\frac{d\chi}{dc} = \alpha + \beta$$

at $c=\frac{1}{2}$. Thus, in order that $T$ should be maximum at $c=\frac{1}{2}$, we must have

$$\alpha=-\beta.$$

In this case, the coefficients $\alpha_1$ and $\alpha_2$ in (2A) are equal and $V_1$ depends symmetrically on A and B atoms. Now,

$$\chi(c)=1+\frac{\alpha c(1-c)z}{z-1},$$

and it is evident that the $T_o-c$ curve is symmetrical with respect to $c=\frac{1}{2}$. It can also be shown that the curve is sharper if $\alpha=-\beta>o$ and flatter if $\alpha=-\beta<o$ Values of $T_o$ (referring to the maximum value as unity) have been calculated for $z=6$ and $z=8$. The results are shown in Fig. 1.

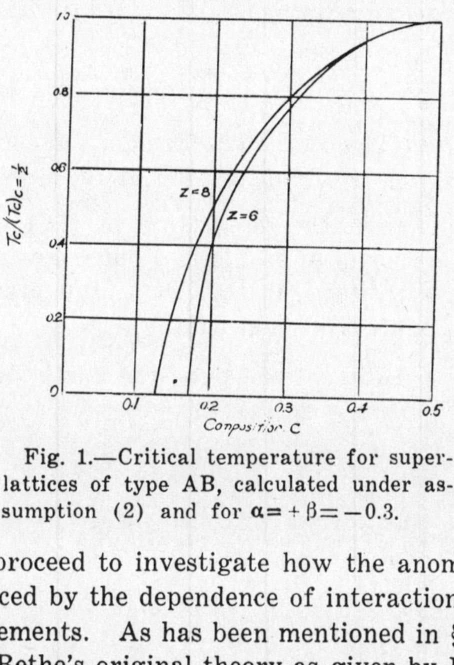


Fig. 1.—Critical temperature for superlattices of type AB, calculated under assumption (2) and for $\alpha=+\beta=-0.3$.

We now proceed to investigate how the anomalous specific heat is influenced by the dependence of interaction energy upon atomic arrangements. As has been mentioned in §2, the energy expression in Bethe's original theory as given by Easthope does not apply here. In order to avoid this difficulty, we seek an approximate energy expression by analogy with the theory of Bragg and Williams. In that theory, the equilibrium condition is

$$\frac{r_a r_b}{W_a W_b} = e^{V/kT} \tag{13}$$

where V is the ordering energy defined by

$$V = -\frac{1}{rN}\left(\frac{\partial E}{\partial r_a}\right)_T \tag{14}$$

If we compare (13) with (5), we see that the ordering energy in our case is implicitly given by

$$e^{V/kT} = \left(\frac{1+\varepsilon x}{\varepsilon + x}\right)^z \left(\frac{1+\eta x}{\eta + x}\right)^z. \tag{15}$$

In this expression, ε, η, and $x$ must be regarded as functions of $r_a$ and $T$, given implicitly by (3), (4) and (6) together with (1) or (2). We now assume that (14) applies also to our case. Then using (15), we obtain

$$E - E_o = -\tfrac{1}{2}NkT \int^{r_a} \log\left(\frac{1+\varepsilon x}{\varepsilon + x}\right)^z \left(\frac{1+\eta x}{\eta + x}\right)^z dr_a + E^*, \tag{16}$$

where $E_o$ is the energy at the absolute zero of temperature and E* is some function of T, unspecified at the present.

In carrying out the integration in (16), $r_a$ and T are regarded as independent of each other, but the equilibrium value of $r_a$ should be substituted in the resulting expression in order to obtain the equilibrium value of the energy.

In order to see how good our approximation (16) is, we compare it with Bethe's expression

$$E - E_o = -zNkT \frac{\varepsilon x}{1+2\varepsilon x+\varepsilon^2} \log x, \tag{17}$$

in the case of constant interaction energy and for $c=\frac{1}{2}$. In this case $\varepsilon = \eta$ and (16) becomes

$$E - E_o = -zNkT \int^{r_a} \log \frac{1+\varepsilon x}{\varepsilon + x}\, dr_a + E^*$$

$$= -\tfrac{1}{2}zNkT \left\{ \frac{1-\varepsilon^2}{1+2\varepsilon x+\varepsilon^2} \log \frac{1+\varepsilon x}{\varepsilon + x} - \log \frac{1+2\varepsilon x+\varepsilon^2}{(1+\varepsilon x)(\varepsilon + x)} \right\}, \tag{18}$$

the function E* being determined such that (17) and (18) agree in the neighborhood of T=0 and give the same change of energy between T=0 and T=∞. (17) and (18) can then be conveniently compared by plotting $\frac{E-E_o}{E_\infty-E_o}$ against $x$, as shown in Fig. 2 for $z=6$ and $z=8$.

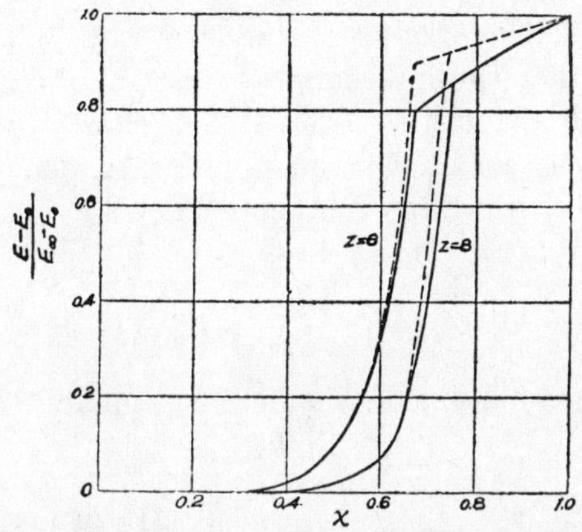


Fig. 2. Comparison of energy expressions according to Bethe (fully drawn) and according to approximation used in this paper (dotted)

In the general case treated here, the integration is too difficult to be carried out explicitly. We shall therefore confine ourselves to the abnormal specific heat at the critical temperature. We regard E given by (16) as a function of $r_a$ and T. The specitic heat is then given by

$$C_v = \frac{dE}{dT} = \left(\frac{\partial E}{\partial T}\right)_{ra} + \left(\frac{\partial E}{\partial r_a}\right)_T \frac{dr_a}{dT}$$

The first term in $C_v$ does not bring about any discontinuity since all the quantities involved in it are continuous. The discontinuity is therefore solely given by the second term, as $\frac{dr_a}{dT}$ is in general discontinuous at the critical temperature.
Hence,

$$\Delta C_v = \lim_{T \to T_o} \left\{ \left(\frac{\partial E}{\partial r_a}\right)_T \frac{dr_a}{dT} \right\}. \tag{19}$$

Write

$$\left(\frac{\partial E}{\partial r_a}\right)_T \frac{dr_a}{dT} = \left(\frac{\partial E}{\partial r_a}\right)_T \frac{dr_a}{dx}\frac{dx}{dT}$$

$$= -\tfrac{1}{2}zNk \log\left(\frac{1+\varepsilon x}{\varepsilon+x}\,\frac{1+\eta x}{\eta+x}\right)\cdot \frac{dr_a}{dx} \times T\frac{dx}{dT}$$

by using (16), The factor before the multiplation sign × is the same whether the interaction energy is constant or not and must approach the same limit at the critical temperature, independent of any assumption on $V_1$. Hence, denoting quantities referring to constant interaction energy by a suffix zero, we have, by using (19),

$$\frac{\Delta C_v}{(\Delta C_v)_o} = T\frac{dx}{dT} \Big/ \left(T\frac{dx}{dT}\right)_o, \tag{20}$$

where the right-hand side should be evaluated at the critical temperature. Now $x = e^{-V_1/kT}$, so that

$$T\frac{dx}{dT} = x \log\frac{1}{x}\left(1 - \frac{T}{V_1}\frac{dV_1}{dx}\frac{dx}{dT}\right).$$

Solving for $T\frac{dx}{dT}$ and substituting in (20), we obtain

$$\frac{\Delta C_v}{(\Delta C_v)_o} = \frac{1}{1 + \frac{x_o}{kT_c}\left(\frac{dV_1}{dx}\right)_o}. \tag{21}$$

If the form (1) for the interaction energy is assumed,

$$\frac{dV_1}{dx} = V_o\beta\frac{dr_a}{dx}.$$

From the case of constant interaction energy, we see that $\frac{dr_a}{dx}$ becomes $-\infty$ as $x$ approaches $x_o$. Hence $\Delta C_v$ will be zero unless $\beta$ vanishes. In the latter case $\Delta C_v$ is the same as in the case of constant interaction energy. If the form (2) for the interaction energy is assumed,

$$\frac{\Delta C_v}{(\Delta C_v)_o} = \frac{1}{1+\beta x_o\frac{V_o}{kT_c}\left(\frac{dm_{AA}}{dx}\right)_o}$$

It can be shown that $\left(\frac{dm_{AA}}{dx}\right)_o$ is finite and positive so that

$$\Delta C_v \gtreqless \left(\Delta C_v\right)_o \text{ according as } \beta \lesseqgtr 0$$

§4. Superlattices of type $AB_3$.

Let us now consider superlattices of the face-centered type. For simplicity, we shall apply the theory of Bragg and Williams.

In this theory, the most important quantity is the ordering energy, which is the increase of energy of the crystal when an atom in a-site and a B atom in b-site are interchanged. By a procedure adopted by Bragg and Williams,[10] which amounts to neglecting local order and regarding all a-sites as equivalent, we obtain the following expression for the ordering energy V:

$$\begin{aligned} V = V_{AA}\,&\{(z_{ba}r_a+z_{bb}w_b)-(z_{aa}r_a+z_{ab}w_b)\}+V_{AB}\{z_{ba}w_a+z_{bb}r_b)\\ &-(z_{aa}w_a+z_{ab}r_b)\}\\ +V_{BB}\,&\{z_{ab}r_b+z_{aa}w_a)-(z_{bb}r_b+z_{ba}w_a)\}+V_{AB}\{z_{ab}w_b+z_{aa}r_a)\\ &-(z_{bb}w_b+z_{ba}r_a)\}, \end{aligned}$$

where $z_{ab}$ is the number of b-sites neighboring to an a-site and the other $z$'s have obviously similar meanings.[4] In terms of $z$ and $z'$ defined by[4]

$$z = z_{aa}+z_{ab} = z_{ba}+z_{bb},$$
$$rz_{ab} = (1-r)z_{ba} = r(1-r)z',$$

V is reduced to

$$V = \frac{2(z'-z)}{1-r}(r_a-c)V_1. \tag{22}$$

With the expression (22) for the ordering energy the total energy of the crystal is obtained by integrating the equation

$$\frac{\partial E}{\partial r_a} = -rNV.$$

Using the assumption (2) for $V_1$ with the help of the formula

$$m_{AA} = C^2 - \frac{r}{1-r}\,\frac{z'-z}{z}\,(r_a-c)^2, \tag{23}$$

which can be easily proved, we find that

10. W. L., Bragg and E. J. Williams, Proc. Roy. Soc. A *151*, 540-566, (1935).

$$E-E_o=-zNV_o\frac{r}{1-r}\frac{z'-z}{z}\left\{1+\alpha c+\beta c^2-\frac{\beta}{2}\frac{r}{1-r}\frac{z'-z}{z}(r_a-c)^2\right\}(r_a-c)^2.$$

For face-centered lattice, $r=\frac{1}{4}$, $z=12$ and $z'=16$, and this becomes

$$E-E_o=-\frac{4}{3}NV_o\left\{1+\alpha c+\beta c^2-\frac{\beta}{18}(r_a-c)^2\right\}(r_a-c)^2. \tag{24}$$

The condition of equilibrium may as usual be expressed in terms of minimum free energy as

$$\frac{\partial E}{\partial r_a}=T\frac{\partial S}{\partial r_a} \tag{25}$$

and the condition for critical temperature as

$$E-E_c=T(S-S_c)\cdot \tag{26}$$

(25) states that the free energy is a minimum at equilibrium and (26) states that thet the transition from ordered phase to disordered phase occurs when the free energy of the two phases are equal.

As the expression for the entropy S, we use Boltzmann's relation between entropy and thermodynamic probability and obtain

$$S=-krN\{r_a\log r_a+(1-r_a)\log(1-r_a)\}-k(1-r)N\{w_b\log w_b+(1-w_b)\log(1-w_b)\}, \tag{27}$$

which is a function of $r_a$ alone on account of the relation $rr_a+(1-r)w_b=c$. In the case of a face-centered lattice, (27) becomes

$$S=-\frac{kN}{4}\left\{r_a\log r_a+(1-r_a)\log(1-r_a)\right\}-\frac{3kN}{4}\left\{w_b\log w_b+(1-w_b)\log(1-w_b)\right\}\cdot \tag{28}$$

Eliminating T between (25) and (26) and writing (28) in the form $S-S_c=-\frac{kN}{4}\varphi(r_a,c)$ for convenience, we obtain the following equation determining $r_a$ at the critical temperature:

$$2\varphi-\left\{1+\frac{\beta(r_a-c)^2/18}{1+\alpha c+\beta c^2-\beta(r_a-c)^2/9}\right\}\frac{\partial\varphi}{\partial r_a}(r_a-c)=o. \tag{29}$$

The critical temperature is then obtained by substituting the value of $r$ from (29) into (27), viz.,

$$T_c=\frac{32}{3}\frac{V_o}{k}\frac{r_a-c}{\dfrac{\partial\varphi}{\partial r_a}}\left\{1+\alpha c+\beta c^2-\frac{\beta}{9}(r_a-c)^2\right\}. \tag{30}$$

The critical temperature has been calculated for $c$ from $o$ to ½ by assuming particular values for $\alpha$ and $\beta$. The results are shown in Fig. 3a.

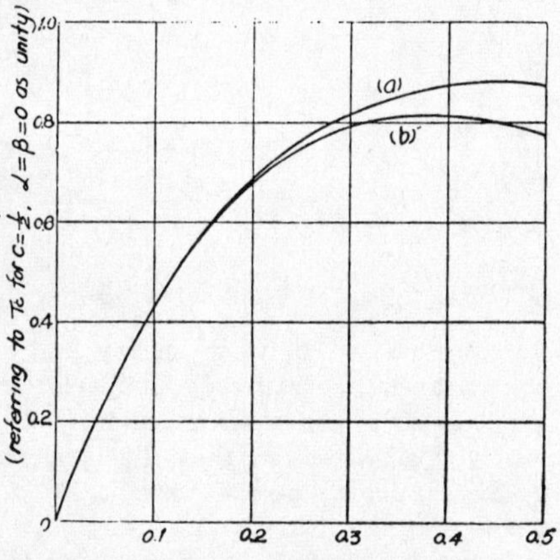


Fig. 3a. Critical temperature for superlattices of type $AB_3$, calculated under assumption (a) $\alpha=-0.3, \beta=0.1$ and (b) $\alpha=\beta=-0.3$.

It is seen from the figures that the maximum of the critical temperature does not occure at $c=½$, In fact, by assigning suitable values to $\alpha$ and $\beta$, the maximum of the critical temperature will occur at any chosen value of $c$. When the maximum position of the critical temperature is fixed, one of the constants $\alpha$ and $\beta$ can still be chosen arbitrarily. They can be so chosen as to fit the experimental data. The values of $\alpha$ and $\beta$ for the maximum critical temperature to occur at $c=\frac{1}{4}$ may be any set of the values (-1.059, -0.719), (-.444, -2.26) and (1.88, -7.69), the last set having the best agreement with experiment[11] (Fig. 3b)

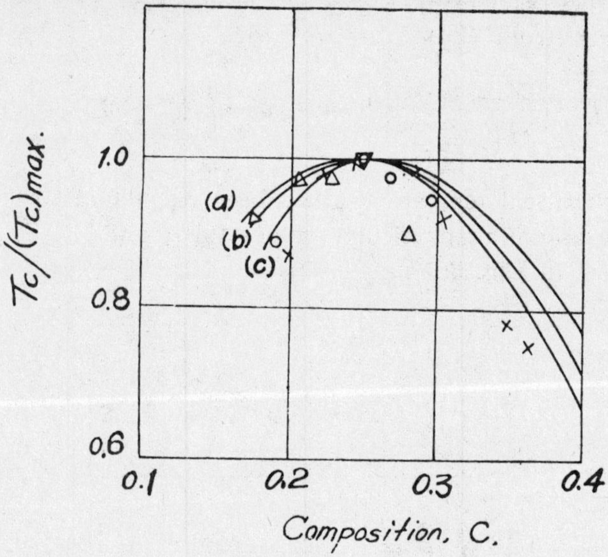


Fig. 3b. Comparison between theory and experiments regarding critical temperature for superlattices of type $AB_3$. The theoretical curves are calculated as in Fig. 3a but with $\alpha, \beta$ adjusted so that $T_0$ is a maximum at $C=\frac{1}{4}$. x: Experimental points for $Cu_3Au$ according to Haughton and Payne, o: Experimental points for $Cu_3Au$ according to Borelius, etc, $\Delta$: Experimental points for $Pb_3Au$ according to Borelius, etc.

The anomalous specific heat can also be easily calculated. We have

$$C_v = \frac{dE}{dT} = \frac{\partial E}{\partial r_a}\frac{dr_a}{dT} \cdot \tag{31}$$

where $\frac{\partial E}{\partial r_a}$ is given by (23). A little calculation will show that

$$C_v = \frac{Nk}{4}\frac{\left(\frac{\partial \varphi}{\partial r_a}\right)^2 (r_a - c)}{(r_a - c)\frac{\partial^2 \varphi}{\partial r_a^2} - \frac{\partial \varphi}{\partial r_a}\left\{1 - \frac{2\beta(r_a - c)^2}{9(1+\alpha c+\beta c^2) - \beta(r_a - c)^2}\right\}} \cdot \tag{32}$$

1. Haughton and Payne, J. Inst. Metals, *46*, 457-480; (1931), and Borelius, Johansson, Linde, Ann. d. Physik *86*, 291-318. (1928).

Regarding (25) and (32) as the parametric equations for $C_v$ $T$ curve, $r_a$ being the parameter, the calculation of $C_v$ as a function of $T$ can be easily carried out for each composition. We are, however, more interested in the discontinuity of $C_v$ at the critical temperature. Since $C_v=0$ above $Tc$, the discontinuity in $C_v$ is given by (32) at the critical temperature. The result of calculation is shown in Fig. 4. It is seen that the maximum value of $\Delta C_v$ does not occur at $c=\frac{1}{2}$.

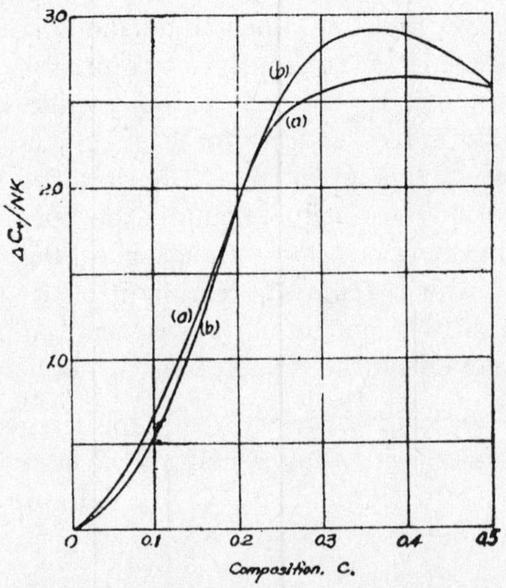


Fig. 4. Anomalous specific heat at critical temperature for superlattices of type $AB_3$ calculated under the same conditions as for Fig. 3a.

If we assume the form (1) for the interaction energy, we obtain results similar to those obtained above. We need only mention that the maximum value of the critical temperature occurs at compositions c=0.4, 0.3 and ¼ for ( $\alpha$, $\beta$ ) equal to (-0.273, 0.141, (-0.926, 0.080) and (-1.568, 0.066) respectively.

It should be remarked that, strictly speaking, these considerations are valid only when $c$ differs not much from ¼, for only then is the division of lattice sites into a- and b-sites legitimate.

It has been shown by Easthope[12] and Shockley[13] that the crystal separates into more than one phase for certain ranges of temperature and composition according to the principle of minimum free energy. In our case, this is still true. For temperatures near the critical temperature calculated above, the crystal separates into two phases, one ordered and the other disordered. We have two curves on the two sides of the $T_o-c$ curve in Fig. 3, between which the ordered and disordered phase coexist. The observed critical temperature ought to be given by the upper curve, and our conclusions should be modified. In order to calculate this separation into two phases, we must know the absolute value (i. e., refered to some value independent of composition) of the free energy for each phase. The absolute value of the entropy is given by (27), but the absolute value of the energy cannot be easily calculated because the energy E at the critical temperature as a function of the composition is not known. However, from the results of Easthope and Shockley this separation will not alter the essential points of our conclusions. Consequently the calculation is not carried out.

I wish to thank Prof. J. S. Wang for suggesting the problem to me and for his invaluable help in the course of investigation. I wish also to thank Prof. C. S. Yeh for many helpful discussions.

12. C. E. Easthope, Proc. Camb. Phil. Soc. *34*, 68-80 (1938).
13. W. Shockley, J. Chem. Phys. *6*, 130-144 (1938).

## Note on the Normalization of Dirac Functions

C. C. Lin
*Department of Applied Mathematics, University of Toronto, Canada*
April 18, 1941

A VERY simple method of normalizing the Dirac functions in the Kepler problem is given in this note. The normalization of Dirac functions in the Kepler problem was first done by Bechert[1] with the help of certain contour integrals in a complex plane. The calculation was, however, rather complicated. With the functions given in Infeld's form, a much simpler method of normalization leading to a neater result is proposed here. The result is, as I verified, the same as that given by Bechert, but it appears in a completely different form.

It has been shown[2] that the radial Dirac functions are given by

$$\chi_1 = C\left\{(\epsilon\kappa-\gamma)^{\frac{1}{2}}\frac{\gamma_2+\gamma_1}{\gamma}\overset{0}{\mathfrak{F}}_{l,\gamma}(x)+(\epsilon\kappa+\gamma)^{\frac{1}{2}}\frac{\gamma_2-\gamma_1}{\gamma}\overset{1}{\mathfrak{F}}_{l,\gamma}(x)\right\} \quad (1)$$

$$\chi_2 = C\left\{(\epsilon\kappa-\gamma)^{\frac{1}{2}}\frac{\gamma_2-\gamma_1}{\gamma}\overset{0}{\mathfrak{F}}_{l,\gamma}(x)+(\epsilon\kappa-\gamma)^{\frac{1}{2}}\frac{\gamma_2+\gamma_1}{\gamma}\overset{1}{\mathfrak{F}}_{l,\gamma}(x)\right\} \quad (2)$$

with the following definitions and formulae ($C$ being the normalization factor):

$$\gamma_1=(\kappa-\alpha Z)^{\frac{1}{2}},\quad \gamma_2=(\kappa+\alpha Z)^{\frac{1}{2}},\quad \gamma=\gamma_1\gamma_2=(\kappa^2-\alpha^2Z^2)^{\frac{1}{2}}, \quad (3)$$

$$\epsilon=\frac{E}{E_0}=\left\{1+\frac{\alpha^2Z^2}{(l+\gamma)^2}\right\}^{-\frac{1}{2}}, \quad (4)$$

where $l$ is a positive integer, or zero (only for $\kappa<0$), and $\kappa$ is an integer, positive or negative, but not zero; $\overset{1}{\mathfrak{F}}_{l,\gamma}$ and $\overset{0}{\mathfrak{F}}_{l,\gamma}$ are the lowest two functions of a ladder of normalized functions defined by

$$\overset{l}{\mathfrak{F}}_{l,\gamma}(x)=\left(\frac{2a}{\gamma+l}\right)^{\gamma+l+\frac{1}{2}}\cdot\frac{1}{[\Gamma(2\gamma+2l+1)]^{\frac{1}{2}}}x^{l+\gamma}e^{-ax/(l+\gamma)}, \quad (5)$$

$$\overset{m}{\mathfrak{F}}_{l,\gamma}(x)=\frac{(l+\gamma)(m+\gamma)}{a[(l+m+2\gamma)(l-m)]^{\frac{1}{2}}}\times\left(\frac{m+\gamma}{x}-\frac{a}{m+\gamma}+\frac{d}{dx}\right)\overset{m+1}{\mathfrak{F}}_{l,\gamma}(x), \quad (6)$$

where $m$ is an integer, $0\leqslant m\leqslant l$, and

$$a=\frac{m_0c}{\hbar}\epsilon\alpha Z. \quad (7)$$

As $C$ is the normalization factor and the functions $\overset{0}{\mathfrak{F}}_{l,\gamma}$, $\overset{1}{\mathfrak{F}}_{l,\gamma}$ are normalized, we obtain from (1), (2) and (3) the condition for $C$ to make $\int_0^\infty(\chi_1^2+\chi_2^2)dx=1$:

$$1=\frac{8C^2}{\gamma^2}\{\epsilon\kappa^2+(\epsilon^2\kappa^2-\gamma^2)^{\frac{1}{2}}\alpha ZI_l(0,1)\}, \quad (8)$$

where

$$I_l(0,1)=\int_0^\infty\overset{0}{\mathfrak{F}}_{l,\gamma}(x)\overset{1}{\mathfrak{F}}_{l,\gamma}(x)dx. \quad (9)$$

It will be shown below that

$$I_l(0,1)=-(\epsilon^2\kappa^2-\gamma^2)^{\frac{1}{2}}/\alpha Z\epsilon, \quad (10)$$

and hence, because of (8), we have

$$C=\epsilon^{\frac{1}{2}}/2\sqrt{2}. \quad (11)$$

This is the normalization factor required.

To calculate $I_l(0,1)$, let us consider

$$I_l(m,m+1)=\int_0^\infty\overset{m}{\mathfrak{F}}_{l,\gamma}\overset{m+1}{\mathfrak{F}}_{l,\gamma}dx. \quad (12)$$

By writing

$$\overset{m\mp}{\mathcal{H}}_{l,\gamma}=\overset{m}{A}_{l,\gamma}\overset{m\mp}{H}_{l,\gamma}=\overset{m}{A}_{l,\gamma}\left(\frac{m+\gamma-1}{x}-\frac{a}{m+\gamma-1}\mp\frac{d}{dx}\right),$$
$$\overset{m}{A}_{l,\gamma}=\frac{(l+\gamma)(m+\gamma-1)}{a[(l+m-1+2\gamma)(l-m+1)]^{\frac{1}{2}}} \quad (13)$$

it is not difficult to verify that

$$H^{(m+1)-}=\xi_mH^{m-}+\eta_mH^{m+}+\zeta_m, \quad (14)$$

where $\xi_m$, $\eta_m$, $\zeta_m$ are constants given by

$$\xi_m+\eta_m=\frac{m+\gamma}{m+\gamma-1},\quad \xi_m-\eta_m=1,$$
$$\zeta_m=a(m+\gamma)\left\{\frac{1}{(m+\gamma-1)^2}-\frac{1}{(m+\gamma)^2}\right\}. \quad (15)$$

Furthermore, we have

$$\int_0^\infty fH^{m-}\varphi dx=\int_0^\infty\varphi H^{m+}fdx \quad (16)$$

provided $f$ and $\varphi$ vanish properly at $x=0$ and $x=\infty$. By making use of (13)–(16), we can easily see that (12) reduces to

$$I_l(m,m+1)=A^{m+1}\left\{(\xi_m+\eta_m)\frac{1}{A^m}I(m-1,m)+\zeta_m\right\}, \quad (17)$$

or

$$\frac{I_l(m-1,m)}{A^m(m+\gamma-1)}+\frac{a}{(m+\gamma-1)^2}=\frac{I_l(m,m+1)}{A^{m+1}(m+\gamma)}+\frac{a}{(m+\gamma)^2}. \quad (18)$$

We note that this means that the quantity on either side of this equation is independent of $m$, and hence is equal to

$$a/(l+\gamma)^2, \quad (19)$$

since $I_l(l,l+1)$ is evidently zero (as $\overset{l+1}{\mathcal{H}}_{l,\gamma}\overset{l}{\mathfrak{F}}_{l,\gamma}=0$). Hence by using the value of $A^{m+1}$ given by (13), we have the value of $I_l(m,m+1)$. Putting $m=0$ in the result, we have

$$I_l(0,1)=-\{1-(\gamma^2/(l+\gamma)^2)\}^{\frac{1}{2}}. \quad (20)$$

Equation (10) is then obtained by solving (4) for $l+\gamma$ and substituting the result into (20).

The author wishes to thank Professor Infeld for suggesting the problem and for his help.

[1] Bechert, Ann. d. Physik 6, 700 (1930). See also Bethe, *Handbuch der Physik* (1933), Vol. 24, p. 315.
[2] L. Infeld, Phys. Rev. 59, 737 (1941). The notations used here are the same as those used in that paper.

# ON THE MECHANICAL BEHAVIOUR OF METALS IN THE STRAIN-HARDENING RANGE*

BY

G. H. HANDELMAN, C. C. LIN AND W. PRAGER

*Brown University*

**1. Introduction.** The present paper is concerned with certain stress-strain relations purporting to describe the mechanical behaviour of quasi-isotropic metals in the strain-hardening range. As a preparation for a more precise characterization of these relations, let us consider the tension test of a metal like copper or aluminum which *does not flow under a constant stress, but exhibits strain hardening.* If the test involves loading only, i.e., if the reduced tensile stress[1] $\sigma$ or the tensile strain $\epsilon$ increase throughout the test, the resulting diagram of reduced stress versus strain will have the general appearance of the curve $OPQ$ in Fig. 1. On the other hand, if the test specimen is unloaded after a certain point, such as $P$, has been reached along this curve, the stress-strain diagram for unloading is found to be very nearly a straight line $PA$ which is parallel to the tangent of the curve $OPQ$ at $O$. After complete unloading, the specimen shows a permanent extension which corresponds to the permanent strain represented by $OA$.

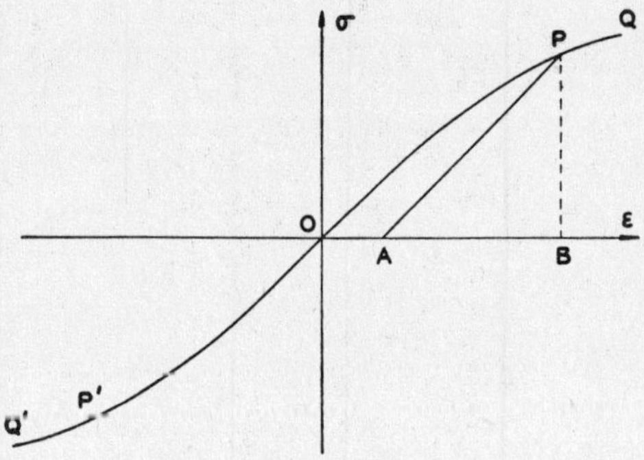

FIG. 1. Typical curve of reduced stress vs. strain.

To simplify the discussion, let us assume at present that the material is incompressible. A longitudinal extension $\epsilon$ of the isotropic specimen is then accompanied by a uniform lateral contraction of the magnitude $\epsilon/2$. If the discussion is restricted to states of stress and strain which can be reached by a *single* loading followed by *one* complete or partial unloading at the most, the mechanical behaviour of the material in simple tension is therefore completely defined by the curve $OPQ$. It will be assumed in the following that for the materials under consideration the stress-strain diagram in simple compression ($OP'Q'$ in Fig. 1) is obtained by reflecting the curve $OPQ$ with respect to the origin $O$, and that the practically important portion of the curve $Q'OQ$, i.e., the portion corresponding to small and moderate strains, is represented with sufficient accuracy by a development of the form

$$\epsilon = \sigma + a_3\sigma^3 + a_5\sigma^5 + \cdots, \tag{1}$$

where $a_3, a_5, \cdots$ are constants. (The coefficient of the linear term on the right-hand side of (1) must be unity since $\sigma$ is the reduced stress. No even powers of $\sigma$ can occur

* Received September 17, 1946.

[1] The reduced stress is defined as the quotient of the stress by Young's modulus.

on the right-hand side of (1), because the stress-strain diagrams for tension and compression are assumed to be congruent.)

In the case of simple tension or compression, the mechanical behaviour of the material during the first loading is readily represented by a *finite* relation of the form (1); the behaviour during the first unloading, however, is most naturally represented by the *differential* stress-strain relation

$$d\epsilon = d\sigma, \tag{2}$$

for this form avoids explicit reference to the state of stress at which the unloading began. Accordingly, it is often convenient to write Eq. (1), too, in differential form:

$$d\epsilon = \alpha(\sigma)d\sigma. \tag{3}$$

Here, $\alpha(\sigma) = d\epsilon/d\sigma = 1 + 3a_3\sigma^2 + 5a_5\sigma^4 + \cdots$ equals the quotient of Young's modulus by the so-called *tangent modulus.* To arrive at a complete analytical description of the mechanical behaviour of the material in simple tension and compression, we must supplement the preceding equations by analytical criteria for loading and unloading. For tension ($\sigma > 0$) loading corresponds to $d\sigma > 0$ and unloading to $d\sigma < 0$; for compression ($\sigma < 0$) these criteria must be reversed. A satisfactory criterion for loading and unloading is therefore furnished by the sign of $\sigma d\sigma = d(\frac{1}{2}\sigma^2)$.

The present paper is concerned with the extension of this analysis to general states of stress and strain which can be reached by a *single* loading followed at most by *one* complete or partial unloading. In the case of simple tension or compression, a differential stress-strain relation of the form (3) which is valid for the first loading can always be integrated under the initial condition $\epsilon = 0$ for $\sigma = 0$ and is thus equivalent to a finite stress-strain relation. For more general states of stress, however, a suitably generalized form of the differential stress-strain relation (3) may be integrable or not. The distinction between differential and finite stress-strain relations for the first loading is therefore no longer a purely formal matter, but acquires physical significance. One of the main results of the following discussion consists in the remark that the assumption of a finite stress-strain relation for the first loading is incompatible with certain postulates concerning the mechanical behaviour under those changes of stress which constitute neither loading nor unloading. This is shown in Section 3. Sections 2 and 4 are devoted to the discussion of finite and differential stress-strain relations, respectively. Section 5 gives a method of correlating experimental results with the present theory. Finally, Section 6 contains a discussion of the limitations of the theory.

**2. Finite stress-strain relations.** Using rectangular Cartesian coordinates $x_i$, ($i = 1, 2, 3$), we denote the displacement from the standard state by $u_i$, the strain by $\epsilon_{ij}$ and the reduced stress by $\sigma_{ij}$. For the small deformations to which the following discussion is restricted, the strain $\epsilon_{ij}$ is given by

$$\epsilon_{ij} = \tfrac{1}{2}(u_{i,j} + u_{j,i}), \tag{4}$$

where $u_{i,j}$ stands for $\partial u_i/\partial x_j$, etc. Adopting the usual summation convention regarding repeated subscripts, we define the *mean normal strain* as

$$e = \tfrac{1}{3}\epsilon_{ii}, \tag{5}$$

and the *strain deviation* as

$$e_{ij} = \epsilon_{ij} - e\delta_{ij}, \tag{6}$$

where $\delta_{ij}$ is the Kronecker delta. Similarly, the *reduced mean normal stress* $s$ and the *deviation* $s_{ij}$ *of the reduced stress* are defined as

$$s = \tfrac{1}{3}\sigma_{ii} \tag{7}$$

and

$$s_{ij} = \sigma_{ij} - s\delta_{ij}. \tag{8}$$

According to the definitions of the deviations $e_{ij}$ and $s_{ij}$, we have

$$e_{ii} = 0, \qquad s_{ii} = 0. \tag{9}$$

The task of generalizing the finite stress-strain relation (1) is simplified by the remark that the first term on the right-hand side represents that part of the total strain $\epsilon$ which is recovered upon *complete* unloading. The remaining terms on the right-hand side of (1) accordingly represent the permanent strain. In Fig. 1 the *total* strain is represented by the segment $OB$, the *recoverable* strain by $AB$, and the *permanent* strain by $OA$.

Setting

$$\epsilon_{ij} = \epsilon'_{ij} + \epsilon''_{ij}, \tag{10}$$

where $\epsilon'_{ij}$ denotes the recoverable and $\epsilon''_{ij}$ the permanent strain, we may assume that the recoverable strain is related to the reduced stress by means of the generalized law of Hooke:

$$\epsilon'_{ij} = (1+\nu)s_{ij} + (1-2\nu)s\delta_{ij}. \tag{11}$$

Here $\nu$ denotes Poisson's ratio. We are then left with the task of supplementing (11) by a relation which expresses the permanent strain occurring during the first loading in terms of the reduced stress. For an *isotropic* material, this relation can only contain scalar constants in addition to the tensors $\epsilon''_{ij}$, $\sigma_{ij}$ and $\delta_{ij}$, and their invariants. Furthermore, the principal axes of $\epsilon''_{ij}$ and $\sigma_{ij}$ must coincide. Under the pressures commonly encountered in the testing of materials, no permanent change of volume is observed, i.e., $\epsilon''_{ii}=0$ and $\epsilon''_{ij}=e''_{ij}$. A state of hydrostatic pressure therefore does not produce any permanent strain, and two states of stress which differ only by a state of hydrostatic pressure may be expected to produce identical permanent strains. The permanent strain $\epsilon''_{ij}$ is thus independent of $s$ and depends only on the deviation $s_{ij}$. Furthermore, if the stress-strain diagrams for simple tension and simple compression are congruent, a reversal of the signs of all stresses may be expected to produce a mere reversal of the signs of all principal strains. Finally, if the ratios of the principal stresses are kept constant during the loading process, the ratios of the principal permanent strains, too, can be expected to remain constant.

In a recent paper,[2] W. Prager established the most general stress-strain relation which is compatible with the preceding postulates. With the notations

$$J_2 = \tfrac{1}{2}s_{ij}s_{ji}, \qquad J_3 = \tfrac{1}{3}s_{ij}s_{jk}s_{ki}, \tag{12}$$

and

$$t_{ij} = s_{ik}s_{kj} - \tfrac{2}{3}J_2\delta_{ij}, \tag{13}$$

Prager's stress-strain relation can be written in the form

[2] W. Prager, *Strain-hardening under combined stresses*, J. Appl. Phys. **16**, 837–840 (1945).

$$\epsilon''_{ij} = F(J_2, J_3^2)[P(J_2, J_3^2)s_{ij} + Q(J_2, J_3^2)J_3\, t_{ij}], \tag{14}$$

where $P$ and $Q$ must be homogeneous in the components of the stress deviation, the degree of $P$ exceeding that of $Q$ by 4. The expressions (12) are second and third order invariants of the stress deviation $s_{ij}$ (the first order invariant $s_{ii}$ vanishes). The tensor (13) is the deviation of the square $s_{ik}s_{kj}$ of the stress deviation $s_{ij}$.

Combining (11) and (14), we obtain the desired generalization of the finite stress-strain relation (1):

$$\epsilon_{ij} = (1 + \nu)s_{ij} + (1 - 2\nu)s\delta_{ij} + F(J_2, J_3^2)[P(J_2, J_3^2)s_{ij} + Q(J_2, J_3^2)J_3\, t_{ij}]. \tag{15}$$

**3. Neutral changes of stress. Inadmissibility of finite stress-strain relations.** In the case of simple tension or compression the sign of $\sigma d\sigma = d(\frac{1}{2}\sigma^2)$ proved to be a satisfactory criterion for loading and unloading. Accordingly, one might consider the possibility of using the sign of $\sigma_{ji}d\sigma_{ji}$ as a criterion in the general case. If, however, the term "loading" is reserved for such changes of stress which are accompanied by a change of the permanent strain, this criterion is not satisfactory. Indeed, on account of (8) and the second Eq. (9), we have

$$\sigma_{ij}d\sigma_{ij} = (s_{ij} + s\delta_{ij})(ds_{ij} + ds\delta_{ij}) = s_{ij}ds_{ij} + 3sds. \tag{16}$$

If loading were to correspond to $\sigma_{ij}d\sigma_{ij} > 0$, a change of stress for which $ds_{ij} = 0$ might therefore constitute loading in spite of the fact that such a change of stress is not accompanied by a change of the permanent strain. To avoid this difficulty, we shall use the sign of $s_{ij}\,ds_{ij} = dJ_2$ as the desired criterion, an increase of $J_2$ corresponding to loading, a decrease to unloading.

Whereas for uniaxial stress any change of stress constitutes either loading or unloading, we have three kinds of change of stress in the general case, according to whether $J_2$ increases, remains constant, or decreases. An infinitesimal change of stress for which $dJ_2 = 0$, will be called a *neutral* change of stress. For instance, any change of stress which affects only the mean normal stress, but leaves the stress deviation untouched, is a neutral change of stress. A more interesting example of a neutral change of stress is given by

$$\sigma_{ij} = \begin{Vmatrix} \sigma & 0 & 0 \\ 0 & 0 & 0 \\ 0 & 0 & 0 \end{Vmatrix}, \qquad d\sigma_{ij} = \begin{Vmatrix} 0 & d\tau & 0 \\ d\tau & 0 & 0 \\ 0 & 0 & 0 \end{Vmatrix}. \tag{17}$$

Equation (17) represents the stress system which arises from a combined tension and torsion test of a thin walled circular cylinder. Specifically, consider such a test piece which is pulled to an arbitrary tensile stress $\sigma$. If the traction is then kept constant and a small torque applied, the resulting systems of stress and increments of stress are represented by Eq. (17).

Let us now suppose that for the first loading ($dJ_2 > 0$) we have the finite stress-strain relation (15) and for unloading ($dJ_2 < 0$) the generalized-law of Hooke in the differential form

$$d\epsilon_{ij} = (1 + \nu)ds_{ij} + (1 - 2\nu)ds\delta_{ij}. \tag{18}$$

The simultaneous use of the stress-strain relations (15) and (18) will lead to obvious difficulties, unless these relations give identical strain increments for neutral changes of stress. We shall show that, in general, this *continuity condition* is not fulfilled. Indeed, if (15) is written in differential form, the first two terms on the right-hand side equal the right-hand side of (18); the continuity condition therefore requires the vanishing of the remaining terms on the right-hand side of the differential form of Eq. (15). Consider, for example, the stress and increment of stress given by Eq. (17). A simple computation will show that

$$s_{ij} = \begin{pmatrix} \frac{2}{3}\sigma & 0 & 0 \\ 0 & -\frac{1}{3}\sigma & 0 \\ 0 & 0 & -\frac{1}{3}\sigma \end{pmatrix}, \qquad t_{ij} = \begin{pmatrix} \frac{2}{9}\sigma^2 & 0 & 0 \\ 0 & -\frac{1}{9}\sigma^2 & 0 \\ 0 & 0 & -\frac{1}{9}\sigma^2 \end{pmatrix},$$

$$ds_{ij} = \begin{pmatrix} 0 & d\tau & 0 \\ d\tau & 0 & 0 \\ 0 & 0 & 0 \end{pmatrix}, \qquad dt_{ij} = \begin{pmatrix} 0 & \frac{1}{3}\sigma d\tau & 0 \\ \frac{1}{3}\sigma d\tau & 0 & 0 \\ 0 & 0 & 0 \end{pmatrix},$$

as well as $dJ_2 = dJ_3 = 0$. For this special case the differential form of Eq. (14) reduces to

$$d\epsilon''_{ij} = F(J_2, J_3^2)[P(J_2, J_3^2)ds_{ij} + Q(J_2, J_3^2)J_3\, dt_{ij}], \tag{19}$$

where $J_2$ and $J_3$ are evaluated for an arbitrary state of pure tension. Since this state of stress satisfies the condition for a neutral change of stress ($dJ_2 = 0$), $d\epsilon''_{ij}$ must vanish. We find then, upon substituting the values of $ds_{ij}$ and $dt_{ij}$ previously computed, that

$$F(J_2, J_3^2)[P(J_2, J_3^2) + \tfrac{1}{3}Q(J_2, J_3^2)J_3\,\sigma] = 0. \tag{20}$$

Now let us return to the finite stress-strain relation, Eq. (15), for the case of pure tension. The first component of the strain tensor (the other non-vanishing terms differ from this only by a constant factor) becomes

$$\epsilon_{11} = \sigma + \tfrac{2}{3}F(J_2, J_3^2)[P(J_2, J_3^2)\sigma + \tfrac{1}{3}Q(J_2, J_3^2)J_3\,\sigma^2]. \tag{21}$$

The invariants appearing in Eq. (20) have been evaluated for an arbitrary state of pure tension. Consequently, Eq. (20) is valid for pure tension and the second term in Eq. (21) equals zero. (A similar remark holds true for each of the other non-vanishing strain components.) Therefore, the stress-strain relation will reduce to Hooke's law for pure tension if the continuity condition is to be fulfilled. On the other hand, we have seen in Section 1 that the stress-strain law for tension need not be linear. Thus the most general *finite* stress-strain law coupled with Hooke's law for unloading will not be sufficiently flexible to represent a tensile test if the continuity condition is to be fulfilled. It is necessary, therefore, to turn to differential stress-strain relations if both loading and unloading are to be adequately represented.

**4. Differential stress-strain relations.** A system of differential stress-strain relations can be obtained from the properties discussed in Section 2 provided certain of these are rewritten in such a way as to be directly applicable in differential form. We shall assume that given the components of the stress tensor $\sigma_{ij}$ and the increments $d\sigma_{ij}$ there correspond unique strain increments $d\epsilon_{ij}$. This implies that the increment in strain, $d\epsilon_{ij}$, depends only on the state of stress at the given instant, $\sigma_{ij}$, and the increment in stress, $d\sigma_{ij}$, and is independent of the way in which this state of stress has been achieved provided only loading has taken place. In particular, we shall assume that this dependence is such that the increments in strain are linear functions of the increments in stress. Thus $d\epsilon_{ij}$ can be written in the form

$$d\epsilon_{ij} = (1 + \nu)ds_{ij} + (1 - 2\nu)ds\delta_{ij} + c_{ijkl}\,d\sigma_{kl}, \tag{22}$$

where the fourth order tensor $c_{ijkl}$ is a function of $\sigma_{ij}$ only. For unloading, the material is assumed to satisfy the differential form of Hooke's law given in Eq. (18). Loading is supposed to take place when $dJ_2 > 0$ and unloading occurs for $dJ_2 < 0$. For a neutral change of stress, $dJ_2 = 0$, the continuity condition requires that Eqs. (18) and (22) coincide. Consequently,

$$c_{ijkl}\,d\sigma_{kl} = 0 \quad \text{whenever} \quad dJ_2 = s_{kl}\,ds_{kl} = 0.$$

Since $s_{kl}$ is a deviator, $dJ_2$ may also be written in the form $dJ_2 = s_{kl}\,d\sigma_{kl}$. Thus the linear form in $d\sigma_{kl}$, $c_{ijkl}\,d\sigma_{kl}$, must vanish whenever $s_{kl}\,d\sigma_{kl}$ vanishes. The coefficients of $d\sigma_{kl}$ in the two forms must be proportional or

$$c_{ijkl} = C_{ij}s_{kl},$$

where the second order tensor $C_{ij}$ is a function of $\sigma_{kl}$ alone. The stress-strain relations then become

$$\left.\begin{aligned} d\epsilon_{ij} &= (1 + \nu)ds_{ij} + (1 - 2\nu)ds\delta_{ij} + C_{ij}dJ_2, && \text{when} \quad dJ_2 \geqq 0; \\ d\epsilon_{ij} &= (1 + \nu)ds_{ij} + (1 - 2\nu)ds\delta_{ij}, && \text{when} \quad dJ_2 \leqq 0. \end{aligned}\right\} \tag{23}$$

In a certain sense, the term $C_{ij}$ measures the permanent deformation. Indeed, let us consider the infinitesimal cycle of stress which results when first $d\sigma_{ij}$ is applied and then $-d\sigma_{ij}$. We assume, in addition, that the material is being loaded when $d\sigma_{ij}$ is applied. The permanent increment in strain $d\epsilon_{ij}''$ will then be

$$d\epsilon_{ij}'' = C_{ij}\,dJ_2. \tag{24}$$

Since the permanent strain is independent of a state of hydrostatic stress for pressures within the range normally encountered in testing of materials, the tensor $C_{ij}$ can only be a function of the components of the stress deviator rather than the stress tensor itself. Furthermore, there can be no permanent change in volume; that is, $d\epsilon_{ii}'' = 0$ or $C_{ii} = 0$. Since the tensors $d\epsilon_{ij}$, $ds_{ij}$, and $\delta_{ij}$ are symmetric, $C_{ij}$ will also be symmetric. In addition, a reversal of the signs of all the stresses is assumed to produce a reversal of sign of all the strain increments. This implies that $C_{ij}$ must be an odd function of the stress components and thus will vanish when all the $s_{ij}$ vanish.

The material is supposed to become orthotropic under the stress $\sigma_{ij}$ in the sense that the $C_{ij}$ can be represented as a power series in the stress deviator $s_{ij}$ with *scalar* coefficients. These coefficients are either constants or else functions of the invariants of $s_{ij}$, i.e., functions of $J_2$ and $J_3$. It is convenient at this point to change from the subscript notation for tensors to Gibbs' notation; the tensor $C_{ij}$ will be denoted by **C** and $s_{ij}$ by **S**. The multiplications indicated below are the usual matrix multiplications. Under the assumptions stated above, the tensor **C** can be written as

$$\mathbf{C} = \sum_{n=0}^{\infty} a_{2n+1}(J_2, J_3)\mathbf{S}^{2n+1}. \tag{25}$$

We note that only odd powers appear in Eq. (25) since **C** is assumed to be an odd function of the stresses. Equation (25) can be simplified further by the Hamilton-Cayley theorem which states that the tensor **S** must satisfy its own characteristic equation.[3] For the stress deviator **S**, this implies that

$$\mathbf{S}^3 = J_2\mathbf{S} + J_3\mathbf{I}, \tag{26}$$

where **I** is the unit tensor. Through Eq. (26), we can reduce[4] any power of **S** greater than the second to a linear combination of **I**, **S**, and $\mathbf{S}^2$ with coefficients which are functions of $J_2$ and $J_3$. For example, consider the reduction of the power $\mathbf{S}^5$. According to Eq. (26),

$$\mathbf{S}^4 = J_2\mathbf{S}^2 + J_3\mathbf{S};$$

thus

$$\mathbf{S}^5 = J_2\mathbf{S}^3 + J_3\mathbf{S}^2 = J_3\mathbf{S}^2 + J_2^2\mathbf{S} + J_2J_3\mathbf{I}.$$

In general, we can rewrite Eq. (25) as

$$\mathbf{C} = a(J_2, J_3)\mathbf{S}^2 + b(J_2, J_3)\mathbf{S} + c(J_2, J_3)J_3\mathbf{I}.$$

We recall that $C_{ii}=0$; since **S** is a deviator, this implies that

$$2a(J_2, J_3)J_2 + 3c(J_2, J_3)J_3 = 0,$$

or

$$c(J_2, J_3) = -\frac{2J_2}{3J_3}a(J_2, J_3).$$

Consequently,

$$\mathbf{C} = a(J_2, J_3)[\mathbf{S}^2 - \tfrac{2}{3}J_2\mathbf{I}] + b(J_2, J_3)\mathbf{S}.$$

The expression appearing in square brackets is just the tensor $t_{ij}$ which was defined in Eq. (13). Returning now to the subscript notation we can write the tensor $C_{ij}$ as

$$C_{ij} = a(J_2, J_3)t_{ij} + b(J_2, J_3)s_{ij}.$$

A further simplification can be made by noting that $C_{ij}$ must be an odd function of the stress components. Since $J_2$ is even, $J_3$ odd, $t_{ij}$ even, and $s_{ij}$ odd, we must have

$$C_{ij} = p(J_2, J_3^2)s_{ij} + q(J_2, J_3^2)J_3t_{ij}.$$

Thus the complete differential stress-strain relations can be written in the form[5]

$$\left.\begin{aligned} d\epsilon_{ij} &= (1+\nu)ds_{ij} + (1-2\nu)ds\delta_{ij} + [p(J_2, J_3^2)s_{ij} + q(J_2, J_3^2)J_3t_{ij}]dJ_2 \quad \text{when } dJ_2 \geqq 0 \\ d\epsilon_{ij} &= (1+\nu)ds_{ij} + (1-2\nu)ds\delta_{ij}, \quad \text{when } dJ_2 \leqq 0. \end{aligned}\right\} \tag{27}$$

**5. Further study of the stress strain relations. Experimental determination.** In this section we shall discuss the relation between the differential form of the stress-strain relation (cf. (27))

[3] M. Bôcher, *Introduction to higher algebra*, The Macmillan Co., New York, 1907, p. 296.

[4] This technique has been used recently by Marcus Reiner, Am. J. Math. **67**, 350–362 (1945) and W. Prager, *loc. cit.*

[5] These relations contain, as special cases, the stress-strain laws developed by W. Prager, Proc. Fifth International Congress of Applied Mechanics, Cambridge, Mass., 1938, pp. 234–237, and by J. H. Laning in an unpublished paper (1942).

$$d\epsilon''_{ij} = \{p(J_2, J_3^2)s_{ij} + q(J_2, J_3^2)J_3t_{ij}\}dJ_2, \qquad dJ_2 \geqq 0 \tag{28}$$

and the integral form

$$\epsilon''_{ij} = F(J_2, J_3^2)[P(J_2, J_3^2)s_{ij} + Q(J_2, J_3^2)J_3t_{ij}], \tag{14}$$

which holds only when the ratios of the principal stresses are kept constant during the loading process, i.e., if

$$s_{ij} = ks_{ij}^{(0)}, \tag{29}$$

where $s_{ij}^{(0)}$ is fixed while $k$ is the scalar variable. We shall then show how a series of tests necessary to establish the Lode diagram will be sufficient to determine the stress-strain relations completely. First of all, it is convenient to bring out the homogeneity properties in the relations (28) and (14) by introducing the symbols

$$\alpha = J_3^2/J_2^3, \qquad \gamma_{ij} = J_3\, t_{ij}/J_2^2, \tag{30}$$

where $\alpha$ is dimensionless, while $\gamma_{ij}$ has the same dimensions as $J_2$. The relation (14) can be written in the form

$$\epsilon''_{ij} = \lambda(J_2, \alpha)\{s_{ij} + \beta(\alpha)\gamma_{ij}\}, \tag{31}$$

where

$$\lambda(J_2, \alpha) \equiv F(J_2, J_3^2)P(J_2, J_3^2), \qquad \beta(\alpha) \equiv J_2^2Q(J_2, J_3^2)/P(J_2, J_3^2). \tag{32}$$

Note that $\beta$ is independent of $J_2$, because of the homogeneity relation between $P$ and $Q$ established in Section 2.

With a similar change of notation, the relation (28) can be written as

$$d\epsilon''_{ij} = G(J_2, \alpha)\{s_{ij} + \beta'(\alpha)\gamma_{ij}\}dJ_2, \qquad dJ_2 \geqq 0 \tag{33}$$

where

$$G(J_2, \alpha) \equiv p(J_2, J_3^2), \qquad \beta'(\alpha) \equiv J_2^2q(J_2, J_3^2)/p(J_2, J_3^2). \tag{34}$$

Since we did not establish a homogeneity relation between $p$ and $q$, we cannot immedilately conclude that $\beta'$ is independent of $J_2$. However, we shall see immediately that this is true and that indeed

$$\beta' \equiv \beta. \tag{35}$$

We shall also show that $G(J_2, \alpha)$ may be obtained from $\lambda(J_2, \alpha)$ by the relation

$$G(J_2, \alpha) = \frac{\lambda}{2J_2} + \frac{\partial\lambda}{\partial J_2}\cdot \tag{36}$$

The relations (35) and (36) will then determine the differential relation (33) *completely* once the integral relation (31) is known by a series of experiments of the *special* type (29). It is to be noted that the functions $G(J_2, \alpha)$ and $\beta(\alpha)$ in (33) determined through the use of (31) will by no means restrict the application of (33) to processes connected in any manner with (29).

To establish the relations (35) and (36), consider the application of (31) and (33) to a process of the type (29). Let $d\epsilon''_{ij}$ be the change in $\epsilon''_{ij}$ corresponding to a change $dk$. Then

$$dJ_2 = 2J_2 \frac{dk}{k}, \qquad d\alpha = 0; \tag{37}$$

and (31) gives

$$d\epsilon''_{ij} = \frac{\partial \lambda}{\partial J_2} dJ_2 \{s_{ij} + \beta\gamma_{ij}\} + \lambda\{s_{ij} + \beta\gamma_{ij}\} \frac{dk}{k},$$

while (33) yields

$$d\epsilon''_{ij} = G(J_2, \alpha)\{s_{ij} + \beta'\gamma_{ij}\} dJ_2.$$

Equating coefficients of $s_{ij}$ and $\gamma_{ij}$, we obtain the relations (35) and (36).

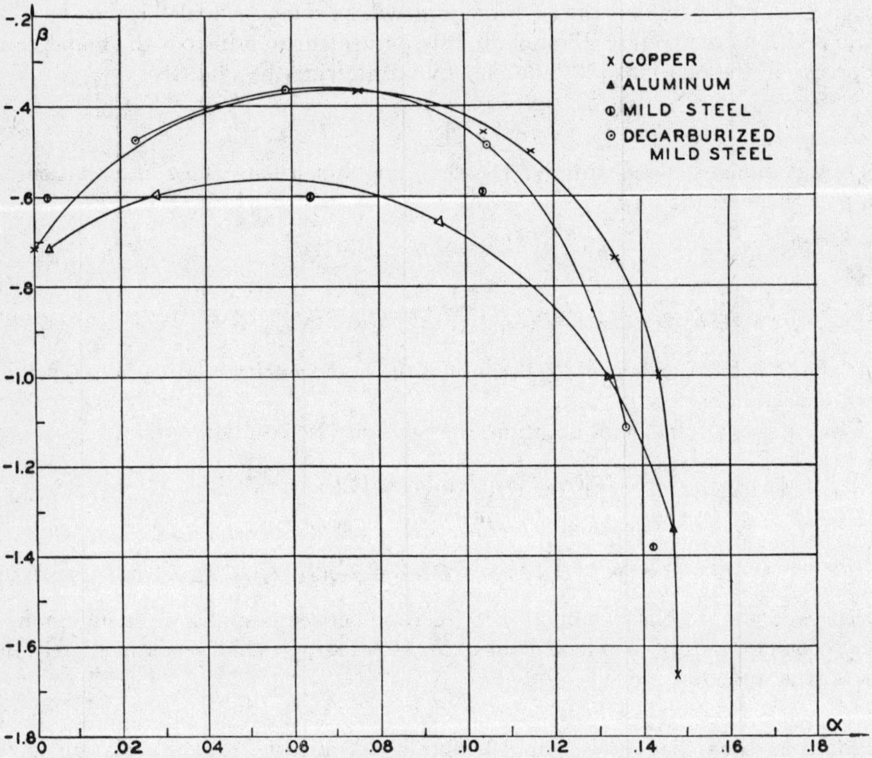


FIG. 2a. The $\alpha-\beta$ diagram (Eq. (38)) of the experimental results of Taylor and Quinney for copper, aluminum, mild steel and decarburized mild steel. The data for mild steel are too scattered for a definite curve to be drawn.

The experimental determination of the stress strain relations can then be reduced to that of (31) alone. This can be done by a series of tests of the type (29), which is of the class described by Lode,[6] Taylor-Quinney[7] and Hohenemser-Prager.[8] Indeed,

[6] W. Lodge, Forschungsarbeiten a.d. Gebiete d. Ingenieurwesens, No. 303, VDI-Verlag, Berlin, 1928.

[7] G. I. Taylor and H. Quinney, Phil. Trans. Roy. Soc. London (A) **230,** 323–362 (1931).

[8] K. Hohenemser and W. Prager, Z. angew. Math. Mech. **12,** 1–14 (1932). An English translation of this paper is available as R.T.P. Translation No. 2468 (Durand Reprinting Committee, in care of California Institute of Technology, Pasadena 4, Calif.).

the relation $\beta(\alpha)$ is merely another presentation of Lode's diagram. It can be easily verified that $\alpha$, $\beta$ are related to Lode's parameters[9] $\mu$ and $\nu$ by the relations

$$\left.\begin{aligned} \alpha &= \frac{4}{27}\,\frac{\mu^2(9-\mu^2)^2}{(3+\mu^2)^3}, \\ \beta &= \frac{9(3+\mu^2)^2}{2(9-\mu^2)}\,\frac{1-\nu/\mu}{\mu^2(1+2\nu/\mu)-3}. \end{aligned}\right\} \tag{38}$$

This new system has the advantage that $\beta$ gives directly the extent of deviation from "von Mises' second hypothesis" discussed by Taylor and Quinney, which is equivalent to putting $\beta\equiv 0$. Indeed, one principal aim of Taylor and Quinney is to find out

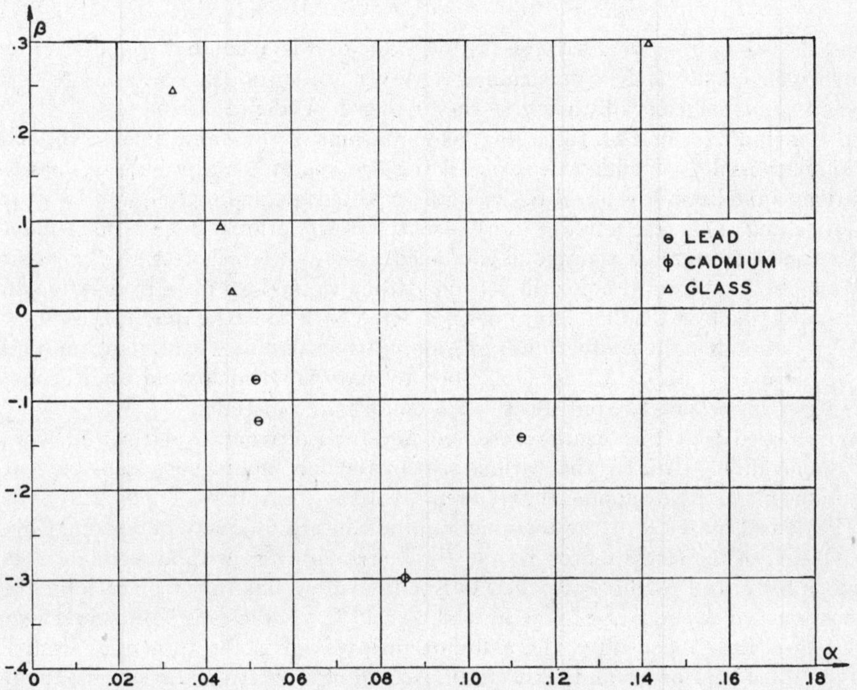


FIG. 2b. The $\alpha-\beta$ diagram (Eq. (38)) of the experimental results of Taylor and Quinney for lead, cadmium and glass. The data are too few to allow any curve to be drawn.

this extent and is therefore to determine the value of $\beta$. Figs. 2(a) and 2(b) show the results of Taylor and Quinney converted into the $(\alpha, \beta)$ diagram. This diagram reveals any experimental error more strongly, since $\beta$ is essentially related to the slope of the $(\mu, \nu)$ curve; e.g.,

[9] W. Lode, *loc. cit.*, pp. 1 and 12.

$$\frac{d\nu}{d\mu} = \frac{9 - 4\beta}{9 + 2\beta} \quad \text{for} \quad \mu \to 1,$$

$$\frac{d\nu}{d\mu} = 1 + \frac{2\beta}{3} \quad \text{for} \quad \mu = 0.$$

It should be noted in passing that $\alpha$ must satisfy the inequality

$$0 < \alpha < 4/27 \tag{39}$$

to insure real values of $\mu$.

Having determined $\beta(\alpha)$ from (38), we may determine $\lambda(J_2, \alpha)$ by noting that (cf. (31))

$$I_2 \equiv \tfrac{1}{2}\epsilon_{ij}''\epsilon_{ij}'' = \lambda^2 J_2\{1 + 3\alpha\beta + \tfrac{1}{3}\alpha^2\beta^2\}. \tag{40}$$

For each loading process given by (29) the value of $\alpha$ is fixed, and (40) gives the dependence of $\lambda^2$ on $J_2$ if $I_2$ is determined for given values of $J_2$. A series of tests with different principal axes will then give the further dependence of $\lambda$ on $\alpha$.

**6. Concluding remarks.** In closing, we note some of the limitations of the stress-strain relations developed in this paper. It has been pointed out previously that these equations have been developed to cover the case of one loading followed by at most *one* unloading. This restriction is quite essential, for relations (27) are not applicable for a second loading. For example, if we consider a simple tensile test, the stress-strain diagram obtained from (27) for the second loading would be a mere translation of the diagram for the first loading. This does not agree with the experimental results. Secondly, we note that these equations apply only to metals which exhibit strain-hardening. They are not applicable, for example, to materials which yield under constant shearing stress or satisfy von Mises' yield condition, $J_2 = \text{const}$.

It is hoped that the results presented here will provoke experimental work to test their validity. Among the various features which should be tested are two assumptions made in developing the differential stress-strain laws. The first hypothesis (cf. Section 4) states that the increments in strain are uniquely determined by the components of the stress tensor $\sigma_{ij}$ and the increments $d\sigma_{ij}$ without reference to the previous history of loading provided only one loading has taken place followed by at most one unloading. The range in which such a hypothesis is valid should be explored empirically. Secondly, the assumption involved in the transition from Eqs. (24) to (25) should be examined carefully. According to these two relations, the principal axes of the increment in permanent strain $d\epsilon_{ij}''$ will coincide with the principal axes of the existing state of stress $s_{ij}$ independent of the increments in stress $d\sigma_{ij}$, provided only loading takes place. This conclusion should be tested by experiment.

Reprinted from the
TRANSACTIONS OF THE AMERICAN MATHEMATICAL SOCIETY
January, 1960, Vol. 94, No. 1
Pp. 24-57

# ON THE ASYMPTOTIC SOLUTIONS OF A CLASS OF ORDINARY DIFFERENTIAL EQUATIONS OF THE FOURTH ORDER

## I. EXISTENCE OF REGULAR FORMAL SOLUTIONS

BY
C. C. LIN AND A. L. RABENSTEIN[1]

1. **Introduction.** This paper is concerned with the asymptotic solutions of the linear differential equation of the fourth order

$$\mathcal{L}(\varphi) = 0 \tag{1.1}$$

when the parameter $\lambda$ is large in the differential operator

$$\mathcal{L}(\varphi) = \frac{d^4\varphi}{dx^4} + \lambda^2\left\{P(x, \lambda)\frac{d^2\varphi}{dx^2} + Q(x, \lambda)\frac{d\varphi}{dx} + R(x, \lambda)\varphi\right\}. \tag{1.2}$$

In the above expression, $P(x, \lambda)$, $Q(x, \lambda)$ and $R(x, \lambda)$ are analytic functions of the complex variable $x$, and they depend on $\lambda$ in such a manner that the following asymptotic expansions hold:

$$P(x, \lambda) = \sum_{n=0}^{\infty} \lambda^{-n} P_n(x), \tag{1.3a}$$

$$Q(x, \lambda) = \sum_{n=0}^{\infty} \lambda^{-n} Q_n(x), \tag{1.3b}$$

$$R(x, \lambda) = \sum_{n=0}^{\infty} \lambda^{-n} R_n(x). \tag{1.3c}$$

Especially we are interested in the behavior of the solution in the neighborhood of a *simple* zero of the function $P_0(x)$. Such a point is called a *turning point* (of the first order). It is of special interest because asymptotic solutions of (1.1) in the form

$$\varphi = e^{\lambda S(x)} \sum_{n=0}^{\infty} \lambda^{-n} \varphi_n(x) \tag{1.4}$$

exhibit singularities at such a point even though the point is a regular point of the differential equation (1.1). Other solution forms must therefore be found, if we wish to find a complete set of asymptotic solutions *uniformly* valid in a neighborhood containing the turning point. It is the principal aim of the present investigation to find such solutions.

---

Received by the editors, October 31, 1958.

(1) This work overlaps in part with the doctoral dissertation of the junior author at the Massachusetts Institute of Technology, June, 1958.

The interest in the equation (1.1) has been especially stimulated by the controversy surrounding the Orr-Sommerfeld equation of hydrodynamic stability, which plays a central role in the dynamics of a viscous fluid and belongs to the type (1.1) under consideration. We shall not go into further details of the discussion of the issues here(2) but merely wish to observe that one principal point in the controversy concerns the role of the *inviscid equation* obtained by formally putting the viscosity coefficient equal to zero. This corresponds to formally letting $\lambda \to \infty$ in (1.1) and leads to the *reduced equation*

$$P_0(x)\frac{d^2\varphi}{dx^2} + Q_0(x)\frac{d\varphi}{dx} + R_0(x)\varphi = 0. \tag{1.5}$$

This equation, in contrast to the full equation (1.1), does have a regular singularity at the zero of $P_0(x)$, except in the most unlikely case that both $Q_0(x)$ and $R_0(x)$ also vanish there. We shall see that the nature of the solutions of the reduced equation also plays an important role in the present mathematical theory. Although the general outline of the theory (cf. §2) is not greatly affected, the actual construction of the solutions (cf. §§6, 7, 8) can vary substantially with the nature of the reduced equation.

There have been many papers published in the literature of hydrodynamics and differential equations that deal with the problem at hand. Especially we wish to mention the theories of Wasow [12] and of Langer [4; 5], with which the present paper shares some common ideas but also has some essential differences. Langer's theory is based on his earlier work for a differential equation of the third order (Langer [2]) which in turn depends on the solution of the special differential equation (Langer [3])

$$v''' + \lambda^2(zv' + \alpha v) = 0, \tag{1.6}$$

where $\alpha$ is a parameter which may depend on $\lambda$ in the manner

$$\alpha = \sum_{n=0}^{\infty} \lambda^{-n}\alpha^{(n)}. \tag{1.7}$$

The present work is based on the special differential equation of the fourth order

$$L(u) \equiv u^{\text{iv}} + \lambda^2(zu'' + \alpha u' + \beta u) = 0 \tag{1.8}$$

where $\alpha$ and $\beta$ may depend on $\lambda$ asymptotically in the following manner:

(2) For a discussion of the physical and mathematical issues involved in the hydrodynamical problem, see Lin [6, Chapter VIII]. Other references to the hydrodynamical problem are given at the end of this section (§1.)

$$\alpha = \sum_{n=0}^{\infty} \lambda^{-n} \alpha^{(n)}, \tag{1.9a}$$

$$\beta = \sum_{n=0}^{\infty} \lambda^{-n} \beta^{(n)}. \tag{1.9b}$$

The differential equation (1.8), which we shall call the *basic reference equation*, as distinct from the approximating *related equation*, is a special case of the given equation (1.1). It is in general not a close approximation to (1.1), but does retain the essential feature of having a turning point. Furthermore, the properties of its solutions can be studied in detail by the method of Laplace transformation, and this has been carried out by one of us (Rabenstein [10]). In this phase of our work, the ideas of Wasow [11] were used, but the greater generality of the equation brings new interesting features. The main differences between the present work and that of Wasow lie in the manner of construction of the solutions of (1.1) (cf. Wasow [13]), and in Wasow's restriction of (1.8) to the case $\alpha = 0$. Certain limitation in Wasow's theory are thereby removed.

To summarize, it is perhaps fair to say(3) that the present work follows more closely the spirit of Langer's theory for equations of the lower orders than either the theory of Wasow or of Langer for the fourth order case. We have also cast the general ideas of the theory in terms of a system of equations of the first order to make it more perspicuous. But by far the most important features of the present work are the following. In the first place, we have carried out the proof of the existence theorem for a *complete* complex neighborhood of the turning point. It is not difficult to prove that the formal solution is asymptotic to a true solution in every fan shaped sector centered at the origin, provided the size of the sector is limited. The difficult part is to prove that, in a *complete* complex neighborhood, there is a *single* solution of the given equation that is asymptotically represented by the *same* formal expansion(4). We shall see the contrast between these two types of theorems in the second part of this paper. The other important feature is its contrast with theories for equations of lower orders. This lies in the central role played by the reduced equation (1.5) and the reduced equation

$$zu'' + \alpha^{(0)} u' + \beta^{(0)} u = 0 \tag{1.10}$$

of the basic reference equation (1.8). Both of them have a regular singularity, and it is important, as we shall see below, to choose $\alpha^{(0)}$ and $\beta^{(0)}$ so that the essential characters of the solution of (1.5) are reproduced in the solutions of (1.10). In the case of equations of the third order (Langer [2]), a similar step has also been taken; but in that case, the reduced equations are of the

(3) An observation made by Professor Norman Levinson.

(4) As far as we know, this has not even been proved for equations of lower orders.

first order, the nature of their solutions are consequently much simpler; and there is no conspicuous effort for such a determination. In the case of second order equations (Langer [1]), the reduced equations become trivial. Thus, the present work also points a way to the development of theories for equations of higher orders: the basic reference equation should be chosen so that the solutions of its reduced equation reproduce the essential characteristics of the solutions of the reduced equation of the given differential equation.

On the basis of the above remarks, a close relationship can already be expected to exist between the theory of Langer and the present work in a fairly wide class of problems. If, in the reduced equation (1.10), $\alpha^{(0)}$ is not an integer, positive, negative or zero, the character of its solution is essentially determined by $\alpha^{(0)}$ and is independent of $\beta^{(0)}$. Consequently, we may put $\beta^{(0)}=0$ in (1.10) and it would then be natural also to try to put $\beta=0$ in the basic reference equation (1.8). This equation then becomes identical with Langer's equation (1.6) provided $u'=v$. The two theories should therefore yield comparable results. On the other hand, if $\alpha^{(0)}$ is an integer, positive, negative, or zero, (as it is in the hydrodynamical problem), the properties of the solution of (1.10) depend essentially on $\beta^{(0)}$, and the results of the two theories may not be directly comparable. In particular, when $\alpha^{(0)}$ is thus specified, the present theory is not essentially modified. On the other hand, there may exist, in Langer's theory, a distinctive class of problems that require special treatment. This class is at the present characterized in terms of certain features that occur in the process of the construction of the solution. As no corresponding features occur in the present theory, it appears that a proper comparison of the two theories can be made only after the distinctive class in Langer's has been characterized directly in terms of the original differential equation (1.1). At the time of this writing, Professor Langer has informed us that he is contemplating such an investigation.

The presentation of the present theory naturally divides itself into three parts:

PART I. Existence of formal asymptotic solutions with regular coefficients.

PART II. Existence of actual solutions approximated by these formal asymptotic solutions.

PART III. Applications to the hydrodynamical problem.

The new features of the present problem, in comparison with cases of lower orders, exhibit themselves most clearly in the development of Part I of the theory. In this sense, it is the most important part. The existence of actual solutions can be proved by following the usual method of constructing a related equation, which goes back to Liouville. The new feature of the present work, as mentioned before, is the existence theorem for a *complete* neighborhood. This is by no means a trivial application of the old ideas. There are certain steps which are quite distinctive for the problem at hand. Furthermore, it is necessary to make certain mathematical investigations which are interesting by themselves, such as a complete discussion of a fundamental system of solutions of the basic reference equation and its adjoint. These studies, made in the spirit of the classical analysis of special functions, are also essential for the final physical applications.

Since the types of mathematical considerations in Parts I and II are very much different from each other, it is thought desirable to present them separately. Accordingly, the present paper deals only with Part I; a sequel will deal with Part II. To provide an overall picture, an outline of the theory is given in §2. Part III will be published in the literature of mechanics(5), but perhaps a word is in order here about the significance of the present work in regard to the studies of the hydrodynamical problem. While the past controversies will no doubt be resolved by any kind of a uniformly valid solution, the extent of the usefulness of such solutions still depend on the solution being given in some convenient form. If our aim is merely to compute the stability characteristics of the flow on the basis of the Orr-Sommerfeld equation, this is still not a very serious point, for the lack of convenience of the solution form can be compensated, at least partly, by speedy modern computing methods. However, the issue goes deeper. The ultimate purpose of the theory of hydrodynamic stability is to clarify the mechanism of transition of the laminar flow into turbulence. Here, recent investigations, both theoretical and experimental, reveal new important roles of the *critical layer*—which corresponds to the turning point in the mathematical theory. It is thus essential that the nature of the solutions in the neighborhood of the turning point be fully understood. Specifically, it is desirable that the theory should yield correct third order derivatives in this neighborhood, while in the stability problem only derivatives of the first order are needed. For more detailed discussion of the hydrodynamic issues, the reader is referred to the earlier reports of the senior author (Lin [7; 8; 9]). In those reports, the inherent significance of the nature of the solutions of the reduced equation (1.8) has not been as explicitly brought out as in the present paper, but their application for the construction of useful solutions for the purpose of the hydrodynamical problem has been emphasized.

ACKNOWLEDGMENT. The authors wish to thank Professor K. O. Friedrichs, Norman Levinson, and especially R. E. Langer and Wolfgang Wasow for their interest in this work and their discussions during various stages of its completion. The main part of this work was completed at the Massachusetts Institute of Technology under the sponsorship of the Office of Naval Research. The final revision of the paper was made at the Mathematics Research Center, University of Wisconsin.

2. **Outline of the theory.** Although the asymptotic theory for the fourth-order equation (1.1) follows the general outline of the theory for lower orders, there are some significant deviations in the manner in which the solution is

(5) Preliminary forms of such work have already been reported (Lin, [7; 8; 9]).

constructed. As mentioned above, this is mainly due to the fact that the reduced equation (1.5) is of the second order. We shall describe in this section the major steps in the theory so that the outline will stand out more clearly.

The usual method of solution is to construct formal solutions of (1.1) in terms of the solutions of (1.8) in the following form

$$\varphi = c_0 u + c_1 u' + c_2 u'' + c_3 u''' \tag{2.1}$$

where the variables $z$ and $x$ are connected by a suitable analytic function, and $c_0(x, \lambda)$, $c_1(x, \lambda)$, $c_2(x, \lambda)$, $c_3(x, \lambda)$ are expected *not* to be all of the same order in $\lambda$. Rather, we might expect $c_m$ to be of the order $O(\lambda^{-m})$, if we take the analogy with equations of lower orders. However, in the present case, it will turn out that

$$c_0, c_1 = O(1), \tag{2.2}$$

$$c_2, c_3 = O(\lambda^{-2}). \tag{2.3}$$

The fact that we need *two* coefficients of the order unity can be expected from the second-order nature of the reduced equation. For if we consider the *two* solutions $u_1^{(0)}$ and $u_2^{(0)}$ of the reduced equation (1.10), we should expect to get *two* solutions $\varphi_1^{(0)}$, $\varphi_2^{(0)}$ of the reduced equation (1.5) in the initial approximation. If we now think of the connection in terms of two first order systems, we should expect these solutions to be related to each other in the form

$$\overset{(0)}{\varphi} = \overset{(0)}{c_0}\,\overset{(0)}{u} + \overset{(0)}{c_1}\,\overset{(0)'}{u}. \tag{2.4}$$

Such a form can follow from (2.1) only if (2.2) is satisfied.

Detailed calculations (cf. Appendix A) show that in the present case, we should actually consider relations between vectors of the forms

$$\begin{aligned} \boldsymbol{\phi} &= (\varphi, \varphi', \lambda^{-1}\varphi'', \lambda^{-2}\varphi'''), \\ \boldsymbol{u} &= (u, u', \lambda^{-1}u'', \lambda^{-2}u'''). \end{aligned} \tag{2.5}$$

(It is immaterial here whether the dashes on $\varphi$ denote differentiation with respect to $z$ or $x$.)

The critical step in the solution is to show that we can indeed find functions

$$c_i(x, \lambda) = \lambda^{-m_i} \sum_{n=0}^{\infty} \lambda^{-n} \overset{(n)}{c_i}(x) \tag{2.6}$$

with analytic coefficients $c_i^{(n)}(x)$. In the case of equations of lower orders, it has always been possible to prove this by explicit quadrature. This is no longer true in the present case. As we shall see in §3, we have to depend on relations of the type (2.4) to overcome this difficulty. Indeed, we have to modify the form of the expansion somewhat for certain special cases (§8).

In order to gain maximum simplicity and to bring out the significance of the reduced equation, we shall deviate somewhat from the above described method of construction of formal solutions. We first introduce a minimum amount of transformation of variables to bring the equation (1.1) into a form having *exactly the same type of reduced equation* as (1.8). This is done in two steps.

FIRST TRANSFORMATION. We introduce into (1.1) the new independent variable

$$z = \left[\frac{3}{2}\int_{x_0}^{x} P_0^{1/2}(x)dx\right]^{2/3} \tag{2.7}$$

and the new dependent variable

$$\psi(z, \lambda) = \varphi(x, \lambda)[P_0(x)/z]^{3/4}. \tag{2.8}$$

Then the equation (1.1) becomes

$$\psi^{\mathrm{iv}} + \lambda^2(p\psi'' + q\psi' + r\psi) = 0, \tag{2.9}$$

where $p$, $q$, $r$ have behaviors identical with that specified in (1.3), i.e.,

$$p(z, \lambda) = \sum_{n=0}^{\infty} \lambda^{-n} p_n(z), \tag{2.10a}$$

$$q(z, \lambda) = \sum_{n=0}^{\infty} \lambda^{-n} q_n(z), \tag{2.10b}$$

$$r(z, \lambda) = \sum_{n=0}^{\infty} \lambda^{-n} r_n(z). \tag{2.10c}$$

The change of independent variable has the purpose of making (2.9) resemble (1.8) more closely in that

$$p_0(z) = z. \tag{2.11}$$

SECOND TRANSFORMATION. Next we consider a transformation of the form

$$\chi = A(z, \lambda)\psi + B(z, \lambda)\psi' + \lambda^{-1}C(z, \lambda)\psi'' + \lambda^{-2}D(z, \lambda)\psi'', \tag{2.12}$$

where $A$, $B$, $C$, $D$ have the behavior (1.3). By formal differentiation, we can then obtain the transformation in the vector form

$$\mathfrak{x} = G\mathfrak{y}, \tag{2.13}$$

where

$$\mathfrak{x} = (\chi, \chi', \lambda^{-1}\chi'', \lambda^{-2}\chi'''), \tag{2.14}$$

$$\mathfrak{y} = (\psi, \psi', \lambda^{-1}\psi'', \lambda^{-2}\psi'''), \tag{2.15}$$

and $G(z, \lambda)$ is a nonsingular matrix with asymptotic dependence on $\lambda$ in the manner

$$G(z, \lambda) = \sum_{n=0}^{\infty} \lambda^{-n} G^{(n)}(z). \tag{2.16}$$

We expect $G^{(n)}(z)$ to be all analytic. It will be seen that this form of $G(z, \lambda)$ is obtained provided certain simple conditions are fulfilled by the coefficients in (2.12).

It is then possible to show that we can *explicitly* determine the functions $A(z, \lambda)$, $B(z, \lambda)$, $C(z, \lambda)$, and $D(z, \lambda)$ as polynomials in $\lambda^{-1}$ in such a manner that $\chi$ satisfies an equation of the form

$$\begin{aligned} L_0(\chi) &= \chi^{\text{iv}} + \lambda^2(z\chi'' + \alpha_0\chi' + \beta_0\chi) \\ &= \lambda(\bar{a}\chi + \bar{b}\chi' + \lambda^{-1}\bar{c}\chi'' + \lambda^{-2}\bar{d}\chi''') \end{aligned} \tag{2.17}$$

where the coefficients $(\bar{a}, \bar{b}, \bar{c}, \bar{d})$ have the behavior (1.3). This is a major step in the theory and will be treated in detail in the next section. The form (2.17) will be referred to as the *normal form* of the equation.

In the process of carrying out this transformation, it will become clear that the constant $\alpha_0$ is indeed completely specified by the given equation if the transformation is to be regular, and that the constant $\beta_0$ is also determined in the cases where $\alpha_0$ is an integer, positive, negative, or zero.

The construction of asymptotic solutions of the normalized equation (2.17) then follows the usual methods. If we introduce the vector

$$\boldsymbol{u} = (u, u', \lambda^{-1}u'', \lambda^{-2}u'''), \tag{2.18}$$

where $u$ is a solution of the equation (1.8), then we may expect it possible, by proper determination of $\alpha$ and $\beta$ to obtain formal solutions of the structure

$$\boldsymbol{\chi} = H\boldsymbol{u} \tag{2.19}$$

and with $H$ expected to be of the form[6]

$$H = I + \sum_{m=1}^{\infty} \lambda^{-m} h^{(m)}(z), \tag{2.20}$$

where $I$ is the identity matrix. Combining (2.12) and (2.19), we may write

$$X = G\Psi = HU, \tag{2.21}$$

or

$$\Psi = G^{-1}HU \tag{2.22}$$

where $U$, $\Psi$, $X$ are fundamental matrix solutions of the systems of equations corresponding to (1.8), (2.9) and (2.17). For convenience of reference, we list below the equations for $\boldsymbol{u}$ and $\boldsymbol{\chi}$ explicitly. The equation for $\boldsymbol{u}$ may be written

(6) Deviations from this expectation that occur in certain special cases will be discussed in §8; but the following outline will not be essentially altered.

$$\frac{du}{dz} = Mu, \tag{2.23}$$

where

$$M = \begin{bmatrix} 0 & 1 & 0 & 0 \\ 0 & 0 & \lambda & 0 \\ 0 & 0 & 0 & \lambda \\ -\beta & -\alpha & -\lambda z & 0 \end{bmatrix}. \tag{2.24}$$

The equation for $\chi$ may be written as

$$\frac{d\chi}{dz} = (M_0 + \epsilon)\chi \tag{2.25}$$

where $M_0$ is obtained from $M$ by putting $\alpha=\alpha_0$, $\beta=\beta_0$ and

$$\epsilon = \lambda^{-1}\begin{bmatrix} 0 & 0 & 0 & 0 \\ 0 & 0 & 0 & 0 \\ 0 & 0 & 0 & 0 \\ \tilde{a} & \tilde{b} & \tilde{c} & \tilde{d} \end{bmatrix}. \tag{2.26}$$

If we terminate the series (2.20) with $h^{(n)}(z)$ as the last term(7) we obtain the $n$th approximation

$$\chi_n = H_n u \tag{2.27}$$

which satisfies the differential equation

$$\frac{d\chi_n}{dz} = (M_0 + \epsilon_n)\chi_n, \tag{2.28}$$

with

$$\epsilon - \epsilon_n = \lambda^{-(n+1)}\Delta, \tag{2.29}$$

where $\Delta$ is a matrix of the same form as $\lambda\epsilon$. The equation (2.28) is the *related equation* of (2.25). To prove that there exists a solution $\chi$ approximated by each solution $\chi_n$ of the related equation, we consider the difference vector

$$\delta = \chi - \chi_n \tag{2.30}$$

which satisfies the differential equation

$$\frac{d\delta}{dz} = M_0\delta + \epsilon\delta + \lambda^{-(n+1)}\Delta\chi_n. \tag{2.31}$$

(7) To be more precise, we do this for the first row of the matrix and let the other rows be derived by differentiation of $\chi_n$.

Thus, $\delta$ is given by the solution of the integral equation(8).

$$\delta(z, \lambda) = \int^z K(z, \zeta; \lambda)\epsilon(\zeta, \lambda)\delta(\zeta, \lambda)d\zeta + \lambda^{-(n+1)} \int^z K(z, \zeta; \lambda)\Delta(\zeta, \lambda)\chi_n(\zeta, \lambda)d\zeta \tag{2.32}$$

where the kernel $K(z, \zeta; \lambda)$ is given by

$$K(z, \zeta; \lambda) = U_0(z, \lambda)U_0^{-1}(\zeta, \lambda), \tag{2.33}$$

and $U_0(z, \lambda)$ is a fundamental matrix solution of the differential system

$$\frac{dU_0}{dz} = M_0U_0. \tag{2.34}$$

This kernel can be calculated explicitly, since we know the matrix functions $U_0(z, \lambda)$ and its inverse. The appraisal of the order of magnitude of $\delta$ then follows the conventional methods of choosing the proper paths of integration and using the known asymptotic behavior of the functions $u(z, \lambda)$.

Variations of the above method might be found desirable. For example, one may wish to consider, instead of

$$\delta = \chi - \chi_n, \tag{2.35}$$

the quantity

$$H_n^{-1}\delta = \tilde{u}_n - u, \qquad \tilde{u}_n = H_n^{-1}\chi. \tag{2.36}$$

In this manner, we are comparing the function $\tilde{u}_n$ (which is explicitly related to the exact solution $\chi$) with the exact solution of the basic reference equation (2.23). Obviously, $\tilde{u}_n$ may be expected to satisfy an equation very close to (2.23), and the usual method of appraisal of errors can be made. The advantage of such an approach is that the solutions of the equation (2.23) itself are well-known. The details of these modified methods will be discussed in Part II when the actual proofs are carried out.

For practical construction of solutions, the method used for developing the theory is not suitable. Rather, once we know the existence of the solutions of the form (2.1), the connection coefficients can be calculated by using the formal asymptotic solutions of $\varphi$ and $u$ of the type (1.4). This method was used in the hydrodynamical applications (Lin [9]).

3. **Reduction of the differential equation into the normal form.** We shall now carry out the second transformation that will bring the given differential equation in the form (2.9) into the form (2.17). The detailed calculations in-

(8) It is sometimes desirable to use a somewhat different approach and obtain an integral equation where the first term on the right-hand side is also small to higher orders of $\lambda^{-1}$. For these discussions, see Part II.

volved are very lengthy and are given in detail in appendices (A) and (B). Here, we shall recapitulate the major steps with frequent reference to the equations derived there.

If we introduce [cf. (A3)]

$$\chi = A\psi + B\psi' + \lambda^{-1}C\psi'' + \lambda^{-2}D\psi''' \tag{3.1}$$

as our new variable, where $A(z, \lambda)$, $B(z, \lambda)$, $C(z, \lambda)$ and $D(z, \lambda)$ have the asymptotic behavior (1.3), then all the derivatives of $\chi$ take on the form [cf. (A4), (A8), etc.]

$$\chi^{(i)} = A_i\psi + B_i\psi' + \lambda^{-1}C_i\psi'' + \lambda^{-2}D_i\psi'''. \tag{3.2}$$

However, the matrix $G$ connecting the vectors $\boldsymbol{\chi}$ and $\boldsymbol{\psi}$, as defined by (2.13) and (2.14), will not be of the order of unity unless we impose the conditions [cf. (A6)]

$$B - pD = \lambda^{-1}E, \quad \text{and} \quad C = \lambda^{-1}F \tag{3.3}$$

with $E$ and $F$ having the asymptotic behavior (1.3). The form of $G$ is then given by [cf. (A12)]

$$G = \sum_{n=0}^{\infty} \lambda^{-n}G^{(n)} \tag{3.4}$$

with [cf. (A13)]

$$G^{(0)} \equiv \begin{bmatrix} A_0^{(0)} & B_0^{(0)} & 0 & D^{(0)} \\ A_1^{(0)} & B_1^{(0)} & C_1^{(0)} & D_1^{(0)} \\ 0 & 0 & B_1^{(0)} - p_0D_1^{(0)} & C_1^{(0)} \\ 0 & 0 & -p_0C_1^{(0)} & B_1^{(0)} - p_0D_1^{(0)} \end{bmatrix}. \tag{3.5}$$

It is necessary that $|G^{(0)}(z)| \neq 0$ and in particular this should hold at $z=0$. For this to be true, we must have [cf. (A14)].

$$A_0^{(0)}(0) \neq 0, \qquad B_1^{(0)}(0) \neq 0. \tag{3.6}$$

If we now form the operator $L_0(\chi)$, with $\alpha_0$ and $\beta_0$ as yet unspecified, we find that an equation of the form (2.17) is satisfied, with the coefficients having the specified orders, provided the following conditions are satisfied [cf. (B10)]:

$$-r_0B_1^{(0)} + zA_1^{(0)\prime} + \alpha_0A_1^{(0)} + \beta_0A^{(0)} = 0, \tag{3.7a}$$

$$-q_0B_1^{(0)} + zA_1^{(0)} + zB_1^{(0)\prime} + \alpha_0B_1^{(0)} + \beta_0B^{(0)} = 0, \tag{3.7b}$$

$$-p_1(B_1^{(0)} - zD_1^{(0)}) - (q_0 + 1)C_1^{(0)} - 2zC_1^{(0)\prime} + \alpha_0C_1^{(0)} = 0, \tag{3.7c}$$

$$-zC_1^{(0)} + 2B_1^{(0)\prime} + B_2^{(0)} - 2(zD_1^{(0)})' + \alpha_0D_1^{(0)} + \beta_0D^{(0)} = 0, \tag{3.7d}$$

where [cf. (A6), (A7) and (A9)]

$$B_0^{(0)} = zD_0^{(0)}, \tag{3.8}$$

$$A_1^{(0)} = A^{(0)'} - r_0 D^{(0)}, \tag{3.9a}$$

$$B_1^{(0)} = B^{(0)'} + A^{(0)} - q_0 D^{(0)}, \tag{3.9b}$$

$$C_1^{(0)} = E^{(0)}, \tag{3.9c}$$

$$D_1^{(0)} = D^{(0)'} + F^{(0)}, \tag{3.9d}$$

and

$$B_2^{(0)} = B_1^{(0)'} + A_1^{(0)} - q_0 D_1^{(0)}. \tag{3.10}$$

Thus, we are faced with the problem of solving the differential equations (3.7)–(3.10), for the variables $A^{(0)}$, $B^{(0)}$, $D^{(0)}$, $E^{(0)}$ and $F^{(0)}$ under the restriction (3.6). The constants $\alpha_0$ and $\beta_0$ are at our disposal. They should obviously be chosen toward making the solution functions regular at $z=0$. If these functions can be found, the conditions (3.6) indeed guarantee that $|G^{(0)}(z)|$ is a constant [cf. (A16) and the discussion following]. We may then write (3.1) in the form

$$\chi = A^{(0)}\psi + [B^{(0)} + \lambda^{-1}(E^{(0)} + p_1 D^{(0)})]\psi' + \lambda^{-2}[F^{(0)}\psi'' + D^{(0)}\psi'''] \tag{3.11}$$

where the higher approxinations in the coefficients are omitted, since they do not enter the conditions for the function (3.11) to satisfy an equation of the form (2.17).

The system of equations to be solved can be divided into two groups. If we consider $D_0$ as eliminated from (3.9a) and (3.9b) by the use of (3.8), then the four equations (3.7a), (3.7b), (3.9a), (3.9b) form a system of four linear differential equations of the first order in the dependent variables $A^{(0)}$, $B^{(0)}$, $A_1^{(0)}$, $B_1^{(0)}$. After this system is solved, the equations (3.7c) and (3.7d) may be regarded as a system of two equations of the first order for the variables $C_1^{(0)}$ and $D_1^{(0)}$, or equivalently the variables $E^{(0)}$ and $F^{(0)}$ (cf. (3.9c), (3.9d)). They take on the form

$$2zE^{(0)'} + (q_0 + 1 - \alpha_0)E^{(0)} - p_1 z F^{(0)} = g_1(z), \tag{3.12}$$

$$2zF^{(0)'} + (q_0 + 2 - \alpha_0)F^{(0)} - zE^{(0)} = g_2(z), \tag{3.13}$$

where $g_1(z)$ and $g_2(z)$ are known analytic functions of $z$.

The system of four equations for $A^{(0)}$, $B^{(0)}$, $A_1^{(0)}$, $B_1^{(0)}$ appears to be very difficult to solve. However, one may anticipate the answers from other considerations and then verify the conjecture. From (3.11), it is clear that the solutions $\chi^{(0)}$ and $\psi^{(0)}$ of the reduced equations

$$z\chi^{(0)''} + \alpha_0\chi^{(0)'} + \beta_0\chi^{(0)} = 0, \tag{3.14}$$

$$z\psi^{(0)''} + q_0\psi^{(0)'} + r_0\psi^{(0)} = 0 \tag{3.15}$$

must satisfy the relation

$$\chi^{(0)} = A^{(0)}\psi^{(0)} + B^{(0)}\psi^{(0)\prime}. \tag{3.16}$$

From this, we can calculate $\chi^{(0)\prime}$ and obtain

$$\chi^{(0)\prime} = A_1^{(0)}\psi^{(0)} + B_1^{(0)}\psi^{(0)\prime}. \tag{3.17}$$

We expect that $A_1^{(0)}$ and $B_1^{(0)}$ will satisfy (3.9a) and (3.9b), if $D^{(0)}$ is eliminated by using (3.8). To verify the above conjecture, we need only show that (3.7a) and (3.7b) are indeed satisfied, provided $A^{(0)}$, $B^{(0)}$, $A_1^{(0)}$, $B_1^{(0)}$ satisfy (3.16) and (3.17). This can be done with a little calculation (§4), if we keep the equations (3.14) and (3.15) in mind. These calculations also show that there are indeed four sets of solutions, which agrees with the fact that equations (3.7a), (3.7b), (3.9a), (3.9b) form a system of four equations of the first order. Moreover, one of the solutions can be shown to be regular with $B^{(0)}(0)=0$ if the parameters $\alpha_0$ and $\beta_0$ are properly chosen. The details of these discussions will be given in the next section.

Having found such regular solutions for $A^{(0)}$, $B^{(0)}$, $A_1^{(0)}$, $B_1^{(0)}$, we may turn back to the discussion of (3.12) and (3.13). It will be found (cf. §4) that

$$\alpha_0 = q_0(0) \tag{3.18}$$

so that we know exactly the values of $E^{(0)}$ and $F^{(0)}$ in (3.12) at $z=0$. It will then not be difficult to verify that a power series solution exists for the system (3.12) and (3.13). That such a formal series solution is a true solution depends on the fact that the system is of the Fuchsian type (Coddington and Levinson, 1955, p. 117).

We may also prove the existence of a regular solution by rewriting the differential system (3.12), (3.13) into an integral system:

$$\hat{E}(z) = [R(z)]^{-1}z^{-1/2}\int_0^z z^{3/2}R(z)[h_1(z) + p_1\hat{F}]dz, \tag{3.19a}$$

$$\hat{F}(z) = [R(z)]^{-1}z^{-1/2}\int_0^z z^2R(z)[h_2(z) + \hat{E}]dz, \tag{3.19b}$$

where

$$R(z) = \exp\left[\int_0^z \frac{1}{2z}(q_0 - \alpha_0)dz\right], \tag{3.20}$$

and

$$\hat{E}(z) = E^{(0)}(z) - E^{(0)}(0), \qquad \hat{F}(z) = F^{(0)}(z) - F^{(0)}(0), \tag{3.21a}$$

$$h_1(z) = \frac{1}{z}[g_1(z) - g_1(0)], \qquad h_2(z) = \frac{1}{z}[g_2(z) - g_2(0)]. \tag{3.21b}$$

It is easy to see that an iteration method would converge and that only regular functions are involved in each step. Consequently, the solution obtained is also regular.

Thus, anticipating the results obtained from the analysis of the reduced equations in the next section, we may state the following lemma.

LEMMA. *The asymptotic theory for the differential equation* (1.1) *is equivalent to that for the special differential equation* (2.17), *in which the constant parameter* $\alpha_0$ *is fixed by the given differential equation, but the constant* $\beta_0$ *is at our disposal except in the cases where* $\alpha_0$ *is an integer, positive, negative, or zero. In the latter case,* $\beta_0$ *is also fixed by the given differential equation. The coefficients* $\bar{a}$, $\bar{b}$, $\bar{c}$, $\bar{d}$ *in* (2.17) *have the asymptotic behavior in* $\lambda$ *similar to those in the given differential equation as specified by* (1.3).

4. **The solutions of the reduced equations.** As we have just seen, the success of the desired transformation depends on obtaining regular solutions of the system of differential equations (3.7a), (3.7b), (3.9a) and (3.9b) for the functions $A^{(0)}$, $B^{(0)}$, $A_1^{(0)}$ and $B_1^{(0)}$. We have also noticed that these functions should be the same as those that appear in the connection formulae

$$\chi^{(0)} = A^{(0)}\psi^{(0)} + B^{(0)}\psi^{(0)\prime}, \tag{4.1}$$

$$\chi^{(0)\prime} = A_1^{(0)}\psi^{(0)} + B_1^{(0)}\psi^{(0)\prime}, \tag{4.2}$$

between the solutions of the equations

$$z\chi^{(0)\prime\prime} + \alpha_0\chi^{(0)\prime} + \beta_0\chi^{(0)} = 0, \tag{4.3}$$

and

$$z\psi^{(0)\prime\prime} + q_0\psi^{(0)\prime} + r_0\psi^{(0)} = 0. \tag{4.4}$$

To verify these conclusions, let us differentiate (4.1) with respect to $z$ to obtain

$$\chi^{(0)\prime} = A^{(0)\prime}\psi^{(0)} + (A^{(0)} + B^{(0)\prime})\psi^{(0)\prime} + B^{(0)}\psi^{(0)\prime\prime},$$

and hence, by using (4.4),

$$\chi^{(0)\prime} = (A^{(0)\prime} - r_0B^{(0)}/z)\psi^{(0)} + (B^{(0)\prime} + A^{(0)} - q_0B^{(0)}/z)\psi^{(0)\prime}. \tag{4.5}$$

Thus, a comparison of (4.2), (4.5) shows that (3.9a) and (3.9b) are satisfied. To verify that (3.7a) and (3.7b) are also satisfied, we have to differentiate (4.2) once more. We obtain (cf. (4.5))

$$\chi^{(0)\prime\prime} = (A_1^{(0)} - r_0B_1^{(0)}/z)\psi^{(0)} + (B_1^{(0)\prime} + A_1^{(0)} - q_0B_1^{(0)}/z)\psi^{(0)\prime}. \tag{4.6}$$

Substituting (4.1), (4.2) and (4.6) into (4.3), we find that the coefficient of $\psi^{(0)}$ is precisely the right-hand side of (3.7a) and the coefficient of $\psi^{(0)\prime}$ is precisely that of (3.7b). The above contentions are therefore verified.

Having established the relations (4.1) and (4.2), we can make use of them to determine the desired solutions in a purely algebraic manner. Consider *any two independent* solutions $\psi_1^{(0)}$ and $\psi_2^{(0)}$ of (4.4) and *any* two solutions $\chi_1^{(0)}$ and $\chi_2^{(0)}$ of (4.3), and *define* the functions $A^{(0)}$ and $B^{(0)}$ by the pair of relations

$$\chi_1^{(0)} = A^{(0)}\psi_1^{(0)} + B^{(0)}\psi_1^{(0)\prime}, \tag{4.7}$$

$$\chi_2^{(0)} = A^{(0)}\psi_2^{(0)} + B^{(0)}\psi_2^{(0)\prime}. \tag{4.8}$$

These functions and their associated functions $A_1^{(0)}$ and $B_1^{(0)}$ will satisfy all the desired requirements. Solving for $A^{(0)}$ and $B^{(0)}$, we obtain

$$A^{(0)} = W_0^{-1}(\chi_1^{(0)}\psi_2^{(0)\prime} - \chi_2^{(0)}\psi_1^{(0)\prime}), \tag{4.9}$$

$$B^{(0)} = -W_0^{-1}(\chi_1^{(0)}\psi_2^{(0)} - \chi_2^{(0)}\psi_1^{(0)}) \tag{4.10}$$

and

$$W_0 = \psi_1^{(0)}\psi_2^{(0)\prime} - \psi_2^{(0)}\psi_1^{(0)\prime}. \tag{4.11}$$

Now, if we let

$$\psi_1^{(0)} = \psi_{\mathrm{I}}^{(0)}, \qquad \psi_2^{(0)} = \psi_{\mathrm{II}}^{(0)} \tag{4.12}$$

be a *specific* pair of *independent* solutions, and $\chi_1^{(0)}$ and $\chi_2^{(0)}$ be an *arbitrary* pair expressed in terms of a *specific* pair of independent solutions,

$$\chi_1^{(0)} = c_{11}\chi_{\mathrm{I}}^{(0)} + c_{12}\chi_{\mathrm{II}}^{(0)}, \qquad \chi_2^{(0)} = c_{21}\chi_{\mathrm{I}}^{(0)} + c_{22}\chi_{\mathrm{II}}^{(0)}, \tag{4.13}$$

then (4.9) and (4.10) gives

$$A^{(0)} = W_0^{-1}\{c_{11}\chi_{\mathrm{I}}^{(0)}\psi_{\mathrm{II}}^{(0)\prime} + c_{12}\chi_{\mathrm{II}}^{(0)}\psi_{\mathrm{II}}^{(0)\prime} - c_{21}\chi_{\mathrm{I}}^{(0)}\psi_{\mathrm{I}}^{(0)} - c_{22}\chi_{\mathrm{II}}^{(0)}\psi_{\mathrm{I}}^{(0)\prime}\}, \tag{4.14}$$

$$B^{(0)} = W_0^{-1}\{c_{11}\chi_{\mathrm{I}}^{(0)}\psi_{\mathrm{II}}^{(0)} + c_{12}\chi_{\mathrm{II}}^{(0)}\psi_{\mathrm{II}}^{(0)} - c_{21}\chi_{\mathrm{I}}^{(0)}\psi_{\mathrm{I}}^{(0)} - c_{22}\chi_{\mathrm{II}}^{(0)}\psi_{\mathrm{I}}^{(0)}\}. \tag{4.15}$$

Thus, we have four independent solutions, since the constants $c_{11}$, $c_{12}$, $c_{21}$, and $c_{22}$ are entirely arbitrary. (They need not be chosen to make $\chi_1^{(0)}$ and $\chi_2^{(0)}$ independent solutions of (4.3). For example, if $\chi_1^{(0)}$ and $\chi_2^{(0)}$ are both zero, we simply obtain the trivial solution $A^{(0)}=0$, $B^{(0)}=0$, which is perfectly valid.)

We now wish to show that *at least one set of the solutions is a nontrivial regular set with the further property* $A_0^{(0)} \neq 0$, $B_1^{(0)} \neq 0$, *and* $B_0^{(0)}=0$ *at* $z=0$. This requires a detailed examination of the solutions of the reduced equations (4.3) and (4.4), and it is here that the coefficients $\alpha_0$ and $\beta_0$ are determined.

The properties of the equations of the type (4.4) are well-known[9]. It has a regular singularity at the point $z=0$, with indices $\rho=0$ and $1-q_0(0)$.

[9] Equation (4.3) is merely a special case of (4.4).

Let us denote the indices by $\rho_1$ and $\rho_2$, with $\rho_2-\rho_1\leqq 0$. Obviously, for our purposes, we must take $\alpha_0=q_0(0)$ so that the indices of (4.3) and (4.4) are the same. Further discussions depend on whether $\rho_2-\rho_1$ is an integer.

If $\rho_2-\rho_1$ *is not an integer*, then two linearly independent solutions of (4.4) may be denoted by [10]

$$\psi_1 = z^{\rho_1}(1 + \cdots), \tag{4.16}$$

$$\psi_2 = z^{\rho_2}(1 + \cdots) \tag{4.17}$$

where the expressions in the parentheses are regular functions. The Wronskian determinant has the value $\rho_2-\rho_1$. If we take $\chi_1$ and $\chi_2$ to be two corresponding solutions of (4.3), we may eliminate the powers $z^{\rho_1}$ and $z^{\rho_2}$ from these equations, and deal only with regular functions. Specifically, we have the equations

$$(z^{-\rho_1}\chi_1) = A^{(0)}(z^{-\rho_1}\psi_1) + (B^{(0)}/z)(z^{-\rho_1+1}\psi_1'), \tag{4.18}$$

$$(z^{-\rho_2}\chi_2) = A^{(0)}(z^{-\rho_2}\psi_2) + (B^{(0)}/z)(z^{-\rho_2+1}\psi_2') \tag{4.19}$$

to solve for $A^{(0)}$ and $B^{(0)}/z$. The coefficient determinant can be easily shown to have the constant value $\rho_2-\rho_1\neq 0$, and the functions $A^{(0)}$ and $B^{(0)}$ thus obtained have the required properties. Indeed, a little calculation will show that $A^{(0)}(0)=B_1^{(0)}(0)=1$, and consequently the restrictions (3.6) are satisfied.

If $\rho_2-\rho_1$ *is an integer*, then the linearly independent solutions of (4.4) take on the form

$$\psi_1 = z^{\rho_1}(1 + \cdots), \tag{4.20}$$

$$\psi_2 = \tilde{\psi}_2 + k\psi_1 \log z, \qquad \tilde{\psi}_2 = z^{\rho_2}(1 + \cdots) \tag{4.21}$$

where $k$ is a constant, which is in general not equal to zero but may vanish under special circumstances. We must then choose not only $\alpha_0$ to make (4.3) have the same indices as (4.4), but also $\beta_0$ to make the two solutions of (4.3) have the same character as (4.20) and (4.21). Thus, if $k$ happens to be zero, we must also take $\beta_0=0$, which is the condition for the special equation (4.3) to have no logarithmic terms in the solutions. Thus, we may write the solutions of (4.3) in the form

$$\chi_1 = z^{\rho_1}(1 + \cdots), \tag{4.22}$$

$$\chi_2 = \tilde{\chi}_2 + k\chi_1 \log z, \qquad \tilde{\chi}_2 = z^{\rho_2}(1 + \cdots), \tag{4.23}$$

where the coefficient of $k$ can be made to be the same as in (4.21).

The relations (4.7) and (4.8) then become

$$\chi_1 = A^{(0)}\psi_1 + B^{(0)}\psi_1',$$

$$\tilde{\chi}_2 = A^{(0)}\psi_2 + B^{(0)}(\tilde{\chi}_2' + k\psi_1/z)$$

or

[10] For convenience of notation, the superscript (0) is temporarily dropped in the following paragraphs of this section.

(4.24) $$(z^{-\rho_1}\chi_1) = A^{(0)}(z^{-\rho_1}\psi_1) + (B^{(0)}/z)(z^{-\rho_1+1}\psi_1'),$$

(4.25) $$(z^{-\rho_2}\tilde{\chi}_2) = A^{(0)}(z^{-\rho_2}\psi_2) + (B^{(0)}/z)(z^{-\rho_2+1}\tilde{\psi}_2' + kz^{-\rho_2}\psi_1).$$

These are equations for $A^{(0)}$ and $B^{(0)}/z$ involving only regular functions. The function

$$z^{-\rho_2} = z^{\rho_1-\rho_2}(1 + \cdots)$$

is regular because $\rho_1 \geqq \rho_2$. The coefficient determinant for the above equations (4.24), (4.25) is a constant, by a consideration of the Wronskian determinant. That it is not equal to zero can also be directly verified at $z=0$. At that point, we obtain the value $\rho_2-\rho_1$ except when this is zero, and the value $(\rho_2-\rho_1)-k$ when $\rho_2-\rho_1=0$. In the latter case, $k$ never vanishes. Thus we have indeed a nonvanishing constant in all nontrivial cases.

We have now completed the proof of the feasibility of the reduction of the given equation (1.1) into the normalized form (2.17), with $\alpha_0$ and $\beta_0$ completely specified in the case $\alpha_0$ is an integer(11). If $\alpha_0$ is not an integer, there is no restriction on $\beta_0$, and we may take $\beta_0=0$. The implications of these differences have been discussed in the introduction.

In the above analysis, we may have arbitrary multiples of $\chi_1$ and $\chi_2$ in (4.18), (4.19) or (4.24), (4.25) and thus obtain answers involving two parameters. Thus, in general there are actually *two* linearly independent regular solutions.

If the equation (4.4) has *two* regular solutions, and this can happen only if the two indices $\rho=0$ and $\rho=1-q_0$ are both integers, positive or zero, then equation (4.3) takes on the form

(4.26) $$z\chi'' - n\chi' = 0,$$

i.e., we have $\alpha_0=-n$, $\beta_0=0$. The solutions of (4.26) are 1 and $z^n$. We may take in this case

$$\chi_1 = c_1 + c_2 z^n,$$
$$\chi_2 = \qquad c_3 z^n$$

and thereby obtain a *three-parameter* family of regular solutions from equations (4.24) and (4.25). These facts will prove useful in later discussions.

5. **Solution of the normalized differential equation.** We have shown in the previous sections that the asymptotic theory of the equation (1.1) can be reduced to that for equation (2.17), which is now redesignated as

(5.1) $$L_0(\chi) = \chi^{\text{iv}} + \lambda^2(z\chi'' + \alpha_0\chi' + \beta_0\chi) = \lambda(\tilde{a}\chi + \tilde{b}\chi' + \lambda^{-1}\tilde{c}\chi'' + \lambda^{-2}\tilde{d}\chi'''),$$

where $\tilde{a}$, $\tilde{b}$, $\tilde{c}$, $\tilde{d}$ have the behavior (1.3). This last equation can be further transformed to eliminate the term in $\chi'''$, but this is not necessary in view of

(11) The difference of the coefficients $\rho_2-\rho_1$ is equal to $-|\alpha_0-1|$.

the concise matrix form (2.25) into which (5.1) can be put. To solve the above equation, we write

$$\chi = \mu u + \nu u' + \lambda^{-1}\sigma u'' + \lambda^{-2}\tau u''', \tag{5.2}$$

where $u$ is a solution of the fundamental reference equation, and $\mu$, $\nu$, $\sigma$, $\tau$ are expected to have the behavior (1.3). This expression has exactly the same character as (3.1), since the equation for $u$ is a special case of the equation for $\psi$. Thus, all the formulae in §3 and in the Appendices apply, except that we should now put

$$p = z, \text{ i.e., } p_0 = z,\ p_1 = 0,\ p_k = 0,\ k = 2, 3, \cdots, \tag{5.3a}$$

$$q = \alpha, \quad \text{and} \quad r = \beta. \tag{5.3b}$$

From (B10) it follows that if $\chi$ is to satisfy (2.17), we must have

$$z\mu_1' + \alpha_0\mu_1 + \beta_0\mu - \beta\nu_1 = \lambda^{-1}\overline{A}, \tag{5.4a}$$

$$z\nu_1' + \alpha_0\nu_1 + \beta_0\nu + z\mu_1 - \alpha\nu_1 = \lambda^{-1}\overline{B}, \tag{5.4b}$$

$$-2z\sigma_1' + [\alpha_0 - \alpha - 1]\sigma_1 + \beta_0\sigma = \lambda^{-1}\overline{C}, \tag{5.4c}$$

$$-2z\tau_1' + [\alpha_0 - \alpha - 2]\tau_1 + \beta_0\tau = \lambda^{-1}\overline{D} - (3\nu_1' + \mu_1) \tag{5.4d}$$

where $\overline{A}$, $\overline{B}$, $\overline{C}$, $\overline{D}$ are given by (C3) with $(A, B, C, D)$ replaced by $(\mu, \nu, \sigma, \tau)$, and

$$\nu - z\tau = \lambda^{-1}\tilde{\sigma}, \qquad \sigma = \lambda^{-1}\tilde{\tau}; \tag{5.5}$$

$$\mu_1 = \mu' - \beta\tau, \tag{5.6a}$$

$$\nu_1 = \nu' + \mu - \alpha\tau, \tag{5.6b}$$

$$\sigma_1 = \tilde{\sigma}, \tag{5.6c}$$

$$\tau_1 = \tau' + \tilde{\tau}. \tag{5.6d}$$

We shall now solve this set of equations in terms of asymptotic expansions in inverse powers of $\lambda$. Then it is clear that the exact forms of the expressions on the right-hand side of (5.4) do not matter. Only the forms of the operator on the left are important. It is also convenient to think of $\sigma_1$ and $\tau_1$ rather than $\tilde{\sigma}$ and $\tilde{\tau}$ as the variable sought, and rewrite (5.5) as

$$\nu - z\tau = \lambda^{-1}\sigma_1, \qquad \sigma = \lambda^{-1}(\tau_1 + \tau'). \tag{5.7}$$

By using (5.5) and (5.7), we may rewrite (5.4) as

$$[z\mu'' + \alpha_0\mu' + (\beta_0 - \beta)\mu] - \beta[2z\tau' + \tau + (\alpha_0 - \alpha)\tau] = O(\lambda^{-1}), \tag{5.8a}$$

$$[z^2\tau'' + (\alpha_0 - 2\alpha + 2)z\tau' + (\alpha_0 - \alpha)(1 - \alpha)\tau + (\beta_0 - \beta)z\tau] + [2z\mu' + (\alpha_0 - \alpha)\mu] = O(\lambda^{-1}), \tag{5.8b}$$

$$-2z\sigma_1' + (\alpha_0 - \alpha - 1)\sigma_1 = O(\lambda^{-1}), \tag{5.8c}$$

$$-2z\tau_1' + (\alpha_0 - \alpha - 2)\tau_1 = O(\lambda^{-1}) - [4\mu' + 3z\tau'' + 3(1 - 2\alpha)\tau' + (\beta_0 - \beta)\tau] \tag{5.8d}$$

where $O(\lambda^{-1})$ denotes terms of that order in the variables $\mu$, $\tau$, $\sigma_1$ and $\tau_1$, and the parameters $\alpha$ and $\beta$. Clearly, in solving for successively higher approximations, the equations (5.8a) and (5.8b) may be first solved for $\mu$ and $\tau$ and then the equations (5.8c) and (5.8d) solved for $\sigma_1$ and $\tau_1$. It is to be noted that $\alpha$ and $\beta$ are also unknowns to be determined, but we may expect $\alpha$ and $\beta$ to be equal to $\alpha_0$ and $\beta_0$ in the limit $\lambda \to \infty$.

The solution of these equations will now be discussed in the next three sections for three different cases, depending on the nature of solutions of the reduced equation

$$z\chi'' + \alpha_0\chi' + \beta_0\chi = 0. \tag{5.9}$$

CASE I. If the indices $\rho = 0$ and $\rho = 1 - \alpha_0$ of equation (5.9) do not differ by an integer, positive, negative or zero, the solutions of (5.9) *do not* contain logarithmic terms. Indeed, their essential characteristics do not depend on $\beta_0$. This is the simplest case.

CASE II. If the indices differ by an integer, positive, negative or zero, and $\beta_0 \neq 0$, one of the solutions of (5.9) definitely contains a logarithmic term. This case is still relatively simple.

CASE III. If the indices differ by an integer, positive, negative or zero, and $\beta_0 = 0$, the solutions of (5.9) are explicitly known to be

$$\chi = 1 \quad \text{and} \quad z^{-\alpha_0+1} \qquad \text{if } \alpha_0 \neq 0 \tag{5.10a}$$

or

$$\chi = 1 \quad \text{and } z \qquad \text{if } \alpha_0 = 0. \tag{5.10b}$$

This is the most complicated case, because the simplicity of the initial approximation makes them less useful for controlling the behavior of the higher approximations. Indeed, it will be found that the coefficients $(\mu, \nu, \sigma, \tau)$ in (5.1) have the usual behavior (1.3) only under very special circumstances. In general, they must be taken to have the form

$$\mu(z, \lambda) = \sum_{p=0}^{\infty} \lambda^{-p/(n+1)}\mu^{(p)}(z). \tag{5.11}$$

It is clear that the general outline of the theory, as presented in §2, will be somewhat modified. But the essential spirit of the analysis will not [cf. footnote 6].

6. **Existence of regular formal solutions**—CASE I. We shall now consider the solution of the system of equations (5.8) in the simplest case when the equation (5.9) has one nonintegral index. The method that we discuss in some detail will be the one that may be used without any essential change in the next two cases. Other methods will be briefly indicated at the end of this section.

We consider expansions of the type

$$\alpha = \sum_{j=0}^{\infty} \lambda^{-j}\alpha^{(j)}, \qquad \beta = \sum_{j=0}^{\infty} \lambda^{-j}\beta^{(j)}, \tag{6.1}$$

$$\mu = \sum_{j=0}^{\infty} \lambda^{-j}\mu^{(j)}, \tag{6.2}$$

and similar equations for $\nu$, $\sigma$ and $\tau$. We naturally attempt to set

$$\alpha^{(0)} = \alpha_0, \qquad \beta^{(0)} = \beta_0. \tag{6.3}$$

The initial approximation for (5.8) then becomes

$$[z\mu^{(0)''} + \alpha_0\mu^{(0)'}] - \beta_0[2z\tau^{(0)'} + \tau^{(0)}] = 0, \tag{6.4a}$$

$$z[z\tau^{(0)''} + (2 - \alpha_0)\tau^{(0)'} + 2\mu^{(0)'}] = 0, \tag{6.4b}$$

$$2z\sigma_1^{(0)'} + \sigma_1^{(0)} = 0, \tag{6.4c}$$

$$2z\tau_1^{(0)'} + 2\tau_1^{(0)} = [4\mu^{(0)'} + 3z\tau^{(0)''} + 3(1 - 2\alpha_0)\tau^{(0)'}]. \tag{6.4d}$$

We expect the initial approximation to be $\mu^{(0)}=1$, $\nu^{(0)}=\sigma^{(0)}=\tau^{(0)}=0$; and this is indeed a possible solution for (6.4). The higher approximations are given by

$$[z\mu^{(j)''} + \alpha_0\mu^{(j)'}] - \beta_0[2z\tau^{(j)'} + \tau^{(j)}] = \beta^{(j)} + R_j(z), \tag{6.5a}$$

$$z[z\tau^{(j)''} + (2 - \alpha_0)\tau^{(j)'} + 2\mu^{(0)'}] = \alpha^{(j)} + R_j(z), \tag{6.5b}$$

$$2z\sigma_1^{(j)'} + \sigma_1^{(j)} = R_j(z), \tag{6.5c}$$

$$2z\tau_1^{(j)} + 2\tau_1^{(j)'} = R_j(z) + [4\mu^{(j)'} + 3z\tau^{(j)''} + 3(1 - 2\alpha_0)\tau^{(j)'}], \tag{6.5d}$$

where $R_j(z)$ is the generic symbol for a regular function which is known when the $(j-1)$th approximations are all known. After regular solutions of the equations (6.5a) and (6.5b) are found, we may integrate (6.5c) and (6.5d) as follows:

$$\sigma_1^{(j)} = z^{-1/2}\int_0^z z^{1/2}R_j(z)dz, \tag{6.6}$$

$$\tau_1^{(j)} = z^{-1}\int_0^z 2^{-1}\tilde{R}_j(z)dz \tag{6.7}$$

where $\tilde{R}_j(z)$ denotes the right-hand side of (6.5d).

To solve the equations (6.5a) and (6.5b), we first adopt the simpler notation $(^-)$ for all unknowns with an index $j$, and use different symbols $R_1(z)$ and $R_2(z)$ for the two regular functions; thus

$$[z\bar{\mu}'' + \alpha_0\bar{\mu}'] - \beta_0[2z\bar{\tau}' + \bar{\tau}] = \bar{\beta} + R_1(z), \tag{6.8}$$

$$z[z\bar{\tau}'' + (2 - \alpha_0)\bar{\tau}' + \bar{\mu}'] = \bar{\alpha} + R_2(z). \tag{6.9}$$

Clearly, to have regularity, we must take

(6.10) $$\bar{\alpha} = -R_2(0)$$

so that (6.9) becomes

(6.11) $$z\bar{\tau}'' + (2 - \alpha_0)\bar{\tau}' + 2\bar{\mu}' = R_3(z).$$

This equation can be integrated once more to give

(6.12) $$z\bar{\tau}' + (1 - \alpha_0)\bar{\tau} + 2\bar{\mu} = R_4(z)$$

where $R_4(z)$ has an additive constant of integration which is at our disposal.

We shall now attempt to solve $\bar{\mu}$ and $\bar{\tau}$ from (6.8) and (6.12) as regular power series. This method is based on the fact that the system is Fuchsian, as can be easily verified by eliminating $z\bar{\tau}'$ from (6.8) by means of (6.12), and consider the system of equations of the first order with $(\bar{\mu}, \bar{\mu}', \bar{\tau})$ as the three dependent variables. Thus, formal power series solutions (possibly including logarithmic terms) are true solutions.

Let us write

(6.13) $$\bar{\mu} = \sum_{k=0}^{\infty} \bar{\mu}_k z^k, \qquad \bar{\tau} = \sum_{k=0}^{\infty} \bar{\tau}_k z^k.$$

Then (6.8) leads to

(6.14) $$-\beta_0(2k + 1)\bar{\tau}_k + (k + \alpha_0)(k + 1)\bar{\mu}_{k+1} = R_{1,k} - \bar{\beta}\delta_{k0}, \quad k = 0, 1, 2, \cdots$$

and (6.12) leads to

(6.15) $$2\bar{\mu}_k + [k + (1 - \alpha_0)]\bar{\tau}_k = R_{4,k}, \qquad k = 0, 1, \cdots .$$

The latter may be solved for $\bar{\mu}_k$ in terms of $\bar{\tau}_k$. To determine $\bar{\tau}_k$, write (6.15) as

(6.16) $$2\bar{\mu}_{k+1} + [k + (2 - \alpha_0)]\bar{\tau}_{k+1} = R_{4,k+1}, \qquad k = -1, 0, 1, \cdots$$

and eliminate $\bar{\mu}_{k+1}$ from (6.4). Thus

(6.17) $$2\beta_0(2k + 1)\bar{\tau}_k + (k + 1)(k + \alpha_0)[k + (2 - \alpha_0)]\bar{\tau}_{k+1} = S_k + 2\bar{\beta}\delta_{k0},$$

for $k=0, 1, 2, \cdots$, where $S_k$ is a known constant. Equation (6.17) enables us to solve all the $\bar{\tau}_k$'s in terms of the leading coefficient $\bar{\tau}_0$, unless the coefficient for one of the unkowns $\bar{\tau}_{k+1}$ vanishes; i.e., when

(6.18) $$(k + \alpha_0)(k + 2 - \alpha_0) = 0$$

for some integral value of $k>0$. This is possible only when

(6.19) $$\alpha_0 = n + 2 \text{ or } -n$$

for some value of $n=0, 1, 2, \cdots$. Thus, if the parameter $\alpha_0$ is not an integer, or if it takes on the special integral value

(6.20) $$\alpha_0 = 1,$$

the above method can be carried through. The condition (6.20) implies that the indices of the reduced equation

$$z\chi'' + \alpha_0\chi' + \beta_0\chi = 0 \tag{6.21}$$

are both equal to zero, and the operator $L_0(\chi)$ is self-adjoint. One of the solutions of (6.21) contains logarithmic terms no matter what the value of $\beta_0$ may be. This is however really an exceptionally simple case of Case II, and will be included in the next section. We may therefore formulate the following theorem.

THEOREM I. *If the indices* $\rho=0$ *and* $\rho=1-\alpha_0$ *of the reduced equation* (6.21) *of the normalized differential equation* (5.1) *do not differ by an integer, then the latter has formal solutions of the form* (5.2) *where the coefficients* $\mu$, $\nu$, $\sigma$, $\tau$ *are given by formulas of the form* (6.2) *and* $u(z, \lambda)$ *is any solution of the differential equation* (1.8) *with coefficients* $\alpha$ *and* $\beta$ *in the form* (1.9). *Furthermore, we have*

$$\mu^{(0)} = 1,\ \nu^{(0)} = 0,\ \sigma^{(0)} = 0,\ \tau^{(0)} = 0,\ \alpha^{(0)} = \alpha_0,\ \beta^{(0)} = \beta_0;$$

*and that all the coefficients* $\alpha^{(j)}$ $(j=1, 2, \cdots)$ *must be chosen in a manner prescribed by the original differential equation; the coefficients* $\beta^{(j)}$ $(j=1, 2, \cdots)$ *are however arbitrary.*

There are at least two other ways of treating the system of equations (6.8) and (6.9). By analogy with §§3, 4, we see that the solutions of the homogeneous system are known. They are simply the connection coefficients of the solutions of the reduced equation (6.1) with themselves. Thus, they are indeed explicitly expressible in terms of Bessel functions. We may then write the solutions of the nonhomogeneous system by the method of variation of parameters and use the disposable constants to make the solution regular.

Another method is to use the form (6.8) and (6.12) and eliminate $z\bar{\tau}'$ from the former by using the latter. It is then easy to reduce this system into an integral system by solving (6.8) for $\bar{\mu}$ and (6.12) for $\bar{\tau}$. The reasoning here runs parallel to that used at the end of §3.

7. **Existence of regular formal solutions**—CASE II. We shall now apply the method used above to consider the case where $\alpha_0$ is an integer and $\beta_0 \neq 0$. The case $\beta_0=0$ will be considered in the next section. We shall prove the following theorem.

THEOREM II. *If* $\alpha_0$ *is an integer, positive, negative, or zero, and* $\beta_0 \neq 0$, *Theorem* I *still holds, but only when the coefficients* $\beta^{(j)}$ $(j=1, 2, 3, \cdots)$ *are also chosen in a manner prescribed by the original differential equation. Only the case* $\alpha_0=1$ *is free from this requirement.*

To prove this theorem, let us write out equation (6.17) explicitly as follows:

$$
(7.1)\quad
\begin{aligned}
&k = 0, && 2\beta_0\bar{\tau}_0 + \alpha_0(2 - \alpha_0)\bar{\tau}_1 = R_0^{(3)} + 2\bar{\beta},\\
&k = 1, && 6\beta_0\bar{\tau}_1 + 2(1 + \alpha_0)(3 - \alpha_0)\bar{\tau}_2 = R_1^{(3)},\\
&k = 2, && 10\beta_0\bar{\tau}_2 + 3(2 + \alpha_0)(4 - \alpha_0)\bar{\tau}_3 = R_2^{(3)},\\
&\cdots\cdots && \cdots\cdots\\
&k = n, && 2\beta_0(2n + 1)\bar{\tau}_n + (n + 1)(n + \alpha_0)(n + 2 - \alpha_0)\bar{\tau}_{n+1} = R_n^{(3)}.
\end{aligned}
$$

If we regard these as a system of equations for $\bar{\tau}_1, \cdots, \bar{\tau}_{n+1}$, the coefficient matrix has the structure in the form of two diagonal terms

$$
\begin{bmatrix}
\alpha_0(2 - \alpha_0) & & & & \\
6\beta_0 & 2(1 + \alpha_0)(3 - \alpha_0) & & & \\
& 10\beta_0 & 3(2 + \alpha_0)(4 - \alpha_0) & & \\
& & \ddots & \ddots & \\
& & & 2\beta_0(2n + 1) & (n + 1)(n + \alpha_0)(n + 2 - \alpha_0)
\end{bmatrix}
$$

The determinant can vanish only if one of the terms in the main diagonal vanishes and this can happen only when (6.19) is satisfied, and only once. We now consider the matrix for which the lower right corner term vanishes, and this will still be denoted by the above matrix. Then, in order to be able to solve for $\bar{\tau}_1, \cdots, \bar{\tau}_{n+1}$, the matrix obtained by replacing the last column by the right-hand side column of the system of equations (7.1) must have a vanishing determinant. This can always be made to be true *if the coefficient of* $2\bar{\beta}$ *does not vanish.* This coefficient is the $n \times n$ determinant at the lower left side of the above matrix (up to a sign), and its value is clearly $(2\beta_0)^n(2n+1)(2n-1) \cdots 5.3$. If $\beta_0 \neq 0$, then, the augumented matrix can be made to have the same rank as the coefficient matrix, and there is a one-parameter family of solutions for $\bar{\tau}_1, \cdots, \bar{\tau}_{n+1}$. The equations for $k = n+1, \cdots$ then determine all the other coefficients. We have thus indeed found regular solutions by proper choice of the coefficients $\beta^{(j)}$, and the theorem is proved.

8. **Existence of regular formal solutions**—CASE III. We shall now consider the apparently simpler but actually more complicated case where $\beta_0 = 0$, and the reduced equation (6.21) takes the form

$$
(8.1)\qquad z\chi'' + \alpha_0\chi' = 0
$$

where $\alpha_0$ is an integer, positive, negative, or zero. We shall restrict our discussion only to cases

$$
(8.2)\qquad \alpha_0 = -n, \qquad n = 0, 1, 2, \cdots,
$$

since the adjoint equation of (5.1) has the corresponding coefficient given by[12]

[12] Cf. Langer [4].

$$\bar{\alpha}_0 = 2 - \alpha_0 = 2, 3, 4, \cdots. \tag{8.3}$$

The gap left by (8.2) and (8.3) is the special case

$$\alpha_0 = \bar{\alpha}_0 = 1, \tag{8.4}$$

when the operator $L_0(\chi)$ is self-adjoint. This case has been covered by Theorems I and II.

Let us first notice the similarity and the differences between the present case and the above cases. The equations (5.8) remain unchanged. Clearly, if a method of successive approximations is used, Equations (5.8c) and (5.8d) can be treated in exactly the same manner as in §6. Thus, the problem again focuses itself on the first two equations (5.8a) and (5.8b). In the special case considered here, these may be written in the following convenient form:

$$z\mu'' - n\mu' = (\beta - \beta_0)(\mu + 2z\tau' + \tau - (\alpha - \alpha_0)\tau) + O(\lambda^{-1}), \tag{8.5}$$

$$\begin{aligned} z[z\tau'' + (n+2)\tau'] = &- 2z\mu' + (\alpha - \alpha_0)(\mu + 2z\tau' + \tau - \alpha\tau) \\ &+ (\beta - \beta_0)z\tau + O(\lambda^{-1}), \end{aligned} \tag{8.6}$$

where we have put $\alpha_0 = -n$ and $\beta_0 = 0$ in some of the expressions, but have retained the differences $\alpha - \alpha_0$, $\beta - \beta_0$ to suggest that they are terms of an order higher. In Equation (8.6), we put $\mu'$ on the right-hand side, because it is supposed to be obtained from Equation (8.5) in a process of successive approximations. As before, the initial approximation is given by $\mu^{(0)} = 1$, $\tau^{(0)} = 0$.

It is easy to see from Equation (8.5) that the usual expansion of $\mu$ and $\tau$ in inverse powers of $\lambda$ would not work, except under extremely special circumstances; for logarithmic terms tend to appear. The successive approximations for $\mu$, beyond the initial one, satisfy the inhomogeneous differential equation

$$z\bar{\mu}'' - n\bar{\mu}' = \bar{R}(z), \tag{8.7}$$

where $\bar{R}(z)$ is a regular function determined by the lower approximations. The solution of (8.7) can be easily written down. It is[13]

$$\bar{\mu}' = z^n \left\{ c - \frac{1}{n} \int z^{-(n+1)} \bar{R}(z) dz \right\}, \tag{8.8a}$$

$$\bar{\mu} = c_1 + c_2 z^{n+1} - \int \frac{z^n}{n} dz \int z^{-(n+1)} \bar{R}(z) dz. \tag{8.8b}$$

Thus, in general, if $\bar{R}(z)$ contains a term $\bar{R}_m z^m$, there is a corresponding term $-\bar{R}_m\{(m-n)(m+1)n\}^{-1} z^{m+1}$ occurring in $\bar{\mu}$. However, if $m = n$, i.e., if $\bar{R}(z)$ should contain a term $\bar{R}_n z^n$, the integral of (8.7) will contain a term of the form

(13) We are treating the case $n \neq 0$ here. The case $n = 0$ requires some obvious minor modifications.

$$\bar{\mu}' = cz^n + \bar{R}_n z^n \log z, \tag{8.9a}$$

$$\bar{\mu} = c_1 + c_2 z^{n+1} \frac{\bar{R}_n}{n+1} z^{n+1}(\log z - 1). \tag{8.9b}$$

The occurrence of logarithmic terms must therefore be prevented, if they tend to occur, by trying to remove terms proportional to $z^n$ on the right-hand side of (8.7).

There is no such difficulty associated with (8.6). As in §6, we can choose $\alpha$ properly to counteract the factor $z$ on the left-hand side of the equation. After this is done, we are faced with solving equations of the type

$$z\bar{\tau}'' + (n+2)\bar{\tau}' = \bar{R}(z), \tag{8.10}$$

whose regular solution is given by

$$\bar{\tau}' = (n+2)^{-1} z^{-(n+2)} \left[ \int_0^z z^{(n+1)} \bar{R}(z) dz \right], \tag{8.11a}$$

and

$$\bar{\tau}' = (n+2)^{-1} \left\{ \int_0^z z^{-(n+2)} \left[ \int_0^z z^{(n+1)} \bar{R}(z) dz \right] dz + c \right\}, \tag{8.11b}$$

where $c$ is an arbitrary constant of integration. Thus, a positive power term $\bar{R}_m z^m$ in $\bar{R}(z)$ would always produce a positive power term

$$\bar{R}_m \{ (m+n+2)(m+1)(n+2) \}^{-1} z^{m+1} \text{ in } \bar{\tau}.$$

*Removal of the logarithmic term by fractional powers.* To be able to remove the logarithmic terms in any scheme of successive approximation it is then necessary to be able to control an explicit term $cz^n$ which enters $\bar{R}(z)$ in (8.7) and whose coefficient $c$ is at our disposal. If we attempt to solve (8.5) and (8.6) by expanding $\mu$ and $\tau$ in the usual form

$$\mu = \sum_{j=0}^{\infty} \lambda^{-j} \mu^{(j)} \quad \text{and} \quad \tau = \sum_{j=0}^{\infty} \lambda^{-j} \tau^{(j)} \tag{8.12}$$

with corresponding expansions for $\alpha$ and $\beta$, we do not have such a control even at the stage of determining $\mu^{(1)}$ and $\tau^{(1)}$. Let us assume, for definiteness, that the logarithmic terms would arise in the first approximation by using the above expansion (8.12).

Clearly, some modification of this expansion is needed. Heuristically, one may reason as follows to arrive at the form of this expansion. The occurrence of the logarithmic term indicates that $\mu$ is *not* of the form

$$\mu = \mu^{(0)} + O(\lambda^{-1}), \qquad \mu^{(0)} = 1,$$

but rather that $\mu - \mu^{(0)}$ must be *larger* than $O(\lambda^{-1})$. This immediately suggests the use of fractional powers. The next question is "Which fraction?"

Let us consider an expansion in the parameter $\epsilon=\lambda^{-(1/N)}$ where $N>1$, but is otherwise unspecified. We note from (8.5) that, before terms of the order of $\lambda^{-1}$ are included, the right-hand sides of (8.7) and (8.10) can only arise from the lower approximations. We have also noticed that a term in $z^m$ on the right-hand sides of these equations gives rise to a term in $z^{m+1}$ in the solution.

Since $\mu^{(0)}=1$, $\tau^{(0)}=1$, we may expect that $z^i$ is precisely the significant power in $\mu^{(i)}$ and $\tau^{(i)}$. Thus, the $n$th approximation should appear in the right-hand side of the equation which also includes the term $O(\lambda^{-1})$ in (8.5), i.e., we should have

$$\epsilon^{n+1} = \lambda^{-1},$$

or

$$\epsilon = \lambda^{-1/(n+1)}. \tag{8.13}$$

Having thus determined $\epsilon$ in a tentative manner, we may try to put

$$\mu = \sum_{p=0}^{\infty} \epsilon^i \mu^{(i)}(z), \qquad \mu^{(0)} = 1, \tag{8.14a}$$

$$\tau = \sum_{p=0}^{\infty} \epsilon^i \tau^{(i)}(z), \qquad \tau^{(0)} = 0, \tag{8.14b}$$

$$\alpha = \sum_{p=0}^{\infty} \epsilon^i \alpha^{(i)}, \qquad \alpha^{(0)} = \alpha_0 = -n, \tag{8.14c}$$

$$\beta = \sum_{p=0}^{\infty} \epsilon^i \beta^{(i)}, \qquad \beta^{(0)} = \beta_0 = 0 \tag{8.14d}$$

in (8.5) and (8.6). Then, we obtain for the first approximation

$$z\mu^{(i)''} - n\mu^{(i)'} = \beta^{(1)}, \tag{8.15}$$

and

$$z\left[z\tau^{(1)''} + (h+2)\tau^{(1)'}\right] = -2z\mu^{(1)'} + \alpha^{(1)}; \tag{8.16}$$

and for the higher approximations, $2 \leqq i \leqq n-1$,

$$z\mu^{(i+1)'} - n\mu^{(i+1)} = \beta_1(\mu^{(i)} - 2z\tau^{(i)'} + \tau^{(i)}) + \beta^{(i+1)} + \cdots \tag{8.17}$$

and

$$\begin{aligned} &z\left[z\tau^{(i+1)''} + (n+2)\tau^{(i+1)'}\right] \\ &= -2z\mu^{(i+1)'} + \alpha_1(\mu^{(i)} + 2z\tau^{(i)'} + (n+1)\tau^{(i)}) + \alpha^{(i+1)} + \beta_1 z\tau^{(i)} + \cdots \end{aligned} \tag{8.18}$$

where the terms not written out explicitly involve only $\mu^{(i-1)}$, $\tau^{(i-1)}$, $\alpha^{(i)}$, $\beta^{(i)}$

and lower approximations. As we have noted before, the terms $\alpha^{(1)}$ in (8.16) and $\alpha^{(i+1)}$ in (8.18) must be chosen to make the solution regular.

Let us now examine the rise of power in the successive approximations. Clearly, we have from (8.15) [cf. (8.8b)]

$$\mu^{(1)} = c_1 + c_2 z^{n+1} - (\beta^{(1)}/n)z \tag{8.19}$$

and, from (8.16) [cf. (8.11b)].

$$\tau^{(1)} = c_3 - \frac{c_2}{(n+1)(n+2)} z^{n+1} + \frac{2\beta^{(1)}}{n(n+2)^2} z. \tag{8.20}$$

Terms with power higher than $n$ will not interfere with the cancellation of the term $z^n$ in the right-hand side of the $(n+1)$th approximation. We may therefore omit such terms by putting $c_2=0$. Thus, the leading powers in $\mu^{(1)}$ and $\tau^{(1)}$ indeed behave in the manner as expected. In the higher approximations, as one can see from (8.17) and (8.18), the leading powers would arise from the terms $\beta_1\mu^{(i)}$ and $-2z\mu^{(i+1)\prime}$ on the right-hand side of these equations. Thus, if we denote by $C_i$ and $\overline{C}_i$ the coefficients of $z^i$ in $\mu^{(i)}$ and $\tau^{(i)}$ respectively, we have the following relations for the determination of these coefficients:

$$(i+1)(i-n)C_{i+1} = \beta_1 C_i, \tag{8.21}$$

$$(i+n+2)\overline{C}_{i+1} = -2C_{i+1}, \tag{8.22}$$

where $2 \leqq i \leqq n-1$. Thus, the coefficient of $z^n$ in $\mu^{(n)}$ is proportional to $\beta_1^n$. This contributes a term proportional to $\beta_1^{n+1}z^n$ in the right-hand side of (8.17) in the $(n+1)$th approximation. In this approximation, the right-hand side contains other terms in $z^n$ through the term $O(\lambda^{-1})$ in (8.5), and $\beta_1^{n+1}$ must be so chosen as to annihilate such terms. Note that these new terms to be annihilated will depend only on the initial approximation $(\mu^{(0)}, \tau^{(0)}) = (1, 0)$, and are not influenced by the intermediate approximations in fractional powers of $\lambda$.

A little calculation will show that as long as $\beta_1 \neq 0$, nothing difficult will arise in the higher approximations. Thus, we have covered the case when the logarithmic terms would appear in the approximation $O(\lambda^{-1})$. If this is not the case, we simply defer the introduction of the fractional powers until the occurrent of logarithmic terms is imminent. Actually, no change of procedure is necessary. If logarithmic terms do not tend to appear until the term $O(\lambda^{-2})$ is considered, the coefficients $\beta^{(1)}, \cdots, \beta^{(n)}$ would automatically be forced to equal to zero by the condition of regularity.

We may summarize our results in the following theorem.

Theorem III. *If the reduced equation of the normalized differential equation* (5.1) *has two regular solutions, i.e., if* $\alpha_0 = -n$ $(n=0, 1, 2, \cdots)$ *and* $\beta_0 = 0$, *then there exists a set of formal solutions in the form*

$$\chi = \mu u + \nu u' + \lambda^{-1}\sigma u'' + \lambda^{-2}\tau u''' \tag{8.23}$$

*where u is any solution of the basic reference equation*

$$u^{\text{iv}} + \lambda^{-2}(zu'' + \alpha u' + \beta u) = 0, \tag{8.24}$$

*and the coefficients* $\mu(z, \lambda)$, $\nu(z, \lambda)$, $\sigma(z, \lambda)$, $\tau(z, \lambda)$ *and* $\alpha(\lambda)$, $\beta(\lambda)$ *have asymptotic expansions of the form* (8.14) *in which* $\epsilon$ *is given by* (8.13) *and the functions* $\mu^{(i)}(z), \cdots, \tau^{(i)}(z)$ *are all regular.*

## Appendix A

In this appendix, we shall give some transformation formulae which will be found useful both for the reduction of the given equation into the normal form and for the solution of the normalized differential equation. We consider the transformation of the dependent variable $\psi(z, \lambda)$ in the differential equation

$$\psi^{\text{iv}} + \lambda^2(p\psi'' + q\psi' + r\psi) = 0 \tag{A 1}$$

of the type (1.1), having in mind, however, that (A 1) is to be regarded as the system of differential equations

$$\frac{d}{dz}\begin{bmatrix} \psi \\ \psi' \\ \lambda^{-1}\psi'' \\ \lambda^{-2}\psi''' \end{bmatrix} = \begin{bmatrix} 0 & 1 & 0 & 0 \\ 0 & 0 & \lambda & 0 \\ 0 & 0 & 0 & \lambda \\ -r & -q & -\lambda p & 0 \end{bmatrix}\begin{bmatrix} \psi \\ \psi' \\ \lambda^{-1}\psi'' \\ \lambda^{-2}\psi''' \end{bmatrix} \tag{A 2}$$

for the vector variable $\boldsymbol{\psi}(\psi, \psi', \lambda^{-1}\psi'', \lambda^{-2}\psi''')$. We introduce a new variable $\chi$ by the relation

$$\chi = A\psi + B\psi' + \lambda^{-1}C\psi'' + \lambda^{-2}D\psi''', \tag{A 3}$$

where $(A, B, C, D)$ are connection coefficients with the type of asymptotic behavior (1.3). If we differentiate (A3) with respect to $z$, we find

$$\chi' = A_1\psi + B_1\psi' + \lambda^{-1}C_1\psi'' + \lambda^{-2}D_1\psi''', \tag{A 4}$$

where

$$A_1 = A' - rD, \tag{A 5a}$$

$$B_1 = B' + A - qD, \tag{A 5b}$$

$$C_1 = C' + \lambda(B - pD), \tag{A 5c}$$

$$D_1 = D' + \lambda C. \tag{A 5d}$$

If we now impose the condition that the connection coefficients $(A_1, B_1, C_1, D_1)$ shall also have the type of behavior (1.3), then we must have

(A 6) $$B - pD = \lambda^{-1}E \quad \text{and} \quad C = \lambda^{-1}F,$$

and the relations (A5) become

(A 7a) $$A_1 = A' - rD,$$

(A 7b) $$B_1 = B' + A - qD,$$

(A 7c) $$C_1 = E + \lambda^{-1}F',$$

(A 7d) $$D_1 = D' + F.$$

The second derivative $\chi''$ is given by

(A 8) $$\chi'' = A_2\psi + B_2\psi' + \lambda^{-1}C_2\psi'' + \lambda^{-2}D_2\psi''',$$

where

(A 9a) $$A_2 = A_1' - rD_1,$$

(A 9b) $$B_2 = B_1' + A_1 - qD_1,$$

(A 9c) $$C_2 = C_1' + \lambda(B_1 - pD_1),$$

(A 9d) $$D_2 = D_1' + \lambda C_1.$$

Note that $C_2$ and $D_2$ are of the order of $\lambda$, and hence the connection coefficients $\lambda^{-1}(A_2, B_2, C_2, D_2)$, which are appropriate for $\lambda^{-1}\chi''$, automatically have the behavior of the type (1.3).

The third derivative $\chi'''$ is given by

(A 10) $$\chi''' = A_3\psi + B_3\psi' + \lambda^{-1}C_3\psi'' + \lambda^{-2}D_3\psi''',$$

where

(A 11a) $$A_3 = A_2' - rD_2 = (A_2' - rD_1) - \lambda rC_1,$$

(A 11b) $$B_3 = B_2' + A_2 - qD_2 = (B_2' + A_2 - qD_1') - \lambda qC_1,$$

(A 11c) $$\begin{aligned} C_3 &= C_2' + \lambda(B_2 - pD_2) \\ &= C_1'' + \lambda[(B_1 - pD_1)' + (B_2 - pD_2')] - \lambda^2 pC_1, \end{aligned}$$

(A 11d) $$D_3 = D_2' + \lambda C_2 = D_1'' + 2\lambda C_1' + \lambda^2(B_1 - pD_1).$$

Note that the order of $\lambda$ is indicated explicitly in the final form of the above expressions, and hence the connection coefficients $\lambda^{-2}(A_3, B_3, C_3, D_3)$, which are appropriate for $\lambda^{-2}\chi'''$, automatically have the behavior of the type (1.3).

Thus, the matrix $G$ connecting the vectors $\mathfrak{x}(\chi, \chi', \lambda^{-1}\chi'', \lambda^{-2}\chi''')$ and $\mathfrak{y}(\psi, \psi', \lambda^{-1}\psi'', \lambda^{-2}\psi''')$ has the form

(A 12) $$G = \sum_{n=0}^{\infty} \lambda^{-n}G_n,$$

where $G_0$ has the form

$$\text{(A 13)}\qquad G_0 \equiv \begin{bmatrix} A_0 & B_0 & 0 & D_0 \\ A_{1,0} & B_{1,0} & C_{1,0} & D_{1,0} \\ 0 & 0 & B_{1,0} - p_0 D_{1,0} & C_{1,0} \\ 0 & 0 & -p_0 C_{1,0} & B_{1,0} - p_0 D_{1,0} \end{bmatrix}.$$

It is essential that $|G_0| = \det G_0 \neq 0$ so that the transformation of variables is nonsingular. In particular, this should hold for the point $z=0$. At that point, we have

$$\text{(A 14)}\qquad | G_0(0) | = A_0(0)\{B_{1,0}(0)\}^3,$$

and the following conditions are necessary and sufficient for the transformation to be nonsingular in a neighborhood(14) of $z=0$:

$$\text{(A 15)}\qquad A_0(0) \neq 0, \qquad B_{1,0}(0) = A_0(0) + D_0(0)[1 - q_0(0)] \neq 0,$$

where $B_{1,0}$ has also been expressed in terms of more elementary quantities by means of (A 7b) and (A 6).

As a matter of fact, we shall require the function $\chi$ to be a solution of a fourth order differential equation of the form (B 1) specified below, which is quite similar to (A 1). Thus, if we consider a fundamental matrix solution $\Psi$ of the system (A 2), then

$$\text{(A 16)}\qquad \mathrm{X} = G\Psi$$

is a fundamental matrix solution of the corresponding system for $\chi(\chi, \chi', \lambda^{-1}\chi'', \lambda^{-2}\chi''')$. The determinant of $G$ thus appears as the ratio of two Wronskian determinants, and the leading term $|G_0|$ must be a constant. Thus, it would take on the constant value at $z=0$ as given by (A 14).

We note that the nature of the matrix $G$ indicates that the above manner of defining the vectors $\chi$ and $\psi$ is the natural one. Other choices of powers of $\lambda$ in the definition would not have been as convenient.

## Appendix B

In this appendix, we shall derive the differential equation satisfied by the function $\chi$ defined in the last section, and then obtain the conditions which must be satisfied by the coefficients $A, B, C, D$ so that $\chi$ satisfies a differential equation of the type

$$\text{(B 1)}\qquad L_0(\chi) = \lambda[\bar{a}\chi + \bar{b}\chi' + \bar{c}\lambda^{-1}\chi'' + \bar{d}\lambda^{-2}\chi''']$$

where

$$\text{(B 2)}\qquad L_0(\chi) \equiv \chi^{\mathrm{iv}} + \lambda^2(z\chi'' + \alpha_0\chi' + \beta_0\chi),$$

as defined in (2.17), and the coefficients $(\bar{a}, \bar{b}, \bar{c}, \bar{d})$ have the behavior (1.3).

(14) The zeros of the analytic function $|G_0(z)|$ are isolated.

To do this, we shall first calculate the fourth derivative of $\chi$. We may write it in the form

(B 3) $$\chi^{\text{iv}} = A_4\psi + B_4\psi' + \lambda^{-1}C_4\psi'' + \lambda^{-2}D_4\psi''',$$

where the coefficients $(A_4, B_4, C_4, D_4)$ are given by

(B 3a) $$A_4 = A_3' - rD_3,$$

(B 3b) $$B_4 = B_3' + A_3 - qD_3,$$

(B 3c) $$C_4 = C_3' + \lambda(B_3 - pD_3),$$

(B 3d) $$D_4 = D_3' + \lambda C_3.$$

After some calculation, we find that

(B 4a) $$A_4 = (A_1' - rD_1')' - rD_1'' - \lambda[(rC_1)' + 2rC_1'] - \lambda^2 r(B_1 - pD_1),$$

(B 4b) $$B_4 = (B_2' + A_2 - qD_1')' + (A_2' - rD_1') - qD_1'' - \lambda[(qC_1)' - rC_1 - 2qC_1'] - \lambda^2 q(B_1 - pD_1),$$

(B 4c) $$C_4 = C_1''' + \lambda[(B_1 - pD_1)'' + (B_2 - pD_1')' + B_2' + A_2 - qD_1' - pD_1''] + \lambda^2[-(pC_1)' - qC_1 - 2pC_1'] - \lambda^3 p(B_1 - pD_1),$$

(B 4d) $$D_4 = D_1''' + 3\lambda C_1'' + \lambda^2[2(B_1 - pD_1)' + (B_2 - pD_1')] - \lambda^3 pC_1.$$

If we now form the differential operator $L_0(\chi)$, we have

(B 5) $$L_0(\chi) = \tilde{A}\psi + \tilde{B}\psi' + \lambda^{-1}\tilde{C}\psi'' + \lambda^{-2}\tilde{D}\psi''',$$

where

(B 6a) $$\tilde{A} = A_4 + \lambda^2[zA_2 + \alpha_0 A_1 + \beta_0 A],$$

(B 6b) $$\tilde{B} = B_4 + \lambda^2[zB_2 + \alpha_0 B_1 + \beta_0 B],$$

(B 6c) $$\tilde{C} = C_4 + \lambda^2[zC_2 + \alpha_0 C_1 + \beta_0 C],$$

(B 6d) $$\tilde{D} = D_4 + \lambda^2[zD_2 + \alpha_0 D_1 + \beta_0 D].$$

Now, *if* the matrix $G$ in the connection formula

(B 7) $$\mathfrak{x} = G\mathfrak{y}$$

for the vectors $\mathfrak{x}(\chi, \chi', \lambda^{-1}\chi'', \lambda^{-2}\chi''')$ and $\mathfrak{y}(\psi, \psi', \lambda^{-1}\psi'', \lambda^{-2}\psi''')$ is nonsingular in the manner specified in part A (cf. (A 10) and (A 11)), we may also write (B 5) in the final form (B 1), and obtain a differential equation for $\chi$, where $(\lambda\bar{a}, \lambda\bar{b}, \lambda\bar{c}, \lambda\bar{d})$ are linear combinations of $(\tilde{A}, \tilde{B}, \tilde{C}, \tilde{D})$ with coefficients of the order of unity. Thus, in order to make the coefficients $(\lambda\bar{a}, \lambda\bar{b}, \lambda\bar{c}, \lambda\bar{d})$ small to a certain order, *all the coefficients* $(\tilde{A}, \tilde{B}, \tilde{C}, \tilde{D})$ *must be made uniformly of that order.*

The detailed expressions for $(\tilde{A}, \tilde{B}, \tilde{C}, \tilde{D})$ are too long to be of interest. We shall merely give their leading terms in the case

(B 8) $$p = z + \sum_{n=1}^{\infty} \lambda^{-n} p_n, \qquad q = \sum_{n=0}^{\infty} \lambda^{-n} q_n, \qquad r = \sum_{n=0}^{\infty} \lambda^{-n} r_n.$$

Then

(B 9a) $\tilde{A} = \lambda^2[-r_0 B_1 + zA_1'' + \alpha_0 A_1 + \beta_0 A] + O(\lambda)$,

(B 9b) $\tilde{B} = \lambda^2[-q_0 B_1 + zA_1 + zB_1'' + \alpha_0 B_1 + \beta_0 B] + O(\lambda)$,

(B 9c) $\tilde{C} = \lambda^2[-p_1(B_1 - zD_1) - (q^{(0)} + 1)C_1 - 2zC_1'' + \alpha_0 C_1] + O(\lambda)$,

(B 9d) $\tilde{D} = \lambda^2[-p_0 C_1 + 2B_1'' + B_1 - 2(zD_1)' + \alpha_0 D_1 + \beta_0 D] + O(\lambda)$.

We notice that in the above expressions, only the ***known*** functions $p$ and $q$ are replaced by their asymptotic forms. The as yet unspecified functions $A, B, C, D; A_1, B_1, C_1, D_1$ are left as they are. This is important for examining the higher approximations. For the initial approximation, we may then make $\tilde{A}$, $\tilde{B}$, $\tilde{C}$, $\tilde{D}$ all of the order $\lambda$ by requiring

(B 10a) $-r_0 B_{1,0} + zA_{1,0}'' + \alpha_0 A_{1,0} - \beta_0 A_0 = 0$,

(B 10b) $-q_0 B_{1,0} + zA_{1,0} + zB_{1,0}' + \alpha_0 B_{1,0} + \beta_0 B_0 = 0$,

(B 10c) $-p_1(B_{1,0} - zD_{1,0}) - (q_0 + 1)C_{1,0} - 2zC_{1,0}' + \alpha_0 C_{1,0} = 0$,

(B 10d) $-p_0 C_{1,0} + 2B_{1,0}' + B_{1,0} - 2(zD_{1,0})' + \alpha_0 D_{1,0} + \beta_0 D_0 = 0$.

If these relations can be satisfied under the restrictions (A 15), then the coefficients $(\tilde{A}, \tilde{B}, \tilde{C}, \tilde{D})$ and hence also $(\tilde{a}, \tilde{b}, \tilde{c}, \tilde{d})$ have the behavior (1.3), as desired in the final equation (B1).

The restrictions (A 15) guarantee the validity of the above conclusions only in a small neighborhood of $z=0$. However, since all the functions involved are generally analytic, the conclusions would hold except at the singularities of the functions involved.

## Appendix C

The above calculations can also be used, with rather minor changes, for expressing the solutions $\chi$ of (B 1) in terms of the solutions of the differential equation (A 1), if we regard the latter as known. In such a case, instead of calculating $(\tilde{a}, \tilde{b}, \tilde{c}, \tilde{d})$ in terms of $(\tilde{A}, \tilde{B}, \tilde{C}, \tilde{D})$, we express the latter in terms of the former. We express the right-hand side of (B 1) in terms of $\psi$, $\psi'$, $\lambda^{-1}\psi''$, $\lambda^{-2}\psi'''$, by means of (A 3), (A 4), (A 8) and (A 10) so that the differential equation (B 1) is satisfied if

(C 1a) $$\tilde{A} = \lambda(\tilde{a}A + \tilde{b}A_1 + \tilde{c}A_2 + \tilde{d}A_3),$$

(C 1b) $$\tilde{B} = \lambda(\tilde{a}B + \tilde{b}B_1 + \tilde{c}B_2 + \tilde{d}B_3),$$

(C 1c) $$\tilde{C} = \lambda(\tilde{a}C + \tilde{b}C_1 + \tilde{c}C_2 + \tilde{d}C_3),$$

(C 1d) $$\tilde{D} = \lambda(\tilde{a}D + \tilde{b}D_1 + \tilde{c}D_2 + \tilde{d}D_3).$$

Equating the corresponding expressions in (B 6) and (C 1), we obtain a set of homogeneous equations for $A$, $B$, $C$, $D$. For convenience of reference, these are listed as follows:

(C 2a) $$zA_1' + \alpha_0 A_1 + \beta_0 A - rB_1 = -rD_1(p-z) + \lambda^{-1}\bar{A},$$

(C 2b) $$zB_1' + \alpha_0 B_1 + \beta_0 B + zA_1 - qB_1 = -qD_1(p-z) + \lambda^{-1}\bar{B},$$

(C 2c) $$(z-3p)C_1' + (\alpha_0 - q - p')C_1 + \beta_0 C = -\lambda(B_1 - pD_1)(p-z) + \lambda^{-1}\bar{C},$$

(C 2d) $$(z-3p)D_1' + (\alpha_0 - q - 2p')D_1 + \beta_0 D = -\lambda C_1(p-z) - (3B_1' + A_1) + \lambda^{-1}\bar{D},$$

in which $\bar{A}$, $\bar{B}$, $\bar{C}$, $\bar{D}$ are given by

(C 3a) $$\bar{A} = \lambda^{-1}\tilde{A} + (rC_1') + 2rC_1' + \lambda^{-1}(A' - rD_1'),$$

(C 3b) $$\bar{B} = \lambda^{-1}\tilde{B} - qC_1' + (q'-r)C_1 + \lambda^{-1}[B_2'' + 2A_2' - 2qD_1'' - (q'+r)D_1'],$$

(C 3c) $$\bar{C} = \lambda^{-1}\tilde{C} + [3B_1'' + 3A_1' - 3pD_1'' - (3p'+3q)D_1' - (p''+2q'+r)D_1] + \lambda^{-1}C_1''',$$

(C 3d) $$\bar{D} = \lambda^{-1}\tilde{D} + 3C_1'' + \lambda^{-1}D_1''',$$

where $\tilde{A}$, $\tilde{B}$, $\tilde{C}$, $\tilde{D}$ are in turn given by (C 1). We have put some of the apparently important terms on the right-hand side of (C 2), because we are interested in the special case when the coefficients in (A 1) are given by

(C 4) $$p = z, \qquad q = \alpha, \qquad r = \beta$$

where $\alpha$ and $\beta$ are given by (1.8).

## References

1. R. E. Langer, *The asymptotic solutions of ordinary linear differential equations of the second order, with special reference to a turning point*, Trans. Amer. Math. Soc. vol. 67 (1949) pp. 461–490.

2. ———, *On the asymptotic forms of the solutions of ordinary differential equations of the third order in a region containing a turning point*, Trans. Amer. Math. Soc. vol. 80 (1955) pp. 93–123.

3. ———, *The solutions of the differential equation* $v''' + \lambda^2 zv' + 3\mu\lambda^2 v = 0$, Duke Math. J. vol. 22 (1955b) pp. 525–542.

4. ———, *On the asymptotic solution of a class of ordinary differential equations of the fourth order, with special reference to an equation of hydrodynamics*, Trans. Amer. Math. Soc. vol. 84 (1957) pp. 144–191.

5. ———, *Formal solutions and a related equation for a class of fourth order differential equations of a hydrodynamic type*, MRC Technical Summary Report No. 22, to appear in Trans. Amer. Math. Soc.

6. C. C. Lin, *Theory of hydrodynamic stability*, Cambridge University Press, 1955.

7. ———, *On uniformly valid asymptotic solutions of the Orr-Sommerfeld equation*, Proceedings of the Ninth International Congress of Applied Mechanics (Brussels) vol. I (1957) pp. 136–148.

8. ———, *On the stability of the laminar boundary layer*, Symposium on Naval Hydro dynamics (National Research Council Publication 515), Washington, D. C., 1958, pp. 353–371.

9. ———, *On the instability of laminar flow and its transition to turbulence*, Proceedings of the Symposium on Boundary Layer Theory (Freiburg), Springer, 1958, pp. 144–160.

10. A. L. Rabenstein, *Asymptotic solutions of the differential equation* $u^{iv}+\lambda^2(zu''+\alpha u'+\beta u)=0$ *for large values of the parameter* $\lambda$, Arch. Rational Mech. Anal. vol. 1 (1958) pp. 418–435.

11. W. Wasow, *A study of the solutions of the differential equation* $y^{(4)}+\lambda^2(xy''+y)=0$ *for large values of* $\lambda$, Ann. of Math. vol. 52 (1950) pp. 350–361.

12. ———, *Asymptotic solution of the differential equation of hydrodynamic stability in a domain containing a transition point*, Ann. of Math. vol. 58 (1953) pp. 222–252.

13. ———, Discussion following paper by Lin, [8], ibid., pp. 363–366.

Massachusetts Institute of Technology,
Cambridge, Massachusetts
Mathematics Research Center, University of Wisconsin,
Madison, Wisconsin

# ERRATA

| Page | Eq. or Line | For | Read |
|---|---|---|---|
| 30 | 2.12 | $\chi = \ldots + \lambda^{-2} D(z,\lambda)\psi''$ | $\chi = \ldots + \lambda^{-2} D(z,\lambda)\psi'''$ |
| 31 | – 8 | Combining (2.12) . . . | Combining (2.13) . . . |
| 34 | 3.7d | $-zC_1^{(0)} + \ldots$ | $-p_1 C_1^{(0)} + \ldots$ |
| 35 | 3.13 | $\ldots - zE^{(0)} = g_2(z)$ | $\ldots + p_1 E^{(0)} = g_2(z)$ |
| 36 | 3.19a | $\hat{E}(z) = [R(z)]^{-1} z^{-1/2} \int_0^z z^{3/2} \times R(z)[h_1(z) + p_1 \hat{F}]\, dz$ | $\hat{E}(z) = [R(z)]^{-1} z^{-1/2} \int_0^z \frac{1}{4} \times z^{1/2} R(z)[2h_1(z) + p_1 g_2(0) - p_1 p_1(0) \times g_1(0) + 2(\alpha_0 - q_0) \times g_1(0)/z + 2p_1 \hat{F}]\, dz$ |
| 36 | 3.19b | $\hat{F}(z) = [R(z)]^{-1} z^{-1/2} \int_0^z z^2 \times R(z)[h_2(z) + \hat{E}]\, dz$ | $\hat{F}(z) = [R(z)]^{-1} z^{-1} \int_0^z \frac{1}{4} \times zR(z)\big[2h_2(z) + [g_2(0) - p_1(0) g_1(0)] \times (\alpha_0 - q_0)/z - 2\hat{E} p_1/z\big]\, dz$ |
| 37 | 4.6 | $\chi^{(0)''} = (A_1^{(0)} + \ldots)$ | $\chi^{(0)''} = (A_1^{(0)'} + \ldots)$ |
| 38 | 4.14 | $A^{(0)} = W_0^{-1}\{\ldots - c_{21}\chi_{\rm I}^{(0)} \times \psi_{\rm I}^{(0)} - \ldots\}$ | $A^{(0)} = W_0^{-1}\{\ldots - c_{21}\chi_{\rm I}^{(0)} \times \psi_{\rm I}^{(0)'} - \ldots\}$ |

| Page | Eq. or Line | For | Read |
|---|---|---|---|
| 38 | 4.15 | $B^{(0)} = W_0^{-1}\{\ldots\}$ | $B^{(0)} = -W_0^{-1}\{\ldots\}$ |
| 39 | − 2 | $\tilde{\chi}_2 = A^{(0)}\psi_2 + B^{(0)} \times (\tilde{\chi}_2' + k\psi_1/z)$ | $\tilde{\chi}_2 = A^{(0)}\tilde{\psi}_2 + B^{(0)} \times (\tilde{\psi}_2' + k\psi_1/z)$ |
| 40 | 4.25 | $\ldots = A^{(0)}(z^{-\rho 2}\psi_2) + \ldots$ | $\ldots = A^{(0)}(z^{-\rho 2}\tilde{\psi}_2) + \ldots$ |
| 40 | 5 | $z^{-\rho 2} = \ldots$ | $\psi_1 z^{-\rho 2} = \ldots$ |
| 41 | 11 | From (B10) | From (C2) |
| 41 | 5.6c | $\sigma_1 = \tilde{\sigma}$ | $\sigma_1 = \tilde{\sigma} + \lambda^{-1}\tilde{\tau}'$ |
| 41 | 5.7 | $\nu - z\tau = \lambda^{-1}\sigma_1$, $\sigma = \lambda^{-1}(\tau_1 + \tau')$ | $\nu - z\tau = \lambda^{-1}\sigma_1 - \lambda^{-2}\tilde{\tau}'$, $\sigma = \lambda^{-1}(\tau_1 - \tau')$ |
| 41 | 5.8d | $\ldots = O(\lambda^{-1}) - [\ldots + 3 \times (1 - 2\alpha)\tau' + \ldots]$ | $\ldots = O(\lambda^{-1}) - [\ldots + 3 \times (2 - \alpha)\tau' + \ldots]$ |
| 43 | 6.4d | $\ldots = [\ldots + 3(1 - 2\alpha_0)\tau^{(0)'}]$ | $\ldots = [\ldots + 3(2 - \alpha_0)\tau^{(0)'}]$ |
| 43 | 6.5b | $z[\ldots + 2\mu^{(0)'}] = \ldots$ | $z[\ldots + 2\mu^{(i)'}] = \ldots$ |
| 43 | 6.5d | $2z\tau_1^{(j)} + 2\tau_1^{(j)'} = R_j(z) + [\ldots + 3(1 - 2\alpha_0) \times \tau^{(j)'}]$ | $2z\tau_1^{(j)'} + 2z\tau_1^{(j)} = R_j(z) + [\ldots + 3(2 - \alpha_0) \times \tau^{(j)'}]$ |

| Page | Eq. or Line | For | Read |
|---|---|---|---|
| 43 | 6.6 | $\sigma_1^{(j)} = z^{-1/2} \int_0^z z^{1/2} \times R_j(z) dz$ | $\sigma_1^{(j)} = z^{-1/2} \int_0^z \frac{1}{2} z^{-1/2} \times R_j(z) \mathrm{d}z$ |
| 43 | 6.9 | $z[\ldots + \bar{\mu}'] = \ldots$ | $z[\ldots + 2\bar{\mu}'] = \ldots$ |
| 44 | 6.14 | $\ldots = R_{1,k} - \bar{\beta}\,\delta_{k0}$ | $\ldots = R_{1,k} + \bar{\beta}\,\delta_{k0}$ |
| 44 | 21 | (6.4). | (6.14). |
| 44 | 6.17 | $\ldots = S_k + 2\bar{\beta}\,\delta_{k0}$ | $\ldots = S_k - 2\bar{\beta}\,\delta_{k0}$ |
| 44 | – 5 | $k > 0$ | $k \geqq 0$ |
| 46 | 1 | $\ldots = R_0^{(3)} + 2\bar{\beta}$ | $\ldots = S_0 - 2\bar{\beta}$ |
| 46 | 7.1 | $R_i^{(3)} \quad i = 0, 1, \ldots, n$ | $S_i \quad i = 0, 1, \ldots, n$ |
| 47 | 8.5 | $\ldots - (\alpha - \alpha_0)$ | $\ldots + (\alpha - \alpha_0)$ |
| 47 | 8.8a | $\bar{\mu}' = z^n \left\{ c - \frac{1}{n} \int z^{-(n+1)} \times \bar{R}(z) dz \right\}$ | $\bar{\mu}' = z^n \left\{ c + \int z^{-(n+1)} \times \bar{R}(z) dz \right\}$ |
| 47 | 8.8b | $\bar{\mu} = c_1 + c_2 z^{n+1} - \int \frac{z^n}{n} dz \int \ldots$ | $\bar{\mu} = c_1 + c_2 z^{n+1} + \int z^n dz \int \ldots$ |
| 47 | – 3 | $-\bar{R}_m \left\{ (m-n)(m+1)n \right\}^{-1} \times z^{m+1}$ | $-\bar{R}_m \left\{ (m-n)(m+1) \right\}^{-1} \times z^{m+1}$ |

| Page | Eq. or Line | For | Read |
|---|---|---|---|
| 48 | 8.11a | $\bar{\tau}' = (n+2)^{-1} z^{-(n+2)} [\ldots]$ | $\bar{\tau}' = z^{-(n+2)} [\ldots]$ |
| 48 | 8.11b | $\bar{\tau}' = (n+2)^{-1} \{\ldots\}$ | $\bar{\tau} = \{\ldots\}$ |
| 48 | 16 | $\bar{R}_m \{(m+n+2)(m+1) \times (n+2)\}^{-1} z^{m+1}$ | $\bar{R}_m \{(m+n+2)(m+1)\}^{-1} \times z^{m+1}$ |
| 49 | 8 | Since $\mu^{(0)} = 1$, $\tau^{(0)} = 1, \ldots$ | Since $\mu^{(0)} = 1$, $\tau^{(0)} = 0, \ldots$ |
| 49 | 8.15 | $z\mu^{(i)''} - n\mu^{(i)'} = \beta^{(1)}$ | $z\mu^{(1)''} - n\mu^{(1)'} = \beta^{(1)}$ |
| 49 | 8.16 | $z[z\tau^{(1)''} + (h+2)\tau^{(1)'}] = \ldots$ | $z[z\tau^{(1)''} + (n+2)\tau^{(1)'}] = \ldots$ |
| 49 | 8.17 | $z\mu^{(i+1)'} - n\mu^{(i+1)} = \beta_1 \times (\mu^{(i)} - 2z\tau^{(i)'} + \ldots$ | $z\mu^{(i+1)''} - n\mu^{(i+1)'} = \beta^{(1)} \times (\mu^{(i)} + 2z\tau^{(i)'} + \ldots$ |
| 49 | 8.18 | $\alpha_1$, $\beta_1$ | $\alpha^{(1)}$, $\beta^{(1)}$ |
| 50 | 8.20 | $\tau^{(1)} = c_3 - \frac{c_2}{(n+1)(n+2)} \times z^{n+1} + \frac{2\beta^{(1)}}{n(n+2)^2} z$ | $\tau^{(1)} = c_3 - \frac{c_2}{(n+1)} z^{n+1} + \frac{2\beta^{(1)}}{n(n+2)} z$ |
| 51 | 8.24 | $u^{\text{iv}} + \lambda^{-2} (\ldots)$ | $u^{\text{iv}} + \lambda^{2} (\ldots)$ |
| 52 | A 11a | $A_3 = \ldots = (A_2' - rD_1) - \lambda r C_1$ | $A_3 = \ldots = (A_2' - rD_1') - \lambda r C_1$ |

| Page | Eq. or Line | For | Read |
|---|---|---|---|
| 52 | A 11c | $C_3 = \ldots = C_1'' + \lambda[\ldots + (B_2 - pD_2')] - \ldots$ | $C_3 = \ldots = C_1'' + \lambda[\ldots + (B_2 - pD_1')] - \ldots$ |
| 54 | B 4a | $A_4 = (A_1' - rD_1')' - \ldots$ | $A_4 = (A_2' - rD_1')' - \ldots$ |
| 54 | B 4b | $B_4 = \ldots - \lambda[(qC_1)' - rC_1 - 2qC_1'] - \ldots$ | $B_4 = \ldots - \lambda[(qC_1)' + rC_1 + 2qC_1'] - \ldots$ |
| 55 | B 9d | $\tilde{D} = \lambda^2[-p_0C_1 + 2B_1' + B_1 - \ldots] + O(\lambda)$ | $\tilde{D} = \lambda^2[-p_1C_1 + 2B_1' + B_2 - \ldots] + O(\lambda)$ |
| 55 | B 10a | $\ldots - \beta_0 A_0 = 0$ | $\ldots + \beta_0 A^{(0)} = 0$ |
| 55 | B 10d | $-p_0C_{1,0} + 2B_{1,0}' + B_{1,0} - \ldots = 0$ | $-p_1C_1^{(0)} + 2B_1^{(0)'} + B_2^{(0)} - \ldots = 0$ |
| 55 | C 1a | $\tilde{A} = \lambda(\tilde{a}A + \tilde{b}A_1 + \tilde{c}A_2 + \tilde{d}A_3)$ | $\tilde{A} = \lambda(\tilde{a}A + \tilde{b}A_1 + \lambda^{-1}\tilde{c}A_2 + \lambda^{-2}\tilde{d}A_3)$ |
| 55 | C 1b | $\tilde{B} = \lambda(\tilde{a}B + \tilde{b}B_1 + \tilde{c}B_2 + dB_3)$ | $\tilde{B} = \lambda(\tilde{a}B + \tilde{b}B_1 + \lambda^{-1}\tilde{c}B_2 + \lambda^{-2}\tilde{d}B_3)$ |
| 55 | C 1c | $\tilde{C} = \lambda(\tilde{a}C + \tilde{b}C_1 + \tilde{c}C_2 + \tilde{d}C_3)$ | $\tilde{C} = \lambda(\tilde{a}C + \tilde{b}C_1 + \lambda^{-1}\tilde{c}C_2 + \lambda^{-2}\tilde{d}C_3)$ |
| 55 | C 1d | $\tilde{D} = \lambda(\tilde{a}D + \tilde{b}D_1 + \tilde{c}D_2 + \tilde{d}D_3)$ | $\tilde{D} = \lambda(\tilde{a}D + \tilde{b}D_1 + \lambda^{-1}\tilde{c}D_2 + \lambda^{-2}\tilde{d}D_3)$ |

| Page | Eq. or Line | For | Read |
|---|---|---|---|
| 56 | C 2c | $\ldots = -\lambda(\ldots)$ | $\ldots = +\lambda(\ldots)$ |
| 56 | C 2d | $\ldots = -\lambda C_1(\ldots)$ | $\ldots = +\lambda C_1(\ldots)$ |
| 56 | C 3a | $\bar{A} = \lambda^{-1}\tilde{A} + (rC_1') + 2rC_1'$ $+\lambda^{-1}(A' - rD_1')$ | $\bar{A} = \lambda^{-1}\tilde{A} + (rC_1)' + 2rC_1'$ $-\lambda^{-1}[(A_2' - rD_1')' - rD_1'']$ |
| 56 | C 3b | $\bar{B} = \lambda^{-1}\tilde{B} - qC_1' + (q' - r)C_1$ $+\lambda^{-1}[\ldots]$ | $\bar{B} = \lambda^{-1}\tilde{B} + (qC_1)' + 2qC_1'$ $+rC_1 - \lambda^{-1}[\ldots]$ |
| 56 | C 3c | $\bar{C} = \lambda^{-1}\tilde{C} + [\ldots] + \lambda^{-1}C_1'''$ | $\bar{C} = \lambda^{-1}\tilde{C} - [\ldots] - \lambda^{-1}C_1'''$ |
| 56 | C 3d | $\bar{D} = \lambda^{-1}\tilde{D} + 3C_1'' + \lambda^{-1}D_1'''$ | $\bar{D} = \lambda^{-1}\tilde{D} - 3C_1'' - \lambda^{-1}D_1'''$ |

In addition, the symbols $A_0$ and $A_{1,0}$ are used through the paper in place of $A^{(0)}$ and $A_1^{(0)}$. This is true also for $B$, $C$, and $D$.

Estratto da *Rendiconti della Scuola Internazionale di Fisica «E. Fermi»* - XXI Corso

# Hydrodynamics of Helium II.

C. C. LIN
*Massachusetts Institute of Technology - Cambridge, Mass.*

## Introduction.

This set of notes was originally prepared for use of the Enrico Fermi International School of Physics held at Varenna, Italy, July 3-5, 1961. It has been slightly revised for purposes of publication. However, it was felt desirable to preserve the informal nature of the lecture notes, especially since there is necessarily a great deal of speculation on the subject of superfluid helium at the present time.

I am indebted to several of my friends for their help and encouragement in the work that forms the basis of these lectures; especially to C. N. YANG for initiating my interest in this subject and for many helpful discussions during my stay at the Institute for Advanced Study, Princeton, for which I wish to thank Professor J. R. OPPENHEIMER for his hospitality. The work was partly supported by the Office of Naval Research.

When Professor CARERI asked me to give a few lectures at this Course, he assigned two duties to me. The first is to present a brief up-to-date account of our knowledge of hydrodynamics of ordinary fluids, especially those aspects that might relate to the current studies of superfluid helium. The second is to discuss some of my thoughts on the subject itself. This dual assignment is a fortunate one, because I cannot make my points clear in the discussion of the hydrodynamics of superfluids without first reviewing some known results in the ordinary case. I shall attempt to do this in the six hours allotted to me with perhaps two hours devoted to the first subject and four hours devoted to the latter. The topics of my discussions are as follows:

I) Hydrodynamics of ordinary fluids – reversible processes.

II) Hydrodynamics of ordinary fluids – irreversible processes.

III) Hydrodynamics of superfluids – reversible processes.

IV) Hydrodynamics of superfluids – irreversible processes.

V) Theoretical basis for the hydrodynamics of superfluid helium.

VI) Experimental evidence for quantization of superfluid flow.

In the last two chapters, I shall attempt to explain why I feel that the concept of quantized vortex lines with *macroscopic* lengths lacks a satisfactory foundation from the theoretical point of view. Questions will also be raised about the experimental evidence for their existence.

Before I begin to discuss the details, I should perhaps make some general points clear about my line of thinking. I wish to elaborate on a point which is obvious in classical hydrodynamics but wich becomes blurred in the study of superfluid helium. This is the possibility of studying macroscopic and microscopic aspects of a problem almost independently of each other, as is usually done in the hydrodynamics of ordinary fluids. To be sure, it would be very satisfying to develop the macroscopic theory from the molecular theory. However, this has not yet been completed even in the case of ordinary liquids. On the other hand, there is no question that the macroscopic hydrodynamics of water is well developed. Indeed, the same theory applies to air, which has an entirely different molecular structure. Thus, the advantage of a macroscopic theory is to avoid the difficulties of the complicated molecular structure and yet at the same time provide a framework for the description of the macroscopic phenomena.

Such a sharp distinction is questionable in the case of superfluid helium, because we are specifically concerned with *macroscopic quantum-mechanical effects*. However, this does not mean that the development of a continuum theory must be handicapped. Rather, the attempt should be made to decide which experimental observations can be described by a unified macroscopic theory and which *must* be described in terms of some molecular model. Clearly, experiments performed with beams of neutrons and ions are usually directly concerned with the internal molecular structure of the liquid, and belong to the second category. Others, in which only macroscopic effects are observed, can hopefully be entirely described by continuum theories. Indeed, the most prominent and basic phenomenon in superfluid helium is the reversible heat flux, which is directly related to the phenomenon of Bose-Einstein condensation—a decidedly quantum-mechanical effect. Yet its macroscopic effect can be handled entirely on the basis of continuum concepts (see Part III). The mechanism for its origin is of course an entirely different matter. If one would then take an optimistic point of view, one might hope that, even in the case of superfluid helium, there is the possibility of separately developing a macroscopic theory of fluid motion and a microscopic theory of molecular structure.

# I. - Hydrodynamics of Ordinary Fluids. Reversible Processes.

## 1. – Fundamental equations. Variational principle.

The conventional hydrodynamic theory is based on fields of macroscopic quantities. The thermodynamic state of the fluid is characterized by the mass density $\varrho$ and the entropy per unit mass $s$. A physical description of the system at any instant of time $t$ is given by the specification, at every point $P(\boldsymbol{x})$ of these quantities together with the macroscopic velocity $\boldsymbol{v}$: $\varrho(\boldsymbol{x}, t)$, $s(\boldsymbol{x}, t)$, and $\boldsymbol{v}(x, t)$. For reversible processes, the conservation of mass and energy can be expressed by the following equations:

$$\frac{\partial \varrho}{\partial t} + \operatorname{div}(\varrho \boldsymbol{v}) = 0\,, \tag{I.1.2}$$

$$\frac{\mathrm{D}s}{\mathrm{D}t} = 0\,, \tag{I.1.2}$$

where $\mathrm{D}/\mathrm{D}t$ is the substantial derivative $\partial/\partial t + (\boldsymbol{v}\cdot\nabla)$. The equations of motion can be derived by means of the variational principle (*) (Hamilton's principle)

$$\delta \iint L(\boldsymbol{v}, \varrho, s)\,\mathrm{d}\tau\,\mathrm{d}t = 0, \tag{I.1.3}$$

with (I.1.1) and (I.1.2) as subsidiary conditions. In (I.1.3), the Lagrangian density function is

$$L = \tfrac{1}{2}\varrho \boldsymbol{v}^2 - \varrho(e + \Omega)\,, \tag{I.1.4}$$

where $e = e(\varrho, s)$ is the internal energy, and $\Omega$ is the potential of the external force. However, in this formulation, the Euler equation associated with $\delta\boldsymbol{v}$ gives

$$\boldsymbol{v} = \operatorname{grad}\varphi + \beta \operatorname{grad} s \tag{I.1.5}$$

and restricts the flow to be irrotational, when the entropy is uniform. To remove this difficulty, we introduce the condition for the conservation of the identity of particles

$$\frac{\mathrm{D}\boldsymbol{X}}{\mathrm{D}t} = 0\,. \tag{I.1.6}$$

(*) For further details, see the discussion of this writer as quoted by SERRIN [1], p. 148 ff.

Then (I.1.5) becomes (*)

$$v = \operatorname{grad} \varphi + \beta \operatorname{grad} s + \operatorname{grad} \boldsymbol{X} \cdot \boldsymbol{\gamma} , \tag{I.1.5a}$$

where $\varphi$, $\beta$, $\boldsymbol{\gamma}$ are Lagrangian multipliers associated with (I.1.1), (I.1.2) and (I.1.6) respectively. It is possible to show that (I.1.3) is equivalent to the set of equations of motion

$$\varrho \frac{\mathrm{D}\boldsymbol{v}}{\mathrm{D}t} = - \operatorname{grad} p , \tag{I.1.7}$$

provided (I.1.1), (I.1.2) and (I.1.6) are assumed. In (I.1.7), the pressure $p$ is defined by the thermodynamic identity

$$T\,\mathrm{d}s = \mathrm{d}e + p\,\mathrm{d}(1/\varrho) , \tag{I.1.8}$$

where $T$ is the absolute temperature.

*Potential motions.* – An important theorem in the hydrodynamics of perfect fluids is the following: a motion will remain irrotational if it is so at a given instant and if the entropy is uniform. Singularities in the form of vortex lines and vortex rings are however permitted in this theory. As we shall note in Chapter II, such singularities cannot exist when irreversible processes are considered.

## 2. – Rotating co-ordinates.

It is well known in particle dynamics that, in terms of quantities relative to a rotating system, with a prescribed angular velocity $\boldsymbol{\omega}$, the acceleration of a particle is given by

$$\boldsymbol{a}_0 = \frac{\mathrm{d}\boldsymbol{v}}{\mathrm{d}t} + 2\,[\boldsymbol{\omega} \times \boldsymbol{v}] + \boldsymbol{\omega} \times [\boldsymbol{\omega} \times \boldsymbol{r}] . \tag{I.2.1}$$

The same principle finds its application in hydrodynamics. It is interesting to note that the centrifugal term can be expressed in terms of a gradient. Specifically, if we take the $z$-axis to coincide with the axis of $\boldsymbol{\omega}$, then

$$\boldsymbol{\omega} \times [\boldsymbol{\omega} \times \boldsymbol{r}] = -\omega^2 \times \boldsymbol{i} - \omega^2 y \boldsymbol{j} = - \nabla \left[ \frac{\omega^2 r^2}{2} - \frac{(\boldsymbol{\omega} \cdot \boldsymbol{r})^2}{2} \right] , \tag{I.2.2}$$

(*) In eq. (I.1.5a), grad $\boldsymbol{X} \cdot \boldsymbol{\gamma}$ has components $(\partial X_k / \partial x_i) \gamma_k$.

and this must hold in general. The Coriolis term is expected to exhibit other interesting properties.

One of the first attempts to study the effect of rotation was made by LORD KELVIN (*) in 1879. He showed that the shallow water waves of small amplitudes are now governed by equations

$$\frac{\partial u}{\partial t} - 2\omega v = -g\frac{\partial \zeta}{\partial x} - \frac{\partial \Omega}{\partial x}, \tag{I.2.3}$$

$$\frac{\partial v}{\partial t} + 2\omega u = -g\frac{\partial \zeta}{\partial y} - \frac{\partial \Omega}{\partial y}, \tag{I.2.4}$$

where $\zeta$ is the elevation of water above the equilibrium height

$$Z_0 = \frac{1}{2}\frac{\omega^2}{g}(x^2 + y^2), \tag{I.2.5}$$

and $\Omega$ is the potential of some disturbing force (*e.g.*, the attraction of the moon). The equation of continuity is

$$\frac{\partial \zeta}{\partial t} + \frac{\partial}{\partial x}(uh) + \frac{\partial}{\partial y}(vh) = 0, \tag{I.2.6}$$

where $h$ denotes the depth from the free surface to the bottom, in the undisturbed condition. If the bottom has the same form as (I.2.5), the depth is uniform. In this case, with $\Omega$ absent, we have a set of equations with constant coefficients. If the solutions have periodic dependence $\exp[i\sigma t]$, then

$$\left\{\begin{aligned} u &= \frac{g}{\sigma^2 - 4\omega^2}\left(i\sigma\frac{\partial}{\partial x} + 2\omega\frac{\partial}{\partial y}\right)\zeta, \\ v &= \frac{g}{\sigma^2 - 4\omega^2}\left(i\sigma\frac{\partial}{\partial y} - 2\omega\frac{\partial}{\partial x}\right)\zeta, \\ &\frac{\partial^2\zeta}{\partial x^2} + \frac{\partial^2\zeta}{\partial y^2} + \frac{\sigma^2 - 4\omega^2}{gh}\zeta = 0. \end{aligned}\right. \tag{I.2.7}$$

It is clear that there is «resonance» at $\sigma = 2\omega$, and that the character of the solution changes as $\sigma$ goes through $2\omega$. This phenomenon will be seen to be relevant to experiments on rotating helium (see Section **3** of Part IV).

---

(*) Sir W. THOMSON: *On gravitational oscillations of rotating water* [2]. See LAMB [3], p. 317.

## 3. – Momentum and impulse.

We include here a discussion of the momentum of irrotational fluid motions because the question frequently comes up in connection with the motion of a « quantized vortex ring ».

Let us first observe that for any kind of continuous motion in a long cylindrical tube, the total momentum of the fluid motion along the tube is zero, provided there is no net flux. This is a simple consequence of the equation of continuity. If the cylinder is along the $x$-axis, then

$$\iint u \, \mathrm{d}y \, \mathrm{d}z = \text{const} = 0 ,$$

and consequently

$$M_x = \iiint \varrho u \, \mathrm{d}x \, \mathrm{d}y \, \mathrm{d}z = 0 .$$

The same is true if the motion is caused by the translation of a sphere along the axis. Indeed, the symmetry of the pattern of the flow would verify this conclusion (there is no question here of the proper convergence of the integrals, as there would be in the case of infinite space). However, LORD KELVIN has introduced the concept of impulse, which behaves in many ways like the momentum of the fluid. It must be remembered, however, that the momentum is strictly zero. The impulse is that required of the external forces to produce the motion, starting from a state of rest. The momentum created by this external force is actually not retained in the fluid but propagated to infinity. With this subtlety involved in the concept of the impulse, one must exercise care in applying it to complex hydrodynamic situations. In the Appendix to this chapter, we discuss the problem in some detail with the intention that it might help the user in forming his judgment.

## 4. – Bound vortices and free vortices.

The prevailing theory of quantized vortex lines in the theory of superfluid helium will be discussed in Parts V and VI. It might be well to recall some aspects of the classical theory of the motion of rectilinear vortices here.

KIRCHHOFF has shown that the motion of a system of isolated rectilinear vortices with circulations $\varkappa_1, \varkappa_2, \ldots, \varkappa_N$ and located at $(x_1, y_1), (x_2, y_2), \ldots, (x_N, y_N)$

respectively is governed by the set of equations

$$\varkappa_{(i)} \frac{\mathrm{d}x_i}{\mathrm{d}t} = + \frac{\partial W}{\partial y_i}, \tag{I.4.1}$$

$$\varkappa_{(i)} \frac{\mathrm{d}y_i}{\mathrm{d}t} = - \frac{\partial W}{\partial x_i}; \qquad i = 1, 2, \ldots, N, \tag{I.4.2}$$

where

$$W = -\frac{1}{2\pi} \sum_{i \neq j} \varkappa_i \varkappa_j \log r_{ij} . \tag{I.4.3}$$

A generalization of this theory to vortices placed in arbitrary cylindrical containers has been developed by LIN [4, 5]. The same formulae (I.4.1) and (I.4.2) are found to hold, but the expression for (I.4.3) is more complicated.

In this theory, the vortices are « *free* » in the sense that they move freely under the influence of *other* vortices with no forces of reaction involved in the process. The vortex line is simply a singularity of the fluid motion and it moves with the velocity of the fluid at the vortex, with the circulatory motion due to the vortex itself substracted. In contrast to this, if we have a fine wire placed in the location of a vortex so as to remove the singularity of the fluid motion, we obtain an irrotational motion with circulation around the wire. The vortex is now « *bound* » to the wire and may be held fixed. In this case, there is a Magnus effect with a cross-wind force given by the Joukowsky formula

$$\boldsymbol{f} = \varrho \boldsymbol{V}' \times \boldsymbol{\varkappa} \tag{I.4.4}$$

per unit length of the cylinder, where $\boldsymbol{V}'$ is the fluid velocity past the wire. Indeed, it is the velocity at which the vortex *would* be moving, had it been *free*.

The significance of this distinction will become clear when we discuss the possible effect of quantized vortex lines in Part VI.

## APPENDIX

### Momentum and impulse.

One of the subtle concepts in classical hydrodynamic theory is the impulse of the fluid motion introduced by Lord KELVIN. We include a brief discussion of this here, since it is often needed in connection with the motion of vortex rings.

The treatment given by LAMB [3], (Sections 119, 152) should be consulted, but we wish to bring out some points not clearly discussed by him. In particular, we wish to bring into discussion any possible effect of an enclosure such as a straight cylindrical tube.

To begin with, let us consider the classical problem of the motion of a sphere described by the velocity potential

$$\Phi = -\frac{Ua^3}{2r^2}\cos\theta\,, \tag{I.A.1}$$

where the conventional notation is used. The $x$-component of the velocity is

$$\Phi_x = -\frac{Ua^3}{2r^3}(1-3\cos^2\theta)\,,$$

and it is clear that the momentum of the fluid in any concentric spherical shell is zero; for

$$\int_0^\pi (1-3\cos^2\theta)\sin\theta\,d\theta = 0\,.$$

However, we also know that to accelerate the sphere instantaneously to the velocity $U$, an impulse of the amount

$$\int F_x\,dt = \frac{1}{2}\left(\frac{4\pi}{3}\varrho a^3\right)U\,, \tag{I.A.2}$$

is needed. What happened to the momentum produced by this force?

The answer to this question is not difficult to provide, if we examine the total momentum transfer through each spherical surface. Take an arbitrary surface of radius $b$. The rate of transfer is composed of two parts: (*a*) the momentum carried by mass flux; (*b*) the effect of the pressure force. It is easy to see that (*a*) gives a zero net effect by fore-aft symmetry. The effect of (*b*) is

$$F_x = \int_0^\pi p\cos\theta\cdot 2\pi b^2\sin\theta\,d\theta\,.$$

The total transfer of momentum due to an impulsive action is

$$\int F_x\,dt = \int_0^\pi \{p\,dt\}\cdot 2\pi b^2\sin\theta\cos\theta\,d\theta\,.$$

Now, by Bernoulli's equation

$$\frac{p}{\varrho} = -\frac{\partial\Phi}{\partial t} - \frac{1}{2}q^2 + F(t)\,.$$

Hence, the impulse of pressure is

$$\tilde{\omega} = \int p\,\mathrm{d}t = -\varrho\Phi\,. \tag{I.A.3}$$

By using (I.A.1), this immediately gives (I.A.2). Thus, *the impulse produced by the external force is immediately transmitted through each spherical surface to infinity; no momentum is retained by any spherical region of the fluid.*

In general, if we associate $\Phi$ with the impulsive pressure, the momentum of fluid enclosed between an inner surface $S_i$ and an outer surface $S_o$ is given by

$$\varrho\int \nabla\Phi\,\mathrm{d}\tau = \int_{S_i} \tilde{\omega}\boldsymbol{n}\,\mathrm{d}S - \int_{S_o} \tilde{\omega}\boldsymbol{n}\,\mathrm{d}S\,, \tag{I.A.4}$$

stating that the momentum generated by the pressure impulse at the inner surface $S_i$ is partly transmitted through the outer surface $S_o$ and partly retained in the fluid in between.

We shall now turn to the discussion of the momentum and impulse of a fluid motion involving a vortex system of finite dimensions in an incompressible fluid which fills an infinitely long cylindrical tube (Fig. 1). We assume that the vorticity lines close upon themselves and do not terminate on the wall. We shall show that the method of Kelvin (LAMB [3], p. 214) can be easily adapted to this case.

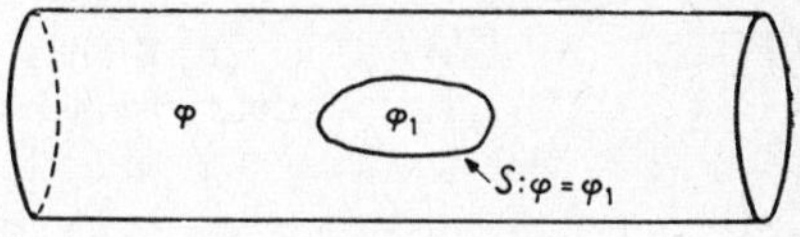

Fig. 1. – Diagram showing two regions of potential flow.

First of all, we notice that *the total x-momentum must be zero.* For

$$\varrho\iiint u\,\mathrm{d}x\,\mathrm{d}y\,\mathrm{d}z = \varrho\int \mathrm{d}x \iint u\,\mathrm{d}y\,\mathrm{d}z\,,$$

and, by continuity

$$\iint u\,\mathrm{d}y\,\mathrm{d}z = \text{const}\,. \tag{I.A.5}$$

If there are no sources, as it is the case in the present problem, then this constant value of flux must be zero, by examining the conditions at infinity. However, the impulse that is required to create the motion from rest is not necessarily zero. Indeed, if we consider the surface $S$ enclosing the vortex system in a simply connected region and denote by $\varphi$ the single-valued velocity potential which obtains outside of $S$, then the impulse required is given by

$$I_x = (\tilde{\omega}_2 - \tilde{\omega}_1)A\,, \tag{I.A.6}$$

where $A$ is the section area and

$$\tilde{\omega}_2 = \lim_{x \to +\infty} (-\varrho\Phi) , \qquad \tilde{\omega}_1 = \lim_{x \to -\infty} (-\varrho\Phi) .$$

for the right-hand side of (6) represents the impulse transmitted to infinity.

If we adopt the method described in LAMB [3] (pp. 214-215), we find that (as the following derivation shows)

$$I_x = \frac{1}{2}\varrho \iiint (y\zeta - z\eta)\,\mathrm{d}\tau , \tag{I.A.7}$$

where $(\xi, \eta, \zeta)$ denotes the components of the vorticity (curl of velocity). It is noteworthy that this formula is *independent* of the presence of the solid boundary, as it is derived by calculations with $S$ alone (*). For a vortex ring symmetrically placed with respect to the tube, we find that

$$I_x = \varkappa \varrho A_R , \tag{I.A.8}$$

where $\varkappa$ is the circulation around the filament, and $A_R$ is the area of the vortex ring (**).

The formula (I.A.8) can be checked against (I.A.6). If we consider the case where the vortex ring nearly touches the tube, the flow may be described by placing an image ring outside of the tube. The flow speed at the axis would be effectively zero and the circulation is

$$\varkappa = \oint \boldsymbol{q}\cdot\mathrm{d}\boldsymbol{s} = \int_{+\infty}^{-\infty} \boldsymbol{q}\cdot\mathrm{d}\boldsymbol{s} , \quad \text{along the } \textit{wall}.$$

Thus, (cf. Fig. 1)

$$\varrho\varkappa = \varrho[\Phi(-\infty) - \Phi(+\infty)] = \tilde{\omega}_2 - \tilde{\omega}_1 .$$

The derivation of (I.A.7) runs as follows: consider the continuation of the potential $\varphi$ *into* the region $S$, not analytically but by a *regular* harmonic function *continuous* with $\varphi$ on $S$. Denote this function by $\varphi_1$.

We now consider a motion produced by the impulsive force $J_i$ per unit mass such that

$$\begin{cases} J_i = u_i - \dfrac{\partial \Phi_1}{\partial x_i} , & \text{inside } S, \\ J_i = 0 , & \text{outside } S, \end{cases} \tag{I.A.9}$$

(*) On the other hand, if we were dealing with the motion of a sphere through a tube, the result would not be the same as (I.A.2). For the « vortex sheet » created at the surface of the sphere would be altered.

(**) Cf. the equivalent magnetic shell of a loop carrying an electric current.

Then, the original flow would be reproduced. For the original flow is compatible with (I.A.9), and the solution should be unique. On the surface $S$ (approaching from inside),

$$J_i = u_i - \frac{\partial \Phi_1}{\partial x_i} = \frac{\partial \Phi}{\partial x_i} - \frac{\partial \Phi_1}{\partial x_i} = J n_i\,,$$

since $\Phi - \Phi_1 = 0$ or $S$.

Consider now the integral

$$K_i = \int \varepsilon_{ijk} x_j \omega_k \,\mathrm{d}\tau\,, \tag{I.A.10}$$

over the volume enclosed by $S$. Then, since

$$\omega_k = \varepsilon_{kpq} \frac{\partial u_q}{\partial x_p} = \varepsilon_{kpq} \frac{\partial J_q}{\partial x_p}\,,$$

we have

$$K_i = \varepsilon_{kij}\, \varepsilon_{kpq} \int x_i \frac{\partial J_q}{\partial x_p} \,\mathrm{d}\tau = \varepsilon_{kij}\, \varepsilon_{kpq} \int \left[\frac{\partial}{\partial x_p}(x_j J_q) - J_q \delta_{jp}\right] \mathrm{d}\tau =$$

$$= \varepsilon_{kij}\, \varepsilon_{kpq} \int x_j J_q n_p \,\mathrm{d}S - (\delta_{ip}\delta_{jq} - \delta_{iq}\delta_{jp})\delta_{jp} \int J_q \,\mathrm{d}\tau\,.$$

Since $J_q = J n_q$, the first term vanishes. Thus,

$$K_i = 2 \int J_i \,\mathrm{d}\tau\,, \tag{I.A.11}$$

which is indeed (I.A.7).

To summarize, *for motions associated with vortices in a long tube, the momentum is always identically zero. But to create such a motion from rest, external forces must be applied to supply a certain amount of momentum to be transmitted to infinity during the transition to the final state from the state of rest.*

## II. - Hydrodynamics of Ordinary Fluids. Irreversible Processes.

### 1. – Navier-Stokes equations.

The principal irreversible processes that occur in fluid motion are *a*) internal friction and *b*) thermal conduction. In the case of homogeneous incompressible fluids, (satisfying an additional restriction to be specified later), the study of fluid motion can be made independently of the equation for thermal energy. For sake of brevity, we shall limit our discussion to such cases. It is important to remember that the internal friction will dissipate mechanical energy into heat, so that the system is not conservative.

Internal friction in many fluids obeys Newton's law of friction

$$\tau = \eta \frac{\mathrm{d}U}{\mathrm{d}y}, \tag{II.1.1}$$

where $\eta$ is the viscosity coefficient, and the case under discussion is one in which the flow velocity is only in the $x$-direction and depends on $y$ only. In general, the velocity gradient can be decomposed in terms of a rate of strain and a rotation:

$$\frac{\partial v_i}{\partial x_j} = e_{ij} + \omega_{ij}, \tag{II.1.2}$$

where

$$e_{ij} = \frac{1}{2}\left(\frac{\partial v_i}{\partial x_j} + \frac{\partial v_j}{\partial x_i}\right) \tag{II.1.3}$$

and

$$\omega_{ij} = \frac{1}{2}\left(\frac{\partial v_i}{\partial x_j} - \frac{\partial v_j}{\partial x_i}\right). \tag{II.1.4}$$

The stress tensor associated with internal friction is *assumed*, as a generalization of (II.1.1), as dependent *linearly* on $e_{ij}$. By a purely mathematical reasoning we can then show that the components of stress tensor are given by

$$\tau_{ij} = 2\eta e_{ij} \tag{II.1.5}$$

for an isotropic incompressible fluid ($e_{kk} = 0$).

The equations of motion for a viscous fluid are then obtained from those for a perfect fluid by the inclusion of the body force (per unit volume)

$$\frac{\partial \tau_{ij}}{\partial x_j} = \eta \Delta v_i. \tag{II.1.6}$$

Here, we have again used the equation of continuity for an incompressible fluid:

$$\frac{\partial v_k}{\partial x_k} = 0. \tag{II.1.7}$$

This equation, together with the Navier-Stokes equation

$$\frac{\mathrm{D}v_i}{\mathrm{D}t} = -\frac{1}{\varrho}\frac{\partial p}{\partial x_i} + \nu \nabla^2 v_i, \qquad \nu = \eta/\varrho, \tag{II.1.8}$$

constitute the basic equations for the four dynamical variables $v_i$ ($i=1, 2, 3$) and $p$. The coefficient of viscosity $\eta$ has been assumed to be a constant (*). In vector notation, we may write (II.1.7) and (II.1.8) in the form

$$\nabla\cdot\boldsymbol{v} = 0\,, \tag{II.1.7$'$}$$

and

$$\frac{\mathrm{D}\boldsymbol{v}}{\mathrm{D}t} = -\frac{1}{\varrho}\nabla p + \nu\nabla^2\boldsymbol{v}\,. \tag{II.1.8$'$}$$

An alternative form of (8′) is

$$\frac{\mathrm{D}\boldsymbol{v}}{\mathrm{D}t} = -\frac{1}{\varrho}\nabla p + \nu\nabla\times[\nabla\times\boldsymbol{v}]\,. \tag{II.1.9}$$

In this form, it can be seen that *the body forces due to viscous friction vanish if the motion is irrotational.* This has often led to the erroneous conclusion that *viscous dissipation is absent in irrotational motion.* We shall explain why this is incorrect in Section **3**.

*The vorticity equation.* – If we take half of the curl of (II. 1.9), we obtain the vorticity equation

$$\frac{\mathrm{D}\boldsymbol{\omega}}{\mathrm{D}t} = (\boldsymbol{\omega}\cdot\nabla)\boldsymbol{v} + \nu\nabla^2\boldsymbol{\omega}\,. \tag{II.1.10}$$

The first term on the right represents the generation of vorticity by stretching of the vortex lines, while the second term represents the diffusion of vorticity. In two-dimensional motions, only the second term is effective. Thus, a two-dimensional vortex gradually diffuses. *The idealized vortex motion consisting of a potential motion with circulation around a singularity cannot exist in a viscous fluid.*

## 2. – Boundary conditions.

In the case of perfect fluids, the condition at a solid boundary is that the fluid cannot penetrate the wall. Thus, if $V_i$ is the velocity of the wall, the relative velocity

$$w_i = v_i - V_i \tag{II.2.1}$$

(*) Otherwise, temperature distribution becomes important in the determination of the flow field.

must not have a normal component:

$$w_i n_i = 0 . \tag{II.2.2}$$

This condition must continue to hold in the present case. However, additional boundary conditions must now hold for the tangential components of velocity

$$w_\alpha = w_i t^i_\alpha \qquad (\alpha = 1, 2), \tag{II.2.3}$$

where $t^i_\alpha$ is the tangent vector (with two components $\alpha = 1, 2$ in the tangent plane). Mathematically, this is dictated by the fact that the equations of motion are of higher orders in the spatial derivatives. Physically, this is suggested by the fact that a spinning cylinder can bring the air around it into motion. Clearly, the viscous stresses at the boundary must enter the formulation of the boundary conditions as well. This has components

$$T_\alpha = (\tau_{ij} n_j) t^i_\alpha . \tag{II.2.4}$$

From the invariance of physical laws with respect to co-ordinate transformations, we should then have

$$T_\alpha = F(w^2) w_\alpha , \qquad w^2 = w_\alpha w^\alpha . \tag{II.2.5}$$

The simplest assumption is that $F =$ constant. This was the condition suggested by NAVIER, who actually derived it from a certain molecular model (1816, 1823), and used it to account for Girard's experiment for flow of liquid through tubes (1813, 1814, 1815). This slip layer has a thickness $\delta = 0(\eta/F)$. Stokes was inclined to the condition that the fluid should stick to the boundary, *i.e.*,

$$w_\alpha = 0 , \tag{II.2.6}$$

which is a special case of (5) when $F \to \infty$. However, the experiments known to Stokes could not bear (II.2.6) out, and he hesitated (1846) between this hypothesis and Navier's, even though the experimental results of HAGAN (1839) and POISEUILLE (1841) were already available at the time. A few years later (1851), STOKES came to the conclusion that the agreement with observation of results obtained on the assumption of no slip was highly satisfactory; and the hypothesis gradually gained ground. Even then, HELMHOLTZ concluded (1860) that there was a considerable slip for water in contact with a gilt surface, though not with glass.

MAXWELL (1879) concluded from kinetic theory of gases, that the slip at the surface was given by Navier's formula, but the thickness of the slip layer $\delta = \eta/F$ is of the order of a few mean free-paths. Thus, we finally have the modern version of our knowledge, that the slip is negligible in ordinary cases, but must be considered under exceptional circumstances. For example, in the case of rarefied gases, the Navier-Stokes equations still hold in the « slip-flow » range, but the boundary conditions take on the form given by NAVIER rather than that given by STOKES. The period of this investigation, which consists of both theoretical and experimental parts, lasted over quite a long period (1816-1879).

The above historical account is based on the article by S. GOLDSTEIN ([6], II, pp. 676-680). The reader is referred to the original article for further details and the exact references to the literature.

## 3. – The rate of energy dissipation.

The viscous stresses represent irreversible processes which dissipate mechanical energy into heat. The amount of dissipation may be calculated as follows: Consider any surface enclosing a given mass of fluid. The viscous surface stress $\tau_{ij}n_j\,\mathrm{d}S$ is doing work at the rate of $u_i\tau_{ij}n_j\,\mathrm{d}S$ over the surface element $\mathrm{d}S$. For the whole surface, the rate of work done is

$$\text{(II.3.1)}\qquad \int u_i\tau_{ij}n_j\,\mathrm{d}S = \int \frac{\partial}{\partial x_j}(u_i\tau_{ij}))\,\mathrm{d}V,$$

showing that the rate of work done per unit volume is

$$\text{(II.3.2)}\qquad \frac{\partial}{\partial x_j}(u_i\tau_{ij}) = u_i\frac{\partial\tau_{ij}}{\partial x_j} + \frac{\partial u_i}{\partial x_j}\tau_{ij} = u_i\frac{\partial\tau_{ij}}{\partial x_j} + \tau_{ij}e_{ij}\,.$$

The first term (cf. (II.1.6) and (II.1.8)) goes to the increase of the kinetic energy while the second term gives the rate of dissipation per unit volume per second:

$$\text{(II.3.3)}\qquad D = \tau_{ij}e_{ij} = 2\eta e_{ij}e_{ij} = 2\eta\left(\omega_{ij}\omega_{ij} + \frac{\partial u_i}{\partial x_j}\frac{\partial u_j}{\partial x_i}\right).$$

In the case of irrotational motions $\omega_{ij} = 0$, we have

$$\text{(II.3.4)}\qquad D = 2\eta\frac{\partial u_i}{\partial x_j}\frac{\partial u_j}{\partial x_i} = 2\eta\left(\frac{\partial u_i}{\partial x_j}\right)^2.$$

Thus, although there is no net body force from viscous friction in the case of irrotational motions, the rate of energy dissipation is not zero.

As an example, consider the flow between rotating cylinders. If the inner cylinder (radius $R_1$) is rotating at an angular speed $\Omega_1$ and the outer cylinder (radius $R_2$) has an angular speed $\Omega_2$, the solution is given by

$$v = Ar + B/r\,, \tag{II.3.5}$$

where

$$\begin{cases} A = (\Omega_2 R_2^2 - \Omega_1 R_1^2)/(R_2^2 - R_1^2)\,, \\ B = -(\Omega_2 - \Omega_1) R_2^2 R_1^2/(R_2^2 - R_1^2)\,. \end{cases} \tag{II.3.6}$$

If $A = 0$, *i.e.*, if the circulation along the two cylinders is the same, then the motion is given by

$$v = B/r\,, \tag{II.3.7}$$

which is an irrotational flow due to a hypothetical vortex. This is also the case if $\Omega_2 = 0$ and $R_2 \to \infty$, *i.e.*, if the inner cylinder is rotating in a viscous fluid of infinite extent. The shear stress is

$$\tau_{r\theta} = \eta\left(\frac{\partial v}{\partial r} - \frac{v}{r}\right) = -2\eta\Omega_1 (R_1/r)^2\,. \tag{II.3.8}$$

The total energy dissipation per unit height of the fluid is

$$\int_{R_1}^{\infty} 4\eta(\Omega_1 R_1^2)^2 \left(\frac{2\pi r\,\mathrm{d}r}{r^4}\right) = 4\pi\eta\Omega_1^2 R_1^2\,. \tag{II.3.9}$$

This is supplied by the work done by the external torque $2\eta\Omega_1 \cdot 2\pi R_1 \cdot R_1$ which has to balance a stress $2\eta\Omega_1$ at the surface of the cylinder (see (II.3.8)). Since the torque is acting on the fluid at the angular velocity $\Omega_1$, it supplies to the fluid exactly the amount of energy required by (II.3.9).

The general case can be discussed by using (II.3.3) or (II.3.4). *When the boundaries are stationary*, one can easily verify, by using (3), that

$$\int D\,\mathrm{d}V = \int 2\eta\omega_{ij}\omega_{ij}\,\mathrm{d}V = 4\eta\int \boldsymbol{\omega}^2\,\mathrm{d}V\,. \tag{II.3.10}$$

There are however no such irrotational motions. Solutions with $e_{ij} = 0$ are also not known, except for the case of uniform rotation. Thus, in general, *the dissipative effect of viscous forces does not vanish.*

## 4. – Instability of laminar motion; turbulence.

It is well known that laminar flow through a pipe can only be maintained at low Reynolds numbers. At higher Reynolds numbers, the motion is in general turbulent. However, it is also known that if the Reynolds number is reduced to below $10^3$, turbulent flow cannot be maintained. Here the Reynolds number is defined by

$$R = \frac{U_m a}{\nu}, \tag{II.4.1}$$

where $U_m$ is the average velocity over a cross-section and $a$ is the radius of the pipe.

For $\nu = 10^{-4}$, and $a = 0.1$ cm, this critical Reynolds number is reached if $U_m = 1$ cm/s.

Flows at higher Reynolds numbers can still be kept in a laminar state, provided extreme care is exercised. Indeed, laminar flows up to Reynolds numbers of the order $10^5$ have been reported by various authors. However, these experiments were performed with extreme caution. It is usually necessary to let the water settle for days before the experiment is started (*).

*a) Instability with respect to infinitesimal disturbances.* Transition to turbulence in a pipe does not seem to be caused by instability of the fully developed Hagen-Poiseuille profile with respect to infinitesimal disturbances. Indeed, such a flow appears to be stable with respect to all infinitesimal disturbances. This is however a rather exceptional case. Most flows are definitely unstable with respect to infinitesimal disturbances at high speeds. The theory for the various types of instability has been fully investigated in recent years (**). The two major types of instability are exemplified by the following two cases, where the basic flows are both solutions of the Navier-Stokes equations:

*a)* pressure flow through a channel (HEISENBERG, LIN);

*b)* flow between rotating cylinders (TAYLOR).

In the first case, oscillations develop after the flow becomes unstable; the instability is said to be of the « overstable » type (after EDDINGTON). In the second case, a *secondary flow* develops, and the instability is said to be of the marginal type. Transition to turbulence seems to follow directly the oscillations in the first case; in the second case, a stationary state is first reached,

---

(*) See GOLDSTEIN [6], chap. VII and VIII for further details regarding turbulent flow in pipes, channels and along flat plates. Applications of the classical theories of turbulent transport are to be found there.

(**) See LIN [7]; CHANDRASEKHAR [8].

and instability of the « overstable » type eventually develops as the speed is further increased (*). Transition then ensues.

*b) Oscillations at finite amplitudes.* Transition to turbulent flow does not occur as soon as the amplitude of the oscillations becomes finite. There is a stage of regular finite amplitude oscillations. Then turbulent spots occur. As more and more of these spots develop, the pressure drop along the channel must be correspondingly increased to maintain the same rate of flow. Eventually, the flow becomes fully turbulent.

The study of nonlinear oscillations is one of the main concerns of workers in this field. The reason for this interest is twofold. First, even in fully developed turbulent flow, the *large scale* motions are more in the nature of a continuation of these regular oscillations than random motions satisfying certain universal laws. Secondly, in geophysical problems, the behavior of finite amplitude oscillations in the atmosphere and in the ocean are decidedly of interest.

*c) Turbulent motion* (**). Fully developed turbulent flow has two prominent characteristics. The large scale motions mentioned above are generally responsible for transport of various physical quantities; the small scale motions (with large rate of strain) are chiefly responsible for energy dissipation. As the energy at small scales is destroyed by viscous dissipation, it is being continually replenished by the breakdown of large eddies by a mechanism of stretching of vortex lines (***). Generally speaking, this is similar to the mechanism of ultraviolet catastrophe in the classical theory of electromagnetic radiation. In the present case, there is no quantum effect to cause an equilibrium spectrum to be formed. Rather a quasi-equilibrium is formed according to the theory of Kolmogoroff.

In the Kolmogoroff theory, the total rate of energy dissipation $\varepsilon$ (per unit mass per second) is regarded as the fundamental quantity. Thus, the small scale eddies are governed only by $\varepsilon$ and the kinetic viscosity $\nu$. From these two quantities, we can form the following scales of length, time, and velocity:

$$\eta_0 = (\nu^3/\varepsilon)^{\frac{1}{4}} , \tag{II.4.2}$$

$$\tau_0 = (\nu/\varepsilon)^{\frac{1}{2}} , \tag{II.4.3}$$

$$v_0 = (\nu\varepsilon)^{\frac{1}{4}} . \tag{II.4.4}$$

---

(*) The instability of flow through curved channels furnishes a good example for the contrast between instabilities of both types. See LIN [7], p. 48.

(**) For an account of the statistical theory, see article by LIN and REID [9].

(***) Cf. Section II.1.

The spatial spectrum at high wave numbers $\varkappa$ is then of the form

$$E(\varkappa) = v_0^2 \eta_0 f(\varkappa\eta_0) , \tag{II.4.5}$$

where $f(\xi_0)$ is a universale function of $\xi_0$. In Kolmogoroff's original theory, the small scale eddies are assumed to be isotropic. Lately, there has been some evidence that this need not be true. Some directional characteristics may then have to be introduced as further parameters for describing a specific field.

At the lower end of the universal spectrum, we have

$$E(\varkappa) = C\varepsilon^{\frac{2}{3}}\varkappa^{-\frac{5}{3}} . \tag{II.4.6}$$

For larger scale, there is no universal theory to cover all cases. For the case of decaying isotropic turbulence, HEISENBERG has proposed the following formula to describe the transfer of energy among the various scales:

$$\frac{\partial}{\partial t}\int_0^{\varkappa} E(\varkappa)\,\mathrm{d}\varkappa = -2(\nu_\varkappa + \nu)\int_{\varkappa}^{\infty} \varkappa^2 E(\varkappa)\,\mathrm{d}\varkappa , \tag{II.4.7}$$

where

$$\nu_\varkappa = \int_{\varkappa}^{\infty} \left[\frac{E(\varkappa)}{\varkappa^3}\right]^{\frac{1}{2}} \mathrm{d}\varkappa . \tag{II.4.8}$$

This formula has also been adapted to the study of shear flow.

Another principal idea in the study of turbulence is to ask how closely the theory of Brownian motion can be applied to the phenomena of turbulent motion. It is known that turbulent motion cannot be a completely random process with joint Gaussian distribution for the velocities at various points of the field. Quasi-Gaussian assumptions have been adopted with success. There are still some basic differences between the Kolmogoroff theory and the quasi-Gaussian theory that urgently require reconciliation.

Despite the recent advances in our understanding of the basic mechanisms of turbulent motion, calculations of mean quantities are still made by the use of the older transport theories. This is partly due to the fact that the larger eddies, which determine the transport properties, do not have universal properties, but depend on the boundary conditions to a large extent. Thus, it is difficult to improve upon the crude theories developed earlier. However, it is now possible to give more detailed descriptions in terms of modern concepts. In particular, the problem of turbulent dispersion has now been quite satisfactorily solved by analogy with the theory of Brownian motion. For example, one can now give a theoretical deduction of Richardson's empirical law, which states that the dispersion coefficient associated with the increase of the size of a cloud of particles should itself be increasing as the $\frac{4}{3}$ power of the size. Specifically, if the mean square separation between two particles in the cloud is $\overline{s^2}$, and the dispersion coefficient is defined by

$$\mathscr{D} = \frac{1}{2}\frac{d\overline{s^2}}{dt}, \tag{II.4.9}$$

then one can show theoretically that

$$\mathscr{D} = \left(\frac{3}{2}\right)^{\frac{2}{3}} B^{\frac{1}{3}} \left(\overline{s^2}\right)^{\frac{2}{3}}, \tag{II.4.10}$$

where $B$ may be identified approximately with the rate of energy dissipation $\varepsilon$.

## III. - Hydrodynamics of Superfluids. Reversible Processes (*).

### 1. – General theoretical considerations.

We shall now attempt to develop the continuum theory of the hydrodynamics of liquid helium. The usefulness of such a theory lies in providing a logical structure of a macroscopic hydrothermodynamics which is not specifically dependent on the molecular structure. As we shall see, it can be used to bring out more sharply the basic points that need to be settled by further theoretical and experimental investigations. Specifically, the two-fluid concept introduced in this approach will be somewhat less restrictive than the picture of excitations in a background fluid, with the latter essentially equivalent to a state of the liquid at 0 °K. Of course, the concept must be generally consistent with those developed from the statistical mechanical theory of mole cular structure.

We take this continuum approach, since there does not, as yet, exist a *complete* hydrodynamic theory *derived* from the statistical mechanics of the quasi-equilibrium state of the liquid. The current hydrodynamic theory of liquid helium II is based on the two-fluid concept developed by LONDON,

(*) A more detailed version of this chapter had been distributed in preprint form from the Institute for Advanced Study, Princeton, N. J., April 1960. It will be published elsewhere.

TISZA, and LANDAU (which we shall introduce in a somewhat different manner), and on the Onsager-Feynman theory of quantized vortex lines (which we shall examine critically). In this theory, the supercomponent of the flow must be pointwise irrotational but there can exist line singularities in the fluid around which there is a quantized circulation (cf. Part V). The supercomponent of the fluid can therefore have angular momentum in bulk. Recently, LEE, HUANG, and YANG [10] studied a dilute system of bosons with hardsphere interaction as a model for liquid helium. LEE and YANG [11] have further developed the theory for application at *finite* temperatures. This furnishes a natural connection between the concept of the degenerate phase in the sense of London's work (see LONDON [12]) and the concept of phonon excitation in the sense of Landau's treatment [13]. They have also developed a quasi-equilibrium theory in which Landau's equations of motion are reproduced (with a slight modification). They stated, however, that the question whether the superfluid flow is irrotational was not resolved in their paper (*).

The difficulty is to answer the question: what is superfluid flow? If one adopts the idea of a background fluid plus excitations as the model of He II, and one thinks of a background fluid as the *single* ground state equivalent to the state of the fluid at 0 °K, it is then difficult to see how the fluid can move without excitations being created in it except for a uniform translation. In other words, it is difficult to tell flow (which is hydrodynamical) from internal excitations (which are thermodynamical). It is indeed because of this confusion that in the earlier days the theoretical critical velocity predicted was more than 1000 times the experimental value, because they refer to different physical phenomena (cf. Part V). One should realize that in hydrodynamics we are dealing with extremely low energies. The flow velocity $v$ is of the order of 1 cm/s, whereas the velocity of sound $c$ (second sound) is of the order of 20 m/s. Thus,

$$v^2/c^2 = \left(\frac{1}{2000}\right)^2 = \frac{1}{4}\cdot 10^{-6}\,.$$

There are thus two theories:

*a*) the theory of fluid structure,

*b*) the theory of flow.

I shall concentrate my discussions on the second aspect.

In spite of the elegance and appeal of the theory of quantized vortex lines, it is indeed very difficult to establish the irrotationality of the superfluid flow,

(*) A more recent paper by T. T. WU [14] takes up this question again, but also leaves it open in the case of interacting bosons.

and the existence of quantized vortex lines so firmly as to exclude other possibilities. Conclusive theoretical arguments have yet to be given (*). Experimental evidence is often indirect, and does not exclude alternative explanations (such as those we shall present in Part IV). Only VINEN has attempted a direct experimental verification of the existence of single quantization, but the results are as yet inconclusive (**). It is thus desirable to attempt to develop the theory of flow of superfluid helium from the usual continuum approach, and to investigate how far such concepts would carry us. This we shall begin in the next section.

## 2. – An outline of the perfect fluid theory.

As mentioned above, our aim is to develop a macroscopic theory which would enable us to describe the various observed phenomena by starting out with a minimum of experimental facts, and by using concepts which are compatible with the microscopic theory. In particular, the application of the concept of a classical thermomechanical continuum to liquid helium II should not be accepted uncritically. This point will be taken up in Part V. As far as specific experimental facts are concerned, we shall start out with the single observation that *liquid helium* II *at rest can transmit heat in a reversible manner*. This reversible heat current we associate with a *flux of order* (*disorder*) *in the momentum space* (***).

We shall show that by introducing this extra degree of freedom into the conventional thermohydrodynamics, we can develop a complete theory of perfect fluid flow, which allows the introduction of a two-fluid model as a purely mathematical step. The resultant equations in the two-fluid language will be found to be compatible with the Landau equations (*•*) in the following sense. Our equations permit a class of solutions for which the motion of the super-

---

(*) In the section on « Rotation of the superfluid », FEYNMAN [15] states (p. 40): « We must therefore still consider it conjectural whether the considerations on rotational flow reported here are actually correct ».

(**) Toward the end of his paper, VINEN [16] stated: « But the fact remains that the present measurements do at least *suggest* that the simplest picture of a free vortex line is in some way inadequate, and that the theory behind this picture should be re-examined ».

(***) Historically, as soon as the fountain effect was discovered by ALLEN and JONES [17], the flux of disorder had been suggested as a primary concept by JONES instead of the two-fluid concept, but the latter prevailed. At that time, JONES was probably thinking in terms of order in the configurational space, as suggested by FRÖHLICH. Subsequent X-ray examination of the structure of liquid helium revealed no such configurational order.

(*•*) See LONDON [12], Section **19**.

fluid is irrotational. If then we restrict our interest to this special class of solutions, the equations of motion are identical with those of LANDAU. However, our equations also permit other classes of solutions, and there appears to be no particular basis for restricting the superfluid motion to the irrotational type. This is only natural, because such a restriction must come from quantum-mechanical considerations and should be beyond the scope of purely continuum concepts. We are thus leaving the question of the irrotationality open at the present stage of the theory. In certain experiments, the superfluid component is known to rotate (at least in bulk motion). The present results may then have to be applied. The discussion of such experiments must await the development of the dynamics of a viscous fluid (Chapter IV), where the question of irrotationality of the supercomponent will be taken up more fully.

In the meantime, while we are dealing with phenomena in which irreversible processes are unimportant, and for which the theory of a perfect fluid is therefore applicable, we are in general restricted to irrotational motions; for the fluid generally starts from rest and cannot gain rotation (cf. Section **4**). Thus, the predictions of our theory would in general be identical with those following Landau's equations (*), which are well verified experimentally. Thus, in developing this part of our theory, we need not examine the experimental data. All we need to study would be its compatibility with our concept of the structure of the liquid. This we shall do in Part V.

## 3. – Fundamental equations. Variational principle.

We shall now proceed to develop a continuum theory of liquid helium II, considering only reversible processes. As mentioned above, we start with the single fact that the fluid is capable of transmitting heat in a reversible manner in the absence of a temperature gradient. To account for this, we introduce the *reversible entropy flux* $\boldsymbol{S}$ *relative to the fluid as a new thermodynamic variable.* The other thermodynamic variables are the mass density $\varrho$, and the entropy per unit mass $s$. Alternatively, we shall define a *velocity of internal convection* $\boldsymbol{c}$ by the relation

$$\boldsymbol{S} = \varrho s \boldsymbol{c} \, . \tag{III.3.1}$$

The internal energy of the fluid, being a scalar, can depend on $\boldsymbol{c}$ only through its magnitude. Thus, the thermodynamic state of the system is given by a

(*) Judging from the paper presented at Toronto by BEKAREVICH and KHALATNIKOV [18], it is clear that people are also concerned with the exact form of the equations of motion.

relation of the form

$$e = e(\varrho, s, \zeta), \quad \zeta = \tfrac{1}{2}\boldsymbol{c}^2 \tag{III.3.2}$$

where $e$ is the internal energy per unit mass.

Next we consider the conservation relations when the liquid is in motion. We adopt the Eulerian description and denote by $\boldsymbol{v}(\boldsymbol{x}, t)$ the velocity field at time $t$ in the Cartesian co-ordinate system $\boldsymbol{x}$. The equation of mass conservation is the well-known one:

$$\frac{\partial \varrho}{\partial t} + \nabla\cdot(\varrho\boldsymbol{v}) = 0 . \tag{III.3.3}$$

The entropy flux through space is $\varrho s(\boldsymbol{v}+\boldsymbol{c})$, and thus the conservation of thermal energy (for reversible processes) may be written as

$$\frac{\partial(\varrho s)}{\partial t} + \nabla\cdot[\varrho s(\boldsymbol{v}+\boldsymbol{c})] = 0 . \tag{III.3.4}$$

The equation for the conservation of momentum is more difficult, as we do not know the role of the convection vector $\boldsymbol{c}$. In fact, one needs more than the equation for momentum conservation. There should be one set of equations each governing the behavior of $\boldsymbol{v}$ and $\boldsymbol{c}$ respectively.

To obtain such equations, we shall assume that the Hamiltonian principle is applicable for treating both degrees of frredom. We write down the Lagrangian density function

$$L = \varrho\left[\frac{1}{2}\boldsymbol{v}^2 - e_L(\varrho, s, \zeta) - \Omega\right], \tag{III.3.5}$$

where $\Omega$ is the potential energy of an external field, and $e_L$ need not be identical with $e$ of (III.3.2), for reasons explained elsewhere. We then impose the variational relation

$$\delta\int_{t_0}^{t_1} \mathrm{d}t \int_V L\,\mathrm{d}V = 0 , \tag{III.3.6}$$

subject to suitable subsidiary conditions. Two of these are (III.3.2) and (III.4.3). There is a third one which relates to the fact that the Hamiltonian principle should be applied in the Lagrangian formulation. This requires the introduction of the Lagrangian co-ordinates $\boldsymbol{X}(\boldsymbol{x}, t)$ of the fluid particles. In the Eulerian description, they satisfy the relationship

$$\text{(III.3.7)} \qquad \frac{\mathrm{D}\boldsymbol{X}}{\mathrm{D}t} = \frac{\partial \boldsymbol{X}}{\partial t} + (\boldsymbol{v}\cdot\nabla)\boldsymbol{X} = 0\,.$$

It may be recalled that in the application of the Hamiltonian principle in particle mechanics, there are constraints at the end points $(t_0, t_1)$ of the interval of time. It is through these constraints that the Lagrangian co-ordinates $\boldsymbol{X}$ enter the variational relation (III.3.6). A fuller discussion of this point will be made elsewhere.

The variational principle yields the equations for $\boldsymbol{v}$ and $\boldsymbol{c}$. After some calculation, one obtains the following equation of motion

$$\text{(III.3.8)} \qquad \frac{\mathrm{D}\boldsymbol{v}}{\mathrm{D}t} = -\nabla\Omega + \frac{1}{\varrho}\nabla\cdot\tau\,,$$

where $\tau$ is a symmetrical stress tensor given by

$$\text{(III.3.9)} \qquad \tau = -\pi\boldsymbol{I} + z\varrho\boldsymbol{c}\boldsymbol{c}\,,$$

and $\pi$ and $z$ are defined by

$$\text{(II.3.10)} \qquad \mathrm{d}e_L = \theta\,\mathrm{d}s + (\pi/\varrho^2)\,\mathrm{d}\varrho + z\,\mathrm{d}\zeta\,,$$

Thus, *the reversible heat flux gives rise to a momentum transfer*, besides the scalar stress $\pi$ (which is to be identified with pressure). The form of the additional terms calls to mind the Reynolds stress in the theory of turbulence. The coefficient $z$ is a kind of correlation coefficient, and the turbulence analogy would suggest that it should be negative, as we shall find it to be indeed the case after detailed investigations.

The equation governing the internal convection vector $\boldsymbol{c}$ is most conveniently left in the form (*)

$$\text{(III.3.11)} \qquad \frac{\mathrm{D}\beta}{\mathrm{D}t} = \theta - (2z\zeta)/s\,, \quad \boldsymbol{c} = (s/z)\nabla\beta\,.$$

Equations (III.3.3), (III.3.4), (III.3.8), (III.3.11) constitute the fundamental equations of the theory of perfect fluids. They govern the behavior of the quantities $\varrho$, $s$, $\boldsymbol{c}$ and $\boldsymbol{v}$.

---

(*) This equation shows that the heat flux $S$ is not subject to Coriolis effect.

## 4. – Sound waves and thermal waves.

Let us now consider approximate solutions for which both $\boldsymbol{v}$ and $\boldsymbol{c}$ are small, so that the quadratic terms can be neglected in the above equations. One notices immediately that the term $z\varrho\boldsymbol{cc}$ should now be neglected from the stress tensor (III.3.9) and that thermal convection can influence the dynamical motion only in an indirect manner, through its thermal effects. The eqs. (III.3.3), (III.3.4), (III.3.8) and (III.3.11) are now approximately given by

$$\frac{1}{\varrho_0}\frac{\partial\varrho}{\partial t}+(\nabla\cdot\boldsymbol{v})=0\,, \tag{III.4.1}$$

$$\frac{1}{s_0}\frac{\partial s}{\partial t}+(\nabla\cdot\boldsymbol{c})=0\,, \tag{III.4.2}$$

$$\frac{\partial\boldsymbol{v}}{\partial t}=-\nabla\Omega+\frac{1}{\varrho_0}\nabla\pi\,, \tag{III.4.3}$$

$$(z_0/s_0)\frac{\partial\boldsymbol{c}}{\partial t}=\nabla\theta\,, \tag{III.4.4}$$

where $\pi$ and $\theta$ are functions of $\varrho$ and $s$. Now for very low temperatures,

$$\left(\frac{\partial s}{\partial\pi}\right)_\theta\to 0\,,$$

and hence

$$\left(\frac{\partial\varrho}{\partial\theta}\right)_\pi=\varrho^2\left(\frac{\partial s}{\partial\pi}\right)_\theta=0\,.$$

If this approximation is adopted, namely,

$$s=s(\theta)\quad\text{and}\quad\varrho=\varrho(\pi)\,, \tag{II.4.5}$$

the systems (III.4.1)–(III.4.4) can be separated into the two sets (III.4.1), (III.4.3) and (III.4.2), (III.4.4). The first set gives a compression wave with a velocity $u_\text{I}$ given by

$$u_\text{I}^2=\left(\frac{\partial\pi}{\partial\varrho}\right)_s, \tag{III.4.6}$$

and the second set gives a thermal wave with velocity $u_\text{II}$ given by

$$u_\text{II}^2=-\frac{s_0^2}{z_0}\left(\frac{\partial\theta}{\partial s}\right)_\varrho. \tag{III.4.7}$$

These are the waves discussed by TISZA before they were experimentally discovered. If we compare (III.4.7) with the usual formula for velocity of the second sound (*), we see that

$$z = -\varrho_n/\varrho_s\,, \tag{III.4.8}$$

where $\varrho_n$ and $\varrho_s$ are respectively the densities of the normal and the super-components. Their meaning in the present theory will be clarified in Section **5**.

## 5. – The two-fluid model.

Current macroscopic theories of liquid helium are formulated in terms of a two-fluid concept, which was initiated by LONDON [19] and TISZA [20] on the basis of the phenomenon of Bose-Einstein condensation. LANDAU developed a two-fluid theory from a different approach and obtained the nonlinear thermohydrodynamic equations. These equations were given a variational formulation by ZILSEL.

It has been shown in various ways that the two approaches to the two-fluid theory are equivalent. Recently, LEE and YANG [11, 21] showed that the two approaches to the two-fluid theory are equivalent by giving the detailed microscopic picture of a dilute Bose system of hard spheres. It has also been repeatedly emphasized that the two-fluid concept must be understood from a formalistic point of view. There are two modes of motions, each of which is associated with its own « effective mass » such that the sum of both of these masses is equal to the total true mass of the liquid.

In our present formulation, this point is clearly presented from the macroscopic point of view. We have yet to introduce the two-fluid concept, although we have been able to derive Tisza's theory of thermal waves. We shall now introduce this concept by a purely mathematical process, and show that the equations of Section **3** agree with Landau's equations, *provided* the superfluid component can move only in an irrotational manner. It is not a part of our theory (as it is with LANDAU) that the superfluid component *must* be irrotational. This is one of the central problems in the hydrodynamics of superfluid helium, and it will be taken up in detail in Part V.

We shall now consider a two-component picture of liquid helium: a normal component with density $\varrho_n$ and entropy $s_n$ per unit mass of the normal component and a supercomponent with density $\varrho_s$ and entropy $s_s$ per unit mass of the supercomponent. Each component has its own velocity field $\boldsymbol{v}_n$ and $\boldsymbol{v}_s$ respectively. This model is to be connected with our above formulation by

(*) See LONDON [12], eq. (11), p. 84.

the identification of the densities of mass, entropy, mass flux and entropy flux. Thus, we have

$$\varrho = \varrho_n + \varrho_s\,, \tag{III.5.1}$$

$$\varrho s = \varrho_n s_n + \varrho_s s_s\,, \tag{III.5.2}$$

$$\varrho \boldsymbol{v} = \varrho_n \boldsymbol{v}_n + \varrho_s \boldsymbol{v}_s\,, \tag{III.5.3}$$

$$\varrho s(\boldsymbol{v} + \boldsymbol{c}) = \varrho_n s_n \boldsymbol{v}_n + \varrho_s s_s \boldsymbol{v}_s\,. \tag{III.5.4}$$

It is customary (though not necessary at this stage) to assume that $s_s = 0$.

With the two-fluid model described above, one can rewrite the basic equations of Section **1** in another form. The equations for the conservation of mass and entropy read

$$\frac{\partial \varrho}{\partial t} + \nabla\cdot\varrho[x\boldsymbol{v}_n + (1-x)\boldsymbol{v}_s] = 0\,, \tag{III.5.5}$$

$$\frac{\partial(\varrho s)}{\partial t} + \nabla\cdot[\varrho s \boldsymbol{v}_n] = 0\,. \tag{III.5.6}$$

The derivation of the equations for $\boldsymbol{v}_n$ and $\boldsymbol{v}_s$ is lengthy; but they may be put into the final form:

$$\frac{\mathrm{D}_n\boldsymbol{v}_n}{\mathrm{D}t} = -\frac{1}{\varrho}\nabla\pi - \frac{1-x}{x}s\nabla\theta - \nabla\Phi - \frac{1-x}{2}\nabla q^2 - (1-x)\boldsymbol{q}\times[\nabla\times\boldsymbol{v}_s]\,, \tag{III.5.7}$$

$$\frac{\mathrm{D}_s\boldsymbol{v}_s}{\mathrm{D}t} = -\frac{1}{\varrho}\nabla\pi + s\nabla\theta - \nabla\Phi + \frac{x}{2}\nabla\boldsymbol{q}^2 + x\boldsymbol{q}\times[\nabla\times\boldsymbol{v}_s]\,. \tag{III.5.8}$$

Equation (III.5.8) can also be written as

$$\frac{\partial \boldsymbol{v}_s}{\partial t} = -\nabla H + \boldsymbol{v}\times[\nabla\times\boldsymbol{v}_s]\,, \tag{III.5.9}$$

where

$$H = g + \Omega + \frac{1}{2}\boldsymbol{v}_s^2 - \frac{x}{2}\boldsymbol{q}^2\,, \qquad g = e + \pi/\varrho - Ts\,. \tag{III.5.10}$$

Thus, *the equations permit the class of solutions satisfying the condition*

$$\nabla\times\boldsymbol{v}_s = 0\,. \tag{III.5.11}$$

For this class, (III.5.7) and (III.5.8) read

$$\frac{\mathrm{D_n}\boldsymbol{v}_\mathrm{n}}{\mathrm{D}t} = -\frac{1}{\varrho}\nabla p - \frac{1-x}{x}s\nabla T - \nabla\Omega - \frac{1-x}{2}\nabla \boldsymbol{q}^2, \tag{III.5.12}$$

$$\frac{\mathrm{D_s}\boldsymbol{v}_\mathrm{s}}{\mathrm{D}t} = -\frac{1}{\varrho}\nabla p + s\nabla T - \nabla\Omega + \frac{x}{2}\nabla \boldsymbol{q}^2\,. \tag{III.5.13}$$

Equations (III5.5), (III.5.6), (III.5.11), (III.5.12) and (III.5.13) are the set of equations given by LANDAU.

The difference between the present theory and Landau's theory is now obvious. In Landau's theory, (III.5.11) is an independent *physical* requirement; here it merely characterizes a possible class of solutions. It is interesting to note that in Zilsel's [22, 23] variation formulation, the condition (III.5.11) has been automatically (though unintentionally) incorporated, because of a defect in the variational principle used. It can be shown that if one had introduced the constraint (III.3.7) into Zilsel's derivation, the equations obtained would have been identical with the present ones, without the restriction (III.5.11).

A brief word should be added here regarding the experimental evidence of the irrotationality of the supercomponent, even though we shall take it up in detail in the next two chapters. As is well known, Landau's condition of irrotationality was verified by the experiments of ANDRONIKASHVILI' who found that only the normal component is moving with an oscillating disc system. In the framework of present theory, this is accountable, as we shall see in Part IV by allowing a slip at the boundary in the systems of basic equations for the motion of a viscous field. Thus, we agree to the fact that the superfluid is indeed not rotating. The reason is, however, not that it cannot rotate, but rather that it does not rotate under the conditions specified (cf. ATKINS [24], p. 111).

## IV. - Hydrodynamics of Superfluids. Irreversible Processes.

### 1. – The basic equations and boundary conditions.

The Landau equations discussed above appear to be quite adequate for the discussion of reversible processes. The behavior of superfluid helium is far more complicated when irreversible processes are present. We shall continue to develop our theory in the spirit of continuum mechanics, in a manner analogous to the theory of ordinary fluids (Part II).

For the « viscous stresses », we shall *assume* their *linear* dependence on the rate of strain. Since there are now two components of fluid motion there must be exchange coefficients connecting the stress acting on one component with the rate of strain of the other. Thus, we should have four *exchange coefficients*, defined as follows:

$$\tau_{ij}^{(n)} = 2\eta^{(nn)} e_{ij}^{(n)} + 2\eta^{(ns)} e_{ij}^{(s)} , \tag{IV.1.1}$$

$$\tau_{ij}^{(s)} = 2\eta^{(sn)} e_{ij}^{(n)} + 2\eta^{(ss)} e_{ij}^{(s)} . \tag{IV.1.2}$$

where $\tau_{ij}^{(n)}$ is the stress tensor acting on the normal component, $e_{ij}^{(n)}$ is the rate of strain of the normal component:

$$e_{ij}^{(n)} = \frac{1}{2}\left(\frac{\partial v_i^{(n)}}{\partial x_j}\right) + \left(\frac{\partial v_j^{(n)}}{\partial x_i}\right), \tag{IV.1.3}$$

and the other symbols have similar meanings. We have used a modified notation for the normal and supercomponents for convenient use of the indicial notation. We have also restricted ourselves to the case of incompressible fluids. The extension to the compressible case can be made in the usual manner.

The equations of motion may be obtained from those for the irreversible case by including the volume forces due to the viscous stresses (IV.1.1) and (IV.1.2). We have the additional force per unit volume

$$\frac{\partial \tau_{ij}^{(n)}}{\partial x_j} = \eta^{(nn)} \Delta v_i^{(n)} + \eta^{(ns)} \Delta v_i^{(s)} ; \tag{IV.1.4}$$

for the normal component, and

$$\frac{\partial \tau_{ij}^{(s)}}{\partial x_j} = \eta^{(sn)} \Delta v_i^{(n)} + \eta^{(ss)} \Delta v_i^{(s)} . \tag{IV.1.5}$$

for the supercomponent. Again, the compressibility of the fluid is neglected.

For boundary conditions, we shall make obvious adaptations of the familiar forms (II.1.2) and (II.1.5). If the solid boundary is impermeable to both components, and there is no supply or withdrawal of heat through the boundary, we have

$$w_i^{(n)} n_i = 0 , \qquad w_i^{(s)} n_s = 0 , \tag{IV.1.6}$$

where $w_i^{(n)}$ and $w_i^{(s)}$ are the velocities relative to the surface of the normal and the supercomponents respectively. If there is heat supply, then the perpendicular component of mass flux must be zero, while the perpendicular component of the heat flux must be continuous.

For the parallel components, eq. (II.1.5) is to be adopted without modification,

(IV.1.7) $$T_\alpha^{(n)} = F^{(n)}(w^{(n)2})w_\alpha^{(n)} ,$$

where $T_\alpha^{(n)} = (\tau_{ij}^{(n)} n_j) t_\alpha^i$ ,

(IV.1.8) $$T_\alpha^{(s)} = F^{(s)}(w^{(s)2})w_\alpha^{(s)} ,$$

where $T_\alpha^{(s)} = (\tau_{ij}^{(s)} n_j) t_\alpha^i$ .

The functions $F^{(n)}$ and $F^{(s)}$ are of course expected to be different from each other. In fact, we shall assume $F^{(n)}$ to be very large, as in the case of an ordinary fluid, and $F^{(s)}$ to be very small at small velocities, so that there is a complete lack of shear interaction between the superfluid and the solid boundary at low speeds. Thus, to a first appxroximation

(IV.1.9) $$w_\alpha^{(n)} = 0 ,$$

(IV.1.10) $$T_\alpha^{(s)} = 0 .$$

We shall have occasion to use (IV.1.8) in a more general form, *e.g.*,

(IV.1.11) $$T_\alpha^{(s)} = \beta \, |w^{(s)}|^2 w_\alpha^{(s)} ,$$

which is obtained from (IV.1.8) by a power series expansion of $F^{(s)}$ and retaining only the first nonvanishing term. (The velocities of flow are usually much smaller compared with the velocities of sound.)

In the above formulation, we have adopted the simplest possible forms. The actual situation can be more complicated. For example, it might be true that $F^{(s)} \equiv 0$ for $|w^{(s)}|$ smaller than a certain « critical value », and then follows approximately the law (IV.1.9). We shall, however, make no attempt to deviate from the form (IV.1.8) in most applications, since the critical value (if it exists) is very small.

The continuum theory has wide latitudes. Although we have chosen to put it in the simplest form, several other effects could have been included. First of all, the concept of mutual friction (specifically the Gorter-Mellink formula) is consistent with the present theory, and could be added to our basic equations. Secondly, we have adopted a linear relationship between the stresses and the rates of strain. There is no reason why the coefficients cannot be dependent on the rates of strain (with suitable rules of covariance followed). We have also assumed that rotation of the supercomponent does not lead to an essential change in *thermodynamical* properties of the medium. However, Bekarevich and Khalatnikov [18] did suggest a continuum theory in which the internal energy of the fluid depends on the magnitude of the superfluid rotation. Indeed, according to their concept, $\varrho_s$ is a tensor when the fluid is in rotation. We shall not indulge in such generalizations until the experimental evidence indicates that it is absolutely necessary to do so. It is to be hoped that such effects (if there be any) should be « small »; otherwise, the general concept of a continuum would appear to be so far different from the usual ones as to make it rather inconvenient.

## 2. – Applications to some simple problems.

We shall now apply the theory to a few flow problems. Since the formulation has been made for the incompressible case, we shall exclude for the moment problems dealing with the attenuation of sound waves, which will be discussed in Section **4**, and Part VI.

*a*) *Flow through a tube.* The first experiment to be examined is the flow through a wide capillary. In this case, the most remarkable feature is the fact that the heat flux is directly proportional to the pressure gradient. In the ordinary theory of mutual friction, this is accounted for by the fact that the pressure gradient is balanced by the viscous forces associated with the normal component alone. In our formulation, the equation of motion for the fluid as a whole contains the viscous force

$$\frac{\partial \tau_{ij}^{(n)}}{\partial x_j} + \frac{\partial \tau_{ij}^{(s)}}{\partial x_j} = [\eta^{(nn)} + \eta^{(sn)}]\Delta v_i^{(n)} + [\eta^{(ns)} + \eta^{(ss)}]\Delta v_i^{(s)} . \tag{IV.2.1}$$

Thus, the same conclusion is obtained if

$$\eta^{(ns)} + \eta^{(ss)} = 0 . \tag{IV.2.2}$$

We shall adopt this relation in our following investigations. If we refer back to (IV.1.1) and (IV.1.2), eq. (IV.2.2) means that the total stress

$$\tau_{ij} = \tau_{ij}^{(n)} + \tau_{ij}^{(s)} = 2\,[\eta^{(nn)} + \eta^{(ns)}]\, e_{ij}^{(n)} = 2\eta^{(n)} e_{ij}^{(n)} , \tag{IV.2.3}$$

*i.e.*, the rate of transfer of mechanical momentum is associated with the normal component alone.

The rate of mean flow (if we adopt (1.13) and put $\eta_{sn} = 0$) is given by a formula similar to that obtained from the Gorter-Mellink formula (ATKINS [24], p. 189), except that the coefficient $A$ is now replaced by a slowly varying function of the diameter [25]. The same dependence of $A$ on channel size was found experimentally by HUNG, HUNT, and WINKEL [26].

Notice that the nonlinear behavior is not caused by a nonlinear term in the equations of motion (as in the Gorter-Mellink theory), but by the nonlinearity in the boundary condition for the superfluid motion.

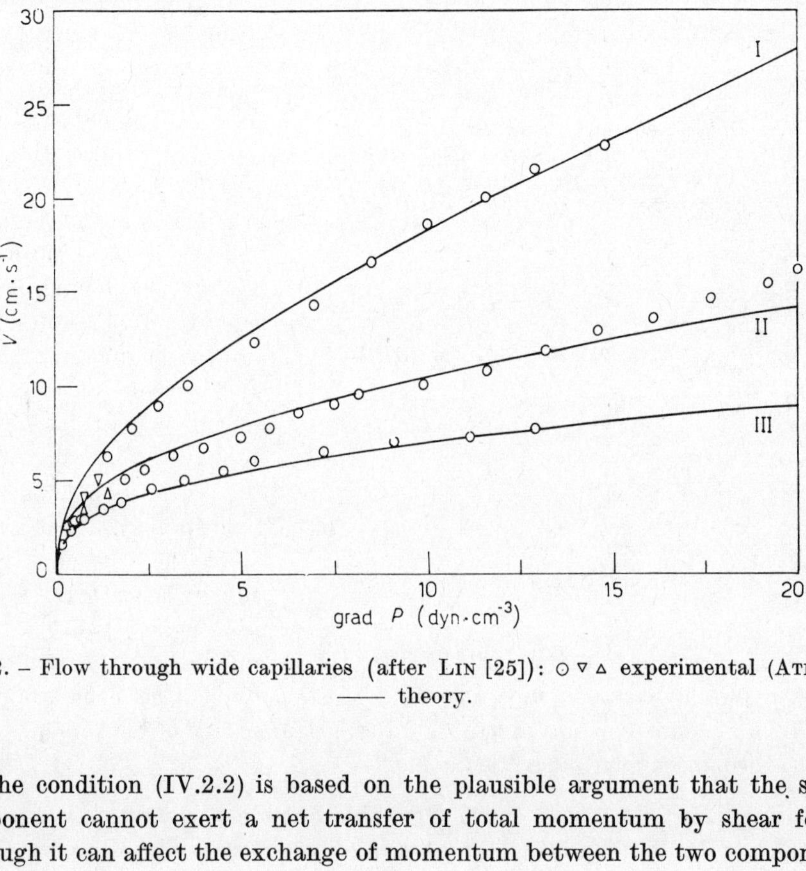


Fig. 2. – Flow through wide capillaries (after LIN [25]): ○ ▽ △ experimental (ATKINS); —— theory.

The condition (IV.2.2) is based on the plausible argument that the supercomponent cannot exert a net transfer of total momentum by shear forces, although it can affect the exchange of momentum between the two components. The assumption that $\eta_{sn} = 0$ again reflects the same idea that the supercomponent is difficult to excite. It appears that the deductions based on both assumptions are well supported by experiments.

*b) Flow between rotating cylinders.* The next simplest geometrical configuration is a set of two concentric cylinders. Under this category of flow phenomena, we shall also consider the rotating bucket (inner cylinder missing) and the rotating rod (outer cylinder missing).

i) The viscometer. Measurements with the viscometer to determine the relation between the angular speed and the torque have been made by HALLET and his collaborators [27-29] (Fig. 3). At first, it was thought there was a nonlinear dependence. Later experiments show that this is a spurious phenomenon associated with secondary flow, and that careful experiments show complete linearity. With the condition (IV.2.2) imposed, the theory gives linear dependence in all cases, despite the nonlinear boundary condition (IV.1.11). This gives further evidence for the condition (IV.2.2) and the underlying physical ideas.

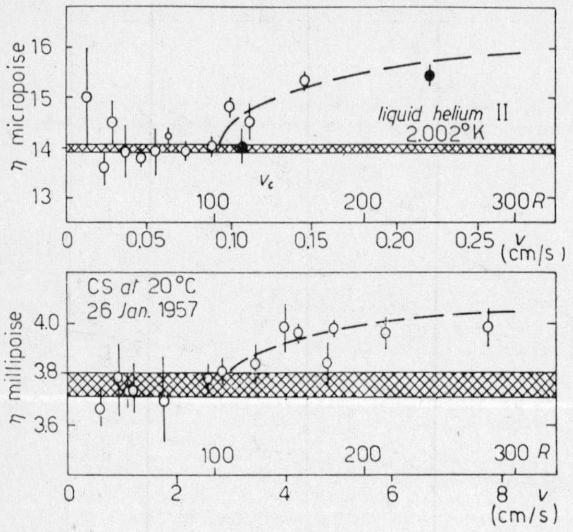


Fig. 3. – « Critical velocity » found with rotating cylinder viscometer in liquid He II and in $CS_2$ (after WOODS and HALLETT [28]).

ii) Uniform rotation. Whether the inner cylinder is present or absent, the final state of fluid motion is one of uniform rotation, if the boundaries are kept rotating at the same uniform angular speed. This is the prediction of the present theory; it is also found to be true from experimental observations, at reasonably high speeds [30].

In the case of the rotating bucket, there seems to be a critical velocity, below which there is no interaction with the supercomponent [31]. This velocity is extremely small. Once the fluid starts moving, the behavior of the transfer of angular momentum can be described by means of the present theory (for detailed discussions, see Part VI).

iii) Rotating rod. The flow field outside of a rotating rod is given by a pure vortex motion, as in the case of an ordinary fluid. The normal fluid follows the rod, while the superfluid lags behind. Thus, the circumferential velocities are given respectively by

$$v^{(n)} = V\left(\frac{a}{r}\right), \qquad v^{(s)} = V'\left(\frac{a}{r}\right), \tag{IV.2.4}$$

while $\beta(V-V')^3$ gives the stress acting on the supercomponent. The total stress acting on the rod is given by (IV.2.3),

$$\tau = 2\eta^{(n)}(V/a), \tag{IV.2.5}$$

This is doing work at the ratio of $\tau \cdot 2\pi a \cdot V$. The separate components $\tau^{(n)}$

and $\tau^{(s)}$ are doing work at the total rate

$$\tau^{(n)} V + \tau^{(s)} V' = \tau V - \tau^{(s)}(V - V') .$$

Thus, there is a dissipation of the amount $\tau^{(s)}(V - V')$ associated with the slip.

iv) General case. In the general case, the flow speed for each component is of the form

$$v = Ar + B/r .$$

The coefficients $A$ and $B$ are determined by certain cubic equations. The interesting point in this problem is a study of its stability. This problem has been studied by CHANDRASEKHAR and DONNELLY [32] on the basis of a set of equations suggested by HALL and VINEN [33]. It would be interesting to repeat the investigation on the basis of the present theory.

## 3. – Oscillating systems.

One of the basic experiments in the study of superfluid helium is Andronikashvili's oscillating disc system. The continuum theory givenab ove would predict the complete slipping of the superfluid component, not because the superflow must normally remain irrotational, but because of the boundary condition (IV.1.10), which holds for low speeds. At higher speeds, HALLET [34] found that the superfluid component was indeed entrained to move with the

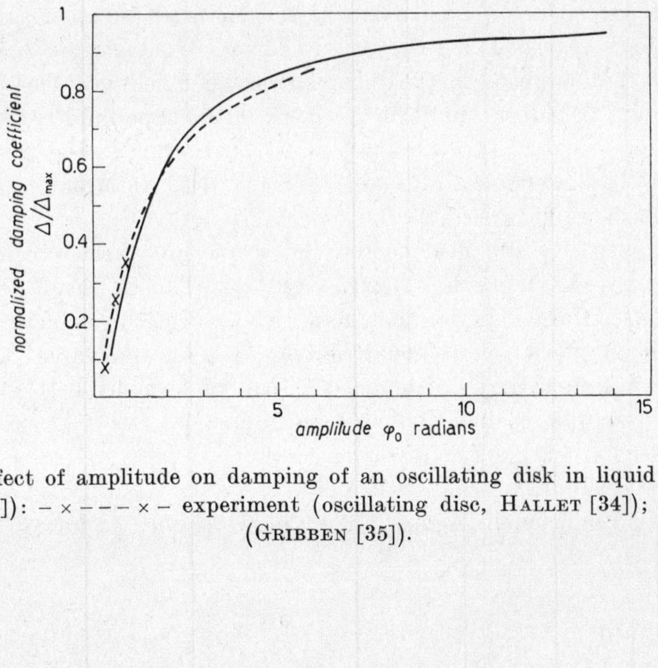


Fig. 4. – Effect of amplitude on damping of an oscillating disk in liquid He II (after GRIBBEN [35]): – × – – – × – experiment (oscillating disc, HALLET [34]); —— theory (GRIBBEN [35]).

normal component. This increasing entrainment with increasing amplitude of oscillation can be accounted for by using the more general boundary condition (IV.1.11).

HALLET also observed the same phenomenon of gradual entrainment with the oscillation of a single disc. The principal features of these observations have been described by using the present theory by R. J. GRIBBEN [35].

In both cases, HALLET found that the mutual friction theory of Gorter and Mellinck does not give a sufficiently large entrainment to account for the observed increase of moment of inertia or damping.

HALL [36] observed a reasonance phenomenon when the above experiments are performed in a rotating system. When the frequency of the oscillation passes through twice the frequency of rotation, there is a change of the nature of the motion. Although HALL was successful in accounting for his observations in terms of « vortex waves », the same phenomenon is expected from the Coriolis effect discussed in the classical case (Section **1**, Part I), and is therefore present also in the continuum theory of superfluid flow. Indeed, the crucial term in Hall's final equations, responsible for this observed change of character, is identical with the Coriolis term in the usual theory. HALL also observed a less pronounced phenomenon of resonance, which we shall discuss in Part VI.

## 4. – Attenuation of sound waves.

The entrainment of the superfluid into a state of rotation has often been spoken of as the breakdown of superfluidity. This was thought, at one time, to be in analogy with the breakdown of the Meissner effect when the critical magnetic field is exceeded. In that case, the superconducting property is lost. Nothing like that happens in the case of superfluid helium. The propagation of second sound is hardly modified at all when the superfluid is in solid body rotation. The study of the attenuation of the thermal wave under various circumstances, including the state of the superfluid rotation, is therefore a matter of considerable interest.

Although viscosity and heat conduction coefficients furnish sources for the attenuation of sound, they have often been found to be insufficient even in ordinary fluids. For example, the measured absorption coefficient of sound in Benzol (at 17 °C) is about 100 *times* larger than the value predicted on the basis of a simple theory. Although this can be formally described in terms of a second viscosity coefficient, the large numerical factor shows that complicated molecular effects are involved.

As a first attempt, we shall still follow the classical approach of using second viscosity coefficients. Again, there could possibly be four of them. We noticed that the condition (IV.2.2) means that the viscosity coefficient associated with the normal component is alone responsible for the momentum transport. Thus, in the propagation of ordinary sound waves, only the normal coefficient of viscosity exerts a damping influence. If a condition similar to (IV.2.2) is satisfied for the second viscosity coefficients, then the absorption coefficient of the first sound is associated with the behavior of the normal component alone, and should rapidly become small as the temperature decreases, even though the attenuation of the second sound is not.

The effect of rotation on the attenuation of second sound has been observed by HALL and VINEN [33] and by DONNELLY and SNYDER (*). There is no obvious interpretation of such a phenomenon on the basis of a simple continuum theory (**) except for the general suggestion that there could be secondary flow in the fluid, whose exact nature would depend on the rotation. (Cf. Hallet's experience with the viscometer experiments). One must remember that, with $\nu = 10^{-4}\ \text{cm}^2/\text{s}$, $\Omega = 9\ \text{rad/s}$, $L = 1\ \text{cm}$, the Reynolds number is of the order of $10^5$. The secondary motion would be mainly along the axis of rotation, because of Coriolis effect. Thus, the fluid would appear to be anisotropic. We shall return to the discussion of this problem in Part VI.

## 5. – Turbulence.

There is, as yet, very little established knowledge regarding turbulent flow in superfluid helium. It is however clear that the critical velocities observed in many of the experiments could be caused by transition to turbulence, because the Reynolds number is usually found to be of the order needed for transition in the case of ordinary fluids.

In the present theory, the turbulent flow of the supercomponent may still be described in terms of classical concepts. One should notice the very small scales attained because of the smallness of $\nu$, for both the normal and the supercomponents. With rate of energy dissipation of the order of $1\ \text{cm}^2/\text{s}^3$ per unit mass, the Kolmogoroff length scale is of the order of $10^{-3}$ cm. Notice that this value is not sensitive to the assumed rate of energy dissipation.

We shall not go into the details of the discussion of possible turbulent flows in the experiments performed so far. Suffice it to say that, in the spirit of the present point of view, we would use classical concepts, including the exchange coefficients, mixture lengths, as well as the more recent statistical concepts, unless the contrary is indicated (***).

(*) Private communication.

(**) One could simply assert, with BEKAREVICH and KHALATNIKOV, that the rotation of the superfluid is a thermodynamic variable.

(***) Cf. Seminar by TACONIS and STAAS given at this Course. The experimental evidence in their paper points toward the soundness of the present point of view.

## 6. – Critical velocities.

In ordinary hydrodynamics, the resistance to the motion of a solid body in the fluid increases sharpy at certain *critical velocities*. There are at least two such velocities:

*a*) the velocity marking substantial conversion of the energy of mechanical motion into internal energy of the fluid (actually characterized by the Mach number), and

*b*) the velocity marking transition from laminar to turbulent flow (actually characterized by the Reynolds number).

In the case of superfluids, the situation is expected to be more complicated. In the usual theory of superfluid structure, the critical velocity calculated (for the production of phonons) is of the same general character as that of type *a*). In the theory of quantized vortex lines, there is a critical velocity associated with the production of these vortex lines, and this is sometimes interpreted as the critical velocity marking the production of turbulence in the superfluid (which is alternatively described as a flow with « tangled vortex lines »).

In terms of the present concepts, we recognize *three* types of critical velocities:

*a*) the critical velocity for internal excitations;

*b*) the critical velocity for transition to turbulence;

*c*) the critical velocity for the incidence of « laminar » friction of the superfluid.

Within each type, there may be several distinct possibilities. For example, within type *a*), there are four possible phenomena and each might be associated with a critical velocity:

*a*) the excitation of first sound;

*b*) the excitation of second sound;

*c*) the production of phonons;

*d*) the destruction of superfluid behavior.

If one considers the analogy between superfluid flow and superconductivity, the velocity *d*) would be an angular velocity which might be compared with the critical magnetic field that destroys superconductivity. It would be interesting to examine these possible velocities experimentally. All of them are expected to be of the order of the velocity of sound.

The critical velocity for transition to turbulence is very much lower, of the general order of $(0.1 \div 1)$ cm/s for apparatus of the sizes commonly used. There are possibly two such velocities, one for the supercomponent, one for the normal component. As in the case of ordinary fluids, one may have to distinguish between critical Reynolds numbers of instability and transition. Experiments related to these phenomena are reported by TACONIS at this Course.

The critical velocity for the incidence of « laminar » friction could refer to a value of $|\boldsymbol{v}_n - \boldsymbol{v}_s|$ in the mutual friction theory or to a value of $|\boldsymbol{w}_s| = |\boldsymbol{v}_s - \boldsymbol{V}|$ in the theory proposed here (cf. Section **1**). In any case, there is a complicating factor for the incidence of interaction between the superfluid with the boundary (or with the normal fluid). The experiments of REPPY and LANE [31] suggest that the incidence of interaction may have to be « triggered » in some manner. An analogy with the initiation of crystallization in a supercooled liquid seems to be suggested.

It is to be noted that, in the present theory, this critical velocity (perhaps of the order of 1 mm/s) is a genuine velocity and should be independent of the geometry of the apparatus. If one adopts the point of view of the quantized vortex lines, the initiation of friction for the supercomponent depends on the product of a velocity and a linear dimension. Such a critical quantity would be difficult to distinguish from a critical Reynolds number for turbulent transition, as far as scale effects are concerned. Thus, *a careful determination of the lowest critical velocity for tubes of suitably chosen sizes would furnish valuable evidence relevant to the theory of quantized vortex lines.* (Cf. VINEN's paper at this conference).

It need hardly be added that critical velocities for film flow and for extremely narrow channels are outside of the scope of the present discussion, for simple continuum concepts are not adequate in such cases (cf. Section **2**, Part V).

## APPENDIX

### Is superfluidity the absence of viscosity?

This question was posed by LONDON ([12], Section **22**) and answered by him as follows: « The apparent absence of viscosity ... would not be the outcome of a mysterious absence of any interactions, which would be quite incredible indeed, but rather the outcome of the absence of vortex motion in the superfluid proper. The term of shearing viscosity $\eta_s \nabla \times \nabla \times \boldsymbol{v}_s$ would be absent not because $\eta_s = 0$ but rather because $\nabla \times \boldsymbol{v}_s = 0$. As we have seen in Part II, the absence of the body force does not mean the absence of shear strain. Thus, there would in general be a shear stress acting on the solid boundary when the superfluid is performing an irrotational motion, *e.g.* in the annular region between two coaxial cylinders. Is this an acceptable state of affairs or should one accept the condition $\eta_s = 0$? The latter contradicts London's strong assertion, and is also at variance with the feeling that superfluid flow is associated with order in the momentum space. A shear strain ought to allow momentum to be transferred in such a case.

## V. - Hydrodynamics of Liquid Helium II. Microscopic Aspects.

### 1. – The molecular basis of a hydrodynamic theory.

We have shown how a continuum theory for the hydrodynamics of superfluid helium can be developed on the basis of the concept of slip of the superfluid at the solid boundary, rather than the concept of basic irrotationality of the supercomponent. This purely phenomenological approach has also the advantage of leaving the molecular picture open. As mentioned before, this approach is adopted because there is as yet no available theory for the study of quasi-equilibrium phenomena of a quantum-mechanical system involving long-range order. Still one can try to check our concepts against the known results in the equilibrium theory. In this chapter, we shall begin to examine this aspect of the theory and compare it with the prevailing theory of quantized vortex lines for the superfluid rotation. Specifically, we shall examine the following questions:

1) Is the continuum concept discussed above justifiable in the light of our knowledge of the structure of the liquid?

2) What are the theoretical arguments in favor of, and against, the basic irrotationality of the supercomponent and the existence of quantized vortex lines?

3) What can one conclude from existing experimental evidence, and what further evidence should be collected to improve the status of our knowledge?

### 2. – The continuum concept.

A continuum theory of fluid motion presupposes the following facts: 1) that there exists a structure length scale $\Lambda$ such that all fluid elements having a larger size have well-defined macroscopic properties, *e.g.*, density and temperature, and 2) that the minimum size of the experimental apparatus is much larger than this scale $\Lambda$. A well-known case where these conditions are violated is the flow of gas through fine tubes at low pressures, first extensively studied by KNUDSEN.

For ordinary media, one might at first think that whenever we take an element of volume with a linear size much larger than the mean interatomic distance (which is of the order of $10^{-8}$ cm in the case of liquid helium), we can safely regard it as a macroscopic element. This is easily seen to be too loose a criterion, for the mean free path in a gas is usually much larger than the interatomic distance, and it is the mean free path that determines the minimum size of the macroscopic element. In the case of liquid helium, the problem is complicated by the existence of a co-operative phenomenon, which leads to a *correlation length.* This is the distance within which the momentum space ordering is strongly effective. Unfortunately, there is no definite knowledge of this correlation length at the present time. In the case of the ideal Bose gas, this scale is infinite. According to the model studied by LEE, HUANG, and YANG [10], in which a dilute system of Bose particles with hard sphere interaction is considered, this correlation length is given by

(V.2.1) $$r_0 = (8\pi an)^{-\frac{1}{2}},$$

where $n$ is the particle number density, and $a$ is the radius of the hard sphere.

If we take

(V.2)2) $$a = 10^{-8}\ \text{cm}, \qquad n = 2\cdot 10^{22}\ \text{cm}^{-3},$$

we get

(V.2.3) $$r_0 = 1.5\cdot 10^{-8}\ \text{cm}.$$

If this figure is trusted, the continuum concept seems to be fully justified in most cases.

However, this estimate appears to be optimistic. Experiments with the propagation of second sound suggest that, at temperatures of the order of 1 °K, there is a strong dispersive effect when the wavelength is of the order $10^{-4}$ cm or smaller. This is also the scale suggested by the length of the mean free path of the phonons, according to the theory of Landau and Khalatnikov [37]. Blatt, Butler, and Schafroth [38] made the conservative estimate of $10^{-2}$ cm or smaller, also on the basis of experiments with propagation of second sound.

If an order of $10^{-4}$ cm is assumed to represent a typical correlation length (around 1 °K), it is clear that flow in films with a thickness of the order of $10^{-6}$ cm is beyond the scope of any hydrodynamic theory. Also, if the correlation length is of the order of the dimensions of the vessel, as it might be expected at 0 °K or very near to it, we have a « completely » macroscopic quantum-mechanical behavior. It would then be unnatural to expect that ordinary continuum concepts should have any validity. The variables we used above are presumably inadequate and insufficient for the description of such flows. In many of the hydrodynamic experiments, however, the continuum concept seems to be well justified.

## 3. – Quantized vortex lines.

The prevailing theory is however not based on a true continuum concept. Although macroscopic quantities are used in the final equations of motion, they are not universal equations but have certain *ad hoc* nature. This is due to the belief of irrotational superflow and the existence of quantized vortex lines of *macroscopic* dimensions in most of the hydrodynamic problems.

It is difficult to trace the origin of the idea that superfluid flow must remain pointwise irrotational. Possibly it was the association of irrotational flows with frictionless flows (which is certainly incorrect in the case of ordinary fluids). Possibly it was by analogy with the Meissner effect in superconductivity (*). Possibly it was associated with the irrotational field of the momentum distribution of a single particle moving in a container with a dimension comparable with the de Broglie wavelength. More likely, the origin can be traced to a combination of these factors. At any rate, all these arguments by analogy are naturally not conclusive. In the discussion of the relationship between irrotationality and lack of resistance, there is even a common misunderstanding (see Appendix, Part IV). Nevertheless, Andronikashvili's disc pile experiment gave experimental support to the idea, and it has become quite generally accepted. When actual rotation of the superfluid component has been experimentally discovered, the idea of quantized vortex lines was introduced by ONSAGER [39, 40] and by FEYNMAN [15]. In this picture, vorticity in the fluid is concentrated along vortex lines as in the classical hydrodynamics of perfect fluids. It might be recalled that in the theory of ordinary fluids, such vortex lines cannot persist (Section **1**, Chapter II). Thus, the natural question to be asked is the following: Why can such vortex lines exist in superfluid helium? In other words, *why are these vortex lines stable structures*? (**) A second question is this: What is the mechanism for the vortex line to be formed, which is after all a mammoth structure involving millions of atoms with correlated motions extending over *macroscopic* distances? Why do the vortex lines (if they must be formed for some reason), not form

(*) Cf. Section IV.**6**.

(**) In the recent analysis of GROSS [41], the question of stability was not discussed.

themselves into tiny rings with a diameter (say $10^{-5}$ cm) much smaller compared with ordinary macroscopic dimensions, but still much larger compared with the diameter of the vortex line (say $10^{-7}$ cm)?

These questions cannot be easily put aside unless one has a conclusive theoretical evidence that the hydrodynamic rotation of a superfluid component is *absolutely impossible*, or *extremely unlikely* from the point of view of statistical considerations. It is not enough to show that irrotational motion, with or without vortex lines, is a *possible* state of motion of the superfluid component, as has often been attempted.

Before we go into the examination of the detailed arguments made on the basis of Schrödinger's equations, let us first clarify one general point. It is sometimes argued that since quantum mechanical laws are followed by liquid He II, some form of discrete quantization involving Planck's constants must be present. Such an argument is likely to be true at 0 °K (provided the effect in question is large enough to be observed). It does not have to hold when *thermodynamic averages* are involved (*). Since the thermal de Broglie wave length for a helium atom is of the order of $10^{-8}$ cm at 1 °K, arguments valid for 0 °K need not be applicable at 1 °K.

Let us now examine the oft-quoted derivation of FEYNMAN [15], which runs roughly as follows: If $\Psi$ is the wave function describing the liquid at rest *at* 0 °K, *i.e.*, it is the wave function representing the ground state of the system, then the wave function

$$\psi_{\text{tr}} = \Psi \exp\left[\frac{im}{\hbar} v \cdot \sum_j \boldsymbol{r}_j\right], \tag{V.3.1}$$

represents a uniform translational motion of the liquid as a whole, where $\sum_j$ represents summation over all the atoms, and $\boldsymbol{v}$ is the uniform translational velocity. Consider now the function

$$\psi = \Psi \exp\left[i \sum_j S(\boldsymbol{r}_j)\right], \tag{V.3.2}$$

where $S$ is a slowly varying function of position and $S(\boldsymbol{r}_j)$ is its value at the location of the $j$-th atom. This function then represents an irrotational flow, with the velocity at any point given by

$$\boldsymbol{v} = \frac{\hbar}{m} \operatorname{grad} S, \tag{V.3.3}$$

(*) The classical example is Planck's law of radiation. The continuous spectrum involves a decidedly quantum effect but no discrete quantization. We wish to emphasize this point because we believe that some effect of discrete quantization should be macroscopically observable in He II at 0 °K.

as a simple quantum-mechanical calculation will show. In a multiply-connected region, such as the annulus between two coaxial cylinders, an irrotational flow can exist with circulation. But since the wave function (V.3.2) must be single-valued, the phase $S$ for each atom must change by an integral multiple of $2\pi$ along a path enclosing the inner cylinder. Hence,

$$\oint \operatorname{grad} S \cdot d\mathbf{s} = 2\pi n ,$$

$n$ being an integer, or

$$\varkappa = \oint \boldsymbol{v} \cdot d\mathbf{s} = 2\pi n \frac{\hbar}{m} . \tag{V.3.4}$$

In the axially symmetrical case,

$$v = n \frac{\hbar}{m} \frac{1}{s} . \tag{V.3.5}$$

This is exactly the velocity distribution for a vortex line in classical hydrodynamics.

If the difficulty with the singularity can be reconciled through a detailed investigation, we may omit the inner cylinder and obtain a solution describing a « quantized vortex line ». The kinetic energy associated with such a vortex line of unit length is

$$\frac{1}{2} \int_{a_0}^{b} \varrho_s v_s^2 \cdot 2\pi r \, dr = \frac{\varrho_s \varkappa^2}{4\pi} \log \frac{b}{a_0} , \tag{V.3.6}$$

if the vortex has a core of radius $a_0$, and it is placed at the center of a cylindrical container of radius $b$. The core size is usually thought to be of the order of the interatomic distance, *i.e.*, $10^{-8}$ cm.

Let us now examine the arguments carefully. First, we notice that, if the temperature had been different from 0 °K, we would have to consider a thermodynamical average. In such a case, the above argument would not hold without a rather elaborate modification, and it is not clear that discrete quantization would be obtained, although macroscopic quantum-mechanical effects can still be present in the same qualitative manner; *e.g.*, the order in momentum space and the resultant superflow of heat.

Secondly, if the above derivation is to be valid, both $\Psi$ and $\psi$ must be the solution of Schrödinger's equation. A little calculation will convince the reader that this is in general impossible, even if we assume the phase angle $S$

to depend on time. The only obvious exception is a pure translation, which is of no interest here (*).

Thirdly, it is clear that the type of argument used above cannot be used to *exclude* other types of motion. The best that can be accomplished is to show that quantized irrotational motion is a *possibility*. We have just seen that even this has not been accomplished.

The above difficulties arise with most of the attempts to establish the theory of quantized vortex lines. To establish the desired theory, it is necessary to prove the *stability* of the proposed state of motion and its statistical likelihood over other rotational motions. This problem still remains to be a challenge to the theoretical physicist.

## VI. - Experimental Evidence

### 1. – Irrotational superflow.

Let us now turn to the discussion of experimental evidence for irrotational superflow and quantized vortex lines. In view of the theoretical difficulties the experiments must be examined critically. This attitude is supported by the unfortunate incidents where some apparent evidence for discrete quantization turned out to be false upon closer examination. After all, mechanical vibrations tend to show discrete behavior, and it is very difficult to avoid small vibrations entirely that might interfere with the necessarily delicate experiments.

We shall divide experiments into the following categories according to the evidence they were designed to provide:

*A*) irrotational superflow,

*B*) quantized circulation,

*C*) individual quantized vortex lines,

*D*) collective effect of a large number of quantized vortex lines.

Experiments of Class *A* will be discussed in this section, Class *B* will be discussed in Section **2**, and Class *D* in Section **3**. There are no experiments which provided direct evidence for the existence of individual quantized vortex lines. Indirect evidence for their existence will be examined in connection with the other experiments mentioned above.

(*) The mathematical difficulty in the present derivation should be contrasted with the case of the derivation of the quantization of flux in the case of superconductivity [42]. This shows clearly a difference in character between superconductivity and superflow.

Irrotational superflow has no doubt been observed, but the main question is whether the cause is the slip at the boundary or whether the superflow must be pointwise irrotational. A related question is the critical velocity of the wall required to excite rotation.

The critical velocity estimated from the mechanism of phonon excitation is far too large to be related to the critical velocity observed by REPPY and LANE [31] in a rotating bucket. Their experimental result is as follows:

« The helium is contained in a glass bucket (radius 1 cm, length 10 cm) which is hung from a Beams' type magnetic suspension, the whole being enclosed in a high vacuum. This arrangement is very nearly frictionless and the empty bucket, once set in rotation, will coast freely for several hours.

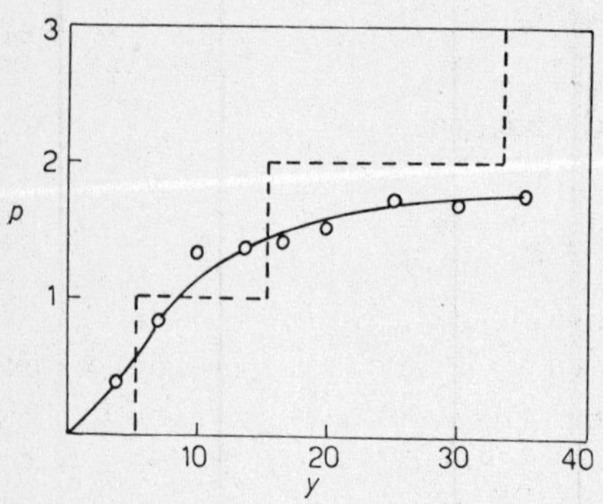


Fig. 5. – Bucket speed as a function of time (after REPPY and LANE [31]).

With the temperature of the helium in the bucket set at 1.2 °K ($\varrho_n/\varrho \approx 3$ per cent) the system is given a short ($\sim \frac{1}{2}$ s) rotational impulse and then permitted to coast freely. The angular velocity of the bucket is then observed as a function of time. In Fig. 5 curve $A$, the bucket speed is seen to decrease, finally reaching a steady value after some 6 ks. This means that the superfluid has derived angular momentum from the bucket and, further this angular momentum turns out to be just that for solid body type rotation at the final speed. In curve $B$, taken at a lower initial speed, the same result obtains except that the superfluid is scarcely accelerated at all for the first few thousand seconds. Finally, in curve $C$, the superfluid remains substantially at rest for the whole course of the experiment which lasts some three hours. The small decline in velocity observed is probably due to the fact that the experiment is not quite isothermal. In the course of 10 ks the helium in the bucket warms up from 1.20 °K ($\varrho_n/\varrho = 3$ per cent) to 1.75 °K ($\varrho_n/\varrho = 27$ per cent) and the normal fluid so created rotates classically thus slowing down the system somewhat ».

REPPY and LANE explained the observed phenomena in terms of vortex lines; but it is just as easy to explain them without a specific vortex line model. All the curious phenomena can be described in terms of a difficulty of interaction between the wall and the superfluid. After the interaction has been initiated, the continuum theory of Part IV is found to be quite adequate to account for the main features of the transition phenomenon, which ends in a state of uniform rotation for the fluid and the bucket alike.

The most conclusive experiment would still be a test of the actual lack of rotation when the bucket speed is extremely small, as originally suggested by LONDON. A specific procedure is as follows. Suppose we have a sealed bucket of superfluid helium suspended in such a manner that it is frictionless with respect to rotation. Let the system be at rest in the laboratory for a *long* time. Since the superfluid helium cannot rotate, the bucket must take on the rotation of the earth by itself. (There is less than one quantum of circulation for a quantized vortex line to be produced for a bucket of ordinary size.) The superfluid helium is then heated up (by radiation, for example). According to the prevailing theory, there should be an observable « spontaneous rotation » of the system, as a calculation of the angular momentum would indicate.

Originally, the apparatus is at rest in the laboratory. It has an angular momentum.

$$J_0 = I_0\omega_0 .$$

After the superfluid becomes ordinary fluid, the whole system has an angular velocity $\omega$ given by

$$I_0\omega_0 = (I_0 + I)\omega ,$$

where $I$ is the moment of inertia of the fluid. The difference in angular velocity

$$\Delta\omega = \omega - \omega_0 = -\frac{I}{I_0 + I}\,\omega_0 ,$$

should be observable as a « spontaneous rotation » in the system of co-ordinates at rest in the laboratory.

There are presumably experimental difficulties to be overcome in the procedure suggested above; but such an experiment or its equivalent is really needed to settle the question completely and to match the observation of Meissner effect in superconductivity. (See also suggestion at end of Section **2**).

## 2. – Experimental evidence. Quantized circulation.

Pending the establishment of irrotational superflow, let us see if quantized circulation can be directly established. VINEN [16] attempted to detect single quantization by performing the following measurements. He filled He II between a pair of concentric cylinders in the form of a wire of length 5 cm and radius $a = 1.3 \cdot 10^{-3}$ cm placed inside of a cylinder with inner radius $b = 0.2$ cm.

The whole instrument system is rotated as various speeds $\omega$ characterized by the dimensionless parameter

$$y = 2\pi b^2\omega/(h/m)$$

which is the number of quanta of the circulation $2\pi b^2\omega$ of the wall speed.

The frequencies of the various modes of vibration of the wire are observed by placing the whole instrument assembly in a stationary magnetic field. It is reasoned that if the wire were circular, there would be a degeneracy of the vibration frequencies; in this case, the circulation around the wire would cause, through the Magnus effect, a split of frequency of

$$\Delta\gamma = \varrho_s \varkappa / 2\pi w ,$$

where $\varrho_s$ is the density of the circulating fluid, and $w$ is the sum of the mass per unit length of the wire and half of the mass of fluid displaced by unit length of wire. It is estimated that

$$\Delta\nu = 0.45 \text{ Hz}$$

per unit quantum of circulation for the wire at hand.

Unfortunately, the experiment is extremely delicate. In the first place, the wire is never symmetrical and has a natural split in frequency amounting to 5 Hz. This has to be corrected by delicate adjustments. Secondly, if the whole instrument assembly were wobbling somewhat during its rotation, such a wobbling frequency would also be registered (*). Thirdly, the damping of the vibration due to the presence of the normal fluid makes the desired measurement barely on the verge of possibility. VINEN treated these and other difficulties very carefully (except for the possible wobbling). The situation is extremely complicated; the explanations given might be satisfactory to an experienced experimentalist, but it is hardly within the confident comprehension of a mere theoretician. For example, it is difficult to understand why the natural splitting of the frequencies should be the observed split when the liquid helium level is just at the top of the wire during the processes of emptying the vessel, a point noted by VINEN himself.

The measured « apparent circulation » is plotted against the speed of rotation (see Fig. 6). It is found that this apparent circulation does not exhibit quantum jumps. By twanging the wire with a current impulse, one could get one quantum of circulation, but two quanta have never been observed, even when the circulation of the outer wall is as high as 35 quanta.

---

(*) Although a number of observations were made when the apparatus was no longer rotating, the wire was usually twanged. Indeed, a part of vortex line was sometimes observed when the apparatus had not been in rotation prior to the observation.

In attempting to explain these phenomena, VINEN considers free vortex lines and partly attached vortex lines. Now it is clear that if the vortex line has a core of $a_0 = 10^{-8}$ cm whereas the wire has a radius of $a = 10^{-3}$ cm there is a tremendous energy advantage

$$\Delta E = -\frac{\varrho \varkappa^2}{4\pi} \log \frac{a}{a_0},$$

in attachment. Indeed, free vortices would tend to be excluded altogether in favor of multiple quantization. VINEN was led to conclude, after elaborate calculation, that the *core* radius of the vortex line must exceed $10^{-4}$ cm. The above formula shows that this value might be anticipated qualitatively, since the free vortex line must be energetically comparable with the state of circulation around the wire of radius $10^{-3}$, if it is to survive at all.

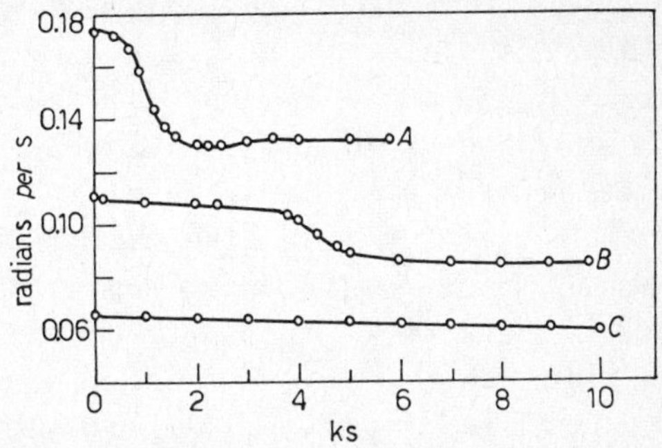


Fig. 6. – Observed apparent circulation around inner cylinder in rotating liquid He II (after VINEN [16]). Solid line connecting experimental points added by the present writer.

If Vinen's estimate of the core radius be accepted, there is either a large hole or a region of distributed vorticity. Either of these would be at variance with the theoretical basis for the existence of vortex lines discussed in Section **3**.

In some of Osborne's experiments, the distance between the vortices is estimated to be of the order of $2 \cdot 10^{-4}$ cm. Thus, according to Vinen's estimate, the vortices are merging into each ither into a continuous distribution of vorticity.

The following quotation from VINEN is noteworthy (*): « But the fact remains that the present measurements do at least *suggest* that the simplest picture of a free vortex line is in some way inadequate, and that the theory behind this picture should be re-examined. It seems possible, for example, that some indeterminacy in the position of a line might appreciably reduce its energy; furthermore, *the value of $a_0$ might vary with the experimental conditions*, and the value that governs the vortex wave velocity might differ from the one that governs the energy per unit length of a stationary line. It should perhaps be stressed that a large value of $a_0$ does not imply the existence of a large hole in the liquid; $a_0$ is to be regarded simply as a parameter that describes the energy per unit length of line, and its value is determined by the

(*) Some italics mine.

detailed conditions near the centre of the line, which may depart considerably from those in the vortex line of classical hydrodynamics ».

The fact that VINEN never observed two quanta of circulation raises the doubt whether the observed apparent single quanta might also be due to extraneous sources. In any case, the extremely large value of $a_0$ is a serious cause for concern. Experiments of the same type, but performed with a different equipment, should help to remove extraneous effects. A different wire size seems to be extremely desirable.

It is noteworthy that the angular speeds of rotation in some of Vinen's experiments are of the same order of magnitude as those in the experiments of Reppy and Lane. In the latter experiments, the equilibrium state is always that of ordinary solid body rotation, whereas the motion is far from that state in Vinen's theoretically calculated results, even when the wall speed amounts to 35 quanta of circulation. It would be interesting to pursue this point further both experimentally and theoretically. Indeed, a general mathematical theory for the application of the theory of quantized vortex lines has never been published.

## 3. – Collective effect of quantized vortex lines.

In most of the experiments dealing with rotating helium, a *large* number of quanta of circulation $h/m$ is involved. Thus, one would expect the net effect to be essentially the same as that for continuous vorticity distribution, even when quantized vortex lines do exist. It is therefore all the more remarkable that attenuation of second sound was observed to increase with rotation [33] (*), even though the amount of increase was not large. (In one experiment, it amounted to about 30 % of the original attenuation when the angular speed of rotation was 9 rad/s.)

As mentioned before, in the idealized situation, no extra attenuation is to be expected from the continuum theory, but that theory is not always reliable for the prediction of sound attenuation. If one insists on using ordinary continuum concepts, one would perhaps suggest some form of secondary motion. This is supported by the observed fact that there is an initial large transient attenuation comparable to that observed by WHEELER, BLAKEWOOD and LANE [43] in a case where secondary motion is definitely present. The possibility therefore remains that some sort of macroscopic secondary motion was present supported partly by rotation and partly by a steady heat current which is present even in the final state of equilibrium.

(*) Also observed by DONNELLY and SNYDER (private communication).

Another approach to the problem is that given by BEKAREVICH and KHALATNIKOV [18]. As mentioned before, in their picture, the internal energy is supposed to depend on the magnitude of superfluid rotation, making it a new thermodynamic variable. In fact, they were able to obtain the same equations proposed by HALL [36], (also lecture at this Course) without *committing themselves to a definite molecular picture* (cf. alternative suggestions below).

If one accepts the idea of quantized vortex lines, one gets a rather natural interpretation for the fact that the extra attenuation is proportional to the speed of the rotation:

$$\alpha = \alpha_0 + \frac{B\Omega}{2u_{\text{II}}},$$

(where $B$ is a constant (varying with temperature) of the order of unity, $\Omega$ is the speed of rotation and $u_{\text{II}}$ is the speed of propagation of second sound), since the number of scattering centers is proportional to $\Omega$. However, in the detailed theory, HALL and VINEN considered a force of interaction on a vortex in the form

$$\boldsymbol{f} = C(\boldsymbol{v}_n - \boldsymbol{v}_L)$$

with the normal fluid, but also in the form

$$\boldsymbol{f} = \varrho_s(\boldsymbol{v}_L - \boldsymbol{v}_s) \times \boldsymbol{\varkappa}$$

with the superfluid, where $\boldsymbol{v}_L$ is the velocity of the vortex line. Now if the vortex line is a singularity in the superfluid, it moves with the local superfluid velocity, and is not subjected to the Magnus effect (Section **4**, Part I). Thus, the explanation given by these authors is also somewhat obscure.

If indeed there be a « *granular structure* », the picture of quantized vortex lines does not represent the only possibility. For one thing, the macroscopic convection cells would have such a structure. Also, as mentioned before, small quantized vortex rings might be easier to form than long vortex lines. These rings, if they are formed, would of course be oriented in some manner with a preferential direction in line with the axis of rotation. Qualitatively, the same results would then be obtained as far as attenuation goes. The vortex ring picture would be more readily reconciled with a continuum picture since we may consider a macroscopic element as one containing many such rings. For example, if the ring diameter is $10^{-5}$ cm, it is still much larger than the probable core diameter $10^{-8}$ cm, but would still be sufficiently small not to modify the continuum concept, except in the more delicate problems of attenuation and the like. Thus, there is a possibility of reconciling the continuum theory with the experimentally observed phenomena. The necessary appeal to detailed « *sub-macroscopic structures* » for the attenuation processes is similar to that in the case of the attenuation of ordinary sound in ordinary fluids.

It might be interesting to try experiments with the scattering of neutron beams or ion beams (*) in rotating He II to see whether there is a granular structure associated with the vortex lines which are presumed to exist according to the prevailing theory.

Another experiment purporting to support the idea of collective effect of quantized vortex lines was performed by HALL [36]. As mentioned before (Section **3**, Part IV), HALL found, with an oscillating disc in rotating liquid helium, a prominent change of the character of the motion as the oscillating frequency passes through twice the frequency of rotation. This is a classical effect. He also found a slight variation of the frequency of the oscillating system as the level of the liquid is changed, and attributed this to a resonance phenomenon. It should be noted that the oscillating system is damped to half its initial period after about $10 \div 20$ oscillations (**) and that the change of frequency observed is only about 1 %. Thus, it does not constitute a conclusive evidence by itself, especially since very few measurements have been made.

## 4. – Conclusion.

The fundamental question whether superflow in He II must have a pointwise irrotational character is not resolved. Conclusive theoretical proof and direct experimental evidence are wanting. Although the theory of quantized vortex lines is ingenious and attractive, there is as yet no conclusive evidence for their existence. If something like these vortex lines should exist, there is as yet no clear picture of their size and structure.

The ultimate solution of these basic theoretical problems must await the development of a systematic theory for the non-equilibrium behavior of many-particle systems obeying the laws of quantum mechanics. On the other hand, the continuum theory of hydrodynamics of He II, as developed in this article, can serve the usual purpose of providing an adequate description for most macroscopic problems. Macroscopic behavior which depends more intimately on the molecular character—*e.g.*, the attenuation of sound—cannot be fully accounted for by a simple macroscopic theory, both in ordinary fluids and in superfluids.

---

(*) Such experiments were reported by Prof. CARERI during this Course.
(**) Private communication from HALL at the conference.

More attention should be paid to the mechanism of interaction between the liquid and the solid boundary, since the boundary conditions are known to have a far reaching significance in fluid motion.

## REFERENCES


[1] J. Serrin: *Handb. d. Phys.*, **8**, 000 (1959).

[2] W. Thomson: *Proc. Roy. Soc. Edin.*, **10**, 92 (1879).

[3] H. Lamb: *Hydrodynamics* (Cambridge, 1932).

[4] C. C. Lin: *Proc. Nat. Acad. Sci.*, **27**, 570 (1941).

[5] C. C. Lin: *On the Motion of Vortices in Two Dimensions* (Toronto, 1943).

[6] S. Goldstein: *Modern Development in Fluid Dynamics* (Oxford, 1938).

[7] C. C. Lin: *Theory of Hydrodynamic Stability* (Cambridge, 1955).

[8] S. Chandrasekhar: *Proc. Roy. Soc.*, **241**, 29 (1957); *Hydrodynamic and Hydromagnetic Stability* (Oxford, 1961).

[9] C. C. Lin and W. H. Reid: *Handb. d. Phys.*, **8**, 000 (1961).

[10] T. D. Lee, K. Huang and C. N. Yang: *Phys. Rev.*, **106**, 1135 (1957).

[11] T. D. Lee and C. N. Yang: *Phys. Rev.*, **112**, 1419 (1958).

[12] F. London: *Superfluids*, vol. **2** (New York, 1954).

[13] L. D. Landau: *Žurn. Ėksp. Teor. Fiz.*, **5**, 71 (1941).

[14] T. T. Wu: *Journ. Math. and Phys.*, **2**, 105 (1961).

[15] R. P. Feynman: *Progress in Low Temp. Phys.*, vol. **1**, ed. by C. J. Gorter (Amsterdam, 1955), chap. II.

[16] W. F. Vinen: *Proc. Roy. Soc.*, **260**, 218 (1960).

[17] J. F. Allen and H. Jones: *Nature*, **141**, 75 (1938).

[18] I. L. Bekarevich and I. M. Khalatnikov: *Proc. VII Intern. Conference on Low Temperature Physics* (1960), p. 459; *Sov. Phys. JETP*, **13**, 643 (1961).

[19] F. London: *Nature*, **141**, 643 (1938); *Phys. Rev.*, **54**, 497 (1938).

[20] L. Tisza: *Nature*, **141**, 913 (1938); *Compt. Rend.*, **207**, 1035 (1938).

[21] T. D. Lee and C. N. Yang: *Phys. Rev.*, **113**, 1406 (1959).

[22] P. R. Zilsel: *Phys. Rev.*, **79**, 309 (1950).

[23] P. R. Zilsel: *Phys. Rev.*, **92**, 1106 (1953).

[24] K. R. Atkins: *Liquid Helium* (Cambridge, 1959).

[25] C. C. Lin: *Phys. Rev. Lett.*, **2**, 245 (1959).

[26] C. S. Hung, B. Hunt and P. Winkel: *Physica*, **18**, 629 (1952).

[27] W. J. Heikkila and A. C. H. Hallett: *Can. Journ. Phys.*, **33**, 420 (1955).

[28] A. D. B. Woods and A. C. H. Hallett: *Proc. V Intern. Conference on Low Temperature Physics and Chemistry* (Madison, 1957), p. 16.

[29] R. J. Donnelly and A. C. H. Hallett: *Ann. Phys.*, **3**, 320 (1958).

[30] D. V. Osborne: *Proc. Phys. Soc.*, A **63**, 909 (1950).

[31] J. D. Reppy and C. T. Lane: *Proc. VII Intern. Conference on Low Temperature Physics* (1960), p. 443.

[32] S. Chandrasekhar and R. J. Donnelly: *Proc. Roy. Soc.*, A **241**, 9 (1957).

[33] H. E. Hall and W. F. Vinen: *Proc. Roy. Soc.*, A **238**, 204, 215 (1956).

[34] A. C. H. Hallett: *Proc. Roy. Soc.*, A **210**, 404 (1952).

[35] R. J. Gribben: *Journ. Math. and Phys.* **15**, 177, 179 (1961).

[36] H. E. Hall: *Phil. Trans.*, A **250**, 359 (1957).
[37] L. D. Landau and I. M. Khalatnikov: *Žurn. Ėksp. Teor. Fiz.*, **19**, 637, 709 (1949).
[38] J. M. Blatt, S. T. Butler and M. R. Schafroth: *Phys. Rev.*, **100**, 476, 481, 495 (1955).
[39] L. Onsager: *Phys. Rev.*, **64**, 114 (1944).
[40] L. Onsager: *Suppl. Nuovo Cimento*, **6**, 249 (1949).
[41] E. P. Gross: preprint (CERN, 1961).
[42] N. Byers and C. N. Yang: *Phys. Rev. Lett.*, **7**, 46 (1961).
[43] R. G. Wheeler, C. H. Blakewood and C. T. Lane: *Phys. Rev.*, **99**, 1667 (1955).

## ERRATA

1. Page 111, line 25 should read: "... of turbulent motion, calculations of distributions of mean quantities are still made by the use ..."

2. Pages 138 and 141: Figures 5 and 6 have been transposed.

# On the Asymptotic Theory of a Class of Ordinary Differential Equations of Fourth Order

## II. Existence of Solutions which are Approximated by the Formal Solutions

*By C. C. Lin and A. L. Rabenstein*

### 1. Introduction

In Part I (Lin and Rabenstein, 1960) a *formal* asymptotic theory was developed for the differential equation

$$\frac{d^4\varphi}{dx^4} + \lambda^2\left[P(x,\lambda)\frac{d^2\varphi}{dx^2} + Q(x,\lambda)\frac{d\varphi}{dx} + R(x,\lambda)\varphi\right] = 0. \tag{1.1}$$

The purpose of this paper is to prove the existence of actual solutions of (1.1) which are represented asymptotically, to a finite order, by the formal solutions which were constructed in Part I. In other words, we shall form a rigorous asymptotic theory from the formal one. This theory will be *uniformly valid* in the sense that the asymptotic behavior of each solution will be known in a complete neighborhood of the turning point.

In part I, it was shown that equation (1.1) could be reduced to the normal form

$$L_0(\chi) \equiv \chi^{(4)} + \lambda^2(z\chi'' + \alpha_0\chi' + \beta_0\chi) = \lambda(a\chi + b\chi' + \lambda^{-1}c'' + \lambda^{-2}d\chi''') \tag{1.2}$$

with independent and dependent variable $z$ and $\chi$ respectively. The quantities $\alpha_0$ and $\beta_0$ are constants. The quantities $a$, $b$, $c$, and $d$ are functions of $z$ and $\lambda$ which are analytic in $z$ in a region containing the origin and are uniformly bounded for $z$ in a closed subset of this region and for $|\lambda| \geq |\lambda_0| > 0$. In this paper we shall deal only with equation (1.2). However, the final results can be expressed in terms of the original variables by reversing the transformations* which lead from (1.1) to (1.2). We shall consider (1.2) only for the case

$$\mathrm{Re}(\alpha_0) > 1. \tag{1.3}$$

When $\mathrm{Re}(\alpha_0) < 1$, the *adjoint* of equation (1.2) can be treated by the theory presented here, and a complete theory for the original equation can be carried through via the adjoint equation.

In Part I, formal approximate solutions of equation (1.2) were constructed. These have the structure

$$\chi_n = \mu(z,\lambda)u + \nu(z,\lambda)u' + \lambda^{-1}\sigma(z,\lambda)u'' + \lambda^{-2}\tau(z,\lambda)u''', \tag{1.4}$$

* These transformations are given by equations (2.7), (2.8), and (2.13) of Part I.

where $u$ is any solution of the basic *reference equation*

$$u^{\mathrm{IV}} + \lambda^2(zu'' + \alpha u' + \beta u) = 0. \tag{1.5}$$

In general, the quantities $\alpha$ and $\beta$ in (1.5) have the forms

$$\alpha = \sum_{j=0}^{n} \alpha^{(j)}\lambda^{-j}, \qquad \beta = \sum_{j=0}^{n} \beta^{(j)}\lambda^{-j}, \tag{1.6}$$

when $\alpha^{(0)} = \alpha_0$, while in (1.4), the function $\mu(z, \lambda)$ has the form

$$\mu(z, \lambda) = \sum_{j=0}^{n} \mu_j(z)\lambda^{-j}, \tag{1.7}$$

the functions $\nu, \sigma$, and $\tau$ having similar expansions. In (1.6) and (1.7), $n$ is an arbitrary, but fixed nonnegative integer. In certain cases, namely when $\beta_0$ is 0 and $\alpha_0$ is one of the integers $\alpha_0 = -N$ or $\alpha_0 = N + 2$, $N = 1, 2, \ldots$ it is necessary to modify the construction somewhat by taking

$$\begin{aligned} \alpha &= \sum_{j=0}^{n} \alpha^{(j)}\varepsilon^j, \qquad \beta = \sum_{j=0}^{n} \beta^{(j)}\varepsilon^j, \\ \mu(z, \lambda) &= \sum_{j=0}^{n(N+1)} \mu_j(z)\varepsilon^j \end{aligned} \tag{1.8}$$

where

$$\varepsilon = \lambda^{-1/(N+1)}. \tag{1.8a}$$

However, these cases require no special consideration in the theory to be presented here. It was shown that the constraints $\alpha^{(j)}$ and $\beta^{(j)}$ and the functions $\mu_j$, $\nu_j$, $\sigma_j$ and $\tau_j$ can be chosen in such a way that the latter are analytic and such that the quantities (1.4) satisfy a *related equation* of the form

$$\begin{aligned} L_0(\chi_n) = {} & \lambda(a\chi_n + b\chi_n' + \lambda^{-1}c\chi_n'' + \lambda^{-2}\, d\chi_n''') \\ & + \lambda^{1-n}(\tilde{a}\chi_n + \tilde{b}\chi_n' + \lambda^{-1}\tilde{c}\chi_n'' + \lambda^{-2}\,\tilde{d}\chi_n'''), \end{aligned} \tag{1.9}$$

where the functions $\tilde{a}$, $\tilde{b}$, $\tilde{c}$, and $\tilde{d}$ have the same properties as the functions $a$, $b$, $c$, and $d$. Thus, corresponding to each solution $u$ of equation (1.5), there is a solution $\chi_n$ of equation (1.9).

We shall show that each solution of equation (1.9) approximates a solution of equation (1.2) in a complete neighborhood of the turning point. A more precise description of our results is given in the next section.

## 2. Results

In order to give a precise statement of our results, we shall need a number of definitions. Let $S$ be a closed disk in the $z$-plane, centered at the origin, and contained in an open region* where the coefficients of equation (1.2) are analytic. Let $C_k$, $k = 1, 2, 3$, be the ray given (mod $2\pi$) by

$$\arg z = \frac{2\pi}{3}(2 - k) - \frac{2}{3}\arg\lambda \equiv \theta_k, \tag{2.1}$$

* The size of this region depends only on the nature of the function $P(x, \infty)$. See equation (2.7), Part I.

and let $S_k$, $k = 1, 2, 3$, be the closed sector of $S$ which lies opposite the ray $C_k$. Let $\xi_k$ be the branch of the function

$$\xi = \tfrac{2}{3}i\lambda z^{3/2}, \tag{2.2}$$

which has a positive real part in the sector $S_k$ and a negative real part in the other two sectors. For any given branch of the function (2.2) let

$$\xi^s = \begin{cases} \xi, & |\xi| > |\xi_0| \\ |\xi_0| e^{i \arg \xi_0}, & |\xi| \le |\xi_0| \end{cases} \tag{2.3}$$

where $|\xi_0|$ is some fixed positive real number, still at our disposal. Let

$$f_k(z, \lambda) = \begin{cases} \left|\dfrac{\xi}{\xi^s}\right| \cos\left\{\dfrac{\pi}{2\varepsilon_1}(\arg z - \theta_k)\right\}, & |\arg z - \theta_k| < \varepsilon_1 \\ 0, \quad \text{elsewhere} \end{cases} \tag{2.4}$$

where $\theta_k$ is given by (2.1) and $\varepsilon_1$ is any fixed real number in the range $0 < \varepsilon_1 \le \pi/3$. Let

$$g_k(z, \lambda) = \begin{cases} \left|\dfrac{\xi}{\xi^s}\right| \cos\left\{\dfrac{\pi}{2\varepsilon_2}(\arg z - \theta_k + \pi)\right\}, & |\arg z - \theta_k + \pi| < \varepsilon_2 \\ 0, \quad \text{elsewhere} \end{cases} \tag{2.5}$$

where $\theta_k + \pi$ is the value (mod $2\pi$) of $\arg z$ on the bisector of the sector $S_k$, and $\varepsilon_2$ is any fixed real number in the range $\pi/3 < \varepsilon_2 \le 2\pi/3$. The function $f_k$ and $g_k$ are bounded, continuous, nonnegative functions which vanish outside of certain sectors of the $z$-plane. The function $f_k$ vanishes outside of a sector which contains the ray $C_k$. The function $g_k$ vanishes outside of a sector which contains the sector $S_k$. Let $E(z, \lambda)$ be a generic symbol for a function which is continuous and uniformly bounded for $z$ in $S$ and $|\lambda| \ge |\lambda_0|$, where $|\lambda_0|$ is a suitably large positive constant. Let $u = B_0, B_k, A_k$, $k = 1, 2, 3$, be the solutions of the reference equation which were defined in the study of that equation. (Rabenstein, 1958.) Let $\chi_n = B_{n,0}, B_{n,k}, A_{n,k}$ be the corresponding solutions of the related equation (1.9).

Then the following statements are valid when $z$ is in the disk $S$, $\operatorname{Re}(\alpha_0) > 1$, and $|\lambda| \ge |\lambda_0|$:

THEOREM I. *Equation* (1.2) *possesses a solution* $B_{n,0}$ *with the property that*

$$\frac{d^m B_{n,0}}{dz^m} = \frac{d^m \tilde{B}_{n,0}}{dz^m} + \lambda^{-(n+1)} E(z, \lambda), \qquad (m = 0, 1, 2, 3) \tag{2.6}$$

where

$$\frac{d^m \tilde{B}_{n,0}}{dz^m} = E(z, \lambda), \qquad (m = 0, 1, 2, 3). \tag{2.6a}$$

THEOREM II. *Equation* (1.2) *possesses solutions* $A_{n,k}$, $k = 1, 2, 3$, *such that*

$$\frac{d^m A_{n,k}}{dz^m} = \frac{d^m \tilde{A}_{n,k}}{dz^m} + \lambda^{-(n+1)} \rho_{m,k}(z, \lambda) E(z, \lambda) \log \lambda, \qquad (m = 0, 1, 2, 3) \tag{2.7}$$

*where*

$$\frac{d^m \tilde{A}_{n,k}}{dz^m} = \rho_{m,k}(z, \lambda) E(z, \lambda), \qquad (m = 0, 1, 2, 3) \tag{2.7a}$$

*and*

$$\rho_{m,k}(z, \lambda) = \lambda^{(2/3)(\alpha_0+m-1)}[|\xi^{[(\alpha_0+m)/3-5/6]s}e^{-\xi_k}| + |\xi^{-(2/3)(\alpha_0+m-1)s}| f_k(z, \lambda)]. \qquad (2.7b)$$

THEOREM III. *Equation* (1.2) *possesses solutions* $B_{n,k}$, $k = 1, 2, 3$, *such that*

$$\frac{d^m B_{n,k}}{dz^m} = \frac{d^m \tilde{B}_{n,k}}{dz^m} + \lambda^{-(n+1)}\eta_{m,k}(z, \lambda)E(z, \lambda)\log\lambda, \qquad (m = 0, 1, 2, 3) \qquad (2.8)$$

*where*

$$\frac{d^m \tilde{B}_{n,k}}{dz^m} = \eta_{m,k}(z, \lambda)E(z, \lambda), \qquad (m = 0, 1, 2, 3) \qquad (2.8a)$$

*and*

$$\eta_{m,k}(z, \lambda) = \lambda^{(2/3)(\alpha_0+m-1)}[|\xi^{-(2/3)(\alpha_0+m-1)s}| + |\xi^{[(\alpha_0+m)/3-5/6]s}e^{\xi_k}|g_k(z, \lambda)]. \qquad (2.8b)$$

The solution $B_{n,0}$ is evidently balanced for $z$ in $S$. The solution $A_{n,k}$ is recessive in the sector $S_k$ and is dominant in the other two sectors. It has a transition line along the ray $C_k$. The solution $B_{n,k}$ is balanced in the sector $S - S_k$ and is dominant in the sector $S_k$.

That fundamental sets of solutions for equation (1.2) can be formed from the solutions described above is easily established. For instance, we have for $z$ in the sector $S - S_3$,

$$W(B_{n,0}, B_{n,3}, A_{n,1}, A_{n,2}) = W(B_{n,0}, B_{n,3}, A_{n,1}, A_{n,2}) + \delta, \qquad (2.9)$$

where

$$\delta = \lambda^4 \xi^s \eta_{0,3}\rho_{0,1}\rho_{0,2}E(z, \lambda)\lambda^{-(n+1)}\log\lambda = \lambda^{2\alpha+2}E(z, \lambda)\lambda^{-(n+1)}\log\lambda. \qquad (2.10)$$

From equations (2.19), (2.20) of Part I, we have

$$W(\tilde{B}_{n,0}, \tilde{B}_{n,3}, \tilde{A}_{n,1}, \tilde{A}_{n,2}) = W(B_0, B_3, A_1, A_2)[1 + o(1)] \qquad (\lambda \to \infty). \qquad (2.11)$$

Since (Rabenstein, 1958, 1959)

$$W(B_0, B_3, A_1, A_2) = 4\pi^2 e^{-2\pi i\alpha}\beta^{\alpha-1}\lambda^{2\alpha+2}, \qquad (2.12)$$

we see that the set of functions $B_{n,0}$, $B_{n,3}$, $A_{n,1}$, $A_{n,2}$ is linearly independent for $z$ in $S$ and $|\lambda|$ sufficiently large. The existence of other fundamental sets of solutions can be established in a similar manner.

In a previous study of fourth order differential equations of the type considered here (Langer, 1957), theorems about the existence and asymptotic behavior of solutions have been proved for $\lambda$ and $z$ in *associated regions*. The one region, $\Lambda$, is a sector of the complex $\lambda$-plane with vertex at the origin, and the other, $R_z$, is a domain of the $z$-plane, which in every case includes the origin. In the case of equation (1.2), every associated pair of regions has an integer $h$ attached to it, since $R_\xi^\Lambda$ the image of $R_z^\Lambda$ under the mapping

$$\xi = \tfrac{2}{3}i\lambda z^{3/2}, \qquad (2.13)$$

is, for every $\lambda$ in $\Lambda$, wholly within the sector

$$(h - \tfrac{3}{2})\pi < \arg\xi < (h + \tfrac{1}{2})\pi. \qquad (2.14)$$

When the argument of $\lambda$ is regarded as fixed, as it is in our theory, the regions of the $z$-plane which map into the sectors (2.13) are the sectors $S - S_k$, $k = 1, 2, 3$. Theorems about the behaviors of solutions in such a sector give no information about the behaviors of the same solutions in the sector $S_k$, and hence they are not uniformly valid results. As we shall show, it is our knowledge of the solutions of the reference equation (1.5) in a complete neighborhood of the turning point which enable us to obtain the uniformly valid theorems stated here.

We shall discuss in detail the three solutions $B_{n,0}$, $B_{n,3}$, $A_{n,1}$ of equation (1.2), which corresponds to the three solutions $B_0$, $B_3$, $A$, of equation (1.5). This set includes the three types of solutions called balanced, partly balanced, and dominant-recessive. Each of the functions $u = B_0, B_k, A_k$, $k = 1, 2, 3$, falls into one of these categories. It is true that neither of the sets of functions $\{A_k\}$, $\{B_k\}$ is completely symmetric, as may be seen from the definitions of these functions. Nevertheless, the transition lines for the functions, and the sectors where they are balanced, dominant, or recessive are symmetrically located. Because of this, the existence proof which corresponds to any one of the above-mentioned functions is similar to one of the proofs given here. The proof of Theorem I is given in Section 7. The proof of Theorem II, for the case $k = 1$, is given in Section 8. The proof of Theorem III, for the case $k = 3$, is given in Section 9.

## 3. Outline of the procedure

In this section we shall describe the general procedure which we use to compare the solutions of equation (1.9) with the solutions of equation (1.2). Detailed comparisons for specific cases will be carried out in Sections 7, 8, 9.

For the purpose of this section, we shall use the formulation of our differential equation as first order systems. Let $\boldsymbol{\chi}, \boldsymbol{\chi_n}, \mathbf{u}$ be the vectors

$$\begin{aligned} \boldsymbol{\chi} &= (\chi, \chi', \lambda^{-1}\chi'', \lambda^{-2}\chi'''), \\ \boldsymbol{\chi_n} &= (\chi_n, \chi_n', \lambda^{-1}\chi_n'', \lambda^{-2}\chi_n'''), \\ \mathbf{u} &= (u, u', \lambda^{-1}u'', \lambda^{-2}u'''). \end{aligned} \tag{3.1}$$

Then equations (1.5), (1.2), and (1.9) are equivalent, respectively, to the systems

$$\frac{d\mathbf{u}}{dz} = M\mathbf{u}, \tag{3.2}$$

$$\frac{d\boldsymbol{\chi}}{dz} = (M_0 + \varepsilon)\boldsymbol{\chi}, \tag{3.3}$$

$$\frac{d\chi_n}{dz} = (M_0 + \varepsilon_n)\boldsymbol{\chi_n}. \tag{3.4}$$

The matrix $M$ is given by

$$M = \begin{bmatrix} 0 & 1 & 0 & 0 \\ 0 & 0 & \lambda & 0 \\ 0 & 0 & 0 & \lambda \\ -\beta & -\alpha & -\lambda z & 0 \end{bmatrix}, \tag{3.5}$$

the matrix $M_0$ is obtained from $M$ by putting $\alpha = \alpha_0$, $\beta = \beta_0$, the matrix $\varepsilon$ has the form

$$\varepsilon = \lambda^{-1}\begin{bmatrix} 0 & 0 & 0 & 0 \\ 0 & 0 & 0 & 0 \\ 0 & 0 & 0 & 0 \\ a & b & c & d \end{bmatrix}, \tag{3.6}$$

and

$$\varepsilon - \varepsilon_n = \lambda^{-(n+1)}\Delta, \tag{3.7}$$

where the matrix $\Delta$ has the form

$$\Delta = \begin{bmatrix} 0 & 0 & 0 & 0 \\ 0 & 0 & 0 & 0 \\ 0 & 0 & 0 & 0 \\ \tilde{a} & \tilde{b} & \tilde{c} & \tilde{d} \end{bmatrix}. \tag{3.8}$$

We wish to compare the solutions of equations (3.3) and (3.4). In the technique adopted here for the purpose, we write

$$\chi - \boldsymbol{\chi}_{\mathbf{n}} = \lambda^{-(n+1)}\delta, \tag{3.9}$$

and consider the vector $\boldsymbol{\delta}$, which satisfies the differential equation

$$\frac{d\boldsymbol{\delta}}{dz} = M_0\boldsymbol{\delta} + \varepsilon\boldsymbol{\delta} + \Delta\boldsymbol{\chi}_{\mathbf{n}}. \tag{3.10}$$

In order to estimate the quantity $\boldsymbol{\delta}$, let us consider the integral equation

$$\begin{aligned} \boldsymbol{\delta}(z, \lambda) = &\int^z U_0(z, \lambda)U_0^{-1}(t, \lambda)\varepsilon(t, \lambda)\boldsymbol{\delta}(t, \lambda)\, dt \\ &+ \int^z U_0(z, \lambda)U_0^{-1}(t, \lambda)\Delta(t, \lambda)\boldsymbol{\chi}_{\mathbf{n}}(t, \lambda)\, dt, \end{aligned} \tag{3.11}$$

where $\boldsymbol{\chi}_{\mathbf{n}}$ is any specific solution of (3.4) and $U_0$ is any fundamental matrix solution of the system

$$\frac{dU_0}{dz} = M_0 U_0. \tag{3.12}$$

We do not necessarily take a single path of integration in (3.11), but may take several paths starting at different points for different terms in the integrands. Multiplying through from the left in (3.11) by the matrix $\lambda\varepsilon(z, \lambda)$, and writing

$$\begin{aligned} \lambda\varepsilon\boldsymbol{\delta} &= \mathbf{w} = (0, 0, 0, w), \\ \Delta\boldsymbol{\chi}_{\mathbf{n}} &= \tilde{\mathbf{w}}_{\mathbf{n}} = (0, 0, 0, \tilde{\mathbf{w}}_n), \end{aligned} \tag{3.13}$$

where

$$w_n = \tilde{a}\chi_n + \tilde{b}\chi_n + \lambda^{-1}\tilde{c}\chi_n'' + \lambda^{-2}\tilde{d}\chi_n''', \tag{3.14}$$

we obtain the equation

$$\mathbf{w}(z, \lambda) = \lambda^{-1} \int^{z} \lambda\varepsilon(z, \lambda) U_0(z, \lambda) U_0^{-1}(t, \lambda) \mathbf{w}(t, \lambda)\, dt$$

$$+ \int^{z} \lambda\varepsilon(z, \lambda) U_0(z, \lambda) U_0^{-1}(t, \lambda) \tilde{\mathbf{w}}_n(t, \lambda)\, dt. \tag{3.15}$$

Taking the last components of the vectors here, we obtain the scalar integral equation

$$w(z, \lambda) = F(z, \lambda) + \lambda^{-1} \int^{z} K(z, t; \lambda) w(t, \lambda)\, dt, \tag{3.16}$$

where

$$F(z, \lambda) = \int^{z} K(z, t; \lambda) \tilde{\mathbf{w}}_n(t, \lambda)\, dt. \tag{3.17}$$

The kernel $K$ is known in terms of the elements of the matrices $\lambda\varepsilon$, $U_0$ and $U_0^{-1}$ and is *the same for any choice of the fundamental matrix* $U_0$. Equation (3.16) is the key to our problem.

So far the development has been formal, since it presupposes the existence of a solution for (3.11). But now suppose that, corresponding to a specific function $\chi_n$ and a specific function $\tilde{w}_n$ defined by (3.14), the integral equation (3.16) can be shown to possess an analytic solution $w(z, \lambda)$. Then (3.15) will have an analytic vector solution $\mathbf{w}(z, \lambda)$ whose first three components are zero. We can then define an analytic vector function $\boldsymbol{\delta}(z, \lambda)$ by means of the formula

$$\boldsymbol{\delta}(z, \lambda) = \lambda^{-1} \int^{z} U_0(z, \lambda) U_0^{-1}(t, \lambda) \mathbf{w}(t, \lambda)\, dt + \int^{z} U_0(z, \lambda) U_0^{-1}(t, \lambda) \tilde{\mathbf{w}}_n(t, \lambda)\, dt. \tag{3.18}$$

It is easy to verify that the function

$$\boldsymbol{\chi} = \boldsymbol{\chi}_n + \lambda^{-(n+1)} \boldsymbol{\delta}, \tag{3.19}$$

will be a solution of equation (3.3). Furthermore, if the asymptotic behavior of the function $w(z, \lambda)$ is known, it will be possible to determine a bound for the function $\boldsymbol{\delta}$ from the formula (3.18).

In view of these remarks, we must now examine the problem of proving the existence of solutions of equation (3.16) for specific choices of the function $F$. Suppose that

$$F(z, \lambda) = \eta(z, \lambda) E(z, \lambda), \tag{3.20}$$

where $\eta$ is continuous and $\eta \neq 0$ for $z$ in $S$ and $|\lambda| \geq |\lambda_0|$ for some $|\lambda_0|$. Let

$$\hat{w}(z, \lambda) = \eta(z, \lambda) w(z, \lambda). \tag{3.21}$$

Then equation (3.16) can be written in the form

$$\hat{w}(z, \lambda) = E(z, \lambda) + \lambda^{-1} \int^{z} \eta^{-1}(z, \lambda) K(z, t; \lambda) \eta(t, \lambda) \hat{w}(t, \lambda)\, dt, \tag{3.22}$$

where the inhomogeneous term is bounded. Suppose that we can select paths of integration in such a way that, for each $z$ in $S$,

$$\int^{z} |\eta^{-1}(z, \lambda) K(z, t; \lambda) \eta(t, \lambda)\, dt| \leq \phi(\lambda), \tag{3.23}$$

where the function $\phi(\lambda)$ is such that

$$|\lambda|^{-1}\phi(\lambda) < 1, \tag{3.24}$$

for $|\lambda| \geq |\lambda_0|$. It will then follow from well-known theory for Volterra integral equations that (3.16) has an analytic solution. Furthermore, this solution will be of the form

$$w(z, \lambda) = F(z, \lambda) + \lambda^{-1}\phi(\lambda)\eta(z, \lambda)E(z, \lambda), \tag{3.25}$$

so that

$$w(z, \lambda) = \eta(z, \lambda)E(z, \lambda). \tag{3.26}$$

In order to obtain bounds for the function $F$ of the type (3.20), and to establish uniform inequalities of the type (3.23), we need a thorough knowledge of the function $K(z, t; \lambda)$. To see how this kernel is formed, let us consider a fundamental matrix for equation (3.2), of the form

$$U_0 = \begin{bmatrix} u_0 & u_1 & u_2 & u_3 \\ u_0' & u_1' & u_2' & u_3' \\ \lambda^{-1}u_0'' & \lambda^{-1}u_1'' & \lambda^{-1}u_2'' & \lambda^{-1}u_3'' \\ \lambda^{-2}u_0''' & \lambda^{-2}u_1''' & \lambda^{-2}u_2''' & \lambda^{-2}u_3''' \end{bmatrix}. \tag{3.27}$$

Here the set of functions $\{u_i\}$ constitutes a fundamental set of solutions for the equation

$$u^{iv} + \lambda^2(zu'' + \alpha_0 u' + \beta_0 u) = 0, \tag{3.28}$$

which is obtained from (1.5) by putting $\alpha = \alpha_0$ and $\beta = \beta_0$. A method for the determination of the inverse matrix has been given (Rabenstein, 1959). The inverse matrix will be of the form

$$U_0^{-1} = \begin{bmatrix} \cdot & \cdot & \cdot & v_0 \\ \cdot & \cdot & \cdot & v_1 \\ \cdot & \cdot & \cdot & v_2 \\ \cdot & \cdot & \cdot & v_3 \end{bmatrix}, \tag{3.29}$$

where the functions $v_i$ are solutions of the adjoint of equation (3.28). The asymptotic behaviors of these functions will be known, since the adjoint equation has the same form as (3.28) but with $\alpha_0$ replaced by $2 - \alpha_0$. Since in (3.15) the first three components of the vectors $\mathbf{w}$ and $\mathbf{w_n}$ are zero, and since the first three rows of the matrix $\lambda\varepsilon$ consist of zeroes, the function $K$ in (3.16) has the form

$$K = ak + bk' + \lambda^{-1}ck'' + \lambda^{-2}dk''', \tag{3.30}$$

where

$$k(z, t; \lambda) = \sum_{i=0}^{3} u_i(z, \lambda)v_i(t, \lambda),$$

the primes in (3.30) denoting differentiation with respect to $z$. Since $K$ and $k$ are the same for every choice of the fundamental set $\{u_i\}$, we can express $K$ in differe forms at our convenience by choosing different fundamental sets. We can also vary the form of $K$ by the use of the relations (4.1), which connect the solutions of (1.5).

In order to carry out the appraisals of the function $F$ and the integral in (3.23), it may be necessary, for some choices of the function $\chi_n$, to take several paths of integration starting at different points, corresponding to different terms in the function $k(z, t; \lambda)$. This is done so that certain exponential functions which appear in the asymptotic forms of some of the terms of $k$ will be bounded on the paths. The details will be shown in the studies of specific cases. The paths corresponding to the various terms in $k$ must start at the same points in (3.18) as they do in (3.16) if the function $\boldsymbol{\chi}$ given by (3.19) is to be a solution of equation (3.3).

Let the components of the vector $\boldsymbol{\delta}$ be given by

$$\boldsymbol{\delta} = (\delta_0, \delta_1, \lambda^{-1}\delta_2, \lambda^{-2}\delta_3). \tag{3.32}$$

Actually $\delta_j = \delta_0^{(j)}$, $j = 1, 2, 3$, since $\boldsymbol{\delta}$ satisfies the equation (3.10). The scalar equations which correspond to (3.18) are

$$\delta_m(z, \lambda) = \lambda^{-1} \int^z k^{(m)}(z, t; \lambda) w(t, \lambda)\, dt,$$

$$+ \int^z k^{(m)}(z, t; \lambda)\tilde{w}_n(t, \lambda)\, dt, \qquad (m = 0, 1, 2, 3). \tag{3.33}$$

These formulae will be used to obtain bounds for the functions $\delta_m$. The scalar equations which correspond to (3.19) are

$$\frac{d^m \chi}{dz^m} = \frac{d^m \chi_n}{dz^m} + \lambda^{-(n+1)}\delta_m. \tag{3.34}$$

Before continuing with the details of the analysis along the lines described above, we might describe briefly one alternative procedure which could be used to obtain the asymptotic behavior of solutions of equation (3.3). In this method, we write

$$\mathbf{v_n} = G_n \boldsymbol{\chi}, \tag{3.35}$$

where the matrix $G_n$ has the form

$$G_n(z, \lambda) = I + \sum_{j=1}^{n} g_j(z)\lambda^{-j}. \tag{3.36}$$

The analytic matrix functions $g_j$ and the quantities $\alpha$ and $\beta$ can be determined in such a way that the function $\mathbf{v_n}$ satisfies a differential equation of the form

$$\frac{d\mathbf{v_n}}{dz} = M\mathbf{v_n} + \lambda^{-(n+1)}\Delta_1 \mathbf{v_n}, \tag{3.37}$$

where the matrix $M$ is given by (3.5) and the matrix $\Delta_1$ has the same form as the matrix $\Delta$. Then $\mathbf{v_n}$ is a solution of the integral equation

$$\mathbf{v_n}(z, \lambda) = \mathbf{u}(z, \lambda) + \lambda^{-(n+1)} \int^z U(z, \lambda)U^{-1}(t, \lambda)\Delta_1(t, \lambda)\mathbf{v_n}(t, \lambda)\, dt, \tag{3.38}$$

where $\mathbf{u}$ is a solution of equation (3.2) and $U$ is a fundamental matrix solution of equation (3.2). The appraisal of the difference $\mathbf{v_n} - \mathbf{u}$ can be carried out in a manner similar to the manner in which the quantity $\boldsymbol{\delta}$ is appraised. The behavior of solutions $\boldsymbol{\chi}$ of equation (3.3) can then be obtained from the formula

$$\boldsymbol{\chi} = G_n^{-1}\mathbf{u} + G_n^{-1}(\mathbf{v_n} - \mathbf{u}). \tag{3.39}$$

## 4. The solutions of the reference equation

The method of procedure described in the last section requires the appraisal of various integrals, of the types (3.17), (3.23), and (3.33). In each of these integrals, the integrands involve the function $k(z, t; \lambda)$ and its derivatives. The function $k$ in turn is expressed in terms of the solutions of equation (3.28) [a special case of (1.5)] and the solutions of its adjoint. In this section we shall list bounds and other formulae for these solutions which will be needed for the analysis which follows. The formulae listed here are either given in the study of the reference equation (Rabenstein, 1958) or else are easily obtained from results given there.

We first of all list the relations which connect the solutions of equation (1.5). These are*

$$\begin{aligned} \bar{B}_2 + A_1 &= B_3, \\ B_3 + A_2 &= B_1 + B_0, \qquad (\bar{B}_2 = e^{-2\pi i\alpha}B_2), \\ B_1 + A_3 &= B_2. \end{aligned} \tag{4.1}$$

We also have the formula

$$\frac{d^n u(z, \alpha)}{dz^n} = u(z, \alpha + n), \qquad (n = 1, 2, \ldots) \tag{4.2}$$

which holds for each of the functions $u = B_0, B_k, A_k, k = 1, 2, 3$. The formulas (4.1) and (4.2) are valid for all values of $\alpha$.

Bounds for the solutions of equation (1.5) and their derivatives when $z$ is in $S$, $\mathrm{Re}(\alpha) > 1$, and $|\lambda| \geq |\lambda_0| > 0$ are given by the formulae

$$\frac{d^m B_0(z, \alpha)}{dz^m} = E(z, \lambda), \qquad (m = 0, 1, 2, 3) \tag{4.3}$$

$$\frac{d^m A_k(z, \alpha)}{dz^m} = \lambda^{(2/3)(\alpha + m - 1)}[|\xi^{[(\alpha + m)/3 - 5/6]s}e^{-\xi_k}| + |\xi^{-(2/3)(\alpha + m - 1)s}| f_k(z, \lambda)]E(z, \lambda),$$
$$(k = 1, 2, 3; m = 0, 1, 2, 3) \tag{4.4}$$

$$\frac{d^m B_k(z, \alpha)}{dz^m} = \lambda^{(2/3)(\alpha + m - 1)}[|\xi^{-(2/3)(\alpha + m - 1)s}| + |\xi^{[(\alpha + m)/3 - 5/6]s}| g_k(z, \lambda)]E(z, \lambda),$$
$$(k = 1, 2, 3; m = 0, 1, 2, 3). \tag{4.5}$$

The definitions of the terms $\xi^s$, $\xi_k$, $f_k$, $g_k$, and $E$ are as given in Section 2. We have omitted the symbol $\lambda$ from the arguments of the functions for the sake of brevity.

The following two formulae are valid for all values of $\alpha$ and for $z$ in the regions specified:

$$A_j(z, \alpha + 3)A_i(z, 1 - \alpha) - e^{z\pi i\alpha}A_i(z, \alpha + 3)A_j(z, 1 - \alpha) = \lambda^{4/3}\xi^{-(4/3)s}E(z, \lambda), \tag{4.6}$$

* The character of these solutions, as given by Rabenstein (1958), are similar to the corresponding solutions described in Theorems I, II, and III.

when $z$ is in the sector $S - S_k, i \neq j \neq k \neq i, j > i$;

$$zA_k'(z, \alpha) + \lambda^{-2}A_k'''(z, \alpha) = \lambda^{(2/3)(\alpha-1)}\xi^{(\alpha/3-5/6)s}e^{-\xi_k}E(z, \lambda), \tag{4.7}$$

when $z$ is in the sector $S - C_k$.

Bounds for the solutions of the adjoint of equation (1.5) when $\mathrm{Re}(\alpha) > 1$ are given by the formulae

$$B_0(t, 2 - \alpha) = E(t, \lambda), \qquad (t \text{ in } S) \tag{4.8}$$

$$A_k(t, 2 - \alpha) = \lambda^{(2/3)(1-\alpha)}\zeta^{-(\alpha/3+1/6)}e^{-\zeta_k}E(t, \lambda), \qquad (t \text{ in } S - C_k, k = 1, 2, 3) \tag{4.9}$$

$$B_k(t, 2 - \alpha) = E(t, \lambda), \qquad (t \text{ in } S - S_k, k = 1, 2, 3) \tag{4.10}$$

$$\left.\begin{aligned} &B_0(t, 2 - \alpha) + (e^{-2\pi i\alpha} - 1)B_3(t, 2 - \alpha) = \lambda^{(2/3)(1-\alpha)}\zeta^{(2/3)(\alpha-1)\sigma}E(t, \lambda), \\ &\qquad\qquad (t \text{ in } S - S_3) \\ &B_0(t, 2 - \alpha) + (1 - e^{2\pi i\alpha})B_j(t, 2 - \alpha) = \lambda^{(2/3)(1-\alpha)}\zeta^{(2/3)(\alpha-1)\sigma}E(t, \lambda), \\ &\qquad\qquad (t \text{ in } S - S_j, j = 1, 2). \end{aligned}\right\} \tag{4.11}$$

When $t$ is the independent variable, we shall write $\zeta$ in place of $\xi$ and $\sigma$ in place of $s$, as we have done here.

## 5. The solutions of the related equation

In this section we shall obtain bounds for some specific solutions of the related equation (1.9) and their derivatives. These bounds will be needed for the appraisal of the inhomogeneous term (3.17) in the integral equation (3.16). We shall suppose that $\mathrm{Re}(\alpha) > 1$.

From equations (5.5)–(5.8) and Appendix A of Part I, we find that the functions $\chi_n$ and their derivatives are given by expressions of the forms

$$\chi_n = \mu u + \nu u' + \lambda^{-1}\sigma u'' + \lambda^{-2}\tau u''', \tag{5.1a}$$

$$\chi_n' = Eu + Eu' + \lambda^{-1}Eu'' + \lambda^{-2}Eu''', \tag{5.1b}$$

$$\chi_n'' = Eu + Eu' + Eu'' + \lambda^{-2}Eu''', \tag{5.1c}$$

$$\chi_n''' = Eu + Eu' + Eu'' + Eu''', \tag{5.1d}$$

where $u$ is a solution of the reference equation (1.5) and $E = E(z, \lambda)$ stands, as usual, for a bounded function. The coefficients in (4.1a) have the forms

$$\nu - z\tau = \lambda^{-2}E(z, \lambda) \tag{5.2a}$$

$$\sigma = \lambda^{-1}E(z, \lambda) \tag{5.2b}$$

$$\mu = 1 + \begin{cases} \lambda^{-1/(N+1)}E(z, \lambda), & \beta_0 = 0, \quad \alpha_0 = N + 2, \quad N = 1, 2, \ldots \\ \lambda^{-1}E(z, \lambda), & \text{otherwise,} \end{cases} \tag{5.2c}$$

$$\tau = \begin{cases} \lambda^{-1/(N+1)}E(z, \lambda), & \beta_0 = 0, \quad \alpha_0 = N + 1 \\ \lambda^{-1}E(z, \lambda), & \text{otherwise.} \end{cases} \tag{5.2d}$$

Taking $u = B_0$ in (5.1) and denoting the corresponding function $\chi_n$ by $B_{n,0}$, we find by the use of the bounds (4.3) that

$$\frac{d^m \tilde{B}_{n,0}}{dz^m} = E(z, \lambda), \qquad (m = 0, 1, 2, 3) \tag{5.3}$$

for $z$ in $S$. This is the same as equation (2.6a) in Theorem I.

Now let $u = A_k$, $k = 1, 2, 3$, in (4.1) and let the corresponding function $\chi_n$ be denoted by $\tilde{A}_{n,k}$. By using the bounds (4.4) and by using (4.7) in connection with (5.1a) and (5.2a), it may be seen that

$$\frac{d^m \tilde{A}_{n,k}}{dz^m} = \lambda^{(2/3)(\alpha+m-1)} \xi^{[(\alpha+m)/3-5/6]s} e^{-\xi_k} E(z, \lambda), \qquad (m = 0, 1, 2, 3) \tag{5.4}$$

for $z$ in the sector $S - C_k$.

If we take $u = B_k$, $k = 1, 2, 3$, in (5.1) and denote the corresponding function $\chi_n$ by $\tilde{B}_{n,k}$, it follows by the use of the bounds (4.5) that

$$\frac{d^m \tilde{B}_{n,k}}{dz^m} = \lambda^{(2/3)(\alpha+m-1)} \xi^{-(2/3)(\alpha_0+m-1)s} E(z, \lambda), \qquad (m = 0, 1, 2, 3) \tag{5.5}$$

for $z$ in the sector $S - S_k$.

Bounds for the function $\tilde{A}_{n,k}^{(m)}$ in a sector which contains the ray $C_k$ and bounds for the function $\tilde{B}_{n,k}^{(m)}$ in a sector $S_k$ may be found by application of the relations (4.1). For instance, from equations (5.1) and the relations (4.1) we find that

$$\tilde{A}_{n,1}^{(m)} = \tilde{B}_{n,0}^{(m)} + (1 - e^{2\pi i\alpha})\tilde{B}_{n,1}^{(m)} - \tilde{A}_{n,2}^{(m)} - e^{2\pi i\alpha}\tilde{A}_{n,3}^{(m)}.$$

Bounds for the functions on the right in a sector which contains the ray $C$, are known from the formulas (5.3)–(5.5). Thus,

$$\tilde{A}_{n,1}^{(m)} = \lambda^{(2/3)(\alpha+m-1)}[|\xi^{[(\alpha+m)/3-5/6]s} e^{-\xi_1}| + |\xi^{-(2/3)(\alpha+m-1)s}|]E(z, \lambda),$$

in such a sector. In this way it may be shown that

$$\frac{d^m \tilde{A}_{n,k}}{dz^m} = \lambda^{(2/3)(\alpha+m-1)}[|\xi^{[(\alpha+m)/3-5/6]s} e^{-\xi_k}| + |\xi^{-(2/3)(\alpha+m-1)s}| f_k(z, \lambda)]E(z, \lambda), \tag{5.6}$$

$$\frac{d^m B_{n,k}}{dz^m} = \lambda^{(2/3)(\alpha+m-1)}[|\xi^{-(2/3)(\alpha+m-1)s}| + |\xi^{[(\alpha+m)/3-5/6]s} e^{\xi_k}| g_k(z, \lambda)]E(z, \lambda), \tag{5.7}$$

for $z$ in $S$ and $\mathrm{Re}(\alpha) > 1$. The function $f_k$ and $g_k$ are given by the definitions (2.4) and (2.5). Since $\alpha = \alpha_0 + o(1)$, these formulae imply the results (2.7a) and (2.8a).

## 6. Lemmas

In this section we shall obtain bounds for certain types of integrals, the results being stated in the form of lemmas. These lemmas will be of frequent use in the remainder of this paper, where we shall give detailed discussions of equations (3.16) and (3.33) for specific cases.

Before we state and prove our lemmas, some definitions and remarks will be in order. Let $S'$ be a closed disk in the $z$-plane, centered at the origin, containing the closed disk $S$ in its interior, and contained in the region where the coefficients in equation (1.2) are analytic. The rays $C_k$ separate $S'$ into three closed sectors

$S_k'$, $k = 1, 2, 3$. The sector $S_k'$ contains the sector $S_k$. Let $z_k$ be the midpoint of the circular arc on the boundary of $S_k'$. Then for $z$ in $S$, we have that $|z| \leq R < |z_k|$, $k = 1, 2, 3$, for some positive constant $R$. The point $\xi^k = \xi_k(z_k)$ will lie on the positive real axis in the $\xi$-plane. Let the symbol $M$ represent a positive constant which may depend on $|\xi_0|$ and $|\lambda_0|$, but not on $z$ and $\lambda$. The constant $M$ may be different in each instance where it occurs. Let $\gamma$ be any real number and let $r$ be any positive real number.

The following will now be proved:

LEMMA I. *If*

$$I_1 = \int_{\gamma_1} |\zeta^{\gamma\sigma} e^{-r\zeta_k} E(t, \lambda)\, dt|, \tag{6.1}$$

*where the path $\gamma_1$ is the straight line segment from $z_k$ to 0, then*

$$I_1 \leq |\lambda|^{-2/3} M. \tag{6.1a}$$

LEMMA II. *Let the path $\gamma_2$ be a line segment from $t = 0$ to any point $z$ in $S$, and let*

$$I_2 = \int_{\gamma_2} |\zeta^{\gamma\sigma} E(t, \lambda)|\, dt. \tag{6.2}$$

*Then*

$$I_2 \leq \begin{cases} |\lambda|^{-2/3} |\xi^{(\gamma + 2/3)s}| M, & \gamma > -\frac{2}{3}, \\ |\lambda|^{-2/3} \log|\xi^s| M, & \gamma = -\frac{2}{3}, \\ |\lambda|^{-2/3} M, & \gamma < -\frac{2}{3}. \end{cases} \tag{6.2a}$$

LEMMA III. *Let $z_*$ be any fixed point on the boundary $S'$ and let $z$ be any point of $S$. Let $\gamma_3$ be the path from $z_*$ to $z$ which consists of a circular arc along the boundary of $S'$ and a line segment along the radius through the point $z$. Then*

$$I_3 \equiv \int_{\gamma_3} |\zeta^{\gamma\sigma} E(t, \lambda)\, dt| \leq \begin{cases} |\lambda|^{-2/3} |\xi^{(\gamma + 2/3)s}| M, & \gamma < -\frac{2}{3}, \\ |\lambda|^{-2/3} \log\left|\dfrac{\lambda}{\xi^s}\right| M, & \gamma = -\frac{2}{3}, \\ |\lambda|^{\gamma} M, & \gamma > -\frac{2}{3}. \end{cases} \tag{6.3}$$

LEMMA IV. *Let $z$ lie in a closed sector which is contained in the region $S - S_j + 0$, $j \neq k$. Then it is possible to choose a path $\gamma_4$ from $z_k$ to $z$, lying within this sector, in such a way that*

$$I_4 \equiv \int_{\gamma_4} |\zeta^{\gamma\sigma} e^{r(\xi_k - \zeta_k)} E(t, \lambda)\, dt| \leq |\lambda|^{-2/3} |\xi^{(\gamma - 1/3)s}| M. \tag{6.4}$$

LEMMA V. *Let $z$ lie in a closed sector of angle $2\pi/3$ which is bisected by the ray $C_k$. Let the path $\gamma_5$ from 0 to $z$ consist of a line segment along $C_k$ and an arc of the circle $|t| = |z|$. Let*

$$I_5 = \int_{\gamma_5} |\zeta^{\gamma\sigma} e^{-r\xi_k} E(t, \lambda)\, dt|. \tag{6.5}$$

*Then*

$$I_5 \leq |\lambda|^{-2/3} |\xi^{(\gamma - 1/3)s} e^{-r\xi_k}| M + |\lambda|^{-2/3} M \begin{cases} |\xi^{(\gamma + 2/3)s}|, & \gamma > -\frac{2}{3}, \\ \log|\xi^s|, & \gamma = -\frac{2}{3}, \\ 1, & \gamma < -\frac{2}{3}. \end{cases} \tag{6.6}$$

*Proof of Lemma* 1. Changing the variable of integration to $\zeta$ in (6.1), we have

$$I_1 \le |\lambda|^{-2/3} M \int_0^{|\xi^k|} |\zeta^{\sigma\gamma - 1/3}| e^{-r|\zeta|}\, d|\zeta|.$$

Considering the parts of the path on which $|\zeta| > |\xi_0|$ and $|\zeta| \le |\xi_0|$ separately, we have

$$I_1 \le |\lambda|^{-2/3} M \left[ \int_0^{|\xi_0|} |\zeta|^{-1/3}\, d|\zeta| + \int_{|\xi_0|}^{\infty} |\zeta|^{\gamma - 1/3} e^{-r|\zeta|}\, d|\zeta| \right].$$

Both integrals in the brackets are bounded, and the lemma follows.

*Proof of Lemma II.* When $|\xi| \le |\xi_0|$, we have

$$I_2 \le |\lambda|^{-2/3} M \int_0^{|\xi|} |\zeta|^{-1/3}\, d|\zeta| \le |\lambda|^{-2/3} M.$$

When $|\xi| > |\xi_0|$, we consider the parts of the path where $|\zeta| \le |\xi_0|$ and $|\zeta| > |\xi_0|$ separately. Then

$$I_2 \le |\lambda|^{-2/3} M \left[ \int_0^{|\xi_0|} |\zeta|^{-1/3}\, d|\zeta| + \int_{|\xi_0|}^{|\xi|} |\zeta|^{\gamma - 1/3}\, d|\zeta| \right].$$

The first integral on the right is bounded. Evaluation of the second integral for the three cases $\gamma > -\frac{2}{3}$, $\gamma = -\frac{2}{3}$, and $\gamma < -\frac{2}{3}$ gives the result (6.2a)

*Proof of Lemma III.* The integral over the circular arc is $O(|\lambda|^{\gamma})$, since $|\zeta^{\sigma}| = \frac{2}{3}|\lambda| \cdot |z_k|^{3/2}$ there. The integral over the line segment may be treated as in the proof of lemmas *R*. The details are omitted.

*Proof of Lemma IV.* The function $\xi_k$ maps the closed sector under consideration into a disk in the $\xi$-plane, excluding a narrow sector which contains either the positive or negative imaginary axis, depending on $j$ and $k$. For the sake of definiteness, let the negative imaginary axis be excluded. The other case can be treated in an analogous fashion. The point $z_k$ goes into a point on the positive real axis.

If $|\xi| \le |\xi_0|$, we may take $\gamma_4$ to consist of two line segments, one from $z_k$ to 0 and the other from 0 to $z$. The function $e^{r(\xi_k - \zeta_k)}$ is bounded on such a path. We may therefore apply Lemmas I and II to show that (6.4) holds in this case.

Now suppose that $|\xi| > |\xi_0|$. We shall consider first the case when the points $\xi_k(z)$ and $\xi^k$ can be joined by a path $\Gamma_4$ in the $\xi$-plane which consists of a circular arc on which $|\zeta| = |\xi^k|$ and a horizontal line segment on which $|\zeta| > |\xi_0|$. This is possible when $\xi_k(z)$ lies in the right half plane and also when $\xi_k(z)$ lies in the second quadrant with $\mathrm{Im}(\xi_k) > |\xi_0|$. Let $\gamma_4$ be the path in the $t$-plane which maps into $\Gamma_4$. The contribution to the circular arc is not greater than

$$|\lambda|^{\gamma} e^{2/3|\lambda|\varepsilon^{3/2}} M, \tag{6.6}$$

when $\xi = |z_k| - R$. The contribution to $I_4$ from the horizontal line segment of $\Gamma_4$ is less than a constant multiple of the quantity

$$J = |\lambda|^{2/3} \int |\zeta^{\gamma - 1/3} e^{r(\xi_k - \zeta_k)}\, d\zeta|, \tag{6.7}$$

the integral being taken over the horizontal line segment from a point $\xi$ on the

image of the boundary of $S'$ to the point $\xi$. Writing

$$\xi = x + iy, \qquad \tilde{\xi} = \tilde{x} + i\tilde{y}, \qquad \zeta = \sigma + i\tau, \tag{6.8}$$

we have

$$J = |\lambda|^{2/3} \int_{\tilde{x}}^{x} (\sigma^2 + y^2)^{\gamma/2 - 1/6} e^{r(x-\sigma)}\, d\sigma. \tag{6.9}$$

Making the change of variable

$$r(x - \sigma) = -|\xi| s, \tag{6.10}$$

we find that

$$J \le |\lambda|^{-2/3(|\xi|^{\gamma} + 1/r)} \int_0^{\infty} |\xi|^{-\gamma} \left[\left(x + \frac{|\xi|}{r} s\right)^2 + y^2\right]^{\gamma/2 - 1/6} e^{-|\xi| s}\, ds. \tag{6.11}$$

An application of Watson's lemma (Copson, 1935, p. 218) to the integral here gives the inequality

$$J \le |\lambda|^{2/3} |\xi^{(\gamma - 1/3)}| M. \tag{6.12}$$

Hence the lemma is proved for this case.

Now suppose that the point $\xi$ cannot be reached by a path of the sort described above, (e.g., $\xi$ in the third quadrant). Let $\gamma_4$ consist of a line segment from $z_k$ to 0, and from 0 to $z$. The image of $\gamma$ in the $\zeta$-plane will consist of line segments from $\xi^k$ to 0 and from 0 to $\zeta$. Let $\xi_*$ be an arbitrary, but fixed point such that $|\xi_*| > |\xi_0|$. If $|\xi_0| \le |\xi| \le |\xi_*|$, the factor $|\xi^{(\gamma - 1/3)s}|$ in the right hand member of (6.4) is bounded, application of Lemmas I and II shows that this lemma holds for $|\xi| \le |\xi_*|$.

When $|\xi| > |\xi_*|$ we can find a point $\hat{\xi} = \theta\xi$ on the line segment from 0 to $\xi$ with $0 < \theta < 1$ and such that $|\hat{\xi}| > |\xi_0|$. We choose $|\hat{\xi}| > \max(|\xi|/2, |\xi_0|)$. Then

$$I_4 \le |\lambda|^{-2/3} M (J_1 + J_2 + J_3), \tag{6.13}$$

where

$$J_1 = \int_0^{|\xi^k|} |\zeta^{\gamma\sigma - 1/3} e^{r(\xi_k - \zeta_k)}| d|\zeta|, \tag{6.14}$$

$$J_2 = |e^{r(1-\theta)\xi_k}| \int_0^{|\hat{\xi}|} |\zeta^{\gamma\sigma - 1/3} e^{r(\hat{\xi} - \zeta_k)}| d|\xi|, \tag{6.15}$$

$$J_3 = \int_{|\hat{\xi}|}^{|\xi|} |\zeta^{\gamma - 1/3} e^{r(\xi_k - \zeta_k)}| d|\zeta|. \tag{6.16}$$

As in the proof of lemma I it follows that

$$J_1 \le M |e^{r\xi_k}|. \tag{6.17}$$

Since the function $e^{r(\hat{\xi} - \zeta_k)}$ is bounded on the path of integration in (5.15), it follows as in the proof of lemma II that

$$J_2 \le M |e^{r(1-\theta)\xi_k}| (|\hat{\xi}|^{\gamma + 2/3} + \log|\hat{\xi}| + 1). \tag{6.18}$$

Since $\frac{1}{2}|\xi| < |\hat{\xi}| < |\xi|$, we have

$$J_3 \le M |\xi|^{\gamma - 1/3} \int_{|\hat{\xi}|}^{|\xi|} e^{r \cos\delta_1 (|\zeta| - |\xi|)}\, d|\zeta|, \tag{6.19}$$

where $\delta_1$ is a suitably small positive constant. Thus

$$J_3 \leq M|\xi|^{\gamma - 1/3}. \tag{6.20}$$

Since $\mathrm{Re}(\xi_k) \leq -|\xi_k| \cos \delta_1$, it follows from (6.13), (6.17), (6.18), and (6.20) that the inequality (6.4) holds for $\xi$ in the region under consideration.

## 7. A balanced solution: proof of Theorem I

In this section we shall show that equation (1.2) possesses a solution $B_{n,0}$ with the properties stated in Theorem I. This solution will be balanced for $z$ in $S$, since the function $\tilde{B}_{n,0}$ has that property.

Following the procedure described in Sect. 3, we must first of all prove the existence of a solution for the integral equation (3.16) in the case when

$$\tilde{w}_n = \tilde{a}\tilde{B}_{n,0} + \tilde{b}\tilde{B}'_{n,0} + \lambda^{-1}\tilde{c}\tilde{B}''_{n,0} + \lambda^{-2}\tilde{d}\tilde{B}''_{n,0}. \tag{7.1}$$

This function $\tilde{w}_n$ is uniformly bounded for $z$ in $S'$ and $|\lambda| \geq |\lambda_0|$, as may be seen from the bounds (4.3). Let $U_0(z, \lambda)$ be the fundamental matrix

$$U_0(z, \lambda) = \begin{bmatrix} B_0(z, \alpha_0) & B_1(z, \alpha_0) & B_2(z, \alpha_0) & B_3(z, \alpha_0) \\ B'_0 & B'_1 & B'_2 & B'_3 \\ \lambda^{-1}B''_0 & \lambda^{-1}B''_1 & \lambda^{-1}B''_2 & \lambda^{-1}B''_3 \\ \lambda^{-2}B'''_0 & \lambda^{-2}B'''_1 & \lambda^{-2}B'''_2 & \lambda^{-2}B'''_3 \end{bmatrix}. \tag{7.2}$$

The inverse matrix (Rabenstein, 1959) will have the form

$$U_0^{-1}(t, \lambda) = \frac{e^{-\pi i \alpha_0}}{2\pi i} \begin{bmatrix} \cdot & \cdot & \cdot & -e^{2\pi i \alpha_0} B_2(t, 2 - \alpha_0) \\ \cdot & \cdot & \cdot & A_1(t, 2 - \alpha_0) \\ \cdot & \cdot & \cdot & A_2(t, 2 - \alpha_0) \\ \cdot & \cdot & \cdot & e^{2\pi i \alpha_0} A_3(t, 2 - \alpha_0) \end{bmatrix}. \tag{7.3}$$

Then the kernel $K(z, t; \lambda)$ in (3.16) and (3.17) will have the form (3.30) with

$$\begin{aligned} k(z, t; \lambda) = \frac{e^{-\pi i \alpha_0}}{2\pi i}[&-e^{2\pi i \alpha_0} B_0(z, \alpha_0) B_2(t, 2 - \alpha_0) + B_1(z, \alpha_0) A_1(t, 2 - \alpha_0) \\ &+ B_2(z, \alpha_0) A_2(t, 2 - \alpha_0) + e^{2\pi i \alpha_0} B_3(z, \alpha_0) A_3(t, 2 - \alpha_0)]. \end{aligned} \tag{7.4}$$

For the reason given in Section 3, we find it necessary to take several different paths of integration starting at different points in the integrals in (3.16) and (3.17). In order to illustrate our choice of paths we shall write

$$k(z, t; \lambda) = \sum_{j=0}^{3} k_j(z, t; \lambda), \tag{7.5}$$

where

$$\begin{aligned} k_0 &= -\frac{e^{\pi i \alpha_0}}{2\pi i} B_0(z, \alpha_0) B_2(t, 2 - \alpha_0), & k_1 &= \frac{e^{-\pi i \alpha_0}}{2\pi i} B_1(z, \alpha_0) A_1(t, 2 - \alpha_0), \\ k_2 &= \frac{e^{-\pi i \alpha_0}}{2\pi i} B_2(z, \alpha_0) A_2(t, 2 - \alpha_0), & k_3 &= \frac{e^{\pi i \alpha_0}}{2\pi i} B_3(z, \alpha_0) A_3(t, 2 - \alpha_0). \end{aligned} \tag{7.6}$$

We shall also write

$$K(z, t; \lambda) = \sum_{j=0}^{3} K_j(z, t; \lambda), \tag{7.7}$$

where

$$K_j = ak_j + bk_j' + \lambda^{-1}ck_j'' + \lambda^{-2}dk_j''', \qquad (j = 0, 1, 2, 3). \tag{7.8}$$

We write (3.16) in the form

$$w(z, \lambda) = F(z, \lambda) + \lambda^{-1} \sum_{j=0}^{3} \int_{z_j}^{z} K_j(z, t; \lambda) w(t, \lambda)\, dt, \tag{7.9}$$

where

$$F(z, \lambda) = \sum_{j=0}^{3} \int_{z_j}^{z} K_j(z, t; \lambda) \tilde{w}_n(t, \lambda)\, dt. \tag{7.10}$$

Here $z_0 = 0$ and the points $z_1$, $z_2$, and $z_3$ are as defined in Section 6. The function $A_j(t, 2 - \alpha_0)$, which appears as a factor in the term $K_j(z, t; \lambda)$, is recessive in the sector $S_j'$. The term $e^{-\zeta_j}$, which appears in the bound (4.9) for the function $A_j$, has its minimum in the sector $S_j$ at the point $z_j$. Although the kernel $K$ appears unsymmetric in its general asymptotic properties, as it is written in the form (7.8), we shall see that actually this is not the case.

We shall now show that the paths of integration may be chosen in such a way that

$$\int^{z} |K(z, t; \lambda)\, dt| \le M, \tag{7.11}$$

for $z$ in $S$ and $|\lambda| \ge |\lambda_0|$. Since $|\tilde{w}_n| \le M$, this will imply that

$$|F(z, \lambda)| \le M. \tag{7.12}$$

The two inequalities (7.11) and (7.12) will imply the existence of a solution for (3.16) in accordance with the remarks following (3.24).

We shall consider separately the three sectors $S - S_j, j = 1, 2, 3$, and shall give a detailed discussion for the sector $S - S_3$ only. The other two sectors may be treated in a symmetric manner.

Let us consider, for $z$ in a closed sector which is contained in the region $S - S_3 + 0$, the quantity

$$\sum_{j=0}^{3} \int_{z_j}^{z} K_j(z, t; \lambda)\, dt.$$

By the use of the relations (4.1) we can write the contribution of the term $a(z, \lambda)k(z, t; \lambda)$ in the integrands of (7.13) as

$$\begin{aligned}\int^{z} ak\, dt = {} & \frac{e^{-\pi i \alpha_0}}{2\pi i} \left\{ \int_{z_1}^{z} aA_2(z, \alpha_0) A_1(t, 2 - \alpha_0)\, dt - e^{2\pi i \alpha_0} \int_{z_2}^{z} aA_1(z, \alpha_0) A_2(t, 2 - \alpha_0)\, dt \right. \\ & \left. + \int_{z_1}^{0} a[B_3(z, \alpha_0) - B_0(z, \alpha_0)] A_1(t, 2 - \alpha_0)\, dt \right\} \\ & + e^{2\pi i \alpha_0} \int_{z_2}^{0} aB_3(z, \alpha_0) A_2(t, 2 - \alpha_0)\, dt + \int_{z_3}^{0} aB_3(z, \alpha_0) A_3(t, 2 - \alpha_0)\, dt \\ & + \int_{0}^{z} a\{B_3(z, \alpha_0)[(1 - e^{2\pi i \alpha_0}) B_3(t, 2 - \alpha_0) + e^{2\pi i \alpha_0} B_0(t, 2 - \alpha_0)] \\ & - B_0(a, \alpha_0) B_3(t, 2 - \alpha_0)\}\, dt. \end{aligned} \tag{7.14}$$

What we have done is to write $k$ in terms of the functions $A_1, A_2, B_0, B_3$, except where $t$ is in the Sector $S_3$ where $A_3$ is recessive. None of these functions have transition lines in the sector $S - S_3$. The paths for the terms in $k_1$ and $k_2$ which involve exponential-type functions of $z$ are chosen as in lemma IV. The paths for all the other terms in $k$ consist of straight line segments. Thus, actually, we have taken two paths from $z_1$ to $z$ and two paths from $z_2$ to $z$ corresponding to the different types of terms in $k_1$ and $k_2$. One path in each case passes through the origin. On the path from $z_k$ to 0, $k = 1, 2, 3$, the function $A_k(t, 2 - \alpha_0)$ is recessive. On the ray from 0 to $z$, the integrand is balanced. The contributions from the terms $bk', \lambda^{-1}ck''$, and $\lambda^{-2}dk'''$ in (7.13) may be found by replacing the function $a$ in (7.14) by $b$, $\lambda^{-1}c$, and $\lambda^{-2}d$ respectively, and by differentiating the functions $A_1(z, \alpha_0)$, $A_2(z, \alpha_0)$, $B_3(z, \alpha_0)$, $B_0(z, \alpha_0)$ the proper number of times.

In the treatment of each of the other two sectors $S - S_j, j = 1, 2$, the kernel is expressed in terms of a set of functions appropriate to that sector. For instance, in considering the sector $S - S_1$, we should express $K$ in terms of the functions $A_2, A_3, B_1, B_0$. In that case, the terms of $k_2$ and $k_3$ would be handled like the terms $k_1$ and $k_2$ were in (7.14).

In the present case, it follows from the bounds (4.3)–(4.5) and (4.8)–(4.11) that the functions $k^{(m)}(z, t; \lambda)$, $m = 0, 1, 2, 3$, have the bounds

$$\lambda^{(2/3)m}\xi^{[(\alpha_0 + m)/3 - 5/6]s}\zeta^{-(\alpha_0/3 + 1/6)\sigma}e^{\xi_j - \zeta_j}E(z, t; \lambda), \tag{7.15a}$$

on the path from $z_j$ to $z, j = 1, 2$;

$$\lambda^{(2/3)m}\xi^{(2/3)(\alpha_0 + m - 1)s}\zeta^{-(\alpha_0/3 + 1/6)\sigma}e^{-\zeta_j}E(z, t; \lambda), \tag{7.15b}$$

on the path from $z_j$ to $0, j = 1, 2, 3$;

$$\lambda^{(2/3)m}\xi^{-(2/3)(\alpha_0 + m - 1)s}\zeta^{(2/3)(\alpha_0 - 1)}E(z, t; \lambda) + E(z, t; \lambda), \tag{7.15c}$$

on the path from 0 to $z$. From these bounds and the relation (7.8), we find that the integral (7.13) has the form

$$\begin{aligned}\int^z K(z, t; \lambda)\, dt = &\sum_{j=1}^{2}\int_{z_j}^{z}\lambda^{2/3}\xi^{(\alpha_0/3 - 1/2)s}\zeta^{-(\alpha_0/3 + 1/6)\sigma}e^{\xi_j - \zeta_j}E\, dt\\ &+ \sum_{j=1}^{3}\int_{z_j}^{z}\lambda^{2/3}\xi^{-(2/3)\alpha_0 s}\zeta^{-(\alpha_0/3 + 1/6)\sigma}e^{-\zeta_j}E\, dt\\ &+ \int_0^z (\lambda^{2/3}\xi^{-(2/3)\alpha_0 s}\zeta^{(2/3)(\alpha_0 - 1)}E + E)\, dt,\end{aligned} \tag{7.16}$$

where the paths of integration are as in the description following (7.14). Application of Lemmas I, II, and IV to the above integrals shows that the left hand member of (7.11) is bounded for $z$ in the closed sector under consideration.

Similarly, it can be shown that the left hand member of (7.11) is bounded for $z$ in any closed sector which is contained in either of the regions $S - S_1 + 0$ or $S - S_2 + 0$. Hence (7.11) and (7.12) hold uniformly for $z$ in $S$.

It now follows that equation (7.9) possesses an analytic solution $w(z, \lambda)$ with the property that

$$w(z, \lambda) = F(z, \lambda) + \lambda^{-1}E(z, \lambda). \tag{7.17}$$

Since the function $F$ is uniformly bounded for $z$ in $S$, the function $w$ is also.

Actually, the solution $w$ exists and is uniformly bounded for $z$ in any closed set which lies in the interior of the disk $S'$. In particular, this is the case for $z$ in a closed disk which contains $S$ in its interior. We shall again denote such a disk by $S'$.

Formula (3.33) now becomes

$$\delta_m(z,\lambda) = \lambda^{-1}\sum_{j=0}^{3}\int_{z_j}^{z} k_j^{(m)}(z,t;\lambda)w(t,\lambda)\,dt + \sum_{j=0}^{3}\int_{z_j}^{z} k_j^{(m)}(z,t;\lambda)\tilde{w}_n(t,\lambda)\,dt, \qquad (m = 0, 1, 2, 3). \tag{7.18}$$

We shall show that

$$\delta_m(z,\lambda) = E(z,\lambda), \qquad (m = 0, 1, 2, 3). \tag{7.19}$$

Actually, we shall first show that (7.19) holds for $z$ in $S$ and $|z| \geq |z_*| > 0$ for some positive constant $|z_*|$. Then, since the left hand members of (7.19) are analytic functions of $z$, it will be a consequence of the maximum principle for analytic functions that (7.19) holds for $z$ in $S$. When $z$ is bounded away from the origin, we have the simplification that $|\xi|$ behaves like $|\lambda|$, since then

$$c_1|\lambda| \leq |\xi| \leq c_2|\lambda|, \tag{7.20}$$

where $0 < c_1 < c_2$.

We again treat separately the sectors $S - S_j, j = 1, 2, 3$. For $z$ in a closed subset of the sector $S - S_3$, we express $k$ in terms of the functions $A_1, A_2, B_3, B_0$ and choose paths of integration corresponding to the various terms of $k$ in (7.18) exactly as we did in (7.14). The bounds for the integrands on the paths have the forms (7.15), since the functions $\tilde{w}_n$ and $w$ are bounded. Using Lemmas I, II, and IV to appraise the integrals, and keeping (7.20) in mind we find that the quantities $\delta_m$ are bounded for $m = 0, 1, 2$. For $m = 3$, an additional step is necessary. In that case we find that

$$\delta_3 = E(z,\lambda) + J(z,\lambda), \tag{7.21}$$

where

$$J = \frac{e^{-\pi i\alpha_0}}{2\pi i}\left[\int_{z_1}^{z} A_2''(z,\alpha_0)A_1(t, 2-\alpha_0)\tilde{w}_n(t)\,dt - e^{2\pi i\alpha_0}\int_{z_2}^{z} A_1''(z,\alpha_0)A_2(t, 2-\alpha_0)\tilde{w}_n(t)\,dt\right]. \tag{7.22}$$

Integrating by parts in (7.22), we have

$$\begin{aligned}2\pi i e^{\pi i\alpha_0}J = {} & [A_2(z,\alpha_0+3)A_1(z, 1-\alpha_0) - e^{2\pi i\alpha_0}A_1(z,\alpha_0+3)A_2(z,1-\alpha_0)]\tilde{w}_n(z)\\ & -[A_2(z,\alpha_0+3)A_1(z_1, 1-\alpha_0)w_n(z_1)\\ & -e^{2\pi i\alpha_0}A_1(z,\alpha_0+3)A_2(z_2,1-\alpha_0)\tilde{w}_n(z_2)]\\ & -\int_{z_1}^{z} A_2(z,\alpha_0+3)A_1(t,1-\alpha_0)\tilde{w}_n'(t)\,dt\\ & +e^{2\pi i\alpha_0}\int_{z_2}^{z} A_1(z,\alpha_0+3)A_2(t,1-\alpha_0)\tilde{w}_n'(t)\,dt.\end{aligned} \tag{7.23}$$

By the use of relation (4.6) we find that the first expression in brackets is bounded. Since $|z| \leq R < |z_1| = |z_2|$, the second expression in brackets tends to 0 as $|\lambda|$

becomes large, and is uniformly bounded for $|\lambda|$ sufficiently large. The function $\tilde{w}_n'$ is bounded for $z$ in $S$, as may be seen by differentiating the expression (7.1) and using the differential equation (1.9) to eliminate the fourth derivation of $B_{n,0}$. From the bounds (4.4) and (4.9) and Lemma IV, it is found that each of the integrals in (7.23) is bounded. Thus $J$ is bounded, and so is the function $\delta_3$. Hence (7.19) holds for $z$ in any closed subset of the sector $S - S_3$. The sectors $S - S_1$ and $S - S_2$ may be treated in a similar manner. In this way, it may be shown that (7.19) holds in an annular region $0 < |z_*| \le |z| \le R$. According to the argument involving the maximum principle, (7.19) holds for $z$ in $S$.

Theorem I now follows from the theory of Section 3 and the equations (3.19), (5.3), and (7.19).

## 8. A dominant-recessive solution: proof of Theorem II

In this section we shall show that equation (1.2) possesses a solution $A_{n,1}$ which is approximated by the function $\tilde{A}_{n,1}$ in a complete neighborhood of the turning point, as stated in Theorem II. This solution, like the function $\tilde{A}_{n,1}$, will be recessive in the sector $S_1$ and dominant in the sectors $S_2$ and $S_3$.

This time we shall consider the integral equation (3.16) when the function $w_n$ is given by the expression

$$\tilde{w}_n = \tilde{a}\tilde{A}_{n,1} + \tilde{b}\tilde{A}_{n,1}' + \lambda^{-1}\tilde{c}\tilde{A}_{n,1} + \lambda^{-2}\tilde{d}\tilde{A}_{n,1}''. \tag{8.1}$$

As may be seen from the bounds (5.6), this function has the behavior

$$w_n(z, \lambda) = \rho_1(z, \lambda)E(z, \lambda), \tag{8.2}$$

where

$$\rho_m(z, \lambda) = \lambda^{(2/3)(\alpha_0 + m - 1)}[|\xi^{[(\alpha_0 + m)/3 - 5/6]s} e^{-\xi_1}| + |\xi^{-(2/3)(\alpha_0 + m - 1)s}| f_1(z, \lambda)] \quad (m = 0, 1, 2, 3). \tag{8.3}$$

We shall choose all paths of integration in (3.16) and (3.17) to start at the point $z$. This is possible for the integral (3.17) because the function $\tilde{w}_n$ is recessive in the sector $S_1$. It is to be expected that a solution $w$ of equation (3.16) will have a similar behavior. In this case, the existence of a solution for equation (3.16) follows from the theory of ordinary differential equations. However, it is still necessary to carry out the appraisals of the types (3.20) and (3.23) in order to obtain a bound for the solution. Let us therefore consider the integral

$$\int_{z_1}^{z} |\rho_1^{-1}(z, \lambda)K(z, t; \lambda)\rho_1(t, \lambda)\, dt|, \tag{8.4}$$

which may depend on the path or paths between the points $z_1$ and $z$.

The functions $\tilde{A}_{n,1}$ and $\tilde{w}_n$ possess a transition line along the ray $C_1$. We shall therefore consider first the case when $z$ is excluded from a narrow sector which contains $C_1$.

Let us suppose then, that $z$ lies in a closed sector which is contained in the region $S - C_1 + 0$. In order to appraise the integral (8.4), it will be necessary to write the kernel $K$ in two different forms, one of which is useful when $z$ is in the sector $S - S_2$ and the other when $z$ is in the sector $S - S_3$. First let us form the fundamental matrix $U_0$ as in the formula (3.27) from the functions $A_1$, $A_3$, $B_2$, $B_0$.

Then the kernel $K$ has the form (3.30) with

$$\begin{aligned} k(z, t; \lambda) = -\frac{e^{-\pi i \alpha_0}}{2\pi i}\{ & e^{2\pi i \alpha_0} A_1(z, \alpha_0) A_3(t, 2-\alpha_0) - A_3(z, \alpha_0) A_1(t, 2-\alpha_0) \\ & + B_2(z, \alpha_0)[(1 - e^{2\pi i \alpha_0}) B_2(t, 2-\alpha_0) + B_0(t, 2-\alpha_0)] \\ & - e^{2\pi i \alpha_0} B_0(z, \alpha_0) B_2(t, 2-\alpha_0)\}. \end{aligned} \tag{8.5}$$

This form of $k$ is suitable when $z$ is in the sector $S - S_2$. Since the functions $A_3$ and $B_2$ have transition lines along the ray $C_3$, this form of $k$ is not suitable when the paths of integration must cross the ray $C_3$ into the sector $S_2$. However, by forming a fundamental matrix $U_0$ from the functions $A_1$, $A_2$, $B_3$, $B_0$, we obtain the alternate form

$$\begin{aligned} k(z, t; \lambda) = -\frac{e^{\pi i \alpha_0}}{2\pi i}\{ & A_1(z, \alpha_0) A_2(t, 2-\alpha_0) - e^{-2\pi i \alpha_0} A_2(z, \alpha_0) A_1(t, 2-\alpha_0) \\ & + B_3(z, \alpha_0)[(1 - e^{-2\pi i \alpha_0}) B_3(t, 2-\alpha_0) - B_0(t, 2-\alpha_0)] \\ & + e^{-2\pi i \alpha_0} B_0(z, \alpha_0) B_3(t, 2-\alpha_0)\}. \end{aligned} \tag{8.6}$$

This form will be appropriate when $z$ is in the sector $S - S_3$.

Let $z$ and $t$ lie in a closed sector which is contained in either of the regions $S - S_j + 0, j = 2, 3$. Then from (8.3), the bounds of Section 4, and the appropriate form of $k$, it follows that

$$\begin{aligned} k^{(m)}(z, t; \lambda)\rho_1(t, \lambda) = {} & \lambda^{(2/3)(\alpha_0+m)} \xi^{[(\alpha_0+m)/3 - 5/6]s} \zeta^{-(2/3)\sigma} e^{\xi_1 - 2\zeta_1} E(z, t, \lambda) \\ & + \lambda^{(2/3)(\alpha_0+m)} \xi^{[(\alpha_0+m)/3 - 5/6]s} \zeta^{-(2/3)\sigma} e^{-\xi_1} E(z, t, \lambda) \\ & + \lambda^{(2/3)(\alpha_0+m)} \xi^{-(2/3)(\alpha_0+m-1)s} \zeta^{(\alpha_0 - 7/6)\sigma} e^{-\zeta_1} E(z, t, \lambda) \\ & + \lambda^{(2/3)\alpha_0} \zeta^{(\alpha_0/3 - 1/2)\sigma} e^{-\zeta_1} E(z, t, \lambda), \qquad (m = 0, 1, 2, 3). \end{aligned} \tag{8.7}$$

From (8.3), (8.7), and (3.30) we find that

$$\begin{aligned} \rho_1^{-1}(z, \lambda) K(z, t; \lambda) \rho_1(t, \lambda) = {} & \lambda^{2/3} \zeta^{-(2/3)\sigma} e^{2(\xi_1 - \zeta_1)} E(z, t, \lambda) + \lambda^{2/3} \zeta^{-(2/3)\sigma} E(z, t, \lambda) \\ & + \lambda^{2/3} \xi^{(1/2 - \alpha_0)s} \zeta^{(\alpha_0 - 7/6)} e^{\xi_1 - \zeta_1} E(z, t, \lambda) \\ & + \xi^{(1/2 - \alpha_0/3)s} \zeta^{(\alpha_0/3 - 1/2)\sigma} e^{\xi_1 - \zeta_1} E(z, t, \lambda). \end{aligned} \tag{8.8}$$

Choosing paths of integration for the terms of the function (8.8) in (8.4) as in Lemmas III and IV, and applying these lemmas, we obtain the inequality

$$\int_{z_1}^{z} |\rho_1^{-1}(z, \lambda) K(z, t; \lambda) \rho_1(t, \lambda)\, dt| \leq M \log|\lambda|. \tag{8.9}$$

This inequality holds for $z$ in any closed sector of $S$ which excludes the ray $C_1$.

In order to treat the case when $z$ lies in a sector which contains the ray $C_1$, we express the quantity $k$ in terms of the functions $A_2$, $A_3$, $B_1$, $B_0$. Thus we obtain the form

$$\begin{aligned} k(z, t; \lambda) = \frac{e^{-\pi i \alpha_0}}{2\pi i}\{ & e^{2\pi i \alpha_0} A_2(z, \alpha_0) A_3(t, 2-\alpha_0) \\ & - A_3(z, \alpha_0) A_2(t, 2-\alpha_0) + e^{2\pi i \alpha_0} B_0(z, \alpha_0) B_1(t, 2-\alpha_0) \\ & + B_1(z, \alpha_0)[(e^{2\pi i \alpha_0} - 1) B_1(t, 2-\alpha_0) - B_0(t, 2-\alpha_0)]\}. \end{aligned} \tag{8.10}$$

We choose a single path of integration in (8.4) which consists of the radius from $z_k$ to 0 and a line segment from 0 to $z$. From (8.3), (8.10) and the bounds of Section 4 we find that the functions $k^{(m)}(z, t; \lambda)\rho_1(t, \lambda)$, $m = 0, 1, 2, 3$, have the forms

$$\begin{aligned} k^{(m)}\rho_1 &= \lambda^{(2/3)(\alpha_0+m)}\xi^{[(\alpha_0+m)/3-5/6]s}\zeta^{-(2/3)\sigma}e^{-\zeta_1}E \\ &+ \lambda^{(2/3)(\alpha_0+m)}\xi^{-(2/3)(\alpha_0+m-1)s}[\zeta^{-2/3}E + \zeta^{(\alpha_0-7/6)\sigma}e^{-\zeta_1}E] \\ &+ \lambda^{(2/3)\alpha_0}\zeta^{(\alpha_0/3-1/2)\sigma}e^{-\zeta_1}E + \lambda^{2/3}\zeta^{-2/3}E, \end{aligned} \tag{8.11a}$$

on the path from $z_1$ to 0, and

$$\begin{aligned} k^{(m)}\rho_1 &= \lambda^{(2/3)(\alpha_0+m)}\xi^{[(\alpha_0+m)/3-5/6]s}\zeta^{-(2/3)\sigma}[|e^{\xi_1-2\zeta_1}|E + |e^{-\xi_1}|E] \\ &+ \lambda^{(2/3)(\alpha_0+m)}\xi^{[(\alpha_0+m)/3-5/6]s}\zeta^{-(\alpha_0+1/6)}[|e^{\xi_1-\zeta_1}|E + |e^{\zeta_1+\xi_1}|E] \\ &+ \lambda^{(2/3)(\alpha_0+m)}\xi^{-(2/3)(\alpha_0+m-1)s}[\zeta^{(\alpha_0-7/6)\sigma}|e^{-\zeta_1}|E + \zeta^{-(2/3)\sigma}E] \\ &+ \lambda^{(2/3)\alpha_0}[\zeta^{(\alpha_0/3-1/2)}|e^{-\zeta_1}|E + \zeta^{-(2/3)\alpha_0\sigma}E], \end{aligned} \tag{8.11b}$$

on the path from 0 to $z$. The behavior of the function $K(z, t; \lambda)\rho_1(t, \lambda)$ is described by expressions like those in the right hand members of (8.11) with $m = 1$. Application of Lemmas I, II, and III shows that

$$\int_{z_1}^{z} |\rho_1^{-1}(z, \lambda)K(z, t; \lambda)\rho_1(t, \lambda)\, dt| \leq M \log|\lambda|, \tag{8.12}$$

for $z$ in the sector which contains $C_1$.

It now follows from (3.17), (8.2), (8.9), and (8.12) that

$$F(z, \lambda) = \rho_1(z, \lambda)E(z, \lambda) \log \lambda \tag{8.13}$$

for $z$ in $S$. It follows from (8.9), (8.12), (8.13), and the theory of Section 3 that the solution $w$ of the integral equation (3.16) has the behavior

$$w(z, \lambda) = F(z, \lambda) + \lambda^{-1}\rho_1(z, \lambda)E(z, \lambda) \log \lambda. \tag{8.14}$$

Thus $w$ has the form

$$w(z, \lambda) = \rho_1(z, \lambda)E(z, \lambda) \log \lambda, \tag{8.15}$$

for $z$ in $S$. The function $w$ also has the form (8.15) for $z$ in a closed disk which contains $S$ in its interior. We again denote this disk by $S'$.

Formula (3.33) becomes

$$\delta_m(z, \lambda) = \lambda^{-1} \int_{z_1}^{z} k^{(m)}(z, t; \lambda)w(t, \lambda)\, dt + \int_{z_1}^{z} k^{(m)}(z, t; \lambda)\tilde{w}_n(t, \lambda)\, dt, \qquad (m = 0, 1, 2, 3), \tag{8.16}$$

In this case it can be shown that the functions $\delta_m$ have the forms

$$\delta_m(z, \lambda) = \rho_m(z, \lambda)E(z, \lambda) \log \lambda, \qquad (m = 0, 1, 2, 3). \tag{8.17}$$

That the forms (8.17) are valid for $z$ in a closed sector which is contained in the region $S - C_1 + 0$ follows from (8.16) by the use of the bounds (8.2), (8.15), (8.7), and Lemmas III and IV. The appraisals of the integrals in (8.16) are straightforward in this case, and the details will be omitted.

When $z$ lies in a sector which contains the ray $C_1$, we may use the bounds (8.2), (8.15), (8.11), and Lemmas I, II, and III to appraise the integrals in (8.16). When

$m = 1, 2, 3$, the forms (8.17) follow easily. When $m = 0$, an additional step is involved. In this case we find that

$$\delta_0(z, \lambda) = \rho_0(z, \lambda)E(z, \lambda) \log \lambda + J, \tag{8.18}$$

where

$$J = \frac{e^{-\pi i \alpha_0}}{2\pi i} \int_{z_1}^{z} \{B_1(z, \alpha_0)[(e^{2\pi i \alpha_0} - 1)B_1(t, 2 - \alpha_0) - B_0(t, 2 - \alpha_0)]$$
$$+ e^{2\pi i \alpha_0} B_0(z, \alpha_0)B_1(t, 2 - \alpha_0)\}\tilde{w}_n(t)\, dt. \tag{8.19}$$

Integrating by parts, we find that

$$2\pi i e^{\pi i \alpha} J = [B_1(z, \alpha_0)[(e^{2\pi i \alpha_0} - 1)B_1(t, 2 - \alpha_0) - B_0(t, 2 - \alpha_0)]\tilde{W}_n(t)$$
$$+ e^{2\pi i \alpha_0} B_0(z, \alpha_0)B_1(t, 2 - \alpha_0)\tilde{W}_n(t)]_{t=z_1}^{t=z}$$
$$- \int_{z_1}^{z} \{B_1(z, \alpha_0)[(e^{2\pi i \alpha_0} - 1)B_1(t, 3 - \alpha_0) - B_0(t, 3 - \alpha_0)]$$
$$+ e^{2\pi i \alpha_0} B_0(z, \alpha_0)B_1(t, 3 - \alpha_0)\}\tilde{W}_n(t)\, dt, \tag{8.20}$$

where

$$\tilde{W}_n(z, \lambda) = \int_{z_1}^{z} \tilde{w}_n(t, \lambda)\, st. \tag{8.21}$$

In order to obtain a bound for the function $\tilde{W}_n$, we write $\tilde{w}_n$ as in (8.1) and integrate by parts. Thus

$$\tilde{W}_n = [\tilde{b}\tilde{A}_{n,1} + \lambda^{-1}\tilde{c}\tilde{A}'_{n,1} + \lambda^{-2}\tilde{d}\tilde{A}''_{n,1}]_{z_1}^{z}$$
$$+ \int_{z_1}^{z} [\tilde{a}\tilde{A}_{n,1} - \tilde{b}'\tilde{A}_{n,1} - \lambda^{-1}c'\tilde{A}'_{n,1} + \lambda^{-2}\, d\tilde{A}''_{n,1}]\, dt. \tag{8.22}$$

From the bounds (5.6), we see that the integrated part is of the form $\rho_0(z, \lambda)E(z, \lambda)$. The functions $\tilde{b}'$, $\tilde{c}'$, $\tilde{d}'$ are bounded for $z$ in $S'$, so the integrand above is of the form $\rho_0(z, \lambda)E(z, \lambda)$. Appraisal of the integral by the use of Lemmas I, II, and IV shows that the integral is also of this form. Thus the function $\tilde{W}_n$ has the form

$$\tilde{W}_n(z, \lambda) = \rho_0(z, \lambda)E(z, \lambda), \tag{8.23}$$

for $z$ in any closed disk contained in the interior of $S'$. Choosing such a disk which contains $S$ in its interior, we again denote it by $S'$. From (8.23) and the bounds of Section 4, we see that the integrated part in (8.20) is of the form $\rho_0(z, \lambda)E(z, \lambda)$. Application of Lemmas II, III, and V to the integrals in (8.20) shows that

$$J = \rho_0(z, \lambda)E(z, \lambda) \log \lambda. \tag{8.24}$$

It now follows from (8.18) and (8.24) that (8.17) holds for $m = 0$.

The existence of a solution $A_{n,1}$ of equation (1.2) which is approximated by the function $\tilde{A}_{n,1}$ follows from the results of this section and the theory of Section 3. The formula (2.7) with $k = 1$, follows from (3.26) and (8.17). As indicated in the introduction, the proofs for the cases $k = 2$ and $k = 3$ in Theorem II are similar to the proof given here for the case $k = 1$.

## 9. A partly balanced solution: proof of Theorem III

In this section, we shall show that equation (1.2) possesses a solution $B_{n,3}$ which is approximated by the function $B_{n,3}$ in a complete neighborhood of the turning point, as stated in Theorem III. This solution is balanced in the sector $S - S_3$ and is dominant in the sector $S_3$, since the function $B_{n,3}$ has this type of behavior.

Let us again consider the integral equation (3.16), this time with the function $w_n$ given by the relation

$$\tilde{w}_n = \tilde{a}\tilde{B}_{n,3} + \tilde{b}\tilde{B}'_{n,3} + \lambda^{-1}\tilde{c}\tilde{B}''_{n,3} + \lambda^{-2}\tilde{d}\tilde{B}'''_{n,3}. \tag{9.1}$$

This function $\tilde{w}_n$ has the form

$$\tilde{w}_n(z, \lambda) = \eta_1(z, \lambda)E(z, \lambda), \tag{9.2}$$

where

$$\eta_m(z, \lambda) = \lambda^{(2/3)(\alpha_0+m-1)}[|\xi^{-(2/3)(\alpha_0+m-1)s}| + |\xi^{[(\alpha_0+m)/3-5/6]s}e^{\xi_3}|g_3(z, \lambda)],$$
$$(m = 0, 1, 2, 3) \tag{9.3}$$

as may be seen from (8.1) and the bounds (5.7). By forming a fundamental matrix $U_0$ from the functions $B_0, B_1, B_2, B_3$ and making use of the first of the relations (4.1), we find that the kernel $K$ in (3.16) has the form (3.30) with

$$k(z, t; \lambda) = \frac{e^{-\pi i\alpha_0}}{2\pi i}\{[B_1(z, \alpha_0) + B_0(z, \alpha_0)]A_1(t, 2 - \alpha_0) + B_2(z, \alpha_0)A_2(t, 2 - \alpha_0)$$
$$+ e^{2\pi i\alpha_0}B_3(z, \alpha_0)A_3(t, 2 - \alpha_0) - B_0(z, \alpha_0)B_3(t, 2 - \alpha_0)\}. \tag{9.4}$$

It is necessary in this case to take paths of integration starting at different points in (3.16) and (3.17) corresponding to different terms in $K$. We therefore define

$$K_j(z, t; \lambda) = ak_j + bk'_j + \lambda^{-1}ck''_j + \lambda^{-2}dk'''_j, \qquad (j = 1, 2, 3, 4) \tag{9.5}$$

where

$$\begin{aligned} k_1 &= \frac{e^{-\pi i\alpha_0}}{2\pi i}[B_1(z, \alpha_0) + B_0(z, \alpha_0)]A_1(t, 2 - \alpha_0), \\ k_2 &= \frac{e^{-\pi i\alpha_0}}{2\pi i}B_2(z, \alpha_0)A_2(t, 2 - \alpha_0), \\ k_3 &= \frac{e^{\pi i\alpha_0}}{2\pi i}B_3(z, \alpha_0)A_3(t, 2 - \alpha_0), \\ k_4 &= -\frac{e^{-\pi i\alpha_0}}{2\pi i}B_0(z, \alpha_0)B_3(t, 2 - \alpha_0), \end{aligned} \tag{9.6}$$

and write (3.16) in the form

$$w(z, \lambda) = F(z, \lambda) + \lambda^{-1}\sum_{j=1}^{4}\int_{z_j}^{z} K_j(z, t; \lambda)w(t, \lambda)\,dt, \tag{9.7}$$

where

$$F(z, \lambda) = \sum_{j=1}^{4}\int_{z_j}^{z} K_j(z, t; \lambda)\tilde{w}_n(t, \lambda)\,dt. \tag{9.8}$$

The points $z_k$, $k = 1, 2, 3$, are as defined in Section 6, and the point $z_4$ is at the intersection of the ray $C_3$ and the boundary of the disk $S'$.

In order to obtain a bound for the function $F$ and to establish the existence of a solution for equation (9.7), let us consider the integral

$$\sum_{j=1}^{4} \int_{z_j}^{z} \eta_1^{-1}(z, \lambda)K(z, t; \lambda)\eta_1(t, \lambda)\, dt, \tag{9.9}$$

for a specific choice of paths. Since the function $\tilde{w}_n$ exhibits different types of asymptotic behaviors in the sectors $S_3$ and $S - S_3$, it will be convenient to consider these two regions separately. Let us first suppose that $z$ is confined to a closed sector which lies in the region $S - S_3 + 0$. By using the relations (4.1), we may write the contribution to the integral (9.9) from the term $ak$ in $K$ as

$$\begin{aligned} \frac{e^{-\pi i\alpha_0}}{2\pi i}\eta_1^{-1}(z)\bigg\{\int_{z_1}^{z} & aA_2(z, \alpha_0)A_1(t, 2 - \alpha_0)\eta_1(t)\, dt \\ & - e^{2\pi i\alpha_0}\int_{z_2}^{z} aA_1(z, \alpha_0)A_2(t, 2 - \alpha_0)\eta_1(t)\, dt + \int_{z_1}^{0} aB_3(z, \alpha_0)A_1(t, 2 - \alpha_0)\eta_1(t)\, dt \\ & + e^{2\pi i\alpha_0}\int_{z_2}^{z} aB_3(z, \alpha_0)A_2(t, 2 - \alpha_0)\eta_1(t)\, dt \\ & + e^{2\pi i\alpha_0}\int_{z_3}^{0} aB_3(z, \alpha_0)A_3(t, 2 - \alpha_0)\eta_1(t)\, dt \\ & + e^{2\pi i\alpha_0}\int_{0}^{z} aB_3(z, \alpha_0)[B_0(t, 2 - \alpha_0) + (e^{2\pi i\alpha_0} - 1)B_3(t, 2 - \alpha_0)]\eta_1(t)\, dt \\ & - \int_{z_4}^{z} aB_0(z, \alpha_0)B_3(t, 2 - \alpha_0)\eta_1(t)\, dt\bigg\}. \end{aligned} \tag{9.10}$$

Here we have expressed $k$ in terms of the functions $A_1$, $A_2$, $B_3$, $B_0$, except for $t$ in the sector $S_3$ where the function $A_3$ is recessive. The paths for the terms in $k_2$ and $k_3$ which involve exponential type functions of $z$ are chosen as in Lemma IV. The path corresponding to $k_4$ is chosen as in Lemma III. The other paths in (9.10) consist of straight line segments. The contributions to the integral (9.10) from the terms $bk$, $\lambda^{-1}ck'$, and $\lambda^{-2}dk'''$ may be written in similar forms.

From (9.3) and the bounds of Section 4, we find that the functions $k^{(m)}(z, t; \lambda)\eta_1(t, \lambda)$, $m = 0, 1, 2, 3$, have the bounds

$$k^{(m)}\eta_1 = \lambda^{(2/3)(\alpha_0 + m)}\xi^{[(\alpha_0 + m)/3 - 5/6]s}\zeta^{-(\alpha_0 + 1/6)\sigma}e^{\xi_j - \zeta_j}E, \tag{9.11a}$$

on the path from $z_j$ to $z, j = 1, 2$;

$$k^{(m)}\eta_1 = \lambda^{(2/3)(\alpha_0 + m)}\xi^{-(2/3)(\alpha_0 + m - 1)s}[\zeta^{-(\alpha_0 + 1/6)\sigma}e^{-\zeta_j}E + \zeta^{-(2/3)\sigma}g_3(t, \lambda)E], \tag{9.11b}$$

on the path from $z_j$ to $0, j = 1, 2, 3$;

$$k^{(m)}\eta_1 = \lambda^{(2/3)(\alpha_0 + m)}\xi^{-(2/3)(\alpha_0 + m - 1)s}\zeta^{-(2/3)\sigma}E, \tag{9.11c}$$

on the path from 0 to $z$;

$$k^{(m)}\eta_1 = \lambda^{(2/3)\alpha_0}\zeta^{-(2/3)\alpha_0}E, \tag{9.11d}$$

on the path from $z_4$ to $z$. Then the function $\eta_1^{-1}(z, \lambda)K(z, t; \lambda)\eta_1(t, \lambda)$ has the bounds

$$\lambda^{(2/3)}\xi^{(\alpha_0 - 1/2)s}\zeta^{-(\alpha_0 + 1/6)\sigma}e^{\xi_j - \zeta_j}E, \tag{9.12a}$$

on the path from $z_j$ to $z, j = 1, 2$;

$$\lambda^{2/3}[\zeta^{-(\alpha + 1/6)\sigma}e^{-\zeta_j}E + \zeta^{-(2/3)\sigma}g_3(t, \lambda)E], \tag{9.12b}$$

on the path from $z_j$ to $0, j = 1, 2, 3$;

$$\lambda^{2/3}\zeta^{-(2/3)\sigma}E, \tag{9.12c}$$

on the path from 0 to $z$;

$$\xi^{-(2/3)\alpha s}\zeta^{-(2/3)\alpha\sigma}E, \tag{9.12d}$$

on the path from $z_4$ to $z$. Use of these bounds and application of Lemmas I, II, III, and IV shows that

$$\sum_{j=1}^{4}\int_{z_j}^{z} |\eta_1^{-1}(z, \lambda)K_j(z, t; \lambda)\eta_1(t, \lambda)\, dt| \leq M \log|\lambda| \tag{9.13}$$

for $z$ in the sector under consideration.

In order to treat the case when $z$ lies in the sector $S_3$, we consider two closed sectors, one of which contains the ray $C_1$ but not $C_2$, and the other contains the ray $C_2$ but not $C_1$. These two sectors may be chosen so that the sector $S_3$ lies in their union.

Suppose first that $z$ lies in the sector which contains the ray $C_1$. By the use of the form (9.4) for $k$ and the relations (4.1), we may write the contribution of the term $ak$ in $K$ to the integral (9.9) as

$$\begin{aligned}
&\frac{e^{-\pi i\alpha_0}}{2\pi i}\eta_1^{-1}(z)\left\{\int_{z_2}^{z} aA_3(z, \alpha_0)A_2(t, 2 - \alpha_0)\eta_1(t)\, dt\right.\\
&\quad + e^{2\pi i\alpha_0}\int_{z_3}^{z} aA_2(z, \alpha_0)A_3(t, 2 - \alpha_0)\eta_1(t)\, dt\\
&\quad + \int_{z_1}^{0} a[B_3(z, \alpha_0) + B_0(z, \alpha_0)]A_1(t, 2 - \alpha_0)\eta_1(t)\, dt\\
&\quad + \int_{z_2}^{0} aB_1(z, \alpha_0)A_2(t, 2 - \alpha_0)\eta_1(t)\, dt\\
&\quad + e^{2\pi i\alpha_0}\int_{z_3}^{0} a[B_1(z, \alpha_0) + B_0(z, \alpha_0)]A_3(t, 2 - \alpha_0)\eta_1(t)\, dt\\
&\quad + \int_{0}^{z} a[B_0(z, \alpha_0) + B_1(z, \alpha_0)][B_0(t, 2 - \alpha_0) + (1 - e^{2\pi i\alpha_0})B_1(t, 2 - \alpha_0)]\eta_1(t)\, dt\\
&\quad + \int_{z_4}^{0} aB_0(z, \alpha_0)A_2(t, 2 - \alpha_0)\eta_1(t)\, dt\\
&\quad \left.- \int_{z_4}^{0} aB_0(z, \alpha_0)[B_1(t, 2 - \alpha_0) + B_0(t, 2 - \alpha_0)]\eta_1(t)\, dt\right\}.
\end{aligned} \tag{9.14}$$

Here we have expressed $k$ in terms of the functions $A_2$, $A_3$, $B_1$, $B_0$, except for $t$ in the sector $S_1$ where the function $A_1$ is recessive. The paths for the terms in $k_1$

and $k_2$ which involve exponential type functions of $z$ are chosen as in Lemmas III and IV. In the integral from $z_4$ to $z$, we take two different paths corresponding to the two terms in the function $\eta_1(t)$. Both paths pass through the point $z_2$ and avoid the sector $S_1$ where the function $B_1$ is dominant. The other paths in (9.14) are straight line segments. The contributions from the terms $bk'$, $\lambda^{-1}ck''$, and $\lambda^{-2}dk'''$ in (9.9) may be written in similar forms.

From (9.3) and the bounds of Section 4, we find that the functions $k^{(m)}(z, t; \lambda)\eta_1(t)$, $m = 0, 1, 2, 3$, have the forms

$$\lambda^{(2/3)(\alpha_0+m)}\xi^{[(\alpha_0+m)/3-5/6]s}[\zeta^{-(\alpha_0+1/6)}e^{\xi_2-\zeta_2}E + \zeta^{-(2/3)\sigma}e^{\xi_2-2\zeta_2}E], \tag{9.15a}$$

on the path from $z_2$ to $z$;

$$\lambda^{(2/3)(\alpha_0+m)}\xi^{[(\alpha_0+m)/3-5/6]s}[\zeta^{-(\alpha_0+1/6)}e^{\xi_3-\zeta_3}E + \zeta^{-(2/3)\sigma}e^{\xi_3}E], \tag{9.15b}$$

on the paths from $z_3$ to $z$;

$$\lambda^{(2/3)(\alpha_0+m)}\xi^{-(2/3)(\alpha_0+m-1)s}[\zeta^{-(\alpha_0+1/6)\sigma}e^{-\zeta_j}E + \eta^{-(2/3)\sigma}g_3(t, \lambda)E], \tag{9.15c}$$

on the path from $z_j$ to 0, $j = 1, 2, 3$;

$$\lambda^{(2/3)(\alpha_0+m)}\xi^{-(2/3)(\alpha_0+m-1)s}[\zeta^{-(2/3)\sigma}E + \zeta^{(\alpha_0-7/6)\sigma}e^{\zeta_3}E], \tag{9.15d}$$

on the path from 0 to $z$;

$$\lambda^{2/3}\zeta^{-(\alpha_0+1/6)\sigma}E, \tag{9.15e}$$

on the path from $z_4$ to 0;

$$\lambda^{(2/3)\alpha_0}\zeta^{-(2/3)\alpha_0\sigma}E + \lambda^{(2/3)\alpha}\zeta^{(\alpha_0/3-1/2)\sigma}e^{-\zeta_2}g_3(t, \lambda)E, \tag{9.15f}$$

on the path from $z_4$ to $z$. The bounds for the function $K(z, t; \lambda)\eta_1(t, \lambda)$ on the paths have the same forms as the bounds for the function $k'(z, t; \lambda)\eta_1(t)$. Thus they are like the expressions (9.15), with $m = 1$. From these bounds and Lemmas I–IV, we find that

$$\sum_{j=1}^{4}\int_{z_j}^{z} |\eta_1^{-1}(t, \lambda)K_j(z, t; \lambda)\eta_1(t, \lambda)\,dt| \le M\log|\lambda|, \tag{9.16}$$

for $z$ in the sector which contains the ray $C_1$.

In order to treat the case when $z$ lies in the sector which contains the ray $C_2$, we use the form (9.4) for $k$ and the relations (4.1) to express the kernel $K$ in terms of the functions $A_1, A_3, B_2, B_0$. In this case, (9.14) is replaced by the expression

$$\begin{aligned}\frac{e^{-\pi i\alpha_0}}{2\pi i}\eta_1^{-1}(z)\Big\{&e^{2\pi i\alpha_0}\int_{z_3}^{z} aA_1(z, \alpha_0)A_3(t, 2-\alpha_0)\eta_1(t)\,dt\\ &-\int_{z_1}^{z} aA_3(z, \alpha_0)A_1(t, 2-\alpha_0)\eta_1(t)\,dt\\ &+\int_{z_1}^{0} a[B_2(z, \alpha_0) + B_0(z, \alpha_0)]A_1(t, 2-\alpha_0)\eta_1(t)\,dt\\ &+\int_{z_2}^{z} aB_2(z, \alpha_0)A_2(t, 2-\alpha_0)\eta_1(t)\,dt + \int_{z_3}^{0} aB_2(z, \alpha_0)A_3(t, 2-\alpha_0)\eta_1(t)\,dt\\ &+\int_{0}^{z} aB_2(z, \alpha_0)[B_0(t, 2-\alpha_0) + (1 - e^{2\pi i\alpha_0})B_2(t, 2-\alpha_0)]\eta_1(t)\,dt\end{aligned}$$

$$- \int_{z_4}^{0} aB_0(z, \alpha_0)A_1(t, 2 - \alpha_0)\eta_1(t)\, dt$$

$$- e^{2\pi i\alpha_0} \int_{z_4}^{z} aB_0(z, \alpha_0)B_2(t, 2 - \alpha_0)\eta_1(t)\, dt \Big\}. \tag{9.17}$$

The paths for the terms in $k_1$ and $k_2$ which involve exponential-type functions of $z$ are chosen as in Lemmas III and IV. In the integral from $z_4$ to $z$, we again take two paths of integration corresponding to the two terms in the function $\eta_1(t)$. These paths pass through the point $z_1$ and avoid the Sector $S_2$ where the function $B_2$ is dominant. The other paths in (9.17) are straight line segments. The bounds for the functions $k^{(m)}(z, t; \lambda)\eta_1(t, \lambda)$ $m = 0, 1, 2, 3$, are similar to the expressions (9.15), except that $z_2$, $\xi_2$, and $\zeta_2$ must be replaced by $z_1$, $\xi_1$, and $\zeta_1$, respectively, in (9.15a), and $\zeta_2$ must be replaced by $\zeta_1$ in (9.15f). By the use of Lemmas I–IV, it may be shown that an inequality of the form (9.16) holds for $z$ in the sector which contains the ray $C_2$.

From these results, it follows that the function (9.8) has the form

$$F(z, \lambda) = \eta_1(z, \lambda)E(z, \lambda) \log \lambda. \tag{9.18}$$

It also follows that equation (9.7) has a solution $w$ such that

$$w(z, \lambda) = F(z, \lambda) + \lambda^{-1}\eta_1(z, \lambda)E(z, \lambda) \log \lambda. \tag{9.19}$$

Thus $w$ has the form

$$w(z, \lambda) = \eta_1(z, \lambda)E(z, \lambda) \log \lambda. \tag{9.20}$$

This function $w$ actually has the form (9.20) in a disk $S'$ which contains $S$ in its interior.

Formula (3.33) becomes

$$\delta_m(z, \lambda) = \lambda^{-1} \sum_{j=1}^{4} \int_{z_j}^{z} k_j^{(m)}(z, t; \lambda)w(t, \lambda)\, dt$$

$$+ \sum_{j=1}^{4} \int_{z_j}^{z} k_j^{(m)}(z, t; \lambda)\tilde{w}_n(t, \lambda)\, dt, \qquad (m = 0, 1, 2, 3). \tag{9.21}$$

In this case, it turns out that the functions $\delta_m$ have the forms

$$\delta_m(z, \lambda) = \eta_m(z, \lambda)E(z, \lambda) \log \lambda, \qquad (m = 0, 1, 2, 3) \tag{9.22}$$

for $z$ in $S$. To show how this may be established, we again treat the sectors $S - S_3$ and $S_3$ separately. Suppose first that $z$ lies in a closed sector which is contained in the region $S - S_3 + 0$. We express the function $k$ in terms of the functions $A_1$, $A_2$, $B_3$, $B_0$, and choose paths of integration corresponding to the various terms of $k$ in (9.21) just as we did in (9.10). Then we may use the bounds (9.2), (9.20) and (9.12), and Lemmas I–IV to appraise the integrals. When $m = 0, 1, 2$, the appraisals are straightforward, and the details will be omitted. When $m = 3$, we find that

$$\delta_3(z, \lambda) = \eta_3(z, \lambda)E(z, \lambda) \log \lambda + J, \tag{9.23}$$

where

$$J = \frac{e^{-\pi i\alpha_0}}{2\pi i}\left[\int_{z_1}^{z} A_2'''(z, \alpha_0)A_1(t, 2 - \alpha_0)\tilde{w}_n(t)\,dt \right.$$

$$\left. - e^{2\pi i\alpha_0}\int_{z_2}^{z} A_1'''(z, \alpha_0)A_2(t, 2 - \alpha_0)\tilde{w}_n(t)\,dt\right]. \tag{9.24}$$

Partial integration yields the relation

$$\begin{aligned} 2\pi i e^{\pi i\alpha_0}J = {} & [A_2'''(z, \alpha_0)A_1(z, 1 - \alpha_0) - e^{2\pi i\alpha_0}A_1'''(z, \alpha_0)A_2(z, 1 - \alpha_0)\tilde{w}_n(z)] \\ & - [A_2'''(z, \alpha_0)A_1(z_1, 1 - \alpha_0)\tilde{w}_n(z_1) \\ & - e^{2\pi i\alpha_0}A_1'''(z, \alpha_0)A_2(z_2, 1 - \alpha_0)\tilde{w}_n(z_2)] \\ & - \int_{z_1}^{z} A_2'''(z, \alpha_0)A_1(t, 1 - \alpha_0)\tilde{w}_n'(t)\,dt \\ & + e^{2\pi i\alpha_0}\int_{z_2}^{z} A_1'''(z, \alpha_0)A_2(t, 1 - \alpha_0)\tilde{w}_n'(t)\,dt. \end{aligned} \tag{9.25}$$

From the bound (9.2) and the relation (4.6), it follows that the first term in brackets on the right is of the form $\eta_3(z, \lambda)E(z, \lambda)$. The second quantity in brackets tends to zero as $|\lambda|$ becomes large. The function $\tilde{w}_n'$ is of the form

$$\tilde{w}_n'(z, \lambda) = \eta_2(z, \lambda)E(z, \lambda), \tag{9.26}$$

as may be shown by differentiating the expression (9.1) and using equation (1.9) to eliminate the fourth derivative of the function $B_{n,3}$. The bounds (4.4) and (9.26) and Lemma IV may be used to show that each of the integrals in (9.25) is of the form $\eta_3(z, \lambda)E(z, \lambda)$. Thus $J = \eta_3(z, \lambda)E(z, \lambda)$, and (9.22) holds for $m = 3$ and $z$ in the sector under consideration.

In the case when $z$ lies in the sector $S_3$, we again consider two sectors, one of which contains the ray $C_1$ and the other the ray $C_2$. When $z$ lies in the sector which contains $C_1$, we express the function $k$ in terms of the functions $A_2, A_3, B_1, B_0$, and choose paths of integration corresponding to the different terms in $k$ for the integrals in (9.21) just as we did in (9.14). Bounds for the integrands may be obtained from (9.2), (9.15), (9.20). The appraisal of the integrals by the use of Lemmas I–IV is again straightforward when $m = 0, 1, 2$. When $m = 3$, we find that

$$\delta_3(z, \lambda) = \eta_3(z, \lambda)E(z, 4)\log\lambda + \tilde{J}. \tag{9.27}$$

## 10. Acknowledgment

The authors wish to thank Dr. Y. Sibuya for discussions with him during the final writing of the manuscript. The work presented here was done at the Massachusetts Institute of Technology and at the United States Army Mathematics Research Center, University of Wisconsin. The work is supported in part by the Office of Naval Research.

## References

COPSON, E. T. Theory of Functions of a Complex Variable, Oxford, (1935).

LANGER, R. E. On the asymptotic solutions of a class of ordinary differential equations of the fourth order, with special reference to an equation of hydrodynamics; Trans. Amer. Math. Soc. **84**, 144–191, (1957).

LIN, C. C. AND RABENSTEIN, A. L. On the asymptotic solutions of a class of ordinary differential equations of the fourth order. Trans. Amer. Math. Soc. **94**, 24–57, (1960).

RABENSTEIN, A. L. Asymptotic solutions of $u^{IV} + \lambda^2(zu'' + \alpha u' + \beta u) = 0$ for large $|\lambda|$; Archive for Rat. Mech. and Anal. **1**, 418–435, (1958).

RABENSTEIN, A. L. The determination of the inverse matrix for a basic reference equation for the theory of hydrodynamic stability; Archive for Rat. Mech. and Anal. **2**, 355–366, (1959).

MASSACHUSETTS INSTITUTE OF TECHNOLOGY

MACALESTER COLLEGE, MINNESOTA

(Received July 18, 1969)

*Offprint from "Archive for Rational Mechanics and Analysis"*
*Volume 1, Number 5, 1958, P. 418—435*

*Springer-Verlag, Berlin · Göttingen · Heidelberg*

# *Asymptotic Solutions of* $u^{iv} + \lambda^2(z u'' + \alpha u' + \beta u) = 0$ *for Large* $|\lambda|$

ALBERT L. RABENSTEIN

*Communicated by* C. C. LIN

## 1. Introduction

The asymptotic theory for the Orr-Sommerfeld equation of hydrodynamic stability has a large literature. The structure of the equation is such that it possesses a *turning point*; and herein lies the main difficulty in the development of such a theory. The equation is of the form

$$\varphi^{iv} + \lambda^2 \sum_{n=0}^{2} P_n(y)\, \varphi^{(2-n)} + \sum_{n=0}^{2} Q_n(y, \lambda)\, \varphi^{(2-n)} = 0 \tag{1.1}$$

where the complex parameter $\lambda$ is supposed to be large in magnitude, the functions $P_n(y)$ are analytic with

$$P_0(y_0) = 0, \qquad P'(y_0) \neq 0 \tag{1.1a}$$

and

$$Q_n(y, \lambda) \sim \sum_{\nu=0}^{\infty} \frac{q_{n\nu}(y)}{\lambda^{\nu}}, \tag{1.1b}$$

where the functions $q_{n\nu}(y)$ are analytic. In the hydrodynamical case, $P_1(y) \equiv 0$. We are interested in the behavior of solutions of (1.1) in a neighborhood of the turning point $y_0$ when $|\lambda|$ is large.

In order to use the method of a comparison equation to obtain such an asymptotic theory, one needs an equation which is sufficiently similar to (1.1) but which can still be solved explicitly. Such an equation is

$$u^{iv} + \lambda^2(z u'' + \alpha u' + \beta u) = 0. \tag{1.2}$$

Here $z$ is the complex independent variable and $\alpha$ and $\beta$ are complex parameters, dependent on $\lambda$ in such a way that they are uniformly bounded for $|\lambda| \geqq |\lambda_0|$ for some $\lambda_0$. This equation has a turning point at $z=0$, and is indeed a special case of (1.1). Furthermore it can be solved using the method of Laplace integrals, since the independent variable appears in only one of the coefficients and appears there linearly.

The complete theory for (1.1), based on the functions defined by (1.2), will be presented later*.

---

* Preliminary forms of this theory have been reported by LIN. See the Bulletin of the Amer. Math. Soc. **62**, 394 (1956), and the Proceedings of the IX[e] Congrès International de Mécanique Appliquée, Université de Bruxelles **1**, 137 (1957).

The present paper concerns the behavior of the functions defined by (1.2), for large $|\lambda|$. This equation is clearly a generalization of that considered by WASOW [*1*]. It turns out that it is necessary to consider (1.2) with general $\alpha$ to make the general theory for (1.1) feasible, even in the case where $P_1(y) \equiv 0$. Also, there are certain general relations that hold for the functions $u(z;\lambda,\alpha,\beta)$, and these are not evident for the functions $u(z;\lambda,0,\beta)$ considered by WASOW. Indeed, one of the important differences between our theory for (1.1) and those of WASOW [*2*] and LANGER [*3*] is the use of the comparison equation (1.2). The present theory follows more closely the earlier work of LANGER for the second order equation [*4*]; it is simpler in structure and more readily applicable.

Series which satisfy (1.2) formally can be found by well known methods. These have the forms

$$\sum_{n=0}^{\infty} \sigma_{1n}(z)\,\lambda^{-2n}, \qquad \sum_{n=0}^{\infty} \sigma_{2n}(z)\,\lambda^{-2n},$$

$$(1.3) \qquad e^{\frac{2}{3}i\lambda z^{\frac{3}{2}}} \sum_{n=0}^{\infty} \sigma_{3n}(z)\,\lambda^{-n}, \qquad e^{-\frac{2}{3}i\lambda z^{\frac{3}{2}}} \sum_{n=0}^{\infty} \sigma_{4n}(z)\,\lambda^{-n}.$$

In these, $\sigma_{1n}(z)$ is an entire function, $\sigma_{20}(z)$ has a branch point at $z=0$, and $\sigma_{30}(z)=\sigma_{40}(z)=z^{\frac{\alpha}{2}-\frac{5}{4}}$. Since all solutions of (1.2) are entire functions of $z$, the multi-valued nature of the expressions (1.3) precludes them from representing actual solutions of (1.2) in a full neighborhood of $z=0$. It is known [*5*] that one of these formal series may represent one solution in a certain region of the $z$ plane and represent another solution in another region. In any case, the representation is known to be valid only for $|z| \geqq |z_0| > 0$, where $z_0$ is a fixed number, so the immediate vicinity of the turning point is not included in the region of representation.

If we put $\eta = \lambda^{\frac{2}{3}} z$, $u(z) = \chi(\eta)$ in (1.2) we obtain the equation

$$(1.4) \qquad \chi^{iv} + \eta\chi'' + \alpha\chi' + \beta\lambda^{-\frac{2}{3}}\chi = 0.$$

Solutions of (1.4) having the form of power series in $\lambda^{-\frac{2}{3}}$ will be valid for $|\eta| < |\eta_0|$ or $|z| < |\lambda|^{-\frac{2}{3}}|\eta_0|$ for an arbitrary but finite $\eta_0$. Thus, for any solution of (1.4), the region of validity in the $z$ plane becomes vanishingly small as $|\lambda|$ becomes large, while the representation of a solution by one of the expressions (1.3) still holds only for $|z| \geqq |z_0|$. Also there remains the problem of connecting the solutions of (1.4) with the solutions represented by the expressions (1.3).

In this article we obtain solutions of (1.2) by Laplace integrals. Certain relations which hold among the solutions defined by these integrals are deduced with the aid of CAUCHY's theorem. These relations enable one to describe fully the asymptotic behavior of each particular solution in a full neighborhood of $z=0$.

## 2. General formulae

Seeking solutions of (1.2) in the form

$$u = \int_C e^{zt} f(t)\,dt$$

where $t$ is a complex variable, we find that $f$ satisfies the first order equation

$$(t^2 f)' = (\lambda^{-2} t^2 + \beta t^{-2} + \alpha t^{-1})\,(t^2 f)$$

where $f$ and $C$ are to be such that $[e^{zt}t^2f]_C=0$*. Then a solution of (1.2) is

$$u=\int_C t^{\alpha-2}\exp\left(\tfrac{1}{3}\lambda^{-2}t^3-\beta t^{-1}+zt\right)dt, \tag{2.1}$$

where $C$ is to be such that

$$\left[t^{\alpha}\exp\left(\tfrac{1}{3}\lambda^{-2}t^3-\beta t^{-1}+zt\right)\right]_C=0. \tag{2.2}$$

If repeated differentiation under the integral sign is permissible, there follows from (2.1) that

$$\frac{d^n u(z,\alpha)}{dz^n}=u(z,\alpha+n). \tag{2.3}$$

It will be seen that this is the case for every solution considered here. Thus, if the asymptotic behavior of a solution (2.1) can be obtained for general $\alpha$, the behavior of its derivatives follows from (2.3). From (1.2) and (2.3) we also obtain the recurrence formula

$$u(z,\alpha+4)=-\lambda^2\left[z\,u(z,\alpha+2)+\alpha\,u(z,\alpha+1)+\beta\,u(z,\alpha)\right]. \tag{2.4}$$

In what follows, we suppose $\lambda$ to be large in magnitude and to have constant argument. Having chosen a particular value for $\arg\lambda$, by $\lambda^p$ we shall mean

$$\lambda^p=\exp\{p(\log|\lambda|+i\arg\lambda)\}$$

for all complex $p$. We assume $\beta\neq 0$, but $\alpha$ and $\beta$ are otherwise arbitrary. As remarked in the introduction, they may depend on $\lambda$ in such a way that they are uniformly bounded for $|\lambda|\geqq|\lambda_0|$. We take for $\arg\beta$ the value which lies in the range

$$\tfrac{2}{3}\arg\lambda\leqq\arg\beta<2\pi+\tfrac{2}{3}\arg\lambda. \tag{2.5}$$

We now define particular solutions of (1.2) by making appropriate choices for the path $C$ of integration in (2.1). Because of the essential singularity of the integrand at $t=0$, paths tending to the origin must do so with direction $\left(\arg\beta+\gamma\frac{\pi}{2}\right)\pmod{2\pi}$, where $-1<\gamma<1$. Because of the factor $t^{\alpha}$, the integrand in (2.1) is in general multiple-valued. The quantity $\lambda^{-2}t^3$ has a negative real part in the sectors

$$T_1:\qquad \frac{\pi}{6}+\frac{2}{3}\arg\lambda<\arg t<\frac{\pi}{2}+\frac{2}{3}\arg\lambda$$

$$T_2:\qquad \frac{5\pi}{6}+\frac{2}{3}\arg\lambda<\arg t<\frac{7\pi}{6}+\frac{2}{3}\arg\lambda$$

$$T_3:\qquad -\frac{\pi}{2}+\frac{2}{3}\arg\lambda<\arg t<-\frac{\pi}{6}+\frac{2}{3}\arg\lambda.$$

These are the unshaded sectors in Figure 1. We also define

$$\overline{T}_2:\qquad -\frac{7\pi}{6}+\frac{2}{3}\arg\lambda<\arg t<-\frac{5\pi}{6}+\frac{2}{3}\operatorname{arc}\lambda.$$

* For a discussion of the use of the Laplace integral in solving differential equations, see CODDINGTON & LEVINSON, Theory of Ordinary Differential Equations, New York, 1955, p. 170.

In Figure 1, $T_2$ and $\overline{T}_2$ have the same location, but the function $t^\alpha$ will have in general* different values in $T_2$ and $\overline{T}_2$. We define three solutions $u = A_k$, $k = 1, 2, 3$, by taking paths $C(A_k)$ going from infinity to infinity in two unshaded sectors. The path $C(A_1)$ goes to infinity in $\overline{T}_2$ and $T_3$, the path $C(A_2)$ goes to infinity in $T_3$ and $T_1$, and the path $C(A_3)$ goes to infinity in $T_1$ and $T_2$.

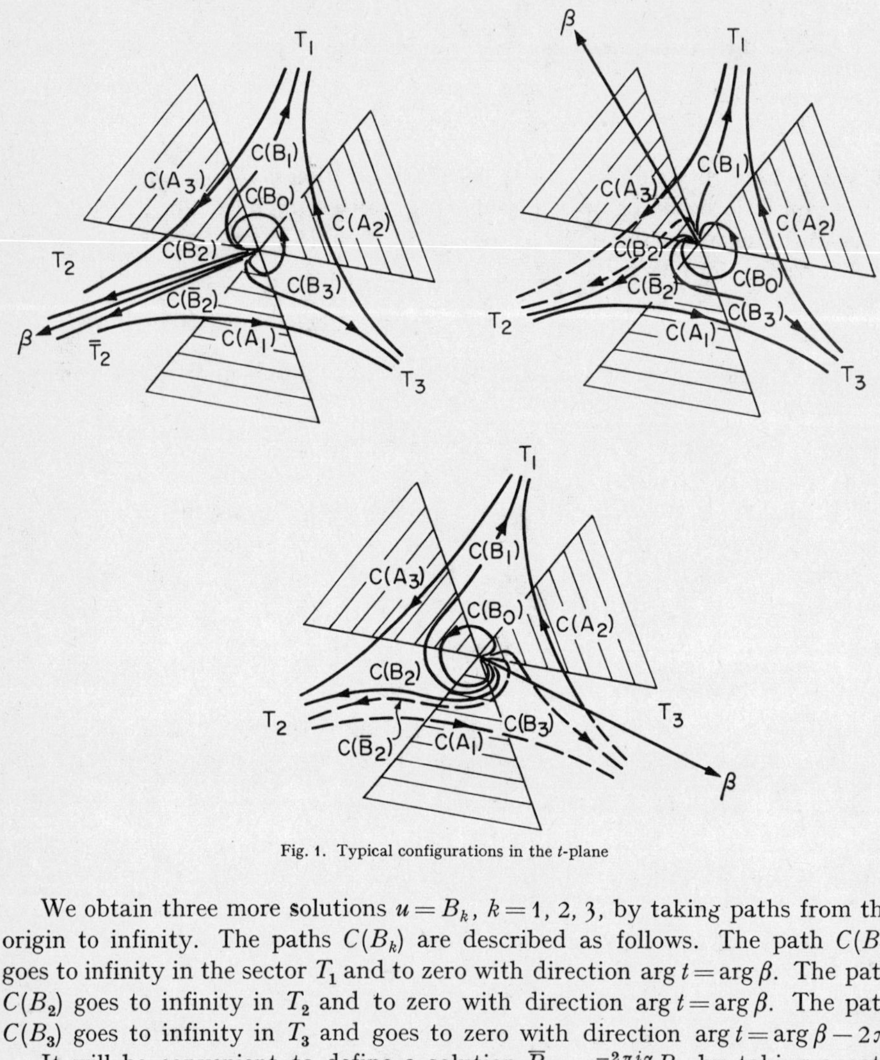


Fig. 1. Typical configurations in the $t$-plane

We obtain three more solutions $u = B_k$, $k = 1, 2, 3$, by taking paths from the origin to infinity. The paths $C(B_k)$ are described as follows. The path $C(B_1)$ goes to infinity in the sector $T_1$ and to zero with direction $\arg t = \arg \beta$. The path $C(B_2)$ goes to infinity in $T_2$ and to zero with direction $\arg t = \arg \beta$. The path $C(B_3)$ goes to infinity in $T_3$ and goes to zero with direction $\arg t = \arg \beta - 2\pi$.

It will be convenient to define a solution $\overline{B}_2 = e^{-2\pi i \alpha} B_2$ by taking a path $C(\overline{B}_2)$ going to infinity in the sector $\overline{T}_2$ and tending to the origin with direction $\arg t = \arg \beta - 2\pi$.

We also define a solution $u = B_0$ corresponding to a path $C(B_0)$ which emerges from the origin with direction $\arg t = \arg \beta - 2\pi$, circles the origin, and tends to zero again with direction $\arg t = \arg \beta$.

* In other words, $T_2$ and $\overline{T}_2$ will lie on different sheets of the Riemann surface for the function $t^\alpha = \exp[\alpha(\log|t| + i \arg t)]$.

Convergence at the origin is not altered if the direction of approach is changed by an angle less than $\frac{1}{2}\pi$, in accordance with the remarks following (2.5). Use will be made of this later. For each path $C$ described above, the integral (2.1) converges uniformly for $z$ in any finite neighborhood $S$ of $z=0$ and for $\alpha$ in any bounded region of the $\alpha$ plane. Also, the condition (2.2) is satisfied. Then differentiation under the integral sign is valid, and (2.3) follows. These paths of integration are shown in Figure 1 for various values of $\arg\beta$. In the visible portion of the Reimann surface, $\arg t$ ranges from $\arg\beta-2\pi$ to $\arg\beta$. The dotted parts of the paths are the parts which lie on other sheets of the surface. The paths which go to infinity do so with directions independent of $\beta$. At the origin, the direction of approach varies continuously with $\arg\beta$, for $\arg\beta$ in the range (2.5). As a consequence of CAUCHY's theorem, the following relations hold among the solutions:

$$\begin{aligned} &\overline{B}_2+A_1=B_3, \\ (2.6)\qquad &B_3+A_2=B_1+B_0 \qquad (\overline{B}_2=B_2e^{-2\pi i\alpha}), \\ &B_1+A_3=B_2. \end{aligned}$$

The paths of integration have been chosen so that these relations hold for all values of $\beta$.

We now seek the asymptotic description of each of the functions defined above in a full neighborhood of the turning point $z=0$. The development follows closely that used by WASOW in [*1*].

### 3. The solution $B_0$

We have

$$B_0=\int_{C(B_0)} t^{\alpha-2}\exp(zt-\beta t^{-1})\exp(\tfrac{1}{3}\lambda^{-2}t^3)\,dt. \tag{3.1}$$

Since the series

$$\exp\left(\frac{1}{3}\lambda^{-2}t^3\right)=\sum_{n=0}^{\infty}\frac{1}{n!}\left(\frac{1}{3}\lambda^{-2}t^3\right)$$

converges uniformly for $t$ in any bounded domain and $|\lambda|\geqq|\lambda_0|>0$, and since the rest of the integrand in (3.1) is uniformly bounded for $t$ on $C(B_0)$, termwise integration is permissible and we obtain

$$B_0=b_0(z,\alpha)+\tfrac{1}{3}\lambda^{-2}b_0(z,\alpha+3)+\cdots \tag{3.2}$$

where

$$b_0(z,\alpha)=\int_{C(B_0)} t^{\alpha-2}\exp(zt-\beta t^{-1})\,dt. \tag{3.3}$$

This may be expressed in terms of a Bessel function. The change of variable $t=e^{-\pi i}\beta\tau^{-1}$ gives

$$b_0(z,\alpha)=e^{-\pi i\alpha}\beta^{\alpha-1}\int_{C'(B_0)}\tau^{-\alpha}\exp\left(\tau-\frac{\beta z}{\tau}\right)d\tau.$$

Comparison with the integral representation formulas for a Bessel function of the first kind ([*6*], p. 176) gives

$$b_0(z,\alpha)=-2\pi i\,e^{-\pi i\alpha}(z^{\frac{1}{2}}\beta^{-\frac{1}{2}})^{1-\alpha}J_{\alpha-1}(2z^{\frac{1}{2}}\beta^{\frac{1}{2}}). \tag{3.4}$$

This may be written

$$(3.5)\qquad b_0(z,\alpha) = -2\pi i\, e^{-\pi i \alpha}\, \beta^{\alpha-1} \sum_{n=0}^{\infty} \frac{(-1)^n (\beta z)^n}{n!\, \Gamma(n+\alpha)}.$$

Thus $B_0$ tends to a single-valued analytic function of $z$ in a full neighborhood of $z=0$. We note here that $b_0(z,\alpha)$ satisfies the second order equation

$$(3.6)\qquad z u'' + \alpha u' + \beta u = 0$$

obtained from (2.1) by formally allowing $\lambda$ to become infinite.

## 4. The solutions $A_k$

We have

$$(4.1)\qquad A_k = \int_{C(A_k)} t^{\alpha-2} \exp\left(\tfrac{1}{3}\lambda^{-2} t^3 + z t - \beta t^{-1}\right) dt.$$

We shall evaluate $A_k$ asymptotically by the method of steepest descents ([7], p. 503). For this purpose let us write

$$(4.2)\qquad t = \nu\tau$$

and

$$\tfrac{1}{3}\lambda^{-2} t^3 + z t = \tfrac{1}{3}\lambda^{-2}\nu^3\tau^3 + \nu z \tau = \mu\left(\tfrac{1}{3}\tau^3 - \tau\right).$$

Then

$$(4.3)\qquad \mu = \lambda^{-2}\nu^3 = -\nu z,$$

whence

$$(4.4)\qquad \nu = -i\lambda z^{\frac{1}{2}}, \qquad \mu = i\lambda (z^{\frac{1}{2}})^3,$$

where the same branch of the square root is to be taken. From the formal solutions (1.3) we expect the quantity

$$(4.5)\qquad \xi = \tfrac{2}{3} i \lambda z^{\frac{3}{2}} = \tfrac{2}{3}\mu$$

to be important. In terms of $\xi$, (4.3) gives

$$(4.6)\qquad \nu = \lambda^{\frac{2}{3}}\left(\tfrac{3}{2}\xi\right)^{\frac{1}{3}}, \qquad z = -\lambda^{-\frac{2}{3}}\left[\left(\tfrac{3}{2}\xi\right)^{\frac{1}{3}}\right]^2,$$

where the same branch of the cube root is taken. The proper branches will be determined in the sequel. Let us now write

$$\tfrac{1}{3}\lambda^{-2} t^3 + z t = \xi\left(\tfrac{1}{2}\tau^3 - \tfrac{3}{2}\tau\right) = |\xi|\, f(\tau)$$

where

$$(4.7)\qquad f(\tau) = e^{i\vartheta}\left(\tfrac{1}{2}\tau^3 - \tfrac{3}{2}\tau\right), \qquad \vartheta = \arg\xi.$$

Then (4.1) becomes

$$(4.8)\qquad A_k = \left[\lambda^{\frac{2}{3}}\left(\tfrac{3}{2}\xi\right)^{\frac{1}{3}}\right]^{\alpha-1} \int_{C'(A_k)} F(\tau,\xi,\lambda)\, e^{|\xi| f(\tau)}\, d\tau,$$

where

$$F(\tau,\xi,\lambda) = \tau^{\alpha-2}\exp\left\{-\beta\lambda^{-\frac{2}{3}}\left(\tfrac{3}{2}\xi\right)^{-\frac{1}{3}}\tau^{-1}\right\}.$$

$f(\tau)$ is given by (4.7) and $C'(A_k)$ is the image of $C(A_k)$ in the $\tau$ plane. Clearly the location of $C'(A_k)$ depends on the choice of a branch for $\nu$ in (4.6).

We consider first the case where the variable $\xi$ is large in magnitude, say $|\xi| \geqq |\xi_0|$ for some $\xi_0$.

The function $\mathrm{Re}\{f(\tau)\}$ has a saddle point at $\tau = 1$ where $f'(\tau) = 0$, and

$$f(\tau) = e^{i\vartheta}\left[-1 + \tfrac{3}{2}(\tau - 1)^2 + \tfrac{1}{2}(\tau - 1)^3\right]. \tag{4.9}$$

The path of steepest descent through $\tau = 1$ is given by

$$\mathrm{Im}\{f(\tau) - f(1)\} = 0. \tag{4.10}$$

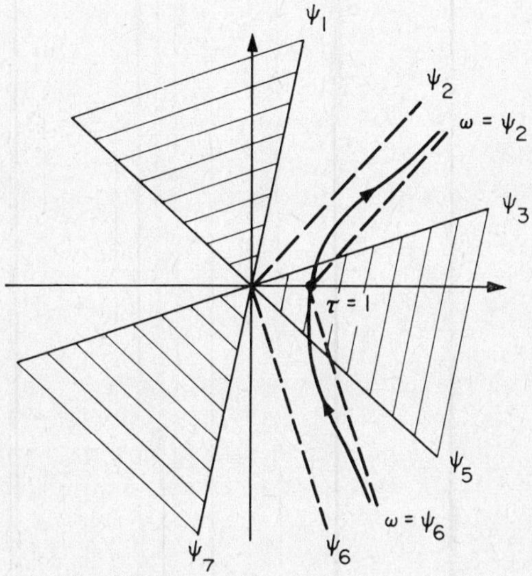


Fig. 2. Path of steepest descent

$\tau$-plane

$\psi_j = \frac{\pi}{2} - (j-1)\frac{\pi}{6} - \frac{\vartheta - 2m\pi}{3}$, $\quad j = 1, 2, 3, 5, 6, 7 - \pi < \vartheta - 2m\pi < \pi$, $\quad \psi = \arg\tau$

To plot this path, put $\tau - 1 = \varrho e^{i\omega}$. Then (4.10) becomes

$$\varrho^2\left[\varrho \sin(3\omega + \vartheta) + 3\sin(2\omega + \vartheta)\right] = 0.$$

For $\vartheta$ in each of the ranges

$$-\pi + 2m\pi < \vartheta < \pi + 2m\pi \qquad (m = 0, \pm 1, \ldots), \tag{4.11}$$

the path of steepest descent is as shown in Figure 2. In the Figure the unshaded sectors are those in which $e^{i\vartheta}\tau^3$ has a negative real part. For each range of $\vartheta$ in (4.11), the path of steepest descent goes to $\infty$ in the two unshaded sectors indicated. As $\vartheta$ passes out of any one of the ranges (4.11), the path changes abruptly and goes to $\infty$ in the third unshaded sector.

The idea now is to choose, for each $A_k$, a branch for $\nu$ in (4.6) so that for $\xi$ in the region specified by (4.11), $C'(A_k)$ goes to infinity in the same two sectors of the $\tau$-plane as does the path of steepest descent. Then we can take for $C'(A_k)$ this path of steepest descent. This requirement is satisfied if we take $m = k - 2$ in (4.11) so that

$$-\pi + 2\pi(k-2) < \vartheta < \pi + 2\pi(k-2) \tag{4.12}$$

and take

$$\nu = \lambda^{\frac{2}{3}} \left|\tfrac{3}{2}\xi\right|^{\frac{1}{3}} e^{\frac{1}{3}i\vartheta} \tag{4.13}$$

for $\vartheta$ in the range (4.12). To see that this is so, let $C(A_k)$ go to infinity in the $t$-plane with the directions

$$\arg t = \pm\frac{\pi}{3} + \frac{2}{3}\arg\lambda + \frac{2\pi}{3}(k-2).$$

Then since $\arg\tau = \arg t - \arg\nu$, we see that $C'(A_k)$ goes to infinity in the $\tau$-plane with directions

$$\arg\tau = \pm\frac{\pi}{3} - \frac{\vartheta - 2\pi(k-2)}{3}.$$

Inspection of Figure 2 with $m$ replaced by $k-2$ shows that the choices (4.12) and (4.13) are satisfactory.

To evaluate $A_k$ by the method of steepest descents, set

$$f(\tau) - f(1) = e^{i\vartheta}\left[\tfrac{3}{2}(\tau-1)^2 + \tfrac{1}{2}(\tau-1)^3\right] = -\tfrac{1}{2}s^2. \tag{4.14}$$

Choose

$$s = (-1)^{k+1} i\, e^{\frac{1}{2}i\vartheta}(\tau-1)(\tau+2)^{\frac{1}{2}} \tag{4.15}$$

with the branch cut along the negative real axis from $\tau = -2$ to $\tau = \infty$, and take the branch of the square root which is real and positive for $\tau$ real and $\tau > -2$. Then as $\tau$ traverses the path of steepest descent in the direction corresponding to that for $C'(A_k)$, $s$ increases through real values. The derivative $ds/d\tau$ is finite and non-zero in the cut plane, so near $s=0$, $\tau$ is an analytic function of $s$. For real $s$, and $|s| \geqq \delta > 0$,

$$|F(\tau(s), \xi, \lambda)| \leqq K_1 |s^{\frac{2}{3}(\alpha-2)}|, \qquad \left|\frac{d\tau}{ds}\right| \leqq K_2.$$

Hence there exist positive constants $K$ and $a$ such that

$$|F(\tau(s), \xi, \lambda)| \cdot \left|\frac{d\tau}{ds}\right| \leqq K\, e^{\frac{1}{2}a s^2}$$

for real $s$. If we expand $F\,d\tau/ds$ in a Taylor series about $s=0$, it follows from WATSON's lemma ([7], p. 501) that the series obtained formally by termwise integration over $(-\infty, \infty)$ represents $A_k$ asymptotically for

$$-\frac{3\pi}{2} + 2\pi(k-2) < \vartheta < \frac{3\pi}{2} + 2\pi(k-2). \tag{4.16}$$

WATSON's lemma extends the range of $\vartheta$ in (4.12) by $\frac{1}{2}\pi$ in each direction. It can be verified from (4.4) that taking $\vartheta$ in the range (4.16) and choosing $\nu$ as in (4.13) is equivalent to restricting $\varphi = \arg z$ to the range

$$\frac{4\pi}{3}(k-2) - \frac{2}{3}\arg\lambda < \varphi < 2\pi + \frac{4\pi}{3}(k-2) - \frac{2}{3}\arg\lambda \tag{4.17}$$

and taking

$$\nu = e^{-\frac{\pi i}{2}} \lambda |z|^{\frac{1}{2}} e^{\frac{1}{2}i\varphi}, \tag{4.18}$$

$$\xi = \tfrac{2}{3} i\lambda |z|^{\frac{3}{2}} e^{\frac{3}{2}i\varphi}. \tag{4.19}$$

29*

The results of the integration are

$$(4.20)\qquad A_k \sim i\sqrt{\pi}\,(-1)^k\,\lambda^{\frac{2}{3}(\alpha-1)}\left(\tfrac{3}{2}\xi\right)^{\frac{1}{3}\alpha-\frac{5}{6}}e^{-\xi}\left[1+\left(\tfrac{2}{3}\beta z-(\alpha-3)^2+\tfrac{1}{6}\right)\xi^{-1}+\cdots\right],$$

where $\vartheta = \arg \xi$ is in the range (4.16). This can be written

$$(4.21)\qquad A_k = i\sqrt{\pi}\,(-1)^k\,e^{-\frac{3\pi i}{2}\left(\frac{\alpha}{3}+\frac{1}{2}\right)}\lambda^{\alpha-\frac{3}{2}}z^{\frac{\alpha}{2}-\frac{5}{4}}e^{-\frac{2}{3}i\lambda z^{\frac{3}{2}}}\left[1+O(\xi^{-1})\right]$$

where $\varphi = \arg z$ is in the range (4.17).

Let us consider again the function $\xi$ in (4.5). In the $z$ plane, $\mathrm{Re}(\xi)$ is zero on three rays, denoted by $C_k$, $k=1, 2, 3$, extending from the origin. These rays are given (mod $2\pi$) by

$$C_k: \qquad \arg z = \frac{2\pi}{3} - (k-1)\frac{2}{3}\pi - \frac{2}{3}\arg\lambda.$$

These divide the finite neighborhood $S$ of $z=0$ into three sectors $S_k$ of angle $\frac{2}{3}\pi$. We take $S_k$ to be the sector opposite $C_k$ and take $S_k$ to be *closed*. It can now be verified that the choice of a branch for $\xi$ in (4.17), (4.19) is equivalent to restricting $z$ to $S - C_k$ and choosing the branch of $\xi$ for which $\mathrm{Re}(\xi) \geqq 0$ in $S_k$ and $\mathrm{Re}(\xi) \leqq 0$ in the other two sectors.

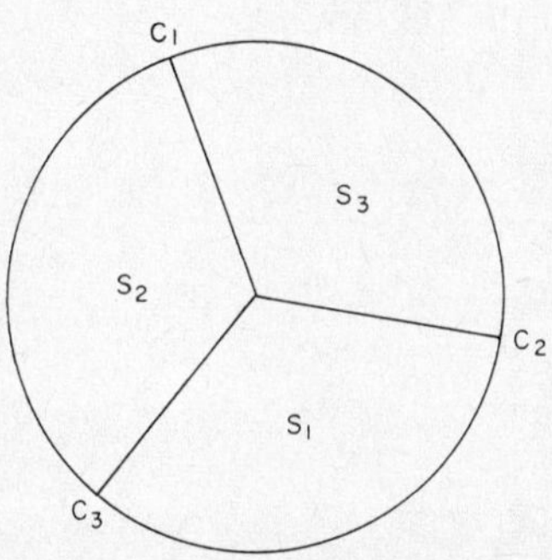

Fig. 3. Diagram for choice of branches

If we examine the relations (4.20) we see that for $z$ in any closed subset of the interior of $S_k$, $A_k$ tends exponentially to zero as $\lambda$ becomes infinite and becomes infinite exponentially for $z$ in a closed subset of the interior of either of the other two sectors. The behavior is described by saying that $A_k$ is *subdominant* or *recessive* in $S_k$ and *dominant* in the other two sectors.

The behavior of $A_k$ near $C_k$ can be obtained from the relations (2.6), once the asymptotic properties of the solutions $B_k$ have been investigated.

For the case when $|\xi| < |\xi_0|$ we put $t = \lambda^{\frac{2}{3}}\sigma$, $z = \lambda^{-\frac{2}{3}}\eta$ in (4.1). We note that $|\xi| < |\xi_0|$ implies $|\eta| < |\eta_0|$. These substitutions give

$$(4.22)\qquad A_k = \lambda^{\frac{2}{3}(\alpha-1)}\int\limits_{C^*(A_k)}\sigma^{\alpha-2}\exp\left(\frac{\sigma^3}{3}+\eta\sigma\right)\exp\left(-\beta\lambda^{-\frac{2}{3}}\sigma^{-1}\right)d\sigma.$$

If we write

$$\exp\left(-\beta\lambda^{-\frac{2}{3}}\sigma^{-1}\right) = \sum_{n=0}^{\infty}\frac{1}{n!}\left(-\frac{\beta\lambda^{-\frac{2}{3}}}{\sigma}\right)^n$$

and integrate term by term, this being valid here, we get

$$(4.23)\qquad A_k = \lambda^{\frac{2}{3}(\alpha-1)}\sum_{n=0}^{\infty}\frac{\left(-\beta\lambda^{-\frac{2}{3}}\right)^n}{n!}\,g_k(\eta, \alpha-n),$$

where

$$(4.24)\qquad g_k(\eta, \alpha) = \int\limits_{C^*(A_k)}\sigma^{\alpha-2}\exp\left(\frac{\sigma^3}{3}+\eta\sigma\right)d\sigma.$$

We observe from (4.24) that

$$\frac{d^n g_k(\eta, \alpha)}{d\eta^n} = g_k(\eta, \alpha + n). \tag{4.25}$$

The functions $g_k(\eta, \alpha)$ satisfy the equation

$$\chi^{iv} + \eta\chi'' + \alpha\chi' = 0 \tag{4.26}$$

obtained from (1.4) by putting formally $\lambda = \infty$. Then because of (4.25),

$$g_k(\eta, \alpha + 4) + \eta g_k(\eta, \alpha + 2) + \alpha g_k(\eta, \alpha + 1) = 0. \tag{4.27}$$

This holds for all $\alpha$, and therefore

$$g_k(\eta, \alpha + 3) + \eta g_k(\eta, \alpha + 1) + (\alpha - 1) g_k(\eta, \alpha) = 0. \tag{4.28}$$

Hence $g_k(\eta, \alpha)$ actually satisfies the third order equation

$$\chi''' + \eta\chi' + (\alpha - 1)\chi = 0. \tag{4.29}$$

For a discussion of this equation see LANGER [8] and HERSHENOV*.

## 5. The solutions $B_k$, $k = 1, 2, 3$

We have

$$B_k = \int_{C(B_k)} t^{\alpha-2} \exp\left(\tfrac{1}{3}\lambda^{-2} t^3 + zt - \beta t^{-1}\right) dt. \tag{5.1}$$

Let

$$b_k(z, \alpha) = \int_{C(B_k)} t^{\alpha-2} \exp(tz - \beta t^{-1})\, dt. \tag{5.2}$$

This integral converges for all $\alpha$ if $\operatorname{Re}(tz) < 0$ for large $|t|$. The requirements $\operatorname{Re}(tz) < 0$ and $\operatorname{Re}(\lambda^{-2} t^3) < 0$ for large $|t|$ can be satisfied by deformation of the path $C(B_k)$ if and only if $z \in S - S_k$. Now

$$e^{\frac{1}{3}\lambda^{-2}t^3} = \sum_{j=0}^{n} \frac{1}{j!}\left(\frac{1}{3}\lambda^{-2} t^3\right)^j + \lambda^{-2(n+1)} t^{3(n+1)} h_n(t, \lambda),$$

where $h_n(t, \lambda)$ is analytic in $t$ for fixed $\lambda$ and is uniformly bounded for $t$ on $C(B_k)$ and $|\lambda| \geqq |\lambda_0| > 0$. Therefore (5.1) gives

$$B_k = b_k(z, \alpha) + \cdots + \frac{1}{n!}\left(\frac{1}{3}\lambda^{-2}\right)^n b_k(z, \alpha + 3n) + R_n, \tag{5.3}$$

where

$$R_n = \lambda^{-2(n+1)} \int_{C(B_k)} t^{\alpha+3(n+1)-2} \exp(tz - \beta t^{-1})\, h_n(t, \lambda)\, dt. \tag{5.4}$$

If $|z| \geqq |z_0| > 0$, then

$$R_n = O(\lambda^{-2(n+1)}). \tag{5.5}$$

In the case where $|\xi| \geqq |\xi_0|$ but $z$ is not bounded away from the origin, put $\tau = tz$ in (5.4). Then

$$R_n = \lambda^{-2(n+1)} z^{-(\alpha+3n+z)} \int_{C'(B_k)} \tau^{\alpha+3n+1} \exp\left(\tau - \frac{\beta z}{\tau}\right) h_n^*(\tau, z, \lambda)\, d\tau. \tag{5.6}$$

---

* Master's thesis, Department of Mathematics, M.I.T. (1957).

For $n > -\frac{2}{3} - \frac{1}{3}\mathrm{Re}(\alpha)$; the integral here is uniformly bounded and

$$R_n = O(z^{1-\alpha}\xi^{-2(n+1)}). \tag{5.7}$$

Later it will be seen that regardless of the value of $\alpha$,

$$B_k = b_k(z, \alpha)\left[1 + O(\xi^{-2})\right]. \tag{5.8}$$

For some values of $\alpha$ the error term $O(\xi^{-2})$ can be replaced by one of still smaller order of magnitude.

To express the functions $b_k(z, \alpha)$ in terms of special functions, we make the change of variable

$$s = t z^{\frac{1}{2}} \beta^{-\frac{1}{2}} \tag{5.9}$$

in (5.2), obtaining

$$b_k = (z^{\frac{1}{2}}\beta^{-\frac{1}{2}})^{1-\alpha} \int\limits_{C^*(B_k)} s^{\alpha-2} \exp\left[z^{\frac{1}{2}}\beta^{\frac{1}{2}}\left(s - \frac{1}{s}\right)\right] ds.$$

Comparison with the integral representation formulae for the Hankel functions (see [6], p. 179) shows that with the proper choice of branches in (5.9),

$$b_1 = \pi i (\beta^{-\frac{1}{2}} z^{\frac{1}{2}})^{1-\alpha} H^{(1)}_{1-\alpha}(2\beta^{\frac{1}{2}} z^{\frac{1}{2}}), \tag{5.10a}$$

$$b_2 = \pi i (\beta^{-\frac{1}{2}} z^{\frac{1}{2}})^{1-\alpha} H^{(1)}_{1-\alpha}(2\beta^{\frac{1}{2}} z^{\frac{1}{2}}), \tag{5.10b}$$

$$b_3 = \pi i (e^{2\pi i}\beta^{-\frac{1}{2}} z^{\frac{1}{2}})^{1-\alpha} H^{(1)}_{1-\alpha}(2\beta^{\frac{1}{2}} z^{\frac{1}{2}}) \tag{5.10c}$$

where $z^{\frac{1}{2}} = |z|^{\frac{1}{2}} e^{\frac{1}{2}i\varphi}$,

$$-\frac{2\pi}{3} - \frac{2\pi}{3}(k-2) - \frac{2}{3}\arg\lambda < \varphi < \frac{2\pi}{3} - \frac{2\pi}{3}(k-2) - \frac{2}{3}\arg\lambda, \tag{5.11}$$

and $\arg\beta^{\pm\frac{1}{2}} = \pm\frac{1}{2}\arg\beta$, with $\arg\beta$ as in (2.5). Examination of these functions for various values of $\alpha$ proves (5.8). We remark that the functions $b_k(z, \alpha)$ satisfy the second order equation (3.6).

We now give a more detailed analysis for the case $k = 2$, using the formula in [6], p. 179. For large $|t|$ we wish to have

$$\frac{5\pi}{6} + \frac{2}{3}\arg\lambda < \arg t < \frac{7\pi}{6} + \frac{2}{3}\arg\lambda$$

according to the choice for $C(B_2)$, and we want

$$\frac{\pi}{2} + 2n\pi < \arg(tz) < \frac{3\pi}{2} + 2n\pi$$

for some integer $n$, so that the integral $b_2$ converges. These will hold if we take

$$\arg t = -\tfrac{1}{4}\arg z + \pi + \tfrac{1}{2}\arg\lambda + \tfrac{1}{2}n\pi,$$

where

$$-\frac{2}{3}\pi + 2n\pi - \frac{2}{3}\arg\lambda < \arg z < \frac{2\pi}{3} + 2n\pi - \frac{2}{3}\arg\lambda,$$

for large $|t|$. For the moment take $\arg\beta^{\frac{1}{2}} = \frac{1}{2}\arg\beta + m\pi$. In the $s$-plane, the path of integration tends to infinity in the direction

$$\arg s = \pi + \tfrac{1}{4}\arg z + \tfrac{1}{2}\arg\lambda + \tfrac{1}{2}n\pi - m\pi - \tfrac{1}{2}\arg\beta = \pi - \omega.$$

(See [6], p. 179 for the significance of $\omega$.) The direction of approach for $C(B_2)$ at the origin in the $t$-plane can be taken as $\arg t = \arg \beta + \gamma \frac{\pi}{2}$, where $-1 < \gamma < 1$. We choose $\gamma$ so that at the origin in the $s$-plane

$$\arg s = \frac{1}{2} \arg z + \frac{1}{2} \arg \beta - m\pi + \gamma \frac{\pi}{2} = \omega .$$

The two expressions for $\omega$ give

$$\gamma = \frac{\pi}{2}\left(2m\pi - \frac{n\pi}{2} - \frac{1}{2} \arg \lambda - \frac{3}{4} \arg z\right).$$

To have $|\gamma| < 1$ we must have $m - n = 0$. Next, we find

$$-\frac{\pi}{2} - 2n\pi < \omega - \arg(\beta^{\frac{1}{2}} z^{\frac{1}{2}}) < \frac{\pi}{2} - 2n\pi ,$$

so we take $n = 0$, and then $m = 0$. Here $-\frac{\pi}{6} < \omega < \frac{7}{6}\pi$. The formula in [6] is given for $|\omega| < \pi$, but extending the range of $\omega$ simply gives an analytic continuation of the Hankel function. The relations (5.10b) and (5.11) with $k = 2$ now follow.

The asymptotic behavior of $B_k$ in $S_k$ can be obtained from the relations (2.6). We note that $B_k$ is dominant in $S_k$. For $z$ in $S - S_k$ and $|z| \geqq |z_0| > 0$, $B_k$ tends to a finite non-zero limit as $\lambda$ becomes infinite. We say that $B_k$ is *balanced* in $S - S_k$.

We now examine the functions $B_k$ in the case where $|\xi| < |\xi_0|$. Here it is impossible to proceed as with the solutions $A_k$ because the path of integration goes to the origin.

When $\alpha$ is not an integer we can solve the equations (2.6) for the three functions $B_k$ in terms of the functions $B_0$ and $A_k$. We get

(5.12a) $$B_1 = \frac{A_1 + A_2 - B_0 + e^{-2\pi i \alpha} A_3}{1 - e^{-2\pi i \alpha}},$$

(5.12b) $$B_2 = \frac{A_1 + A_2 + A_3 - B_0}{1 - e^{-2\pi i \alpha}},$$

(5.12c) $$B_3 = \frac{A_1 + e^{-2\pi i \alpha}(A_2 + A_3 - B_0)}{1 - e^{-2\pi i \alpha}}.$$

The asymptotic expansions for the functions $B_0$ and $A_k$ when $|\xi| < |\xi_0|$ are given by (3.2) and (4.23).

Since all solutions here are entire functions of $\alpha$, we can pass to the limit as $\alpha$ tends to an integer $n$, obtaining

(5.13a) $$B_1 = \frac{1}{2\pi i}\left[\frac{\partial}{\partial \alpha}(A_1 + A_2 + A_3 - B_0) - 2\pi i A_3\right]_{\alpha = n},$$

(5.13b) $$B_2 = \frac{1}{2\pi i}\left[\frac{\partial}{\partial \alpha}(A_1 + A_2 + A_3 - B_0)\right]_{\alpha = n},$$

(5.13c) $$B_3 = \frac{1}{2\pi i}\left[\frac{\partial}{\partial \alpha}(A_1 + A_2 + A_3 - B_0) + 2\pi i A_1\right]_{\alpha = n}.$$

It is now necessary to examine the quantity

$$W_n = \left[\frac{\partial}{\partial\alpha}(A_1 + A_2 + A_3 - B_0)\right]_{\alpha=n}. \tag{5.14}$$

Let $K = C(A_1) + C(A_2) + C(A_3)$, $\Gamma = K - C(B_0)$. Then

$$W_n = \left[\frac{\partial}{\partial\alpha}\int_\Gamma t^{\alpha-2}\exp\left(\frac{1}{3}\lambda^{-2}t^3 + zt - \beta t^{-1}\right)dt\right]_{\alpha=n},$$
$$= \int_\Gamma t^{n-2}\log t\exp\left(\frac{1}{3}\lambda^{-2}t^3 + zt - \beta t^{-1}\right)dt.$$

If we put

$$t = \lambda^{\frac{2}{3}}\sigma, \qquad z = \lambda^{-\frac{2}{3}}\eta, \tag{5.15}$$

this becomes

$$W_n = \lambda^{\frac{2}{3}(n-1)}\left[\frac{2}{3}(\log\lambda)\int_{\Gamma^*}\sigma^{n-2}\exp\left(\frac{\sigma^3}{3} + \eta\sigma - \frac{\beta}{\lambda^{\frac{2}{3}}\sigma}\right)d\sigma + \right.$$
$$\left. + \int_{\Gamma^*}\sigma^{n-2}\log\sigma\exp\left(\frac{\sigma^3}{3} + \eta\sigma - \frac{\beta}{\lambda^{\frac{2}{3}}\sigma}\right)d\sigma\right].$$

The first integral here vanishes because of CAUCHY's theorem. Then

$$W_n = \lambda^{\frac{2}{3}(n-1)}\left[\sum_{m=0}^{\infty}\frac{1}{m!}\left(\frac{-\beta}{\lambda^{\frac{2}{3}}}\right)^m\int_{K^*}\sigma^{n-2}\log\sigma\exp\left(\frac{\sigma^3}{3} + \eta\sigma\right)d\sigma - \right.$$
$$\left. - \int_{C^*(B_0)}\sigma^{n-2}\log\sigma\exp\left(\frac{\sigma^3}{3} + \eta\sigma - \frac{\beta}{\lambda^{\frac{2}{3}}\sigma}\right)d\sigma\right].$$

Reversing the transformation (5.15) in the last integral here, we find its contribution to be

$$\frac{2}{3}(\log\lambda)\int_{C(B_0)} t^{n-2}\exp\left(\frac{1}{3}\lambda^{-2}t^3 + zt - \beta t^{-1}\right)dt -$$
$$- \int_{C(B_0)} t^{n-2}\log t\exp\left(\frac{1}{3}\lambda^{-2}t^3 + zt - \beta t^{-1}\right)dt$$
$$= \frac{2}{3}(\log\lambda)\,B_0 - \left[\frac{\partial B_0}{\partial\alpha}\right]_{\alpha=n}.$$

Finally, then,

$$W_n = \lambda^{\frac{2}{3}(n-1)}\sum_{k=1}^{3}\sum_{m=0}^{\infty}\frac{1}{m!}\left(\frac{-\beta}{\lambda^{\frac{2}{3}}}\right)^m\left[\frac{\partial}{\partial\alpha}g_k(\eta, \alpha - m)\right]_{\alpha=n} +$$
$$+ \frac{2}{3}(\log\lambda)\,B_0 - \left[\frac{\partial B_0}{\partial\alpha}\right]_{\alpha-n}. \tag{5.16}$$

The functions $g_k(\eta, \alpha)$ are defined by (4.24). The asymptotic expansions of the functions $B_k$ when $\alpha$ is an integer $n$ are then given by (5.16) and (5.13).

By taking $\alpha = 0$, $\beta = 1$, these results may be shown to agree with those of WASOW ([2], p. 227, equations (2.8a, b, c, d)). The relations connecting his solutions $U_1$, $U_2$, $U_3$ with those considered here are

$$U_1 = B_3 - B_0,$$
$$U_2 = B_1,$$
$$U_3 = B_2.$$

## 6. A set of independent solutions

We consider here the set of solutions $B_0$, $B_3$, $A_1$, $A_2$. Similar results will hold for any set $B_0$, $B_i$, $A_j$, $A_k$, where no two of $i$, $j$, $k$ are equal. The solution $B_0$ is balanced everywhere in $S$. The solution $B_3$ is balanced in $S - S_3$ and is dominant in $S_3$. The solution $A_j$ is subdominant in $S_j$ and is dominant in the other two sectors. This set will be shown to be linearly independent for $|\lambda|$ sufficiently large.

Let
$$C(\alpha) = i\sqrt{\pi}\, e^{-\frac{3\pi i}{2}\left(\frac{\alpha}{3}+\frac{1}{2}\right)}.$$

Then for $z$ in $S - S_3$ and $|\xi| \geqq |\xi_0|$ we have

$$B_0 = b_0(z, \alpha) + O(\lambda^{-2}), \tag{6.1}$$

$$B_3 = b_3(z, \alpha)\left[1 + O(\xi^{-2})\right], \tag{6.2}$$

$$A_1 = -C(\alpha)\, \lambda^{\alpha - \frac{3}{2}} z^{\frac{\alpha}{2} - \frac{5}{4}} e^{-\frac{2}{3} i \lambda z^{\frac{3}{2}}} \left[1 + O(\xi^{-1})\right], \tag{6.3}$$

$$A_2 = -i e^{\pi i \alpha} C(\alpha)\, \lambda^{\alpha - \frac{3}{2}} z^{\frac{\alpha}{2} - \frac{5}{4}} e^{\frac{2}{3} i \lambda z^{\frac{3}{2}}} \left[1 + O(\xi^{-1})\right], \tag{6.4}$$

where in (6.2), (6.3), (6.4)

$$-\frac{4\pi}{3} - \frac{2}{3}\arg\lambda < \arg z < -\frac{2}{3}\arg\lambda.$$

Let $W(f_1, \ldots, f_n)$ be the Wronskian of $f_1, \ldots, f_n$ with respect to $z$. Using (2.3) to evaluate the derivatives of the above functions, by calculation we find

$$\begin{aligned} W(B_0, B_3, A_1, A_2) &= 2\pi i\, \lambda^{2\alpha+2} z^{\alpha} W(b_0, b_3)\left[1 + O(\xi^{-1})\right], \\ &= 4\pi^2 e^{-2\pi i \alpha} \beta^{\alpha-1} \lambda^{2\alpha+2}\left[1 + O(\xi^{-1})\right]. \end{aligned} \tag{6.5}$$

In the case where $|\xi| < |\xi_0|$ we have

$$B_0 = -2\pi i\, e^{-\pi i \alpha} \beta^{\alpha-1} \frac{1}{\Gamma(\alpha)} + O(\lambda^{-\frac{2}{3}}), \tag{6.6}$$

$$A_j = \lambda^{\frac{2}{3}(\alpha-1)}\left[g_j(\eta, \alpha) + O(\lambda^{-\frac{2}{3}})\right] \qquad (j = 1, 2, 3). \tag{6.7}$$

Using (5.12c) we find

$$W(B_0, B_3, A_1, A_2) = \frac{e^{-2\pi i \alpha}}{1 - e^{-2\pi i \alpha}} W(B_0, A_1, A_2, A_3). \tag{6.8}$$

In the case where $\alpha$ is an integer $n$, we must take the limiting value of this expression as $\alpha$ tends to $n$. By use of (6.6) and (6.7), this becomes

$$\frac{-2\pi i \beta^{\alpha-1}}{\Gamma(\alpha)} \frac{e^{-3\pi i \alpha}}{1 - e^{-2\pi i \alpha}} \lambda^{2\alpha+2} W_{\eta}(g_1', g_2', g_3')\left[1 + O(\lambda^{-\frac{2}{3}})\right], \tag{6.9}$$

where $W_\eta$ denotes the Wronskian with respect to the variable $\eta$ and the primes denote differentiation with respect to $\eta$. With the aid of (4.29) there follows

$$W_\eta(g_1', g_2', g_3') = (1-\alpha)\, W_\eta(g_1, g_2, g_3). \tag{6.10}$$

Since the Wronskian on the right is independent of $\eta$, it may be evaluated at $\eta = 0$. The quantities $g_k(0, \alpha)$ may be expressed in terms of gamma functions ([9], p. 225). In fact,

$$g_k(0,\alpha) = \int\limits_{C^*(A_k)} \sigma^{\alpha-2} e^{\frac{\sigma^3}{3}}\, d\sigma = 2\pi i \frac{3^{\frac{\alpha-4}{3}} e^{\frac{2\pi i}{3}(\alpha-1)(k-2)}}{\Gamma\left(\frac{4-\alpha}{3}\right)}.$$

By the identity ([9], p. 225)

$$\Gamma\left(\frac{2-\alpha}{3}\right)\Gamma\left(\frac{3-\alpha}{3}\right)\Gamma\left(\frac{4-\alpha}{3}\right) = \frac{2\pi\,\Gamma(2-\alpha)}{3^{\frac{3}{2}-\alpha}}$$

and (4.25) there follows

$$W_1(g_1, g_2, g_3) = \frac{-4\pi^2}{\Gamma(2-\alpha)}. \tag{6.11}$$

From (6.9), (6.10), (6.11) and the familiar identity

$$\Gamma(2-\alpha)\,\Gamma(\alpha) = \frac{\pi(1-\alpha)}{\sin\pi\alpha}$$

we obtain finally

$$W(B_0, B_3, A_1, A_2) = 4\pi^2 e^{-2\pi i\alpha}\beta^{\alpha-1}\lambda^{2\alpha+2}\left[1 + O(\lambda^{-\frac{2}{3}})\right]. \tag{6.12}$$

Then from (6.5) and (6.12) we see that if $|\lambda|$ is sufficiently large, the solutions $B_0, B_3, A_1, A_2$ are linearly independent.

A summary of the asymptotic properties of these functions is given below. The behavior of the other functions may be calculated from this information by using the connection formulas (2.6).

(*i*) *The case* $|\xi| \geqq |\xi_0|$.

Let $\xi = \frac{2}{3} i\lambda z^{\frac{3}{2}}$, and in the formulae below let $\arg z$ lie in the appropriate part of the range

$$-2\pi - \frac{2}{3}\arg\lambda < \arg z < \frac{2\pi}{3} - \frac{2}{3}\arg\lambda.$$

$$B_0 = b_0(z,\alpha) + O(\lambda^{-2}) \quad \text{for} \quad z \text{ in } S, \tag{6.13}$$

where

$$b_0(z,\alpha) = -2\pi i\, e^{-\pi i\alpha}(\beta^{-\frac{1}{2}} z^{\frac{1}{2}})^{1-\alpha} J_{\alpha-1}(2\beta^{\frac{1}{2}} z^{\frac{1}{2}}),$$

$$A_1 = -C(\alpha)\,\lambda^{\alpha-\frac{3}{2}} z^{\frac{\alpha}{2}-\frac{5}{4}} e^{-\xi}\left[1 + O(\xi^{-1})\right] \qquad \text{for } z \text{ in } S - C_1, \tag{6.14}$$

$$A_2 = -i\lambda^{\alpha-\frac{3}{2}} e^{\pi i\alpha} C(\alpha)\, z^{\frac{\alpha}{2}-\frac{5}{4}} e^{\xi}\left[1 + O(\xi^{-1})\right] \qquad \text{for } z \text{ in } S - C_2, \tag{6.15}$$

$$B_3 = \pi i (e^{2\pi i}\beta^{-\frac{1}{2}} z^{\frac{1}{2}})^{1-\alpha} H^{(1)}_{1-\alpha}(2\beta^{\frac{1}{2}} z^{\frac{1}{2}})\left[1 + O(\xi^{-2})\right] \qquad \text{for } z \text{ in } S - S_3, \tag{6.16}$$

$$B_3 = -C(\alpha)\,\lambda^{\alpha-\frac{3}{2}} z^{\frac{\alpha}{2}-\frac{5}{4}} e^{-\xi}\left[1 + O(\xi^{-1})\right] + \tag{6.17}$$
$$+ \pi i\, e^{-2\pi i\alpha}(\beta^{-\frac{1}{2}} z^{\frac{1}{2}})^{1-\alpha} H^{(1)}_{1-\alpha}(2\beta^{\frac{1}{2}} z^{\frac{1}{2}})\left[1 + O(\xi^{-2})\right] \quad \text{for } z \text{ in } S_3 - C_1.$$

The expressions for $A_1$ and $B_3$ near $C_1$ and for $A_2$ near $C_2$ are more complicated but may be found easily from the relations

$$
\begin{aligned}
&B_3 = B_0 + B_1 - A_2,\\
(6.18)\qquad &A_1 = B_0 + (1 - e^{-2\pi i\alpha})\,B_1 - A_2 - e^{-2\pi i\alpha}A_3,\\
&A_2 = B_0 + (1 - e^{-2\pi i\alpha})\,B_2 - A_1 - A_3.
\end{aligned}
$$

These follow from (2.6).

(*ii*) *The case* $|\xi| < |\xi_0|$ (for $z$ in $S$).

$$(6.19)\quad B_0 = b_0(z,\alpha) + O(\lambda^{-2}) = -2\pi i\,e^{-\pi i\alpha}\beta^{\alpha-1}\left[\frac{1}{\Gamma(\alpha)} - \frac{\beta\eta}{\Gamma(\alpha+1)}\lambda^{-\frac{2}{3}} + \cdots\right],$$

$$(6.20)\quad A_j = \lambda^{\frac{2}{3}(\alpha-1)}\left[g_j(\eta,\alpha) + O(\lambda^{-\frac{2}{3}})\right],$$

$$(6.21)\quad B_3 = \frac{e^{-\pi i\alpha}}{2i\sin\pi\alpha}\left\{\lambda^{\frac{2}{3}(\alpha-1)}\left[(e^{2\pi i\alpha}g_1 + g_2 + g_3) + O(\lambda^{-\frac{2}{3}})\right] - b_0(z,\alpha) + O(\lambda^{-2})\right\}$$

when $\alpha$ is not an integer.

$$
\begin{aligned}
B_3 = {}&\frac{1}{2\pi i}\lambda^{\frac{2}{3}(n-1)}\left\{\left[2\pi i\,g_1(\eta,\alpha) + \sum_{k=1}^{3}\frac{\partial g_k(\eta,\alpha)}{\partial\alpha}\right]_{\alpha=n} + O(\lambda^{-\frac{2}{3}})\right\} +\\
(6.22)\qquad &+ \frac{1}{3\pi i}(\log\lambda)\left[b_0(z,n) + O(\lambda^{-2})\right] - \frac{1}{2\pi i}\left[\frac{\partial b_0(z,\alpha)}{\partial\alpha}\right]_{\alpha=n} + O(\lambda^{-2})
\end{aligned}
$$

when $\alpha$ is an integer $n$.

In these formulae

$$g_k(\eta,\alpha) = \int\limits_{C^*(A_k)} \sigma^{\alpha-2}\exp\left(\frac{\sigma^3}{3} + \eta\sigma\right)d\sigma,$$

$$g_k(0,\alpha) = 2\pi i\,\frac{3^{\frac{\alpha-4}{3}}\,e^{\frac{2\pi i}{3}(\alpha-1)(k-2)}}{\Gamma\left(\frac{4-\alpha}{3}\right)}.$$

For $z = 0$, these formulae (*ii*) yield

$$(6.23)\quad B_0(0) \sim -2\pi i\,e^{-\pi i\alpha}\beta^{\alpha-1}\frac{1}{\Gamma(\alpha)},$$

$$(6.24)\quad A_k(0) \sim 2\pi i\,\frac{3^{\frac{\alpha-4}{3}}\,e^{\frac{2\pi i}{3}(\alpha-1)(k-2)}}{\Gamma\left(\frac{4-\alpha}{3}\right)}\lambda^{\frac{2}{3}}(\alpha-1),$$

$$
\begin{aligned}
B_3(0) \sim {}&\pi\frac{3^{\frac{\alpha-4}{3}}}{\Gamma\left(\frac{4-\alpha}{3}\right)}\,\frac{e^{-\pi i\alpha} + 2\cos\frac{\pi}{3}(\alpha+2)}{\sin\pi\alpha}\lambda^{\frac{2}{3}}(\alpha-1) +\\
(6.25)\qquad &+ \pi\frac{e^{-2\pi i\alpha}\beta^{\alpha-1}}{\Gamma(\alpha)\sin\pi\alpha}\qquad(\alpha \neq n),
\end{aligned}
$$

$$
\begin{aligned}
B_3(0) \sim {}&\lambda^{\frac{2}{3}(n-1)}\frac{3^{\frac{n-4}{3}}}{\Gamma\left(\frac{4-n}{3}\right)}\times\\
(6.26)\qquad &\times\left[2\pi i\,e^{\frac{2\pi i}{3}(1-n)} - \frac{4\pi}{3}\sin\frac{2\pi}{3}(n-1) + \frac{1}{3}\log 3 + \frac{1}{3}\psi\left(\frac{4-n}{3}\right)\right] -\\
&- \frac{2}{3}e^{-n\pi i}\beta^{n-1}\frac{1}{\Gamma(n)}\log\lambda + \frac{d}{d\alpha}\left[e^{-\pi i\alpha}\beta^{\alpha-1}\frac{1}{\Gamma(\alpha)}\right]_{\alpha=n}\qquad(\alpha = n).
\end{aligned}
$$

*(iii) Bounds.*

Let $\xi_j$ be the branch of $\xi = \frac{2}{3} i \lambda z^{\frac{3}{2}}$ which has a positive real part in $S_j$ and a negative real part in the other two sectors. Let $E = E(z, \lambda)$ be a generic symbol for a uniformly bounded function of $z$ and $\lambda$. Then the following relations hold.

For $|\xi| \geqq |\xi_0|$ and $z \in S$,

$$B_0 = E. \tag{6.27a}$$

For $|\xi| < |\xi_0|$ and $z \in S$,

$$B_0 = \begin{cases} E, & \alpha \neq 0, -1, -2, \ldots \\ \lambda^{\frac{2}{3}(\alpha-1)} E, & \alpha = 0, -1, -2, \ldots. \end{cases} \tag{6.27b}$$

For $|\xi| \geqq |\xi_0|$,

$$A_j e^{\operatorname{Re} \xi_j} = \lambda^{\frac{2}{3}(\alpha-1)} \xi^{\frac{\alpha}{3} - \frac{5}{6}} E, \qquad z \in S - C_j \qquad (j = 1, 2, 3). \tag{6.28a}$$

For $z \in C_j$,

$$A_j e^{\operatorname{Re} \xi_j} = \begin{cases} E, & \operatorname{Re} \alpha \leqq 1, \ \alpha \neq 0, -1, -2, \ldots \\ \lambda^{\frac{2}{3}(\alpha-1)} \xi^{-\frac{2}{3}(\alpha-1)} E, & 1 \leqq \operatorname{Re} \alpha \leqq \frac{3}{2}, \text{ or } \alpha = 0, -1, -2, \ldots \\ \lambda^{\frac{2}{3}(\alpha-1)} \xi^{\frac{\alpha}{3} - \frac{5}{6}} E, & \operatorname{Re} \alpha \geqq \frac{3}{2}. \end{cases} \tag{6.28b}$$

For $|\xi| < |\xi_0|$ and $z \in S$,

$$A_j e^{\operatorname{Re} \xi_j} = \lambda^{\frac{2}{3}(\alpha-1)} E. \tag{6.28c}$$

For $|\xi| \geqq |\xi_0|$ and $z \in S$ $(j = 1, 2, 3, \ j \neq k \neq l \neq j)$,

$$B_j e^{\frac{1}{2} \operatorname{Re} \{\xi_k + \xi_l\}} = \begin{cases} E, & \operatorname{Re} \alpha \leqq 1, \ \alpha \neq 1 \\ \lambda^{\frac{2}{3}(\alpha-1)} \xi^{-\frac{2}{3}(\alpha-1)} E, & 1 < \operatorname{Re} \alpha < \frac{3}{2} \\ \lambda^{\frac{2}{3}(\alpha-1)} \xi^{\frac{\alpha}{3} - \frac{5}{6}} E, & \operatorname{Re} \alpha \geqq \frac{3}{2} \\ \left(\log \frac{\lambda}{\xi}\right) E, & \alpha = 1. \end{cases} \tag{6.29a}$$

For $|\xi| < |\xi_0|$ and $z \in S$,

$$B_j e^{\frac{1}{2} \operatorname{Re} \{\xi_k + \xi_l\}} = \begin{cases} E, & \operatorname{Re} \alpha \leqq 1, \ \alpha \neq 1 \\ (\log \lambda) E, & \alpha = 1 \\ \lambda^{\frac{2}{3}(\alpha-1)} E, & \operatorname{Re} \alpha > 1. \end{cases} \tag{6.29b}$$

**Acknowledgement.** This work was sponsored in part by the Office of Naval Research. The author wishes to express his thanks to Professor C. C. LIN for suggesting this problem and for many helpful discussions with him during its development.

**Note added in proof.** Professor TRUESDELL has kindly pointed out the connection between equation (2.3) and his own work on the equation $\frac{\partial}{\partial z} F(z, \alpha) = F(z, \alpha + 1)$. (Annals of Mathematics Studies, Number 18, Princeton University Press, 1948.) Use of the theory for the $F$-equation would have shortened several of the formal steps in this paper.

## References

[1] Wasow, W.: A study of the solutions of the differential equation $y^{(4)} = \lambda^2(xy'' + y) = 0$ for large values of $\lambda$. Ann. of Math. **52**, 350–361 (1950).

[2] Wasow, W.: Asymptotic solution of the differential equation of hydrodynamic stability in a domain containing a transition point. Ann. of Math. **58**, 222–252 (1953).

[3] Langer, R. E.: On the asymptotic solutions of a class of ordinary differential equations of the fourth order, with special reference to an equation of hydrodynamics. Trans. Amer. Math. Soc., **84**, 144–191 (1957).

[4] Langer, R. E.: The asymptotic solutions of ordinary linear differential equations of the second order, with special reference to a turning point. Trans. Amer. Math. Soc. **67**, 461–490 (1949).

[5] Wasow, W.: The complex asymptotic theory of a fourth order differential equation of hydrodynamics. Ann. of Math. **49**, 852–871 (1948).

[6] Watson, G. N.: A Treatise on the Theory of Bessel Functions, second edit. Cambridge 1944.

[7] Jeffreys, H., & B. S. Jeffreys: Methods of Mathematical Physics, third edit. Cambridge 1956.

[8] Langer, R. E.: The solutions of the differential equation $v''' + \lambda^2 z v' + 3\mu\lambda^2 v = 0$. Duke Math. Journ. **22**, 525–541 (1955).

[9] Copson, E. T.: Theory of Functions of a Complex Variable. Oxford 1935.

Massachusetts Institute of Technology
Cambridge, Massachusetts

*(Received March 14, 1958)*

Reprinted from ADVANCES IN MATHEMATICS
All Rights Reserved by Academic Press, New York and London
Vol. 19, No. 3, March 1976

# On the Role of Applied Mathematics

C. C. LIN

*Massachusetts Institute of Technology, Cambridge, Massachusetts*

TO MY GOOD FRIEND STANISLAW MARCIN ULAM IN HONOR OF HIS 65TH BIRTHDAY

This paper concerns itself with the nature of applied mathematics, and how to make its role effective. Its differences from pure mathematics and from theoretical sciences are pointed out with the help of illustrative examples. Recommendations are made for supporting a program of education and research in applied mathematics to make it an independent scientific discipline and to create a community of applied mathematicians.

## I. INTRODUCTION

### 1. *The Role of Applied Mathematicians*

It is a pleasure for me to be given this opportunity to speak to this distinguished audience on the role of applied mathematics—even though I would rather, like everybody else, speak about my own research work on stellar systems, on galaxies, on hydrodynamics and on turbulence—because I do have a few thoughts to share with you regarding the profession of applied mathematics as a whole. Every applied mathematician here knows what he is doing in his own line of specialty, and it would be presumptive for me to comment on that. However, I feel that there are things which are left unattended and which should be done by all applied mathematicians as a group to enhance the role of our profession. There are also several rather philosophical issues which, I feel, deserve our continued attention, because they would influence the long-term outlook of applied mathematics. It is toward these issues that I shall address my discussion. I hope that my remarks are also of interest to the pure mathematicians and to the theoretical and empirical scientists, with whom the applied mathematicians often develop close working relationships.

I should add that, like anybody else, my opinions will be biased by my own experience, which does not include any extensive stay in a laboratory such as this one. I hope, therefore, that I shall be excused and corrected if my remarks do not quite fit the conditions here.

It is almost tautological to say that the role of the applied mathematician is to promote the most effective use of mathematics in the natural and social sciences, in applied sciences, in engineering, in industry, in government, and in all other kinds of human activity. But this is not the complete list of his professional activities. As an intelligent scientist, he can also ask intelligent questions and originate new ideas which are only remotely or partly mathematical but which are nevertheless extremely important. I have in mind such ideas and practical inventions as those pioneered by our friend Stan Ulam at this laboratory.

However, in this talk, I shall restrict myself more narrowly to the effective use of mathematics in the sciences and to the stimulation of pure mathematical research through such efforts. One of the focal points of my discussion will be *how* to make this role effective.

In the course of this discussion, I shall comment on a few philosophical issues for which there seem to be, as yet, no generally accepted opinion. This is not surprising: philosophy is a subject in which it is easy to ask good questions and difficult to find good answers. The questions I would like to ask are the following: (a) Applicability: Why is mathematics so broadly applicable?[1] (b) Universality: Are the mathematics usable in one science also most likely to be usable in another? (c) Probability of applicability: (i) Can we identify a body of knowledge known as "applicable mathematics"? (ii) Is mathematics ever to be applicable to such subjects as history?

I shall attempt to give answers to some of these questions.

To continue with this discussion, we must first agree on an answer to the question "What is applied mathematics?" Different people will probably give answers with different emphases. I hope that my answer will be broad enough to receive general acceptance and be useful as a basis for our discussion.

[1] Cf. [1].

## II. What is Applied Mathematics?

### 2. *The Nature and Scope of Applied Mathematics*

Generally speaking, applied mathematics is a disciplined activity which lies between the empirical sciences and pure mathematics. In essence, it is characterized by an attitude, an approach, a way of thinking. The principal theme is the *interdependence* of mathematics and the sciences. In common with the pure mathematician, the applied mathematician is interested in the stimulation of the development of new mathematics (see [2]),—but with primary emphasis on those aspects *directly* or at least very strongly motivated by scientific problems. In common with the theoretical scientists, the applied mathematician seeks knowledge and understanding of scientific facts and real world phenomena through the use of mathematical methods.

A particularly challenging type of activity is to find *new* ideas for the *application* of mathematics and indeed to develop mathematical theories in those scientific subjects which have not hitherto been subjected to systematic mathematical treatment. (Social sciences and biology are the oftcited examples.) These efforts may in turn lead to the creation of new mathematical ideas and theories (by abstraction, generalization, or otherwise), which are interesting in their own right as a part of pure mathematics. The recognition of this duality is essential to the spirit of applied mathematics. By stressing this duality, there is obviously a *difference in emphasis* between applied mathematics and either pure mathematics or the empirical sciences.

The scope of applied mathematics is very broad, and can be best described by borrowing the following words from Albert Einstein [3]:

> "Its realm is accordingly defined as
> that part of the sum total of our
> knowledge which is capable of being
> expressed in mathematical terms."

These words were originally used to define physics. Taken literally, the statement clearly includes the mathematical theories of economics, biology, communication, etc., and is perhaps a more adequate description of applied mathematics.

Clearly, the activity of every branch of science includes major efforts *not* deeply mathematical. For example, the *act* of observation of the behavior of sun spots and solar winds, isolation of radio-active elements, taking of opinion polls (not the analysis), and cultivation of bacteria. Thus, applied mathematics does not *include* all the sciences. It merely *overlaps* with all the sciences. The extent of this overlap may be large (as in physics) or small (as in biology, for the moment). It is indeed the extension of this overlap that should concern the applied mathematician. At the same time, the *wisdom* of the applied mathematician, generated by his intimate knowledge of mathematical methods (or "applicable mathematics," if you wish) and his experience with a *variety* of mathematical applications, puts him in the best position to judge in which areas or problems of science the power of a particular branch of mathematics would be most effective, would have only limited effectiveness, or would have no effectiveness at all. His efforts must obviously be devoted to those aspects of science where mathematical methods can be effectively used. The practitioner must *continue* to improve upon his own mathematical knowledge[2] and new mathematics must be created when necessary. Fortunately, our experience has shown that there is indeed simplicity and order in the fundamental aspects of the sciences. All fundamental laws of physics can be stated in mathematical form on one sheet of paper. At the same time, it appears to be true that human intelligence insists on conceptual simplicity in the formulation of the fundamentals of any scientific subject. *Thus, the mathematical approach is also the basic approach.* At the other extreme, mathematical methods also can be effectively used in detailed scientific or technological calculations.

To summarize, a well-educated applied mathematician must have a clear appreciation of the above-mentioned spirit of duality and a comprehensive knowledge of the total picture of the mathematical sciences. Let us now turn to a more specific discussion of his creative activities.

## 3. *Creative Activities of Applied Mathematicians, Part A*[3]

Applied mathematicians are often found to be engaged in the following efforts: (a) the formulation of scientific concepts and problems in mathematical terms, (b) the solution of the resultant mathematical problems, and (c) the discussion, interpretation and the evaluation of the results of his analysis, including the making of specific predictions. But the solution of specific problems often serves merely as a focus and

[2] See, for example, Theodore von Kármán's remarks at the age of 80 in [4, Preface].
[3] Cf. [5].

an aid in reaching a deeper understanding. The final goal of these efforts of an applied mathematician lies in (d) the creation of ideas, concepts, and methods that are of basic significance and general applicability to the subject in question, including the formulation of general principles. As mentioned above, these efforts may lead to (e) the creation of new mathematical ideas and theories. It might be added that, because of the origin of such mathematical theories, they are more likely to be applicable to other branches of the sciences.[4]

*The basic difference in motivation between pure and applied mathematicians* is reflected in the habits and practices of their activities. Although the applied mathematician understands and appreciates the nature of a rigorous demonstration, he cannot be made inactive by these considerations. In phase (b) of his activities, his primary emphasis is always directed toward the ultimate solution, and he frequently uses heuristic scientific reasoning to achieve this end. In this way, an analysis might be made more amenable to solution, approximations introduced, or arguments made plausible. His treatment is at all times responsible and disciplined, but he is not a deductive logician interested solely in the beauty of form and the power of abstraction. He should, however, have enough background in pure mathematics to be able to distinguish between rigorous proof, clear demonstration, plausible arguments, and hopeful speculation; all of which are used in the study of applied mathematics. In particular, when he is engaged in the creation of new mathematics, the applied mathematician should follow the practices (including the degree of rigor) used in pure mathematics in the *formulation* of his results. Because of his background, heuristic reasoning leading to the final form of the theory will doubtlessly be emphasized.

In many cases, the development of a rigorous form of the mathematical method or theory cannot be accomplished, yet the method has to be used with only plausible arguments to support its reliability. Most work in numerical solution of scientific problems belongs to this category.

In the other two phases of his activity, (a) and (c), the applied mathematician should follow the practices of a theoretical scientist. *The construction of an idealized mathematical model is indeed the most important and the most difficult phase*, especially in a new field of application of

[4] Cf. Hirsh Cohen [6]. In this article, he pointed out the similarity of the nature of these mathematical problems and their solution to those in mechanics and classical applied mathematics. Also, Lawrence Klein, in [7], stated ". . . but there are few examples in which it can be said that our subject has called forth its own branch of mathematics or given inspiration for great new mathematical discoveries."

mathematics. It usually requires a comprehensive knowledge and a deep understanding of the empirical facts related to the particular phenomenon under consideration as well as penetrating insight and mature judgement. These are also required in phases (c) and (d) of his activities. In these phases, the applied mathematician should examine his results to reach a deeper understanding of the problem at hand, and attempt to abstract the essentials to form concepts which are of more general applicability. At the same time, his conclusions must, of course, be checked against the existing body of knowledge. Any new inferences or predictions from his results are also subject to verification by further experimentation and observation, since their truth cannot be determined by purely logical means. With complementary theoretical and empirical efforts, a deeper and more penetrating understanding may be achieved.

Despite these similarities of activity, there are *subtle differences in attitude between a theoretical scientist and an applied mathematician*. The theoretical physicist, for example, has his primary interest in the discovery of new physical laws. Indeed, a correct theory can be formulated only after a great deal of trial and error. (Consider the current efforts in high-energy physics.) In contrast, the applied mathematician places more emphasis on the use of mathematical methods for the description of physical phenomena in terms of known physical laws. (Consider the current efforts in numerical weather forecasting.) He is also greatly interested in the stimulation of new mathematical ideas. One might say that the difference lies in the relative extent to which *inductive* and *deductive* reasoning are emphasized in each discipline.

However, these subtle differences in motivation can lead to substantial differences in the choice of the educational program and in the attitude developed in a person, which in turn color his activities. For example, classical mechanics (including particle mechanics and continuum mechanics) is seldom studied in detail by modern students of physics, yet it remains one of the principal subjects for the education of applied mathematicians. The fundamental concepts and the mathematical methods which have been developed through such studies still constitute the essential basis for the application of mathematics, whether the particular subject in question is atmospheric turbulence, radio astronomy, or demography. Indeed, the applied mathematician does not draw sharp lines along traditional boundaries of subject matter, and it is desirable for him to adapt his interests to the present and future vitality of the subject matter if he expects his research efforts to have an impact beyond the development of applicable mathematical methods (cf. Appendix).

*The desire and ability to cut across traditional scientific disciplines, through the medium of mathematics, are perhaps the unique characteristics of an applied mathematician.* Indeed, the education of an applied mathematician must provide him with a breadth of knowledge in both mathematics and the fundamentals of the sciences. He is then in the best position to advance a specific subject by the creation of suitable mathematical models, by his critical and precise thinking, and by transferring the mathematical knowledge gained through the study of another scientific discipline.

The above description of an applied mathematician almost compels him to stay at the general fundamental level of a braod spectrum of sciences. Of course, an applied mathematician could, if he so chooses, become a specialist in a given scientific subject. In doing this, he should perhaps subtly change his attitude and adopt entirely the spirit of the specialists in that field.

In order to educate applied mathematicians, a program of study must be provided both in the undergraduate and the graduate levels. Opportunities for post-doctoral research should also be available. In particular, there must be a course dealing with the fundamentals of the five-step processes described at the beginning of this section, with special emphasis on the first step, the construction of suitable mathematical models in a variety of problems. This last part is usually lacking in most of the curricula in mathematics.

## 4. *Creative Activities of Applied Mathematicians, Part B*

Applied mathematicians are sometimes also found in the resolution of mathematical issues that arise from their scientific problems. This interest they share with the willing pure mathematicians. Although this is a less frequent type of activity than that described above, it would be unfortunate to ignore this aspect of interaction between mathematics and the sciences. To quote from Chapter II of the COSRIMS Report [8], "Fourier considered mathematics as a tool for describing nature. But the impact of Fourier, important as it was in physics and engineering, was particularly felt in some of the purest branches of mathematics."

It is also to be recalled that when Fourier submitted his mathematical theory of heat to the French Academy for the Grand Prize, the referees,

Laplace, Lagrange, and Legendre pointed out that the mathematical treatment left much to be desired in the way of rigor, even though they admitted the novelty and importance of Fourier's work. More than a century was to elapse before all the subtle mathematical issues were satisfactorily resolved.[5] (To be sure, some of these issues were raised later.)

Here is a case from which much can be learned on the nature of the interaction. *The type of mathematical theory which is developed for the solution of certain scientific problems can often find application in other scientific problems.* For, when formulated in abstraction, i.e., in mathematical terms, they become similar or even identical. Secondly, in the early stages of the development of such work, its implication in pure mathematics is frequently unclear, and cannot be fully appreciated. The evaluation of such activities, therefore, must be made by the professionals as in any other discipline, to avoid unfair judgment.

In order to carry on with this tradition of Fourier, it is clear that the applied mathematician should be educated in the modern developments in pure mathematics. Even if his aim is less ambitious, he should be well informed in pure mathematics in order to make use of the latest accomplishments. In a university, special courses should be planned to introduce the essentials of pure mathematics to students in applied mathematics (cf. Sections 7 and 8 in Part III). As mentioned before, the applied mathematician also must be prepared to learn new mathematics throughout his career whenever it becomes necessary.

The contribution of pure mathematicians, who show interest in applications, can be extremely great in this area of interaction between science and mathematics. They can frequently contribute to the abstraction, consolidation, and generalization of special developments into a more elegant and a more powerful structure. Naturally, they would also contribute to *other mathematical developments which might appear to be useless at the beginning, but would later turn out to have important scientific applications.* However, I must disqualify myself from pursuing the discussion further along these lines. I hope that my friends in pure mathematics would expound on this theme further and furnish us with impressive examples.

From the standpoint of an applied mathematician, I would like, however, to add one comment. The theorems we need are often not the most general. We would be perfectly happy to have theorems proved

[5] See [9, 10].

under suitable restrictions. For Fourier analysis, piecewise smooth functions are usually sufficient. The extension to functions of bounded variation, Lebesque measurable functions, etc., are decidedly of lower importance. (For a modern problem in control theory, see [11]. Piecewise analytic functions are found to be most suitable.)

## 5. *Examples*

We shall now examine a few examples in physical mathematics to see how the above idealized description of the creative activities of applied mathematics fits realistic developments in specific instances.

### (a) *Fourier Analysis, a Case of Unqualified Success*

First, consider the case of Fourier analysis. This is perhaps the best example of a complete "cycle" of development. A relatively simple class of problems in mathematical physics, heat conduction, was solved according to steps (a)–(c) outlined in Section 3. The mathematical theories of Fourier series, Fourier integral, and generalized harmonic analysis are very impressive developments. The theory of Lebesque measure is easily motivated by asking a simple question: Is the Converse to Parseval's theorem a true statement? Modern radio astronony, with its multimillion dollar equipments and multimillion dollar annual budgets, depends on Fourier analysis for its very operation. The process of Fourier analysis is indeed built into the automatic online data processing equipment. The spectral analysis of the signals from distant astronomical objects is actually done by calculating the autocorrelation function and its Fourier transform (see [12, Fig. 5]).

The spatial Fourier analysis of the sky is also essential to radio astronomy. For this purpose, an array of radio telescopes is needed. A very large array, shaped in the form of the letter *Y* has been planned for completion in this country as a 10-year program, with a total budget exceeding one hundred million dollars. Figure 1 shows a linear array in Holland. It has a system of 12 telescopes spaced over a distance of 1600m. While the signal received by each of the telescopes can be analyzed in time by the method described above, the use of pairs of telescopes as interferometers enables the system to give a spatial Fourier analysis of the sky along the array. As the earth spins around its axis and revolves in its orbit, different directions in the sky may be covered. The Dutch astronomers are now in the process of doubling the length of the telescope system and the number of interferometer pairs in it.

FIG. 1. Part of a row of 12 radio-telescopes, each 25 m in diameter. The system, extending over a distance of 1600 m, provides 20 interferometers for the analysis of the structure of the radio sources. Present plans call for a doubling of the distance and the number of interferometers.

At microscales, the study of molecular structure is done by the study of X-ray diffraction patterns. The analysis is again based on Fourier methods. On the theoretical side, fundamental concepts in quantum mechanics, the wave-particle duality, can be best understood by using Fourier integrals.

### (b) *Turbulence, an Incomplete Picture*

However, complete success of this type is rare. In many instances, success is only partial.

Consider the problem of turbulence. Fourier analysis is a necessary part of the theory, and indeed, provides the basic language for the subject. However, there are essential difficulties elsewhere. Turbulence is basically a nonlinear random process, in which the dissipative effect plays a paramount role as the stochastic process. There is as yet no mathematical theory for turbulence despite many heroic efforts over several decades.

Even on the level of a single field of fluid motion, which is governed by a well-established set of partial differential equations (the Navier-Stokes equations) the rigorous mathematical theory provides little help. The equations are quasilinear. The existence theory for this set of partial differential equations has been established only for very small values of the Reynolds number, a characteristic dimensionless parameter of the problem. To study the theory of turbulence, one must however deal with the fluid motion at extremely large Reynolds numbers.

In the more limited problem of hydrodynamic stability, where we may linearize the equations in the first approximation, rigorous mathematical theories have been found to be more helpful. There have also been very interesting new mathematical developments stimulated by the needs in the solution of the physical problem.

### (c) *Numerical Analysis, a Case of Acquiescence*

The difficulty of dealing with nonlinear partial differential equations shows up even more clearly in the need to proceed with large-scale numerical calculations, despite the lack of mathematical theorems that would guarantee the convergence of such procedure. Examples that easily come to mind include much of the work done in this laboratory on the numerical solution of fluid dynamical problems and numerical weather forecasting. Logical-minded persons have to acquiesce to such decisions to go ahead for the lack of a better solution. Applied mathematicians are of course quite at home with such approaches, because there are various ways to ensure the plausible correctness of such procedures.

### (d) *Plasma Physics and Stellar Dynamics, a Case of Mutual Support Without Firm Mathematical Foundation*

Plasma physics is a subject of great interest to this laboratory: The problem of plasma containment is still an unsolved problem, Professor Harold Grad would be better qualified than I to speak on the mathematical issues. But I am also not aware of any significant contributions to this science from new mathematical theorems. Most of the theoretical progress has been made in the spirit of applied mathematics. Indeed, in some of the exploratory calculations being made now, the procedures followed would seriously worry theoretical hydrodynamicists. The substantiation of such theoretical analysis depends heavily on experimental verification. At the same time, I am informed by my M.I.T. colleagues, who are theoretical plasma physicists working closely with experimental programs, that most of the desired experiments, with very few exceptions, are yet to be done. Here is a challenging opportunity for applied mathematicians to make substantial contributions in consolidating and improving upon the methods of analysis to make the predictions more reliable.

At the same time, these theoretical calculations can derive indirect support from distant places. The study of spiral structure in galaxies has led to the investigation of collective modes in stellar systems which are essentially plasmoidal in nature. These modes have been observed in terms of their effects on star formation. To be sure, the analysis in these cases has been done quite carefully, but even modes calculated with similar care in plasma physics have not yet been observed.

I mention this case to stress the importance of reasoning by analogy in the work of applied mathematicians. Confidence and insight can be gained in our understanding of plasma physics even though the work on galaxies contains no logical deduction in the context of electromagnetic plasmas.

Indeed, the concept of collective modes might be applicable (perhaps already applied) to social sciences. Each person is analogous to a star. At least in economics, a person interacts more with the institutions than with individuals. The latter kind of interaction plays a rather minor role to influence the economics of the country as a whole.

(e) *Transonic Flow, a Case of Mismatch*

When engineers began to design airplanes to fly at very high subsonic speeds, the problem of the existence of a local region of supersonic flow became important. People were concerned that shocks might develop to influence the behavior of the boundary layer, and thereby produce seriously increased drag. However, it was later realized that if reasonable and ingenious care were exercised to avoid strong shocks, the problem of highly increased resistance could be overcome.

In the meantime, the problem has stimulated a mathematical investigation of the possibility of obtaining smooth solutions of partial differential equations of mixed elliptic and hyperbolic types (parabolic on the line of transition). This is a very interesting mathematical development in itself, but its value for the practical problem is limited. The stringent mathematical conditions required for guaranteeing a smooth flow would exclude very *weak* shocks, whose occurrence does not really matter in the practical problem.

I cite this example to warn our colleagues against investing too much effort in problems posed by scientists and engineers before finding out the real nature and implications of the solution of the mathematical problem. In the example cited, the effort is not fruitless, because the mathematical problem is of interest in itself. In other cases, the loss could be almost total.

## 6. *Applied Mathematics as an Independent Discipline*

I now come to the central point of my talk, which is a plea for the establishment of a healthy community of applied mathematicians and for increased support of education and research in applied mathematics, especially the support of academic activities. We cannot expect the important role of applied mathematics to be fulfilled without a steady supply of applied mathematicians of very high caliber. It is not enough to have only pure mathematicians who would devote a small fraction of their efforts to applied mathematics. Neither is it sufficient to have only theoretical scientists who are unaware of mathematical methods used in other sciences. Applied mathematicians must be educated with a broad knowledge of applicable mathematics and an extensive exposure to the application of mathematics.

A common educational program for applied mathematicians is also needed to create a community spirit, without which applied mathematics cannot remain a healthy profession.

I think it is time that we should recognize applied mathematics as an independent discipline, fairly distinct from pure mathematics.

The difference in spirit between pure and applied mathematics is clear and often very great. In a very interesting and important article [13], Professor J. Schwartz warned mathematicians against the danger of single-mindedness, literal-mindedness, and simple-mindedness in dealing with scientific problems. The above examples also show clearly the difference between the approach of pure and applied mathematicians. Yet the need for the education of applied mathematicians in a comprehensive spirit described in this article has been generally neglected. It was recommended in [8, Recommendation 18]. Nevertheless, only a few schools have established such programs.

One might ask whether it is realistic to adopt this comprehensive approach. From my own examination of this matter, the answer is an unqualified "yes." I have recently read papers and books in ecology, mathematical theories of population, economics, combinatorial analysis, etc. I did not find the mathematics used to be *that* different one from the other. The mathematical methods used in biological sciences also do not differ from those in physical sciences (see [6]). Certainly, people use similar methods in numerical analysis and similar computing machines.

If the mathematical methods used in the various sciences were widely divergent, the establishment of a community spirit among applied mathematicians would of course be more difficult.

In this country, there had not been a tradition of applied mathematics. People now in this profession come from a variety of backgrounds. There is thus a definite need for practicing applied mathematicians to do something important for this community. This is to educate each other in the subject of his own special line of research in a well-motivated manner. This would create mutual understanding, establish mutual confidence, and build up the community spirit.

In the following sections, I present a discussion of the educational program and conclude with a number of recommendations. (In the oral presentation, this part was omitted.)

## III. Education

### 7. *Education of Applied Mathematicians*

The ultimate aim of any liberal education program must include the following:

(1) the cultivation of an attitude, points of view, and value judgements;

(2) the acquirement of a vast body of coherent knowledge which deserves to be handed down from one generation to another, and which is best understood and special to this particular area of pursuit; and

(3) the development of certain talents, especially creative ability in the particular area in question.

From these, it is expected that the person will develop, through his personal experience in creative activity, a wisdom which enables him to act as an advisor to his co-workers and a teacher to the younger generation. In the latter capacity, he must be able to present a perspective of the total activity so that the younger generation will be able to decide how and where to devote their efforts, and to carry on their work with confidence. The lack of this perspective would often lead to wasteful efforts.

How can we arrange an educational program in applied mathematics to achieve the objectives described above?

First of all, the education must be started in the undergraduate years, during the formative period of the youth. If a person has formed an attitude to consider it far more valuable to prove an elegant theorem in algebra than to explain a phenomenon observed in the atmosphere or in the galaxies with the help of mathematical methods (or to consider the latter as "somebody else's business"), it would be very difficult to convert him to another set of beliefs. If a person has been impressed with a zeal to use his mathematical talent to elucidate a point in high energy physics, he might not be as enthusiastic in attacking a problem in biology or in economics.

The educational material (or the primary content of the knowledge of a student in applied mathematics) then must be based on an account of the typical impressive past accomplishments of mathematics selected from *all* the sciences. The method of case study must be used. At the same time, the students should also be made to understand that, while there would be a continuous opportunity for the applied mathematician in established areas (such as mechanics, or more broadly, physics) dramatic accomplishments of the future might lie in some other areas, where mathematics has not, as yet, been fully utilized. One must *avoid the danger of sterility by an overemphasis of traditional subject matter.*

It is the lack of survey courses of this broad nature that makes it difficult to educate future comprehensive applied mathematicians. A young student with an aspiration to become one usually finds himself in the almost impossible position of having to sift out for himself the mathematical aspects of the basic sciences from the courses that are available to him from the various scientific departments at a university. While this attempt itself is of great educational value, the total effort needed is overwhelming. The usual course of action for the student is then to attach himself to one of the traditional disciplines, either in mathematics or in one of the sciences. This specialization tends to defeat the major purpose of the education of comprehensive applied mathematicians.

*The job of abstracting and synthesizing the total picture of the interplay between mathematics and the sciences must be done for these inspired young students by the experienced scientists.* Indeed, this type of effort is the primary responsibility of the teachers in any profession. They must provide the introductory or survey courses to present the (undergraduate) students with a *total* picture of the *status quo* in the whole field so that future research efforts can be pursued with the greatest effectiveness.

Courses of another type, best taught by professional applied mathematicians, are those usually designated "methods of applied mathematics." These should be available at various levels. Besides these two types of "professional courses," the students must have a good foundation in basic pure mathematics, and some depth of education in at least one branch of science.

We may thus summarize a basic educational program in comprehensive applied mathematics[6] as follows:

(a) an education in the attitude of an applied mathematician,

(b) an education in the usual working ability (methods of applied mathematics),

(c) a survey in applied mathematics with emphasis on the total picture,

(d) a basic education in pure mathematics,

(e) An education in at least one branch of science in depth.

The subprograms (a), (b), and (c) should be normally taught by applied mathematicians; subprogram (d), by pure mathematicians; and subprogram (e), by professional scientists in various departments.

Graduate education should naturally be an extension and a continuation of the basic program discussed above; it must naturally include the development of the ability of the student to do research. The broad outlines of such abilities have been given in Section III. Here, as in all subjects, specialization is unavoidable, and the available lines of research at a given institution must depend on the faculty members available. It would, therefore, be useless to go into further details in this general discussions.

One should note the *central role of courses for subprogram* (c) (survey in applied mathematics), which should be approximately a two-year sequence in the junior and senior years. (It should be started in the

[6] In practice, this might mean three semester courses in pure mathematics, four semester courses in scientific subjects, and six semester courses in applied mathematics [(b) and (c)].

sophomore year, if possible.) Without these courses, we tend to educate only (i) pure mathematicians who might be occasionally interested in applications, or (ii) highly specialized applied mathematicians and theoretical scientists who would individually be very competent in a relatively narrow area of mathematics in his particular field of interest. These people do contribute to applied mathematics; but their presence alone is not sufficient for the healthy development of applied mathematics. The propagation of a comprehensive educational discipline of applied mathematics is essential to its success, since we must attract enterprising and brilliant young students in large numbers into the profession, and offer them a promising academic career. This last point is perhaps the most important point to the success of the development of applied mathematics. Applied mathematicians must be judged as such in terms of their total ability in promoting the interaction. They must also have the opportunity to educate their own future professionals.

Clearly, we do not advocate the education of applied mathematicians by first training them as pure mathematicians. Such a training has the danger of introducing a frame of mind which is disadvantageous to their creative activities. Those already educated in pure mathematics may wish to practice applied mathematics; but they should keep in mind the warnings of Professor Schwartz mentioned earlier in this paper.

## 8. *Some Practical Aspects of the Educational Program*

There are at least three important practical reasons for providing an undergraduate educational program in applied mathematics at a university.

(A) If only a graduate program is provided, one has to concentrate on education toward research ability in a certain specialized area; and there will not be time for education in breadth. This results in a splintering of activities in applied mathematics, making the total effect less effective. The usual sympton of difficulty is the small number of excellent students entering the graduate school to study applied mathematics.

(B) The maintenance of an undergraduate program would give the faculty in applied mathematics a common focus of interest and activity. The subject matter in an undergraduate program will evolve, but comparatively slowly (Newton mechanics still has its role, quantum mechanics will not be changed overnight). This permits a relatively stable joint effort on the part of the faculty. Joint effort creates stability.

In a graduate program, there is necessarily a diversification of interest among the faculty members. Furthermore, their interests often have to change quite rapidly in accordance with the current trends of scientific progress and the vitality of the subject matter. There cannot be real communication except among the specialists themselves. The complete lack of communication among colleagues is not healthy for any group.

(C) The courses in mathematical methods, and even the survey course, have tremendous educational value to many other students in pure mathematics, science, and engineering. Indeed, by a systematic offering of courses in mathematical methods, one may eliminate the need to provide special "service courses," which have often been the source of great discontent at a university.

A faculty member is usually required to fulfill a certain minimum teaching obligation. With the undergraduate program, the teaching obligations of an applied mathematician would be essentially to offer courses for the education of future applied mathematicians. The usefulness of these courses for the other purposes mentioned above is a natural beneficial side effect. If the applied mathematicians must teach mostly "service courses," they cannot help the feeling of being "second-class citizens" in the academic community.

## IV. Recommendations

### 9. *Support of Academic Activities in Applied Mathematics*

To prevent mathematics from drifting further and further away from science and technology, there is a need to develop academic applied mathematics: There is a need to have highly qualified scientists specifically devoted to such activities. I wish to make the following recommendations toward accomplishing this goal.

*Recommendation* 1. *The university administration and the scientific community, including the pure mathematicians, should be made keenly aware of the above-mentioned need.*

*Recommendation* 2. *Programs in academic applied mathematics should be offered in the universities, at both the undergraduate and graduate levels, aiming at the highest intellectual caliber.*[7]

At first, because of the shortage of applied mathematicians, these programs could be offered by the cooperative efforts of scientists and engineers together with pure mathematicians who have an interest in applications. But it is clear that the service teaching mentality would be detrimental to the morale. Persons of highest scientific caliber, whose primary efforts are elsewhere and who must work in a highly competitive atmosphere, rarely can be depended upon to offer their time and energy to "other people's business" for an extended period of time. We therefore need professional applied mathematicians to educate future professional applied mathematicians, whose ability and contributions are to be judged in terms of the creative activities described above and in terms of promoting the interaction between mathematics and the sciences.

We therefore arrive at the following most important recommendation.

*Recommendation* 3. *The universities should create faculty positions specifically in Applied Mathematics and appoint to these positions only people of the highest caliber.*

Their research activities should be well respected in the scientific community. Judgement of their work should be based on their contributions to the effective interaction between mathematics and science.

Since the National Science Foundation is awarding research grants to support all branches of pure science, the academic aspect of applied mathematics (AAM) must not be an exception. Otherwise, due to the pressure of the needs of mission-oriented support,[8] the academic nature of such work, with all its long-range benefits, would be jeopardized. We therefore offer

*Recommendation* 4. *The National Science Foundation should continue to expand its explicit support to academic activities in applied mathematics*, as such (not under the guise of research in some subject matter). Those activities, whose usefulness is easily identifiable, may continue to be supported by mission-oriented agencies. A plurality of pattern of support should be maintained.

[7] A secondary beneficial effect of such programs is the elimination of the need to offer service courses of low caliber to students in engineering and science.

[8] It must be said that, at present, the mission-oriented agencies have been very generous in their support of basic research whose usefulness may not be directly apparent. But this is a policy susceptible to pressure from the U.S. Congress.

## V. Appendix

### *Toward New Horizons with Great Traditions*

While the newly developed and developing subjects in applied mathematics would naturally attract greater attention, the role of applied mathematics developed in the classical spirit should also be fully appreciated in future research, and especially as a vehicle for education. The usual danger of emphasizing all kinds of traditional effects whether in science or in other fields is to fall into a pit of sterility. It must be avoided by accepting new challenges with a flexible attitude. With this in mind, one can indeed make the best of a wealth of knowledge made available to us through efforts of the great minds of past generations.

(a) *Education*

While excitement is often created by new developments in any scientific subject, traditional subject matter, sifted through many generations of critical thinking and brought up to its modern form, can often serve as a core of knowledge on which future developments can be based. It is indeed these crystalized thoughts, ideas, and reasoning, which must be taught in the undergraduate years as a source of stimulation and as a basis for future research efforts. Indeed, a common background of undergraduate education would serve as a basis of establishing rapport among applied mathematicians who might later specialize in different lines of research.

Needless to say, the pattern of education and detailed choice of subject matter must evolve in the course of time as our perspective over the total picture of applied mathematics changes in view of new developments.

(b) *Research*

The nature of research efforts in the "traditional" line actually evolves rapidly and substantially in the course of time. These research efforts also bring out clearly unsolved classical problems whose implications are obviously important but yet not fully understood. As an example in the latter category, the study of "turbulence" in the context of hydrodynamics is still an unfinished task. It is now clear that the existing studies and the still unresolved problems both relate to the broader problem of "nonlinear random processes." In the case of incompressible viscous fluids, the physical processes are very well understood from a general point of view, and many detailed comparisons of theoretical and observed results have become possible; but the fundamental mathematical theory remains unformulated.

Stability and instability, wave motion, linear versus nonlinear processes, reversible versus irreversible processes, entropy, and one can name some more, are certainly classical topics whose implications are general. The concept of entropy has already found its new application in information theory. It is quite possible that the concept of wave motion and the description of the process of shock wave formation might turn out to be useful in the mathematical description of social processes. On the other hand, one can easily trace the change of direction and find new vitality in an old subject like fluid mechanics. Around the turn of the century, aerodynamics, as related to aeronautical engineering, was one of the great challenges of the times. It remained strong as a subject for research of applied mathematicians for a long time. Nowadays, aerodynamics is so well understood that it is no longer the major concern of the applied mathematician. In its place, other aspects of fluid mechanics, those related to geophysical and astrophysical problems, are attracting greater attention. Certain concepts developed primarily in aerodynamics, such as the boundary layer, however, remain important even in these newer studies.

Continuum mechanics is of course but one aspect of classical applied mathematics, although it does have a special position, at least for the following three reasons: (i) the physical concepts are simple, (ii) the mathematical problems involved are interesting and of wide applicability, and (iii) a variety of natural phenomena fall within its scope. Particle dynamics also occupied and continues to occupy an important position in applied mathematics. There is currently a great deal of activity based on the theory of dynamical systems (integrals of motion, adiabatic invariants, etc.) in the construction of models of galaxies of stars. Even the whole subject of mechanics and statistical mechanics is but one part of "classical" mathematical physics. There will no doubt remain many problems to be explored in the whole broad area of such activities for years to come. The solid foundation of traditional efforts will always help us to reach new horizons if we adopt the progressive point of view.

## References

1. S. Ulam, The applicability of mathematics, *in* "Mathematical Sciences, a Collection of Essays," MIT Press, Cambridge, 1969.
2. H. Weyl, Relativity theory as a stimulus in mathematical research, *Proc. Amer. Phil. Soc.* **93** (1949), 535.

3. ALBERT EINSTEIN, "Out of my Later Years," Philosophical Library, New York, 1950.
4. T. VON KÁRMÁN AND C. C. LIN, On the existence of an exact solution of the equations of Navier–Stokes, *Comm. Pure Appl. Math.* **14** (1961), 645–655.
5. H. P. GREENSPAN, Applied mathematics as a science, *Amer. Math. Monthly* **68** (1961), 872–880.
6. H. COHEN, Mathematics and the biological sciences, *in* "Mathematical Sciences, A Collection of Essays," MIT Press, Cambridge, 1969.
7. L. KLEIN, The role of mathematics in economics, *in* "Mathematical Sciences, A Collection of Essays," MIT Press, Cambridge, 1969.
8. Report of the Committee on Support of Research *in* the Mathematical Sciences. National Academy of Sciences, Washington, D.C. 1969.
9. E. T. BELL, "Men of Mathematics," pp. 197–198. Simon and Schuster, New York, 1937.
10. G. BIRKHOFF, "Source Book in Classical Analysis," Harvard Univ. Press, Cambridge, 1974.
11. N. LEVINSON, Minmax, leapunov and "bang–bang," *J. Differential Equations* **2** (1966), 218.
12. R. WIELEBINSKI, The Effelsberg 100-m radio telescope, *Naturwissenschaften* **58** (1971), 109–116.
13. J. SCHWARTZ, The pernicious influence of mathematics on science, *in* "Proceedings of 1960 International Congress on Logic, Methodology, and Philosophy of Science," pp. 356–360, (Nagel, Suppes, and Tarske, Eds.), Stanford Univ. Press, Stanford, 1962.

Printed by the St Catherine Press Ltd., Tempelhof 37, Bruges, Belgium

SIAM REVIEW
Vol. 20, No. 4, October 1978


# EDUCATION OF APPLIED MATHEMATICIANS*

C. C. LIN†

**Abstract.** The role of an educational program for applied mathematicians is discussed in terms of the nature of applied mathematics and the development of a community of applied mathematicians with mutual understanding. A specific sample program—the one just adopted at the Massachusetts Institute of Tehcnology—is described.

**1. Introductory remarks.** You just heard from Dr. Burton Colvin on the use of mathematics in a variety of research programs. As you can see, the mathematician who participates in such work would need quite a different kind of knowledge from that taught in an educational program suitable for future workers in pure mathematics. This situation presents a challenging problem in designing educational programs in mathematics at a university. There are limitations to what can be done at a university where the students have their freedom of choice. Can we design an educational program in applied mathematics

a. which is coherent from an academic point of view,
b. which is intellectually challenging and attractive to the talented students, and
c. which is, at the same time, serving the needs of students who wish to find employment in industry and government?

I think we can.

I speak here of students both in the undergraduate and in the graduate years, but I feel (and my feelings have been growing stronger in recent years) that an important element in our discipline, like many others, is its ability to attract undergraduates. I shall explain this more fully later.

When I speak of a discipline of education, I am of course not referring to a program of technical training. I am speaking primarily of the preparation for a long-term academic endeavor; a program of education would enable the student eventually to *think* for himself. There must be a coherent philosophy, a clear perspective, and a concrete set of common objectives toward which all applied mathematicians will find it compatible to devote their efforts. These basic principles should be sufficient to sustain the career of an individual for his whole lifetime. They should also be able to sustain the profession indefinitely.

Diverse opinions have been expressed regarding the nature and role of applied mathematics. Basically, there are three approaches: one is centered on pure mathematics, with the field of application as an extension or specialization (and hence occupying a secondary role); another is essentially to leave the subject to the theorists in the individual scientific disciplines and not to think of applied mathematics as a separate profession (obviously, this position cannot be advocated at a meeting of SIAM); third is to look at all branches of applied mathematics as interrelated—similar to the various branches of pure mathematics—and to encourage a comprehensive understanding of these subjects by talented young students, with the hope that the interrelationship between all these subjects will become strengthened through cooperative efforts of all applied mathematicians. At least, there should be mutual understanding and mutual respect among the various specializations. This approach is

* Received by the editors September 21, 1977. Retiring President's address, Presented on December 4, 1975 at the meeting of the Society for Industrial and Applied Mathematics, San Francisco, California.

† Department of Mathematics, Massachusetts Institute of Technology, Cambridge, Massachusetts 02139.

what I call the "comprehensive approach" and I have advocated it on several occasions. It is, I believe, the only approach that will hold the profession—in particular this professional society—together.

**2. What is applied mathematics?** Let me first describe what I mean by *applied* mathematics by examining the relationship between *pure* mathematics and the sciences. This is not a simple matter. Indeed, the very term *pure* mathematics has sometimes caused controversy; some mathematicians feel that it is artificial to separate pure and applied mathematics. When I served on the COSRIMS under the Chairmanship of Lipman Bers, the current President of the AMS, the Committee agreed on the term *core* mathematics. The relationship between mathematics and the sciences is then as shown in Fig. 1. Here, at the center, is the core of basic mathematical principles—the inner citadel of pure mathematics (or core mathematics); on the outer fringes, there are the various sciences; between is applied mathematics. I do not think there is much disagreement on this picture.

From this picture you can see why Professor Joseph Keller once defined applied mathematics as a science of which pure mathematics is a part. One simply regards the whole inner area as one discipline. I sympathize with him, but do not wish to wipe out the identity of pure mathematics by such a sweeping statement.

There is also not much disagreement on the process of applied mathematics. It consists of at least three steps; (1) formulation; (2) solution; and (3) interpretation.

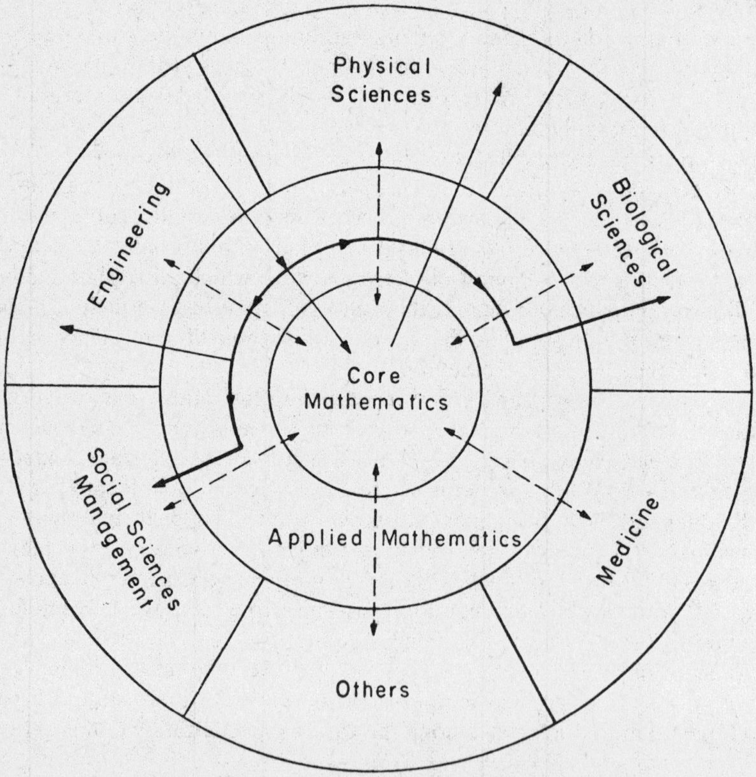


FIG. 1. *Relationship between mathematics and other disciplines. The arrow leading into the area of core mathematics from engineering represents Fourier analysis. It was invented by a mechanical engineer. One can branch off on the level of applied mathematics, and apply Fourier analysis to various disciplines; but one can also go deeply into the core. The choice depends on one's objectives.*

Some people would add an introductory step of assessment of the general nature of the scientific problem and a concluding step of examining the ramifications.

It is also generally agreed that the formulation process is often the most difficult. Plausible models for a given scientific problem may have to be examined for months or years before the correct one is obtained. One must examine the scientific facts in great detail. Sometimes, there is not even a "best model." All models may turn out to be quite imperfect, and each would contribute only to a partial understanding. Clearly, a deep understanding of the scientific problem is needed in judging the usefulness (or lack thereof) of the various models.

Pure and applied mathematicians would have their differences, particularly on the educational process. Many pure mathematicians would insist that a person must first become a pure mathematician before learning about applications. However, mathematics is not an easy subject. Learning applications is another difficult matter, especially the processes of formulation and interpretation, as we have noted. Thus, this approach leads to a gloomy picture and some people have indeed argued that one might as well give up attempts to recruit applied mathematicians because these people are rare geniuses.

As an applied mathematician, I cannot agree with the view that the applied mathematician is a superman. We simply do not know as much pure mathematics as the professional pure mathematician (and we do not have to know as much). All we need to know well in pure mathematics is the foundation of the *relevant* mathematics we are using. A statistician should know probability theory; a physical mathematician should know theory of differential equations (ordinary and partial) etc., but neither *has* to know algebraic topology.

We would rather start with the middle ground in Fig. 1 (and perhaps even from the periphery); i.e., we start with scientific problems and adopt the method of *case study* in order to motivate the mathematical theory. In this way, one can get to the interesting applications more directly.

This is indeed consistent with the development of applied mathematics. Consider the Fourier series as an example. Starting with the problem of heat conduction, one can easily demonstrate its usefulness in representing arbitrary functions. The pure mathematician would then ask the fundamental question as to what kinds of functions would allow Fourier representation, and go deeply into pure analysis (cf. Fig. 1). On the other hand, the applied mathematician may stop with piecewise smooth functions and go a long way toward Fourier transforms, Laplace transforms, generalized harmonic analysis, and the applications of these methods to a large number of scientific problems, such as radio astronomy and molecular structure. There is simply a difference in *objectives*.

Another example which shows the difference in character between applied mathematics and pure mathematics is numerical analysis and machine computation. Since one cannot analytically solve the partial differential equations necessary for numerical weather forecasting, one has to do it by numerical methods. Furthermore, one does not have existence theorems for these systems of partial differential equations; one cannot prove that the process converges in the asymptotic limit of infinitesimal steps. But one must continue with a confidence based on the use of empirical tests for assuring the validity of the procedures. Otherwise, one should immediately stop a large fraction of computations now processed by the machines in this country. And I was told that computing is a very large industry; of the order of a billion dollars a year.

The difficulty with educating applied mathematicians is therefore not as great as it is sometimes described to be.

**3. Education of applied mathematicians.** Now, let me discuss the educational program, beginning with some general comments and concluding with the presentation of a specific educational program.

As mentioned above, I believe in a comprehensive approach; i.e., there should be a common educational program—at least during the undergraduate years—to create a community spirit of applied mathematicians.

One might ask whether it is realistic to adopt this comprehensive approach. From my own examination of this issue, the answer is an unqualified "yes." I have recently read papers and books in ecology, mathematical theories of population, economics, combinatorial analysis, etc. I did not find the mathematics used to be *that* different, one from the other. I found that mathematical methods used in biological sciences do not differ substantially from those in physical sciences (see, e.g., article by Hirsch Cohen [1]). People use similar methods in numerical analysis and similar computing machines.

If mathematical methods used in various sciences were widely divergent, the establishment of a community spirit among applied mathematicians would, of course, be more difficult; perhaps even meaningless.

At this meeting of SIAM, there is also a very good example. The joint work of Gordon Kaufman and Eytan Barouch and their collaborators on the exploration of oil fields is a problem in operations research. Barouch was trained as a mathematical physicist with specialization in statistical mechanics (which of course is quite different from statistics). However, the similarity of basic concepts enabled Barouch to move quite easily into applied probability.

Certainly, it requires special effort for professors to organize elementary courses in such a manner as to bring out the similarities and the differences, but that is exactly what professors, as teachers, should do.

Let us now turn to the discussion of an educational *program*.

The ultimate aim of any liberal education program must include the following:

a. Cultivation of an attitude, points of view, and value judgements.
b. Acquisition of a vast body of coherent knowledge which deserves to be handed down from one generation to another, and which is best understood and special to this particular area of pursuit.
c. Development of certain talents, especially creative ability in the particular area in question.

From these, it is expected that the person will develop, through his personal experience in creative activity, a wisdom which will enable him to act as an advisor to his co-workers and a teacher to younger generations. In the latter capacity, he should be able to present a perspective of the *total* activity so that the younger generation will be able to decide how and where to apply their efforts and to carry on their work with confidence. The lack of this perspective would often lead to wasted efforts.

How can we arrange an educational program in applied mathematics to achieve the objectives described above?

First, this education process must be started in the undergraduate years, during the formative period of the youth. If a person has formed an attitude to consider it far more valuable to prove an elegant theorem in algebra than to study the exploration of oil fields, or to explain a phenomenon observed in the atmosphere or in the galaxies with the help of mathematical methods (or to consider the latter as "somebody else's business"), it would be very difficult to convert him to another set of beliefs. If a person has been impressed with a zeal to use his mathematical talent to elucidate a point in high energy physics, he might not be as enthusiastic in attacking a problem in biology or economics.

The educational material (or the primary content of the knowledge of a student in applied mathematics) must then be based on an account of the typical impressive past accomplishments of mathematics *selected from all the sciences*. The method of case study must be used. At the same time, the students should be made to understand that, while there would be a continuous opportunity for the applied mathematician in established areas (such as mechanics, or more broadly, physics), dramatic accomplishements of the future might lie in other areas; areas where mathematics has not, as yet, been fully utilized. One must *avoid the danger of sterility by an exclusive emphasis on traditional subject matter.*

It is a lack of survey courses of this broad nature that makes it difficult to educate future comprehensive applied mathematicians. A young student with an aspiration to become one usually finds himself in the almost impossible position of having to sift out for himself the mathematical aspects of the basic sciences from the courses that are available to him from the various scientific departments at a university. While this attempt itself is of great educational value, the total effort needed is overwhelming. The usual course of action for the student is then to attach himself to one of the traditional departments, either in mathematics or in one of the sciences. This specialization defeats the major purpose of the education of comprehensive applied mathematicians.

*The job of abstracting and synthesizing the total picture of the interplay between mathematics and the sciences must be done for these inspired young students by experienced scientists.* Indeed, this type of effort is the primary responsibility of teachers in any profession. They must provide the introductory or survey courses in an attempt to present the [undergraduate] students with a *total* picture of the *status quo* in the whole field in order that future research efforts can be pursued with greatest effectiveness.

Courses of another type, best taught by professional applied mathematicians, are those usually designated "methods of applied mathematics." They should be available at various levels. Other than these two types of "professional courses," the students should have a good foundation in basic pure mathematics, and some depth of education in at least one branch of science.

We may thus summarize a basic educational program in comprehensive applied mathematics[1] as follows. It should include:

(a) an education in the attitude of an applied mathematician, with emphasis on solving real-world problems;
(b) an education in the usual working tools (methods of applied mathematics);
(c) a survey in applied mathematics with emphasis on the total picture;
(d) a basic education in pure mathematics; and
(e) an in-depth education in at least one branch of science.

Subprograms (a), (b) and (c) should be normally taught by applied mathematicians; subprogram (d) by pure mathematicians; and subprogram (e) by professional scientists in various professional departments.

Graduate education should naturally be an extension and continuation of the basic program discussed above; naturally it must include the development of the

[1] See below for one specific program.

ability for the student to do research and to think for himself. The broad outlines of such abilities have been given in § 2. Here, as in all subjects, specialization is unavoidable, and the available lines of research, at a given institution, must depend on faculty members available. Therefore, it would be useless to prolong this general discussion in this article.

One should note the *central role of courses for subprogram* (c) (survey in applied mathematics), which should be approximately a two-year sequence during the sophomore and junior years. Without these courses, we tend to educate only (i) pure mathematicians who might be occasionally interested in applications, or (ii) highly specialized applied mathematicians and theoretical scientists who would individually be very competent in a relatively narrow area of mathematics in their particular fields of interest. These people do contribute to applied mathematics but their presence alone is not sufficient for the healthy development of applied mathematics as a profession. The propagation of a comprehensive educational discipline of applied mathematics is essential to its success. We must attract enterprising and brilliant young students in large numbers into the profession and offer them promising academic careers. This last point is perhaps most important to the success and development of applied mathematics. Applied mathematicians must be judged as such in terms of their total ability in promoting the interaction. They must also have the opportunity to educate future professionals in applied mathematics.

Clearly, we do not advocate the education of applied mathematicians by first training them as pure mathematicians. Such training bears the danger of introducing a frame of mind which is disadvantageous to their creative activities. Those already educated in pure mathematics may wish to practice applied mathematics; but they should keep in mind the warnings of Professor Schwartz [2] in one of his most stimulating papers: *The pernicious influence of mathematics in the sciences.*

Let me now present a sample undergraduate program in applied mathematics; i.e., the one adopted at M.I.T.

**4. An undergraduate degree program in applied mathematics.** The Department of Mathematics at M.I.T., at the recommendation of its Committee on Applied Mathematics, has adopted a program leading to the degree of B.Sc. in Mathematics, with emphasis in applied mathematics. Besides fulfilling the general stipulations of the Department, students in this Option must fulfill the following requirements:

a. *Basic program* (all required)
   Calculus
   Differential Equations
   Linear and Modern Algebra
   Complex Variables
   Introduction to Discrete Mathematics
   Introduction to Continuum Mathematics

b. *Specialization*

At least four courses from the following two groups; of these, at least one course must be chosen from each group.

Group I: (1) Probability and Random Variables, or Probability
(2) Statistical Inference, or Statistics for Applications
(3) Applied Combinatorial Analysis
(4) Introduction to the Mathematical Theory of Computation

Group II: (1) Numerical Analysis
(2) Introduction to Physical Mathematics I

(3) Introduction to Physical Mathematics II
(4) Introduction to Fluid Mechanics

c. Students planning to continue graduate school are strongly advised to study a course in Analysis.

*General Comments*

1. It is believed that the establishment of this option provides several advantages for both students and faculty. Specifically, it would make the applied mathematics program more visible to the students.

2. In formulating the above program, attempts have been made to include *very few* restrictive requirements in order to accommodate the diverse intertests of the faculty members in the Department.

3. The course requirements in this program should (and presumably would) be revised from time to time; but the general nature of the program would be expected to remain essentially unchanged over a substantial period of time. The courses available at M.I.T. make it possible to offer choices; e.g., a course of Statistical Inference or Statistics for Applications. This is unnecessary. Any elementary course in statistics would be adequate.

**5. Some practical aspects of the educational program.** There are at least three important practical reasons for providing an undergraduate educational program in applied mathematics at a university.

a. If a graduate program only is provided, one has to concentrate on education toward research ability in a certain specialized area; there will be insufficient time for education in breadth. This results in a splintering of activities in applied mathematics and consequently a weakening of the total effort. The usual symptom of difficulty is the small number of excellent students entering the graduate school to study applied mathematics.

b. The maintenance of an undergraduate program would give the applied mathematics faculty a common focus of interest and activity. The subject matter in an undergraduate program will evolve, but comparatively slowly. (Newtonian mechanics still has its role. Quantum mechanics will not change overnight.) This permits a relatively stable joint effort on the part of the faculty. Joint effort creates stability.

In a graduate program, there is necessarily a diversification of interest among the faculty members. Furthermore, their interests often have to change quite rapidly in accordance with current trends of scientific progress and vitality of the subject matter. There cannot be real communication except among specialists themselves. Complete lack of communication among colleagues is unhealthy for any group.

c. The courses in mathematical methods, including the survey course, have tremendous educational value to many other students in pure mathematics, science, and engineering. Indeed, by a systematic offering of courses in mathematical methods, one might even be able to eliminate the need to provide special "service courses"—which have often been the source of great discontent at a university.

A faculty member is usually required to fulfill a certain minimum teaching obligation. With the undergraduate program, the teaching obligation of an applied mathematician would be essentially to offer courses for the education of future applied mathematicians. The usefulness of these courses for the other purposes mentioned above is a natural beneficial side effect. If applied mathematicians must teach primarily "service courses," they cannot help the feeling of being "second-class citizens" in the academic community.

**Concluding remarks.** It is not my purpose today to discuss research in applied mathematics, which I had occasion to discuss elsewhere [3]. I shall just make a few concise statements.

(1) Emphasis should be placed on the solution of real-world problems. The applied mathematician must be able to produce a real *impact* on science [4] and be ready to compete with theorists otherwise educated in this effort.

(2) In the development of general mathematical methods and theories, the applied mathematician should place emphasis on those directly related to applications. Leave most (if not all) pure mathematical issues to pure mathematicians interested in applications.

(3) Be venturesome. Nothing ventured, nothing done. Attack new problems. Avoid too much introspection and narrowing of interests.

I do not deny that there are other students who would prefer to study pure mathematics first and then learn a specialized line of application. These people can make important contributions to applied mathematics. But I think the survival of a professional society such as SIAM, as a coherent body, requires the existence of the type of applied mathematicians with comprehensive understanding.

## REFERENCES

[1] H. COHEN, *Mathematics and the biological sciences*, COSRIMS Essays, M.I.T. Press, Cambridge, MA, 1969.

[2] J. SCHWARTZ, *The pernicious influence of mathematics in the sciences*, Proceedings of 1960 International Congress on Logic, Methodology and Philosophy of Science, Nagel, Suppes and Tarski, eds., Stanford University Press, Stanford, CA, 1962, pp. 356–360.

[3] C. C. LIN, *On the role of applied mathematics*, Invited paper presented at the First Los Alamos Workshop on Mathematics in the Natural Sciences, June 12–18, 1974; Advances in Math., 19 (1976), pp. 267–288.[2]

[4] FRED WAN, *Director's Statement of Policy*, Institute of Applied Mathematics, University of British Columbia, Canada.

[2] It has been pointed out to me that the example cited in § (e) on p. 278 is not good. I apologize to Cathleen Morawetz for not appreciating her important contributions properly. But I am sure that readers of the paper can think of their own examples.

# SELECTED PAPERS
*of*
# C. C. LIN

*C. C. Lin speaking at Steward Observatory in Tucson, Arizona (March, 1967)*

# SELECTED PAPERS
*of*
# C. C. LIN

— Volume 2 —

**Astrophysics**

*Editors*
David J. Benney, Frank H. Shu and Chi Yuan

**World Scientific**
*Singapore • New Jersey • Hong Kong*

*Published by*

**World Scientific Publishing Co. Pte. Ltd.**
P.O. Box 128, Farrer Road, Singapore 9128

*U.S.A. office*: World Scientific Publishing Co., Inc.
687 Hartwell Street, Teaneck NJ 07666, USA

Library of Congress Cataloging-in-Publication data is available.

**SELECTED PAPERS OF C. C. LIN**



ISBN 9971-50-318-2

Printed in Singapore by Kim Hup Lee Printing Co. Pte. Ltd.

# TABLE OF CONTENTS

## Volume II: Astrophysics

# RECOLLECTIONS AND COMMENTS

## II

Papers in this volume largely concern the development of the density theory for the spiral structure in galaxies. In a series of three papers ([70], [72], [81]), the density wave theory for spiral structure has been developed to a point where comparison with observations can be made with confidence. This is made possible by adopting a semi-empirical approach centered around the hypothesis of quasi-stationary spiral structure rather than a deductive approach with emphasis on mechanisms. Such studies were left to a second stage. Thus, after the initial success, many dynamical issues remain to be examined, including the basic processes for the formation and the maintenance of such quasi-stationary spiral structures and the conditions under which such spiral structures may be expected to exist. The early phase of the theoretical developments in this semi-empirical approach are summarized in [84] and [85]. A continuation of this work is described in [99], which dealt with the observation of the Oort constants that are basic to the studies of the Milky Way.

The completion of this phase of the work on density wave theory is naturally the most exciting, since the correctness of a concept and a theory has been demonstrated through comparison with empirical data. There was great excitement everytime my discussions met with the approval of eminent astronomers such as J. Oort, B. Strömgren, B. Bok, and L. Woltjer. In particular, I recall the summer of 1968 when I gave a series of lectures in a course of Galactic Astronomy, alternatively with another series given by G. Westerhout in the same course. Oort and Bok also gave a series of lectures. Westerhout described the data from radio observations in the Milky Way, and I described the analysis of such data by C. Yuan and myself. The two series of lectures fit into each other perfectly even though they were not planned that way in advance. There was great satisfaction in seeing theoretical predictions confirmed by observations. Similar excitement was evoked by the observations of the concentration of the magnetic field by Mathewson et al in the whirlpool galaxy M51.

Nevertheless, questions continue to be raised on the dynamical basis of the QSSS hypothesis. The natural reply is the modal approach, and the study of the relevant dynamical mechanisms. Beginning with [86], the investigation turned towards the study of dynamical mechanisms. A fluid dynamical model was formulated for the sake of simplicity, but with careful inclusion of the modifications needed for the simulation of the stellar system which the fluid model intends to simulate. By this time dynamical studies by other authors have become quite well-advanced. We persisted in the dynamical studies through the use of steady wave trains and modes. It is hoped that the use of modes would make it easier to approach the study of the morphology of galaxies.

The first calculation of spiral modes were published in the 1976 paper ([94]). It is an asymptotic theory which allows the mechanisms for the excitation and the maintenance of spiral structure to be clearly described through the use of an ordinary differential equation of the second order, which allows two waves at each location. At first, the theory was restricted to tightly-wound normal spirals. Detailed development of such an approach may be found in [98], [103] and [106]. More recently, the theory of open modes appropriate to bar spirals have also been developed ([112]).

The study of spiral modes, through the computation of the eigenfunctions and the study of the dynamical mechanisms for their excitation and maintenance, has been described in a series of papers ([108] – [113]). Specifically, attempts were made, successfully, to cover all morphological types in the Hubble classification. The dynamical mechanisms can be described in terms of the propagation of wave trains (which is sometimes quite complicated). A full account of this research effort will soon be published.

It is interesting to compare the dynamical processes in a galactic disk with those in the classical studies of shear flows. In both cases, the excitation mechanism is closely related to the "comoving zone", i.e., a neighborhood where the waves propagate at approximately the same phase velocity (and group velocity) as that of the mean material motion in the direction of its motion. This is the "corotation zone" in galactic dynamics and the "critical layer" in shear flows.

Indeed, in the classical theory of hydrodynamic stability, Lord Raleigh took the modal approach whereas Lord Kelvin took the wave packet approach in the neighborhood of the critical layer. The shearing wavelet in the Kelvin approach, however, has only transient growth, but no net over-reflection, because of the lack of gravitational energy. Naturally, the spectral approach ([113]) in the galactic problem has a counterpart in the Kelvin problem. The detailed evolution of the wavelet was worked out by Orr (1907) in that simple case.

C. C. Lin
*Massachusetts Institute of Technology*
*May 1987*

# SELECTED PAPERS
*of*
# C. C. LIN

— Volume 2 —

**Astrophysics**

# ON THE SPIRAL STRUCTURE OF DISK GALAXIES

C. C. LIN AND FRANK H. SHU
Department of Mathematics, Massachusetts Institute of Technology
*Received March 20, 1964*

## ABSTRACT

It is shown that gravitational instability is a plausible basis for the formation of the spiral pattern in disk galaxies. An explicit asymptotic formula is obtained for the form of the spiral. It gives reasonable numerical results for the galaxy, and qualitatively satisfactory trends for normal spirals of various types.

## I. INTRODUCTION

The mechanism for the formation of the spiral patterns observed in most disk-shaped galaxies has not yet been fully understood. There is little doubt, from the observational data available, that these magnificent manifestations are associated with the interstellar gas and the brilliant young stars born in them. But could the old stars also play an important role in the formation of the spiral structure?

To construct a theory of the spiral structure, one must bear in mind the following important components of a galaxy:

*a*) *The stars*—with their gravitational forces, circular velocity, and velocity dispersion
*b*) *The interstellar gas*—with *its* gravitational field and pressure
*c*) *The magnetic field*—which exerts its influence through the highly conducting interstellar gas.

A complete theory should take all these components and forces into account, and put their relative importance into perspective. Such a theory is not yet available.

There are at least two possible types of spiral theories. The first alternative is to associate every spiral arm with a *given body of matter;* e.g., such an arm might essentially be a tube of gas primarily constrained by the interstellar magnetic field. The difficulty with the disrupting influence of differential rotation in such a theory is well known. The various issues associated with this point of view have been thoroughly discussed recently by Oort (1962). The second alternative is to regard the spiral structure as a *wave pattern*, which either remains stationary, or at least quasi-stationary, in a frame of reference rotating around the center of the galaxy at a proper angular speed (possibly zero).

Three years ago, through discussions with Professors B. Strömgren and L. Woltjer, one of us (C. C. L.) became interested in the possibility of spiral structure from the point of view of gravitational instabilities of the galactic disk, and several of us, including A. Toomre and C. Hunter, began to examine the problem in some detail. Our emphasis and conjectures are not yet in complete agreement. Since A. Toomre's (1964) point of view has been published, it seems desirable to publish our point of view even though the work is not yet as complete as the present writers would wish to have it. Furthermore, we feel that, in any case, the role of gravitational force, which is by far the largest in a galaxy, deserves to be carefully examined for a complete understanding of the problem, whether or not it eventually turns out to be the predominant mechanism.

Toomre tends to favor the first of the possibilities described above. In his point of view, the material clumping is periodically destroyed by differential rotation and regenerated by gravitational instability. It is somewhat difficult to see how this mechanism alone can account for the *relatively regular spiral pattern over the whole disk* in most of the flat galaxies. The present authors favor the second point of view, i.e., that the matter in the galaxy (stars and gas) can maintain a density wave through gravitational interaction in the presence of the differential rotation of the various parts of the disk. This density wave provides a spiral gravitational field which *underlies* the observable concentration of young stars and the gas. In this way, an observable spiral pattern can be maintained over the whole disk. Indeed, an explicit formula, equation (14), for its approximate description has been found.[1]

It is almost certain that the basic mechanism to be proposed here is related to the idea of "density waves" discussed by P. O. Lindblad (1960) and B. Lindblad (1961). Our analysis will, on the whole, tend to support their general conclusions. It is, however, somewhat difficult to make a detailed comparison between their ideas and ours, since the methods used are very different from each other. There is considerably more emphasis on individual stellar orbits in the Lindblads' theory, while the interaction of the stars over the disk as a whole is directly considered in our work through the use of the distribution function. This approach will enable us to exhibit certain spiral patterns very easily from general dynamical considerations (Sec. II). Some attempts at comparison will be made in Section IV. An investigation that further clarifies the relationship between these two approaches would no doubt enhance our understanding of the whole problem.

## II. THE STABILITY PROBLEM OF THE GALACTIC DISK

It is natural to ask whether distribution of mass in the form of a thin rotating disk, such as that in our own Galaxy, is in a state of *stable* equilibrium. Instability can take the form of a warping of the disk or of motions in its plane. We shall concentrate on the latter mode as being more relevant to the problem at hand. In our idealized model, the mass is concentrated in an infinitesimally thin disk, with a surface density that is approximately equal to the projected density in the galactic plane. The gravitational forces are in balance with appropriate circular and random velocities. Such a representation necessarily involves a high surface density in the central region, where a typical galaxy has a bulge due to the large three-dimensional velocity dispersion of the stars. Thus, it will often be found convenient to have a singularity of the density distribution at the center, provided the total mass is finite.

To investigate the stability of such a disk, we may start with the basic equations of stellar dynamics expressed in terms of the distribution function in phase space (Chandrasekhar 1960). But it is sometimes convenient to use the infinite set of equations for the (tensorial) moments of the velocity components, and to terminate them by imposing a suitable approximation. As a first step, we shall neglect the velocity dispersion altogether.[2] One then has the following set of equations:

$$\mu_t + r^{-1}[(r\mu u)_r + (\mu v)_\theta] = 0\,, \tag{1a}$$

$$u_t + uu_r + (v/r)u_\theta - v^2/r = \phi_r\,, \tag{1b}$$

$$v_t + uv_r + (v/r)v_\theta + uv/r = \phi_\theta/r\,, \tag{1c}$$

$$\phi_{rr} + \phi_r/r + \phi_{\theta\theta}/r^2 + \phi_{zz} = -4\pi G\mu(r, \theta)\delta(z)\,, \tag{1d}$$

where the cylindrical system of coordinates $(r, \theta, z)$ is used, $\mu$ is the surface density, and $\phi$ is the negative of the gravitational potential. The first three equations are restricted to the plane $z = 0$, while equation (1d) is for the *three*-dimensional space. For the initial state of equilibrium, we consider a solution of the form $\mu = \mu_0(r)$, $u = 0$, $v = V(r) = r\Omega(r) > 0$. We then consider a disturbance from the equilibrium state; e.g., $\mu = \mu_0(r) + \mu'(r, \theta, t)$. As in all stability problems, we begin by considering the case of small dis-

[1] Preliminary results obtained by detailed numerical calculations carried out by Mr. Ronald Rehm confirm the validity of the formula and the general conclusions.

[2] In this approximation, the gaseous component of the galaxy is naturally included. We simply regard the density, $\mu$, below as the *total* density of matter including both the stars and the gas.

turbances. The equations governing the quantities $\mu'(r, \theta, t)$, etc., can then be linearized.[3] We obtain

$$\mu'_t + \Omega\mu'_\theta + \frac{1}{r}(r\mu_0 u')_r + (\mu_0/r)v'_\theta = 0\,, \tag{2a}$$

$$u'_t + \Omega u'_\theta - 2\Omega v' = \phi'_r\,, \tag{2b}$$

$$v'_t + \Omega v'_\theta + (\kappa^2/2\Omega)u' = \phi'_\theta/r\,, \tag{2c}$$

$$\phi'_{rr} + \phi'_r/r + \phi'_{\theta\theta}/r^2 + \phi'_{zz} = -4\pi G\mu'\,\delta(z)\,. \tag{2d}$$

In the above equations, $\kappa$ is the epicyclic frequency given by

$$\kappa^2 = 4\Omega^2[1 + (r/2\Omega)(d\Omega/dr)]\,. \tag{3}$$

The set of linearized equations (2a–d) admits solutions of the type

$$\mu' = Re\{\mu^{(1)}(r)\exp[i(\omega t - n\theta)]\}\,, \qquad \omega = \omega_r + i\omega_i\,, \tag{4}$$

where $n$ is an integer (which may be taken to be positive without loss of generality), and $\omega_r$ and $\omega_i$ are the real and imaginary parts of the parameter $\omega$. Unstable modes are given by $\omega_i < 0$.

Under such a general formulation, $\omega$ takes on characteristic values for the solution of a set of linear integro-differential equations in the single independent variable $r$, subjected to suitable boundary conditions at $r = 0$ and as $r \to \infty$. However, since we would admit a singularity at $r = 0$ in the solution, if we adopt the disk model right to the center of the galaxy, and the range of $r$ is infinite, the set of characteristic values may be expected to have a *continuous spectrum* (in addition to any discrete part). Physically, this means that we are looking for a representation of the situation in the disklike part of the galaxy, and leaving the conditions near the central bulge, where the random velocities are large and three-dimensional, to adjust themselves to almost any requirement of the disk part. We also expect that all the disturbances would die off at infinity in a reasonable manner.

The solution (4) is clearly of the nature of a *density wave.* Indeed, it generally has a *spiral form.* To see this, let us write (as we can always do),

$$\mu^{(1)}(r) = S(r)\exp[i\Phi(r)]\,, \tag{5}$$

where $S(r)$ and $\Phi(r)$ are *real.* Then equations (4) give

$$\mu'(r, \theta, t) = S(r)e^{-\omega_i t}\cos[\omega_r t - n\theta + \Phi(r)]\,. \tag{6}$$

If $S(r)$ varies *slowly* with the radial distance $r$ while $\Phi(r)$ varies *quickly*, then equation (6) gives a spiral impression in the density distribution at any instant of time, the form of the spiral being

$$\theta = \frac{1}{n}[\Phi(r) + \text{const.}]\,. \tag{7}$$

There are $n$ arms in the spiral. These are *trailing spiral arms* if $\Phi'(r) < 0$, and leading ones if $\Phi'(r) > 0$. Note that we have taken $\Omega(r)$ and $n$ to be positive by convention. By comparing equation (7) with observed two-armed galactic spirals, it is easy to see that $\Phi$ should change by an order of $4\pi$ over a typical radial distance in such cases as the whirlpool nebula.

[3] It is suggested that this is justifiable for our present purposes, even though the random velocities neglected might be comparable to the plausible amplitude of the disturbance, since we are dealing with the *co-operative* effect of grouping of stars, which acts in a manner quite different from the effect of velocity dispersion (see also Sec. IV).

### III. AN ASYMPTOTIC SOLUTION

The stability problem formulated above is very similar to those in ordinary hydrodynamics.[4] It has, however, the additional mathematical difficulty that, after the variable $z$ is eliminated, $\phi'(r, \theta, 0)$ and $\mu'(r, \theta)$ are connected by an integral relationship. Fortunately, this difficulty disappears in an asymptotic approximation based on the rapid change of the phase factor $\Phi(r)$ mentioned at the end of the last section. In this case, we may attempt to integrate the Poisson equation

$$\phi_{rr}^{(1)} + \phi_r^{(1)}/r - n^2\phi^{(1)}/r^2 + \phi^{(1)}_{zz} = -4\pi G\mu^{(1)}(r)\delta(z) , \tag{8}$$

by an asymptotic process.[5] We write

$$\mu^{(1)}(r) = S(r) \exp [i\lambda f(r)], \text{ and} \tag{9}$$

$$\phi^{(1)}(r, z) = \psi(r, z, \lambda) \exp [i\lambda h(r, z)] , \tag{10}$$

where all the functions involved may be complex, $\lambda$ is a large real parameter introduced for the convenience of the formal asymptotic procedure, and $\psi(r, z, \lambda)$ is assumed to have an expansion in inverse powers of $\lambda$. In effect, we are looking for asymptotic solutions of the Laplace equation for $z > 0$ and $z < 0$, bounded at infinity (in a suitable sense for $\lambda$ large), continuous at $z = 0$, and fulfilling the jump condition $[\phi_z] = -4\pi G\mu^{(1)}(r)$ at $z = 0$ that can be easily derived from equation (8). Such a solution leads to the simple result that, to a first approximation,

$$\phi_r^{(1)} = 2\pi i G\epsilon\mu^{(1)} , \qquad \phi_\theta^{(1)} = O(\mu^{(1)}/\lambda) , \tag{11}$$

where $\epsilon = \pm 1$ according to the sign of the real part of $f'(r)$.

Within this approximation, it is very easy to find solutions of the type (5) for the system of equations (2a–d). Our first aim is to determine the phase factor, which turns out to be

$$\lambda f'(r)\epsilon = [\kappa^2 - (\omega - n\Omega)^2]/2\pi G\mu_0 . \tag{12}$$

Thus, a solution of the type considered is possible only if the real part of $\kappa^2 - (\omega - n\Omega)^2$ is positive, i.e., if

$$\kappa^2 + \omega_i^2 - (\omega_r - n\Omega)^2 > 0 . \tag{13}$$

The sign of $\Phi'(r) = Re[\lambda f'(r)]$ can, however, be positive or negative. Thus, *both leading and trailing arms are permitted*. The choice between the two might be resolved only after the solution is found (e.g., by numerical integration) in the ranges where the inequality (13) is violated. In those ranges, the solution is expected to have a behavior different from that indicated above.[6] In particular, the condition at the center of the galaxy deserves special attention.

Another effect that might distinguish between leading and trailing arms is differential rotation. Although density waves are propagated primarily by gravitational forces, they would be modified by differential rotation, when the non-linear terms, omitted from equations (2a–d), are included. The effect might be analogous to that of fluid motion

[4] The similarity might hold not only for problem of initial instability, but also for the subsequent process of development into a quasi-stationary final state, with the random velocities of the stars and of the interstellar-gas clouds supplying the smoothing effects that limit the growth of the waves. Differential rotation corresponds to the shear flow which can supply the energy to the oscillations and random motions (cf. Sec. IV).

[5] For details see Appendix.

[6] Toomre (1964) carried out numerical calculations for axisymmetrical disturbances over the whole range of disk radii and found that the solution drops off roughly in an exponential manner at infinity.

that leads to the distortion of acoustic waves. In that case, a density decrease in the direction of wave propagation tends to be accentuated into a compression shock, whereas a density decrease would tend to be smoothed out by the motion of the fluid. Thus, it is conceivable that only trailing waves are stable in the presence of non-linear effects.

In the case of axisymmetrical disturbances, $n = 0$, the formula (12) reduces to one already obtained by Toomre (1964) for a local wavenumber in terms of an assumed frequency. The present asymptotic approach can also yield the slowly varying amplitude distribution over the disk, which is not available from the local theory. The importance of this approximate analysis over the whole disk will be seen in Section IV.

The present study shows that non-axisymmetrical disturbances can propagate around the disk without change of shape even *in the presence of differential rotation.* Indeed, for the range $r_1 < r < r_2$ in which expression (13) applies, the geometrical form of the spiral pattern is found from equations (7) and (12) to be given by

$$n(\theta - \theta_0) = -\int_{r_0}^{r} (2\pi G\mu_0)^{-1}[\kappa^2 + \omega_i^2 - (\omega_r - n\Omega)^2]\,dr\,. \tag{14}$$

The end points of this range, $r = r_1$ and $r_2$, correspond to the points of *local gravitational resonance* in the neutral case (cf. Sec. IV). As a typical example, if we refer to the density data given by Schmidt (1956) and the velocity data given by P. O. Lindblad (1960, Table 2) on the basis of Schmidt's model, and take $\omega_r = 20$ km/sec kpc and $\omega_i = 50$ km/sec kpc (which is somewhat less than the value of $\kappa$ at 5 kpc from galactic center), we would have a wavelength of the order of 2–3 kpc in our neighborhood. The lower limit $r_1$ would be at about 2 kpc from the galactic center, while the upper limit $r_2$ would disappear altogether. Such a mode amplifies rather rapidly in the absence of velocity dispersion. Its expected behavior in the presence of velocity dispersion will be discussed in Section IV.

The contrast between the spiral patterns of Sa, Sb, and Sc galaxies can also be brought out analytically by equation (14). If there is a comparatively greater concentration of mass in the center, the density $\mu_0$ of the disk part is relatively smaller. Equation (14) then predicts tighter spirals, as indeed are observed in Sa galaxies. More even distribution of matter corresponds to loosely wound spirals, as observed in Sc galaxies.

## IV. HYPOTHESIS OF QUASI-STATIONARY SPIRAL STRUCTURE IN THE SPATIAL DISTRIBUTION OF STARS

The above analysis suggests that, in the absence of velocity dispersion of the stars, there are many possible spiral patterns of density modifications in a basically axisymmetrical disk of gravitating matter. Those modes which are strongly unstable may be expected to bring themselves into prominence. On the other hand, there is considerable velocity dispersion among the stars, and it is to be expected that all gravitational instabilities would tend to be smoothed out by their effects. Two possibilities then exist. The first is that all the modes of gravitational instability are suppressed. The second possibility is that the two tendencies might balance each other for certain modes, or a *group* of modes, in such a manner that a stable or nearly stable pattern might be maintained over long periods of time. Modes which are sufficiently different from this select group would, however, be suppressed.

It is not easy to examine the various factors that would enter into the decision between these two possibilities. A first attempt could be made by examining whether the velocity dispersion present is sufficient to suppress all the waves according to a *local* theory. Such a calculation has been carried out by Toomre (1964) for the solar neighborhood, and he reached the conclusion that an rms velocity dispersion $\sigma_u$ of the order of 20–30 km/sec would be needed for the suppression of all the waves. This amount of velocity dispersion is actually present, and he concluded that the solar neighborhood is locally stable. However, if one would adopt the same method of analysis for a location at 4–5 kpc from the galactic center, the value of $\sigma_u$ required would be about 70–90 km/sec (if we adopt Schmidt's model [1956] for our Galaxy). If such a high-velocity dispersion were actually present, it would imply that a considerable number of stars with high radial velocities would reach our neighborhood from the interior part of the Galaxy, contrary to observational evidence. Thus, that part of the disk is locally unstable.

This picture of a galactic disk, which is in part stable and in part unstable according to a *local* theory, would suggest the possibility of a balance resulting in a neutral density wave extending over the *whole* disk and having a scale of the order of (but smaller than) the distance between the stable and the unstable regions.[7]

The question may be raised as to why the velocity dispersion might cease to grow to the extent needed for complete stabilization. One possible reason is that large-scale waves are ineffective in producing local velocity dispersion. Furthermore, it may be conjectured that these large-scale waves are stabilized at small but finite amplitudes[8] by the joint effects of velocity dispersion and by the reaction of the gas together with its associated magnetic field (Chandrasekhar and Fermi 1953).

We now venture to suggest that there are indeed such large-scale neutral (or nearly neutral) waves for most of the disk galaxies, and formulate our ideas in the form of the following hypothesis:

The total stellar population, which has various degrees of velocity dispersion, forms a *quasi-stationary spiral structure* in space of the general nature discussed above. This is primarily due to the effect of gravitational instability as limited by velocity dispersion (and secondarily to the influence of the gas and the magnetic field). The extent of density variation in the spiral pattern may be only a small fraction of the symmetrical mean density distribution.

In advancing this hypothesis, we are not ignoring the fact that transient local patterns of a general spiral form are relatively easy to produce in a system in rotation. However, it seems rather difficult to account for a *coherent pattern over the whole disk* without bringing into account the co-operative effects of long-range gravitational forces.

If we accept the above hypothesis and follow the line of reasoning that led to it, the following inferences and conjectures may be made.

1. Due to the prevailing spiral gravitational field, all components of the galaxy, including the gas and the young stars, should form similar spiral patterns on the scale of the radius of the disk, whatever other secondary processes there might be. Different stellar components, with different extent of velocity dispersion, should lie on somewhat different spirals (cf. Zwicky 1957) and exhibit different degrees of unevenness. As an extreme case, one may even attempt to analyze the behavior of the gas, which has very little pressure, as a separate system (including possibly an appropriate magnetic field), which is subjected to the action of a spiral gravitational field produced by the rest of the galactic population. Its density contrast may therefore be expected to be far larger than that in the stellar components, as is indeed known from observations. These discussions should be supplemented by a consideration of the reaction of the gas on the stellar population, as mentioned above.

2. It is known from observations that galaxies devoid of gas do not show prominent spiral patterns. This is consistent with the present proposal. The velocity dispersion in such galaxies is usually thought to be comparatively large, and the mechanism of instability discussed here may indeed be suppressed completely. Even if this were not so,

[7] Although such waves cannot be analyzed by a strictly local theory, it is still feasible to apply the asymptotic method described in this paper; for the method does allow for the interrelationship between the various regions in the higher approximations. An asymptotic analysis that includes the effect of velocity dispersion would be extremely desirable.

[8] Cf. Landau's theory of stable oscillations at small but finite amplitudes in the problem of hydrodynamic stability (in Landau and Lifschitz 1959).

and the old stars did form a spiral structure, the lack of gas and the young stars stemming therefrom might so impair the manifestation of the spiral structure that it would hardly be observable.

3. One simple method to account for the effect of velocity dispersion and the reaction of the gas is to adopt the Lindblads' concept of a fixed smooth disk of stars belonging to the disk population and Population II and a mobile disk of Population I (gas and young stars). Then the basic equations (1a–d) should be modified by the addition of a force $F(r)$ in equation (1b) to represent the fixed field, and the surface density $\mu$ would refer to that of Population I alone. Except for this change of the interpretation of $\mu$, the above theory will remain unchanged.[9] The spiral pattern is still given by equation (14), but the wavelength is much shorter, since $\mu$ is reduced. However, there is no sharp demarcation between the fixed and the mobile components. One could include in the mobile part also some older stars with relatively low-velocity dispersion. Thus, if we choose $\omega_r = 20$ km/sec kpc, $\omega_i = 0$, the mobile component in our neighborhood should be set at about 25 per cent of the total mass[10] in order to get a wavelength of approximately 2 kpc. In contrast to Section III, we consider neutral waves here, since the stabilizing influence of velocity dispersion has largely been taken into account.

4. Both the Lindblads emphasized the approximate constancy of $\Omega - \kappa/2$ over a large part of the Galaxy. Our theory does not seem to be dependent on it. Instead, we have a certain kind of local resonance only at two distinct radial locations in the galaxy where $\kappa^2 - (\omega - n\Omega)^2 = 0$ in the case of neutral waves (and also approximately for slightly amplifying and damped waves). This condition clearly means that the traveling spiral gravitational field is in step with the local epicyclic motion. Indeed, near these points, one would expect the gas, even more than the stars, to react very strongly and possibly to acquire very large radial velocities. These radial velocities should be outward and inward over alternate sectors of extent $\pi/n$ each. It would be very interesting to pursue this point further both theoretically and in relation to observational facts (e.g., the 3-kpc arm).[11]

We are indebted to Professors B. Strömgren and L. Woltjer for introducing this subject to one of us (C. C. L.) at a conference in Princeton, N.J., in 1961 and for many helpful discussions during the course of this study. We had the pleasure of collaborating with Dr. A. Toomre. Frequent discussions with him have been to our mutual benefit, although we have placed somewhat different emphasis in our work and have arrived at somewhat different conjectures. We also had the pleasure of discussing the problem with Professors M. Krook and D. Layzer and Mr. A. J. Kalnajs.

This work is partially supported by a grant from the National Science Foundation.

## APPENDIX

### ASYMPTOTIC SOLUTION OF POISSON'S EQUATION

We now consider, in some detail, the solution of the equation for gravitational potential corresponding to a surface distribution of density in the plane $z = 0$. Especially we wish to demonstrate an asymptotic method used in the text.

[9] Dr. Kevin Prendergast first mentioned to me (C. C. L.) the idea that the present analysis might be made in the presence of the gravitational field of a fixed disk during a discussion at the early stages of this study.

[10] This is not unreasonable, especially since we expect this region to be locally stable and forced to oscillate by the inner regions. It might even be too low an estimate in view of the possible existence of some molecular hydrogen as yet undetected.

[11] See Oort (1962), pp. 3–22, and esp. his reference to B. Lindblad on p. 12.

Since we are working with a cylindrical coordinate system, the convenient form for the basic equation is

$$\frac{\partial^2\phi}{\partial r^2}+\frac{1}{r}\frac{\partial\phi}{\partial r}+\frac{\partial^2\phi}{\partial z^2}+\frac{1}{r^2}\frac{\partial^2\phi}{\partial\theta^2}=-4\pi G\sigma(r,\theta)\delta(z). \tag{A1}$$

In other words, we look for solutions of Poisson's equation for $z>0$ and for $z<0$ such that, at $z=0$, there is a discontinuity in $\phi_z$ given by

$$\left[\frac{\partial\phi}{\partial z}\right]=\phi_z(r,\theta,0+)-\phi_z(r,\theta,0-)=-4\pi G\sigma(r,\theta). \tag{A2}$$

Since the problem is linear, we may consider the individual harmonic components in

$$\sigma=\sigma_n(r)\cos n\theta\,,\qquad \phi=F_n(r,z)\cos n\theta \tag{A3}$$

and obtain the general solution by superposition. We thus have the partial differential equation

$$\frac{\partial^2F_n}{\partial r^2}+\frac{1}{r}\frac{\partial F_n}{\partial r}+\frac{\partial^2F_n}{\partial z^2}-\frac{n^2F_n}{r^2}=-4\pi G\sigma_n(r)\delta(z) \tag{A4}$$

to be solved under the condition

$$\left[\frac{\partial F_n}{\partial z}\right]=-4\pi G\sigma_n\,. \tag{A5}$$

We consider the case where the variation of $\sigma_n(r)$ is sufficiently rapid in some sense. That is, we consider a representation

$$\sigma_n(r)=Re[\sigma^{(1)}(r)e^{i\lambda f(r)}], \tag{A6}$$

where $\lambda$ is large, while both $\sigma^{(1)}(r)$ and $f(r)$ vary but slowly in $r$. We may then expect a solution of the form

$$F_n(r,z)=Re[A(r,z,\lambda)e^{i\lambda\Phi(r,z)}], \tag{A7}$$

where $A(r,z,\lambda)$ depends on $\lambda$ in the following asymptotic manner:

$$A(r,z,\lambda)=A^{(0)}(r,z)+\lambda^{-1}A^{(1)}(r,z)+\lambda^{-2}A^{(2)}(r,z)+\ldots \tag{A8}$$

The functions $\Phi(r,z)$, $A^{(0)}(r,z)$, . . . , etc., are slowly varying functions of $r$ and $z$. The condition (A5) becomes the pair of relations

$$\left[\frac{\partial A}{\partial z}\right]+i\lambda A(r,0)\left[\frac{\partial\Phi}{\partial z}\right]=-4\pi G\sigma^{(1)}(r),\qquad \Phi(r,0)=f(r). \tag{A9}$$

We have imposed the conditions that $\Phi(r,z)$ and $A(r,z,\lambda)$ are continuous at $z=0$.

*Differential equations.*—We now substitute

$$F=A(r,z,\lambda)e^{i\lambda\Phi(r,z)} \tag{A10}$$

into the equation (A4); we get

$$\begin{aligned}&\frac{\partial^2A}{\partial r^2}+\frac{\partial^2A}{\partial z^2}+\frac{1}{r}\frac{\partial A}{\partial r}-\frac{n^2A}{r^2}\\&+i\lambda\left[A\left(\frac{\partial^2\Phi}{\partial r^2}+\frac{\partial^2\Phi}{\partial z^2}+\frac{1}{r}\frac{\partial\Phi}{\partial r}\right)+2\left(\frac{\partial A}{\partial r}\frac{\partial\Phi}{\partial r}+\frac{\partial A}{\partial z}\frac{\partial\Phi}{\partial z}\right)\right]\\&-\lambda^2A\left[\left(\frac{\partial\Phi}{\partial r}\right)^2+\left(\frac{\partial\Phi}{\partial z}\right)^2\right]=0\,.\end{aligned} \tag{A11}$$

This leads immediately to the condition

$$\left(\frac{\partial\Phi}{\partial r}\right)^2+\left(\frac{\partial\Phi}{\partial z}\right)^2=0\,, \tag{A12}$$

which is satisfied by

$$\Phi=f_1(r+iz) \qquad \text{or} \qquad f_2(r-iz)\,, \tag{A13}$$

but not a combination of both. The second of the boundary conditions (A9) imposes the requirement that

$$f_1(r)=f_2(r)=f(r)\,. \tag{A14}$$

The choice of solution for $z>0$ and $z<0$ must be such that the solution decreases with $|z|$ increasing. Thus, we should have

$$\Phi=f(r+i\epsilon z) \qquad \text{for} \qquad z>0\,, \qquad \text{and} \qquad \Phi=f(r-i\epsilon z) \tag{A15}$$

for $z<0$, where $\epsilon=\pm 1$ accordingly as $f'(r)\gtrless 0$. The first condition (A9) becomes

$$\left[\frac{\partial A}{\partial z}\right]+i\lambda A\,(\epsilon 2i)\,f'(r)=-4\pi G\sigma^{(1)}(r)$$

or

$$\left[\frac{\partial A}{\partial z}\right]=-4\pi G\sigma^{(1)}(r)+2\lambda A\,|\,f'(r)\,|\,. \tag{A16}$$

Equation (A16) suggests that $A(r,z,\lambda)$ should be of the form

$$A(r,z,\lambda)=\lambda^{-1}[\,A^{(1)}(r,z)+\lambda^{-1}A^{(2)}(r,z)+\ldots]\,, \tag{A17}$$

and that $A^{(1)}(r,z)$ satisfies the condition

$$0=-4\pi G\sigma^{(1)}(r)+2A^{(1)}(r,0)\,|\,f'(r)\,|\,. \tag{A18}$$

With $\Phi(r,z)$ given in the form of expression (A13), equation (A11) can be somewhat simplified. Indeed, we get

$$\begin{aligned}&\left(\frac{\partial^2 A}{\partial r^2}+\frac{\partial^2 A}{\partial z^2}+\frac{1}{r}\frac{\partial A}{\partial r}-\frac{n^2 A}{r^2}\right)\\ &+i\lambda f'(r\pm i\epsilon z)\left[\frac{A}{r}+2\left(\frac{\partial A}{\partial r}\pm i\,\frac{\partial A}{\partial z}\right)\right]=0\,.\end{aligned} \qquad (z\gtrless 0) \tag{A19}$$

Using the form (A17), the equation for $A^{(1)}(r,z)$ becomes

$$\frac{A_\pm^{(1)}}{2r}+\left[\frac{\partial A_\pm^{(1)}}{\partial r}\pm i\epsilon\,\frac{\partial A_\pm^{(1)}}{\partial z}\right]=0$$

or

$$\frac{\partial}{\partial r}[\,r^{1/2}A_\pm^{(1)}]\pm i\epsilon\,\frac{\partial}{\partial z}[\,r^{1/2}A_\pm^{(1)}]=0\,.$$

The solution of this is immediately seen to be

$$[\,r^{1/2}A_\pm^{(1)}]=g(r\pm i\epsilon z)\,, \qquad (z\gtrless 0) \tag{A20}$$

where $g$ is to be determined from the boundary condition (A18).

We have

$$r^{-1/2}g(r)=2\pi G\sigma^{(1)}(r)/\,|\,f'(r)\,|\,. \tag{A21}$$

Thus, to a first approximation, the surface distribution $\sigma^{(1)}(r) \exp [i\lambda f(r) - in\,\theta]$ gives rise to the potential

$$\phi(r,z) = A_{\pm}(r,z)\exp[i\lambda f(r \pm iz\epsilon) - in\theta], \qquad (z \gtrless 0) \quad \text{(A22)}$$

where $\epsilon = sgn\, f'(r)$,

$$A_{\pm}(r,z) = \lambda^{-1}r^{-1/2}g(r \pm iz\epsilon) \quad \text{(A23)}$$

and

$$r^{-1/2}g(r) = 2\pi G\sigma^{(1)}(r)/|f'(r)|. \quad \text{(A24)}$$

The force per unit mass in the plane $z = 0$ is given by

$$\phi_r(r, 0) = 2\pi i G\sigma\epsilon, \qquad [\epsilon = \pm 1 \text{ according as } f'(r) \gtrless 0] \quad \text{(A25)}$$

where

$$\sigma = \sigma^{(1)}(r)\exp[i\lambda f(r) - in\theta].$$

Notice that $n$ does not appear in the initial approximation for the amplitude functions. It appears only in the higher approximations.

It is understood in the above reasoning that $\sigma^{(1)}(r)$ and $f(r)$ are analytic functions. Otherwise, a more elaborate process is needed to justify the same results. See Shu (1963).

The higher approximations can be easily found by using the asymptotic representation (A17) for $A(r, z)$. It appears from the form of the solution that it holds only for a finite wedge ($|z/r| <$ const.) in the $(z, r)$-plane with the further restrictions that $r$ should be finite, and that $f'(r)$ does not vanish over the range in question.

## REFERENCES

Chandrasekhar, S. 1960, *Principles of Stellar Dynamics* (New York: Dover Publications), p. 184.
Chandrasekhar, S., and Fermi, E. 1953, *Ap. J.*, **118**, 113, 116.
Landau, L. D., and Lifschitz, E. M. 1959, *Fluid Mechanics* (Reading, Mass.: Addison Wesley), p. 104.
Lindblad, B. 1961, *Stockholm Obs. Ann.*, **21**, 8, esp. Sec. 5, "Spiral Structure."
Lindblad, P. O. 1960, *Stockholm Obs. Ann.*, **21**, 4.
———. 1962, *Interstellar Matter in Galaxies*, ed. L. Woltjer (New York: W. A. Benjamin, Inc.), pp. 222–233.
Oort, J. H. 1962, *Interstellar Matter in Galaxies*, ed. L. Woltjer (New York: W. A. Benjamin, Inc.), pp. 234–244.
Schmidt, M. 1956, *B.A.N.*, **13**, No. 468, 15.
Shu, Frank, H. 1963, "Gravitational Instability and Spiral Structure in Disk Galaxies," thesis for B.Sc. in Physics, Massachusetts Institute of Technology.
Toomre, A. 1964, *Ap. J.*, **139**, 1217.
Zwicky, F. 1957, *Morphological Astronomy* (Berlin: Springer-Verlag), pp. 198–201.

THE ASTROPHYSICAL JOURNAL
Vol. 142, No. 4, November 15, 1965


# THE GRAVITATIONAL COLLAPSE OF A UNIFORM SPHEROID

C. C. LIN, L. MESTEL,* AND F. H. SHU†
Department of Mathematics, Massachusetts Institute of Technology
*Received March 29, 1965*

## ABSTRACT

A uniform, non-rotating, pressure-free spheroid is supposed to collapse gravitationally from rest. It is shown that the initial eccentricity is steadily increased by the anisotropic gravitational field: an initially oblate spheroid tends toward a disk, and an initially prolate spheroid toward a spindle. Numerical results are computed for a series of initial eccentricities.

## I. INTRODUCTION

In cosmogonical studies there arises naturally the problem of the collapse from rest of a cool, self-gravitating cloud. In the simplest idealization, with the cloud density initially spherically symmetric, the collapse clearly takes place through a series of spherically symmetric states, because of the isotropy of the gravitational field. A convenient Lagrangian coordinate is the radius $r(m,t)$ at time $t$ of the subsphere containing mass $m$. With pressure negligible, the equation of motion is

$$\frac{d^2 r}{dt^2} = -\frac{Gm}{r^2}; \tag{1}$$

this integrates to

$$r(m,t) = r_0(m)\cos^2\theta\,, \tag{2}$$

where

$$\theta + \tfrac{1}{2}\sin 2\theta = t\left[\frac{8\pi}{3}G\bar{\rho}_0(m)\right]^{1/2}, \tag{3}$$

and $\bar{\rho}_0(m) = m/(4\pi r_0^3/3)$ is the mean density of the mass sphere, $m$, at the initial state of rest ($t = 0$). If in particular the sphere has a uniform density $\rho_0$ at $t = 0$, then $\theta$ and so also $r/r_0$ are functions of time only; a uniform sphere contracts homologously, and so stays uniform. The same result holds if at $t = 0$ each shell is not at rest but is contracting with a speed proportional to $r_0$.

The instability of such a sphere against the formation of subcondensations—the "fragmentation" problem (Hoyle 1953)—is currently exciting much interest (Hunter 1962, 1964; Layzer 1963). In this paper, however, we are concerned with the more elementary question—the stability of the cloud as a whole against a change of shape that leaves the density uniform. It is shown that the state of spherical collapse is singular: the slightest over-all departure from sphericity is systematically magnified. The significance of this result will be discussed elsewhere (Mestel 1965): it is relevant to the asymptotic state achieved by a fragmenting cloud and to the contraction, with or without breakup, of a strongly magnetic cloud. In this paper, however, we shall assume that no breakup occurs, and ignore all forces other than gravitation.

## II. THE EQUATIONS TO THE PROBLEM

Although the detailed calculations below are performed just for axially symmetric bodies, the basic result is valid for all uniform ellipsoids and is most easily established in

* Department of Applied Mathematics and Theoretical Physics, University of Cambridge, Cambridge, England.

† Now at Department of Astronomy, Harvard University.

symmetric notation. In a study of the flattening of a *uniformly rotating* uniform sphere, Lynden-Bell (1962, 1964) pointed out that collapse from an initial state of pure rotation will be through a series of uniform spheroids of steadily increasing eccentricity. The essence of the proof depends on the gravitational potential within a uniform spheroid being quadratic in Cartesian coordinates, and it is immediately applicable to the present problem with zero rotation, but with the initial anisotropy due not to centrifugal force but to an anisotropic mass distribution.

For collapse to occur through a series of uniform ellipsoids, a particle with coordinates $(x_0,y_0,z_0)$ at time $t = 0$ must subsequently be at $(x,y,z)$ with

$$x = x_0X(t)\,, \qquad y = y_0Y(t)\,, \qquad z = z_0Z(t)\,; \tag{4}$$

i.e., $X$, $Y$, and $Z$ must be independent of $(x_0,y_0,z_0)$, so that $x$, $y$, and $z$ are individually subjected to *homologous* transformations that maintain both the ellipsoidal shape and a uniform density

$$\rho = \frac{\rho_0}{XYZ} \tag{5}$$

but alter the ratios of axes (except in the degenerate case $X = Y = Z$). The potential inside a uniform ellipsoid is (Lyttleton 1953)

$$V = -\pi G\rho(\alpha x^2 + \beta y^2 + \gamma z^2 - \delta)\,, \tag{6}$$

where

$$\alpha = abc\int_0^\infty \frac{d\lambda}{(a^2+\lambda)\Delta}\ \text{etc.}\,, \qquad \delta = abc\int_0^\infty \frac{d\lambda}{\Delta}\,, \tag{7}$$

and

$$\Delta^2 = (a^2 + \lambda)(b^2 + \lambda)(c^2 + \lambda)\,. \tag{8}$$

The quantities $\alpha$, $\beta$, and $\gamma$ depend only on the ratios $a:b:c$. Thus, the equations of motion of a point mass at $(x,y,z)$

$$\frac{d^2x}{dt^2} = -2\pi G\rho\alpha x\,, \qquad \frac{d^2y}{dt^2} = -2\pi G\rho\beta y\,, \qquad \frac{d^2z}{dt^2} = -2\pi G\rho\gamma z \tag{9}$$

reduce, by the relations (4) and (5), to

$$\frac{d^2X}{dt^2} = -\left(\frac{2\pi G\rho_0}{XYZ}\right)\alpha X\,, \tag{10}$$

etc. At $t = 0$,

$$X = Y = Z = 1\,, \qquad \frac{dX}{dt} = \frac{dY}{dt} = \frac{dZ}{dt} = 0\,. \tag{11}$$

Thus, since both equation (10) and initial conditions (11) are independent of $(x_0,y_0,z_0)$, so are also the functions $X(t)$, $Y(t)$ and $Z(t)$, and the collapse is through a series of uniform ellipsoids. The same result clearly holds under the relaxed condition with $dX/dt$, $dY/dt$, $dZ/dt$ initially not zero but merely independent of $(x_0,y_0,z_0)$. Further, if, for definiteness,

$$a > b > c\,, \tag{12}$$

then by equations (7)

$$\alpha < \beta < \gamma\,, \tag{13}$$

and from equations (10)

$$-\frac{1}{X}\frac{d^2X}{dt^2} < -\frac{1}{Y}\frac{d^2Y}{dt^2} < -\frac{1}{Z}\frac{d^2Z}{dt^2} \tag{14}$$

*the ratios* a:b:c *steadily increase with time*, when the initial conditions $dX/dt = dY/dt = dZ/dt = 0$ are imposed. In particular, an oblate spheroid increases in eccentricity and approaches an infinitely thin disk (provided pressure can always be neglected); equally, a prolate spheroid becomes an infinitely thin cylinder.

### III. SOLUTION OF THE EQUATIONS

For computation of the spheroidal cases, we use cylindrical coordinates $(\varpi,\phi,z)$, which are related to initial values $(\varpi_0,\phi_0,z_0)$ by

$$\varpi = \varpi_0 R(t)\,, \qquad z = z_0 Z(t)\,, \qquad \phi = \phi_0\,. \tag{15}$$

The potential (6) can be written

$$V = \text{constant} - \frac{G\rho}{2}\,[A(e)\varpi^2 + C(e)z^2]\,, \tag{16}$$

where $A$ and $C$ are known functions of the instantaneous eccentricity $e$. The analogues of equation (10) are

$$\frac{d^2R}{a t^2} = -\frac{G\rho_0 A(e)}{RZ} \tag{17}$$

and

$$\frac{d^2Z}{dt^2} = -\frac{G\rho_0 C(e)}{R^2} \tag{18}$$

with

$$Z^2 = R^2\,\frac{(1+e^2)}{(1-e_0{}^2)} \tag{19a}$$

for an oblate spheroid, and

$$R^2 = Z^2\,\frac{(1-e^2)}{(1-e_0{}^2)} \tag{19b}$$

for a prolate spheroid. Equations (17) and (18) must be solved subject to the initial condition at $t = 0$:

$$R = Z = 1\,, \qquad dR/dt = dZ/dt = 0\,. \tag{20}$$

For the oblate case

$$A(e) = \frac{2\pi(1-e^2)}{e^3}\,[\sin^{-1} e - e(1-e^2)^{1/2}]\,, \tag{21}$$

$$C(e) = \frac{4\pi}{e^2}\left[1 - \frac{(1-e^2)^{1/2}}{e}\sin^{-1} e\right]; \tag{22}$$

for the prolate case

$$A(e) = \frac{2\pi}{e^2}\left[1 - \frac{1-e^2}{2e}\log\frac{(1+e)}{(1-e)}\right], \tag{23}$$

$$C(e) = \frac{4\pi(1-e^2)}{e^2}\left[-1 + \frac{1}{2e}\log\frac{(1+e)}{(1-e)}\right] \tag{24}$$

(McMillan 1958).

#### *a) Solution for the Case of Oblate Spheroids*

To fix our ideas, let us first consider the case of the oblate spheroid. In this case, the equations to be integrated are (17) and (18), with $A(e)$ and $C(e)$ defined by equations

(21) and (22), and $e$ defined by equation (19a). The initial conditions are given by (20).

In order to get analytical solutions, we adopt variables suggested by the spherical case, in which

$$R = Z\,, \qquad e = 0\,, \qquad A(e) = C(e) = 4\pi/3\,, \tag{25a}$$

and the solution is given by

$$R = \cos^2\theta\,, \qquad V = -(8\pi/3)^{1/2}\tan\theta\,, \qquad \tau = (8\pi/3)^{-1/2}(\theta + \tfrac{1}{2}\sin 2\theta) \tag{25b}$$

where

$$\tau \equiv (G\rho_0)^{1/2}\,t\,, \qquad V \equiv dR/d\tau\,. \tag{26}$$

Thus, for oblate spheroids, we use the same definitions (26) and introduce $\theta$ and $x$ by

$$R = \cos^2\theta\,, \qquad x = \tan\theta \tag{27}$$

as the more convenient variables. We also introduce a parameter $\alpha$ through the relation

$$V = \frac{dR}{d\tau} = -\,\alpha\tan\theta\,. \tag{28}$$

We further define $E$, $\psi$, and $Q$ by (cf. eq. [19a])

$$E = \frac{Z}{R} = \frac{(1-e^2)^{1/2}}{(1-e_0^2)^{1/2}} = \frac{\cos\psi}{\cos\psi_0}\,, \tag{29a}$$

$$Q = R^2\,\frac{dE}{d\tau}\,. \tag{29b}$$

The system of equations to be integrated then becomes

$$x\,\frac{d\alpha^2}{dx} + 2\alpha^2 = \frac{4A}{E}\,, \tag{30}$$

$$\frac{dE}{dx} = \frac{2Q}{\alpha}\,, \tag{31}$$

$$(1+x^2)\,\frac{dQ}{dx} = \frac{2(A-C)}{\alpha}\,, \tag{32}$$

where $E$ is given by equation (29a), and

$$A = 2\pi - \frac{C}{2}\,, \tag{33a}$$

$$C = 4\pi\,(\sin\psi)^{-3}\,(\sin\psi - \psi\cos\psi)\,. \tag{33b}$$

Note that in the spherical case ($A = C = 4\pi/3$), the variables to be sought in equations (30)–(32) are all constants:

$$Q = 0\,, \qquad E = 1\,, \qquad \text{and} \qquad \alpha^2 = \frac{2A}{E} = \frac{8\pi}{3}\,. \tag{34}$$

The solution then takes on the form (25) given above.

Corresponding to equation (20), equations (30)–(32) are to be integrated under the following initial conditions at $x = 0$:

$$a = a_0 = (2A_0)^{1/2}\,, \qquad E = E_0 = 1\,, \qquad Q = 0\,, \tag{35}$$

where $A_0$ and $E_0$ are the values of $A$ and $E$ evaluated at $\psi = \psi_0$ according to equations (29) and (33). The initial eccentricity $e_0$ remains a parameter of the problem.

We shall attempt to express the solutions in powers of $x$. If this does not carry the solution far enough in any particular case, the numerical method[1] will be used to give more accurate results. It actually turns out that two or three terms of the series solution give extremely accurate answers right up to the collapse of the oblate spheroids into a disk, if the spheroid has a fair amount of initial oblateness. Even when the initial flattening is as little as 10 per cent (minor axis:major axis = 0.90), the results are moderately accurate. In this case, the time of collapse obtained by the series solution is 0.536 (when the series is used up to three terms for $A$, $C$, and $E$) while numerical integration gives 0.538. The ratio "collapse radius":"initial radius" is 0.140 by series method and 0.134 by numerical integration. (For further details, see Table 1 and Fig. 1.)

TABLE 1*

NUMERICAL VALUES AT COLLAPSE FOR NATURAL TIME AND RADIUS FOR OBLATE CASE

| $z_0/\varpi_0$ | $e_0$ | $\tau_c$ (Series Solution) | $\tau_c$ (Numerical Integration) | $\varpi_c/\varpi_0$ (Series Solution) | $\varpi_c/\varpi_0$ (Numerical Integration) |
|---|---|---|---|---|---|
| 0.95...... | 0.312 | 0.539 | 0.541 | 0.0868 | 0.0744 |
| .90...... | .436 | .536 | .538 | .140 | .134 |
| .80...... | .600 | .527 | .529 | .252 | .251 |
| .70...... | .714 | .516 | .518 | .358 | .360 |
| .60...... | .800 | .504 | .505 | .461 | .464 |
| .50...... | .866 | .489 | .492 | .560 | .563 |
| .40...... | .917 | .474 | .477 | .655 | .658 |
| .30...... | .954 | .457 | .462 | .747 | .750 |
| .20...... | .980 | .439 | .447 | .835 | .838 |
| 0.10...... | 0.995 | 0.419 | 0.428 | 0.920 | 0.917 |

* Here $\tau_c$ is the dimensionless time (eq. [26]) at which collapse into a disk occurs. The parameter $e_0$ denotes the initial eccentricity of the spheroid and $Z_0/\varpi_0$ denotes the ratio of its minor axis to radius (major axis). The ratio $\varpi_c/\varpi_0$ is the disk radius as a fraction of that of the initial spheroid.

The power series for the various quantities are calculated as follows. First, we note that the initial conditions require that $a$, $A$, $C$, $E$ are all even in $x$, and that $Q$ is odd in $x$, when they are calculated from equations (30)–(32). Thus, we may write

$$a = a_0 + a_2x^2 + a_4x^4 + \ldots\,, \tag{36}$$

$$A = A_0 + A_2x^2 + A_4x^4 + \ldots\,, \tag{37}$$

$$C = C_0 + C_2x^2 + C_4x^4 + \ldots\,, \tag{38}$$

$$E = E_0 + E_2x^2 + E_4x^4 + \ldots\,, \qquad (E_0 = 1) \tag{39}$$

$$Q = Q_1x + Q_3x^3 + \ldots\,. \tag{40}$$

[1] Kutta-Runge procedure, as extended by Gill (1953).

Note that the variables $A$, $C$, and $E$ must be related to each other, since they are all defined in terms of $\psi$ (see eqs. [29] and [33]). By introducing these relations and by making use of equations (30)–(32), the coefficients can all be determined in terms of the initial eccentricity $e_0$. The details of these calculations are given in Appendix I. The more important coefficients are given explicitly in equations (A5a)–(A5e).

After the series (36) and (39) are calculated, we can easily integrate equation (28) for the time $\tau$ in terms of $x = \tan\theta$. We obtain

$$\alpha_0 \tau = (\theta + \tfrac{1}{2}\sin 2\theta) - \frac{\alpha_2}{\alpha_0}(\theta - \tfrac{1}{2}\sin 2\theta). \quad (41)$$

In this form, it is apparent that, when $\alpha_2/\alpha_0$ and $\theta$ are moderately small, the change in $\tau(\theta)$ is given essentially by the scale factor $\alpha_0$.

*Numerical results.*—The quantity $E = Z/R$ is expressed in terms of $\theta$ via equation (39), where $x = \tan\theta$ according to definition (27). The dimensionless time $\tau = (G\rho_0)^{+1/2}\, t$

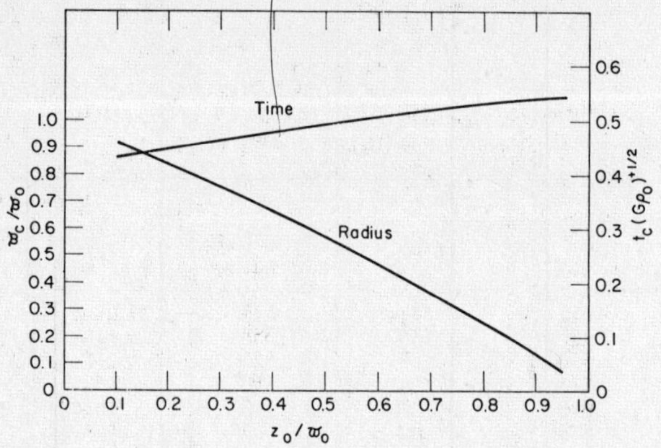


FIG. 1.—"Collapse" values for oblate spheroids as computed by numerical integration. The abscissa is the initial ratio of the minor axis to the major axis. The radius of the final disk as a fraction of the initial radius of the oblate spheroid is shown by the curve marked "radius," with vertical scale marked on the left. The "natural" time $\tau_c = t_c(G\rho_0)^{+1/2}$ at collapse into a disk is shown by the curve marked "time," with vertical scale marked on the right. Corresponding plots of the collapse values as given by series solution would be virtually indistinguishable from the ones given here.

is also expressed in terms of $\theta$ by expression (41). Thus, equations (27), (39), and (41) give the parametric representation of $Z(\tau)$ and $R(\tau)$. Only the terms that appear explicitly in these equations are needed for sufficient accuracy. All of the coefficients needed can be determined in terms of the initial eccentricity $e_0$ by means of equation (A5) in Appendix I.

Two quantities of general interest are (*a*) the time required for complete collapse of the spheroid into a disk, and (*b*) the radius of this disk as a fraction of the initial radius of the spheroid. These are given in Table 1.

Corresponding results for these "collapse values" were also obtained by numerical integration, and they are shown for comparison. These were obtained by the Kutta-Runge method with $z(M,t)/\varpi(M,t) = E(t)\cos\psi_0$ as the independent variable, where $\varpi(M,t)$ is the radius of the entire spheroid of mass $M$ at time $t$ and $z(M,t)$ is the semi-minor axis. It was found necessary to start the calculation by using the series solution for a small initial interval (e.g., from $E\cos\psi_0 = 0.95$ to $0.94$). The machine calculation was then carried out to $E\cos\psi_0 = 0.01$, and the "collapse values" themselves were obtained by extrapolation. It is seen that the agreement is close between results obtained by the two methods. These results are also shown in Figure 1.

The history of the collapse is shown in Figure 2. Almost identical results are obtained by series solution and by numerical integration.

### *b) Solution for the Case of Prolate Spheroids*

Although the following discussion for the prolate case will be found to be very similar to that in the previous section, it is thought desirable to give a complete treatment, precisely because a cross-reference would cause endless confusion. In this case, the equations to be integrated are equations (17) and (18), with $A(e)$ and $C(e)$ given by equations (23)

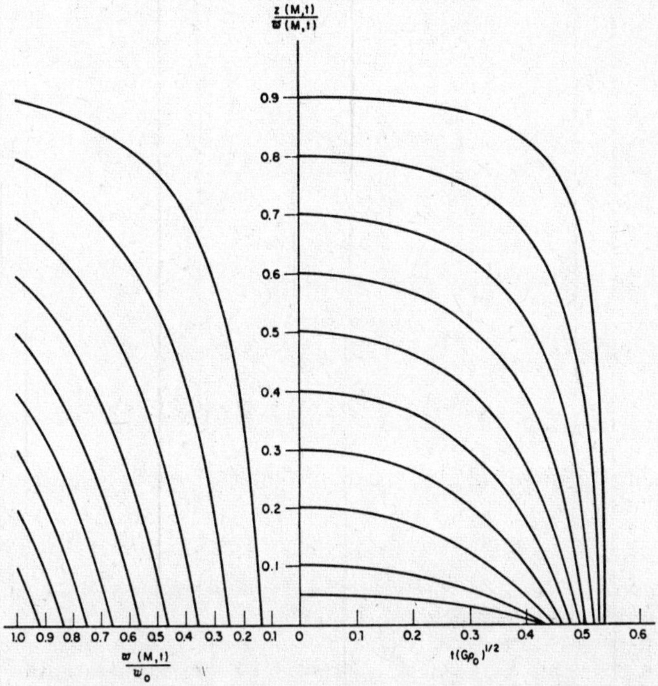


FIG. 2.—History of collapse of oblate spheroids as computed by numerical integration. The ordinate is the instantaneous ratio of the minor axis to the major axis. The abscissa running to the right is the "natural" time $\tau = t(G\rho_0)^{+1/2}$ and the abscissa running to the left is the instantaneous radius as a fraction of the initial radius of the oblate spheroids. Curves are shown for various values of initial oblateness (which may be read off the vertical axis for the initial conditions $t = 0$, $\varpi/\varpi_0 = 1$). Corresponding plots of the history of collapse as given by series solution would be virtually indistinguishable from the ones given here.

and (24), and $e$ defined by equation (19b). The initial conditions are given by equation (20).

In analogy with the oblate case, we define $\theta$ and $x$ by

$$Z = \cos^2\theta\,, \qquad x = \tan\theta \tag{42}$$

and define $W$ and $\alpha$ through

$$W = \frac{dZ}{d\tau} = -\alpha\tan\theta\,, \tag{43}$$

where $\tau \equiv (G\rho_0)^{+1/2} t$. We further define $E$ by

$$E \equiv \frac{R}{Z} = \frac{(1-e^2)^{1/2}}{(1-e_0^2)^{1/2}}. \tag{44}$$

Note that although $E$ in equation (44) stands in the same relation to $e$ as it does in equation (29), its relation to the semimajor axis and the radius is reversed.

By introducing the variable

$$Q \equiv Z^2 \frac{dE}{d\tau}, \tag{45}$$

the system of equations to be integrated becomes

$$x\frac{d\alpha^2}{dx} + 2\alpha^2 = \frac{4C}{E^2}, \tag{46}$$

$$\frac{dE}{dx} = \frac{2Q}{\alpha}, \tag{47}$$

$$(1+x^2)\frac{dQ}{dx} = \frac{2(C-A)}{\alpha E}, \tag{48}$$

where $E$ is given by expression (44) and

$$A = 2\pi - \frac{C}{2}, \tag{49a}$$

$$C = 4\pi\frac{1-e^2}{e^2}\left(-1 + \frac{1}{2e}\log\left|\frac{1+e}{1-e}\right|\right). \tag{49b}$$

Corresponding to equation (20), equations (46)–(48) are to be integrated under the initial conditions

$$\alpha = \alpha_0 = (2C_0)^{1/2}, \qquad E = E_0 = 1, \qquad Q = 0 \tag{50}$$

at $x = 0$, where $C_0$ and $E_0$ are the values of $C$ and $E$ evaluated at $e = e_0$. The initial eccentricity $e_0$ remains a parameter of the problem.

Upon comparison with the oblate case, we notice that all equations (42)–(48) can be obtained from their oblate case counterparts through the formal transformation

$$\text{Oblate} \rightarrow \text{Prolate}, \quad R \rightarrow Z, \quad Z \rightarrow R, \quad V \rightarrow W, \quad A \rightarrow C/E, \quad C \rightarrow A/E. \tag{51}$$

The last two transformations lead to the slight disparity between the formal results obtained for the prolate case and the oblate case. These differences arise from the fact that the original equations of motion (17) and (18) are themselves unsymmetrical in $R$ and $Z$.

A power series solution in $x$ may be developed with $\alpha$, $A$, $C$, and $E$ all even in $x$ and $Q$ odd in $x$ as required by initial condition (50). Thus, we have

$$\alpha = \alpha_0 + \alpha_2 x^2 + \alpha_4 x^4 + \ldots, \tag{52}$$

$$A = A_0 + A_2 x^2 + A_4 x^4 + \ldots, \tag{53}$$

$$C = C_0 + C_2 x^2 + C_4 x^4 + \ldots, \tag{54}$$

$$E = E_0 + E_2 x^2 + E_4 x^4 + \ldots, \tag{55}$$

$$Q = Q_1 x + Q_3 x^3 + \ldots. \tag{56}$$

The set of equations (44), (46), (47), (48), and (49) forms a complete set for the determination of the coefficients in equations (52)–(56) in terms of the initial eccentricity $e_0$. The details of this calculation are given in Appendix II.

The recovery of time is effected in the analogous way to the oblate case by integration of equation (43), and leads to the same formal relation (41). The coefficients $\alpha_0, \alpha_2, \ldots,$ of course, depend on the initial eccentricity $e_0$ in a manner different from that in the oblate case.

*Numerical results.*—It turns out that the series solutions for prolate spheroids are somewhat less accurate than their oblate counterparts. The convergence is less rapid because the ratios of successive coefficients in the expansions for $E$, $\alpha$, and $Q$ in powers of $x$ are larger in the prolate case for the same initial eccentricity. This difference is accentuated for larger initial eccentricities. An indication of this behavior is seen upon comparison of the values of the important ratios $E_4/E_2$ and $\alpha_2/\alpha_0$ for the prolate and oblate cases at initial eccentricity $e_0 = 1$ (cf. Tables A1 and A2).

TABLE 2*

NUMERICAL VALUES AT COLLAPSE FOR NATURAL TIME AND SEMIMAJOR AXIS FOR PROLATE CASE

| $\varpi_0/z_0$ | $e_0$ | $\tau_c$ (Series Solution) | $\tau_c$ (Numerical Integration) | $Z_c/Z_0$ (Series Solution) | $Z_c/Z_0$ (Numerical Integration) |
|---|---|---|---|---|---|
| 0.95...... | 0.312 | 0.527 | 0.542 | 0.200 | 0.0842 |
| .90...... | .436 | .523 | .541 | .263 | .150 |
| .80...... | .600 | .518 | .539 | .368 | .276 |
| .70...... | .714 | .514 | .536 | .464 | .397 |
| .60...... | .800 | .511 | .532 | .558 | .511 |
| .50...... | .866 | .508 | .529 | .649 | .620 |
| .40...... | .917 | .505 | .526 | .739 | .722 |
| .30...... | .954 | .502 | .524 | .824 | .816 |
| .20...... | .980 | .500 | .523 | .902 | .900 |
| 0.10...... | 0.995 | 0.498 | 0.521 | 0.966 | 0.966 |

* Notation here similar to that in Table 1.

Nevertheless, the series results for the collapse values do not suffer badly upon comparison with those obtained by numerical integration, provided the prolate spheroid is initially sufficiently elongated. Thus, when the initial elongation is 30 per cent (minor axis:major axis = 0.70), the time of collapse obtained by the series solution is 0.514 as compared to the value 0.536 by numerical integration; the ratio "collapse major axis": "initial major axis" is 0.464 by series solution and 0.397 by numerical integration. Other collapse values for different initial eccentricities are given in Table 2, and are shown graphically in Figure 3.

The numerical integration uses a four-step Runge-Kutta method (see Gill 1953) with $\varpi/z = (1 - e_0^2)^{1/2} E$ as the independent variable. The same procedure used before is followed in starting the integration process with the help of the series solution, and in terminating it before "collapse."

The history of the collapse obtained by series solution and by numerical integration is shown in Figure 4. The agreement between the two calculations in the initial stages justifies the use of the series solution itself to start the numerical integration.

## IV. DISCUSSION

It must be emphasized that the systematic increase in eccentricity is a purely gravitational effect: it is due not to a force that is essentially anisotropic (such as centrifugal force) but to the gravitational field of a non-isotropic distribution of sources. The effect

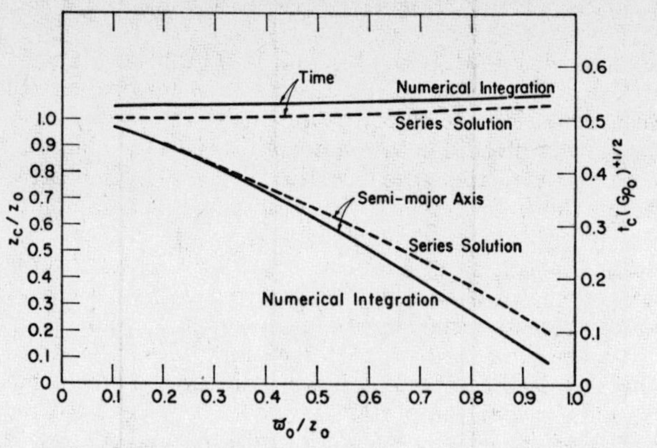


FIG. 3.—"Collapse" values for prolate spheroids. The abscissa is the initial ratio of the minor axis to the major axis. The major axis of the final spindle as a fraction of the initial major axis of the prolate spheroid is shown by the curve marked "semimajor axis," with vertical scale marked on the left. The "natural" time $\tau_c = t_c(G\rho_0)^{+1/2}$ at collapse into a disk is shown by the curve marked "time," with vertical scale marked on the right. Values as computed by numerical integration are shown in solid lines; those computed by series solution are shown in dotted lines.

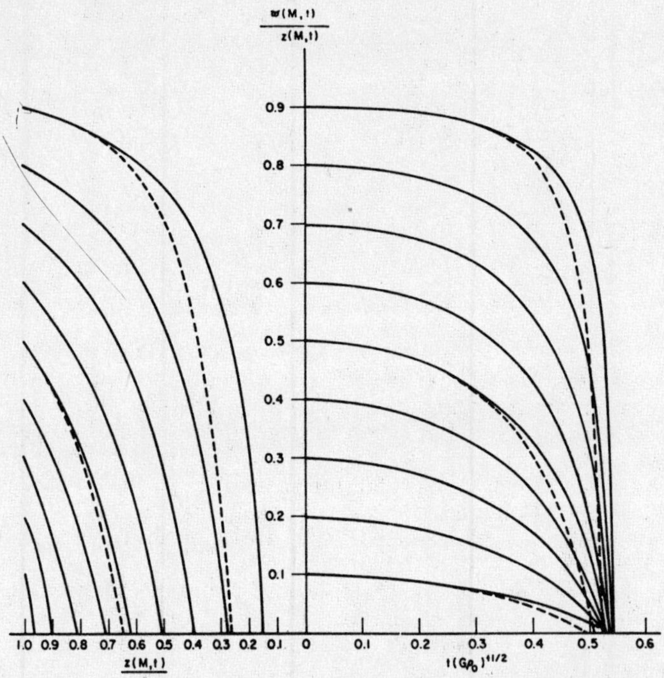


FIG. 4.—History of collapse of prolate spheroids. The ordinate is the instantaneous ratio of the minor axis to the major axis. The abscissa running to the right is the "natural" time $\tau$, and the abscissa running to the left is the instantaneous major axis as a fraction of the initial major axis of the prolate spheroids. Curves are shown for various values of initial prolateness (which may be read off the vertical axis for the initial conditions $t = 0$, $\varpi/\varpi_0 = 1$). Those in solid lines are computed by numerical integration. Three cases of varying initial prolateness ($\varpi_0/z_0 = 0.9$, 0.5, and 0.1) as computed by series solution are plotted in dotted lines and offered for comparison. The left curves for $\varpi_0/z_0 = 0.1$ as computed by series solution and numerical integration are virtually indistinguishable throughout the collapse process.

would not occur in a star, which by definition is a body maintained very close to hydrostatic equilibrium, with the isotropic thermal pressure the main force opposing gravity: a slight oblateness in a non-rotating star, for example, would be reversed by the excess pressure gradient along the axis of oblateness, and the star would execute non-radial oscillations.

The assumption of zero pressure gradients is only an approximation. Even in a uniform gas cloud, the density gradient is large at the edge, and there will be an isothermal pressure gradient that is locally large. A more realistic model would have a density that is nearly uniform over the bulk of the cloud, but decreases monotonically toward the edge. If the cloud is strictly spherical and contracts from an initial state with negligible pressure gradients, then since $d\bar{\rho}/dm < 0$, solution (3) shows that the outer shells lag behind the inner ones, and the isothermal pressure gradient becomes a still smaller fraction of the gravitational force density. However, if the cloud is initially oblate, then once it has become highly flattened, with a semithickness $\bar{z}$, the ratio of the $z$-pressure gradient to the $z$-component of gravity is $(c^2/\lambda)/4\pi G\rho\bar{z}$, where $\lambda$ is a density scale height, $c$ is the isothermal sound speed, and the value $4\pi$ has been taken from the limit of equation (22) as $e \rightarrow 1$. Thus, since $\rho\bar{z}$ stays constant by mass conservation, an isothermal pressure is able to halt indefinite flattening, and the spheroid will oscillate. When the kinetic energy of the oscillation has all been dissipated and radiated away, the cloud (still assumed unfragmented) will achieve hydrostatic $z$-equilibrium, with a semithickness $\bar{z}$ given by

$$\bar{z} \cong \left(\frac{c^2}{2\pi G\rho}\right)^{1/2} \tag{57}$$

(Spitzer 1942; Ledoux 1951). The $\varpi$-component of gravity will be effectively unopposed, and free fall will continue in two dimensions, with $\bar{z}$ adjusting itself so as to satisfy (57) approximately. If $\bar{\varpi}$ is the instantaneous equatorial radius, then the cloud mass $M \cong 4\pi\rho\bar{z}\bar{\varpi}^2/3$, and

$$\frac{\bar{z}}{\bar{\varpi}} \cong \frac{Mc^2}{3GM^2/2\bar{\varpi}} \propto \bar{\varpi}\,. \tag{58}$$

Thus the oblateness continues to increase, though at a rate determined by free fall in the $\varpi$-direction only.

If, however, the cloud is initially prolate, then once it has become quasi-cylindrical, the ratio of pressure to gravitational force becomes (by eq. [23]) $\cong (c^2/\lambda)/2\pi G\rho\bar{\varpi}$, which behaves like $\bar{\varpi}/\lambda$, since in cylindrical motion $\rho \propto \bar{\varpi}^{-2}$. Thus an initially weak isothermal pressure gradient stays weak during isothermal cylindrical collapse (McCrea 1957): indefinite collapse will be halted only by departure from adiabatic process or by nonthermal forces such as magnetic force (Chandrasekhar and Fermi 1953; Mestel 1965; Strittmatter 1965).

## APPENDICES

### I. DETERMINATION OF THE COEFFICIENTS IN EQUATIONS (36)–(40)

The most important quantities to be calculated are $E$ and $\alpha$. When equations (36)–(40) are substituted into equations (30)–(32), we obtain (noting that $E_0 = 1$)

$$E_2 = 1 - C_0/A_0\,, \tag{A1}$$

$$E_4 = (6A_0^2)^{-1}[A_2C_0 - A_0C_2 - C_0(A_0 - C_0)]\,, \ldots, \tag{A2}$$

and

$$\alpha_0 = (2A_0)^{1/2}\,, \tag{A3}$$

$$\frac{\alpha_2}{\alpha_0} = (4A_0)^{-1}(A_2E_0 - A_0E_2)\,. \tag{A4}$$

The above relations are insufficient for determining all the coefficients until the relationships among the coefficients for $A$, $C$, and $E$ are introduced. This step is carried out below. The results for some of the more important coefficients are as follows:

$$E_0 = 1\,, \qquad E_2 = 3 - 4\pi/a_0\,, \tag{A5a}$$

$$E_4/E_2 = \tfrac{1}{3}(\chi - \sin\chi)^{-2}[\chi^2 + \chi\sin\chi - 4(1 - \cos\chi)]\,, \tag{A5b}$$

$$\alpha_0 = (2a_0)^{1/2}, \qquad \frac{\alpha_2}{\alpha_0} = \frac{E_2{}^2}{4}\frac{1+\cos\chi}{1-\cos\chi}, \tag{A5c}$$

TABLE A1

NUMERICAL VALUES FOR CERTAIN IMPORTANT COEFFICIENTS IN OBLATE CASE

| $\psi_0$ | $\alpha_0$ | $\alpha_2/\alpha_0$ | $E_2$ | $E_1/E_2$ |
|---|---|---|---|---|
| 0 | 2.895 | 0 | 0 | 0.0333 |
| $\pi/4$ | 2.678 | 0.0635 | −0.504 | .039 |
| $\pi/3$ | 2.437 | .126 | −1.230 | .043 |
| $5\pi/12$ | 1.955 | .245 | −3.58 | .052 |
| $\pi/2$ | 0 | 0.405 | $-\infty$ | 0.0632 |

TABLE A1—*Continued*

| $Z_0/\varpi_0$ | $E_2$ | $E_2/E_2$ | $\alpha_0$ | $\alpha_2/\alpha_0$ | $Q_1$ | $Q_3/Q_1$ |
|---|---|---|---|---|---|---|
| 1.00 | 0 | 0.0333 | 2.895 | 0 | 0 | 0.0667 |
| 0.95 | − 0.0682 | .0375 | 2.86 | 0.00986 | − 0.195 | .0849 |
| 0.90 | − 0.134 | .0351 | 2.83 | .0191 | − 0.379 | .0894 |
| 0.80 | − 0.303 | .0371 | 2.76 | .0407 | − 0.835 | .115 |
| 0.70 | − 0.522 | .0392 | 2.67 | .0653 | − 1.39 | .144 |
| 0.60 | − 0.816 | .0416 | 2.57 | .0935 | − 2.09 | .177 |
| 0.50 | − 1.23 | .0442 | 2.44 | .126 | − 3.00 | .215 |
| 0.40 | − 1.86 | .0472 | 2.27 | .164 | − 4.22 | .258 |
| 0.30 | − 2.91 | .0505 | 2.06 | .209 | − 5.99 | .310 |
| 0.20 | − 5.02 | .0542 | 1.77 | .262 | − 8.88 | .370 |
| 0.10 | −11.4 | .0583 | 1.32 | .326 | −15.0 | .443 |
| 0.05 | −24.1 | .0607 | 0.963 | .364 | −23.2 | .485 |
| 0.00 | $-\infty$ | 0.0632 | 0 | 0.405 | $-\infty$ | 0.531 |

where

$$a_0 = 2\pi\sin\chi(\chi - \sin\chi)/(1 - \cos\chi)^2\,, \tag{A5d}$$

and

$$\chi = 2\psi_0 = 2\sin^{-1} e_0\,. \tag{A5e}$$

Numerical values for these coefficients are given in Table A1.

The final formulae used for the numerical calculations are as follows:

$$E = 1 + E_2(\tan\theta)^2 + E_4(\tan\theta)^4\,, \tag{A6a}$$

$$\alpha_0\tau = (\theta + \tfrac{1}{2}\sin 2\theta) - (\alpha_2/\alpha_0)(\theta - \tfrac{1}{2}\sin 2\theta)\,, \tag{A6b}$$

$$R = \cos^2\theta\,, \tag{A6c}$$

and

$$Z = ER\,. \tag{A6d}$$

*Expansion in Powers of $\psi - \psi_0$*

To relate $A$, $C$, and $E$, we expand them in powers of

$$\delta = \psi - \psi_0\,. \tag{A7}$$

We have

$$A = a_0 + a_1\delta + a_2\delta^2 + \ldots, \tag{A8a}$$

$$C = c_0 + c_1\delta + c_2\delta^2 + \ldots, \tag{A8b}$$

$$E = \epsilon_0 + \epsilon_1\delta + \epsilon_2\delta^2 + \ldots; \tag{A8c}$$

and (since $\delta$ is even in $x$ and vanishes at $x = 0$)

$$\delta = \delta_2 x^2 + \delta_4 x^4 + \ldots. \tag{A9}$$

The coefficients in the above expansions for $A$, $C$, and $E$ are obviously functions of $\psi_0 = \sin^{-1} e_0$ only. The coefficients $\delta_2, \delta_4, \ldots$, have to be determined from the differential equations, or alternatively, from the formulae (A1)–(A4).

Clearly, by comparing equations (37), (38), (39), and (A8), we have

$$A_0 = a_0\,, \qquad C_0 = c_0\,, \qquad E_0 = \epsilon_0 = 1\,. \tag{A10}$$

By substituting equation (A9) into equations (A8), we get, by comparison with equations (37)–(39),

$$a_1\delta_2 = A_2\,, \qquad a_1\delta_4 + a_2\delta_2^2 = A_4\,, \ldots, \tag{A11a}$$

$$c_1\delta_2 = C_2\,, \qquad c_1\delta_4 + c_2\delta^2\delta_2^2 = C_4\,, \ldots, \tag{A11b}$$

$$\epsilon_1\delta_2 = E_2\,, \qquad \epsilon_1\delta_4 + \epsilon_2\delta_2^2 = E_4\,, \ldots. \tag{A11c}$$

The successive coefficients in equations (A1)–(A4) can then be calculated explicitly by using them in alternate steps with equations (A11). Thus, $E_2$ is determined from equation (13), which in turn defines $\delta_2$ from equation (A11c); $A_2$, $C_2$ are then obtained from equations (A11a) and (A11b), enabling $E_4$ to be calculated. The results are as follows:

$$E_2 = 1 - c_0/a_0 \tag{A12a}$$

$$E_4 = (6a_0^2\epsilon_1)^{-1}\,(a_1c_0 - a_0c_1 - a_0b_0\epsilon_1)\,, \ldots; \tag{A12b}$$

$$\alpha_0 = (2a_0)^{1/2} \tag{A13a}$$

$$\alpha_2/\alpha_0 = (4a_0^2\epsilon_1)^{-1}\,(a_1 - a_0\epsilon_1)(a_0 - c_0) \ldots. \tag{A13b}$$

[Note that since $A + C/2 = 2\pi$, we can also write $E_4$ as

$$E_4/E_2 = (6a_0^2\epsilon_1)^{-1}\,(1 - 2\pi c_1 - a_0c_0\epsilon_1)\,. \tag{A12b'}$$]

The coefficients $a_0, a_1, \ldots, c_0, c_1, \ldots, \epsilon_0, \epsilon_1, \ldots$, are as follows:

$$a_0 = 2\pi - \frac{c_0}{2}, \qquad c_0 = 4\pi(\sin\psi_0)^{-3}(\sin\psi_0 - \psi_0\cos\psi_0), \tag{A14a}$$

$$a_1 = -\frac{c_1}{2}, \qquad c_1 = 4\pi(\sin\psi_0)^{-3}\left[\frac{\psi_0}{\sin\psi_0}(\sin^2\psi_0 + 3\cos^2\psi_0) - 3\cos\psi_0\right] \tag{A14b}$$

$$a_2 = -\frac{c_2}{2},$$

$$c_2 = 4\pi(\sin\psi_0)^{-3}[6\cot^2\psi_0(\sin\psi_0 - \psi_0\cos\psi_0) + 2\sin\psi_0 - 4\psi_0\cos\psi_0], \quad \ldots, \tag{A14c}$$

$$a_n = -\frac{c_n}{2}, \tag{A14d}$$

$$\epsilon_0 = 1, \qquad \epsilon_1 = -\tan\psi_0, \qquad \epsilon_2 = -\tfrac{1}{2}, \ldots. \tag{A14e}$$

By substituting these explicit expressions into equations (A12) and (A13), we get finally equations (A5).

## II. DETERMINATION OF THE COEFFICIENTS IN EQUATIONS (52)–(56)

We may obtain the coefficients in power series expansion in $x$ of $\alpha$, $E$, and $Q$ in terms of those of $A$ and $C$ either by direct substitution of equations (52)–(56) into the differential equations (46)–(48), or by comparison with the results of the oblate case with the help of the substitution rules given by equation (51). Both methods yield the same results, of course, and the corresponding relations to equations (A1)–(A4) are

$$E_0 = 1, \tag{A15a}$$

$$E_2 = 1 - A_0/C_0, \tag{A15b}$$

$$E_4 = (6C_0)^{-1}[A_0C_2 - C_0A_2 - A_0(C_0 - A_0)] \ldots; \tag{A15c}$$

$$\alpha_0 = (2C_0)^{1/2}, \tag{A16a}$$

$$\alpha_2/\alpha_0 = (4C)^{-1}(C_2 - 2C_0E_2) \ldots; \tag{A16b}$$

$$Q_1 = \alpha_0 E_2, \tag{A17a}$$

$$Q_3 = 2\alpha_0 E_4 + \alpha_2 E_2, \ldots. \tag{A17b}$$

The relations between the coefficients $A$, $C$, and $E$ as required by equations (44), (49a), and (49b) can be introduced to give the functional dependence of the above coefficients on the initial eccentricity $e_0$. This step is carried out later. The results for some of the more important coefficients are

$$E_0 = 1, \qquad E_2 = \tfrac{1}{2}\left(3 - \frac{4\pi}{c_0}\right); \tag{A18a}$$

$$\frac{E_4}{E_2} = (6c_0^2)^{-1}\left(2\pi\frac{c_1}{\epsilon_1} - \frac{a_0c_0}{\epsilon_1}\right); \tag{A18b}$$

$$\alpha_0 = (2c_0)^{1/2}, \qquad \frac{\alpha_2}{\alpha_0} = (4c_0^2)^{-1}(c_0 - a_0)\left(\frac{c_1}{\epsilon_1} - 2c_0\right); \tag{A18c}$$

$$Q_1 = \alpha_0 E_2, \qquad \frac{Q_3}{Q_1} = \frac{2c_4}{c_2} + \frac{\alpha_2}{\alpha_0}; \tag{A18d}$$

where

$$a_0 = 2\pi - \frac{c_0}{2}, \qquad c_0 = 4\pi \frac{1-e_0^2}{e_0^2}\left(-1 + \frac{1}{2e_0}\log\frac{1+e_0}{1-e_0}\right), \tag{A18e}$$

$$c_1 = \frac{4\pi}{\epsilon_0^2}\left[3 - (3-e_0^2)\frac{1}{2e_0}\log\frac{1+e_0}{1-e_0}\right], \qquad \epsilon_1 = -\frac{e_0}{1-e_0^2}. \tag{A18f}$$

TABLE A2

NUMERICAL VALUES FOR CERTAIN IMPORTANT COEFFICIENTS IN PROLATE CASE

| $\varpi_0/z_0$ | $E_2$ | $E_4/E_2$ | $a_0$ | $a_2/a_0$ | $Q_1$ | $Q_3/Q_1$ |
|---|---|---|---|---|---|---|
| 1.00 | 0 | $+\infty$ | 2.895 | 0 | 0 | $+\infty$ |
| 0.95 | − 0.0637 | 0.726 | 2.83 | 0.0188 | − 0.181 | 1.47 |
| 0.90 | − 0.136 | 0.582 | 2.77 | 0.0392 | − 0.376 | 1.20 |
| 0.80 | − 0.312 | 0.502 | 2.63 | 0.0863 | − 0.821 | 1.09 |
| 0.70 | − 0.548 | 0.500 | 2.48 | 0.144 | − 1.36 | 1.14 |
| 0.60 | − 0.881 | 0.541 | 2.30 | 0.218 | − 2.02 | 1.30 |
| 0.50 | − 1.38 | 0.633 | 2.09 | 0.318 | − 2.88 | 1.58 |
| 0.40 | − 2.20 | 0.810 | 1.84 | 0.461 | − 4.05 | 2.08 |
| 0.30 | − 3.74 | 1.18 | 1.55 | 0.693 | − 5.79 | 3.04 |
| 0.20 | − 7.46 | 2.12 | 1.18 | 1.16 | − 8.83 | 5.39 |
| 0.10 | −23.1 | 6.34 | 0.714 | 2.71 | −16.5 | 15.4 |
| 0.05 | −72.6 | 20.2 | 0.412 | 6.60 | −29.9 | 47.1 |
| 0.00 | − ∞ | $+\infty$ | 0 | $+\infty$ | $-\infty$ | $+\infty$ |

Numerical values for the coefficients $E_2$, $E_4/E_2$, $a_0$, $a_2/a_0$, $Q_1$, and $Q_3/Q_1$ are given in Table A2.

The final formulae used for numerical computations are

$$E = 1 + E_2(\tan\theta)^2 + E_4(\tan\theta)^4\,, \tag{A19a}$$

$$Z = \cos^2\theta\,, \tag{A19b}$$

$$R = EZ\,, \tag{A19c}$$

$$a_0\tau = \theta + \tfrac{1}{2}\sin 2\theta - \frac{a_2}{a_0}(\theta - \tfrac{1}{2}\sin 2\theta). \tag{A19d}$$

*Expansion in* $(e - e_0)$

To relate $A$, $C$, and $E$, we expand them in powers of

$$\delta = e - e_0\,. \tag{A20}$$

We have

$$A = a_0 + a_1\delta + a_2\delta^2 + \ldots\,, \tag{A21a}$$

$$C = c_0 + c_1\delta + c_2\delta^2 + \ldots\,, \tag{A21b}$$

$$E = \epsilon_0 + \epsilon_1\delta + \epsilon_2\delta^2 + \ldots\,. \tag{A21c}$$

It is clear that the coefficients above are functions only of $e_0$ and can be obtained from the expressions (44), (49a), and (49b). Indeed some of these are given explicitly by equations (A18e) and (A18f). Since $A$, $C$, and $E$ are given in $x$, it is clear that $\delta$ must be even in $x$ and vanishes at $x = 0$; thus

$$\delta = \delta_2 x^2 + \delta_4 x^4 + \ldots\,, \tag{A22}$$

Upon substitution of equation (A22) into equation (A21) and comparison with equations (53)–(55), we get

$$A_0 = a_0\,, \qquad A_2 = a_1\delta_2\,, \qquad A_4 = a_1\delta_4 + a_2\delta_2^2\,, \ldots\,, \tag{A23a}$$

$$C_0 = c_0\,, \qquad C_2 = c_1\delta_2\,, \qquad C_4 = c_1\delta_4 + c_2\delta_2^2\,, \ldots\,, \tag{A23b}$$

$$E_0 = \epsilon_0 = 1\,, \qquad E_2 = \epsilon_1\delta_2\,, \qquad E_2 = \epsilon_1\delta_4 + \epsilon_2\delta_2^2\,, \ldots\,. \tag{A23c}$$

The coefficients in equations (A15)–(A17) can be obtained explicitly in terms of $a_0, a_1, \ldots$; $b_0, b_1, \ldots$; $c_0, c_1, \ldots$, which are known functions of the initial eccentricity, by using equations (A23) in alternate steps with equations (A15)–(A17). Thus, the first equations of (A23a) and (A23b) coupled with equation (A15b) determine $E_2$. The second relation of (A23c) serves to define $\delta_2$ once $E_2$ is determined, from which $A_2$ and $C_2$ are obtained from the second equations of (A23a) and (A23b). This enables the calculation of $E_4$ from equation (A15c). Partial results are summarized in equations (A18).

One of us (L. M.) wishes to thank Professor C. C. Lin for arranging a visit to the Department of Mathematics, Massachusetts Institute of Technology as research associate during July and August, 1964. The cosmogonical importance of the present problem first arose during discussions with Professor E. E. Salpeter during his visit to Cambridge, England, in 1961. The authors are grateful to the Computation Center at Massachusetts Institute of Technology for the use of its facilities.

This work is supported in part by the National Aeronautics and Space Administration through the Center for Space Research at Massachusetts Institute of Technology.

## REFERENCES

Chandrasekhar, S., and Fermi, E. 1953, *Ap. J.*, **118**, 113.
Gill, S. 1953, *Proc. Cambridge Phil. Soc.*, **47**, 96.
Hoyle, F. 1953, *Ap. J.*, **118**, 513.
Hunter, C. 1962, *Ap. J.*, **136**, 594.
———. 1964, *ibid.*, **139**, 570.
Layzer, D. 1963, *Ap. J.*, **137**, 351.
Ledoux, P. 1951, *Ann. d'ap.*, **14**, 438.
Lynden-Bell, D. 1962, *Proc. Cambridge Phil. Soc.*, **58**, 709.
———. 1964, *Ap. J.*, **139**, 1195.
Lyttleton, R. A. 1953, *The Stability of Rotating Liquid Masses* (Cambridge: Cambridge University Press).
McCrea, W. H. 1957, *M.N.*, **117**, 562.
McMillan, W. D. 1958, *The Theory of the Potential* (New York: Dover Publications).
Mestel, L. 1965, *Q.J.R.A.S.* (in press).
Spitzer, L., Jr. 1942, *Ap. J.*, **95**, 329.
Strittmatter, P. A. 1965, *M.N.* (in preparation).

Reprinted from the PROCEEDINGS OF THE NATIONAL ACADEMY OF SCIENCES
Vol. 55, No. 2, pp. 229–234. February, 1966.

# ON THE SPIRAL STRUCTURE OF DISK GALAXIES, II. OUTLINE OF A THEORY OF DENSITY WAVES*

BY C. C. LIN† AND FRANK H. SHU

INSTITUTE FOR ADVANCED STUDY, PRINCETON, NEW JERSEY,
AND HARVARD COLLEGE OBSERVATORY

*Communicated December 3, 1965*

1. *Introduction.*—In a previous paper[1] (hereafter referred to as Paper I or the elementary theory), we discussed a theory in which the *spiral patterns* observed in disklike galaxies are described in terms of density waves. Specifically, we stated our ideas in the form of a hypothesis of the quasi-stationary spiral structure of the stars (hereafter referred to as the QSSS hypothesis).[2] As already emphasized in that paper, the primary problem we wish to resolve is the persistence of the spiral *pattern* or the grand design over the whole disk, on the scale of 10 kpc. In such a problem, hydromagnetic forces are estimated to be negligible, to a first approximation. On the other hand, when we consider the structure of an individual spiral arm, on the scale of 1 kpc, the effect of the magnetic field must be considered.

The purpose of this note is to outline a gravitational theory that provides a full dynamical basis of the QSSS hypothesis. In particular, we wish to determine, *quantitatively*, the relative importance of the gas and the stars in providing the spiral gravitational field. Although the stars would obviously provide the predominant part of the symmetrical gravitational field, their considerable dispersive motion tends to smooth out any spiral distribution that might be present. This reduction of the effectiveness of the stars in forming a spiral structure has now been calculated on the basis of the principles of stellar dynamics (cf. §4 and Fig. 1; see also Remark 3, p. 652, Paper I). In view of this smoothing effect, we shall find that the *spiral* gravitational field of the gas is *not* negligible, although the stars still play the more important role.

2. *Self-Sustained Density Waves.*—We consider a small deviation from a symmetrical disk of stars and gas, and attempt to demonstrate that density waves of a spiral structure will be self-sustained at a small but finite amplitude. Suppose that such a wave were maintained; then there must be an associated gravitational field of a generally spiral form. We shall start with this field and carry out the analysis as indicated in the following diagram:

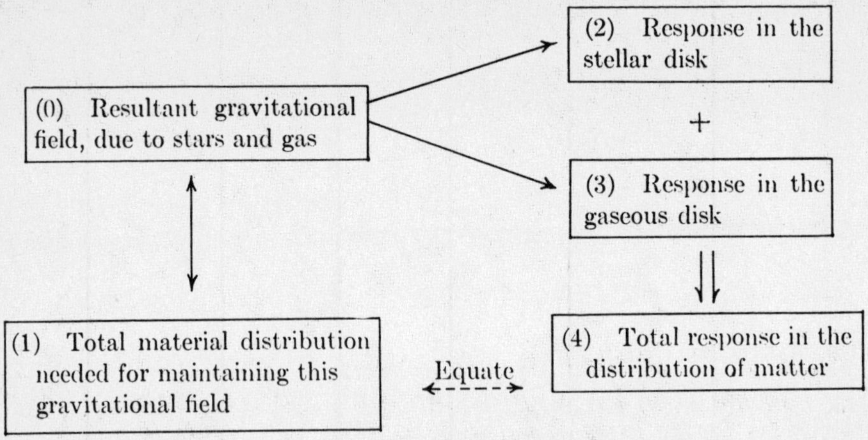


Under the influence of the resultant spiral-like gravitational field (0) [which will eventually be considered as due to a total material distribution (1) consisting partly of gas and partly of stars], the laws of stellar dynamics would dictate a certain redistribution of the stars (2), and the laws of gas dynamics would dictate a certain redistribution of the gas (3). The sum of these two distributions yields a total distribution of matter (4) that must be equal to the density distribution (1) which we needed to give rise to the gravitational field (0) according to Poisson's equation. This last condition is the equation to be solved for the unspecified functions and parameters that occur in the problem.

The analysis has so far been carried out only for the linear theory. Although the nonlinear effects may be expected to tend to prefer trailing patterns, we shall see that, already in the linear theory, trailing spiral waves will be found to be amplified. It may then be expected that this wave will grow to a small but finite amplitude and then attain an equilibrium through the nonlinear effects (see §6 below).

3. *The Asymptotic Theory.*—To carry out the analysis, we adopt the natural cylindrical coordinate system $(\varpi, \theta, z)$ such that the galactic disk is in the plane $z = 0$, with its center at the origin. In the linear theory, the resultant gravitational potential may be assumed to be given by a superposition of spiral modes, and the response to these individual modes may be treated separately. Let the potential of each of these modes be given by the real part of

$$\mathcal{V}_1(\varpi, \theta, t) = A(\varpi) \exp\left\{i[\omega t - m\theta + \Phi(\varpi)]\right\}, \tag{3.1}$$

where $A(\varpi)$ and $\Phi(\varpi)$ are real functions of the axial distance, $\omega$ is a complex parameter, and $m$ is a *positive* integer (or zero in the case of rings). The function $A(\varpi)$ is slowly varying with $\varpi$, whereas the function $\Phi(\varpi)$ is of the form $\epsilon^{-1}f(\varpi)$, where $f(\varpi)$ is slowly varying (and monotonic), and $\epsilon$ is a small parameter of the order of the angle of inclination of the spiral arms. Indeed, the function (3.1) clearly has a spiral structure described by the family of curves $m(\theta - \theta_0) = \Phi(\varpi) - \Phi(\varpi_0)$. The angle of inclination $i$ of any one of these curves is given by

$$(k\varpi)^{-1} = (1/m) \tan i,$$

where $k(\varpi) = \Phi'(\varpi)$. We note that $|k(\varpi)|$ is essentially the wave number in the radial direction. Thus, a natural approach is to adopt an asymptotic solution based on a rapidly varying phase angle, with the aid of the small parameter $\epsilon$ mentioned above. For example, if the basic symmetrical state is described by the distribution function $\Psi = \Psi_0(\varpi, c_{\tilde{\omega}}, c_\theta)$, where $c_{\tilde{\omega}}$ and $c_\theta$ are the peculiar velocity components of the stars relative to a state of circular motion $\Theta_c = \varpi\Omega(\varpi)$, then the distribution function in the disturbed state is given by the real part of

$$\Psi = \Psi_0(1 + \psi) \tag{3.2}$$

with

$$\psi = \varphi(\varpi, c_{\tilde{\omega}}, c_\theta) \exp\left\{i[\omega t - m\theta + \Phi(\varpi)]\right\}, \tag{3.3}$$

and

$$\varphi = \sum_{n=0}^{\infty} \epsilon^n \varphi^{(n)}(\varpi, c\ , c_\theta). \tag{3.4}$$

4. *Neutral Waves.*—The method used by the present writers in the elementary theory (Paper I) is essentially a special case of the scheme outlined above. Only gas and stars with zero dispersion were considered. It is expected that the high-dispersion stars would not participate in the spiral pattern in full, and that therefore the "effective" mass density of the stars must be smaller than its actual value by a suitable factor. Indeed, by following the scheme outlined in section 3 to the initial approximation, we can work out this reduction factor explicitly. If we assume that, in the basic symmetrical state, the dispersion velocities follow the Schwarzschild law of distribution (with a magnitude of dispersion sufficiently large to prevent local gravitational collapse), *neutral* density waves of the spiral form can be found. The radial wave number of such waves is given by

$$|k(\varpi)| = \frac{\kappa^2 - (\omega - m\Omega)^2}{2\pi G[\sigma_0 + \sigma_* \mathfrak{F}_\nu(x)]}, \tag{4.1}$$

or the range of values of $\varpi$ such that

$$\kappa^2 - (\omega - m\Omega)^2 > 0. \tag{4.2}$$

In the above formulae, $\kappa(\varpi)$ is the epicyclic frequency corresponding to the circular velocity $\varpi\Omega(\varpi)$, $\sigma_0(\varpi)$ is the basic density distribution of the gas, $\sigma_*(\varpi)$ is the basic density distribution of the stars, and $\mathfrak{F}_\nu(x)$ is the *reduction factor* defined as follows:

$$\mathfrak{F}_\nu(x) = \frac{1 - \nu^2}{x}\left\{1 - \frac{\nu\pi}{\sin \nu\pi} \cdot \frac{1}{2\pi} \int_{-\pi}^{\pi} e^{-x(1 + \cos s)} \cos \nu s\, ds\right\}, \tag{4.3}$$

where $\nu = (\omega - m\Omega)/\kappa$, $x = k^2 \langle c_{\tilde{\omega}}^2\rangle/\kappa^2$, and $\langle c_\omega^2\rangle$ is the mean square value of the radial component of the dispersion velocity of the stars in the basic symmetrical state.

Equation (4.1) may be rewritten into the pair of relations

$$\frac{\sigma_p}{\sigma_*} = \frac{|k|}{k_*}(1 - \nu^2), \tag{4.4a}$$

$$\frac{\sigma_p}{\sigma_*} = \frac{\sigma_0}{\sigma_*} + \mathfrak{F}_\nu(x), \tag{4.4b}$$

where $k_* = \kappa^2/2\pi G\sigma_*$. We may describe $\sigma_p/\sigma_*$ as the total "mobile component" (measured in terms of the stellar mass) which determines, according to (4.4*a*), the wave number $|k|$ for a given frequency $\nu$. It consists, according to (4.4*b*), of a gaseous component and a stellar component.

Equation (4.1) is one of the central relationships in the present theory. By setting $x = 0$, it can be readily compared with equation (12) in Paper I. In general, $\mathfrak{F}_\nu(x)$ is real for real values of $\nu$ and $x$ and its values are plotted in Figure 1 against $\nu^2$ for a series of values of $x$.

5. *Comments on the Neutral Waves.*—(*a*) We note that equation (4.1) holds only for a range of values of $\varpi$ such that (4.2) is satisfied. We shall refer to this range as the *principal part of the spiral pattern.* If we now use the Schmidt model of 1956 (for which the sun is located at 8.2 kpc) and if we adopt the value 20 km/sec-kpc as the angular velocity of the pattern $\Omega_p = \omega/m$ $(m = 2)$, we find that the principal part of the spiral pattern extends over the range 2 kpc $< \varpi <$ 12 kpc. The inner radius is somewhat uncertain; the outer radius is well beyond the solar neighborhood. For other integral values of $m$, one finds a very limited principal part.

(*b*) We note that equation (4.1) does not give any indication of the sign of $k(\varpi)$. Thus, up to the present approximation, there is no way to differentiate between leading waves $(k > 0)$ and trailing waves $(k < 0)$. This is indeed expected from general considerations of symmetry. Trailing waves will be found to be preferred in the next approximation, which includes terms of the order of

$$\epsilon = (1/m) \tan i.$$

(*c*) By considering the special case $\nu = 0$ in equation (4.1), one obtains the marginal condition for neutral oscillations at a given location. For a given value of $\sigma_0/\sigma_*$, there is a relationship between the wavelength $2\pi/|k|$ and the velocity dispersion $\langle c_{\tilde{\omega}}{}^2\rangle$. However, realistic results can be obtained only if we apply a similar reduction factor to the gas. The result of such a calculation is shown in Figure 2, where we have taken the mean square turbulent motion to be 1/10 of $\langle c_{\tilde{\omega}}{}^2\rangle$ for the stars.[3] Conditions above the marginal curve correspond to neutral oscillations, those below to instability.

The marginal curve for zero gas content has been obtained earlier by Toomre[4] and Kalnajs.[5] Toomre also discussed the effect of the presence of the gas, but he did not examine the effect of its turbulent motion.

(*d*) With the help of Figure 1, one may easily determine the values of the parameters corresponding to any neutral wave. Consider $\sigma_*$, $\sigma_0$, $\kappa^2$, $\langle c_{\tilde{\omega}}{}^2\rangle$ (and hence $k_*$) as known. Then for a given wavelength, i.e., a chosen value of $|k|/k_*$, there is an associated value of $x$. If we now plot a straight line to represent (4.4*a*) for this value of $|k|/k_*$ (shown dashed in Fig. 1 for $k_* = 0.2|k|$) and associate it with the corresponding curve for $\mathfrak{F}_\nu(x)$ for $x = 7$ (see ref. 6), the value of $\nu$ at which the two curves separate by $\sigma_0/\sigma_*$ is the proper value, and the ordinates on the two curves give, respectively, the total mobile component and the part due to the stellar disk. For example, the neutral wave with $\nu = 0.7$ requires a total mobile component of 11 per cent of the stellar mass, with 9 per cent from the stars and 2 per cent from the gas.

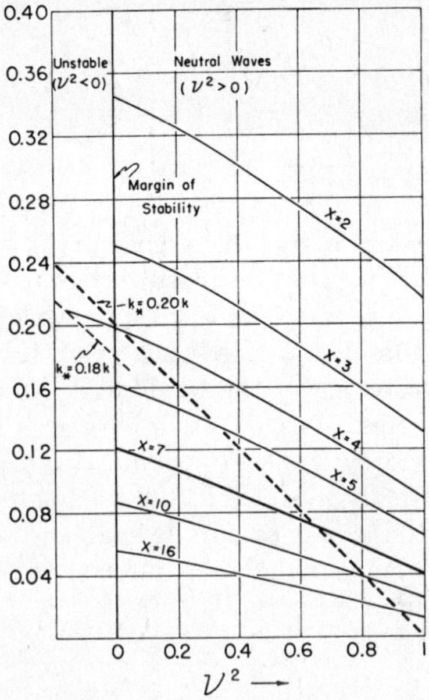


FIG. 1.—Diagram showing the reduction factor $\mathfrak{F}_\nu(x)$ and the relationship (4.4).

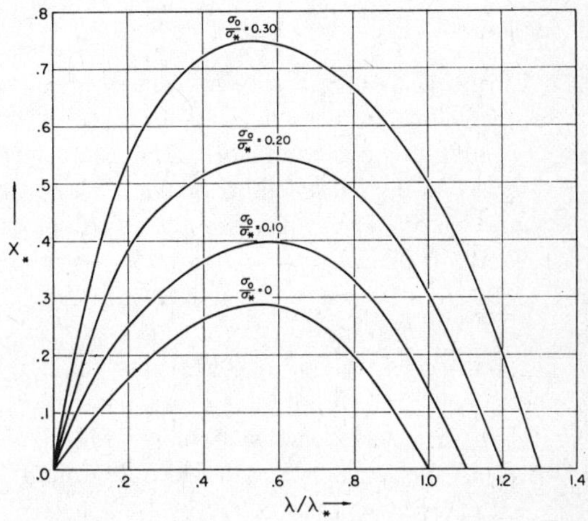


FIG. 2.—Margin of stability when there is gas in turbulent motion. ($\lambda = 2\pi/|k|$, $\lambda_* = 2\pi/k_*$, $x_* = k_*^2 \langle c_{\bar{\omega}}^2\rangle/\kappa^2$.)

In particular, the conditions of marginal stability are given by the real axis $\nu^2 = 0$. Unstable conditions correspond to $\nu^2 < 0$.

6. *Amplification of Trailing Waves.*—All the above results are stated for the initial approximation in the series (3.4). In the next approximation, in which we include terms of the order of the angle of inclination of the spiral arms, we shall find indications of *instability*. This small amplification rate is then expected to be balanced by the nonlinear effects to produce neutral waves at amplitudes of the order of $\sqrt{\epsilon}$, where $\epsilon$ is of the order of $(1/m) \tan i$.

An indication of instability may already be found in the fore-aft asymmetrical deviation from the Schwarzschild distribution in the axially symmetrical basic state, even if the spiral disturbance (3.3) is calculated only to the initial approximation. But the more important effect is found only if we carry out the calculations to the order of $\epsilon$ in equation (3.4). *Amplification of trailing waves* is then found for a range of values of the radial distance for which

$$-1 < m(\Omega_p - \Omega)/\kappa < 0. \tag{6.1}$$

For the Schmidt model, with $\Omega = 20$ km/sec-kpc, this yields the range 2 kpc $< \varpi <$ 10 kpc, which extends well beyond the solar neighborhood. The less massive part still farther out from the galactic center is expected to be driven into a trailing pattern by the more massive interior part. These self-excited trailing spiral waves may also be expected to produce a small barlike singularity near the center.

The amplification index $-\omega_i$ is given by

$$\frac{\omega_i}{\kappa} \operatorname{sign}(k) = \frac{\langle c_{\bar{\omega}}^2 \rangle}{\kappa \varpi} \cdot f(\zeta) \cdot D \tag{6.2}$$

to a first approximation for large values of $x$, where $\zeta = m(\Omega_p - \Omega)/\kappa$,

$$D = -d \log \langle c_{\varpi}^2 \rangle / d \log \varpi + {}^1/_2,$$

and

$$f(\zeta) = \sqrt{\frac{2}{\pi}} \frac{\sin^2 \zeta}{\sin \zeta \cos \zeta - \zeta} > 0 \qquad (\text{for } -\pi < \zeta < 0). \tag{6.3}$$

7. *Concluding Remarks.*—Observations of the phenomena over a galactic disk show transient features as well as features which appear to be quasi-permanent. It is not difficult to account for transient features with general spiral-like appearance in a system with differential rotation. The central problem is indeed to account for the more *permanent* features such as the spiral pattern over the whole disk. In this note, we outlined such a theory in terms of quasi-stationary self-sustained density waves. Such a gravitational theory has some obvious consequences. For example, pure circular motion is usually assumed in the reduction of data in 21-cm radio observations for the purpose of locating the spiral arm. The present theory would suggest the need to re-examine such data reduction since there are expected some systematic radial and circular motions of the gas distributed in the form of a spiral-like flow field.

One of us (C. C. L.) would like to thank Professors R. Oppenheimer and B. Strömgren for their hospitality during his current stay at the Institute for Advanced Study, and to thank the Director of the Harvard College Observatory and the Chairman of the Astronomy Department for the use of its facilities. The authors wish also to thank several of their friends for various helpful discussions during the course of this work. Especially to be noted are discussions with A. Toomre and A. J. Kalnajs, who were conducting related research work at the same time, and discussions with Professor B. Strömgren, who gave continuous encouragement to the point of view of quasi-stationary density waves.

* This work was largely done at the Massachusetts Institute of Technology with the financial support of the National Science Foundation and the National Aeronautics and Space Administration.

† On leave of absence from Massachussetts Institute of Technology, Cambridge, Mass.

[1] Lin, C. C., and F. H. Shu, *Astrophys. J.*, **140**, 646–655 (1964).

[2] B. Lindblad [*Stockholm Obs. Ann.*, **22**, 3–20 (1963)] arrived at the same hypothesis from somewhat different considerations, with emphasis on individual orbits (theory of dispersion orbits). The present theory is based on statistical considerations, but is doubtless related to his theory. For example, the relations (4.1) and (4.2) are obviously related to Lindblad resonance which occurs at locations where $\kappa^2 - (\omega - m\Omega)^2 = 0$. There are also significant differences. Thus, in the absence of the gas, the condition for resonance is obtained from (4.1) for a *finite* extent in the galactic disk, as suggested by B. Lindblad, only if we set the reduction factor $\mathfrak{F}_\nu(x)$ equal to zero, i.e., if we neglect the spiral gravitational field due to the "cooperative" effect of the stars. See Paper I and later communications for further comments on the theories of B. Lindblad and P. O. Lindblad.

[3] The reduction factor applied to the gas may be taken either from the stellar dynamical theory or the gas dynamical theory. The results are substantially the same. Curves of similar nature are obtained when the mean square turbulent velocity is changed roughly in proportion to the amount of gas present.

[4] Toomre, A., *Astrophys. J.*, **139**, 1217 (1964).

[5] Kalnajs, A. J., Ph.D. thesis, Department of Astronomy, Harvard University (1965).

[6] The value $x = 7$ chosen is based on marginal stability in a pure stellar sheet for the most unstable waves in the sense of Toomre. If we make use of Fig. 2 to allow for the effect of the gas, a slightly higher value of $x$ (perhaps $x = 8$) should be chosen. This does not modify the conclusions to any significant extent.

J. SIAM Appl. Math.
Vol. 14, No. 4, July, 1966

# ON THE MATHEMATICAL THEORY OF A GALAXY OF STARS*

C. C. LIN†

**1. Introduction.** Recent progress in radio astronomy has enabled us to make more accurate observations on the physical processes involved in the structure of our galaxy and of other galaxies of stars. It is therefore more fruitful now to make further theoretical studies of hydromagnetics and stellar dynamics, and of their roles in the theory of galactic structure. It is the purpose of this paper to bring these problems to the attention of the mathematicians at this time, and to report on our recent efforts in these studies.

I take great pleasure in having this opportunity to present this paper at this particular symposium in honor of Professor Richard Courant, because he has been, for many years, the nucleus of a galaxy of stars in mathematics, who might be interested in the behavior of stars of nature. Indeed, the subject matter is closely related to the work of several of the stars of this Institute. I have in mind in particular the work in plasma physics by Harold Grad, Cathleen Morawetz, and their associates, and the work on dynamical systems by Jürgen Moser and his associates. In fact, Moser is the leading expert in the mathematical aspects of one of the famous problems in this subject: the problem of the third integral[1]. Unfortunately, I shall have no occasion to refer to that problem here, for it will be seen that *two* integrals are sufficient for the problem at hand.

Before I go on with the description of the mathematical theory, I think it would be desirable to make a few further remarks on the general nature of the subject matter and on the purpose of my paper. In view of the publicity that astronomy is currently receiving in the newspapers and news magazines, especially on the esoteric and speculative aspects, there is a natural tendency for a person with only a limited knowledge of astronomy to feel uneasy. He would feel that all the problems in astrophysics and cosmology involve speculations on the distant past or on the validity of fundamental physical laws. Let me assure you from the outset that I shall be dealing with a relatively well-defined problem in astrophysics—not in cosmology, which is a subject of a different type—and that the problem is

* Received by the editors July 6, 1965. Presented by invitation at the Courant Symposium, held at the Courant Institute of Mathematical Sciences, New York, June 8–9, 1965.

† Department of Mathematics, Massachusetts Institute of Technology, Cambridge, Massachusetts. This research was partially supported by the National Science Foundation and the National Aeronautics and Space Administration. The author wishes to thank Mr. Frank H. Shu for his help in the preparation of the paper.

[1] A proof of the existence of this integral in some general sense was given by George Contopoulos, who was in the audience.

one which can be treated on the basis of well-known laws of physics. Furthermore, it concerns a process which is going on at the present, and the theoretical deductions can be checked by observations.

The purpose of this paper is then to pose the problems that arise from observations on the structure of the galaxy, and to formulate them into a suitable mathematical framework, in order that they may be better appreciated by (applied) mathematicians with experience in other fields. (It is of course my hope that they will be tempted to join our efforts in trying to understand certain aspects of the mysteries of the universe.) I shall also report, toward the end of this paper, some of the results we have obtained so far in order that one may see the relationship between theory and observations. The detailed steps leading to these conclusions will however be published elsewhere.

The main theme of this paper will be the suggestion that the principal mechanisms of galactic structure can be treated by the principles of stellar dynamics. In 1942 S. Chandrasekhar [1] published a monograph on this subject. In the same year he also published, in collaboration with John von Neumann, an extensive joint paper [3]. On the problem of spiral structure, however, he emphasized, in a joint paper with Enrico Fermi [2], the role of hydromagnetic forces for the maintenance of spiral *arms*. This is certainly an important process, but it appears that the problem of spiral *patterns* can be resolved by studying the dynamical processes on the basis of Newtonian mechanics.

I shall now briefly describe the problem of the spiral structure which presents, at first sight, some paradoxical features.

The Earth we live on is one of the nine major planets revolving around the sun. The sun is one of the two hundred billion stars revolving around the nucleus of our galaxy—the Milky Way. In Fig. 1, we show a picture of the whirlpool galaxy, which is a close relative of our own. There is also a smaller companion whose shape is rather irregular. The main galaxy has a rather striking spiral structure over a disklike distribution. The side view of another galaxy (Fig. 2) shows how thin the disk is. Clearly, the disk shape is associated with a large angular momentum in the system. Otherwise, the gravitational forces would have pulled the galaxy together.

The classification of regular galaxies into various types is shown schematically in Fig. 3. We shall not go into the detailed discussions of the characteristics and the relative abundance of galaxies of the various types. Suffice it to say that a large majority of them (70%) are normal spirals like the whirlpool. A few per cent are irregulars, like its companion (not represented in Fig. 3). One of the main problems in galactic structure is: *why do most of them have the regular spiral structure?* A second question is: *how can these structures persist?*

The urgency of the second question can be appreciated if one realizes

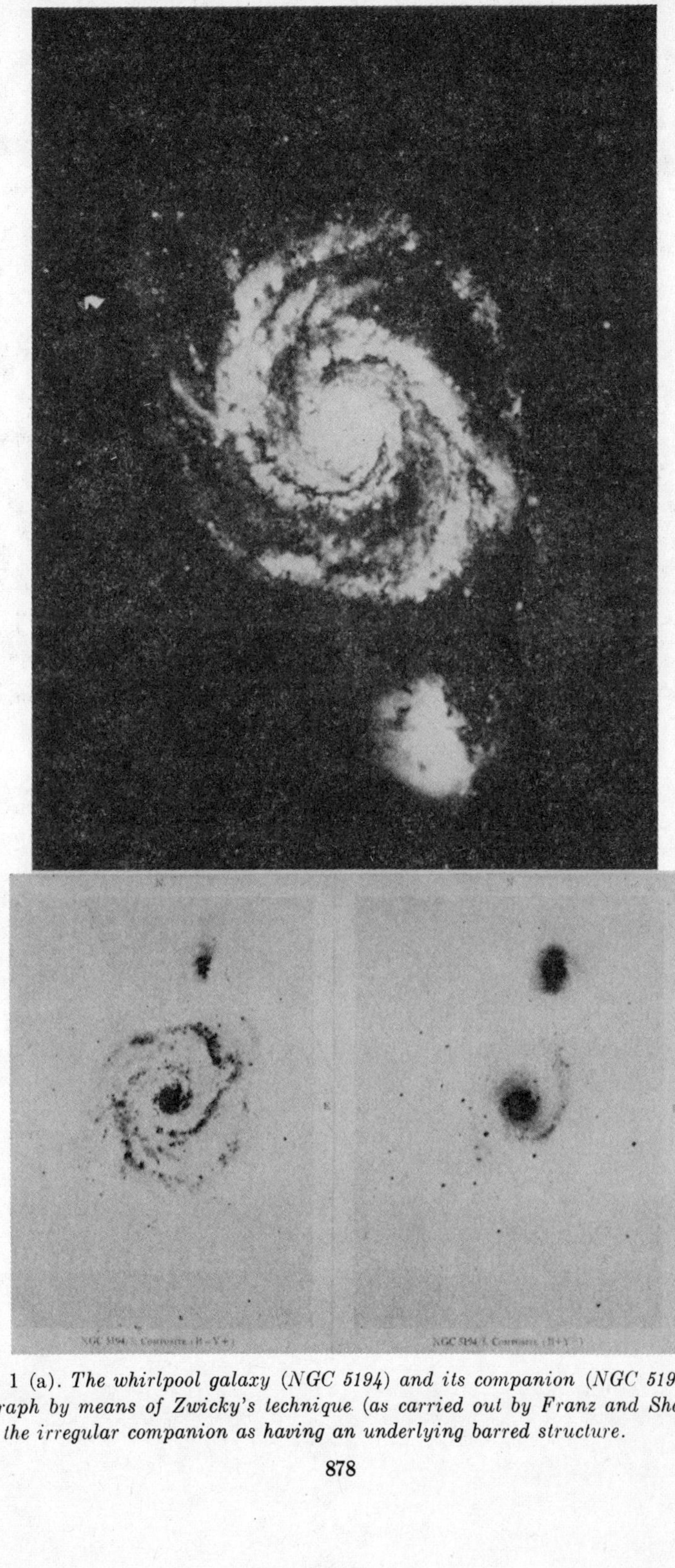

FIG. 1 (a). *The whirlpool galaxy (NGC 5194) and its companion (NGC 5195).* (b). *Photograph by means of Zwicky's technique (as carried out by Franz and Sharpless) reveals the irregular companion as having an underlying barred structure.*

FIG. 2. *The side view of a galaxy (NGC 4565)*

that there is strong *differential rotation*; i.e., the galaxy is not rotating as a solid body, but the inner part rotates much faster than the outer part. Indeed, in our galaxy the linear velocity is nearly the same for a substantial range of the radial distance. The problem is very well stated by Oort [13]:

"In systems with strong differential rotation, such as is found in all non-barred spirals,[2] spiral features are quite natural. Every structural irregularity is likely to be drawn out into a part of a spiral. But *this* is not the phenomenon we must consider. We must consider a spiral structure extending over the whole galaxy, from the nucleus to its outermost part, and consisting of two arms starting from diametrically opposite points. Although this structure is often hopelessly irregular and broken up, the general form of the large-scale phenomenon can be recognized in many nebulae."

In other words, the primary problem is to explain the grand design over the whole disk on the scale of 10 kpc. There is also a secondary problem of

[2] The problem of barred spirals has been studied by Kevin Prendergast [14]. Many of the features can be understood; but the mechanism for maintaining the bar cannot yet be described in detail.

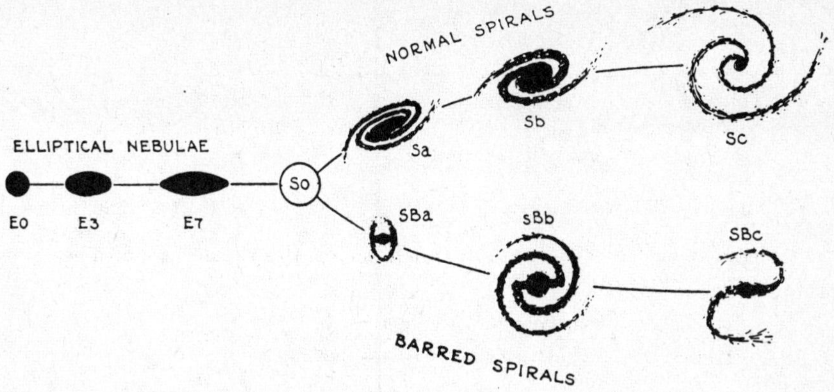


FIG. 3. *Hubble's classification of galaxies*

describing the structure of an individual *arm*, which might involve somewhat different mechanisms.

In looking for mechanisms for creation and maintenance of regular spiral patterns, it should be borne in mind that one could also use essentially the same mechanism to account for spiral patterns which are less perfect and comparatively transitory. For the superposition of several patterns rotating relative to each other would have the desired effect.

In this paper, I shall attempt to develop a stellar dynamical theory for the *persistence of the spiral pattern in the presence of differential rotation.* But first of all, let us establish the physical basis of the theory by examining the physical processes in the galaxy, and formulating the equations governing them.

## 2. A brief outline of the physical basis.[3]

**2a. Physical processes in the galaxy.** The physical processes in a galaxy of stars are multivarious. First of all, there are the nuclear reactions within the stars themselves. The energy coming out of the stars (especially the bright, young, massive stars of the O and B types) will ionize the surrounding hydrogen gas, forming the Strömgren spheres (also called H II regions). But an individual star is only one of 200 billion in our galaxy and the nuclear processes from them can stimulate but cannot control the large-scale dynamical processes that lead to the relatively regular spiral pattern over the whole disk in most of the flat galaxies. Indeed, most of the stars are like the sun, which is relatively old and dim. It is well-known that the magnificent manifestations of the spiral patterns are associated with the interstellar gas and the relatively few brilliant young stars born

[3] For a more detailed account of the physical facts, see Appendix 1.

TABLE 1

*Energy densities in the galaxy* [12]

| | |
|---|---|
| Total radiation (star light) | $0.7 \times 10^{-12}$ erg cm.$^{-3}$ |
| Turbulent gas motion | $0.5 \times 10^{-12}$ erg cm.$^{-3}$† |
| Total energy of galactic rotation | $1300 \times 10^{-12}$ erg cm.$^{-3}$ |
| Cosmic rays | $1 \times 10^{-12}$ erg cm.$^{-3}$ |
| Magnetic field ($10^{-5}$ gauss) | $4 \times 10^{-12}$ erg cm.$^{-3}$ |

† Remark by present writer: For certain purposes, this should be multiplied by a factor equal to

$$2 \times \frac{\text{linear velocity of galactic rotation}}{\text{turbulent velocity of gas motion}} \cong 50\text{—}70$$

in them. But could the old stars also play an important role in the formation of the spiral structure?

To deal with the large-scale structure of the galaxy, one must bear in mind the following important components:

(i) *the stars*—with their gravitational forces, circular velocity and peculiar velocity;

(ii) *the interstellar gas*—with its gravitational field and pressure;

(iii) *the magnetic field*—which exerts its influence through the highly conducting interstellar gas.

Two other components of the galaxy are the radiation from the stars and the cosmic ray particles.

The relative importance of the various components can be seen from Table 1. Clearly, the kinetic energy of galactic rotation far outweighs any other component. Furthermore, this table is based on a mean gas density of about $2 \times 10^{-24}$g./c.c. (1.2 H atoms/c.c.). The total mass density of gas and stars is about 5 times larger in the solar neighborhood, and about 50 times larger in the interior of the galaxy. It is thus clear that even an extremely small deviation of the stars from circular motion would involve a level of energy comparable to the rest. A one-per-cent deviation would be associated with substantially more energy density than, e.g., the total magnetic energy density.

While we wish to emphasize this predominance of stellar mass in our galaxy, the situation may not be as favorable in other galaxies, such as those in the Sc type which has considerably more gas. We shall therefore formulate the basic equations to include all the three components named above. The effect of star light and cosmic rays will not be explicitly considered, even though they are acknowledged as the source for ionizing the gas.

**2b. Dynamical equations.** Consider first the motion of the stars. Each star moves under the joint gravitational field of the other stars and the gas. Clearly, we wish to formulate a statistical treatment of the 200 billion stars regarding each one of them as a molecule, since we know that stars are very far apart from each other. Indeed, the star next to the sun ($\alpha$-Centauri) is about 4 light years away, or about $2.5 \times 10^5$ times the earth's orbit (1 A.U.). Thus, even when the "orbital electrons" (the planets) of the "stellar atom" (solar system) are included in defining its size, we still have a highly rarefied gas. The frequency of collisions is even smaller than these linear dimensions would suggest, for the relative velocity among the stars is rather small, amounting only to the order of 30 km./sec. If we would consider the "effective collision size" of the sun as given by the diameter of the orbit of Pluto (80 A.U.), the mean free time between collisions works out to be of the order of $10^{13}$ years. A more careful analysis as given by Chandrasekhar [1] leads to the value of $10^{14}$ years. Since we are only interested in processes lasting over a period of the order of $10^{10}$ years, the "gas of stellar atoms" may be regarded as "collision-free." We may thus describe the motion of the stars by the Boltzmann equation without the "collision term," but with the collective gravitational field as the source of interaction between the stars themselves and between the stars and the gas. A more refined treatment should of course include the "fluctuations" of the gravitational field due to the nearby stars (cf. [3]), but this is not necessary for present purposes.

*Equations of stellar dynamics.* Thus, if we consider an idealized model with stars all having the same mass $m_*$ and if we introduce a distribution function $\Psi_3$ in the phase space $(x_i, v_i)$, $i = 1, 2, 3$, the equation governing the development of $\Psi_3$ in time $t$ is as follows:

$$\frac{\partial \Psi_3}{\partial t} + v_i \frac{\partial \Psi_3}{\partial x_i} + a_i \frac{\partial \Psi_3}{\partial v_i} = 0, \tag{2.1}$$

where $v_i$ is the velocity of the individual stars, and $a_i$ is acceleration due to the combined gravitational field of the gas and the stars. The latter is thus derivable from a gravitational potential $\mathcal{V}(x_i, t)$:

$$a_i = -\frac{\partial \mathcal{V}}{\partial x_i}, \tag{2.2}$$

and the gravitational potential is in turn determined by the gas and the stars in the galaxy (and perhaps the effect of neighboring galaxies):

$$\frac{\partial^2 \mathcal{V}}{\partial x_k^2} = 4\pi G \rho_t, \qquad \rho_t = \rho + \rho_*. \tag{2.3}$$

In this equation, $G$ is the gravitational constant, $\rho$ is the mass density of the gas, and $\rho_*(x_i, t)$ is the mass density of the stars given by

$$\rho_*(x_i, t) = m_* \int \Psi_3(x_i, v_i, t)\, d\tau(v_i), \tag{2.4}$$

where the integral extends over the infinite velocity space. The mass density of the gas is one of the variables in the hydromagnetic equations.

*Equations of hydromagnetics.* Because of the large linear dimensions involved, the effect of viscosity and resistivity are expected to be small, for the proper measure of diffusive effects is always the dimensionless parameter

$$\mathfrak{D} = Dt_0/L^2, \tag{2.5}$$

where $D$ is the diffusion coefficient, $t_0$ is a typical time scale, and $L$ is a typical linear dimension. If we take $L = 10^3$ l.y. $= 10^{21}$ cm., $t_0 = 10^{10}$ yrs. $= 3 \times 10^{17}$ sec., then

$$\mathfrak{D} = D \times 3 \times 10^{-25}, \qquad D \text{ in cm.}^2/\text{sec.} \tag{2.6}$$

The resistivity of fully ionized hydrogen with one atom per cubic centimeter is (within a factor of 1.4)

$$\eta = 10^{14} \times T^{-3/2} \tag{2.7}$$

for temperature $T$ in the range $10^2$–$10^4$ °K [19, (5.37), p. 84, and Table 5.1, p. 73]. Thus, even if the gas is only ionized to a small extent, the parameter $\mathfrak{D}$ is still very small, which justifies the approximation of infinite conductivity. Similarly, viscous forces may be neglected. The transfer of thermal energy by radiation and thermal conductivity is a more complicated problem which we shall not go into in the present discussion.

With this provision, the hydromagnetic equations are as follows.[4] For the dynamics of the gas, we have respectively the following equations of state, mass conservation, motion, and energy:

$$p = \rho RT, \tag{2.8}$$

$$\frac{\partial \rho}{\partial t} + \frac{\partial}{\partial x_k}(\rho u_k) = 0, \tag{2.9}$$

$$\rho \frac{Du_i}{Dt} = -\frac{\partial p}{\partial x_i} + \rho a_i + \frac{1}{4\pi} \frac{\partial}{\partial x_j}\left\{B_i B_j - \frac{1}{2} B^2 \delta_{ij}\right\}, \tag{2.10}$$

$$\frac{Dp}{Dt} = \gamma \frac{p}{\rho} \frac{D\rho}{Dt} + \epsilon(\gamma - 1). \tag{2.11}$$

The symbols in the above equations are as follows: the thermodynamic state of the gas is specified by the pressure $p$, the density $\rho$, and the temperature $T$; these are related through the parameter $R$, the gas constant

[4] Gaussian (c.g.s.) units are used throughout this paper.

per unit mass; the motion of the gas is described by the Eulerian velocity field $u_i(\mathbf{x}, t)$, and its substantial derivative $Du_i/Dt$, which is the acceleration. The force terms on the right-hand side of (2.10) contain not only the pressure force, but also (i) the acceleration due to the combined gravitational field of the gas and the stars and (ii) the force due to the Maxwell stresses of the magnetic field $B_i$. These last two items are the agents through which the gas interacts with the other two important components of the galaxy: the stars and the magnetic field. The last equation for energy contains the unspecified term $\epsilon$ which represents the rate of energy transfer into a unit volume per second by heat conduction, radiative transfer, etc.

The displacement current is always neglected in a hydromagnetic problem. In the case of infinite conductivity, it is well-known that the Maxwell equations lead to the following two equations for the magnetic field:

$$\frac{\partial B_k}{\partial x_k} = 0, \tag{2.12}$$

$$\frac{D}{Dt}\left(\frac{B_i}{\rho}\right) = \left(\frac{B_k}{\rho}\right)\frac{\partial u_i}{\partial x_k}. \tag{2.13}$$

The second equation above guarantees the conservation of magnetic flux associated with a given piece of fluid surface. (The magnetic field is "frozen" into the fluid).

In view of the numerical values indicated in Table 1, we need not consider, to a first approximation, the effect of the gas and the magnetic field when we deal with the stellar-dynamical problem, unless the change in the density in the gas is to a much greater extent than the change in the density of the stars. Such a condition could happen in our neighborhood, but it is unlikely to happen in the more massive interior part of our galaxy. The motion of the gas, on the other hand, would be heavily dependent on the gravitational field of the stars. In Sc galaxies, the consideration of stellar dynamics should in general be made together with the hydromagnetic problem. But even there it might be reasonable to adopt an iteration process starting with the consideration of the stellar dynamical problem by itself.

Since the disk galaxies are extremely thin, it is possible to simplify our analysis by adopting a limiting model of zero thickness. To derive the proper stellar dynamical equations in that case, we integrate (2.1) with respect to the coordinate $x_3$ and the velocity $v_3$ in the direction normal to the disk. This is carried out in Appendix 2. Clearly, the central bulge is not well-approximated in this process. Thus, *singularities in the solution should be allowed at the center (and along the axis) in the theory based on the thin disk model*, for the solution obtained is only asymptotically valid when the axial distance is much larger than the radius of the nuclear region.

**3. Basic equations for an infinitesimally thin disk of stars.** Consider an infinitesimally thin disk of stars in differential rotation. This might be coexisting with a sheet of gas and a magnetic field. Let us consider only motions in the plane of the disk. We shall write down the *fundamental equation of stellar dynamics with reference to an arbitrarily chosen differentially rotating system* (see Appendix 2, (16)). It reads:

$$\frac{\partial \Psi}{\partial t} + c_\varpi \frac{\partial \Psi}{\partial \varpi} + \left(\Omega + \frac{c_\theta}{\varpi}\right)\frac{\partial \Psi}{\partial \theta} + \left(a_\varpi + \Omega^2\varpi + 2\Omega c_\theta + \frac{1}{\varpi} c_\theta^2\right)\frac{\partial \Psi}{\partial c_\varpi} + \left(a_\theta - \frac{\kappa^2}{2\Omega} c_\varpi - \frac{1}{\varpi} c_\varpi c_\theta\right)\frac{\partial \Psi}{\partial c_\theta} = 0, \tag{3.1}$$

where the symbols are defined as follows. The function $\Psi(\varpi, \theta, c_\varpi, c_\theta, t)$ is the distribution function in two-dimensions, such that

$$dN = \Psi\varpi \, d\varpi \, d\theta \, dc_\varpi \, dc_\theta \tag{3.2}$$

is the number of stars (each of mass $m_*$) lying in the element of volume

$$d\tau_c = \varpi \, d\varpi \, d\theta \, dc_\varpi \, dc_\theta \tag{3.3}$$

of the phase space with the spatial cylindrical coordinate system $(\varpi, \theta)$, and the *peculiar velocity* components $(c_\varpi, c_\theta)$. The "peculiar velocity" is defined with respect to an *arbitrarily chosen* circular motion with angular velocity $\Omega(\varpi)$; there is no radial velocity in the reference system. Thus, the components of the total velocity of the stars are given by

$$\Pi = c_\varpi, \qquad \Theta = \varpi\Omega(\varpi) + c_\theta . \tag{3.4}$$

The mean values of $c_\varpi$ and $c_\theta$ are not necessarily equal to zero, even in an axially symmetrical state of motion.

The *surface density* of the stellar system is given by

$$\sigma_*(\varpi, \theta\, t) = m_* \iint \Psi \, dc_\varpi \, dc_\theta , \tag{3.5}$$

whereas the mean values $(V_\varpi + v_\varpi, V_\theta + v_\theta)$ of the total velocity components $(\Pi, \Theta)$ are given by

$$\frac{\sigma_*}{m_*}(V_\varpi + v_\varpi) = \iint \Pi\Psi \, dc_\varpi \, dc_\theta , \tag{3.6a}$$

$$\frac{\sigma_*}{m_*}(V_\theta + v_\theta) = \iint \Theta\Psi \, dc_\varpi \, dc_\theta , \tag{3.6b}$$

where $V_\varpi = 0$, $V_\theta = \varpi\Omega(\varpi)$.

The components of acceleration $(a_\varpi, a_\theta)$ (in the plane of the disk $z = 0$) might be only partly attributable to the surface density distribution $\sigma_*$, but we shall usually regard them as derivable from a gravitational potential $\mathcal{V}(\varpi, \theta, z, t)$ in the plane $z = 0$. The part due to $\sigma_*$ is the self-gravitation of the stars and will be designated[5] with a superscript $s$. Thus, the self-field $(a_\varpi^{(s)}, a_\theta^{(s)})$ is given by

$$a_\varpi^{(s)} = -\frac{\partial}{\partial\varpi}\mathcal{V}^{(s)}(\varpi, \theta, 0, t) \tag{3.7a}$$

and

$$a_\theta^{(s)} = -\frac{1}{\varpi}\frac{\partial}{\partial\theta}\mathcal{V}^{(s)}(\varpi, \theta, 0, t) \tag{3.7b}$$

in terms of the self-gravitational potential $\mathcal{V}^{(s)}(\varpi, \theta, z, t)$, which satisfies the three-dimensional Poisson equation

$$\frac{\partial^2\mathcal{V}^{(s)}}{\partial\varpi^2} + \frac{1}{\varpi}\frac{\partial\mathcal{V}^{(s)}}{\partial\varpi} + \frac{1}{\varpi^2}\frac{\partial^2\mathcal{V}^{(s)}}{\partial\theta^2} + \frac{\partial^2\mathcal{V}^{(s)}}{\partial z^2} = 4\pi G\sigma_*\delta(z). \tag{3.8}$$

In the last equation, $\delta(z)$ is the Dirac delta function.

The quantity $\kappa(\varpi)$ is the usual epicyclic frequency [1, (4.327), p. 156] associated with $\Omega(\varpi)$. It is defined by

$$\kappa^2 = (2\Omega)^2\left\{1 + \frac{\varpi}{2\Omega}\frac{d\Omega}{d\varpi}\right\}. \tag{3.9}$$

Since $\Omega(\varpi)$ is "arbitrarily" chosen, there is no direct interpretation for $\kappa$ in the physical system in question. However, for a "reasonable" choice of $\Omega(\varpi)$, as we shall insist upon in all the following discussions, we also have the usual interpretation for $\kappa$.

The characteristic equations of the partial differential equation (3.1) are

$$\begin{aligned} dt = \frac{d\varpi}{c_\varpi} = \frac{d\theta}{\Omega + c_\theta/\varpi} &= \frac{dc_\varpi}{a_\varpi + (\Omega\varpi + c_\theta)^2/\varpi} \\ &= \frac{dc_\theta}{a_\theta - (\kappa^2/2\Omega + c_\theta/\varpi)c_\varpi}. \end{aligned} \tag{3.10}$$

This system possesses the following two integrals:

$$\tfrac{1}{2}\{c_\varpi^2 + (\varpi\Omega + c_\theta)^2\} + \mathcal{V}(\varpi, \theta, 0) = E \tag{3.11}$$

for *time-independent* potentials, and

$$\varpi(\varpi\Omega + c_\theta) = J \tag{3.12}$$

for *axially symmetric* potentials (cf. Appendix 2, (22a), (23a)).

[5] The superscript $s$ will be used alternatively with an asterisk in the subscript.

**4. Equations for small disturbances from a certain symmetrical distribution.** Consider first a certain state of axially symmetric distribution in equilibrium, which we shall describe in terms of a distribution function

$$\Psi = \Psi_0(\varpi, c_\varpi, c_\theta). \tag{4.1}$$

The components of acceleration are

$$(a_\varpi, a_\theta) = (a_{\varpi 0}(\varpi), 0). \tag{4.2}$$

The basic equation (3.1) becomes

$$\begin{aligned} &c_\varpi \frac{\partial \Psi_0}{\partial \varpi} + \left\{a_{\varpi 0} + \Omega^2 \varpi + 2\Omega c_\theta + \frac{c_\theta^2}{\varpi}\right\} \frac{\partial \Psi_0}{\partial c_\varpi} \\ &\qquad + \left\{0 - \left(\frac{\kappa^2}{2\Omega} + \frac{c_\theta}{\varpi}\right) c_\varpi\right\} \frac{\partial \Psi_0}{\partial c_\theta} = 0. \end{aligned} \tag{4.3}$$

For convenience, we shall choose $\Omega(\varpi)$ such that the condition

$$a_{\varpi 0} + \Omega^2 \varpi = 0 \tag{4.4}$$

is satisfied. With this choice, (4.3) is simplified into

$$c_\varpi \frac{\partial \Psi_0}{\partial \varpi} + \left\{2\Omega + \frac{c_\theta}{\varpi}\right\} c_\theta \frac{\partial \Psi_0}{\partial c_\varpi} - \left\{\frac{\kappa^2}{2\Omega} + \frac{c_\theta}{\varpi}\right\} c_\varpi \frac{\partial \Psi_0}{\partial c_\theta} = 0. \tag{4.5}$$

Since both of the integrals (3.11) and (3.12) exist in this case, the general solution of (4.5) is of the form

$$\Psi_0 = \Psi_0\{\varpi(\varpi\Omega + c_\theta), \tfrac{1}{2}[c_\varpi^2 + (\varpi\Omega + c_\theta)^2] + \mathcal{V}_0(\varpi)\}, \tag{4.6}$$

where

$$\mathcal{V}_0(\varpi) = \int_{\varpi_0}^{\varpi} \Omega^2 \varpi \, d\varpi. \tag{4.7}$$

Note that $\mathcal{V}_0(\varpi)$ need not be due to the stellar distribution (4.1) alone, as there might be, for example, a symmetrical distribution of gas in the galaxy.

Consider now a slight deviation from $\Psi_0$ associated with a small change in $(a_\varpi, a_\theta)$. The distribution function is now

$$\Psi = \Psi_0(1 + \psi), \tag{4.8}$$

and the field (which is not necessarily wholly determined by (4.8)) is

$$a_\varpi = a_{\varpi 0} + a_{\varpi 1}, \qquad a_\theta = a_{\theta 1}. \tag{4.9}$$

We are thus considering the change in $\Psi$ due to an imposed field $(a_{\varpi 1}, a_{\theta 1})$ while leaving it open whether the field is due to self-gravitation alone. To be sure, there is a self-field $(a_\varpi^{(s)}, a_\theta^{(s)})$ associated with (4.8), and this may be subtracted from $(a_\varpi, a_\theta)$ to determine the contributions of other agencies (e.g., gas and other nearby galaxies).

To derive the differential equation for $\psi$, it is convenient to write the fundamental equation (3.1) in the form

$$\mathfrak{L}(\Psi) = 0, \tag{4.10}$$

where $\mathfrak{L}$ is the linear differential operator defined by

$$\begin{aligned}\mathfrak{L} \equiv \frac{\partial}{\partial t} + c_\varpi \frac{\partial}{\partial \varpi} + \left(\Omega + \frac{c_\theta}{\varpi}\right)\frac{\partial}{\partial \theta} + \left\{a_\varpi + \Omega^2 \varpi + 2\Omega c_\theta + \frac{c_\theta^2}{\varpi}\right\}\frac{\partial}{\partial c_\varpi} \\ + \left\{a_\theta - \left(\frac{\kappa^2}{2\Omega} + \frac{c_\theta}{\varpi}\right)c_\varpi\right\}\frac{\partial}{\partial c_\theta}.\end{aligned} \tag{4.11}$$

We shall denote the corresponding operator with $(a_\varpi, a_\theta) = (a_{\varpi 0}, a_{\theta 0})$ by $\mathfrak{L}_0$ and write

$$\mathfrak{L} = \mathfrak{L}_0 + \mathfrak{L}_1, \tag{4.12}$$

where

$$\mathfrak{L}_1 \equiv a_{\varpi 1}\frac{\partial}{\partial c_\varpi} + a_{\theta 1}\frac{\partial}{\partial c_\theta} \tag{4.13}$$

and $a_{\varpi 1}, a_{\theta 1}$ (defined by (4.9)) are "small" in some sense (to be clarified below). If we now substitute (4.8) into (4.10), we obtain, with the help of these new symbols and the relation $\mathfrak{L}_0\{\Psi_0\} = 0$, the following nonlinear equation for the small disturbance $\psi$:

$$\mathfrak{L}_0\{\psi\} = \mathfrak{L}_1\{Q_0\} + (\psi\mathfrak{L}_1\{Q_0\} - \mathfrak{L}_1\{\psi\}), \tag{4.14}$$

where

$$Q_0 = -\log \Psi_0 . \tag{4.15}$$

Since $\psi$ is expected to be of the same order as the perturbation $(a_{\varpi 1}, a_{\theta 1})$ that causes it, the last two terms in (4.14) are of smaller order than the term $\mathfrak{L}_1\{Q_0\}$. Thus, the linearized differential equation for the small disturbance $\psi$ is

$$\mathfrak{L}_0\{\psi\} = \mathfrak{L}_1\{Q_0\} = a_{\varpi 1}\frac{\partial Q_0}{\partial c_\varpi} + a_{\theta 1}\frac{\partial Q_0}{\partial c_\theta}, \tag{4.16}$$

where $\mathfrak{L}_0$ is defined by (4.11) with $(a_\varpi, a_\theta) = (a_{\varpi 0}, a_{\theta 0})$. With the particular choice of $\Omega(\varpi)$ given by (4.4), the explicit form of (4.16) is as follows:

$$\begin{aligned}\frac{\partial \psi}{\partial t} + c_\varpi \frac{\partial \psi}{\partial \varpi} + \left(\Omega + \frac{c_\theta}{\varpi}\right)\frac{\partial \psi}{\partial \theta} + \left(2\Omega + \frac{c_\theta}{\varpi}\right)c_\theta \frac{\partial \psi}{\partial c_\varpi} \\ - \left(\frac{\kappa^2}{2\Omega} + \frac{c_\theta}{\varpi}\right)c_\varpi \frac{\partial \psi}{\partial c_\theta} = a_{\varpi 1}\frac{\partial Q_0}{\partial c} + a_{\theta 1}\frac{\partial Q_0}{\partial c_\theta}.\end{aligned} \tag{4.17}$$

This is then the differential equation serving as the basis of much of our subsequent investigations. After $\psi$ is obtained, the density perturbation $\sigma_1^{(s)}$ produced by the disturbance field $(a_{\varpi 1}, a_{\theta 1})$ may be obtained from the formula

$$\sigma_1^{(s)} = m_* \iint \psi \Psi_0 \, dc_\varpi \, dc_\theta = m_* \iint \psi e^{-Q_0} \, dc_\varpi \, dc_\theta . \tag{4.18}$$

This produces an induced field $(a_{\varpi 1}^{(s)}, a_{\theta 1}^{(s)})$, which forms a part of the *resultant impressed field* $(a_{\varpi 1}, a_{\theta 1})$. If there is no other field than the self-induced field, we are dealing with the problem of a self-maintained *collective mode* of oscillation, whose nature will be discussed in subsequent sections.

**5. Methods for solving the equation for small disturbances.** While the calculation of self-consistent symmetrical models with velocity dispersion is an interesting and challenging problem in itself, we shall proceed with the main task of this paper, which is to search for solutions of the spiral form. This is not difficult to find from (4.17), for the method of separation of variables immediately yields the answer.

Clearly, the variables $t$ and $\theta$ are cyclic in the left-hand side of (4.17). We may thus look for solutions of the form

$$\psi = \hat{\psi}(\varpi, c_\varpi, c_\theta) e^{i(\omega t - m\theta)}, \tag{5.1}$$

where $m$ is an integer and $\omega$ is a complex parameter, $\omega = \omega_r + i\omega_i$. As usual, the real part of the solution is to be taken, and thus $\omega_i < 0$ indicates exponentially growing modes.

Since (4.17) is not homogeneous, a solution of the form (5.1) requires that $(a_{\varpi 1}, a_{\theta 1})$ should also be of the same general form, i.e.,

$$(a_{\varpi 1}, a_{\theta 1}) = [\hat{a}_{\varpi 1}(\varpi), \hat{a}_{\theta 1}(\varpi)] e^{i(\omega t - m\theta)}. \tag{5.2}$$

This may be a consequence of the nature of the imposed field, but it is also consistent with the self-field associated with the distribution function (5.1), for (4.18) leads to the density distribution

$$\sigma_1 = \hat{\sigma}_1(\varpi) e^{-i(\omega t - m\theta)}, \tag{5.3}$$

which in turn yields a field of the form (5.2) according to the Poisson equation.

There are at least three standard ways open to us to proceed further:

(i) a consideration of the normal modes of self-sustained oscillations and the solutions obtained by superposition,

(ii) a consideration of the initial value problem by using a Fourier series in $\theta$ and a Laplace transform with respect to $t$, and

(iii) a consideration of the "response" of the stellar sheet due to an imposed field of the form (5.2).

While all of these methods are essentially equivalent to each other, they tend to emphasize slightly different aspects of the problem. We shall adopt the last method for the present, because of its simplicity. It also leaves open the possibility of including a gravitational field of gas. To obtain self-sustained oscillations, one has to solve the Poisson equation to demonstrate the self-consistency of the gravitational fields.

Since we anticipate the relatively rapid variation of phase with radial distance in (5.1) and (5.2), we shall exhibit the phase angle explicitly by introducing it via the gravitational potential. We write

$$a_{\varpi 1} = -\frac{\partial \mathcal{V}_1}{\partial \varpi}, \qquad a_{\theta 1} = -\frac{1}{\varpi}\frac{\partial \mathcal{V}_1}{\partial \theta}, \tag{5.4a}$$

where

$$\mathcal{V}_1 = A(\varpi) \exp\{i[\omega t - m\theta + \Phi(\varpi)]\}. \tag{5.4b}$$

In (5.4b), $A(\varpi)$ and $\Phi(\varpi)$ are both real, being respectively the amplitude and phase of the gravitational potential. We then write

$$\psi = \phi(\varpi, c_\varpi, c_\theta) \exp\{i[\omega t - m\theta + \Phi(\varpi)]\}, \tag{5.5}$$

where $\phi$ is expected to vary *slowly* with $\varpi$. Note that $\phi$ is in general complex·

All the quantities in our problem now have a *spiral appearance*: the spiral being defined by lines of constant phase of the gravitational potential (5.4b):

$$m(\theta - \theta_0) = \Phi(\varpi) - \Phi(\varpi_0). \tag{5.6}$$

The spiral arms are leading if $\Phi'(\varpi) > 0$ and trailing if $\Phi'(\varpi) < 0$. The circular velocity of the stars is always taken in the counterclockwise sense.

A crude approximation to the spiral pattern in our galaxy is given by taking

$$\Phi(\varpi) = -20 \log \varpi. \tag{5.7}$$

Indeed, for any smooth spiral which is relatively tightly wound,[6] we should have

$$\Phi(\varpi) = \epsilon^{-1} f(\varpi), \qquad \epsilon^{-1} \gg 1, \tag{5.8}$$

where $\epsilon^{-1}$ is a conveniently chosen real parameter. Thus, a natural method for obtaining simple approximate answers is to adopt an asymptotic series

[6] Smoothness insures that the higher derivatives of the phase function $\Phi(\omega)$ are of the same order of magnitude as its first derivative.

of the type

$$(5.9)\qquad \phi(\varpi, c_\varpi, c_\theta, \epsilon) = \phi^{(0)}(\varpi, c_\varpi, c_\theta) + \epsilon\phi^{(1)}(\varpi, c_\varpi, c_\theta) + \cdots .$$

**6. The equation for the amplitude function.** When the expression (5.5) is substituted into (4.17), we obtain the following equation for $\phi(\varpi, c_\varpi, c_\theta)$:

$$(6.1)\qquad \begin{aligned} &c_\varpi \frac{\partial\phi}{\partial\varpi} + \left(2\Omega + \frac{c_\theta}{\varpi}\right) c_\theta \frac{\partial\phi}{\partial c_\varpi} - \left(\frac{\kappa^2}{2\Omega} + \frac{c_\theta}{\varpi}\right) c_\varpi \frac{\partial\phi}{\partial c_\theta} \\ &\qquad + i\left[kc_\varpi - m\frac{c_\theta}{\varpi} + (\omega - m\Omega)\right]\phi = g_\varpi \frac{\partial Q_0}{\partial c_\varpi} + g_\theta \frac{\partial Q_0}{\partial c_\theta}, \end{aligned}$$

where (cf. (5.4)) we have written

$$(6.2)\qquad k(\varpi) = +\Phi'(\varpi),$$

so that $k(\varpi) < 0$ for trailing arms, and

$$(6.3)\qquad (a_{\varpi 1}, a_{\theta 1}) = (g_\varpi, g_\theta) \exp\{i[\omega t - m\theta + \Phi(\varpi)]\}.$$

Thus,

$$(6.4a)\qquad g_\varpi = -[A'(\varpi) + ik(\varpi)A(\varpi)],$$

$$(6.4b)\qquad g_\theta = \operatorname{Im} A(\varpi)/\varpi.$$

Equation (6.1) is the partial differential equation for the amplitude function. Let us first examine the nature of this equation and its solution without the use of asymptotic integration. Two points deserve special attention: (a) the characteristics and integrals of (6.1), and (b) the singularity of the equation at $(c_\varpi, c_\theta) = (0, 0)$, where the coefficients of all the partial derivatives vanish.

**6a. The characteristics and integrals of (6.1).** The characteristics of (6.1) are determined by the following autonomous system of differential equations:

$$(6.5)\qquad \frac{d\varpi}{ds} = c_\varpi,$$

$$(6.6)\qquad \frac{dc_\varpi}{ds} = \left(2\Omega + \frac{c_\theta}{\varpi}\right) c_\theta,$$

$$(6.7)\qquad \frac{dc_\theta}{ds} = -\left(\frac{\kappa^2}{2\Omega} + \frac{c_\theta}{\varpi}\right) c_\varpi,$$

$$(6.8)\qquad \frac{d\phi}{ds} = -i\left[kc_\varpi - m\frac{c_\theta}{\varpi} + (\omega - m\Omega)\right]\phi + \left[g_\varpi \frac{\partial Q_0}{\partial c_\varpi} + g_\theta \frac{\partial Q_0}{\partial c_\theta}\right].$$

Clearly, (6.5)–(6.7) are the same as those occurring in the symmetrical case (cf. (3.10)), and we thus have the same integrals:

$$J = \varpi(\varpi\Omega + c_\theta), \tag{6.9}$$

and

$$E = \tfrac{1}{2}[c_\varpi^{\,2} + (\varpi\Omega + c_\theta)^2] + \mathcal{V}_0(\varpi). \tag{6.10}$$

If we consider the problem geometrically in the three-dimensional space $(\varpi, c_\varpi, c_\theta)$, then each of the equations $J$ = const. and $E$ = const. represents a two-dimensional surface. Along their intersecting curve, (6.8) represents the ordinary differential equation for the determination of $\phi$. A little reflection will show that this is equivalent to the following statement. If we introduce a change of variables and adopt $J$ and $E$ as two of the three new independent variables by (6.9) and (6.10), then the third variable $s = s(\varpi, c_\varpi, c_\theta)$ may be so chosen that the transformation functions

$$\varpi = \varpi(J, E, s), \qquad c_\varpi = c_\varpi(J, E, s), \qquad c_\theta = c_\theta(J, E, s), \tag{6.11}$$

satisfy the conditions

$$\frac{\partial \varpi}{\partial s} = c_\varpi, \qquad \frac{\partial c_\varpi}{\partial s} = \left(2\Omega + \frac{c_\theta}{\varpi}\right) c_\theta, \qquad \frac{\partial c_\theta}{\partial s} = -\left(\frac{\kappa^2}{2\Omega} + \frac{c_\theta}{\varpi}\right) c_\varpi. \tag{6.12}$$

Indeed, with this choice of variables, the partial differential equation (6.1) becomes

$$\frac{\partial \phi}{\partial s} = -i\left[kc_\varpi - \frac{mc_\theta}{\varpi} + (\omega - m\Omega)\right]\phi + \left[g_\varpi \frac{\partial Q_0}{\partial c_\varpi} + g_\theta \frac{\partial Q_0}{\partial c_\theta}\right] \tag{6.13}$$

in which $(\varpi, c_\varpi, c_\theta)$ are regarded as functions of $(J, E, s)$ via (6.11).

Instead of introducing this transformation into characteristic variables all at once, we shall first introduce the system of variables $(J, c_\varpi, c_\theta)$ in order to bring out the nature of the solution near the singularity of the equation $(c_\varpi, c_\theta) = (0, 0)$.

**6b. The singularity of the equation at $(c_\varpi, c_\theta) = (0, 0)$.** The general solution of the partial differential equation (6.1) contains certain arbitrary functions. Indeed, if $\phi_1(\varpi, c_\varpi, c_\theta)$ is a solution, then we may obtain another by adding to it a constant multiple of $\phi_0(\varpi, c_\varpi, c_\theta)$ which satisfies the homogeneous part of (6.1). However, if we require regularity of the solutions at $(c_\varpi, c_\theta) = 0$, it becomes clear that

$$\phi_0(\varpi, 0, 0) = 0$$

and that the same is true for all the derivatives of $\phi_0(\varpi, c_\varpi, c_\theta)$ with respect to $c_\varpi$ and $c_\theta$. Thus,

$$\phi_0(\varpi, c_\varpi, c_\theta) \equiv 0$$

if it is to be regular at $(c_\varpi, c_\theta) = 0$ for all values of $\varpi$. There is therefore reason to expect a unique regular solution of (6.1), as is indeed anticipated from physical considerations.[7]

To bring out the nature of the solution of (6.1) and the nature of the characteristic surface (6.10) near this singularity, it is convenient to introduce $J$ as an independent variable in place of $\varpi$. Thus, with reference to (6.9), we consider a change of variables $(\varpi, c_\varpi, c_\theta) \to (J, c_\varpi, c_\theta)$, where $c_\varpi$ is left unchanged. It can be easily verified, as it may be expected from the general theory, that (6.1) reduces to the form

$$\begin{aligned} \left(2\Omega + \frac{c_\theta}{\varpi}\right) c_\theta \frac{\partial \phi}{\partial c_\varpi} - \left(\frac{\kappa^2}{2\Omega} + \frac{c_\theta}{\varpi}\right) c_\varpi \frac{\partial \phi}{\partial c_\theta} \\ + i \left[ k c_\varpi - \frac{m c_\theta}{\varpi} + (\omega - m\Omega) \right] \phi \\ = g_\varpi \frac{\partial Q_0}{\partial c_\varpi} + g_\theta \left( \frac{\partial Q_0}{\partial c_\theta} + \varpi \frac{\partial Q_0}{\partial J} \right), \end{aligned} \tag{6.14}$$

whose left-hand side may be formally obtained from (6.1) by the suppression of the term in $\partial\phi/\partial\varpi$. However, in (6.14), $\varpi$ must be regarded as a function of $c_\theta$ and $J$, and the partial derivatives of $Q_0$ are defined in terms of the new set of variables.

In a plane of $J = J_0$, we now consider the characteristics of (6.14). Clearly, near $(c_\varpi, c_\theta) = 0$, the approximate equation for them is

$$\frac{dc_\varpi}{2\Omega \cdot c_\theta} = -\frac{dc_\theta}{(\kappa^2/2\Omega) \cdot c_\varpi}. \tag{6.15}$$

To the same order of approximation, $J = J_0$ may be regarded as $\varpi = \varpi_0$. Thus, the integration of (6.15) leads to

$$\left(\frac{\kappa}{2\Omega}\right)^2 c_\varpi^2 + c_\theta^2 = C(\varpi), \tag{6.16}$$

i.e., for various choices of $C$, we have a set of ellipses around the origin. Clearly, this qualitative nature of closed contour curves around the origin will not be changed by the inclusion of higher order terms in the exact calculation of the characteristics of (6.14), since there will still be an energy integral.

*Remark on choice of variables.* If desired, the variable $J$ may be replaced by any function of $J$. Since it is essentially used to replace $\varpi$ in the above

[7] The same conclusion can also be reached quite readily from (6.13), by referring to (6.12).

transformation, it might be found desirable to think in terms of a characteristic parameter which is close to $\varpi$. Indeed, if we introduce the *characteristic radial distance* $r$ by the relation

$$r^2\Omega(r) = J = \varpi[\varpi\Omega(\varpi) + c_\theta], \tag{6.17}$$

this purpose can be accomplished. For small values of $c_\theta$, $r$ is nearly identical with $\varpi$. We may thus find it convenient to write

$$\varpi = r + \delta(r, c_\theta), \tag{6.18}$$

where $\delta(r, c_\theta) \to 0$ as $c_\theta \to 0$. We might have used $r$ in place of $J$ in the above discussion. The last term in (6.14) would have been slightly altered. We have

$$\varpi \frac{\partial Q_0}{\partial J} = \varpi \frac{\partial Q_0/\partial r}{dJ/dr} = \frac{\varpi}{r}(r\Omega' + 2\Omega)^{-1} \frac{\partial Q_0}{\partial r}.$$

**7. The intrinsic form of the amplitude equation.** The approximate form of the energy integral (6.16) suggests that we should introduce dimensionless variables $(\xi, \eta)$ in place of $(c_\varpi, c_\theta)$ such that the first approximation to the energy integral is

$$\xi^2 + \eta^2 = \text{function of } \varpi.$$

To accomplish this purpose, let us define $(\xi, \eta)$ by

$$c_\varpi = \xi V_1(\varpi), \qquad c_\theta = \eta V_2(\varpi),$$

where $V_1(\varpi)$ and $V_2(\varpi)$ are as yet to be chosen, but they must conform to the relationship $V_2/V_1 = \kappa/2\Omega$. It would be tempting to set these scales as $2\Omega\varpi$ and $\kappa\varpi$, but this did not turn out to be the best choice. Rather, noting the fact that $\varpi\Omega + c_\theta$ occurs in the integrals for angular momentum and energy, (6.9) and (6.10), we choose $V_1 = 2\varpi\Omega^2/\kappa$, $V_2 = \varpi\Omega$:

$$c_\varpi = \xi(2\varpi\Omega^2/\kappa), \qquad c_\theta = \eta(\varpi\Omega). \tag{7.1}$$

We apply this transformation directly to (6.1), i.e., we take $(\varpi, \xi, \eta)$ as the independent variables. A direct calculation then leads to the following equation:

$$\begin{aligned} &\frac{1}{2}\gamma^2\xi\varpi \frac{\partial\phi}{\partial\varpi} + \left(\eta + \frac{\eta^2}{2} - \beta\xi^2\right)\frac{\partial\phi}{\partial\xi} - \xi(1 + \eta)\frac{\partial\phi}{\partial\eta} \\ &\qquad + i\left(\nu + \alpha\xi - \frac{m}{2}\gamma\eta\right)\phi = \frac{1}{2}\left[f_\varpi \frac{\partial Q_0}{\partial\xi} + \gamma f_\theta \frac{\partial Q_0}{\partial\eta}\right]. \end{aligned} \tag{7.2}$$

In the above equation, we have introduced the following new symbols:

$$\gamma^2 = \left(\frac{2\Omega}{\kappa}\right)^2, \qquad \beta = \frac{\gamma^2}{2}\frac{\varpi}{V_1}\frac{dV_1}{d\omega}, \tag{7.3a}$$

(where $V_1 = 2\varpi\Omega^2/\kappa$) which depend only on $\Omega(\varpi)$, the rotation curve;

$$\alpha = \frac{\gamma^2}{2} k\varpi = \frac{\gamma^2}{2} \varpi\Phi'(\varpi), \qquad \nu = \frac{\omega - m\Omega}{\kappa}, \tag{7.3b}$$

which depend further on the nature of the spiral wave[8] and

$$\begin{aligned} f_\varpi &= \frac{g_\varpi}{\varpi\Omega^2} = -\frac{A'(\varpi)}{\varpi\Omega^2} - 2i\alpha \frac{A(\varpi)}{V_1^2}, \\ f_\theta &= \frac{g_\theta}{\varpi\Omega^2} = +\frac{imA(\varpi)}{\varpi^2\Omega^2}, \end{aligned} \tag{7.3c}$$

which are proportional to the amplitude of the impressed gravitational field. We recall that $A(\varpi)$ and $\Phi(\varpi)$ are both real, being the magnitude and phase of the amplitude function for the spiral gravitational potential (cf. (5.4b)).

We shall call (7.2) the *intrinsic form* of the equation for the amplitude function of the disturbance in distribution function, since the velocity components are referred to typical velocities inherent to a given location. The equation is also essentially dimensionless. (The only variable that still carries dimension is $\varpi$, which can be trivially made dimensionless.)

*Remark.* It should be mentioned here that, on the right-hand side of (7.2), the function $Q_0 = Q_0(\varpi, c_\varpi, c_\theta)$ is now a function of $\xi$ and $\eta$ through (7.1), and that the partial derivatives are taken with $\varpi = \text{const}$.

It is very easy to recover the integral (6.9) for angular momentum from (7.2). Indeed, we have, along a characteristic,

$$\frac{d\varpi}{\frac{1}{2}\gamma^2\varpi\xi} = -\frac{d\eta}{(1+\eta)\xi}, \tag{7.4}$$

which yields

$$J = \varpi^2\Omega(1+\eta). \tag{7.5}$$

The energy integral is not so easy to find from (7.2). However, an especially useful approximate form is obvious from the relation

$$(1+\eta)\xi\, d\xi + [\eta + \tfrac{1}{2}\eta^2 - \beta\xi^2]\, d\eta = 0 \tag{7.6}$$

along the characteristic, if we are willing to approximate[9] $\beta(\varpi)$ as a function of $J$ by using (7.5). This amounts to neglecting a certain quantity of the order $\xi^2 O(\eta)$ in (7.7). Equation (7.6) is now linear in $\xi^2$ and can be integrated in terms of quadratures:

$$\frac{1}{2}\xi^2(1+\eta)^{-\beta} + \int_0^\eta (1+\eta)^{-(\beta+1)}\left(1+\frac{\eta}{2}\right)\eta\, d\eta = \mathcal{E}, \tag{7.7}$$

[8] Note that $(k\omega)^{-1} = (1/m)\tan i$, where $i$ is the inclination of the spiral arm to the circular arc.

[9] Note that $\beta$ is a numerical constant if $V_1(\omega)$ varies as a power of $\omega$. In particular, $\beta = 0$ if $\omega\Omega = \text{const}$.

where $\beta$ is a function of $J$, and $\mathcal{E}$ is the new characteristic variable corresponding to $E$. For small values of $\eta$, (7.7) reduces to (6.16), as it should. It is however quite involved to verify directly that (7.7) is actually an approximation to (6.10).

*Remark.* Equation (7.7) is obviously also an approximate characteristic integral for the symmetrical distribution function. Indeed, a reasonable form for $\Psi_0$ is

$$\Psi_0 = \exp(-\mu\mathcal{E}), \tag{7.8}$$

where $\mu$ is an absolute constant, and $\beta = \beta(\varpi)$.

We shall now re-write (7.2) in two other useful forms.

*Change of scale.* Clearly, for the significant[10] range of the variables $(c_\varpi, c_\theta)$, the dimensionless variables $\xi$ and $\eta$ are small. To make use of this fact, and to bring out the important terms in (7.2), it is convenient to introduce a change of scale, as follows. If the wave number $\alpha$ is to be important in (7.2), the term $\alpha\xi$ must not be negligible. Thus, we are essentially concerned with *short* wave-lengths, or spirals of small inclination. We now introduce a small parameter $\epsilon > 0$ such that (cf. (5.8))

$$\bar{\alpha} = \alpha\xi \tag{7.9}$$

is of the order of unity. We would then expect

$$\bar{\xi} = \xi/\epsilon \quad \text{and} \quad \bar{\eta} = \eta/\epsilon \tag{7.10}$$

to be also of the order of unity, and we should be able to separate the important terms from the unimportant ones. Indeed, (7.2) becomes

$$\begin{aligned} \frac{1}{2}\xi\gamma^2\bar{\xi}\varpi\frac{\partial\phi}{\partial\varpi} + \left[\bar{\eta} + \epsilon\left(\frac{\bar{\eta}^2}{2} - \beta\bar{\xi}^2\right)\right]\frac{\partial\phi}{\partial\bar{\xi}} - \bar{\xi}(1 + \epsilon\bar{\eta})\frac{\partial\phi}{\partial\bar{\eta}} \\ + i\left(\nu + \bar{\alpha}\bar{\xi} - \epsilon\frac{m}{2}\gamma\bar{\eta}\right)\phi = \frac{1}{2}\epsilon\left(f_\varpi\frac{\partial Q_0}{\partial\bar{\xi}} + \gamma f_\theta\frac{\partial Q_0}{\partial\bar{\eta}}\right), \end{aligned} \tag{7.11}$$

where

$$Q_0 = \bar{Q}_0(\varpi, \bar{\xi}, \bar{\eta}, \epsilon) \tag{7.12}$$

and

$$\begin{aligned} \epsilon f_\varpi &= -\epsilon A'/\varpi\Omega^2 - 2i\bar{\alpha}A/V_1^2, \\ \epsilon f_\theta &= +\operatorname{Im}\epsilon A/\varpi^2\Omega^2. \end{aligned} \tag{7.13}$$

[10] The specification of the significant range may be seen from (4.8) and (4.18). For a distribution function $\Psi_0$ similar to the Maxwellian function, the exact behavior of the function $\phi(\varpi, c_\varpi, c_\theta)$ for $c_\varpi^2 \gg \langle c_\varpi^2\rangle$ and $c_\theta^2 \gg \langle c_\theta^2\rangle$ would appear to be unimportant.

Thus, if we write the perturbation series

$$\epsilon^2\phi = \phi^{(0)}(\varpi, \bar{\xi}, \bar{\eta}) + \epsilon\phi^{(1)}(\varpi, \bar{\xi}, \bar{\eta}) + \cdots, \tag{7.14}$$

we obtain the following equation for $\phi^{(0)}$:

$$\bar{\eta}\frac{\partial\phi^{(0)}}{\partial\bar{\xi}} - \bar{\xi}\frac{\partial\phi^{(0)}}{\partial\bar{\eta}} + i(\nu + \bar{\alpha}\bar{\xi})\phi^{(0)} = -i\bar{\alpha}\frac{A}{V_1^2}\frac{\partial Q}{\partial\bar{\xi}}. \tag{7.15}$$

*The modified intrinsic form.* Equation (7.2) also suggests the adoption of $\alpha\xi$ (and hence also $\alpha\eta$) as the new independent variables. Indeed, these variables

$$p = \alpha\xi = kc_\varpi/\kappa, \qquad q = \alpha\eta = \gamma kc_\theta/\kappa$$

can be easily interpreted in terms of the ratio of the epicyclic radius to the wave length of the imposed gravitational field. The precise form of the resultant equation is rather awkward, but the initial approximation has essentially the same form as (7.14). In view of the simple interpretation of $(p, q)$, this form might be convenient as a basis for a perturbation calculation. The parameter $\epsilon$ should still be introduced as above.

**8. Density waves.** One of our principal purposes is to construct solutions which represent self-sustained density waves. That is, starting from the *imposed gravitational field* (5.4), which is characterized by the two real functions $A(\varpi)$ and $\Phi(\varpi)$, we calculate the *induced density variation* (4.18) by means of (5.5), where the amplitude $\phi(\varpi, c_\varpi, c_\theta)$ of the disturbance in the distribution function is to be solved from (6.1). The condition for self-sustained density waves is then that the *induced gravitational field* due to this induced density variation should be equal to the original imposed gravitational field.

Clearly, at each stage of the calculation, the relationship is linear in the amplitude. Thus, we have eventually a linear functional equation for the unknown function $A(\varpi)$, but with $\Phi(\varpi)$ as a second unknown function, entering in a complicated nonlinear fashion. We also have the parameters $m$ (an integer) and $\omega$ (a complex number). The characteristic value problem is a rather unusual one, as already stated in a previous study [8] using the moment method:

"Under such a general formulation, $\omega$ takes on characteristic values for the solution of a set of linear integro-differential equations in the single independent variable $\omega$, subjected to suitable boundary conditions at $\varpi = 0$ and as $\varpi \to \infty$. However, since we would admit a singularity at $\varpi = 0$ in the solution, if we adopt the disk model right to the center of the galaxy, and the range of $\varpi$ is infinite, the set of characteristic values may be expected to have a continuous spectrum (in addition to any discrete part).

Physically, this means that we are looking for a representation of the situation in the disklike part of the galaxy, and leaving the conditions near the central bulge, where the random velocities are large and three-dimensional, *to adjust themselves to almost any requirement of the disk part*. We also expect that all the disturbances would die off at infinity in a reasonable manner."

As mentioned in the same paper, the concept of density waves was first introduced by B. Lindblad, and was studied by him and his son P. O. Lindblad; but the methods they used are quite different. The role of individual stellar orbits is more prominent in their work, whereas the statistical treatment is used here. But there is no doubt that our theories are basically related. Indeed, independently of each other, B. Lindblad [9] and the present writer came to the same suggestion of a *quasi-stationary spiral structure of the stars* in a disk galaxy. This idea was formulated by the present writer in the joint paper with Shu [8] in the form of the following hypothesis:

"The total stellar population, which has various degrees of velocity dispersion, forms a quasi-stationary spiral structure in space of the general nature discussed above. This is primarily due to the effect of gravitational instability as limited by velocity dispersion (and secondarily to the influence of the gas and the magnetic field). The extent of density variation in the spiral pattern may be only a small fraction of the symmetrical mean density distribution."

It is clear that the type of solutions discussed here would help to decide whether this hypothesis can be borne out by the principles of stellar dynamics.

Up to now, only the linear theory has been considered in detail. If the linear theory shows the existence of neutral or unstable density waves of a spiral form, it tends to support the abovementioned hypothesis. It is expected that the amplitude of the unstable density waves will be limited by nonlinear effects (or possibly by gas through its gravitational influence and hydromagnetic processes).

The theory is in a more advanced stage of development when the wavelength is relatively short (compared with a typical radial distance from the galactic center) so that the asymptotic method may be used. But the asymptotic approach is by no means "local," for it can be applied, in principle, to a finite range over the disk extending over a few wave lengths. Up to now, studies of the theory for the whole disk have essentially been limited to the case of zero dispersion ("zero temperature")[11]. The results in both cases will now be briefly described under the two headings: asymptotic theory, finite dispersion; whole disk, zero dispersion.

[11] Kalnajs [7] has studied the problem of the whole disk, with the peculiar velocities of the stars included. Further clarification of his results seems desirable.

**8a. Asymptotic theory, finite dispersion.** As one may see from §7, in the first step of an asymptotic theory based on short wave lengths, we approximate (7.2) by (7.15), in which the $\varpi$-derivative does not appear. This may also be regarded as taking $\varpi = \varpi_0$ as characteristics instead of $J = J_0$. Equation (7.15) can be solved rather easily if we assume that the basic distribution function follows a modified Schwarzschild law;[12] e.g.,

$$\Psi_0 = A_0(1 + h_0\,\eta)\exp\left\{-\frac{\mu_0}{2}(\xi^2 + \eta^2)\right\}, \tag{8.1}$$

where $A_0$, $\mu_0$ and $h_0$ are all functions of $\varpi$. Indeed, we have

$$m_* \iint_{-\infty}^{\infty} \Psi_0\, dc_\varpi\, dc_\theta = m_* A_0\, V_1\, V_2(2\pi/\mu_0) = \sigma_0^{(s)}(\varpi), \tag{8.2}$$

where $\sigma_0^{(s)}$ is the surface density of matter in the basic state of equilibrium.

With the solution thus obtained, we may calculate the density disturbance $\sigma_1^{(s)}$ from (4.18); then the induced gravitational field can be approximated [8, (11)] by

$$\begin{aligned} g_\varpi^{(\text{ind.})} &= 2\pi i G \sigma_1^{(s)}\,\operatorname{sgn}\{\Phi'(\varpi)\}, \\ g_\theta^{(\text{ind.})} &= 0. \end{aligned} \tag{8.3}$$

From the linear nature of all the steps, it can be seen that the induced field is proportional to $A(\varpi)$, as is the impressed field. Indeed, to this approximation, we have

$$g_\varpi^{(\text{imp.})} = -ikA, \qquad g_\theta^{(\text{imp.})} = 0. \tag{8.4}$$

Thus, we may speak of a *response ratio*

$$R = g_\varpi^{(\text{ind.})}/g_\varpi^{(\text{imp.})}, \tag{8.5}$$

which works out to be

$$R = \frac{2\pi G\,|\,k\,|\,\sigma_e^{(s)}}{\kappa^2 - (\omega - m\Omega)^2}, \qquad k = \Phi'(\varpi). \tag{8.6}$$

The symbols are defined as follows: the quantity $\sigma_e^{(s)}$ is the *effective surface density* which is related to the true surface density $\sigma_0^{(s)}$ by a *reduction factor* $F_\nu(x)$, $(\nu = (\omega - m\Omega)/\kappa)$,

$$\sigma_e^{(s)} = \sigma_0^{(s)} F_\nu(x), \tag{8.7}$$

---

[12] It should be noted that $\Psi_0$ was determined via (4.5), which is independent of the approximation of small inclination of the spiral patterns (or short radial wave lengths). It has also been verified by Shu that the final result (8.8), when expressed in terms of physical variables, is not sensitive to the exact form (8.1) assumed here.

where

$$(8.8)\qquad F_\nu(x) = \frac{1-\nu^2}{x}\left\{1 - \frac{\nu\pi}{\sin \nu\pi} e^{-x}[\mathfrak{s}_\nu(x) + ih\{\mathfrak{s}_\nu(x) - \mathfrak{s}_\nu'(x)\}]\right\},$$

$$(8.9a)\qquad x = \alpha^2/\mu_0 = k^2\langle c_\varpi^2\rangle/\kappa^2,$$

$$(8.9b)\qquad h = \alpha h_0/\mu_0 = k\langle c_\theta\rangle/(\kappa^2/2\Omega),$$

and

$$(8.10)\qquad \mathfrak{s}_\nu(x) = \frac{1}{2\pi}\int_{-\pi}^{\pi} e^{-x\cos s} \cos \nu s \, ds.$$

Clearly, $\mathfrak{s}_\nu(x)$ reduces to the modified Bessel function $I_\nu(x)$ for the integral values of $x$, up to a factor $(-1)^\nu$. For self-sustained waves, the response factor should be set equal to unity.

The term "reduction factor" is justified for $F_\nu(x)$ since it becomes unity for the limiting case of zero velocity dispersion (with $h = 0$, $x = 0$) and is generally less than unity for $x > 0$. Indeed, the response ratio (8.6) with $\sigma_e^{(s)}$ replaced by $\sigma_0^{(s)}$, is identical with that calculated for the case of zero dispersion by the hydrodynamic equations.[13] It is through the use of a reduction factor (unproven at that time) that Lin and Shu [8, Remark 3, p. 652] considered the effect of velocity dispersion in their elementary theory.

A number of interesting conclusions can be drawn from the formulas (8.6)–(8.8).

(i) *Schwarzschild distribution in the equilibrium state.* First, we note that, for the ellipsoidal distribution, $h_0 = 0$, and hence the reduction factor is real for real $\nu$. Fig. 4 gives the graphs of $F_\nu(x)$, for $\nu = 0$ and $\nu = \pm 1$, which are

$$(8.11)\qquad F_0 = \frac{1}{x}\{1 - e^{-x}I_0(x)\}$$

and

$$(8.12)\qquad F_1(x) = \frac{2}{x}e^{-x}I_1(x) = F_{-1}(x).$$

Their numerical values are also shown in Table 2.

The response ratio $R$ is also always real. By setting $R = 1$, one gets a relation for the determination of $k(\varpi)$ for each real value of $\varpi$, within certain limits set by

[13] Compare [8, (12), p. 649], noting that the response ratio should be set equal to unity for self-sustained waves.

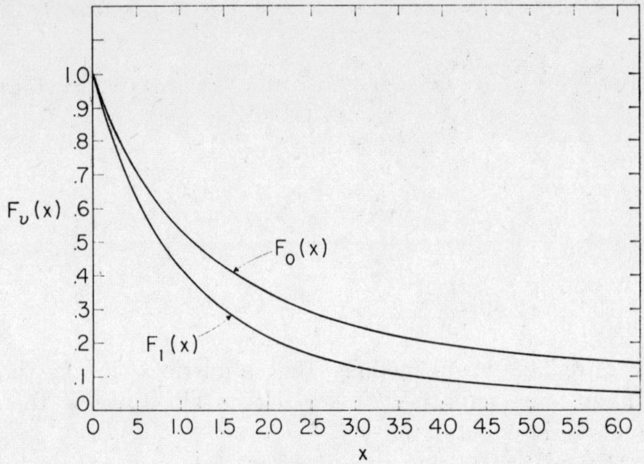


FIG. 4. *Reduction factor for Schwarzschild law of ellipsoidal distribution*

TABLE 2
*Reduction factor*

| $x$ | $F_0(x)$ | $F_1(x)$ |
|---|---|---|
| 0.00 | 1. | 1. |
| 0.50 | 0.7099294 | 0.6256832 |
| 1.00 | 0.5342404 | 0.4158208 |
| 1.50 | 0.4311233 | 0.2920260 |
| 2.00 | 0.3457458 | 0.2152693 |
| 2.50 | 0.2919814 | 0.1652677 |
| 3.00 | 0.2523332 | 0.1312178 |
| 3.50 | 0.2220565 | 0.1070857 |
| 4.00 | 0.1982495 | 0.0893754 |
| 4.50 | 0.1790670 | 0.0759817 |
| 5.00 | 0.1632918 | 0.0655889 |
| 5.50 | 0.1500939 | 0.0573458 |
| 6.00 | 0.1388904 | 0.0506838 |

$$\kappa^2 - (\omega - m\Omega)^2 > 0. \tag{8.13}$$

The endpoints of this range are the points of *Lindblad resonance.*

The solution represents a density wave traveling around the disk at an angular frequency $\omega_r/m$ and having a spiral structure given by (5.6). There is clearly a range of such frequencies. To this approximation, one cannot discriminate them relative to each other. There is also found no difference in character between trailing and leading waves.

*Remark.* One natural question that would come up is the following:

Suppose we consider a combination of parameters such that $R > 1$, would this indicate instability? The answer is no! It merely means that the imposed frequency is different from the natural frequency. The reader will easily see this by considering the response of a simple pendulum to an applied periodic force with the restoring force temporarily suspended. Stability or instability can be discussed in terms of the response to an applied force only when the applied frequency is zero.

*Special case of axially symmetrical disturbances.* In this case, Kalnajs [7] obtained, independently of the work of the author, a dispersion relationship of the form $R(\omega, k) = 1$ by a quite different method with an apparently quite different form for $R$. The result can however be identified with the present one for the case of Schwarzschild distribution of velocities.[14] Toomre [20] and Kalnajs [21] also obtained conditions for marginal stability of axially symmetric modes. This may be reproduced from the present work by putting $m = 0$ and $\nu = 0$. The relation $R = 1$ then reads

$$(8.14) \qquad 1 = \frac{|k|}{|k_c|} \cdot \frac{1}{x}\{1 - e^{-x} I_0(x)\}, \quad \text{where} \quad |k_c| = \frac{\kappa^2}{2\pi G\sigma_0}.$$

This relationship yields a marginal value $x_m$ for all $|k| > |k_c|$, i.e., for all waves shorter than a certain critical wave-length. For conditions below the boundary curve, there is strong instability. For points above the boundary curve, Toomre marked the condition "stable". This has to be interpreted as "neutrally stable", permitting oscillations (which correspond to the neutral waves in the nonsymmetrical case). It must be left open whether there might be mild instabilities which remain undetected in this particular approximation. Indeed, as we shall see immediately, (8.8) shows that there are such instabilities (*over-stable* modes) for the nonsymmetrical modes, provided that $h_0 \neq 0$.

(ii) *Nonsymmetrical distribution in the equilibrium condition.* If the response ratio is complex while the imposed force varies in a strictly periodical manner, there is damping or amplification of the self-sustained oscillations. This can again be seen by considering a simple pendulum with damping or amplification. It can also be seen directly from (8.6) and (8.8). If we put $R = 1$, this condition can be fulfilled only if $\nu$ (i.e., $\omega$) is complex, if all other quantities are real. Since the imaginary factor

$$ih = i\,\frac{k\langle c_\theta\rangle}{(\kappa^2/2\Omega)}$$

is small, we may expect $\omega_i$ to be proportional to $h$. Thus, for a given asymmetry of the basic distribution $\Psi_0$, which determines the sign of $\langle c_\theta\rangle$, trailing

[14] This may be done by regarding $F_\nu(x)$ as an analytic function of $\nu$ and expanding it in terms of partial fractions according to the theorem of Mittag-Leffler.

($k < 0$) and leading ($k > 0$) patterns are differentiated.[15] This differentiation is inherent in the asymmetry of $\Psi_0$, and the results would therefore remain qualitatively the same even if the form (8.1) for $\Psi_0$ is altered. From general considerations and observational facts, it is known that $\langle c_\theta \rangle < 0$.

The detailed calculations of $\omega_i$ are rather complicated but it can be shown that trailing patterns are preferred from the consideration of the range of values of $\varpi$ with $\omega_r < 0$, i.e., in the relatively massive part of the galaxy. This is the first indication of a mechanism for preferring trailing waves. Even more important, it is an indication of a mechanism for differentiating the two types of waves. Thus, *the pattern always has a spiral form.*

It is to be noted that we have not yet considered the higher approximations in the asymptotic series (5.9) or (7.14). (Note added in proof: See [8a] published after the lecture was delivered.) Detailed examination of the scale of the pattern observed shows that the frequency $\omega$ should be near the range given by the resonance condition $\kappa^2 - (\omega - m\Omega)^2 = 0$. In view of the fact that $\Omega - \kappa/2$ is nearly constant in the interior part of the galaxy,[16] two-armed spirals are preferred.

**8b. Whole disk, zero dispersion.** The study of density waves over the whole disk is obviously a very complicated problem. Apart from the attempts of Kalnajs [7] mentioned above, all calculations have been made for the case of zero dispersion. Toomre [20] considered the possible instability of axially symmetrical disturbances, imposing a condition of regularity at the center ($\omega = 0$) for both the disturbance and the basic distribution of density. He found that short waves are unstable. Rehm [15] calculated the nonsymmetrical modes, again imposing the condition of regularity at the center. He found that if the basic system has differential rotation of the usual type, the amplified spiral patterns are leading.[17] Both authors used numerical methods, and it is therefore difficult to conclude whether this is a general feature, but it is presumably so. Still, it would be desirable to examine the problem by the more flexible analytical methods, and to determine whether one can find trailing patterns if one allows singularity of a vaguely barred shape at the center, as suggested by the continuation of two arms into the center. This was done by Rehm for a certain special model studied also by Hunter [5]. Hunter previously considered oscillations over a uniformly rotating disk. He found that the disturbances

---

[15] These remarks refer to the case $m > 0$. For $m = 0$, the corresponding conditions are outgoing ($k < 0$) and incoming ($k > 0$) waves ($\omega_r > 0$).

[16] This is a fact strongly emphasized by the Lindblads; cf. Fig. D, Appendix 1.

[17] It is significant that P. O. Lindblad [10a] obtained leading patterns in the early stages of the development in his numerical investigations of the stability of a material disk.

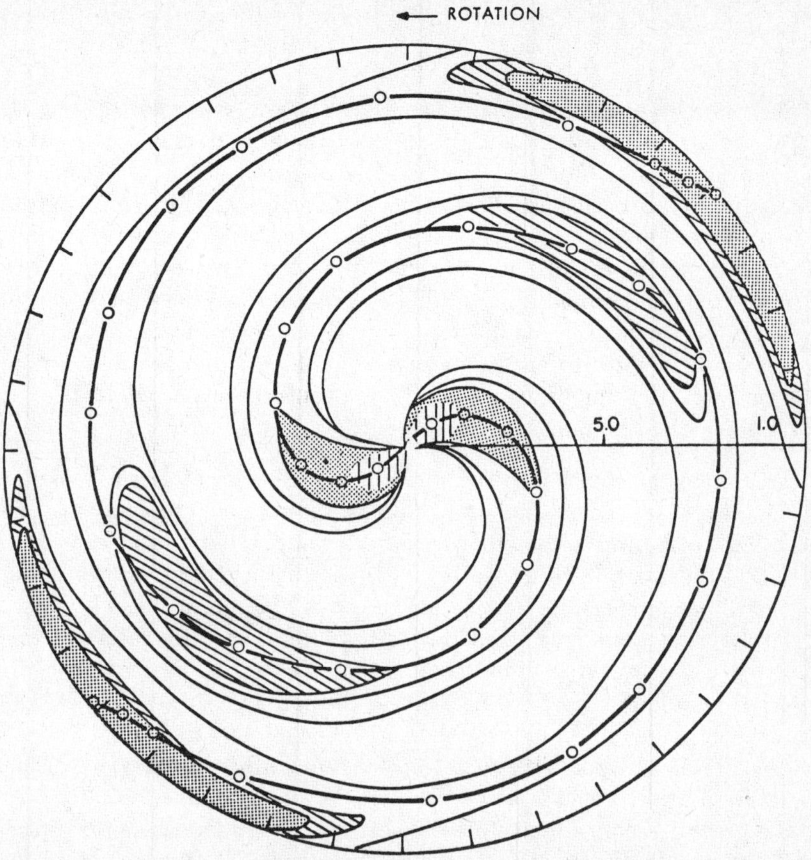


FIG. 5. *Trailing spiral patterns with singularity at center (after Rehm)*

regular at the center showed neither leading nor trailing character. He also suggested the consideration of disturbances over a class of differentially rotating disks which permits analytical treatment. Rehm found that the regular solutions over these disks with differential rotation again showed leading spiral patterns. However,[18] he was also able to show the existence of both trailing and leading spirals if the solution is allowed to have a bar-like singularity at the center. (These are solutions expressible in terms of series of Legendre functions $P_n$ and $Q_n$, with a few terms of each kind.) Fig. 5 shows a pattern obtained by him. He also found that, in the middle

[18] Although in this instance Rehm's calculations were made only in the case of uniform rotation, it is safe to conjecture that the result would not be greatly changed by differential rotation.

part of the radial range (neither close to the center nor close to the rim of his disk), the pattern formula given by Lin and Shu [8] yielded a reasonably good approximation.

The bar-like singularity might be regarded as a source for driving the spiral wave. On the other hand, if the disk proper prefers a spiral pattern (as mentioned above for the cases with velocity dispersion), this bar-like singularity could be regarded as a sink. After all, if a two-armed pattern is continued right to the center of the disk, some bar-like pattern is almost unavoidable. Rehm's calculations would at least give a crude indication as to how dynamical compatibility could be maintained. In the actual galaxies, the large dispersion velocity of the stars in the three-dimensional nuclear region would be able to adjust to any requirement from the disk proper. Furthermore, the amplitude of the wave would be expected to be very small outside of the range specified by (8.13).

**9. Concluding remarks.** In the study of the behavior of a galaxy of stars in the form of a disk in differential rotation, a primary problem is that of the spiral structure. Although various possibilities have been considered (see [13]), the present work, together with other earlier studies, suggests that the spiral patterns can be understood in terms of a quasi-stationary spiral structure of the stars maintained primarily according to the laws of stellar dynamics. The next step to be taken is then to describe the motion of the gas and the young stars born in them in the spiral gravitational field predominantly determined by the stars of all ages, but including the self-gravitation of the gas. The migration of young stars, as they move in a spiral gravitational field is an especially interesting problem.

The present theory (quantitatively) justifies the usual concept of (qualitatively) describing the stellar disk as composed of younger stars of low peculiar velocities moving in a smooth background of old stars of high dispersion speeds. The fraction of stars that should be assigned to each population is explicitly given in terms of a reduction factor, which of course varies somewhat with the purpose at hand. In terms of this reduction factor, one can justify (§8) the suggestion of Lin and Shu [8] for the use of their formula for the spiral pattern with a suitably reduced surface density of matter.

The theory also leads to the conclusion that trailing and leading arms are differentiated from each other by mild instabilities[19] which depend on the asymmetry in the distribution functions. This suggests that only spiral patterns are possible. Rings can occur only under exceptional circumstances. Whether trailing or leading arms should prevail depends on more detailed

[19] In the nature of an "over-stability."

considerations; but closer examination does show that trailing patterns are preferred.

The mild instability just described can only be balanced either by the effect of finite amplitudes or through the gas together with its own gravitational field and the associated magnetic field. Much work remains to be done to form a complete picture (beginning perhaps with a search for exactly periodic waves of small but finite amplitudes); but the present work does show a possible mechanism for causing spiral patterns to develop in a disk of stars with peculiar velocities.

Even within the scope of the linear theory, there are various interesting features regarding the basic processes. First of all, the mechanism of Landau damping is obviously important for these collective modes. Lynden-Bell [11] considered this mechanism for the case of uniform rotation (by solving an initial-value problem), and found that there is little of it for the type of modes that are relevant here. It would be extremely desirable to extend the study of this problem to the case of differential rotation. By analogy with the problems in plasma dynamics, with uniform and nonuniform magnetic fields, one would expect more pronounced Landau damping when the epicyclic frequency is variable. Some investigations of related problems have been carried out by Julian and Toomre [6].

Another very interesting mathematical problem is the study of the asymptotic procedure near a condition of resonance. Here, it might be important to avoid the approximation of the exact characteristics (specified by the condition of constant angular momentum) by the approximate characteristics (constant radial distance). A familiar analogy is that occurring in acoustics, where the exact characteristics must be used when resonance condition is approached.

The conditions at the center of the galaxy can only be conjectured at the present time. The three-dimensional nuclear bulge is certainly a complicated system by itself. In an idealized disk model, one might admit singularities of a bar-like shape to accommodate the extension of a two-armed pattern into the center. Only the case of the zero-dispersion disk has been studied in detail. It is a relatively urgent task to consider the problem over a whole disk of stars including their peculiar velocities.

**Appendix 1. Elementary facts about galaxies.**[20] We collect below a number of elementary facts about galaxies in general, and normal spirals in particular. More detailed information is given about our own galaxy.

**A1a. Classification of galaxies.** Galaxies are usually classified into the

---

[20] For more detailed information about various galaxies, see [16].

following four categories according to their appearance:

| | *Relative abundance of each type (approx.)* |
|---|---|
| elliptical galaxies | 10% |
| normal spirals | 70% |
| barred spirals | 15% |
| irregulars | 5% |

Fig. 3 of the text shows the scheme of this Hubble classification. The irregulars are not shown in that diagram. Some more examples of galaxies of various types are shown in Fig. A.

We are concerned mainly with normal spirals. Here it is interesting to note the increase of gas content as we pass through the sequence Sa, Sb, Sc to irregulars:

| | *Approximate gas content* |
|---|---|
| Sa spirals | 1% |
| Sb spirals | 2–3% |
| Sc spirals | 8% |
| Irregulars | 15% |

This estimation of gas content is based on atomic hydrogen alone. If there is molecular hydrogen, the amount might even be twice as great.

Irregulars are listed here for comparison, because it has been reported that an underlying spiral structure can be detected for the companion of the whirlpool galaxy and a companion of our own galaxy, the Large Magellanic Cloud. It would be interesting to investigate whether all of the irregulars (or which types of these) have an underlying spiral structure.

**A1b. Properties of individual galaxies, especially our own.**

(a) The linear dimensions of all spiral galaxies are of the order of 10,000 parsecs (1 pc. = 3.26 l.y. = $3.08 \times 10^{18}$ cm.), some smaller, some larger. Our own galaxy is among the largest of all spirals. The sun is at a distance of about 10 kpc. from the galactic center. About 80% of the mass of the galaxy lies within the solar orbit around the center.

(b) The mass of our galaxy is about $2 \times 10^{11}$ solar masses. The solar mass is about $2 \times 10^{33}$ g., making the mass of our galaxy $4 \times 10^{44}$ g.

(c) The circular velocity is approximately constant at 250 km./sec. Since (1 km./sec.) $\times 10^6$ yrs. = 1.02 pc., the period of revolution is about 200 million years at a radial distance of 8 kpc. from the center, and about 100 million years at 4 kpc. The detailed distribution of rotation velocities is given in Fig. B. (See the Schmidt model in §A1c.)

(d) For our galaxy, the spiral arms have a width of the order of 500 pc.; the inter-arm regions have a width of about 1500 pc.;—corresponding to a

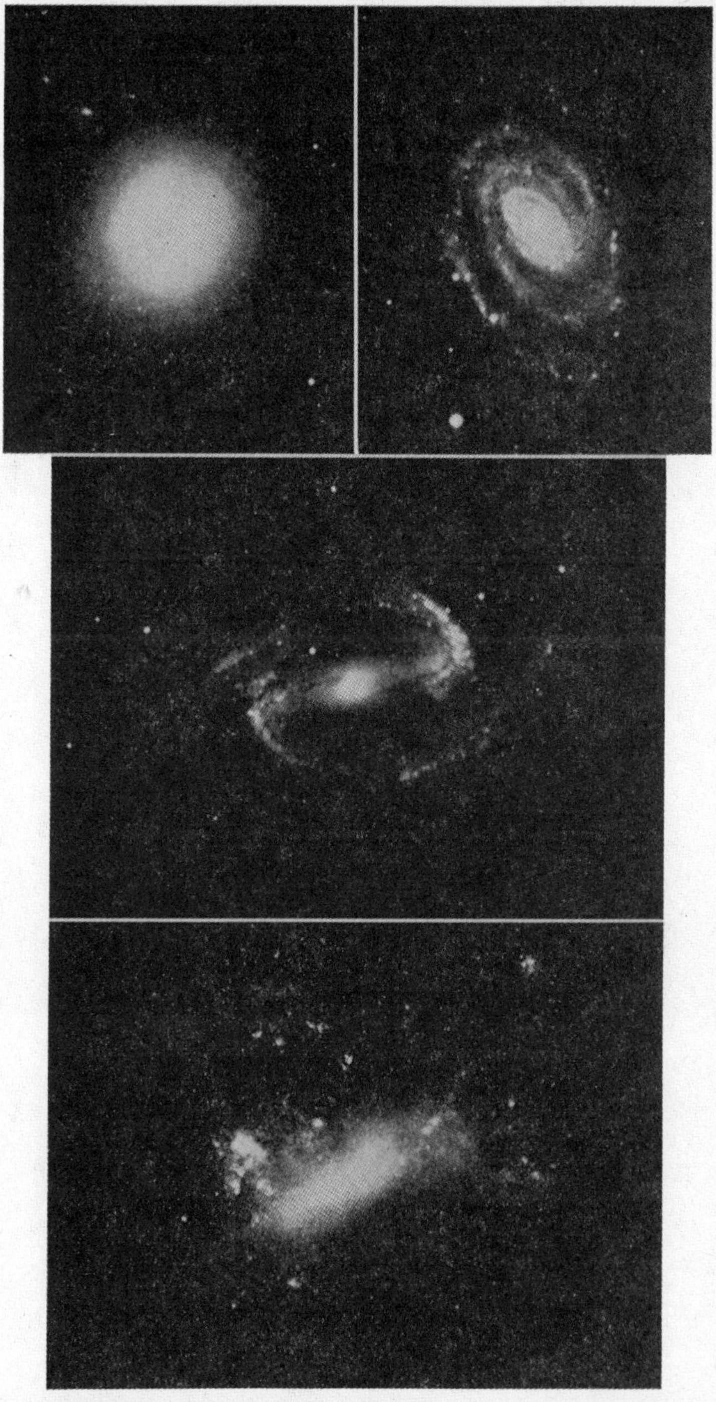

FIG. A. *Examples of galaxies of various types. Upper left: Elliptical galaxy EO pec. NGC 4486 Upper right: Normal spiral SC NGC 5364 Center: Barred galaxy SBb(s), NGC 1300 Lower: Irregular galaxy Irr I, the Large Magellanic Cloud.*

*Motion of Interstellar Gas* 7

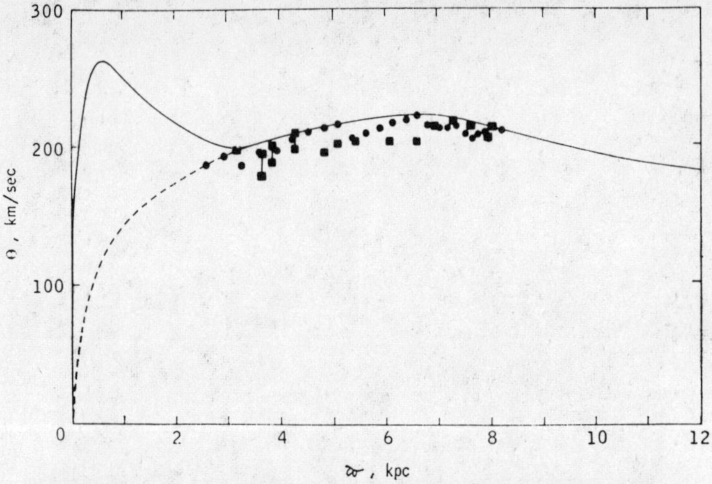


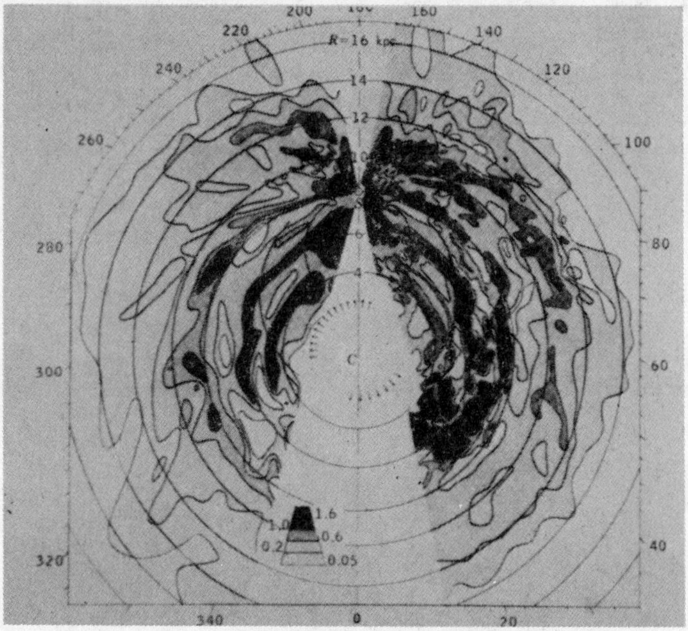


Fig. B. *Variation of circular velocity with distance from center of our own galaxy (after Oort* [*13*]).

Fig. C. *Distribution of hydrogen in the galactic plane (after Oort* [*13*])

radial "wave-length" of the pattern of the order of 2 kpc. The grand design of the spiral pattern extends over a scale of the order of 10 kpc. For NGC 5364 (Fig. A), the linear width of the outer spiral arms is about 700 pc. The spacing between the center of the outer arms is about 3300 parsecs on the left side.

(e) There are more gas and young stars (of types O and B) in the bright arms. Thus, the contrast in brightness does not represent a proportional contrast in total mass density (which is very small). Contrast in the density of the gas may however amount to about three or four to one.

*All the following is for our own galaxy.*

(f) The distribution of stars in the direction perpendicular to the galactic disk extends over a thickness of 600 pc. between layers where the density is one-half of the maximum. The corresponding figure for the gas is about 170 pc. Thus, the disk is indeed very thin. The distribution of gas over the galactic plane is shown in Fig. C.

(g) The fraction of matter that is gas (atomic hydrogen) is about 20% in our neighborhood, but only about 2% at a location, say, 5 kpc. from the center.

TABLE A

*Angular velocity and epicyclic frequency for the Schmidt model (after P. O. Lindblad)*

| $\varpi$ | $\Omega$ | $\kappa$ | $\Omega - \kappa/2$ | $\varpi$ | $\Omega$ | $\kappa$ | $\Omega - \kappa/2$ |
|---|---|---|---|---|---|---|---|
| kpc | km/sec. kpc | km/sec. kpc | km/sec. kpc | kpc | km/sec. kpc | km/sec. kpc | km/sec. kpc |
| 2.0 | 88.2 | 141.7 | 17.3 | 12.0 | 14.5 | 11.5 | 8.8 |
| .5 | 74.1 | 120.0 | 14.1 | .5 | 13.5 | 10.2 | 8.4 |
| 3.0 | 65.7 | 103.9 | 13.8 | 13.0 | 12.6 | 9.6 | 7.9 |
| .5 | 58.4 | 91.0 | 12.8 | .5 | 11.8 | 9.3 | 7.2 |
| 4.0 | 52.5 | 81.4 | 11.8 | 14.0 | 11.2 | 9.2 | 6.6 |
| .5 | 47.7 | 73.3 | 51.1 | .5 | 10.5 | 9.2 | 5.9 |
| 5.0 | 43.7 | 66.4 | 10.5 | 15.0 | 10.0 | 8.9 | 5.5 |
| .5 | 40.3 | 61.1 | 9.8 | .5 | 9.4 | 8.5 | 5.2 |
| 6.0 | 37.4 | 56.2 | 9.3 | 16.0 | 9.0 | 8.0 | 5.0 |
| .5 | 34.8 | 50.1 | 9.8 | .5 | 8.6 | 7.4 | 4.8 |
| 7.0 | 32.2 | 43.2 | 10.6 | 17.0 | 8.2 | 7.1 | 4.6 |
| .5 | 29.7 | 36.3 | 11.6 | .5 | 7.8 | 6.9 | 4.3 |
| 8.0 | 27.3 | 29.7 | 12.4 | 18.0 | 7.4 | 6.7 | 4.1 |
| .5 | 25.0 | 24.5 | 12.7 | .5 | 7.1 | 6.6 | 3.8 |
| 9.0 | 22.8 | 21.3 | 12.2 | 19.0 | 6.8 | 6.5 | 3.6 |
| .5 | 21.0 | 19.8 | 11.1 | .5 | 6.6 | 6.4 | 3.4 |
| 10.0 | 19.4 | 18.5 | 10.2 | 20.0 | 6.3 | 6.0 | 3.3 |
| .5 | 18.0 | 17.1 | 9.4 | | | | |
| 11.0 | 16.7 | 15.2 | 9.1 | | | | |
| .5 | 15.6 | 13.0 | 9.0 | | | | |

(h) Except for the "3 kpc. arm," where the radial motion of the gas is observed to be 53 km./sec., this motion is believed to have a velocity of no more than 15 km./sec.

(i) The mean density of gas (atomic hydrogen) is about $2 \times 10^{-24}$ g./c.c. (1.2 H atoms/c.c.). This corresponds to a kinetic energy of galactic

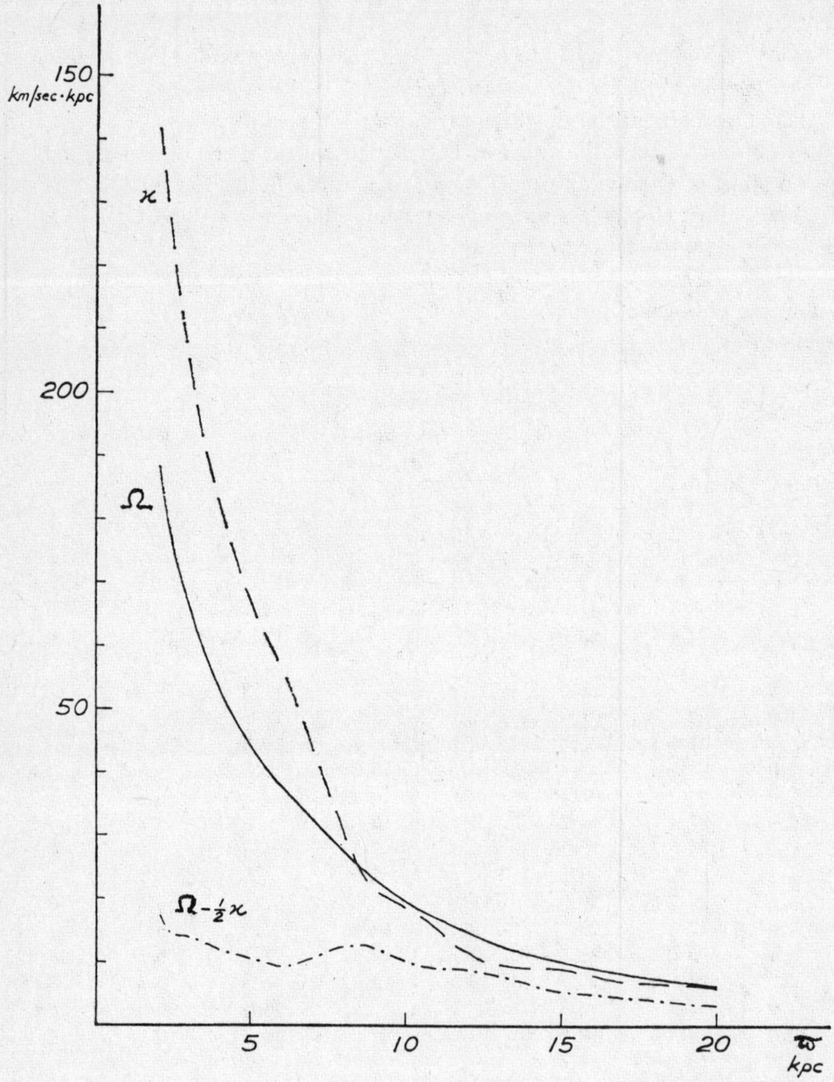


FIG. D. *Circular velocity and epicyclic frequency according to Schmidt model (after P. O. Lindblad).*

rotation of 1300 × $10^{-12}$ erg cm.$^{-3}$. It is not yet known whether there might be molecular hydrogen of comparable mass density (see [4]).

(j) The gas is actually in the form of clouds with a diameter of the order of 10 pc. The density in the cloud is 3-10 times that in the surroundings.

(k) The turbulent motion of the gas has a velocity of the order of 7 km./sec., giving an energy density of the order of 0.5 × $10^{-12}$ erg cm.$^{-3}$.

(l) Near bright stars, there is an ionized region in the gas called H II region (or Strömgren spheres). The un-ionized regions are called H I regions, which are estimated to have a temperature of $10^2$ °K.

(m) The magnetic field is estimated to have a magnitude of the order of $10^{-5}$ gauss (3 times larger or smaller). This corresponds to an energy density of 4 × $10^{-12}$ erg cm.$^{-3}$.

(n) The energy density of cosmic ray particles is of the order of 1 × $10^{-12}$ erg cm.$^{-3}$.

(o) There are dust particles in the interstellar regions. They might play an important role for the combination of atomic hydrogen into molecular form.

**A1c. The Schmidt model.** The distribution or rotational velocity $\Omega(\varpi)$ and epicyclic frequency $\kappa(\varpi)$ are given in Table A and Fig. D, according to the model constructed by Schmidt [17]. These data are taken from [10a, pp. 10, 11].

**Appendix 2. Mathematical theory of stellar dynamics for an infinitesimally thin disk in differential rotation: general principles.**

**A2a. Notation.**

$t$ — time.

$\varpi, \theta, z$ — cylindrical coordinates.

$\Pi, \Theta, Z$ — components of total stellar velocity in cylindrical coordinates.

$\Psi_3(\varpi, \theta, z, \Pi, \Theta, Z, t)$ — distribution function in three-dimensional case.

$\Pi_{00}(\varpi, \theta, z, t)$, $\Theta_{00}(\varpi, \theta, z, t)$, $Z_{00}(\varpi, \theta, z, t)$ — mean stellar velocity at a given locality.

$\Pi_0(\varpi, \theta, z, t)$, $\Theta_0(\varpi, \theta, z, t)$, $Z_0(\varpi, \theta, z, t)$ — similar quantities (but *not* necessarily mean values) for an arbitrarily chosen reference system.

$c_\varpi^{(0)} = \Pi - \Pi_{00}$, $c_\theta^{(0)} = \Theta - \Theta_{00}$, $c_z^{(0)} = Z - Z_{00}$ — components of dispersion velocity.

$c_\varpi = \Pi - \Pi_0$, $c_\theta = \Theta - \Theta_0$, $c_z = Z - Z_0$ — components of dispersion velocity relative to an arbitrary reference system ($\Pi_0 = Z_0 = 0$, $\Theta_0 = \varpi\Omega(\varpi)$ in the present paper).

In the two-dimensional case, the following notation is adopted:

$\Psi_2(\varpi, \theta, \Pi, \Theta, t)$
distribution function in two-dimensional case.

$\sigma = m \iint \Psi_2 \, d\Pi \, d\Theta$
surface density, where $m$ is the mass of an individual star.

$\Pi = \Pi_{00}(\varpi, \theta, t) + c_\varpi^{(0)} = \Pi_0(\varpi, \theta, t) + c_\varpi,$

$\Theta = \Theta_{00}(\varpi, \theta, t) + c_\theta^{(0)} = \Theta_0(\varpi, \theta, t) + c_\theta,$
with definition of symbols similar to those above.

$\langle q \rangle = \iint q\Psi_2 \, d\Pi \, d\Theta \Big/ \iint \Psi_2 \, d\Pi \, d\Theta$
local mean value for $q$.

$\langle \Pi \rangle = \Pi_0 + \langle c_\varpi \rangle = \Pi_0 + v_\varpi = V_\varpi + v_\varpi = \Pi_{00},$

$\langle \Theta \rangle = \Theta_0 + \langle c_\theta \rangle = \Theta_0 + v_\theta = V_\theta + v_\theta = \Theta_{00}$
($V_\varpi$ usually taken to be zero, $V_\theta = \varpi\Omega(\varpi)$).

$\mathcal{V}(\varpi, \theta, z, t)$
gravitational potential.

$a_\varpi, a_\theta$ acceleration components in the plane $z = 0$.

$\Psi(\varpi, \theta, c_\varpi, c_\theta, t) \equiv \Psi_2(\varpi, \theta, \Pi, \Theta, t)$
representation of the two-dimensional distribution function in terms of the dispersion velocities for a given choice of the reference system.

$c_\varpi' = c_\varpi - \langle c_\varpi \rangle, \quad c_\theta' = c_\theta - \langle c_\theta \rangle$
components of dispersion velocity relative to the instantaneous mean motion (same as $c_\varpi^{(0)}, c_\theta^{(0)}$).

**A2b. Basic equations.** The basic equation of stellar dynamics, when written in terms of cylindrical coordinates ($\varpi$, $\theta$, $z$) reads as follows [1, p. 187]:

$$\begin{aligned}(1) \quad & \frac{\partial \Psi_3}{\partial t} + \Pi \frac{\partial \Psi_3}{\partial \varpi} + \frac{\Theta}{\varpi} \frac{\partial \Psi_3}{\partial \theta} + Z \frac{\partial \Psi_3}{\partial z} - \left( \frac{\partial \mathcal{V}}{\partial \varpi} - \frac{\Theta^2}{\varpi} \right) \frac{\partial \Psi_3}{\partial \Pi} \\ & \qquad - \left( \frac{1}{\varpi} \frac{\partial \mathcal{V}}{\partial \theta} + \frac{\Pi\Theta}{\varpi} \right) \frac{\partial \Psi_3}{\partial \Theta} - \frac{\partial \mathcal{V}}{\partial z} \frac{\partial \Psi_3}{\partial Z} = 0,\end{aligned}$$

where $\Psi_3$ is the distribution function such that

$$(2) \qquad dN = \Psi_3(\varpi, \theta, z; \Pi, \Theta, Z; t) \varpi \, d\varpi \, d\theta \, dz \, d\Pi \, d\Theta \, dZ$$

gives the number of stars (all of mass $m$) in the element of phase space

$$(3) \qquad d\tau = \varpi \, d\varpi \, d\theta \, dz \, d\Pi \, d\Theta \, dZ.$$

The mean velocity of the stars $\Pi_{00}$, $\Theta_{00}$, $Z_{00}$ are defined, as for all mean quantities, by relations of the type

$$\Pi_{00} = \frac{\iiint \Pi\Psi_3 \, d\Pi \, d\Theta \, dZ}{\iiint \Psi_3 \, d\Pi \, d\Theta \, dZ.} \tag{4}$$

Note that $\Pi_{00}$ is a function of $(\varpi, \theta, z, t)$ and is therefore fundamentally different from $\Pi$, which is an independent variable.

We now define the dispersion velocity components $(c_\varpi^{(0)}, c_\theta^{(0)}, c_z^{(0)})$ by

$$\Pi = \Pi_{00} + c_\varpi^{(0)}, \qquad \Theta = \Theta_{00} + c_\theta^{(0)}, \qquad Z = Z_{00} + c_z^{(0)}. \tag{5}$$

Note that the components of the dispersion velocities $(c_\varpi^{(0)}, c_\theta^{(0)}, c_z^{(0)})$ are functions of the variables $(\varpi, \theta, z, \Pi, \Theta, Z, t)$, since $\Pi_{00}$, $\Theta_{00}$, and $Z_{00}$ are functions of $\varpi$, $\theta$, $z$, $t$. This point is especially important to note when we eliminate $\Pi$, $\Theta$, $Z$ in terms of the components of dispersion velocities as independent variables. It means that we must think in terms of the simultaneous transformation from the six variables $(\varpi, \theta, z, \Pi, \Theta, Z)$ to $(\varpi, \theta, z, c_\varpi^{(0)}, c_\theta^{(0)}, c_z^{(0)})$. The Jacobian of this transformation can, however, be shown to be identically equal to unity.

**A2c. The infinitesimally thin disk.** We shall now specialize to the problem of a "flat" disk. We introduce the two-dimensional distribution function

$$\Psi_2(\varpi, \theta, \Pi, \Theta, t) = \int_{-\infty}^{\infty}\int_{-\infty}^{\infty} \Psi_3 \, dz \, dZ, \tag{6}$$

and apply the same integration process to (1). Clearly, the terms involving $Z$ disappear if $\Psi_3$ is such that (6) is absolutely convergent. Most of the other terms yield corresponding expressions involving $\Psi_2$, but the following terms

$$-\frac{\partial \mathcal{V}}{\partial \varpi}\frac{\partial \Psi_3}{\partial \Pi}, \qquad -\frac{1}{\varpi}\frac{\partial \mathcal{V}}{\partial \theta}\frac{\partial \Psi_3}{\partial \Theta}$$

deserve special attention. We treat them as follows: Define

$$a_\varpi = -\left(\frac{\partial \mathcal{V}}{\partial \varpi}\right)_{z=0}, \qquad a_\theta = -\frac{1}{\varpi}\left(\frac{\partial \mathcal{V}}{\partial \theta}\right)_{z=0}; \tag{7}$$

then we may write

$$-\iint_{-\infty}^{\infty} \frac{\partial \mathcal{V}}{\partial \varpi}\frac{\partial \Psi_3}{\partial \Pi} \, dz \, dZ = a_\varpi \frac{\partial \Psi_2}{\partial \Pi} - \iint_{-\infty}^{\infty} \left(a_\varpi + \frac{\partial \mathcal{V}}{\partial \varpi}\right)\frac{\partial \Psi_3}{\partial \Pi} \, dz \, dZ, \tag{8}$$

where the last integration should be carried out first with respect to $z$. The integrand vanishes for $z = 0$ and $z = \pm\infty$, and the integral approaches zero as the thickness of the disk approaches zero. A similar argument holds for the other term. Thus, we finally obtain the following equation for an infinitesimally thin disk:

$$\frac{\partial \Psi_2}{\partial t} + \Pi \frac{\partial \Psi_2}{\partial \varpi} + \frac{\Theta}{\varpi} \frac{\partial \Psi_2}{\partial \theta} + \left( a_\varpi + \frac{\Theta^2}{\varpi} \right) \frac{\partial \Psi_2}{\partial \Pi} + \left( a_\theta - \frac{\Pi \Theta}{\varpi} \right) \frac{\partial \Psi_2}{\partial \Theta} = 0. \tag{9}$$

The surface density of this disk is

$$\sigma_*(\varpi, \theta, t) = m_* \iint_{-\infty}^{\infty} \Psi_2 \, d\Pi \, d\Theta. \tag{10}$$

Due to a given surface density, such as (10), the components of acceleration $(a_\varpi, a_\theta)$ are given by

$$a_\varpi = -\left( \frac{\partial \mathcal{V}}{\partial \varpi} \right)_{z=0}, \qquad a_\theta = -\frac{1}{\varpi} \left( \frac{\partial \mathcal{V}}{\partial \theta} \right)_{z=0}, \tag{11}$$

where $\mathcal{V}$ satisfies the Poisson equation

$$\frac{\partial^2 \mathcal{V}}{\partial \varpi^2} + \frac{1}{\varpi} \frac{\partial \mathcal{V}}{\partial \varpi} + \frac{1}{\varpi^2} \frac{\partial^2 \mathcal{V}}{\partial \theta^2} + \frac{\partial^2 \mathcal{V}}{\partial z^2} = 4\pi G \sigma \delta(z). \tag{12}$$

In (12), $G$ is the gravitational constant and $\delta(z)$ is Dirac's delta function.

We are usually interested in a system in differential rotation. This means that we have a state of motion with

$$\Pi_{00} \cong 0, \qquad \Theta_{00} \cong \varpi \Omega(\varpi), \qquad (Z_{00} \cong 0). \tag{13}$$

Thus, it is convenient to introduce the following dispersion velocity relative to an arbitrarily chosen reference state: $(\Pi_0 = 0, \Theta_0 = \omega \Omega(\omega))$:

$$c_\varpi = \Pi - 0, \qquad c_\theta = \Theta - \varpi \Omega(\varpi). \tag{14}$$

The components of mean dispersion velocities $\langle c_\varpi \rangle$ and $\langle c_\theta \rangle$ are no longer necessarily zero; but it is implied that they are "small" compared with $\varpi \Omega(\varpi)$. We shall now introduce $(c_\varpi, c_\theta)$ as the independent variables in place of $(\Pi, \Theta)$ in (9), and write

$$\Psi_2(\omega, \theta, \Pi, \Theta, t) \equiv \Psi(\omega, \theta, c_\varpi, c_\theta, t). \tag{15}$$

Then, we have transformations of derivatives typified by

$$\frac{\partial \Psi_2}{\partial \varpi} = \frac{\partial \Psi}{\partial \varpi} + \frac{\partial \Psi}{\partial c_\theta} \left\{ -\frac{d}{d\varpi} (\varpi \Omega) \right\}, \qquad \frac{\partial \Psi_2}{\partial \Theta} = \frac{\partial \Psi}{\partial c_\theta},$$

and (9) becomes

$$\begin{aligned} &\frac{\partial \Psi}{\partial t} + c_\varpi \frac{\partial \Psi}{\partial \varpi} + \left( \Omega + \frac{c_\theta}{\varpi} \right) \frac{\partial \Psi}{\partial \theta} + a_\varpi + \Omega^2 \varpi + 2\Omega c_\theta + \frac{1}{\varpi} c_\theta^2 \Big) \frac{\partial \Psi}{\partial c_\varpi} \\ &\qquad + \left( a_\theta - \frac{\kappa^2}{2\Omega} c_\varpi - \frac{1}{\varpi} c_\varpi c_\theta \right) \frac{\partial \Psi}{\partial c_\theta} = 0, \end{aligned} \tag{16}$$

where $\Psi$ is associated with the volume element

$$d\tau_c = \varpi \, d\varpi d\theta \, dc_\varpi \, dc_\theta$$

of the phase space, and

$$\kappa^2 = (2\Omega)^2 \left\{1 + \frac{\varpi}{2\Omega}\frac{d\Omega}{d\varpi}\right\}. \tag{17}$$

The (local) averages of the components of the dispersion velocity are in general different from zero, and these will be respectively denoted by $v_\varpi$ and $v_\theta$. We have

$$\frac{\sigma_*}{m_*} v_\varpi = \iint c_\varpi \Psi \, dc_\varpi \, dc_\theta \, , \qquad \frac{\sigma_*}{m_*} v_\theta = \iint c_\theta \Psi \, dc_\varpi \, dc_\theta \, , \tag{18}$$

where $\sigma_*$ is the surface density:

$$\sigma_* = m_* \iint \Psi \, dc_\varpi \, dc_\theta \, , \tag{19}$$

and $m_*$ is mass per star. The local mean of the total stellar velocity $(\Pi, \Theta)$ is easily verified to be given by $(\varpi\Omega(\varpi) + v_\varpi, v_\theta)$, as expected.

Equation (16) is the final equation for stellar dynamics with reference to a conveniently chosen differentially rotating system, for the configuration of a thin disk with infinitesimal thickness. The acceleration $(a_\varpi, a_\theta)$ derives part of its contributions from the surface density $\sigma_*$ given by (19) (according to (11) and (12)), but it may also have contributions from any "external" agency (e.g., gas or another neighboring galaxy).

**A2d. Characteristics and integrals.** The characteristics of the differential equation (9) are defined by the system of differential equations

$$dt = \frac{d\varpi}{\Pi} = \frac{d\theta}{\Theta/\varpi} = \frac{d\Pi}{a_\varpi + \Theta^2/\varpi} = \frac{d\Theta}{a_\theta - \Pi\Theta/\varpi}. \tag{20}$$

From the last two equations, we can derive

$$dt = \frac{\Pi \, d\Pi + \Theta \, d\Theta}{a_\varpi \Pi + a_\theta \Theta/\varpi}. \tag{21}$$

If the potential $\mathcal{V}$ is *time-independent*, this leads to the energy integral

$$\tfrac{1}{2}(\Pi^2 + \Theta^2) + \mathcal{V}(\varpi, \theta, 0) = E, \quad \text{a constant.} \tag{22}$$

If the potential is *axially symmetric* ($a_\theta = 0$), then (20) yields

$$\frac{d\varpi}{\Pi} = \frac{d\Theta}{-\Pi\Theta/\varpi},$$

and we have

$$\varpi\Theta = J, \quad \text{a constant.} \tag{23}$$

This is the integral for angular momentum.

In terms of the variables $(c_\varpi, c_\theta)$, these integrals become

$$\tfrac{1}{2}\{c_\varpi^2 + (\varpi\Omega + c_\theta)^2\} + \mathcal{V}(\varpi, \theta, 0) = E \tag{22a}$$

for time-independent potentials, and

$$\varpi(\varpi\Omega + c_\theta) = J \tag{23a}$$

for axially-symmetric potentials. These should, of course, be derivable from the equations for the characteristics of (16):

$$\begin{aligned} dt = \frac{d\varpi}{c_\varpi} = \frac{d\theta}{\Omega + c_\theta/\varpi} &= \frac{dc_\varpi}{a_\varpi + (\varpi\Omega + c_\theta)^2/\varpi} \\ &= \frac{dc_\theta}{a_\theta - (\kappa^2/2\Omega + c_\theta/\varpi)c_\varpi}. \end{aligned} \tag{20a}$$

Indeed they are. But the derivation is difficult to discover without the guide given above. For example, in the axially symmetrical case, the second differential in (20a) and the last one give

$$\frac{d\varpi}{c_\varpi} = \frac{dc_\theta}{-(\kappa^2/2\Omega + c_\theta/\varpi)c_\varpi}. \tag{24}$$

Although this can be verified to lead to (23a), the process required is somewhat camouflaged. A similar remark applies to the energy integral.

**A2e. Symmetric distribution in equilibrium.** Consider now the basic state, which is an axially symmetric distribution in equilibrium:

$$\Psi = \Psi_0(\varpi, c_\varpi, c_\theta). \tag{25}$$

We then have

$$(a_\varpi, a_\theta) = (a_{\varpi 0}(\varpi), 0). \tag{26}$$

The basic equation (16) becomes

$$\begin{aligned} c_\varpi \frac{\partial \Psi_0}{\partial \varpi} &+ \left(a_{\varpi 0} + \Omega^2\varpi + 2\Omega c_\theta + \frac{c_\theta^2}{\varpi}\right)\frac{\partial \Psi_0}{\partial c_\varpi} \\ &+ \left(0 - \left(\frac{\kappa^2}{2\Omega} + \frac{c_\theta}{\varpi}\right)c_\varpi\right)\frac{\partial \Psi_0}{\partial c_\theta} = 0. \end{aligned} \tag{27}$$

The characteristics of (27) are given by (22a) and (23a), and thus we have the general integral of (27) in the form

$$\Psi_0 = F(J_1, J_2), \tag{28}$$

where $F$ is an arbitrary function of the two arguments, and $J_1$ and $J_2$ are defined by

$$J_1 \equiv \varpi(\varpi\Omega + c_\theta), \tag{29}$$

$$J_2 \equiv \tfrac{1}{2}[c_\varpi^{\ 2} + (\varpi\Omega + c_\theta)^2] + \mathcal{V}(\varpi). \tag{30}$$

Clearly, $J_1$ and $J_2$ are to be identified with $J$ and $E$ above.

**A2f. Choice of the reference state.** It is clear that (27) can be somewhat simplified if we introduce the special choice of $\Omega(\varpi)$ to make

$$a_{\varpi 0} + \Omega^2\varpi = 0, \tag{31}$$

i.e., if we take $\Omega(\varpi)$ to be the angular velocity that a particle would have if it were moving in a purely circular orbit. This is not to be confused with the mean angular velocity $\Omega(\varpi) + \langle c_\theta\rangle/\varpi$ in a symmetrical system, but the two are expected to be close. To calculate the difference, let us evaluate $\langle c_\theta\rangle$ from (27), after (31) is imposed.

This is done by multiplying (27) with $c_\varpi$ and integrating over the infinite velocity space. We get (cf. (38) below)

$$2\Omega\langle c_\theta\rangle = \frac{1}{\varpi_*}\left\{\frac{1}{\sigma_*}\frac{\partial}{\partial\varpi}[\varpi\langle c_\varpi^{\ 2}\rangle\sigma_*] - \langle c_\theta^{\ 2}\rangle\right\} - (a_{\varpi 0} + \Omega^2\varpi). \tag{32}$$

If (31a) is imposed, the last term on the right vanishes. This shows that $\langle c_\theta\rangle$ is indeed smaller compared with $\langle c_\theta^{\ 2}\rangle^{1/2}$ if the latter is smaller compared with the circular velocity $\Theta_0 = \varpi\Omega$.

This special choice of the reference state will often—but not always—be adopted. It might be convenient, sometimes, to choose $\Omega(\varpi)$ such that $\langle c_\theta\rangle = 0$. In that case, we have

$$a_{\varpi 0} + \Omega^2\varpi = \frac{1}{\varpi}\left\{\frac{1}{\sigma_*}\frac{\partial}{\partial\varpi}[\varpi\sigma_*\langle c_\varpi^{\ 2}\rangle] - \langle c_\theta^{\ 2}\rangle\right\}, \tag{33}$$

which is of the order of the mean square dispersion velocities.

**A2g. Moment equations: hydrodynamic equations.** In some studies, it is convenient to use the moment equations. We shall write these down for convenience of reference. If we multiply $mc_\varpi^{\ \alpha}c_\theta^{\ \beta}$ into the fundamental equation (16) for the distribution function relative to a differentially rotating system, and then integrate with respect to $c_\varpi$ and $c_\theta$, we obtain the following equation:

$$\frac{\partial}{\partial t}[\sigma_*\langle c_\varpi^{\ \alpha}c_\theta^{\ \beta}\rangle] + \frac{\partial}{\partial\varpi}[\sigma_*\langle c_\varpi^{\alpha+1}c_\theta^{\ \beta}\rangle] + \frac{\partial}{\partial\theta}[\sigma_*\Omega\langle c_\varpi^{\ \alpha}c_\theta^{\ \beta}\rangle - \frac{\sigma_*}{\varpi}\langle c_\varpi^{\ \alpha}c_\theta^{\ \beta+1}\rangle]$$

$$
\begin{aligned}
&= \alpha\sigma_* \left\{ (a_\varpi + \Omega^2\varpi)\langle c_\varpi^{\alpha-1} c_\theta^{\beta}\rangle + 2\Omega\langle c_\varpi^{\alpha-1} c_\theta^{\beta+1}\rangle + \frac{1}{\varpi}\langle c_\varpi^{\alpha-1} c_\theta^{\beta+2}\rangle \right\} \\
(34)\qquad &+ \beta\sigma_* \left\{ a_\theta\langle c_\varpi^{\alpha} c_\theta^{\beta-1}\rangle - \frac{\kappa^2}{2\Omega}\langle c_\varpi^{\alpha+1} c_\theta^{\beta-1}\rangle \right\} \\
&+ (\beta+1)\sigma_* \frac{1}{\varpi}\langle c_\varpi^{\alpha+1} c_\theta^{\beta}\rangle,
\end{aligned}
$$

where

$$
\sigma_*\langle c_\varpi^{\alpha} c_\theta^{\beta}\rangle = m_* \iint_{-\infty}^{\infty} \Psi c_\varpi^{\alpha} c_\theta^{\beta}\, dc_\varpi\, dc_\theta . \tag{35}
$$

The terms on the right are obtained by making use of the following relations:

$$
\begin{aligned}
\frac{\partial\Psi}{\partial c_\varpi} c_\varpi^{\alpha} c_\theta^{\beta} &= \left[ \frac{\partial}{\partial c_\varpi}(c_\varpi^{\alpha}\Psi) - \alpha c_\varpi^{\alpha-1}\Psi \right] c_\theta^{\beta}, \\
(36)\qquad \frac{\partial\Psi}{\partial c_\theta} c_\varpi^{\alpha} c_\theta^{\beta'} &= \left[ \frac{\partial}{\partial c_\theta}(c_\theta^{\beta'}\Psi) - \beta' c_\theta^{\beta'-1}\Psi \right] c_\varpi^{\alpha}.
\end{aligned}
$$

We note that when $\alpha = 0$, the first term on the right-hand side vanishes, and thus the embarrassing terms involving $c_\varpi^{\alpha-1} = c_\varpi^{-1}$ do not actually occur.

*Hydrodynamic equations.* The equations with $(\alpha, \beta) = (0, 0), (1, 0), (0, 1)$, respectively, correspond to those usually occurring in hydrodynamics. They are therefore explicitly written down here for convenience of reference.

(A) Equation of continuity, $(\alpha, \beta) = (0, 0)$:

$$
\frac{\partial\sigma_*}{\partial t} + \frac{\partial}{\partial\varpi}[\sigma_*\langle c_\varpi\rangle] + \frac{\partial}{\partial\theta}\left[\sigma_*\Omega + \frac{\sigma_*}{\varpi}\langle c_\varpi\rangle\right] + \frac{\sigma_*}{\varpi}\langle c_\varpi\rangle = 0. \tag{37}
$$

(B) Equation of motion in the radial direction, $(\alpha, \beta) = (1, 0)$:

$$
\begin{aligned}
&\frac{\partial}{\partial t}[\sigma_*\langle c_\varpi\rangle] + \frac{\partial}{\partial\varpi}[\varpi\sigma_*\langle c_\varpi^2\rangle] + \frac{\partial}{\partial\theta}\left[\sigma_*\Omega\langle c_\varpi\rangle + \frac{\sigma_*}{\varpi}\langle c_\varpi c_\theta\rangle\right] \\
(38)\qquad &\qquad = \sigma_*\left\{(a_\varpi + \Omega^2\varpi) + 2\Omega\langle c_\theta\rangle + \frac{1}{\varpi}\langle c_\theta^2\rangle\right\} - \frac{\sigma_*}{\varpi}\langle c_\varpi^2\rangle.
\end{aligned}
$$

(C) Equation of motion in the angular direction, $(\alpha, \beta) = (0, 1)$:

$$
\begin{aligned}
&\frac{\partial}{\partial t}[\sigma_*\langle c_\theta\rangle] + \frac{\partial}{\partial\varpi}[\sigma_*\langle c_\varpi c_\theta\rangle] + \frac{\partial}{\partial\theta}\left[\sigma_*\Omega\langle c_\theta\rangle + \frac{\sigma_*}{\varpi}\langle c_\theta^2\rangle\right] \\
(39)\qquad &\qquad = \sigma_*\left\{a_\theta - \frac{\kappa^2}{2\Omega}\langle c_\varpi\rangle - \frac{2}{\varpi}\langle c_\varpi c_\theta\rangle\right\}.
\end{aligned}
$$

These equations cannot yet be compared with ordinary hydrodynamic equations without the introduction of the dispersion velocity relative to the mean motion. We shall now write

$$\langle c_\varpi \rangle = v_\varpi\,, \qquad \langle c_\theta \rangle - v_\theta\,, \tag{40}$$

and define

$$c_\varpi' = c_\varpi - \langle c_\varpi \rangle, \qquad c_\theta' = c_\theta - \langle c_\theta \rangle. \tag{41}$$

We also introduce the symbols

$$\tau_{\varpi\varpi} = -\sigma_*\langle c_\varpi'^2 \rangle, \qquad \tau_{\varpi\theta} = \tau_{\theta\varpi} = -\sigma_*\langle c_\varpi' c_\theta' \rangle, \qquad \tau_{\theta\theta} = -\sigma_*\langle c_\theta'^2 \rangle, \tag{42}$$

which represent the "stress" components due to the peculiar velocity $(c_\varpi', c_\theta')$ over the mean motion. The equations (37)–(39) can then be rewritten in the following form:

$$\frac{\partial \sigma_*}{\partial t} + \frac{1}{\varpi}\frac{\partial}{\partial \varpi}(\sigma_* \varpi v_\varpi) + \frac{1}{\varpi}\frac{\partial}{\partial \theta}[\sigma_*(\Omega\varpi + v_\theta)] = 0, \tag{37a}$$

$$\begin{aligned} &\sigma_*\left[\frac{\partial v_\varpi}{\partial t} + v_\varpi \frac{\partial v_\varpi}{\partial \varpi} + \left(\Omega + \frac{v_\theta}{\varpi}\right)\frac{\partial v_\varpi}{\partial \theta}\right] - \frac{\sigma_*}{\varpi}(\varpi\Omega + v_\theta)^2 \\ &\qquad = \sigma_* a_\varpi + \left[\frac{\partial \tau_{\varpi\varpi}}{\partial \varpi} + \frac{\partial \tau_{\varpi\theta}}{\partial \theta} + \frac{1}{\varpi}(\tau_{\theta\theta} - \tau_{\varpi\varpi})\right]. \end{aligned} \tag{38a}$$

$$\begin{aligned} &\sigma_*\left[\frac{\partial v_\theta}{\partial t} + v_\varpi \frac{\partial v_\theta}{\partial \varpi} + \left(\Omega + \frac{v_\theta}{\varpi}\right)\frac{\partial v_\theta}{\partial \theta}\right] + \sigma_* v_\varpi \left(\frac{\kappa^2}{2\Omega} + \frac{v_\theta}{\varpi}\right) \\ &\qquad = \sigma_* a_\theta + \left(\frac{\partial \tau_{\theta\varpi}}{\partial \varpi} + \frac{1}{\varpi}\frac{\partial \tau_{\theta\theta}}{\partial \theta} + \frac{2\tau_{\varpi\theta}}{\varpi}\right). \end{aligned} \tag{39a}$$

These equations are now familiar to the hydrodynamicists: the total "hydrodynamic" velocity in the angular direction being $\Omega\varpi + v_\theta$. The term with $\kappa^2/2\Omega$ requires a little calculation to verify. The terms in the "stresses" are identical with those occurring in turbulent motion. (See, e.g., [22, p. A3].)

If the "stresses" are neglected, we get the equations used by Lin and Shu [8]. Equations (1a)–(1c) in that paper can be obtained from (37a)–(39a) by the following changes in notation:

$$\begin{gathered} (t.\ \varpi.\ \theta) \Rightarrow (t, r, \theta), \\ \sigma_* \Rightarrow \mu, \\ (v_\varpi\,, \Omega\varpi + v_\theta) \Rightarrow (u, v), \\ (a_\varpi\,, a_\theta) \Rightarrow (\phi_r\,, \phi_\theta/r), \\ \tau_{\varpi\varpi}\,, \tau_{\varpi\theta} = \tau_{\theta\omega}\,, \tau_{\theta\theta} \Rightarrow 0. \end{gathered}$$

## REFERENCES

[1] S. CHANDRASEKHAR, *Principles of Stellar Dynamics*, Dover, New York, 1942, 1960.

[2] S. CHANDRASEKHAR AND E. FERMI, *Magnetic fields in spiral arms*, Astrophys. J., 118 (1953), pp. 113–116.

[3] S. CHANDRASEKHAR AND JOHN VON NEUMANN, *The statistics of the gravitational field arising from a random distribution of stars*, Ibid., 95 (1942), pp. 489–531.

[4] R. J. GOULD, T. GOLD, AND E. E. SALPETER, *The interstellar abundance of the hydrogen molecule II. Galactic abundance and distribution*, Ibid., 138 (1963), pp. 408–425.

[5] C. HUNTER, *The structure and stability of self-gravitating disks*, Monthly Notices Roy. Astronom. Soc., 126 (1963), pp. 299–315.

[6] W. H. JULIAN AND A. TOOMRE, *On the non-axisymmetric stability and responses of a differentially rotating disk of stars*, Astrophys. J., to appear.

[7] A. J. KALNAJS, *The stability of highly flattened galaxies*, Ph.D. Thesis, Department of Astronomy, Harvard University, Cambridge, 1965.

[8] C. C. LIN AND F. H. SHU, *On the spiral structure of disk galaxies*, Astrophys. J., 140 (1964), pp. 646–655.

[8a] C. C. LIN AND F. H. SHU, *On the spiral structure of disk galaxies, II. Outline of a theory of density waves*, Proc. Nat. Acad. Sci. U.S.A., 55 (1966), pp. 229–234.

[9] B. LINDBLAD, *On the possibility of a quasi-stationary spiral structure in galaxies*, Stockholm Obs. Ann., 22 (1963), pp. 3–20.

[10a] P. O. LINDBLAD, *The development of spiral structure in a galaxy approached by numerical computations*, Ibid., 21 (1960), pp. 3–73.

[10b] ———, *Gravitational resonance effects in the central layer of a galaxy*, Interstellar Matter in Galaxies, L. Woltjer, ed., Benjamin, New York, 1962, pp. 222–233.

[11] D. LYNDEN-BELL, *The stability and vibrations of a gas of stars*, Monthly Notices Roy. Astronom. Soc., 124 (1962), pp. 279–296.

[12] P. MORRISON, *On the origins of cosmic rays*, Rev. Modern Phys., 29 (1957), pp. 235–243.

[13] J. H. OORT, *Spiral structure*, Interstellar Matter in Galaxies, L. Woltjer, ed., Benjamin, New York, 1962, pp. 234–244.

[14] K. H. PRENDERGAST, *The motion of gas in barred spiral galaxies*, Ibid., pp. 217–221.

[15] R. G. REHM, Ph.D. Thesis, Department of Mathematics, Massachusetts Institute of Technology, Cambridge, 1965.

[16] A. SANDAGE, *The Hubble atlas of galaxies*, Carnegie Institute of Washington, Washington, D. C., 1961.

[17] M. SCHMIDT, *A model of the distribution of mass in the galactic system*, Bull. Astronom. Inst. Netherlands, 13 (1956), pp. 15–41.

[18] S. SHARPLESS AND O. G. FRANZ, *Composite photography of galaxies*, P.A.S.P., 75 (1963), pp. 219–225.

[19] L. SPITZER, *Physics of Fully Ionized Gases*, Interscience, New York, 1956.

[20] A TOOMRE, *On the gravitational stability of a disk of stars*, Astrophys. J., 139 (1964), pp. 1217–1235.

[21] F. ZWICKY, *Morphological Astronomy*, Springer-Verlag, Berlin, 1957, pp. 198–201.

[22] S. GOLDSTEIN, *Modern Developments in Fluid Dynamics*, vol. I, Oxford University Press, Oxford, 1938.

I.A.U. Symposium no. 31, 1967

# 52. DENSITY WAVES IN DISK GALAXIES

C. C. LIN and F. H. SHU

*(Massachusetts Institute of Technology and Harvard College Observatory, Cambridge, Massachusetts, U.S.A.)*

ABSTRACT

Density waves in the nature of those proposed by B. Lindblad are described by detailed mathematical analysis of collective modes in a disk-like stellar system. The treatment is centered around a hypothesis of quasi-stationary spiral structure. We examine (*a*) the mechanism for the maintenance of this spiral pattern, and (*b*) its consequences on the observable features of the galaxy.

## 1. GENERAL OBSERVATIONS

In discussing the spiral features of disk galaxies, it is important to note the difference between these two questions: (*a*) the organization of a spiral pattern over the whole disk, and (*b*) the detailed structure of a stretch of an individual arm. This point was clearly made by Oort (1962) in the following words:

> 'In systems with strong differential rotation, such as is found in all non-barred spirals, spiral features are quite natural. Every structural irregularity is likely to be drawn out into a part of a spiral. But *this* is not the phenomenon we must consider. We must consider a spiral structure extending over the whole galaxy, from the nucleus to its outermost part, and consisting of two arms starting from diametrically opposite points. Although this structure is often hopelessly irregular and broken up, the general form of the large-scale phenomenon can be recognized in many nebulae.'

The primary purpose of our present theory is indeed an attempt to explain the grand design over the whole disk. We believe that the mechanism for maintaining this grand design can be described essentially in terms of *density waves* of the general nature discussed by the late Bertil Lindblad.* Gravitational forces are predominant, while hydromagnetic forces can at most play a secondary role. Specifically, we believe that the density waves form a quasi-stationary spiral structure.

A more detailed description of our ideas and conclusions has been published elsewhere (Lin and Shu 1964, 1966). Here we shall only indicate a few high points of the theory, and report on the latest numerical results. The density waves are primarily collective modes in a stellar system, which may be described as a 'gravitational plasma'. Numerical results reported below are based on the simplest form of the theory, but they serve well the purpose of demonstrating the relevance of these density waves to the observed spiral features. Briefly, we conclude that the principal spiral pattern in our Galaxy can be

*For further discussions of the relationship between the present theory and that of Lindblad, see Lin and Shu (1966).

associated with a density wave propagating around the center at an angular speed of about 11 km $\sec^{-1}$ $\text{kpc}^{-1}$ in the direction of the general galactic rotation of stars.

## 2. SELF-SUSTAINED DENSITY WAVES OF A SPIRAL FORM

Spiral patterns in the plane of a disk galaxy may be described by functions of the form

$$F(\tilde{\omega}, \theta, t) = A(\tilde{\omega}) \exp \{i [\omega t - m\theta + \Phi(\tilde{\omega})]\}, \quad (1)$$

where $t$ is the time, $(\tilde{\omega}, \theta)$ are the polar coordinates in the plane, $A(\tilde{\omega})$ is a slowly varying function of $\tilde{\omega}$, $\Phi(\tilde{\omega})$ is a slowly varying (monotonic) function multiplied by a large parameter, $\omega$ is a real constant (for neutral waves), and $m$ is an integer. As usual, the real part of the above expression is to be taken. The lines of constant $\text{Re}(F)$ are approximately given by the equation

$$m(\theta - \theta_0) = \Phi(\tilde{\omega}) - \Phi(\tilde{\omega}_0), \quad (2)$$

which represents a spiral pattern with $m$ arms. The pattern given by (1) rotates at an angular velocity

$$\Omega_p = \omega/m. \quad (3)$$

The radial wave number of the spiral pattern is given by $|k(\tilde{\omega})|$, where

$$k(\tilde{\omega}) = \Phi'(\tilde{\omega}). \quad (4)$$

If the motion of the stars is in the direction of increasing $\theta$, trailing waves correspond to $k(\tilde{\omega}) < 0$ and leading waves to $k(\tilde{\omega}) > 0$.

It can be shown (see Lin and Shu 1966) that a self-sustained density wave of the general nature of (1) can be obtained through an analysis of small disturbances over a symmetrical disk of stars and gas in differential rotation. These waves have the following properties.

(*a*) They extend essentially over a range of the galactic disk where the condition

$$\Omega - \frac{\kappa}{m} < \Omega_p < \Omega + \frac{\kappa}{m} \quad (5)$$

is satisfied, where $\Omega(\tilde{\omega})$ is the angular speed of the stars, and $\kappa(\tilde{\omega})$ is the epicyclic frequency. We shall refer to this range as the *principal part* of the spiral pattern. Figure 1 shows that, for $m = 2$, the pattern would extend from $\tilde{\omega} = 4$ kpc outwards without limit, if $\Omega_p = 11$ km $\sec^{-1}$ $\text{kpc}^{-1}$. For other values of $m$, this principal part would be quite limited in extent.

(*b*) The dispersion relation for such waves is expressible in terms of the wave number $\lambda = 2\pi/|k|$ and the frequency $\nu = m(\Omega_p - \Omega)/\kappa$ at which the stars see the gravitational field. This relationship includes an additional parameter, which measures the velocity dispersion of the stars. If the velocity dispersion is barely enough to stabilize the disk against gravitational collapse (Toomre 1964), the dispersion relationship is as shown in Figure 2, where $\lambda_* = 4\pi^2 G\sigma_*/\kappa^2$, $G$ being the gravitational constant and $\sigma_*$ being the projected stellar density.

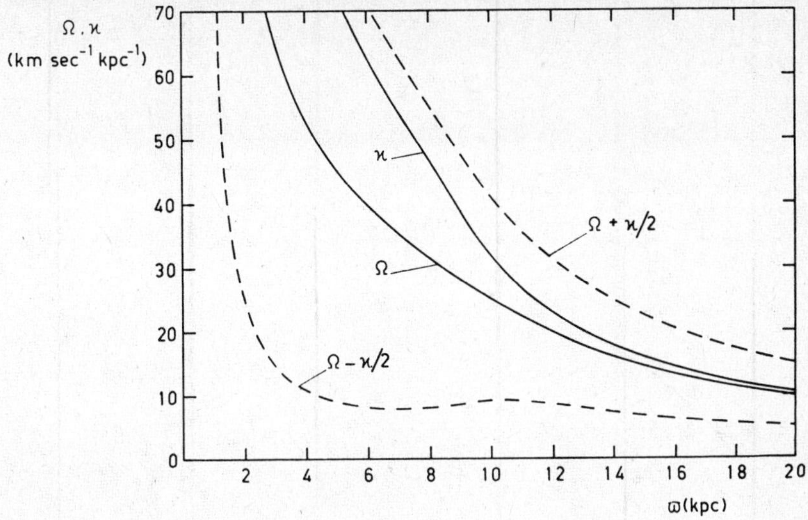


FIG. 1. Angular velocity $\Omega$, epicyclic frequency $\kappa$, and related quantities for the 1965 Schmidt model of the Galaxy.

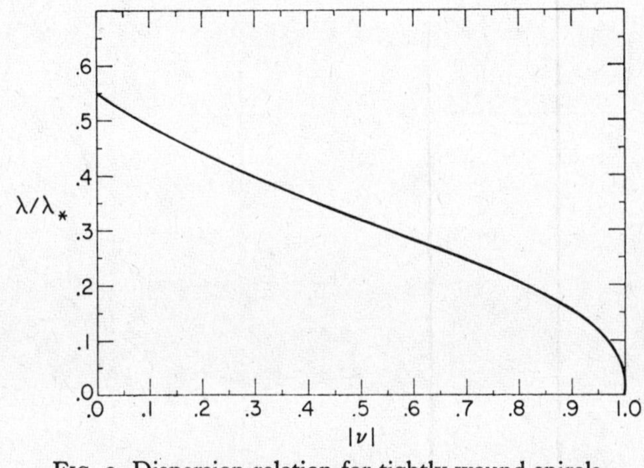


FIG. 2. Dispersion relation for tightly wound spirals.

(*c*) Trailing patterns are preferred over leading patterns, if the velocity dispersion of stars increases toward the center. (Several other effects would also tend to prefer trailing patterns.)

### 3. AN EXAMPLE OF A COMPUTED PATTERN

By taking $\Omega_p = 11$ km sec$^{-1}$ kpc$^{-1}$, we get a pattern that starts with a dispersion ring at the Lindblad resonant point, $\tilde{\omega} = 3{\cdot}75$ kpc, and continues with a spiral as shown in Figure 3. This may be compared with the spiral pattern shown by Mrs Burbidge in her Introductory Report (W. Becker 1964, Figure 4).

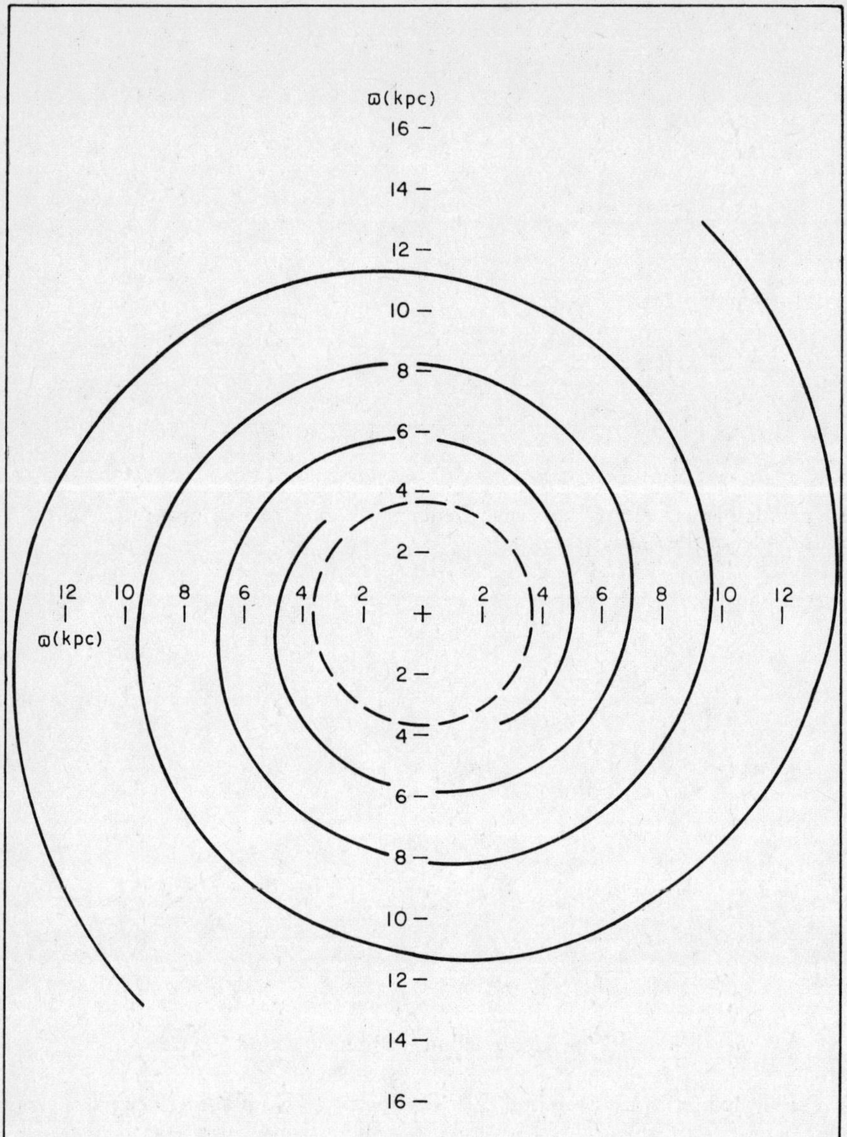


FIG. 3. Spiral pattern calculated for the 1965 Schmidt model, on the assumption that the pattern rotates at an angular velocity of 11 km sec$^{-1}$ kpc$^{-1}$ in the general direction of the rotation. Dispersion ring for Lindblad resonance taken at 3·75 kpc from the center.

No criterion based on the present theory is yet available for determining the pattern frequency. The above value of $\Omega_p$ was chosen so that the location of the dispersion ring would correspond to that of the '3-kpc Arm'.* In general, there may be a superposition of a group of waves, and the spiral pattern becomes quasi-stationary. There may also be other spurious spiral arms superposed on the overall pattern. But density waves of the type described appear to be the most natural candidate for the explanation of the existence of a grand design.

REFERENCES

Becker, W. 1964, *Z. Astrophys.*, **58**, 202.

Lin, C. C., Shu, F. H. 1964, *Astrophys. J.*, **140**, 646.

Lin, C. C., Shu, F. H. 1966, *Proc. nat. Acad. Sci. USA*, **55**, 229.

Oort, J. H. 1962, in *Interstellar Matter in Galaxies*, Ed. L. Woltjer, W. A. Benjamin, Inc., New York, p. 234.

Schmidt, M. 1965, in *Galactic Structure*, Eds. A. Blaauw and M. Schmidt, Vol. V in the series: *Stars and Stellar Systems*, Univ. of Chicago Press, Chicago, p. 513.

Toomre, A. 1964, *Astrophys. J.*, **139**, 1217.

## Discussion

*S. B. Pikel'ner:* I should like to know how you avoid the winding of spiral arms in your picture. You have a stable pattern for the density waves, a potential wheel which moves with constant angular velocity. The gas flows across the potential wheel and its density increases in the wheel. The difference between the velocities of gas and wheel must be rather large, of the same order as the rotational velocity, in order to avoid the winding. If the gas density inside the arms is considerably higher than that outside, the velocity of the rarefied gas between arms must be much higher than the differential velocity of gas and wheel. Is this consistent with theory and observations?

*C. C. Lin* answers: The gas flow velocity required is only a few km/sec in the direction perpendicular to the arms. This is largely because the inclination of the arms with respect to the circular direction is small only: of the order of 5°, or a tenth of a radian. Thus, the required radial velocities are of the order of 1/10 of the velocity differences in the circular direction. Detailed estimation of these velocities, on the basis of reasonable assumptions for the density contrasts, does not indicate any contradiction with theory or observation.

It might be of interest to bring up a related point, namely, the relative motion of the young stars with respect to the gas concentrations. In our neighborhood the stars are moving with a circular velocity of 250 km/sec, whereas the pattern is moving at 110 km/sec (if we adopt the value used for our pattern calculations). This difference in speed would cause a separation of only 140 pc (in the direction perpendicular to the arms, i.e., almost in the radial direction) between the gas concentration and the stellar arm as delineated by stars of the age 10 million years. Older stars would merge into the general population and cannot clearly mark out a spiral arm.

* This is, however, not a very accurate determination, and the value of $\Omega_p$ could be as high as 20 km $sec^{-1}$ $kpc^{-1}$.

This work was supported in part by the National Science Foundation

# Density Wave Theory of Spiral Structure

**C. C. LIN**

*Massachusetts Institute of Technology*

**F. H.-S. SHU**

*State University of New York at Stony Brook*

## Contents

## 1. GENERAL BACKGROUND

### 1.1. Introduction—The Problem of Spiral Structure

The discovery of spiral galaxies was already made during the last century. It is now known that nearly all the disk-shaped galaxies have a spiral structure. The problem of explaining these spiral structures has also occupied the interest of prominent astronomers for many years. Until recently, however, there has always existed the *dilemma of differential rotation*, which shall be the starting point for our subsequent discussions. The main question is whether the spiral patterns are composed of *material arms* or *density waves*. Although the late Bertil Lindblad proposed the theory of *density waves*, and devoted many years to its development, he was unable to put the theory into a quantitative form because the limited mathematical method used by him was not adequate for treating *collective modes in stellar systems*. The theory therefore did not impart confidence to the readers. Feelings of reservation were expressed by most astronomers and astrophysicists.

It is the purpose of this article to present the modern version of the *density wave* theory as developed over the past few years by myself and my collaborators Frank H. Shu, Chi Yuan, and William W. Roberts. We are indebted to many friends for their helpful discussions and information. Expecially I should mention Professors Bengt Strömgren, L. Woltjer, Kevin H. Prendergast, Alar Toomre, and Dr. A. Kalnajs for their contributions and criticisms during the early and crucial phases of this work.

I shall first briefly describe the problem of the spiral structure which presents, at first sight, some paradoxical features. In Fig. 1.1, we show a picture of the whirlpool galaxy, which is a close relative of our galaxy—the Milky Way. There is also a smaller companion whose shape is rather irregular. The main galaxy has a rather striking structure over a disklike distribution. The side view of another galaxy (Fig. 1.2) shows how thin the disk is. Clearly, the disk shape is associated with a large angular momentum in the system. Otherwise, the gravitational forces would have pulled the galaxy together.

The classification of regular galaxies into various types is shown schematically in Fig. 1.3. We shall not go into the detailed discussions of the characteristics and the relative abundance of galaxies of the various types.† Suffice it to say that a majority of them are normal spirals like the whirlpool. A few percent are irregulars, like its companion (not represented in Fig. 1.3). One of the main problems in galactic structure is: *why do most of them have the regular spiral structure?* A second question is: *how can these structures persist?*

The urgency of the second question can be appreciated if one realizes that there is an apparent contradiction between two sets of observational facts. On the one hand, there is evidence (see Sec. 2 on QSSS hypothesis) for us to believe that the spiral spacing in a given galaxy does not change in the course of time. Yet, on the other hand, there is strong *differential rotation*; i.e., the galaxy is not rotating as a solid body, but the inner part rotates much faster than the outer part. Indeed, in our galaxy the linear velocity is nearly the same for a substantial range of the radial distance. The problem is very well stated by Oort (1962).

"In systems with strong differential rotation, such as is found in all nonbarred spirals,‡ spiral features are quite natural. Every structural irregularity is likely to be drawn out into a part of a spiral. But *this* is not the phenomenon we must consider. We must consider a spiral structure extending over the whole galaxy, from the nucleus to its outermost part, and consisting of two arms starting from diametrically opposite points. Although this structure is often hopelessly irregular and broken up, the general form of the large-scale phenomenon can be recognized in many nebulae".

In other words, the primary problem is to explain the grand design over the whole disk on the scale of 10 kpc. There is also a secondary problem of describing the structure of an individual *arm*, which might involve somewhat different mechanisms.

But one immediate difficulty stands out if the spiral arm were a material arm, because there would be winding or unwinding of the spiral

† Some general information about galaxies and the Milky Way System is given in Appendix A.

‡ The problem of barred spirals has been studied by Kevin Prendergast (1962) and by K. C. Freeman (1965). Many of the features can be understood; but the mechanism for maintaining the bar cannot yet be described in detail.

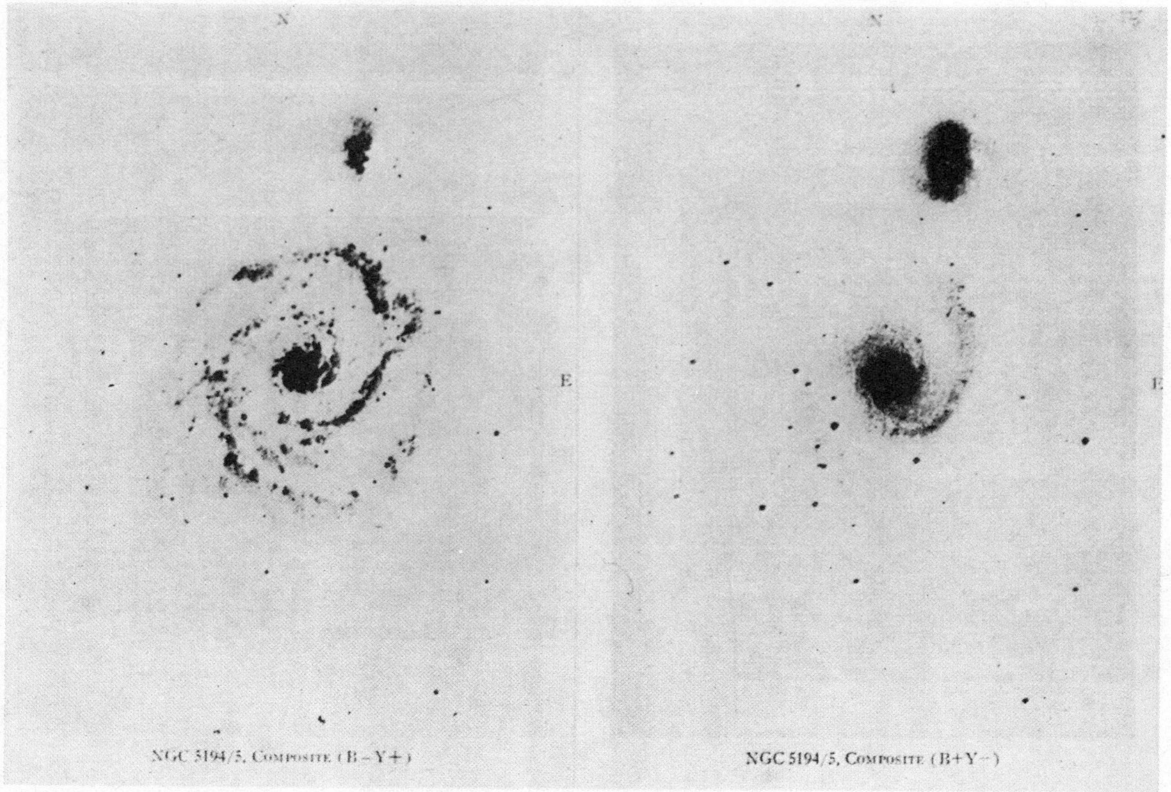


a b

Fig. 1.1. (a) The whirlpool galaxy (NGC 5194) and its companion (NGC 5195). (b) Photograph by means of Zwicky's technique (as carried out by Franz and Sharpless) reveals the irregular companion as having an underlying barred structure.

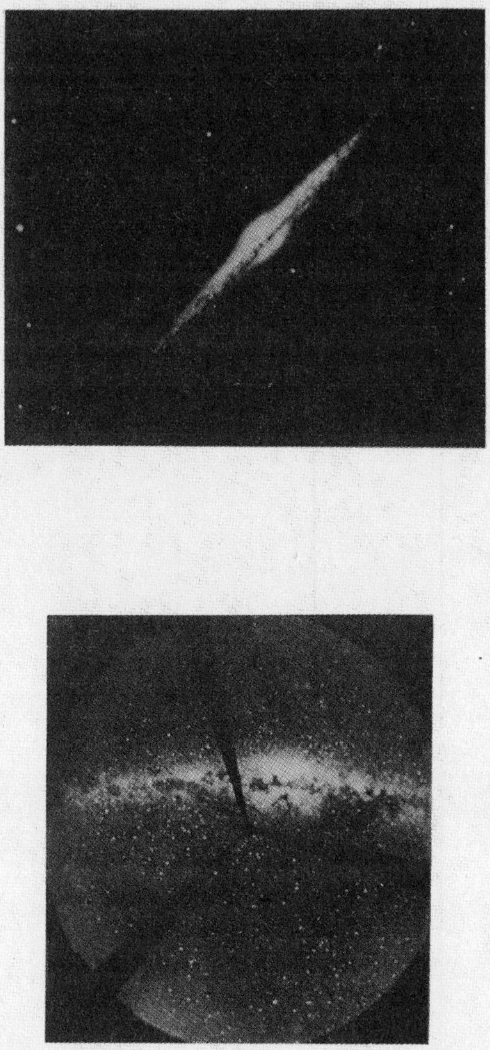

FIG. 1.2. The side view of a galaxy (NGC 4565).

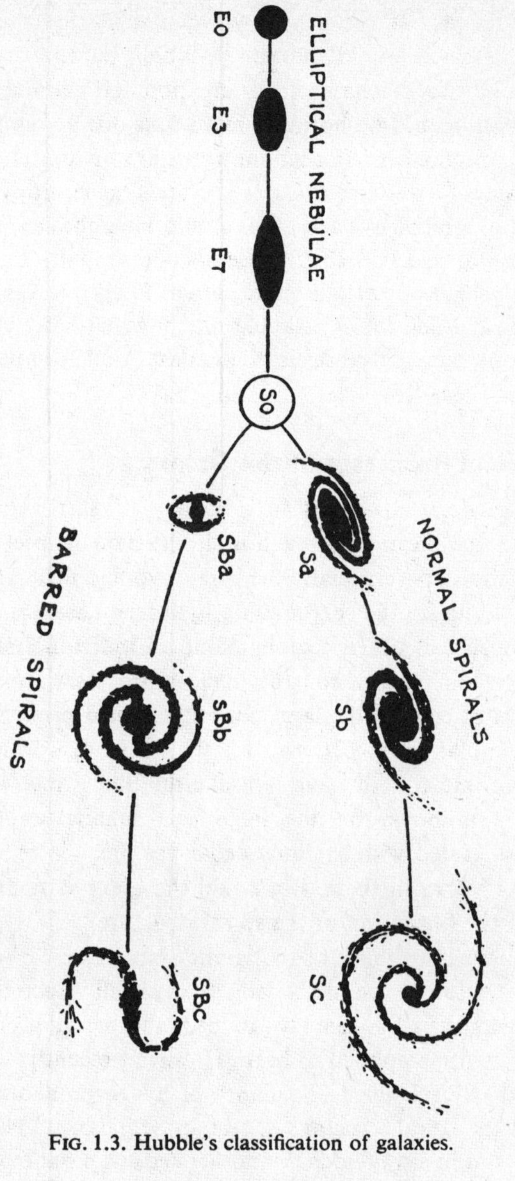


FIG. 1.3. Hubble's classification of galaxies.

is forced to think in terms of the density-wave pattern. This concept has other apparent difficulties, which we shall examine later (see Sec. 2.3).

It is also to be noted that the winding of material arms also leads to the perennial increase of the magnetic field, whereas a stationary value pattern. Since there is evidence (see Sec. 2) that this is not happening, one

of the field is reached in the density-wave concept discussed in Sec. 2. In both cases, the large-scale field lines run essentially parallel to the spiral arms.

In looking for mechanisms for creation and maintenance of regular spiral patterns, it should be borne in mind that one could also use essentially the same mechanism to account for spiral patterns which are less perfect and comparatively transitory. For the superposition of a group of patterns rotating relative to each other would have the desired effect.

In this article, we shall present an essentially stellar dynamical theory for the *persistence of the spiral pattern in the presence of differential rotation.* But first of all, let us establish the physical basis of the theory by examining the physical processes in the Galaxy, and formulating the equations governing them.

## 1.2. Physical Processes in the Galaxy‡

The physical processes in a galaxy of stars are multivarious. First of all, there are the nuclear reactions in the stars themselves. The energy coming out of the stars (especially the bright, young, massive stars of the O and B types) will ionize the surrounding hydrogen gas, forming the Strömgren spheres (also called H II regions). But an individual star is only one of 200 billion in our galaxy and the nuclear processes from them can influence but cannot control the large-scale dynamical processes that lead to the relatively regular spiral pattern over the whole disk in most of the flat galaxies. Indeed, most of the stars are like the Sun, which is relatively old and dim. It is well-known that the magnificent manifestations of the spiral patterns are associated with the interstellar gas and the relatively few brilliant young stars born in them. But could the older stars also play an important role in the formation of the spiral structure?

It is natural to think of the young stars as being born in a region of high concentration of atomic hydrogen, especially since they are usually distributed together in the Galaxy. Indeed, the designation Population I (due to Baade) is often applied to both. However, recently Morton Roberts (1967) found that the highest abundance of hydrogen actually occurs in a ring *outside* of the spiral structure in certain Sc galaxies. This is a remarkable discovery and in a way adds to another serious paradox to the problem of the spiral structure. But we shall see that it will be naturally resolved in our theory of density waves.

Another very intersting new line of investigation made possible by radio astronomy is the study of the distribution of H II regions by spectral

‡ For a more detailed account of the physical facts, see Appendix to Sec. 1.

lines of recombination at radio frequencies (109α). This enables one to penetrate further than by optical means. One remarkable fact noted is that there are no H II regions within 4 kpc from the center of the Milky Way system. (Burke *et al.*, 1968.) We shall find that this is in agreement with earlier predictions from the density-wave theory.

To deal with the large-scale structure of the Galaxy, one must bear in mind the following important components:

(i) *the stars*—with their gravitational forces, circular velocity, and peculiar velocity;

(ii) *the interstellar gas*—with its gravitational field and pressure;

(iii) *the magnetic field*—which exerts its influence through the highly conducting interstellar gas.

Two other components of the Galaxy are the radiation from the stars and the cosmic-ray particles.

The relative importance of the various components can be seen from Table I. Clearly, the kinetic energy of galactic rotation far outweighs any other component. Furthermore, this table is based on a mean gas density of about $2 \times 10^{-24}$ g/cc (1.2 H atoms/cc). The total mass density of gas and stars is about 5 times larger in the solar neighborhood, and about 50 times larger in the interior of the Galaxy. It is thus clear that even an extremely small deviation of the stars from circular motion would involve a level of energy comparable to the rest. A one-per-cent deviation would be associated with substantially more energy density than, e.g., the total magnetic energy density.

TABLE I. Energy densities in the Galaxy in erg $cm^{-3}$ [see Morrison (1957)].

| | |
|---|---|
| Total radiation (star light) | $0.7 \times 10^{-12}$ |
| Turbulent gas motion | $0.5 \times 10^{-12}$ [a] |
| Total energy of galactic rotation | $1300 \times 10^{-12}$ |
| Cosmic rays | $1 \times 10^{-12}$ |
| Magnetic field ($10^{-5}$ gauss) | $4 \times 10^{-12}$ |

[a] Remark by present writer: For certain purposes, this should be multiplied by a factor equal to

$$2 \times \frac{\text{linear velocity of galactic rotation}}{\text{turbulent velocity of gas motion}} \cong 50\text{–}70.$$

While we wish to emphasize the predominance of stellar mass in our galaxy, the situation may not be as favorable in other galaxies, such as those in the Sc type which have considerably more gas. We shall therefore

16*

formulate the basic equations to include all the three components named above. The effect of star light and cosmic rays will not be explicitly considered, even though they are acknowledged as the source for ionizing the gas.

## 1.3. Mathematical Formulation of the Basic Principles

Consider first the motion of the stars. Each star moves under the joint gravitational field of the other stars and the gas. Clearly, we wish to formulate a statistical treatment of the 200 billion stars regarding each one of them as a molecule, since we know that stars are very far apart from each other.‡ Indeed, the star next to the Sun ($\alpha$-Centauri) is about 4 light years away, or about $2.5 \times 10^5$ times the Earth's orbit (1 a.u.). Thus, even when the "orbital electrons" (the planets) of the "stellar atom" (solar system) are included in defining its size, we still have a highly rarefied gas. The frequency of collisions is even smaller than these linear dimensions would suggest, for the relative velocity among the stars is rather small, amounting only to the order of 30 km/sec. If we would consider the "effective collision size" of the Sun as given by the diameter of the orbit of Pluto (80 a.u.), the mean free time between collision works out to be of the order of $10^{13}$ years. A more careful analysis as given by Chandrasekhar (1942) leads to the value of $10^{14}$ years. Since we are only interested in processes lasting over a period of the order of $10^{10}$ years, the "gas of stellar atoms" may be regarded as "collision-free". We may thus describe the motion of the stars by the Boltzmann equation without the "collision term", but with the collective gravitational field as the source of interaction between the stars themselves and between the stars and the gas. A more refined treatment should of course include the "fluctuations" of the gravitational field due to the nearby stars (Chandrasekhar, 1942), but this is not necessary for present purposes.

*Equations of stellar dynamics*. Thus, if we consider an idealized model with stars all having the same mass $m_*$ and if we introduce a distribution function $\Psi_3$ in the phase space $(x_i, v_i)$, $i = 1, 2, 3$, the equation governing the development of $\Psi_3$ in time $t$ is as follows:

$$\frac{\partial \Psi_3}{dt} + v_i \frac{\partial \Psi_3}{\partial x_i} + a_i \frac{\partial \Psi_3}{\partial v_i} = 0, \tag{1.1}$$

where $v_i$ is the velocity of the individual stars, and $a_i$ is acceleration due to the combined gravitational field of the gas and the stars. The latter is

‡ A binary system is treated as a single star.

thus derivable from a gravitational potential $\mathscr{V}(x_i, t)$:

$$a_i = -\frac{\partial \mathscr{V}}{\partial x_i}, \tag{1.2}$$

and the gravitational potential is in turn determined by the gas and the stars in the galaxy (and perhaps the effect of neighboring galaxies):

$$\frac{\partial^2 \mathscr{V}}{\partial x_k^2} = 4\pi G \varrho_t, \quad \varrho_t = \varrho + \varrho_*. \tag{1.3}$$

In this equation, $G$ is the gravitational constant, $\varrho$ is the mass density of the gas, and $\varrho_*(x_i, t)$ is the mass density of the stars given by

$$\varrho_*(x_i, t) = m_* \int \Psi_3(x_i, v_i, t)\, d\tau(v_i), \tag{1.4}$$

where the integral extends over the infinite velocity space. The mass density of the gas is one of the variables in the hydromagnetic equations.

*Equations of hydromagnetics.* Because of the large linear dimensions involved, the effects of viscosity and resistivity are expected to be small, for the proper measure of diffusing processes is always the dimensionless parameter

$$\mathscr{D} = Dt_0/L^2, \tag{1.5}$$

where $D$ is the diffusion coefficient, $t_0$ is a typical time scale, and $L$ is a typical linear dimension. If we take $L = 10^3$ lt-yr $= 10^{21}$ cm, $t_0 = 10^{10}$ yr $= 3 \times 10^{17}$ sec, then

$$\mathscr{D} = D \times 3 \times 10^{-25}, \quad D \text{ in cm}^2/\text{sec}. \tag{1.6}$$

The resitivity of fully ionized hydrogen with one atom per cubic centimer is (within a factor of 1.4)

$$\eta = 10^{14} \times T^{-3/2} \tag{1.7}$$

for temperature $T$ in the range $10^2$–$10^4$ °K [Spitzer (1956), Eq. (5.37), p. 84, Table 5.1, p. 73]. Thus, even if the gas is only ionized to a small extent, the parameter $\mathscr{D}$ is still very small, which justifies the approximation of infinite conductivity. Similarly, viscous forces may be neglected. The transfer of thermal energy by radiation and thermal conductivity is a more complicated problem which we shall not go into in the present discussion.

With this provision, the hydromagnetic equations are as follows.‡ For the dynamics of the gas we have, respectively, the following equations of state, mass conservation, motion, and energy:

$$p = \varrho RT, \tag{1.8}$$

$$\frac{\partial \varrho}{\partial t} + \frac{\partial}{\partial x_k}(\varrho u_k) = 0, \tag{1.9}$$

‡ Gaussian (cgs) units are used throughout this paper.

$$\varrho \frac{Du_i}{Dt} = -\frac{\partial p}{\partial x_i} + \varrho a_i + \frac{1}{4\pi}\frac{\partial}{\partial x_j}\left\{B_i B_j - \frac{1}{2}B^2 \delta_{ij}\right\}, \tag{1.10}$$

$$\frac{Dp}{Dt} = \gamma \frac{p}{\varrho}\frac{D\varrho}{Dt} + \varepsilon(\gamma - 1). \tag{1.11}$$

The symbols in the above equation are as follows: the thermodynamic state of the gas is specified by the pressure $p$, the density $\varrho$, and the temperature $T$; these are related through the equation of state; the motion of the gas is described by the Eulerian velocity field $u_i(x, t)$, and its substantial derivative $Du_i/Dt$, which is the acceleration. The force terms on the right-hand side of (3.10) contain not only the pressure force, but also (i) the acceleration due to the combined gravitational field of the gas and the stars and (ii) the force due to the Maxwell stresses of the magnetic field $B_i$. These last two items are the agents through which the gas interacts with the other two important components of the Galaxy: the stars and the magnetic field. The last equation for the energy contains the unspecified term $\varepsilon$ which represents the rate of energy transfer into a unit volume per second by heat conduction, radiative transfer, etc.

The displacement current is always neglected in a hydromagnetic problem. In the case of infinite conductivity, it is well-known that the Maxwell equations lead to the following two equations for the magnetic field:

$$\frac{\partial B_k}{\partial x_k} = 0, \tag{1.12}$$

$$\frac{D}{Dt}\left(\frac{B_i}{\varrho}\right) = \left(\frac{B_k}{\varrho}\right)\frac{\partial u_i}{\partial x_k}.$$

The second equation above guarantees the conservation of magnetic flux associated with a given piece of fluid surface. (The magnetic field is "frozen" into the fluid.)

In view of the numerical values indicated in Table I, we need not consider, to a first approximation, the effect of the gas and the magnetic field when we deal with the stellar-dynamical problem, unless the change in the density in the gas is to a much greater extent than the change in the density of the stars. Such a condition could happen in our neighborhood, but it is unlikely to happen in the more massive interior part of our galaxy. The motion of the gas, on the other hand, would be heavily dependent on the gravitational field of the stars. In general, especially in Sc galaxies, the consideration of stellar dynamics should be made together with the hydromagnetic problem; both the gas and the stars move in the resultant gravitational field produced by the combined effect of the gas and the stars.

Since the disk galaxies are extremely thin, it is possible to simplify our analysis by adopting a limiting model of zero thickness. To derive the proper stellar dynamical equations in that case, we integrate (3.1) with respect to the coordinate $x_3$ and the velocity $v_3$ in the direction normal to the disk. This is carried out in the Appendix to Sec. 4. Clearly, the central bulge is not well approximated in this process. Thus, *singularities in the solution should be allowed at the center (and possibly along the axis) in the theory based on the thin-disk model,* for the solution obtained is only asymptotically valid when the axial distance is much larger than the radius of the nuclear region.

## 2. QSSS HYPOTHESIS

### 2.1. Physical Picture

Let us now review briefly some of the observational facts in order to form a clearer picture of what the problems are.

1. The spiral arms are associated with the gas and the young stars. The contrast of total mass density between the brighter and darker regions is small.

2. There is a typical spacing for spiral galaxies of various types as determined by other physical characteristics as shown in the following table:

| | Sa → Sb → Sc |
|---|---|
| Nuclear concentration | decreasing |
| Gas content | increasing |
| Tom Spacing | increasing |
| Artal mass | decreasing |

3. There appears to be a considerable degree of regularity of the pattern *extending over the whole disk*. In the case of some of the galaxies, such as NGC 5364 or the whirlpool galaxy, this regularity is remarkably strong. At the same time, irregular features on small scales are clearly present.

There are at least two possible types of spiral theories. The first alternative is to associate every spiral arm with a *given body of matter*; e.g., such an arm might essentially be a tube of gas primarily constrained by the interstellar magnetic field. The difficulty with the disrupting influence of differential rotation in such a theory is well known. The various issues associated with this point of view have been thoroughly discussed recently by Oort (1962). The second alternative is to regard the spiral structure as a *wave pattern*, which either remains stationary, or at least quasi-stationary, in a frame of reference rotating around the center of the galaxy at a proper speed (possibly zero).

In this article, we shall show that we are led to favor the second point of view, i.e., that the matter in the Galaxy (stars and gas) can maintain a density wave through gravitational interaction in the presence of the differential rotation of the various parts of the disk. This density wave provides a spiral gravitational field which *underlies* the observable concentration of young stars and the gas. In this way, an observable spiral pattern, including small scale irregularities, can be maintained over the whole disk.

Specifically, we venture to suggest that there are indeed large-scale neutral (or nearly neutral) waves of a spiral form for most of the disk galaxies, and formulate our ideas in the form of the following hypothesis.

The total galactic population, which consists primarily of the gas and the stars with various degrees of velocity dispersion, forms a *quasi-stationary spiral structure* in space of the general nature discussed above. This is primarily associated with a balance between the effects of gravitational instability and velocity dispersion (and secondarily to the influence of the gas and the magnetic field). The extent of density variation in the spiral pattern may be only a small fraction of the symmetrical mean density distribution.

In advancing this hypothesis, we are not ignoring the fact that transient local patterns of a general spiral form are relatively easy to produce in a system in rotation. However, it seems rather difficult to account for a *coherent pattern over the whole disk* without bringing into account the cooperative effects of long-range gravitational forces.

If we accept the above hypothesis and follow the line of reasoning that led to it, the following inferences and conjectures may be made.

(a) In the coordinate system rotating with the pattern, all phenomena are stationary. Both the stream lines and the magnetic field lines form closed nearly circular loops coinciding with each other. Such a situation is shown in Fig. 2.1 (W. W. Roberts, 1968).

(b) Due to the prevailing spiral gravitational field, all components of the Galaxy, including the gas and the young stars, should form similar spiral patterns on the scale of the radius of the disk, whatever other secondary processes there might be. Different stellar components, with different extent of velocity dispersion, should lie on somewhat different spirals (cf. Zwicky, 1957) and exhibit different degrees of unevenness. As an extreme case, one may even attempt to analyze the behavior of the gas, which

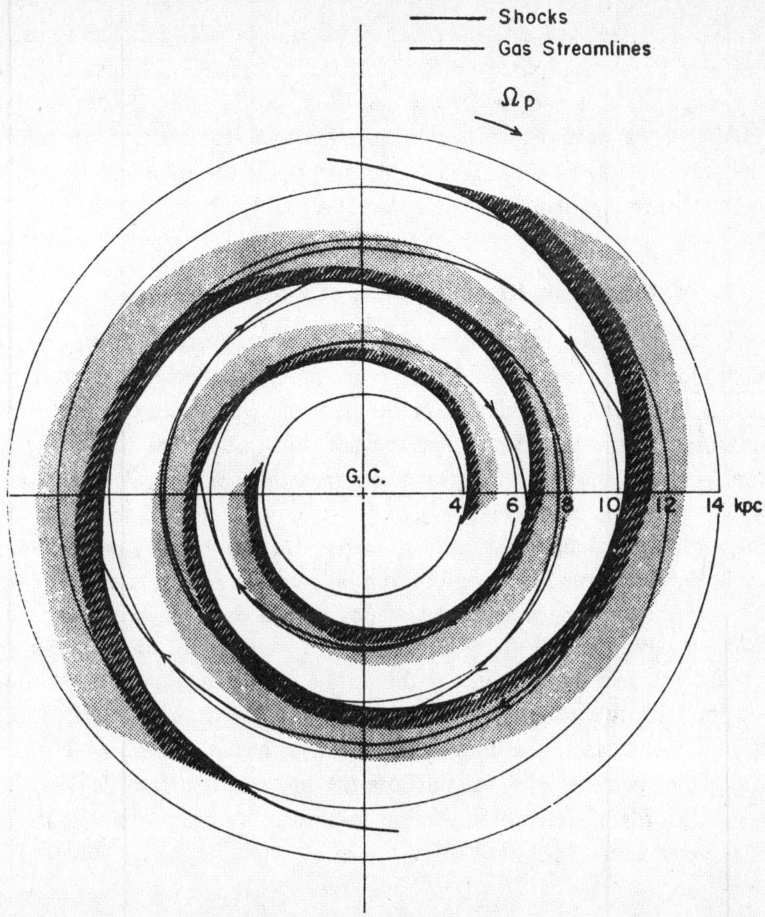


FIG. 2.1. A theoretical spiral pattern in the density wave theory, showing stream lines, magnetic-field lines, and regions of concentration of neutral hydrogen gas (lightly shaded) and young stars (heavily shaded). After W. W. Roberts.

has very little pressure, as a separate system (including possibly an appropriate magnetic field), which is subjected to the action of a spiral gravitational field produced by the rest of the galactic population. Its density contrast may therefore be expected to be far larger than that in the stellar components, as is indeed known from observations. These discussions should be supplemented by a consideration of the reaction of the gas on the stellar population, as mentioned above.

(c) It is known from observations that galaxies devoid of gas do not show prominent spiral patterns. This is consistent with the present proposal. The velocity dispersion of the stars in such galaxies is usually thought to be comparatively large, and the mechanism of instability discussed here may indeed be suppressed completely. Even if this were not so, and the old stars did form a spiral structure, the lack of gas and the young stars stemming therefrom might so impair the manifestation of the spiral structure that it would hardly be observable.

## 2.2. Mathematical Description of a Spiral Pattern

Spiral patterns are also relatively easy to describe from a mathematical point of view. Indeed, the elementary method of separation of variables naturally gives rise to spiral patterns when one considers small disturbances from an axi-symmetric basic state. In order to see this, let us first consider such small disturbances in a thin gas sheet in the galactic disk. We adopt a cylindrical system $(\varpi, \theta, z)$ with the plane $z = 0$ coinciding with the galactic disk, and the origin at the galactic center. In the undisturbed state, the disk is in differential rotation with a circular velocity $\varpi\Omega(\varpi)$, where the angular velocity $\Omega(\varpi)$ decreases with increasing radial distance $\varpi$. We denote the components of velocity of the gas by $(u_{\tilde{\omega}}, \varpi\Omega + u_\theta, u_z)$, respectively, in the directions of increasing $(\varpi, \theta, z)$, where $(u_{\tilde{\omega}}, u_z, u_\theta)$ are functions of $(\varpi, \theta, z, t)$, $t$ being the time. We shall restrict ourselves to the case of an infinitesimally thin sheet so that $z = 0$, $u_z = 0$. In the plane sheet, let $\sigma(\varpi, \theta, t)$ denote the projected surface density.

In order to see some general features of spiral waves, let us consider the equation of continuity:

$$\frac{\partial \sigma}{\partial t} + \frac{1}{\varpi}\left[\frac{\partial}{\partial \varpi}(\sigma\varpi u_\varpi) + \frac{\partial}{\partial \theta}(\sigma u_\theta) + \frac{\partial}{\partial \theta}(\sigma\varpi\Omega)\right] = 0. \tag{2.1}$$

If we write $\sigma = \sigma_0(\varpi) + \sigma'$ and linearize, we have

$$\frac{\partial \sigma'}{\partial t} + \frac{1}{\varpi}\left[\frac{\partial}{\partial \varpi}(\sigma_0\varpi u_\varpi) + \frac{\partial}{\partial \theta}(\sigma_0 u_\theta) + \frac{\partial}{\partial \theta}(\sigma'\varpi\Omega)\right] = 0. \tag{2.2}$$

This equation, linear in the quantities $({}_{\tilde{\omega}}\sigma', u, u_\theta)$, has coefficients dependent only on $\varpi$. Thus, solutions may be sought with a typical perturbation term in the form

$$q'(\varpi, \theta, t) = \operatorname{Re}\{\hat{q}(\varpi)\exp[i(\omega t - m\theta)]\}, \tag{2.3}$$

where $\omega$ is a constant and $m$ is an integer. The quantity $\hat{q}(\varpi)$ is in general complex, and has the form

$$\hat{q}(\varpi) = A(\varpi)\,e^{i\Phi(\varpi)}. \tag{2.4}$$

Now, if $A(\varpi)$ is slowly varying while $\Phi(\varpi)$ is of the form $\Phi(\varpi) = \Lambda f(\varpi)$, where $f(\varpi)$ is also slowly varying and $\Lambda$ is a large parameter, we have *tightly wound spirals*. To see this, let us note that lines of constant $q'$ are approximately given by

$$\omega t - m\theta + \Phi(\varpi) = \text{const}, \tag{2.5}$$

with maximum value reached when the constant is an integral multiple of $2\pi$. Equation (2.3) represents a spiral pattern for each instant $t$ described by a typical curve

$$m(\theta - \theta_0) = \Phi(\varpi) - \Phi(\varpi_0). \tag{2.6}$$

The pattern has $m$ arms. In the course of time, the pattern rotates with a uniform angular velocity

$$\Omega_p = \omega/m \tag{2.7}$$

when $\omega$ is real. The pattern is tightly wound, if $\Lambda$ is large. For example, if $\Phi(\varpi) = \Lambda \log \varpi$, Eq. (2.5) gives equiangular spirals with pitch angle

$$i = \tan^{-1}(m/\Lambda). \tag{2.8}$$

In general, the pitch angle $i$ is given by

$$\tan i = m(k\varpi)^{-1}, \quad \text{where} \quad k(\varpi) = \Phi'(\varpi). \tag{2.9}$$

The spacing between spiral arms is given by

$$\lambda = 2\pi/|k|.$$

If $\Omega < 0$, trailing patterns are obtained when $k < 0$.

Since the pitch angle $i$ is of the order of 5° for our galaxy, the parameter $\Lambda$ is of the order of 20 for $m = 2$. Even for angles three times as large, the parameter $\Lambda$ is still appreciable for asymptotic calculations of a rapidly varying phase to be very good approximation. If we apply such an approximation to (2.1), we obtain

$$i(\omega - m\Omega)\,\sigma' + ik\sigma_0 u_\varpi = 0,$$

or

$$\frac{\sigma'}{\sigma_0} = \frac{k u_\varpi}{m(\Omega - \Omega_p)},$$

or

$$u_\varpi = \frac{\sigma_{\max} - \sigma_{\min}}{\sigma_{\max} + \sigma_{\min}} \cdot \varpi(\Omega - \Omega_p)\tan i. \tag{2.10}$$

This allows us to calculate the radial systematic motion of the gas from observed density contrasts.

## 2.3. Kinematical Tests of the Concept of Density Waves

Since the first difficulty that we face is a kinematical one, let us begin by examining the consistency of our concept with the observed kinematical behavior of the various components of our galaxy. The concept of a density wave pattern must pass two "kinematical tests": the predictions on (i) the radial motion of the gas and (ii) the winding of young stellar arms, must both be consistent with observations.

The behavior of the older stars in a galaxy can be readily observed only in terms of a statistical distribution and therefore does not present any immediate difficulties in a physical picture involving a density wave. A similar remark applies to the gas, but here we have already a more serious point to be reconciled. The *material* velocity of the gas can be observed by the Doppler effect and it is necessary to show that the radial motion of the gas required by a density wave, circulating around the galactic center, does not exceed any upper limit that might be imposed by observations. We shall see that this is indeed the case in our own galaxy.

The arrangement of young stars can also be followed in greater detail. It is possible, at least approximately, to draw "isochronic lines" for young stars of the same age. If the winding process of differential rotation is operative, then these arms of young stars should be rapidly wound up in the course of one ot two revolutions. However, the "life time" of these brilliant young stars of the O and B types are of the order of ten million years, and we shall see that the result of the winding process over such a short period is not impressive.

Consider therefore an observer moving with the angular velocity $\Omega_p$ of the spiral pattern. The density pattern of the gas will then be stationary, but the gas particles will be moving at a different speed. Let us denote the gas velocity by $(u_\theta, V + u_{\tilde{\omega}})$; then it has been shown [Eq. (2.10)], that the radial motion of the gas has an upper limit $u_{\max}$ given by

$$\frac{u_{\max}}{V(\varpi)} = \frac{1}{\pi}\frac{\Delta\varpi}{\varpi}\frac{1-r}{1+r}\left(1 - \frac{\Omega_p}{\Omega}\right), \tag{2.11}$$

where $\Delta\varpi$ is the spacing between two successive arms around the location $\varpi$, $r$ is the radio of minimum surface density of the gas to its maximum surface density, and $\Omega$ is the angular velocity at $\varpi$. If we take

$$r = \tfrac{1}{3}, \quad \Delta\varpi = 2\text{ kpc at } \varpi = 8\text{ kpc},$$

$$\Omega = 30\text{ km/sec-kpc}, \quad \Omega_p = 14\text{ km/sec-kpc},$$

$$V = 250\text{ km/sec},$$

we obtain

$$u_{max} = 10 \text{ km/sec}. \tag{2.12}$$

This is safely within the upper limit of 15 km/sec set by observations.

Consider now the behavior of the young stars. They are presumed to be born in the gas concentration $ACB$ (Fig. 2.2). After a time interval $\Delta t$, the stars born at $C$ would move *ahead* of the density wave (which is still at $ACB$) to the point $C'$ given by

$$C'C = \varpi(\Omega - \Omega_p)\Delta t.$$

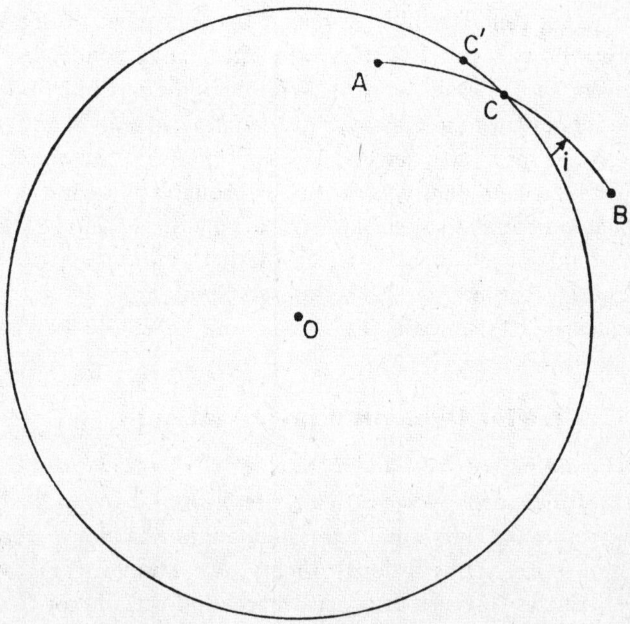


FIG. 2.2. Diagram illustrating the negligible winding of an arc of young stars.

The stars born at $A$ would advance further while the stars born at $B$ would not advance as much, so that the arc $A'B'C'$ would be inclined at a somewhat smaller angle $i'$ ($<i$). But the stars in the *arc* $A'C'B'$ do not deviate very much from the *arc* $ACB$. Indeed, the deviation of $C'$ from $ACB$ is given by

$$\Delta\varpi = \varpi(\Omega - \Omega_p)(\Delta t) \tan i. \tag{2.13}$$

If we take the above values of $\varpi$, $\Omega$, and $\Omega_p$, and $\Delta t = 10 \times 10^6$ yr, $\tan i = 0.11$, we get

$$\Delta\varpi = 240 \text{ pc}. \tag{2.14}$$

Thus, the stellar arm is now only *slightly* displaced from its place of formation. There is indeed observational evidence to give general support to this kind of displacement. More detailed comparative study of observational data and theoretical predictions may be found in the discussions by W. W. Roberts (1968).

The change of the angle of inclination is clearly very small over ten-million or even twenty-million years. For longer periods of time, the stars merge into the general population and can no longer be distinguished as isochronic lines by age groups. In this way, the winding dilemma is avoided as far as the young stellar arms are concerned.

There is a similar difficulty related to the problem of star formation. Since the gas flows through a spiral arm, it must be condensed into stars in a relatively short period of time. Why does the gas remain in the free state while it circulates the Galaxy, but becomes suddenly ready to collapse into stars in the spiral arm region? We shall find that the answer lies in a sudden increase of its density as it flows through the region of minimum gravitational potential to form a "shock" on the galactic scale. Such a "shock" would indeed predict star formation along a narrow organized "front" causing the bright H II regions to lie along a narrow curve like "beads on a string" (Morgan, Osterbrock, and Sharpless, 1951).

## 2.4. Outline for Dynamical Investigations

Having passed the "kinematical tests", we now turn to the dynamical basis of these density waves. They are primarily due to the "cooperative" behavior of the stars due to gravitational interaction. Before we proceed with a purely gravitational theory, we should consider all the components of the Galaxy: stars, gas, magnetic field, photons, cosmic-ray particles, etc. and make an assessment of their relative importance in a dynamic process. This may be done by using the basic equations of stellar dynamics and hydromagnetics. I shall not go into the details here. Suffice it to say (See Appendix to Sec. 5) that the hydromagnetic effect becomes appreciable for a scale of 1 kpc only if the field is as high as $5 \times 10^{-5}$ gauss (which is uncomfortably high). Since we are dealing with the problem of the *spiral pattern*, on the scale of 10 kpc, it will never be as large as the gravitational forces involved.

We shall therefore consider a purely gravitational theory, including both the gravitational field of the stars and of the gas. We attempt to demonstrate that density waves of a spiral structure will be self-sustained at a small but finite amplitude. Suppose such a wave were maintained, then there must be an associated gravitational field of a generally spiral form. We shall start with this field and carry out the analysis as indicated in the following diagram:

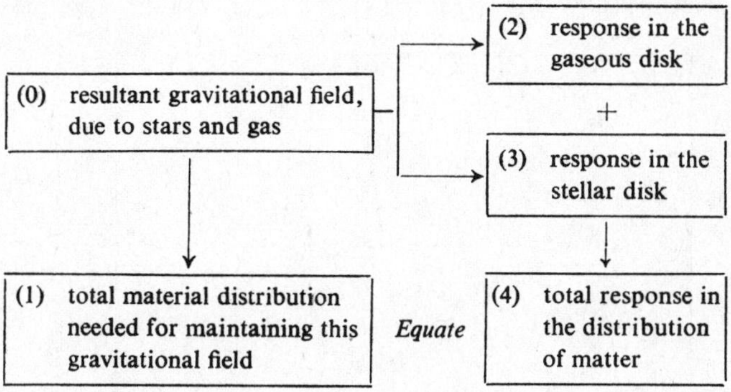


The resultant gravitational field (0) must be associated, according to Poisson's equation, with a certain distribution of matter (1), which may consist partly of gas and partly of stars. The distribution of gas (2) may be calculated in terms of the *resultant* gravitational field without any *further* reference to the distribution of the stars. Similarly, the distribution of the stars (3) may be calculated without any *further* reference to the distribution of the gas. The sum of these two distributions yields a total distribution of matter (4) that must be identical with the density distribution (1) which is needed to give rise to the field. This last condition is the equation to be solved for the unspecified functions and parameters that occur in the resultant gravitational field (0) initially assumed.

The analysis outlined has so far been carried out only for the linear theory. Clearly, it is in the nature of a dispersion relationship, i.e., a relationship between a wavelength and a frequency. The detailed nature of such a relation will be clarified by considering a simplified model (Sec. 3).

## 3. AN ELEMENTARY THEORY

As mentioned above, the purpose of this section is to give a simplified theory for the dispersion relationship. The model we shall adopt is a self-gravitating gaseous disk in differential rotation; the gas is assumed to have pressure, simulating turbulence. We now shall find that the more complete dynamical theory to be presented in Sec. 5 can be simulated by a gas dynamical theory almost completely, except near resonance. (See Fig. 3.1.) This is not entirely unexpected since the "moment equations" for stellar systems are essentially hydrodynamical equations in general nature (see Appendix B).

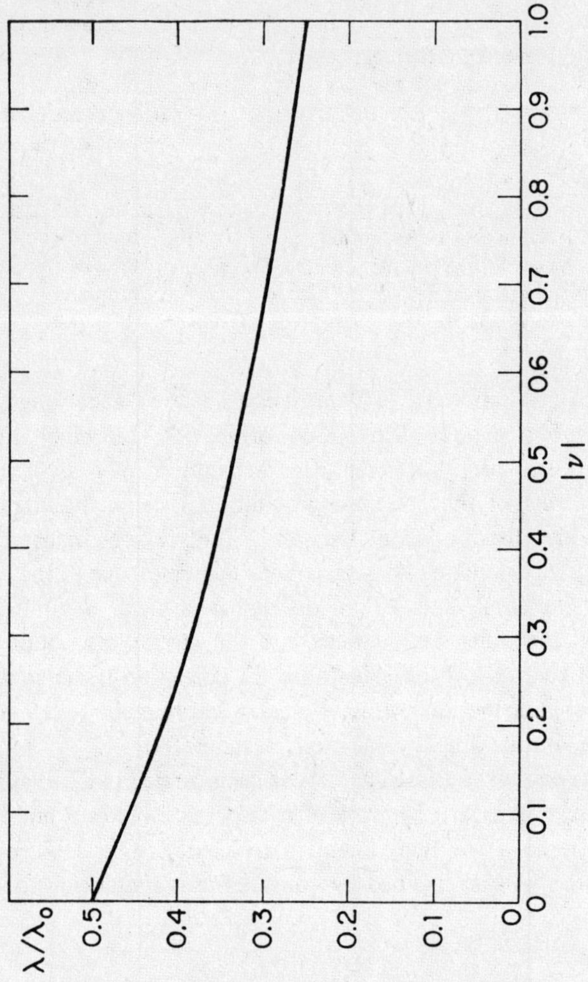


FIG. 3.1. Dispersion relationship for density waves in the gas-dynamical theory.

## 3.1. Formulation of the Mathematical Problem

We adopt a cylindrical coordinate system $(\varpi, \theta, z)$ with the plane $z = 0$ coinciding with the galactic disk, and the origin at the galactic center. In the undisturbed state, the disk is in differential rotation with a circular velocity $\varpi\Omega(\varpi)$, where the angular velocity $\Omega(\varpi)$ decreases with increasing radial distance $\varpi$. We denote the velocity components of the gas by $(u, \varpi\Omega + v, w)$ in the directions of increasing $(\varpi, \theta, z)$ respectively, where $(u, v, w)$ are functions of $(\varpi, \theta, z, t)$, $t$ being time. We shall restrict ourselves to the cases $w = 0$. Then the dependence on $z$ disappears upon integration. The surface density will be denoted by

$$\sigma_{\text{total}} = \sigma_{\text{fixed}}(\varpi) + \sigma(\varpi, \theta, t), \tag{3.1}$$

where $\sigma_{\text{fixed}}(\varpi)$ is the surface density of the stars with such large dispersion velocities that they cannot be expected to participate in the formation of spiral patterns.

In the plane $z = 0$, the equation of continuity and the equations of motion may be written as follows:

$$\frac{\partial\sigma}{\partial t} + \frac{1}{\varpi}\left\{\frac{\partial}{\partial\varpi}(\varpi\sigma u) + \frac{\partial}{\partial\theta}\left[\sigma(\varpi\Omega + V\right]\right\} = 0, \tag{3.2}$$

$$\frac{\partial u}{\partial t} + u\frac{\partial u}{\partial\varpi} + \left(\Omega + \frac{v}{\varpi}\right)\frac{\partial u}{\partial\theta} - \frac{(\varpi\Omega + v)^2}{\varpi} = -\frac{a_0^2}{\sigma}\frac{\partial\sigma}{\partial\varpi} - \frac{\partial\mathscr{V}}{\partial\varpi}, \tag{3.3}$$

$$\frac{\partial v}{\partial t} + u\frac{\partial}{\partial\varpi}(\varpi\Omega + v) + \left(\Omega + \frac{v}{\varpi}\right)\frac{\partial v}{\partial\theta} + u\left(\Omega + \frac{v}{\varpi}\right)$$
$$= -\frac{1}{\varpi}\left(\frac{a^2}{\sigma}\frac{\partial\sigma}{\partial\theta} - \frac{\partial\mathscr{V}}{\partial\theta}\right), \tag{3.4}$$

where $\mathscr{V}$ is the gravitational potential (in the plane $z = 0$), and $a_0$ is a measure of the turbulent velocity, taken to be constant in our analysis. The potential $\mathscr{V}(\varpi, \theta, z, t)$ satisfies the Poisson equation

$$\frac{\partial^2\mathscr{V}}{\partial\varpi^2} + \frac{1}{\varpi}\frac{\partial\mathscr{V}}{\partial\varpi} + \frac{1}{\varpi^2}\frac{\partial^2\mathscr{V}}{\partial\theta^2} + \frac{\partial^2\mathscr{V}}{\partial z^2} = 4\pi G\sigma_{\text{total}}\,\delta(z). \tag{3.5}$$

In the basic state, $u = v = 0$, and the density and potential distributions are axi-symmetric:

$$\begin{aligned}\sigma &= \sigma_0(\varpi),\\ \mathscr{V} &= \mathscr{V}_0(\varpi).\end{aligned} \tag{3.6}$$

Equations (3.2) and (3.4) are trivially satisfied, while equation (3.3) gives the balance in the radial direction

$$-\varpi\Omega^2 = -\frac{a_0^2}{\sigma_0}\frac{d\sigma_0}{d\varpi} - \frac{d\mathscr{V}_0}{d\varpi}. \tag{3.7}$$

### 3.2. Small Disturbances

If we now consider small disturbances over the basic state, and write

$$\begin{aligned} \sigma &= \sigma_0(\varpi) + \sigma_1(\varpi, \theta, t), \\ \mathscr{V} &= \mathscr{V}_0(\varpi, z) + \mathscr{V}_1(\varpi, \theta, z, t), \end{aligned} \tag{3.8}$$

we may neglect terms quadratic in $(u, v, \sigma_1)$ and obtain the following linearized equations from (3.2), (3.3), (3.4), and (3.5):

$$\frac{\partial \sigma_1}{\partial t} + \Omega \frac{\partial \sigma_1}{\partial \theta} + \frac{1}{\varpi}\left\{\frac{\partial}{\partial \varpi}(\varpi \sigma_0 u) + \frac{\partial}{\partial \theta}(\sigma_0 v)\right\} = 0, \tag{3.9}$$

$$\frac{\partial u}{\partial t} + \Omega \frac{\partial u}{\partial \theta} - 2\Omega v = -\frac{a_0^2}{\sigma_0}\frac{\partial \sigma_1}{\partial \varpi} - \frac{\partial \mathscr{V}_1}{\partial \varpi}, \tag{3.10}$$

$$\frac{\partial v}{\partial t} + \Omega \frac{\partial v}{\partial \theta} + \frac{\varkappa^2}{2\Omega} u = -\frac{1}{\varpi}\left\{\frac{a_0^2}{\sigma_0}\frac{\partial \sigma_1}{\partial \theta} + \frac{\partial \mathscr{V}_1}{\partial \theta}\right\}, \tag{3.11}$$

$$\frac{\partial^2 \mathscr{V}_1}{\partial \varpi^2} + \frac{1}{\varpi}\frac{\partial \mathscr{V}_1}{\partial \varpi} + \frac{1}{\varpi^2}\frac{\partial^2 \mathscr{V}_1}{\partial \theta^2} + \frac{\partial^2 \mathscr{V}_1}{\partial z^2} = 4\pi G \sigma_1 \delta(z), \tag{3.12}$$

where

$$\varkappa^2 = (2\Omega)^2\left\{1 + \frac{\varpi}{2\Omega}\frac{d\Omega}{d\varpi}\right\}. \tag{3.13}$$

### 3.3. Spiral Waves

Clearly, the above equations allow solutions where each of the small disturbance quantities is of the form

$$q_1 = \text{real part of } A(\varpi) \exp\{i[\omega t - m\theta + \Phi(\varpi)]\}, \tag{3.14}$$

where $\omega$ is a parameter (real or complex), $m$ is an integer, and $A(\varpi)\, e^{i\Phi(\varpi)}$ is a complex function of the real variable $\varpi$. As explained in Sec. 2.2, these are spiral waves. If we write

$$\{\sigma_1, u, v\} = \{\hat{\sigma}(\varpi), \hat{u}(\varpi), \hat{v}(\varpi)\} \exp\{i[\omega t - m\theta + \Phi(\varpi)]\}, \tag{3.15}$$

to show the same spiral structure in all these quantities, then $\hat{\sigma}(\varpi)$, $\hat{u}(\varpi)$, $\hat{v}(\varpi)$ are not necessarily real, but they must be slowly varying. The gravitational field for a potential $\mathscr{V}_1$ of the form (3.14) has components

$$\begin{aligned} -\frac{\partial \mathscr{V}_1}{\partial \varpi} &= -\{A' + ikA\} \exp\{-\}, \\ -\frac{1}{\varpi}\frac{\partial \mathscr{V}_1}{\partial \theta} &= +\frac{imA}{\varpi} \exp\{-\}. \end{aligned} \tag{3.16}$$

if $k\varpi$ is large, then

$$ikA \gg \frac{imA}{\varpi} \quad \text{or} \quad A'. \tag{3.17}$$

We may therefore neglect the $\theta$-component of the gravitational field. We denote the amplitude of the $\varpi$-component by $g_\varpi$. Thus,

$$g_\varpi = -(A' + ikA) \cong -ikA. \tag{3.18}$$

## 3.4. Asymptotic Theory for Tightly Wound Spirals

By introducing asymptotic approximations of the type (3.17), we may derive from (3.9), (3.10), and (3.11) the following system of linear equations for the amplitude functions defined in (3.15):

$$\begin{bmatrix} \nu & k\sigma_0/\varkappa & 0 \\ 0 & i(\nu^2 - k^2a_0^2/\varkappa^2)\nu & -2\Omega/\varkappa \\ 0 & \varkappa/2\Omega & i\nu \end{bmatrix} \begin{bmatrix} \hat{\sigma} \\ \hat{u} \\ \hat{v} \end{bmatrix} = \begin{bmatrix} 0 \\ g_\varpi/\varkappa \\ 0 \end{bmatrix} \tag{3.19}$$

where

$$\nu = (\omega - m\Omega/\varkappa = m(\Omega_p - \Omega)/\varkappa. \tag{3.20}$$

In case $\omega$ is real, $\Omega_p$ is the pattern speed, and $\nu$ is the frequency at which the gas particles encounter a periodic gravitational field, the frequency being expressed in dimensionless form in terms of the epicyclic frequency $\varkappa$.

The first and the third rows of (3.19) lead to

$$\frac{\hat{\sigma}}{\sigma_0} : \frac{\hat{u}}{\varpi\Omega} : \frac{\hat{v}}{\varpi\Omega} = (-k\varpi) : m\left(\frac{\Omega_p}{\Omega} - 1\right) : i\frac{\varkappa^2}{2\Omega^2}. \tag{3.21}$$

They are essentially the consequences of the conservation of mass and of angular momentum. (Angular momentum is conserved, since $-\partial\mathscr{V}/\partial\theta$ has been neglected). The second equation of (3.19) leads to

$$\frac{\hat{\sigma}}{\sigma_0} = -\frac{ikg_\varpi}{\varkappa^2(1 - \nu^2) + k^2a_0^2}. \tag{3.22}$$

Equations (3.21) and (3.22) give the response of the gaseous disk to the spiral gravitational field (3.16).

*Poisson's equation.* The asymptotic solution of the Poisson equation (3.12) for tightly wound spiral distributions (3.15) may be readily obtained. It is found (see Lin and Shu, 1964) that

$$g_{\tilde{\omega}} = 2\pi iG\hat{\sigma} \times \text{sign of } (k). \tag{3.23}$$

17*

By combining (3.22) and (3,23) (cf. scheme described in Sec. 2), we obtain the dispersion relation

$$\frac{|k|}{k_0} = (1 - \nu^2) + \frac{k^2 a_0^2}{\varkappa^2}, \tag{3.24}$$

where

$$k_0 = \varkappa^2/2\pi G\sigma_0. \tag{3.25}$$

It is slightly more convenient to introduce the wavelengths

$$\lambda = 2\pi/|k| \quad \text{and} \quad \lambda_0 = 2\pi/k_0, \tag{3.26}$$

and rewrite (3.24) in the form

$$x_0 = \frac{\lambda}{\lambda_0} - (1 - \nu^2)\left(\frac{\lambda}{\lambda_0}\right)^2, \quad \text{where} \quad x_0 = \frac{k_0^2 a_0^2}{\varkappa^2}. \tag{3.27}$$

The relation (3.24) or (3.27) represents a dispersion relation between $\lambda/\lambda_0$ and $\nu$, which includes the parameter $x_0$. We shall now see that $x_0$ can be determined from stability considerations. These stability considerations were first made by Toomre (1964) for the problem of the stellar disk‡ and later by Lynden–Bell (1967) for the problem of the gaseous disk. Both authors, however, were skeptical of the existence of spiral waves of the type discussed by Lin and Shu (1964, 1966). [See Lynden–Bell and Ostriker (1966) and the discussions at the end of this section]. Clearly, (3.27) imples that $\nu^2$ must be real. Neutral waves are obtained for $\nu^2 > 0$. If $\nu^2 < 0$, the solutions (3.15) must be interpreted as having an instability, since the frequency $\omega$ now has an imaginary part, by (3.20). The marginal condition between these cases is given by $\nu^2 = 0$.

In this case, Eq. (3.27) represents a parabola in the plane $(\lambda/\lambda_0, x_0)$ with vertex at (1/2, 1/4). Points under the parabola correspond to unstable conditions ($\nu^2 < 0$); points above correspond to neutral waves. If all wavelengths are to be stable, we must have $x_0 \geqq 1/4$. If turbulence is assumed to be produced only by instabilities, then we would expect to have

$$x_0 = \frac{1}{4}, \tag{3.28}$$

since there is no reason for it to grow much further. In this case, we have

$$\frac{\lambda}{\lambda_0} = \frac{1}{2}\frac{1}{(1 \pm \nu)}. \tag{3.29}$$

‡ In the stellar case, one could speculate that Landau damping might prevail after the peculiar velocity of the stars ecxeeds that required for stability. In a rotating system under consideration here, this is found not to be the case in the asymptotic approximation. Neutral waves are found.

There are two branches to our solution; the case $(1 + \nu)$ is shown in Fig. 3.1, this being the case of the shorter waves. Both short and long waves have also been noted in the stellar theory by Shu (1968).

We note, by comparing Fig. 3.1 with Fig. 5.2, that the stellar theory and the gas theory give quite similar results, except in the vicinity of $|\nu| = 1$. The condition $|\nu| = 1$ corresponds to a type of resonance pointed out by Lindblad: the material particles see a gravitational field at the natural frequency of epicyclic motions of the stars, and hence resonance is expected. This leads to an infinitesimally small scale in the stellar case. In the gaseous case, the effect of pressure blunts this resonance effect. In the case of a combination of both sheets, the stellar effect prevails, and the scale is again zero at resonance.

The absence of sharp resonance is a rather serious weakness when one attempts to apply a gas dynamical theory to the whole galactic disk. The pattern is no longer restricted to a "principal range" [see Lin and Shu (1966)] defined by the Lindblad resonance points, but extends throughout the whole galactic disk. The restrictions due to boundary conditions at the galactic center may then exclude the possibility of spiral solutions in the linear theory. Such restrictions do not apply in the case of stellar systems, since the spiral patterns essentially terminate at the inner point of Lindblad resonance. Neither would the anti-spiral theorem apply in a nonlinear theory.

In the next section, we shall consider the dynamics of an infinitesimally thin disk of stars as a basis for developing a theory of spiral structure based on density waves involving both the gaseous and the stellar components.

## 4. DYNAMICS OF AN INFINITESIMALLY THIN STELLAR DISK

### 4.1. Basic Equations

Consider an infinitesimally thin disk of stars in differential rotation. This might by coexisting with a sheet of gas and a magnetic field. Let us consider only motions in the plane of the disk. We shall write down the *fundamental equation of stellar dynamics with reference to an arbitrarily chosen differentially rotating system* [see Appendix to Sec. 4, (B16)]. It reads:

$$\frac{\partial \Psi}{\partial t} + c_\varpi \frac{\partial \Psi}{\partial \varpi} + \left(\Omega + \frac{c_\theta}{\varpi}\right)\frac{\partial \Psi}{\partial \theta} + \left(a_\varpi + \Omega^2 \varpi + 2\Omega c_\theta + \frac{1}{\varpi} c_\theta^2\right)\frac{\partial \Psi}{\partial c_\varpi} + \left(a_\theta - \frac{\varkappa^2}{2\Omega} c_\varpi - \frac{1}{\varpi} c_\varpi c_\theta\right)\frac{\partial \Psi}{\partial c_\theta} = 0, \tag{4.1}$$

where the symbols are defined as follows. The function $\Psi(\varpi, \theta, c_\varpi, c_\theta, t)$ is the distribution function in two dimensions, such that

$$dN = \Psi \varpi \, d\varpi \, d\theta \, dc_\varpi \, dc_\theta \tag{4.2}$$

is the number of stars (each of mass $m^*$) lying in the element of volume

$$d\tau_c = \varpi \, d\varpi \, d\theta \, dc_\varpi \, dc_\theta \tag{4.3}$$

of the phase space with the spatial cylindrical coordinate system $(\varpi, \theta)$, and the *peculiar velocity* components $(c_\varpi, c_\theta)$. The "peculiar velocity" is defined with respect to an *arbitrarily chosen* circular motion with angular velocity $\Omega(\varpi)$; there is no radial velocity in the reference system. Thus, the components of the total velocity of the stars are given by

$$\Pi = c_\varpi, \qquad \Theta = \varpi\Omega(\varpi) + c_\theta. \tag{4.4}$$

The mean values of $c_\varpi$ and $c_\theta$ are not necessarily equal to zero, even in an axially symmetrical state of motion.

The *surface density* of the stellar system is given by

$$\sigma_*(\varpi, \theta, t) = m_* \iint \Psi \, dc_\varpi \, dc_\theta, \tag{4.5}$$

whereas the mean values $(V_\varpi + v_\varpi, V_\theta + v_\theta)$ of the total velocity components $(\Pi, \Theta)$ are given by

$$\frac{\sigma_*}{m_*}(V_\varpi + v_\varpi) = \iint \Pi\Psi \, dc_\varpi \, dc_\theta, \tag{4.6a}$$

$$\frac{\sigma_*}{m_*}(V_\theta + v_\theta) = \iint \Theta\Psi \, dc_\varpi \, dc_\theta, \tag{4.6b}$$

where $V_\varpi = 0$, $V_\theta = \varpi\Omega(\varpi)$.

The components of the acceleration $(a_\varpi, a_\theta)$ (in the plane of the disk $z = 0$) might be only partly attributable to the surface density distribution $\sigma_*$, but we shall usually regard them as derivable from a gravitational potential $\mathcal{V}(\varpi, \theta, z, t)$ in the plane $z = 0$. The part due to $\sigma_*$ is the self-gravitation of the stars and will be designated‡ with a superscript $s$. Thus, the self-field $(a_\varpi{}^{(s)}, a_\theta{}^{(s)})$ is given by

$$a_\varpi^{(s)} = -\frac{\partial}{\partial\varpi}\mathcal{V}^{(s)}(\varpi, \theta, 0, t) \tag{4.7a}$$

and

$$a_\theta^{(s)} = -\frac{1}{\varpi}\frac{\partial}{\partial\theta}\mathcal{V}^{(s)}(\varpi, \theta, 0, t) \tag{4.7b}$$

‡ The superscript $s$ will be used alternatively with an asterisk in the subscript.

in terms of the self-gravitational potential $\mathscr{V}^{(s)}(\varpi, \theta, z, t)$, which satisfies the three-dimensional Poisson equation

$$\frac{\partial^2 \mathscr{V}^{(s)}}{\partial \varpi^2} + \frac{1}{\varpi}\frac{\partial \mathscr{V}^{(s)}}{\partial \varpi} + \frac{1}{\varpi^2}\frac{\partial^2 \mathscr{V}^{(s)}}{\partial \theta^2} + \frac{\partial^2 \mathscr{V}^{(s)}}{\partial z^2} = 4\pi G \sigma_* \delta(z). \quad (4.8)$$

In the last equation, $\delta(z)$ is the Dirac delta function.

The quantity $\varkappa(\varpi)$ is the usual epicyclic frequency [Chandrasekhar (1960), Eq. (4.327), p. 156] associated with $\Omega(\varpi)$. It is defined by

$$\varkappa^2 = (2\Omega)^2 \left\{1 + \frac{\varpi}{2\Omega}\frac{d\Omega}{d\varpi}\right\}. \quad (4.9)$$

Since $\Omega(\varpi)$ is "arbitrarily" chosen, there is no direct interpretation for $\varkappa$ in the physical system in question. However, for a "reasonable" choice of $\Omega(\varpi)$, as we shall insist upon in all the following discussions, we also have the usual interpretation for $\varkappa$ as the epicyclic frequency.

The characteristic equations of the partial differential equation (4.1) are

$$dt = \frac{d\varpi}{c_\varpi} = \frac{d\theta}{\Omega + c_\theta/\varpi} = \frac{dc_\varpi}{a_\varpi + (\Omega\varpi + c_\theta)^2/\varpi}$$

$$= \frac{dc_\theta}{a_\theta - (\varkappa^2/2\Omega + c_\theta/\varpi)\, c_\varpi}. \quad (4.10)$$

This system possesses the following two integrals:

$$\frac{1}{2}\{c_\varpi^2 + (\varpi\Omega + c_0)^2\} + \mathscr{V}(\varpi, \theta, 0) = E \quad (4.11)$$

for *time-independent* potentials and

$$\varpi(\varpi\Omega + c_\theta) = J \quad (4.12)$$

for *axially symmetric* potentials [cf. Appendix to Sec. 4, (B22a), (B23a)].

## 4.2. The Axially Symmetrical State

Consider now an equilibrium state with axial symmetry described by the distribution function

$$\Psi = \Psi_0(\varpi, c_\varpi, c_\theta). \quad (4.13)$$

The components of the acceleration are

$$(a_\varpi, a_\theta) = (a_{\varpi 0}(\varpi), 0). \quad (4.14)$$

The basic equation (4.1) becomes

$$c_\varpi \frac{\partial \Psi_0}{\partial \varpi} + \left[ a_{\varpi 0} + \Omega^2 \varpi + 2\Omega c_\theta + \frac{c_\theta^2}{\varpi} \right] \frac{\partial \Psi_0}{\partial c_\varpi} + \left[ 0 - \left( \frac{\varkappa^2}{2\Omega} + \frac{c_\theta}{\varpi} \right) c_\varpi \right] \frac{\partial \Psi_0}{\partial c_\theta} = 0. \tag{4.15}$$

For convenience, we shall choose $\Omega(\varpi)$ such that the condition

$$a_{\varpi 0} + \Omega^2 \varpi = 0 \tag{4.16}$$

is satisfied. With this choice, (4.15) is simplified into

$$c_\varpi \frac{\partial \Psi_0}{\partial \varpi} + \left( 2\Omega + \frac{c_\theta}{\varpi} \right) c_\theta \frac{\partial \Psi_0}{\partial c_\varpi} - \left( \frac{\varkappa^2}{2\Omega} + \frac{c_\theta}{\varpi} \right) c_\varpi \frac{\partial \Psi_0}{\partial c_\theta} = 0. \tag{4.17}$$

Since both of the integrals (4.11) and (4.12) exist in this case, the general solution of (4.17) is of the form

$$\Psi_0 = \Psi_0 \left\{ \varpi(\varpi\Omega + c_\theta),\ \frac{1}{2}[c_\varpi^2 + (\varpi\Omega + c_\theta)^2] + \mathscr{V}_0(\varpi) \right\}, \tag{4.18}$$

where

$$\mathscr{V}_0(\varpi) = \int_{\varpi_0}^{\varpi} \Omega^2 \varpi \, d\varpi. \tag{4.19}$$

Note that $\mathscr{V}_0(\varpi)$ need not be due to the stellar distribution (3.13) alone, as there might be, for example, a symmetrical distribution of gas in the Galaxy.

### 4.3. The Modified Schwarzschild Distribution

It is known from observations that the dispersion velocities satisfy an ellipsoidal distribution, to a first approximation. A distribution function of the form

$$\Psi_s = P_0(\varpi) \exp \left\{ -\frac{c_\varpi^2}{2\langle c_\varpi^2 \rangle} - \frac{c_\theta^2}{2\langle c_\theta^2 \rangle} \right\}, \tag{4.20}$$

is known as the Schwarzschild distribution, where $P_0(\varpi)$, $\langle c_\varpi{}^2 \rangle$, and $\langle c_\theta{}^2 \rangle$ are functions of the radial distance $\varpi$. Such a function (4.20) is a solution of (4.17) only if we neglect the terms $c_\theta/\varpi$ against $2\Omega$ and $\varkappa^2/2\Omega$. In that case, the angular momentum integral $J = \varpi(\varpi\Omega + c_\theta)$ becomes

$$J = \varpi^2 \Omega, \tag{4.21}$$

and (4.17) yield a characteristic integral

$$\frac{1}{2}\left[c_\varpi^2 + \left(\frac{2\Omega}{\varkappa}\right)^2 c_\theta^2\right] = C(\varpi). \tag{4.22}$$

Thus, the form (4.20) may be constructed from the two integral invariants (4.21) and (4.22).

How does one obtain an exact solution of (4.17) which would immediately reduce to (4.20) in the case of small dispersion velocities? This question was answered by Shu (1968) in the following manner.

First, for the integral of angular momentum, define a distance $r$ by the relation

$$J = \varpi(\varpi\Omega + c_\theta) = r^2\Omega(r). \tag{4.23}$$

Then a star in circular motion at a distance $r$ has precisely the given value of $J$. Next, for the energy integral, introduce the "epicyclic energy" $\mathscr{E}$ defined by

$$\mathscr{E} = E - E_c, \tag{4.24}$$

where $E_c$ is the energy of a star in the above-mentioned circular orbit, i.e.,

$$E_c = \frac{1}{2}r^2\Omega^2(r) + \mathscr{V}_0(r). \tag{4.25}$$

Clearly such a star has no epicyclic motion and no epicyclic energy.

Now $r$ and $\mathscr{E}$ are well defined integrals of motion. Thus, any function of them is an exact solution of (4.17) (cf. Appendix Be). It is to be expected that $\mathscr{E}$ would be proportional to the quantity $C(\varpi)$ in (4.22), for clearly the latter is also a measure of the extent of epicyclic motion. A little calculation shows that indeed

$$\mathscr{E} \cong \frac{1}{2}\left[c_\varpi^2 + \left(\frac{2\Omega}{\varkappa}\right)^2 c_\theta^2\right] \tag{4.26}$$

when the dimensionless parameter $\sqrt{2\mathscr{E}}/r\varkappa(r)$ is small. Thus, if we write

$$\Psi_S = P_0(r)\exp\{-\beta(r)\mathscr{E}\}, \tag{4.27}$$

then we have an exact solution of the Eq. (4.17) for a stationray distribution function, which, at the same time, is approximately the same as the function (4.20). Shu has shown that the physical interpretation of $P_0(r)$ and $\beta(r)$ may be obtained as follows:

$$\beta(r) = \{\langle c_\varpi^2\rangle(r)\}^{-1} \tag{4.28}$$

$$P_0(r) = \frac{2\Omega(r)}{\varkappa(r)}\,\frac{\sigma_{*0}(r)}{2\pi m_*\langle c_\varpi^2\rangle(r)}, \tag{4.29}$$

where $\langle c_\varpi{}^2 \rangle (\varpi)$ is the mean-square value of the radial dispersion speed at radial distance $\varpi$, and it is to be evaluated at the "circular distance" $r$ defined by (4.23) for a given angular momentum $J$. The results (4.28) and (4.29) are accurate to the order $\{1 + O[\sqrt{2\mathscr{E}}/r\varkappa(r)]\}$. Shu calls the distribution function defined by (4.27), (4.28), and (4.29) the *modified Schwarzschild distribution.*

## 4.4. Construction of a Galactic Model

Clearly we can now construct a model of a galaxy with the modified Schwarzschild distribution everywhere if we are given $\Omega(\varpi)$, $\sigma_{*0}(\varpi)$, and $\langle c_\varpi{}^2 \rangle (\varpi)$. We note that the two former quantities are related to each other through (4.19) and

$$\mathscr{V}_0(\varpi) = \int_0^\infty \varkappa(\varpi, \varpi')\, \sigma_{*0}(\varpi')\, d\varpi' \tag{4.30}$$

for a pure disk model, where $K(\varpi, \varpi')$ is expressible in terms of an elliptic integral:

$$K(\varpi, \varpi') = -G \oint \frac{d\varphi}{\sqrt{1 + \alpha^2 + 2\alpha \cos\varphi}}, \qquad \alpha = \frac{\varpi}{\varpi'}. \tag{4.31}$$

This is the problem of the mass model of a disk galaxy after the rotation curve is known (from observations, for example), and it has been treated by a number of authors in a variety of ways. We shall see later that there is also a criterion for obtaining $\langle c_\varpi{}^2 \rangle$ from stability considerations, once the mass model is given. Thus, in principle, we can construct a model for a pure stellar disk once the rotation curve is known. Such a model can only be expected to be approximate since the effects of gas and of the thickness of the disk are found to be very significant for the determination of the dispersion velocity $\langle c_\varpi{}^2 \rangle^{1/2}$.

# 5. THEORY OF DENSITY WAVES IN A COMPOSITE SYSTEM OF STELLAR AND GASEOUS DISKS

## 5.1. Theory of Density Waves

The density-wave theory was developed especially to resolve the problems of (i) the persistence of the spiral structure in the face of differential rotation, and (ii) the large-scale nature of the phenomenon, i.e., the existence of a grand design over the whole galactic disk. We therefore start out with the QSSS hypothesis (hypothesis of the existence of density waves with a quasi-stationary spiral structure), rather than attempt to study the problem of the origin of the spiral structure. As outlined in Sec. 2, the theory is based on the calculation of the responses of an infinitesimally thin disk of stars and gas to a *resultant* gravitational field of a spiral form. The effect of finite thickness has been recently examined by Shu (1968), but we shall not go into the details in this section (see Shu's lectures).

*Response of the stellar disk.* The stellar disk is described by a distribution function $\Psi(\varpi, \theta, c_\varpi, c_\theta, t)$, where $t$ is the time, $(\varpi, \theta)$ are polar coordinates in the plane of the disk, and $(c_\varpi, c_\theta)$ are the peculiar velocities relative to a (differentially) rotating observer moving at the circular speed $\varpi\Omega(\varpi)$. In the absence of spiral modes, $\Psi$ is given by a function $\Psi_0(\varpi, c_\varpi, c_\theta)$ which is time-independent and axially symmetric in the configurational space.

We now make some brief observations on the purely kinematical aspects of smooth but tightly wound spiral patterns in the plane of a disk galaxy. First of all, we recall that these patterns may be described by functions of the form

$$F(\varpi, \theta, t) = \mathscr{A}(\varpi) \exp\{i[\omega t - m\theta + \Phi(\varpi)]\}, \tag{5.1}$$

where $\varpi$, $\theta$, $t$ are defined as before, $A(\varpi)$ is a slowly varying function of $\varpi$, $\Phi(\varpi)$ is a slowly varying (monotonic) function multiplied by a large parameter, $\varpi$ is a constant (real for neutral waves), and $m$ is an integer. As usual, the real part of the above expression is to be taken. The lines of constant Re $(F)$ are approximately given by the equations

$$m(\theta - \theta_0) = \Phi(\varpi) - \Phi(\varpi_0), \tag{5.2}$$

which represents a spiral pattern with $m$ arms.

The pattern given by (5.1) rotates with an angular velocity

$$\Omega_p = \omega_r/m, \tag{5.3}$$

where $\omega_r$ is the real part of $\omega$. The radial wave number of the spiral pattern is given by $k(\varpi)$ where

$$k(\varpi) = \Phi'(\varpi). \tag{5.4}$$

If the motion of the stars is in the direction of increasing $\theta$, trailing waves correspond to $k(\varpi) < 0$, and leading waves to $k(\varpi) > 0$.

In the presence of a small spiral gravitational potential

$$\mathscr{V}_1 = A(\varpi) \exp\{i[\omega t - m\theta + \Phi(\varpi)]\}, \tag{5.5}$$

of the form (5.1), the response $\Psi_1 = \Psi - \Psi_0$ in the distribution function should also have a spiral distribution in the configurational space. When the dispersion velocities of the stars are not large, this response $\Psi_1$ can be easily worked out. In this case, the basic distribution function $\Psi_0$ is of the form

$$\Psi_0 = \Psi_0(\xi^2 + \eta^2, \varpi), \tag{5.6}$$

where $\xi$ and $\eta$ are defined by

$$c_\varpi = \xi(2\Omega/\varkappa)(\varpi\Omega) = \xi V_1(\varpi) \tag{5.7a}$$

and

$$c_\theta = \eta(\varpi\Omega) \tag{5.7b}$$

in which $\varkappa$ is the epicyclic frequency defined by

$$\varkappa^2 = (2\Omega)^2\left\{1 + \frac{2\varpi}{\Omega}\frac{d\Omega}{d\varpi}\right\}. \tag{5.8}$$

Within the approximation of small dispersion speeds, $\varpi$ is a parameter related to the integral for angular momentum, and for given $\varpi$, $\xi^2 + \eta^2$ is the integral of epicyclic energy. The response $\Psi_1$ to the resultant potential (5.5) is found to be

$$\Psi_1 = \Psi_0'(\xi^2 + \eta^2)\{1 - q(\alpha\xi, \alpha\eta, \nu)\}\frac{\mathscr{V}_1}{V_1^2} \tag{5.9}$$

in the case of tightly wound spirals, where $\Psi_0'$ denotes the derivative of $\Psi_0$ with respect of its argument,‡ $V_1$ is defined in (5.7a), and the parameters $\alpha$ and $\nu$ are defined by

$$\alpha = (k\varpi)(2\Omega^2/\varkappa^2), \tag{5.10}$$

and

$$\nu = (\omega - m\Omega)/\varkappa. \tag{5.11}$$

The function $q$ is defined by the integral

$$q = \frac{\nu\pi}{\sin\nu\pi}\cdot\frac{1}{2\pi}\int_{\pi}^{\pi}\exp\{i[\nu s - \alpha\xi\sin s + \alpha\eta(1 + \cos s)]\}\,ds. \tag{5.12}$$

The derivation of the result (5.9) may be found in the Appendix to Sec. 5.

‡ Here, as in other quantities, the dependence on the galacto-centric $\varpi$ is understood but not explicitly exhibited.

In the case of the Schwarzschild distribution,‡ the perturbation in density, as well as those in the components of the mean velocity of the stars, can be very readily calculated to exhibit their dependence on the dispersion velocity of the stars (see Appendix to Sec. 5). One obtains

$$\frac{\hat{\sigma}}{\sigma_0} = -\frac{k^2 A}{\varkappa^2} \frac{1}{1-\nu^2} \mathscr{F}_\nu(x), \tag{5.13a}$$

$$\frac{\hat{v}_\varpi}{\varpi\Omega} = \frac{k^2 A}{\Omega^2} \frac{(k\varpi)^{-1}\nu}{1-\nu^2} \mathscr{F}_\nu^{(1)}(x), \tag{5.13b}$$

$$\frac{\hat{v}_\theta}{\varpi\Omega} = \frac{1}{2} \frac{\varkappa^2 A}{\Omega^2} \frac{(k\varpi)^{-1}}{1-\nu^2} \mathscr{F}_\nu^{(2)}(x), \tag{5.13c}$$

where

$$x = k^2 \langle c_\varpi^2 \rangle / \varkappa^2, \tag{5.14}$$

and the functions $\mathscr{F}_\nu(x)$, $\mathscr{F}_\nu^{(1)}(x)$, and $\mathscr{F}_\nu^{(2)}(x)$ are "reduction factors" defined by Eqs. (C37), (C42), and (C45) in the Appendix to Sec. 5. They all take on the value unity for $x = 0$, i.e., in the case of zero dispersion. The function $\mathscr{F}_\nu(x)$ is most important for our subsequent discussions. Its values are already tabulated and presented in graphical form (Lin and Shu, 1966; Lin, 1967*b*). One form of the graph is reproduced in the Appendix for convenience of reference.

We note that since all the above results are given for tightly wound spirals, they are subject to an uncertainty of a factor of $1 + O[(k\varpi)^{-1}]$.

*Response of the gaseous disk.* The response of the gaseous disk can be much more readily obtained, if the turbulent motion of the gas is simulated by pressure. We then introduce the square of the equivalent acoustic velocity $a$, and define the quantity

$$x_g = k^2 a^2 / \varkappa^2 \tag{5.15}$$

analogous to $x$ defined by (5.10). Then all the three *reduction factors* are found to be of the same form, i.e.,

$$\mathscr{F}_\nu^{(g)}(x_g) = \frac{1}{1 + x_g/(1-\nu^2)}. \tag{5.16}$$

‡ In earlier papers, we have only presented results for the case of the Schwarzschild distribution [for which $\Psi_0$ is an exponential function of $(\xi^2 + \eta^2)$]. Clearly, this is an important case, and we shall continue to adopt it in following sections of this paper. But the generalization given here enables one to study, separately, the role played by stars with different epicyclic energies; this may be done simply by taking $\Psi_0$ to be of the form of the Dirac delta-function. Such studies will be carried out in a separate investigation.

For convenience of reference, we shall record the explicit formulae as follows:

$$\frac{\hat{\sigma}_g}{\sigma_{g0}} = -\frac{k^2 A}{\varkappa^2}\frac{1}{1-\nu^2}\mathscr{F}_\nu^{(g)}(x_g), \tag{5.17a}$$

$$\frac{\hat{u}_\omega}{\varpi\Omega} = \frac{k^2 A}{\Omega^2}\frac{(k\varpi)^{-1}\nu}{1-\nu^2}\mathscr{F}_\nu^{(g)}(x_g), \tag{5.17b}$$

$$\frac{\hat{u}_\theta}{\varpi\Omega} = \frac{i}{2}\frac{k^2 A}{\Omega^2}\frac{(k\varpi)^{-1}}{1-\nu^2}\mathscr{F}_\nu^{(g)}(x_g), \tag{5.17c}$$

where $\hat{u}_\varpi$ and $\hat{u}_\theta$ are the amplitude functions for the components of the gaseous velocity. Let us also not that, irrespective of pressure, the following ratios hold:

$$\frac{\hat{\sigma}_g}{\sigma_{g0}} : \frac{\hat{u}_\omega}{\varpi\Omega} : \frac{\hat{u}_\theta}{\varpi\Omega} = (-k\varpi) : \frac{m(\Omega-\Omega_p)}{\Omega} : \frac{\varkappa^2}{(2\Omega)^2}, \tag{5.18}$$

where $\Omega_p$ is the pattern speed. This follows from the fact that the pressure gradient plays no role in the equation of continuity and in the $\theta$-component of the equation of motion (in the case of tightly wound spirals).

## 5.2. The Dispersion Relation for Density Waves

The central relationship for any type of wave motion is that between wavelength and frequency. In the theory of self-sustained density waves, one must equate the density required to sustain the gravitational potential (5.1) according to Poisson's equation, i.e.,

$$\hat{\sigma}_p = -|k|\, A/2\pi G, \tag{5.19}$$

with the *sum* of the responses in density variation in the stellar and gaseous disks. Thus, by using (5.13a) and (5.16a), we get the *dispersion relationship*‡

$$\frac{k_*}{|k|}(1-\nu^2) = \mathscr{F}_\nu(x) + \frac{\sigma_0}{\sigma_*}\mathscr{F}_\nu^{(g)}(x_g) \tag{5.20}$$

which connects $k$ and $\nu$ for a given galactic model. In the above equation,

$$k_* = \varkappa^2/2\pi G\sigma_*, \tag{5.21}$$

‡ In the limit $|\nu| - 1$, and $k - \infty$, the contribution from the gas is negligible- We also have $\lambda/\lambda_* \sim \{1-|\nu|\}^{1/2}$ in this vicinity.

and

$$1 - \nu^2 > 0, \quad \text{i.e.,} \quad \Omega - \frac{\varkappa}{m} < \Omega_p < \Omega + \frac{\varkappa}{m}. \qquad (5.22)$$

The limiting cases $\nu = \pm 1$, correspond to Lindblad resonance $\Omega_p = \Omega \pm \varkappa/m$; they deserve special treatment.

The waves described by the dispersion relationship (5.20) have the following properties.

(a) They extend essentially over a range of the galactic disk for which the conditions

$$\Omega - \frac{\varkappa}{m} < \Omega_p < \Omega + \frac{\varkappa}{m} \qquad (5.23)$$

are satisfied, where $\Omega(\varpi)$ is the angular speed of the stars, and $\varkappa(\varpi)$ is the epicyclic frequency, as defined before. We shall refer to this range as the principal part of the spiral pattern. In Fig. 5.1, we show the graphs of $\Omega(\varpi)$, $\varkappa(\varpi)$, and $\Omega \pm \varkappa/2$ on the basis of the model given by Schmidt (1965).

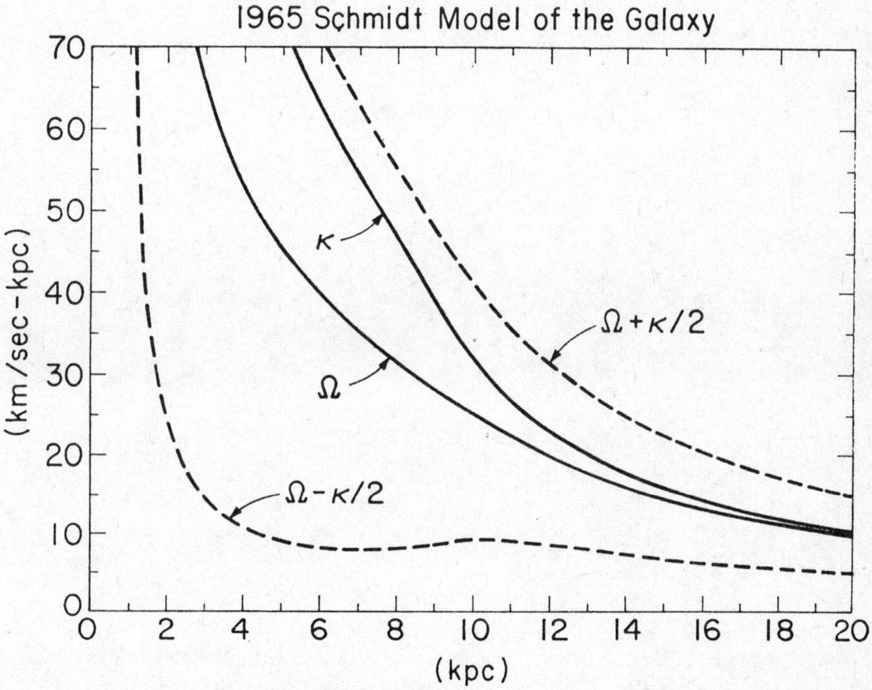


FIG. 5.1. The graphs of $\Omega$, $\varkappa$, and $\Omega \pm \varkappa/2$ at various galactocentric distances $\varpi$; where $\Omega$ is the circular velocity in angular measure and $\varkappa$ is the epicyclic frequency.

According to this diagram, for $m = 2$, the pattern would extend from $\varpi = 4$ kpc outwards without limit, if $\Omega_p = 11$ km/sec$^{-1}$ kpc$^{-1}$. For other values of $m$, this principal part would be quite limited in extent, whatever the value of $\Omega_p$ may be. This is the reason why all spirals have two arms, for the rotation curves are qualitatively the same.

(b) We notice that the dispersion relation for such waves is expressed in terms of the wavelength $\lambda = 2\pi/|k|$ and the frequency $\nu = m(\Omega - \Omega_p)/\varkappa$ at which the stars see the gravitational field. This relationship includes two additional parameters, $x_*$ and $x_g$, which measure the velocity dispersion of the stars and the turbulence in the gas. As a first attempt to appreciate the significance of this dispersion relationship, it is convenient to work out explicitly the results in the case of a pure stellar disk. In Fig. 5.2, we show the result for the case where the velocity dispersion is barely enough to stabilize the idsk against gravitational collapse (Toomre, 1964).‡ The margin of instability is obtained by setting $\nu^2 = 0$ in (5.20); conditions with $\nu^2 < 0$ correspond to instability, while those

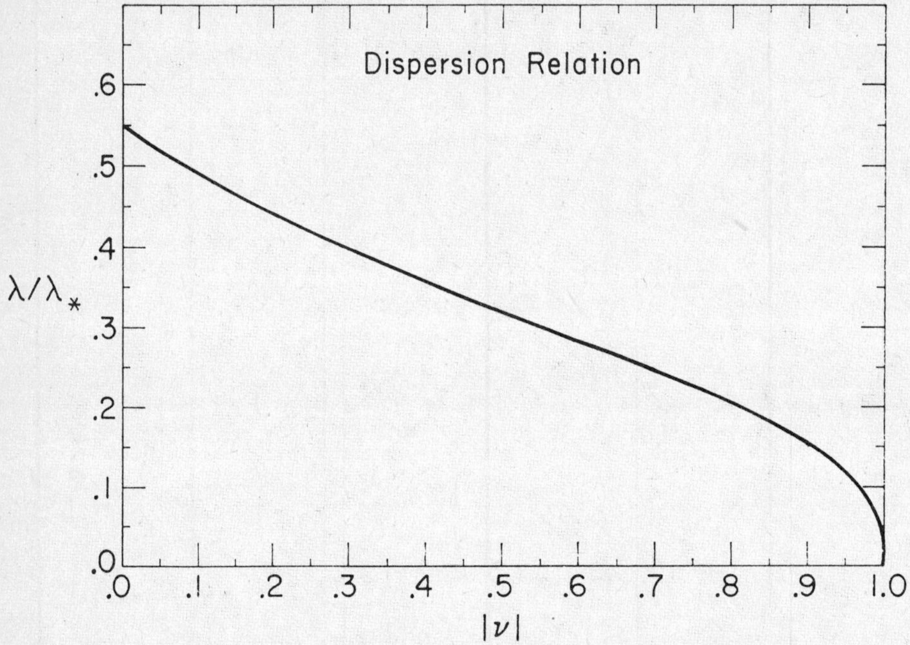


FIG. 5.2. Standard dispersion relationship for density waves in the stellar dynamical theory; where $\nu = m(\Omega - \Omega_p)/\varkappa$ ($m$ = number of arms, $\Omega_p$ = pattern frequency), $\lambda$ is the spacing between two neighboring arms, and $\lambda_*$ is a typical length scale defined by $\lambda_* = 4\pi^2 G\sigma_*/\varkappa^2$, where $\sigma_*$ is the projected surface density of the stars.

with $\nu^2 > 0$ correspond to neutral waves. The condition $\nu^2 = 0$ leads to a minimum dispersion velocity $\langle c_\varpi{}^2 \rangle$ given by (see Toomre, 1964).‡

$$\frac{k_*^2 \langle c_\varpi^2 \rangle}{\varkappa^2} = 0.2857 \tag{5.24}$$

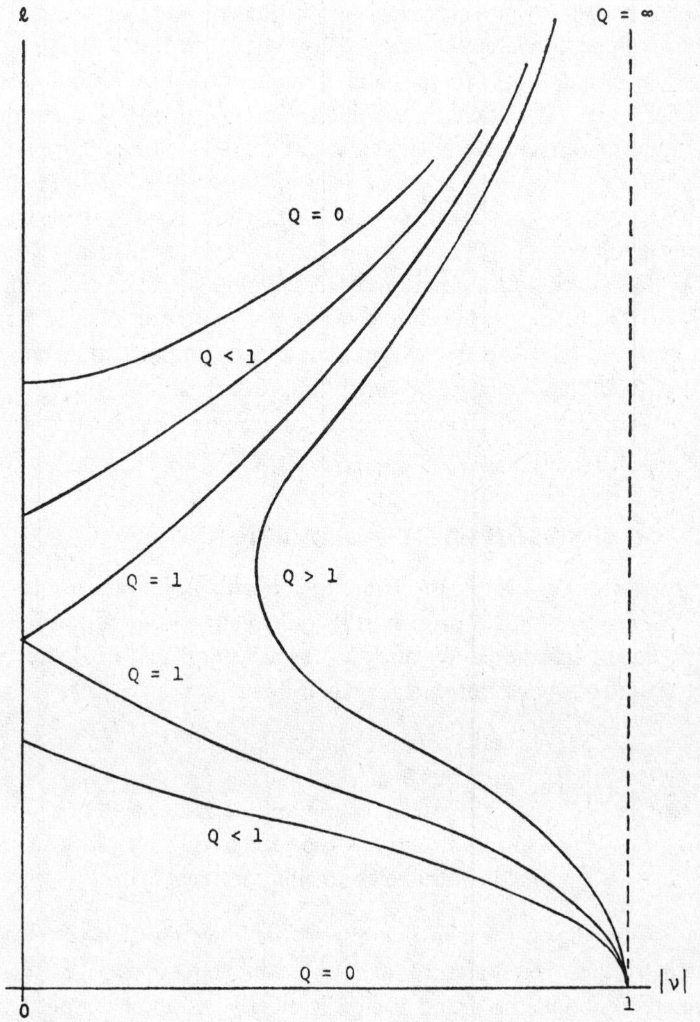


FIG. 5.3. General dispersion relationship ($\mathfrak{l} = \lambda/\lambda^*$) for density waves in the stellar dynamical theory, for various amounts of stellar peculiar velocities. The case shown in Fig. 5.2 is that marked $Q = 1$ (after Frank Shu).

‡ Toomre considered exponentially damped or growing rings not, neutral spirals; but the mathematical relationships are related. See Lin and Shu (1966).

below which there would be gravitational collapse. The corresponding wavelengths of the unstable waves if $\langle c_\varpi^2 \rangle$ is slightly lowered from these critical values, are however usually so large that their significance in the spiral theory is not clear. Thus, one might be inclined to think that even when $\langle c_\varpi^2 \rangle$ is not as large as that given by (5,24), the dispersion speeds may not increase any further. For slightly lower values of $\langle c_\varpi^2 \rangle$, the dispersion relationship yields the other curves shown in Fig. 5.3.

As noted above, these calculations are based on an infinitesimally thin stellar sheet. The effect of a fair amount of gas and the modification due to finite thickness have been studied in detail by Shu (1968) and found to be not critical for the particular relationship in Fig. 5.3. Some further discussion of these points will be given when our results are used in the following sections.

(c) To the approximation discussed here, there are no indications that either trailing waves or leading waves are preferred, one to the other. However, other effects, especially nonlinear terms representing convection of matter, would also tend to prefer trailing patterns.

We are now in a position to compare our theoretical results with observed features in our own galaxy.

## 6. COMPARISON WITH OBSERVATIONS

Before we discuss the comparison with observations, let us examine a few terms whose proper use would help to put proper emphasis on the relevant questions. We use the term "spiral features" for general purposes and subdivide this category as follows:

Spiral features {
(1) Spiral arms—a piece, e.g., the Orion arm or the Sagittarius arm
(2) Spiral pattern—the grand design (visible)
(3) Spiral structure—the underlying spiral gravitational field (invisible controlling influence)

Referring to the two pictures of M51, we see that (3) is perhaps indicated by the "red" picture, and is the most regular. Even (2) is not as regular, although the grand design is clearly visble. The spiral pattern might be described as composed of pieces of spiral arms sprinkled over an underlying spiral structure. (Imagine a spiral magnetic end piece placed under a plate of glass. We now sprinkle iron fillings on the glass, knock the glass, and we obtain a spiral pattern. There need not be continuous spiral arms.)

We note that even a piece of gaseous arm, complete with O-associations would move under the influence of the gravitational field, as if the spiral arm were regular, at least to a very close approximation. Thus, a disrupted distribution in H I and H II regions does not necessarily indicate a serious disruption in the spiral structure of the gravitational field.

Finally, we note that in a coordinate system rotating at the pattern speed, the gas flows in closed stream tubes. The large-scale structure of the magnetic field follows the stream lines.

## 6.1. Distribution of Atomic Hydrogen in the Galaxy

On the basis of the basic model shown in Fig. 5.1, and the dispersion relationship shown in Fig. 5.2, we can calculate a spiral pattern as soon as the pattern speed $\Omega_p$ is decided upon. A cursory examination of the numerical values obtained will show that $\Omega_p$ must be in the range of 11–13 km/sec-kpc to arrive at spacings between spiral arms that are reasonable compared with observations. One must also keep in mind that there is no reason why there should be only a single pattern associated with a single angular frequency $\Omega_p$. A group of waves should in general be present. For simplicity, we shall however try our best to compare the observational data with the theoretical results for a *single* pattern. It is therefore too much to expect very close agreement; yet we shall see that the results are surprisingly good.

In Fig. 6.1, we reproduce the spiral pattern presented at the Noordwijk Symposium (Lin and Shu, 1967) based on a pattern speed

$$\Omega_p = 11 \text{ km/sec-kpc}.$$

The related basic data used are given in Table A in Appendix A. When this pattern is compared with the observed disrtibution of atomic hydrogen (see Fig. A3), it appears to be in good agreement, although an even better agreement can probably be obtained by taking $\Omega_p$ slightly higher. As it is, the spiral pattern is perhaps a little too tight; but it does show the correct number of arms from the solar vicinity inwards, up to the "3-kpc arm". The latter is tentatively associated with the strong motions that might be expected at inner Lindblad resonance, where

$$\Omega_p = \Omega - \varkappa/2.$$

According to the theory, the pattern ends here as a ring; there is no prominent pattern inside of this ring. Indeed, no strong indications of a spiral structure have been obtained from observations. A detailed study of the conditions

18*

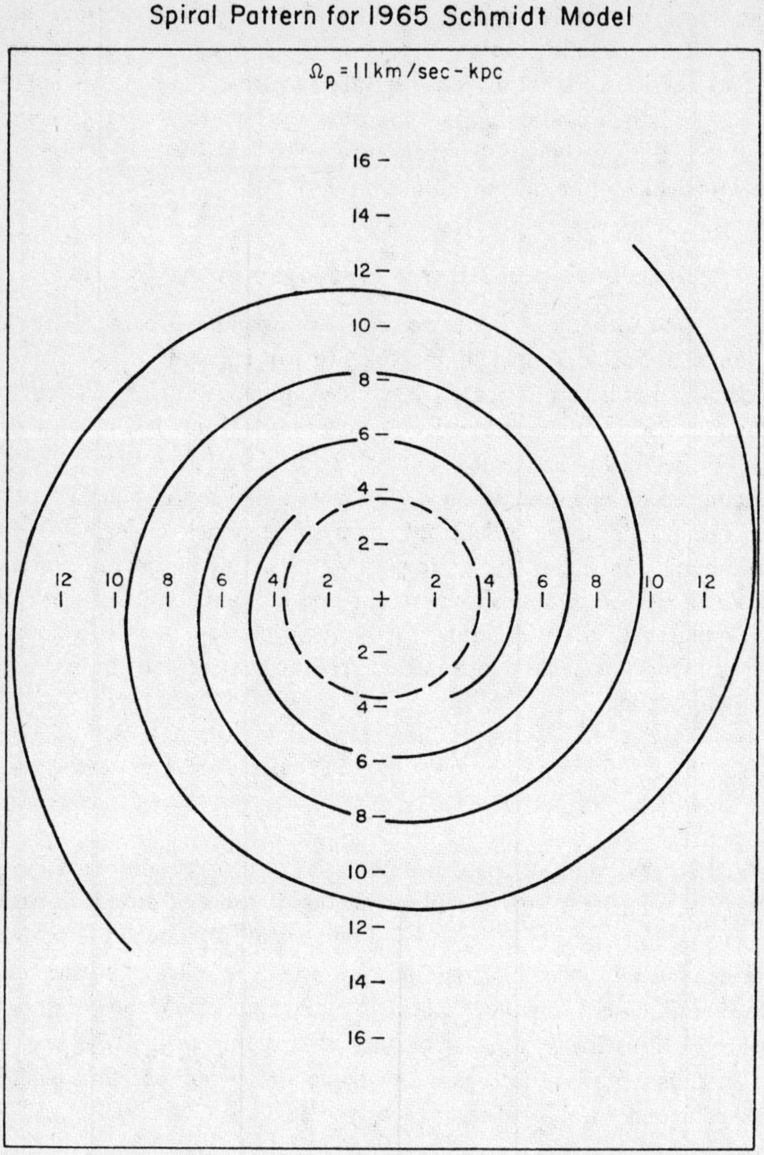


FIG. 6.1. A sprial pattern for the 1965 Schmidt model, for pattern speed $\Omega_p = 11$ km/sec-kpc.

near Lindblad resonance requires an extensive reformulation of the theory for dealing with a stellar sheet. This has been carried out by Shu (1968).

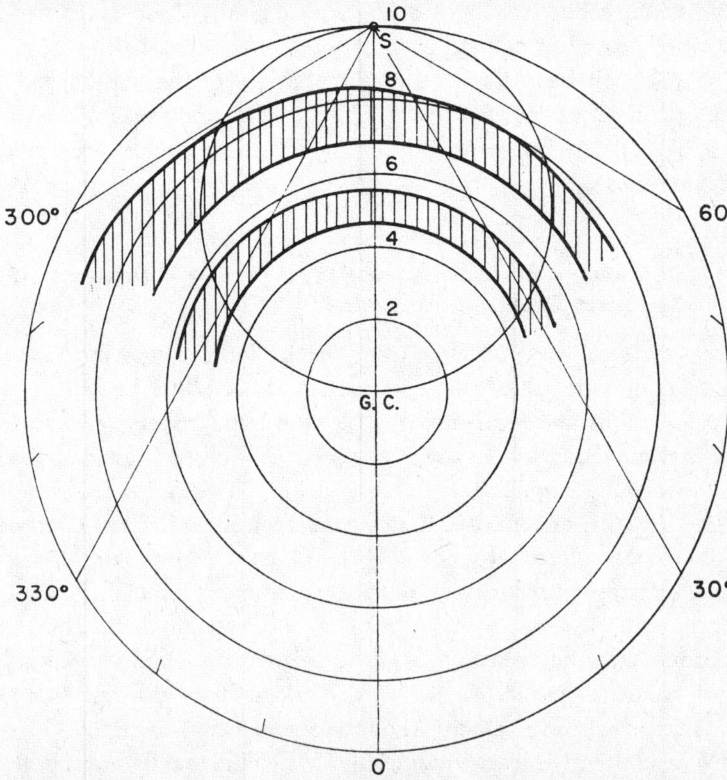


FIG. 6.2. Locating gaseous arms from variations in the rotation curves.

The Perseus arm and the Sagittarius arm can be fitted into an overall spiral pattern with a skeleton shown by Fig. 6.1, if each arm is assigned a width equal to about one quarter of the spacing. The distribution of the young optical objects will be discussed in Sec. 7; no serious problems arise. The inner arm (the one inside of the Sagittarius arm) also fits well into the picture. The Orion arm, however, presents somewhat of a problem. It is known to be inclined at about 30° (cf. section 8 on magnetic field), from observations of neutral hydrogen, optical objects, and magnetic field. If an arm with such an inclination were to be fitted into a two-armed pattern in our vicinity, the interarm spacing would be about 18 kpc, since the pitch angle $i$ is related to this spacing $\lambda$ by the simple relationship $\lambda = \pi\varpi_0 \tan i$, where $\varpi_0$ is the distance to the center of the Galaxy

($\varpi_0$ = 10 kpc in the present case). Such a spacing is clearly inadmissible, and we are therefore forced to conclude that the Orion arm (in which the Sun is located) is an interarm branch. Such branches, of course, occur quite frequently in the outer parts of disk-shaped spiral galaxies, and therefore we need not be unduly alarmed.

We shall leave the details of the choice of the pattern speed $\Omega_p$ and the initial point $(\varpi_0, \theta_0)$ in Eq. (5.2) to a separate discussion (Yuan, to be published). Suffice it to mention here that the choices are limited to very narrow ranges.

### 6.2. Dispersion Velocities of Stars, Finite Thickness of the Galactic Disk

One subtle but rather crucial check of the theory is to examine the dispersion velocities of the stars predicted according to (5.24). In the above calculations, the disk is taken to be a pure stellar sheet of infinitesimal thickness. This leads to an rms value $\langle c_\varpi{}^2 \rangle^{1/2}$ of the radial dispersion velocity of about 52 km/sec for the solar vicinity. This is decidedly too high by at least 25%. If the effect of the gas had been included, the discrepancy would have been larger. However, Shu (1968) has shown that all the discrepancies disappear when we take into consideration the effect of the finite thickness of the disks of stars and of gas. It is found that the pattern speed should also be increased slightly, to about 13.5 km/sec-kpc. (This value will be used in Sec. 7). Shu's calculations show that the rms value of the radial dispersion velocity should be somewhat less than 37 km/sec in our vicinity, if it is barely sufficient to sustain gravitational collapse. This value is in good agreement with observations.

In the same study, Shu found the relative importance of gas and of stars to be about comparable. Here it is important to keep in mind that the gaseous disk is much thinner than the stellar disk. Consequently, the projected mass density of gas is not as large as indicated by the volume density.

### 6.3. Systematic Motion of the Gas

The possible existence of systematic motions was suspected quite early. Kerr (1962) suggested it as a possible source of difference between northern and southern observations. More recently, Burton (1966) considered the possible existence of a systematic motion of the gas near the outer edge of the Sagittarius arm, and concluded that "the observations are consistent with a situation in which the high-velocity stream is hydrogen flowing along the spiral arm, parallel to its axis". But he also suggested other possibilities. Shane and Bieger-Smith (1966) also discussed the role of systematic motions as part of their extensive effort in constructing alternative galactic models.

None of these authors, however, stressed the dynamical mechanism by which these systematical motions are produced. To be sure, whenever there is a circular arm (or nearly circular arm) of concentrated matter, its associated gravitational field would produce a systematic motion of the gas along its edges. It is easy to see this from considerations of angular momentum; indeed, the motion must be *with* the general rotation on the *outside* edge, and *against* the general rotation on the inside edge. But the density-wave theory goes deeper in the following two respects.

(a) Since the pattern speed must in general differ from the speed of the material objects, which are in differential rotation, there *must* by systematic radial motion of the gas to maintain *conservation* of matter.

(b) The observation of the gaseous arm clearly *indicates* the presence of a minimum in gravitational potential; but the gravitational potential is expected to be *jointly* maintained by the gas and by the stars with dispersion speeds comparable to the turbulent speeds in the gas. In many situations, *the stellar component would be dominant in supplying the gravitational field.* Thus, there is no need to be concerned about having sufficient contrast in gaseous density in order to *maintain* the field.

The systematic motion of the gas is estimated to have velocity components of the order of 10 km/sec. The radial component‡ cannot be easily singled out from observational data, but its magnitude is certainly compatible with the range of theoretically predicted values. The component in the circular direction would cause variations in the observed rotation curve. These variations have long been observed, but they were thought to be possibly the consequence of missing gas over interarm regions. A detailed study by Yuan has conclusively shown that the latter effect does not give significant contributions to the variation in velocity.

A comparison of theory with observation is shown in Figs. 6.2, 6.3(a) and 6.3(b). In Fig. 6.2, we show the location of the gaseous arms as deduced from the location of the maxima and minima in the rotation curves

‡ For convenience for camparison with observational features, let it be noted that this component is directed towards the galactic center when the spiral pattern is trailing and travelling at a pattern speed lower than the local circular velocity. See Miller (1967) for relevant data in the Perseus arm.

shown in Fig. 6.3. It is in very good agreement with direct observations of the location of atomic hydrogen. In Fig. 6.3, we show the theoretical curve (dashed) assuming a certain "true rotation curve" on each side. There is good agreement in major features. The sudden dip in the northern curve at about 53° in galactic longitude $l^{\text{II}}$ may be traced to the interaction of the Orion arm with the Sagittarius arm. The northern and southern rotation curves are still different, suggesting an oval distortion of the Galaxy as a

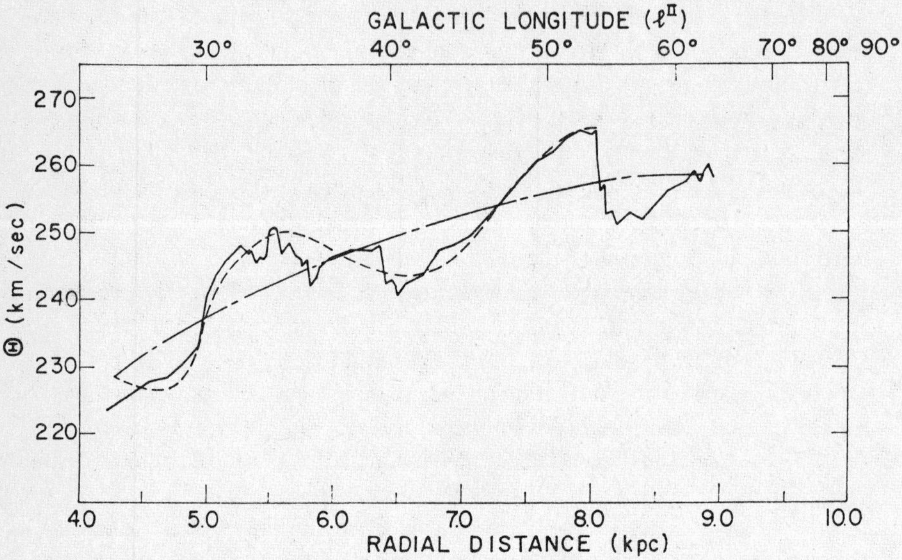


FIG. 6.3a. Rotation curves. (a) Northern hemisphere.

whole. Such a distortion was suggested by Kerr. For a uniformly rotating disk, Hunter's (1963) theoretical work also indicates that it is an important mode of instability; more recently, Kalnajs (private communication) showed that this mode was particularly difficult to stabilize by stellar dispersion velocities. Work remains to be done to correlate the observational data with these general theoretical conclusions.

In the quantitative study of systematic motion, it is neceseary to keep the role of the hydromagnetic forces in the picture. Yuan (to be published) has discussed this point in great detail. The amplitudes of the velocity variation shown in Figs. 6.2 and 6.3 are all about 8 km/sec. This could be caused by a spiral gravitational field equal to about 5% of the symmetrical field (cf. Sec. 7) acting on a gas with turbulent velocity equal to 7 km/sec (rms value of one component) and with a magnetic field of about 5 × $10^{-6}$ gauss. The amplitude of the total variation in mass is about 10% of the mean, with gas and stars each contributing about half of the amount.

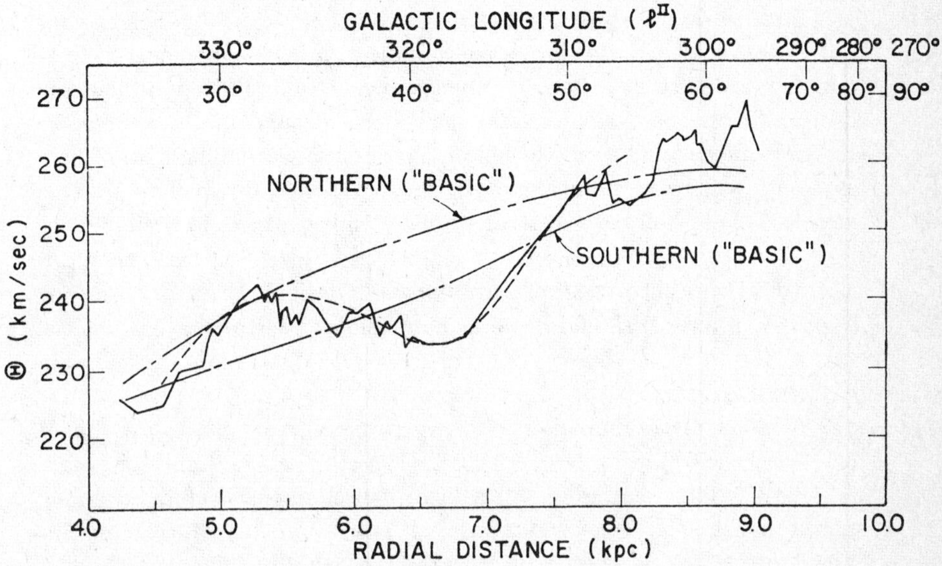


Fig. 6.3b. Rotation curves Southern hemisphere. Solid lines, observation; dashed lines, theory; dash-dot lines, mean motion, theory.

## 6.4. The Distribution of Young Stars and Other Optical Objects

It is well known that young O, B stars are indicators of the spiral arm. It is also generally agreed that they are formed at places of the highest gaseous concentration.† In the density-wave theory, this has a very important implication: the newly formed stars must migrate out of the arms since the former moves at the material circular speed (say 250 km/sec) while the latter moves at the pattern speed (say 125 km/sec). Over a period of 10 million years, there would be a separation of the stars from the gaseous arm by something like 1.25 kpc. However, since the inclination of the spiral arms is small, the *radial* separation is only about one-tenth of the above

† Recent work by William W. Roberts (1968) shows that the process of star formation is intimately related to the formation of shocks in the gaseous flow, but the discussions in this section are not altered.

amount, and becomes difficult to detect. (For a more detailed discussion, see Sec. 2.3).

With the spiral pattern discussed in Sec. 6.1, it is found that the young optical objects indeed lie within the spiral arms, with the notable exception of the Orion arm (already discussed in Sec. 6.1). Although the optical objects in the Perseus arm seem to indicate a fairly large local inclination, their radial spread is not so large as to cause any difficulty in locating them within an arm of small overall inclination. There seems to be however some difficulty with the optical objects in the Vela-Carina direction. There may very well be some local irregularities special to this region. Further theoretical and observational studies are certainly much to be desired.

At this point, it should be emphasized again that the theory primarily deals with the overall pattern, the grand design, rather than with the detailed structure. It should however be possible to introduce local effects, such as the radial expansion of stars from an O association, to explain local irregularities.

A closer examination of the structure of the spiral arm should then reveal the following three "lanes": (1) a dust lane indicating the place of highest gas concentration, (2) a lane of brilliant newly formed young stars with large H II regions—the string of beads noted by Morgan and his collaborators (see Sharpless, 1965), and (3) a lane of dying O stars with much smaller H II regions.‡

Since the dust is suspended in the gas, it is plausible that it should be the thickest where the gas density is the highest. This is known to be true on a large scale. As the dust particles pass through the spiral arm, they will evaporate under the effect of the radiation of the brilliant young stars. (On a smaller scale, there must also be very little dust in the immediate vicinity of bright stars.) One might therefore conceive of a process of dust formation after the gas (moving in a nearly circular orbit) has left the bright part of the sprial arm. This process continues to produce more dust resulting in the highest dust concentration just before the particles are again evaporated at the next arm. Assuming that the pattern speed is about one-half of the material circular speed (as is expected to be approximately true in our vicinity), the time between destruction and re-congregation into a dust lane is about one galactic year. This should be compared with a typical time-scale for dust formation, and a figure of $2 \times 10^8$ years is by no means unreasonable.

‡ Specifically, A. Sandage pointed out to me the possibility of making these detailed observations and analyses in connection with M33. He also mentioned the plans of B. Lynds in this work.

## 6.5. The Migration of Moderately Young Stars

Ever since the method for the determination of stellar ages was greatly improved by Strömgren and his collaborators‡ (Crawford, Kelsall, Perry, *et al.*), it has become meaningful to look for the birth places of the moderately young stars by following the history of their migration. It is expected that these stars would all be found to be formed inside of spiral arms. For these calculations, one has of course to know the present location of the stars and to obtain their velocities from observations of their proper motions. All these are now obtainable with sufficient accuracy. Indeed, Contopoulos and Strömgren (1965) constructed tables for the easy calculation of such orbits, and had some success with the program. Some preliminary results were reported by Strömgren at the Noordwijk Symposium (Strömgren, 1966); and it was concluded that these are generally compatible with the density-wave theory.

An important factor not included in this earlier analysis was the effect of the spiral gravitational field. Preliminary explorations made by the present writer, with the help of S. C. Wang, showed that even a small field could be quite significant. For example, a spiral field of a strength equal to 5% of the symmetrical field can cause displacements to the extent of 1–2 kpc over a period of 100–200 million years. This is not entirely surprising, since the star has migrated over a distance of the order of 25–50 kpc during that period, and a deviation of 5% would yield the calculated displacements.

It is thus necessary first to determine the spiral gravitational field for our galaxy. This is done by using the density-wave theory and by checking the predictions on gas concentrations as "tracers" and also partial contributors of minima in gravitational potential. By experimenting with different choices of pattern frequencies and field strengths, one of us Yuan (to be published) found that a good choice would be a pattern speed of 13.5 km/sec-kpc and a field strength of about 5% of the symmetrical field. The results of his analysis are shown in Figs. and 6.4 6.5. In Fig. 6.4, the places of formation of 25 stars are shown by using the ages determined by Strömgren *et al.* Clearly, they appear to lie on one spiral arm which is *leading* and does not fit into any known structure from radio observations. In Fig. 6.5, the places of formation of the same 25 stars are shown when

‡ See Strömgren (1963–1966), Strömgren and Perry (1965), Kelsall and Strömgren (1966), Crawford (1963, 1966), Crawford and Perry (1966), Crawford and Barnes (1966), and Crawford and Strömgren (1966).

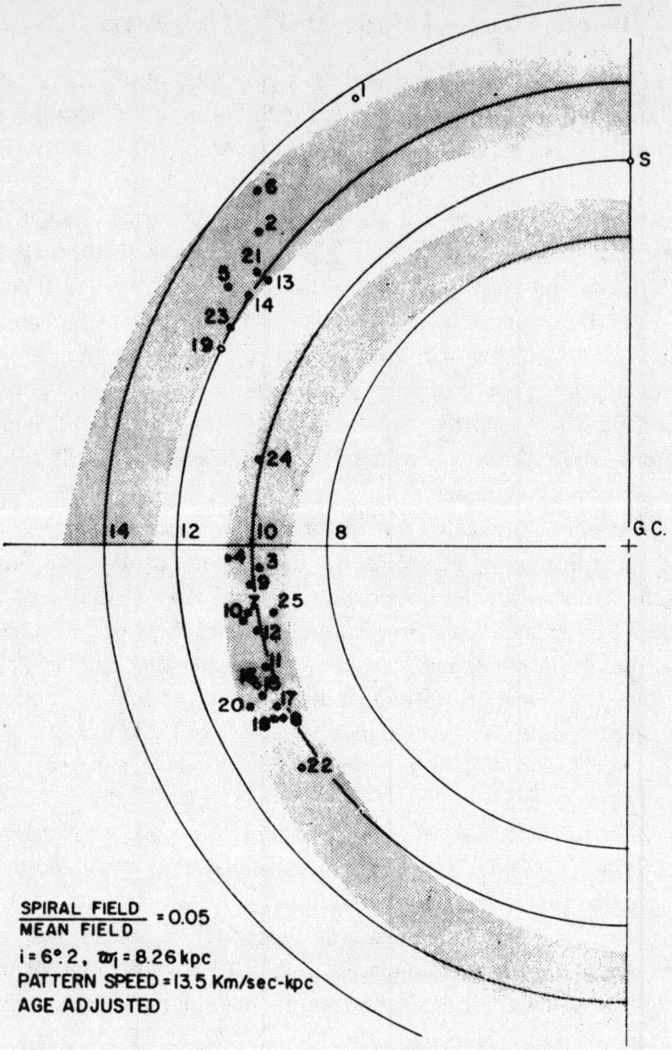


FIG. 6.4. Places of formation of stars as determined without a spiral field.

the analysis includes the field described above. It is seen that all of them lie in spiral arms.

The age determination suffers from an uncertainty of 10–15 percent. We have taken the age recommended by Strömgren *et al.*, with a few exceptions. For instance, for star No. 23, the age was taken to be 168 instead of 191 with a difference of less than 13%. The reason is that, with this choice, the dispersion velocity components are most reasonable "at birth". In all other cases, the dispersion velocity components are also all reasonable

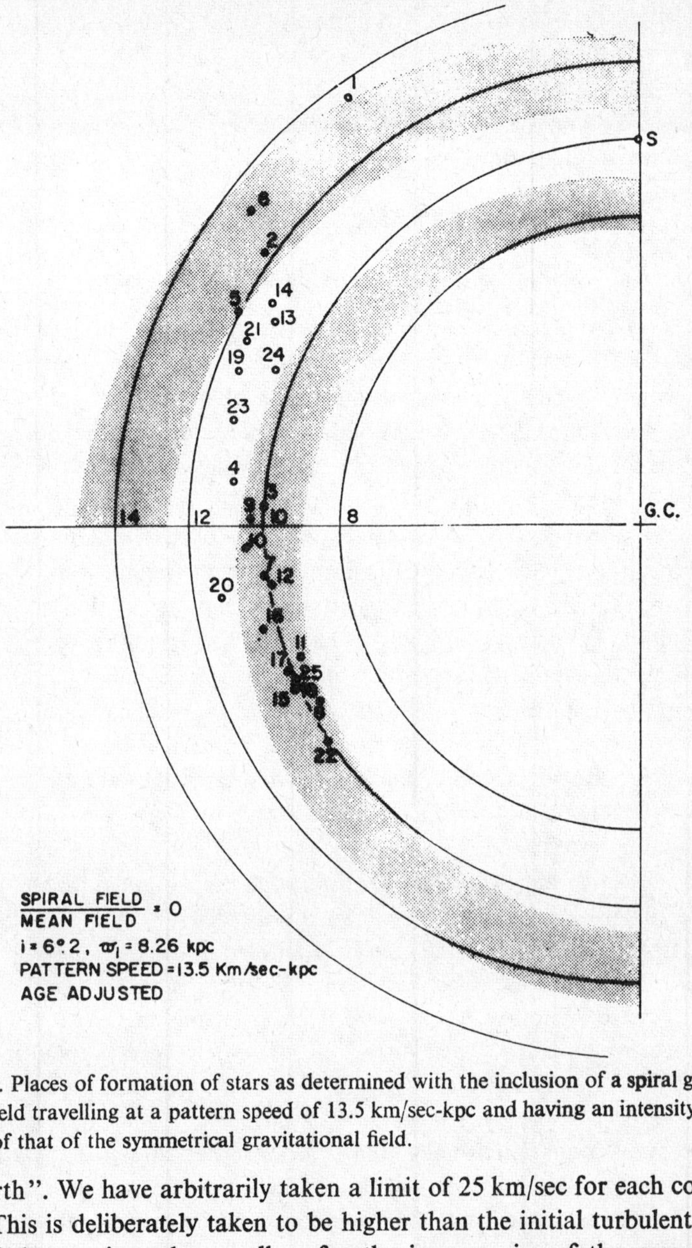


FIG. 6.5. Places of formation of stars as determined with the inclusion of a spiral gravitational field travelling at a pattern speed of 13.5 km/sec-kpc and having an intensity equal to 5% of that of the symmetrical gravitational field.

"at birth". We have arbitrarily taken a limit of 25 km/sec for each component. This is deliberately taken to be higher than the initial turbulent velocity of the gas in order to allow for the inaccuracies of the parameters involved in our calculations. Furthermore, it is to be expected that the dispersion velocity will increase considerably during the period of the formation process, as the star migrates for something of the order of 10 million years. If we lower this limit, we would be led to a slightly lower field strength (about 4% of the symmetric field), and to somewhat different age assignments, which however still lie well within the error limits given by Strömgren *et al.* The major conclusions are not altered; all stars were found to be born inside of spiral arms.

## 7. CONCLUDING REMARKS

### 7.1. An Appraisal of the Density-Wave Theory

(I) The density-wave theory has many attractive features. At the present time, there is as yet no definite criterion for the precise determination of the pattern speed $\Omega_p$. The observational check of the theory therefore lies in the comparison with a multitude of data. One set of data may be regarded as a basis for determining the pattern speed, while the others are then used as checks. Indeed, the value of $\Omega_p$ turns out to be *rather sharply defined* by the observational data. Other comparisons between theory and observations are also extremely satisfactory. For example, in the problem of the migration of moderately young stars, there are so many factors that could cause the results to go astray; and yet we have obtained impressive agreement. Further continuation and extension of the observational program would be extremely desirable.

In general, it might be noted that numerical values determined in this paper are subject to changes with the observational data and theoretical models used; but we can have reasonable confidence in them perhaps to within 20 percent.

(II) The range of values 11–13 km/sec-kpc obtained for the pattern speed $\Omega_p$ is also very satisfactory from the theoretical point of view. With reference to Fig. 5.1, it is clear that Lindblad resonance is very nearly realized with $\Omega_p$ in this range, but *there is a definite deviation*. In the theory of Lindblad, cooperative effects were neglected; thus, he insisted on exact resonance, which can happen only if an extremely special rotation curve $\Omega(\varpi)$ holds; i.e., one for which $\Omega - \varkappa/2 = \text{const}$. We include the cooperative effects, and *the deviation is a measure of the scale of the spiral spacing* in the theory of the infinitesimally thin disk. A finite scale must be allowed; otherwise, the thickness effect would suppress the waves. Thus, by taking the pattern speed $\Omega_p$ to be in the range indicated by present investigations, we are "as close to resonance as possible", without giving up density waves altogether. Such is indeed the condition expected of any mechanical system.

(III) The strength of the spiral field has now been estimated from the study of the migration of moderately young stars. It would be interesting to see how much contrast in gas density would ensue as a consequence. Calculations of this type had been made by Fujimoto (1966) and by Roberts (1968). Since the parameters chosen by Fujimoto do not correspond to those for a two-arms pattern discussed in this paper, only the results of Roberts are useful as a basis for checking the linear theory. A comparative study shows that estimates made on the basis of an extrapolation of the linear theory give generally satisfactory results.

(IV) The calculations of Fujimoto and of Roberts indicate the formation of a shock. This might indeed be an essential element in the process of star formation. One might imagine that the gas in turbulent motion has "clouds" before the shock, which are on the verge of gravitational collapse in the presence of turbulent stresses; and that the sudden compression would trigger off the collapse of the clouds, leading to star formation. After the gas leaves the shock region, it is again decompressed, and the process of star formation ceases. Superficially, such a process appears to be in such excellent agreement with our general concepts that it deserves further careful study in the immediate future.

An implication of this discussion is that a relatively high concentration of gas need not be sufficient to imply formation of new stars. The other prerequisite of a shock must be fulfilled in many cases. This conclusion appears to be in general agreement with the abundance distribution of neutral hydrogen found by Morton Roberts (1967) in certain galaxies. The highest abundance of hydrogen in certain galaxies is found in a ring *outside* of the spiral structure. This is readily explained on the basis of the present theory if it is supposed that shocks have not been able to form in these outer parts, because the conditions there (with too few stars and too much gas) are not favorable to the propagation of density waves.

Another corroborative evidence is the absence of HII region within 4 kpc of the galactic center. Our theory suggests that this should be the case, since there cannot be any density waves inside of the radius of inner Lindblad resonance (cf. Fig. 6.1). Indeed, both the continuous spectrum (Kerr and Westerhout, 1965) and the 109$\alpha$ (Burke, 1968) radio observations reveal no HII regions in such localities (outside of the dense nucleus).

## 7.2. Further Theoretical Developments

The density-wave theory, as outlined above, needs extension in several directions to complete the theoretical understanding of basic mechanisms and to develop its implications. We shall outline some of these considerations here.

(I) Thickness effect.

The galactic disk has a finite thickness. Indeed, there are two thickness scales, one for the gas (about 200 pc between half-density points) and another for the stars (about 600 pc). These thickness scales could be the final deciding factors for the determination of the most prominent scale observed. In any case, they enter into the determination of the dispersion velocity, as mentioned before. Without taking them into consideration, the value obtained is about 30% too high. The detailed discussions may be found in the lecture by Frank Shu, Appendix F.

(II) Complete formulation of the linear theory.

In the asymptotic treatment of the theory, we have not considered the boundary conditions to be satisfied at the center of the Galaxy and at the outer fringes. Our confidence in this approach is largely based on the fact that the spirals are actually observed. To a lesser degree, we rely on the existence of a principal part bounded by points of Lindblad resonance. It is of course necessary, as the next step in our investigations, to study the complete problem. This turns out to be extremely complicated; even the formulation of the problem is by no means simple.

The details of this formulation have been given by Frank Shu‡ (see Appendix F, where preliminary investigations are also given.) Two points deserve to be mentioned here. (a) The asymptotic calculations given above are inadequate near resonance, and it is necessary to recalculate the response of the stars to a given gravitational field by using exact characteristics of the symmetrical disk. (b) Without trying to solve the problem, it is possible to show that neutral spiral-like solutions cannot exist if resonance points are absent. This is a very important point. It shows that the so-called anti-spiral theorems can easily be misleading. If certain restrictions are inadvertently imposed, spiral solutions cannot exist.

The exact nature of the boundary conditions is also a point requiring great care. Since there is a nucleus, singularities in the solution might be permissible in the disk model. The condition near the outer edge is largely controlled by gas, and turbulent dissipation might be important in removing energy from the large-scale motions. Thus one could easily have a boundary condition that includes energy absorption.

(III) Origin of galactic spirals.

The problem of the origin of galactic spirals cannot be solved without using the complete formulation mentioned above. Only prelimin-

‡ An equivalent formulation using another (approximate) representation was given by A. Kalnajs (1965).

ary investigations are available (Kalnajs, 1965; Shu, 1968). However, the asymptotic theory can also give indications of preference, although the arguments are not conclusive. It appears that trailing spirals are favored by the fact (predicted from stability considerations) that the dispersion velocity of stars decreases outwards. (Lin and Shu, 1966). Much work remains to be done.‡

(VI) Nonlinear theory.

If one looks beyond the developments of the complete linear theory, one still has to consider nonlinear effects, which might again show a preference for trailing spirals.

The calculations of Fujimoto and of Roberts are investigations in nonlinear theory, but not complete ones, since the condition of self-consistency was not imposed in these studies. One could project a similar investigation into the behavior of stars under an imposed spiral gravitational field. It would then be possible to look into the question of self-consistency and to decide whether leading waves are possible in the nonlinear theory. Trailing waves are presumably only slightly altered by nonlinear effects.

## 7.3. Further Applications of the Density-Wave Theory

(I) In Sec. (7.1) we have already discussed the implications of the density-wave theory on the process of star formation. Such studies require a detailed understanding of the dynamics of interstellar matter. With our present concepts, one should re-examine all aspects of the dynamical problems. It would be very interesting to make use of existing facilities (such as the Haystack Antenna system) to study the turbulent motion of the interstellar clouds. This will contribute greatly to our understanding of the process of star formation.

(II) Similar efforts should be made on conditions of the galactic magnetic field, and its role in the containment of cosmic-ray particles.

(III) Turning our attention to external galaxies ,we see the possibility of examining them in greater details. Indeed, it in easy to conceive of a program to construct a catalogue of theoretical models for galaxies on the basis of the present theory, with a definite spiral pattern shown for each of the models. This might facilitate the identification of galaxies from their superficial appearance.

---

‡ Toomre (private communication) has also carried out studies involving the propagation of a group of waves and their initiation by external agents.

(IV) Our galaxy has relatively little gas. Studies should now be focussed on cases where the gas content is appreciable. It may be that the dynamical theory can explain why spirals cannot be formed when the gas content is appreciable, and to describe the correlation between spiral spacing and gas content. A superficial argument can easily be advanced to predict the increase of spiral spacing with gas content on the basis of the present theory; but careful examination of such reasoning is needed.

(V) Since we now have the beginning of the understanding of the process of star formation, we can look toward the future when we may be able to study the evolution of a galaxy as a whole. Presumably, larger galaxies evolve faster and the gas content in them would be more nearly depleted by *now* than in a smaller galaxy, if we assume all the galaxies were formed in the same epoch. Thus, in a future epoch, a less massive galaxy might take on some of the present appearances of a more massive galaxy, even though the two objects are *never* the same during any epoch of their evolution. Much work remains to be done on such problems.

## APPENDIX A. (TO SECTION 1): ELEMENTARY FACTS ABOUT GALAXIES‡

We collect below a number of elementary facts about galaxies in general, and normal spirals in particular. More detailed information is given about our own galaxy.

### Aa. Classification of Galaxies

Galaxies are usually classified into the following four categories according to their appearance:

elliptical galaxies,
normal spirals,
barred spirals,
irregulars.

Fig. 1.3 of the text shows the scheme of this Hubble classification. The irregulars are not shown in that diagram. Some more examples of galaxies of various types are shown in Fig. A1.

We are concerned mainly with normal spirals, which are the most abundant among the brighter galaxies. Here it is interesting to note the

‡ For more detailed information about various galaxies, see Sandage (1961) and Woltjer (1967).

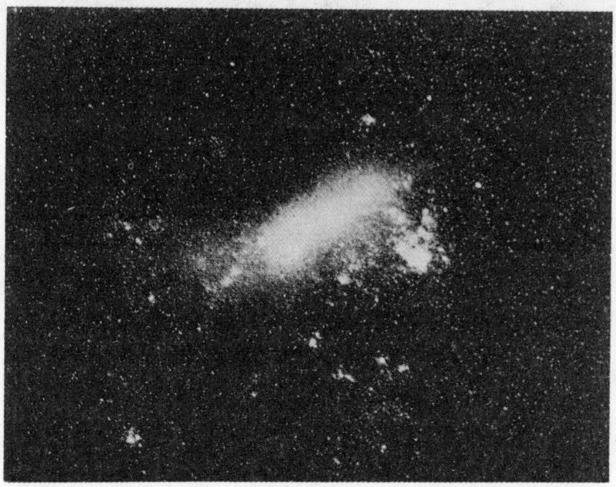

FIG. A1. Examples of galaxies of various types. Upper left: Elliptical galaxy EO pec- NGC 4486. Upper right: Normal spiral SC NGC 5364. Center: Barred galaxy SBb(s), NGC 1300. Lower: Irregular galaxy Irr I, the Large Magellanic Cloud.

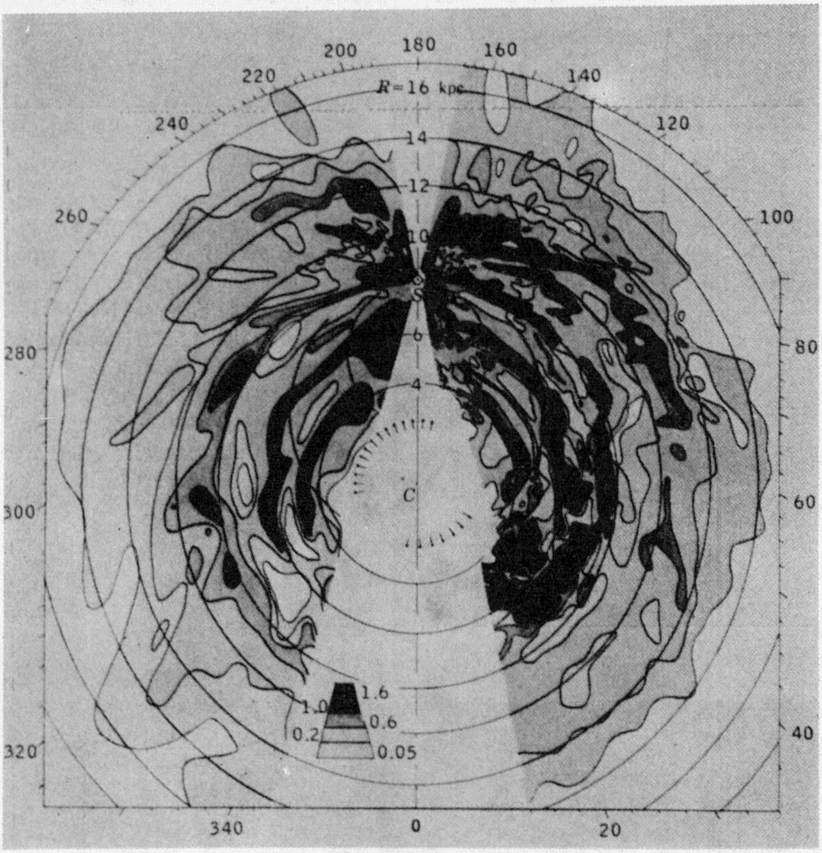


FIG. A3. Distribution of hydrogen in the galactic plane [after Oort (1962)]. The *maximum* densities in the $z$ direction are projected on the galactic plane, and contours are drawn through the points. The density scale indicated is in atoms per $cm^3$.

increase of gas content as we pass through the sequence Sa, Sb, Sc to irregulars:

| | Approximate gas content (%) |
|---|---|
| Sa spirals | 1 |
| Sb spirals | 2–3 |
| Sc spirals | 8 |
| Irregulars | 15 |

This estimation of gas content is based on atomic hydrogen alone. If there is molecular hydrogen, the amount might even be twice as great.

Irregulars are listed here for comparison, because it has been reported that an underlying spiral structure can be detected for the companion of the whirlpool galaxy and a companion of our own galaxy, the Large Magellanic Cloud. It would be interesting to investigate whether all of the irregulars (or which types of these) have an underlying spiral structure.

## Ab. Some Properties of the Milky Way System

Our own Galaxy is among the largest of all spirals. The Sun is at a distance of about 10 kpc from the galactic center. About 80% of the mass of the Galaxy lies within the solar orbit around the center. The mass of our Galaxy is about $2 \times 10^{11}$ solar masses. The solar mass is about $2 \times 10^{33}$ g, making the mass of our Galaxy $4 \times 10^{44}$ g. The circular velocity is approximately constant at 250 km/sec. Since (1 km/sec) $\times 10^6$ yr $= 1.02$ pc, the period of revolution is about 200 million years at a radial distance of 8 kpc from the center, and about 100 million years at 4 kpc. The detailed distribution of rotation velocities is given in Fig. A2. (See the Schmidt model in Sec. Ac.)

There are more gas and young stars (of types O and B) in the bright spiral arms. Thus, the contrast in brightness does not represent a proportional contrast in total mass density (which is very small). Contrast in the density of the gas may, however, amount to about three or four to one. Near bright stars, there is an ionized region in the gas called H II region (or Strömgren spheres). The un-ionized regions, called H I regions, are estimated to have a temperature of the order of $10^2$ °K.

The distribution of stars in the direction perpendicular to the galactic disk extends over a thickness of about 600 pc between layers where the density is one-half of the maximum. The corresponding figure for the gas is about 170 pc. Thus, the disk is indeed very thin. The distribution of gas over the galactic plane is shown in Fig. A3.

19*

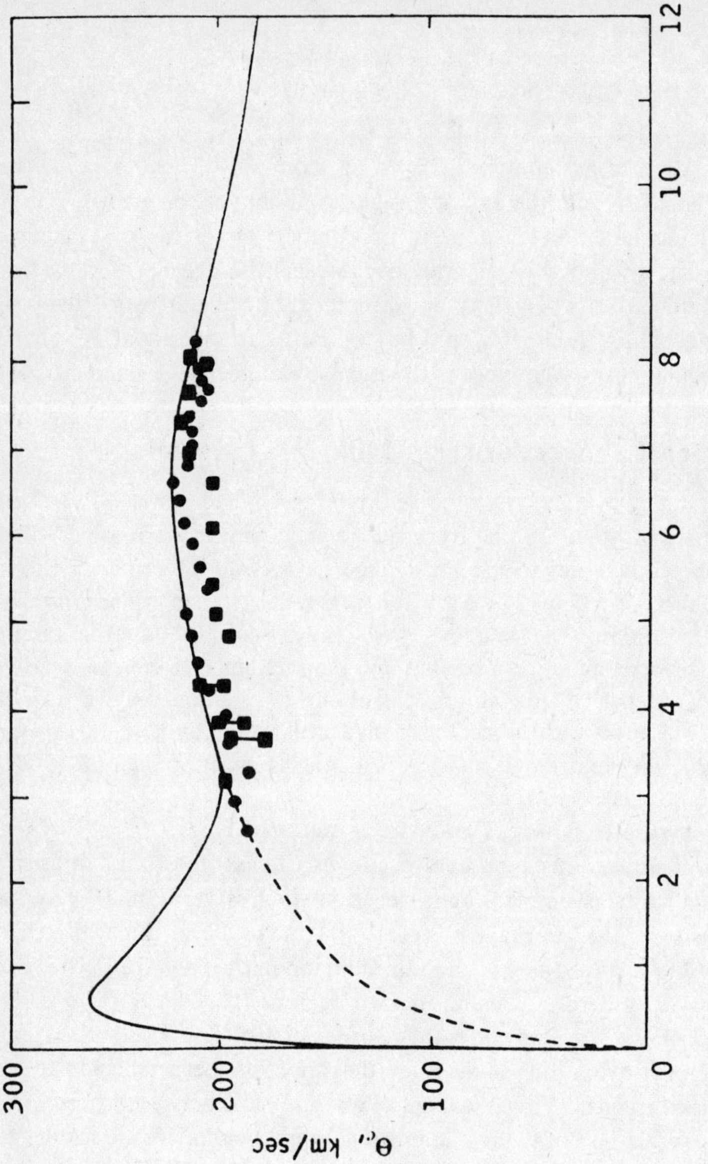


FIG. A2. Variation of circular velocity with distance from center of our own galaxy [after Oort (1962)].

The fraction of matter that is gas (atomic hydrogen) is about 20% in our neighborhood, but only about 2% at a location, say 5 kpc from the center. A typical density of gas (atomic hydrogen) is about $2 \times 10^{-24}$ g/cc (1.2 H atoms/cc). This corresponds to a kinetic energy of galactic rotation of $1300 \times 10^{-12}$ erg cm$^{-3}$. It is not yet known whether there might be molecular hydrogen of comparable mass density [see Gould, Gold, and Salpeter (1963)]. The gas is actually in the form of clouds with a diameter of the order of 10 pc. The density in the cloud is 3–10 times that in the surroundings.

Except in the central regions and at the "3-kpc arm", where the radial motion of the gas is observed to be 53 km/sec, this motion is believed to have a velocity of no more than 15 km/sec. The turbulent motion of the gas has a velocity of the order of 7 km/sec, giving an energy density of the order of $0.5 \times 10^{-12}$ erg cm$^{-3}$.

The magnetic field is estimated to have a magnitude of the order of $10^{-5}$ gauss. This corresponds to an energy density of $4 \times 10^{-12}$ erg cm$^{-3}$. The energy density of cosmic-ray particles is of the order of $1 \times 10^{-12}$ erg cm$^{-3}$. There are dust particles in the interstellar regions. They might play an important role for the recombination of atomic hydrogen into molecular form.

## Ac. The Schmidt Model

The distribution or rotational velocity $\Omega(\varpi)$ and epicyclic frequency $\varkappa(\varpi)$ are given in Table A and Fig. 5.1, according to the model constructed by Schmidt (1965).

TABLE A. 1965 Schmidt Model of the Galaxy

(a) Rotation parameters

| $\varpi$ (kpc) | $\varpi\Omega$ (km/sec) | $\Omega$ (km/sec-kpc) | $\varkappa$ (km/sec-kpc) | $\Omega - \varkappa/2$ (km/sec-kpc) | $\Omega + \varkappa/2$ (km/sec-kpc) |
|---|---|---|---|---|---|
| 1 | 200 | 200 | 245 | 77.6 | 323 |
| 2 | 187 | 93 | 136 | 25.4 | 161 |
| 3 | 198 | 66.0 | 103 | 14.5 | 118 |
| 4 | 213 | 53.2 | 85.0 | 10.7 | 95.7 |
| 5 | 227 | 45.3 | 72.6 | 9.0 | 81.6 |
| 6 | 238 | 39.7 | 62.8 | 8.3 | 71.1 |
| 7 | 247 | 35.2 | 54.4 | 8.0 | 62.4 |
| 8 | 252 | 31.4 | 46.7 | 8.1 | 54.8 |
| 9 | 253 | 28.1 | 39.2 | 8.5 | 47.7 |

| $\varpi$ (kpc) | $\varpi\Omega$ (km/sec) | $\Omega$ (km/sec-kpc) | $\varkappa$ (km/seckpc) | $\Omega-\varkappa/2$ (km/sec-kpc) | $\Omega+\varkappa/2$ (km/sec-kpc) |
|---|---|---|---|---|---|
| 10 | 250 | 25.0 | 31.6 | 9.2 | 40.8 |
| 11 | 244 | 22.2 | 26.5 | 8.9 | 35.4 |
| 12 | 238 | 19.8 | 22.8 | 8.4 | 31.2 |
| 13 | 231 | 17.8 | 20.0 | 7.8 | 27.7 |
| 14 | 224 | 16.0 | 17.7 | 7.2 | 24.9 |
| 15 | 218 | 14.6 | 15.9 | 6.6 | 22.5 |
| 16 | 213 | 13.3 | 14.3 | 6.1 | 20.4 |
| 17 | 207 | 12.2 | 13.0 | 5.7 | 18.7 |
| 18 | 202 | 11.2 | 11.9 | 5.3 | 17.2 |
| 19 | 197 | 10.4 | 10.9 | 4.9 | 15.9 |
| 20 | 193 | 9.6 | 10.1 | 4.6 | 14.7 |

(b) Structure parameters

| $\varpi$ (kpc) | $\sigma(\varpi)$ $(M_\odot/\mathrm{pc}^2)$ | $M(\varpi)$ [$(10^9 M_\odot)$ | $\lambda_{tot}$(kpc) |
|---|---|---|---|
| 1 | 1097 | 11 | 3.1 |
| 2 | 817 | 20 | 7.5 |
| 3 | 646 | 31 | 10.4 |
| 4 | 521 | 44 | 12.3 |
| 5 | 421 | 57 | 13.6 |
| 6 | 338 | 70 | 14.5 |
| 7 | 267 | 82 | 15.3 |
| 8 | 206 | 93 | 16.0 |
| 9 | 155 | 103 | 17.1 |
| 10 | 114 | 111 | 19.3 |
| 11 | 86 | 117 | 20.7 |
| 12 | 65.9 | 123 | 21.5 |
| 13 | 51.8 | 127 | 22.1 |
| 14 | 41.5 | 131 | 22.5 |
| 15 | 33.7 | 135 | 22.8 |
| 16 | 27.8 | 138 | 23.0 |
| 17 | 23.2 | 140 | 23.2 |
| 18 | 19.5 | 143 | 23.4 |
| 19 | 16.6 | 145 | 23.5 |
| 20 | 14.2 | 147 | 23.7 |

## APPENDIX B (TO SECTION 4): MATHEMATICAL THEORY OF STELLAR DYNAMICS FOR AN INFINITESIMALLY THIN DISK IN DIFFERENTIAL ROTATION—GENERAL PRINCIPLES

### Ba. Notation

$t$, time.

$\varpi, \theta, z$, cylindrical coordinates.

$\Pi, U, Z$, components of total stellar velocity in cylindrical coordinates.

$\Psi_3(\varpi, \theta, z, \Pi, H, Z, t)$, distribution function in the three-dimensional case.

$\Pi_{00}(\varpi, \theta, z, t)$, $\theta_{00}(\varpi, \theta, z, t)$, $Z_{00}(\varpi, \theta, z, t)$, mean stellar velocity at a given locality.

$\Pi_0(\varpi, \theta, z, t)$, $\theta_0(\varpi, \theta, z, t)$, $Z_0(\varpi, \theta, z, t)$, similar equantities (but *not* necessarily mean values) for an arbitrarily chosen reference system.

$c_\varpi{}^{(0)} = \Pi - \Pi_{00}$, $c^{(0)} = H - H_{00}$, $c_z{}^{(0)} = Z - Z_{00}$, components of dispersion velocity.

$c_\varpi = \Pi - \Pi_0$, $c_\theta = \theta - \theta_0$, $c_z = Z - Z_0$, components of dispersion velocity relative to an arbitrary reference system [$\Pi_0 = Z_0 = 0$, $H_0 = \varpi\Omega(\varpi)$ in the present paper].

In the two-dimensional case, the following notation is adopted:

$\Psi_2(\varpi, \theta, \Pi, H, t)$, distribution function in two-dimensional case.

$\sigma = m \iint \Psi_2 \, d\Pi \, dH$, surface density, when $m$ is the mass of an individual star.

$$\Pi = \Pi_{00}(\varpi, \theta, t) + c_\varpi^{(0)} = \Pi_0(\varpi, \theta, t) + c_\varpi,$$

$$\Theta = \Theta_{00}(\varpi, \theta, t) + c_\theta^{(0)} = \Theta_0(\varpi, \theta, t) + c_\theta,$$

with definition of symbols similar to those above.

$\langle q \rangle = \iint q\Psi_2 \, d\Pi \, dH / \iint \Psi_2 \, d\Pi \, dH$, local mean value for $q$.

$$\langle \Pi \rangle = \Pi_0 + \langle c_\varpi \rangle = \Pi_0 + v_\varpi = V_\varpi + v_\varpi = \Pi_{00},$$

$$\langle \Theta \rangle = \Theta_0 + \langle c_\theta \rangle = \Theta_0 + v_\theta = V_\theta + v_\theta = \Theta_{00}$$

[$V_\varpi$ usually taken to be zero, $V_\theta = \varpi\Omega(\varpi)$].

$V(\varpi, \theta, z, t)$, gravitational potential.

$a_\varpi$, $a_\theta$, acceleration components in the plane $z = 0$.

$\Psi(\varpi, \theta, c_\varpi, c_\theta, t) \equiv \Psi_2(\varpi, \theta, \Pi, H, t)$, representation of the two-dimensional distribution function in terms of the dispersion velocities for a given hoice of the reference system.

$c'_\varpi = c_\varpi - \langle c_\varpi \rangle$, $c'_\theta = c_\theta - \langle c_\theta \rangle$, components of dispersion velocity relative to the instantaneous mean motion (same as $c_\varpi{}^{(0)}$, $c_\theta{}^{(0)}$).

## Bb. Basic Equations

The basic equation of stellar dynamics, when written in terms of cylindrical coordinates $(\varpi, \theta, z)$ reads as follows [Chandrasekhar (1960), p. 187]:

$$\frac{\partial \Psi_3}{\partial t} + \Pi \frac{\partial \Psi_3}{\partial \varpi} + \frac{\Theta}{\varpi} \frac{\partial \Psi_3}{\partial \theta} + Z \frac{\partial \Psi_3}{\partial z} - \left( \frac{\partial \mathscr{V}}{\partial \varpi} - \frac{\Theta^2}{\varpi} \right) \frac{\partial \Psi_3}{\partial \Pi}$$
$$- \left( \frac{1}{\varpi} \frac{\partial \mathscr{V}}{\partial \theta} + \frac{\Pi\Theta}{\varpi} \right) \frac{\partial \Psi_3}{\partial \Theta} - \frac{\partial \mathscr{V}}{\partial z} \frac{\partial \Psi_3}{\partial Z} = 0, \tag{B1}$$

where $\Psi_3$ is the distribution function such that

$$dN = \Psi_3(\varpi, \theta, z; \Pi, \Theta, Z; t)\, \varpi\, d\varpi\, d\theta\, dz\, d\Pi\, d\Theta\, dZ \tag{B2}$$

gives the number of stars (all of mass $m$) in the element of phase space

$$d\tau = \varpi\, d\varpi\, d\theta\, dz\, d\Pi\, d\Theta\, dZ. \tag{B3}$$

The mean velocity of the stars $\Pi_{00}, \theta_{00}, Z_{00}$ are defined, as for all mean quantities, by relations of the type

$$\Pi_{00} = \frac{\iiint \Pi \Psi_3\, d\Pi\, d\Theta\, dZ}{\iiint \Psi_3\, d\Pi\, d\Theta\, dZ}. \tag{B4}$$

Note that $\Pi_{00}$ is a function of $(\varpi, \theta, z, t)$ and is therefore fundamentally different from $\Pi$, which is an independent variable.

We now define the dispersion velocity components $(c_\varpi{}^{(0)}, c_\theta{}^{(0)}, c_z{}^{(0)})$ by

$$\Pi = \Pi_{00} + c_\varpi^{(0)}, \quad \Theta = \Theta_{00} + c_\theta^{(0)}, \quad Z = Z_{00} + c_z^{(0)}. \tag{B5}$$

Note that the omponents of the dispersion velocities $(c_\varpi{}^{(0)}, c_\theta{}^{(0)}, c_z{}^{(0)})$ are functions of the variables $(\varpi, \theta, z, \Pi, \theta, Z, t)$ ,since $\Pi_{00}, \theta_{00}$, and $Z_{00}$ are functions of $\varpi, \theta, z, t$. This point is especially important to note when we eliminate $\Pi, \theta, Z$ in terms of the components of dispersion velocities as independent variables. It means that we must think in terms of the simultaneous transformation from the six variables $(\varpi, \theta, z, \Pi, \theta, Z)$ to $(\varpi, \theta, z, c_\varpi{}^{(0)}, c_\theta{}^{(0)}, c_z{}^{(0)})$. The Jacobian of this transformation can, however, be shown to be identically equal to unity.

## Bc. The Infinitesimally Thin Disk

We shall now specialize to the problem of a "flat" disk. We introduce the two-dimensional distribution function

$$\Psi_2(\varpi, \theta, \Pi, \Theta, t) = \int_{-\infty}^{\infty} \int_{-\infty}^{\infty} \Psi_3\, dz\, dZ, \tag{B6}$$

and apply the same integration process to (B1). Clearly, the terms involving $Z$ disappear if $\Psi_3$ is such that (B6) is absolutely convergent. Most of the other terms yield corresponding expressions involving $\Psi_2$, but the following terms

$$-\frac{\partial \mathscr{V}}{\partial \varpi}\frac{\partial \Psi_3}{\partial \Pi}, \quad -\frac{1}{\varpi}\frac{\partial \mathscr{V}}{\partial \theta}\frac{\partial \Psi_3}{\partial \Theta}$$

deserve special attention. We treat them as follows: Define

$$a_\varpi = -\left(\frac{\partial \mathscr{V}}{\partial \varpi}\right)_{z=0}, \quad a_\theta = -\frac{1}{\varpi}\frac{\partial \mathscr{V}}{\partial \theta}\Big)_{z=0}; \tag{B7}$$

then we may write

$$-\iint_{-\infty}^{\infty} \frac{\partial \mathscr{V}}{\partial \varpi}\frac{\partial \Psi_3}{\partial \Pi}\, dz\, dZ = a_\varpi \frac{\partial \Psi_2}{\partial \Pi} - \iint_{-\infty}^{\infty} a_\varpi + \frac{\partial \mathscr{V}}{\partial \varpi}\frac{\partial \Psi_3}{\partial \Pi}\, dz\, dZ, \tag{B8}$$

where the last integration should be carried out first with respect zo $z$. The integrand vanishes for $z = 0$ and $z = \pm\infty$, and the integral approaches zero as the thickness of the disk approaches zero. A similar argument holds for the other term. Thus, we finally obtain the following equation for an infinitesimally thin disk:

$$\frac{\partial \Psi_2}{\partial t} + \Pi \frac{\partial \Psi_2}{\partial \varpi} + \frac{\Theta}{\varpi}\frac{\partial \Psi_2}{\partial \theta} + \left(a_\varpi + \frac{\Theta^2}{\varpi}\right)\frac{\partial \Psi_2}{\partial \Pi} + \left(a_\theta - \frac{\Pi\Theta}{\varpi}\right)\frac{\partial \Psi_2}{\partial \Theta} = 0. \tag{B9}$$

The surface density of this disk is

$$\sigma_*(\varpi, \theta, t) = m_* \iint_{-\infty}^{\infty} \Psi_2\, d\Pi\, d\Theta. \tag{B10}$$

Due to a given surface density, such as (10), the components of the acceleration $(a_\varpi, a_\theta)$ are given by

$$a_\varpi = -\left(\frac{\partial \mathscr{V}}{\partial \varpi}\right)_{z=0}, \quad a_\theta = -\frac{1}{\varpi}\left(\frac{\partial \mathscr{V}}{\partial \theta}\right)_{z=0}, \tag{B11}$$

where $\mathscr{V}$ satisfies the Poisson equation

$$\frac{\partial^2 \mathscr{V}}{\partial \varpi^2} + \frac{1}{\varpi}\frac{\partial \mathscr{V}}{\partial \varpi} + \frac{1}{\varpi^2}\frac{\theta^2 \mathscr{V}}{\partial \theta^2} + \frac{\partial^2 \mathscr{V}}{\partial z^2} = 4\pi G \sigma \delta(z). \tag{B12}$$

In (B12), $G$ is the gravitational constant and $\delta(z)$ is Dirac's delta function.

We are usually interested in a system in differential rotation. This means that we have a state of motion with

$$\Pi_{00} \cong 0, \quad \Theta_{00} \cong \varpi\Omega(\varpi), \quad (Z_{00} \cong 0). \tag{B13}$$

Thus, it is convenient to introduce the following peculiar velocity relative to an arbitrarily chosen reference state $[\Pi_0 = 0, \Theta_0 = \varpi\Omega(\varpi)]$:

$$c_\varpi = \Pi - 0, \quad c_\theta = \Theta - \varpi\Omega(\varpi). \tag{B14}$$

The components of mean dispersion velocites $\langle c_\varpi \rangle$ and $\langle c_\theta \rangle$ are no longer necessarily zero; but it is implied that they are "small" compared with $\varpi\Omega(\varpi)$. We shall now introduce $(c_\varpi, c_\theta)$ as the independent variables in place of $(\Pi, \theta)$ in (B9), and write

$$\Psi_2(\varpi, \theta, \Pi, \Theta, t) \equiv \Psi(\varpi, \theta, c_\varpi, c_\theta, t). \tag{B15}$$

Then, we have transformations of derivatives typified by

$$\frac{\partial \Psi_2}{\partial \varpi} = \frac{\partial \Psi}{\partial \varpi} + \frac{\partial \Psi}{\partial c_\theta}\left[-\frac{d}{d\varpi}(\varpi\Omega)\right], \quad \frac{\partial \Psi_2}{\partial \Theta} = \frac{\partial \Psi}{\partial c_\theta},$$

and (B9) becomes

$$\frac{\partial \Psi}{\partial t} + c_\varpi \frac{\partial \Psi}{\partial \varpi} + \left(\Omega + \frac{c_\theta}{\varpi}\right)\frac{\partial \Psi}{\partial \theta} + \left(a_\varpi + \Omega^2 \varpi + 2\Omega c_\theta + \frac{1}{\varpi} c_\theta^2\right)\frac{\partial \Psi}{\partial c_\varpi}$$
$$+ \left(a_\theta - \frac{\varkappa^2}{2\Omega} c_\varpi - \frac{1}{\varpi} c_\varpi c_\theta\right)\frac{\partial \Psi}{\partial c_\theta} = 0, \tag{B16}$$

where $\Psi$ is associated with the volume element

$$d\tau_c = \varpi \, d\varpi \, d\theta \, dc_\varpi \, dc_\theta$$

of the phase space, and

$$\varkappa^2 = (2\Omega)^2\left(1 + \frac{\varpi}{2\Omega}\frac{d\Omega}{d\varpi}\right). \tag{B17}$$

The (local) averages of the components of the dispersion velocity are in general different from zero, and these will be respectively denoted by $v_\varpi$ and $v_\theta$. We have

$$\frac{\sigma_*}{m_*} v_\varpi = \iint c\Psi_\varpi \, dc_\varpi \, dc_\theta, \quad \frac{\sigma_*}{m_*} v_\theta = \iint c_\theta \Psi \, dc_\varpi \, dc_\theta, \tag{B18}$$

where $\sigma_*$ is the surface density:

$$\sigma_* = m_* \iint \Psi \, dc_\varpi \, dc_\theta, \tag{B19}$$

and $m^*$ is mass per star. The local mean of the total stellar velocity $(\Pi, \theta)$ is easily verified to be given by $[\varpi\Omega(\varpi) + v_\varpi, v_\theta]$, as expected.

Equation (B16) is the final equation for stellar dynamics with reference to a conveniently chosen differentially rotating system, for the configuration of a thin disk with infinitesimal thickness. The acceleration $(a_\varpi, a_\theta)$ derives part of its contributions from the surface density $\sigma_*$ given by (B19) [according to (B11) and (B12)], but it may also have contributions from any "external" agency (e.g., gas or another neighboring galaxy).

## Bd. Characteristics and Integrals

The characteristics of the differential Eq. (9) are defined by the system of differential equations

$$dt = \frac{d\varpi}{\Pi} = \frac{d\theta}{\Theta/\varpi} = \frac{d\Pi}{a_\varpi + \Theta^2/\varpi} = \frac{d\Theta}{a_\theta - \Pi\Theta/\varpi}. \tag{B20}$$

From the last two equations, we can derive

$$dt = \frac{\Pi\, d\Pi + \Theta\, d\Theta}{a_\varpi \Pi + a_\theta \Theta/\varpi}. \tag{B21}$$

If the potential $\mathscr{V}$ is the *time-independent*, this leads to the energy integral

$$\frac{1}{2}(\Pi^2 + \Theta^2) + \mathscr{V}(\varpi, \theta, 0) = E, \quad a \text{ constant}. \tag{B22}$$

If the potential is *axially symmetric* ($a_\theta = 0$), then (B20) yields,

$$\frac{d\varpi}{\Pi} = \frac{d\Theta}{-\Pi\Theta/\varpi},$$

and we have

$$\varpi\Theta = J, \quad a \text{ constant}. \tag{B23}$$

This is the integral for angular momentum.

In terms of the variables ($c_\varpi$, $c_\theta$), these integrals become

$$\frac{1}{2}\{c_\varpi^2 + (\varpi\Omega + c_\theta)^2\} + \mathscr{V}(\varpi, \theta, 0) = E \tag{B22a}$$

for time-independent potentials, and

$$\varpi(\varpi\Omega + c_\theta) = J \tag{B23a}$$

for axially-symmetric potentials. These should, of course, be derivable from the equations for the characteristics of (B16):

$$dt = \frac{d\varpi}{c_\varpi} = \frac{d\theta}{\Omega + c_\theta/\varpi} = \frac{dc_\varpi}{a_\varpi + (\varpi\Omega + c_\theta)^2/\varpi}$$

$$= \frac{dc_\theta}{a_\theta - (\varkappa^2/2\Omega + c_\theta/\varpi)\, c_\varpi}. \tag{B20a}$$

Indeed they are. But the derivation is difficult to discover without the guide given above. For example, in the axially symmetrical case, the second differential in (B20a) and the last one give

$$\frac{d\varpi}{c_\varpi} = \frac{dc_\theta}{-(\varkappa^2/2\Omega + c_\theta/\varpi)\, c_\varpi}. \tag{B24}$$

Although this can be verified to lead to (B23a), the process required is somewhat camouflaged. A similar remark applies to the energy integral.

## Be. Symmetric Distribution in Equilibrium

Consider now the basic state, which is an axially symmetric distribution in equilibrium:

$$\Psi = \Psi_0(\varpi, c_\varpi, c_\theta). \tag{B25}$$

We then have

$$(a_\varpi, a_\theta) = (a_{\varpi 0}(\varpi), 0). \quad \text{(B25)}$$

The basic equation (B16) becomes

$$c_\varpi \frac{\partial \Psi_0}{\partial \varpi} + \left(a_{\varpi 0} + \Omega^2 \varpi + 2\Omega c_\theta + \frac{c_\theta^2}{\varpi}\right) \frac{\partial \Psi_0}{\partial c_\varpi} + \left[0 - \left(\frac{\varkappa^2}{2\Omega} + \frac{c_\theta}{\varpi}\right) c_\varpi\right] \frac{\partial \Psi_0}{\partial c_\theta} = 0. \tag{B27}$$

The characteristics of (B27) are given by (B22a) and (B23a), and thus we have the general integral of (B27) in the form

$$\Psi_0 = F(J_1, J_2), \tag{B28}$$

where $F$ is an arbitrary function of the two arguments, and $J_1$ and $J_2$ are defined by

$$J_1 \equiv \varpi(\varpi\Omega + c_\theta), \tag{B29}$$

$$J_2 \equiv \tfrac{1}{2}\left[c_\varpi^2 + (\varpi\Omega + c_\theta)^2\right] + \mathscr{V}(\varpi). \tag{B30}$$

Clearly, $J_1$ and $J_2$ are to be identified with $J$ and $E$ above.

## Bf. Choice of the Reference State

It is clear that (B27) can be somewhat simplified if we introduce the special choice of $\Omega(\varpi)$ to make

$$a_{\varpi 0} + \Omega^2 \varpi = 0, \tag{B31}$$

i.e., if we take $\Omega(\varpi)$ to be the angular velocity that a particle would have if it were moving in a purely circular orbit. This is not to be confused with the mean angular velocity $\Omega(\varpi) + \langle c_\theta \rangle/\varpi$ in a symmetrical system, but the two are expected to be close. To calculate the difference, let us evaluate $\langle c_\theta \rangle$ from (B27), after (B31) is imposed.

This is done by multiplying (B27) with $c_\varpi$ and integrating over the infinite velocity space. We get [c.f. (B38) below]

$$2\Omega\langle c_\theta \rangle = \frac{1}{\varpi_*}\left\{\frac{1}{\sigma_*}\frac{\partial}{\partial\varpi}[\varpi\langle c_\varpi^2\rangle\,\sigma_*] - \langle c_\theta^2\rangle\right\} - (a_{\varpi 0} + \Omega^2\varpi). \tag{B32}$$

If (B31) is imposed, the last term on the right vanishes. This shows that $\langle c_\theta \rangle$ is indeed smaller compared with $\langle c_\theta{}^2\rangle^{1/2}$ if the latter is smaller compared with the circular velocity $\theta_0 = \varpi\Omega$.

This special choice of the reference state will often—but not always—be adopted. It might be convenient, sometimes, to choose $\Omega(\varpi)$ such that $\langle c_\theta \rangle = 0$. In that case, we have

$$a_{\varpi 0} + \Omega^2\varpi = \frac{1}{\varpi}\left\{\frac{1}{\sigma_*}\frac{\partial}{\partial\varpi}[\varpi\sigma_*\langle c_\varpi^2\rangle] - \langle c_\theta^2\rangle\right\}, \tag{B33}$$

which is of the order of the mean square dispersion velocities.

## Bg. Moment Equations: Hydrodynamic Equations

In some studies, it is convenient to use the moment equations. We shall write these down for convenience of reference. If we multiply $m_* c_\varpi{}^\alpha c_\theta{}^\beta$ into the fundamental equation (B16) for the distribution function relative to a differentially rotating system, and then integrate with respect to $c_\varpi$ and $c_\theta$, we obtain the following equation:

$$\begin{aligned}
&\frac{\partial}{\partial t}[\sigma_*\langle c_\varpi^\alpha c_\theta^\beta\rangle] + \frac{\partial}{\partial\varpi}[\sigma_*\langle c_\varpi^{\alpha+1} c_\theta^\beta\rangle] + \frac{\partial}{\partial\theta}\left[\sigma_*\Omega\langle c_\varpi^\alpha c_\theta^\beta\rangle - \frac{\sigma_*}{\varpi}\langle c_\varpi^\alpha c_\theta^{\beta+1}\rangle\right] \\
&= \alpha\sigma_*\left\{(a_\varpi + \Omega^2\varpi)\langle c_\varpi^{\alpha-1} c_\theta^\beta\rangle + 2\Omega\langle c_\varpi^{\alpha-1} c_\theta^{\beta+1}\rangle + \frac{1}{\varpi}\langle c_\varpi^{\alpha-1} c_\theta^{\beta+2}\rangle\right\} \\
&\quad + \beta\sigma_*\left\{a_\theta\langle c_\varpi^\alpha c_\theta^{\beta-1}\rangle - \frac{\varkappa^2}{2\Omega}\langle c_\varpi^{\alpha+1} c_\theta^{\beta-1}\rangle\right\} \\
&\quad + (\beta + 1)\,\sigma_*\frac{1}{\varpi}\langle c_\varpi^{\alpha+1} c_\theta^\beta\rangle,
\end{aligned} \tag{B34}$$

where

$$\sigma_*\langle c_\varpi^\alpha c_\theta^\beta\rangle = m_* \iint_{-\infty}^{\infty} \Psi c_\varpi^\alpha c_\theta^\beta dc_\varpi dc_\theta. \tag{B35}$$

The terms on the right are obtained by making use of the following relations:

$$\begin{aligned} \frac{\partial\Psi}{\partial c_\varpi} c_\varpi^\alpha c_\theta^\beta &= \left[\frac{\partial}{\partial c_\varpi}(c_\varpi^\alpha \Psi) - a c_\varpi^{\alpha-1}\Psi\right] c_\theta^\beta, \\ \frac{\partial\Psi}{\partial c_\theta} c_\varpi^\alpha c_\theta^{\beta'} &= \left[\frac{\partial}{\partial c_\theta}(c_\theta^{\beta'}\Psi) - \beta' c_\theta^{\beta'-1}\Psi\right] c_\varpi^\alpha. \end{aligned} \tag{B36}$$

We note that when $\alpha = 0$, the first term on the right-hand side vanishes, and thus the embarrassing terms involving $c_\varpi{}^{\alpha-1} = c_\varpi{}^{-1}$ do not actually occur.

*Hydrodynamic equations.* The equations with $(\alpha, \beta) = (0, 0)$, $(1, 0)$, $(0, 1)$, respectively, correspond to those usually occurring in hydrodynamics. They are therefore explicitly written down here for convenience of reference.

(I) Equation of continuity, $(\alpha, \beta) = (0, 0)$:

$$\frac{\partial\sigma_*}{\partial t} + \frac{\partial}{\partial\varpi}[\sigma_*\langle c_\varpi\rangle + \frac{\partial}{\partial\theta}\left[\sigma^*\Omega + \frac{\sigma_*}{\varpi}\langle c_\varpi\rangle\right] + \frac{\sigma_*}{\varpi}\langle c_\varpi\rangle = 0. \tag{B37}$$

(II) Equation of motion in the radial direction, $(\alpha, \beta) = (1, 0)$:

$$\begin{aligned} &\frac{\partial}{\partial t}[\sigma_*\langle c_\varpi\rangle] + \frac{\partial}{\partial\varpi}[\varpi\sigma_*\langle c_\varpi^2\rangle + \frac{\partial}{\partial\theta}\left[\sigma_*\Omega\langle c_\varpi\rangle + \frac{\sigma_*}{\varpi}\langle c_\varpi c_\theta\rangle\right] \\ &= \sigma_*\left\{(a_\varpi + \Omega^2\varpi) + 2\Omega\langle c_\theta\rangle + \frac{1}{\varpi}\langle c_\theta^2\rangle\right\} - \frac{\sigma_*}{\varpi}\langle c_\varpi^2\rangle. \end{aligned} \tag{B38}$$

(III) Equation of motion in the angular direction, $(\alpha, \beta) = (0, 1)$:

$$\begin{aligned} &\frac{\partial}{\partial t}[\sigma_*\langle c_\theta\rangle] + \frac{\partial}{\partial\varpi}[\sigma_*\langle c_\varpi c_\theta\rangle] + \frac{\partial}{\partial\theta}\left[\sigma_*\Omega\langle c_\theta\rangle + \frac{\sigma_*}{\varpi}\langle c_\theta^2\rangle\right] \\ &= \sigma_*\left\{a_\theta - \frac{\varkappa^2}{2\Omega}\langle c_\varpi\rangle - \frac{2}{\varpi}\langle c_\varpi c_\theta\rangle\right\}. \end{aligned} \tag{B39}$$

These equations cannot yet be compared with ordinary hydrodynamic equations without the introduction of the dispersion velocity relative to the mean motion. We shall now write

$$\langle c_\varpi\rangle = v_\varpi, \quad \langle c_\theta\rangle = v_\theta, \tag{B40}$$

and define

$$c'_\varpi = c_\varpi - \langle c_\varpi \rangle, \quad c'_\theta = c_\theta - \langle c_\theta \rangle. \tag{B41}$$

We also introduce the symbols

$$\tau_{\varpi\varpi} = -\sigma_* \langle c'^2_\varpi \rangle, \quad \tau_{\varpi\theta} = \tau_{\theta\varpi} = -\sigma_* \langle c'_\varpi c'_\theta \rangle, \quad \tau_{\theta\theta} = -\sigma_* \langle c'^2_\theta \rangle, \tag{B42}$$

which represent the "stress" components due to the peculiar velocity $(c_\varpi', c_\theta')$ over the mean motion. The Eqs. (B37)–(B39) can then be rewritten in the following form:

$$\frac{\partial \sigma_*}{\partial t} + \frac{1}{\varpi}\frac{\partial}{\partial \varpi}(\sigma_* \varpi v_\varpi) + \frac{1}{\varpi}\frac{\partial}{\partial \theta}[\sigma_*(\Omega\varpi + v_\theta)] = 0, \tag{B37a}$$

$$\sigma_*\left[\frac{\partial v_\varpi}{\partial t} + v_\varpi \frac{\partial v_\varpi}{\partial \varpi} + \left(\Omega + \frac{v_\theta}{\varpi}\right)\frac{\partial v_\varpi}{\partial \theta}\right] - \frac{\sigma^*}{\varpi}(\varpi\Omega + v_\theta)^2$$
$$= \sigma_* a_\varpi + \left[\frac{\partial \tau_{\varpi\varpi}}{\partial \varpi} + \frac{\partial \tau_{\varpi\theta}}{\partial \theta} + \frac{1}{\varpi}(\tau_{\theta\theta} - \tau_{\varpi\varpi})\right], \tag{B38a}$$

$$\sigma_*\left[\frac{\partial v_\theta}{\partial t} + v_\varpi \frac{\partial v_\theta}{\partial \varpi} + \left(\Omega + \frac{v_\theta}{\varpi}\right)\frac{\partial v_\theta}{\partial \theta}\right] + \sigma_* v_\varpi\left(\frac{\varkappa^2}{2\Omega} + \frac{v_\theta}{\varpi}\right)$$
$$= \sigma_* a_\theta + \left(\frac{\partial \tau_{\theta\varpi}}{\partial \varpi} + \frac{1}{\varpi}\frac{\partial \tau_{\theta\theta}}{\partial \theta} + \frac{2\tau_{\varpi\theta}}{\varpi}\right). \tag{B39a}$$

These equations are now familiar to the hydrodynamicists: the total "hydrodynamic" velocity in the angular direction being $\Omega\varpi + v_\theta$. The term with $\varkappa^2/2\Omega$ requires a little claculation to verify. The terms in the "stresses" are identical with those occurring in turbulent motion. [See, e.g., Goldstein (1938), p. A3].

If the "stresses" are neglected, we get the equations used by Lin and Shu (1964). Equations (1a)–(1c) in that paper can be obtained from (B37a)–(B39a) by the following changes in notation:

$$(t, \varpi, \theta) \to (t, r, \theta),$$
$$\sigma_* \to \mu,$$
$$(v_\varpi, \Omega\varpi + v_\theta) \to (u, v),$$
$$(a_\varpi, a_\theta) \to (\phi_r, \phi_\theta/r),$$
$$\tau_{\varpi\varpi}, \tau_{\varpi\theta} = \tau_{\theta\varpi}, \tau_{\theta\theta} \to 0.$$

## APPENDIX C (TO SECTION 5)

### Ca. Response of a Stellar Disk to a Spiral Gravitational Field

We give here an outline of the calculation of the response of a stellar disk to a spiral gravitational field.

Let $t$ be the time, and $(\Pi, \Theta, Z)$ be the components of total stellar velocity in the cylindrical coordinates $(\varpi, \theta, z)$. We shall restrict ourselves to the two-dimensional case and introduce the peculiar velocity components $(c_\varpi, c_\theta)$ by the relations

$$c_\varpi = \Pi, \quad c_\theta = \Theta - \varpi\Omega(\varpi), \tag{C1}$$

where $\Omega(\varpi)$ is the angular velocity of circular motion defined by balancing the centrifugal acceleration $\varpi\Omega^2$ with the symmetrical gravitational field. The distribution function $\Psi(\varpi, \theta, c_\varpi, c_\theta, t)$ then satisfies the differential equation

$$\begin{aligned}\frac{\partial\Psi}{\partial t} + c_\varpi\frac{\partial\Psi}{\partial\varpi} + \left(\Omega + \frac{c_\theta}{\varpi}\right)\frac{\partial\Psi}{\partial\theta} \\ + \left(a_\varpi + \Omega^2\varpi + 2\Omega c_\theta + \frac{c_\theta^2}{\varpi}\right)\frac{\partial\Psi}{\partial c_\varpi} \\ + \left(a_\theta - \frac{\varkappa^2}{2\Omega}c_\varpi - \frac{c_\varpi c_\theta}{\varpi}\right)\frac{\partial\Psi}{\partial c_\theta} = 0,\end{aligned} \tag{C2}$$

where $\varkappa(\varpi)$ is the epicyclic frequency defined by

$$\varkappa^2 = (2\Omega)^2\left\{1 + \frac{\varpi}{2\Omega}\frac{d\Omega}{d\varpi}\right\}, \tag{C3}$$

and $(a_\varpi, a_\theta)$ are the components of acceleration of the prevailing gravitational field. In conformity to our scheme (Lin and Shu, 1966), we do not impose a condition of self-consistency at this point. Instead, we assume that the resultant field is of the form

$$(a_\varpi, a_\theta) = (\Omega^2\varpi, 0) + (a_{\varpi 1}, a_{\theta 1}), \tag{C4}$$

where $(a_{\varpi 1}, a_{\theta 1})$ are derivable from a spiral gravitational potential

$$\mathscr{V}_1 = A(\varpi)\exp\{i[\omega t - m\theta + \Phi(\varpi)]\}; \tag{C5}$$

that is

$$a_{\varpi 1} = -\frac{\partial\mathscr{V}_1}{\partial\varpi}, \quad a_{\theta 1} = -\frac{1}{\varpi}\frac{\partial\mathscr{V}_1}{\partial\theta}. \tag{C6}$$

In the equation (C5), both $A(\varpi)$ and $\Phi(\varpi)$ are real and $\omega$ is a complex parameter. (Cf. Eq. (2.1) in the text].

If $\Psi_0$ is the symmetrical distribution function corresponding to the case $A = 0$, and if we write

$$\Psi = \Psi_0(1 + \psi) = e^{-Q_0}(1 + \psi), \tag{C7}$$

then $\psi$ satisfies the linearized equation

$$\begin{aligned} &\frac{\partial\psi}{\partial t} + c_\varpi \frac{\partial\psi}{\partial\varpi} + \left(\Omega + \frac{c_\theta}{\varpi}\right)\frac{\partial\psi}{\partial\theta} \\ &+ \left(2\Omega + \frac{c_\theta}{\varpi}\right)c_\theta \frac{\partial\psi}{\partial c_\varpi} - \left(\frac{\varkappa^2}{2\Omega} + \frac{c_\theta}{\varpi}\right)c_\varpi \frac{\partial\psi}{\partial c_\theta} \\ &= a_{\varpi 1}\frac{\partial Q_0}{\partial c_\varpi} + a_{\theta 1}\frac{\partial Q_0}{\partial c_\theta}. \end{aligned} \tag{C8}$$

With the field components (C6), $\psi$ is naturally expected to be of the form

$$\psi = \phi(\varpi, c_\varpi, c_\theta) \exp\{i[\omega t - m\theta + \Phi(\varpi)]\}. \tag{C9}$$

The amplitude function $\Phi$ satisfies the equation

$$\begin{aligned} &c_\varpi \frac{\partial\phi}{\partial\varpi} + \left(2\Omega + \frac{c_\theta}{\varpi}\right)c_\theta \frac{\partial\phi}{\partial c_\varpi} - \left(\frac{\varkappa^2}{2\Omega} + \frac{c_\theta}{\varpi}\right)c_\varpi \frac{\partial\phi}{\partial c_\theta} \\ &+ i\left[kc_\varpi - m\frac{c_\theta}{\varpi} + (\omega - m\Omega)\right]\phi \\ &= g_\varpi \frac{\partial Q_0}{\partial c_\varpi} + g_\theta \frac{\partial Q_0}{\partial c_\theta}, \end{aligned} \tag{C10}$$

where

$$k(\varpi) = \Phi'(\varpi) \tag{C11}$$

and

$$(a_{\varpi 1}, a_{\theta 1}) = (g_\varpi, g_\theta) \exp\{i[\omega t - m\theta + \Phi(\varpi)]\}. \tag{C12}$$

*Asymptotic solution of* (C10).

If we are dealing with the case of tightly wound spirals, we may attempt an asymptotic series of the form

$$\varphi = \varphi^{(0)} + \varepsilon\varphi^{(1)} + \cdots, \tag{C13}$$

where $\varepsilon$ is a parameter of the order of $(k\varpi)^{-1}$. Consistent with this approximation, we shall first introduce dimensionless velocities referred to local values; thus, we shall write

$$c_\varpi = \xi V_1 = \xi(2\Omega/\varkappa)(\varpi\Omega), \tag{C14a}$$

$$c_\theta = \eta(\varpi\Omega). \tag{C14b}$$

(The scales are suggested by our experience with the theory of epicycles.) It is then not difficult to verify that the equation for $\varphi^{(0)}$ is

$$\eta\frac{\partial\varphi^{(0)}}{\partial\xi} - \xi\frac{\partial\varphi^{(0)}}{\partial\eta} + i(\nu + \alpha\xi)\varphi^{(0)} = -\frac{i\alpha A}{2V_1^2}\frac{\partial Q_0}{\partial\xi}, \tag{C15}$$

where $\alpha$ and $\nu$ are defined by

$$\alpha = (k\varpi)(2\Omega^2/\varkappa^2), \tag{C15a}$$

and

$$\nu = (\omega - m\Omega)/\varkappa. \tag{C15b}$$

Notice that a characteristic integral of (C15) is

$$\xi^2 + \eta^2 = C(\varpi) \tag{C16}$$

thanks to the choice of scales adopted. For our present purposes, we shall also take the basic distribution to be a function of $\xi^2 + \eta^2$ and $\varpi$; i.e.,

$$Q_0 = Q_0(\xi^2 + \eta^2, \varpi). \tag{C17}$$

Such a distribution function holds, when only stars of small dispersion speeds are considered. Since only a small fraction of stars have large dispersion speeds and they are not expected to be greatly influenced by the spiral gravitational field, (C17) is a good approximation for our present purposes. It is also compatible with the usually adopted Schwarzschild distribution. In that case, $Q_0$ is indeed of the form

$$Q_0 = \frac{1}{2}\left[\frac{c_\varpi^2}{\langle c_\varpi^2\rangle} + \frac{c_\theta^2}{\langle c_\theta^2\rangle}\right] + Q_{00}(\varpi), \tag{C18}$$

where $\langle c_\varpi{}^2\rangle$ and $\langle c_\theta{}^2\rangle$ are the mean-square values of the components of dispersion velocities for the location $\varpi$. For the moment, we shall consider general distribution functions of the form (C17). Note that $\varpi$ is a parameter in all these calculations.

To integrate (C15), we introduce the characteristic integral (C16) as one of the variables, i.e., we introduce the polar coordinate system $(\tau, s)$ in the $(\xi, \eta)$ plane:

$$\xi = \tau\cos s, \quad \eta = \tau\sin s. \tag{C19}$$

Then (C14) becomes

$$-\frac{d\varphi^{(0)}}{ds} + i(\nu + \alpha\tau\cos s)\,\varphi^{(0)} = -\frac{i\alpha A}{V_1^2} Q_0'(\tau^2)\,\tau\cos s, \qquad \text{(C20)}$$

where $Q_0'(\tau^2)$ denotes the derivative of $Q_0$ with respect to $\tau^2$, the dependence on $\varpi$ being understood. Both $\tau^2$ and $\varpi$ are of course to be treated as parameters for the purposes of the differential equation (C20). As we shall demonstrate below, the solution of this differential equation is

$$\varphi^{(0)} = -\frac{A}{V_1^2} Q_0'(\tau^2)\,\{1 - q(\alpha\xi, \alpha\eta, \nu)\}, \qquad \text{(C21)}$$

where

$$q = \frac{\nu\pi}{\sin\nu\pi}\,\frac{1}{2\pi}\int_{-\pi}^{\pi} \exp\{i[\nu s - \alpha\xi\sin s + \alpha\eta(1+\cos s)]\}\,ds. \qquad \text{(C22)}$$

It is readily verified by direct substitution that (C22) is indeed a solution of the original equation (C15).

The derivation of (C21) from (C19) is also not difficult. The general solution of (C21), which involves one "constant" of integration depending both on $\tau$ (to be exhibited) and on $\varpi$ (not to be exhibited), is

$$\frac{V_1^2}{AQ_0'(\tau^2)}\,\varphi^{(0)} = e^{iF(s,\tau)}\left\{C(\tau) - i\alpha\int_0^s e^{-iF(s',\tau)}\,\tau\cos s'\,ds'\right\}, \qquad \text{(C23)}$$

where

$$F(s,\tau) = \nu s + \alpha\tau\sin s. \qquad \text{(C24)}$$

To determine $C(\tau)$, we apply the condition of periodicity, equating $\varphi_0(s)$ with $\varphi_0(s + 2\pi)$. We then consider the integral in $\varphi_0(s + 2\pi)$ in its alternative form given by the right-hand side of the identity

$$\int_0^{s+2\pi} = \int_0^{2\pi} + \int_{2\pi}^{2\pi+s}. \qquad \text{(C25)}$$

The change of variable $s' = 2\pi + s''$ in the last integral of (C25) reduces it to the form

$$\int_0^s e^{-iF(s''+2\pi,\tau)}\,\tau\cos s''\,ds''$$

$$= e^{i\nu(2\pi)}\int_0^s e^{-iF(s'',\tau)}\,\tau\cos s''\,ds''. \qquad \text{(C26)}$$

20*

In the form on the right-hand side of (C25), the integral is the same as that occurring in $\varphi_0(s)$. Some obvious calculations would then yield the desired determination of $C(\tau)$; and thence the result (C21).

*Some remarks on the solution* (C21)

The solution (C21) may also be written in the following form. If we recall the definitions of $\psi$ and $\varphi$ [Eqs. (C7) and (C9)], we see that the "response", in terms of the distribution function itself, is

$$\Psi_1 = \Psi_0 \psi = \frac{\mathscr{V}_1}{V_1^2} \Psi_0'(\tau^2) \{1 - q\}, \qquad \text{(C27)}$$

where $q(\alpha\xi, \alpha\eta, \nu)$ is defined by (C22), $(\xi, \eta)$ are defined by (C14), and $\tau^2 = \xi^2 + \eta^2$. The parameters $\alpha$ and $\nu$ are defined by (C16) and (C17). For the Schwarzschild distribution, we have

$$\frac{\mathscr{V}_1}{V_1^2} \Psi_0'(\tau^2) = \frac{\mathscr{V}_1}{\langle c_\varpi^2 \rangle} \Psi_0(\tau^2). \qquad \text{(C28)}$$

It is important to notice that the density response, obtained by integrating (C27) with respect to $c_\varpi$ and $c_\theta$, is always *real for real* $\varpi$. To see this, we first note that $\Psi_0'(\tau^2)$ is even in $\eta$. The imaginary part of $q$ comes in terms of the two integrals

$$I_1 = \int_{-\pi}^{\pi} \cos[\nu s - \alpha\xi \sin s] \sin[\alpha\eta(1 + \cos s)]\, ds,$$

and

$$I_2 = \int_{-\pi}^{\pi} \sin[\nu s - \alpha\xi \sin s] \cos[\alpha\eta(1 + \cos s)]\, ds.$$

Now, $I_2$ is clearly zero, since the integrand is odd in $s$. The integral $I_1$ does not vanish, but it is clearly odd in $\eta$. Thus, when the integration in $\eta$ (i.e., in $c_\theta$) is performed, the net contribution vanishes. Indeed, all even moments of $c_\theta$, weighted with an arbitrary power of $c_\varpi$, vanish for the perturbation (C27).

## Cb. Calculation of Responses in Density and in Components of Mean Stellar Velocity

We consider the case where the basic distribution function is the Schwarzschild distribution

$$\Psi_0 = P_0(\varpi) \exp\left\{-\frac{\mu_0}{2}(\xi^2 + \eta^2)\right\}, \qquad \text{(C29)}$$

where $\mu_0^{-1}$ is a measure of the mean-square value of the dispersion velocities,

$$\mu_0^{-1} = \langle c_\varpi^2 \rangle / V_1^2, \quad V_1 = (2\Omega/\varkappa)(\varpi\Omega), \tag{C29a}$$

and $P_0$ is the normalization factor, so that

$$\sigma_{*0} = m_* \iint \Psi_0 \, dc_\varpi \, dc_\theta . \tag{C30}$$

The response of the distribution function to the spiral gravitational field is now of the form

$$\Psi_1 = \frac{1}{\langle c_\varpi^2 \rangle} \Psi_0 (1 - q), \tag{C31}$$

where $q$ si defined by (C21).

For the state $\Psi = \Psi_0 + \Psi_1$, we have

$$\sigma_*(\varpi, \theta, t) = m_* \iint \Psi \, dc_\varpi \, dc_\theta \tag{C32a}$$

and also the following relations for the mean stellar velocity components $v_\varpi$ and $v_\theta$:

$$\sigma_* v_\theta = m_* \iint \Psi c_\varpi \, dc_\varpi \, dc_\theta , \tag{C32b}$$

$$\sigma_* v_\theta = m_* \iint \Psi c_v \, dc_\varpi \, dc_\theta . \tag{C32c}$$

It is easy to verify that, within the approximation of the linear theory,

$$\frac{\sigma^*{}_1}{\sigma^*{}_0} = \frac{\iint \Psi_1 \, dc_\varpi \, dc_\theta}{\iint \Psi_0 \, dc_\varpi \, dc_\theta} = \frac{\mathscr{V}_1}{\langle c_\varpi^2 \rangle} \langle (1 - q) \rangle , \tag{C33a}$$

$$\frac{v_\varpi}{V_1} = \frac{\mathscr{V}_1}{\langle c_\varpi^2 \rangle} \langle \xi (1 - q) \rangle , \tag{C33b}$$

$$\frac{v_\theta}{\varpi\Omega} = \frac{\mathscr{V}_1}{\langle c_\varpi^2 \rangle} \langle \eta (1 - q) \rangle , \tag{C33c}$$

where $\langle \; \rangle$ denotes the weighted average with respect to $\Psi_0$, or the function $\exp\{-\frac{1}{2}\mu_0(\xi^2 + \eta^2)\}$.

The calculation of these averages depends on well-known relationships of the type

$$\langle e^{i\lambda\xi} \rangle = e^{-\lambda^2/2\mu_0}, \tag{C34a}$$

$$\langle \xi e^{i\lambda_1\xi} \rangle = \frac{i\lambda_1}{\mu_0} e^{-\lambda_1^2/2\mu_0}, \tag{C34b}$$

$$\langle \eta e^{i\lambda_2\eta} \rangle = \frac{i\lambda_2}{\mu_0} e^{-\lambda_2^2/2\mu_0}, \tag{C34c}$$

for Gaussian distribution functions. It is then easy to verify, by using (C34a), that

$$\langle(1-q)\rangle = 1 - \frac{\nu\pi}{\sin\nu\pi}\mathscr{G}_\nu(x), \tag{C35}$$

where $x = k^2\langle c_\varpi^2\rangle/\varkappa^2$, and

$$\mathscr{G}_\nu(x) = \frac{1}{2\pi}\int_{-\pi}^{\pi} e^{-x(1+\cos s)}\cos\nu s\, ds. \tag{C36}$$

The reduction factor $\mathscr{F}_\nu(x)$ that occurs in (5.13a) in the text may then be introduced by defining it with the relation

$$\mathscr{F}_\nu(x) = \frac{1-\nu^2}{x}\left\{1 - \frac{\nu\pi}{\sin\nu\pi}\mathscr{G}_\nu(x)\right\}, \tag{C37}$$

so that (5.13a) takes on a form directly comparable with the case of zero dispersion speeds.

The evaluation of the integral in (C33b) can be done with the help of (C34b), where $\lambda_1 = \alpha \sin s$. We find that an integral of the form

$$\mathscr{H}_\nu(x) = \frac{1}{2\pi}\int_{-\pi}^{\pi} e^{-x(1+\cos s)}\sin\nu s\sin s\, ds \tag{C38}$$

is involved. By partial integration, this can be transformed into

$$\mathscr{H}_\nu(x) = \frac{\sin\nu\pi}{\pi x}\left\{1 - \frac{\nu\pi}{\sin\nu\pi}\mathscr{G}_\nu(x)\right\}. \tag{C39}$$

Thus, the result can be best summarized by writing

$$\langle\xi(1-q)\rangle : \langle(1-q)\rangle = -\frac{\nu}{\alpha} \tag{C40}$$

or

$$\frac{v_\varpi}{\Omega\varpi} : \frac{\sigma_{*1}}{\sigma_{*0}} = -\frac{\omega - m\Omega}{\Omega(k\varpi)}. \tag{C41}$$

It is worth noting that Eq. (C41) may be obtained directly by using the equation of continuity in the continuum form. The reduction factor $\mathscr{F}_\nu^{(1)}(x)$ in (2.13b) is thus identical with $\mathscr{F}_\nu(x)$:

$$\mathscr{F}_\nu^{(1)}(x) = \mathscr{F}_\nu(x). \tag{C42}$$

The numerical values of $\mathscr{F}_\nu(x)$ are shown in Table C and Fig. C1.

TABLE C. Numerical values of the reduction factor $\mathscr{F}(\nu, x)$.

| $\nu$ \ $x$ | 0 | 1 | 2 | 3 | 4 | 5 | 6 | 7 | 8 |
|---|---|---|---|---|---|---|---|---|---|
| 0 | 1 | 0.5342 | 0.3457 | 0.2523 | 0.1982 | 0.1633 | 0.1389 | 0.1209 | 0.1068 |
| 0.1 | 1 | 0.5333 | 0.3447 | 0.2514 | 0.1974 | 0.1625 | 0.1381 | 0.1202 | 0.1066 |
| 0.2 | 1 | 0.5305 | 0.3416 | 0.2484 | 0.1946 | 0.1600 | 0.1359 | 0.1182 | 0.1046 |
| 0.3 | 1 | 0.5258 | 0.3363 | 0.2434 | 0.1901 | 0.1558 | 0.1321 | 0.1147 | 0.1013 |
| 0.4 | 1 | 0.5191 | 0.3286 | 0.2362 | 0.1835 | 0.1499 | 0.1267 | 0.1097 | 0.0966 |
| 0.5 | 1 | 0.5100 | 0.3185 | 0.2266 | 0.1748 | 0.1420 | 0.1195 | 0.1032 | 0.0907 |
| 0.6 | 1 | 0.4984 | 0.3056 | 0.2145 | 0.1638 | 0.1321 | 0.1105 | 0.0950 | 0.0833 |
| 0.7 | 1 | 0.4838 | 0.2894 | 0.1994 | 0.1503 | 0.1199 | 0.0995 | 0.0849 | 0.0739 |
| 0.8 | 1 | 0.4658 | 0.2695 | 0.1810 | 0.1337 | 0.1050 | 0.0861 | 0.0726 | 0.0629 |
| 0.9 | 1 | 0.4435 | 0.2452 | 0.1586 | 0.1136 | 0.0871 | 0.0700 | 0.0581 | 0.0494 |
| 1.0 | 1 | 0.4158 | 0.2153 | 0.1312 | 0.0894 | 0.0656 | 0.0507 | 0.0408 | 0.0341 |

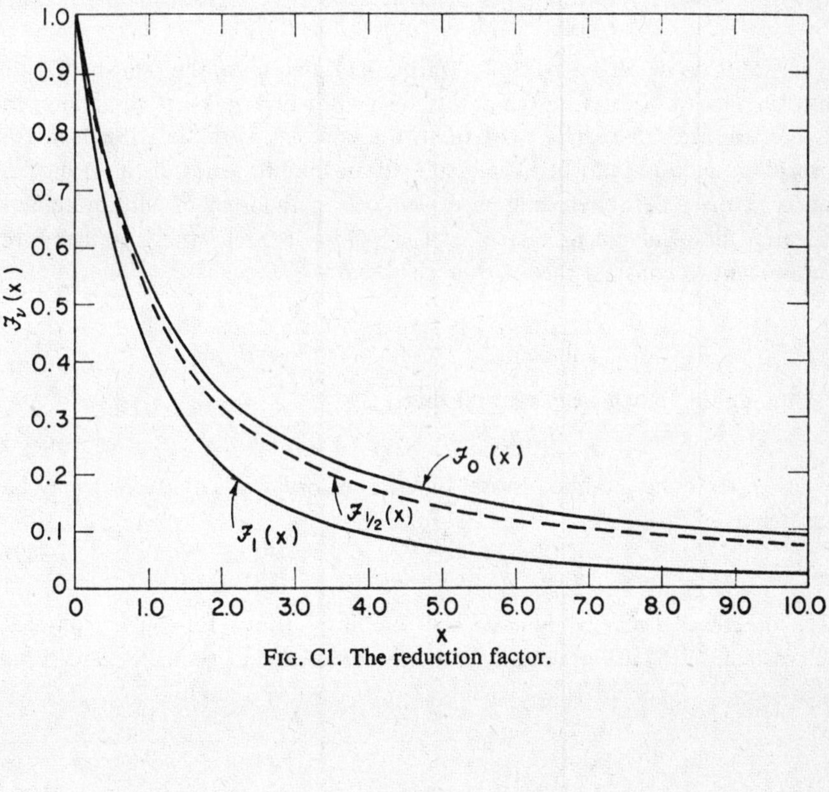


FIG. C1. The reduction factor.

To evaluate (C32c), we follow a similar method by using (C34c) with $\lambda_2 = \alpha(1 + \cos s)$. The result can be best summarized by writing

$$\langle \eta(1 - q)\rangle : \langle \xi(1 - q)\rangle = -i\mathscr{G}_\nu'(x) : \mathscr{H}_\nu(x), \tag{C43}$$

or

$$\frac{\langle c_\theta \rangle}{\varpi\Omega} : \frac{\langle c_\varpi \rangle}{(\varpi\Omega)(2\Omega/\varkappa)} = -i\mathscr{G}_\nu'(x) : \mathscr{H}_\nu(x). \tag{C44}$$

In the form used in Eq. (5.13c) in text, we then have

$$\mathscr{F}_\nu^{(2)}(x) = \frac{(1 - \nu^2)\,\nu\pi}{\sin \nu\pi}\{-\mathscr{G}_\nu'(x)\}, \tag{C45}$$

where $\mathscr{G}_\nu'(x)$ is the derivative, with respect to its own argument $x$, of the function $\mathscr{G}_\nu(x)$ defined by (C36).

## Cc. An Alternative Form of the Dispersion Relation

The dispersion relationship (5.20), in the case of a purely stellar disk, will now be presented in a form that reminds us of the similarity between our problem and that of waves in electromagnetic plasma. In this case, the dispersion relation (5.20) may also be written as

$$\frac{k^*}{|k|} = \frac{1}{x}\left\{1 - \frac{\nu\pi}{\sin(\nu\pi)}\mathscr{G}_\nu(x)\right\}, \tag{C46}$$

where $\mathscr{G}_\nu(x)$ is defined by (C36). Clearly, the function on the right-hand side has singularities (poles) at the points $\nu = \pm n$, where $n$ is a positive integer. Otherwise, the function is analytic in the complex variable $\nu$. Such meromorphic functions can be expanded into an infinite series of the principal parts at the poles, according to a well-known theorem of Mittag-Leffler. Clearly, the principal parts at $\nu = +n, -n$ are related, and it is therefore convenient to consider the series in the form

$$\frac{\nu\pi}{\sin(\nu\pi)}\mathscr{G}_\nu(x) = 1 + \sum_{n=1}^{\infty} \frac{B_n(x)}{\nu^2 - n^2}. \tag{C47}$$

By taking the limit $\nu \to n$, we find that

$$B_n(x) = 2n^2 e^{-x} I_n(x), \tag{C48}$$

where $I_n(x)$ is the modified Bessel function of order $n$. Thus, (C46) may be written as

$$\frac{k_*}{k} = -\frac{2e^{-x}}{x}\sum_{n=1}^{\infty} \frac{I_n(x)}{(\nu/n)^2 - 1}\Bigg\}. \tag{C49}$$

The difference between this and the familiar formula for plasma physics (Bernstein, 1958) lies essentially in (i) the difference in geometry and (ii) a reversal of sign caused by the difference between gravitational and electromagnetic forces.

The present form of the theory, as embodied in (5.20) is, however, much more perspicuous for the sprial problem, especially when there is also gas present. The concept of the reduction factor has been found to be particularly useful.

## Cd. A Comparison between the Present Problem and Bernstein's Problem

It is noted that there is some similarity between the present dispersion relationship and that obtained by Bernstein for plasma waves in the presence of a magnetic field. It would be interesting to point out the basis for this similarity and for the differences.

The similarity between the gravitational problem and the plasma problem was already pointed out by Lynden-Bell, and is shown in the following table.

| | Stellar System (Gravitational) | Plasma (Electromagnetic) |
|---|---|---|
| (1) | inverse square law for attraction | inverse square law for repulsion |
| (2) | Coriolis force $2\Omega \times \mathbf{v}$ | Lorentz force $(e/c)(\mathbf{v} \times \mathbf{H})$ |

The difference between the present problem and that studied by Lynden-Bell (which is almost a complete analogue of Bernstein's problem) lies in the difference in geometry. We are dealing with a disk geometry, and they are dealing with volume distributions. Thus, the individual Fourier components of the field in the two cases are given, respectively, by

(3) $g_{\varpi} = 2\pi i G\sigma \times \text{sign of } k, \qquad E_k = 4\pi i G\varrho_k/k,$

The factor $k$ causes great a difference in the dispersion relation.

As a consequence of (1) and (2), there are also differences between these two cases, independent of geometry. Propagating waves are possible when

(4) $\lambda < \lambda_J, \qquad \lambda > \lambda_D,$

(5) $\nu < 1, \qquad \nu > 1.$

That is, in the stellar (plasma) case, the propagating waves have a wavelength shorter (longer) than the Jeans (Debye) length and a frequency lower (higher) than the basic particle frequency.

## APPENDIX D. SOME COMMENTS ON DISPERSION VELOCITIES AND LENGTH SCALES

There are three kinds of dispersion velocities that are important to us:

(1) stellar dispersion,
(2) molceular velocity, and
(3) turbulent motion of the gas.

Their typical magnitudes are (1) 30 km/sec, (2) 1 km/sec, and (3) 8 km/sec. But they have similar behavior towards resisting gravitational collapse. Indeed they enter the virial theorem in a similar manner, as contributions to the kinetic energy. Thus, the density distribution in a plane stellar sheet with uniform dispersion velocity is given by

$$\varrho(z) = \varrho_0 \operatorname{sech}^2 (z/z_0),$$

with

$$z_0^2 = \langle c_z^2 \rangle / 2\pi G \varrho_0,$$

where $\varrho(z)$ is density at height $z$, $\varrho_0$ is the density at $z = 0$, and $\langle c_z^2 \rangle$ is the mean-square value of the $z$-component of stellar dispersion. Stellar collisions are neglected. A similar relation holds for an isothermal gaseous sheet, although molecular collisions are present in this case.

There is a similar relationship with spherical geometry. Since the scale maintained by the stellar dispersion velocities is essentially the Jeans length, a spherical system is expected to be entirely stable: all modes are "Landau-damped". For a disk geometry, particularly one maintained by rotation, there are three scales. They are, in the order of increasing size, (1) the thickness, (2) the Jeans length for dispersion velocities in the plane of the disk, and (3) the overall size of the disk. The waves we are dealing with have scales intermediate between (2) and (3).

## APPENDIX E. THE EFFECT OF FINITE DISK THICKNESS‡

### Ea. Theoretical Development

Let me point out at the outset that we are not particulartly interested in solving the full three-dimensional problem. The spiral disturbances under consideration have characteristic wavelengths which are appreciably longer than the thickness of the galactic disk. Limiting ourselves to the discussion of such waves results in a substantial simplification of the mathematics.

‡ This part is the contribution of F. H. S. Shu.

For thin disks possessing reflection symmetry about the plane $z = 0$, motion in the $z$ direction is essentially decoupled from that in the plane. Let us suppose that a resultant (perturbation) spiral gravitational potential of the form

$$\mathscr{V}_1(\varpi, \theta, z, t) = \mathscr{A}(\varpi, |z|) \exp\{i[\omega t - m\theta + \Phi(\varpi)]\} \qquad \text{(E1)}$$

is superposed onto the basic time-indpendent axisymmetric potential $\mathscr{V}_0(\varpi, z) = \mathscr{V}_0(\varpi, 0) + \mathscr{B}_0(\varpi, z)$. Further, we shall suppose the variation of $\Phi$ with $\varpi$ to be rapid in comparison with the variation of $\mathscr{A}$ with $\varpi$ (but not rapid in comparison with the variation of $\mathscr{A}$ with $|z|$).

Under these conditions, it is easy to demonstrate from the linearized dynamical equations that the surface density responses of the stars and gas are given by the relations

$$\sigma_{*1} = -\frac{k^2}{\varkappa^2(1-\nu^2)}\sigma_{*0}\mathscr{F}_\nu(x_*)\langle\mathscr{V}_1\rangle_{*z}, \qquad \text{(E2a)}$$

$$\sigma_{g1} = -\frac{k^2}{\varkappa^2(1-\nu^2)}\sigma_{g0}\mathscr{F}_\nu(x_g)\langle\mathscr{V}_1\rangle_{gz}, \qquad \text{(E2b)}$$

where $\langle\mathscr{V}_1\rangle_{*z}$ and $\langle\mathscr{V}_1\rangle_{gz}$ are the mass-weighted averages of the gravitational potential:

$$\langle\mathscr{V}_1\rangle_{*z} = \frac{1}{\sigma_{*0}}\int_{-\infty}^{\infty} \varrho_{*0}\mathscr{V}_1\, dz, \qquad \text{(E3a)}$$

$$\langle\mathscr{V}_1\rangle_{gz} = \frac{1}{\sigma_0}\int_{-\infty}^{\infty} \varrho_{g0}\mathscr{V}_1\, dz. \qquad \text{(E3b)}$$

The responses given in Eqs. (E2) are exactly of the same form as those obtained in the theory for a disk of zero thickness except that the averages (E3) replace $\mathscr{V}_1$ evaluated in the central plane $z = 0$. There is a *weakening* of the response to a given gravitational field because in the course of executing finite $z$ motions, the matter in the Galaxy samples fields which are weaker than those in the central plane. A mass-average is involved in Eqs. (E2) because the total time spent by all the stars and gas at a given level of $z$ is directly proportional to their respective basic volume densities there.

For self-sustained waves, apart from the weakening of the response due to finite $z$ motions, there is also a *dilution* of the resultant gravitational field because the distribution is spread out rather than concentrated in a central sheet. For a thin disk, the spiral gravitational potential is related to the total *volume density* perturbation through

$$\mathscr{V}_1(\varpi, \theta, z, t) = -\frac{2\pi G}{|k|} \int_{-\infty}^{\infty} \varrho_{\text{tot},1}(\varpi, \theta, c, t)\, e^{-|k||z-c|}\, dc, \tag{E4a}$$

$$\varrho_{\text{tot}1} = \varrho_{*1} + \varrho_{g1}. \tag{E4b}$$

A simple derivation of this relation results from the realization that the function $-e^{-|k|z}/2|k|$ is the appropriate Green's function for Poisson's equation in this case.

We wish to estimate the $z$ variation of $\mathscr{V}_1$ from Eq. (E4a) by approximating the $z$ variation of $\varrho_{\text{tot}1}$. Dynamical considerations reveal that the $z$ variation of $\varrho_{*1}$ and $\varrho_{g1}$ to be proportional to the basic mass distribution modified by the $z$ variation of the perturbation potential. Suppressing all dependences except that in $z$, we may heuristically write

$$\varrho_{*1}(z) = \varrho_{*0}(z)\, F_*[\mathscr{V}_1(z)], \tag{E5a}$$

$$\varrho_{g1}(z) = \varrho_{g0}(z)\, F_g[\mathscr{V}_1(z)], \tag{E5b}$$

where $F_*$ and $F_g$ are dimensionless smoothly varying functionals of $\mathscr{V}_1$. For disks which are thin compared to the wavelength of the disturbance, the $z$ variation of $V_1$ (and hence the $z$ variation of $F_*$ and $F_g$) is slow compared to that of $\varrho_{*0}$ and $\varrho_{g0}$. Hence, it suffices to replace $F_*$ and $F_g$ by some average constant values ("constant" only in reference to variations with $z$). These constant values are chosen so that integration of $\varrho_{*1}$ and $\varrho_{g1}$ over all $z$ recover $\sigma_{*1}$ and $\sigma_{g1}$, respectively. Thus, we may write

$$\varrho_{*1} = \varrho_{*0}\frac{\sigma_{*1}}{\sigma_{*0}}, \tag{E6a}$$

$$\varrho_{g1} = \varrho_{g0}\frac{\sigma_{g1}}{\sigma_{g0}}, \tag{E6b}$$

as the essential parts of the volume density perturbations which contribute to the gravitational potential (E4a).

If we now substitute Eqs. (E6) into Eqs. (E4) and put the resulting expression into Eqs. (E2), we obtain the following homogeneous matrix equation to solve for $\sigma_{*1}$ and $\sigma_{g1}$:

$$\begin{bmatrix} \dfrac{\lambda_T\sigma_{*0}/\sigma_{\text{tot}}}{\lambda(1-\nu^2)}\mathscr{F}_\nu T_{**} - 1 & \dfrac{\lambda_T\sigma_{*0}/\sigma_{\text{tot}}}{\lambda(1-\nu^2)}\mathscr{F}_\nu T_{*g} \\ \dfrac{\lambda_T\sigma_{g0}/\sigma_{\text{tot}}}{\lambda(1-\nu^2)}\mathscr{F}_\nu^{(g)} T_{*g} & \dfrac{\lambda_T\sigma_{g0}/\sigma_{\text{tot}}}{\lambda(1-\nu^2)}\mathscr{F}_\nu^{(g)} T_{gg} - 1 \end{bmatrix} \begin{bmatrix} \sigma_{*1} \\ \sigma_{g1} \end{bmatrix} = 0 \tag{E7}$$

where we have defined as usual,

$$\lambda = \frac{2\pi}{|k|} = \frac{2\pi}{|\Phi'(\varpi)|}, \quad \nu = \frac{\omega - m\Omega(\varpi)}{\varkappa(\varpi)}, \tag{E8a}$$

$$\lambda_T = \frac{4\pi^2 G\sigma_{\text{tot}}}{\varkappa^2}, \quad \sigma_{\text{tot}} = \sigma = \sigma_{*0} + \sigma_{g0}, \tag{E8b}$$

and where we have introduced the dimensionless "reduction factors" due to thickness:

$$T_{**} = \iint_{-\infty}^{\infty} \frac{\varrho_{*0}(\varpi, z)}{\varrho_{*0}(\varpi)} \frac{\varrho_{*0}(\varpi, c)}{\sigma_{*0}(\varpi)} e^{-|k||z-c|}\, dz\, dc, \tag{E9a}$$

$$T_{*g} = \iint_{-\infty}^{\infty} \frac{\varrho_{*0}(\varpi, z)}{\sigma_{*0}(\varpi)} \frac{\varrho_{g0}(\varpi, c)}{\sigma_{g0}(\varpi)} e^{-|k||z-c|}\, dz\, dc, \tag{E9b}$$

$$T_{gg} = \iint_{-\infty}^{\infty} \frac{\varrho_{g0}(\varpi, z)}{\sigma_{g0}(\varpi)} \frac{\varrho_{g0}(\varpi, c)}{\sigma_{g0}(\varpi)} e^{-|k||z-c|}\, dz\, dc. \tag{E9c}$$

The functions $\mathscr{F}_\nu(x_*)$ and $\mathscr{F}_\nu^{(g)}(x_g)$ appearing in Eq. (E7) are the reduction factors associated with the random motions of the stars and the gas in the plane of the Galaxy. They were introduced on page 269.

In order that Eq. (E7) have nontrivial solutions, the determinant of the coefficient matrix must be zero. This condition yields the dispersion relation which relates the wavelength to the dimensionless frequency once the basic parameters are given. I won't write down the general form taken by this dispersion relation but shall confine the discussion to the simplified case when the term

$$\frac{\sigma_{*0}\sigma_{g0}}{(\sigma_{\text{tot}})^2}\,[T_{**}T_{gg} - T_{*g}^2] \ll 1. \tag{E10}$$

Numerically, this condition is almost always satisfied for disk galaxies whose gas content is small. Under conditions when Eq. (E10) holds, the eigenvalue matrix equation (E7) yields the relations

$$\frac{\lambda}{\lambda_T}(1 - \nu^2) = \frac{\sigma_{*0}}{\sigma_{\text{tot}}} \mathscr{F}_\nu(x_*)\, T_{**} + \frac{\sigma_{g0}}{\sigma_{\text{tot}}} \mathscr{F}_\nu^{(g)}(x_g)\, T_{gg}, \tag{E11}$$

$$\frac{\sigma_{*1}}{\sigma_{g1}} = \frac{\sigma_{g0}\mathscr{F}_\nu^{(g)} \sqrt{T_{gg}}}{\sigma_{*0}\mathscr{F}_\nu \sqrt{T_{**}}}. \tag{E12}$$

We have written the dispersion relation (E11) in the usual form where the right-hand side represents the "effective mass" which responds to the gravitational disturbance compared with that which would have responded if the material components only had purely circular motions. The reduction of the stellar response due to peculiar motions in the plane of the Galaxy and perpendicular to it is given by the functions $\mathscr{F}_\nu^{(g)}(x_*)$ and $T_{**}$ whereas the corresponding reduction for the gas is given by $F_\nu(x_g)$ and $T_{gg}$.

The self-gravitating basic state of thin disk galaxies is well approximated by the mass distributions

$$\varrho_{*0}(\varpi, z) = \varrho_{*0}(\varpi, 0)\,\mathrm{sech}^2\,(z/z_{*0}), \tag{E13a}$$

$$\varrho_{g0}(\varpi, z) = \varrho_{g0}(\varpi, 0)\exp\,(-z^2/z_{g0}^2), \tag{E13b}$$

with the scale heights given by

$$z_{*0} = \sqrt{\frac{2\langle c_z^2\rangle}{\mu^2}}, \tag{E14a}$$

$$z_{g0} = \sqrt{\frac{2a_0^2}{\mu^2}}, \tag{E14b}$$

TABLE E

| $y_*$ | $T(y_*)$ | $y_*$ | $T(y_*)$ |
|---|---|---|---|
| 0.000 | 1.0000 | 0.50 | 0.4672 |
| 0.002 | 0.9333 | 0.55 | 0.4542 |
| 0.004 | 0.9108 | 0.60 | 0.4425 |
| 0.006 | 0.8941 | 0.65 | 0.4317 |
| 0.008 | 0.8804 | 0.70 | 0.4217 |
| 0.01 | 0.8686 | 0.75 | 0.4125 |
| 0.02 | 0.8247 | 0.80 | 0.4039 |
| 0.03 | 0.7935 | 0.85 | 0.3959 |
| 0.04 | 0.7686 | 0.90 | 0.3884 |
| 0.05 | 0.7478 | 0.95 | 0.3814 |
| 0.06 | 0.7298 | 1.00 | 0.3747 |
| 0.07 | 0.7138 | 1.10 | 0.3625 |
| 0.08 | 0.6995 | 1.20 | 0.3514 |
| 0.09 | 0.6864 | 1.30 | 0.3414 |
| 0.10 | 0.6745 | 1.40 | 0.3322 |
| 0.15 | 0.6261 | 1.50 | 0.3238 |
| 0.20 | 0.5897 | 1.60 | 0.3160 |
| 0.25 | 0.5606 | 1.70 | 0.3088 |
| 0.30 | 0.5363 | 1.80 | 0.3021 |
| 0.35 | 0.5156 | 1.90 | 0.2958 |
| 0.40 | 0.4975 | 2.00 | 0.2899 |
| 0.45 | 0.4815 | — | — |

TABLE E continued.

| $\sqrt{y}$ | $T(y)$ | $\sqrt{y}$ | $T(y)$ | $\sqrt{y}$ | $T(y)$ |
|---|---|---|---|---|---|
| 0.00 | 1.0000 | 0.17 | 0.8338 | 0.34 | 0.7079 |
| 0.01 | 0.9888 | 0.18 | 0.8253 | 0.35 | 0.7015 |
| 0.02 | 0.9778 | 0.19 | 0.8172 | 0.36 | 0.6952 |
| 0.03 | 0.9670 | 0.20 | 0.8090 | 0.37 | 0.6889 |
| 0.04 | 0.9564 | 0.21 | 0.8010 | 0.38 | 0.6828 |
| 0.05 | 0.9460 | 0.22 | 0.7932 | 0.39 | 0.6768 |
| 0.06 | 0.9357 | 0.23 | 0.7854 | 0.40 | 0.6708 |
| 0.07 | 0.9257 | 0.24 | 0.7778 | 0.41 | 0.6649 |
| 0.08 | 0.9157 | 0.25 | 0.7703 | 0.42 | 0.6591 |
| 0.09 | 0.9060 | 0.26 | 0.7629 | 0.43 | 0.6534 |
| 0.10 | 0.8965 | 0.27 | 0.7557 | 0.44 | 0.6478 |
| 0.11 | 0.8871 | 0.28 | 0.7486 | 0.45 | 0.6423 |
| 0.12 | 0.8778 | 0.29 | 0.7415 | 0.46 | 0.6368 |
| 0.13 | 0.8687 | 0.30 | 0.7346 | 0.47 | 0.6314 |
| 0.14 | 0.8597 | 0.31 | 0.7278 | 0.48 | 0.6261 |
| 0.15 | 0.8510 | 0.32 | 0.7210 | 0.49 | 0.6209 |
| 0.16 | 0.8423 | 0.33 | 0.7210 | 0.50 | 0.6157 |

where $\mu = \sqrt{\partial^2 B_0(\varpi, 0)/\partial z^2}$ is the $z$ oscillation frequency and $a_0$ is the rms turbulent velocity in the interstellar gas.

With such mass distributions, the thickness reduction factors take the forms

$$T_{**} = T_*(y_*) = \int_0^1 dp \int_{-p}^{p} \left[\frac{(1+q)}{(1-q)}\frac{(1-p)}{(1+p)}\right]^{\sqrt{y_*/2}} dq, \qquad \text{(E15a)}$$

$$T_{gg} = T_g(y_g) = e^{y_g}\,\mathrm{erfc}\left(\sqrt{y_g}\right), \qquad \text{(E15b)}$$

where we have defined

$$y_* = \frac{k^2\langle c_z^2\rangle}{\mu^2}, \qquad \text{(E16a)}$$

$$y_g = \frac{k^2 a_0^2}{\mu^2}, \qquad \text{(E16b)}$$

as a measure of the square of the average $z$ excursion of the stars and gas, respectively. In Table E we have tabulated the numerical behavior of the functions defined by Eqs. (E15).

## Eb. Comparison with Observations

For comparison with observations, we shall adopt the following values as being representative of the local conditions in our own galaxy:

$$\begin{aligned} 2z_g(\tfrac{1}{2}) &= 280\text{ pc}, \quad 2z_*(\tfrac{1}{2}) = 620\text{ pc}, \\ \varrho_{g0}/\varrho_{*0} &= 0.26, \quad \sigma_{\text{tot}} = 94\ \text{M}_\odot/\text{pc}^2, \\ \Omega &= 25\text{ km/sec-kpc}, \quad \varkappa = 31\text{ km/sec-kpc}, \end{aligned} \tag{E17}$$

where $2z_*(\frac{1}{2})$ and $2z_g(\frac{1}{2})$ refer to the total thickness between the half-density points in the disks of the stars and gas, respectively. With these figures, the ratio of gas to stellar surface density is

$$\sigma_{g0} : \sigma_{*0} = 1 : 9 \tag{E18}$$

and the ratios of the scale heights to $\lambda_T$ are

$$z_{g0}/\lambda_T = 0.010, \quad z_{*0}/\lambda_T = 0.021. \tag{E19}$$

## Ec. Stability Criterion

For axisymmetric disturbances ($m = 0$), the dispersion relation allows both oscillatory ($\omega$ real) and exponential ($\omega$ imaginary) time dependences. The curve of marginal stability is obtained by setting $\omega = 0$ (i.e. $\nu = 0$) in the dispersion relation (E11). In Fig. E.1 we present for comparison the curves of marginal stability obtained for the case considered and that for a pure stellar disk of zero thickness.

Finite thickness has primarily two effects: it lowers the amount of minimum velocity dispersion required in the galactic plane to stabilize

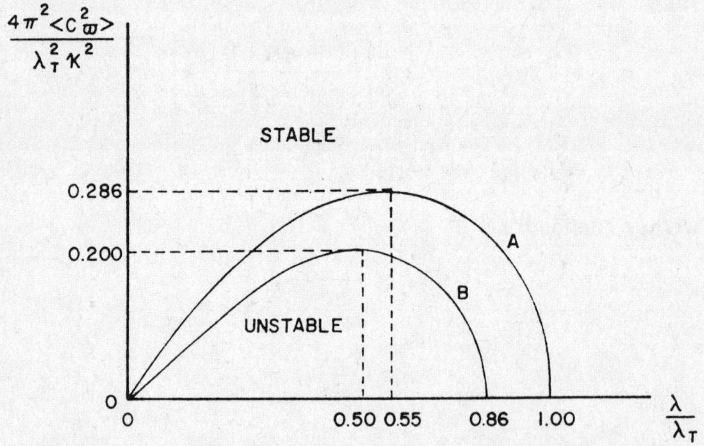


FIG. E.1. The curves marginal stability drawn schematically for the cases: A—No gas, zero thickness; B—10% ($\sigma_{g0}/\sigma_{\text{tot}} = 0.10$), finite thickness ($z_{g0}/\lambda_T = 0.010$, $z_{*0}/\lambda_T = 0.021$).

all axisymmetric disturbances, and it helps rotation to stabilize the long waves at a shorter scale than rotation would have been able to do by itself without the assistance of velocity dispersion either in the plane or perpendicular to it. The effect of the presence of gas is to raise the amount of stellar dispersion required for stability.

With the values adopted previously for the basic parameters, the minimum amounts of velocity dispersion required to stabilize the disk are

$$\langle c_\varpi^2\rangle^{1/2} = 37\ \text{km/sec}, \quad \langle c_\theta^2\rangle^{1/2} = 23\ \text{km/sec}, \qquad \text{(E20)}$$
$$\langle c_z^2\rangle^{1/2} = 20\ \text{km/sec}, \quad a_0 = 8.7\ \text{km/sec}.$$

If we had neglected the effect of finite thickness in both the disks of stars and gas, we would have obtained stellar dispersions in the galactic plane of amounts which are 30% higher than those predicted above. The present values are in reasonable agreement with observations although the velocities required are still slightly too high (by about 10%).

## Ed. The Spiral Pattern: Comparison with the Simple Theory

When we include the effects of gas and finite thickness, the dispersion relation is somewhat modified from the simple one appropriate for a stellar disk of zero thickness. In particular, in the neighborhood of a resonance point, finite thickness combined with the considerations of exact characteristics prevents the scale of the disturbances which can propagate from becoming too small. Elsewhere, the form of the dispersion relationship is relatively insensitive to these modifications. The effects of finite disk thickness and the presence of a small amount of gas have the tendency to balance one another resulting in a geometric form for the pattern which is not substantially altered.

It is found that an increase of the pattern frequency $\Omega_p$ of only 1.5 km/sec-kpc from the value required by the simple theory proves to be sufficient to reproduce the correct arm spacing in the solar neighborhood. Such a slight increase of $\Omega_p$ is found to be favorable for Yuan's star migration calculations.

## Ee. Relative Participation of Stars and Gas

For a choice of the pattern frequency $\Omega_p = 14$ km/sec-kpc, the total reduction factors associated with the gas and stars have the values

$$\mathcal{F}_\nu^{(g)}(x_g)\, T_g(y_g) = (0.571)\,(0.789) = 0.451, \qquad \text{(E21)}$$
$$\mathcal{F}_\nu(x^*)\, T_*(y_*) = (0.115)\,(0.571) = 0.066.$$

Because of their smaller random motions, the gas plays an enhanced role in determining the scale of the spiral structure. The "effective participation" of the gas is only reduced to slightly less than one-half of the participation it would have had if it had no turbulence, whereas the "effective participation" of the stars is reduced to less than 7% of its corresponding value. Only the stars which have peculiar velocities of comparable smallness as the turbulent velocity of the gas will respond with effectiveness to the spiral wave. The older disk stars merely serve to provide a background field, their participation in the spiral structure is negligible.

Thus, although the basic surface densities were taken to be in the ratio

$$\sigma_{g0}/\sigma_{*0} = 0.11$$

the perturbation density responses, as given by Eq. (E12), were found to be in the ratio

$$\sigma_{g1}/\sigma_{*1} = 0.65.$$

The participation in the spiral structure between stars and gas is nearly equal at distances of 10 kpc from the galactic center. At points interior to this distance, where the relative amount of gas is smaller, the stars would be dominant in maintaining the spiral field.

## APPENDIX F. THE FORMULATION OF THE COMPLETE PROBLEM ‡

We shall now formulate the problem of the normal modes of oscillation which occur in the place of a differentially rotating collisionless stellar sheet possessing velocity dispersion only in the plane of the sheet. Such a sheet, of course, has zero thickness.

We shall reduce our task to solving an integral equation which implicitly contains appropriate boundary conditions at the origin and at infinity. Although our treatment is a normal mode treatment using the method of separation of variables, it will become necessary to appeal to the initial-value problem to resolve certain ambiguities which arise when there exist stars in resonance with the wave. It is presisely the possibility for the occurence of important stellar resonances that distinguishes the stellar dynamical theory from the gas dynamical theory. In particular, "anti-spiral theorems" (cf. Lynden-Bell and Ostriker, 1967) are now seen in the proper perspective when dealing with stellar rather than gaseous disks.

‡ This part is the contribution of F. H.-S. Shu.

## Fa. The Mathematical Development

We suppose the basic state of our model galaxy to be one in which the mass distribution is time-independent and axisymmetric. Superposed onto the basic state are small perturbations in the surface density which in turn give rise to corresponding perturbations in the gravitational potential:

$$\sigma(\varpi, \theta, t) = \sigma_0(\varpi) + \sigma_1(\varpi, \theta, t), \tag{F1a}$$

$$\mathscr{V}(\varpi, \theta, t) = \mathscr{V}_0(\varpi) + \mathscr{V}_1(\varpi, \theta, t). \tag{F1b}$$

In the linearized theory, we need consider only disturbances of the form

$$\sigma_1(\varpi, \theta, t) = S(\varpi) \exp\{i[\omega t - m\theta]\}, \tag{F2a}$$

$$\mathscr{V}_1(\varpi, \theta, t) = V(\varpi) \exp\{i[\omega t - m\theta]\}, \tag{F2b}$$

with $m$ restricted to be an integer or zero while $\omega$ can be a complex constant.

The perturbation potential is obtained from the imposed perturbation density through potential theory which gives

$$V(\varpi) = -2\pi G \int_0^\infty P_m(\varpi, a)\, aS(a) \frac{da}{\varpi + a}, \tag{F3a}$$

$$P_m(\varpi, a) = \frac{2}{\pi} \int_0^{\frac{\pi}{2}} \frac{\cos 2m\chi}{\sqrt{1 - \zeta \cos^2 \chi}}\, d\chi, \quad \zeta = \frac{4\varpi a}{(\varpi + a)^2}. \tag{F3b}$$

The kernel $P_m(\varpi, a)$ has a logarithmic singularity at $a = \varpi$. For $m = 0$, $P_m$ assumes the form of the complete elliptic integral of the first kind.

The calculation for the response of the stars to the imposed gravitational field is facilitated if we introduce the characteristic variables which are the integrals of motion in the basic state, namely, the unperturbed energy and angular momentum (per unit mass) of a star:

$$E_0 = \tfrac{1}{2}[\Pi^2 + \Theta^2] + \mathscr{V}_0(\varpi), \tag{F4a}$$

$$J = \varpi\Theta. \tag{F4b}$$

The use of these variables rather than the velocity components $(\Pi, \Theta)$ is essential for a substantial improvement of the theory near resonance.

The phase (number) density of stars in the perturbed state is written as

$$\psi = F_0(E_0, J) + \frac{\partial F_0}{\partial E_0}(E_0, J)\mathscr{V}_1(\varpi, \theta, t) + F_1(\varpi, \theta, E_0, J, t), \tag{F5}$$

where $F_1$ has the same angle and time dependence as $V_1$:

$$F_1(\varpi, \theta, E_0, J, t) = f(\varpi, E_0, J) \exp\{i[\omega t - m\theta]\}. \tag{F6}$$

The linearized Liouville equation can be integrated along the characteristics of the unperturbed orbits to give

$$f(\varpi, E_0, J) = -\frac{\omega\, \partial F_0/\partial E_0 + m\, \partial F_0/\partial J}{2 \sin(\omega\tau_0 - m\theta_0)} \int_{-\tau_0}^{\tau_0} V(\bar{\varpi}_0(\tau)) e^{i[\omega\tau - m - \bar{\theta}_0(\tau)]}\, d\tau. \tag{F7}$$

This formula contains two parameters $\tau_0$ and $\theta_0$ and two functions $\bar{\varpi}_0(\tau)$ and $\bar{\theta}_0(\tau)$. They are defined as follows. The parameter $2\tau_0(E_0, J)$ is the period of radial oscillation for a (bound) star of energy $E_0$ and angular momentum $J$, and the other parameter $2\theta_0(E_0, J)$ is the angular displacement in azimuth suffered by a star in time $2\tau_0$. They are related to $E_0$ and $J$ as follows:

$$\tau_0(E_0, J) = \int_{\varpi_1(E_0, J)}^{\varpi_2(E_0, J)} \frac{d\varpi}{|\Pi_0(\varpi, E_0, J)|}, \tag{F8a}$$

$$\theta_0(E_0, J) = \int_{\varpi_1(E_0, J)}^{\varpi_2(E_0, J)} \frac{J\, d\varpi}{\varpi^2 |\Pi_0(\varpi, E_0, J)|}, \tag{F8b}$$

where $\varpi_1(E_0, J)$ and $\varpi_2(E_0, J)$ are the zeroes (i.e. the "turning points of the motion") of the radial velocity

$$\Pi_0(\varpi, E_0, J) = \left\{2[E_0 - \mathscr{V}_0(\varpi)] - \frac{J^2}{\varpi^2}\right\}^{1/2}. \tag{F9}$$

The functions $\bar{\varpi}_0(\tau)$ and $\bar{\theta}_0(\tau)$ are so defined as to describe the orbit in time of a star whose motion satisfies the equations of motion

$$\frac{d\bar{\varpi}_0}{d\tau}(\tau) = \Pi_0(\bar{\varpi}_0, E_0, J), \quad \bar{\varpi}_0(\pm\tau_0) = \varpi, \tag{F10a}$$

$$\frac{d\bar{\theta}_0}{d\tau}(\tau) = \frac{J}{\bar{\varpi}_0^2(\tau)}, \qquad \bar{\theta}_0(\pm\tau_0) = \pm\theta_0. \tag{F10b}$$

A typical orbit described by these equations is illustrated schematically in Fig. F1.

For $\omega$ real, resonance in Eq. (F7) can arise whenever $\omega\tau_0 - m\theta_0$ is an integral multiple of $\pi$. The resonant stars are those which, in the linear theory, always see the *same* phase of the wave in each cycle of their peculiar motions. This phenomenon is very important for the discussion of self-sustained spiral waves.

Integration over all velocity space of Eq. (F5) yields the surface density response to the applied gravitational potential. In terms of $E_0$ and $J$, this integration gives the radial dependence of the perturbation surface density as

$$\varpi S_*(\varpi) = m_* \iint \frac{dE_0\, dJ}{|\Pi_0(\varpi, E_0, J)|} \left\{ \frac{\partial F_0}{\partial E_0}(E_0, J)\, V(\varpi) + f(\varpi, E_0, J) \right\}. \tag{F11}$$

For simplicity we have taken all stars to be of the same mass $m_*$.

For self-sustained oscillations the imposed surface density $S(\varpi)$ must be equal to the surface density response $S_*(\varpi)$. This condition com-

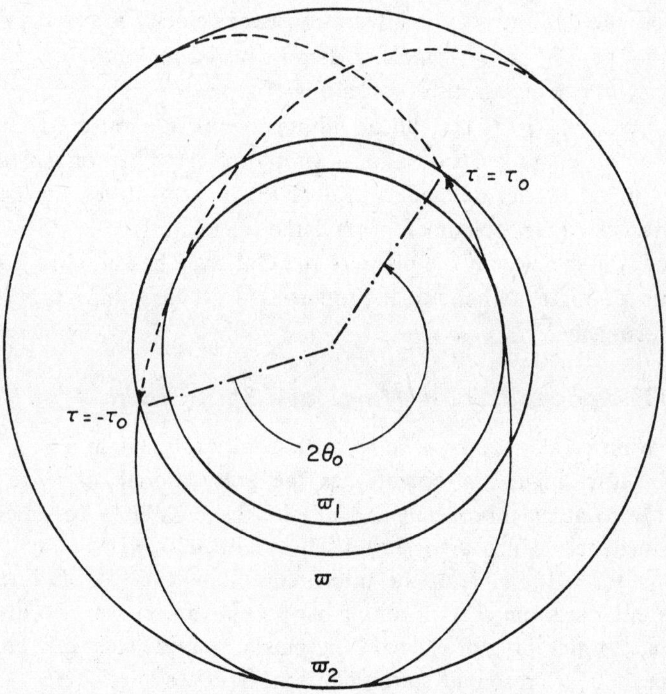


FIG. F1. The solid part of the orbit represents one complete period of radial oscillation, the particle having started and returned to radial distance $\varpi$ after having reached the maximum radial distance $\varpi_2$ and the minimum $\varpi_1$.

bined with Eqs. (F1), (F7), and (F3) leads to the following integral equation satisfied by the eigenfunction $S(\varpi)$:

$$\varpi S(\varpi) = \int_0^\infty k_m(\varpi, a)\, aS(a) \frac{da}{(\varpi + a)}, \tag{F12}$$

where the kernel $K_{m\omega}(\varpi, a)$ is given by the successive integrations of $P_m(\varpi, a)$:

$$K_{m\omega}(\varpi, a) = -4\pi G \iint_{\Pi_0(\varpi, E_0, J) \geqq 0} \frac{dE_0\, dJ}{\Pi[\varpi, E_0, J)}$$

$$\times \left\{ \frac{\partial F_0}{\partial E_0} P_m(\varpi, a) - \frac{\omega\, \partial F_0/\partial E_0 + m\, \partial F_0/\partial J}{2 \sin(\omega\tau_0 - m\theta_0)} \right.$$

$$\left. \int_{-\tau_0}^{\tau_0} P_m(\bar{\varpi}_0(\tau), a) \cos[\omega\tau - m\bar{\theta}_0(\tau)]\, d\tau \right\}. \tag{F13}$$

It was possible to contract the integration over velocity space in Eq. (F13) to only that part where $\Pi_0(\varpi, E_0, J)$ is positive by making use of the time reversibility of the equations of motion (F10).

Equation (F12) is a linear homogeneous integral equation to be solved for the characteristic value $\omega$ and the eigenfunction $\varpi S(\varpi)$. Together with the kernel defined by Eq. (F13), it constitutes the complete formulation of the problem of normal modes in the whole disk of stars with velocity dispersion in the plane of the Galaxy. The kernel in Eq. (F13) has certain peculiar mathematical properties which are important for the physical problem.

## Fb. The possibility for Waves of a Spiral Form

First of all, $K_{m\omega}(\varpi, a)$ is logarithmically singular at $a = \varpi$. While integrable, this singularity accentuates the gravitational influence of the stars in the local neighborhood and admits the possibility for short-scale wavy disturbances which are (nearly) self-sustained *locally*.

Second, for $\omega$ real, the interpretation of the $(E_0, J)$ integration may present the same sort of ambiguity as encountered in the discussion of the Landau damping of electrostatic plasma waves. This difficulty arises whenever there are resonant stars such that the poles of $[\sin(\omega\tau_0 - m\theta_0]^{-1}$ lie along the path of integration. The dilemma to be resolved in this case is whether the integrations are to be interpreted in a principal-value sense or not.

As is well-known, the resolution of this difficulty is unambigupusly provided by the initial-value treatment. When applied to our problem, the involves recipe interpreting Eq. (F13) as defining $K_{m\omega}(\varpi, a)$ only for *amplified* disturbances [i.e. only for $\mathrm{Im}(\omega) < 0$]. For neutral or damped disturbances, $K_{m\omega}(\varpi, a)$ is to be obtained by *analytically continuing* the integral expression on the right-hand side of Eq. (F13) from the region where $\mathrm{Im}(\omega) < 0$ into the region $\mathrm{Im}(\omega) \geqq 0$. In the presence of resonant stars, the contribution of the poles of $[\sin(\omega\tau_0 - m\theta_0)]^{-1}$ in this analytic continuation will generally lead to a nonzero imaginary part for $K_{m\omega}(\varpi, a)$ *even if $\omega$ is real.*

If $K_{m\omega}(\varpi, a)$ is complex, the integral equation (F12) must allow complex-valued solutions for $S(\varpi)$. Such solutions, especially those whose phase variations are rapid compared with their amplitude variations, will display a *spiral-like* appearance.

However, the above described situation need not arise. For instance, it is possible for $K_{m\omega}(\varpi, a)$ to be purely real if neutrales modes (i.e. $\omega$ real) can be found which involve no resonant stars. For such cases, by taking the complex conjugate of Eq. (F12), we easily show that if $S(\varpi)$ is a solution of Eq. (F12), so is its complex conjugate $S^*(\varpi)$. If $\omega$ is not degenerate, $S^*(\varpi)$ can differ from $S(\varpi)$ only by a constant complex multiplicative factor of modulus unity, i.e.,

$$S^*(\varpi) = e^{-2i\chi}S(\varpi), \tag{F14}$$

where $\chi$ is some real constant. Equating the arguments of both sides results in the relation:

$$\arg\{S(\omega)\} = \chi. \tag{F15}$$

In this case, the phase of $S(\varpi)$ is a strict constant and $S(\varpi)$ shows no spiral structure.

If $\omega$ is degenerate and $S(\varpi)$ and $S^*(\varpi)$ are linearly independent, they must show spiral structure of opposite sense, one leads wherever the other trails. Even more, we could take

$$\frac{1}{2}[S(\varpi) + S'^*(\varpi)], \quad \frac{1}{2i}[S(\varpi) - S'^*(\varpi)], \tag{F16}$$

rather than $S(\varpi)$ and $S^*(\varpi)$ themselves, as the two linearly independent solutions. Both these solutions are purely real and consequently do not exhibit spiral structure.

In summary, we have shown that an "anti-spiral theorem", analogous to that given by Linden–Bell and Ostriker (1967) for a gaseous disk, applies to all neutral modes for which resonance does not occur. We do not, of course, interpret this as proof of the impossibility for spirals, but rather as a convincing argument for either the existence of strong resonances or weak overstabilities or a combination of both to account for the observed spiral structure of disk galaxies.

## BIBLIOGRAPHY


Becker, W. (1964), Z. Astrophys. **58**, 202.

Bernstein, I. B. (1958), Phys. Rev. **109**, 10.

Bok, B. J. (1967), Am. Scientist **55**, 375–399.

Bok, B. J., and George Contopoulos (1967), Report to the I.A.U. (Commission 33).

Burke, B. F. (1968), Presented at Charlottesville Conferences on HII regions.

Burton, W. B. (1966), Bull. Astron. Inst. Neth. **18**, 247–255.

Chandrasekhar, S. (1942), *Principles of Stellar Dynamics* (Dover, New York, 1960).

Contopoulos, G., and Bengt Strömgren (1965), *Tables of Plane Galactic Orbits* (Institute for Space Studies, New York).

Crawford, D. L. (1963), Astrophys. J. **137**, 523–529.

Crawford, D. L. (1966), Proceedings of the I.A.U. Symposium No. 24, pp. 170–179 (Contribution No. 110 from Kitt Peak National Observatory).

Crawford, D. L., and C. L. Perry (1966), Astron. J. **71**, 206–215 (Contribution No. 147 from Kitt Peak National Observatory).

Crawford, D. L., and J. V. Barnes (1966), Astron. J. **71**, 610–614 (Contribution No. 192 from Kitt Peak National Observatory).

Crawford, D. L., and Bengt Strömgren (1966), Vistas Astron. **8**, 149–157. (Contribution No. 82 from Kitt Peak National Observatory).

Fujimoto, M. (1966), Columbia University Preprint (to be published in the Proceedings of the I.A.U. Symposium No. 29).

Hunter, C. (1963), Monthly Notices Roy. Astron. Soc. **126**, 299–315.

Hunter, C. (1965), Monthly Notices Roy. Astron. Soc. **129**, 321–343.

Kalnajs, A. J. (1965), Ph. D. dissertation, Harvard University.

Kelsall, T., and Bengt Strömgren (1966), Vistas Astron. **8**, 159–178.

Kerr, F. J. (1962), Monthly Notices Roy. Astron. Soc. **123**, 327. See also article by F. J. Kerr and G. Westerhout in *Galactic Structure* (University of Chicago Press, 1965), Chap. 9.

Lin, C. C. (1966), J. SIAM Appl. Math **14**, 876–921.

Lin, C. C. (1967*a*), Ann. Rev. Astron. Astrophys. **5**, 453–464.

Lin, C. C. (1967*b*), *Relativity Theory and Astrophysics* (1965), Am. Math. Soc. Lectures in Applied Mathematics, Vol. 2, pp. 66–97.

Lin, C. C. (1968), *Galaxies and the Universe*, Vetlesen Symposium, 1966, edited by L. Woltjer (Columbia University, New York), pp. 33–51.

Lin, C. C., and F. H. Shu (1964), Astrophys. J. **140**, 646–655.

Lin, C. C., and F. H. Shu (1966), Proc. Natl. Acad. Sci. U.S. **55**, 229–234.

Lin, C. C., and F. H. Shu (1967), Proceedings of the I.A.U. Symposium No. 31, Noordwijk, 1966.

Lin, C. C., F. H. Shu, and Chi Yuan (1968).

LINDBLAD, B. (1963), Stockholm Obs. Ann. **22**, 3–20.

LINDBLAD, P. O. (1960), Stockholm Obs. Ann. **21**, 3–73.

LINDBLAD, P. O. (1962), *Interstellar Matter in Galaxies*, edited by L. Woltjer (W. A. Benjamin, New York), pp. 222–233.

LYNDEN-BELL, D. (1967), AMS Lectures.

LYNDEN-BELL, D., and J. D. OSTRIKER (1967), Monthly Notices Roy. Astron. Soc. **136**, 293.

LYNDS, B. (1967), Sky and Telescope **33**, 343–346; **34**, 18–21.

MILLER, JOSEPH S. (1968), Astrophys. J. **151**, 473–489.

MILLER, R. H., and K. H. PRENDERGAST (1968), Astrophys. J. **151**, 699–709.

MORRISON, P. (1957), "On the Origins of Cosmic Rays", Rev. Mod. Phys. **29**, 235–243.

OORT, J. H. (1962), in *Interstellar Matter in Galaxies*, edited by L. Woltjer (W. A. Benjamin, Inc., New York), pp. 234–244.

PRENDERGAST, K. H. (1962), "The Motion of Gas in Barred Spiral Galaxies" in *Interstellar Matter in Galaxies*, edited by L. Woltjer (W. A. Benjamin, New York), pp. 217–221.

ROBERTS, M. S. (1967), Presented at the Prague meeting of the I.A.U. (Commission 33).

ROBERTS, W. W. (1968), Ph. D. Thesis, Massachusetts Institute of Technology.

SCHMIDT, M. (1965), in *Galactic Structure* (University of Chicago Press), pp. 513–530.

SHANE, W. W., and G. P. BIEGER-SMITH (1966), Bull. Astron. Inst. Neth. **18**, 263–292.

SHARPLESS, S. (1965), in *Galactic Structure* (University of Chicago Press), pp. 131–156.

SHU, F. H. (1968), "The Dynamics and Large Scale Structure of Spiral Galaxies", Ph. D. Thesis, Harvard University.

SPITZER, L. (1956), *Physics of Fully Ionized Gases* (Interscience, New York).

STRÖMGREN, BENGT (1963), George Darwin Lecture (October, 1962), Quart. J. Roy. Astron. Soc. **4**, 8.

STRÖMGREN, BENGT (1963), *Stars and Stellar Systems*, Vol. III, Chap. 9, pp. 123–191.

STRÖMGREN, BENGT (1964), *Astrophysica Norvegica*, (Contribution No. 59 from Kitt Peak National Observatory), Vol. IX, pp. 333–343.

STRÖMGREN, BENGT (1966), *Stellar Evolution* (Plenum Press), pp. 391–394.

STRÖMGREN, BENGT (1966), Ann Rev. Astron. Astrophys. **4**, 433–472.

STRÖMGREN, BENGT (1967), Proceedings of the I.A.U. Symposium No. 31, Noordwijk, 1966.

STRÖMGREN, BENGT, and C. PERRY (1965), Photoelectric *UVBY* Photometry for 1217 Stars Brighter than $V = 6\ 5$, mostly of spectral classes A, F, and G. (second version.)

TOOMRE, ALAR (1964), Astrophys. J. **139**, 1217.

YUAN, CHI (to be published), "Application of the Density-Wave Theory to the Spiral Structure of the Milky Way".

THE ASTROPHYSICAL JOURNAL, Vol. 155, March 1969

# ON THE SPIRAL STRUCTURE OF DISK GALAXIES
# III. COMPARISON WITH OBSERVATIONS*

C. C. LIN AND C. YUAN
Massachusetts Institute of Technology
AND
FRANK H. SHU
Harvard College Observatory

*Received June 21, 1968; revised August 16, 1968*

ABSTRACT

The density-wave theory of galactic spirals is developed in a form slightly more general than that outlined by Lin and Shu in an earlier short communication. Only self-sustained waves are studied in this paper, and the problem of the origin of the spiral structure is barely discussed. The implications of the theory are examined, both in general terms and in detail. The conclusions are compared with observations. Specifically, we consider (1) the distribution of atomic hydrogen in the Galaxy, (2) the systematic motion of the gas, (3) the distribution of young stars and other optical objects, and (4) the migration of moderately young stars. Good agreements are obtained in all cases if we adopt a pattern speed of about 13 km $sec^{-1}$ $kpc^{-1}$, and a spiral gravitational field equal to about 5 per cent of the symmetrical field. General discussions are also given on (*a*) the structure of the magnetic field and its role on the systematic motion of the gas, (*b*) the role of the density wave in the process of star formation, and (*c*) the distribution of H II regions as revealed by the 109$\alpha$ radio observations.

## I. INTRODUCTION

The density-wave theory of galactic spirals was proposed by the late B. Lindblad a long time ago. An evaluation of the early phase of the theory was made by Chandrasekhar (1942). Many subsequent papers on this subject published by Lindblad and his co-workers may be traced through his last paper in the *Stockholm Observatory Annalen* (Lindblad 1963). Feelings of reservation were expressed by astronomers and astrophysicists about Lindblad's ideas, largely because the method adopted by him, which leans heavily on the properties of individual stellar orbits, cannot yield quantitative predictions for comparison with observations. Without making use of the statistical theory of stellar dynamics, he could hardly treat the behavior of stellar collective modes in a quantitative manner. Yet these modes indeed constitute the essence of the density waves. His discussions were therefore largely qualitative and did not convey sufficient conviction to the readers.

P. O. Lindblad (1960, 1962) attempted to study these collective modes by an extensive calculation of the orbits of a number of stars with the help of large computing machines. His model consists of a smooth background representing Population II stars, together with 192 stars representing Population I. Density waves of a spiral form were indeed found from these calculations, but they were rather transient and were not quasi-stationary, as suggested by B. Lindblad. In view of the limitations inherent in direct computational methods, the results obtained are perhaps to be described as semiquantitative, when comparison with observations is attempted.[1]

The explicit adoption of stellar collective modes as a basis for a density-wave theory

* A condensed version of this paper was reported at a meeting of Commission 33 at the General Assembly of the International Astronomical Union in Prague, August 22–31, 1967 (not to be published).

[1] More recently, Miller and Prendergast (1968) carried out extensive calculations for a disk model consisting of 120000 stars. Others have dealt with cases involving cylindrical symmetry.

of galactic spirals was first published[2] by two of the present writers (Lin and Shu 1964, 1966, hereinafter referred to as Papers I and II, respectively). A description of the history and the evolution of ideas may be found in a survey article by Lin (1967*a*) on the dynamics of disk-shaped galaxies. In Paper I, an elementary theory was presented in which the dispersion of stellar velocities and the pressure (or turbulent motion) of the gas were neglected. Only speculative discussions were given concerning their effects. These speculations were confirmed by subsequent investigations, and the results were summarized in Paper II, which also contains an outline of the methods used for dealing with a galactic system composed of stars with velocity dispersion and of gas with pressure simulating the effect of turbulent motion. In the present paper, we shall also include brief discussions (in §§ I, V, and VIII) of the magnetic field and its influence on the gaseous component. The details of these investigations will be presented elsewhere (Yuan 1969).

As emphasized before (Oort 1962; Lin and Shu 1967; Lin 1968), the primary purpose of our theory is to attempt to explain the *grand design* over the whole disk, and in particular to explain its persistence. As one can see upon a little reflection, the problem of the origin of the spiral structure is mathematically more difficult, since we have to deal with a combined boundary-value and initial-value problem. These studies remain a challenge for future investigations.[3] In contrast, to explain persistence, one need only produce solutions representing self-sustained waves. The method outlined in Paper II is especially formulated for this purpose. Even here, there are some subtle points on the basic mechanism which should be clarified by future investigations. We shall bring out these points in the subsequent discussions.

### *a) Some General Implications of the Density-Wave Theory*

The winding dilemma of material spiral arms implies also a difficulty in the associated magnetic field. As the winding process goes on, a material arm is stretched, and the magnetic field, being frozen into the gas, also continues to increase in its strength. These difficulties are all avoided if we associate the spiral pattern with a *density wave* that has a *quasi-stationary spiral structure.*

Consider the ideal case of a single wave. To an observer rotating with the pattern speed, the gas flow is steady and in *closed* stream lines[4] (which deviate only slightly from circular shape). Thus, one visualizes a situation in which the magnetic field always runs *approximately along* the spiral arm in the over-all pattern. There may also exist material arms which are being distorted by differential rotation. In such an arm, there might be even a fair amount of deviation between the field lines and the spiral arm, but it would still be a good first approximation to think of the field as primarily oriented along an arm. Such general conclusions appear to be in agreement with observations.

### *b) Objectives of the Present Paper*

In the absence of a complete theory for the mechanism of density waves, the need for observational support is urgent. Accordingly, the main purpose of the present paper is to describe the applications of the theory to several observable features in our own Galaxy, in order to enable the reader to form an over-all picture of the extent to which the theory may yield results that agree with observations.[5] Some further details of the

[2] Kalnajs (1965) described a more ambitious scheme in attacking the same problem along similar lines. He envisaged the importance of collective modes of a very large scale. These are mathematically more difficult to treat, and his results have not yet been published. See also n. 3.

[3] See Shu (1968) for a fuller discussion of the complete mathematical formulation of the problem and of the possibility of tracing the origin of spiral structure to "gradient instability," as suggested in Paper II.

[4] Such flow patterns, including shock formation, have been recently calculated by Roberts (1969).

[5] An article by Lynds (1967*a*, *b*) gave a general account of these comparisons. The article also contains discussions of other data not discussed in detail in this paper.

analyses used and the comparisons made in each case will be published separately in later communications. Specifically, we shall consider (1) the distribution of atomic hydrogen, (2) the systematic motion of the gas, (3) the distribution of young stars and their place of formation, (4) the migration of moderately young stars and their place of formation, and, finally, (5) the role of the magnetic field in item (2), its alignment with the spiral arm having been mentioned above. We shall also examine the problems related to the local structure in the solar vicinity, in the light of these studies. Some comments will be made toward the end of this paper on the general status of the theory and on some general implications, especially on the process of star formation and on the evolution of the galaxies.

A secondary purpose of this paper is to give an orderly presentation of the theory in a form somewhat more general than that presented before. The principal results will be described in § 2 to pave the way for the discussion of the comparison of the theory with observations. The mathematical details of these developments may be found in the Appendices. The exposition of the present paper will be organized essentially as a continuation of Papers I and II cited above; only occasional references will be made to the other papers (e.g., Lin and Shu 1967; Lin 1966, 1967*a*, *b;* the reader is also referred to Prendergast 1967).

It should be mentioned from the outset that the detailed calculations in this work are based on an asymptotic analysis of the WKBJ type, valid for reasonably tightly wound spirals (pitch angle less than 15°). For larger inclinations, the results can be only qualitatively correct; accurate results can only be obtained from the type of approach formulated by Kalnajs (1965) and Shu (1968).

Our future plans include further applications of our theory to the structure of the Milky Way, and the study of other galaxies. A comprehensive discussion of the existing observable features that bear on the spiral structure of galaxies, especially our own, was given by Bok (1967) and by Bok and Contopoulos (1967). The reader should refer to them for further details, and for future observations recommended.

## II. THEORY OF DENSITY WAVES

The density-wave theory was developed especially to resolve the problems of (i) the persistence of the spiral structure in the face of differential rotation and (ii) the large-scale nature of the phenomenon, i.e., the existence of a grand design over the whole galactic disk. We therefore start out with the QSSS hypothesis (hypothesis of the existence of density waves with a quasi-stationary spiral structure), rather than attempt to study the problem of the origin of the spiral structure.[6] As outlined in Paper II, the theory is based on the calculation of the responses of an infinitesimally thin disk of stars and gas to a *resultant* gravitational field of a spiral form. The effect of finite thickness has been recently examined by Shu (1968), but we shall not go into the details in this paper.

### *a*) *Response of the Stellar Disk*

The stellar disk is described by a distribution function $\Psi(\varpi,\theta,c_\varpi,c_\theta,t)$, where $t$ is the time, $(\varpi,\theta)$ are polar coordinates in the plane of the disk, and $(c_\varpi,c_\theta)$ are the peculiar velocities relative to a (differentially) rotating observer moving at the circular speed $\varpi\Omega(\varpi)$. In the absence of spiral modes, $\Psi$ is given by a function $\Psi_0(\varpi,c_\varpi,c_\theta)$ which is time-independent and axisymmetric in the configurational space.

We now make some brief observations on the purely kinematical aspects of a smooth but tightly wound spiral pattern in the plane of a disk galaxy. First of all, we notice that these patterns may be described by functions of the form

$$F(\varpi,\theta,t) = \mathfrak{A}(\varpi) \exp \{i[\omega t - m\theta + \Phi(\varpi)]\} \,, \tag{2.1}$$

[6] The reader is referred to Lin (1967*b*) for a discussion of the "kinematical tests" of the feasibility of the QSSS hypothesis as the basis for a theory.

where $\varpi,\theta,t$ are defined as before; $\mathfrak{A}(\varpi)$ is a slowly varying function of $\varpi$; $\Phi(\varpi)$ is a slowly varying (monotonic) function multiplied by a large parameter; $\omega$ is a constant (real for neutral waves); and $m$ is an integer. As usual, the real part of the above expression is to be taken. The lines of constant $\mathrm{Re}(F)$ are approximately given by the equation

$$m(\theta - \theta_0) = \Phi(\varpi) - \Phi(\varpi_0) , \tag{2.2}$$

which represents a spiral pattern with $m$ arms.

The pattern given by equation (2.1) rotates with an angular velocity

$$\Omega_p = \omega_r/m , \tag{2.3}$$

where $\omega_r$ is the real part of $\omega$. The radial wavenumber of the spiral pattern is given by $|k(\varpi)|$, where

$$k(\varpi) = \Phi'(\varpi) . \tag{2.4}$$

If the motion of the stars is in the direction of increasing $\theta$, trailing waves correspond to $k(\varpi) < 0$, and leading waves, to $k(\varpi) > 0$.

In the presence of a small spiral gravitational potential,

$$\mathcal{V}_1 = A(\varpi) \exp \{i[\omega t - m\theta + \Phi(\varpi)]\} , \tag{2.5}$$

of the form of equation (2.1), the response $\Psi_1 = \Psi - \Psi_0$ in the distribution function should also have a spiral distribution in the configurational space. When the dispersion velocities of the stars are not large, this response $\Psi_1$ can be easily worked out. In this case, the basic distribution function $\Psi_0$ is of the form

$$\Psi_0 = \Psi_0(\xi^2 + \eta^2, \varpi) , \tag{2.6}$$

where $\xi$ and $\eta$ are defined by

$$c_\varpi = \xi(2\Omega/\kappa)(\varpi\Omega) = \xi V_1(\varpi) , \tag{2.7a}$$

and

$$c_\theta = \eta(\varpi\Omega) , \tag{2.7b}$$

in which $\kappa$ is the epicyclic frequency defined by

$$\kappa^2 = (2\Omega)^2\left(1 + \frac{\varpi}{2\Omega}\frac{d\Omega}{d\varpi}\right) . \tag{2.8}$$

Within the approximation of small dispersion speeds is a parameter related to the integral for angular momentum, and (for given $\varpi$) $\xi^2 = \eta^2$ is the integral of epicyclic energy. The response $\Psi_1$ to the resultant potential (2.5) is found to be

$$\Psi_1 = \Psi_0'(\xi^2 + \eta^2)[1 - q(a\xi, a\eta, \nu)]\frac{\mathcal{V}_1}{V_1^2} , \tag{2.9}$$

in the case of *tightly wound spirals*, where $\Psi_0'$ denotes the derivative of $\Psi_0$ with respect to its argument,[7] $V_1$ is defined in equation (2.7a), and the parameters $a$ and $\nu$ are defined by

$$a = (k\varpi)(2\Omega^2/\kappa^2) \tag{2.10}$$

and

$$\nu = (\omega - m\Omega)/\kappa . \tag{2.11}$$

[7] Here, as in other quantities, the dependence on the galactocentric distance $\varpi$ is understood but not explicitly exhibited.

The function $q$ is defined by the integral

$$q = \frac{\nu\pi}{\sin \nu\pi} \cdot \frac{1}{2\pi} \int_{-\pi}^{\pi} \exp \{i[\nu s - a\xi \sin s + a\eta(1 + \cos s)]\} ds . \tag{2.12}$$

The derivation of the result (2.9) may be found in Appendix A. In all previous papers, we have presented results only for the case of the Schwarzschild distribution (for which $\Psi_0$ is an exponential function of $\xi^2 + \eta^2$). Clearly, this is an important case, and we shall continue to adopt it in following sections of this paper. But the generalization given here enables one to study, separately, the roles played by stars with different epicyclic energies; this may be done simply by taking $\Psi_0$ to be of the form of the Dirac $\delta$-function. Such studies will be carried out in a separate investigation.

In the case of the Schwarzschild distribution, the perturbations in density and in the components of the mean velocity of the stars can be very readily calculated to exhibit their dependence on the dispersion velocity of the stars (see Appendix B). One obtains

$$\frac{\sigma}{\sigma_*} = -\frac{k^2 A}{\kappa^2} \cdot \frac{1}{1 - \nu^2} \cdot \mathfrak{F}_\nu(x) , \tag{2.13a}$$

$$\frac{v_\varpi}{\varpi\Omega} = \frac{k^2 A}{\kappa\Omega} \cdot \frac{(k\varpi)^{-1}\nu}{1 - \nu^2} \cdot \mathfrak{F}_\nu^{(1)}(x) , \tag{2.13b}$$

$$\frac{v_\theta}{\varpi\Omega} = \frac{i}{2} \frac{k^2 A}{\Omega^2} \cdot \frac{(k\varpi)^{-1}}{1 - \nu^2} \cdot \mathfrak{F}_\nu^{(2)}(x) , \tag{2.13c}$$

where

$$x = k^2 \langle c_\varpi^2 \rangle / \kappa^2 \tag{2.14}$$

and the functions $\mathfrak{F}_\nu(x)$, $\mathfrak{F}_\nu^{(1)}(x)$, and $\mathfrak{F}_\nu^{(2)}(x)$ are "reduction factors" defined by equations (B9), (B14), and (B17) in Appendix B. They all take on the value unity for $x = 0$, i.e., in the case of zero dispersion. The function $\mathfrak{F}_\nu(x)$ is most important to our subsequent discussions. Its values have already been tabulated and presented in graphical form (Lin and Shu 1966; Lin 1967*b*). One form of the graph is reproduced in Figure 8 for convenience of reference.

We note that, since all the above results are given for tightly wound spirals, they are subject to an uncertainty of a factor of $1 + O(1/k\varpi)$.

### *b) Response of the Gaseous Disk*

The response of the gaseous disk can be much more readily obtained if the turbulent motion of the gas is simulated by pressure. We then introduce the square of the equivalent acoustic velocity $a$, and define the quantity

$$x_g = k^2 a^2 / \kappa^2 , \tag{2.15}$$

analogous to $x$ defined by equation (2.14). Then all the three *reduction factors* are found to be of the same form, i.e.,

$$\mathfrak{F}_\nu^{(g)}(x_g) = \frac{1}{1 + x_g/(1 - \nu^2)} . \tag{2.16}$$

For convenience of reference, we shall record the explicit formulae as follows:

$$\frac{\sigma_g}{\sigma_0} = -\frac{k^2 A}{\kappa^2} \cdot \frac{1}{1 - \nu^2} \cdot \mathfrak{F}_\nu^{(g)}(x_g) , \tag{2.17a}$$

$$\frac{u_\varpi}{\varpi\Omega} = \frac{k^2 A}{\kappa\Omega} \cdot \frac{(k\varpi)^{-1}\nu}{1 - \nu^2} \mathfrak{F}_\nu^{(g)}(x_g) , \tag{2.17b}$$

$$\frac{u_\theta}{\varpi\Omega} = \frac{i}{2} \cdot \frac{k^2 A}{\Omega^2} \cdot \frac{(k\varpi)^{-1}}{1-\nu^2} \mathfrak{F}_\nu^{(g)}(x_g) , \tag{2.17c}$$

where $u_\varpi$ and $u_\theta$ are the amplitude functions for the components of the gaseous velocity. Let us also note that, irrespective of pressure, the following ratios hold:

$$\frac{\delta_g}{\sigma_0} : \frac{u_\varpi}{\varpi\Omega} : \frac{u_\theta}{\varpi\Omega} = -k\varpi : m\left(\frac{\Omega_p}{\Omega} - 1\right) : i \frac{\kappa^2}{2\Omega^2} , \tag{2.18}$$

where $\Omega_p$ is the pattern speed. This follows from the fact that the pressure gradient plays no role either in the equation of continuity or in the $\theta$-component of the equation of motion (in the case of tightly wound spirals).

## III. THE DISPERSION RELATION FOR DENSITY WAVES

The central relationship for any type of wave motion is that between wavelength and frequency. In the theory of self-sustained density waves, one must equate the density required to sustain the gravitational potential (eq. [2.1]) according to Poisson's equation, i.e.,

$$\delta_p = -|k| A/2\pi G , \tag{3.1}$$

with the *sum* of the responses in density variation in the stellar and gaseous disks. Thus, by using equations (2.13a) and (2.17a), we get the *dispersion relationship*,

$$\frac{k_*}{|k|} (1 - \nu^2) = \mathfrak{F}_\nu(x) + \frac{\sigma_0}{\sigma_*} \mathfrak{F}_\nu^{(g)}(x_g) , \tag{3.2}$$

which connects $|k|$ and $\nu$ for a given galactic model. In the above equation, we define $k_*$ by

$$k_* = \kappa^2/2\pi G\sigma , \tag{3.3}$$

and we must have

$$1 - \nu^2 > 0 . \tag{3.4}$$

The limiting cases $\nu = \pm 1$ correspond to Lindblad resonance $\Omega_p = \Omega \pm \kappa/m$; they deserve special treatment (see item *b* in § VIII).

The waves described by the dispersion relationship (eq. [3.2]) have the following properties:

*a*) They extend essentially over a range of the galactic disk for which the conditions

$$\Omega - \frac{\kappa}{m} < \Omega_p < \Omega + \frac{\kappa}{m} \tag{3.5}$$

are satisfied, where $\Omega(\varpi)$ is the angular speed of the stars, and $\kappa(\varpi)$ is the epicyclic frequency, as defined before. We shall refer to this range as the principal part of the spiral pattern. In Figure 1, we show the graphs of $\Omega(\varpi)$, $\kappa(\varpi)$, and $\Omega \pm \kappa/2$ on the basis of the model given by Schmidt (1965); the relevant numerical values may be found in Table 1. According to this diagram, for $m = 2$, the pattern would extend from $\varpi = 4$ kpc outward beyond 20 kpc, if $\Omega_p = 11$ km sec$^{-1}$ kpc$^{-1}$. For other values of $m > 2$ this particular part would be quite limited in extent, whatever the value of $\Omega_p$ may be. This is the reason why *all spirals have two arms* (or only one arm), for the rotation curves are qualitatively the same. Note that the argument no longer holds for the outer part of a galaxy, leaving the possibility open for the existence of "multiple-armed" galaxies such as M101, which has still a well-defined two-armed structure in the interior parts. The possibility that one-armed galaxies exist cannot be ruled out by the present discussion, but it requires a more complicated study which will be presented later.

*b*) We notice that the dispersion relation for such waves is expressed in terms of the wavelength $\lambda = 2\pi/|k|$ and the frequency $\nu = m(\Omega - \Omega_p)/\kappa$ at which the stars see the gravitational field. This relationship includes two additional parameters, $x$ and $x_g$, which measure the velocity dispersion of the stars and the turbulence in the gas. As a first attempt to appreciate the significance of this dispersion relationship, it is convenient to work out explicitly the results in the case of a pure stellar disk. In Figure 2, we show the result for the case where the velocity dispersion is barely enough to stabilize the disk against gravitational collapse (Toomre 1964). The margin of instability is obtained by setting $\nu^2 = 0$ in equation (3.2); conditions with $\nu^2 < 0$ correspond to instability, while

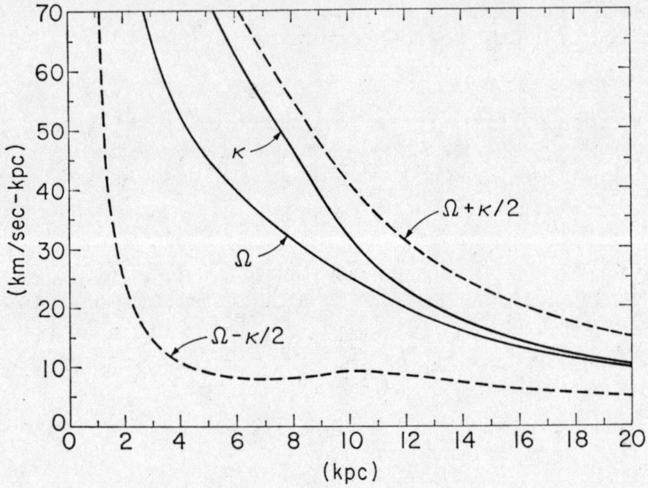


FIG. 1.—Rotation curve, etc., of our own Galaxy according to Schmidt model (symbols defined in Fig. 2).

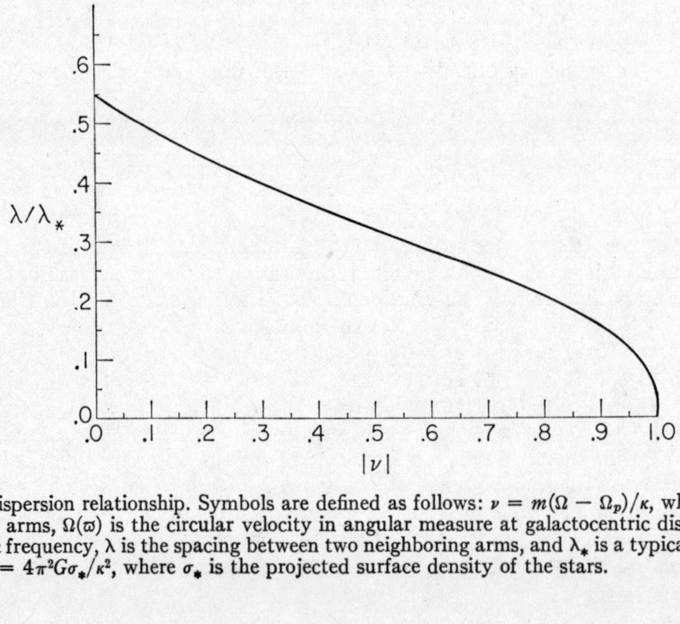


FIG. 2.—Dispersion relationship. Symbols are defined as follows: $\nu = m(\Omega - \Omega_p)/\kappa$, where $m = 2$ is the number of arms, $\Omega(\varpi)$ is the circular velocity in angular measure at galactocentric distance $\varpi$, $\kappa(\varpi)$ is the epicyclic frequency, $\lambda$ is the spacing between two neighboring arms, and $\lambda_*$ is a typical length scale defined by $\lambda_* = 4\pi^2 G\sigma_*/\kappa^2$, where $\sigma_*$ is the projected surface density of the stars.

TABLE 1

1965 SCHMIDT MODEL OF THE GALAXY

A. ROTATION PARAMETERS

| $\varpi$ (kpc) | $\varpi\Omega$ (km sec$^{-1}$) | $\Omega$ (km sec$^{-1}$ kpc$^{-1}$) | $\kappa$ (km sec$^{-1}$ kpc$^{-1}$) | $\Omega-\kappa/2$ (km sec$^{-1}$ kpc$^{-1}$) | $\Omega+\kappa/2$ (km sec$^{-1}$ kpc$^{-1}$) |
|---|---|---|---|---|---|
| 1 | 200 | 200.0 | 245.0 | 77.6 | 323.0 |
| 2 | 187 | 93.0 | 136.0 | 25.4 | 161.0 |
| 3 | 198 | 66.0 | 103.0 | 14.5 | 118.0 |
| 4 | 213 | 53.2 | 85.0 | 10.7 | 95.7 |
| 5 | 227 | 45.3 | 72.6 | 9.0 | 81.6 |
| 6 | 238 | 39.7 | 62.8 | 8.3 | 71.1 |
| 7 | 247 | 35.2 | 54.4 | 8.0 | 62.4 |
| 8 | 252 | 31.4 | 46.7 | 8.1 | 54.8 |
| 9 | 253 | 28.1 | 39.2 | 8.5 | 47.7 |
| 10 | 250 | 25.0 | 31.6 | 9.2 | 40.8 |
| 11 | 244 | 22.2 | 26.5 | 8.9 | 35.4 |
| 12 | 238 | 19.8 | 22.8 | 8.4 | 31.2 |
| 13 | 231 | 17.8 | 20.0 | 7.8 | 27.7 |
| 14 | 224 | 16.0 | 17.7 | 7.2 | 24.9 |
| 15 | 218 | 14.6 | 15.9 | 6.6 | 22.5 |
| 16 | 213 | 13.3 | 14.3 | 6.1 | 20.4 |
| 17 | 207 | 12.2 | 13.0 | 5.7 | 18.7 |
| 18 | 202 | 11.2 | 11.9 | 5.3 | 17.2 |
| 19 | 197 | 10.4 | 10.9 | 4.9 | 15.9 |
| 20 | 193 | 9.6 | 10.1 | 4.6 | 14.7 |

B. STRUCTURE PARAMETERS

| $\varpi$ (kpc) | $\sigma(\varpi)$ ($M_\odot$ pc$^{-2}$) | $M(\varpi)$ ($10^9 M_\odot$) | $\lambda_{tot} = 4\pi^2(\sigma_* + \sigma_0)/\kappa^2$ (kpc) |
|---|---|---|---|
| 1 | 1097.0 | 11 | 3.1 |
| 2 | 817.0 | 20 | 7.5 |
| 3 | 646.0 | 31 | 10.4 |
| 4 | 521.0 | 44 | 12.3 |
| 5 | 421.0 | 57 | 13.6 |
| 6 | 338.0 | 70 | 14.5 |
| 7 | 267.0 | 82 | 15.3 |
| 8 | 206.0 | 93 | 16.0 |
| 9 | 155.0 | 103 | 17.1 |
| 10 | 114.0 | 111 | 19.3 |
| 11 | 86.0 | 117 | 20.7 |
| 12 | 65.9 | 123 | 21.5 |
| 13 | 51.8 | 127 | 22.1 |
| 14 | 41.5 | 131 | 22.5 |
| 15 | 33.7 | 135 | 22.8 |
| 16 | 27.8 | 138 | 23.0 |
| 17 | 23.2 | 140 | 23.2 |
| 18 | 19.5 | 143 | 23.4 |
| 19 | 16.6 | 145 | 23.5 |
| 20 | 14.2 | 147 | 23.7 |

those with $\nu^2 > 0$ correspond to neutral waves. The condition $\nu^2 = 0$ leads to a minimum dispersion velocity $\langle c_\varpi^2 \rangle$ given by (see Toomre 1964)[8]

$$\frac{k_*^2 \langle c_\varpi^2 \rangle}{\kappa^2} = 0.2857 \cong \tfrac{2}{7}\,, \tag{3.6}$$

below which there would be gravitational collapse. However, the corresponding wavelengths of the unstable waves, if $\langle c_\varpi^2 \rangle$ is slightly lowered from these critical values, are usually so large that their significance in the spiral theory is not clear. Thus, one might be inclined to think that even when $\langle c_\varpi^2 \rangle$ is not as large as that given by equation (3.6), the dispersion speeds may not increase any further.

As noted above, these calculations are based on an infinitesimally thin stellar sheet. The effect of a fair amount of gas and the modification due to finite thickness have been studied in detail by Shu (1968) and found to be not critical for the particular relationship in Figure 2. Some further discussion of these points will be given when our results are used in the following sections.

*c*) To the approximation discussed here, there are no indications that either trailing waves or leading waves are preferred, one to the other. However, such indications (though not conclusive) are found in the next approximation: Trailing patterns are preferred over leading patterns, if the velocity dispersion of stars increases toward the center. Other effects, especially non-linear terms representing convection of matter, would also tend to prefer trailing patterns. We are now in a position to compare our theoretical results with observed features in our own Galaxy.

## IV. DISTRIBUTION OF ATOMIC HYDROGEN IN THE GALAXY

On the basis of the basic model shown in Figure 1, and the dispersion relationship shown in Figure 2, we can calculate a spiral pattern as soon as the pattern speed $\Omega_p$ is decided upon. A cursory examination of the numerical values obtained will show that $\Omega_p$ must be in the range of 11–13 km sec$^{-1}$ kpc$^{-1}$ to arrive at spacings between spiral arms that are reasonable compared with observations. One must also keep in mind that there is no reason why there should be only a single pattern associated with a single angular frequency $\Omega_p$. A group of waves should in general be present. For simplicity, however, we shall try our best to compare the observational data with the theoretical results for a *single* pattern. It is therefore too much to expect very close agreement; yet we shall see that the results are surprisingly good.

In Figure 3, we reproduce the spiral pattern presented by Lin and Shu (1967), based on a pattern speed

$$\Omega_p = 11 \text{ km sec}^{-1} \text{ kpc}^{-1} \,.$$

The related basic data used are given in Table 1. When this pattern is compared with the observed distribution of atomic hydrogen, it appears to be in good agreement, although an even better agreement can probably be obtained by taking $\Omega_p$ slightly higher. As it is, the spiral pattern is perhaps a little too tight, but it does show the correct number of arms from the solar vicinity inward, up to the "3-kpc arm." The latter is tentatively associated with the strong motions that might be expected at inner Lindblad resonance, where

$$\Omega_p = \Omega - \kappa/2 \,.$$

According to the theory, the pattern ends here as a ring; there is no prominent pattern inside this ring. Indeed, no strong indications of a spiral structure have been obtained from observations. A detailed study of the conditions near Lindblad resonance requires

[8] Toomre considered exponentially damped or growing rings, not neutral spirals; but the mathematical theory is related to that in the present paper (see Lin and Shu 1966).

an extensive reformulation of the theory for dealing with a stellar sheet. This has been carried out by Shu (1968).

The Perseus arm and the Sagittarius arm can be fitted into an over-all spiral pattern with a skeleton shown by Figure 3, if each arm is assigned a width equal to about one-half the spacing. The distribution of the young optical objects in these arms will be discussed in § VI; no serious problems arise. The inner arm (the one inside the Sagittarius arm) also fits well into the picture. The Orion arm, however, presents somewhat of a problem. It is known to be inclined at about 30°, from observations of neutral hydrogen, optical objects, and magnetic field. If an arm with such an inclination were to be fitted into a

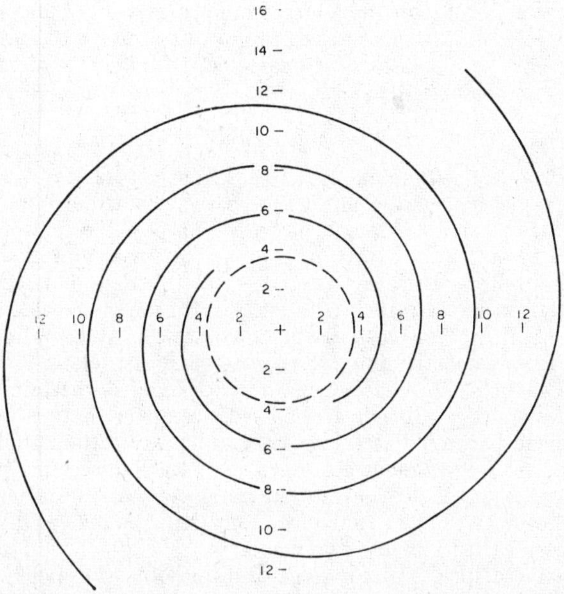


FIG. 3.—Spiral patterns at $\Omega_p = 11$ km sec$^{-1}$ kpc$^{-1}$

two-armed pattern in our vicinity, the interarm spacing would be about 18 kpc, since the pitch angle $i$ is related to this spacing $\lambda$ by the simple relationship $\lambda = \pi\varpi_0 \tan i$, where $\varpi_0$ is the distance to the center of the Galaxy ($\varpi_0 = 10$ kpc in the present case). Such a spacing is clearly inadmissible, and we are therefore forced to conclude that the Orion arm (in which the Sun is located) is an interarm branch. Such branches, of course, occur quite frequently in the outer parts of disk-shaped spiral galaxies, and therefore we need not be unduly alarmed.

We shall leave the details of the choice of the pattern speed $\Omega_p$ and the initial point $(\varpi_0,\theta_0)$ in equation (2.2) to a separate discussion (Yuan 1969). Suffice it to mention here that the choices are limited to very narrow ranges.

*Dispersion Velocities of Stars, Finite Thickness of the Galactic Disk*

One subtle but rather crucial check of the general concepts is to examine the dispersion velocities of the stars predicted according to equation (3.6). In the above calculations, the disk is taken to be a pure stellar sheet of infinitesimal thickness. This leads to an rms value $\langle c_{\varpi}^2\rangle^{1/2}$ of the radial dispersion velocity of about 52 km sec$^{-1}$ for the solar vicinity, which is decidedly too high by at least 25 per cent. If the effect of the gas had been included, the discrepancy would have been larger. However, Shu (1968) has shown that all the discrepancies disappear when we take into consideration the effect of the finite thickness of the disks of stars and of gas. It is found that the pattern speed should also be increased slightly, to about 13.5 km sec$^{-1}$ kpc$^{-1}$. (This value will be used in § VII.) Shu's calculations show that the rms value of the radial dispersion velocity should be about 37 km sec$^{-1}$ in our vicinity (but presumably somewhat lower), if it is barely sufficient to sustain gravitational collapse. This value is in reasonable agreement with observations.

In the same study, Shu found the relative importance of gas and of stars to be about comparable. Here it is important to keep in mind that the gaseous disk is much thinner than the stellar disk. Consequently, the projected mass density of gas is not as large as indicated by the local volume density.

## V. SYSTEMATIC MOTION[9] OF THE GAS

The possible existence of systematic motions was suspected quite early. Kerr (1962) suggested it as a possible source of difference between northern and southern observations. More recently, Burton (1966) considered the possible existence of a systematic motion of the gas near the outer edge of the Sagittarius arm, and concluded that "the observations are consistent with a situation in which the high-velocity stream is hydrogen flowing along the spiral arm, parallel to its axis." But he also suggested other possibilities. Shane and Bieger-Smith (1966) also discussed the role of systematic motions as part of their extensive effort in constructing alternative galactic models.

None of these authors, however, stressed the dynamical mechanism by which these systematical motions are produced. To be sure, whenever there is a circular (or nearly circular) arm of concentrated matter, its associated gravitational field would produce a systematic motion of the gas along its edges. It is easy to see this from considerations of angular momentum; indeed, the motion must be *with* the general rotation on the *outside* edge, and *against* the general rotation on the inside edge. But the density-wave theory goes deeper in the following two respects:

*a*) Since the pattern speed must in general differ from the speed of the material objects, which are in differential rotation, there *must* be systematic motion of the gas to maintain *conservation of matter*.

*b*) The existence of the gaseous arm clearly *indicates* the presence of a minimum in gravitational potential, but the gravitational potential is expected to be *jointly* maintained by the gas and by the stars with dispersion speeds comparable to the turbulent speeds in the gas. In many situations, *the stellar component would be dominant in supplying the gravitational field*. Thus, there is no need to be concerned about having sufficient contrast in gaseous density in order to *maintain* the field.

The systematic motion of the gas is estimated to have velocity components of the order of 10 km sec$^{-1}$. The radial component[10] cannot be easily singled out from observational data, but its magnitude is certainly compatible with the range of theoretically predicted values. The component in the circular direction would cause variations in the observed rotation curve. These variations have long been observed, but they were thought to be possibly the consequence of missing gas over interarm regions. A detailed study by Yuan (1969) has conclusively shown that the latter effect does not give significant contributions to the variation in velocity.

[9] By systematic motion, we mean the bulk motion of the gas *in addition to* the circular motion of the basic galactic model.

[10] For convenience of comparison with observational features, let it be noted that this component is directed toward the galactic center when the spiral pattern is trailing and traveling at a pattern speed lower than the local circular velocity. Cf. Miller (1968) for relevant data in the Perseus arm.

A comparison of theory with observation (Kerr 1964) is shown in Figures 4, 5, *a*, and 5, *b*. In Figure 4, we show the location of the gaseous arms as deduced from the location of the maxima and minima in the rotation curves shown in Figure 5. The figure is in very good agreement with the location of atomic hydrogen determined from direct observations. In Figure 5, we show the theoretical curve (dashed), assuming a certain "true rotation curve" on each side. There is good agreement in major features. The sudden dip in the northern curve at about 53° in galactic longitude $l^{\rm II}$ may be traced to the intersection of the Orion arm with the Sagittarius arm. The northern and southern rotation curves are still different, suggesting an oval distortion of the Galaxy as a whole. Such a distortion was suggested by Kerr. For a uniformly rotating disk, Hunter's (1963) theoretical work also indicates that it is an important mode of instability; more recently, Kalnajs (private communication) showed that this mode was particularly difficult to

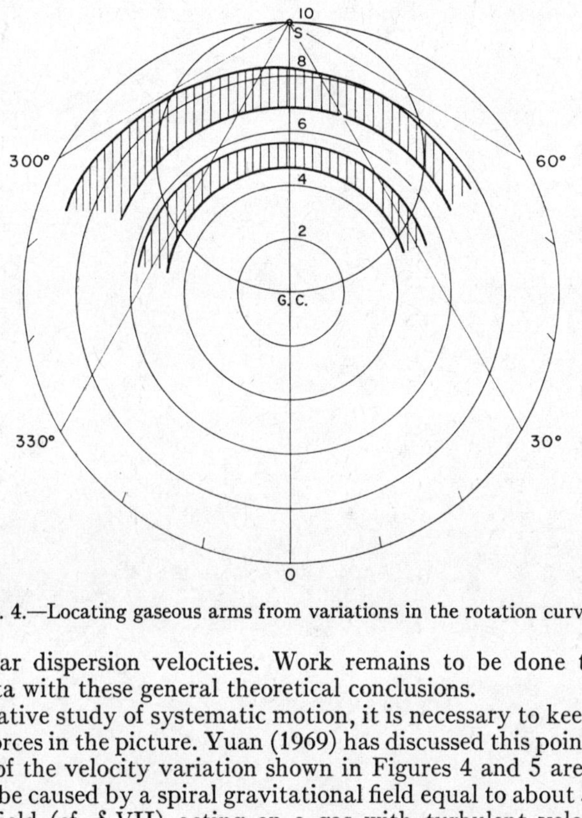


FIG. 4.—Locating gaseous arms from variations in the rotation curves

stabilize by stellar dispersion velocities. Work remains to be done to correlate the observational data with these general theoretical conclusions.

In the quantitative study of systematic motion, it is necessary to keep the role of the hydromagnetic forces in the picture. Yuan (1969) has discussed this point in great detail. The amplitudes of the velocity variation shown in Figures 4 and 5 are all about 8 km sec$^{-1}$. This could be caused by a spiral gravitational field equal to about 5 per cent of the axisymmetrical field (cf. § VII) acting on a gas with turbulent velocity equal to 7 km sec$^{-1}$ (rms value of one component) and with a magnetic field of about $5 \times 10^{-6}$ gauss. The amplitude of the total variation in mass is about 10 per cent of the mean, with gas and stars each contributing about half of the amount.

## VI. THE DISTRIBUTION OF YOUNG STARS AND OTHER OPTICAL OBJECTS

It is well known that young O and B stars are indicators of the spiral arm. It is also generally agreed that they are formed at places of the highest gaseous concentration.[11]

[11] Recent work by Roberts (1968) shows that the process of star formation is intimately related to the formation of shocks in the gaseous flow, but the discussions in this section are not altered (see § VIII).

In the density-wave theory, this has a very important implication: The newly formed stars must eventually migrate out of the arms, since the former revolve at the angular velocity of the material (say 25 km sec$^{-1}$ kpc$^{-1}$), while the latter move at the speed of the pattern (say 13 km sec$^{-1}$ kpc$^{-1}$). Over a period of 10 million years, there would be a separation of the stars from the gaseous arm by something like 1.2 kpc in the solar vicinity. However, since the inclination of the spiral arms is small, the *radial* separation is only about one-tenth of the above amount, and becomes difficult to detect. (For a more detailed discussion, see Lin 1967*b*.)

With the spiral pattern discussed in § IV, it is found that the young optical objects indeed lie within the spiral arms, with the notable exception of the Orion arm (already discussed in § IV). Although the optical objects in the Perseus arm seem to indicate a fairly large local inclination, their radial spread is not so large as to cause any difficulty in locating them within an arm of small over-all inclination. There seems to be, however, some difficulty with the optical objects in the Vela-Carina direction. There may very

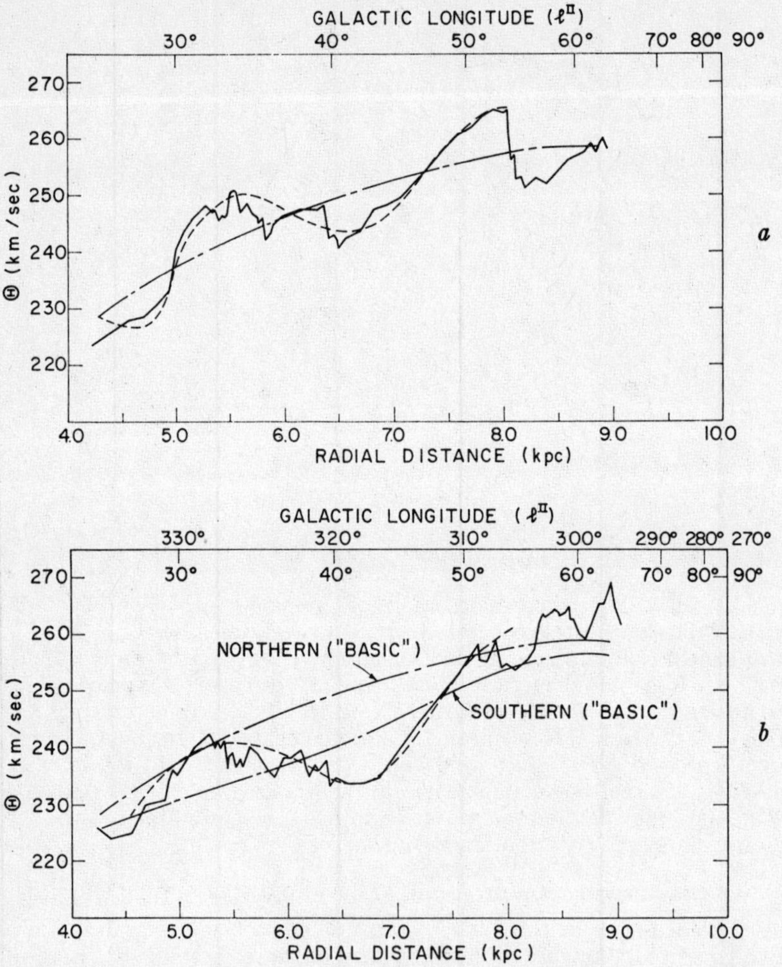


FIG. 5.—*a*, Northern Hemisphere. *b*, Southern Hemisphere. Rotation curves. Observation (Kerr 1964) ——; theory - - - -; mean motion, theory — - — - —.

well be some local irregularities special to this region. Further theoretical and observational studies are certainly much to be desired.

At this point, it should be emphasized again that the theory primarily deals with the over-all pattern, the grand design, rather than with the detailed structure. It should, however, be possible to introduce local effects, such as the radial expansion of stars from an O-association, to explain local irregularities.

A closer examination of the structure of the spiral arm should then reveal the following three "lanes": (1) a dust lane indicating the place of highest gas concentration, (2) a lane of brilliant newly formed young stars with large H II regions—the string of beads noted by Morgan and his collaborators (see Sharpless 1965)—and (3) a lane of dying O stars with much smaller H II regions.[12]

Since the dust is suspended in the gas, it is plausible that it should be the thickest where the gas density is the highest. This is known to be true on a large scale. As the dust particles pass through the spiral arm, they will evaporate under the effect of the radiation of the brilliant young stars. (On a smaller scale, there must also be very little dust in the immediate vicinity of bright stars.) One might therefore conceive of a process of dust formation after the gas (moving in a nearly circular orbit) has left the bright and hot part of the spiral arm. This process continues to produce more dust, so that the highest concentration of dust is reached just before the particles are again evaporated at the next arm. If it is assumed that the pattern speed is about one-half the circular speed of the material (as is expected to be approximately true in our vicinity), the time between destruction and recongregation into a dust lane is about 1 galactic year. This should be compared with a typical time scale for dust formation, and a figure of $2 \times 10^8$ years is by no means unreasonable.

## VII. THE MIGRATION OF MODERATELY YOUNG STARS

Ever since the method for the determination of stellar ages was greatly improved by Strömgren and his collaborators,[13] it has become meaningful to look for the birthplaces of the moderately young stars by following the history of their migration. It is expected that these stars would all be found to be formed inside spiral arms. For these calculations, one has, of course, to know the present location of the stars and to obtain their velocities from observations of their proper motions. All these are now obtainable with sufficient accuracy. Indeed, Contopoulos and Strömgren (1965) constructed tables for the easy calculation of such orbits, and had some success with the program. Some preliminary results were reported by Strömgren (1967), and it was concluded that these are generally compatible with the density-wave theory.

An important factor not included in this earlier analysis was the effect of the spiral gravitational field. Preliminary explorations made by Lin with the help of S. C. Wang, showed that even a small spiral field (equal to 5 per cent of the symmetrical field, for example) could be quite significant. It is thus necessary first to determine the spiral gravitational field for our Galaxy by using the results of § V. By experimenting with different choices of pattern frequencies and field strengths, Yuan (1969) found that a good choice would be a pattern speed of about 13.5 km $sec^{-1}$ $kpc^{-1}$ and a field strength of about 5 per cent of the symmetrical field. The results of his analysis are shown in Figures 6 and 7. In Figure 6, the places of formation of twenty-five stars are shown by using the ages determined by Strömgren *et al.* (private communication). Clearly, they appear to lie on one spiral arm which is *leading* and does not fit into any known struc-

[12] Specifically, A. Sandage pointed out to Lin the possibility of making these detailed observations and analyses in connection with M33. He also mentioned the plans of B. Lynds in this work.

[13] See Crawford (1963, 1966); Strömgren (1963*a*, *b*, 1964, 1965, 1966*a*, *b*); Strömgren and Perry (1965); Kelsall and Strömgren (1966); Crawford and Barnes (1966); Crawford and Perry (1966); and Crawford and Strömgren (1966).

ture from radio observations. In Figure 7, the places of formation of the same twenty-five stars are shown when the analysis includes the field described above. It is seen that all of them now lie in spiral arms.

The age determination suffers from an uncertainty of 10–15 per cent. We have taken the ages recommended by Strömgren *et al.*, with a few exceptions, but all well within the error limits given. In these few cases, the age was chosen to minimize the components of the dispersion velocity "at birth." In all cases, each component of the dispersion velocity is within 25 km sec$^{-1}$. This is deliberately taken to be higher than the initial turbulent velocity of the gas in order to allow for the increase during the formation process and for the inaccuracies of the parameters involved in our calculations.

### VIII. CONCLUDING REMARKS

*a*) The density-wave theory has many attractive features. At the present time, there is as yet no definite criterion for the precise determination of the pattern speed $\Omega_p$. The observational check of the theory therefore lies in the comparison with a multitude of data. One set of data may be regarded as a basis for determining the pattern speed, while the others are then used as checks. Indeed, the value of $\Omega_p$ turns out to be *rather sharply*

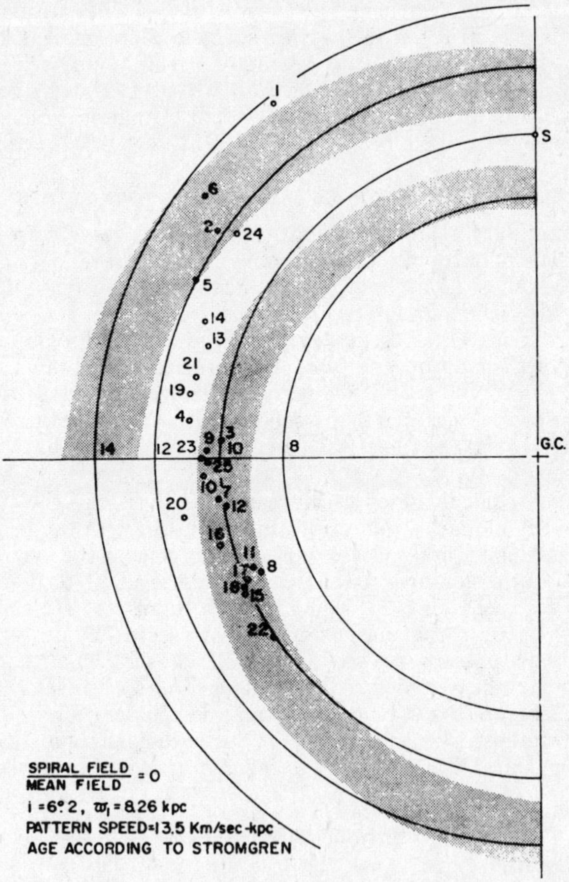


FIG. 6.—Places of formation of stars as determined without a spiral field

*defined* by the observational data. Other comparisons between theory and observations are also extremely satisfactory. For example, in the problem of the migration of moderately young stars, there are a great many factors that could cause the results to go astray, and yet we have obtained impressive agreement. Further continuation and extension of the observational program would be extremely desirable.

In general, it might be noted that the numerical values determined in this paper are subject to changes with the observational data and the theoretical models used, but we can have reasonable confidence in them, perhaps to within 20 per cent.

*b*) The range of values 11–13 km sec$^{-1}$ kpc$^{-1}$ obtained for the pattern speed $\Omega_p$ is also very satisfactory from the theoretical point of view. With reference to Figure 1, it is clear that Lindblad resonance is very nearly realized with $\Omega_p$ in this range, but *there is a definite deviation*. In the theory of Lindblad, co-operative effects were neglected; thus, he insisted on exact resonance, which can happen only if an extremely special rotation curve $\Omega(\varpi)$ holds—i.e., one for which $\Omega - \kappa/2 =$ const. We include the co-operative effects, and *the deviation is a measure of the scale of the spiral spacing* in the theory of the infinitesimally thin disk. A finite scale must be allowed; otherwise, the peculiar motions of the stars, both in the galactic plane and perpendicular to it, would

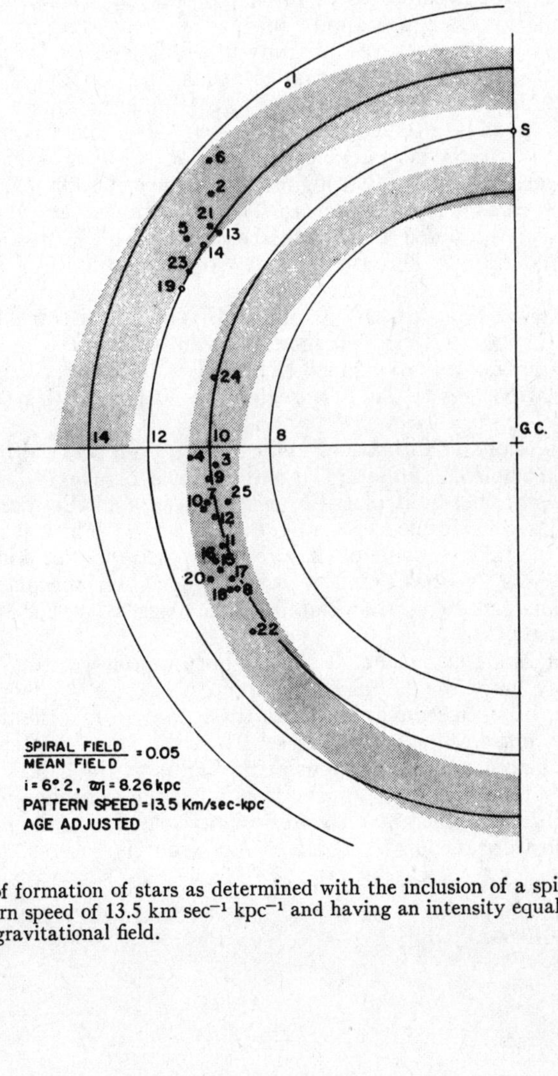


FIG. 7.—Places of formation of stars as determined with the inclusion of a spiral gravitational field travelling at a pattern speed of 13.5 km sec$^{-1}$ kpc$^{-1}$ and having an intensity equal to 5 per cent of that of the symmetrical gravitational field.

suppress the waves. Thus, by taking the pattern speed $\Omega_p$ in the range indicated by present investigations, we are "as close to resonance as possible," without giving up density waves altogether. Such is indeed the condition expected of any mechanical system.

*c*) The strength of the spiral field has now been estimated from the study of the migration of moderately young stars. It would be interesting to see how much contrast in gas density would ensue as a consequence. Calculations of this type have been made by Fujimoto (1966) and by Roberts (1969). Since the parameters chosen by Fujimoto do not correspond to those for a two-armed pattern discussed in this paper, only the results of Roberts are useful as a basis for checking the linear theory. A comparative study shows that estimates made on the basis of an extrapolation of the linear theory give generally satisfactory results.

*d*) The calculations of Fujimoto and of Roberts indicate the formation of a shock. This might indeed be an essential element in the process of star formation. One might imagine that the gas in turbulent motion has "clouds" before the shock, which are on the verge of gravitational collapse; the sudden compression would then trigger off the collapse of the clouds, which would lead to star formation. After the gas left the shock region, it would again be decompressed, and the process of star formation would cease. Superficially, such a process appears to be in such excellent agreement with our general concepts that it deserves further careful study in the immediate future.

An implication of this discussion is that the concentration of gas, though relatively high, need not be sufficient to imply extensive formation of new stars. The other prerequisite[14] of a shock must be fulfilled in many cases. This conclusion appears to be in general agreement with the abundance distribution of neutral hydrogen found by Roberts (1967) in certain galaxies. The highest abundance of hydrogen in certain galaxies is found in a ring *outside* the spiral structure. This is readily explained on the basis of the present theory, if it is supposed that shocks have not been able to form in these outer parts, because the conditions there (with too few stars and too much gas) are not favorable to the propagation of density waves.

Other corroborative evidence is the absence of H II regions within 4 kpc of the galactic center. Our theory suggests that this should be the case, since there cannot be any density waves inside the radius of inner Lindblad resonance (cf. Fig. 3). Indeed, the 109$\alpha$ radio observations reveal no H II regions in such localities (outside the dense nucleus).[15]

*e*) Once the process of star formation is clarified, one can proceed to study the evolution of a galaxy as a whole. Presumably, larger galaxies evolve faster, and the gas content in them would be more nearly depleted *by now* than in a smaller galaxy, if we assume that all the galaxies were formed around the same epoch. Thus, in a future epoch, a smaller galaxy might take on some of the present appearances of a larger galaxy, even though the two objects are *never* the same at any time. These speculations must be confirmed by careful analysis and substantiated by the observed stellar and gas contents in galaxies of various types.

*f*) The points of Lindblad resonance play important roles in the present theory. In the first place, they mark the extent of the spiral pattern in a theory involving stars. The scale of the wavelength reduces to zero at these points when the theory is developed on the basis of an infinitesimally thin disk. Thus, the spiral pattern is substantially "insulated" from the boundary conditions at the center of the galaxy. If the conditions are such that both the resonance points are absent, the existence of a *spiral* pattern becomes more doubtful. (This remark is presumably relevant to the anti-spiral theorem of Lynden-Bell and Ostriker (1967) in the case of a purely gaseous disk with pressure.

[14] Still another favorable condition is a high degree of turbulence, as will be discussed elsewhere.

[15] See Burke (1968) and the older work of Westerhout on background radiation, as quoted on page 196 in Kerr and Westerhout (1965).

In the gaseous case, the resonance condition described above does not exist; there is no case in which the wavelength reduces to zero.)

The behavior of stars near resonance has been studied thoroughly by Shu (1968). The behavior of gas has been studied by Roberts (1969), who found that there are "free modes" near resonance, i.e., there are solutions to equations of gas dynamics which correspond to non-circular motion even when the spiral component of the gravitational field is absent.

Clearly, the initiation of a spiral structure by mild instability of the type described in Paper II (caused by the gradient of dispersion velocities) is intimately related to the existence of a neutral spiral mode satisfying boundary conditions. Thus, the total absence of resonance points makes it unlikely for this instability mechanism to become operative. Consequently, conclusions obtained from the studies of a uniformly rotating disk need not apply to real, differentially rotating, disk-shaped galaxies.

During the course of this work, various discussions were held with numerous friends, too many to be acknowledged individually. We wish to register our gratitude collectively for their contributions to the theoretical ideas and to the proper interpretation of the observational data. We should, however, mention especially Professors B. Strömgren, L. Woltjer, and K. Prendergast for their invaluable help during the early and crucial phases of this work. Discussions with Professors J. H. Oort, Bart J. Bok, Bernard F. Burke, Frank Kerr, Morton S. Roberts, Allan R. Sandage, and Maarten Schmidt have given us valuable information on various observational facts. We have also benefited from discussions of the theoretical ideas with Professor D. Layzer at the early stages and from extensive discussions with Professor A. Toomre and Dr. A. Kalnajs over a period of the past few years.

This work is supported in part by grants from the National Science Foundation and from the National Aeronautics and Space Administration. Computations were carried out at the Computation Center of the Massachusetts Institute of Technology and at the Institute for Space Studies NASA in New York. We wish to express our thanks to Dr. Robert Jastrow for making the facilities available and to Dr. Kenneth Grossman for carrying out certain calculations related to the problem of star migration.

## APPENDIX A

### RESPONSE OF A STELLAR DISK TO A SPIRAL GRAVITATIONAL FIELD

We give here an outline of the calculation of the response of a stellar disk to a spiral gravitational field.

Let $t$ be the time, and $(\Pi, \Theta, Z)$ be the components of total stellar velocity in the cylindrical coordinates $(\varpi,\theta,z)$. We shall restrict ourselves to the two-dimensional case and introduce the peculiar velocity components $(c_\varpi,c_\theta)$ by the relations

$$c_\varpi = \Pi\,, \qquad c_\theta = \Theta - \varpi\Omega(\varpi)\,, \tag{A1}$$

where $\Omega(\varpi)$ is the angular velocity of circular motion defined by balancing the centrifugal acceleration $\varpi\Omega^2$ with the symmetrical gravitational field. The distribution function $\Psi(\varpi,\theta,c_\varpi,c_\theta t)$ then satisfies the differential equation

$$\begin{aligned} \frac{\partial\Psi}{\partial t} + c_\varpi\frac{\partial\Psi}{\partial\varpi} + \left(\Omega + \frac{c_\theta}{\varpi}\right)\frac{\partial\Psi}{\partial\theta} + \left(a_\varpi + \Omega\varpi^2 + 2\Omega c_\theta + \frac{c_\theta^2}{\varpi}\right)\frac{\partial\Psi}{\partial c_\varpi} \\ + \left(a_\theta - \frac{\kappa^2}{2\Omega}c_\varpi - \frac{c_\varpi c_\theta}{\varpi}\right)\frac{\partial\Psi}{\partial c_\theta} = 0\,, \end{aligned} \tag{A2}$$

where $\kappa(\varpi)$ is the epicyclic frequency defined by

$$\kappa^2 = (2\Omega)^2\left(1 + \frac{\varpi}{2\Omega}\frac{d\Omega}{d\varpi}\right), \tag{A3}$$

and $(a_\varpi,a_\theta)$ are the components of acceleration of the prevailing gravitational field. In conformity with our scheme (Lin and Shu 1966), we do not impose a condition of self-consistency at this point. Instead, we assume that the resultant field is of the form

$$(a_\varpi,a_\theta) = (\varpi\Omega^2,0) + (a_{\varpi 1},a_{\theta 1}) , \tag{A4}$$

where $(a_{\varpi 1},a_{\theta 1})$ are derivable from a spiral gravitational potential,

$$\mathcal{V}_1 = A(\varpi) \exp \{i[\omega t - m\theta + \Phi(\varpi)]\} ; \tag{A5}$$

that is,

$$a_{\varpi 1} = -\frac{\partial \mathcal{V}_1}{\partial \varpi}, \quad a_{\theta 1} = -\frac{1}{\varpi}\frac{\partial \mathcal{V}_1}{\partial \theta} . \tag{A6}$$

In equation (A5), both $A(\varpi)$ and $\Phi(\varpi)$ are real, and $\omega$ is a complex parameter (cf. eq. [2.1] in the text).

If $\Psi_0$ is the symmetrical distribution function corresponding to the case $A = 0$, and if we write

$$\Psi = \Psi_0(1 + \psi) = e^{-Q_0}(1 + \psi) , \tag{A7}$$

then $\Psi$ satisfies the linearized equation,

$$\begin{aligned}\frac{\partial \psi}{\partial t} + c_\varpi \frac{\partial \psi}{\partial \varpi} + \left(\Omega + \frac{c_\theta}{\varpi}\right)\frac{\partial \psi}{\partial \theta} + \left(2\Omega + \frac{c_\theta}{\varpi}\right)c_\theta \frac{\partial \psi}{\partial c_\varpi}\\ - \left(\frac{\kappa^2}{2\Omega} + \frac{c_\theta}{\varpi}\right)c_\varpi \frac{\partial \psi}{\partial c_\theta} = a_{\varpi 1}\frac{\partial Q_0}{\partial c_\varpi} + a_{\theta 1}\frac{\partial Q_0}{\partial c_\theta} .\end{aligned} \tag{A8}$$

With the field components (eq. [A6]), $\Psi$ is naturally expected to be of the form

$$\psi = \phi(\varpi, c_\varpi, c_\theta) \exp \{i[\omega t - m\theta + \Phi(\varpi)]\} . \tag{A9}$$

The amplitude function $\phi$ satisfies the equation

$$\begin{aligned}c_\varpi \frac{\partial \phi}{\partial \varpi} + \left(2\Omega + \frac{c_\theta}{\varpi}\right)c_\theta \frac{\partial \phi}{\partial c_\varpi} - \left(\frac{\kappa^2}{2\Omega} + \frac{c_\theta}{\varpi}\right)c_\varpi \frac{\partial \phi}{\partial c_\theta}\\ + i\left[k c_\varpi - m\frac{c_\theta}{\varpi} + (\omega - m\Omega)\right]\phi = g_\varpi \frac{\partial Q_0}{\partial c_\varpi} + g_\theta \frac{\partial Q_0}{\partial c_\theta} ,\end{aligned} \tag{A10}$$

where

$$k(\varpi) = \Phi'(\varpi) \tag{A11}$$

and

$$(a_{\varpi 1},a_{\theta 1}) = (g_\varpi, g_\theta) \exp \{i[\omega t - m\theta + \Phi(\varpi)]\} . \tag{A12}$$

*Asymptotic solution of equation (A10).*—If we are dealing with the case of tightly wound spirals, we may attempt an asymptotic series of the form

$$\varphi = \varphi^{(0)} + \epsilon\varphi^{(1)} + \ldots , \tag{A13}$$

where $\epsilon$ is a parameter of the order of $(k\varpi)^{-1}$. Consistent with this approximation, we shall first introduce dimensionless velocities referred to local values; thus, we shall write

$$c_\varpi = \xi V_1 = \xi(2\Omega/\kappa)(\varpi\Omega) , \tag{A14a}$$

$$c_\theta = \eta(\varpi\Omega) . \tag{A14b}$$

(The scales are suggested by our experience with the theory of epicycles.) It is then not difficult to verify that the equation for $\varphi^{(0)}$ is

$$\eta \frac{\partial \varphi^{(0)}}{\partial \xi} - \xi \frac{\partial \varphi^{(0)}}{\partial \eta} + i(\nu + a\xi)\varphi^{(0)} = - \frac{iaA}{V_1^2} \frac{\partial Q_0}{\partial \xi}, \tag{A15}$$

where $a$ and $\nu$ are defined by

$$a = (k\varpi)(2\Omega^2/\kappa^2), \tag{A15a}$$

and

$$\nu = (\omega - m\Omega)/\kappa. \tag{A15b}$$

Notice that a characteristic integral of equation (A15) is

$$\xi^2 + \eta^2 = C(\varpi), \tag{A16}$$

thanks to the choice of scales adopted. For our present purposes, we shall also take the basic distribution to be a function of $\xi^2 + \eta^2$ and $\varpi$; i.e.,

$$Q_0 = Q_0(\xi^2 + \eta^2, \varpi). \tag{A17}$$

Such a distribution function holds, when only stars of small dispersion speeds are considered. Since only a small fraction of stars have large dispersion speeds, and since they are not expected to be greatly influenced by the spiral gravitational field, equation (A17) is a good approximation for our present purposes. It is also compatible with the usually adopted Schwarzschild distribution. In that case, $Q_0$ is indeed of the form

$$Q_0 = \tfrac{1}{2}\left[\frac{c_\varpi^2}{\langle c_\varpi^2\rangle} + \frac{c_\theta^2}{\langle c_\theta^2\rangle}\right] + Q_{00}(\varpi), \tag{A18}$$

where $\langle c_\varpi^2\rangle$ and $\langle c_\theta^2\rangle$ are the mean-square values of the components of dispersion velocities for the location $\varpi$. For the moment, we shall consider general distribution functions of the form (A17). Note that $\varpi$ is a parameter in all these calculations.

To integrate equation (A15), we introduce the characteristic integral (A16) as one of the variables; i.e., we introduce the polar coordinate system $(\tau, s)$ in the $(\xi,\eta)$-plane:

$$\xi = \tau \cos s, \quad \eta = \tau \sin s. \tag{A19}$$

Then equation (A15) becomes

$$-\frac{d\varphi^{(0)}}{ds} + i(\nu + a\tau \cos s)\varphi^{(0)} = -\frac{2iaA}{V_1^2} Q'(\tau^2)\tau \cos s, \tag{A20}$$

where $Q_0'(\tau^2)$ denotes the derivative of $Q_0$ with respect to $\tau^2$, the dependence on $\varpi$ being understood. Both $\tau^2$ and $\varpi$ are, of course, to be treated as parameters for the purposes of the differential equation (A20). As we shall demonstrate below, the solution of this differential equation is

$$\varphi^{(0)} = -\frac{2A}{V_1^2} Q_0'(\tau^2)[1 - q(a\xi, a\eta, \nu)], \tag{A21}$$

where

$$q = \frac{\nu\pi}{\sin(\nu\pi)} \cdot \frac{1}{2\pi} \int_{-\pi}^{\pi} \exp\{i[\nu s - a\xi \sin s + a\eta(1 + \cos s)]\} ds. \tag{A22}$$

It is readily verified by direct substitution that equation (A22) is indeed a solution of the original equation (A15).

The derivation of equation (A21) from equation (A19) is also not difficult. The general solution of equation (A21), which involves one "constant" of integration depending both on $\tau$ (to be exhibited) and on $\varpi$ (not to be exhibited), is

$$\frac{V_1^2}{2AQ_0'(\tau^2)}\varphi^{(0)} = e^{iF(s,\tau)}\left[C(\tau) + ia\int_0^s e^{-iF(s',\tau)}\tau \cos s' ds'\right], \tag{A23}$$

where

$$F(s,\tau) = \nu s + a\tau \sin s . \tag{A24}$$

To determine $C(\tau)$, we apply the condition of periodicity, equating $\varphi_0(s)$ with $\varphi_0(s + 2\pi)$. We then consider the integral in $\varphi_0(s + 2\pi)$ in its alternative form given by the right-hand side of the identity,

$$\int_0^{s+2\pi} = \int_0^{2\pi} + \int_{2\pi}^{2\pi+s} . \tag{A25}$$

The change of variable $s' = 2\pi + s''$ in the last integral of equation (A25) reduces it to the form

$$\int_0^s e^{-iF(s''+2\pi,\tau)}\tau \cos s'' ds'' = e^{2\pi i\nu}\int_0^s e^{-F(s'',\tau)}\tau \cos s'' ds'' . \tag{A26}$$

In the form on the right-hand side of equation (A26), the integral is the same as that occurring in $\varphi_0(s)$. Some obvious calculations would then yield the desired determination of $C(\tau)$, and thence the result (A21).

*Some remarks on the solution (A21)*.—The solution (A21) may also be written in the following form. If we recall the definitions of $\psi$ and $\varphi$ (eqs. [A7] and [A9]), we see that the "response," in terms of the distribution function itself, is

$$\Psi_1 = \Psi_0\psi = \frac{2\mathcal{V}_1}{V_1^2}\Psi_0'(\tau^2)(1-q) , \tag{A27}$$

where $q(a\xi,a\eta,\nu)$ is defined by equation (A22), $\xi$, $\eta$ are defined by equation (A14), and $\tau^2 = \xi^2 + \eta^2$. The parameters $a$ and $\nu$ are defined by equations (A15a) and (A15b). For the Schwarzschild distribution, we have

$$\frac{-2\mathcal{V}_1}{V_1^2}\Psi_0'(\tau^2) = \frac{\mathcal{V}_1}{\langle c_\varpi^2\rangle}\Psi_0(\tau^2) . \tag{A28}$$

It is important to notice that the density response, obtained by integrating equation (A27) with respect to $c_\varpi$ and $c_\theta$, is always *real for real* $\omega$. To see this, we first note that $\Psi_0'(\tau^2)$ is even in $\eta$. The imaginary part of $q$ appears in terms of the two integrals,

$$\mathfrak{J}_1 = \int_{-\pi}^{\pi} \cos(\nu s - a\xi \sin s) \sin[a\eta(1 + \cos s)]ds$$

and

$$\mathfrak{J}_2 = \int_{-\pi}^{\pi} \sin(\nu s - a\xi \sin s) \cos[a\eta(1 + \cos s)]ds .$$

Now, $\mathfrak{J}_2$ is clearly zero, since the integrand is odd in $s$. The integral $\mathfrak{J}_1$ does not vanish, but it is clearly odd in $\eta$. Thus, when the integration in $\eta$ (i.e., in $c_\theta$) is performed, the net contribution vanishes. Indeed, all even moments of $c_\theta$, weighted with an arbitrary power of $c_\varpi$, vanish for the perturbation (A27).

## APPENDIX B

## CALCULATION OF RESPONSES IN DENSITY AND IN COMPONENTS OF MEAN STELLAR VELOCITY

We consider the case where the basic distribution function is the Schwarzschild distribution,

$$\Psi_0 = P_0(\varpi) \exp\left[-\frac{\mu_0}{2}(\xi^2+\eta^2)\right], \tag{B1}$$

where $\mu_0^{-1}$ is a measure of the mean-square value of the dispersion velocities

$$\mu_0^{-1} = \langle c_\varpi^2 \rangle / V_1^2, \qquad V_1 = (2\Omega/\kappa)(\varpi\Omega), \tag{B1a}$$

and $P_0$ is the normalization factor, so that

$$\sigma_{*0} = m_* \int\int \Psi_0 dc_\varpi dc_\theta . \tag{B2}$$

The response of distribution function to the spiral gravitational field is now of the form

$$\Psi_1 = \frac{-\mathcal{V}_1}{\langle c_\varpi^2 \rangle} \Psi_0(1-q), \tag{B3}$$

where $q$ is defined by equation (A21).

For the state $\Psi = \Psi_0 + \Psi_1$ we have

$$\sigma_*(\varpi,\theta,t) = m_* \int\int \Psi dc_\varpi dc_\theta , \tag{B4a}$$

and also the following relations for the mean stellar velocity components $v_\varpi$ and $v_\theta$:

$$\sigma_* v_\varpi = m_* \int\int \Psi c_\varpi dc_\varpi dc_\theta , \tag{B4b}$$

$$\sigma_* v_\theta = m_* \int\int \Psi c_\theta dc_\varpi dc_\theta . \tag{B4c}$$

It is easy to verify that, within the approximation of the linear theory,

$$\frac{\sigma_{*1}}{\sigma_{*0}} = \frac{\int\int \Psi_1 dc_\varpi dc_\theta}{\int\int \Psi_0 dc_\varpi dc_\theta} = \frac{-\mathcal{V}_1}{\langle c_\varpi^2 \rangle} \langle (1-q) \rangle , \tag{B5a}$$

$$\frac{v_\varpi}{V_1} = \frac{-\mathcal{V}_1}{\langle c_\varpi^2 \rangle} \langle \xi(1-q) \rangle , \tag{B5b}$$

$$\frac{v_\theta}{\varpi\Omega} = \frac{-\mathcal{V}_1}{\langle c_\varpi^2 \rangle} \langle \eta(1-q) \rangle , \tag{B5c}$$

where $\langle\ \rangle$ denotes the weighted average with respect to $\Psi_0$ or the function $\exp[-\frac{1}{2}\mu_0(\xi^2+\eta^2)]$.

The calculation of these averages depends on well-known relationships of the type

$$\langle e^{i\lambda\xi} \rangle = e^{-\lambda^2/2\mu_0} , \tag{B6a}$$

$$\langle \xi e^{i\lambda_1\xi} \rangle = \frac{i\lambda_1}{\mu_0} e^{-\lambda_1^2/2\mu_0} , \tag{B6b}$$

$$\langle \eta e^{i\lambda_2\eta} \rangle = \frac{i\lambda_2}{\mu_0} e^{-\lambda_2^2/2\mu_0} , \tag{B6c}$$

for Gaussian distribution functions. It is then easy to verify, by using equation (B5a), that

$$\langle (1-q) \rangle = 1 - \frac{\nu\pi}{\sin(\nu\pi)} \mathfrak{G}_\nu(x) , \tag{B7}$$

where $x = k^2\langle c_\varpi^2\rangle/\kappa^2$, and

$$\mathfrak{G}_\nu(x) = \frac{1}{2\pi}\int_{-\pi}^{\pi} e^{-x(1+\cos s)} \cos \nu s ds \, . \tag{B8}$$

The reduction factor $\mathfrak{F}_\nu(x)$ that occurs in equation (2.13a) in the text may then be introduced by defining it with the relation

$$\mathfrak{F}_\nu(x) = \frac{1-\nu^2}{x}\left[1 - \frac{\nu\pi}{\sin(\nu\pi)}\mathfrak{G}_\nu(x)\right], \tag{B9}$$

so that equation (2.13a) takes on a form directly comparable with the case of zero dispersion speeds.

TABLE 1A

NUMERICAL VALUES OF THE REDUCTION FACTOR $\mathfrak{F}(\nu,x)$

| $\nu$ | $x$ | | | | | | | | |
|---|---|---|---|---|---|---|---|---|---|
| | 0 | 1 | 2 | 3 | 4 | 5 | 6 | 7 | 8 |
| 0.0 | 1 | 0.5342 | 0.3457 | 0.2523 | 0.1982 | 0.1633 | 0.1389 | 0.1209 | 0.1068 |
| 0.1 | 1 | .5333 | .3447 | .2514 | .1974 | .1625 | .1381 | .1202 | .1066 |
| 0.2 | 1 | .5305 | .3416 | .2484 | .1946 | .1600 | .1359 | .1182 | .1046 |
| 0.3 | 1 | .5258 | .3363 | .2434 | .1901 | .1558 | .1321 | .1147 | .1013 |
| 0.4 | 1 | .5191 | .3286 | .2362 | .1835 | .1499 | .1267 | .1097 | .0966 |
| 0.5 | 1 | .5100 | .3185 | .2266 | .1748 | .1420 | .1195 | .1032 | .0907 |
| 0.6 | 1 | .4984 | .3056 | .2145 | .1638 | .1321 | .1105 | .0950 | .0833 |
| 0.7 | 1 | .4838 | .2894 | .1994 | .1503 | .1199 | .0995 | .0849 | .0739 |
| 0.8 | 1 | .4658 | .2695 | .1810 | .1337 | .1050 | .0861 | .0726 | .0629 |
| 0.9 | 1 | .4435 | .2452 | .1586 | .1136 | .0871 | .0700 | .0581 | .0494 |
| 1.0 | 1 | 0.4158 | 0.2153 | 0.1312 | 0.0894 | 0.0656 | 0.0507 | 0.0408 | 0.0341 |

The evaluation of the integral in equation (B5b) can be done with the help of equation (B6b), where $\lambda_1 = a \sin s$. We find that an integral of the form

$$\mathfrak{H}_\nu(x) = \frac{1}{2\pi}\int_{-\pi}^{\pi} e^{-x(1+\cos s)} \sin \nu s \sin s ds \tag{B10}$$

is involved. By partial integration, this can be transformed into

$$\mathfrak{H}_\nu(x) = \frac{\sin(\nu\pi)}{\pi x}\left[1 - \frac{\nu\pi}{\sin(\nu\pi)}\mathfrak{G}_\nu(x)\right]. \tag{B11}$$

Thus, the result can be best summarized by writing

$$\langle \xi(1-q)\rangle : \langle (1-q)\rangle = -\frac{\nu}{a} \tag{B12}$$

or

$$\frac{v_\varpi}{\varpi\Omega} : \frac{\sigma_{*1}}{\sigma_{*0}} = -\frac{\omega - m\Omega}{\Omega(k\varpi)} \, . \tag{B13}$$

It is worth noting that equation (B13) may be obtained directly by using the equation of continuity in the continuum form. The reduction factor $\mathfrak{F}_\nu^{(1)}(x)$ in equation (2.13b) is thus identical with $\mathfrak{F}_\nu(x)$:

$$\mathfrak{F}_\nu^{(1)}(x) = \mathfrak{F}_\nu(x) \, . \tag{B14}$$

The numerical values $\mathfrak{F}_\nu(x)$ are shown in Table 1A and Figure 8.

To evaluate equation (B4c), we follow a similar method by using equation (B6c) with $\lambda_2 = \alpha(1 + \cos s)$. The result can be best summarized by writing

$$\langle \eta(1 - q) \rangle : \langle \xi(1 - q) \rangle = -i\mathfrak{G}_\nu'(x) : \mathfrak{H}_\nu(x) \,, \tag{B15}$$

or

$$\frac{\langle c_\theta \rangle}{\varpi\Omega} : \frac{\langle c_\varpi \rangle}{(\varpi\Omega)(2\Omega/\kappa)} = -i\mathfrak{G}_\nu'(x) : \mathfrak{H}_\nu(x) \,. \tag{B16}$$

In the form used in equation (2.13c) in text, we then have

$$\mathfrak{F}_\nu^{(2)}(x) = \frac{(1 - \nu^2)(\nu\pi)}{\sin(\nu\pi)} [-\mathfrak{G}_\nu'(x)] \,, \tag{B17}$$

where $\mathfrak{G}_\nu'(x)$ is the derivative, with respect to its own argument $x$, of the function $\mathfrak{G}_\nu(x)$ defined by equation (B8).

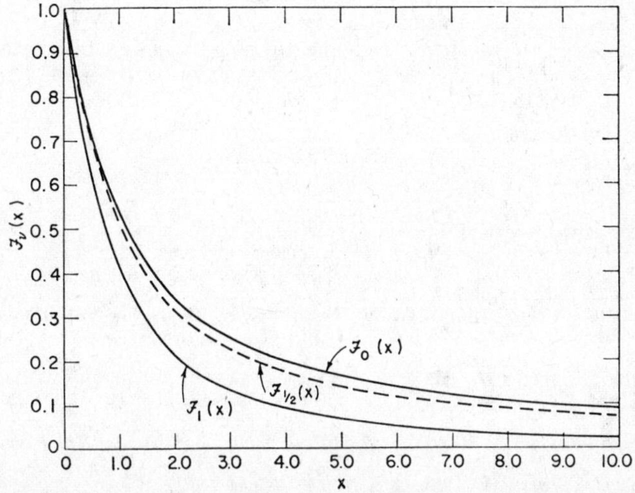


FIG. 8.—The reduction factor. Abscissa is a measure of the mean-square value of the radial component of the peculiar velocity of the stars.

## APPENDIX C

## AN ALTERNATIVE FORM OF THE DISPERSION RELATION

The dispersion relationship (3.2), in the case of a purely stellar disk, will now be presented in a form that reminds us of the similarity between our problem and that of waves in electromagnetic plasma. In this case, the dispersion relation (3.2) may also be written as

$$\frac{k_*}{|k|} = \frac{1}{x}\left[1 - \frac{\nu\pi}{\sin(\nu\pi)}\mathfrak{G}_\nu(x)\right] , \tag{C1}$$

where $\mathfrak{G}_\nu(x)$ is defined by equation (B8). Clearly, the function on the right-hand side has singularities (poles) at the points $\nu = \pm n$, where $n$ is a positive integer. Otherwise, the function is analytic in the complex variable $\nu$. Such meromorphic functions can be expanded into an infinite series of the principal parts at the poles, according to a well-known theorem of Mittag-Leffer. Clearly, the principal parts at $\nu = +n, -n$ are related, and it is therefore convenient to consider the series in the form

$$\frac{\nu\pi}{\sin(\nu\pi)} \mathfrak{G}_\nu(x) = \sum_{n=1}^{\infty} \frac{B_n(x)}{\nu^2 - n^2} . \tag{C2}$$

By taking the limit $\nu \to n$, we find that

$$B_n(x) = (-)^n n^2 I_n(x) , \tag{C3}$$

where $I_n(x)$ is the modified Bessel function of order $n$. Thus, equation (C1) may be written as

$$\frac{k_*}{k} = \frac{1}{x}\left[1 - \sum_{n=1}^{\infty} \frac{(-)^n I_n(x)}{(\nu/n)^2 - 1}\right] . \tag{C4}$$

The difference between this and the familiar formula for plasma physics (Bernstein 1958) lies essentially in (i) the difference in geometry and (ii) a reversal of sign caused by the difference between gravitational and electromagnetic forces.

The present form of the theory, as embodied in equation (3.2), is, however, much more perspicuous for the spiral problem, especially when there is also gas present. The concept of the reduction factor has been found to be particularly useful.

## REFERENCES


Becker, W. 1964, *Zs. f. Ap.*, **58**, 202.
Bernstein, I. B. 1958, *Phys. Rev.*, **109**, 10.
Bok, B. J. 1967, *Am. Scientist*, **55**, 375.
Bok, B. J., and Contopoulos, G. 1967, *Rept. I.A.U. Commission*, No. 33.
Burke, B. F. 1968, presented at the Charlottesville Conference on H II regions.
Burton, W. B. 1966, *B.A.N.*, **18**, 247–255.
Chandrasekhar, S. 1942, *Principles of Stellar Dynamics* (rev. ed. 1960; New York: Dover Publications).
Contopoulos, G., and Strömgren, B. 1965, *Tables of Plane Galactic Orbits* (New York: Institute for Space Studies).
Crawford, D. L. 1963, *Ap. J.*, **137**, 523.
———. 1966, in *Proc. I.A.U. Symp. No. 24*, p. 170 (*Contributions from the Kitt Peak National Observatory*, No. 110).
Crawford, D. L., and Barnes, J. V. 1966, *A.J.*, **71**, 610 (*Contributions from the Kitt Peak National Observatory*, No. 192).
Crawford, D. L., and Perry, C. L. 1966, *A.J.*, **71**, 206 (*Contributions from the Kitt Peak National Observatory*, No. 147).
Crawford, D. L., and Strömgren, B. 1966, in *Vistas in Astronomy*, ed. A. Beer, **8** (New York: Pergamon Press), 149–157 (*Contributions from the Kitt Peak National Observatory*, No. 82).
Fujimoto, M. 1966, in *Proc. I.A.U. Symp. No. 29* (to be published).
Hunter, C. 1963, *M.N.R.A.S.*, **126**, 299.
———. 1965, *ibid.*, **129**, 321.
Kalnajs, A. J. 1965, unpublished Ph.D. dissertation, Harvard University.
Kelsall, T., and Strömgren, B. 1966, in *Vistas in Astronomy*, ed. A. Beer, **8** (New York: Pergamon Press), 159–178.
Kerr, F. J. 1962, *M.N.R.A.S.*, **123**, 327.
———. 1964, *I.A.U. Symp. No. 20*, p. 81.
Kerr, F. J., and Westerhout, G. 1965, in *Galactic Structure*, ed. A. Blaauw and M. Schmidt (Chicago: University of Chicago Press), p. 167.
Lin, C. C. 1966, SIAM *J. Appl. Math.*, **14**, 876.
———. 1967*a*, *Ann. Rev. Astr. and Ap.*, **5**, 453.
———. 1967*b*, *Lectures in Applied Mathematics (1965)*, **2**: *Relativity Theory and Astrophysics* (New York: American Mathematical Society), p. 66.
———. 1968, in *Galaxies and the Universe*, ed. L. Woltjer (New York: Columbia University Press), p. 33.
Lin, C. C., and Shu, F. H. 1964, *Ap. J.*, **140**, 646.
———. 1966, in *Proc. Nat. Acad. Sci.*, **55**, 229.
———. 1967, in *Proc. I.A.U. Symp. No. 31, Noordwijk, 1966*, p. 313.
Lindblad, B. 1963, *Stockholm Obs. Ann.*, **22**, 3.
Lindblad, P. O. 1960, *Stockholm Obs. Ann.*, **21**, 3.

Lindblad, P. O., 1962, in *Interstellar Matter in Galaxies*, ed. L. Woltjer (New York: W. A. Benjamin), p. 222.
Lynden-Bell, D., and Ostriker, J. D. 1967, *M.N.R.A.S.*, **136**, 293.
Lynds, B. 1967*a*, *Sky and Tel.*, **33**, 343.
———. 1967*b*, *ibid.*, **34**, 18.
Miller, Joseph S. 1968, *Ap. J.*, **151**, 473.
Miller, R. H., and Prendergast, K. H. 1968, *Ap. J.*, **151**, 699.
Oort, J. H. 1962, in *Interstellar Matter in Galaxies*, ed. L. Woltjer (New York: W. A. Benjamin), p. 234.
Prendergast, K. H. 1967, in *Proc. I.A.U. Symp. No. 31, Noordwijk, 1966*, p. 303.
Roberts, M. S. 1967, presented at the Prague meeting of the I.A.U. (Commission 33).
Roberts, W. W. 1969, unpublished Ph.D. thesis, Massachusetts Institute of Technology.
Schmidt, M. 1965, in *Galactic Structure*, ed. A. Blaauw and M. Schmidt (Chicago: University of Chicago Press), p. 513.
Shane, W. W., and Bieger-Smith, G. P. 1966, *B.A.N.*, **18**, 263.
Sharpless, S. 1965, in *Galactic Structure*, ed. A. Blaauw and M. Schmidt (Chicago: University of Chicago Press), p. 131.
Shu, F. H. 1968, unpublished Ph.D. thesis, Harvard University.
Strömgren, B. 1963*a*, *Quart. J.R.A.S.*, **4**, 8.
———. 1963*b*, in *Basic Astronomical Data*, **3**, ed. K. Aa. Strand (Chicago: University of Chicago Press), p. 123.
———. 1964, *Astrophysica Norvegica*, **9**, 333 (*Contributions from the Kitt Peak National Observatory*, No. 59).
———. 1966*a*, in *Stellar Evolution*, ed. R. F. Stein and A. G. W. Cameron (New York: Plenum Press), p. 391.
———. 1966*b*, *Ann. Rev. Astro. and Ap.*, **4**, 433.
———. 1967, in *Proc. I.A.U. Symp. No. 31, Noordwijk, 1966*.
Strömgren, B., and Perry, C. 1965, *Photoelectric uvby Photometry for 1217 Stars Brighter than* $V = 6^{\mathrm{m}}.5$, *Mostly of Spectral Classes A, F, and G* (2d version).
Toomre, Alar. 1964, *Ap. J.*, **139**, 1217.
Yuan, Chi. 1969 (to be published).



## ERRATUM

Equations (C2), (C3), and (C4), in Appendix C, should be corrected to read as follows:

$$\frac{\nu\pi}{\sin(\nu\pi)}\,\mathfrak{G}_\nu(x) = 1 + \sum_{n=1}^{\infty} \frac{B_n(x)}{\nu^2 - n^2}\,. \tag{C2}$$

$$B_n(x) = 2n^2 I_n(x) e^{-x}\,, \tag{C3}$$

$$\frac{k_*}{k} = -\frac{2e^{-x}}{x} \sum_{1}^{\infty} \frac{I_n(x)}{(\nu/n)^2 - 1}\,. \tag{C4}$$

# 72. INTERPRETATION OF LARGE-SCALE SPIRAL STRUCTURE

C. C. LIN

*Massachusetts Institute of Technology, Cambridge, Mass., U.S.A.*

**Abstract.** The present paper consists of three parts: (1) A general explanation, from a semi-empirical point of view, of the density wave theory and its ramifications, with only a few remarks on those detailed features which have not been stressed before; (2) a statement of our conclusions about the Milky Way System; and (3) a discussion of the problem of the origin of density waves of spiral form.

## 1. Density Wave Theory of Galactic Spirals: A Semi-Empirical Approach

### A. EXISTENCE OF 'GRAND DESIGN'

During this conference, several speakers have pointed out that the spiral features in a galaxy are complicated, but there is also no doubt about the existence of a 'grand design', usually two-armed in the inner parts, but often multiple-armed in the outer parts. As properly stressed by Oort (1962) at the Princeton conference, the explanation of the existence of the grand design needed urgent consideration. At the present stage of the development of the theory, now that this grand design is reasonably well understood, it is perhaps more appropriate, if one looks into the future, to pay more attention to the more complicated spiral features on a smaller scale, for here we may learn more about the physical processes in the galaxy. Indeed, Oort has placed slightly greater emphasis on these features in his introductory lecture at this symposium (see p. 1). One might indeed paraphrase his present statement in terms of his 1961 words as follows: Although the general form of the large-scale phenomenon can be recognized in many nebulae, this structure is often hopelessly irregular and broken up.

For the past few years, Shu, Yuan and I myself have been trying to ascertain the grand design of our galaxy by calculating a spiral pattern from the density wave theory (Lin and Shu, 1964, 1966, 1967), in the spirit of the late B. Lindblad, and comparing the theoretical deductions with observational data (Lin *et al.*, 1969; Yuan, 1969) We began our studies by adopting a semi-empirical approach. We followed Oort in dividing the problem of spiral structure into two parts; (a) the persistence of the spiral structure in the presence of differential rotation (or the question of material arms versus density waves), and (b) the origin of this structure. This turns out to be a very fortunate decision, for problem (b) turns out to be a much more difficult theoretical problem than problem (a). Indeed, it is still largely unsolved (cf. Section 3). We shall therefore continue to stress this semi-empirical approach in the present paper.

### B. GENERAL FEATURES TO BE EXPLAINED

I list below ten general observational features which one must consider in dealing with spiral features in galaxies. Their explanation in terms of the density wave theory will be considered in the next section.

*Becker and Contopoulos (eds.), The Spiral Structure of Our Galaxy, 377–390.* 

(1) Existence of *grand design*
The spiral structure extends as a 'pattern' over the 'whole' galactic disk.

(2) Persistence
(a) The spacing between spiral arms, which is used to classify normal spirals into Sa, Sb, Sc types, is correlated with other physical characteristics of the galaxy.
(b) In particular, a smaller nucleus is associated with wider spacing.

(3) *Two*-armed, trailing
The spiral pattern is in general two-armed and, as far as is known, always trailing.

(4) *Multiple*-armed structure
Multiple-armed structure is often observed in the outer regions of many galaxies.

(5) *Ring* structure possible
A ring structure is often observed in the inner part of many galaxies. This structure is very clearly seen in NGC 5364.

(6) *Slender strings of beads*
The newly formed stars and young H II regions are arranged neatly like a slender string of beads. This indicates star formation in restricted regions, but simultaneously over a wide front. (It is perhaps this feature that led people to think of magnetic containment as a way to pinch the gas along narrow tubes.)

(7) *Dust lanes*
The *principal* dust lanes lie on the *inside* edge of the bright optical arm, although there are other features associated with dust.

(8) *Abundance distribution of* H II
Both the continuum survey of Westerhout and the recombination line observations of Burke, Mezger, and their collaborators Reifenstein and Wilson indicate a marked deficiency of H II region inside of the '3 kpc circle'. This may be compared with the absence of H II regions within the ring of NGC 5364.

(9) *Abundance distribution of* H I
M. Roberts has shown that the peak distribution of H I extends *beyond* the bright spiral structure in M 33 and other galaxies. This agrees in general trend with that known in our galaxy: the peak H I distribution lies outside of the H II peak (Westerhout).

(10) *Magnetic field*
The magnetic field is relatively weak and, in the case of the Orion arm, it lies roughly along the arm (Hiltner).

### C. QSSS HYPOTHESIS

It is clear that items (1) and (2) – the persistence feature mentioned above – would strongly suggest the density wave concept. Indeed, based on these features, we postulated the QSSS hypothesis (hypothesis of quasi-stationary spiral structure), which enabled us to explain a large body of observational data. It also raised a number of interesting questions on the dynamical mechanisms in stellar systems; e.g., the behaviors of material arms and of density waves, the origin of density waves, the uniqueness of the spiral pattern, and the behavior of stars, collectively and individually, near the points of co-rotation and Lindblad resonance*.

We can expect the existence of density waves also to have deep implications on the physical processes in the interstellar medium, and in particular on the formation of new stars. We can also expect the success of the theory, as applied to the Milky Way System, to embolden us to apply the theory to external galaxies. These are indeed fruitful avenues for future research work.

We therefore take the existence of density waves, and in particular, the QSSS hypothesis as our central theme. On the one hand, we examine its implications in our own galaxies and in other galaxies. On the other hand, we look for the mechanism for its origin, and use this as a starting point to examine many interesting behaviors of the stellar system.

One important theme to be kept in mind is COEXISTENCE. The complicated spiral structure of the galaxies indicates the coexistence of material arms and density waves, – and indeed of the possible coexistence of several wave patterns. When conflicting results appear to be suggested by observations, the truth might indeed lie in the co-existence of several patterns. Before taking this 'easy way out', one should of course try to examine each interpretation of the observational data as critically as possible.

There is also coexistence in the problem of origin of spiral structure. From our experience with plasma physics, we learned that there are many types of instabilities. Since a stellar system is basically a plasmoidal system, various types of instability can also occur in the problem of the galactic disk. The instability mechanism suggested in Section 3 is therefore only one of several possibilities. It has the advantage of being 'gross' or 'ponderous' instead of being 'delicate'. Such 'gross' instabilities, whenever they exist, usually occur prominently (as our experience in problems of hydrodynamic instabilities tells us). Another 'gross' mechanism for the origin of spiral structure is the existence of a short bar at the center, whose formation may or may not be related to the mechanism discussed in Section 3.

### D. DISCUSSION OF THE GENERAL FEATURES

To go beyond items (1) and the first part of (2), we need to work out the detailed dispersion relationship for density waves. A description of the explanation of the second part of item (2), items (3), (4), and (5) may be found in the paper by Lin *et al.* (1969). We note especially that the ring structure is associated with Lindblad resonance. Items (6), (7), (8), and (9) can be explained in terms of a sudden compression of the gas, as it passes through a density wave, as will be discussed in some detail by W. W. Roberts (this volume, p. 415; see also Roberts, 1969). Clearly, the role of magnetic field (item (10)) is secondary in our picture.

* When the stars meet the density wave at a frequency equal to its epicyclic frequency, there is a resonance behavior. This is referred to as Lindblad resonance (cf. the lecture of Contopoulos, this volume, p.303). There is of course also 'resonance' behavior when the stars move with the w ave. This is however called specifically co-rotation to avoid confusion.

The detailed discussion, as presented at the conference, will not be given here. I wish only to record the following remarks.

1. The existence of the primary dust lane on the *inner* side of the bright spiral arm is compatible with a *trailing* spiral pattern travelling at a pattern speed *lower* than the material speed (cf., Lynds, this volume, p. 26).

2. The separation between the dust lane and the bright spiral arm should be, according to the theory, a measure of the *difference* between the pattern speed and the material speed (cf., Roberts, this volume, p. 415). It should therefore be *larger in the interior part* of the galaxy (cf., Morgan, this volume, p. 9) where the angular velocity of the material motion is larger, the angular velocity of the pattern being independent of radial distance.

3. As we move away from the center of the galaxy, formation of new stars and hence the bright spiral pattern should terminate where the material speed and the pattern speed become nearly the same. This comment is very useful for determining the pattern speed in external galaxies.

4. External galaxies should be studied in great detail for confirmation of these ideas. In particular, the ring structure deserves special attention, since extraordinary behavior may be expected near resonance.

5. Secondary dust lanes across a spiral arm ('feathers') or on the outer edge of an arm are probably related to the visibility of the dust and other secondary dynamical behavior in an arm. Dust is probably present throughout an arm. These points deserve further attention, as they will cast light on the mechanisms occurring within a spiral arm.

## 2. Spiral Structure of the Galaxy

### A. A BRIEF DESCRIPTION OF THE SPIRAL STRUCTURE

Having stated my general position, I shall proceed directly to a description of our conclusions. In Figure 1, we show an old picture already published in the *Proceedings of the Noordwijk Symposium*. It shows a theoretical pattern of our galaxy that is in reasonable agreement with observations. This diagram gives the *skeleton* of the spiral structure calculated according to the density wave theory, as applied to a basic galactic model given by Schmidt (1965). The pattern is trailing, and rotating around the Galaxy at a pattern speed of 11 km $s^{-1}$ $kpc^{-1}$. Thus, the material motion is *faster* than the wave pattern. According to the calculations of W.W. Roberts (which are qualitatively in agreement with the earlier calculations of Fujimoto for other values of the parameters involved), this leads to a sharp rise of the gas density (on the *inside* of the optical spiral arm) followed by a sudden expansion (towards the *outside* of the spiral arm). We could infer from this that the dust lane would be expected to occur close to the inner edge of the bright optical arm.

The pattern speed of 11 km $s^{-1}$ $kpc^{-1}$ is now considered a lower limit; a more accurate value is perhaps 11.5 km $s^{-1}$ $kpc^{-1}$, if we continue to adopt a model of an infinitesimally thin disk. The detailed calculations were made by Yuan, and will be discussed by him in detail. I shall merely mention here that the spiral pattern derived by him from the observational data differs only slightly from the theoretical pattern. It must be recognized that the agreement should *not* be perfect, since the galactic disk is perhaps not perfectly circular and the actual structure may not be a pure mode in the theory.

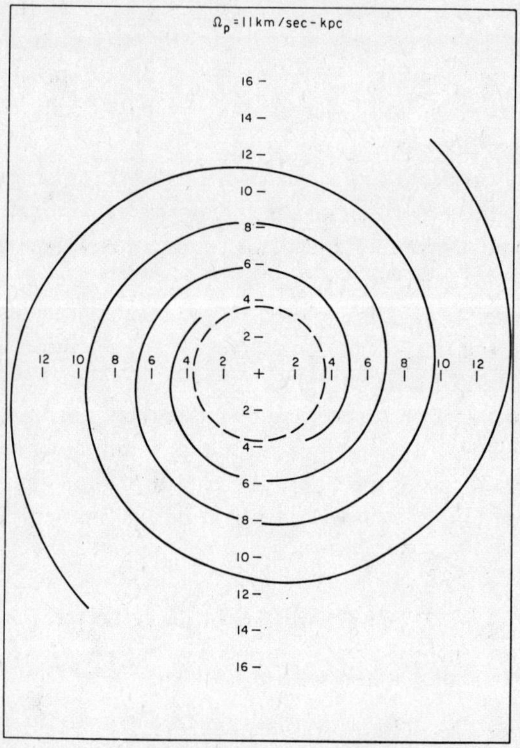


Fig. 1. Spiral pattern for 1965 Schmidt model.

When the thickness of the galactic disk is considered, Shu found that the pattern speed should be approximately 13.5 km $s^{-1}$ $kpc^{-1}$ in order to reproduce the observed spiral pattern. These calculations were made on the basis of his approximate evaluation of the thickness effect. At this symposium, Vandervoort (p. 341) reported further and more careful investigations of the effect of thickness of the galactic disk. The results agree very well with the earlier results obtained by Shu.

Other parameters that characterise the nature of the primary density wave are also shown in Table I. For a complete discussion of the manner in which these parameters are determined, the reader is referred to the paper by Lin *et al.* cited above.

The pattern speed we obtained is in sharp contrast with the value of 32 km $s^{-1}$ $kpc^{-1}$ obtained by Kalnajs. One can reconcile the two different results only if we accept the idea that there may be more than one mode present in the galaxy. The mode we obtained can be compared with a number of observational data, such as (a) the distribution of neutral hydrogen and its correlation with the oscillations in the rotation curve, (b) the distribution of young stars, (c) the origin of the local stars, and (d) the vertex deviation of the velocity distribution of the local stars. The discussion of items (a), (b) and (c) may be found in the paper by Lin *et al.*; item (d) will be discussed in a forthcoming paper by Strömgren, Lin and Yuan. One principal result is that the velocity distribution of the A stars, as observed by Eggen, can be very well reproduced by theoretical calculations based on the parameters shown in Table I.

TABLE I
Certain dynamical parameters

| | |
|---|---|
| 1. rms radial velocity predicted for stars in the solar neighborhood | $\leqslant 37$ km s$^{-1}$ |
| 2. Spiral pattern (primary component), pattern speed | 13.5 km s$^{-1}$ kpc$^{-1}$ |
| 3. In the solar vicinity | |
| (a) arm spacing (between Perseus and Sagittarius arms) | 3.5 kpc |
| (b) amplitude of spiral gravitational field | 5% of mean field |
| (c) amplitude of variation of projected mass density | 10% of mean |
| (i) in stars | 5% |
| (ii) in gas | 5% |
| (d) rms turbulent velocity of the gas (adopted) | 7 km s$^{-1}$ |
| (e) magnetic field (adopted for dynamical consistency) | 5 $\mu$G |

## B. SOME SPECIAL PROBLEMS

I wish now to call attention to three special problems which have been raised from time to time. The answers to these questions are not entirely clear, but I shall offer some opinions and comments.

(1) Do stars and gas move together?

(2) What is the nature of the local arms:
(i) the Carina arm, (ii) the Orion arm,
(iii) the Perseus arm, (iv) the Sagittarius arm?

(3) What are the 3 kpc arms?

(1) During this Symposium, Kerr has offered observational evidence that the H I and H II regions move at approximately the same velocity, within a scatter of about 10 km s$^{-1}$. According to the gravitational picture, the stars and gas should move together, in the mean, provided that they are located at the same point in space, and that their velocity dispersions (turbulence in one case, peculiar velocity of stars in the other) exert a similar influence. Neither condition is exactly satisfied when one observes in a given direction. Thus, a small amount of deviation does not give evidence to the existence of other types of forces in action.

(2) The local arms are not necessarily all of the same nature. This is a godsent

opportunity for us to understand the various physical processes, through theoretical considerations and the complementary use of a variety of observations. The Perseus and the Sagittarius arms fit into our large scale structure. The Orion arm is presumably a material arm (possibly formed out of material separated from the Sagittarius arm about 80–100 × $10^6$ yr ago). The Carina arm could be composed of young objects migrated out of the Sagittarius arm. It may also be a part of a trailing pattern travelling at about 15 km $s^{-1}$ $kpc^{-1}$. There is also the possibility that it might be a part of a trailing pattern, travelling at a speed *higher* than the local circular speed. Such an arm would have its dust lane on the outside of its bright objects.

(3) The '3-kpc arm' is a deep mystery, as discussed in some detail in the forthcoming paper by W. W. Roberts (1969). One more suggestion might be made here. If indeed the mechanism suggested in Section 3 is correct, then the '3-kpc arm' is a part of a reflected leading wave, of an evanescent type (i.e., one whose amplitude decreases exponentially in space as it leaves the ring of reflection). Such a wave carries no energy flux, the reflected energy being carried outwards by the long waves described in Section 3. This picture is necessarily tentative. A careful study of the reflection mechanism is therefore an urgent matter to prove or disprove the conjecture.

## 3. On the Origin and the Maintenance of Galactic Spirals*

### A. OVERVIEW

Having provided a possible mechanism for a grand design over the whole galactic disk, we must now investigate the origin and the long term maintenance of the density waves that underlie the galactic spiral structure. In his survey lecture at the beginning of this Symposium, Oort raised this question. The consideration of the propagation of a group of density waves (Toomre, 1969) brings the problem even into sharper focus. In the following discussions, we suggest a mechanism for the initiation of trailing spiral waves and for the maintenance of a two-armed spiral pattern in a quasi-stationary manner. In our picture, trailing spiral arms, co-rotating with the general mass motion, are produced by stretching of irregularities at the outer parts of a galaxy through differential rotation. Owing to resonance, the two-armed structure will prevail as the disturbances propagate inwards as a group of waves, which extracts energy from the basic rotation of the galaxy. The pattern speed expected from this mechanism in our Galaxy is thus around 11–13 km $s^{-1}$ $kpc^{-1}$, consistent with that determined from observational data (Lin *et al.*, 1969).

The reflection of the waves from the central region then stabilizes the wave pattern into a quasi-stationary form by transmitting the signal, via long-range forces, back to the outer regions where the waves originated. Thus, there is necessarily the co-existence of a very loose spiral structure and a tight spiral structure. Population I objects stand out sharply in the tight pattern while stars with large dispersive motion would primarily participate in the very loose pattern.

* The author is grateful to F. Shu and A. Toomre for several helpful discussions.

### B. A STRETCH OF A TRAILING SPIRAL ARM PRODUCED BY DIFFERENTIAL ROTATION

Since the galactic disk is in a state of strong differential rotation, any structural irregularity is likely to be stretched into a part of a trailing spiral arm, moving approximately at the local circular velocity, and inclined to the circular direction at various and varying angles. This material arm exerts its own induced effect and it may be expected that it eventually becomes a roughly self-sustained entity, somewhat like the self-sustained density waves (Lin and Shu, 1964, 1966, 1967) with inherent frequency $\nu=0$ (co-rotating waves). The process of such an evolution may be followed by the theoretical studies* of Goldreich and Lynden-Bell (1965) and of Julian and Toomre (1966).

Where would such structural irregularities most likely occur? The answer seems to be: in the outer regions of the galaxy. There is observational evidence for this statement: connecting links between major spiral arms are frequently observed in the outer parts of many galaxies. In our opinion, the Orion arm is such an interarm branch (Lin *et al.*, 1969). There is also a theoretical basis for this conjecture. The Jeans instability of the galactic disk (Toomre, 1964) tends to produce irregularities unless it is stabilized by the peculiar velocities of the stars and the turbulent motion of gas. Since the stellar content is smaller in the outer parts, and the turbulent motion of the gas may be dissipated, these outer parts of a galaxy are likely to be gravitationally unstable. Structural irregularities develop and are stretched out into short spiral arms, co-rotating with the prevailing local circular motion. Such a disturbance $q(\varpi, \theta, t)$, at a distance $\varpi=\varpi_0$ from the galactic center may be represented by a formula of the general form

$$q = F(t/\tau_0, \quad \theta - \Omega_0 t), \tag{1}$$

where $\Omega_0$ is the angular velocity at $\varpi_0$, and $\tau_0$ is a time scale much longer than $2\pi/\Omega_0$. We shall henceforth neglect the slow dependence on time indicated by the argument $t/\tau_0$.

### C. RESONANCE

The disturbance in the form of a piece of a co-rotating spiral arm would naturally exert its influence on other parts of the galaxy, and could induce density waves. But in general its effect would be expected to be limited, unless there is some form of *resonance*. A 'near-resonance' mechanism** becomes apparent when we consider the individual harmonic components of (1) and concentrate on the component with $2\theta$ dependence.

* The analysis made by these authors becomes inconclusive for application to waves on a finite disk when the angle of inclination becomes small; for the radial distance of the waves from the galactic center was effectively taken to be infinite. We therefore do not believe that the waves would eventually be wiped out by Landau damping, as indicated by Julian and Toomre.

** As pointed out by G. Contopoulos, the 'near resonance' considered here is, strictly speaking, still quite far away from the resonance condition defined by Equation (5). The term as used here merely indicates that the waves can propagate over a *substantial* part of the disk where $|2(\Omega - \Omega_0)| < \kappa$, which defines the "principal part of the spiral pattern" (Lin and Shu, 1966).

Let us write

$$q = \sum_{m=0}^{\infty} q_m \, e^{im(\theta - \Omega_0 t)} \tag{2}$$

and consider the effect of the individual components. The stars at a galacto-concentric distance $\varpi$ will feel a gravitational field at an angular frequency

$$f = m\left[\Omega(\varpi) - \Omega_0\right] \tag{3}$$

If this angular frequency $f$ is nearly equal to the epicyclic frequency $\kappa(\varpi)$, then there is near-resonance, and we may expect the disturbance (2) to have a large effect. The condition of resonance is

$$\kappa(\varpi) = m\left(\Omega(\varpi) - \Omega_0\right) \tag{4}$$

or

$$\Omega_0 = \Omega(\varpi) - \kappa(\varpi)/m\,. \tag{5}$$

Clearly, this can be satisfied only for one particular value of $\varpi$. However, near-resonance may be expected for a wide range of values of $\varpi$ if the right-hand side of (5) is nearly constant. In the outer parts of a galaxy, the variations in $\Omega(\varpi)$ and $\kappa(\varpi)$ are small. Thus, the required condition may be nearly satisfied for a substantial range of the radial distance $\varpi$ for a few values of $m$. In the inner parts of a galaxy, the required condition can be satisfied only for $m=2$. (In the case of uniform rotation, $\Omega-\kappa/2$ is exactly equal to zero.) This is a fact noted by B. Lindblad, and it has played an important role in the theory of density waves.

It is thus possible to conclude that the component of (2) with $m=2$ could produce a substantial effect in the whole galaxy, provided it occurs at such a galacto-centric distance that $\Omega_0$ is nearly equal to the nearly constant value of $\Omega(\varpi)-\kappa(\varpi)/2$. In our galaxy, this means that

$$\Omega_0 = 11 - 13 \text{ km s}^{-1} \text{ kpc}^{-1}\,. \tag{6}$$

and that the disturbances should originate around $\varpi_0 = 15$ kpc. This is indeed a region where the stars are less abundant, and hence the system is gravitationally unstable.

### D. PROPAGATION OF THE DISTURBANCE

There is therefore the possibility of a multiple-armed structure at the outer reaches of a galaxy, but all prominent spiral structures are expected to be two-armed. The spiral structure consists of a group of trailing waves with phase velocity (in angle)

$$\Omega_p = 11 - 13 \text{ km s}^{-1} \text{ kpc}^{-1}\,. \tag{7}$$

Let us now consider how these waves would behave as they propagate in the radial direction.

There are two possible groups of trailing waves (Toomre, 1969): in the range $\Omega > \Omega_p$, they are (i) the short waves with a negative group velocity, and (ii) the long waves with a positive group velocity. Clearly, only the *short trailing* waves will prevail*, if the disturbances are initiated in the manner described above.

It is of course too much to expect that these short trailing waves would naturally produce a quasi-stationary spiral pattern. In the first place, the energy supplied by the disturbances, randomly created at the outer regions, must be limited. Furthermore, it will disperse during propagation. Secondly, the 'slow' dependence on the time scale $\tau_0$ is bound to show up fairly soon, and a well-organized pattern is therefore unlikely to emerge. How can these two difficulties be met?

The first difficulty, we shall see, will be resolved by the feeding of energy from the differentially rotating galactic disk. The second difficulty will be resolved by a reflection of the wave from the central regions of the galaxy, either at a resonance ring or at the galactic center, where a more-or-less bar-shaped structure will be formed.

### E. ENERGY FEEDING DURING PROPAGATION

It is easy to imagine that a *differentially* rotating disk can exchange energy with waves propagating through it, but one must determine whether the waves gain or lose energy. Also, one must consider the associated re-adjustment of the distribution of the rotation of the galaxy; the time scale of this adjustment should be very long. This latter point is easy to settle. A rough estimate of this time scale is $\varepsilon^{-2}$ times a typical period of revolution, where $\varepsilon$ is the fractional variation of the various quantities in the disturbance. If we take $\varepsilon$ to be of the order of 10%, the time scale for this perennial adjustment would be of the order $10^{10}$ yr.

The feeding of energy can be calculated quantitatively. It is known that, in many cases of wave propagation, there is a principle of conservation of density of action (Whitham, 1965; Bretherton and Garrett, 1967). In the case of density waves under consideration, this principle has also been found to be true by Toomre (1969)** based on results obtained by Shu (1968).

Briefly, this principle may be stated as follows. If the energy density of the wave is denoted by $\mathscr{E}$, we define the density of action by

$$\mathscr{A} = \mathscr{E}/f\,, \tag{8}$$

where $f$ is the angular frequency seen by an observer co-moving with the general stream. It then follows that

$$\frac{\partial \mathscr{A}}{\partial t} + \nabla\cdot(\mathscr{A}\mathbf{c}_g) = 0\,, \tag{9}$$

where $\mathbf{c}_g$ is the group velocity.

* The waves outside the co-rotation point would also be short trailing waves, since they propagate outwards where $\Omega < \Omega_p$. Note that $\Omega(\varpi)$ decreases with increasing values of $\varpi$.

** Toomre's calculations were made for axisymmetric modes, for which the effect of change in frequency is not evident. The extension to non-axisymmetric modes, although natural to anticipate, does involve rather non-trivial calculations. This was completed by Shu. During this Symposium, Kalnajs pointed out that the energy density defined here is that associated with the rotating coordinate system. This does not alter the main line of reasoning here, especially that part related to the vanishing of the group velocity near resonance.

Following the motion of the group of waves, we then have, by (9),

$$\left(\frac{\mathrm{d}\mathscr{A}}{\mathrm{d}t}\right)_{\mathbf{c}_g} = \frac{\partial \mathscr{A}}{\partial t} + (\mathbf{c}_g \cdot \nabla)\mathscr{A} = \mathscr{A}(\nabla \cdot \mathbf{c}_g), \tag{10}$$

which may also be written as

$$\left\{\frac{\partial}{\partial t} + (\mathbf{c}_g \cdot \nabla)\right\} \log \mathscr{A} = -\nabla \cdot \mathbf{c}_g, \tag{11}$$

or

$$\left\{\frac{\partial}{\partial t} + (\mathbf{c}_g \cdot \nabla)\right\} \log \mathscr{E} = \left\{\frac{\partial}{\partial t} + (\mathbf{c}_g \cdot \nabla)\right\} \log f - \nabla \cdot \mathbf{c}_g. \tag{12}$$

If the right-hand side of the above equation is positive, we have an amplification of the wave group during propagation. If it is negative, we have an attenuation. In the present case, we shall find that the right-hand side is indeed *positive*, and we have consequently an amplification of the disturbance energy as the wave propagates inwards, when the waves are stationary ($\partial(\ \ )/\partial t = 0$). This is very easy to verify since

$$c_g < 0, \quad \text{and} \quad \frac{\mathrm{d}\Omega}{\mathrm{d}\varpi} < 0. \tag{13}$$

The form (9) yields a more vivid description of the mechanism. In the stationary case, we have

$$\varpi c_g \mathscr{A} = \text{constant}. \tag{14}$$

Near the point of origin, the angular frequency $f$ (Equation (3)) is very small. Consequently (Equation (8)), it requires only a negligible amount of *energy* to produce a substantial amount of *action*. As the group of waves propagates inwards, the angular frequency $f$ steadily increases, and the energy density $\mathscr{E}$ must therefore increase proportionally. In addition, there is further enhancement of $\mathscr{E}$ due to the decrease of $|c_g|$ and $\varpi$. Indeed, if there is a point of inner Lindblad resonance, $|c_g| \to 0$ there*, and the energy density becomes very large. The increase of energy density with decreasing $\varpi$ is of course to be expected from the cylindrical geometry of the galactic system.

We note however that the mechanism will not be operative unless there is a pre-existing mechanism for wave propagation. Furthermore, if the propagation mechanism is imperfect, i.e., if energy may be partly lost by dispersion, the effect of this gradient instability may be somewhat obscured. In the present case, we are fortunate to have a condition of near-resonance, which tends to insure a favorable propagation of the waves.

### F. REFLECTION OF WAVES NEAR THE CENTRAL REGION

From the discussions of the last section, there appear two possibilities for the action of the wave group to become infinite, if we insist on using Equation (14). Either we

* Indeed, it vanishes as $(1 - |\nu|)^{3/2}$.

have (i) $\varpi=0$, or we have (ii) $c_g=0$. The first condition refers to the galactic center, and the second condition refers to Lindblad resonance. These are the places where the propagation of the group of waves must undergo a change, and it is natural to think of them as places where the waves, having reached a very large (but not infinite) magnitude, are reflected. Let us examine the consequences of this reflection process, beginning with the galactic center.

As the group of waves approaches the galactic center, its greatly increased amplitude will be sufficient to cause the galactic nucleus to be slightly distorted* into a 'short bar'. This 'bar' configuration will be rotating at an angular velocity $\Omega_p$, appropriate to the forcing incoming wave. There is thus an associated gravitational field also rotating at the same angular frequency $\Omega_p$, propagating outwards. Its effect will be particularly strongly felt by the outer rings of the galaxy where the circular velocity is $\Omega_p$; i.e., where the waves originated. Thus, the cycle *does* complete itself and a stationary state may be maintained.

The completion of the cycle must be done with the right phase in time and right distribution in space. An indication for this possibility may be seen by examining the outward propagation. If we persist in applying the results obtained for the short waves to the relatively long waves, we find that the two branches coincide for the condition of co-rotation $\nu=0$. Thus, if the energy is carried out by the long trailing waves, it would be forcing the outer parts of the galactic disk with an appropriate 'mode' (i.e., an appropriate distribution in space and variation in time) provided one numerical value characterizing the phase is correctly specified. Since we are dealing with a group of waves whose 'wave length' is quite long, and not really well defined, this is not a critical requirement. At the same time, one cannot expect the re-enforcement to be perfect, but this is again not a critical matter. A *partial feedback* would be able to sustain a relatively stationary pattern, since there is a substantial gain of energy as the short waves move inwards, but not a corresponding loss of energy as the long waves, with length scale on the order of the galactic radius, propagate outwards.

### G. INNER RESONANCE RING

In cases where a sharp Lindblad resonance exists (e.g. in NGC 5364 or the Milky Way), the waves cannot penetrate into the center, but would be reflected already at the resonance ring. The violent response of the stars suggests that they are likely to group themselves into certain orbits, collectively showing an oval structure (cf., Contopoulos, this volume, p. 303), which takes the place of the bar in the above described reflection mechanism; but the general points made above still hold. The precise mechanism of reflection is obviously a very important and very interesting process deserving much future attention.

There is the distinct possibility that the waves would be *absorbed* at the inner resonance ring, and that there is no reflection of the group of waves. If this were to be found to be the case, there would be difficulty in accounting for the regular orga-

* In a pure disk model, the 'short bar' thus produced at the center would presumably be represented by a singularity. See Lin and Shu (1964), Rehm (1965), or as quoted by Lin and Shu (1967).

nization of a grand design. It would then be more likely that we would see a rather chaotic superposition of two-armed waves, unless certain nonlinear effects related to resonant stars (Contopoulos, 1970), would prevail.

### H. CONCLUSION

We have thus arrived at a picture involving the co-existence of a relatively tightly wound pattern of short waves and a rather loosely wound pattern of long waves. Clearly, objects with small peculiar velocities – particularly Population I objects – would participate in the tight spiral pattern. On the same length scale, the loosely wound pattern, with at most one winding for the whole galaxy would present itself as a distortion of the whole galaxy. It would be less visible to casual optical and radio observations. It could conceivably be mapped out via very careful analysis of optical and radio data.

All the above discussions are still fairly speculative. It would be extremely desirable to follow through the detailed development of these ideas with more careful mathematical analysis.

## 4. Concluding Remarks: A Bird's Eye View of Theoretical Developments

As mentioned before, we use the QSSS hypothesis as the focal point for our theoretical developments. On the side of basic mechanism, we raise various dynamical issues whose clarification may also help us to understand the behavior of electro-magnetic plasmas. For the study of our own galaxy, and perhaps other galaxies as well, there appear to be a better opportunity for the understanding of the physical processes, including such microscopic behavior as the formation of molecules and dust grains, induced by macroscopic compression via density waves. The effect of the density wave can also influence the cosmic ray particles via the galactic magnetic field. Here it is important to keep alive the general concept of density waves, and not to attach it *only* to a two-armed grand design.

We have also the opportunity to explore the external galaxies, since here certain types of observations can be made more easily*, and we have a larger sample of objects for our study. The relationship between normal spirals and barred spirals is a fascinating subject. It is perhaps sufficient, but not necessary, to have a short barred structure at the center of a normal galaxy, to initiate a two-armed grand design. The coexistence of several origins for spiral structure should be stressed. Simultaneously, there is the possibility of the coexistence of several kinds of spiral structures and their actual revelation as observable features. Indeed, it is the feeling of the present writer that relatively loose spiral patterns tend to be obtained by computational or analytical methods directed at the whole galactic disk. The asymptotic method yields relatively tight spirals when near resonance is approached, but it can also yield fairly loose spirals.

* One must beware of possibilities of misinterpretation, for we can only measure the velocity in the line of sight. Since the material in a galactic disk has two other components of motion besides that in the circular direction, inferences on the small deviations in circular motion must be drawn with care.

## Acknowledgement

The work reported here has been supported in part by grants from the National Science Foundation and the National Aeronautics and Space Administration.


## References


Bretherton, F. P. and Garrett, C. J. R.: 1967, *Proc. Roy. Soc. London* **A302**, 529.
Contopoulos, G.: 1970, *Astrophys. J.*, in press.
Goldreich, P. and Lynden-Bell, D.: 1965, *Monthly Notices Roy. Astron. Soc.* **130**, 97, 125.
Julian, W. H. and Toomre, A. 1966, *Astrophys. J.* **146**, 810.
Lin, C. C.: 1967, in *Relativity Theory and Astrophysics*, 2: *Galactic Structure*, Amer. Math. Soc., Providence, p. 66.
Lin, C. C. and Shu, F. H.: 1964, *Astrophys. J.* **140**, 646.
Lin, C. C. and Shu, F. H.: 1966, *Proc. Nat. Acad. Sci. U.S.A.* **55**, 229.
Lin, C. C. and Shu, F. H.: 1967, IAU-URSI Symposium No. 31, p. 313.
Lin, C. C., Yuan, C., and Shu, F. H.: 1969, *Astrophys. J.* **155**, 721.
Oort, J. H.: 1962, in *Interstellar Matter in Galaxies* (ed. by L. Woltjer), Benjamin, New York, p. 234.
Roberts, W. W.: 1969, *Astrophys. J.* **158**, 123.
Schmidt, M.: 1965, *Stars and Stellar Systems* **5**, 513.
Shu, F. H.: 1968, Ph.D. Thesis, Harvard University.
Toomre, A.: 1964, *Astrophys. J.* **139**, 1217.
Toomre, A.: 1969, *Astrophys. J.* **158**, 899.
Whitham, G. B.: 1965, *J. Fluid Mech.* **22**, 273.
Yuan, C.: 1969, *Astrophys. J.* **158**, 871, 889.

# THEORY OF SPIRAL STRUCTURE*

C. C. LIN
*Massachusetts Institute of Technology, Cambridge, Mass., U.S.A*

It is an honor for me to continue the discussion of galactic spirals following Professor Bok, who has contributed to the subject for many years. My experience has been relatively short. My first encounter with the study of galactic spirals occurred in 1961, when I was invited by Professor Bengt Strömgren to attend a conference on interstellar matter at Princeton, N.J. There I first learned about the winding dilemma from Professor Jan Oort. After the conference, Professor Lodewijk Woltjer, who edited the Proceedings, visited me for about a month. Thus, I began slowly to learn about spiral galaxies and to work on the subject. It turned out to be an extremely rewarding experience, for the observational data were already ripe for theoretical analysis, and the hydromagnetic theory of spiral arms was clearly encountering great difficulties. Since I was not educated as an astronomer, I owe my gratitude to Professors Strömgren, Woltjer, and Kevin Prendergast, on whom I depended for correct astronomical facts as I started my work. Without their help, I would not be standing here today. Later on, I was to receive help from many other distinguished astronomers, including Professor Bok from whom you just heard. I am greatly impressed with the community of astronomers as a dedicated group of scientists.

I wish to dedicate my talk this evening to the memory of a former President of the IAU – who was a good friend of Professor Strömgren's and of many of you present here – the late Bertil Lindblad. As is well known, he first suggested the concept of density waves as a basis for the spiral structure in disk-shaped galaxies. Over a number of years, he and his collaborators attempted to establish this concept by showing that stars in epicyclic motion *tend* to aggregate into a spiral gravitational well, which is then maintained in turn by the excess of stars gathered there. Unfortunately, the mathematical method he used did not enable him to calculate the *collective* behavior of the stars in a convenient manner, and he could not produce the necessary *quantitative* conclusions for comparison with observations in order to substantiate his reasoning. His ideas were therefore not widely accepted.

Modern computing machinery provides one approach to the quantitative treatment of stellar systems. This method was first adopted by Lindblad (1960, 1962) for the study of spiral structure. More recently, other investigators (Miller and Prendergast, 1968; Hohl and Hockney, 1969; Hohl, 1970) have carried out more extensive 'numerical experiments' along similar lines. The most extensive of these, just carried out by Miller *et al.* (1970) involves the consideration of both a gaseous component and a

* The preparation of this paper and much of the work reported here were done with the partial support of the National Aeronautics and Space Administration and the National Science Foundation.

*De Jager (ed.), Highlights of Astronomy, 88–121.* 

stellar component, the latter consisting of approximately $10^5$ stars. Their general conclusions support the concept of density waves. The numerical method has the advantage of providing at least a qualitative description of the process of evolution. However, it has not, as yet, yielded any quantitative results for specific comparison with observations. (A short section of their movie was shown and briefly discussed during the discourse.)

Quantitative results can be more readily obtained by analytical methods more suitable for the study of *collective modes*. In order to stay close to comparison with observations, Frank Shu and I formulated, a few years ago, (Lin and Shu, 1964, 1966, 1967) the *hypothesis of quasi-stationary spiral structure* (QSSS hypothesis); that is, we adopt as a working hypothesis the statement that a density wave pattern of spiral form, however it was originated, does exist in a galaxy, simply because it is directly observable. We then work in two different directions from this central position. On the one hand, we examine its consequences and compare them with observational data. On the other hand, we examine the basic dynamical mechanisms to see how such patterns can be initiated and maintained in an almost permanent manner.

The study of basic mechanisms turns out to be – as one would expect – close to the study of *inhomogeneous* electromagnetic plasmas, with magnetic field replaced by rotation. Various aspects of these problems have by now been studied by a number of investigators in the gravitational case. (See Section 3A for references.) Most of my presentation will be devoted to the discussion of the *consequences* of the QSSS hypothesis, but I shall also briefly refer to the problem of the origin and permanence of galactic spirals.

In order to maintain continuity with the lecture by Dr. Bok, let it be mentioned now that one of the first theoretical results obtained is the pattern of the Milky Way System presented by Frank Shu and myself at the *IAU Symposium No. 31* held at Noordwijk in 1966. This is the first theoretical spiral pattern ever worked out on the basis of dynamical principles. We believe that it still remains essentially correct, although minor refinements of the model have since been introduced. Frank Shu has since worked out the spiral pattern of three other galaxies. The significance of his results will be discussed later.

Many other theoretical predictions have by now been worked out and found to be in satisfactory agreement with observations by myself and my collaborators Frank Shu, William Roberts, and Chi Yuan. These I shall report briefly this evening in the following four sections:

1. Nature of the problem: material arms or density waves?
2. General spiral features observed
3. Theory of density waves, including applications to external galaxies
4. Application to the Milky Way system

Toward the end of this paper, the reader will find a discussion of future prospects and some remarks on other observable features and physical processes that we might examine in view of recent developments.

**Table of Contents**



## 1. Material Arms or Density Waves?

As is well known, the first problem that faces us is the winding dilemma; i.e., whether the spiral arms observed each contain the same material over many revolutions of the stellar system. Superficially, the answer appears to be a definite 'yes', for the brilliant young stars marking the spiral arms are definitely material objects. However, the spiral pattern for many galaxies must then change appreciably over a period of time of the order of two or three revolutions of the system; for the inner parts of a galaxy are generally rotating at a rate several times that of the outer parts, as exemplified by the Milky Way system. It is most unlikely that such a rapid change of appearance is actually taking place, for the classification of spiral galaxies into Sa, Sb, and Sc types is not only based on geometry but also on other physical characteristics; e.g., the gas content and the mass concentration in the nuclear region. A more subtle difficulty is the implication that the galactic magnetic field must steadily increase in the course of time, if indeed the material arms wind up more tightly. Clearly, both difficulties would be avoided if the spiral structure were associated with a *wave pattern*.

As it turns out, if we adopt the concept of density waves (or density waves modified by hydromagnetic effects), we can explain not only the 'winding dilemma' just discussed, but also a number of other observational features (Section 2). But before we turn to this discussion, let us take up another issue: *the nature of the intergalactic bridge*, such as that connecting M 51 (NGC 5194) and its companion (NGC 5195). Many such interconnected galaxies have been examined by Arp (1969). Such a bridge is doubtless a material arm, and yet it usually joins into one of the principal spiral arms in the wave pattern. The question again arises: are spiral arms material objects or wave patterns?

Obviously, this difficulty disappears if the wave pattern *co-rotates* with the material somewhere in the outerparts of the galaxy; i.e., if they have the same angular velocity. Within the co-rotating radius, the observed spiral structure is a wave pattern. Beyond this radial distance, the spiral arm may well consist of essentially the same material. In fact, it is difficult to expect wave propagation in these outer parts. Notice that there is no observational evidence against a change of the geometrical shape of the intergalactic bridge*. Indeed, as long as a material bridge persists, its shape is expected to change, as the galaxies move relative to each other.

As we shall see later, there are other pieces of supporting evidence for the co-rotation of the spiral pattern with the material objects in the outer parts of a galaxy. These will be discussed below in connection with the process of star formation and the origin of galactic spirals. (Section 3E.)

## 2. General Spiral Features Observed

We shall adopt a semi-empirical approach in presenting the concept of density waves, beginning with the consideration of a number of general features observed in galactic spirals. In the study of these features, there is one important theme to be kept in mind: *coexistence***. The complicated spiral structure of the galaxies indicates the coexistence of material arms and density waves – and indeed of the possible coexistence of several wave patterns. These features influence but do not destroy one another. When apparently conflicting conclusions are indicated by observations, the truth might indeed lie in the coexistence of material arms and several wave patterns. To be sure, before taking this 'easy way out', one should examine each interpretation of observational data as critically as possible.

Furthermore, as is well known, the different categories of optical objects defining essentially the same spiral arm usually appear displaced relative to one another (cf. Morgan, 1970). The radio features again do not necessarily coincide with the optical features.

Eleven prominent spiral features, including those briefly discussed above, will first be described. It will then be shown that these features can all be explained in terms of the concept of density waves. (See Table I for a brief outline including the theoretical interpretations.) Other support for the density wave theory may be found from detailed observations in the Milky Way system. These data will be discussed in Section 4.

* A. Toomre and J. Toomre have recently demonstrated (in a paper presented at the June 1970 meeting of the American Astronomical Society) that an intergalactic bridge can be produced by the close encounter of a galaxy with its satellite at a suitable relative velocity.

** Zwicky (1957) used the term 'coexistence' to describe the 'blue' and the 'red' components of M 51. We are using it in a broader context.

TABLE I

Observed features in spiral galaxies and their theoretical explanation in terms of density waves

| Observation | Theory (QSSS hypothesis) |
|---|---|
| (1) existence of a grand design | (1) wave pattern |
| (2) persistence of spiral pattern | (2) quasi-stationary spiral structure |
| (3) intergalactic bridge joining smoothly into spiral pattern | (3) co-rotation of wave pattern and material in outer regions |
| (4) spiral pattern usually two-armed | (4) $\Omega - \kappa/2$ nearly constant (Lindblad) |
| (5) multiple-armed in outer regions | (5) $\Omega - \kappa/m$ nearly constant in outer regions |
| (6) ring structure | (6) $\lambda \approx 0$ at resonance |
| (7) H II regions arranged like a string of beads | (7) galactic shock triggering star formation |
| (8) dust lanes on inner side of bright spiral arms | (8) gas compressed at galactic shock before stars form |
| (9) abundance distribution of ionized hydrogen varies greatly over the disk; none inside ring, except at center | (9) wave amplitude varies (according to definite laws); becomes tightly wound as resonance is approached |
| (10) peak of abundance distribution of neutral hydrogen outside of H II distribution | (10) shock mechanism needed for star formation; not available in outer regions |
| (11) magnetic field generally weak | (11) absence of perennial stretching by differential rotation |

(1) *Grand Design*

The first impressive feature of disk-shaped galaxies (already mentioned above) is the existence of a grand design in the form of a spiral pattern extending over the whole galactic disk, usually with other fragmentary spiral features superimposed*.

(2) *Persistence*

The persistence of spiral structure against differential rotation may be inferred from the fact that the spacing between spiral arms, which is used to classify normal spirals into Sa, Sb, Sc types, is correlated with other physical characteristics of the galaxy, such as the total mass, gas content, and concentration of nuclear mass. In particular, a smaller nucleus is generally associated with wider spacing (Sc galaxies).

(3) *Intergalactic Bridge*

In many pairs of galaxies, there is observable an intergalactic bridge, which joins smoothly with a major spiral arm in the main pattern.

(4) *Two-armed Trailing Pattern*

The spiral pattern is generally two-armed, especially in the central regions. As far as is known, the spiral arms are always trailing. (See also footnote to (1).)

* Sometimes these extra spiral features suggest the existence of a secondary spiral pattern, e.g., in M 51. The blue objects in M 51 clearly delineate one two-armed pattern. When this pattern is subtracted out from the ordinary photograph, there appears to remain another two-armed pattern displaced relative to the first by a quarter of a turn. Cf. NGC 6946, often labelled a four-armed spiral (Lynds, 1967).

(5) *Multiple-armed structure*

Multiple-armed structures are often observed in the outer regions of many galaxies. A well-known multiple-armed Sc spriral is M 101 (NGC 5457). These galaxies still have a two-armed structure in the central regions.

(6) *Ring Structure*

A ring structure is often found in the central regions of many normal galaxies (and also in the outer regions of barred galaxies). This inner ring structure is very clearly seen in NGC 5364, the Sc galaxy described by Sandage (1961) in the *Hubble Atlas* as 'one of the most regular galaxies in the sky'.

(7) *Narrowness of Spiral Arms*

The newly formed stars and young H II regions are neatly arranged like beads on a string, forming spiral arms much narrower than the spacing between the arms. This indicates that the process of star formation goes on in restricted regions, but simultaneously over a wide front.

(8) *Dust Lanes*

The principal dust lanes always lie on the inner side of the bright optical arm, although secondary dust features may appear in other forms. Indeed, the dust lanes often delineate the spiral pattern better than the bright stars or H II regions (Lynds, 1970).

(9) *Abundance Distribution of Ionized Hydrogen*

Both the continuum survey of Westerhout (1958) and the recombination line observations of Burke, Mezger and their collaborators indicate a marked deficiency in the abundance of H II regions inside of the '3 kpc circle' in the Milky Way. The abundance increases rather abruptly outside of this circle, and then declines with increasing distance from the center, to negligible amounts at about 13 kpc from the center. This may be compared with the absence of H II regions within the ring of NGC 5364. Extensive data on H II regions are now available from the work of Courtès, Hodge and their collaborators.

(10) *Abundance Distribution of Neutral Hydrogen*

Roberts (1966, 1968) found, by studying a number of external galaxies, that the peak of the distribution of neutral hydrogen usually appears to lie well *beyond* the bright optical structure. A prominent example is the galaxy M 33 (Figure 1). It is also well known that our galaxy conforms to this description. This fact suggests that the formation of stars requires more than the mere availability of hydrogen gas. Some triggering mechanism appears to be essential.

(11) *Magnetic Field*

The magnetic field in the Milky Way appears to be relatively weak, generally of the

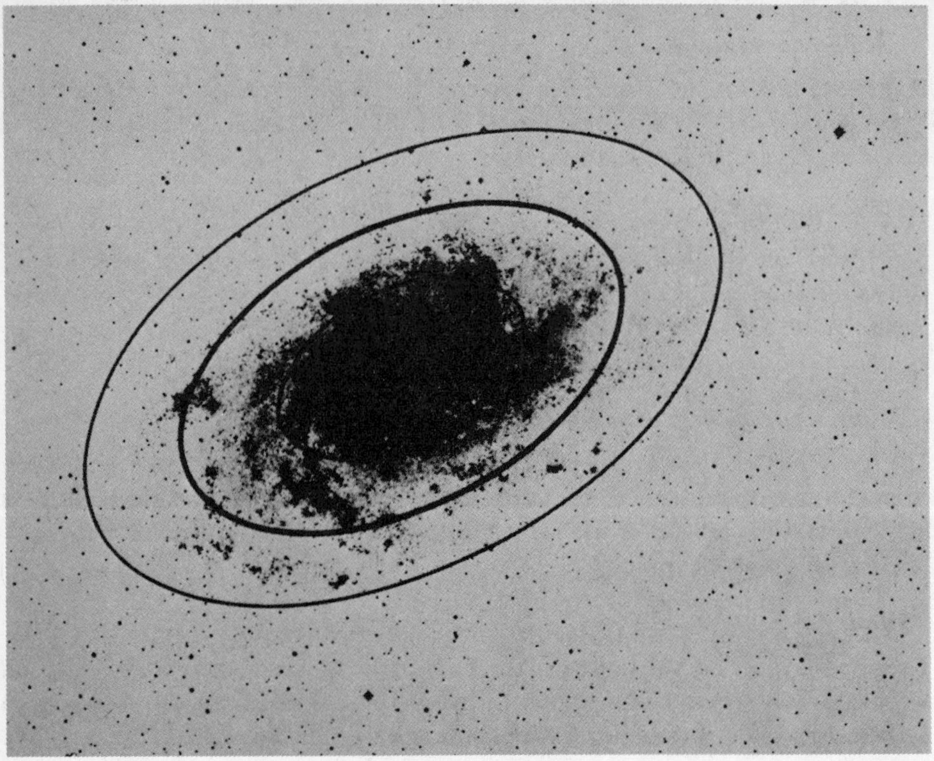

Fig. 1. The galaxy M 33, with region of maximum abundance of neutral hydrogen marked by two solid lines (after Morton Roberts). Notice that this region is entirely exterior to the optical spiral structure.

order of 1–5 $\mu$G. The structure of the magnetic field, however, remains largely unknown.

Because of the weakness of the magnetic field, we must look toward gravitational forces for a theory of spiral structure.

## 3. Theory of Density Waves

### A. A SURVEY OF THEORETICAL DEVELOPMENTS BY ANALYTICAL METHODS

As mentioned above, it is desirable to develop analytical methods to obtain useful results from the density wave theory for comparison with observations. In particular, it is found convenient to adopt the *hypothesis of quasi-stationary spiral structure*. The first theoretical calculations are naturally based on a linearized theory, assuming small density variations in a spiral pattern extending over a large portion of the galactic disk. But it is reassuring to find that, in the case of the Milky Way, one can obtain a spiral pattern that is in general agreement with that observed if one adopts a pattern speed of about one-half of the circular speed in the solar vicinity. To account for the process of star formation, it is necessary to show that the gas increases in density by a large factor (of the order of ten) as it goes into the spiral arm. Such calculations were made by Roberts (1969, 1970) for a two-armed spiral pattern determined by Lin *et al.* (1969) from an extensive study of observational data in our own Galaxy. These results will be discussed in some detail below.

The analytic approach is also more flexible for the exploration of the dynamical processes in order to reach a deeper understanding. Various aspects of these problems have been studied by Contopoulos (1970a, b, c), Goldreich and Lynden-Bell (1965), Julian and Toomre (1966), Kalnajs (1965, 1970), Lynden-Bell (1970), Lynden-Bell and Ostriker (1967), Marochnik and Suchkov (1969), Shu (1968, 1970a, b, c), and Toomre (1964, 1969, 1970). Subjects such as the antispiral theorem, the origin and permanence of galactic spirals are considered in these papers. They are very challenging and not yet completely solved. I regret that I shall not be able to do justice to all these works in this review, but I shall have occasion to refer to some of them in the subsequent discussions. At this point, I only wish to refer you to the survey paper by Contopoulos (1970a) and Shu's papers (1970a, b) for the latest status of the theory. A complete formulation of the linear theory may be found in these papers. I should also call your attention to studies of the dynamics of the whole galactic disk by using self-gravitating models either in the 'cold' case (Hunter, 1963, 1965, 1969a, 1970; Miyamoto, 1969; Rehm, 1965; Yabushita, 1968, 1969a, b) or in the case with dispersive velocities by the use of moment equations for stellar dynamics (Hunter, 1969b). In particular, Hunter noted the tendency for a galactic disk to deform into an oval shape (as did Kalnajs for the case of the uniformly rotating stellar disk, unpublished). Nonlinear spiral waves have been studied in the asymptotic theory by Berry and Vandervoort (1970) in the 'cold' case and by Vandervoort (preprint) for a gaseous disk with pressure. Another important direction for research is the study of stellar orbits by Contopoulos (1970a, b, c), Barbanis (1968a, b), Barbanis and Woltjer (1969), and others in a spiral gravitational field; for these will eventually form the basis for a deeper theory. In particular, they may be used in the future to connect the original approach of Lindblad and Langebartel (1953) with the modern approach of *collective* modes. For the rest of the paper, I shall follow the approach centered on the QSSS hypothesis.

### B. IMPLICATIONS OF A SPIRAL PATTERN OF DENSITY WAVES

The QSSS hypothesis is suggested by the existence and the persistence of a grand design. To explain the correlation of the spacing between spiral arms with the concentration of nuclear mass, however, requires a quantitative calculation related to the mechanism for sustaining the density waves. Similarly, a detailed analysis is needed to explain why a two-armed structure is preferred, why a multiple-armed structure often occurs on the outer parts of certain spiral galaxies, and why the spiral arms often wind up tightly around a ring structure, which is at a substantial distance from the center of a galaxy. These calculations will be presented in the written text in the next section. I regret that, because of limitation of time, I shall have to omit this part from my oral presentation. On the other hand, it is possible to give at least a qualitative explanation of the other five observational features [(7)–(11)] described above, provided that we are willing to accept certain simple physical pictures obtained from calculated results (Roberts, 1969; see also earlier work of Fujimoto discussed in Roberts' paper).

Consider an observer in a coordinate system co-rotating with the wave pattern. To this observer, the pattern is fixed , and material objects are moving through its gravitational field. We may visualize them as moving along imaginary stream tubes.

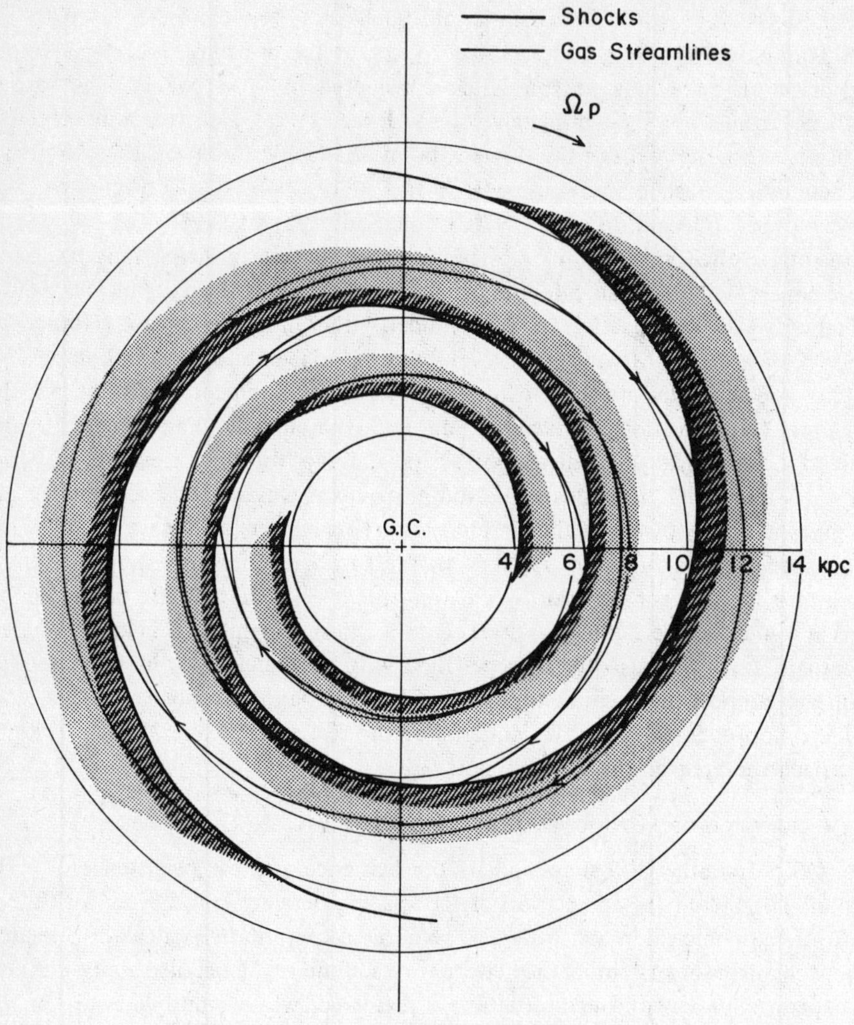


Spiral Pattern in the Galaxy

Fig. 2. Theoretical diagram showing the spiral structure of a galaxy (approximately the Milky Way) and the associated kinematical behavior of the gas and the stars (after W. W. Roberts). Notice the large scale galactic shocks that precede regions of star formation.

Calculations show that the gas is suddenly compressed near the minimum of the gravitational potential. We shall call this compression along a wide front of many kiloparsecs a *galactic shock* (Figure 2). At a shock front, a streamline suddenly changes its direction, and eventually closes on itself after another similar turn at the other shock. The sudden compression of the gas collects the existing dust particles into a prominent *dust lane*, whose strength may be further enhanced by the formation of more molecules and dust particles, induced by increased gas density. This dust lane is very *narrow*, since the calculated results show that the sudden compression of the gas at the galactic shock is followed by a rather rapid decompression.

This same compression process also provides a triggering mechanism for star formation, by bringing the individual gaseous clouds into a state of continuing gravitational collapse. Once begun, this process will continue for each cloud complex and for the individual clouds of the complex, even after decompression sets in on the scale of the spiral arm. Brilliant stars are therefore formed almost simultaneously *over a wide front* of many kiloparsecs. These stars and the associated H II regions are aligned as a narrow spiral arm, like beads on a string, for their lifetime is on the order of $10^7$ yr, and there is very little radial displacement over such a short period of time. *The observed spiral arm is therefore the brilliant manifestation of very young objects whose location and arrangement are controlled by the invisible gravitational field determined by the older stars*, – an establishment behind the scenes.

The formation of these bright young objects is triggered off by compression on a large scale. However, this compression does not guarantee the formation of stars, since the intial conditions of the gas also play an important role. Thus, the distribution of H II regions is often found to be *patchy*, and the spiral arms are often better delineated by the dust lanes (Lynds, 1970).

The bright stellar arm is expected to be somewhat separated from the dust lane where the compression occurs, for there is a time of the order of $30–50 \times 10^6$ yr required for the collapse of the gaseous clouds into stars. It may be shown that this separation is generally of the order of several hundred parsecs. The exact extent of the separation depends, among other things, upon the component of the velocity at which the gas rushes through the dusty region in the direction normal to the shock. As it turns out, this relative velocity decreases with increasing distance from the galactic center, and hence the separation is minimal towards the end of the spiral pattern. As we move inwards, greater separation is expected. Such a change can indeed be seen in M 51. At the inner resonance ring, where the spiral features are tightly wrapped together, there may again be very little separation. However, at least in M 51, up to the point where a dust lane may be traced, the winding is not yet so tight that a substantial separation is still noticeable.

The process of star formation described above also explains why the *abundance* distribution of neutral hydrogen does not always match that of ionized hydrogen, for the latter can be readily produced only by density waves, and the strength of such waves varies greatly over the galactic disk. However, the theory enables us to calculate these variations. In particular, a spiral density wave may terminate by winding into a ring, and hence there are no H II regions inside*. In the outer parts of a galaxy, even though neutral hydrogen may be present beyond the tip of the observed spiral arm, there are no strong density waves, and hence star formation on a large scale cannot take place and no spiral structure is observed. In this picture, the tip of the spiral arm is roughly the location where the material objects co-rotate with the wave pattern.

In the coordinate system co-rotating with the spiral pattern, the gaseous motion is essentially steady. The magnetic field lines are expected to coincide with the stream line, and run in a direction nearly parallel to the spiral arm (except where the field is very weak). There is no perennial stretching of the lines of the magnetic field to increase its strength. The field remains weak (if it is initially so) and plays only a secondary role in the dynamical processes (e.g., in determining the rolling motion inside of a spiral arm; cf. Fujimoto and Miyamoto, 1970).

One may also speculate, from our dynamical picture, about the magnetic lines associated with a local spur of material produced by differential rotation of a clump of gas. The field lines would lie roughly in the direction of the material arm, but could point in opposite directions, perhaps above and below the galactic plane (cf. Roberts and Yuan, 1970). This is compatible with certain observed features in the Orion arm.

Recently, Mathewson and his collaborator (Mathewson, 1968, 1969; Mathewson and Nicholls, 1968) suggested the existence of a local helical magnetic field in addition to a field on a larger scale. I do not feel competent to comment on these suggestions, but would refer you to the papers by Woltjer (1970) and Fujimoto and Miyamoto (1970) at the Basel Symposium.

### C. SPIRALS WITH MODERATELY SMALL PITCH ANGLE**

Most of the spirals, – even the Sc's – have their principal spiral arms inclined at moderately small pitch angles. For such spirals, the asymptotic method of WKBJ (see Morse and Feshbach, 1953) may be applied, and simple relationships can be obtained. We shall discuss some of the general conclusions obtained from such an approach in the present section. In the next section, we shall apply the results to a general discussion of the spiral patterns in the Milky Way and three other external galaxies. The detail treatment of observational data in the Milky Way will be presented in Section 4.

1. The theory predicts the possible existence of a density wave pattern propagating around the galaxy with a pattern speed $\Omega_p$. The pattern can only extend over the part of the galactic disk for which

$$\Omega - \kappa/m < \Omega_p < \Omega + \kappa/m,$$

where $\Omega(\varpi)$ is the circular velocity (in angular measure) of the material around the galactic center, at a distance $\varpi$ therefrom, $\kappa(\varpi)$ is the epicyclic frequency, and $m$ is an integer giving the number of arms.

* Except possibly at the center, where the H II regions may be formed by different mechanisms.
** Omitted from oral presentation.

This relationship implies that only two-armed spirals are expected to occur in prevalence while multiple-armed structure can be expected only in the outer parts of a galaxy. This can be easily seen if we examine the trend of the curves for $\Omega(\varpi) \pm \kappa(\varpi)/m$ in the Galaxy ($m$=integer). The flatness of the curve $\Omega - \kappa/2$ (Figure 3) is at the root of the conclusion. We may expect all other galaxies to have a similar behavior simply because $\Omega - \kappa/2 = 0$ for a galactic disk in uniform rotation. In the outer parts, patterns with other values of $m$ are permitted. Thus a composite structure with two arms inside and more arms outside is expected. The Sc galaxy M 101 is an excellent example for a composite structure of this kind.

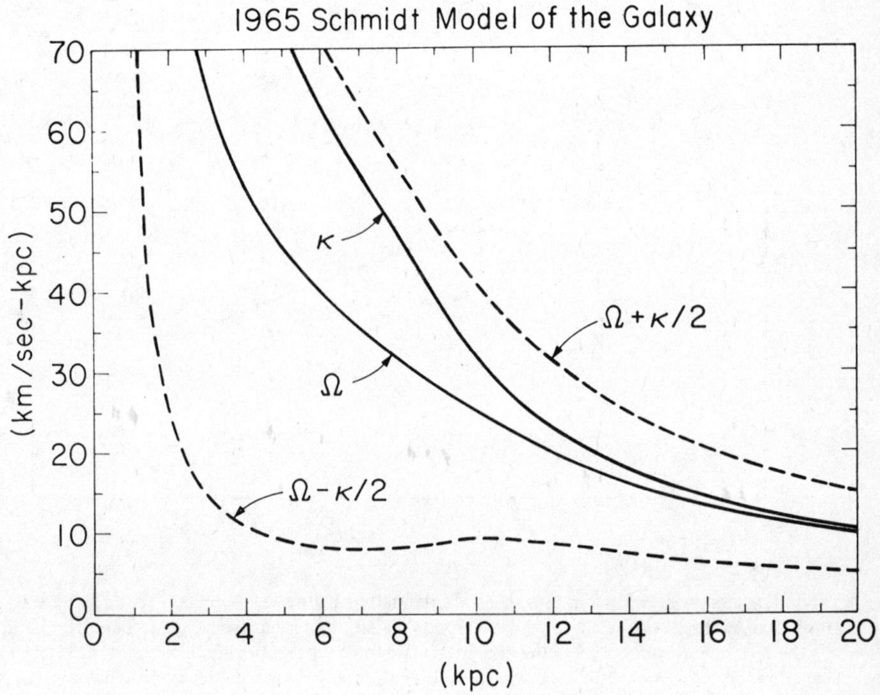


Fig. 3. Rotation curve, etc., of our own Galaxy according to the Schmidt model of 1965. (Symbols defined in Figure 4). Notice the flatness of the curve of $\Omega - \kappa/2$ for a substantial part of the Galaxy, as noted by Bertil Lindblad.

2. A dispersion relationship can be obtained connecting the spacing (or wavelength) between the spiral arms with the frequency at which the material moves through the wave pattern. In dimensionless form, we have the curve shown in Figure 4 (after Lin and Shu, 1967). The typical length scale is $\lambda_* = 4\pi^2 G\sigma_*/\kappa^2$, where $G$ is the gravitational constant, $\sigma_*$ is the mass surface density, and $\kappa$ is the epicyclic frequency. The parameter $\nu$ is the frequency in dimensionless form: $\nu = m(\Omega_p - \Omega)/\kappa$, where $\Omega(\varpi)$ and $\Omega_p$ are defined above.

3. Note that the spacing $\lambda$ approaches zero as the points of resonance (if they exist)

are approached and the spiral pattern should terminate as a ring both inside and outside. The inner ring is observed in many disk-shaped galaxies but not the outer ring, since it is often too far out for the theory to be applicable*. In these outer regions, gaseous clumps are formed by gravitational collapse, and regular density waves are not favored. Within the inner resonance ring, there are no strong density waves, and hence no H II regions. A possible exception is the very center of the galactic nucleus, where H II regions can be formed by other means of condensation.

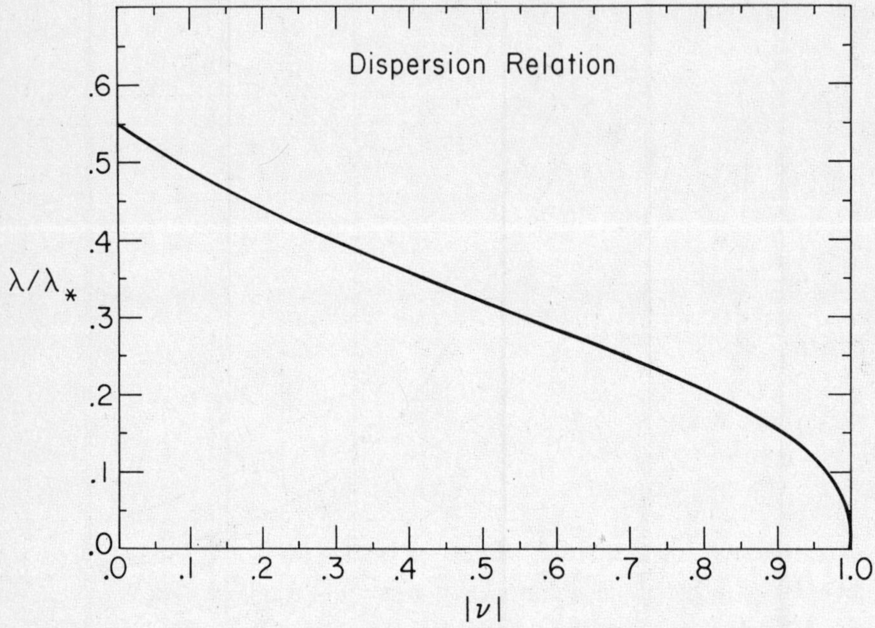


Fig. 4. The Dispersion relationship. Symbols are defined as follows: $\nu = m(\Omega - \Omega_p)/\kappa$, where $m = 2$ is the number of arms, $\Omega(\varpi)$ is the circular velocity in angular measure at galactocentric distance $\varpi$, $\kappa(\varpi)$ is the epicyclic frequency, $\lambda$ is the spacing between two neighboring arms, and $\lambda_*$ is a typical length scale defined by $\lambda_* = 4\pi^2 G\sigma_*/\kappa^2$, where $\sigma_*$ is the projected surface density of the stars, and $G$ is the constant of universal gravitation. (After Lin and Shu, 1967). Note that a ring structure ($\lambda = 0$) is associated with resonance ($|\nu| = 1$).

4. The ratio of the scale $\lambda_*$ to the diameter of the galactic disk can be shown to be roughly proportional to the ratio of the mass of the galactic disk to the total mass, including the galactic nucleus. Thus, Sc galaxies with smaller nuclei tend to be loosely wound.

We have thus explained why a two-armed structure is preferred, why a multiple-armed structure can only occur in the outerparts of certain spiral galaxies, and why the spiral arms often wind up tightly around a ring structure which is at a substantial distance from the galactic center. We have also given one reason for the loose structure

* The situation may be different for barred spirals.

of Sc spirals. Another reason will be found from a discussion of the pattern speed at which quasi-stationary spirals are maintained.

All the eleven general observational features have been accounted for. We shall now turn toward more specific discussions.

### D. SPIRAL PATTERNS OF THE MILKY WAY, M 33, M 51 AND M 81: DETERMINATION OF PATTERN SPEED

For spirals with moderately small pitch angles, the theory permits us to determine the spiral pattern in a simple manner once the pattern speed is known. In addition to the Milky Way, three other galaxies, M 33, M 51 and M 81, have been studied by Shu *et al.* (1970). In each case, after the mass model is known*, there remains only one adjustable parameter; namely, the pattern speed $\Omega_p$. However, if we wish to make the calculated spiral pattern agree with observations, there is practically no freedom in its choice, for the spiral structure depends very sensitively on the pattern speed. In particular, whenever there is an inner resonance ring where the spiral structure terminates, a convenient determination presents itself; for the theory asserts that, near such a ring, the stars are encountering the spiral gravitational field at the local epicyclic frequency. Such a ring appears to exist in the cases of the Milky Way, M 51, and M 81.

The results are shown in Figures 5, 6, 7, 8; there is good agreement between theory and observations. The case of the Galaxy was briefly mentioned above, and will be treated in greater detail below. In the case of M 51, there is an inner resonance ring, within which giant H II regions are deficient. Good agreement is obtained with the primary spiral pattern. There also appears to exist a long-wave pattern and a secondary two-armed pattern which is displaced by an angle of 90° from the first. Their significance will be discussed in Section 3E.

The general features of M 81 are similar to those in M 51. In the case of M 33, there is no inner resonance ring, and the spiral pattern terminates at the nucleus where there is a short bar. In both cases, there are also indications of secondary spiral patterns, although they are not as clear as in M 51.

In all three cases, Frank Shu found that the pattern is *co-rotating* with the material in the outer parts of the galaxy. This is a very encouraging indication that the mechanism proposed earlier (Lin, 1970) to explain the origin and permanence of spiral patterns is essentially correct. We shall consider this presently. The result is also compatible with the processes of star formation considered in Section 3B, and our description of the intergalactic bridge in Section 1.

### E. ORIGIN AND PERMANENCE OF GALACTIC SPIRALS

So far, we have adopted a semi-empirical approach. The QSSS hypothesis has led to a number of deductions in agreement with the major features observed in galactic spirals. Just how the observed spiral patterns originated is not yet fully explained. They

* To be sure, there is uncertainty in the observed rotation curve, and hence of the mass model. But these variations are not large enough to alter the following general conclusions.

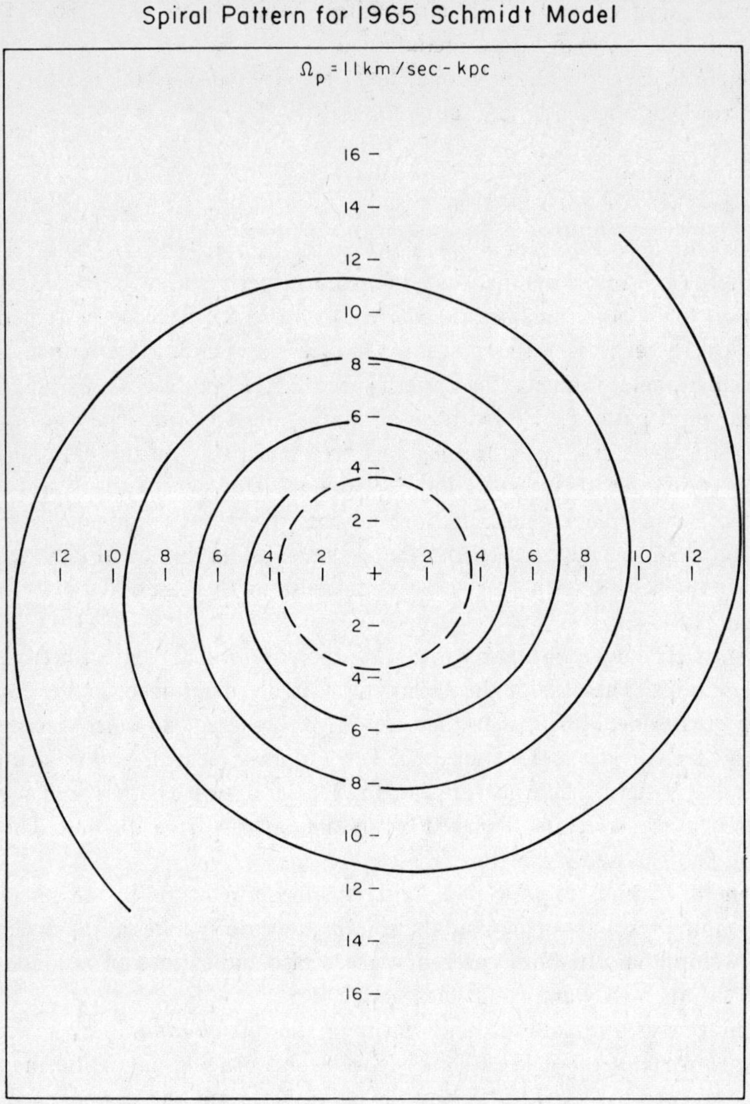


Fig. 5. Spiral pattern of the Milky Way.

might be caused simultaneously by several mechanisms. Both excitation external to the galaxy and instability internal to the system may be present. However, from the prevalence of spiral galaxies without nearby companions, it appears that *there must be a mechanism inherent to the galactic system itself.*

Search for an unstable spiral normal mode via the solution of the boundary-value problem (Hunter, 1965, 1969, 1970; Kalnajs, 1965, 1970; Miyamoto, 1969; Shu, 1970b) is not yet wholly successful. Indeed, the problem of setting realistic boundary condi-

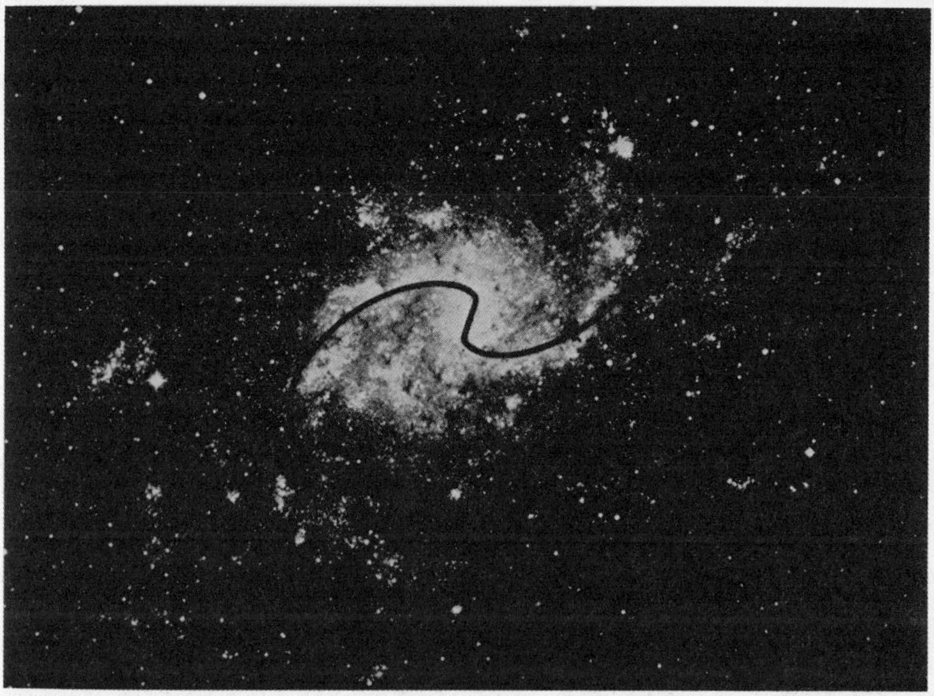

Fig. 6. Spiral pattern of M 33, comparison of theory and observations (after Shu).

tions in the outer parts of the galaxy may be very complicated, because of the clumpy structure of the gaseous component.

One of the mechanisms proposed (Lin, 1970) for the initiation and the maintenance of density waves is the Jeans instability in the outer regions of the galaxy*. Once the gaseous condensations are produced by the process of gravitational clumping and stretched into material arms (Goldreich and Lynden-Bell, 1965) – aided by the excitation of density 'wakes' in the stellar sheet (Julian and Toomre, 1966) – short trailing waves may be initiated that propagate both inwards and outwards as a group (Toomre, 1969). The outward-propagating waves cannot be strong, for the difference is small between the wave speed and the material speed. The clumpy and irregular distribution of gas in this area is also expected to prevent the waves from being well organized. For the inward-propagating waves, the energy density tends to pile up in the interior where the 'group velocity' of such waves is small (Toomre, 1969; Shu, 1970a). If there is no inner resonance ring, the waves would reach the center and be 'reflected' via the resultant bar-like structure (e.g., in the case of M 33). If there is a resonance ring, the

* Jeans (1929, p. 349) specifically stated "that condensations cannot form in elliptical nebulae – or in the central masses of spiral nebulae, but they must inevitably form in the equatorial extensions of the spiral nebulae". However, in the work of Jeans, the effect of differential rotation of the galactic disk was not included. Toomre (1964) did the first work including such an effect for a stellar system. The simpler case of the gaseous system was then given by Lynden-Bell (1967).

Fig. 7. Spiral pattern of M 51, comparison of theory and observations (after Shu). The dotted curve shows the 'long-wave' pattern.

waves would be 'reflected' at the ring by a two-step process (e.g., in the cases of the Milky Way, M 51, and M 81). First, the waves would be absorbed there by Landau damping according to the linear theory (James Mark, private communication). This damping process would be 'saturated' and cease after about a billion years (Contopoulos, 1970c) and the ring would be deformed into two oval structures with their major axes at right angles to each other (Contopoulos, 1970a, 1970c). The gravitational influence of these oval structures is *in step* with the waves at the outer ring of initiation, and thereby constitutes an *effective* feedback mechanism, which maintains the permanence of the wave pattern. It has been suggested (Lin, 1970) that this 'reflection' process may take the form of a 'long-wave pattern'. Indications of such a pattern have been found by Shu *et al.* in M 51 (cf. Figure 7).

The mechanism just discussed can be subjected to two observational tests. First, a correct spiral pattern must be obtained if the wave pattern is assumed to be co-rotating with the material in the outer parts of the galactic disk. Secondly, the 'piling up' of the wave energy would lead to an increase in the number density of H II regions towards the galactic center even when there is little increase of H I. As mentioned above, the first test is satisfied, as shown by Shu *et al.* in their investigations of the Milky Way and of the external galaxies M 33, M 51 and M 81. The second (partial)

Fig. 8. Spiral pattern of M 81, comparison of theory and observations (after Shu).

test will be taken up (Section 4A) as part of the comparison of the theory with observations in the Milky Way system. This system has the fortunate feature that its H I abundance changes very little over a distance in which its H II abundance changes by about one order of magnitude. For other galaxies such as those studied by Hodge (1969), a more detailed knowledge of their H I distribution is required for a complete analysis.

## 4. The Milky Way

The theory discussed above can be tested against more detailed observations in the Milky Way system. Indeed, many such data clearly suggest the existence of density waves. However, it should be made clear at the very outset that the determination of the spiral structure of the Milky Way is intrinsically a very difficult task. The observer, being inside of the system, can be easily distracted by local features. Under such circumstances, it is difficult to distinguish a double two-arm spiral pattern, such as that seen in M 51, from a single two-arm spiral pattern which is doubly as tight. What I shall present is therefore based on a self-consistent dynamical model which is in reasonable agreement with a number of observations of various kinds. Any other suggestion should also be checked against the observational data discussed below.

These include the distribution and motion of (1) neutral hydrogen, (2) ionized hydrogen, and (3) stars.

### A. NEUTRAL HYDROGEN

On the basis of the density wave concept, it is expected that, together with associated radial motions, there exists a higher circular speed of gaseous motion on the outer side of a concentration of hydrogen gas and a lower speed on the inner side of such a concentration, than would have been observed in its absence. Lin *et al.* (1969) used this to correlate the oscillations in the rotation curve with the location of the Sagittarius arm and the Norma-Scutum arm. (Figure 9, Figure 10. See Kerr, 1969, for a survey of the relevant observational data). Earlier, Shane and Bieger-Smith (1966)

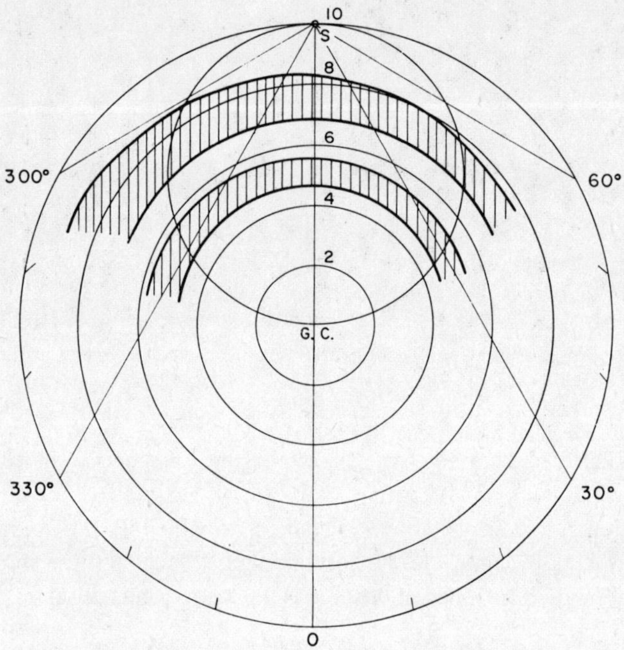


Fig. 9. Locating gaseous arms for variations in the rotation curves shown in Figure 10.

represented apparent irregularities in the measured rotation curve by attributing them to non-circular motions. A high stream of gas outside of the Sagittarius arm was noted by Burton (1966). More recently, Burton (1970), Burton and Shane (1970) found that much of their data could be satisfactorily interpreted only by invoking streaming motions similar to those implied by the concept of density waves.

Once noncircular motions are admitted, the analysis of the observational data becomes very complicated, and a process of successive approximation (or trial-and-error) has to be adopted. From a theoretical standpoint, one would then begin by calculating a spiral pattern and then attempt to reproduce the observational data with

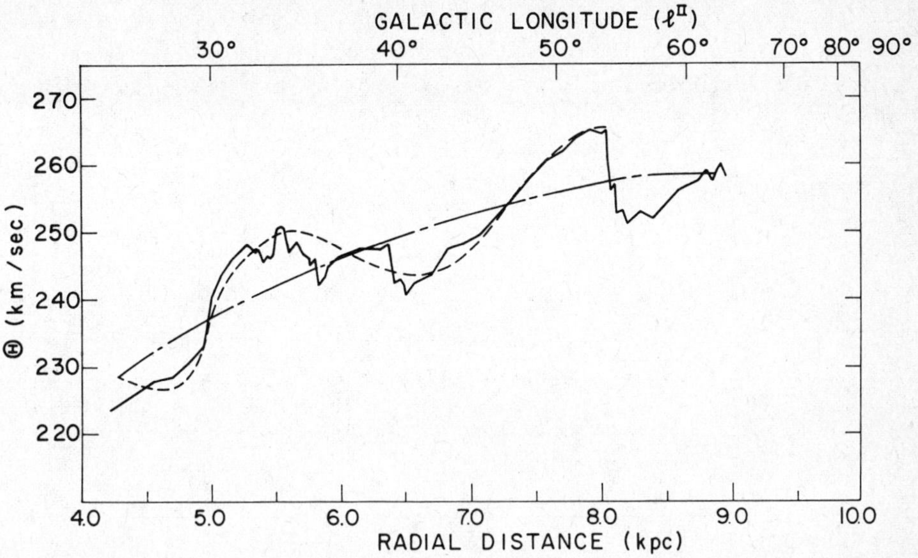


Fig. 10a.

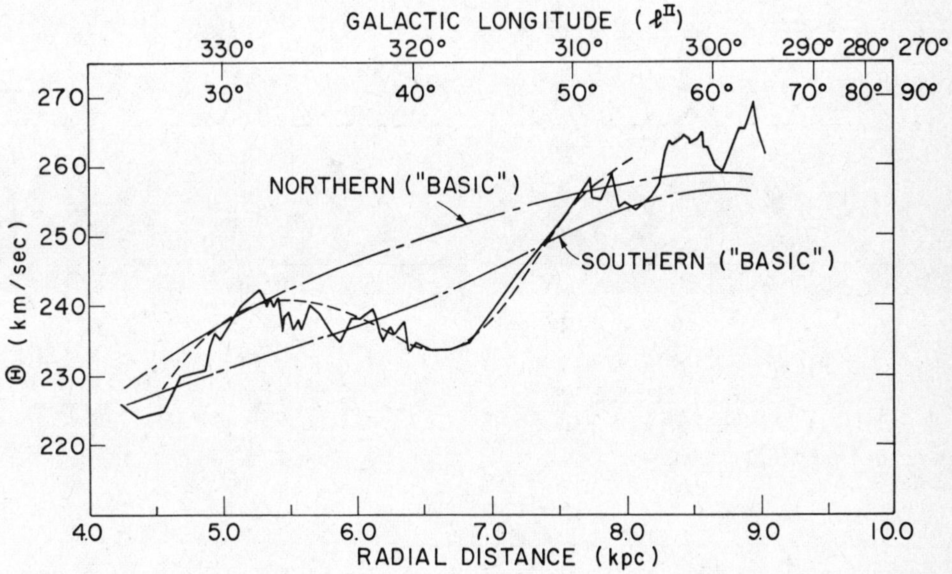


Fig. 10b.

Figs. 10a, b. Rotation curves: (a) 'Northern' Hemisphere; (b) 'Southern' Hemisphere. Observation (Kerr, 1964) ———; Theory — — — —; Mean motion (theory) ·—·—·

theoretical profiles. Such calculations were started by Yuan about two years ago and a preliminary report was given by him at Basel (Yuan, 1970a). His current results are reproduced in Figure 11. The overall comparison between theory and observations for all galactic longitudes is reasonably satisfactory – at least to the extent that the grand design may be said to be properly represented – but refinement of the theoretical model is certainly needed in order to account for detailed features*, especially in the central regions, ($|l^{II}|<30°$).

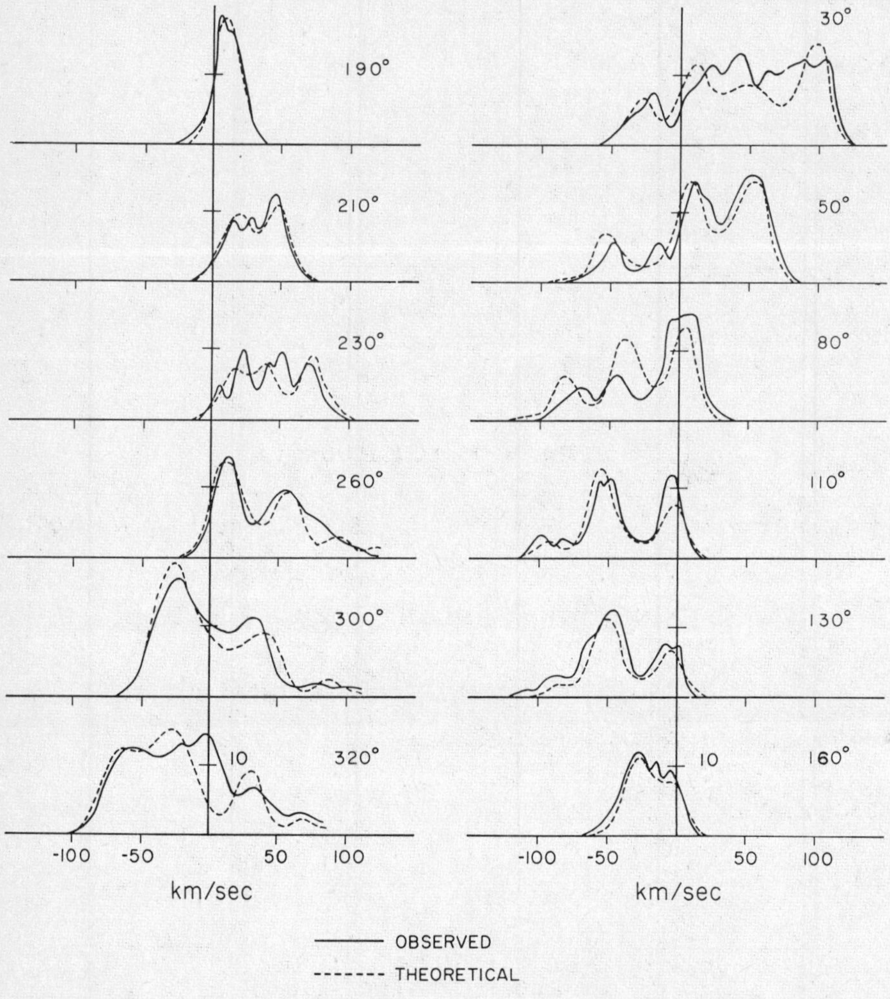


Fig. 11. Theoretical velocity-intensity profiles and their comparison with observations (after Yuan). Data for $|l^{II}|<30°$ are not analyzed since they are dominated by features in the central regions not related to a simple spiral structure.

* For example, in the direction of the galactic center, Yuan (1969a, p. 881) pointed out that the observed motion of the gas is compatible with a theoretically expected inwards motion of 7 km/sec, provided that there is an overturning motion (Henderson, 1967) with an amplitude of 5 km/sec.

Independently of Yuan's work, Burton (1970), Burton and Shane (1970) used the density wave theory to analyze their observational data by constructing theoretical contour maps in the longitude-velocity plane. These are restricted to certain northern directions, but more detailed than Yuan's analysis. The results obtained are more satisfactory than those obtained by using a circular model. It is found that an additional expansion field is needed within 5 kpc at least in the region around $l^{\mathrm{II}}=30°$. This may well be associated with the oval shaped distortion of the mass distribution near the resonance ring, as suggested by the nonlinear analysis of Contopoulos.

### B. IONIZED HYDROGEN

The continuum survey by Westerhout (1958) shows that the bulk of the galactic continuum radiation comes from a narrow range of latitude centered about the galactic plane and that most of the ionized hydrogen must be concentrated in a ring somewhat outside of the 4 kpc radius ($l=\pm 26°$). From this distance outwards, there is a decline of the density of ionized hydrogen to practically nothing at 13 kpc, while the amount of neutral hydrogen increases to a peak value. As mentioned above, this feature is consistent with that discovered by Roberts (1966, 1968) in a number of external galaxies.

The theoretical curve shown in the figure (Figure 12) is a rough estimate of the abundance ratio of ionized hydrogen to neutral hydrogen at various galactocentric distances. The estimate is based only on the strength of the gravitational field in the density wave pattern (See Figure 3 in Shu, 1970b). The calculations were made by Stuart Feldman (private communication), based on data analyzed by Mezger (1969). Other factors* that may influence the amount of ionized hydrogen produced are not taken into account. It is remarkable that there is still general agreement between theory and observations. Within the ring of resonance at a radial distance of about 4 kpc, there is little ionized hydrogen expected, and indeed very little is observed. The rather sudden decline of the abundance of ionized hydrogen in the transition region from 5 kpc inwards is also well described by the theoretical results of James Mark (private communication).

The data analyzed by Mezger (1969) are based on the work of Reifenstein *et al.* (1970), and of Wilson *et al.* (1970). In Mezger's original presentation of the data, the five giant H II regions near or inside the 4 kpc circle were evenly distributed inside the circle, for their distances are not well known. Actually, four of these may well be associated with the very center of the galaxy**. A fifth one (G347.6+0.2) is not far

* One important factor is the relative velocity between the gas and the density wave. This factor is especially important in the outer regions where the relative velocity is small and hence the shock is weak. In particular, there should be little star formation near co-rotation. For most part of the data shown in Figure 12, this factor is however not very important.

** In a private communication, dated May 11, 1970, Dr Mezger kindly informed me that these H II regions have physical characteristics different from those in the spiral arms. The author wishes also to thank Drs B. F. Burke and W. W. Shane for the discussion of the data and their interpretations.

from features associated with the '3 kpc arm'. Thus, the giant H II regions inside of the resonance ring are not relevant to the problem at hand.

A similar analysis could be made from the data of Hodge (1969) for external galaxies. Since the distribution of neutral hydrogen may not be as uniform in the region in question as in the Milky Way, definite conclusions cannot be drawn until more detailed H I observations become available.

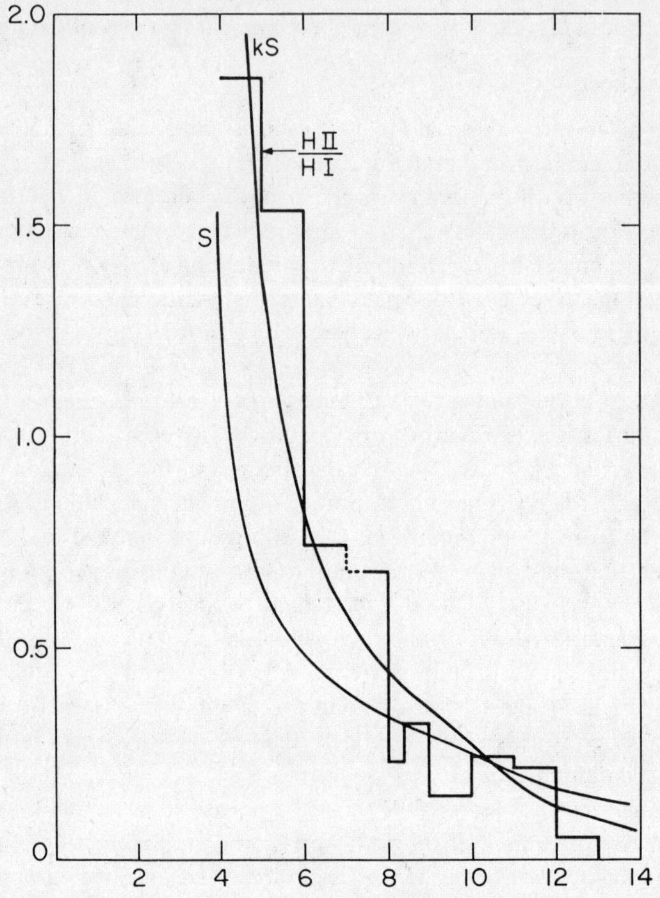


Fig. 12. Relative abundance distribution of ionized hydrogen to neutral hydrogen (arbitrary scale). Horizontal scale is galacto-centric distance in kpc. The theoretical curve (marked kS) is a rough estimate of the relative abundance, based only on the strength of the gravitational field in the spiral wave.

The kinematics of ionized and neutral hydrogen are essentially identical, based on the observations made and analyzed by Dieter (1967), by Kerr *et al.* (1968), and by Mezger *et al.* (1970). Their data suggest that the very young stars are co-moving with the gas and that they are essentially distributed along the spiral arms defined by neutral hydrogen, in agreement with the gravitational theory of spiral structure.

C. STARS

The very young stars* are associated with their H II regions (Figure 13). For somewhat older stars in the solar vicinity, Strömgren (1967) has discussed their migration on the basis of their current kinematical data and their age, in order to determine the places of their origin. The stars are expected to be formed inside of spiral arms. The program had not been entirely successful until the density wave concept was brought into the

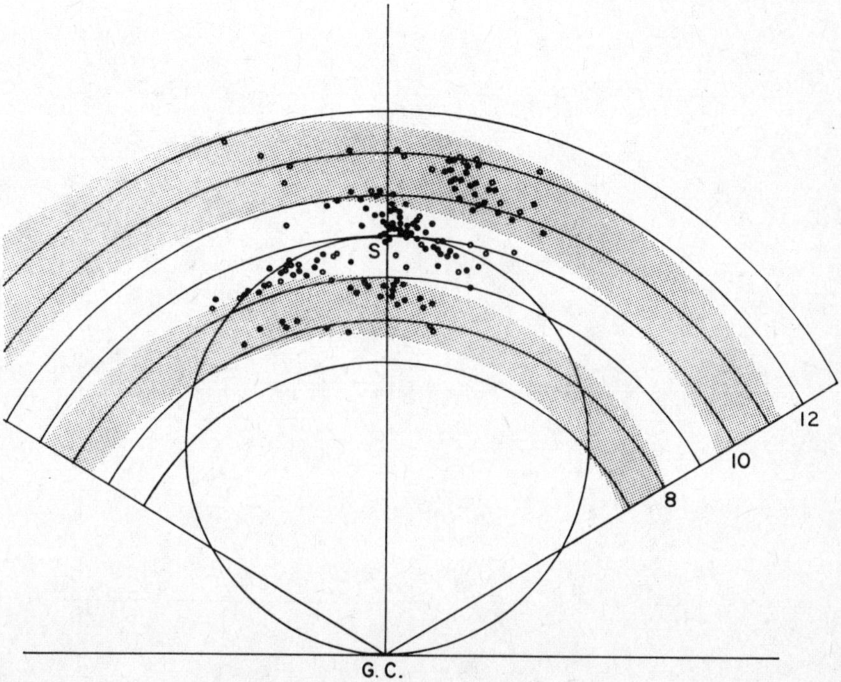


Fig. 13. Distribution of young stellar populations. The Orion arm might be a material arm. The Carina arm is probably part of a secondary feature.

picture. A comparison of the analyses for B 8–B 9 stars with and without the gravitational field (Figure 14; see Lin *et al.*, 1969; Yuan, 1969b), shows clearly the existence of a spiral field with an amplitude of about 5% of the mean field. A larger field of 7% or a smaller field of 4% would yield considerably worse agreement (Cf. Section D).

Another well known phenomenon is the 'vertex deviation' of the velocity distribution of local stars, particularly the A stars. Asymmetries in the field and in the origin of these stars have been suggested as the cause for this vertex deviation. With the help of the density wave theory, we can actually make detailed calculations for stars of one

* In a study of the space distribution and kinematics of supergiant stars, Humphreys (1969) found that "in the direction $l = 285°–300°$, systematic motions of 10 km/sec were found between the two sides of the arm as expected from Lin's density wave theory of spiral structure".

particular age group. In Figure 15, we show the comparison of the theoretical results (Yuan, 1970b) with the observational data (Eggen, 1965) for stars in the age group of 300–500 million years. It is seen that the density wave theory gives the desired explanation of the observed phenomena: the A stars show a more pronounced vertex deviation than both the B stars and the F stars (Eggen, 1969). (Another approach to this problem has been adopted by Mayor (1970), but it would take us too far afield to discuss all the issues involved.)

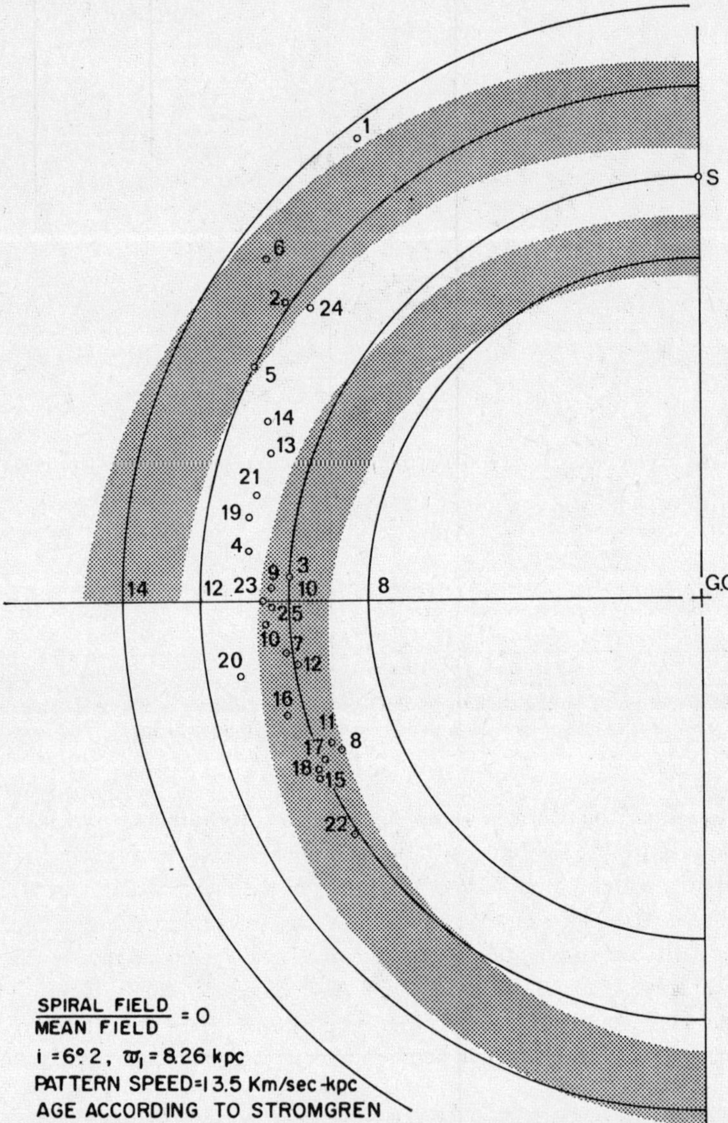


Fig. 14a.

D. SUMMARY OF PHYSICAL PARAMETERS

At this point, it is important to emphasize that we have used the same spiral pattern with the same physical parameters in our analysis of each of the various phenomena. These parameters are shown in Table II. In particular, we note that the dispersion of

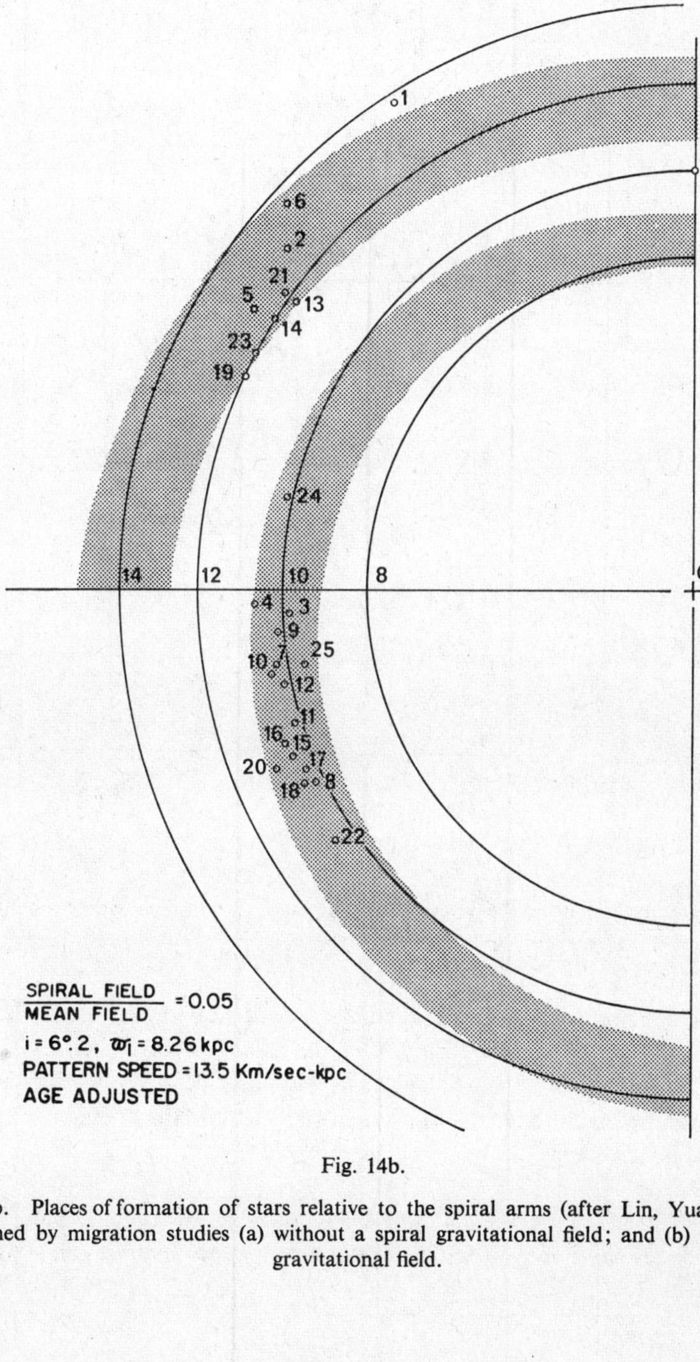


Fig. 14b.

Figs. 14a, b. Places of formation of stars relative to the spiral arms (after Lin, Yuan and Shu), as determined by migration studies (a) without a spiral gravitational field; and (b) with a spiral gravitational field.

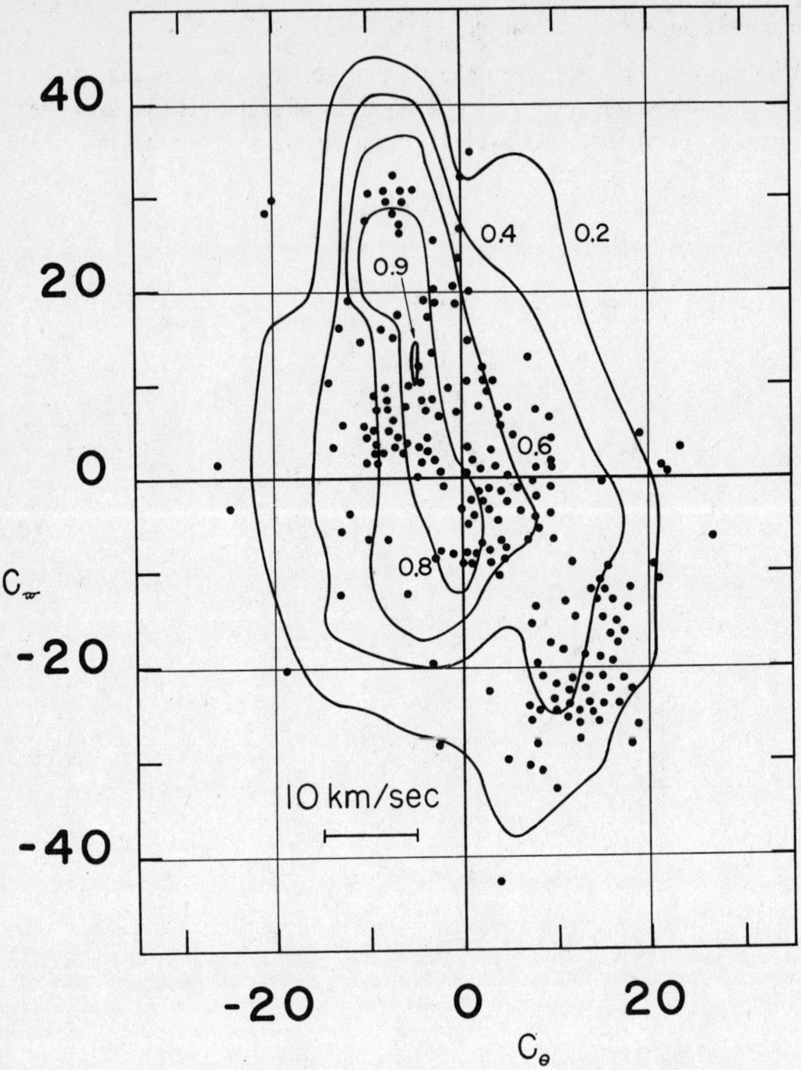


Fig. 15. 'Vertex deviation' in the velocity distribution of A stars compared with theoretically expected velocity distribution for stars in the age group of 300–500 × $10^6$ yr (after Yuan).

stellar velocities predicted from the theory is in good agreement with observational data. This agreement is reached only after the thickness of the disk is considered, following Shu's analysis (1968). Shu's theory has since been verified by Vandervoort (1970a) who developed a theory of galactic disks of finite thickness based on the use of adiabatic invariants.

## E. LOCAL STRUCTURE

All the successes cited above should not obscure the fact that we can only suggest a

TABLE II

Certain physical parameters in the Milky Way (mostly near the Sun)

| | |
|---|---|
| 1. rms radial velocity predicted for stars in the solar neighborhood | $\leqslant$ 37 km sec$^{-1}$ |
| 2. Spiral pattern (primary component), pattern speed | 13.5 km sec$^{-1}$ kpc$^{-1}$ |
| 3. In the solar vicinity | |
| (a) arm spacing (between Perseus and Sagittarius arms) | 3.5 kpc |
| (b) amplitude of spiral gravitational field | 5% of mean field |
| (c) amplitude of variation of projected mass density | 10% of mean |
| (i) in stars | 5% |
| (ii) in gas | 5% |
| (d) rms turbulent velocity of the gas (adopted) | 7 km sec$^{-1}$ |
| (e) magnetic field (adopted for dynamical consistency) | 5 $\mu$G |

tentative picture for the various spiral features in the solar vicinity. We suggest the following picture*.

(1) The Sagittarius and the Perseus arms belong to the primary spiral pattern. (So does the Norma-Scutum arm).

(2) The Carina-Cygnus arm (Bok, 1932, 1969; Bok *et al.*, 1970; Kerr 1969; Dickel *et al.* 1970) belongs to a secondary spiral pattern. (See detailed explanation below).

(3) The Orion arm is essentially a material arm at high inclination, or, as suggested by Frank Shu (private communication), it might be part of the 'long-wave' pattern.

The most important observational support for the above point of view is the absence of giant H II regions in the Orion and the Carina arms. There are some giant H II regions in the Carina direction, but they are so far out that they may very well belong to another spiral arm. Yuan (1969a) has also presented other evidence to show that the Orion arm differs from an ordinary major spiral arm.

It must be stressed again that, in these outer parts of the Milky Way, there could easily be coexistence of many features, as suggested by the photographs of many external galaxies. It is therefore not surprising that the solar vicinity is so difficult to analyze, since we can observe many minor features clearly. Alternative interpretations should be examined by more detailed studies.

The opportunities offered by such local structures should not be forgotten. We can examine the detailed structure of each spiral arm by using diverse methods – optical, radio, infrared, etc. – and by studies of various components – young stars, older stars, neutral hydrogen, ionized hydrogen, magnetic field and various molecules. The fact that the various arms have different characteristics would enable us to learn more about the physical processes in the galaxy. A prominent case that has received considerable attention is the mystery in the Perseus arm (Münch, 1965; Miller, 1967; Rickard, 1967). Here, two components of gaseous motion have been detected, one agreeing quite well with the motion of the stellar association, the other appearing to

* Cf. Prof. Bok's lecture. See also Kerr and Kerr (1970) for additional evidence for a gap between Sagittarius and Carina arms.

expand away from it. It is not yet known for sure whether this is a special phenomenon or a consequence of the density waves*.

## 5. General Remarks: Future Outlook

I have presented a broad survey of the comparison of observational features with the deductions from the density wave theory, – both for the Milky Way and for some external galaxies. I hope that this will convince everyone that the density wave concept introduced by Lindblad, including the material concentration of both gas and stars, is the essential basis for the spiral structure of disk-shaped galaxies. At this point, it seems appropriate to summarize some of the remaining points to be clarified and some of the natural implications to be considered in order to provide a guide for future research.

### A. UNDERSTANDING THE DYNAMICAL MECHANISM

The theory of density waves in stellar systems is very similar to the theory of certain plasma waves – but there are also essential differences. (See Lin *et al.*, 1969, Appendix C). Thus, the understanding of electromagnetic plasma waves and the density waves go hand in hand. Indeed, many of the problems we face are those associated with inhomogeneous plasmas, for which the theory is still not fully developed at this time. Yet it is precisely here that we might find the key to the containment problem of thermo-nuclear reactors. Thus, the understanding of natural phenomena could lead to useful applications. But this is not our main concern here. We must attempt to understand better some of the observed phenomena in nature.

We have not yet fully understood the origin and the development of spiral patterns in the galaxies. I have reported briefly on the work done up to now on (1) numerical experiments, (2) the existence of an unstable mode over the whole galactic disk, and (3) the initiation of the density waves by Jeans instability. Further work appears to be essential and challenging. Various possible mechanisms for exciting the waves should be studied.

In connection with the last possibility, the reflection of the density wave from the inner Lindblad resonance ring presents a very interesting problem. So far, there is only James Mark's linear analysis of such waves. Total absorption by Landau damping was found. Going beyond the linear theory, Contopoulos (1970) has made calculations of the nonlinear response of the stellar system to a given spiral field near resonance, and Roberts (1969), of the gaseous component; but self-sustained waves have not yet been considered in either case. Nonlinear self-sustained waves in gaseous disks have been recently treated by Berry and Vandervoort (1970), extending an earlier work in the linear approximation by Fujimoto (1968).

There should perhaps be more work done to connect the density wave concept with our knowledge of the nature of the individual orbits in a spiral gravitation field. It is

* In a paper presented Aug. 25, 1970, W. W. Roberts showed that the latter is the case.

possible that one can construct self-sustained solutions from the orbital studies in a spiral gravitational field without linearizing the Boltzmann equation. Such an approach would also clarify the relationship of the modern theories with the original Lindblad concepts (Lindblad and Langebartel, 1953).

### B. MILKY WAY

Since Prof. Bok has discussed the desiderata for future observational studies, I shall be rather brief on this subject.

(1) The study of large-scale distribution and motion of neutral hydrogen deserves continued effort. Even with the incorporation of the density wave concept (Yuan, 1969, 1970a; Burton, 1970; Burton and Shane, 1970), the analysis still needs improvement. At the same time, the complication of the local features needs to be resolved. In a sense, this complication is a blessing in disguise, since we are perhaps offered a variety of different features for close scrutiny. The proximity of such features allows us to examine them by diverse methods of observation. The clarification of their nature, with the help of the density-wave concept, would lead to deeper understanding. I would particularly like to call attention to the extensive work done on the Carina features and the mystery in the Perseus arm. In this connection, I wish again to stress *coexistence*. Besides the grand design, with both its short-wave and long-wave patterns, there might also be material arms, spiral patterns of density waves at higher pattern speeds, patterns with more than two arms, or even isolated spiral arms of density waves.

(2) I have spoken very little about the central region of the Milky Way, partly because we believe that there are no density waves of the type we discussed (as indicated by the deficiency of H II regions), but largely because there are many observed features lacking proper interpretation. Another class of problems lacking a full dynamical explanation is that of the high latitude features (Oort, 1970). Both classes of problems are fruitful areas for future research.

(3) Finally, I should mention the subject of the interstellar medium, including gas, dust particles, magnetic field, and cosmic ray particles. Clearly the nature of the nearby spiral arms, and especially the nature of the Orion arm, greatly influence our concept of the interstellar medium; for observational data are often derived from such regions. One should remember the special nature of the Orion arm. It is also important to keep in mind that the interstellar medium is extremely inhomogeneous on the scale of the spacing between spiral arms, because of the existence of a spiral gravitational field. There has as yet been very little discussion in the literature on the implications of such inhomogeneities.

### C. EXTERNAL GALAXIES

We should perhaps now pay more attention to external galaxies for further work on general spiral features, for their structure can be more readily observed. The spiral patterns of only three such galaxies have been analyzed in terms of the density wave theory (Shu *et al.*, 1970). This work should obviously be extended.

A closer examination of the distribution of the various components of the galaxies would be very instructive. (Cf. Morgan, 1970). We mentioned the problem of the position of the dust lane relative to the stellar arm. Zwicky (1957) has stressed the different distributions of the 'blue' and the 'red' components in M 51. Sharpless has done extensive work in isolating the two components in many external galaxies. A great deal of data have been made available by Courtès and his collaborators (Carranza *et al.*, 1968; Carranza *et al.*, 1969; Crillon and Monnet, 1969). I look forward with great expectation to the time when the distribution of neutral hydrogen in many external galaxies can be resolved into spiral arms. A comparison of the spiral structure of neutral hydrogen, ionized hydrogen, blue supergiants, non-blue stellar population and dust lanes would be very instructive. So far, such a comparison is rather incomplete both in the Milky Way and in external galaxies.

With high resolution instruments, it would be very desirable to examine the conditions near the ring structure in external galaxies. This may be expected to throw light on the mechanism just discussed in Section A.

I shall close my remarks with some brief comments on barred spiral galaxies. A theory attributing spiral structure to material arms has been studied by Prendergast (1962, also unpublished notes) and by Freeman (1970). This may represent one class of bar-shaped spirals. I share the feeling expressed by many astronomers that there is a gradual transition from bar-shaped galaxies with a very open structure (e.g., NGC 1300) to disk-shaped galaxies with a short bar structure near the center or no bar structure at all. The spiral structure may then also have a gradual transition from essentially material arms to a pattern of density waves. I look forward to the complete understanding of such problems through further theoretical and observational investigations.

## Bibliography

GENERAL REFERENCES

1. *Galactic Structure*
   *Stars and Stellar Systems;* vol. V
   (ed. by A. Blaauw and M. Schmidt), University of Chicago Press, Chicago, 1965.
2. *Galactic Structure* (Ithaca)
   *Lectures in Applied Mathematics*, vol. IX
   (ed. by Jürgen Ehlers), American Mathematical Society, Providence, R.I., 1967.
3. *Proceedings of IAU Symposium No. 38*, held at Basel, Switzerland, August 29–September 4, 1969
   (ed. by W. Becker and G. Contopoulos), Reidel, The Netherlands, 1970.

SPECIFIC REFERENCES TO INDIVIDUAL ARTICLES

References to articles in the above volumes will be made with the following abbreviations:
1. *Galactic Structure;*
2. *Galactic Structure (Ithaca);*
3. *Basel Symposium.*

Other references are cited in the conventional manner.

Arp, H.: 1969, *Astron. Astrophys.* **3**, 418.
Barbanis, B.: 1968a, *Astrophys. J.* **153**, 71.
Barbanis, B.: 1968b, *Astron. J.* **73**, 784.

Barbanis, B.: 1970, *Basel Symposium*, p. 343.
Barbanis, B. and Woltjer, L.: 1967, *Astrophys. J.* **150**, 461.
Berry, C. L. and Vandervoort, P. O.: 1970, *Basel Symposium*, p. 336.
Bok, B. J.: 1932, Ph.D. Thesis, Harvard University.
Bok, B. J.: 1959, *Observatory* **74**, 61.
Bok, B. J.: 1970, *Basel Symposium*, p. 457.
Bok, B. J., Hine, A. A., and Miller, E. W.: 1970, *Basel Symposium*, p. 246.
Burton, W. B.: 1966, *B.A.M.* **18**, 247.
Burton, W. B.: 1970, *Astron. Astrophys., Suppl.* **2**, 261, 291, 339.
Burton, W. B. and Shane, W. W.: 1970, *Basel Symposium*, p. 397.
Carranza, G., Courtès, G., Georgelin, Y., Monnet, G., and Pourcelot, A.: 1968, *Ann. Astrophys.* **31**, 63.
Carranza, G., Crillon, R., et Monnet, G.: 1969, *Astron. Astrophys.* **1**, 479.
Contopoulos, G.: 1970a, *Basel Symposium*, p. 303.
Contopoulos, G.: 1970b, Report of IAU Commission 33.
Contopoulos, G.: 1970c, *Astrophys. J.* **160**, 113.
Crillon, R. et Monnet, G.: 1969, *Astron. Astrophys.* **2**, 1.
Dickel, H. R. *et al.*: 1970, Basel Symposium, p. 213.
Dieter, N. H.: 1967, *Astrophys. J.* **150**, 435.
Eggen, O. J.: 1965, *Galactic Structure*, p. 117.
Eggen, O. J.: 1969, *Astrophys. J.* **155**, 701.
Feldman, Stuart: 1970, private communication.
Freeman, K. C.: 1966, *Monthly Notices Roy. Astron. Soc.* **133**, 47; **134**, 1, 15.
Freeman, K. C.: 1970, *Basel Symposium*, p. 351.
Fujimoto, M.: 1968, *Astrophys. J.* **152**, 391.
Fujimoto, M.: 1969, *Publ. Astron. Soc. Japan* **21**, 288.
Fujimoto, M. and Miyamoto, M.: 1969, *Publ. Astron. Soc. Japan* **21**, 194.
Fujimoto, M. and Miyamoto, M.: 1970, *Basel Symposium*, p. 440.
Goldreich, P. and Lynden-Bell, D.: 1965, *Monthly Notices Roy. Astron. Soc.* **130**, 125.
Henderson, A. P.: 1967, Unpublished Ph.D. Thesis, University of Maryland.
Hodge, P. W.: 1969, *Astrophys. J.* **155**, 417.
Hohl, F.: 1970, *Basel Symposium*, p. 368.
Hohl, F. and Hockney, R. W.: 1969, *J. Comput. Phys.* **4**, 306.
Humphreys, Roberta M.: 1969, Ph.D. Thesis, University of Michigan.
Hunter, C.: 1963, *Monthly Notices Roy. Astron. Soc.* **126**, 299.
Hunter, C.: 1965, *Monthly Notices Roy. Astron. Soc.* **129**, 321.
Hunter, C.: 1969a, *Astrophys. J.* **157**, 183.
Hunter, C.: 1969b, *Studies Appl. Math.* **48**, 55.
Hunter, C.: 1970, *Basel Symposium*, p. 326.
Hunter, C. and Toomre, A.: 1969, *Astrophys. J.* **155**, 747.
Julian, W. H. and Toomre, A.: 1966, *Astrophys. J.* **146**, 810.
Kalnajs, A. J.: 1965, Ph.D. Thesis, Harvard University.
Kalnajs, A. J.: 1970, *Basel Symposium*, p. 318.
Kerr, F. J.: 1964, *The Galaxy and the Magellanic Clouds, IAU Symposium* No. 20, p. 81.
Kerr, F. J.: 1969, *Ann. Rev. Astron. Astrophys.* **7**, 39.
Kerr, F. J.: 1970, *Basel Symposium*, p. 95.
Kerr, F. J., Burke, B. F., Reifenstein, E. C., Wilson, T. L., and Mezger, P.: 1968, *Nature* **220**, 1210.
Kerr, F. J. and Kerr, Maureen: 1970, *Astrophys. Letters*, in press.
Lin, C. C.: 1966, *SIAM J. Appl. Math.* **14**, 876.
Lin, C. C.: 1967a, *Ann. Rev. Astron. Astrophys.* **5**, 453.
Lin, C. C.: 1967b, *Galactic Structure (Ithaca)*, p. 66.
Lin, C. C.: 1968, in *Galaxies and the Universe (Columbia)*, p. 33.
Lin, C. C.: 1970, *Basel Symposium*, p. 377.
Lin, C. C. and Shu, F. H.: 1964, *Astrophys. J.* **140**, 646.
Lin, C. C. and Shu, F. H.: 1966, *Proc. Nat. Acad. Sci.* **55**, 229.
Lin, C. C. and Shu, F. H.: 1967, *IAU Symposium* No. 31 (Noordwijk), p. 313.

Lin, C. C., Yuan, C., and Shu, F. H.: 1969, *Astrophys. J.* **155**, 721.
Lindblad, B.: 1963, *Stockholm Obs. Ann.* **22**, No. 5.
Lindblad, B. and Langebartel, R.: 1953, *Stockholm Obs. Ann.* **17**, No. 6.
Lindblad, P. O.: 1960, *Stockholm Obs. Ann.* **21**, No. 4.
Lindblad, P. O.: 1962, in *Interstellar Matter in Galaxies*, Benjamin, New York, p. 222.
Lynden-Bell, D.: 1967, *Galactic Structure (Ithaca)*, p. 150.
Lynden-Bell, D.: 1970, *Basel Symposium*, p. 331.
Lynden-Bell, D. and Ostriker, J. D.: 1967, *Monthly Notices Roy. Astron. Soc.* **136**, 273.
Lynds, B. T.: 1967, *Sky Telescope* **34**, 18.
Lynds, B. T.: 1970, *Basel Symposium*, p. 26.
Mark, James W-K.: 1970, private communication.
Marochnik, L. S. and Suchkov, A. A.: 1969, *Astrophys. Space Sci.* **4**, 317.
Mathewson, D. S.: 1968, *Astrophys. J.* **153**, L47.
Mathewson, D. S.: 1969, *Proc. Astron. Soc. Australia* **1**, 209.
Mathewson, D. S. and Nicholls, D. C.: 1968, *Astrophys. J.* **154**, L11.
Mayor, M.: 1970, *Astron. Astrophys.* **6**, 60–66.
Mezger, P. G.: 1970, *Basel Symposium*, p. 106.
Mezger, P. G., Wilson, T. L., Gardner, F. F., and Milne, D. K.: 1969, *Astron. Astrophys.* **4**, 96.
Miller, J. S.: 1968, *Astrophys. J.* **151**, 473.
Miller, R. H. and Prendergast, K. H.: 1968, *Astrophys. J.* **151**, 699.
Miller, R. H., Prendergast, K. H., and Quirk, W. J.: 1970, *Astrophys. J.* **161**, 903.
Miyamoto, M.: 1969, *Publ. Astron. Soc. Japan* **21**, 319.
Morgan, W. W.: 1970, *Basel Symposium*, p. 9.
Morse, P. M. and Feshbach, H.: 1953, *Methods of Theoretical Physics*, p. 1092.
Munch, G.: 1965, *Galactic Structure*, p. 203.
Oort, J. H.: 1967, *IAU Symposium* No. 31, p. 279.
Oort, J. H.: 1970, *Basel Symposium*, pp. 1, 142.
Prendergast, K. H.: 1962, in *Interstellar Matter in Galaxies*, Benjamin, New York, p. 217.
Reifenstein, E. C., Wilson, T. L., Burke, B. F., Mezger, P. G., and Altenhoff, W.: 1970, *Astron. Astrophys.* **4**, 357.
Rickard, J.: 1968, *Astrophys. J.* **152**, 1019.
Roberts, M. S.: 1966, *Astrophys. J.* **144**, 639.
Roberts, M. S.: 1968, in *Interstellar Ionized Hydrogen*, Benjamin, New York, p. 617.
Roberts, W. W.: 1969, *Astrophys. J.* **158**, 123.
Roberts, W. W.: 1970, *Basel Symposium*, p. 415.
Roberts, W. W. and Yuan, C.: 1970, *Astrophys. J.* **161**, 877.
Sandage, A.: 1961, *The Hubble Atlas of Galaxies*, Carnegie Institution, Washington, D.C.
Shane, W. W. and Bieger-Smith, G. P.: 1966, *Bull. Astron. Inst. Neth.* **18**, 263.
Shu, F. H.: 1968, Ph.D. Thesis, Harvard University.
Shu, F. H.: 1969, *Astrophys. J.* **158**, 505.
Shu, F. H.: 1970a, *Astrophys. J.* **160**, 89.
Shu, F. H.: 1970b, *Astrophys. J.* **160**, 99.
Shu, F. H.: 1970c, *Basel Symposium*, p. 323.
Shu, Frank H., Stachnik, Robert V., and Yost, Jonathan C.: 1970, *Astron. J.* (Abstract only), in press.
Strömgren, Bengt: 1967, *IAU Symposium* No. 31, p. 323.
Toomre, A.: 1964, *Astrophys. J.* **139**, 1217.
Toomre, A.: 1969, *Astrophys. J.* **158**, 899.
Toomre, A.: 1970, *Basel Symposium*, p. 334.
Vandervoort, P. O.: 1970a, *Astrophys. J.* **161**, 187.
Vandervoort, P. O.: 1970b, *Basel Symposium*, p. 337.
Vaucouleurs, G. de: 1970, *Basel Symposium*, p. 18.
Weaver, H.: 1970, *Basel Symposium*, p. 126.
Westerhout, G.: 1958, *Galactic Structure*, p. 196.
Wilson, T. L.: 1969, Ph.D. Thesis, M.I.T.
Wilson, T. L.: 1970, *Basel Symposium*, p. 140.
Wilson, T. L., Mezger, P. G., Gardner, F. F., and Milne, D. K.: 1970, to be published.

Woltjer, L.: 1970, *Basel Symposium*, p. 439.
Yabushita, S.: 1968, *Monthly Notices Roy. Astron. Soc.* **140**, 109.
Yabushita, S.: 1969a, *Monthly Notices Roy. Astron. Soc.* **142**, 201.
Yabushita, S.: 1969b, *Monthly Notices Roy. Astron. Soc.* **143**, 231.
Yuan, C.: 1969a, *Astrophys. J.* **158**, 871.
Yuan, C.: 1969b, *Astrophys. J.* **158**, 889.
Yuan, C.: 1970a, *Basel Symposium*, p. 391.
Yuan, C.: 1970b, *Astron. J.* (Abstract only), in press.
Zwicky, F.: 1957, *Morphological Astronomy*, Springer-Verlag, Berlin, p. 198.

*Studies in Applied Mathematics, Vol. LII, No. 1, March 1973. Copyright © 1973 by The Massachusetts Institute of Technology*

# A Forcing Mechanism for Spiral Density Waves in Galaxies

*By S. I. Feldman and C. C. Lin*

In this paper, we consider gas dynamical models of spiral galaxies in which there is a rigidly rotating, weakly barlike structure in the central regions. It is found that, in the neighborhood of the corotation circle, this barlike structure forces a *trailing spiral wave*. Such a driven wave could then propagate inwards to the bar and complete a feedback loop to maintain the spiral structure. For basic distributions typical of our Galaxy, if a ten-per-cent oval distortion is assumed, the strength of the induced spiral field is found to be of the order of a few per cent of the axisymmetric field, in agreement with observational data.

## 1. Introduction

It is by now widely accepted that density waves underlie the spiral patterns observed in disk-shaped galaxies. The origin and maintenance of such spiral waves still present challenging problems. Specifically, Toomre [1] has shown that short trailing spiral waves propagate away from corotation. Furthermore, his calçulations show that, in a certain model, the inward propagating waves are absorbed at the location of the inner Lindblad resonance. This latter conclusion was confirmed by Mark [2] in a more general analysis. There is thus a need for a mechanism for replenishing these spiral waves.

One possible mechanism for the maintenance of a wave pattern was proposed by one of us [3, 4]. It is visualized that the inward propagating spiral wave pattern causes an oval distortion of the mass distribution in the central region of the galaxy, and that this oval distortion, rotating at the pattern speed, reinforces the spiral wave system. Corotation is the natural place to look for this feedback process since the short trailing waves propagate away from there. (The problems of origin and of maintenance should be treated as distinct. It *was* visualized that the wave pattern might have originated from a Jeans instability in the outer parts of the galactic disk where the velocity is low [3]. It should be made clear that this is only one of the possible mechanisms. The initial irregularity of the system could also be the origin of spiral structure. For further details, see [3].)

It should also be noted that Holmberg [6] believes that all observed galaxies have at least a small bar-shaped structure in their central regions. In all galaxies for which measurements are available, the spiral arms trail the direction or rotation [7].

In this paper, we focus our attention on the driving mechanism. We consider models in which a small imposed rotating gravitational field perturbs a gaseous disk or infinite cylinder, originally in equilibrium. We shall show that a bar-like central distortion induces a stationary trailing spiral wave in the neighborhood of the ring of corotation. The feedback loop in the proposed mechanism will be closed as this wave system propagates inwards and reinforces the original distortion.

Our models consist of three components: a spherically symmetric "stellar" nucleus, a rigidly rotating lightweight "stellar" bar, and a "gaseous" disk. The axis of rotation of the bar passes through the center of the nucleus. In this paper, we wish to discuss the general nature of the forcing process, so we avoid consideration of the internal dynamics of the nucleus and bar by assuming that these components are self-consistent (i.e., that the Poisson's equation and the equilibrium equations are individually satisfied) and that the stars and gas interact only through their gravitational fields. To simplify the calculations, we assume that the bar is much less massive than the nucleus and perturbs the flow of the gas only slightly. In the inertial frame, we erect a cylindrical coordinate system, $(\varpi, \theta, z)$, with the origin at the center of the nucleus and with the positive $z$-axis along the axis of rotation. Consistent with a permanent wave pattern, we shall assume the flow to be stationary in a frame of reference rotating with the bar. Thus, all the physical quantities (the velocities, density, pressure, and gravitational potential) must be independent of time in terms of the coordinates $(\varpi, \phi, z)$, where

$$\phi = \theta - \Omega_p t, \tag{1.1}$$

and $\Omega_p$ is the angular frequency of rotation of the bar.

From observations, most of the stars appear to move in the galactic plane. Since the galaxy is quite thin, it seems reasonable to ignore "rolling" and so to assume a purely two-dimensional flow.

If no bar is present, the basic flow can be axisymmetric and circular; then

$$\underline{u} = (u_{\varpi}, u_{\phi}, u_z) = (0, \varpi[\Omega - \Omega_p], 0), \tag{1.2}$$

where the $u$'s are the cylindrical components of the velocity measured in the rotating frame and $\varpi\Omega(\varpi)$ is the unperturbed circular velocity as measured in the inertial frame. The corotation radius, $\varpi_{co}$, is defined by

$$\Omega(\varpi_{co}) = \Omega_p \tag{1.3}$$

It can be shown [5] that gas and stars act in a similar fashion in the corotation region, even though this approximation breaks down near the Lindblad resonance circles. This fact justifies our considering a purely gaseous disk as a model for the combined disk of stars and gas, even though real galaxies contain little gas. For simplicity, we assume that the gas is inviscid and satisfies a polytrope law with constant index, $\gamma$:

$$p = K_p \rho^{\gamma} \quad \text{where } K_p > 0 \text{ and } \gamma > 1. \tag{1.4}$$

## 2. Gas dynamics

The dynamics of a gravitating, two-dimensional, barytropic gaseous system are determined by five equations: the equation of state, the continuity equation, the radial momentum equation, the tangential momentum equation, and Poisson's equation. The Euler equations in the rotating coordinate system are [8]:

$$\frac{\partial}{\partial \varpi}(\rho \varpi u_{\varpi}) + \frac{\partial}{\partial \phi}(\rho u_{\phi}) = 0, \tag{2.1}$$

$$u_{\varpi}\frac{\partial}{\partial \varpi}u_{\varpi} + \frac{1}{\varpi}u_{\phi}\frac{\partial}{\partial \phi}u_{\varpi} - \frac{1}{\varpi}u_{\phi}^2 - 2\Omega_p u_{\phi} - \Omega_p^2 \varpi = -\left[\frac{\partial \mathscr{V}}{\partial \phi} + \frac{1}{\rho}\frac{\partial p}{\partial \varpi}\right], \tag{2.2}$$

and

$$u_{\varpi}\frac{\partial}{\partial \varpi}u_{\phi} + \frac{1}{\varpi}u_{\phi}\frac{\partial}{\partial \phi}u_{\phi} + 2\Omega_p u_{\varpi} + \frac{1}{\varpi}u_{\varpi}u_{\phi} = -\frac{1}{\varpi}\left[\frac{\partial \mathscr{V}}{\partial \phi} + \frac{1}{\rho}\frac{\partial p}{\partial \phi}\right], \tag{2.3}$$

where $\rho$ is the mass density of the gas, $p$ is the pressure of the gas, and $\mathscr{V}$ is the gravitational potential due to the nucleus, bar, and gas.

We can use the equation of state to simplify the momentum equations by introducing the enthalpy, $\eta$, defined by:

$$d\eta = \frac{dp}{\rho}, \tag{2.4}$$

$$\eta = \frac{\gamma}{\gamma - 1}K_p \rho^{\gamma - 1}. \tag{2.5}$$

The speed of sound, $a_s$, is given by

$$a_s^2 = \frac{dp}{d\rho} = (\gamma - 1)\eta. \tag{2.6}$$

It is convenient to work in terms of dimensionless variables. We therefore introduce, based on a suitable length scale $\varpi_0$ and a frequency scale $\Omega_0$ to be specified below, the dimensionless variables $(r, \omega, \omega_p, u, v, h, \psi)$ defined as follows:

$$\varpi = \varpi_0 r, \tag{2.7}$$

$$\Omega = \Omega_0 \omega, \tag{2.8}$$

$$\Omega_p = \Omega_0 \omega_p, \tag{2.9}$$

$$u_{\varpi} = (\varpi_0 \Omega_0)u, \tag{2.10}$$

$$u_{\phi} = (\varpi_0 \Omega_0)v, \tag{2.11}$$

$$\eta = (\varpi_0 \Omega_0)^2 h, \tag{2.12}$$

and

$$\mathscr{V} = (\varpi_0 \Omega_0)^2 \psi. \tag{2.13}$$

The Euler equations then become:

$$(\gamma - 1)h\left[u + r\frac{\partial u}{\partial r} + \frac{\partial v}{\partial \phi}\right] + \left[ru\frac{\partial h}{\partial r} + v\frac{\partial h}{\partial \phi}\right] = 0, \tag{2.14}$$

$$u\frac{\partial u}{\partial r} + \frac{1}{r}v\frac{\partial u}{\partial \phi} - \frac{v^2}{r} - 2\omega_p v - \omega_p^2 r = -\frac{\partial}{\partial r}(h + \psi), \tag{2.15}$$

and

$$u\frac{\partial v}{\partial r} + \frac{1}{r}v\frac{\partial v}{\partial \phi} + \frac{uv}{r} + 2\omega_p u = -\frac{1}{r}\frac{\partial}{\partial \phi}(h + \psi). \tag{2.16}$$

We assume that the bar changes the basic axisymmetric circular flow only slightly. The perturbations due to the bar can all be expanded in Fourier series of the form

$$\sum_{m=-\infty}^{\infty} c_m(r)\, e^{-im\phi}. \tag{2.17}$$

After linearizing the equations of motion on the assumption that the perturbations are small, we find that the equations for the Fourier components are independent. Therefore, we need only consider perturbations with a single Fourier component:

$$h = h_0(r) + h_1(r)\, e^{-im\phi}, \tag{2.18}$$

$$\psi = \psi_0(r) + \psi_1(r)\, e^{-im\phi}, \tag{2.19}$$

$$u = 0 + u_1(r)\, e^{-im\phi}, \tag{2.20}$$

and

$$v = v_0(r) + v_1(r)\, e^{-im\phi}; \tag{2.21}$$

where the real part is to be taken, and

$$v_0(r) = r(\omega - \omega_p). \tag{2.22}$$

In the following sections, we shall consider the potential as the sum of two terms, one due to the stars, one due to the gas. Thus,

$$\psi = \psi_{\text{stars}} + \psi_{\text{gas}}, \tag{2.23}$$

$$\psi_0 = \psi_{0,\text{nucleus}} + \psi_{0,\text{gas}}, \tag{2.24}$$

and

$$\psi_1 = \psi_b + \psi_g, \tag{2.25}$$

where $\psi_b$ is due to the bar and $\psi_g$ is due to the perturbed gas density.

In the following, let a prime denote differentiation with respect to $r$:

$$f' = \frac{df}{dr}. \tag{2.26}$$

Upon linearizing the Euler equations we find that:

$$\omega^2 = \frac{1}{r}(h_0' + \psi_0'), \tag{2.27}$$

$$(\gamma - 1)h_0(u_1 + ru_1' - imv_1) + [rh_0'u_1 - im(\omega - \omega_p)rh_1] = 0, \tag{2.28}$$

$$im(\omega - \omega_p)u_1 + 2\omega v_1 = (h_1' + \psi_1'), \tag{2.29}$$

and

$$(2\omega + r\omega')u_1 - im(\omega - \omega_p)v_1 = \frac{im}{r}(h_1 + \psi_1). \tag{2.30}$$

To simplify the calculations, we introduce several dimensionless parameters:

$$E^2 = \frac{1}{(\gamma - 1)h_0} = (\text{scaled speed of sound})^{-2}, \tag{2.31}$$

$$H = \frac{1}{\gamma - 1}\frac{d \ln h_0}{d \ln r} = \frac{d \ln \rho_0}{d \ln r}, \tag{2.32}$$

and

$$\nu = -\frac{m\Omega_0}{\kappa}(\omega - \omega_p), \tag{2.33}$$

where the dimensional epicyclic frequency, $\kappa$, is given by

$$\frac{\kappa^2}{\Omega_0^2} = 4\omega^2\left[1 + \frac{r\omega'}{2\omega}\right], \tag{2.34}$$

and $\nu$ is the same symbol as used in previous papers on density waves. The momentum equations permit us to express the perturbation velocities in terms of the perturbation enthalpy and potential; thus,

$$u_1 = \frac{im}{m^2\kappa^2(1 - \nu^2)/\Omega_0^2}\left[\frac{2\omega}{r}(h_1 + \psi_1) + (\omega - \omega_p)(h_1' + \psi_1')\right], \tag{2.35}$$

$$v_1 = \frac{1}{m^2\kappa^2(1 - \nu^2)/\Omega_0^2}\left[\frac{m^2(\omega - \omega_p)}{r}(h_1 + \psi_1) + (2\omega + r\omega')(h_1' + \psi_1')\right] \tag{2.36}$$

Elimination of the velocity components from the continuity equation (2.28) yields

$$h_1'' + Ah_1' + (B + C)h_1 + \psi_1'' + A\psi_1' + B\psi_1 = 0, \tag{2.37}$$

where

$$A(r) = \frac{1 + H}{r} - \frac{d}{dr}\ln[(1 - \nu^2)\kappa^2], \tag{2.38}$$

$$B(r) = -\frac{m^2}{r^2} + \frac{2\omega}{r^2(\omega - \omega_p)}\left\{H - r\frac{d}{dr}\ln[\kappa^2(1 - \nu^2)/\omega]\right\}, \tag{2.39}$$

and

$$C(r) = -E^2m^2\kappa^2(1 - \nu^2)/\Omega_0^2. \tag{2.40}$$

If we define the function $D$ by the equation

$$D(r) = -[\psi_b'' + A\psi_b' + B\psi_b], \tag{2.41}$$

then we can write (2.37) as:

$$h_1'' + Ah_1' + (B + C)h_1 + \psi_g'' + A\psi_g' + B\psi_g = D. \tag{2.42}$$

The functions $A, B, C$ depend only on the zero-order model and the pattern frequency, and are independent of the perturbations. The quantity $D$ depends on the perturbing bar, and represents the driving agent. These functions may be singular at the origin, infinity, the corotation radius, and the Lindblad resonances. The corotation radius is where

$$\omega(r) = \omega_p \quad \text{and} \quad \nu(r) = 0. \tag{2.43}$$

The Lindblad resonances occur at those radii where

$$\nu(r) = \pm 1. \tag{2.44}$$

### 3. The neighborhood of corotation

The coefficients $B$ and $D$ of equation (2.42) are singular at corotation due to the factor

$$\frac{1}{\omega(r) - \omega_p}. \tag{3.1}$$

In order to integrate the differential equation through this pole, we use the usual method of analytic continuation and consider "slightly complex" values of $r, \Omega_p, \omega_p$, and $\omega$, $\omega(r)$ being assumed to be a locally analytic function. Eventually, we find that we must deform the path of integration into the complex $r$-plane when $\omega_p$ is real. We shall see that the proper choice of the contour is as shown in Figure 1b, and that this choice forces the generated waves to *trail behind* the direction of rotation.

For convenience, we choose the length scale, $\varpi_0$, to be the corotation radius and the frequency scale, $\Omega_0$, to the real pattern frequency, $\Omega_p$. Then we have $\omega(1) = 1$, and we may write

$$\Omega_p = \Omega_0(1 + is) \quad \text{and} \quad \omega_p = 1 + is, \tag{3.2}$$

where

$$s = 0^+ \quad \text{or} \quad 0^- \tag{3.3}$$

Now consider a typical perturbation term like

$$h_1(r)\, e^{-im\phi} = h_1(r)\, e^{-im(\theta - \Omega_p t)}\, e^{-ms\Omega_p t}. \tag{3.4}$$

If we were solving an initial value problem for the modes, the expansion in Fourier series in the rotating angle $\phi$ would be obtained by a Fourier–Laplace transform in $(\theta, t)$. The usual rule for the inversion contour would then tell us that *neutral modes are to be regarded as limits of amplified modes* (see Appendix B).

If we permit $s$ to be small but non-zero, the amplified modes are those satisfying

$$ms < 0. \tag{3.5}$$

For $\omega_p$ complex, the pole of (3.1) lies slightly off the real $r$-axis, at the point $1 + i\varepsilon$ such that

$$\omega(1 + i\varepsilon) = 1 + is, \tag{3.6}$$

or

$$\varepsilon = s/\omega'(1). \tag{3.7}$$

For any reasonable mass distribution and choice of $\Omega_p$, the rotation curve $\Omega(\varpi)$ decreases near corotation, so

$$\omega'(1) < 0. \tag{3.8}$$

Thus, $m$ and $\varepsilon$ must have the same sign. For convenience, in the rest of this paper, we consider only positive modes $m$. Thus, the pole is *above* the real $r$-axis, as shown in Figure 1a. If we now take the limit $s \to 0^-$, $\varepsilon \to 0^+$, we see that we must integrate the differential equation along the contour shown in Figure 1b. In the rest of this paper, we shall consider $s$ to be zero, so

$$\Omega_0 = \Omega_p, \tag{3.9}$$

and

$$\omega_p = 1. \tag{3.10}$$

Note that, for $m$ positive, if a solution behaves like

$$h_1(r) \sim e^{ikr}, \tag{3.11}$$

then the result is a trailing spiral if $k < 0$ and a leading spiral if $k > 0$.

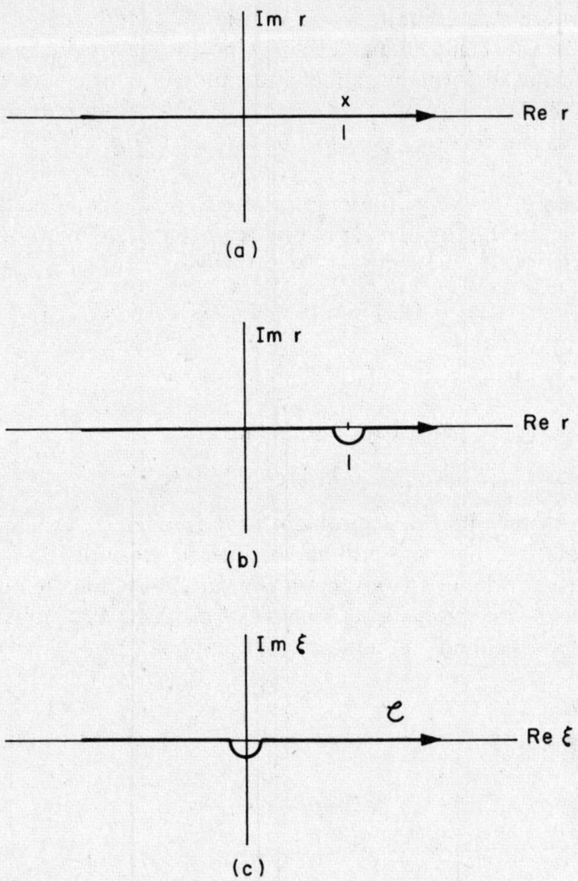


Figure 1. Proper choice of the contour in the case $m > 0$ if the angular velocity of galactic rotation decreases outwards at corotation.

(a) Contour with pole slightly above the real $r$-axis

(b) Contour deformed below pole at $r = 1$

(c) Contour $\mathscr{C}$ in the $\xi$-plane

In order to make further analysis possible, we must make an appraisal of the size of the various coefficients in the differential equations. We shall evaluate them for values for the solar neighborhood and assume that the order relationship (3.16) thus found is valid near corotation. For the solar neighborhood,

$$\varpi_0 = 10\,\text{kpc}, \tag{3.12}$$

$$\Omega_0 = 10\,\text{km/sec/kpc}, \tag{3.13}$$

$$a_s = 10\,\text{km/sec}, \tag{3.14}$$

and thus

$$E = 10. \tag{3.15}$$

In place of the large parameter $E$, let us define the small parameter $\delta$ by

$$\frac{\kappa(1)}{\Omega_0}\,\delta = \frac{1}{E(1)} = \frac{a_s(1)}{\varpi_0 \Omega_0} \ll 1, \tag{3.16}$$

where $\delta$ is the ratio of the acoustic velocity at corotation to the unperturbed circular velocity of galactic rotation. We can now write the coefficients of equation (2.42) in the form:

$$A = a_1(r), \tag{3.17}$$

$$B = \frac{a_2(r)}{r-1}, \tag{3.18}$$

$$C = -\delta^{-2} a_3(r), \tag{3.19}$$

and

$$D = \frac{a_4(r)}{r-1}; \tag{3.20}$$

where all of the $a_i$ are of order unity, regular, and non-zero in the neighborhood of $r = 1$. The size of $C$ suggests balancing the term $(h_1'' + \psi_g'')$ in equation (2.42) by the term $Ch_1$. We therefore introduce a new coordinate, $\xi$:

$$\xi = \frac{r-1}{\delta} \tag{3.21}$$

The coefficients of the different equation in $\xi$ are regular power series in $\delta$. Hence, we may look for solutions $h_1$ and $\psi_g$ in power series of $\delta$:

$$h_1 = h_1^{(0)} + \delta h_1^{(1)} + \cdots \tag{3.22}$$

$$\psi_g = \psi_g^{(0)} + \delta \psi_g^{(1)} + \cdots \tag{3.23}$$

By noting that

$$a_3(1) = 1, \tag{3.24}$$

we see that the first two powers of the gas dynamics equation (2.42) are:

$$\frac{d^2\psi_g^{(0)}}{d\xi^2} + \frac{d^2 h_1^{(0)}}{d\xi^2} - h_1^{(0)} = 0 \tag{3.25}$$

$$\frac{d^2\psi_g^{(1)}}{d\xi^2} + \frac{d^2h_1^{(1)}}{d\xi^2} - h_1^{(1)}$$
$$= \frac{a_4(1)}{\xi} - a_1(1)\frac{d}{d\xi}(h_1^{(0)} + \psi_g^{(0)}) - \frac{a_2(1)}{\xi}(h_1^{(0)} + \psi_g^{(0)}) + \xi a_3'(1)h_1^{(0)} \quad (3.26)$$

Figure 1c shows the path of integration $\mathscr{C}$.

Because of singularities in the coefficients and secularities in the solutions, the expansions are only valid for:

$$\delta < 0(\xi) < \frac{1}{\delta} \quad (3.27)$$

In the following discussions, we shall find it important to consider the quantity $a_4(r)$ at the corotation point $r = 1$, since it measures the effect of the bar as a driving agent. From (3.20), (2.39) and (2.41), it is clear that $a_4(1)$ is given by the product of $\psi_b(1)$ and the quantity

$$B_1(r) = \frac{-2\omega}{r^2(d\omega/dr)}\left\{\frac{1}{\gamma - 1}\frac{d\log h_0}{d\log r} - \frac{d\ln[\kappa^2(1 - \nu^2)/\omega]}{d\ln r}\right\} \quad (3.28)$$

evaluated at $r = 1$ and $\nu = 0$. If we adopt power law approximations for the density and the rotational velocities, then

$$\frac{1}{\gamma - 1}\frac{d\ln h_0}{d\ln r} = \frac{d\ln\rho_0}{d\ln r} = -\mu_1, \quad (3.29)$$

$$\frac{d\ln\omega}{d\ln r} = \frac{d\ln\kappa}{d\ln r} = -\mu_2; \quad (3.30)$$

and $a_4(1)$ becomes

$$a_4(1) = 2\left(1 - \frac{\mu_1}{\mu_2}\right)\psi_b(1) \quad (3.31)$$

where $\psi_b(1)$, as we may recall, is the dimensionless form of the gravitational potential due to the barlike structure [Eq. (2.25)].

## 4. Gravitation

Now we introduce the role of self-gravitation. By assumption, the stellar components are self-consistent, so the gas must satisfy Poisson's equation:

$$\nabla^2\mathscr{V}_{\text{gas}} = 4\pi G\rho. \quad (4.1)$$

We have also assumed that the flow is exactly two-dimensional. There are two models in which this requirement is satisfied: the infinitely thick vertically homogeneous cylinder and the infinitesimally thin disk. We shall show that, in both extreme cases, the bar drives a trailing spiral wave in the neighborhood of corotation.

We shall first (Section 5) consider the cylinder model because of its analytic simplicity. As just mentioned, we assume that the density is independent of $z$ and that there are no vertical motions. By symmetry, there will be no vertical forces. A very thick cylinder will be approximately represented by this model.

In Section 6, we shall consider the disk model, in which all of the mass is constrained to the plane $z = 0$. This model is used in most of the previous papers on the density wave theory to describe approximately the behavior of a very thin disk.

When the driving term due to the bar is absent in equation (2.42), we obtain free spiral solutions as found in earlier work on density waves. In this paper, we are particularly interested in solutions due to the driving.

We do not suggest that an infinite cylinder provides an adequate model for a galaxy, but present it for its mathematical simplicity and because it provides the opposite case from the disk. It is significant that both models give the same qualitative result, a trailing wave driven at corotation. In Section 7, we shall interpret the result of Section 6 by using a simple galactic model.

## 5. The cylinder model

In this model, there are no vertical forces or density gradients so that

$$\frac{\partial^2 \mathscr{V}_{\text{gas}}}{\partial z^2} = 0. \tag{5.1}$$

We can write (4.1) as:

$$\frac{\partial^2 \psi_{\text{gas}}}{\partial r^2} + \frac{1}{r}\frac{\partial \psi_{\text{gas}}}{\partial r} + \frac{1}{r^2}\frac{\partial^2 \psi_{\text{gas}}}{\partial \phi^2} = \frac{4\pi G\rho}{\Omega_0^2}. \tag{5.2}$$

Linearizing, we find

$$\psi''_{0,\text{gas}} + \frac{1}{r}\psi'_{0,\text{gas}} = \frac{4\pi G\rho_0}{\Omega_0^2}, \tag{5.3}$$

and

$$\psi''_g + \frac{1}{r}\psi'_g - \frac{m^2}{r^2}\psi_g = \Lambda h_1, \tag{5.4}$$

where

$$\rho_0 = \left[\varpi_0^2\Omega_0^2\frac{\gamma-1}{\gamma K_p}h_0\right]^{1/(\gamma-1)}, \tag{5.5}$$

and

$$\Lambda = \frac{4\pi G\rho_0}{(\gamma-1)h_0\Omega_0^2} = \frac{4\pi G\rho_0\varpi_0^2}{(a_s)_0^2} = (k_J\varpi_0)^2 = \left(\frac{2\pi\varpi_0}{\lambda_J}\right)^2. \tag{5.6}$$

In the above formula, $\lambda_J$ is the Jeans length, $k_J$ is the Jeans wavenumber, and $(a_s)_0$ is the speed of sound in the unperturbed medium.

All of the dynamics of the linearized problem for the infinite cylinder are contained in the two linear second-order ordinary differential equations (2.42) and (5.4). These could be combined into a single fourth-order linear inhomogeneous ordinary differential equation for $\psi_g$, but it is easier to work with the system of second-order equations.

To determine the size of $\Lambda$, let us assume that the mass density $\rho_0$ is that given by the Oort limit in the solar neighborhood,

$$\rho_0 = 10^{-23}\ \text{g/cm}^3 \tag{5.7}$$

Assuming the other solar neighborhood values (3.12–15), we find

$$\Lambda = 8100 \tag{5.8}$$

We shall assume that $\Lambda$ is of order $\delta^{-2}$ [cf. (3.16)], so we define the parameter $L$ by the relation

$$\Lambda = \delta^{-2} L(r), \tag{5.9}$$

where $L$ is slowly varying, of order unity, and where $L(1) > 1$.

Using this assumption and the expansions (3.22–23), we find the first two orders of Poisson's equation (5.4) to be:

$$\frac{d^2\psi_g^{(0)}}{d\xi^2} = L(1)h_1^{(0)}, \tag{5.10}$$

and

$$\frac{d^2\psi_g^{(1)}}{d\xi^2} = L(1)h_1^{(1)} + \xi L'(1)h_1^{(0)} - \frac{d\psi_g^{(0)}}{d\xi}. \tag{5.11}$$

Combining the gas dynamics equations (3.25–26) with Poisson's equations (5.10–11) then yields

$$\frac{d^2 h_1^{(0)}}{d\xi^2} + \alpha^2 h_1^{(0)} = 0, \tag{5.12}$$

and

$$\frac{d^2 h_1^{(1)}}{d\xi^2} + \alpha^2 h_1^{(1)} = \frac{a_4(1)}{\xi} + F_{\text{cyl}}^{(0)}; \tag{5.13}$$

where

$$F_{\text{cyl}}^{(0)} = \xi[a_3'(1) - L'(1)]h_1^{(0)} + [1 - a_1(1)]\frac{d\psi_g^{(0)}}{d\xi} - a_1(1)\frac{dh_1^{(0)}}{\xi} - \frac{a_2(1)}{\xi}[h_1^{(0)} + \psi_g^{(0)}], \tag{5.14}$$

and

$$\alpha = [L(1) - 1]^{1/2} \tag{5.15}$$

These equations have solutions (see Appendix B)

$$h_1^{(0)} = k_L^{(0)} e^{i\alpha\xi} + k_T^{(0)} e^{-i\alpha\xi} \tag{5.16}$$

and

$$h_1^{(1)} = k_L^{(1)} e^{i\alpha\xi} + k_T^{(1)} e^{-i\alpha\xi} + k_L^{(0)} f_L(\xi) + k_T^{(0)} f_T(\xi) + k_D f_D(\xi). \tag{5.17}$$

The terms involving $f_L$ and $f_T$ can be written in terms of integrals of elementary functions. (The subscripts $L$ and $T$ denote leading and trailing terms, respectively.) These terms, due to the term $F^{(0)}_{\text{cyl}}$ in (5.13), are the first-order corrections to the free spiral solutions, so we write:

$$h_1 = k_L e^{i\alpha\xi}[1 + O(\delta)] + k_T e^{-i\alpha\xi}[1 + O(\delta)] + \delta k_D f_D(\xi) + O(\delta^2), \qquad (5.18)$$

where $k_D$ is given by

$$k_D = \frac{\pi}{\alpha} a_4(1), \qquad (5.19)$$

which is proportional to the strength of the bar, and $f_D(\xi)$ is given by

$$f_D(\xi) = e^{-i\alpha\xi} I(\xi, -\alpha) - e^{i\alpha\xi} I(\xi, \alpha), \qquad (5.20)$$

which has the asymptotic behaviors

$$\begin{cases} 0, & \text{as } \xi \to +\infty, \\ e^{-i\alpha\xi}, & \text{as } \xi \to -\infty. \end{cases}$$

The function $I$ is defined in Appendix B.

Thus, we can write the density perturbation as the sum of a leading spiral ($k_L e^{i\alpha\xi}$), a trailing spiral ($k_T e^{-i\alpha\xi}$), and a wave due to and proportional to the strength of the bar ($k_D f_D$). Furthermore, the driven solution dies away at plus infinity and behaves like a *trailing* spiral at minus infinity; the driven leading spiral solution dies away at minus infinity. Figures 2 and 3 show spirals of these forms for $m = 2$ and $k_L = 0$. In Figure 2, $k_T$ is chosen to put the spiral inside of corotation. In Figure 3, $k_T$ is chosen to make the spiral die away inside corotation. Note the similarity between Figure 2 and a normal spiral galaxy and between Figure 3 and a barred spiral galaxy.

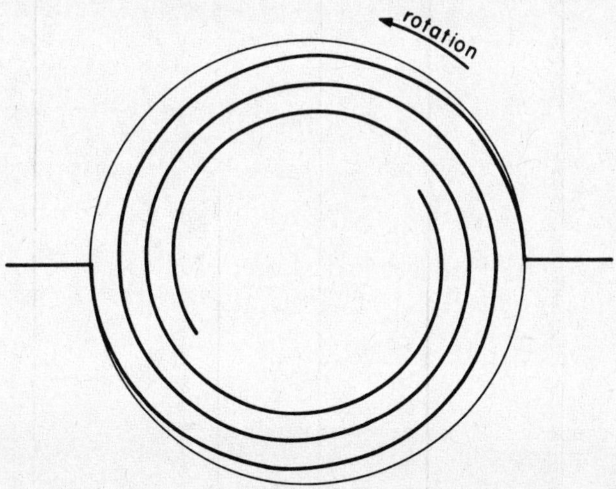


Figure 2. Spiral solution of infinite cylinder model, decaying outside of corotation.

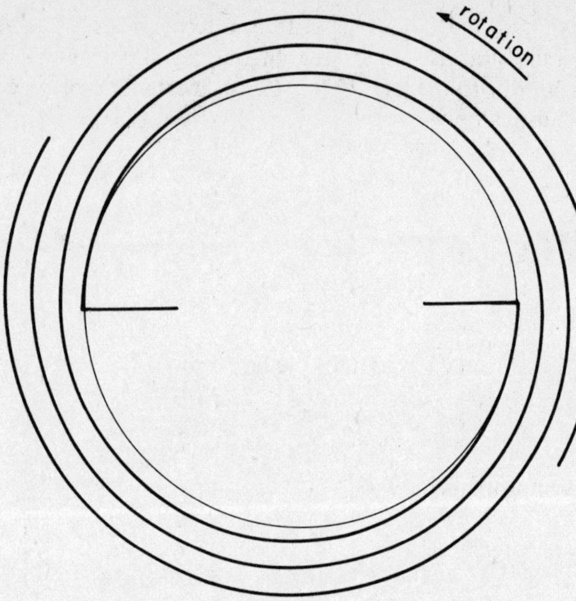


Figure 3. Spiral solution of infinite cylinder model, decaying inside of corotation.

## 6. The disk model

In this model, we project all of the mass into a surface density, $\sigma$; thus,

$$\rho(\omega, \phi, z) = \sigma(\overline{\omega}, \phi)\delta(z), \tag{6.1}$$

where $\delta$ is the Dirac function. From Appendix C, we see that, in the limit of tightly wound spirals,

$$\frac{d\psi_g}{dr} = -i\,\Sigma(r)(\operatorname{sgn} k)h_1, \tag{6.2}$$

where (sgn $k$) is $+1$ if the spirals are leading and $-1$ if the spirals are trailing, and

$$\Sigma = \frac{2\pi G\sigma_0}{(\gamma - 1)h_0\overline{\omega}_0\Omega_0^2}. \tag{6.3}$$

Following [5], we adopt the following value for the solar neighborhood:

$$\sigma_0 = 114\, M_{\text{sun}}/pc^2 = 0.024\ \text{g/cm}^2. \tag{6.4}$$

Using the solar neighborhood values (3.12–15), we find

$$\Sigma = 310, \tag{6.5}$$

so we assume

$$\Sigma = O(\delta^{-1}). \tag{6.6}$$

We therefore define a new dimensionless parameter $S$ by

$$(\operatorname{sgn} k)\Sigma = \frac{2}{\delta}S(r), \tag{6.7}$$

where $S$ is slowly varying, of order unity, but

$$|S| > 1. \tag{6.8}$$

Then (6.2) becomes:

$$\frac{d\psi_g}{d\xi} = -2iSh_1 \tag{6.9}$$

Substituting the series (3.22–23), we find the first two orders of the derivative of Poisson's equation (6.9):

$$\frac{d^2\psi_g^{(0)}}{d\xi^2} = -2iS(1)\frac{dh_1^{(0)}}{d\xi}, \tag{6.10}$$

and

$$\frac{d^2\psi_g^{(1)}}{d\xi^2} = -2iS(1)\frac{dh_1^{(1)}}{d\xi} - 2iS'(1)\left[h_1^{(0)} + \xi\frac{dh_1^{(0)}}{d\xi}\right]. \tag{6.11}$$

Combining these equations with the gas dynamics equations (3.25–26) yields

$$\frac{d^2h_1^{(0)}}{d\xi^2} - 2iS(1)\frac{dh_1^{(0)}}{d\xi} - h_1^{(0)} = 0, \tag{6.12}$$

and

$$\frac{d^2h_1^{(1)}}{d\xi^2} - 2iS(1)\frac{dh_1^{(1)}}{d\xi} - h_1^{(1)} = \frac{a_4(1)}{\xi} + F_{\text{disk}}^{(0)}; \tag{6.13}$$

where

$$F_{\text{disk}}^{(0)} = \left[2iS'(1) + \xi a_3'(1) - \frac{a_2(1)}{\xi}\right]h_1^{(0)} - \frac{a_2(1)}{\xi}\psi_g^{(0)} + [2i\xi S'(1) - a_1(1)]\frac{dh_1^{(0)}}{d\xi} - a_1(1)\frac{d\psi_g^{(0)}}{d\xi}. \tag{6.14}$$

Now let us define the quantities

$$\alpha_\pm = S(1) \pm [S(1)^2 - 1]^{1/2}. \tag{6.15}$$

Note that both $\alpha_+$ and $\alpha_-$ have the same sign as $S(1)$. This formula is consistent with the dispersion relation found in [9, eq. (3.2)]. The solutions of (6.12–13) are:

$$h_1^{(0)} = c_+^{(0)}e^{i\alpha_+\xi} + c_-^{(0)}e^{i\alpha_-\xi}, \tag{6.16}$$

and

$$h_1^{(1)} = c_+^{(1)}e^{i\alpha_+\xi} + c_-^{(1)}e^{i\alpha_-\xi} + c_+^{(0)}g_+(\xi) + c_-^{(0)}g_-(\xi) + c_Dg_D(\xi). \tag{6.17}$$

The terms involving $g_+$ and $g_-$ can be written in terms of integrals of elementary functions. These terms, due to the term $F_{\text{disk}}^{(0)}$ in (6.13), are the first-order corrections to the free spiral solutions, so we write:

$$h_1 = c_+e^{i\alpha_+\xi}[1 + O(\delta)] + c_-e^{i\alpha_-\xi}[1 + O(\delta)] + \delta c_Dg_D(\xi), \tag{6.18}$$

where $c_D$ is given by

$$c_D = \frac{\pi}{[S(1)^2 - 1]^{1/2}} a_4(1), \tag{6.19}$$

which is proportional to the strength of the bar, and $g_D(\xi)$ is given by

$$g_D(\xi) = e^{i\alpha_-\xi} I(\xi, \alpha_-) - e^{i\alpha_+\xi} I(\xi, \alpha_+), \tag{6.20}$$

which has the following asymptotic behaviors:

$$\begin{cases} 0 & \text{as } \xi \to +\infty, \\ 0 \quad \text{for } S > 0, & \text{as } \xi \to -\infty, \\ e^{i\alpha_-\xi} - e^{i\alpha_+\xi} \quad \text{for } S < 0, & \text{as } \xi \to -\infty. \end{cases} \tag{6.21}$$

Thus, the enthalpy perturbation can be written as the sum of three terms, two free spirals and a driven wave. If $S(1) > 0$, the driven waves lead, but die away at plus and minus infinity. If $S(1) < 0$, the solution is the superposition of two trailing waves, a short wave $\exp\{i(\alpha_-\xi - m\varphi)\}$ and a long wave $\exp\{i(\alpha_+\xi - m\varphi)\}$. From earlier work, we should focus our attention on the short wave.

Although the above solution is not highly oscillatory for positive values of $\xi$, it can be shown that this failure to conform to the asymptotic approximation (6.2) does not appreciably affect the solution near the corotation ring.

## 7. Physical interpretation

We have considered galactic models in which a gas is perturbed by a rigidly rotating bar. In the two classes of models considered, infinite cylinders and infinitesimally thin disks, we have found essentially the same result. Without a bar, the models will support both leading and trailing spiral waves. The bar forces both a leading and a trailing wave near corotation. However, the leading wave dies away from corotation but the trailing wave does not.

In order to evaluate the quantitative significance of the results, let us consider the strength of the force field in the short driven wave as a fraction of axisymmetric field. In terms of dimensionless variables, this ratio is

$$f(r) = \mathrm{Re}\left\{-\frac{1}{r\omega^2}\frac{d\psi_g}{dr}\right\} \tag{7.1}$$

For the disk model, by using (6.2), (6.7), (6.18), (6.19) and (6.21), we find that

$$f(r) = \mathrm{Re}\left\{\frac{2\pi i}{r\omega^2}\frac{S(1)}{[S(1)^2 - 1]^{1/2}} a_4(1) g_D(\xi)\right\} \tag{7.2}$$

If we now consider the root-mean-square value $F$ of $f(r)$ just inside of corotation ($r = \omega = 1$), we find that

$$F = \pi\sqrt{2}\left|\frac{S(1)}{[S(1)^2 - 1]^{1/2}}\right| \cdot |a_4(1)|. \tag{7.3}$$

where the averaging process is taken with respect to $\xi$. By using values typical of the solar neighborhood, we find $S(1) = -15{\cdot}5$. Thus, we may approximate the relevant factor in $F$ by unity. Now we use the formula (3.31) for $a_4(1)$, and finally obtain

$$F = 2\pi\sqrt{2}\,.\left|1 - \frac{\mu_1}{\mu_2}\right|.\,|\psi_b(1)|, \tag{7.4}$$

where $\mu_1$ and $\mu_2$ are defined by (3.29) and (3.30). Since $\psi_b(r, \phi)$ has a $2\phi$-dependence, $|\sqrt{2}\psi_b(1)|$ is the root-mean-square value of the tangential component of the driving field due to the bar, as a fraction of the axi-symmetric field. Thus, for a given basic distribution (i.e., for given values of $\mu_1$ and $\mu_2$), *the strength of the induced spiral field in the gas is* $2\pi \cdot |1 - \mu_1/\mu_2|$ *times the tangential component of the driving field due to the bar*. For $\mu_1 = 2$, and $\mu_2 = 1$, this factor is $2\pi$. Note that it is independent of the amount of gas present.

*An example.* To give a more vivid physical picture, let us assume a specific form for the bar. The simplest model is an elliptic disk of uniform surface density. Let the density be $\sigma_{\text{bar}}$, the semi-major axis be $a_{\text{bar}}$, and the semi-minor axis be $\beta a_{\text{bar}}$. The mass of the bar is then $M_{\text{bar}} = \pi\beta a_{\text{bar}}^2\sigma_{\text{bar}}$. The $(2\phi)$-component of the potential can be approximated by its quadruple moment

$$\mathscr{V} = -\frac{GM_{\text{bar}}}{8\varpi^3}a_{\text{bar}}^2\{(3\cos^2\varphi - 1) + \beta^2(3\sin^2\varphi - 1)\}, \tag{7.5}$$

in the plane $z = 0$. Thus,

$$\psi_b(1) = -\frac{3}{16}(1 - \beta^2)\left(\frac{a_b}{\varpi_0}\right)^2\frac{GM_{\text{bar}}}{\varpi_0^3\Omega_0^2} \tag{7.6}$$

If all of the important mass were concentrated in a nucleus of mass $M_{\text{nuc}}$, then

$$\varpi_0^3\,\Omega_0^2 = GM_{\text{nuc}}, \tag{7.7}$$

and we obtain the following form for $F$:

$$F = \frac{3\sqrt{2}\pi}{8}\left|1 - \frac{\mu_1}{\mu_2}\right|\cdot(1 - \beta^2)\cdot\left(\frac{a_b}{\varpi_0}\right)^2\frac{M_{\text{bar}}}{M_{\text{nuc}}} \tag{7.8}$$

As in the above calculations, we choose $\sigma_0 \sim \tilde{\omega}^{-2}$, $\Omega \sim \tilde{\omega}^{-1}$, so that $\mu_1 = 2$ and $\mu_2 = 1$. We then have

$$F = 1{\cdot}67(1 - \beta^2)\left(\frac{a_{\text{bar}}}{\varpi_0}\right)^2\frac{M_{\text{bar}}}{M_{\text{nuc}}}. \tag{7.9}$$

If we consider the bar to be a large, massive but only slightly eccentric disk, we may choose $M_{\text{bar}} = M_{\text{nuc}}$, and assume a value of $\beta$ close to unity. Then (7.9) takes on the approximate form

$$F = 3(1 - \beta)(a_{\text{bar}}/\varpi_0)^2 \tag{7.10}$$

For $\beta = 0.90$ and $a_{\text{bar}} = \varpi_0/3$, we find that

$$F \approx 0.033 \tag{7.11}$$

Thus, a slightly central distortion is capable of producing, near corotation, a spiral force field that amounts to several per cent of the axi-symmetric field. This is of the right order of magnitude for the field strength in our Galaxy, as inferred from a number of observational data (see Table II in [4]). The variation of this field strength as the wave propagates inwards is governed by the principle of conservation of density of wave action, as shown in previous studies of Toomre [1] and Shu [10].

The wave system should connect up with the oval distortion in the central regions of the galaxy. The condition of matching could then lead to an eigenvalue problem for the pattern frequency $\Omega_p$.

## 8. Acknowledgement

The work reported here was done with the partial support of the National Science Foundation. The authors wish to thank Drs. J. W-K. Mark, F. H. Shu, and A. Toomre for helpful comments.

## Appendix A. Laplace transforms

We can solve the partial differential equation

$$\frac{\partial y}{\partial t} = \mathscr{P} y \tag{A1}$$

by modal analysis. We set

$$y(x, t) = \sum_m y_m(x)\, e^{p_m t}, \tag{A2}$$

so

$$(\mathscr{P} - p_m) y_m = 0. \tag{A3}$$

If these equations are singular for real values of $x$, we must deform the path of integration into the complex $x$-plane. To determine the contour, we consider the solution of the initial value problem by Laplace transforms. The solution is given in the form:

$$y(x, t) = \frac{1}{2\pi i} \int_{p_0 - i\infty}^{p_0 + i\infty} \tilde{y}(x, p)\, e^{pt}\, dp, \tag{A4}$$

where $\tilde{y}$ satisfies the equation

$$(\mathscr{P} - p)\tilde{y} = -y(x, 0) \tag{A5}$$

as a consequence of (A1). Thus, (A4) may also be written as

$$y(x, t) = -\frac{1}{2\pi i} \int_{p_0 - i\infty}^{p_0 + i\infty} [(\mathscr{P} - p)^{-1} y(x, 0)]\, e^{pt}\, dp, \tag{A6}$$

where we must integrate to the right of the poles of $(\mathscr{P} - p)^{-1} y(x, 0)$.

If the rightmost pole has imaginary part zero, then we can integrate along the imaginary $p$-axis, indenting the contour toward amplified modes ($\mathrm{Re}\, p > 0$) around each pole on the axis. We then apply this rule to determine the path of integration in the $x$-plane for the equations (A3).

## Appendix B. Solutions of certain differential equations

Consider the second-order constant coefficient inhomogeneous differential equation

$$\frac{d^2y}{d\xi^2} + g_1\frac{dy}{d\xi} + g_2 y = \frac{1}{\xi}, \tag{B1}$$

Assume that the roots of the indicial equation are pure imaginary, so that both solutions of the homogeneous equation are oscillatory:

$$y_1 = e^{i\alpha_1\xi}, \tag{B2}$$

$$y_2 = e^{i\alpha_2\xi}. \tag{B3}$$

($\alpha_1$ and $\alpha_2$ real). If we integrate along the contour $\mathscr{C}$ (Figure 4), then a particular solution, $y_p$, of the inhomogeneous equation is given by

$$y_p = \frac{2\pi}{\alpha_1 - \alpha_2}[e^{i\alpha_2\xi}I(\xi, \alpha_2) - e^{i\alpha_1\xi}I(\xi, \alpha_1)], \tag{B4}$$

where

$$I(\xi, \alpha) = \frac{1}{2\pi i}\int_\xi^\infty \frac{e^{-i\alpha\zeta}}{\zeta}\, d\zeta. \tag{B5}$$

The integral is taken along $\mathscr{C}$.

Clearly,

$$I(\xi, \alpha) = O(1/\xi) \quad \text{as} \quad \xi \to +\infty. \tag{B6}$$

It can easily be shown that

$$I(\xi, \alpha) = \begin{Bmatrix} O(1/\xi) & \text{for } \alpha > 0 \\ \frac{1}{2} + O(1/\xi) & \text{for } \alpha = 0 \\ 1 + O(1/\xi) & \text{for } \alpha < 0 \end{Bmatrix} \quad \text{as } \xi \to -\infty. \tag{B7}$$

## Appendix C. Results for an infinitesimally thin disk

If we assume that

$$\rho = \sigma(\varpi, \phi)\delta(z), \tag{C1}$$

where

$$\sigma = \sigma_0(r) + \sigma_1(r)\, e^{-im\phi}, \tag{C2}$$

then the usual Poisson inversion formula for the potential in terms of the density gives:

$$\psi_g e^{-im\phi} = -\int_0^\infty \frac{\Sigma(r')}{2\pi} h_1(r')r'\, dr' \int_0^{2\pi} \frac{e^{-im\phi'}\, d\phi'}{[r^2 + r'^2 - 2rr'\cos(\phi' - \phi)]^{1/2}} \tag{C3}$$

where

$$\Sigma(r) = \frac{2\pi G\sigma_0}{(\gamma - 1)h_0\Omega_0^2\varpi_0}. \tag{C4}$$

Following Shu [9], we define the function $\hbar_m(x)$

$$\hbar_m(x) = \frac{2}{\pi}\int_0^{\pi/2} \frac{\cos 2mt\, dt}{[1 - x\cos^2 t]^{1/2}}. \tag{C5}$$

We then define the kernel, $Q$:

$$Q(r, r') = -\frac{r}{r + r'}\Sigma(r')\hbar_m\left(\frac{4rr'}{(r + r')^2}\right); \tag{C6}$$

then (C3) becomes:

$$\psi_g(r) = \int_0^\infty Q(r, r')h_1(r')\, dr' \tag{C7}$$

If we now define the operator

$$\mathscr{L} = \frac{d^2}{dr^2} + A\frac{d}{dr} + B \tag{C8}$$

then the gas dynamics equation (2.42) can be written:

$$\mathscr{L}h_1 + Ch_1 + \mathscr{L}\int_0^\infty Q(r, r')h_1(fr')\, dr' = D, \tag{C9}$$

All of the dynamics of the infinitesimally thin gaseous disk are contained in this single singular, integro-differential equation for the perturbed enthalpy. If we remove the bar from the problem by setting $D$ to zero, we get a homogeneous integro-differential equation for locally self-sustained spiral density waves.

If we write $h_1$ in the form

$$h_1(r) = \chi(r)\exp\{i\Phi(r)\} \tag{C10}$$

and assume that $\mu$ is slowly varying and $\Phi$ is a large constant times a slowly varying function, the method of stationary phase yields the dispersion relation

$$k^2 - \Sigma|k| - C = 0, \tag{C11}$$

where

$$k = \frac{d\Phi}{dr} \tag{C12}$$

is the dimensionless wavenumber.

The method of stationary phase allows us to approximate (C7) by:

$$\psi'_g = -i\Sigma\,(\operatorname{sgn} k)h_1. \tag{C13}$$

Equations (C11) and (C13) are consistent with equations (3.2) and (3.1) of [5].

## References

1. Toomre, Alar, "Group Velocity of Spiral Waves in Galactic Disks", *Astrophys. J.* **158**, (1969), 899
2. Mark, James W-K., "The Spiral Wave of Our Galaxy Near Inner Lindblad Resonance", *Proc. Nat. Acad. Sci.* **68**, No. 9 (1971), 2095

3. LIN, C. C., "Interpretation of Large-Scale Spiral Structure", in IAU Symposium No. 38, pp. 377–390. ("The Spiral Structure of our Galaxy", edited by Becker and Contopoulos, Reidel, 1970)
4. LIN, C. C., "Theory of Spiral Structure", *Highlights of Astronomy*, Vol. **2** (1970), 103 (Dordrecht-Holland).
5. LIN, C. C., YUAN, C., and SHU, FRANK H., "On the Spiral Structure of Disk Galaxies. III. Comparison with Observations", *Astrophys. J.* **155** (1969), 721
6. HOLMBERG, ERIK, "A Photographic Photometry of Extragalactic Nebulae", *Medd. Lund.* Ser. **2**, No. 136 (1958), Section 12
7. DE VAUCOULEURS, G., "Tilt Criteria and Direction of Rotation of Spiral Galaxies", *Astrophys. J.* **127** (1958), 487
8. ROBERTS, W. W., JR. and YUAN, C., "Application of the Density-Wave Theory to the Spiral Structure of the Milky Way System. III. Magnetic Field: Large-Scale Hydromagnetic Shock Formation", *Astrophys. J.* **161** (1970), 887
9. SHU, FRANK H., "On the Density-Wave Theory of Galactic Spirals. I. Spiral Structure as a Normal Mode of Oscillation", *Astrophys. J.* **160** (1970), 95
10. SHU, FRANK H., "On the Density-Wave Theory of Galactic Spirals. II. The Propagation of the Density of Wave Action", *Astrophys. J.* **160** (1970), 99

MASSACHUSETTS INSTITUTE OF TECHNOLOGY

(Received November 13, 1972)

Reprinted from
*Proc. Natl. Acad. Sci. USA*
Vol. 73, No. 5, pp 1379–1381, May 1976
Astronomy

# Unstable spiral modes in disk-shaped galaxies

(density waves/galactic dynamics/spiral arms)

Y. Y. LAU, C. C. LIN, AND JAMES W.-K. MARK

Department of Mathematics, Massachusetts Institute of Technology, Cambridge, Mass. 02139

*Contributed by C. C. Lin, February 9, 1976*

**ABSTRACT** The mechanisms for the maintenance and the excitation of trailing spiral modes of density waves in disk-shaped galaxies, as proposed by Lin in 1969 and by Mark recently, are substantiated by an analysis of the gas-dynamical model of the galaxy. The self-excitation of the unstable mode is caused by waves propagating outwards from the corotation circle, which carry away angular momentum of a sign opposite to that contained in the wave system inside that circle. Specifically, a simple dispersion relationship is given as a definite integral, which allows the immediate determination of the pattern frequency and the amplification rate, once the basic galactic model is known.

## 1. Introduction

The development of the theory of normal galactic spirals has been, for a long time, based on the hypothesis that there exists a pattern of density waves with a quasi-stationary spiral structure (QSSS hypothesis) (1, 2). On the other hand, Toomre's study (3) of the radial propagation of waves with a relatively tight winding shows that a pure short trailing wave system would move away from the corotation region (i.e., the region where the angular velocities of the wave pattern and of the material are about the same). Since the propagation time of these waves over the whole galactic disk is found to be of the order of one billion years, Toomre pointed out that the quasi-stationary hypothesis was questionable, unless a mechanism for replenishment could be provided.

Lin (4) responded to this challenge by suggesting the following picture. When these short trailing density waves reach the central regions of the galactic disk, they could induce a feedback process either by outward-propagating long trailing waves or directly by a bar-like structure formed through the action of these short trailing waves. The latter process has been calculated by Feldman and Lin (5) and by Lin and Lau (6). The driving mechanism due to long waves has been calculated by Mark (7–10) in the stellar case and confirmed by Lin and Lau (6) in the gas dynamical model. It is found that there is in general an amplification of these waves by "stimulated emission" at the corotation circle. Mark (10–13) then proposed that unstable modes can actually be obtained by combining the feedback and the amplification mechanisms. Estimations were given of the rates of amplification.

In this note, we shall present the principal results of one simple analysis, following closely the mechanisms proposed. It yields, in the asymptotic theory, explicit solutions that represent unstable modes. These solutions exist in broad classes of disk models with reasonable distributions of density and rotational velocity. If the disk is locally stable everywhere and the corotation circle is in a *marginally* stable region, then the $e$-folding time $T$ for energy is given by the formula

$$T = \tau/\ln 2 = 1.44\tau \qquad [1.1]$$

Abbreviation: kpc, kiloparsec ($= 3.08 \times 10^{16}$ m).

where $\tau$ is the time required for the waves to travel one complete cycle. In a typical cycle, a packet of short trailing waves begins at the corotation circle. It propagates inwards to a central turning point (to be defined more precisely below), where it is turned back outwards (as long trailing waves) to return to the corotation circle. The mechanism for this amplification process will now be described before we present the analytical theory.

Other mechanisms related to the generation and maintenance of spiral modes will be briefly discussed in *Section 4*.

## 2. The models and the mechanisms

For our present purposes, we shall adopt a gas-dynamical model, since our primary aim is to demonstrate the basic mechanism, and the gas-dynamical model has the advantage of simplicity. Indeed, a similar theory has also been developed by Mark (manuscript submitted) for the stellar dynamical model under certain reasonable simplifying approximations, some of which are not needed here. The fundamental dispersion relationship is then slightly different from [**3.6**]. However, a careful comparison shows that, at least for the type of models under consideration, there is no qualitative difference between the results for a model based on gas dynamics and one based on stellar dynamics. Thus, a consideration of the gas dynamical model also helps to clarify the role of the resonant regions: the behavior of individual corotating (or resonant) stars is not important.

To provide a feedback process from central regions, we shall, for the moment, consider galactic disk models without inner Lindblad resonance. We shall further assume that the stability parameter (14, 15) $Q$ (defined in Eq. **3.4** below) increases steadily towards the center (12), so that the disk is essentially "immobile" there. This is a good dynamical simulation of the central regions in real galaxies where there may be a nuclear bulge. It can then be shown that the energy of the incoming short waves is essentially fully turned back by refraction. (See *Section 4* below for other possibilities.)

The amplification process occurs essentially near the corotation region. Consider neutral waves for the moment. As the waves complete the cycle described above, the (radially) outgoing wave, at unit energy flux, induces an inward propagating wave from the corotation circle with twice as much energy flux, since there is also an outgoing wave induced, which carries away energy of an opposite sign, at unit flux, to the outer regions of the galactic disk (6–10). Thus, there is doubling of energy of the waves over a period of time $\tau$, and hence the formula [**1.1**] given above. Naturally, this formula has also been deduced by straightforward mathematical analysis. (See *Section 3* below.) Somewhat different numerical values will be obtained if the disk is not marginally stable in the corotation region (see ref. 10, which also contains results for cases $Q \neq 1$ in the corotation region).

Thus, if we consider the wave system in the whole region

within the corotation circle, it carries negative total energy. As a mode, it is amplified by the outward transfer of energy of an opposite sign across the corotation circle. In terms of angular momentum, this amounts to the transfer of angular momentum outwards by the trailing wave system, in line with the general trend suggested by Lynden-Bell and Kalnajs (16). However, they attributed the generation mechanism to the resonant stars. In our analysis, this effect is neglected, since it has been shown (7–10) that it only leads to a minor modification of the mechanism discussed here.

The fate of the outward propagating waves is easily understood in the stellar dynamical model. They are simply absorbed at the (first-order) *outer* Lindblad resonance. In the gas-dynamical model, it is less clear. However, in a real galaxy, there are even other ways for disposing of the *positive* wave energy (e.g., by turbulent dissipation) besides absorption at the outer Lindblad resonance. The adoption of the radiation boundary condition in the gas dynamical model is therefore just a mathematical device to simulate the real galaxy without explicitly introducing dissipative mechanisms.

## 3. Analytical results

The basic equations governing density waves in a self-gravitating gaseous disk may be found in a convenient form in the paper by Lin and Lau (6). If we retain only the dominant terms in the Euler equations of motion and apply an asymptotic formula to the Poisson equation, we obtain the following differential equation for the enthalpy perturbation $h$ (see ref. 6, p. 362):

$$\frac{d^2u}{dr^2} + k_3^2 u = 0 \qquad [3.1]$$

where

$$u = h \exp\left\{ i \int^r dr \Sigma/2 \right\} \qquad [3.2]$$

$$k_3^2 = \frac{\Sigma^2}{4} - \left(\frac{\kappa}{a}\right)^2 (1 - \nu^2) \qquad [3.3]$$

$$\Sigma = \left(\frac{\kappa}{a}\right)\frac{2}{Q}, \text{ and } Q = \frac{\kappa a}{\pi G \sigma} \qquad [3.4]$$

In these equations, $r$ is the distance from the galactic center, $\kappa$ is the epicyclic frequency, $\sigma$ is the surface density, $a$ is the acoustic speed, $\nu = (\omega - m\Omega)/\kappa$ where $\omega$ is the mode frequency, $m$ is the number of the spiral arms, and $\Omega$ is the circulation frequency (circular velocity in angular measure). Note that $\Sigma$ is proportional to the surface density. At large distances $r$, where we assume $Q = 1$, we have $k_3^2 = (\kappa/a)^2\nu^2$. Note that $k_3^2$ has a *double* zero at the corotation radius $r_{co}$, defined by $\nu(r_{co}) = 0$.

As mentioned in *Section 2*, an unstable normal mode may be expected, if there is a mechanism for turning incoming waves back near the galactic center. We see here that this type of "reflection" near the galactic center is indeed possible if there is a location $r = r_{ce}$ when $k_3^2$ has a simple zero. This simple zero can occur in a model in which $Q$ steadily increases as $r \rightarrow 0$. In that case $k_3^2$ has the appearance shown in Fig. 1.

We consider a mode which satisfies the boundary conditions

$$u = 0 \text{ at } r = 0, \text{ and} \qquad [3.5a]$$

$$u \sim \exp\left[ -i \int_{r_{co}}^{r} dr \frac{\kappa}{a} \nu \right], \text{ for } r \gg r_{co}. \qquad [3.5b]$$

The first condition follows from the condition of regularity, and the second is one of outgoing waves. These conditions then lead to the following dispersion relation for the determination of $\omega$:

$$\exp\left[ -i \oint k dr \right] = -\sqrt{2} \qquad [3.6]$$

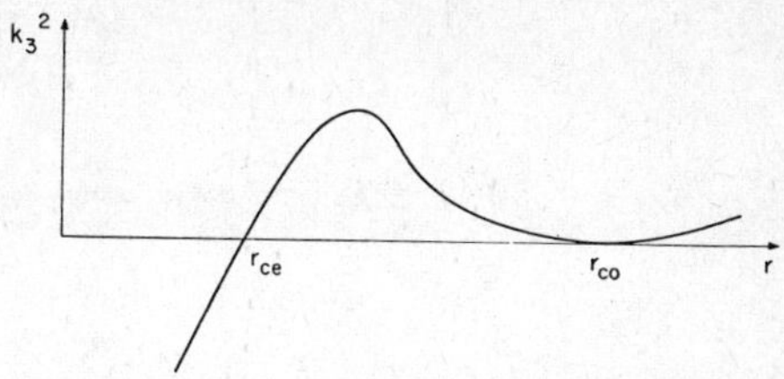


FIG. 1. Schematic diagram showing the variation of $k_3^2$ (the square of the reduced wave number defined in Eq. [3.3] in the *text*) as a function of radial distance $r$.

where the closed path denotes the cycle ($r_{ce} \rightarrow r_{co} \rightarrow r_{ce}$). In [3.6], $k = -\Sigma/2 \pm k_3$, where the branch of $k_3$ is so defined that $Re\, k_3 > 0$. The positive sign is used as $r$ varies from $r_{ce}$ to $r_{co}$, and the negative sign is used in completing the integral from $r_{co}$ back to $r_{ce}$. Equating the magnitude and phase separately in Eq. [3.6], we obtain the following pair of approximate dispersion relations, in the limit of small growth rate:

$$\oint k dr = (2n + 1)\pi, (n = 0, 1, 2 \ldots) \qquad [3.7a]$$

and

$$e^{\omega_i \tau} = \sqrt{2} \qquad [3.7b]$$

where

$$\tau = \oint \frac{dr}{|c_g|} \equiv \oint \frac{dr}{|\partial\omega/\partial k|} \qquad [3.7c]$$

is the transit time for a group of waves to complete a cycle ($r_{ce} \rightarrow r_{co} \rightarrow r_{ce}$). Eq. [3.7b] clearly indicates that the $e$-folding time is on the order of the time required for a group of waves to propagate across the galactic disk, which is of the order of one billion years.

Solutions have been found for the eigenvalues of $\omega$ on the basis of Eq. [3.7] for certain specific models. Clearly, the rate of amplification calculated depends on the amplitude distribution of the wave, which is not properly described by Eq. [3.1]. However, this does not invalidate the above formulas [3.7]. The principle of conservation of density of wave action holds (3, 17, 18) both for the exact theory and for the approximate form [3.1]. Since the propagation of the waves is well described by Eq. [3.1], the same amplification rate is obtained in both cases. The accurate amplitude distribution must be obtained by going to the next approximation. Better still, it should be calculated by numerical integration, by using the exact linearized differential equations of motion and the integral form for the Poisson equation.

It should be emphasized that the dispersion relationship [3.6] is approximate. The accurate treatment of the eigenvalue problem might actually lead to a very narrow band of continuous spectrum.

**Example.** If we take a rotation curve given by $\Omega = 250/(r + 1)$ $km/sec\text{-}kpc$, where $r$ is in kiloparsecs (kpc), $a$ = 45 km/sec, and $Q = 1$ for $r \geq 5$ but increases roughly like $5/r$ for smaller values of $r$, we have a reasonable model for our purposes. For this model, there is a solution with $n = 0$ in Eq. [3.7], with $r_{co}$ = 8.5 kpc and $r_{ce}$ = 3.5 kpc, approximately. The short wavelength approximation is well satisfied even at $r$ = 3–4 kpc, since $|kr| = |\Sigma/2 \pm k_3|r \gtrsim 4$. It is even better satisfied at outer parts.

For spiral waves with $|kr|$ of the order of 4, we have also performed numerical calculations to check the accuracy of our asymptotic solution of the Poisson equation by comparing it with the potential obtained by an exact calculation based on the

Poisson integral formula. The results are found to be in agreement with each other to within 5% as long as $|kr| > 4$.

## 4. Discussion

(*a*) **Modal Analysis.** As mentioned above, the asymptotic theory demonstrated in this note can be improved and put on a more solid foundation by numerical calculations based on the exact linearized equations.

Another improvement to be made is the inclusion of *inner Lindblad resonance*. Certainly, the return of wave energy from the central region would no longer be perfect in this case. But this would presumably only lead to a reduction of the amplification, at least in many cases. Basically, we still have the same physical picture: the waves propagating outwards from the corotation circle amplify the system of waves inside it by carrying away angular momentum of an opposite sign. There is of course the possibility that the energy may be completely absorbed at the inner Lindblad resonance. In that case, the above mechanism would not work, and we would need a different process. [See (*b*) below.]

Zang and Toomre (Ph.D. Dissertation of Thomas Zang, 1976, Massachusetts Institute of Technology) have found self-excited modes for galactic models with uniform linear velocity and various distributions of the mobile component of mass density. Their models do exhibit an inner Lindblad resonance. The underlying mechanism is not explicitly displayed by their analysis. It is suggested here that one component of the sustaining mechanism could be of the type described in this note. However, the length scales and the amplification rates obtained by Zang and Toomre seem both to be too large for our calculations to be properly applied. There are probably other amplifying mechanisms in their particular galactic models.

(*b*) **Some Other Mechanisms.** We would now briefly refer to two other mechanisms for generating unstable modes. One is the interaction with a halo whenever it exists (13), the other is Jeans instability in the outer regions (19). In case Jeans instability is present one could have a distinct mode, even when the mechanism discussed in this paper is absent. If both mechanisms were present, there would be a question of proper phase matching. In general, the spiral pattern may not be perfect. Since real galaxies do not usually have perfect patterns, one should regard our theoretical studies as providing the various possible mechanisms, and leave the choice of the preferred mechanisms and their relative importance to natural forces in action.

(*c*) **Reflection from Galactic Center.** In the previous section, we considered the reflection of waves from the galactic central regions as a result of the increase of the parameter $Q$ near the galactic center. *If* the differential equation developed for the asymptotic treatment of these density waves *remains* essentially valid near the center, then we find that there is a simple modification of our theory possible in order to include the effect of cylindrical geometry. We replace the coefficient $k_3^2$ in [**3.1**] by

$$\overline{k_3}^2 = k_3^2 - \frac{m^2}{r^2} \qquad [4.1]$$

This function is definitely of the form shown in Fig. 1 (with a slight modification near corotation), whether $Q$ increases with decreasing $r$. The feedback process always works. Indeed, certain dynamical factors may even further enhance this geometrical effect. However, the whole discussion is uncertain because the differential equation used may not be valid at the origin, since the usual asymptotic solution of Poisson's equation is not accurate there. Further investigation is needed.

(*d*) **Analogy with Plasma Physics.** The mechanism discussed in this paper is similar to a type of instability known in plasmas. In 1969, Berk *et al.* (20) found that instability may be excited by an exchange of energy between different regions in a nonuniform plasma in a mirror machine geometry. The consideration of the suppression of the unstable modes of this type led to a criterion on the critical length of the mirror device (21). The same critical length scale has been approximately reproduced by Lau (manuscript submitted) in a simplified form of the theory, which is mathematically similar to that presented in this note.

(*e*) **The Role of Resonant Stars.** Even though we stressed the role of the resonant stars in the framework of the linear theory, they may not play a crucial role in the maintenance mechanism of spiral structure in real galaxies. For one thing, after the spiral structure is generated, these stars might be displaced from the condition of resonance after a few hundred million years. If so, the absorption of waves at outer Lindblad resonance in the analytical theory must then be taken over by physically significant dissipative processes (see *Section 2*). The lack of absorption at inner Lindblad resonance can only aid the reflection process from central regions described above.

We wish to thank Dr. A. Toomre for helpful discussions on several aspects of the problem. This work is supported in part by a grant from the National Science Foundation.

1. Lin, C. C. & Shu, F. H. (1964) *Astrophys. J.* **140,** 646–655.
2. Lin, C. C. & Shu, F. H. (1968) in *Astrophysics and General Relativity*, ed. Chretian, M. (Gordon and Breach, New York), Vol. 2, pp. 235–328.
3. Toomre, A. (1969) *Astrophys. J.* **158,** 899–913.
4. Lin, C. C. (1970) in *Proc. International Astronomical Union Symposium No.* 38 (Basel), pp. 377–390.
5. Feldman, S. I. & Lin, C. C. (1973) *Stud. Appl. Math.* **52,** 1–20.
6. Lin, C. C. & Lau, Y. Y. (1975) *SIAM J. Appl. Math.* **29,** 352–370.
7. Mark, J. W.-K. (1974) in *The Formation and Dynamics of Galaxies*, ed. Shakeshaft, J. R. (D. Reidel Publ. Co., Boston, Mass.), pp. 417–423.
8. Mark, J. W.-K. (1975) in *La Dynamique des Galaxies Spirales*, ed. Weliachew, L. (Centre National de la Recherche Scientifique, Paris), pp. 67–79.
9. Mark, J. W.-K. (1976) *Astrophys. J.* **203,** 81–96.
10. Mark, J. W.-K. (1976) *Astrophys. J.*, in press (April 15 issue).
11. Mark, J. W.-K. (1974) *Bull. Am. Astron. Soc.* **6,** 462.
12. Mark, J. W.-K. (1975) *Bull. Am. Astron. Soc.* **7,** 395.
13. Mark, J. W.-K. (1976) *Astrophys. J.*, in press (May 15 issue).
14. Toomre, A. (1964) *Astrophys. J.* **139,** 1217–1238.
15. Julian, W. H. & Toomre, A. (1966) *Astrophys. J.* **146,** 810–830.
16. Lynden-Bell, D. & Kalnajs, A. J. (1972) *Mon. Not. R. Astron. Soc.* **157,** 1–30.
17. Shu, F. H. (1970) *Astrophys. J.* **160,** 99–112.
18. Mark, J. W.-K. (1974) *Astrophys. J.* **193,** 539–559.
19. Lau, Y. Y., Lin, C. C., Mark, J. W.-K. & LaBudde, R. A. (1975) presented at the *Seventh Annual Meeting of the Division of Dynamical Astronomy*, American Astronomical Society (Pasadena, Calif., Dec. 11, 12).
20. Berk, H. L., Pearlstein, L. D., Callen, J. D., Horton, C. W. & Rosenbluth, M. N. (1969) *Phys. Rev. Lett.* **22,** 876–879.
21. Berk, H. L., Pearlstein, L. D. & Cordey, J. (1972) *Phys. Fluids* **15,** 891–910.

*Reprinted from*
*Proc. Natl. Acad. Sci. USA*
Vol. 74, No. 11, pp. 4726–4729, November 1977
Astronomy

# Discrete spiral modes in disk galaxies: Some numerical examples based on density wave theory

(galactic dynamics/spiral arms)

G. BERTIN*, Y. Y. LAU, C. C. LIN, JAMES W.-K. MARK, AND L. SUGIYAMA

Department of Mathematics, Massachusetts Institute of Technology, Cambridge, Massachusetts 02139

*Contributed by C. C. Lin, August 19, 1977*

**ABSTRACT** Discrete growing spiral modes are calculated, on the basis of principles of stellar dynamics, according to the recently developed asymptotic theory of density waves for spiral structure in galaxies. From the growth rates obtained, it appears likely that several spiral modes coexist in a given disk galaxy. General properties of these modes are described for two galaxy models, and some of their important observational implications are discussed.

It has recently been shown that *discrete unstable spiral modes* can be obtained from an *asymptotic* analysis of the density wave problem for disk galaxies. The theory is based on either a gas dynamical model (1, 2) or a stellar dynamical model (3). Some preliminary numerical results have been presented in the papers cited. It is now desirable to apply the stellar dynamical approach to an examination of these unstable spiral modes for some typical galaxy models. This has become possible after Bertin and Mark (unpublished) extended the theory to the regime of small but finite inclination of the spiral arms. In this report, we present a summary of such results for two galaxy models (Fig. 1), hereafter referred to as models C1 and C2, which respectively resemble the galaxies M31 and M101 in several important aspects. Model C1 has a small but sizable nucleus and a rather concentrated bulge-halo system. Model C2 has no nucleus and has a relatively more diffuse bulge-halo system. In both cases, the total mass of the spheroidal system is about 20–22% of the total mass. (See *Appendix* for more detailed descriptions.) Similar numerical results for spiral modes have been obtained earlier by Lau and Bertin (unpublished data) on the basis of the gas dynamical model.

We shall now make two brief comments to relate our results to observable features in galaxies. Detailed discussions will be presented below.

1. Judging from the growth rates observed, we find it plausible to believe that the observed spiral structure of a given galaxy is the superposition of several spiral modes. Such a conclusion has also been hinted at by detailed studies of observational features which often exhibit asymmetries suggestive of the presence of a superposed one-armed structure.

2. For some galaxies, the density structure in the stellar disk has been measured (4, 5). In this connection, the radial position of the corotation circle and the appearance of the density distribution nearby have attracted much interest [for example, see Schweizer's review article (4) and Oort's comments (6)]. We shall show below that this observed radial structure might provide a convenient method for the determination of the corotation radius of the dominant spiral pattern, and we would like to encourage more extensive observations and more careful analysis of the kind of data provided by Schweizer (5). Indeed, some of his data on density distribution are qualitatively similar to those obtained theoretically (cf. figure 7*a* of ref. 5 and our Fig. 4).

The costs of publication of this article were defrayed in part by the payment of page charges. This article must therefore be hereby marked "*advertisement*" in accordance with 18 U. S. C. §1734 solely to indicate this fact.

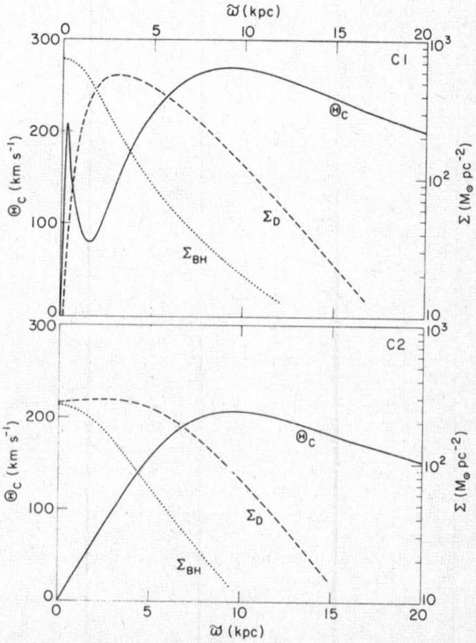


FIG. 1. (*Upper*) Rotation curve $\theta_c$ and the behavior of the spherical ($\Sigma_{BH}$, projected density) and disc ($\Sigma_D$) components for model C1. (*Lower*) Similar illustration of model C2.

## SPIRAL MODES

For the models shown in Fig. 1, we have calculated a few unstable spiral modes with relatively high growth rates. The spiral structures of some typical modes are given in Fig. 2. In this figure, the corotation circle with radius $r_c$ is indicated by a dotted curve in each of the diagrams. The azimuthal mode number $m$ gives the number of spiral arms, whereas the radial mode number $n$ governs the radial structure.

In Fig. 3, we show an example of the contour curves of constant perturbation density.

In Table 1, we list the pattern frequencies $\Omega_p$ and the growth rates $\gamma$ for the modes in Fig. 2 and some other modes. In the

* Present address: Scuola Normale Superiore, 56100 Pisa, Italy.

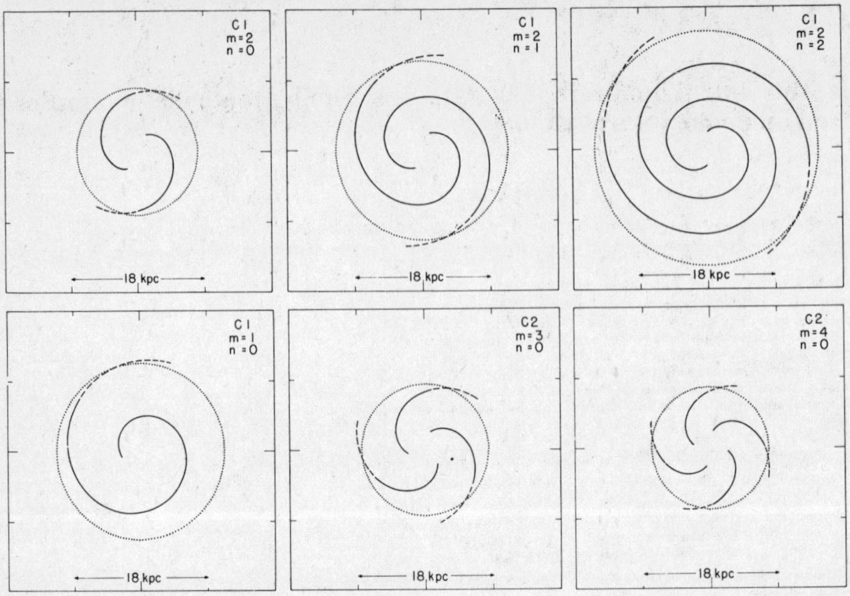


FIG. 2. Spatial structure of some of the modes calculated on the models C1 and C2. The relevant parameters of these modes are reported in Table 1, which includes data for other modes not illustrated here.

determination of $\gamma$, we have not included all known amplification effects, such as that due to the interaction between the disk and the bulge-halo stars (7). The reason is as follows. At the present epoch, the wave patterns in galaxies have probably reached a quasi-stationary state through a balance between dissipative influences (e.g., in the galactic shock) and the tendency for the mode to grow. Modes with higher $\gamma$ are more likely to achieve a higher amplitude and play more important roles. However, the exact value of $\gamma$ is less important, at least when it is small. We need only note that it will generally take a few billion years for the modes to reach observable amplitudes.

For model C1, the important spiral structure is essentially two-armed ($m = 2$); there is often also superposed a one-armed ($m = 1$) mode. Three-armed and four-armed modes are mostly suppressed by wave absorption at Lindblad resonances. In particular, for $m \geq 3$ the outer resonance circle is spatially

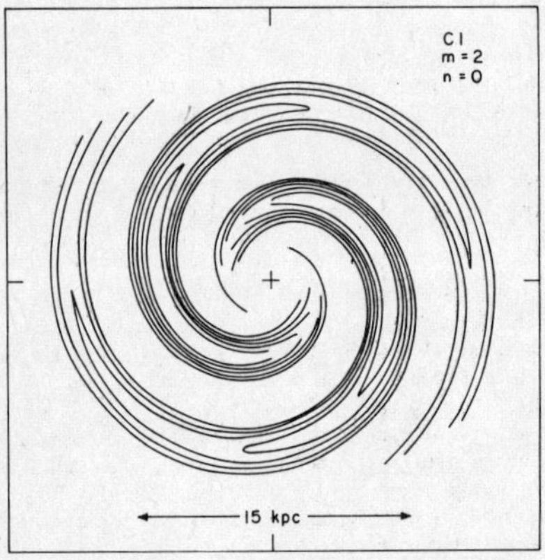


FIG. 3. Contour curves of constant perturbation density for the mode $(m,n) = (2,0)$ in galaxy model C1.

Table 1. Pattern frequencies ($\Omega_p$) and growth rates ($\gamma$) for various modes ($m$)

| $m$ | $n$ | $r_c$, kpc | $\Omega_p$, km s$^{-1}$ kpc$^{-1}$ | $\gamma$, $10^9$ yr$^{-1}$ |
|---|---|---|---|---|
| | | Model Cl | | |
| 1 | 0 | 11.4 | 23.7 | 3.0 |
| 1 | 1 | 17.3 | 12.6 | 0.85 |
| 2 | 0 | 8.0 | 32.8 | 9.3 |
| 2 | 1 | 11.6 | 22.2 | 5.3 |
| 2 | 2 | 15.0 | 15.9 | 2.8 |
| | | Model C2 | | |
| 1 | 0 | 11.7 | 17.4 | 2.2 |
| 2 | 0 | 8.9 | 23.2 | 5.9 |
| 2 | 1 | 12.2 | 16.2 | 3.9 |
| 2 | 2 | 15.2 | 11.9 | 2.2 |
| 3 | 0 | 7.8 | 25.7 | 10.0 |
| 4 | 0 | 7.2 | 27.2 | 14.2 |
| 4 | 1 | 9.4 | 22.1 | 10.9 |
| 4 | 2 | 11.1 | 18.4 | 9.1 |

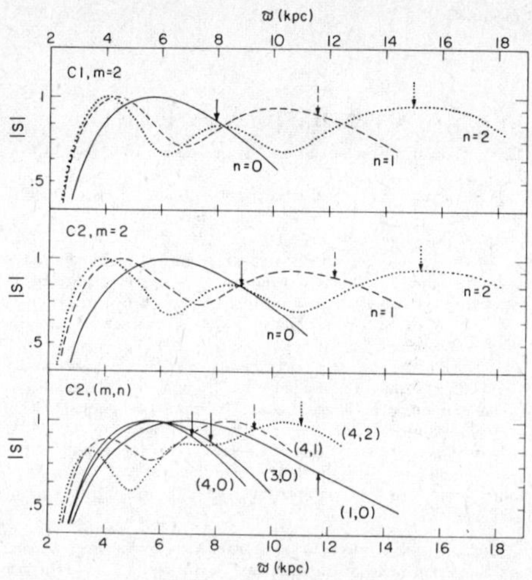


FIG. 4. For several modes (cf. Table 1), the magnitude of the density perturbation $|S|$ is plotted in arbitrary units (logarithmic scale) vs. the radial distance $\tilde{\omega}$. The $n$-dependence of the modes is clearly shown by the radial density profile. Arrows indicate the location of the corotation radius for each mode.

rather close to the corotation circle. This is typical of galaxies with rather concentrated nuclear bulge.

Model C2 does not have a nucleus. It also has a relatively diffuse bulge-halo system. In this case, even three-armed and four-armed structures can coexist on a comparable footing with the two-armed modes.

The radial distribution of the amplitude of the density perturbation is shown in Fig. 4. The nature of this distribution is governed by the radial mode number $n$: there are $n$ local minima and $(n + 1)$ local maxima. Such amplitude modulation has been interpreted by Mark (3) as the result of interference between the long and short waves that superpose to form the mode. Indeed, this reasoning gives the correct dependence on $n$ of the number of local maxima and minima in amplitude distribution. Mark also suggested that nonlinear effects might result in separately observable long and short waves.

The structures of the modes reported here were obtained through a straightforward integration of an ordinary differential equation of the second order. This simple governing equation was derived by using an asymptotic approximation to simplify the stellar dynamic equations for density waves. In the present asymptotic analysis, we have adopted an improved treatment of the potential theory and of the relatively large inclination of the spiral arms. A similar equation based on the gaseous model (equation 1 in ref. 2), has been used by Lau and Bertin (unpublished) to obtain spiral modes. For larger values of $n$, the procedure developed earlier (3) might be more accurate in the stellar case. There is no need to change procedure in the gaseous case.

Both the present asymptotic approach and other existing numerical model calculations (e.g., ref. 8) have their own merits and limitations. Some of the advantages of the former are as follows: (*i*) the method enables one to reach a clear understanding of the underlying physical mechanisms for formation and growth of the mode; (*ii*) it applies to wide classes (although not all) of realistic galaxy models; and (*iii*) for a given galactic model, the computation required to exhibit observable consequences is easy and simple. Especially the calculations of modes with high $n$ values are expected to be much simpler by the present asymptotic approach than by others. These modes with high $n$ values are important for interpretation of observational data.

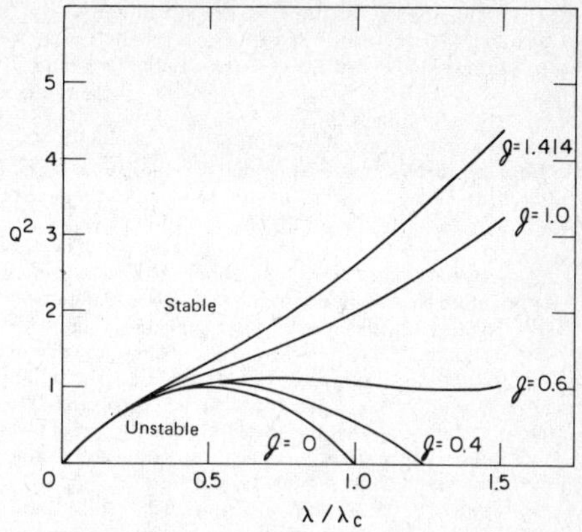


FIG. 5. The marginal stability curve for various values of $\mathcal{J}$.

## COROTATION RADII AND MULTIPLE SPIRAL STRUCTURES

From Figs. 3 and 4, it is seen that the spiral structure of each mode extends smoothly through the corotation circle. Of particular interest is the amplitude of the wave density perturbation as shown in Fig. 4. The curves are plotted on a logarithmic scale so that they can be compared with the surface brightness of the spiral arms as shown, for example, by figure 7*a* of ref. 5. The resemblance is quite striking.

Without attempting an interpretation now, we wish to point out that, for the *higher* modes, the corotation radii typically occur at or near the outermost peak in the wave density amplitude. If this characteristic holds for an overwhelming majority of galaxy models—a fact to be verified by further investigations—then there is possibly another method for the observational determination of the corotation radius $r_c$ (and consequently of the pattern frequency $\Omega_p$). We may simply identify the location of the outermost intensity peak from density observations and assign the corotation radius of the dominant mode to be approximately there (provided that the observed spiral structure is dominated by a mode with higher $n$).

In order to apply these considerations to the observed data for a given galaxy, it must be remembered that there is in general a mixture of several spiral modes in a galaxy at the present epoch. Thus, at the very least, we suggest that the spiral arm brightness results such as those of Schweizer (5) should be Fourier analyzed and presented separately for each $m$ number. In the resultant radial profile of the spiral arms brightness for each $m$, we must still realize that the outermost peak in brightness only determines the corotation radius of the (radially) most extensive of the prominent spiral modes. Due to the nature of the growth rates, the prominent modes are expected to be the ones with low values of the radial mode number $n$ (e.g., $n$

= 0,1,2). It is nevertheless likely for the mode with $(m,n) = (2,2)$ to be the prominent one in a galaxy similar to model C1, because the lower modes with somewhat higher growth rates are more restricted in their radial extent. (Nonlinear effects might also help to determine which mode becomes the most prominent in the observed spiral structure.)

The coexistence of several modes has another important implication. There is now no reason why some HII regions or observable spiral structures might not be found outside the corotation radius of the most prominent spiral mode. The outer spiral structure in the stars could be the smooth extension of the inner structure beyond the corotation radius, or a manifestation of modes with much higher radial mode number $n$. These latter modes have larger radial extent but relatively low growth rates, so that they are less prominent. To separate these possibilities, a detailed examination of observational data is required. Preliminary studies show that the amplitude distribution of these higher modes has a concentration near corotation and a rather tight spiral winding. Such modes might thus show up in the extreme outer part of a galaxy as a nearly ring-like structure which exhibits a spiral feature upon closer examination.

## GROWTH OF SPIRAL WAVES IN DISK GALAXIES

The origin of the high growth rate discussed in the above calculated modes was discussed in ref. 2. It has also been examined in the context of a local theory. Thus, we derived a local dispersion relationship for spiral waves in a gaseous disk. In addition to the well-known parameter $Q = \kappa a/\pi G\sigma_0$, another dimensionless parameter $\mathcal{J} = 2m(\pi G\sigma_0/\kappa^2\tilde{\omega})/\sqrt{(1/s - 1/2)}$ is found to be important in characterizing the growth of spiral waves. Here, $a$ is the acoustic speed, $\sigma_0$ is the surface density, $m$ is the number of spiral arms, $\kappa$ and $\Omega$ are the epicyclic and circular frequencies, respectively, $s = -d\ln\Omega/d\ln\tilde{\omega}$, and $G$ is the gravitational constant. Specifically, for stability against disturbances of local wavelength $\lambda$, we must have the condition $Q^2 > 4[\lambda/\lambda_c - (\lambda/\lambda_c)^2/(1 + \mathcal{J}^2\lambda^2/\lambda_c^{\ 2})]$ where $\lambda_c = 4\pi^2 G\sigma_0/\kappa^2$. The marginal stability curves are shown in Fig. 5 for various values of $\mathcal{J}$. For axisymmetric disturbances, $\mathcal{J} = 0$ and the curve resembles that of Toomre (9). The diagram indicates that nonaxisymmetric waves are more difficult to suppress than the axisymmetric ones, in general agreement with the work by Goldreich and Lynden-Bell (10) and by Julian and Toomre (11).

The details of this work, together with the role of $\mathcal{J}$ in the growth of spiral modes, will be given in a forthcoming paper by Lau and Bertin.

## APPENDIX: The Basic Models

The important parameters of models C1 and C2 are shown in Fig. 1.

The total mass of galaxy model C1 is $M_T \approx 1.6 \times 10^{11} M_\odot$, of which about 20% is in the form of a bulge-halo system. There is an additional small nuclear sphere with mass $M_N \approx 3 \times 10^9 M_\odot$ (not shown in Fig. 1). This may also be regarded as representing a disk in the central region. The value of $Q$, the ratio of the dispersion speed to that required to suppress axisymmetric Jeans instability, is taken to be equal to unity for $\tilde{\omega} \geq 7$ kpc. Toward the galactic center, $Q$ steadily increases (it reaches a value of 3.7 at $\tilde{\omega} = 1$ kpc).

The total mass of galaxy model C2 is $M_T \approx 9.6 \times 10^{10} M_\odot$, of which about 22% is in the form of a bulge-halo system. There is no nuclear sphere in this model. The parameter $Q$ is assumed to be equal to 1 for $\tilde{\omega} \geq 3$ kpc but to rise to 1.75 near the galactic center.

In both models, the disk density decays almost exponentially in the outer regions, and the bulge-halo system does not have a very large radial extent. The mass concentration toward the center is stronger in model C1 than in model C2.

The authors are grateful to Drs. Ronald Allen, François Schweizer, and Frank Shu for helpful discussions. This work was supported in part by the European Space Agency and in part by a grant from the National Science Foundation.

1. Lau, Y. Y., Lin, C. C., & Mark, J. W.-K. (1976) *Proc. Natl. Acad. Sci. USA* **73**, 1379–1381.
2. Lau, Y. Y. & Mark, J. W.-K. (1976) *Proc. Natl. Acad. Sci. USA* **73**, 3785–3787.
3. Mark, J. W.-K. (1977) *Astrophys. J.* **212**, 645–658.
4. Schweizer, F. (1975) in *La Dynamique des Galaxies Spirales*, ed. Weliachew, L. (CNRS, Paris), pp. 337–355.
5. Schweizer, F. (1976) *Astrophys. J. Suppl. Ser.* **31**, 313–332.
6. Oort, J. H. (1975) in *La Dynamique des Galaxies Spirales*, ed. Weliachew, L. (CNRS, Paris), p. 356.
7. Mark, J. W.-K. (1976) *Ap. J.* **206**, 418–434.
8. Zang, T. A. (1976) Ph.D. Dissertation, Massachusetts Institute of Technology, Cambridge, MA.
9. Toomre, A. (1964) *Astrophys. J.* **139**, 1217–1238.
10. Goldreich, P. & Lynden-Bell, D. (1965) *Mon. Not. R. Astron. Soc.* **130**, 125–158.
11. Julian, W. H. & Toomre, A. (1966) *Astrophys. J.* **146**, 810–830.

Astron. Astrophys. 69, 181—198 (1978)

# On the Stellar Streaming Motions and the Observational Determination of the Structural Constants of the Galaxy

C. C. Lin[1]*, C. Yuan[2]** and W. W. Roberts, Jr.[3]*

[1] Massachusetts Institute of Technology, Cambridge, Massachusetts 02139, USA
[2] Department of Physics, City College of New York, Convent Avenue at 138th Street, New York, N.Y. 10031, USA
[3] University of Virginia

Received June 14, 1977; revised January 11, 1978

**Summary.** The observational determination of the Oort constants $A$ and $B$ and of other structural parameters in the Miky Way system is analyzed in the light of the density wave theory. To investigate the nature and the extent of the influence of the spiral gravitational field on the observational determination of Galactic structure, we examined several models of the solar vicinity, including a spiral gravitational field and the accompanying shocks. These models are first tested by comparing the theoretically predicted and the observed fields of stellar motions. Good agreement is found for models similar to the one proposed by Lin et al. in 1969. Specifically, the two "dips" in the data for the determination of the rotation curve are found to correspond to the locations of the shocks. For the models thus chosen, the radial velocities and the proper motions that would be observed from the solar vicinity are then treated by the conventional methods of optical astronomy for obtaining the Oort constants and other structural parameters. In this way, we have a process to recover the galactic parameters originally adopted in our models. The internal consistency of this process is very good, the correction required being quite small in the case of the Oort constants $A$ and $B$, and especially so in $A-B$. This analysis helps us to understand the limitations in and the improvements on the conventional determination of the galactic parameters.

**Key words:** galactic structure — Oort constants — stellar streaming motion — density-wave theory — large-scale shocks

## I. Introduction

In the conventional approach to the determination of galactic structure, a strictly circular model is adopted. However, in the course of such investigations, it became apparent that this model could not be completely accurate. For example, Kraft and Schmidt (1963) stated: "Our failure to derive the curvature of the angular velocity curve from material that, at first sight, would seem quite suitable for the purpose points to the importance of localized systematic deviations from circular motion." Other optical astronomers (Rubin et al., 1962; Humphreys, 1970) agree with this statement. The importance of streaming motions in radio-astronomical work has also been stressed by many observers (Burton, 1966; Shane and Bieger-Smith, 1966), and discussed quite extensively by Burton (1970, 1971, 1972), by Shane (1971, 1972) and by Yuan and his collaborator (Yuan, 1969; Yuan and Liebovitch, 1971) in terms of the streaming motions predicted by the density wave theory. The theory of nonlinear streaming motions of the gas has been quite thoroughly studied in recent years by Roberts (1969), Roberts and Yuan (1970), Shu et al. (1972) and by Shu et al. (1973). In particular, Roberts (1972) applied the theory to the Perseus region and was able to account for the well-known apparent discrepancies discovered in the absorption features (see Munch, 1965).

It appears that, with our current knowledge and understanding of these streaming motions, it might be possible to make better use of the observed data to arrive at a more accurate determination of the structural parameters of the Galaxy. However, this is intrinsically difficult; for the streaming motions caused by the density waves depend on the local spiral structure assumed, and the determination of the parameters specifying the spiral structure can only be made after a basic circular model is adopted (see Lin et al., 1969; also Lin, 1971). Thus, the only convenient approach appears to be a procedure of successive approximations. We accept the conventional determination of the Oort constants as a first approximation, and apply corrections to account for the influence of the spiral field.

In a brief preliminary version of this paper (Lin et al., 1974), one such approach is described and followed. We

*Send offprint requests to:* C. Yuan

* Supported in part by the National Science Foundation
** Supported in part by CUNY FRAF No. 11456 and by the National Science Foundation

have since improved on the analysis and arrived at a deeper understanding of the situation. We now know that the corrections needed are minimal because of a fortunate combination of circumstances. Had we not been fortunately located about midway between two major spiral arms, or had the distribution of young optical objects shown a different preferential distribution in longitudinal directions, we might have had greater difficulty in analyzing the observational data to obtain the essential parameters in the structure of our Galaxy. These details will be explained in Section V of this paper.

Let us now return to a discussion of our plan of theoretical investigations.

### *a) Plan of Investigation – A Theoretical Procedure of "Recovery"*

We adopt the following procedure for the study of radial velocities. Proper motions are treated in a similar manner.

(1) We begin with the circular model of the Galaxy given by Schmidt (1965), and impose on it a spiral structure with a given geometry and given gravitational field. (Several models were attempted in our study. Our experience shows that the solar location must be carefully specified.)

(2) We calculate the streaming motions according to the *nonlinear* theory, including the formation of galactic shocks (see Roberts, 1969; Shu et al., 1972; Shu et al., 1973).

(3) From step (2), we calculate the line-of-sight velocity in heliocentric coordinates. (See Section III for a more detailed explanation).

(4) *We treat these theoretically calculated data by the conventional methods used for analyzing observational data;* i.e., we adopt a circular model and attempt to fit our "data" to such a model in order to determine the structural parameters of the Galaxy.

(5) The difference between the values of the parameters assumed in step (1) and those obtained in step (4) gives the *correction* to be applied to the usual observationally determined value of the Oort constants. To be specific, let us denote by $A_1$ the value of the Oort constant $A$ assumed in step (1), and by $A_4$ the value obtained in step (4), and let us define

$$\Delta A = A_1 - A_4$$

Then we except that the true value of the Oort constant $A$ for our Galaxy is given approximately by

$$A = A_c + \Delta A ,$$

where $A_c$ is obtained by a conventional analysis of the observational data assuming a circular model, in the manner done by Kraft and Schmidt, by Rubin et al. and by Humphreys.

It is hoped that the amount of correction $\Delta A$ would be in error by a small amount if the original model adopted in step (1) were in error to a small extent. We would then have a fairly reliable process for "recovering" the value of $A$ from under the masking effect of the spiral gravitational field.

### *b) Observational Justification of This Approach*

Before the procedure proposed above can be justified, we must examine whether the field of flow calculated is at least qualitatively of the same nature as the field of the motions of young objects. As it turns out, the two fields of motion appear to be in general agreement for the model already adopted (Lin et al., 1969) on the basis of an analysis of a number of other observational data. In this model, the Sun is placed approximately midway between two major spiral arms. The observed stellar motion of young objects will be reviewed in Section II, and the comparison of these data with the calculated flow field will be presented in Section IV.

We have also considered slightly different models. As expected, small variations in our models do not produce significant changes. Major modifications will be shown, however, to lead to qualitatively different results. (See discussions of Model S 3 below.) Such models should be rejected.

The analysis of this paper could be repeated for a wide variety of models, some of which should include secondary spiral features which are known to exist in the solar vicinity. We feel, however, that the simple models adopted in this paper are sufficient for clarifying the major issues, and warrant the following conclusions.

### *c) Principal Conclusions*

(i) The determination of the Oort constants is not significantly influenced by the presence of a spiral gravitational field, provided that the spiral pattern and its associated gravitational field are similar to those prescribed according to the discussions in subsection (b) just presented. For the specific models calculated, our results indicate the following estimate of accuracy in the *theoretical* recovery process: 10% deviation in Oort constant $A$, 20% in Oort constant $B$, about 2% in the angular velocity $\Omega = A - B$. These deviations are in general below the uncertainties caused by errors in observations. The presence of a spiral gravitational field tends to give a higher value for $A$ and a lower absolute value for $B$. We also show that the nonlinear streaming motions and the biased distribution of young stars in longitude do not have strong effects on the recovery of the assumed Oort constant $A$.

It should be noted that, in the usual analysis, observational data used are from all longitude directions and at distances of 1–3 kpc from the Sun, because local

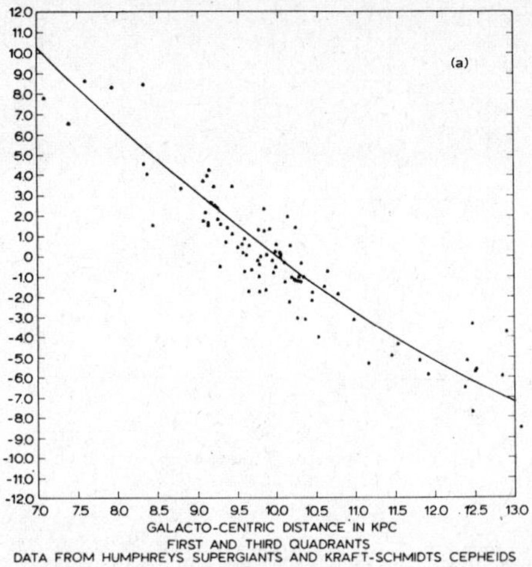


**Fig. 1a.** The angular velocity rotation curve for the cepheids (Kraft and Schmidt, 1965) and the supergiants (Humphreys, 1970) in the first and third quadrants. The ordinate is for the value $V_r/\tilde{\omega}_0 \sin l \cos b$ in $\mathrm{km\,s^{-1}\,kpc^{-1}}$ and the abscissa for $\tilde{\omega}$ in kpc. The solid line is the 1965 Schmidt curve. Stars with galactic longitude $|l| < 25°$ are excluded from the samples

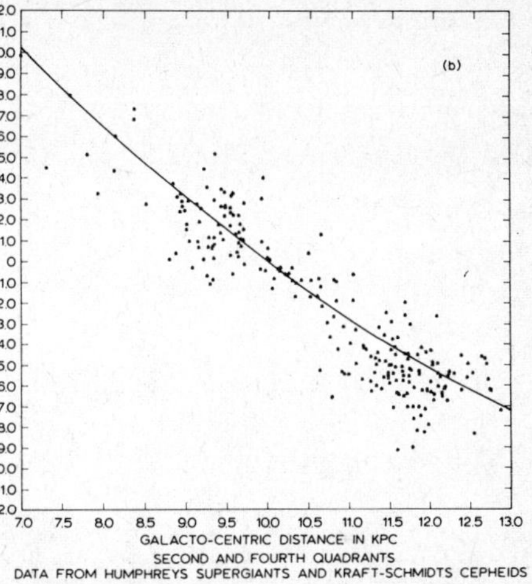


**Fig. 1b.** The angular velocity rotation curve for the cepheids (Kraft and Schmidt, 1965) and the supergiants (Humphreys, 1970) in the second and fourth quadrants. Note the concentrations of the data points hanging below the 1965 rotation curve at about $\tilde{\omega} = 9$ kpc and $\tilde{\omega} = 11.5$ kpc, both of which are interpreted as the regions of star formation in the galactic shock

data show too much fluctuation. This is an important fact because such a procedure tends to reduce and to average out the effect of the asymmetrical field of the streaming motions associated with the density waves. However, as we shall see in Section V, there is still a strong element of luck in this averaging process.

(ii) The recovery of the rotation curve is fairly accurate, provided we assume a reasonable value for $\tilde{\omega}_0$, the galacto-centric distance of the Sun. It is, however, not accurate enough to yield a reliable determination of its curvature at the solar vicinity. This agrees with the experience of astronomers using observational data. Similarly, the $K$-term cannot be consistently determined from data at various radial distances from the Sun. Apparent contradictions have often been reported by the analysis of observational data. On the basis of present studies, it is not difficult to understand the reasons behind these discrepancies.

(iii) The recovery of $A\tilde{\omega}_0$ (and therefore of $\tilde{\omega}_0$) is also quite good. It probably suffers an inaccuracy of less than 10%, the exact error depending on the models used.

(iv) It was noted during the course of investigation that the local standard of rest has really been determined relative to the *young* objects, which are in general moving relative to the very old objects. Data for the latter set of objects are difficult to obtain. It would seem desirable that an extensive observational effort be made to ascertain this relative motion, whose magnitude is estimated to be anywhere between 0 and $10\,\mathrm{km\,s^{-1}}$ according to our theoretical studies (cf. Section IV).

## II. Observational Determination of the Rotation Curve and the Oort Constants

The structure of the Milky Way system in the neighborhood of the Sun is traditionally determined by optical observations of the kinematics of the young objects. Of course the sample of young objects in the *immediate* neighborhood of the Sun may reflect locally perturbed kinematics and dynamics which might not be typical of the mean state of dynamics on the large scale in our Galaxy. Consequently, the conventional approach for the observational determination of galactic structure generally covers a region of a few kiloparsecs about the Sun. For such an extended region, bright young stars are the only objects suitable for optical observations. It has been recognized (Kraft and Schmidt, 1963) that among the brightest stars, the cepheids are ideal for the determination of the structural constants. The O–B stars and supergiants, however, are also frequently used (Rubin et al., 1962, 1964; Humphreys, 1970; Balona and Feast, 1974). Unfortunately, the surveys of O–B stars beyond two kiloparsecs from the Sun are hardly complete and uniform with respect to longitudes. Furthermore, with intrinsic observational difficulties, their individual radial

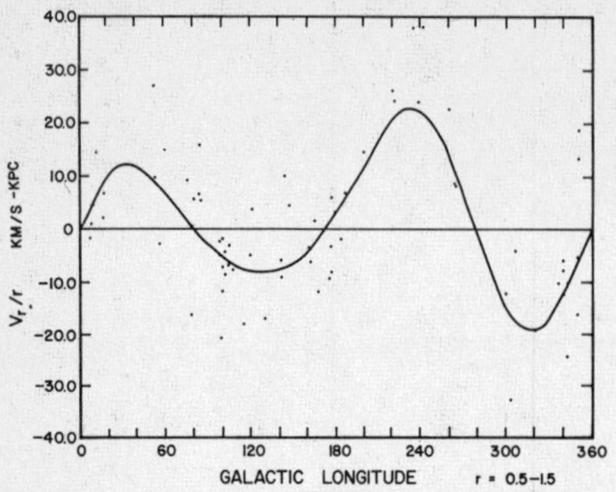


**Fig. 2a.** The "double-sine" curve for supergiants (Humphreys, 1970) with distances between 0.5 kpc and 1.5 kpc from the Sun. The ordinate represents the value $V_r/r\cos^2 b$, viz radial velocity corrected for projection divided by distance projected from the Sun, and the abscissa is the galactic longitude. Since all $b$ values are small and also we try to keep a uniform format between the theoretical and the observed figure, we have omitted $\cos^2 b$ in labelling the ordinate. The solid curve is the least square fit of the data points

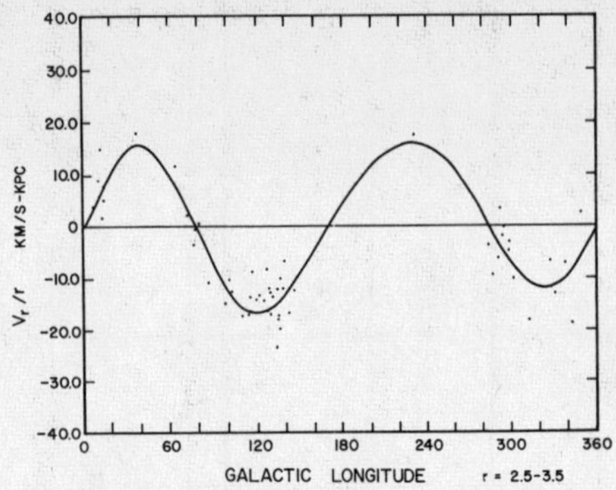


**Fig. 2c.** The "double sine" curve for the supergiants with distances between 2.5 kpc and 3.5 kpc from the Sun

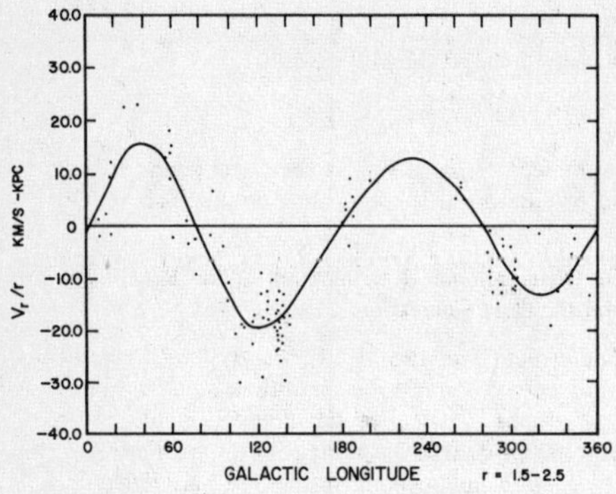


**Fig. 2b.** The "double sine" curve for the supergiants with distances between 1.5 kpc and 2.5 kpc from the Sun

velocities and distances cannot be accurately determined. Although the supergiants also suffer a shortcoming in completeness, they are observed much more uniformly in longitude than the O–B stars. Their sharp spectral lines also lead to a better determination of radial velocity. Therefore, for the present purpose of comparison, we shall only use the data of the cepheids and the supergiants, compiled respectively by Kraft and Schmidt (1963) and by Humphreys (1970). (The samples of Balona and Feast, 1974, were not available at the time when our analysis was made.)

We should point out too that the long period cepheids and the supergiants are particularly suitable for the purpose of comparison with our results from the non-linear calculations of gas flow. Their ages are believed to be of the order of $10^7$ years. Within that age range, the mean velocity of the stars will not significantly deviate away from that of the gas clouds and their location will remain close enough to the shock front where they are formed for all practical purposes.

Conventionally, the observational data are presented in the following three different ways which will lead to the determination of the structural parameters for the Milky Way.

### *a) Angular Velocity Curves*

The radial velocities with respect to the local standard of rest, $V_r$, of young stellar objects divided by $\tilde{\omega}_0 \sin l$ are plotted against $\tilde{\omega} - \tilde{\omega}_0$, where $\tilde{\omega}_0$ is the galacto-centric distance to the Sun. If the Galaxy is in pure circular motion, the configuration will be the angular-velocity curve for the solar vicinity. Clearly $\tilde{\omega}_0$ must be determined from other methods.

It is most illustrative to view the angular velocity curve through the observational data in the first and third quadrants and in the second and fourth quadrants separately. Figures 1a and 1b contain the observational data of Kraft and Schmidt (1963) and of Humphreys (1970) in these quadrants respectively. The general trend of decreasing angular velocity with increasing radius exhibited by the stellar data is shown in both figures. The slightly curved thick line running through the data represents the angular velocity characteristic of the Schmidt model for purely circular motion. The deviations of the stellar data from the Schmidt curve are most evident in the second and fourth quadrants in Figure 2.1b. Here prominent dips in the stellar data occur near 9 kpc and 11.5 kpc. The characteristic streaming motions reflected in these dips are indeed intriguing characteristics which must be accounted for in any complete theory of galactic structure. An understanding of these streaming motions will be provided through the theoretical developments in Sections III and IV of this paper.

*b) Double-sine Curve*

The radial velocities, $V_r$, of young stellar objects divided by their individual stellar distances are plotted against galactic longitude $l$. The result is the famous "double-sine" curve first noticed by Oort, from which the Oort constant $A$ and the constant of local expansion $K$ may be determined (Rubin and Burley, 1964). Figures 2a, 2b, and 2c illustrate this type of plot for the observational data for supergiants (Humphreys, 1970) in three annuli about the Sun, covering the regions lying at 0.5 – 1.5, 1.5 – 2.5, and 2.5 – 3.5 kpc respectively. The solid curve in each diagram represents a least square fit to the data determined through a method of Fourier analysis to be discussed in Sections IId and III. The numerical coefficients which are obtained by this method to prescribe the curve are given in Table 1.

Particularly evident in Figures 2b and 2c is the concentration of the young stellar objects within the troughs of the curve over the regions $l = 100°$ to $l = 140°$ and $l = 280°$ to $l = 320°$. Insight into the underlying mechanism which can trigger the formation of the young stellar objects in these locations will be provided through the theoretical developments in Section IV.

*c) Double-cosine Curve*

A plot of the projected component of the proper motions, corrected for the motion of the sun can be made with respect to galactic longitude. For differential circular motion, such a plot yields a double-cosine curve with the horizontal zero axis displaced downward uniformly. The amount of the downward displacement is a measure of the Oort constant $B$ (Rubin and Burley, 1964). However, it is well known that the proper motions of the more distant stars (1 – 2 kpc) are very difficult to determine to the accuracy needed in the analysis. For this reason, we shall not reproduce any double-cosine curves from observations in this paper. The reader is referred to Figure 3 in the paper by Rubin and Burley.

*d) The Oort Constants*

From the observational data presented in Figures 1 and 2, it is possible to determine the Oort constant $A$, the curvature of the angular velocity curve, and several other parameters of galactic structure. In contrast to the conventional approach (see Kraft and Schmidt, 1963; Humphreys, 1970), we adopt a more general approach

**Table 1.** Fourier coefficients obtained from observations (Humphreys' data)

| distance (kpc) | 1 | 2 | 3 | 0.5–3.5 |
|---|---|---|---|---|
| $\alpha_0$* | 1.37 | −0.87 | 1.31 | 0.63 |
| $\alpha_1$ | 4.40 | 1.25 | −0.74 | −0.28 |
| $\beta_1$ | −2.46 | −3.40 | −4.18 | −4.11 |
| $\alpha_2$ | −0.87 | 0.76 | 0.44 | 0.22 |
| $\beta_2$ | 14.77 | 14.41 | 14.15 | 14.66 |
| $\alpha_3$ | 3.29 | −2.00 | −1.49 | −0.96 |
| $\beta_3$ | 2.44 | 2.60 | 3.13 | 2.47 |

Fourier coefficients of the supergiants (Humphreys, 1970). The values for distance $r = 1$ kpc are obtained from all the stars in the sample between $r = 0.5$ kpc and $r = 1.5$ kpc. The values for other distances are also obtained similarly with 1 kpc intervals. The quantity $\alpha_0$ is the dilatation constant $K$ and the quantity $\beta_2$ is the Oort constant $A$

* For definition of symbols, see Equations, (3.3) and (3.4)

via Fourier analysis in Section III of this paper to identify these parameters. Strictly speaking, this approach provides a definite advantage only when the available data are uniformly distributed in longitude. In the analysis of observational data, we first fit them with Chebyshev polynomials through a standard least square procedure and then find the Fourier coefficients by analyzing the curves represented by the polynomials. This method is convenient for treating data of the angular velocity curve, although it becomes somewhat indirect for treating those of the double-sine curve. When data $V_r/r$ are fitted with the least square method, a proper weight is assigned to each $V_r/r$ to account for the unequal accuracy of the observations in $V_r$ and $r$. The formula for the weight is given in Section V. The general Fourier analysis presented in Section III may or may not represent a substantial improvement over the conventional method. In Table 1, we summarize the principal results of our approach as applied to the observational data of Humphreys (1970).

## III. Theoretical Analysis

When the Oort constants are obtained from a theoretical analysis of flow fields (or fields of mean motion of young stars), it is of course most convenient and even necessary to consider relative motions. It is therefore fortunate that *the conventionally adopted local standard of rest is essentially co-moving which the young* O–B *stars* [1]. The velocity data usually presented, relative to the local standard of rest, are essentially the data for the *relative motions* of young objects.

There are two steps in the analysis of the field of motion in the solar vicinity. The first is a discussion of the *local field* of motion, focussed on the Oort constants.

[1] The relative velocity between the two systems is about 1 km s$^{-1}$. (See Delhaye, 1965)

This part of the analysis is independent of the distance $\tilde{\omega}_0$ of the Sun from the center of the Galaxy. The second is the determination of the *rotation curve* of the Galaxy, and this must be preceded by the determination of the galacto-centric distance $\tilde{\omega}_0$ of the Sun. We shall first examine step one.

### a) Determination of Oort Constants and Other Local Derivatives by Fourier Analysis

Strictly speaking, the Oort constants are defined by the derivatives of the velocity field at the immediate vicinity of the Sun. However, since the velocities of nearby objects are sensitive to local effects, we must be prepared to estimate these derivatives by using data obtained at a few (e.g., two) kiloparsecs of the Sun. We should therefore formulate a method for dealing with the flow field at such distances.

The most systematic approach to the problem is to consider the Fourier series of the two ratios

$$F_r(r,l)=V_r/r\,, \tag{3.1}$$

$$F_t(r,l)=V_t/r\,, \tag{3.2}$$

where $F_t$ is the "proper motion" in the galactic plane in the positive direction of $l$ relative to the local standard of rest and $F_r$ is a similar quantity for the radial motion.

In general, we may write

$$F_r(r,l)=\sum_0^\infty \{\alpha_n(r)\cos nl+\beta_n(r)\sin nl\}\,, \tag{3.3}$$

$$F_t(r,l)=\sum_0^\infty \{\mu_n(r)\cos nl+\nu_n(r)\sin nl\}\,, \tag{3.4}$$

where $\{\alpha_n(r),\beta_n(r)\}$ and $\{\mu_n(r),\nu_n(r)\}$ may be thought of as Taylor series in $r$. For $r=0$, we can prove *exactly* that all the Fourier coefficients vanish except possibly the following:

$$\alpha_0(0)=K\,,\quad \alpha_2(0)=C\,,\quad \beta_2(0)=A\,; \tag{3.5}$$

$$\mu_0(0)=B\,,\quad \mu_2(0)=A\,,\quad \nu_2(0)=-C\,, \tag{3.6}$$

where the notation follows usual astronomical convention: $A$, $B$ are the Oort constants, $K$ is the rate of local expansion (the $K$-term), and $C$ is also the symbol commonly used [see (3.9) below].

The method of Fourier analysis is essentially a more exact mathematical formulation for analyzing the observed radial and proper motions than the conventional method of considering the data as a function of longitude just for the determination of the $(2l)$-component.

### b) Influence of Streaming Motions

We can make a fairly general assessment of the error that might be introduced in the determination of the Oort constants due to the influence of streaming motions. In particular, we shall see that the error in the determination of the angular velocity $\Omega=A-B$ is expected to be small.

In galacto-centric coordinates $(\tilde{\omega},\theta)$, with velocity components denotes by $(\Pi,\Theta)$, the coefficients of local relative motion are as follows[2]:

$$A=-\frac{1}{2}\left(\frac{\partial\Theta}{\partial\tilde{\omega}}-\frac{\Theta}{\tilde{\omega}}+\frac{1}{\tilde{\omega}}\frac{\partial\Pi}{\partial\theta}\right), \tag{3.7}$$

$$B=-\frac{1}{2}\left(\frac{\partial\Theta}{\partial\tilde{\omega}}+\frac{\Theta}{\tilde{\omega}}-\frac{1}{\tilde{\omega}}\frac{\partial\Pi}{\partial\theta}\right), \tag{3.8}$$

$$C=\frac{1}{2}\left(\frac{\partial\Pi}{\partial\tilde{\omega}}-\frac{\Pi}{\tilde{\omega}}-\frac{1}{\tilde{\omega}}\frac{\partial\Theta}{\partial\theta}\right), \tag{3.9}$$

$$K=\frac{1}{2}\left(\frac{\partial\Pi}{\partial\tilde{\omega}}+\frac{\Pi}{\tilde{\omega}}+\frac{1}{\tilde{\omega}}\frac{\partial\Theta}{\partial\theta}\right). \tag{3.10}$$

In Oort's (1927) original discussion, a circular model was adopted. Thus,

$$A=A_c=-\frac{1}{2}\left(\frac{d\Theta_c}{d\tilde{\omega}}-\frac{\Theta_c}{\tilde{\omega}}\right)=-\frac{\tilde{\omega}}{2}\frac{d}{d\tilde{\omega}}\left(\frac{\Theta_c}{\tilde{\omega}}\right), \tag{3.11}$$

$$B=B_c=-\frac{1}{2}\left(\frac{d\Theta_c}{d\tilde{\omega}}+\frac{\Theta_c}{\tilde{\omega}}\right). \tag{3.12}$$

After $A$ and $B$ are determined, we are interested in calculating $A-B$ and $A+B$ in order to get the circular velocity and its radial gradient. In the circular case, we have

$$A_c\ \ B_c=\Theta_c/\tilde{\omega}\,, \tag{3.13}$$

$$A_c+B_c=-d\Theta_c/d\tilde{\omega}\,. \tag{3.14}$$

In general, we have to make suitable corrections. Let $(u_{\tilde{\omega}},u_\theta)$ be the deviation of gaseous motion from circular motion; then the corrections $(\Delta A,\Delta B)$ are given by

$$\Delta A=-\frac{1}{2}\left(\frac{\partial u_\theta}{\partial\tilde{\omega}}-\frac{u_\theta}{\tilde{\omega}}+\frac{1}{\tilde{\omega}}\frac{\partial u_{\tilde{\omega}}}{\partial\theta}\right), \tag{3.15}$$

$$\Delta B=-\frac{1}{2}\left(\frac{\partial u_\theta}{\partial\tilde{\omega}}+\frac{u_\theta}{\tilde{\omega}}-\frac{1}{\tilde{\omega}}\frac{\partial u_{\tilde{\omega}}}{\partial\theta}\right); \tag{3.16}$$

and the corrections for Equations (3.13) and (3.14) are

$$\Delta(A-B)=+\frac{1}{\tilde{\omega}}\left(u_\theta-\frac{\partial u_{\tilde{\omega}}}{\partial\theta}\right), \tag{3.17}$$

$$\Delta(A+B)=-\frac{\partial u_\theta}{\partial\tilde{\omega}}\,. \tag{3.18}$$

Clearly, if $\tilde{\omega}$ is larger than the spacing between spiral arms, $\Delta(A-B)$ is smaller than $\Delta(A+B)$; that is, the calculation of angular velocity in Oort's original dis-

[2] See Chandrasekhar (1942), p. 19 and p. 24. Note that in his notation $\Theta_c=-250\ \mathrm{km\,s^{-1}}$. Hence there are some differences in signs with our notation

**Table 2.** The Models. Five theoretical models were investigated. These are designated S-1, S-2, S-3 and E-1, E-2. The spirals in the S-series have a pitch angle of seven degrees; the E-series, eight degrees. The location of the Sagittarius and the Perseus arms are given in the following table in kpc from the galactic center

| | Sagittarius arm | Perseus arm |
|---|---|---|
| S-1 | 8.26 | 12.3 |
| S-2 | 8.5 | 12.55 |
| S-3 | 7.0 | 10.3 |
| E-1 | 7.9 | 12.4 |
| E-2 | 7.4 | 11.5 |

cussion requires the least correction, whereas that of $\partial\Theta/\partial\tilde{\omega}$ requires the most, and the corrections in $A$ and in $B$ are approximately of the same magnitude. [As it turns out, because of the signs of the two quantities in Equations (3.17) and (3.18), the percentage correction required in $A$ is smaller.]

### c) Rotation Curve

The conventional procedure[3] used for the determination of the rotation curve (see Kraft and Schmidt, 1963; Humphreys, 1970) is to start with a circular model with an angular velocity curve $\Omega(\tilde{\omega})$. Through our approach with a more general model containing streaming motions the formula for relative velocities in the direction of line-of-sight is

$$\begin{aligned} v_r = {} & \tilde{\omega}_0(\Omega(\tilde{\omega})-\Omega(\tilde{\omega}_0))\sin l\cos b \\ & + \left[\left(u_\theta\frac{\tilde{\omega}_0}{\tilde{\omega}}-u_{\theta_0}\right)\sin l-\left(u_\omega\frac{\tilde{\omega}_0}{\tilde{\omega}}-u_{\omega_0}\right)\cos l \right. \\ & \left. + \frac{r\cos b}{\tilde{\omega}}u_\omega\right]\cos b\,, \end{aligned} \tag{3.19}$$

where $(l, b)$ are respectively the longitude and the latitude in the direction of observation, $\tilde{\omega}$ is the galactocentric distance of the material in question and $\tilde{\omega}_0$ is its value for the solar location; $(u_{\tilde{\omega}_0}, u_{\theta_0})$ denote the deviation of the local standard of rest from circular motion. We can write $v_r$ in terms of the coefficients of local relative motion as follows:

$$\begin{aligned} v_r = {} & \tilde{\omega}_0[\Omega(\tilde{\omega})-\Omega(\tilde{\omega}_0)]\sin l\cos b \\ & + r(K+\Delta A\sin 2l+C\cos 2l)\cos^2 b+\dots, \end{aligned} \tag{3.20}$$

where $C$ and $K$ are defined by Equations (3.9) and (3.10) respectively and where

$$\Delta A = -\frac{1}{2}\left(\frac{\partial u_\theta}{\partial\tilde{\omega}}-\frac{u_\theta}{\tilde{\omega}}+\frac{1}{\tilde{\omega}}\frac{\partial u_{\tilde{\omega}}}{\partial\theta}\right). \tag{3.21}$$

[3] Rubin and Burley (1964) adopted a more general field of motion—but not as general as that used in this paper—and evaluated the correction terms from an analysis of their data

Often it is desirable to expand the first term in (3.19) or (3.20) in the form of a truncated Taylor series. We have

$$\Omega-\Omega_0 = -\frac{2}{\tilde{\omega}_0}[A(\tilde{\omega}-\tilde{\omega}_0)+q(\tilde{\omega}-\tilde{\omega}_0)^2]\,, \tag{3.22}$$

where

$$A = -\tfrac{1}{2}\tilde{\omega}_0\Omega'(\tilde{\omega}_0)\,, \tag{3.23}$$

$$q = -\tfrac{1}{4}\tilde{\omega}_0\Omega''(\tilde{\omega}_0)\,. \tag{3.24}$$

The parameter $q$ represents the curvature exhibited by the curve of angular velocity.

For a strictly circular model, it is also possible to determine the product $A\tilde{\omega}_0$ through the following formula (Schmidt, 1965):

$$V_{max} = 2A\tilde{\omega}_0\sin l(1-|\sin l|)\,, \tag{3.25}$$

where $V_{max}$ is the maximum velocity of all objects in the direction of longitude $l$. Clearly, this formula is sensitive to the effect of local streaming motions.

### d) The Local Constant of Expansion K

The constant of strictly local expansion $K$, as defined by Equation (3.10), may be calculated through the Fourier analysis of the double-sine curve to obtain $\alpha_0(r)$, followed by an application of Equation (3.5). Alternatively, $K$ may be determined from the equation of continuity for the theoretical flow field by the following formula:

$$K = \tfrac{1}{2}\mathrm{div}.\,\underline{V} = -\frac{|V|}{2\varrho}\frac{d\varrho}{ds} \tag{3.26}$$

where $\varrho$ is the gas density, $\underline{V}$ is the velocity and $s$ is the space coordinate along the streamline. Since the density $\varrho$ decreases along the stream line as the gas is leaving the shock region, the value of $K$ must be positive for the general region between the spiral arms. This theoretical calculation from the gas dynamical equations provides a value for $K$ of $5\,\mathrm{km\,s^{-1}\,kpc}$, which is in good agreement with the value obtained from the Fourier analysis of the theoretical double-sine curve for $r = 1$ kpc.

## IV. Theoretical Results; Insight into the Observational Data

In this section we shall present our theoretical results calculated from the procedure outlined in Section I and from the analysis described in Section III. The calculations are based on the *non-linear* theory[4] of the streaming gas motions under the influence of a spiral gravitational field. The line-of-sight velocity and the "proper motions" of the gas for the solar vicinity obtained from the above calculations are treated *as if*

[4] An earlier study of the influence of a spiral gravitational field on the determination of the Oort constants was made by Rohlfs (1971) on the basis of a *linear* theory. Extending Rohlfs' results, Feitzinger discussed the density-wave effect on the $K$-term in a more recent study (1977)

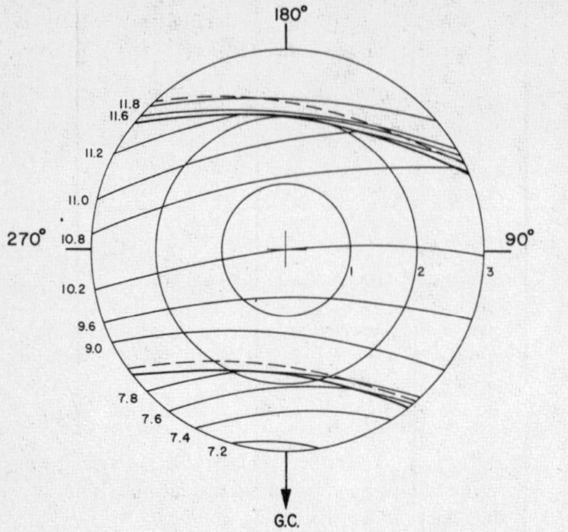


**Fig. 3.** The flow field for model S-1 in the solar vicinity. The broken lines are the locations of the potential minimum of the assumed spiral gravitational field. The numerical values on the left side of the $r = 3$ kpc circle indicate the mean galacto-central distance for each individual streamline in the diagram. A heavy solid line lying inside and also parallel to the potential minimum is the location of the shock front. The streamlines change their direction at the shock front

they were observational data and are subjected to the conventional analysis for the determination of galactic structure. Finally, the structural parameters thus obtained will be compared with the values of the assumed basic model of circular motion, in order to determine the influence of the spiral gravitational field.

For the basic circular model, the 1965 Schmidt model is adopted. As there is as yet no general agreement among astronomers about the spiral structure for the solar vicinity, five theoretical models were used in the calculation of the velocity fields of the gas. The models with pitch angle of the spiral arms equal to 7° are designated as S-1, S-2 and S-3; those with pitch angle equal to 8° are designated as E-1 and E-2. The relative orientations of these five theoretical models with respect to the Sun are described in Table 2. It should be mentioned that the spiral pattern of model S-1 is a close approximation of the theoretical spiral pattern which Lin et al. (1969) adopted for the Milky Way System. For purposes of illustration, the model S-1 will be used most often as the sample model, although some of the other models considered, as well as others not cited, appear just as desirable and adequate in elucidating the observational data.

The computations for the non-linear velocity along a streamline are similar to those of Roberts (1969). The entire flow pattern is obtained by interpolation from a sequence of streamlines. In order to preserve the accuracy, a very fine scale in distance has been used between two subsequent streamlines. The flow pattern of model S-1 within a distance of 3 kpc from the Sun is shown in Figure 3. The Sun is situated between the Sagittarius Arm and the Perseus Arm. The locations of the sinusoidal potential minima are indicated by the heavy curves in the diagram. The streamlines make a sudden turn along the galactic shock front slightly inside of a potential minimum. The theoretical line-of-sight velocity $V_r$ and the theoretical transverse velocity $V_t$, within 3 kpc from the Sun, are computed from the numerical results of the flow pattern. Since the gas velocity at the location of the Sun is more complicated than that of the Schmidt model for purely circular rotation, it is necessary to subtract also the velocity of the gas at the position of the sun in order to consider the relative motions $V_r$ and $V_t$.

This treatment is essentially tantamount to the statement that *the conventionally adopted local standard of rest is co-moving with the young objects.* The amount of velocity subtracted is usually not large, but significant. For model S-1, the tangential component of the streaming motion is 6 km s$^{-1}$ and the component in the anti-center direction is 6.25 km s$^{-1}$.

In what follows, we shall present our theoretical results of the nonlinear gas flow field precisely in the same manner as the observational data are presented. These data will then be treated through our more general analysis for the determination of the constants of the galactic structure.

### *a) Angular Velocity Curve*

As usual, we shall assume $\tilde{\omega}_0$ to be known. The presence of the spiral gravitational field causes the quantity $V_r/\tilde{\omega}_0 \sin l$ to deviate systematically from the angular velocity curve for any originally assumed model with purely circular motion. This deviation, which varies gradually with galactic longitude, changes drastically when the shock is encountered. The situation is shown in Figures 4a and 4b for the model S-1. The heavy curve in each figure is the angular-velocity curve for the 1965 Schmidt model. *Because of the streaming motion, the angular velocity curves in the first and third quadrants* ($l = 0$–90° and 180–270°) *have steeper slopes than that* of the Schmidt model, *and hence a larger value of A, whereas the angular-velocity curves in the second and fourth quadrants* ($l = 90$–180° and 270 – 360°) *have flatter slopes and hence a smaller value of A.*

The jumps in angular velocity which occur at about $\tilde{\omega} = 8.5$ kpc and 12 kpc in Figures 4a and 4b correspond respectively to the galactic shocks calculated for the Sagittarius Arm and the Perseus Arm. Since the bright young stars are expected to be found in concentrations behind the shock and also to share in the nonlinear streaming motion of the gas there, the shock phenomenon is expected to be reflected in the observed angular-velocity curve. Consequently, we view the two

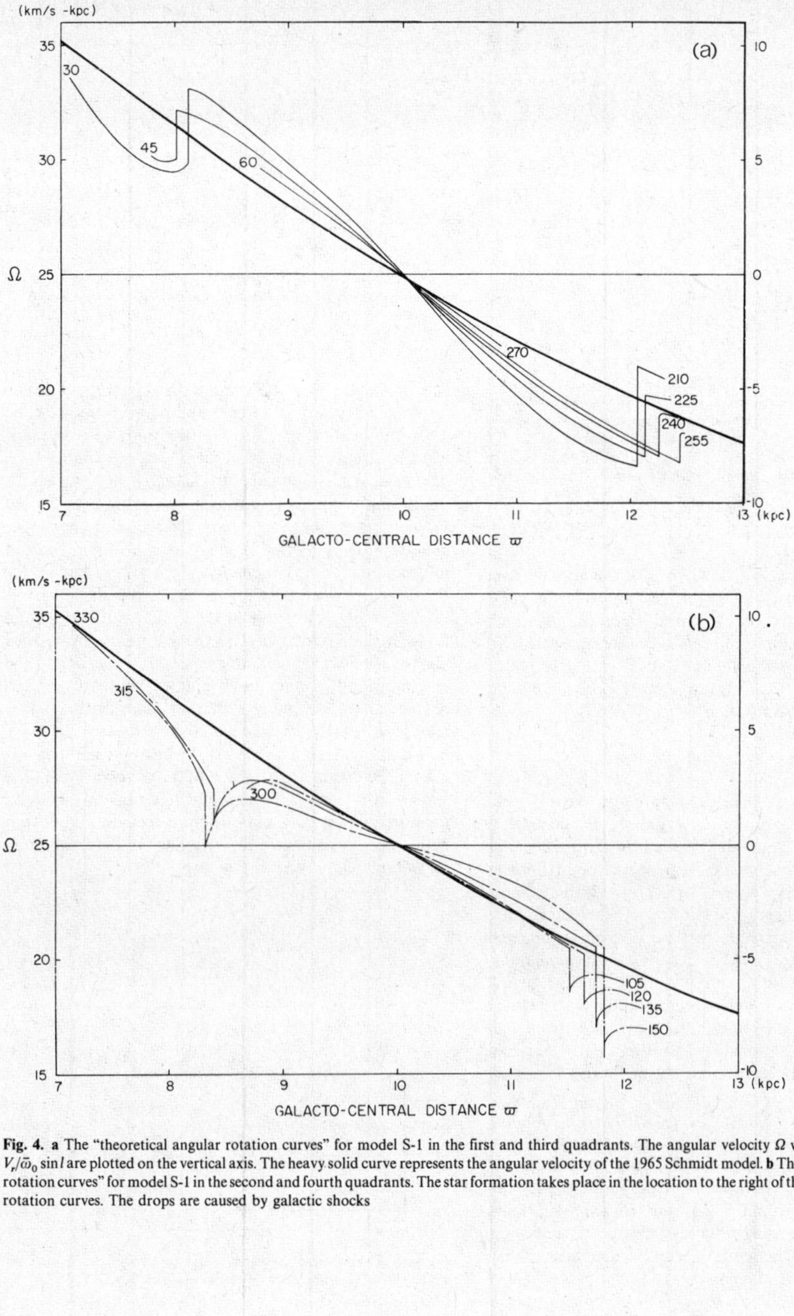


**Fig. 4. a** The "theoretical angular rotation curves" for model S-1 in the first and third quadrants. The angular velocity $\Omega$ which is calculated as $V_r/\bar{\varpi}_0 \sin l$ are plotted on the vertical axis. The heavy solid curve represents the angular velocity of the 1965 Schmidt model. **b** The "theoretical angular rotation curves" for model S-1 in the second and fourth quadrants. The star formation takes place in the location to the right of the sudden drops of the rotation curves. The drops are caused by galactic shocks

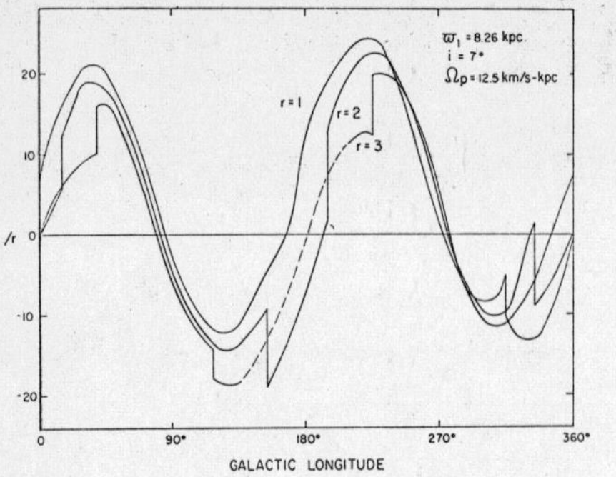


**Fig. 5.** The theoretical "double-sine" curve for model S-1. Three curves in the diagram represent velocity fields at $r = 1, 2$ and 3 kpc. Since the $r = 1$, 2 and 3 kpc. Since the $r = 1$ kpc circle does not intersect any shock in this model, the curve therefore is a smooth sinuisoidal-like curve. For the curves $r = 2$ kpc and $r = 3$ kpc, shocks cause the sudden drops on the curves, although the general double-sine nature is still recognizable. The broken part of the $r = 3$ kpc curve marks out the locations where the doubly periodic shock solution ceases to exist. In general, the mean location for those streamlines has exceeded 12.5 kpc from the galactic center for model S-1

prominent *dips* exhibited by the observational data discussed in Section II (cf. Fig. 1b) as important manifestations of the presence of streaming motions which characterize the nature of the *shock* phenomenon. This similarity between the theoretical and observed flow fields provides the justification of the approach adopted in this paper (cf. Section V).

### b) "Double-sine" Curve

We follow Oort's original procedure for analyzing observational data. We plot, in Figure 5, the theoretical line-of-sight velocity divided by the heliocentric distance $r$ for model S-1 as a function of galactic longitude. For simplicity, only three curves corresponding to distances of $r = 1, 2$ and 3 kpc, are shown. These curves are analyzed with the general method of Fourier analysis described in Section III. The results of the calculations for models S-1 and E-1 are tabulated in Tables 3 and 4. The curve for $r = 1$ is indeed a smooth curve, since the heliocentric circle does not intersect the shock front. For the curves of $r = 2$ kpc and 3 kpc, the overall "double-sine" nature is still preserved, although shock discontinuities set in.

Figures 6 and 7 illustrate the fact that the "double-sine" nature is also preserved by other models as well, such as S-2 and E-1 respectively. This preservation of the "double-sine" characteristics despite the presence and the influence of an underlying spiral gravitational field and corresponding streaming motions is an important result to which we will devote further attention in the discussion of Section V.

**Table 3.** Fourier coefficients for theoretical values of $v_r/r$ for Model S-1

| Distance (kpc) | 1 | 2 | 3 |
|---|---|---|---|
| $\alpha_0$ | 5.00 | 1.81 | −0.16 |
| $\alpha_1$ | −0.96 | 1.88 | 0.31 |
| $\beta_1$ | −1.53 | −3.33 | −4.10 |
| $\alpha_2$ | 4.26 | −0.78 | −1.13 |
| $\beta_2$ | 16.57 | 16.13 | 15.05 |
| $\alpha_3$ | 0.17 | 1.46 | −0.38 |
| $\beta_3$ | 0.01 | −0.50 | 0.88 |

Fourier coefficients of the theoretical "double sine" curves for model S-1. The quantity $\alpha_0$ is the dilatation constant $K$ and the quantity $\beta_2$ is the Oort constant $A$

Although shocks occur in model E-1 (Fig. 7) just as in models S-1 and S-2 (Fig. 5 and 6), we have sketched smoothed curves in Figure 7 for the purpose of simulating the situation in which the shock discontinuity is averaged and smoothed out. Actually, the real shock discontinuity in velocity is expected to be pronounced for the gas in the intercloud phase of the interstellar medium (Shu et al., 1972), but somewhat smoothed out in the cloud phase. The smoothed curves of Figure 7 are therefore provided here to emphasize this distinction.

Extremely important in these figures is the feature that a certain proportion of the young, newly formed stars are expected to be triggered into existence in the neighborhood of the shock and to share to various degrees in the nonlinear streaming velocity there. This feature, suggestive of the theoretical curves in Figures 5, 6, and 7, seems to provide insight into our understanding for the overall distribution of the observational data shown in Figure 2. This insight has not only made a great impact on our analysis, but it has also led us to a better understanding of the meanings of the fundamental parameters of spiral structure.

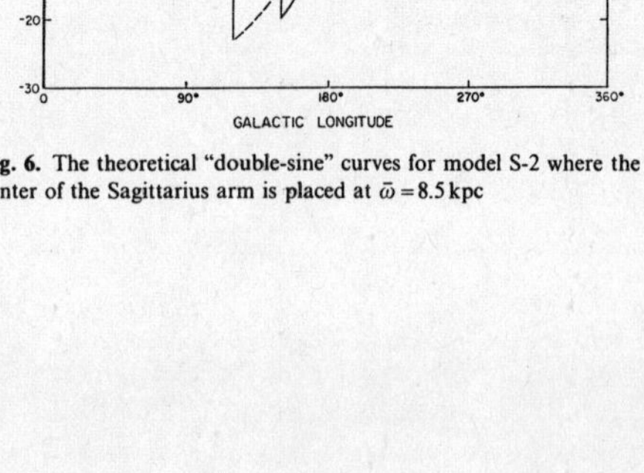


**Fig. 6.** The theoretical "double-sine" curves for model S-2 where the center of the Sagittarius arm is placed at $\bar{\omega} = 8.5$ kpc

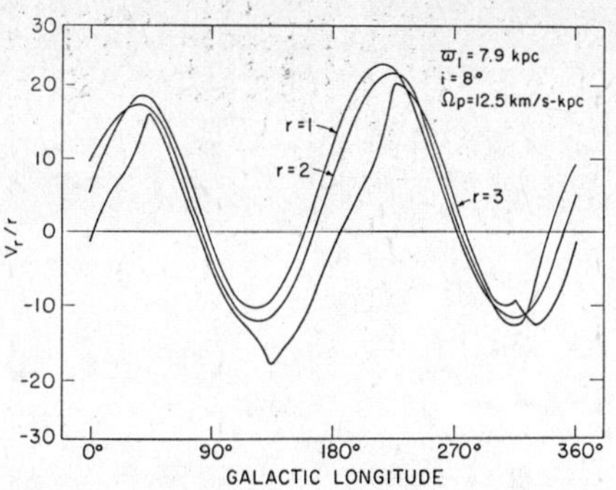


**Fig. 7.** The theoretical "double-sine" curves for model E-1. It is also similar to those of S-1 and S-2, except that the shocks on the curves are connected smoothly

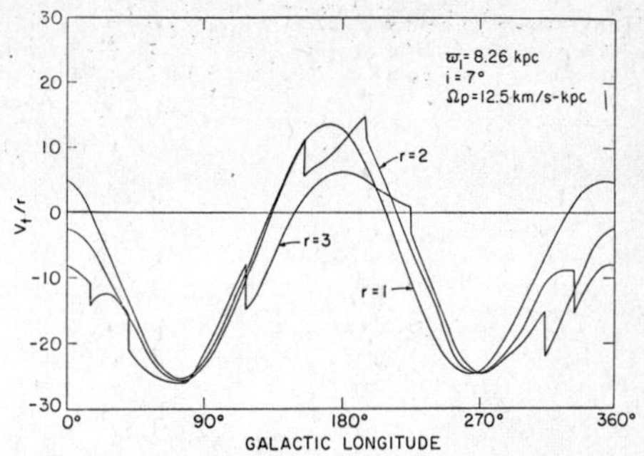


**Fig. 8.** The theoretical "double cosine" curve for model S-1. The vertical axis is tangential velocity divided by distance from the Sun, $V_t/r$. The situation here is still somewhat similar to those of the double-sine cases with a phase shift. Please notice that all the curves are also shifted downward with an amplitude about equal to $10\,\mathrm{km\,s^{-1}\,kpc}$, which, of course, represents the Oort constant $B$ adopted in the calculations

**Table 4.** Fourier coefficients for theoretical values of $v_r/r$ for Model E-1

| Distance (kpc) | 1 | 2 | 3 |
|---|---|---|---|
| $\alpha_0$ | 4.42 | 4.03 | −0.12 |
| $\alpha_1$ | −2.79 | −0.34 | 0.36 |
| $\beta_1$ | −1.05 | −2.37 | −3.66 |
| $\alpha_2$ | 4.10 | 4.06 | 1.81 |
| $\beta_2$ | 15.24 | 15.08 | 14.69 |
| $\alpha_3$ | −0.60 | 1.17 | 0.27 |
| $\beta_3$ | 0.48 | 0.39 | 0.90 |

Fourier coefficients of the theoretical "double sine" curves for model E-1

**Table 5.** Fourier coefficients for theoretical values of $v_t/r$ (Model S-1)

| distance (kpc) | 1 | 2 | 3 |
|---|---|---|---|
| $\mu_0$ | −7.92 | − 9.91 | −11.08 |
| $\mu_1$ | −4.12 | −10.48 | − 7.73 |
| $\nu_1$ | 0.59 | 0.53 | − 0.37 |
| $\mu_2$ | 15.81 | 12.64 | 11.96 |
| $\nu_2$ | −5.59 | − 3.30 | − 2.04 |
| $\mu_3$ | 0.08 | − 0.83 | 0.21 |
| $\nu_3$ | 0.45 | − 0.32 | − 0.99 |

Fourier coefficients of the theoretical "double cosine" curves for model S-1. The quantity $\mu_0$ is the Oort. Constant $B$ and the quantity $\mu_2$ is the Oort constant $A$

### *c) The "Double-cosine" Curve*

Similar to the previous cases, the double-cosine curves may be constructed by calculating the "proper motion" of theoretical gas flows. These curves are also analyzed with the general Fourier analysis to determine the Oort constants and other parameters of galactic structure. In reality, because of the great uncertainty involved in determining the proper motion for stars with distances beyond 1 kpc from the Sun, the analysis of the "double-cosine" curves is less frequent in the literature. The theoretical double-cosine curves for model S-1 are shown in Figure 8 and the Fourier coefficients are listed in Table 5.

### *d) The Oort Constants and the K-term*

Now we summarize our results on the determination of the Oort constants $A$ and $B$ in Table 6. Only models S-1 and E-1 are presented in comparison with the 1965 Schmidt model. Clearly, the values depend on $r$, although the dependence is quite weak. In comparison, we shall confine ourselves to each column in Table 6, i.e. for the same $r$. In doing so, we note that the Oort constant $A$ is recovered within 10% of its assumed value, and the Oort constant $B$, within 20% of its assumed value. Both errors are below that of the inaccuracy in the observations. The determination of $\Omega = A - B$ is especially blessed with extremely good accuracy for the reason explained in the preceding section.

We note that the coefficient $\alpha_0$ in Tables 3 and 4 is the $K$-term in the usual terminology: $\alpha_0 > 0$ implies that the solar vicinity is expanding. Since the value of $\alpha_0$ varies significantly with the radial distance $r$ at which it is determined, the situation is obviously quite complicated. The detailed discussion will be given in the next section, where a more penetrating analysis will be presented.

It should be mentioned that the extrapolated values of $\alpha_0$ at $r=0$ have been independently checked by using Equation (3.26); good agreement has been obtained.

**Table 6.** Tabulation of numerical values of $A$, $B$, $A-B$, $A+B$

1. Oort constant $A$; assumed value $=15\,\mathrm{km\,s^{-1}\,kpc^{-1}}$

| | $r=1$ | $r=2$ | $r=3$ |
|---|---|---|---|
| S | 14.7 | 14.3 | 13.9 |
| E-1 | 15.2 | 15.1 | 14.7 |
| S-1 | 16.6 | 16.1 | 15.1 |

2. Oort constant $B$; assumed value $=-10\,\mathrm{km\,s^{-1}\,kpc^{-1}}$

| | $r=1$ | $r=2$ | $r=3$ |
|---|---|---|---|
| S | −10.3 | −10.6 | −10.9 |
| E-1 | − 8.8 | − 9.5 | −11.1 |
| S-1 | − 7.9 | − 9.9 | −11.1 |

3. Angular velocity $A-B$; assumed value $=25\,\mathrm{km\,s^{-1}\,kpc^{-1}}$

| | $r=1$ | $r=2$ | $r=3$ |
|---|---|---|---|
| S | 25.0 | 24.9 | 24.8 |
| E-1 | 24.0 | 24.6 | 25.8 |
| S-1 | 24.5 | 26.2 | 26.2 |

4. Radial gradient of circular velocity $A+B$; assumed value $=5\,\mathrm{km\,s^{-1}\,kpc^{-1}}$

| | $r=1$ | $r=2$ | $r=3$ |
|---|---|---|---|
| S | 4.4 | 3.7 | 3.0 |
| E-1 | 6.4 | 5.0 | 3.6 |
| S-1 | 8.7 | 6.4 | 4.0 |

Summary of the Oort constants $A$ and $B$ and their differences from the theoretical calculations. $S$ stands for the 1965 Schmidt model. Since all the structural constants vary with $r$, the comparison should be confined to each column.

Thus, the variation of $\alpha_0$ with $r$ is genuine and not caused by numerical inaccuracies.

### e) The Product $A\tilde{\omega}_0$

We have applied Equation (3.25) of Section III to determine the product $A\tilde{\omega}_0$. For both models S-1 and E-1, the values of $A\tilde{\omega}_0$ are somewhat greater than $150\,\mathrm{km\,s^{-1}}$ on the "north" side of the Galaxy and smaller on the "south" side. The average $A\tilde{\omega}_0$, say at $l=75°$ and 285° for model S-1 is $163\,\mathrm{km\,s^{-1}}$, which differs only by 8.3% from the assumed value. The value for model E-1 is almost exactly equal to the assumed value of $150\,\mathrm{km\,s^{-1}}$.

In a recent article, Oort and Plaut (1975) also noticed the tendency to over-estimate $AR_0$ due to the streaming motion of the gas. This is consistent with our analysis. However, since Oort and Plaut used only data of streaming motions observable from the northern hemisphere, the actual correction should perhaps be somewhat lower than they estimated.

## V. Discussion

The results in Section IV are purely theoretical. Their usefulness as a basis for judging the observational determination of the Oort constants depends on the similarity of the observed field of motion to the theoretical flow field. It was already noted that there are indeed similarities between the observed diagrams in Figures 1a and 1b on the one hand and the theoretical diagrams in Figures 4a and 4b on the other.

### a) The Curvature Parameter $q$

It is also clear from these diagrams that the curvature parameter $q$ of the rotation curve cannot be accurately determined. Indeed, Kraft and Schmidt (1963) noted that "in the distance intervals 9–11 kpc and 8–12 kpc, the value of $\alpha$ ($q$ in this paper) is seen to be $+2$ and $+3\,\mathrm{km\,s^{-1}\,kpc^{2}}$ while in the range 7–13 kpc it is about $-2$ in the same units". Their findings were generally confirmed by Humphreys.

From our theoretical analysis, it is clear that the cause of this difficulty is largely the effect of the strong streaming motions associated with the shocks (e.g., in the Perseus arm), which in turn determines the location of the young stars, rather than with the presence of the local arm (as Kraft and Schmidt originally suspected). Indeed, this effect of streaming motions is quite apparent in Figure 4b, in which the theoretical angular velocity curve has a definitely negative curvature in the solar vicinity, opposite to that of the general curve of angular velocity. This effect is clearly largest as a local effect, but diminishes when averaged over a wider range of radial distances. Thus, it is understandable that the value of $q$ may be as large as $+4$ for data in the range $9<\tilde{\omega}<11$, but tends to change over to a negative value as the range is increased to $7<\tilde{\omega}<13$ kpc. In either case, one may expect the "true" value of $q$ to be still more negative, but the accurate determination of this value becomes very difficult.

### b) Is the Solar Vicinity Expanding?

We shall now turn our attention to the dilatation constant $K$, and make a closer analysis of the flow field to see if the solar vicinity is expanding. In so doing, we shall also gain some insight into the determination of the Oort constant $A$.

#### i) The Issues

Considerable attention had been given to the determination of the dilatation coefficient, "the $K$-term", in the solar vicinity. On the basis of the motion of cepheids, Kraft and Schmidt (1965) found that the value of $K$ is negative and about two to three times its mean error. They concluded that there is a general contraction in the solar neighborhood.

On the other hand, somewhat different results were obtained by Rubin and Burley (1964) and by Humphreys (1970). They showed, on the basis of the motion of O–B stars and supergiants, that the value of $K$ is *positive* in the region close to the Sun, but becomes negative as the region under consideration is extended. Since the stars in their samples are several times more numerous than the classical cepheids used by Kraft and Schmidt, especially in the neighborhood of the Sun, the results obtained by Rubin and Burley and by Humphreys are likely to reflect the local picture better. We repeated the analysis by using the data in Humphreys' sample (excluding stars whose data she regarded as of poor quality as well as those whose spectral types are later than $A$) and found the values listed in Table 7. The first value is positive for $r=1$ in model S-1, implying that there exists a general expansion of the flow field in the immediate neighborhood of the Sun. The fact that the observed value is somewhat smaller than the theoretical value indicates that there is a certain degree of departure between the motions of the gas and the young stars, as they move into the interarm region. For the ring-shaped region $1.5<r<2.5$ kpc, the observations, however, suggest a general contraction which is contrary to the theoretical result of a mild expansion. In fact, the contraction appears to be even stronger for Rubin's data of O–B stars in the same ring-shaped region.

We shall attempt to understand why the expansion decreases in the theoretical models as the radial distance increases and why the value of $K$ based on observational data can actually turn negative. The method will be described in the next subsection.

Before doing so, let us note that Lindblad and his coworkers (Lindblad et al., 1973) have used the theoretically predicted *expanding* model to explain the observed features in the 21-cm line profiles of the local gas in his discussion of the observational data of the Gould belt.

**Table 7.** Summary of the dilatation $K$ for two theoretical models, in comparison with those determined from Humphrey's observational data

| | $r=1$<br>0.5–1.5 | $r=2$<br>1.5–2.5 | $r=3$<br>2.5–3.5 | |
|---|---|---|---|---|
| Humphreys | 1.37 | −0.87 | 1.31 | km/s–kpc |
| S-1 | 5.00 | 1.81 | 1.31 | |
| E-1 | 4.41 | 4.03 | −0.12 | |

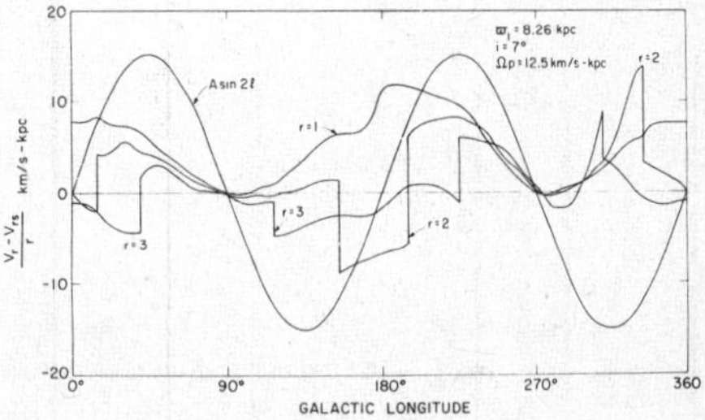


**Fig. 9.** The theoretical streaming motions for model S-1. The ordinate represents the difference between theoretical radial velocity divided by the distance and the radial velocity of the 1965 Schmidt model divided by the distance. The results are the streaming motions divided by the distance in response to the adopted spiral gravitational field. Three curves are respectively for $r=1$, 2 and 3 kpc. The "double-sine" curve is presented to indicate the locations of ups and downs

### ii) The Method

In order to examine the expanding (contracting) motions in the solar vicinity more closely, we calculate the contribution to the term $V_r/r$ from the streaming motions alone for the theoretical flow field in model S-1 (cf. Fig. 5). The results are shown in Figure 9. The quantity plotted is

$$\phi(r,l)=(V_r-V_{rs})/r\,, \quad \text{for} \quad r=1,2,3\,kpc\,, \tag{5.1}$$

where $v_{rs}$ is the line-of-sight velocity associated with the basic Schmidt model.

Clearly, the quantity $\phi(r,l)$ will yield the same value for $K$ as $v_r/r$ for each radial distance $r$, since there is no net (volume) flow across any circle (indeed any closed curve) in the basic circular model.

The Fourier analysis of $\phi(r,l)$ would give essentially the same information as given in Tables 1 and 3 for the quantity $V_r/r$, but an examination of the curve $\phi(r,l)$ can give us new insight into the nature of the motions that contribute directly to expansive motion and to differential rotation. (By referring to the first part of Table 6, we see that, if $\phi(r,l)$ is Fourier analyzed, the coefficient of the $\sin 2l$ term will have the value $16.6-14.7=1.9$ in $\mathrm{km\,s^{-1}\,kpc^{-1}}$.)

### iii) Effect of Streaming Motions

There are clearly *two* effects worth noting in the determination of $K$. (a) From the theoretical curves (Figs. 5 and 9), we see that in the nonlinear streaming motions the effect of the shocks appears prominent. (b) From the observational data (Fig. 2), we note the selective concentration of the young optical objects behind the shock. We have already noted in Section IV the correlation of these two effects.

Both of these effects are decisive in the determination of the precise value of $K$. In particular, the motion associated with the shocks tend to *lower* its value. On the other hand, their effects on the Oort constant $A$ are only minor, because the magnitude of $\phi(r,l)$ is only a few km s kpc, and the curves show no strong $\sin 2l$ component. Indeed, the effect is less than $2\,\mathrm{km\,s^{-1}\,kpc^{-1}}$ as we shall see in subsection 5(c), where the shock effect and the selection effect will be discussed.

It thus remains for us to explain the negative value of the dilatation parameter ($K=-0.87\,\mathrm{km\,s^{-1}\,kpc^{-1}}$) for

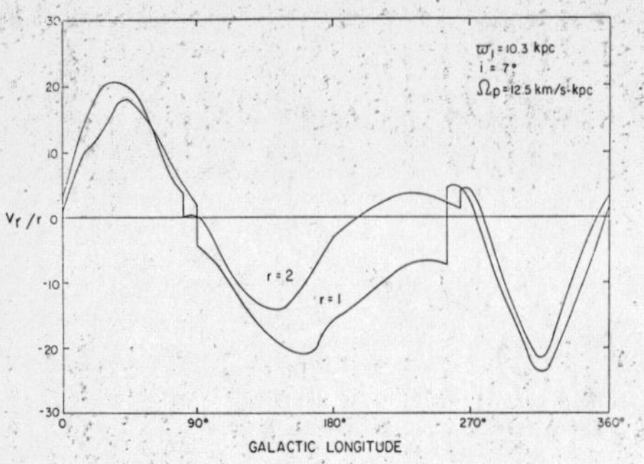

**Fig. 10.** The theoretical "double-sine" curves for model S-3. This shows how much the double-sine nature is distorted when the Sun is placed next to a major arm

Humphreys' data for $r=1.5$–$2.5$ kpc, even though the corresponding value for the theoretical model S-1 is $+1.81$ km s$^{-1}$ kpc$^{-1}$. Briefly, this may be attributed to the lack of star formation in the third quadrant and the unusually active star formation in certain directions (Perseus) in the second quadrant.

For the ring $0.5<r<1.5$ (in kpc), there is no effect from the primary shocks. Despite the scatter in data because of small $r$, we can see from Figure 2a that the net observed motion is expansive, since the average value of $V_r/r$ in longitude is clearly positive. The same is true of the theoretical motion at $r=1$ kpc shown in Figure 5. As we move out to $r=2$ kpc, there is a decline of $K$ even in the theoretical picture, because of the effect of shocks. The quantity $\phi(r,l)$ is simply lowered, on the whole. This trend continues to $r=3$ kpc where more of the flow is influenced by the shocks. In the observational data for $r=1.5$–$2.5$ kpc, this lowering is exaggerated not only by the selective distribution of stars at the shock, but also by the lack of star formation in both sides of the anti-center direction, especially in the whole third quadrant. In the second quadrant, where the shock would produce strong negative values of $\phi(r,l)$, there is a great deal of star formation, thus biasing the trend still further. The value of $K$ becomes negative. In the range $r=2.5$–$3.5$ kpc, although the situation is still similar, the stars in the second quadrant are less numerous, compared with those in the previous case. Furthermore, their velocities are generally less negative. Hence, the final value of $K$ turns slightly positive.

### *c) The Oort Constant A*

In subsection (a), we showed that the nonlinear streaming motions can prevent the curvature parameter $q$ from being accurately determined. These same streaming motions, however, do not prevent reliable values of the Oort constant $A$ from being obtained; why?

#### i) The Solar Location Relative to Spiral Arms

The first obvious reason is that the Oort constant $A$ is a large parameter, and thus the "double-sine" component is not obscured by the streaming motions (cf. theoretical diagrams in Figs. 5, 6, and 7 and observational diagrams in Fig. 2). However, it should be stressed that this is not enough. We are also fortunate to have been located roughly midway between two major spiral arms. Had the Orion arm been a major spiral arm, the situation would have been entirely different. In Figure 10, we show a plot of $(V_r/r, l)$ for Model S-3, in which the Sun is located close to the inside of a major spiral arm. Clearly, the double-sine nature of the curve is highly obscured by the streaming motions. The fact that Oort was able to detect the double-sine nature of the radial motions, despite the the existence of the local streaming motion, is of course a matter of our lucky fate.

If the situation of Model S-3 were the case, we would be so dazzled by the complexity of the large number of young objects nearby that the task of determining the basic structure of our galaxy would have been much more difficult. We show in Figure 11 the rotation curve which would have been obtained for Model S-3. It is clearly very different from Figures 1a and 1b.

However, once we are in our fortunate situation, it is also easy to understand why the correction term $\Delta A$ needed is so small. A crude estimate of the magnitude of $\Delta A$ due to the effect of streaming motions can be made as follows. If we refer to Figure 5 of the paper by Roberts (1969), it is clear that the behavior of $w_{\|}$, the total velocity parallel to the spiral equipotential curve of the density wave, would have the largest effect on $A$ and on $q$. If the left-hand shock in Robert's diagram is identified with the Sagittarius arm and the right-hand shock with the Perseus arm, and if the Sun is located about midway between the shocks (see above), the slope of the curve remains small if we take average values over regions centered around the Sun. The contribution to $A$ is perhaps no larger than the order of 2 km s$^{-1}$ kpc$^{-1}$. By assuming the distance between the arms to be 3.5 kpc, the range of velocity variation to be 14 km s$^{-1}$, the average contribution to $A$ is approximately $\frac{1}{2}(14/3.5)$ $=2$ km s$^{-1}$ kpc$^{-1}$.

#### ii) The Biased Distribution of Young Stars

We already noticed that the young optical objects tend to cluster behind the shock, and to share the nonlinear streaming motion of the interstellar medium. Furthermore, the longitudinal distribution of these young objects is highly nonuniform. They are heavily concentrated in the second quadrant and there are hardly any in the third quadrant. As mentioned above, this phenomenon has effectively lowered the value of the dilatation constant $K$ determined from the observational method. Now, we shall examine whether the

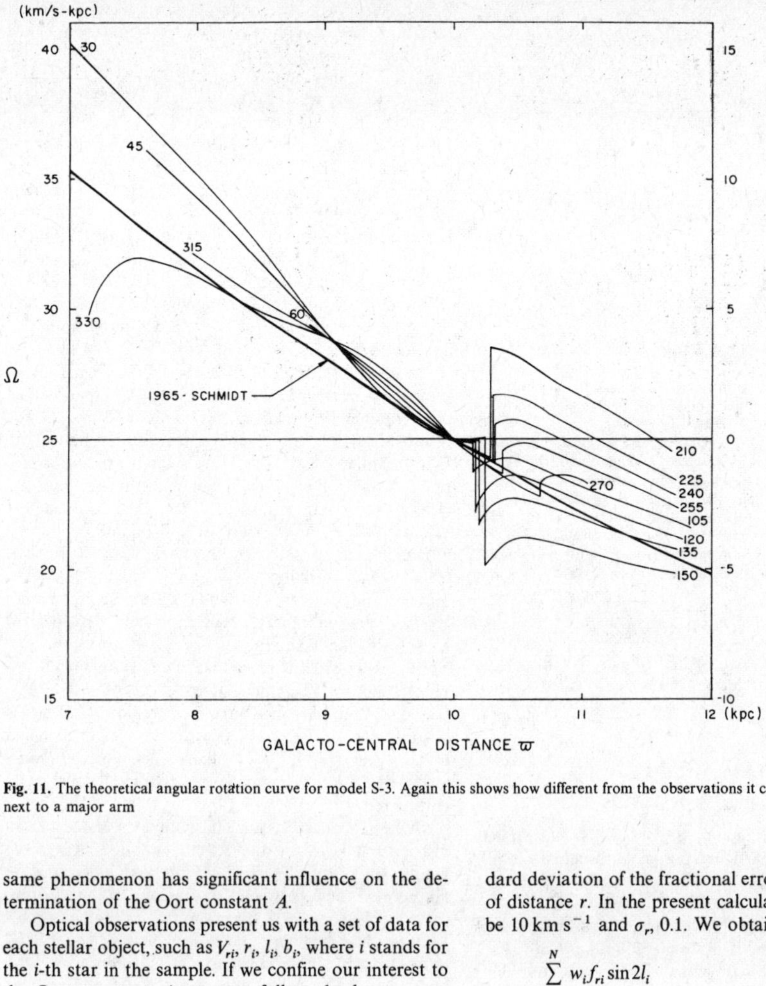


**Fig. 11.** The theoretical angular rotation curve for model S-3. Again this shows how different from the observations it can be if the Sun is located next to a major arm

same phenomenon has significant influence on the determination of the Oort constant $A$.

Optical observations present us with a set of data for each stellar object, such as $V_{ri}$, $r_i$, $l_i$, $b_i$, where $i$ stands for the $i$-th star in the sample. If we confine our interest to the Oort constant $A$, we may follow the least-square procedure by minimizing the quantity[5]

$$\sum_{i=1}^{N} w_i(f_{ri} - A \sin 2l_i)^2 ,$$

for its determination. In the above expression, $f_{ri} = V_{ri}/r_i \cos^2 b_i$ for each star, $N$ is the number of stars in the sample, and $w_i$ is a weighting factor, proportional to $r_i^2 \cos^4 b_i(\sigma_r^2 + V_{ri}^2\sigma_r^2)^{-1}$, where $\sigma_v$ is the standard deviation of error in the observations of $v_r$ and $\sigma_r$ is the standard deviation of the fractional error in the observation of distance $r$. In the present calculations, $\sigma_v$ is taken to be 10 km s$^{-1}$ and $\sigma_r$, 0.1. We obtain the formula

$$A_{\min} = \frac{\sum_{i=1}^{N} w_i f_{ri} \sin 2l_i}{\sum_{i=1}^{N} w_i \sin^2 2l_i} \tag{5.1}$$

If the stars in the sample are essentially at the same distance away, and uniformly distributed over longitudes and the number of stars is sufficiently numerous, the above expression is essentially

$$A_{\min} = \beta_2 = \frac{1}{\pi} \int_0^{2\pi} F_r(r, l) \sin 2l dl ,$$

where $F_r(r, l)$ is defined by (3.1), since $f_{ri}$ is essentially equal to $F_r(r_i, l_i)$. On the other hand, if the data points are grouped for specific narrow ranges of the radial

[5] See, for example, *Applied Regression Analysis* by N. R. Draper and H. Smith, pp. 77–81, John Wiley, 1966. The weighting factor is inversely proportional to the square of the standard deviation of $f_{ri}$

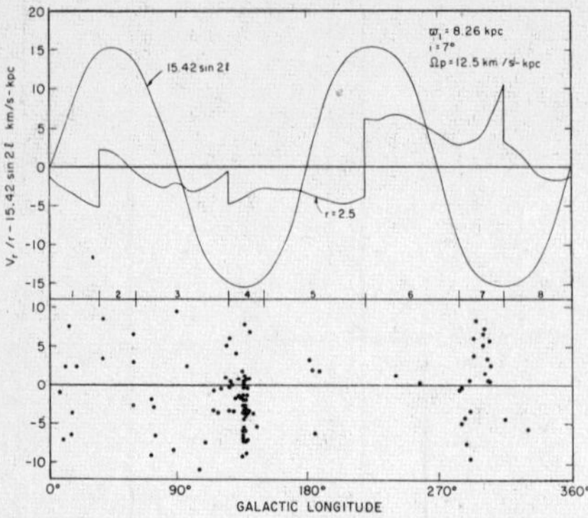


**Fig. 12.** Streaming motions with $(V_r/r - A\sin 2l)$ plotted against the galactic longitude. The upper part: A theoretical curve for model S-1 at $r=2.5$ kpc when $A$ is taken to be 15.42 km s$^{-1}$ kpc, which is the least square value for theoretical $V_r/r$. The lower part: A similar plot of the supergiants with distances between $r=2$ kpc and $r=3$ kpc (Humphreys, 1970). The ordinate of the lower part should be labelled as $V_r/r\cos^2 b - 15.42\sin 2l$. Since the $b$ values are small, in order to maintain the same format of the theoretical figure, we have omitted $\cos^2 b$ in labelling it. The whole range of longitudes is divided into eight regions. Assuming that star formation takes place in the post-shock region up to 1 kpc from the shock, we denote regions 2, 4, 5 and 7 as those regions. In reality, there is no observed star formation in region 5

distance, but aggregated in a highly biased manner in longitude, there is a correction term

$$\delta A = \frac{\sum_{i=1}^{N} w_i \psi_i \sin 2l_i}{\sum_{i=1}^{N} w_i \sin^2 2l_i}, \qquad (5.2)$$

where $\psi_i \equiv f_{ri} - \beta_2 \sin 2l_i$, and $A_{\min} = \beta_2 + \delta A$. The value of $\beta_2$ refers to that particular range of radial distance, and the quantity $\delta A$ is a measure of the effect of biased distribution.

Just as before, we shall first apply this observational procedure to the theoretical flow field and evaluate $\delta A$ for various biased distributions. Then, we shall determine the effect of the biased distribution by applying Fourier analysis to the quantity $\psi_i$ defined above, which is biased in longitude distribution exactly according to the observational data.

For the flow field at $r=2.5$ kpc in Model S-1, the least square procedure with one-degree intervals gives $A_{\min} = 15.42$ km s$^{-1}$ kpc$^{-1}$[6]. The quantity, $\psi_i(l) = F_r(l) - 15.42\sin 2l$, is plotted in Figure 12 together with the "double-since" curve. In the above expression, the weighting factor $w_i$ is assigned to be 1 in this theoretical case. Assuming that stars are formed uniformly in a region 1 kpc behind the shock, we may group all the longitudes into several regions (see Table 8). Some have active star formation and others have none. These regions are also marked out in Figure 12. Regions 2($l$=35–62°), 4($l$=125–150°), 5($l$=150–220°) and 7($l$=285–315°) correspond respectively to the "northern" Sagittarius arm, the Perseus arm, the southern extension of the Perseus arm and the southern Sagittarius-Carina arm (compare Figure 3). Note that we have decided to list region 5 separately from region 4, because the rates of star formation are quite different in these two regions. There is essentially no star formation in region 5, even though it is in the postshock area. On the other hand, there is an abundance of star formation in region 4. The regions designated with other numbers are inter-arm regions.

In order to bring out the effect of the biased distribution, different weights are assigned to different regions in a number of possible combinations. Some of the results are presented in Table 8. It is interesting to note that the values $\delta A$ are all relatively small even in some of the rather strongly biased cases. The reason is quite clear: The contribution to the right-hand side of Equation (5.2) from the heavy star formation in region 4 is usually compensated by the negative contributions from the star formation in region 7, while little contribution came from regions 5 and 6. Of course, our proper choice of the spiral pattern is essential, and it happens to lead to this lucky situation.

We list in Table 8 also a case of an arbitrarily chosen set of weights to show the extreme effects of unfavorable bias in longitude distribution. If region 4 alone is assigned the weight of unity, with zero weight assigned to all other regions, the value of $\delta A$ would become +3.9 km s$^{-1}$ kpc$^{-1}$, much larger than in all other cases considered.

A more realistic test of the influence of the biased distribution is to use the actual longitudes where the young stars are observed. Using the longitudes in Humphreys' data between 2 and 3 kps and the theoretical velocities for $r=2.5$, the quantity $\delta A$ obtained equals 1.65, which is the highest deviation obtained so far in all our reasonable thought experiments. Humphreys' data, $(f_{ri} - 15.42\sin 2l_i)$, are plotted versus $l_i$ in the lower part of Figure 12. Although most of her stars (whose longitude locations are all in good agreement with the theory) lie in region 4 and region 7, the unusually high concentration around $l=135°$ has made $\delta A$ more positive than all the corresponding theoretical cases.

On the other hand, if we apply the same analysis directly to the longitudes and velocities in Humphreys' data, $\delta A$ is equal to 0.69 km s$^{-1}$ kpc$^{-1}$. In other words, the value of $A$ would be 16.11 km s$^{-1}$ kpc$^{-1}$. Since the abovediscussed effect of biased distribution amounts to 1.65 km s$^{-1}$ kpc$^{-1}$, a correction of the true value of $A$ would probably be 14.16 km s$^{-1}$ kpc$^{-1}$, which is well with in the limits of the uncertainty in optical observa-

[6] Compare with Humphreys' value 15.1 ± 0.6 km s$^{-1}$ kpc for stars with heliocentric distance $r$ in the range 2–3 kpc

**Table 8.** Theoretical results on accuracy in the determination of the Oort constant $A$ if the data are based in the distribution in longitude. The division of regions is marked in Figure 12

| Weight Assignment for Each Region (regions according to longitude $l$) | | | | | | | | |
|---|---|---|---|---|---|---|---|---|
| 1 | 2 | 3 | 4 | 5 | 6 | 7 | 8 | $\delta A$ |
| $l=0$–35° | 35–62° | 62–125° | 125–150° | 150–220° | 220–285° | 285–315° | 315–360° | |
| 1 | 1 | 1 | 1 | 1 | 1 | 1 | 1 | 0.09 |
| 0 | 1 | 0 | 1 | 1 | 0 | 1 | 0 | −0.88 |
| 0 | 1 | 0 | 2 | 2 | 0 | 1 | 0 | −0.40 |
| 0.2 | 0.185 | 0.23 | 2 | 0.057 | 0.03 | 0.767 | 0.022 | 0.87[a] |
| 0 | 1 | 0 | 3 | 0 | 0 | 1 | 0 | 1.39 |
| 0 | 1.5 | 0 | 3 | 0 | 0 | 1.5 | 0 | 0.75 |
| 0 | 2 | 0 | 3 | 0 | 0 | 2 | 0 | 0.30 |
| 0 | 0 | 0 | 1 | 0 | 0 | 0 | 0 | 3.90 |

[a] In this case, the weights are assigned according to the actual "number density" for each region ("number density" being number of stars per unit longitude range)

**Table 9.** Tabulated evaluation of the effects of biased observational data on the determination of the Oort constant $A$ and $B$, the dilatation constant $K$ and the curvature term $q$

| | $q$ | $A$ and $B$ | $K$ |
|---|---|---|---|
| Effect of biased distribution | moderate | weak | moderate |
| Effect of shock (streaming motion) | strong | weak | moderate |

tions. Therefore, we may conclude that the combined effect of streaming motions and of biased distribution of star formation is unlikely to be substantial in the determination of the Oort constant $A$, provided we adopt a spiral pattern similar to the one suggested by Lin et al. (1969).

### *d) Concluding Remarks*

From the above discussions (cf. also Table 9), it is clear that, despite the various complications, the observational determination of the Oort constant $A$ is perhaps accurate up to 10%. Although no detailed analysis is made for the Oort constant $B$, the situation is expected to be similar. As emphasized before, this happy conclusion depends on a fortunate combination of circumstances.

On the other hand, the dilatation constant $K$ is a different matter. This dilatation is of course entirely due to the streaming motions, since it is absent in the case of pure rotation. The very local stars (within 1 kpc) will (theoretically) tend to give a "true" value of $K$ close to $+5\,\mathrm{km\,s^{-1}}$ kpc, although the accuracy of such data may be low. Stars further away will not give as large an average value because of the nature of the streaming motions. Finally, the presence of streaming motions makes it very difficult to determine the curvature parameter $q$ which is well defined only for the circular model. However, since the value of $A\tilde{\omega}_0$ can be well determined, one can use the value of $\tilde{\omega}_0$ thus (and otherwise) obtained to construct the rotation curve. Such procedures have been followed in practice. Thus, the usual reconstruction of the rotation curve (see Schmidt, 1965) is very good despite the presence of streaming motions and longitudinal bias in star formation.

*Acknowledgement.* We would like to thank Drs. A. Blaauw, D. L. Crawford, R. M. Humphreys, Frank Kerr, R. P. Kraft, P. O. Lindblad, J. H. Oort, K. Rohlfs, M. Schmidt, T. Schmidt-Kaler, W. W. Shane, Frank Shu, B. Strömgren, A. R. Upgren and R. Wielen for helpful discussions on various aspects of the problem. We wish especially to thank Dr. Shane for his careful reading of the manuscript and the various suggestions that resulted in an improvement of this paper. A part of the study was carried out while one of us (C.Y.) was visiting Sterrewacht, Leiden on a fellowship sponsored by the Netherlands Organization for Pure Science (Z.W.O.). He wishes to thank Prof. H. van der Laan and other members of the department for their hospitality. Besides the support of the National Science Foundation and of the CUNY, we would also like to acknowledge, with thanks, the use of the computing facilities at NASA's Goddard Institute for Space Studies (C.Y.).

## References

Balona, L. A., Feast, M. W.: 1974, *M. N.* **167**, 621
Burton, W. B.: 1966, *Bull. Astron. Inst. Neth.* **18**, 247
Burton, W. B.: 1970, *Astron. Astrophys. Suppl.* **2**, 261, 291
Burton, W. B.: 1971, *Astron. Astrophys.* **10**, 76
Burton, W. B.: 1972, *Astron. Astrophys.* **19**, 51
Burton, W. B., Shane, W. W.: 1970, IAU *Symp.* **38**, 397
Chandrasekhar, S.: 1942, Principles of Stellar Dynamics, Chicago: University of Chicago Press
Delhaye, J.: 1965, in *Stars and Stellar Systems*, Vol. V, Ed. A. Blaauw and M. Schmidt, p. 513, Chicago: University of Chicago Press
Feitzinger, J. V.: 1977, *Astron. Astrophys.* **55**, 277
Humphreys, R. M.: 1970, *Astron. J.* **75**, 602
Kraft, R., Schmidt, M.: 1963, *Astrophys. J.* **137**, 249
Lin, C. C.: 1971, in *Highlights of Astronomy 2*, Ed. C. de Jager, p. 88, Dordrecht, The Netherlands: D. Reidel Publishing Co.

Lin, C.C., Yuan, C.: 1975, *Bull. Am. Astron. Soc.* **7**, 344
Lin, C.C., Yuan, C., Roberts, W.W.: 1974 in *Highlights of Astronomy 3*, Ed. G. Contopoulos, p. 441, Dordrecht, The Netherlands: D. Reidel Publishing Co.
Lin, C.C., Yuan, C., Shu, F.H.: 1969, *Astrophys. J.* **155**, 721
Lindblad, P.O., Grape, K., Sandqvist, Aa., Schober, J.: 1973, *Astron. Astrophys.* **24**, 309
Munch, G.: 1965, in *Stars and Stellar Systems*, Vol. 5, Ed. A. Blaauw and M. Schmidt, p. 203, Chicago: University of Chicago Press
Oort, J.H., Plaut, L.: 1975, *Astron. Astrophys.* **41**, 71
Roberts, W.W.: 1969, *Astrophys. J.* **158**, 123
Roberts, W.W.: 1972, *Astrophys. J.* **173**, 259
Roberts, W.W., Yuan, C.: 1970, *Astrophys. J.* **161**, 877
Rolhfs, K.: 1971, *Astron. Astrophys.* **17**, 246
Rubin, V.C., Burley, J., Kiasatpoor, A., Klock, B., Pease, G., Rutscheidt, E., Smith, C.: 1962, *Astron. J.* **69**, 491
Rubin, V.C., Burley, J.: 1964, *Astron. J.* **69**, 80
Shane, W.W.: 1971, *Astron. Astrophys. Suppl.* **4**, 1, 315
Shane, W.W.: 1972, *Astron. Astrophys.* **16**, 118
Shane, W.W., Bieger-Smith, G.P.: 1966, *Bull. Astron. Inst. Neth.* **18**, 263
Shu, F.H., Milione, V., Roberts, W.W.: 1973, *Astrophys. J.* **183**, 819
Shu, F.H., Milione, V., Gebel, W., Yuan, C., Goldsmith, D.W., Roberts, W.W.: 1972, *Astrophys. J.* **172**, 557
Yuan, C.: 1969, *Astrophys. J.* **158**, 871
Yuan, C.: 1970, IAU *Symp.* **38**, 391
Yuan, C., Leibovitch, L.: 1971, *Bull. Am. Astron. Soc.* **3**, 440; **4**, 268

# Density Wave Theory of Spiral Structure of Galaxies*

*By C. C. Lin and Y. Y. Lau*



The density wave theory of spiral structure in galaxies is studied as a dynamical problem: the gravitational instability of a galactic disk with respect to spiral modes. The asymptotic theory for tightly wound spirals is presented in detail, although the problem has been formulated in a more general form to open the way to future research work in other directions. Some preliminary results based on the general formulation are included; they confirm those based on the

---

Address for correspondence: Professor C. C. Lin, Room 2-330, or Professor Y. Y. Lau, Room 2-332, M.I.T., Cambridge, MA 02139.

*An abridgement of this paper has been presented by the authors at a symposium in honor of Professor Eric Reissner held at La Jolla, California on June 23, 1978. We wish to thank Professor David J. Benney, the Managing Editor of this journal, for his kind permission to publish an abbreviated version [55] in advance of the publication of this full paper. The preparation of this paper and the research work on which it is based were done with the partial support of the National Science Foundation.

*STUDIES IN APPLIED MATHEMATICS* 60, 97–163 (1979) 97

Published by Elsevier North Holland, Inc.

asymptotic theory in cases where the latter should be applicable. The fluid dynamical model is adopted; reference to the stellar dynamical model is made when necessary. Besides providing a coherent account of the present status of the theory, it is hoped that this article will further stimulate workers in continuum mechanics to play their important role in future research work in the astrophysics of galaxies.

---

## Introductory remarks

During the past fifteen years, great progress has been made in the density wave theory of spiral structure in galaxies, both through mathematical analysis of the dynamical processes and through comparison with observations to confirm the theoretical predictions. Several reviews or expository articles have been written on this subject, with different emphases and from different points of view. Some of them describe the extensive observational confirmation of the theoretical predictions; others present or survey the theoretical developments towards better understanding of the mechanisms; still others contain some fairly detailed description of the background and the astronomical issues that motivated the density wave theory. (See the general references [1]–[4] at the end of this article and a brief guide to the type of information contained in some of them.)

The purpose of this paper is to present a coherent exposition of the dynamical theory so that the deductive framework of the theory can be clearly perceived. For this purpose, we have chosen to represent the galactic disk by a fluid model, even though the complete stellar dynamical theory is also available. A brief description of the current status of the dynamical theory is given in Sec. 5, and also in Sec. 20. An assessment of the general theoretical developments is given in Part V.

The implications of the theory concern not only the large scale optical characteristics in various galaxies, including our own, but also the interstellar medium, the detailed physical processes of formation of stars out of the interstellar medium, the formation of cool dense clouds, the molecules in them, the dust particles, and the distribution of the magnetic field in a galaxy. Not all these physical implications have been studied in detail, and we shall not focus our attention on these issues in this paper.

Instead, we shall draw attention to the dynamical aspects of the theory. Can one, for example, classify all the types of galaxies on the basis of their dynamical characteristics? How? Much work remains to be done to answer these questions. Speculations will be made in the final section of this paper.

Another point we shall emphasize is the intrinsic interest of the study of a galaxy as a dynamical system. It is a system similar in nature to electromagnetic plasmas. In both galaxies and in plasmas, we may adopt either a description of the physical system in the continuum model, or a description in terms of a system of discrete particles. In the latter case, we may either deal with the distribution function in phase space or treat the motion of a large number of particles by direct machine calculation. These large scale numerical experiments have yielded much useful information, and will continue to do so in the future. In contrast, in the present paper, we shall adopt the continuum model as the primary basis, and our treatment will be largely analytical.

The analytical theory has the advantage of bringing out the physical mechanisms more clearly. It has so far been limited to the linear approximation: the stability of a galactic disk with respect to spiral modes. The usual approaches —the asymptotic approximation, the initial value approach, the modal approach, etc.—have been used. These reveal all the interesting dynamical processes such as resonances and critical layers. In developing the nonlinear theory, one would expect to have challenging problems similar to those encountered in the study of turbulence. Such nonlinear theories are as yet not available.

In the light of these unresolved issues, the present paper serves merely as the initial phase of a great deal of interesting astronomical and dynamical investigations. The basic physical concepts of the theory have, however, been firmly established over the past fifteen years through the efforts of many theoretical and observational workers. We hope that this paper will help to make it easier for other interested persons, especially the dynamicists, to participate in this effort to push forward the frontiers of our knowledge.

## Part I
## Physical basis of the density wave theory

### *1. The Hubble classification of galaxies: the winding dilemma*

Galaxies are usually classified into ellipticals, normal spirals, and barred spirals, in a scheme initiated by Hubble (Fig. 1), based primarily on their morphological appearance. Although there have been refinements of this classification over the years, the basic scheme remains essentially unchanged.[1] One primary concern in this paper is an attempt to understand the dynamical basis for the spiral structure of the *normal* spirals. Speculations will be made on the *barred* spirals towards the end of this paper (Sec. 23).

The Hubble classification of normal spirals into types Sa, Sb, Sc is based not only on the pitch angle or the tightness of winding of the spiral structure, but also on other physical characteristics. The most obvious of these is the size of the nuclear bulge. It is found that galaxies with larger nuclear bulge usually have spirals with smaller pitch angle; these are classified as Sa galaxies. As one follows the sequence Sa, Sb, Sc, the pitch angle of the spiral arms increases while the size of the nuclear bulge decreases. In Fig. 2, we show a picture of the Sc galaxy NGC 5364, which has features similar to those of our own Milky Way system.

Since it is difficult to imagine that the mass distribution in a galaxy can undergo significant changes over a period of a few galactic revolutions, the Hubble type of a galaxy would be expected to be essentially permanent, perhaps undergoing a change only over a period which amounts to a significant fraction of the Hubble time scale ($10^{10}$ yr). A similar argument can be made for the gas

[1]For example, it was pointed out by Vaucouleurs that the division between normal spirals and barred spirals is not that sharp. Another classification scheme is the luminosity classification initiated by van den Bergh. See Roberts, Roberts, and Shu [84] for a theoretical understanding of the latter scheme on the basis of the density wave theory.

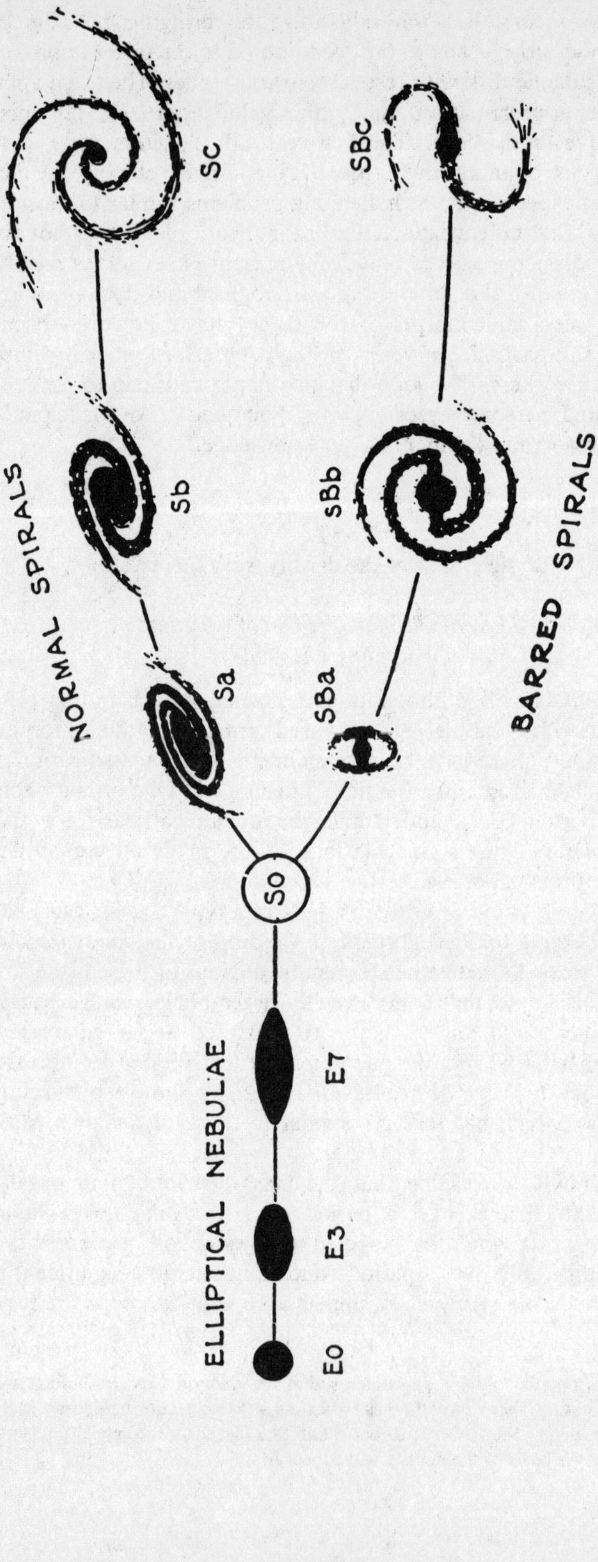


Figure 1. The Hubble classification of galaxies.

Figure 2. The spiral galaxy NGC 5364.

content in the galaxy. Herein lies the source of the well-known *winding dilemma*.

The distribution of stars in a galaxy may be roughly divided into a spheroidal component (consisting of the nuclear bulge and an essentially spherical halo) and a very flat disk component. The spiral structure is observed over the disk (Figs. 2, 3). It is known that the disk material is rotating rather rapidly, the period of revolution being on the order of $10^8$ years. A very important feature is that the angular velocity is not uniform. Indeed, in many cases, the linear velocity in the optically prominent part of a galaxy is nearly constant. Thus, the inner parts rotate much faster, and two different parts of a galaxy may easily differ in angular speed by a factor of two.

**Figure 3. The spiral galaxy NGC 4565.**

Such a pattern of *differential rotation* can be very effective for producing spiral arms like those observed in spiral galaxies. Any material clump can be stretched into a piece of "trailing" spiral structure. However, such a stretching process will *continually* reduce the pitch angle of the trailing spiral arm formed from the original material. Thus, the reasoning suggests that over a period of one or two revolutions, there will be a noticeable shift of a galaxy towards Hubble type Sa. (If the spiral arms are "leading," the shift will occur in the opposite direction.) This is the *winding dilemma*.

### 2. *The existence of grand design*

In addition to the winding dilemma, there is another difficulty in understanding the spiral structure of normal spirals. This is best described in the following words of Jan Oort [77]:

> In systems with strong differential rotation, such as is found in all nonbarred spirals, spiral features are quite natural. Every structural irregularity is likely to be drawn out into a part of a spiral. But *this* is not the phenomena we must consider. We must consider a spiral structure extending over the whole galaxy, from the nucleus to its outermost part, and consisting of two arms starting from diametrically opposite points. Although this structure is often hopelessly irregular and broken up, the general form of the large-scale phenomena can be recognized in many nebulae.

The first part of the statement describes briefly the role of differential rotation discussed above. The second part asserts the existence of a *grand design* over a galactic disk stretching over distances of the order of thousands of light years. It is difficult to see how material arms can be organized in this manner. Thus, we would have two serious difficulties if the spiral arms were always associated with the *same* material objects.

### 3. *The density wave theory*

Both difficulties are resolved if there exists a *spiral wave pattern* of some kind over the galactic disk. In the course of time, the objects marking out the spiral structure do *not*, in general, remain the same.

Dynamically, it is easy to visualize stationary wave patterns over a field of differential rotation. Hydrodynamic waves over shear flows have been known to exist, and indeed they often exist in the form of *self-excited modes* (see, e.g., [46]).

In the present context, there are several specific questions to be answered. The primary question is this:

What is the physical nature of the wave pattern? Is it associated with the old stars, the young stars, the gas, or all of these components?

Once this primary question has been answered, we still have to answer two other questions of a more detailed nature:

(A) How can one explain various observed astronomical phenomena in terms of such a wave pattern? For example, the brilliant blue stars which mark out the observed spiral structure are certainly material objects. How are they associated with the waves?

(B) What is the dynamical process that maintains the spiral wave pattern?

The primary question was given an answer by the late Bertil Lindblad (see [61] and his earlier papers), who first proposed the concept of density waves of stellar distributions. He was led to this concept by the study of individual stellar objects, their "epicyclic motion," and the displacement of their "guiding centers" caused in turn by the gravitational field of the density concentrations. Thus, the density concentrations appear to have a strong tendency to be self-sustained, and will propagate as *waves* over the galactic disk.

In the absence of a clear quantitative description of these density waves, it is obviously impossible to give any detailed description of the physical processes that underlie the observed phenomena (question A above). However, one can make some general plausible speculations. For example, *the blue stars marking the spiral structure may be imagined to be associated with the peak wave amplitude* —to be born and to die out rather quickly at or near that location, like the familiar whitecaps at the crest of water waves in a lake on a windy day. But any attempt to go deeper must depend on a detailed mathematical description of the nature of these waves.

Clearly question B must also await the development of a dynamical theory. Because of the lack of the accompanying mathematical analysis, the original ideas of Lindblad were not widely accepted.

### 4. *QSSS hypothesis*

Of the two questions A and B, the first is obviously the more fundamental. Thus, right from the beginning of the development of the density wave theory of spiral galaxies, there have been two aspects of research work: (1) comparison of theoretical predictions with observations, and (2) attempts to understand the dynamical mechanisms underlying the spiral structure of galaxies.

Even for comparison with observations, one must have the rudiments of the properties of the density waves. The minimum that one needs is a *local dispersion relationship* connecting the wavelength (or spacing between spiral arms) with the frequency of periodic variation of these waves. Such a dispersion relationship was given in 1964–1966 for *spiral waves* by Lin and Shu [2, 56, 57], who also constructed a theoretically derived wave pattern for our Galaxy. In the study of *axisymmetric oscillations*, Kalnajs [33] obtained in 1965 an equivalent form (but entirely different in appearance) of this dispersion relationship. The equivalence of these two forms can be demonstrated by using the Mittage-Leffler theorem, well known in the theory of analytic functions (see [58]).[2]

The theory naturally develops as follows. We first make the postulate that a *stationary spiral structure can exist* as a wave pattern over the disk. However, it is obvious that the observed features over a galactic disk are too complicated to be described in terms of a single density wave pattern, and that we must admit the possible *coexistence of many spiral features*, e.g., a multiplicity of normal modes, transient wave patterns, and even material arms, i.e., spiral arms composed of essentially the same material. We further postulate that the density wave pattern exists in the older stars as well as in the young objects. Indeed, we assume that

[2]For sake of historical accuracy, the dates given in these references include those when the lectures were delivered in the summer schools. The publication of such notes is often delayed by two or three years.

the major part of the nonsymmetric gravitational field is associated with the distribution of density in the older stars, even though such a density variation is rather difficult to detect and was not detected until very recently. The manifestation of the spiral structure is essentially provided by the brilliant young stars and the gas, whose behavior is profoundly influenced (behind the scenes, as it were) by the spiral gravitational field of the older stars.

Such a postulate was formulated and called the hypothesis of *quasistationary spiral structure* (*QSSS hypothesis*). The local dispersion relationship is then derived and made the basis of the discussion of observational data (see, e.g., [51, 59]).

The degree of departure from a *strictly* stationary structure naturally varies from one galaxy to another. Recent developments of the dynamical theory seem to suggest that, for galaxies with large surface mass density in the disk [more specifically, with large values of the parameter $\mathcal{J}$ defined by Eq. (70) below], the spiral structure might never attain a nearly stationary state but would rather evolve continually and might possibly be completely wiped out eventually if the stellar dispersion velocity continually increases during the evolutionary process. The spiral structure of the galaxy would be rather irregular, "hopelessly irregular and broken up," or nonexistent. The last case is perhaps that of an S0 galaxy, in which the stellar velocity dispersion is believed to be quite large.

Once the QSSS hypothesis is accepted and the parameters of the spiral pattern worked out, one can attempt to work out the physical consequences in greater detail. Specifically, the process of star formation out of the interstellar medium may be examined (see [83], [96]). Indeed, the formation of young stars is expected to occur in the neighborhood of the peak density concentration of the older stars. But this in itself is difficult to verify observationally. Other predictions can, however, be checked against observations. For example, one can explain why the dust lane always lies in the immediate vicinity of the bright spiral arms; one can also expect a maximum of magnetic field in the galaxy to occur in the vicinity of the dust lane [86]. The latter prediction was corroborated by the observations of continuum emission at 20 cm ([72]; see Fig. 4). Following this analysis, one can also provide an explanation of the luminosity classification of galaxies [84], which is also at least semiquantitatively compatible with the Hubble classification. Other observational confirmations have been reported (see, for example, [51, 14, 95, 98, 107]). In particular, Strom and Strom [98] surveyed the reports on the direct observation of density variation in the older stars by Schweizer [90] and by Strom et al. [99]. Visser [107], on the other hand, presented a detailed analysis of the flow field and mass distribution of H I gas in M81 and showed the agreement with theory. It is not the purpose of this paper to go into all these observational confirmations. The reader should consult the survey reports cited.

*Physical implications of the density wave theory.* In the process of discussing these observational confirmations, it becomes clear that the density wave concept has a very important impact on the understanding of the behavior of the interstellar medium, particularly on the processes of formation of dense clouds, of molecules in these cool dense clouds, and of stars themselves. On galaxies, we already mentioned the initial steps taken by Roberts, Roberts, and Shu on the

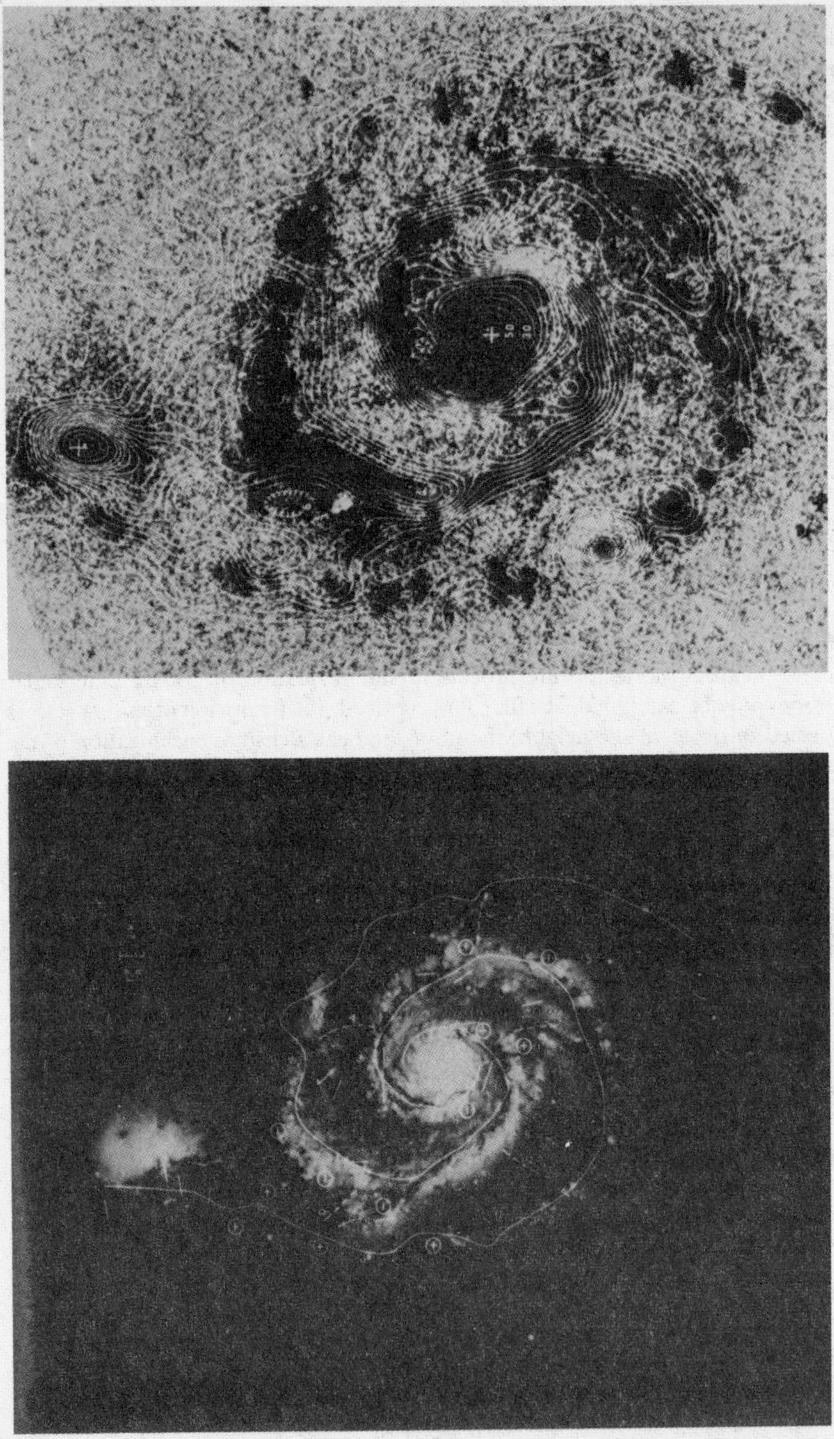

Figure 4. Radio map of the galaxy M51.

physical basis of the classification of normal spirals. It would of course be desirable to predict the spiral appearance of a galaxy from its mass model—including both the mass distribution and the characteristics of the mass. We may hope to accomplish this first for the normal spirals, and eventually to include the barred spirals, and the type of galaxies in transition between the two categories. For these purposes, much theoretical work remains to be done (cf. Sec. 23). In addition, there are purely dynamical issues to be resolved, some of which will now be briefly described.

### 5. *Existence of unstable spiral modes*

It is clear that the key to the understanding of possible mechanisms for quasistationary spiral structure lies in the establishment of the existence of *unstable spiral modes*. Unless *boundary conditions* are included in the study, even strong initial instability does not guarantee the eventual establishment of the *modes*, for the disturbances may increase initially and subsequently die away.[3]

The principal purpose of this paper is indeed to give an up to date account of one aspect of the theory for such *modes*: the theory of unstable spiral modes in the fluid model. Emphasis will be placed on the asymptotic theory, although this restriction has now been removed [79, 80]; the conclusions of the asymptotic theory were essentially confirmed. The theory has also been developed in the stellar model (see Bertin et al. [11] and papers by Mark cited there), but it is much clearer to present the theory in the fluid model in a self-contained account. Existing reviews of the dynamics of unstable spiral modes were written before the general theory was brought to a conclusion, and even before the numerical results based on the asymptotic theory were worked out and published (cf. [11, 12, 40, 80]). The impressions given in these reviews are therefore not up to date.

*Understanding dynamical mechanisms*. The general form of the theory requires a more elaborate numerical approach based on the formulation given in Sec. 10; on the other hand, the asymptotic theory has the advantage of yielding simple results, which enable one to understand the dynamical mechanisms more perspicuously, despite certain limitations in accuracy and in the class of problems covered. For example, the unstable mode is maintained by two kinds of *trailing* waves of different wavelengths traveling in opposite directions, and the instability is based in part on a mechanism of "wave amplification by stimulation of emitted radiation" [69].

The existence of the unstable modes according to the linear theory merely marks the beginning of the dynamical studies. We may list the following items for future consideration:

(1) *The nonlinear theory*. If the growth rate is slow, one might hopefully conjecture that the density wave pattern will be limited to a small but finite amplitude in a galaxy by some nonlinear mechanism such as the shock wave in the gaseous component (see, e.g., [85, 35]). If the growth rate is rather high, one

[3]It has been well known since Orr's classical study of the instability of shear flows [76] that there are "disturbances of pure laminar flow of uniform vorticity [i.e., a simple uniform shearing motion] which increase very greatly initially and subsequently die away." (See [21, p. 195.])

may indeed question whether the galaxy would ever reach a quasistationary state, especially when the coexistence of several modes is recognized.

(2) *The coexistence of several modes*. If, indeed, several modes coexist, is there a dominant one to justify the continued application of the concept of QSSS? Would several strongly unstable modes lead to a situation similar to that of turbulent flow in conventional hydrodynamics? The structure may be continually evolving and rather chaotic, but still may give the impression of having a spiral structure of some general scale over a range covering that of all the unstable modes. Is this indeed the reason why the situation is as Oort described in the statement quoted above (cf. [2, p. 68])?

(3) *The surface density over the galactic disk*. It is for this issue just raised that we feel that the surface density, as measured by the parameter $\mathcal{J}$ mentioned earlier, is extremely important for the determination of the final general appearance of the normal spirals. In addition, there is also the effect of the finite thickness of the galactic disk. As explained above, the resultant amount of velocity dispersion in the stars might also be determined from the process of evolution.

But these are the topics for future investigations. Indeed, all these issues are challenging general problems in applied mathematics. They are extremely similar to the kinds of problems that arise in plasma dynamics and in fluid dynamics. We shall return to some of the discussions of these issues in this paper, particularly in Part V.

## Part II
## Mathematical formulation of the dynamical problem

### 6. *The mass distribution in a galaxy*

For the sake of definiteness, we shall begin our discussions by considering a type of galaxy which presumably corresponds to the normal spirals.

The galaxy has a stellar component and a gaseous component. The distribution of matter can be roughly divided into a spheroidal component (the nuclear bulge and the halo) and a disk component. Figure 5 shows this distribution for our own galaxy. One can see that it is similar to the impression one gets from the photograph of the galaxy NGC 4565 (Fig. 3).

The spheroidal component is composed of stars with large velocity dispersion; the disk component is composed of stars with relatively small velocity dispersion, both perpendicular to the disk and in the plane of the disk. The gas lies essentially in the disk. Most of the gas has such a low temperature that its molecular velocity is on the order of 1 km/sec or less. Even its turbulent motion has a velocity only on the order of 10 km/sec, which is still much smaller than the typical velocity dispersion of the disk stars.

We may therefore treat the spheroidal component as an essentially *rigid* body, which is fixed in space or rotating at a small angular velocity, considerably smaller than that of the disk component. Clearly, if the spheroidal component is to play any role in the dynamics of the disk, beyond providing an axisymmetric field, we must consider its response to the spiral gravitational field of the disk stars. This problem has been investigated by Mark [70]. For the

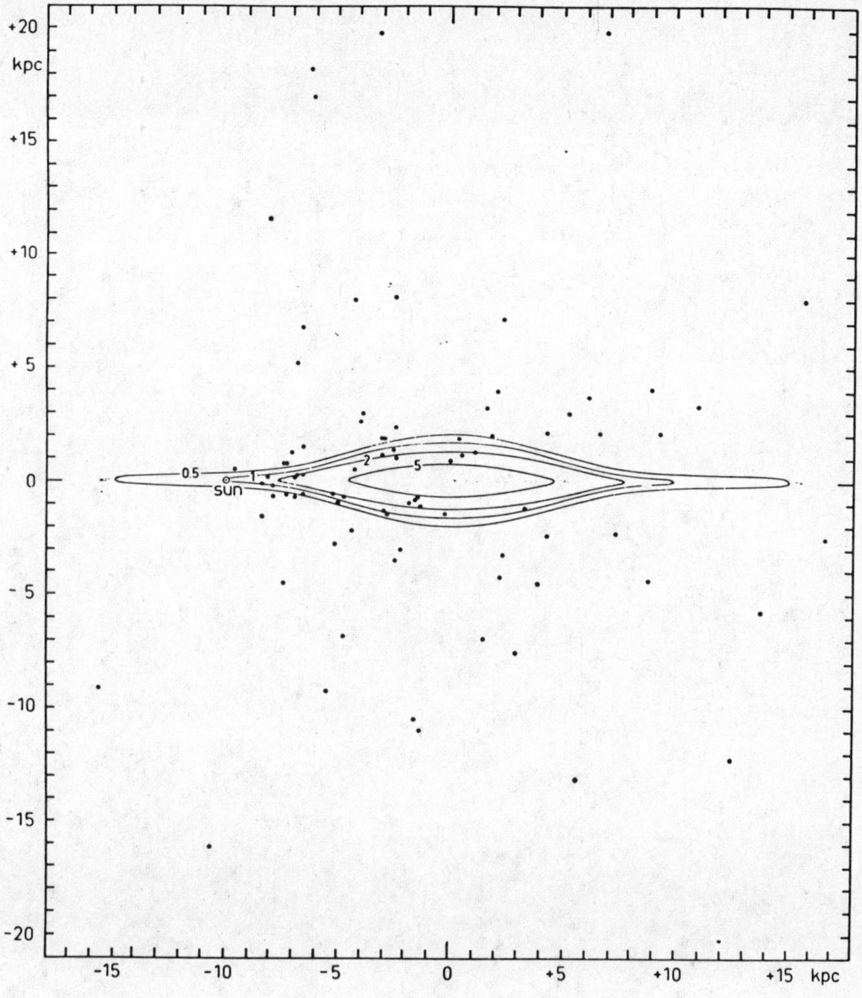


Figure 5. Mass distribution in the Milky Way system.

purposes of this paper, we shall use the simpler model of a rigid spheroidal component.

A given spiral gravitational field in the disk will influence the gaseous component much more strongly than the stellar component, perhaps by an order of magnitude, since the gaseous component has an equivalent pressure (due to turbulent motion) which is about 10 times smaller than that in the disk stars. Thus, even when the gravitational field is so small that the behavior of the stellar component can be treated on the basis of a *linear* theory, the behavior of the gas must be treated on the basis of a *nonlinear* theory, perhaps with shock waves included. The dynamical problem governing the behavior of galactic material is

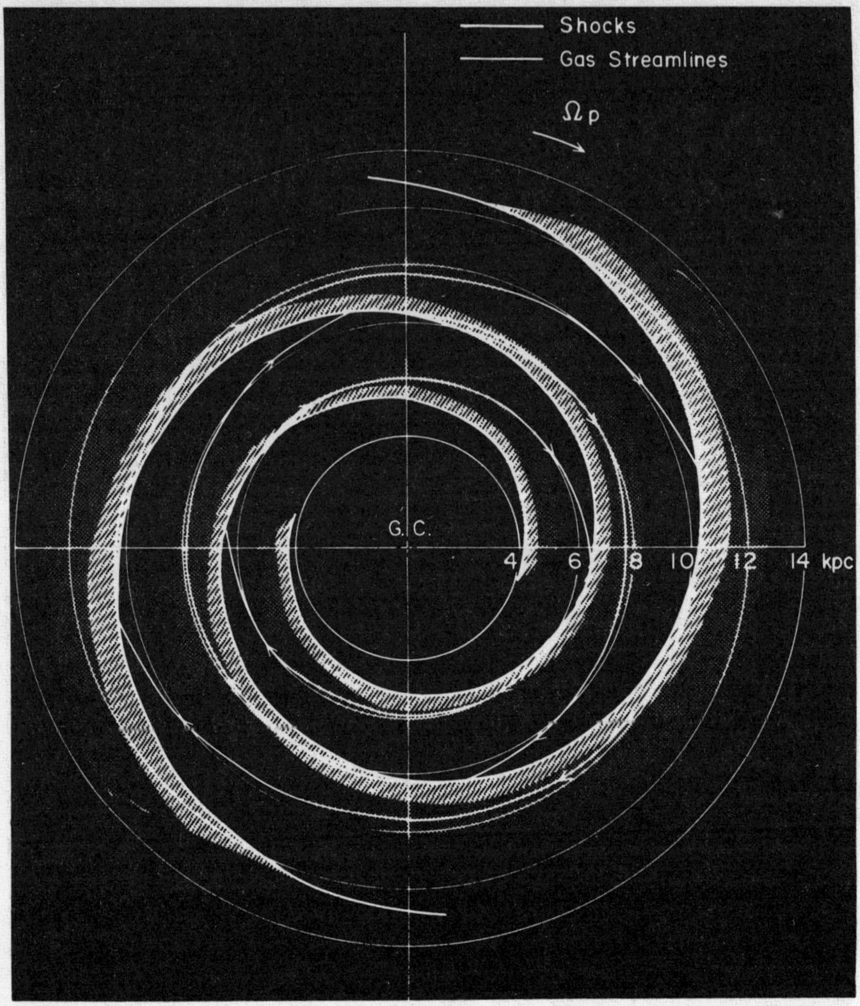


Figure 6. The spiral structure of our galaxy and the process of star formation in such a galaxy.

therefore essentially very complicated. Fortunately the gaseous component usually has such a small mass that we may at first neglect its effect in an approximate treatment of the stellar component. In the next step, we may consider the nonlinear behavior of the gas under the influence of the stellar gravitational field. (See Fig. 6, which is reproduced from a paper by W. W. Roberts [83].) A process of successive approximation may then be used to get the desired answers.

## *7. The stellar component treated by a fluid dynamical model*

The basic equations governing the behavior of the stellar component are well known. They are the Boltzmann equation (usually with close encounters between the stars omitted), and the Poisson equation for the gravitational potential. For simplicity of analysis, however, we shall adopt a *fluid dynamical model for the disk stars*. The justification of this step is not simple; we shall return to its discussion at the end of this paper (Secs. 22, 23). Here we shall merely note that it depends on the characteristics of the waves, which in turn depend on the characteristics of the mass distribution. We shall, as mentioned above, consider at first a type of mass distribution which presumably covers all the normal spiral galaxies, with the possible exception of certain Sc spirals like M33. For galaxies like M33, it might be still possible to use the fluid dynamical model; but the reason for its justification is perhaps somewhat different.

For the galaxies under consideration, there is a sizable nuclear mass, so that in the central regions there is only a relatively small amount of mass that can be assigned to the disk component; i.e., there is presumably only a small fraction of the stars which move essentially in the plane of the disk and which have small velocity dispersion. If we examine a picture of the galaxy NGC 4565 and compare it with a model of our own galaxy, we see that this situation is presumably typical of all Sb galaxies and some Sc galaxies. It holds even more strongly for Sa galaxies.

The total mass in the spheroidal component need not be excessively large. The requirement is actually on the projected surface density in the galactic disk. It should not be too high. (More specifically, the parameter $\mathcal{J}$ to be defined below in Eq. (70) should not be too large.) Experience shows that there are a variety of models which satisfy this requirement. In some cases, the spheroidal component may contain as little as 25% of the total mass.

The mass distribution and the angular velocity of a typical model are shown in Fig. 7.

We shall further simplify our problem by assuming that the fluid (which simulates the "variable part" of the stars) satisfies a barotropic relationship between the pressure $p$ and the density $\rho$ (which are both functions of the time $t$

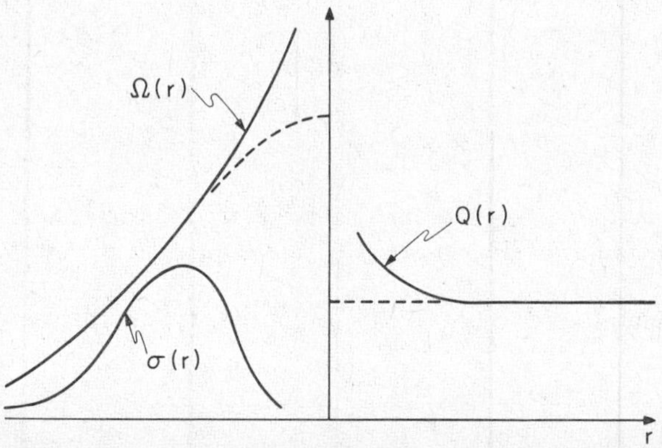


Figure 7. Typical distributions of rotational velocity, surface mass density, and velocity dispersion (as measured by $Q$) in the class of galaxies considered in this paper.

and of the positional vector. Then the force per unit mass is

$$-\frac{1}{\rho}\nabla p = -\nabla h, \tag{1}$$

where

$$h = h(\rho) = \int \frac{dp}{\rho}.$$

Thus, the basic equations governing our model are as follows.

Let the gravitational potential $\mathcal{V}$ be divided into a part $\mathcal{V}_0$, due to the fixed mass distribution $\rho_0$, and a part $\psi$, due to the variable mass distribution $\rho$. Then

$$\nabla^2 \mathcal{V}_0 = 4\pi G \rho_0, \tag{2a}$$

and

$$\nabla^2 \psi = 4\pi G \rho, \tag{2b}$$

where $G$ is the constant of universal gravitation. The variable part $\rho$ satisfies the following gas dynamical equations:

$$\frac{\partial \rho}{\partial t} + \nabla \cdot (\rho \mathbf{u}) = 0, \tag{3}$$

and

$$\frac{\partial \mathbf{u}}{\partial t} + (\mathbf{u} \cdot \nabla)\mathbf{u} = -\nabla(\mathcal{V} + h), \tag{4}$$

where $\mathbf{u}$ is the velocity vector. It is convenient to introduce the acoustic speed $a$ by the relationship

$$a^2 = \rho \frac{dh}{d\rho} \tag{5}$$

This quantity $a$ is to be identified with the root mean square value of the velocity dispersion. (For details, see [10].) In the solar vicinity, its value is about 30 km/sec.

The equation of continuity (3) may be put into the alternative form

$$\frac{D}{Dt}\ln \rho + \nabla \cdot \mathbf{u} = 0, \tag{3a}$$

where $D/Dt$ is the substantial derivative. In terms of the acoustic speed $a$ and the enthalpy $h$, Eq. (3a) may be written as

$$\frac{1}{a^2}\frac{Dh}{Dt}+\nabla\cdot\mathbf{u}=0. \tag{3b}$$

When we attempt to deal with the case of the infinitesimally thin disk (next section), it is convenient to start with Eqs. (4) and (3b).

### 8. *The infinitesimally thin disk*

The mass distribution in a galaxy, including the spheroidal and the disk components, is axisymmetrical. Thus, it is natural to try to describe it in the cylindrical coordinate system $(r,\theta,z)$ with the $z$-plane coinciding with the plane of symmetry in the disk. The $z$-axis is then the axis of rotational symmetry.

We shall be dealing only with the dynamics of a disk, which we shall idealize as being infinitesimally thin. We then speak of a surface density $\sigma(r,\theta,t)$. The corresponding gravitational potential $\psi(r,\theta,z,t)$ then satisfies the differential equation

$$\nabla^2\psi=4\pi G\sigma\delta(z), \tag{6}$$

where $\delta(z)$ is the delta function of Dirac, and

$$\nabla^2\psi=\frac{\partial^2\psi}{\partial r^2}+\frac{1}{r}\frac{\partial\psi}{\partial r}+\frac{1}{r^2}\frac{\partial^2\psi}{\partial\theta^2}+\frac{\partial^2\psi}{\partial z^2}. \tag{7}$$

Let $\{u(r,\theta,t),v(r,\theta,t)\}$ be the components of velocity in the $r$- and $\theta$-directions respectively. Then the equation of continuity is

$$\frac{\partial\sigma}{\partial t}+\frac{1}{r}\frac{\partial}{\partial r}(r\sigma u)+\frac{1}{r}\frac{\partial}{\partial\theta}(\sigma v)=0, \tag{8}$$

or

$$\frac{1}{\sigma}\left[\frac{\partial\sigma}{\partial t}+u\frac{\partial\sigma}{\partial r}+\frac{v}{r}\frac{\partial\sigma}{\partial\theta}\right]+\frac{1}{r}\left[\frac{\partial(ru)}{\partial r}+\frac{\partial v}{\partial\theta}\right]=0. \tag{9}$$

The equations of motion are

$$\frac{\partial u}{\partial t}+u\frac{\partial u}{\partial r}+\frac{v}{r}\frac{\partial u}{\partial\theta}-\frac{v^2}{r}=-\frac{\partial}{\partial r}(\mathcal{V}_0+\psi+h), \tag{10}$$

$$\frac{\partial v}{\partial t}+u\frac{\partial v}{\partial r}+\frac{v}{r}\frac{\partial v}{\partial\theta}+\frac{uv}{r}=-\frac{1}{r}\frac{\partial}{\partial\theta}(\psi+h), \tag{11}$$

where

$$dh=a^2(d\sigma/\sigma). \tag{12}$$

The boundary conditions to be imposed must now be carefully examined. There should certainly be (i) a condition of *regularity* at the center. At infinity, the correct boundary condition for spiral modes turns out to be (ii) a *radiation condition* of *outgoing* (trailing) spiral waves. *This latter condition is crucial*, if the fluid dynamical model is to simulate a stellar dynamical system, as we shall explain below in Sec. 10 and in Sec. 22. It also turns out that, because of this boundary condition and the related physical arguments, *all unstable spiral patterns must be trailing*. If the boundary conditions are not properly imposed, there is the danger that "antispiral theorem" will operate in one form or another, and we cannot obtain spiral modes. One might note in passing that this radiation boundary condition can be readily imposed in a cylindrical coordinate system. In the spheroidal coordinate system, which is also convenient for describing disks, it is not as easy to impose the radiation condition at infinity.

### 9. *The state of equilibrium*

The state of equilibrium has rotational symmetry. Thus,

$$\mathcal{V} = \mathcal{V}_0(r,z) + \psi_0(r,z), \tag{13a}$$

$$\sigma = \sigma_0(r), \tag{13b}$$

$$u = 0, \qquad v = v_0(r). \tag{13c}$$

The equation of continuity is automatically satisfied; so is the $v$-component of the equations of motion. The $u$-component reduces to

$$\frac{v_0^2}{r} = \frac{d}{dr}(\mathcal{V}_0 + \psi_0 + h_0), \tag{14}$$

where $\mathcal{V}_0$ and $\psi_0$ are evaluated at $z=0$. It is convenient to introduce the angular velocity $\Omega(r)$ associated with the circular velocity $v_0(r)$, i.e.,

$$\Omega(r) = v_0(r)/r. \tag{15}$$

In the solar vicinity, $v_0$ is about 250 km/sec, and $\Omega$ is about $25/10^9\cdot$yr. We shall also have occasion to introduce the parameter $\kappa(r)$ defined by

$$\kappa^2 = (2\Omega)^2\left\{1+\frac{1}{2}\frac{d\ln\Omega}{d\ln r}\right\}$$

$$= \frac{1}{r^3}\frac{d}{dr}(r^2\Omega)^2, \tag{16}$$

which is proportional to the gradient of the square of the angular momentum of the circular motion. The equilibrium may be expected to be stable only if $\kappa^2>0$.

Strictly speaking, the circular velocity $V(r)$ and the corresponding angular velocity $\Omega_0(r)$ are, *by convention*, defined by

$$V = r\Omega_0, \qquad r\Omega_0^2 = \frac{d}{dr}(\mathcal{V}_0 + \psi_0). \tag{17}$$

In that case, the corresponding parameter $\kappa_0$ is called the *epicyclic frequency*. Stars which deviate slightly from strict circular motion will go into an *epicyclic motion* at an angular frequency $\kappa_0$ around the *guiding center* which moves at the circular frequency $\Omega_0$ around the galactic center. The differences between $(v_0, \Omega, \kappa)$ and $(V, \Omega_0, \kappa_0)$ are so small that we shall apply the terminology to the former quantities, which alone will appear in the following analysis.

It should be noted that although the epicyclic parameter $\kappa$ appears in the theory, the epicyclic *motion* itself is not properly simulated in the fluid dynamical model. It would therefore be expected that defects in the fluid model will appear whenever resonance is important.

### *10. Spiral modes—linear theory*

We now consider small disturbances in the equilibrium state. We neglect nonlinear terms of the second order from the equations of motion and continuity. The resultant linear equations are cyclic in the variables $t$ and $\theta$, and hence we may seek solutions in the form of normal modes

$$\sigma = \sigma_0(r) + \sigma_1(r)e^{i(\omega t - m\theta)}, \tag{18a}$$

$$h = h_0(r) + h_1(r)e^{i(\omega t - m\theta)}, \tag{18b}$$

$$\psi = \psi_0(r) + \psi_1(r)e^{i(\omega t - m\theta)}, \tag{18c}$$

with similar expressions for the other variables. Here, we have put $z=0$ in $\psi$; the amplitude functions $\sigma_1, h_1, \psi_1$ are in general complex; $\omega$ is the complex frequency; and $m$ must be an integer. For spiral modes, $m$ is the number of spiral arms. The general solution is a linear superposition of these modes of the form (18), properly combined to satisfy the initial conditions.

It is convenient, when $m \neq 0$, to write the exponential factor in an alternative form

$$e^{i(\omega t - m\theta)} = e^{\gamma t}e^{im(\Omega_p t - \theta)}, \qquad \omega = m\Omega_p - i\gamma. \tag{19}$$

Then $\gamma$ is the rate of growth ($\gamma > 0$) or decay ($\gamma < 0$), and $\Omega_p$ is the angular velocity of the wave pattern.

From Eq. (12) it follows that

$$h_1(r) = a_0^2(r)\sigma_1(r)/\sigma_0, \tag{20}$$

where $a_0^2(r)$ is given, being a condition of the basic state. Clearly, the gravitational potential over the disk can be expressed in terms of an integral of the surface mass density. Thus,

$$\psi_1(r) = -2\pi G\int_0^\infty K(r,r')\sigma_1(r')\,dr', \tag{21}$$

where $K(r,r')$ is known. As a matter of fact (see [93, p. 95]),

$$K(r,r') = \frac{r'}{r+r'}P_m(r,r'), \tag{22}$$

where

$$P_m(r,r') = \frac{2}{\pi}\int_0^{\pi/2}\frac{\cos 2mx}{\sqrt{1-\zeta\cos^2 x}}\,dx,$$

$$\zeta = \frac{4rr'}{(r+r')^2}. \tag{22a}$$

The kernel function $K(r,r')$ has a logarithmic singularity at $r=r'$. Its general form is shown in Fig. 8, for $m=2$.

It is possible to eliminate the variables $u_1(r)$ and $v_1(r)$ from the linearized equations of motion and continuity in order to get a single differential equation of the second order in the variables $h_1(r)$ and $\psi_1(r)$. Indeed, we get an equation of the form[4]

$$\mathfrak{L}(h_1+\psi_1) = -Ch_1, \tag{23}$$

where

$$\mathfrak{L} = \frac{d^2}{dr^2} + A\frac{d}{dr} + B, \tag{24}$$

and the coefficients $A$, $B$, and $C$ are given as follows:

$$A = -\frac{1}{r}\frac{d\ln\mathcal{Q}}{d\ln r}, \qquad \mathcal{Q} = \frac{\kappa^2(1-\nu^2)}{\sigma_0 r}; \tag{25a}$$

$$B = -\frac{m^2}{r^2} - \frac{4m\Omega(r\nu')}{\kappa r^2(1-\nu^2)} + \frac{2\Omega m}{r^2\kappa\nu}\frac{d\ln(\kappa^2/\sigma_0\Omega)}{d\ln r}; \tag{25b}$$

$$C = -\frac{\kappa^2(1-\nu^2)}{a_0^2}. \tag{25c}$$

[4]The physical meaning of this equation has been examined by Lau and Bertin [40]; a detailed derivation is also included on p. 516 of that paper.

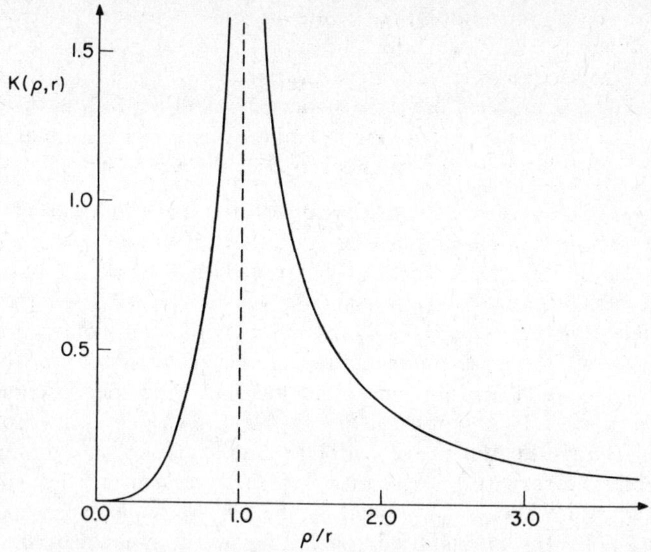


**Figure 8.** The influence function of surface density distributions. (After Pannatoni [79].)

In the above equations, we have introduced an important parameter

$$\nu = \frac{\omega - m\Omega}{\kappa} \tag{26}$$

whose real and imaginary parts are

$$\nu_r = \frac{m(\Omega_p - \Omega)}{\kappa} \quad \text{and} \quad \nu_i = -\frac{\gamma}{\kappa}. \tag{27}$$

In the optically prominent inner part of the galaxy, it turns out that $\nu_r$ is usually negative. Thus, $-\nu_r$ is the dimensionless measure of the frequency at which the material, traveling at angular velocity $\Omega$, experiences the periodic effect of the spiral pattern of $m$ arms, traveling at angular velocity $\Omega_p$. Thus we may call $-\nu_r$ the *dimensionless frequency of encounter* of the material with the wave pattern, and $-\nu_i$ the *dimensionless rate of growth*.

The importance of the parameter $\nu$ can be seen from the fact that the coefficients of the differential equation (23) have singularities and other special features at

$$\nu = 0, +1, -1. \tag{28}$$

These are usually designated as the *corotation resonance*, the *outer Lindblad resonance*, and the *inner Lindblad resonance*. At the Lindblad resonances, the encounter frequency is equal to the epicyclic frequency $\kappa$.

Equations (20), (21), and (23) constitute the basic equations, from which $\sigma_1(r)$ can be eliminated. Thus we have one integral equation and one differential equation of the second order for the two variables $\psi_1(r)$ and $h_1(r)$. The complete formulation of the problem of spiral modes must still depend on (a) the proper formulation of the boundary conditions, and (b) the proper understanding of the behavior of the solution in the neighborhood of the singularities at $\nu=0, \pm 1$.

(a) *Boundary conditions*. As discussed before, the condition of regularity should be imposed at the origin. The condition of outgoing waves at infinity (radiation condition) can be explicitly imposed only after the nature of the solution is better understood, as we shall develop in Part III (cf. also Sec. 22).

(b) *Singularities at resonances*. In common with many other problems of instability theory, the governing equations are expected to hold for the *unstable modes* along the real axis of the independent variable (regarded as complex). The singularities $\nu=0, \pm 1$ are points off the real axis. In this case, the explicit proof has been given by Feldman and Lin [19]. "Stable modes" are unimportant in any instability theory; in the solution of the problem by the initial value approach, they are always dominated by the contributions associated with the initial conditions. In the modal approach, they are dominated by the contributions from a continuous spectrum of neutral solutions with discontinuities at the singular points.

There is a deeper problem in our special case: can we expect the fluid dynamical theory to model the stellar dynamical system properly at these resonances? The answer is *no* if we take the equations at face value. The proper treatment of these singularities can only be understood by comparison with results obtained from the stellar dynamical theory. Fortunately, these are available through the work of Mark [66, 68]. We refer the reader to Sec. 22, where these issues are examined in greater detail.

## Part III
## Asymptotic theory of spiral density waves: rudimentary theory

In the previous part, we have outlined the formulation of the basic equations and the boundary conditions for determining the spiral wave patterns over a galactic disk. The difficulty in solving these equations is mostly attributable to the fact that the description of the gravitational potential is in the form of an integral equation, a fundamental difficulty caused by the long range nature of the force of gravitation. In this part and the next, we shall develop an asymptotic method for simplifying this mathematical problem. The advantage of an asymptotic theory is well known. It is simple and therefore convenient for bringing out the essential physical mechanisms intrinsic to the problem. As we shall see below, we can even arrive at a simple *quantum condition* in the form of a phase integral for the determination of the eigenfrequency.

Admittedly, the accuracy of such an asymptotic theory is expected to be somewhat limited; but we have found the results to be surprisingly accurate for most of the purposes of our problem. The basic form to be developed in this part will be found to be insufficient for the determination of the growth rate, but the improved form to be developed in Part IV will be found to be fairly accurate even for that purpose. In any case, the growth rate has been shown to be so sensitive to the distributions of a number of parameters that its accurate theoretical determination is of doubtful practical value.

Before we go on with the development of the theory of spiral modes, let us first consider *free spiral waves*. We consider waves of such short wave length that the WKBJ method can be applied; we postpone, for the moment, the examination of the boundary conditions.

## 11. *Spiral waves*

To apply the WKBJ method, we write the amplitude functions in the form

$$h_1 = \hat{h}_1 \exp\left\{ i \int k\, dr \right\}, \qquad \psi_1 = \hat{\psi}_1 \exp\left\{ i \int k\, dr \right\}. \tag{29}$$

Here $k$ is presumed to be of the form $\Lambda f(r)$, where $\Lambda$ is a large parameter, and $f(r)$, as well as $\hat{\psi}_1(r)$, is a smooth, slowly varying function of the radial distance $r$; i.e., $d\ln k/d\ln r = O(1)$. As one can easily verify, the combination

$$\psi_1(r) e^{i(\omega t - m\theta)} = \hat{\psi}_1(r) \exp\{\gamma t + i\phi(r,\theta,t)\}, \tag{30}$$

where

$$\phi(r,\theta,t) = m(\Omega_p t - \theta) + \int k(r)\, dr, \tag{30a}$$

represents spiral waves whose crests are located approximately along the curve $\phi = 0$ (mod $2\pi$). Thus, the spiral form of an equipotential curve, at any instant $t$, is given by

$$m(\theta - \theta_0) = \int_{r_0}^{r} k(r)\, dr, \tag{31}$$

and $|k|$ is the wave number. The waves are short, since

$$|k| r \gg 1. \tag{32}$$

The waves are called *trailing* if $k < 0$ and *leading* if $k > 0$.

The validity of the asymptotic approximation depends on the condition (32). This can be easily seen to be equivalent to a requirement of *small pitch angle* $i$ for the spiral arms. Indeed, from (31), it follows easily that

$$|\tan i| = |m/kr|. \tag{31a}$$

For a two armed spiral ($m = 2$) with pitch angle of 10°, $|kr|$ turns out to be approximately 11.

If we introduce solutions of the type (29) into the basic equations (21) and (23), and consider $\Lambda$ as the principal large parameter, we get a series of algebraic relations. As is familiar in any WKBJ procedure, in the initial approximation we should get the dispersion relation; in the next approximation, we expect to get the law of distribution of the amplitude of the waves. The unique situation in the present case is that the integral representation of the Poisson equation should be solved, not by direct substitution, but by a different procedure. (See Appendix A.) To the first two orders, this asymptotic approximation of the Poisson equation yields the differential relationship

$$\frac{d\psi_1}{dr} = -\frac{\psi_1}{2r} - i\Sigma(r)h_1\,\mathrm{sgn}(k), \tag{33}$$

where

$$\Sigma(r) = \frac{2\pi G\sigma_0}{a_0^2}. \tag{34}$$

For simplicity and clarity, we shall first consider *trailing waves*, and write (33) in the form

$$\frac{d\psi_1}{dr} = -\frac{\psi_1}{2r} + i\Sigma(r)h_1. \tag{33a}$$

For *leading waves*, $\Sigma(r)$ should simply be replaced by $-\Sigma(r)$.

The derivation of (33) can be justified if $|kr| \gg 1$. Indeed, during this process of derivation, it also becomes clear that $|k| = O(\Sigma)$, $\psi_1 = O(h_1)$ and that $\Sigma$ is a large parameter in the sense that

$$r\Sigma \gg 1. \tag{35}$$

These statements will also be confirmed *a posteriori* [cf. Eq. (42) below]. In passing, we note that $\psi_1$ and $h_1$ have the same physical dimension.

Let us now turn our attention to the dynamical equation (23), and examine the orders of magnitude of the coefficients $A$, $B$, and $C$. Apart from the singularities, we see that

$$rA = O(1) \quad \text{and} \quad r^2B = O(1). \tag{36}$$

The order of magnitude of $C$ is more difficult to discuss. For this purpose, we write

$$C = -(1-\nu^2)\Sigma/2\delta, \tag{37}$$

where

$$\delta = \pi G\sigma_0/\kappa^2 \tag{38}$$

is a length scale dependent only on the *equilibrium* state, whose significance will be discussed below; we now only note the fact that in many galaxies

$$\frac{\delta}{r} \ll 1. \tag{39}$$

Indeed, we shall restrict ourselves to this class of galaxies for most of our discussions. Thus, the combination of (35) and (39) shows that

$$Cr^2 \simeq (r\Sigma)(r/\delta) \gg 1, \tag{40}$$

which implies that

$$Cr^2 \gg r\Sigma \text{ or } r/\delta, \tag{40a}$$

Clearly, the existence of the small basic linear scales $\delta$ and $\Sigma^{-1}$ underlies the possibility for the existence of solutions in the form of tightly wound spirals.

### *12. The dispersion relation*

The asymptotic procedure described above leads, in the lowest approximation, to the following quadratic equation for the wave number $k$ $(k<0)$:

$$k^2 + \Sigma k - C = 0, \tag{41}$$

or, in terms of physical quantities,

$$(\omega - m\Omega)^2 = \kappa^2 + (ka_0)^2 - 2\pi G\sigma_0|k|. \tag{41a}$$

The solution of (41) is

$$k = -k_0\left\{1 \pm \sqrt{1 - Q^2(1-\nu^2)}\right\}, \tag{42}$$

where

$$k_0 = \frac{\Sigma}{2}, \tag{43a}$$

$$Q^2 = (k_0\delta)^{-1} = \left(\frac{\pi G\sigma_0}{\kappa a_0}\right)^{-2}. \tag{43b}$$

These are the trailing waves $(k<0)$. There are two other roots with $k_0$ replaced by $-k_0$; those are the leading waves $(k>0)$. Thus, we have a total of *four kinds of waves*; of these, two are leading and two are trailing. The dispersion relation (42) is shown graphically in Fig. 9.

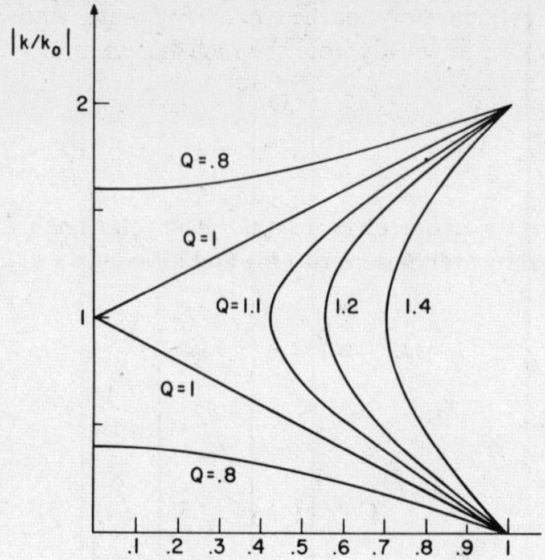


Figure 9. The dispersion relationship of fluid disks.

The angular propagation of the wave is nondispersive; the phase (or group) velocity is $\Omega_p$. In the radial direction, the *group velocity* of propagation of these waves can be easily obtained from the phase angle $\phi(r,\theta,t)$ defined by (30a). It is

$$c_g^{-1} = -\frac{\partial}{\partial\omega}\left(\frac{\partial\phi}{\partial r}\right) = -\frac{\partial k}{\partial\omega} \tag{44}$$

or

$$c_g = -\frac{a_0^2}{\kappa^2}(k+k_0)/\nu. \tag{44a}$$

The factor $k+k_0$ is positive for long waves and negative for short waves. Thus we have the results in Table 1 [53].

TABLE 1
Group Propagation for Four Types of Density Waves

| Waves for | | |
|---|---|---|
| $k<0$: | Short trailing | Long trailing |
| $k>0$: | Long leading | Short leading |
| Propagation: | Departing corotation | Approaching corotation |

These results were first obtained by Toomre [102] in the stellar case, based on the dispersion relation previously obtained by Lin and Shu [57].

### *13. Axisymmetric waves*

The asymptotic theory developed above applies to the axisymmetric waves. Specifically, the dispersion relation (41a) now takes on the form

$$\omega^2 = \kappa^2 + (ka_0)^2 - 2\pi G\sigma_0|k|. \tag{45}$$

This relationship can be understood in terms of oscillations associated with epicyclic motion [$\kappa^2$], gaseous pressure [$(ka_0)^2$, acoustic waves], and the modification of such oscillations by self-gravity [$-2\pi G\sigma_0|k|$]. The negative sign in the last term suggests the possibility of instability if this term dominates the other two. For a given real frequency $\omega$, this dispersion relation suggests a propagation of neutral oscillations whose wavelength $k$ is locally determined by Eq. (45). On the other hand, for spiral waves [$m\neq 0$ in Eq. (42)], we have a given pattern frequency $\Omega_p$ which leads to a variable "Doppler shifted frequency" $\omega - m\Omega$ or $m(\Omega_p - \Omega)$, which in turn determines the spacing between spiral arms in accordance with the dispersion relation (42).

The condition of marginal stability is given by $\omega = 0$. In that case, we have the relationship

$$Q^2 = \frac{1}{4}\left\{\frac{\lambda}{\lambda_c} - \left(\frac{\lambda}{\lambda_c}\right)^2\right\}, \tag{46}$$

where

$$\lambda_c = 4\pi^2 G\sigma/\kappa^2 = 4\pi\delta, \text{(47a)}$$

$$Q = a_0\left(\frac{\pi G\sigma_0}{\kappa}\right)^{-1} = \frac{a_0}{\kappa\delta}, \tag{47b}$$

and $\lambda = 2\pi/|k|$. This diagram is sketched in Fig. 10. We see that, as $\lambda$ changes,

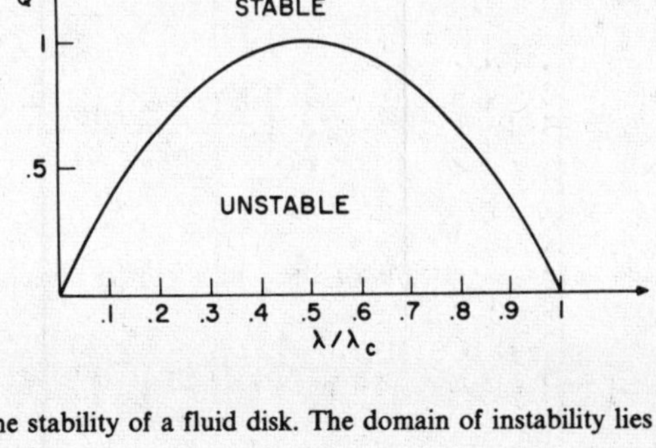


Figure 10. The stability of a fluid disk. The domain of instability lies below the curve.

the peak value of $Q^2$ is unity; i.e., axisymmetric instability is removed if

$$Q \geqslant 1, \quad \text{or} \quad a_0 \geqslant \kappa\delta = \frac{\pi G \sigma_0}{\kappa}. \tag{48}$$

Thus, for acoustic speeds in excess of $\kappa\delta$, the disk is completely *stable with respect to axisymmetric disturbances*; i.e., only neutral waves can propagate. This criterion was first obtained by Toomre [101] in the stellar case; the fluid case was given by Lynden-Bell [64].

The above relationship gives an interpretation for the scale $\delta$. It is known, in stellar dynamics, that the diameter of the epicycle in the radial direction is $2c_r/\kappa$, where $c_r$ is the maximum radial velocity of the star. Thus, $\delta$ is the "radius" of the epicycle of those stars with a maximum radial speed which is barely sufficient to maintain marginal stability according to the fluid model.

### 14. Spiral modes: formulation

We now proceed to the study of spiral modes, first with the primary purpose of understanding the mechanism, even though at some sacrifice of accuracy. In Part IV, we shall return to the study of effects which are necessarily neglected in the present simple analysis, in which only the lowest order terms are retained in the asymptotic approach. It is to be hoped that this lowest approximation will retain the essential features of the problem. Fortunately it does, as will be seen after the more complete analysis in Part IV.

We now consider the equation for the spiral modes. If we eliminate the derivatives $\psi_1''$ and $\psi_1'$ from the dynamical equation (23) by using the differential form (33) of the Poisson equation, we obtain (see [44], [40]), for trailing spirals,

$$h_1'' + (A + i\Sigma)h_1' + \left[ -i\Sigma\left(\frac{1}{2r} - A - \frac{\Sigma'}{\Sigma}\right) + B + C \right] h_1 + \left(B - \frac{A}{2r} + \frac{3}{4r^2}\right)\psi_1 = 0. \tag{49}$$

Since $\psi_1$ is of the same order as $h_1$, and the coefficient $C$ is two orders larger than the coefficients $A$ and $B$, we need not worry about the term in $\psi_1$ in the lowest approximation. We reduce the equation (49) for $h_1$ to the normal form by using the standard transformation

$$h_1 = u \exp\left\{ -\frac{i}{2} \int \Sigma\, dr \right\} \cdot \left( \frac{\kappa^2(1-\nu^2)}{\sigma_0 r} \right)^{1/2}. \tag{50}$$

Then, if we explicitly keep only terms of the lowest order, we have

$$\frac{d^2u}{dr^2} + \left(\frac{\kappa}{a_0}\right)^2 \left\{ (Q^{-2} - 1 + \nu^2) + O\left(\left(\frac{a_0}{\kappa}\right)^{-1}\right) \right\} u = 0. \tag{51}$$

We note that $\kappa/a_0$ is of the order of $\delta^{-1}$, since $Q=a_0/\kappa\delta$, and $Q$ is of the order of unity.

Before we proceed with the investigation of the simplified differential equation

$$\frac{d^2u}{dr^2}+k_3^2u=0,$$

$$k_3^2=\left(\frac{\kappa}{a_0}\right)^2(Q^{-2}-1+\nu^2), \tag{52}$$

let us first note that it does yield the same dispersion relationship as Eq. (42).

We consider a type of galaxy in which the rotation curve and the mass density are as shown in Fig. 7. Furthermore, we shall assume that the parameter $Q$ has a distribution as shown in the same figure. Note that in our model we have assumed $Q$ to be essentially equal to unity for the outer parts, while its value rises towards the central regions. The reason is as follows. The disk material was originally collected there because it has relatively small random motions. However, the gravitational instability discussed above would tend to increase the random motions of the stars and to bring it up to the level required for stability. There is no reason to exceed the marginal value. (For spiral instabilities, see Part IV.) On the other hand, in the inner parts of the galaxy, there is material in the nuclear bulge with high velocity dispersion. Thus, the combination of low disk mass density and relatively high velocity dispersion would combine to yield relatively high values of $Q$ [cf. Eq. (47b)].

For the kind of model chosen, the quantity $k_3^2$ in (52) has the general characteristic shown in Fig. 11 when $\omega$ is real. It has a simple zero at some inner point $r_{ce}$ and a double zero at the corotation point $r_{co}$ where $\nu=0$. These are called *turning points* in the asymptotic theory of differential equations. For $\omega$ complex but nearly real, the general nature of the problem is not greatly changed.

The *boundary conditions* for our problem can now be formulated. For the inner region, $r<r_{ce}$, $k_3^2<0$, and therefore it is natural to specify the condition

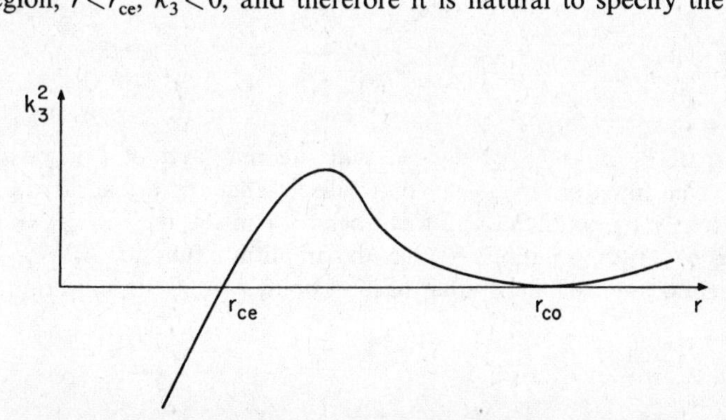


**Figure 11. Distribution of the square of the reduced wave number in the rudimentary theory [42].**

that the solution should *attenuate* as $r$ decreases. For $r>r_{co}$, the solution is still in the nature of propagating waves. For trailing waves, the short wave propagates outwards, the long wave inwards. Based on physical considerations given in the next paragraph, we should specify a *radiation condition* at infinity, since the inward signal from infinity amounts to an external disturbance.

The radiation condition leads to growing modes of trailing spirals. This will be shown mathematically in the next section. Here, we describe the physical processes. In a real galaxy, the outward propagating waves would be absorbed at the outer Lindblad resonance or be dissipated by turbulence, since the energy density of the waves turns out to be positive. On the other hand, the interior waves possess a negative amount of energy (or angular momentum), and so they gain amplitude when they lose angular momentum through an outward transport as modeled by this radiation condition [42].

The use of the radiation condition means that we have the *short waves* at large distances if the waves are *trailing*. For leading spirals, the outward propagating waves are the long waves, which are not absorbed at Lindblad resonances. Neither are they efficiently dissipated by turbulence. The radiation condition therefore gives preference to *trailing spirals* (see Secs. 21C and 22B).

## *15. Spiral modes: solution*

The solution of Eq. (52) by the method of uniformly valid asymptotic solutions was made by Lin and Lau [54] and extended by Li et al. [44] to the next approximation. The essence of this work is reproduced here by using the notation of the latter paper.

In a neighborhood of the simple turning point $r=r_{ce}$, the solution is similar to that in conventional quantum mechanics. We have

$$u_{ce}=\left(\frac{\zeta}{k_3}\right)^{1/2} Z_{1/3}(\zeta), \tag{53}$$

where

$$\zeta=\int_{r_{ce}}^{r} k_3\,dr. \tag{53a}$$

The branch of $k_3$ is to be so chosen that the real part of $\zeta$ is positive for $r_{ce}<r<r_{co}$. The function $Z_{1/3}(\zeta)$ is that Bessel function which has a decaying amplitude for $r<r_{ce}$, and standing wave behavior in the region $r_{ce}<r<r_{co}$. For large values of $|\zeta|$, the solution (53) has the amplitude function $k_3^{-1/2}$.

In a neighborhood of the double turning point $r=r_{co}$, the solution takes on the form

$$u_{co}=\left(\frac{z}{k_3}\right)^{1/2} Z_{1/4}(z), \tag{54}$$

where

$$z = -\int_{r_{co}}^{r} k_3 dr. \tag{54a}$$

In (54), $Z_{1/4}(z)$ is that Hankel function whose phase behaves as $e^{-iz}$ for $r > r_{co}$ in order to satisfy the radiation condition. The branch of $k_3$ chosen must be compatible with that above. Clearly, the real part of $z$ is again essentially positive for $r_{ce} < r < r_{co}$. Again, for large values of $|z|$, the amplitude function is $k_3^{-1/2}$.

In the interval $(r_{ce}, r_{co})$, the solutions given above have the following representations in terms of the conventional Hankel functions. For $u_{ce}$, we have

$$u_{ce} = u_{ce}^{+} + u_{ce}^{-}, \tag{55a}$$

where

$$u_{ce}^{-} = e^{-\frac{1}{6}\pi i}\left(\frac{\zeta}{k_3}\right)^{1/2} H_{1/3}^{(2)}(\zeta), \tag{55b}$$

$$u_{ce}^{+} = e^{+\frac{1}{6}\pi i}\left(\frac{\zeta}{k_3}\right)^{1/2} H_{1/3}^{(1)}(\zeta). \tag{55c}$$

For $u_{co}$, we have

$$u_{co} = u_{co}^{+} + u_{co}^{-}, \tag{56a}$$

where

$$u_{co}^{-} = \sqrt{2}\, e^{\frac{3}{4}\pi i}\left(\frac{z}{k_3}\right)^{1/2} H_{1/4}^{(1)}(z), \tag{56b}$$

$$u_{co}^{+} = e^{\frac{1}{2}\pi i}\left(\frac{z}{k_3}\right)^{1/2} H_{1/4}^{(2)}(z). \tag{56c}$$

In the above representation, we have identified the outward (+) and the inward (−) propagating waves. For large values of $|z|$ and $|\zeta|$, these waves have the wave numbers $\pm k_3$, and the combined wave number $k = -k_0 \pm k_3$ satisfies the local dispersion relationship. Thus, we may write

$$h_1 = A_s e^{i\int k_s dr} + A_l e^{i\int k_l dr}, \tag{57}$$

where $s$ and $l$ designate short and long waves, and we have

$$k_s = -k_0 - k_3, \qquad k_l = -k_0 + k_3. \tag{58}$$

However, when the solution is in the form (55) and (56), we may also write it in the same form (57); for example,

$$u_{ce}^{-} = A_s e^{i\int k_s dr}, \qquad u_{ce}^{+} = A_l e^{i\int k_l dr}. \tag{59}$$

But this does not imply that (58) holds, for the variable $\zeta$ can be quite small. Thus, *the failure of the local dispersion relationship does not imply the failure of the concepts of incoming trailing short waves or outgoing long trailing waves.*

Since there are overlapping domains where the above solutions $u_{ce}$ and $u_{co}$ both hold, they can only differ by numerical constants. Thus, we have

$$au_{ce}^{-} = bu_{co}^{-}, \qquad au_{ce}^{+} = bu_{co}^{+}, \tag{60}$$

where $a$ and $b$ are constants. In the part of the complex $r$-plane where both $|\zeta|$ and $|z|$ are large, the asymptotic forms of the functions in (60) hold, and we obtain the following relations:

$$e^{i\pi} = \sqrt{2}\,\exp\left\{2i\int_{r_{ce}}^{r_{co}} k_3\,dr\right\}, \tag{61a}$$

$$2^{1/4}a^2 = b^2 e^{-\frac{5}{4}\pi i}. \tag{61b}$$

Equation (61b) is not so important, but Eq. (61a) is the all-important *quantum condition* for the determination of the eigenfrequency $\omega$. It may be written in the form

$$\int_{r_{ce}}^{r_{co}} k_3\,dr = \left(n+\tfrac{1}{2}\right)\pi + \tfrac{1}{4}i\ln 2, \qquad n = 0,1,\ldots. \tag{62}$$

This quantum condition was first given by Lau, Lin, and Mark [42]. The rigorous mathematical theory that underlies these calculations was developed by Nishimoto [75].

Having determined the eigenvalue by the quantum condition (62), one may calculate the eigenfunction by using the solutions given above. Alternatively, one may integrate the original differential equation numerically by using the value of $\omega$ just obtained as the initial approximation. As we shall see later, the numerical integration process can be more easily adapted to investigate the effects of other terms that have been ignored so far. For an understanding of the dynamical mechanisms, however, there is as yet no substitute for the asymptotic theory just developed. For example, when the growth rate $\gamma = -\omega_i$ is small, Eq. (62) leads to the approximate relationships

$$\int_{r_{ce}}^{r_{co}} k_3\,dr = \left(n+\tfrac{1}{2}\right)\pi \qquad (\omega = \omega_r \text{ in } k_3) \tag{63a}$$

and

$$\gamma\tau = \tfrac{1}{4}\ln 2, \tag{63b}$$

where

$$\tau = \int_{r_{ce}}^{r_{co}} \frac{dr}{|c_g|},$$

$$|c_g| = \left|\frac{\partial k_3}{\partial \omega}\right|^{-1} \qquad (\omega = \omega_r \text{ in } k_3). \tag{63c}$$

Physically, $|c_g|$ is the magnitude of the group velocity of the waves at a given location (same for incoming short waves and outgoing long waves), and $\tau$ is the time it takes to propagate between $r_{ce}$ and $r_{co}$. Thus, the growth of the amplitude of these waves over a round trip is [cf. (63b)]

$$e^{2\gamma\tau} = \sqrt{2}, \tag{64}$$

which is the factor appearing in (56b). More specifically, the outgoing wave $u_{co}^{+}$ is turned back at $r_{co}$ with a reinforcement (because there is another wave leaving $r_{co}$ to go to infinity), which corresponds to a *doubling* of energy. This incoming wave (56b) propagates to $r_{ce}$ and is turned back at the *same* energy level[cf. Eqs. (55b) and (55c)]. After a period of $2\tau$, the process is complete with a doubling of energy. This is precisely the assertion of Eq. (64).

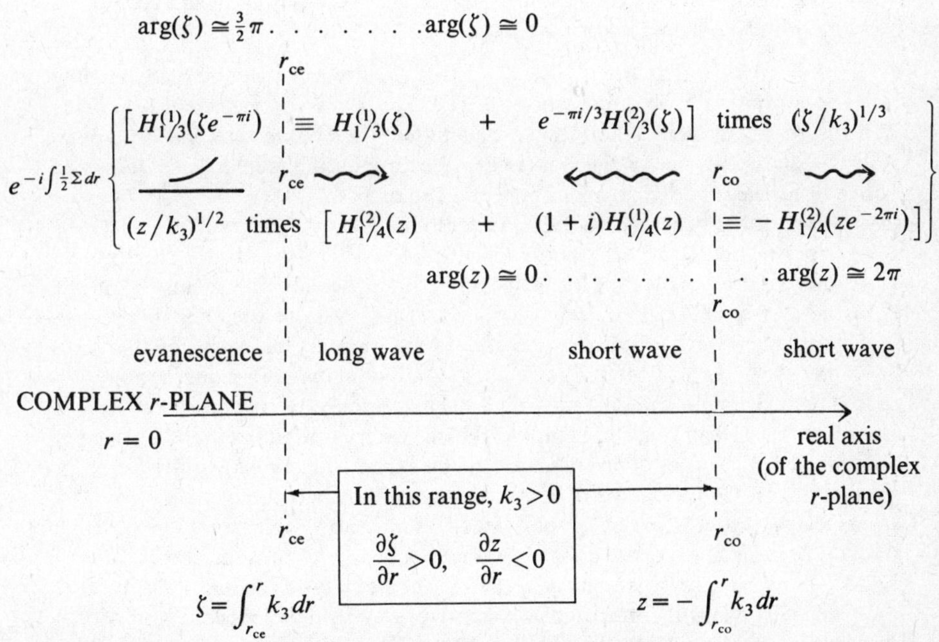


Figure 12. Progressive waves in a galactic disk.

The outgoing wave beyond $r_{co}$ must have a single unit of energy of an opposite sign to those just inside of $r_{co}$. This is indeed the case. (Since energy is not always well defined, we should perhaps speak of the angular momentum carried by the waves.) The solution $u_{co}$ outside of $r_{co}$ is simply

$$u_{co} = \left(\frac{z}{k_3}\right)^{1/2} H^{(2)}_{1/4}(z). \tag{65}$$

[Note that $H^{(2)}$ corresponds to short waves, both here and in (55). The direction of group propagation is, however, reversed, because $\nu$ changes sign across the corotation circle.]

The above solutions and the interpretation of their implications are schematically summarized in the diagram in Fig. 12. The numerical results will be presented for the more accurate calculations to be discussed in the next section.

## Part IV
## Unstable spiral modes

In this part, we shall attempt to remove the uncertainties in the previous rudimentary analysis by examining the remaining issues both mathematically and physically.

### *16. A closer examination*

The above treatment is successful in bringing out the essential mechanisms, but it is not entirely satisfactory on two accounts:

(1) Are we sure that the answers are trustworthy? The double zero turning point at corotation is obtained when all the other terms are neglected. It appears to be a precarious step. When they are included, the double zero will disappear. Will this disappearance of the double zero destroy all the mechanisms described above or at least modify them in important manners?

(2) Even if the basic mechanisms are not changed, the accuracy of the results is certainly in doubt. To what extent would the pattern speed, the growth rate, and other characteristics of the solution be modified by the inclusion of the terms neglected so far? If so, what is the physical basis for such changes?

To answer these two questions is indeed the primary objective of this part. We study the effects of the terms neglected individually or in combination with each other through both analytical and numerical approaches. We further identify the physical mechanism to which each term corresponds, thereby obtaining an idea what their effects might be expected to be like under different circumstances. Confirmation of these arguments has become available recently after the exact calculation of spiral modes was made by Pannatoni [79]. We shall show in Sec. 18 that spiral modes obtained from the asymptotic theory indeed agree well with those obtained from the exact calculation. (See also [80].)

In this part, we will consider two specific galaxy models: Model I and Model II. These models are illustrated in Figs. 13 and 14, their analytic expressions are given in Appendix B. Model II has a lower surface density and its spiral modes would be expected to have a lower rate of growth than Model I.

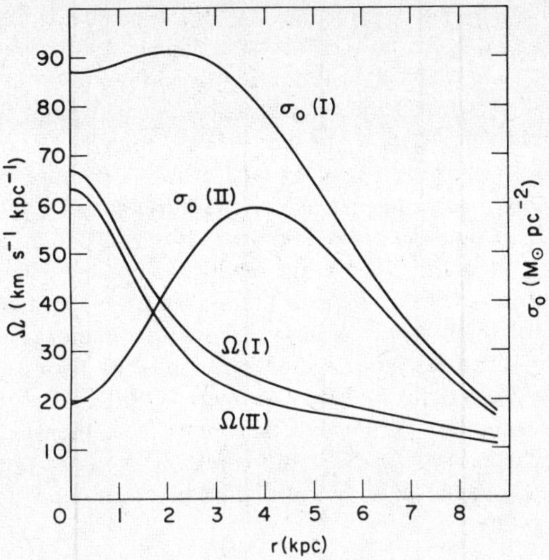


Figure 13. The surface density $\sigma_0$ and the rotation curves of the equilibrium models I and II.

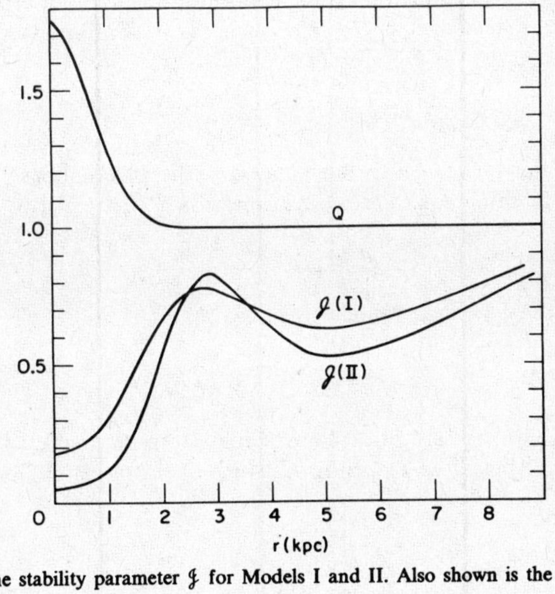


Figure 14. The stability parameter $\mathcal{J}$ for Models I and II. Also shown is the assumed $Q$ profile.

To give another perspective of the instability mechanism, it is useful to reexamine the local dispersion relation to a higher approximation than that given in Part III, and indeed from a slightly different point of view. This alternative approach (cf. Sec. 19) is not based on a small pitch angle, but based on the smallness of $\delta/r$, i.e., of the size of the epicycle. (In both cases, we assume that quantities like $d\ln\Omega/d\ln r$ and $d\ln\sigma_0/d\ln r$ are of the order of unity.) It is therefore similar to that used by Goldreich and Lynden-Bell [20] in the gaseous case, and by Julian and Toomre [32] in the stellar case. Our results developed in Sec. 19 can then be more readily compared with theirs. Specifically, we bring out the effect of the gravitational or pressure forces tangential to the circular motion in terms of a new stability parameter $\mathcal{J}$ (which can be formed out of two of the parameters used by Julian and Toomre [32]). We shall show that this new parameter and the parameter $Q$ introduced by Toomre are the only two dimensionless parameters that enter the dimensionless dispersion relation. Indeed, the parameter $Q$ is sufficient for the study of axisymmetric disturbances in the galactic disk. Spiral disturbances are governed by an additional parameter $\mathcal{J}$ [39, 40].

### *17. The effects of terms of higher orders*

We now attempt to include the effects of the higher order terms in the calculation of spiral modes. First of all, we need to improve on the differential equation for the Poisson equation. This has been provided by Bertin and Mark [13], who showed that the improved relationship is (for trailing waves)

$$\frac{d\psi_1}{dr}+\frac{1}{2r}\psi_1-\frac{m^2-\frac{1}{4}}{i\Sigma r^2}\psi_1=i\Sigma h_1. \tag{66}$$

By introducing the variable $u$ through the same transformations (50), we find that the new differential equation is [see Appendix C, Eq. (C4)]

$$\frac{d^2u}{dr^2}+K^2u=0, \tag{67}$$

where

$$K^2=k_3^2+R, \qquad R=\Sigma R_i, \tag{68}$$

that is, $R$ represents the sum of all the residual terms $R_i$ neglected before. The explicit form of $R$ is given in Appendix C; some of the more important terms $R_i$ are as follows:

$$R_1=-i\frac{\Sigma}{2}\frac{d}{dr}\ln\left[Q^2(1-\nu^2)\right], \tag{69a}$$

$$R_2=\left(\frac{\kappa}{a_0}\right)^2\cdot\tfrac{1}{4}\mathcal{J}^2Q^2, \tag{69b}$$

$$R_3=-\frac{2\Omega/\kappa}{\Omega_p-\Omega-i\gamma}\frac{1}{r^2}\frac{d\ln(\kappa^2/\sigma_0\Omega)}{d\ln r}. \tag{69c}$$

In (69b), the quantity $\mathcal{J}$ is defined by

$$\mathcal{J} = 2m\frac{\delta}{r}\left(\frac{1}{s}-\frac{1}{2}\right)^{-1/2}$$

$$= m\frac{\delta}{r}\frac{4\Omega}{\kappa}\sqrt{s}\,. \tag{70}$$

where

$$s = -\frac{d\ln\Omega}{d\ln r}, \tag{71}$$

and $\delta$ is defined in Eq. (38).

Clearly, if $k \sim \Sigma$,

$$k^{-2}R_1 = O\big((kr)^{-1}\big), \tag{72a}$$

$$k^{-2}R_2 = O\big((m/kr)^{-2}\big), \tag{72b}$$

$$k^{-2}R_3 = O\big((1/kr)^{-2}(\Omega_p-\Omega-i\gamma)^{-1}\big). \tag{72c}$$

The inclusion of $k^{-2}R_1$ thus appears to be natural. The inclusion of $k^{-2}R_2$ is also needed if $m^2/kr \sim 1$. This is indeed the case, even for $m=2$, if $kr$ is only as large as four. The term $k^{-2}R_3$ is needed near the corotation circle where $\Omega_p-\Omega=0$, even though it is only of the order of $(kr)^{-2}$. We refer the readers to Appendix C for a more reliable description of the corotation resonance. See also Sec. 22 on this issue.

The effect of each of the terms requires a careful study in order to be understood. The ultimate decision on the magnitude of its significance can best be checked by numerical integration of the final equations. This has been done by Lau and Bertin [40].

The effect of the out of phase term $R_1$ has been studied by Li et al. [44]. By carrying out the asymptotic analysis to the next approximation, it is found that for the type of models considered in the previous section, there is no change in the growth rate, but a slight change in the pattern speed. A numerical integration of this problem for the specific model described in Sec. 16 has been done by Lau and Pannatoni (see [40]). The result supports this conclusion; there is just a slight increase in the growth rate.

The effect of the term $R_2$ (which is a measure of the angle dependent effect) has been studied by Lau and Mark [43], by Nishimoto [75], and by Lau and Bertin [40]. This term turns out to have a very important effect on the growth rate though only a small one on the pattern speed. We shall focus our attention on this term in the next two sections.

The effect of the term $R_3$ has also been examined in the paper by Lau and Bertin [40]. Its effect is small except possibly for the lowest mode ($n=0$). In this case, it actually requires a modification of the gas dynamical model by considering stellar dispersion near corotation to get its true magnitude.[5] Such an investigation shows that there is a moderate lowering of the growth rate due to this resonant term, but only a small change in the pattern speed.

[5]Otherwise, as also shown by Hu [27] in a somewhat different context, the resonance effect can cause significant differences. See also Sec. 22.

TABLE 2
Pattern Speeds and Growth Rates of Two-Armed Modes According to Various Approximations (Model I, $n=0$, $n=1$)

| $R_i$ terms retained | $n=0$ | | $n=1$ | |
|---|---|---|---|---|
| | $\Omega_p$ | $\gamma$ | $\Omega_p$ | $\gamma$ |
| None | 27 | 0.9 | 19 | 0.4 |
| Only $R_2$ | 30.9 | 5.5 | 21.9 | 2 |
| $R_1$ and $R_2$ | 33 | 6.8 | 23.4 | 2.3 |
| $R_2$ and $R_3$ | 26 | 0.3 | 20 | 2.2 |
| $R_2$ and $R_3'$ [a] | 29.9 | 2.7 | 22.4 | 2.4 |

[a] $R_3'$ is the modified form of $R_3$, as described in Appendix C.

In Table 2 we list the results of numerical integrations of Eq. (67) obtained for Model I (cf. Figs. 13, 14; also Appendix B) under a variety of conditions. The data support the statements made above. We shall discuss below only the effect of the term $R_2$ and refer the reader to the original papers for the effect of the other terms. In all cases studied, the quantum condition is well satisfied. (See [40] for details.)

*18. The effect of tangential forces*

Let us now examine the effect of the term $R_2$ in some detail. In the next section, it will be shown that this term is related to the effect of the tangential forces which have been neglected in Part III. We postpone the discussion of the resonance effect to Sec. 22.

First of all, let us note that $R_2$ is positive. Near the corotation circle ($\nu^2 \approx 0$), it actually dominates the term $K^2$, which is now

$$\hat{k}_3^2 = k_3^2 + R_2$$
$$= \left(\frac{\kappa}{a_0}\right)^2 \left[\frac{1}{Q^2} - 1 + \nu^2 + \tfrac{1}{4}\mathcal{J}^2 Q^2\right]. \tag{71}$$

We may also rewrite the equation in the form

$$\hat{k}_3^2 = \left(\frac{\kappa}{a_0}\right)^2 \left[\frac{1}{Q_e^2} - 1 + \nu^2\right], \tag{72}$$

where

$$Q_e^{-2} = Q^{-2} + \tfrac{1}{4}\mathcal{J}^2 Q^2. \tag{72a}$$

We may call $Q_e$ an equivalent $Q$-parameter. Clearly, $Q_e < 1$ if $Q = 1$ in the

MARK

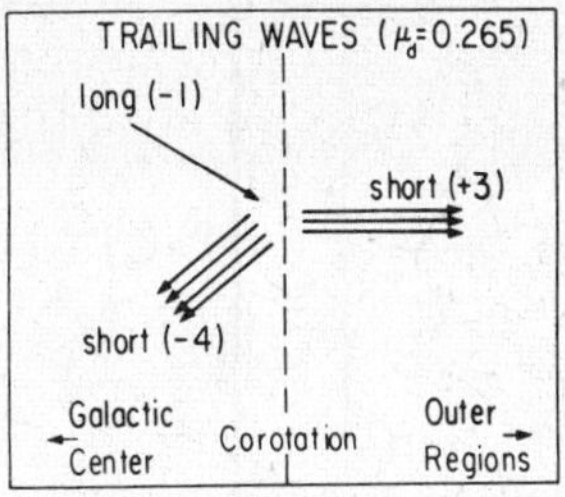


Figure 15. Wave amplification by stimulated emission (after Mark).

model. Thus, we might expect some kind of gravitational instability of *non*-axisymmetric disturbances.

A. *Remark on the numerical value of* $Q_e$. It is easy to verify that when $\mathcal{J}=1$ (which is a realistic value in many galactic models), the following table of values of $Q_e^{-2}$ holds:

| $Q^2$: | 1 | 2 | 4 |
|---|---|---|---|
| $Q_e^{-2}$: | 1.25 | 1 | 1.25 |

In all these cases, there is a wave (neutral or growing) according to the simple dispersion relationship. Indeed, the value of $Q_e$ is within 11% of unity. This may have a bearing on the success of the applications of the simple theory of spiral structure with $Q=1$ assumed (cf. [84]).

It is indeed remarkable that the general character of the solution near the corotation region remains unchanged even with such a serious modification of the local conditions. Actually, this can be understood by referring to the work of Mark [69] concerning conditions near the corotation region (cf. Fig. 15). The long trailing wave approaching the corotation region induces a much *stronger* response when $Q<1$ than the response in the case $Q=1$; but the general mechanisms remain the same. The stronger response is associated with the basic instability just discussed. The process shown in Fig. 15 was compared by Mark with the laser mechanism.

As far as the calculation of modes is concerned, the primary influence of the $R_2$ term can be seen from a comparison of the curve for $k_3^2$ (Fig. 11) with that for $\hat{k}_3^2$ (Fig. 16). The minimum at corotation is now lifted up to a positive value. There is free wave propagation in the region, but the group velocity for short trailing waves still changes sign across the corotation. In Fig. 16, we also show the eigenfunction of the $n=0$ mode of Model I. The contour plot of this function (Fig. 17) shows that the spiral wave appears to be dominated by a short wave.[6] This is also found to be the case by Lau and Haass [41], who applied the present theory to the Milky Way system.

[6]This perhaps is also the reason for the apparent success of Roberts, Roberts, and Shu [84], who used only the short wave branch in their mapping of 24 galaxies.

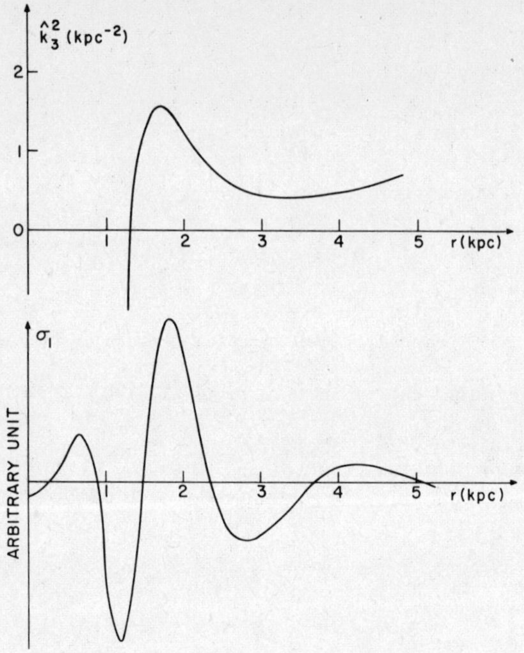


Figure 16. The lowest normal mode in Model I: (a) square of the wave number; (b) amplitude variation along one radius.

The pattern frequency and the growth rate are listed in Table 2. The effect of the term $R_2$ may be seen from a comparison of the first two rows of this table. Clearly, there is a substantial enhancement of the growth rate $\gamma$.

B. *The exact modal calculation of Pannatoni* [*79, 80*]. Here we report some recent work of exact calculations. Pannatoni [79] has recently devised a numerical scheme to calculate spiral modes as specified in the formulation in Sec. 10. The procedure is free from the restriction of asymptotics. It incorporates a radiation condition as one of the boundary conditions along the lines suggested in Secs. 14 and 22. For Model I, without taking into account the corotation resonance effects [i.e., upon dropping the singular term in Eq. (25b)], the $n=0$ mode has a pattern frequency $\Omega_p=29.8$ and a growth rate $\gamma=8.5$ in the exact theory. The contours of constant perturbation density of this mode is shown in Fig. 18. These numbers compare well with those given in the second row in Table 2, namely $\Omega_p=30.9$, $\gamma=5.5$. Note also the striking resemblance between the contour plots shown in Figs. 17 and 18. Keeping the corotation resonance term in the form given by the fluid dynamical theory, the pattern speed of the $n=0$ mode becomes $\Omega_p=27.7$ in the exact theory and $\Omega_p=26$ in the asymptotic theory (fourth row of Table 2); the corresponding growth rates are $\gamma=0.46$ and $\gamma=0.3$. There is thus close agreement in this case.

An outline of the numerical scheme and discussions on these results may be found in [80].

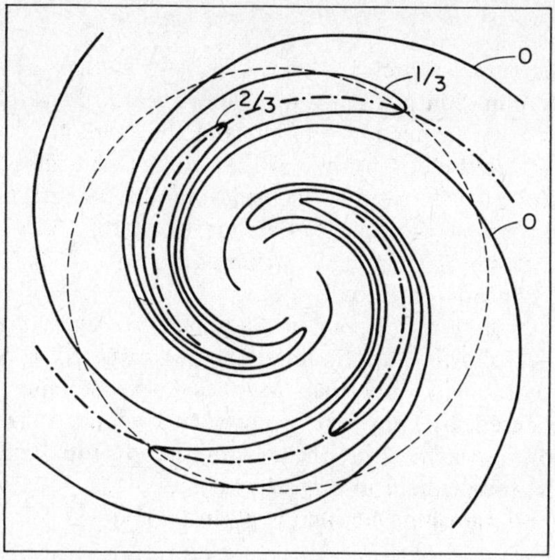


Figure 17. Density contour map for the $n=0$ mode of Model I based on the asymptotic theory. The numbers denote the fraction of maximum perturbation density on the contours. The dash-dot curves denote the constant phase contours of the short wave. The dashed circle is the corotation circle, at 2.7 kpc.

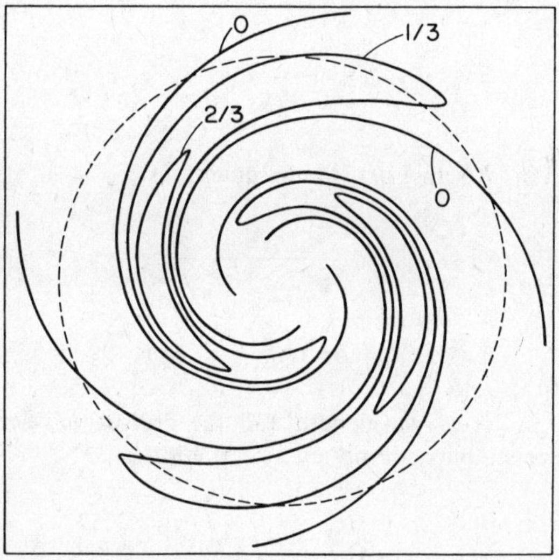


Figure 18. Density contour map for the $n=0$ mode of Model'I obtained from the exact theory. Same explanation and scale as in Fig. 17.

### *19. The local dispersion relation*

From the discussions in the previous section, it is clear that the parameter $\mathcal{J}$ is crucial to the determination of the growth rate of spiral modes. In this section, we shall present a local dispersion relation and the marginal stability diagram for spiral waves, where the importance of the stability parameters $Q$ and $\mathcal{J}$ will be explicitly illustrated. This study will bring out two important general characteristics of spiral waves: (1) open spirals are likely to possess a high rate of growth; and (2) a value of $Q$ larger than unity is required for stability against spiral disturbances, locally or globally.

The derivation of the local dispersion relationship is given in Appendix D. It is made under the following approximations: (a) wave fluid corotation resonance is ignored; and (b) only the "in phase" components in the density response are considered. By "in phase" component, we mean that component of the density response which is in phase with the potential minimum. These approximations can be justified in a *local* analysis.

The dispersion relationship obtained is given by [Eq. (D15)]

$$(\omega - m\Omega)^2 = \kappa^2 + \left(k^2 + \frac{m^2}{r^2}\right)a_0^2(1+\chi)$$

$$-2\pi G\sigma_0\left(k^2 + \frac{m^2}{r^2}\right)^{1/2}(1+\chi), \tag{73}$$

where

$$\chi = \mathcal{J}^2(\lambda/\lambda_c)^2,$$

and $\mathcal{J}$ is defined in (70). In Eq. (74), the quantities

$$\lambda = \frac{2\pi}{(k^2 + m^2/r^2)^{1/2}}, \tag{74a}$$

$$\lambda_c = 4\pi^2 G\sigma_0/\kappa^2 = 4\pi\delta \tag{74b}$$

are respectively the local wavelength and the critical wavelength defined by Toomre [101]. For stability, we obtain the criterion

$$Q^2 > 4\left[\frac{\lambda}{\lambda_c} - \frac{(\lambda/\lambda_c)^2}{1+\mathcal{J}^2(\lambda/\lambda_c)^2}\right], \tag{75}$$

by requiring the right hand side of Eq. (73) to be nonnegative. This marginal stability condition is presented graphically in Fig. 19.

The dispersion relationship (73) may be better understood by examining the cases with $m=0$ and $m\neq 0$. For $m=0$ (or even when $\mathcal{J}=0$), we have $\chi=0$. Equation (73) then reduces to Eq. (41a) [cf. Fig. 10]. Again, the terms involving

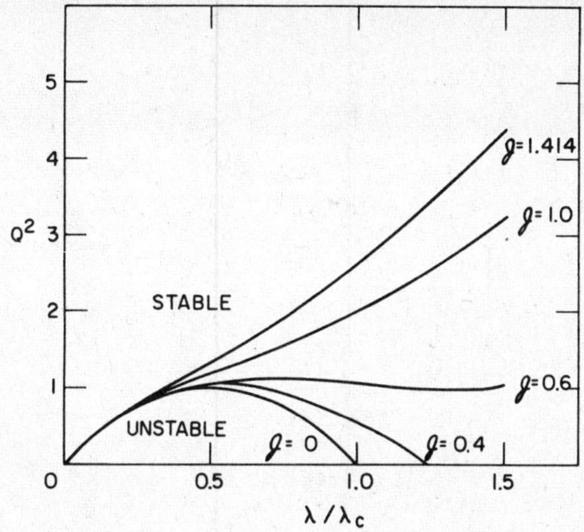


Figure 19. The marginal stability curves for various values of $\mathcal{J}$. The $\mathcal{J}=0$ curve corresponds to the axisymmetric case (cf. Fig. 10).

$a_0$ and $\kappa$ signify the stabilizing influences due to finite pressure and to the restraint imposed by the conservation of angular momentum. For stability, $Q>1$; this is the situation where these two stabilizing factors overcome the tendency toward Jeans collapse. The latter is represented by the last term in Eq. (73). For spiral waves, $m\neq 0$. Besides the obvious replacement of $k^2$ by $k^2+m^2/r^2$, there is an extra term in the last two terms of Eq. (73). This extra term is responsible for the enhanced growth rate discussed above. It has the following simple interpretation. The density response to a spiral force field (gravitational or pressure force) is greater than that to the corresponding axisymmetric field of the *same field strength* by a factor $1+\chi$. This enhancement in the response depends crucially on differential rotation (shear). This is in line with the conclusions of Goldreich and Lynden-Bell [20] and of Julian and Toomre [32], who observed that self-gravitating disks are very responsive to *spiral* disturbances.

Let us supplement the above description of the effect of tangential forces on the density response with graphical illustrations. In order to single out the role of the parameter $\mathcal{J}$, let us consider models in which the net effect of the corotation resonance is *absent* from the density change. One such model is the self-similar disk, where $\Omega\propto 1/r$, $\kappa\propto 1/r$, and $\sigma_0\propto 1/r$. Thus, in such a disk, we have, for Eq. (D12), $d(\kappa^2/\sigma_0\Omega)/dr=0$.

In Figs. 20 and 21, we present the linear, steady state response of test particles in the abovementioned self-similar disk to a spiral potential of the form $A\exp[i\omega t-i2\theta-i\alpha\ln r]$. We assume that in the absence of this spiral field the test particles are presented by three rings, each containing 36 particles which are equally spaced in angle. The unperturbed locations of the three rings correspond to $\nu=-0.67, -0.49, -0.3$ from the innermost to the outermost. The orbits are calculated from Eqs. (D10) and (D11). The response due *only* to the *radial* force

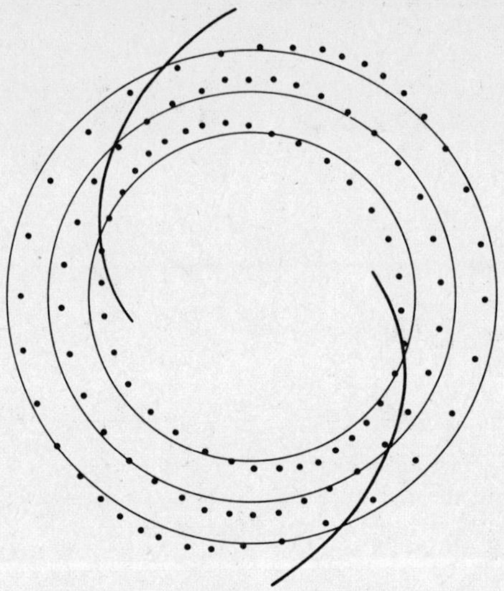

Figure 20. The response due only to the radial component of a spiral field. The strength of the radial force is equal to $0.1r\Omega^2$. Also shown is the spiral potential minimum (solid curves).

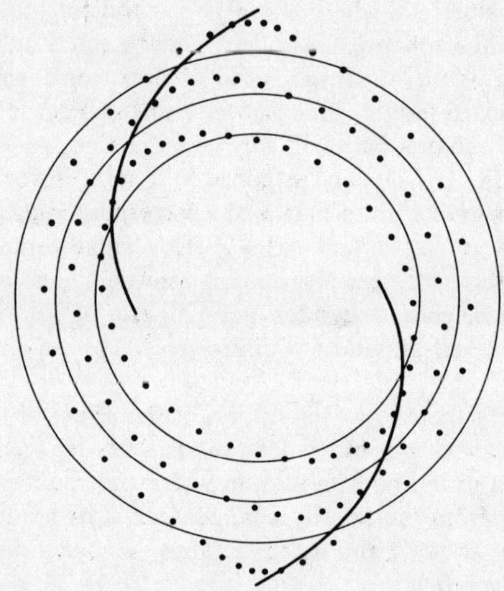

Figure 21. The response due to the combined (radial and tangential) spiral force of field strength $0.1r\Omega^2$. In both Figs. 20 and 21, the corotation resonance is absent [i.e., $d\ln(\kappa^2/\sigma_0\Omega)/d\ln r = 0$].

field is shown in Fig. 20, and that to the combined field (radial and tangential) in Fig. 21. In both figures, we set $\alpha = |kr| = 4$. The coefficient $A$ has been adjusted so that the *total* field strength used in the calculations equals to 10% of the equilibrium force field $r\Omega^2$ in both cases. It is readily seen that the density concentration along the spiral potential minimum in Fig. 21 is higher than that in Fig. 20.

Note that the extra clumping perceived in Fig. 21 is associated with $\mathcal{J}$. This conclusion is reached by examining the expressions for the density clumping resulting from the orbital displacements [cf. Eq. (D12) and the discussion following Eq. (D13)]. Indeed, the extra clumping due to tangential forces is estimated to be of the order of 40% of the total density clumping induced by the spiral field in these figures. Even if the density contrast between Figs. 20 and 21 does not appear as overwhelmingly impressive, we should stress that this extra clumping amounts to a moderate reduction in $Q$ [cf. Eq. (72a)]. Note that a slight decrease of $Q$ below unity (say from $Q=1$ to $Q=0.8$) could result in a substantial enhancement of the growth rates of modes. We have discussed this aspect already in the previous section.

The way in which this extra clumping factor $\chi$ enhances the local growth rate is also seen in the marginal stability diagram (Fig. 19). In this diagram, the curve $\mathcal{J} = 0$ corresponds to axisymmetric waves. When $\mathcal{J}$ is not zero, a higher value of $Q$ is required for stability of spiral waves. This is especially true when the waves are more open (larger $\lambda$ at fixed $m$). Naturally, if the tangential force is to have a destabilizing influence, as is commonly agreed, then the more open the spiral, the larger will be the relative influence of the tangential forces and hence the higher the growth rate. As pointed out to us by Toomre [103], this destabilizing influence is closely related to the one responsible for the transient growth of the shearing wavelets studied by Goldreich and Lynden-Bell [20]. See Appendix D and Ref. [40] for more details.

## *20. Recapitulation of the mathematical analysis*

We have developed an analytical theory for galaxies with an appreciable amount (e.g. 25%) of the total mass in the spheroidal component. The surface density $\sigma_0$ of the mobile disk component is such that the quantity $\delta$, the *reference scale of epicycles*, is small. More precisely, we have

$$\frac{\delta}{r} \ll 1, \qquad \text{where} \quad \delta = \frac{\pi G \sigma_0}{\kappa^2}. \tag{76}$$

We also restrict ourselves to cases where the actual epicycles have scales of the order of this reference scale; i.e., we want

$$Q = \frac{a_0}{\kappa\delta} \approx 1. \tag{77}$$

Under these circumstances, the *reference wave number* of the spiral waves, defined by

$$k_0 = Q^{-2}\delta^{-1}, \tag{78}$$

is large; i.e., $k_0 r \gg 1$. We also make the tentative conclusion that the actual wave number $|k|$ is large, i.e.,

$$|kr| \gg 1. \tag{79}$$

This leads to the formula (42), which states that $|k|$ is indeed of the order of $k_0$, and the original assumption (79) is confirmed. We have thus restricted ourselves to the cases where the pitch angle $i$ is small, since

$$|\tan i| = |m/kr|. \tag{80}$$

Actually, we need to require this pitch angle to be only moderately small; specifically, we should have

$$|\tan^2 i| = |m/kr|^2 \ll 1. \tag{81}$$

Experience shows that the pitch angle should be no larger than 30° for $m=2$ modes, if all the asymptotic relations are to work with sufficient accuracy.

A further requirement is that the parameter $\mathcal{J}$ defined in Eq. (70) should not be too large; i.e., we should have

$$\mathcal{J} \lesssim 1. \tag{82}$$

From the definition of $\mathcal{J}$, it is clear that this condition is relatively easy to fulfill where (76) is valid. In many cases, it is essential to restrict $\mathcal{J}$ to low values in the inner regions in order to ensure the existence of the inner turning point. In general, if the surface density is small and the shear is low in the central regions, then the parameter $\mathcal{J}$ will be rather low and the parameter $Q$ rather high, and the existence of the inner turning point is favored.

Even when the parameter $\mathcal{J}$ is fairly small, the outgoing and incoming waves should be viewed in a broad context: *the local dispersion relationship need not be accurately satisfied for the progressive waves* (cf. Fig. 12). We have also noted in several calculated examples that the combination of short and long waves yields an *impression* of a trailing spiral with a wave length comparable to that of the *short* branch. (See, e.g., Fig. 17; also [11].) It would not be surprising, then, if a spectral analysis should show a predominance of short waves. With still higher values of $\mathcal{J}$, the wave pattern (as determined by the calculations described in Sec. 17) must be regarded as an analytic continuation of the solutions for small $\mathcal{J}$. It may then be desirable to treat the problem by the use of the third order differential equation. Or, better yet, it should be treated with the complete integro-differential system, in the manner described by Pannatoni [79] and by Pannatoni and Lau [80].

When the value of $\mathcal{J}$ is still higher (e.g., when $\mathcal{J}^2 \simeq 2$, as in the models of Refs. [20, 32, 110]), the growth rate is also expected to be large, and the dispersion speed will probably continue to increase as a result of this instability, as if the system were attempting to restore stability (cf. Sec. 19). In such a case, it is unlikely that the original QSSS hypothesis will be satisfied. We shall return to a discussion of such galactic models in Sec. 23.

The asymptotic approach described here, which is based on (79), is to be carefully distinguished from the *local* approximation in which the limit $r \to \infty$ is taken. All lengths are then based on the scale $\delta$. This is the approach used by Goldreich and Lynden-Bell [20] and by Julian and Toomre [32], and by Toomre [102] in part of his paper. It is implicitly used by Lau and Bertin [40] in their paper, as described in Sec. 19. The two methods often lead to similar but not identical conclusions.

Finally, it should be mentioned that the asymptotic theory developed, especially in terms of the procedure of numerical integration in Sec. 18, does not logically depend on the interpretation described in the diagram of Fig. 12. It is still natural, however, to continue to use the interpretation described in the diagram even in the more complicated cases for the sake of a general appreciation of the mechanisms.

*Summary of the present studies.* Within the scope of the asymptotic theory described above, we have reached the following conclusions:

(1) The maintenance of the spiral pattern is essentially accomplished by a feedback mechanism conjectured before [50]. To this earlier concept, one should add the following two amplifying mechanisms: (a) the three wave interaction in the corotation region [69], and (b) the role of the tangential forces [43, 103, 40]. In the first mechanism, there is an outgoing wave from the corotation circle. This wave is to be absorbed at the outer Lindblad resonance. As a consequence, the mechanism can work only if the spiral structure is *trailing*. As a matter of fact, we do not need any leading waves at all for the mechanism to work. Indeed, the preliminary calculation without the restriction of asymptotics [80] appears to have corroborated these arguments, especially for galaxies with low surface density and a sizable fixed central structure.

A mode is usually a standing wave pattern formed by two wave systems propagating in opposite directions. This is also true in the present case. The unusual feature is that both of these wave systems are trailing and that they have *different* wave lengths. The angular momenta carried by these waves are also opposite in sign. The net effect of a trailing spiral mode is to transfer angular momentum to the outer parts of a galaxy.

(2) The calculations usually reveal the possible existence of a few unstable modes. The growth rate for each mode depends so sensitively on the conditions in the basic state that it might be meaningless to quibble over its exact theoretical value. Still, it is safe to conclude that in some cases one predominant mode stands out, while in other cases several modes may *coexist*. In the former cases, the stationary state might be reached when the growth is restricted by several possible effects such as (i) shock dissipation in the interstellar medium [35, 85], (ii) increase of the $Q$-parameter (velocity dispersion in the stars), and (iii) nonlinear terms neglected in the present analysis.

Perhaps it is appropriate to comment here on those cases where several modes of sufficiently high instability are present. The final pattern will be similar to that of *turbulent* flow in ordinary hydrodynamics. Such a flow is often dominated by a small number of interacting modes at large scales, with superposed motions at smaller scales. In turbulent jets, the larger scale motions can be relatively easily detected; in turbulent boundary layers over a flat plate or a convex surface, it is difficult to discern clear patterns. In spiral galaxies, we may also expect to find a variety of situations. For example, one may imagine the existence of a slowly evolving spiral pattern on a certain general length scale, as a result of the coexistence of two competing patterns on similar scales. At any given moment, there is a grand design which may be evolving at a rather slow rate on the order of the "beat frequency" of the two patterns.

The amplitude of a fully developed turbulent motion is determined by a rather complicated process of energy balance involving nonlinear processes and energy dissipation. It is difficult to trace this process in detail, but a general outline is known. In a galaxy, similar work should perhaps be pursued.

## Part V
## An assessment of the theory

### 21. Comparison with other approaches

The basic dynamical issues in the theory are outlined in Sec. 5. In the present paper, we give a coherent account of an attempt for their resolution. We have (1) adopted the *fluid model*, (2) given primary emphasis to the search for *unstable* modes, and (3) used an analytical approach based on an *asymptotic* approximation valid for relatively tightly wound spirals. There are alternative approaches to all three of these steps, and we shall examine some of the advantages and disadvantages of these choices.

A. *Choice of models*. Since we are dealing with a stellar system, the most natural thing to do is to adopt a *stellar model*. In this model, the stellar system is described by a combination of the Poisson equation and the collisionless form of the Boltzmann equation. This model is basic, and we have referred to it in order to supplement the imperfections in the *fluid model* used in this paper. We shall return to a discussion of these points in greater detail later on in this part.

Another model may be called the *particle model*, in which the motions of the individual stars are followed, and their collective gravitational field is then calculated by a self-consistent evaluation of the field. This has been done with the aid of high speed computing machines. It was initiated by P. O. Lindblad [62, 63] on a modest scale. More extensive calculations have been carried out by other investigators [8, 9, 23–26, 73, 74]. This approach gives very valuable information. In particular, it helps to clarify the following issues: (i) How does a stellar system *evolve* (into a spiral structure) when left to itself? (ii) Would the system settle down to a *quasistationary spiral structure*, and if so, under what conditions? (iii) Would the system develop a barlike structure, and if so, under what conditions? On the other hand, because of the limitations of accuracy in such "numerical experiments," it is difficult to obtain quantitative answers for specified galaxy models. Refinements of the "experimental" techniques appear to be highly desirable. We shall therefore not discuss them much further in this paper, but be content to refer to some of their conclusions as needed. Readers interested in further details should consult the original papers.

B. *Description of the evolutionary process as an initial value problem.* In the "numerical experiments," the evolutionary process is directly followed in the course of time. In the analytical approach, we have a choice. We may follow the evolutionary process either directly or by a superposition of normal modes. This is of course a general point in all studies of stability problems, and it is well known that the two procedures can be shown to be equivalent in the end. However, the two approaches can lead to very different *emphases* of the various dynamical processes. Indeed, the first approach tends to give the impression that the system (and the spiral structure in particular) will continue to evolve and does not settle down into a quasistationary mode, even if it actually does. It may also give the impression of great instability when no unstable modes exist, because the growth of the disturbances could be transient (cf. the classical problem of hydrodynamic stability cited in Sec. 5). The second approach tends to give the opposite impression. One must be reminded of the fact that several (perhaps as few as three) competing spiral modes of comparable growth rates may lead to a final state which is fairly transitory and quite irregular in appearance. The existence of unstable spiral modes must not be construed, by itself, to be a guarantee for an eventual attainment of a regular stationary spiral structure, even though it is a hopeful sign. This point will be elaborated below (Sec. 23).

C. *Use of asymptotic approximation.* One may choose to use a suitable asymptotic approximation or dispense with it. With either choice, the solution of the problem is best carried out by an analytical reduction of the problem followed by numerical calculation at a suitable stage. However, the adoption of the asymptotic approach was a crucial decision to be made. In general, such a step leads (i) to a better physical understanding and (ii) to a considerable simplification of the analytical effort, and hence (iii) to flexibility in application. Of course, there will also be some restriction and some sacrifice of accuracy.

The study of the fluid dynamical model without using asymptotic approximation has been carried out by Pannatoni [79, 80], using the approach formulated in this paper. We note that in the discussion of natural modes, the *radiation condition* at infinity is extremely important. Earlier, the study of spiral modes (or other modes) *without using an asymptotic approximation* was made by a number of researchers beginning with Hunter's work on flattened spheroids [28]. In this early work, the spheroidal coordinate system is the natural one to use, and others have followed this approach with some success [82, 6, 31]. However, the radiation condition presents a rather serious limitation on the use of the spheroidal coordinate system. It is rather difficult to impose. (Hunter and Toomre [30] also used the same coordinate system to study the warping of galactic disks either as a natural mode of oscillation or as a result of the passage of a satellite. The problem is three dimensional and beyond the scope of this article.)

D. *Stellar dynamical theories*. We now return to describe briefly the literature on the stellar dynamical approach. The modal problem was formulated by Kalnajs [34, 36] and by Shu [93] in slightly different forms. Shu pointed out that the presence of resonances would invalidate the "antispiral theorem," and Kalnajs [34] announced some preliminary results of modal calculations obtained by such an approach. The most recent modal calculation was made by Toomre and Zang (see [110])*, who studied a special one parameter set of galaxy models with a great degree of *self-similarity*. They are modifications of a rather peculiar galaxy model in which there is no linear scale in the distributions: the linear velocity or rotation is strictly constant, the velocity dispersion is also strictly constant, and the surface density is inversely proportional to the radial distance $r$. In the modified model, the surface density contains an extra factor $(1+r'^{-N})^{-1}$, where $N=1,2,\ldots$, and $r'$ is a dimensionless form of $r$ based on a length scale of the model. For this particular model, they can exhibit a stellar distribution function and carry through a calculation based on the formulation of Kalnajs. The total mass of the galaxy is obviously infinite; but this does not seem crucial, since the spiral waves were found to be absorbed at the outer Lindblad resonance. It is difficult, however, to see how one can extend this unusual feat to other galaxies with different rotation curves. Nevertheless, one can learn a great deal from these calculations, and the results can be used as a reference for evaluating other approximate theories.

In the asymptotic approach, the stellar dynamical theory of the spiral modes has been developed by Mark in a series of papers [66–71]. In these papers, he demonstrated the absorption of short waves at the Lindblad resonances [66] and clarified the mechanism of wave amplification by stimulated emission near the corotation region [69]. This latter phenomenon was also noted in a general manner by Toomre [102], and attributed by him to the "softness" of this region. Mark gave quantitative answers; he also clarified the corotation resonance phenomenon referred to above. Mark's work culminated in a paper by Bertin and Mark [12], whose major conclusions have also been reported earlier in the literature [11]. The stellar dynamical theory of Bertin and Mark can now be applied to the study of spiral modes in galaxies just as easily as the fluid dynamical theory. The results are also similar in most cases, but it tends to give higher growth rates (which appear to be in better agreement with the results from the Toomre-Zang theory in their special case).

Another asymptotic approach is to obtain moment equations from the stellar dynamical equations, primarily in cases where the epicycles are small. This was done by Hunter [29] and by Berman and Mark [10]. In the latter work, it is shown that the gas dynamical equations can be justified if the resonances are examined separately. Indeed, this work may be regarded as consolidating the basis of the present paper. Hunter's approach promised richer results and is being extensively studied by him (private communication).

Clearly, one of the urgent problems is to apply these theories to realistic galactic models. Comparison with observations may then be made.

---

*Note added in proof: See also A. J. Kalnajs in International Astronomical Union Symposium 77 (E. M. Berkhijsen, Ed.) Reidel, Dorchrecht, Holland, 1978, p. 113.

### 22. *Some dynamical issues: the resonances*

In the present linear theory, the resonances exhibit themselves mathematically as singularities in the governing equations of motions. As we have emphasized from the outset, it is their presence which invalidates the antispiral theorem. The latter denies the existence of a spiral structure in galaxies as a neutral mode. Yet paradoxically, the studies of spiral modes reported here do not seem to have taken these resonances explicitly into account. In what follows, we shall reexamine these issues, and provide a description of the role these resonances play in the maintenance of spiral structures and their proper modeling.

There are three resonances: the inner Lindblad resonance, where $\nu = -1$; the corotation resonance, where $\nu = 0$; and the outer Lindblad resonance, where $\nu = 1$. We shall consider them individually.

A. *The inner Lindblad resonance* ($\nu = -1$). The inner Lindblad resonance does not always exist. This statement holds especially for galaxies with small differential rotation. When this resonance does exist, it serves as a "mode selector." That is, it restricts the admissible values of $m, n$ (the azimuthal and radial mode numbers respectively) for a given galaxy model.

Let us recall [102, 66] the fate of an inward-propagating group of density waves of short wavelengths in a disk of *stars* as it reaches the inner Lindblad resonance. These short waves are totally absorbed. The transmission and reflection coefficients, if one chooses to speak in these terms, are practically zero. Now if a normal mode is to be maintained by a feedback process, it is preferable that this incoming short wave should not "see" the inner Lindblad resonance. The latter should be shielded from the incoming short waves by a rising $Q$ barrier.

Now consider the possible restriction on the values of $m$ allowable by the inner Lindblad resonance. If $m$ is large, $\nu$ changes rapidly. As a result, the corotation and the inner Lindblad resonance will be very close to each other (their spacing being roughly proportional to $1/m$). In these cases, the feedback process described earlier cannot operate, and no unstable normal mode can be formed.

Note that these higher $m$ waves could experience a high rate of growth in a transient manner. This is true because the stability parameter $\mathcal{J}$ is proportional to $m$, and is particularly true near the outer part of the galaxies, where differential rotation is expected to be high. The featherlike features often observed there may thus be possibly associated with these high $m$ wavelets, which are transient in time and are restricted in space to a rather small region in the radial direction.

Now let us consider the restriction on the total number of unstable radial modes (for a fixed $m$) by the inner Lindblad resonance. Higher $n$ modes would contain a greater number of wave crests within the corotation radius, and would thus in general require a corotation radius further away from the galactic center. As a result, the inner Lindblad resonances of these modes would also be placed at greater radii from the galactic center. In the event that these resonances lie outside the refracting $Q$-barrier, no unstable normal modes are expected to be formed, for reasons already explained. Thus, only the lower $n$ modes ($n = 0, 1, 2$) are expected to be unstable.

The above description applies to a disk of stars. For a disk of rotating fluid, the situation is more complicated. For one thing, the epicyclic motion of the individual stars cannot be accounted for in the equilibrium of any fluid model. Therefore, the latter would at best, if at all, provide a weaker effect of the Lindblad resonances than a disk of stars. This is indeed found to be the case. If one evaluates the fate of an inward-propagating short trailing wave as it approaches the inner Lindblad resonance in the *fluid* model, one finds that these waves are almost totally absorbed. Typically, the transmission coefficient is about 20% in terms of wave energy. More importantly, the reflection coefficient is practically zero. (Incidentally, resonant absorption of waves of a similar type has played a central role in the heating of a fusion plasma by microwaves, or by a laser. See Stix [97, p. 245]; a similar effect in a geophysical fluid is also noted by Bretherton and Booker [15].) As the short wave is transmitted through the inner Lindblad resonance, it may suffer a phase shift of 45°.

Even though the fluid model cannot provide a very accurate simulation of the inner Lindblad resonance, this shortcoming perhaps is not serious in numerical calculations of spiral modes. First, as described above, the inner Lindblad resonance indeed absorbs a major portion of the incoming short wave and does not provide any reflection, just as in the disk of stars. Secondly, the existence of unstable spiral modes demands the shielding of these resonances from the short wave by the $Q$-barrier. The net effect of this defect in the fluid model at the inner Lindblad resonance would therefore be considerably reduced.

B. *The outer Lindblad resonance* ($\nu = 1$). Similar to what is happening at the inner Lindblad resonance, short trailing waves are also perfectly absorbed at the outer Lindblad resonance in a disk of *stars*, i.e., the transmission and the reflection coefficient are both vanishingly small [66, 102]. Thus in stellar dynamical calculations of spiral modes, homogeneous boundary conditions can be imposed at the outer boundary of the (finite) disk as long as the latter lie *outside* the outer Lindblad resonance by a distance on the order of several epicyclic radii.

In the fluid model, the situation is quite different. Along the same line of reasoning as given above for the inner Lindblad resonance, one finds that, in the fluid model, the outgoing short trailing waves are only partially absorbed at the outer Lindblad resonance (though the reflection coefficient there is again zero). Typically, half (sometimes 70% or higher) of the wave energy is transmitted. This rather low level of wave absorption at the outer Lindblad resonance naturally presents a serious difficulty in imposing boundary conditions at the outer edge of a (finite) disk. If the amplitude is required to vanish at the outer edge, a substantial amount of short *leading* waves would be generated from the boundary. These incoming leading waves would further complicate the picture as they encountered the outer Lindblad resonance. If unstable modes were obtained through the use of this homogeneous boundary condition in the fluid theory, the growth mechanism of these modes would be something more than the outward transport of angular momentum, and the eigenfrequencies so obtained would be anticipated to be sensitive to the exact location where the boundary condition is imposed.

*Outward transport of angular momentum:* Carried along with the outgoing short trailing waves is an *outward transport* of angular momentum. Outward transport of angular momentum has long been thought of as the maintenance mechanism of spiral structure in disk galaxies (e.g. [65, 42, 71]). This angular momentum is readily absorbed at the outer Lindblad resonance in a stellar disk, or by dissipation in the turbulent motion of the gas in the outer part of galaxies. It ought to be emphasized here again that such absorption processes (either by resonance or by dissipation) work only for trailing spirals. For leading spirals, the only waves which have outward group velocity beyond the corotation circle are the long waves. The latter are not easily absorbed. This perhaps explains *why we observe only trailing spirals*.

One can think of two ways to simulate the perfect absorption of short trailing waves in a fluid model. One way is to introduce artificial viscosity in the outer region. It has been attempted by Pannatoni [79], but so far without much success. The other method is to introduce a radiation condition, allowing the short trailing wave to propagate to infinity and at the same time ruling out leading components from the outer region. This radiation condition destroys the self-adjointness of the system and therefore admits unstable spiral modes as a possible solution.[7] It forms the backbone of the present asymptotic theory. It has also been used recently by Pannatoni [79], who numerically solved the complete system of the Euler's equations of motion and the integral form of the Poisson equation. The eigenfrequencies obtained, either in the asymptotic theory or in the more exact treatment of Pannatoni, are not sensitive to the exact location at which the radiation condition is applied. Pannatoni's result agrees well with those obtained from the asymptotic analyses. (See also Lau and Bertin [40, Note Added in Proof]; also Pannatoni and Lau [80].)

C. *The corotation resonance* ($\nu=0$). At the corotation resonance, the fluid comoves with the wave pattern. The strength of the effect of this resonance is proportional to $D=d\ln(\kappa^2/\sigma_0\Omega)/d\ln r$. When this quantity vanishes, the governing equation is regular at $\nu=0$. When $D$ is positive (negative), the corotating fluid gives rise to an outward (inward) transport of angular momentum across the corotation circle. Thus, if $D$ is positive at the corotation radius, $r_{co}$, the growth rate of the mode is enhanced by the resonance effect, whereas if $D$ is negative at $r_{co}$, the growth rate could be reduced, as the growth of spiral modes is related to the outward transport of angular momentum. This indeed has been found to be the case from our numerical studies [40].

While the fluid model underestimates the effect of Lindblad resonances because of the lack of epicyclic motion, for the same reason it overestimates the effect of corotation resonance in comparison with a stellar disk model. In the equilibrium of a fluid disk, it has been assumed that all fluid elements are moving concentrically, and corotating resonance effects would be very pronounced, as such models would not allow for the smearing effects associated with the epicyclic motion of stars. It turns out that the latter effects can be incorporated readily by replacing the $1/\nu$ factor in the governing equation by a "plasma dispersion function," which in the limit of zero velocity dispersion reduces back to $1/\nu$. (See [40] and also [68] for details.) Such differences in the corotation resonance have also been noted in an earlier study of the generating mechanism of spiral waves by a bar. Specifically, it is found that a disk of fluid is more responsive to a bar than a disk of stars in the excitation of spiral waves near the region of corotation.

[7]The use of a radiation condition to simulate perfect absorption at resonances has been widely used in plasma physics. (See, e.g., [7, 81, 38, 5, 105, 87].)

The above discussions pertain to the linear theory. Let us comment briefly on the nonlinear aspects of these resonances. It is generally believed that the corotation resonance is easily saturated by trapping of orbits over a time scale of the order of one galactic year [17, 37, 10]. For the Lindblad resonances, opinions diverge as to whether they would saturate as readily. The answer, one way or the other, does not seem to be as crucial in the case of the outer Lindblad resonance, for short trailing waves can always be absorbed by the turbulence in gas, and the maintenance mechanism of spiral structure remains basically unchanged. With regard to the inner Lindblad resonance, it would be very desirable to have a more definitive answer. As mentioned before, in certain cases the inner Lindblad resonance can act as a "mode selector." At the same time, it is more accessible to observations. A further understanding of the resonance phenomenon may help to resolve a number of mysteries (e.g., 3 kpc arm, ring structure).

### *23. Physical implications: dynamical basis for the classification of galaxies*

A. *Application to specific galaxies*. Attempts have been made to reproduce the spiral patterns of several specific galaxies. Lau and Haass [41] computed the spiral pattern for our own galaxy as a normal mode. The pattern frequency obtained is approximately the same as that found before (see [60]). In this calculation, it has been found important to attribute the inner part of the mass distribution to the immobile spheroidal component, and to consider the effect of the finite thickness of the galactic disk [92, 106].

Preliminary results are also available for the galaxies M31, M101, and M81. The first two galaxies are approximately modeled in the paper by Bertin et al. [11]. In these calculations, several unstable modes are usually found. This is welcome. Indeed, the concept of multiple modes may already be forcing itself upon us through the observations in M31 [91]. In this case, a definite asymmetry is noted in the flow pattern. Superficially, there appears to be a one armed spiral pattern of gaseous motion superposed on the two armed optical pattern. Detailed investigations have not yet been carried out. On the other hand, in the case of M81, the data seem to suggest the predominance of a single pattern [107]. In many galaxies, such as our own Milky Way system, there appears to be a predominant pattern; but there also appear to be coexisting spiral features of other kinds. Indeed, Roberts, Roberts, and Shu [84] classified 24 galaxies according to a predominant pattern identified in each. The pattern frequency was determined essentially by locating the corotation circle near the tip of the prominent optical spiral arms.

From the dynamical point of view, it would be interesting to see whether, in a given galactic model, the pattern thus determined through observational data can be identified with any of the *modes*. If so, is it the mode with $n=0$, or $n=1$? Is it some other mode? Why? These questions can probably be answered in some cases even though there are uncertainties in modeling spiral structure.

The galaxy M51 is obviously complicated by the presence of the satellite,[8] but

[8]It is unlikely that the satellite is responsible for producing the spiral structure. Toomre [104] stated: "...I must here repeat sadly that such hopes seem to be fading." See also Lin's reply to Schatzman in the discussion at the end of the paper by Lin and Yuan [60].

even here it appears possible to identify the dominant mode. The spiral structure of the galaxy M33 could be a complicated superposition of two (or three) modes. It would be interesting to sort that out, if possible. This galaxy has a rather small nucleus and is therefore possibly akin to the barred spirals to be discussed in the next section.

B. *Barred spirals and bar formation in normal spirals.* We have so far focused our attention on the discussion of models of disk galaxies with a sizable nuclear bulge (e.g., 25%) and a relatively small surface density in the disk mass. We believe these are essentially the models for the normal spirals with fairly tight winding (e.g., with pitch angle less than 30°). What are the appropriate dynamical models for the other types of galaxies: the S0's and the barred spirals? Can one develop the theory in a systematic manner so that we may classify the galaxies on the basis of their basic mass distribution? Our discussions below are of course conjectural; they are, however, based on our experience from the present analysis and from other sources, e.g., the "numerical experiments." Suggestions for further investigations, both theoretical and observational, will be made. Let us now consider barred spirals.

There are at least two possible types of barred spirals. In one type, the galaxy may have a substantial amount of mass in a barlike (or triaxial ellipsoidal) distribution. In another type, the galaxy has a basically axisymmetric distribution, but develops a barlike instability in the central regions, possibly to a small extent similar to that of density waves in normal spirals. We shall focus our attention on the latter. Clearly, this is the type which forms a smooth transition into normal spirals in the classification scheme proposed by de Vaucouleurs.

The development of barlike instability in an originally axisymmetric system has been studied in the "numerical experiments," as already mentioned above. Indeed, in many such numerical experiments, the "normal" spiral structure is eventually wiped out because the velocity dispersion in the stars tends to increase rapidly under the influence of the barlike potential field. (The disk becomes "very hot.")

Analytical studies of open spiral modes in a *differentially* rotating system are obviously more difficult, and few conclusive results have been obtained so far. It is generally believed that, if the system does not contain sufficient mass in the spheroidal system, the mass distribution in the central regions will have a strong tendency to develop a barlike distribution similar to those in uniformly rotating disks (see, e.g., [28, 36]). In such a differentially rotating disk, this barlike configuration is presumably rotating at a slower rate than the material there. We have a pattern with a corotation circle somewhere in the outer parts of the galaxy. The calculation of such open modes in the linear theory is one of the urgent problems we face. Such a calculation has just been shown to be possible through the careful analysis of Pannatoni [79].

The barlike instability in the central regions is so strong that it can be avoided only if there is little mass associated with the disk distribution. This is not unreasonable, since the nuclear bulge can take up most of the mass, leaving very little in the "disk." Indeed, it appears reasonable that a spiral galaxy preferentially belongs to the "normal" type if the mass distribution of the disk component has a "hole" in the central regions. Otherwise, there is expected to be at least a superposed barlike structure even if this structure is not sufficiently strong to wipe out the Jeans instability in the disk. The galaxy will then contain a superposition of both instability modes and perhaps falls into the type SAB in the de Vaucouleurs classification.

If the disk mass is substantial in the central regions (e.g., a Gaussian distribution), and if the differential rotation is not very strong, there will be a rather strong barlike distortion which could in turn induce a very strong spiral structure in the differentially rotating disk. This spiral structure would cause the disk to become very hot, and the galaxy would belong to the type SB.

The above ideas possibly contain the germ of a dynamical classification scheme that corresponds to de Vaucouleurs's classification of galaxies into SA, SAB, SB. An idea for the dynamical classification of galaxies into SAa, SAb, SAc has already been presented by Roberts, Roberts and Shu [84].

The limitation of the growth of the amplitude of such wave patterns is perhaps through the same three possible mechanisms discussed above: (i) natural limitation due to nonlinear terms, (ii) shock formation in the gas, and (iii) increase of velocity dispersion in the stars under the influence of the open spiral mode. Flows with shocks have been calculated by several research workers [88, 45], but the "damping" effect may be estimated in a manner similar to those available in normal spirals. A complete understanding of (i) and (iii) must first await the solution of the linear problem. Much work remains to be done.

*Unrelated spiral structures:* From the photograph of NGC 5364, it appears that the small spiral in the central regions is unrelated to the principal normal spiral structure in the main part of the optically prominent disk. Thus, it appears that there are, in this galaxy, *two distinct classes of instability modes* which are not necessarily related to each other (at least not intimately).

Why is the normal spiral structure in NGC 5364 not wiped out? The obvious answer is that the barlike structure in the central regions is underdeveloped. But why is this so?

A clue to the answer may be found from the side view of our own galaxy, and from a picture of NGC 5364. The central part of such galaxies appears to be similar to an elliptical galaxy E7. A disk structure exists only in the outer parts. Thus, the disk component has a mass distribution similar to that shown in Fig. 7; the paucity of disk mass in the central regions implies only a weak barlike instability. The spheroidal component, being very "hot," has relatively small angular momentum and does not undergo any substantial oval distortion. In this case, the only major instability is that associated with the "normal" spiral instability such as that studied in this paper. (However, the distribution of mass in the central regions might be a little more complicated. For example, there is apparently a mass concentration on a rather small scale near the center of NGC 5364. This mass concentration apparently has an oval distortion and drives, in the distorted state, a small spiral structure largely restricted to the interior of the ring.)

C. S0 *galaxies*. The most striking feature of S0 galaxies is their lack of a spiral structure, even though observers find them to be similar to Sa galaxies in many respects. There is, however, also the conspicuous lack of gas, which suggests a resemblance to the ellipticals. This has often been mentioned as a possible reason for the lack of observed spiral structure, because the lack of gas dictates a lack of newly formed brilliant young stars. In this approach, the question of the actual presence of a spiral gravitational field is sidestepped.

On the other hand, there are indeed two dynamical considerations which suggest the actual lack of a spiral mode in S0 galaxies: (i) the excessive thickness of the galactic disk, and (ii) the large velocity dispersion in the stars. Both are natural if S0's are indeed to be linked with the flat elliptical system E7.

(i) *Effect of thickness of the galactic disk*. Shu [92] and Vandervoort [106] have shown that the finite thickness of the galactic disk reduces Jeans instability. Thus, even if the mass distribution in an S0 galaxy is very similar to that in an Sa galaxy, this single modification could make a decisive difference in the instability of spiral modes.

(ii) *Effect of large velocity dispersion*. Elliptical galaxies were found in recent observations to have rather small rotational velocities. One may thus infer that the system is partially supported by "pressure," and that there is considerable velocity dispersion. This property is "passed on" to S0's. Indeed, the large thickness of the galactic disk is also consistent with this trend. We would therefore again expect it to be difficult for the spiral structure to develop in the disk.

D. *Smooth arm spiral galaxies*. One of the most exciting observational discoveries in recent years is the direct observation of stellar density variations of the type envisaged in the density wave theory [89, 90, 99]. In the smooth arm spiral galaxies observed by Strom and his collaborators [109], they found density waves of rather large amplitudes. They believed that these systems have been stripped of gas, perhaps through galactic encounters. "The results of a preliminary study of the 25 galaxy sample suggest that the spiral waves of largest amplitude are found in disks having the bluest colors, as might be expected if these systems were stripped more recently. In systems with weak arm amplitudes, the disk colors appear reddest, a result consistent with the belief that systems stripped relatively long ago should show lower arm amplitudes." (See [98].) They also conjectured that these galaxies are perhaps in the initial stages in the transmutation of a spiral into an S0 galaxy. They suggested the determination of gas content as an observational test of their conjecture.

All these observational results and conjectures are consistent with dynamic evolution, if there is a continual increase of velocity dispersion in the stars after the gas is stripped. The stripping of the gas has to be followed initially by an *increase* of the amplitude of the spiral density waves. But it is precisely this increase (beyond some conceivable critical value) which may lead to an explosive increase of stellar velocity dispersion, which eventually wipes out the spiral structure and leaves the galaxy in the S0 stage.

These conjectures are very tentative; very careful theoretical treatments should follow. Again, the study of the three mechanisms of stabilization mentioned earlier in this section must be examined.

E. *Concluding remarks*. Over the past fifteen years, we have come a long way towards the understanding of density wave patterns in spiral galaxies. Especially well understood are galactic spirals with small pitch angles. But there are still large areas of challenging theoretical investigations to be made and many rewarding observational studies suggested. Our total knowledge of the structure and evolution of galaxies remains to be cultivated.

**Acknowledgments**

The authors wish to thank Drs. G. Bertin, James Mark, R. F. Pannatoni, Frank Shu, and Alar Toomre for many helpful discussions during the period when much of the material in this article was developed. They wish also to thank Frank Shu for a careful reading of an earlier version of the article and for his helpful comments. The support of the National Science Foundation is gratefully acknowledged.

## Appendices

### *Appendix A*
### *Asymptotic solution of the Poisson equation*[9]

The Poisson equation connecting the potential $\psi_1(r,z)e^{i(\omega t - m\theta)}$ with the surface density $\sigma_1(r)e^{i(\omega t - m\theta)}$ is

$$\left(\frac{\partial^2}{\partial r^2} + \frac{1}{r}\frac{\partial}{\partial r} + \frac{\partial^2}{\partial z^2} - \frac{m^2}{r^2}\right)\psi_1 = 4\pi G\sigma_1\delta(z). \tag{A1}$$

Integration of (A1) across the galactic plane gives

$$\sigma_1(r) = \frac{1}{4\pi G}\frac{\partial}{\partial z}\psi_1(r,z)\bigg|_{z=0_-}^{z=0_+} \tag{A2}$$

From the reflection symmetry of the problem, $\psi_1(r,z)$ must depend only on the absolute value of $z$. We shall therefore focus our attention on the *half space* $z>0$.

To accommodate cylindrical symmetry, we introduce

$$W = r^{1/2}\psi_1. \tag{A3}$$

In terms of $W$, Eqs. (A1) and (A2) become

$$\frac{\partial^2 W}{\partial r^2} + \frac{\partial^2 W}{\partial z^2} - \frac{m^2 - \frac{1}{4}}{r^2}W = 0 \tag{A4}$$

and

$$\sigma_1(r) = \frac{r^{-1/2}}{2\pi G}\left\{\frac{\partial W}{\partial z}\right\}_{z=0_+} \tag{A5}$$

If we now refer to Eq. (32), we see that the last term in (A4) is of the order

[9]After Shu [94].

$(kr)^{-2}$ compared with the first term. We neglect this term and get

$$\frac{\partial^2 W}{\partial r^2}+\frac{\partial^2 W}{\partial z^2}=0 \tag{A6}$$

which is to be solved under the boundary condition $W\to 0$ as $z\to\infty$, and

$$W=r^{1/2}\hat{\psi}_1(r)e^{i\Phi(r)} \qquad \text{at} \quad z=0 \tag{A7}$$

where $\Phi(r)=\int k\,dr$ [cf. Eq. (29)]. The solution is clearly

$$W=(r\pm iz)^{1/2}\hat{\psi}_1(r\pm iz)e^{i\Phi(r\pm iz)}. \tag{A8}$$

The choice of sign must be made according to the sign of $k$ so that $W$ vanishes as $z\to\infty$. By substituting (A8) into (A5), we get

$$\sigma_1(r)=-\frac{|k|\psi_1(r)}{2\pi G}\left\{1-\frac{i}{kr}\frac{d\ln\left[r^{1/2}\hat{\psi}_1(r)\right]}{d\ln r}\right\}. \tag{A9}$$

By using (29), this may be written into the form (33). A more refined form of (33) was obtained by Bertin and Mark [13], accurate to two orders of $(kr)^{-1}$. It reads

$$\frac{d\psi_1}{dr}+\frac{\psi_1}{2r}-\frac{m^2-\frac{1}{4}}{i\Sigma r^2}\psi_1=-i\Sigma h_1\,\mathrm{sgn}(k). \tag{A10}$$

## *Appendix B*
## *The Models*

The two galaxy models used in this paper are based on the one used by Mark [71]. Both have a rigid, spherical component of mass $M_s=5.4\times 10^9 M_\odot$. The total mass within radius $r$ of this spherical component is

$$M_0(r)=M_s\frac{(r/a_s)^3}{\left[1+(r/a_s)^2\right]^{3/2}}$$

$$a_s=1.75\ \text{kpc}. \tag{B1}$$

The active thin disk is constructed out of two Toomre disks of index $n=5$. Let $a_1=12$ kpc, $a_2=8$ kpc, and $x_j=[1+(r/a_j)^2]^{1/2}$, $j=1,2$. The surface density $\sigma_0(r)$ of the thin disk is given by

$$\sigma_0(r)=\frac{4.5M_1}{\pi a_1^2}\frac{1}{x_1^{11}}-\frac{4.5M_2}{\pi a_2^2}\frac{1}{x_2^{11}}. \tag{B2}$$

For Model I,

$$M_1=2\times 10^{10}M_\odot, \qquad M_2=5\times 10^9 M_\odot.$$

For Model II,

$$M_1 = 2\times 10^{10}M_\odot, \qquad M_2 = 8\times 10^9 M_\odot.$$

The total mass $M_d$ of the active disk equals $M_1 - M_2$.

The rotation curve can be obtained from the formula

$$\Omega^2 = \frac{16}{105}\left[\frac{M_1 G}{a_1^3}H(x_1) - \frac{M_2 G}{a_2^3}H(x_2)\right] + \frac{M_s G}{a_s^3}\frac{1}{\left[1+\left(\frac{r}{a_s}\right)^2\right]^{3/2}}, \tag{B3}$$

where

$$H(x_j) = \frac{59.0625}{x_j^{11}} + \frac{26.25}{x_j^9} + \frac{16.875}{x_j^7} + \frac{11.25}{x_j^5} + \frac{6.5625}{x_j^3}.$$

In this paper, we assume

$$Q(r) = 1 + 0.75e^{-r^2}, \tag{B4}$$

where $r$ is in kiloparsecs.

## *Appendix C*
## *Derivation of the governing equation (67)*

The linearized Euler's equations of motion give Eq. (23):

$$Ch_1 = C\frac{\sigma_1}{\sigma_0}a_0^2 = -\mathfrak{L}(h_1+\psi_1) \tag{C1}$$

where $C$ and $\mathfrak{L}$ have been defined in Eq. (24) and (25). This equation, written as it appears above, expresses the density response $\sigma_1$ to some imposed potential $\phi_1 = h_1 + \psi_1$. The derivation and interpretation of this equation are postponed to Appendix D.

Upon substituting Eq. (66) into (C1), we obtain, to two orders in $(kr)^{-1}$,

$$h_1'' + (A+i\Sigma)h_1' + \left[C+B-i\Sigma\left(\frac{1}{2r} - A - \frac{\Sigma'}{\Sigma}\right) - \frac{\left(m^2-\frac{1}{4}\right)}{r^2}\right]h_1$$

$$= -\left[-\frac{A}{2r}+B+\frac{3}{4r^2}\right]\psi_1$$

$$\simeq +\left[-\frac{A}{2r}+B+\frac{3}{4r^2}\right]2h_1, \tag{C2}$$

where in the right hand side we have approximated $\psi_1 \simeq -\Sigma h_1/|k| \simeq -2h_1$. For

$m=3,4$, the order $m^2=O(kr)$ would be more appropriate and the result would be qualitatively the same. See [12].

We introduce the standard transformation of variables

$$h_1 = u\sqrt{\frac{\kappa^2(1-\nu^2)}{\sigma_0 r}}\; e^{-i\int^r(\Sigma/2)dr} \tag{C3}$$

to write the last equation as

$$\frac{d^2u}{dr^2} + K^2u = 0, \tag{C4}$$

$$K^2 = k_3^2 + R, \tag{C5}$$

$$k_3^2 = \left(\frac{\kappa}{a_0}\right)^2\left(\frac{1}{Q^2} - 1 + \nu^2\right), \tag{C6}$$

$$R = \sum_{i=1}^{4} R_i, \tag{C7}$$

$$R_1 = -i\frac{\Sigma}{2}\frac{d}{dr}\ln\left[Q^2(1-\nu^2)\right], \tag{C8}$$

$$R_2 = \left(\frac{\kappa}{a_0}\right)^2 \tfrac{1}{4}\mathcal{J}^2 Q^2, \tag{C9}$$

$$R_3 = -\frac{2\Omega m}{r^2\kappa\nu}\,\frac{d\ln(\kappa^2/\sigma_0\Omega)}{d\ln r}, \tag{C10}$$

$$R_4 = \frac{3}{4r^2} - \sqrt{\frac{r\kappa^2(1-\nu^2)}{\sigma_0}}\,\frac{d^2}{dr^2}\sqrt{\frac{\sigma_0}{r\kappa^2(1-\nu^2)}} - \frac{4m\left(\frac{\Omega}{\kappa}\right)\nu\left(m\nu r\frac{\Omega'}{\kappa} + \frac{r\kappa'}{\kappa}\right)}{r^2(1-\nu^2)}. \tag{C11}$$

In these equations $k_3^2$ already appears in the rudimentary theory. The other terms have the following interpretation: $R_1$ represents the "out of phase" component of the density response. It appears to modify only slightly the phase relationship and the growth rate, but vanishes if $Q$=constant and $\nu=0$. The term $R_2$ is the all-important quantity which includes the parameter $\mathcal{J}$ defined in Eq. (70) in the main text. Its role is discussed extensively in Secs. 17–19. The next quantity $R_3$ represents corotation resonance. To account for the smearing effects associated with the epicyclic motions of stars as discussed in Sec. 22, we should replace the singular factor $1/\nu$ in (C10) by $-b\bar{Z}(b\nu)$, where

$$\bar{Z}(\zeta) = -2ie^{-\zeta^2}\int_{-\infty}^{-i\zeta} dt\, e^{-t^2} \tag{C12}$$

is the well-known "plasma dispersion function" and $b$ is given by

$$b = \frac{\kappa}{\sqrt{2}\, \nu' a_0}. \tag{C13}$$

Note that in the limit of vanishingly small acoustic speed ($a_0 \to 0$), $-b\bar{Z}(b\nu) \to 1/\nu$. With such a replacement, $R_3$ is denoted by $R_3'$. This way of accounting for the corotation resonance is basically a simplified version of that introduced by Mark [69]. The remaining term $R_4$ is small and is thus ignored. [Actually, this remaining term includes a contribution proportional to $(\Omega')^2$, which has a stabilizing influence in the fluid case. The corresponding term in the stellar dynamic approach is destabilizing (see [12]).]

## *Appendix D*
## *Derivation of the local dispersion relation*

In this Appendix, we derive the local dispersion relation (73) for spiral waves. As an attempt to gain a better understanding of the motion of a fluid element under the influence of a spiral field, we use a Lagrangian description to calculate the density response to a spiral field. The result is just Eq. (C1).

The instantaneous coordinates $(r,\theta)$ of a fluid element is governed by

$$\ddot{r} - r\dot{\theta}^2 = -\phi_r(r,\theta,t), \tag{D1}$$

$$\frac{d}{dt}(r^2\dot{\theta}) = -\phi_\theta(r,\theta,t), \tag{D2}$$

where $\phi$ is the total potential (combining the gravitational potential $\psi$ and enthalpy $h$) and the subscripts $(r,\theta)$ denote partial derivatives. In the unperturbed state, $r = r_0 =$ constant, and $\theta = \theta_0 = \Omega t$, where $r_0\Omega^2(r_0) = \phi_0'(r_0)$, with $\phi_0$ being the total unperturbed potential and a prime denoting differentiation with respect to $r_0$. In the presence of a small perturbation $\phi_1 = \phi_1(r_0,\theta_0,t)$, we have

$$r = r_0 + r_1(r_0,\theta_0,t), \tag{D3}$$

$$\theta = \theta_0 + \theta_1(r_0,\theta_0,t). \tag{D4}$$

To first order, the radial and azimuthal perturbations $r_1$ and $r_0\theta_1$ are governed by

$$\ddot{r}_1 + 2r_0\Omega\Omega' r_1 - 2r_0\Omega\dot{\theta}_1 = -\phi_{1r}(r_0,\theta_0,t), \tag{D5}$$

$$r_0^2\ddot{\theta}_1 + 2r_0\Omega_0\dot{r}_1 = -\phi_{1\theta}(r_0,\theta_0,t). \tag{D6}$$

We have considered a stationary spiral potential

$$\phi_1(r_0,\theta_0,t) = \Phi_1(r_0)\exp[\,i\omega t - im\theta_0\,]. \tag{D7}$$

This potential produces a force field $\mathbf{F} = \hat{r}F_r + \hat{\theta}F_\theta = -\hat{r}\,\partial\Phi_1/\partial r + \hat{\theta}\, im\Phi_1/r$.

We now obtain an equation for $r_1$ by differentiating Eq. (D5) with respect to time and using Eq. (D6) to eliminate $\ddot{\theta}_1$:

$$\ddot{r}_1 + \kappa^2 \dot{r}_1 = 2\Omega F_\theta + \dot{F}_r$$

$$= \left[ \frac{2\Omega im}{r_0} \Phi_1(r_0) + i(\omega - m\Omega) \frac{d\Phi_1(r_0)}{dr_0} \right] e^{i(\omega - m\Omega)t}. \tag{D8}$$

It is clear from this equation that the desired solution for $\dot{r}_1$ is simply the steady state response to a sinusoidal force of frequency $\omega - m\Omega(r_0)$, with intrinsic spring constant $\kappa^2(r_0)$. Thus $\dot{r}_1$, and hence $r_1$ and $\theta_1$, would in general be functions $r_0$, since the spring constant $\kappa$, the driving frequency $\omega$-$m\Omega(r_0)$, and the amplitude $\Phi_1(r_0)$ are radius dependent. The (Eulerian) density perturbation $\sigma_1$, given by

$$\frac{\sigma_1}{\sigma_0} = -\frac{\partial r_1}{\partial r_0} + im\theta_1 - r_1 \frac{d}{dr_0} \ln(r_0\sigma_0), \tag{D9}$$

can be evaluated after we solve for $r_1$ and $\theta_1$ from (D8) and (D6). We obtain

$$r_1 = \frac{F_r}{\kappa^2 - (\omega - m\Omega)^2} - \frac{2i\Omega F_\theta}{(\omega - m\Omega)[\kappa^2 - (\omega - m\Omega)^2]}, \tag{D10}$$

$$\theta_1 = \frac{2i\Omega F_r}{r(\omega - m\Omega)[\kappa^2 - (\omega - m\Omega)^2]} + \frac{F_\theta[(\omega - m\Omega)^2 - 2r\Omega\Omega']}{r(\omega - m\Omega)^2[\kappa^2 - (\omega - m\Omega)^2]}, \tag{D11}$$

where the subscript of $r_0$ has been dropped, as we are now considering linearized quantities. Upon substituting (D10) and (D11) into (D9), we obtain (C1).

Let us now make a local analysis of Eq. (C1). Then $d/dr$ is replaced by $ik$ and (C1) reads

$$\frac{\sigma_1}{\sigma_0} = \frac{h_1 + \psi_1}{\kappa^2 - (\omega - m\Omega)^2} \left\{ -k^2 - \frac{m^2}{r^2} - T_1 + ikA - \frac{2\Omega m \, d\ln(\kappa^2/\sigma_0\Omega)}{r^2(\omega - m\Omega)\, d\ln r} \right\}, \tag{D12}$$

where

$$T_1 = -\left( \frac{2m\Omega}{\kappa r} \right)^2 \frac{d\ln\Omega}{d\ln r}.$$

The last term in (D12) corresponds to wave-fluid resonance whose role is fairly well understood and will be ignored henceforth. The second to last term represents the density response which is 90° out of phase with the potential minimum. This term will also be ignored in the present *local* analysis.

The dropping of the "out of phase" terms is presumably valid in the consideration of the growth of waves. In particular, we refer to Li et al. [44].

These authors have shown, in their modal analysis, that the "out of phase" terms do not affect the growth rates of spiral modes. The high rate of growth observed in [20] appears to be attributed to the "in phase" components also. We note that this "out of phase" component vanishes identically when $Q$ is constant and $\nu=0$. Keeping only the first three terms in (D12), we write

$$\left(\frac{\sigma_1}{\sigma_0}\right)_{\text{in phase}} = \frac{h_1+\psi_1}{\kappa^2-(\omega-m\Omega)^2}\left[-k^2-\frac{m^2}{r^2}-T_1\right]. \tag{D13}$$

Thus, for a given field strength $\sqrt{k^2+m^2/r^2}\,(h_1+\psi_1)$, the density response is higher for $m\neq 0$, the extra amount being proportional to $T_1$ (or $\mathcal{J}^2$). Furthermore, it is clear that the ratio $T_1/(k^2+m^2/r^2)$, which is a measure of the extra clumping associated with the azimuthal force, is largest when $k=0$ (for fixed $m$). This perhaps is an interpretation of the high growth noted by Goldreich and Lynden-Bell [20] as their wavelets become open.

For the Poisson equation, again keeping only the "in phase" components, we have

$$-\sqrt{k^2+m^2/r^2}\,\psi_1 = 2\pi G\sigma_1. \tag{D14}$$

Combining (D13) and (D14), we obtasin the local dispersion relation

$$(\omega-m\Omega)^2 = \kappa^2+\left(k^2+\frac{m^2}{r}\right)a_0^2(1+\chi)$$

$$-2\pi G\sigma_0\sqrt{k^2+\frac{m^2}{r^2}}\,(1+\chi), \tag{D15}$$

which is reproduced as Eq. (73) in the text, where $\chi$ is defined. This dispersion relationship was reported by Lau and Bertin [39, 11] in 1977.

## Bibliographical notes

References [1]–[4] are general references. Some articles in these books are cited individually in the succeeding references.

For reviews of the theory, see Contopoulos [16], Lin [52], Lin and Shu [58], Toomre [103], and Wielen [108]. For discussion of observations, see Bok [14], Lin [50, 52], and Shu [95]. For a general introduction for nonastronomers, see Lin [49] and Lin and Shu [58]. An abridgement of the present article has been presented by Lin and Lau [55] at a symposium in honor of Eric Reissner on June 23, 1978 at La Jolla, California.

## References

1. *Stars and Stellar Systems* (Univ. of Chicago Press, Chicago), Vol. V. Galactic Structure (A. A. Blaauw and M. Schmidt, Eds.); Vol. IX. Galaxies and the Universe (A. Sandage, M. Sandage and J. Kristian, Eds.).
2. *Galactic Structure* (Summer Seminar at Cornell University, Ithaca 1965), Lectures in Applied Mathematics, Vol. IX, Amer. Math. Soc., Providence, R.I., 1967.

3. *Astrophysics and General Relativity* (Summer Institute in Theoretical Physics, Brandeis Univ., 1968; M. Chrétian, S. Deser, and J. Goldstein, Eds.), Gordon and Breach, New York, 1971, Vol. 2, pp. 239–329.
4. *Proceedings of IAU Symposium 38* (Basel, Switzerland, August 29–September 4, 1969; W. Becker and G. Contopoulos, Eds.), Reidel, The Netherlands, 1970.
5. T. M. Antonsen, *Phys. Rev. Lett.* 41:33 (1978).
6. J. M. Bardeen, *Proc. Internat. Astron. Union Symposium* 69:297 (1975).
7. H. L. Berk, L. D. Pearlstein, J. D. Callen, C. W. Horton, and M. N. Rosenbluth, *Phys. Rev. Lett.* 22:876 (1969).
8. R. H. Berman, D. R. K. Brownrigg, and R. W. Hockney, Numerical models of galaxies, Dept. of Computer Science, University of Reading, Report No. RCS 99, Mar. 1978.
9. R. H. Berman and J. W-K. Mark, *Astron. and Astrophys.*, in press (1979).
10. R. H. Berman and J. W-K. Mark, *Astrophys. J.* 216:257 (1977).
11. G. Bertin, Y. Y. Lau, C. C. Lin, J. W-K. Mark, and L. Sugiyama, *Proc. Nat. Acad. Sci. U.S.A.* 74:4729 (1977).
12. G. Bertin and J. W-K. Mark, *Astronom. and Astrophys.* 64:389 (1978).
13. G. Bertin and J. W-K. Mark, *SIAM J. Appl. Math.*, Vol. 36, No. 3, in press (1979).
14. B. J. Bok, *Amer. Scientist* 60:708 (1972).
15. F. P. Bretherton and J. R. Booker, *J. Fluid Mech.* 27:513 (1967).
16. G. Contopoulos [4, p. 303].
17. G. Contopoulos, *Astrophys. J.* 181:657 (1973).
18. S. A. Ericson, Ph.D. Thesis, M.I.T., 1974.
19. S. I. Feldman and C. C. Lin, *Studies in Appl. Math.* 52:1 (1973).
20. P. Goldreich and D. Lynden-Bell, *Mon. Not. Roy. Astronom. Soc.* 130:125 (1965).
21. S. Goldstein, *Modern Developments in Fluid Dynamics*, Oxford U. P., New York, 1938.
22. R. W. Hockney and D. R. K. Brownrigg, *Mon. Not. Roy. Astronom. Soc.* 167:351 (1974).
23. F. Hohl [4, p. 368].
24. F. Hohl, *Astrophys. J.* 154:353 (1973).
25. F. Hohl, *Astronom. J.* 81:30 (1976).
26. F. Hohl and R. W. Hockney, *J. Comput. Phys.* 4:306 (1969).
27. W. R. Hu, *Sci. Sinica* 20:325 (1977).
28. C. Hunter, *Mon. Not. Roy. Astronom. Soc.* 129:321 (1965).
29. C. Hunter, *Studies in Appl. Math.* 49:59 (1970).
30. C. Hunter and A. Toomre, *Astrophys. J.* 155:747 (1969).
31. M. Iye, *Publ. Astronom. Soc. Japan* 30:223 (1978).
32. W. H. Julian and A. Toomre, *Astrophys. J.* 146:810 (1966).
33. A. J. Kalnajs, The stability of highly flattened galaxies, Ph.D. Thesis, Harvard Univ., 1965.
34. A. J. Kalnajs [4, p. 318].
35. A. J. Kalnajs, *Astrophys. Lett.* 11:41 (1972).
36. A. J. Kalnajs, *Astrophys. J.* 175:63 (1972).
37. A. J. Kalnajs, *Proc. Astronom. Soc. Australia* 2:174 (1973).
38. Y. Y. Lau, *Phys. Fluids* 19:1644 (1976).
39. Y. Y. Lau and G. Bertin, *Bull. Amer. Astronom. Soc.* 9:430 (1977).
40. Y. Y. Lau and G. Bertin, *Astrophys. J.*, 226:508 (1978).
41. Y. Y. Lau and J. Haass, paper presented in the International Astronomical Symposium No. 84, 1978.
42. Y. Y. Lau, C. C. Lin, and J. W-K. Mark, *Proc. Nat. Acad. Sci. U.S.A.* 73:1379 (1976).
43. Y. Y. Lau and J. W-K. Mark, *Proc. Nat. Acad. Sci. U.S.A.* 73:3785 (1976).
44. C. B. Li, N. K. Han, and C. C. Lin, *Sci. Sinica* 19:665 (1976).
45. L. Liebovitch, Ph.D. Thesis, Harvard Univ., 1978.
46. C. C. Lin, *The Theory of Hydrodynamic Stability*, Cambridge U. P., 1955.
47. C. C. Lin [2, p. 66].
48. C. C. Lin, *SIAM J. Appl. Math.* 14:876 (1966).
49. C. C. Lin, in *Proceedings of the 4th U.S. National Congress of Applied Mechanics*, Univ. of Minnesota, 1966, p. 3.
50. C. C. Lin [4, p. 377].
51. C. C. Lin, in *Highlights of Astronomy*, (C. de Jager, Ed.) Internat. Astronom. Union, Reidel, 1971, p. 88.

52. C. C. Lin, Theory of spiral structure, in *Structure and Evolution of Galaxies* (Lectures presented at the NATO Advanced Study Institute held in Erice, Italy, June 22–July 9, 1974; G. Setti, Ed.), Reidel, Boston, 1975, p. 119.
53. C. C. Lin and Y. Y. Lau, *SIAM J. Appl. Math.* 29:352 (1975).
54. C. C. Lin and Y. Y. Lau, *Advances in Math.* 22:120 (1976).
55. C. C. Lin and Y. Y. Lau, in *Mechanics Today*, Vol. 5, 1978.
56. C. C. Lin and F. H. Shu, *Astrophys. J.* 140:646 (1964).
57. C. C. Lin and F. H. Shu, *Proc. Nat. Acad. Sci. U.S.A.* 55:229 (1966).
58. C. C. Lin and F. H. Shu [3, p. 239].
59. C. C. Lin, C. Yuan, and F. H. Shu, *Astrophys. J.* 155:721 (1969).
60. C. C. Lin and C. Yuan, in *Astronomical Papers Dedicated to B. Strömgren* (A. Reiz and T. Andersen, Eds.), Copenhagen Univ. Observatory, 1978, p. 369.
61. B. Lindblad, *Stockholm Obs. Ann.* 22, No. 5 (1963) and his earlier work cited therein.
62. P. O. Lindblad, *Stockholm Obs. Ann.* 21, No. 4 (1960).
63. P. O. Lindblad, in *Interstellar Matter in Galaxies* (L. Woltjer, Ed.) Benjamin, New York, 1962, p. 222.
64. D. Lynden-Bell, in *Lectures in Applied Mathematics*, Vol. 9 (J. Ehlers, Ed.) Amer. Math. Soc., Providence, R.I., 1967.
65. D. Lynden-Bell and A. J. Kalnajs, *Mon. Not. Roy. Astronom. Soc.* 157:1 (1972).
66. J. W-K. Mark, *Proc. Nat. Acad. Sci. U.S.A.* 68:2095 (1971).
67. J. W-K. Mark, *Astrophys. J.* 193:539 (1974).
68. J. W-K. Mark, *Astrophys. J.* 203:81 (1976).
69. J. W-K. Mark, *Astrophys. J.* 205:363 (1976).
70. J. W-K. Mark, *Astrophys. J.* 206:418 (1976).
71. J. W-K. Mark, *Astrophys. J.* 212:645 (1977).
72. D. S. Mathewson, P. C. van der Kruit, and W. N. Brouw, *Astronom. and Astrophys.* 17:468 (1972).
73. R. H. Miller, *Astrophys. J.* 190:539 (1974).
74. R. H. Miller, K. H. Prendergast, and W. J. Quirk, *Astrophys. J.* 161:903 (1970).
75. T. Nishimoto, *Studies in Appl. Math.* 60:11 (1979).
76. W. McF. Orr, *Proc. Roy. Irish Acad. Sect. A* 27:69 (1907).
77. J. H. Oort, in *Stellar Matter in Galaxies* (L. Woltjer, Ed.), Benjamin, New York, 1962, p. 234.
78. J. P. Ostriker and P. J. E. Peebles, *Astrophys. J* 186:467 (1973).
79. R. F. Pannatoni, Ph.D. Thesis, M.I.T. (1979).
80. R. F. Pannatoni and Y. Y. Lau, *Proc. Nat. Acad. Sci. USA* 76:4 (1979).
81. L. D. Pearlstein and H. L. Berk, *Phys. Rev. Lett.* 23:220 (1969).
82. R. G. Rehm, Ph.D. Thesis, M.I.T., Cambridge, Mass., 1965.
83. W. W. Roberts, *Astrophys. J.* 158:123 (1969).
84. W. W. Roberts, M. S. Roberts, and F. H. Shu, *Astrophys. J.* 196:381 (1975).
85. W. W. Roberts and F. H. Shu, *Astrophys. Lett.* 12:49 (1972).
86. W. W. Roberts and C. Yuan, *Astrophys. J.* 161:877 (1970).
87. D. W. Ross and S. M. Mahajan, *Phys. Rev. Lett.* 40:324 (1978).
88. R. H. Sanders and J. M. Huntley, *Astrophys. J.* 209:53 (1976).
89. F. Schweizer, in *La Dynamique des Galaxies Spirales* (L. Weliachew, Ed.), CNRS, Paris, 1975, p. 337.
90. F. Schweizer, *Astrophys. J. Suppl. Ser.* 31:313 (1976).
91. W. W. Shane and J. Bystedt, in International Astronomical Union Symposium 77 (E. M. Berkhuijsen, Ed.), Reidel, Dorchrecht, Holland, 1978, p. 97.
92. F. H. Shu, Ph.D Thesis, Harvard Univ., 1968.
93. F. H. Shu, *Astrophys. J.* 160:89 (1970).
94. F. H. Shu, *Astrophys. J.* 160:99 (1970).
95. F. H. Shu, *Amer. Scientist* 61:524 (1973).
96. F. H. Shu, V. Milione, and W. W. Roberts, *Astrophys. J.* 183:819 (1973).
97. T. H. Stix, *Theory of Plasma Waves*, McGraw-Hill, New York, 1962, p. 245.
98. S. E. Strom and K. M. Strom, in *International Astronomical Union Symposium 77*,(E. M. Berkhuijsen, Ed.), Reidel, Dorchrecht, Holland, 1978, p. 69.
99. S. E. Strom, K. M. Strom, J. W. Goad, F. J. Urba, and W. Rice, *Astrophys. J.* 204:684 (1976).

100. A. TOOMRE, *Astrophys. J.* 138:385 (1963).
101. A. TOOMRE, *Astrophys. J.* 139:1217 (1964).
102. A. TOOMRE, *Astrophys. J.* 158:899 (1969).
103. A. TOOMRE, *Ann. Rev. Astron. Astrophys.* 15:437 (1977).
104. A. TOOMRE, in *The Large Scale Structure of the Universe* (M. S. LONGAIR and J. EINASTO, Eds.), Reidel, 1978, p. 109.
105. K. T. TSANG, P. J. CATTO, J. C. WHITSON, and G. SMITH, *Phys. Rev. Lett.* 40:327 (1978).
106. P. VANDERVOORT, *Astrophys. J.* 161:87 (1970).
107. H. C. D. VISSER, Ph.D. Thesis, Univ. of Groningen, 1978.
108. R. WIELEN, *Publ. Astronom. Soc. Pacific* 86:341 (1974).
109. M. S. WILKERSON, S. E. STROM, and K. M. STROM, *Bull. Amer. Astronom. Soc.* 9:649 (1977).
110. T. A. ZANG, Ph.D. Thesis, M.I.T., 1976.

MASSACHUSETTS INSTITUTE OF TECHNOLOGY

(Received September 29, 1978).

ADVANCES IN APPLIED MECHANICS, VOLUME 24

# Galactic Dynamics and Gravitational Plasmas*

C. C. LIN

*Massachusetts Institute of Technology*
*Cambridge, Massachusetts*

G. BERTIN

*Massachusetts Institute of Technology*
*Cambridge, Massachusetts*
*and*
*Scuola Normale Superiore*
*Pisa, Italy*



* This paper is reproduced from *Plasma Astrophysics* [(T. D. Guyenne and G. Lévy, eds.) ESA SP-161, pp. 191–205. European Space Agency, Noordwijk, The Netherlands]. Minor changes were made. For further discussion of related issues see Lin and Bertin (1984).




ISBN 0-12-002024-6

## I. Introduction

A normal galaxy is primarily a collisionless, gravitationally bound stellar system, although there are other constituents such as interstellar gas, cosmic ray particles, and magnetic field. The study of the stellar system may be made on the basis of Liouville's partial differential equation for the distribution function of point masses (the stars) in phase space and of Poisson's equation for gravitational potential. This combined system of equations is the counterpart of the system of Vlasov and Maxwell equations in electromagnetic plasmas. Hence the term "gravitational plasma" is appropriate for the description of stellar systems. In this discussion, we mention some specific similarities and differences between these two kinds of plasmas, but our main objective is to give a report of current developments in the dynamics of stellar systems, especially in the applications to spiral galaxies.

According to a scheme initiated by Hubble and refined by Sandage and de Vaucouleurs, regular galaxies are classified according to their *morphology* into ellipticals, normal spirals, and barred spirals. Within each category there are subdivisions. For example, the normal spirals are subdivided into Sa, Sb, and Sc galaxies in the order of decreasing disk/bulge mass ratio, increasing pitch angle, and increasing gas content. There are also transition types, such as SO and SAB. One of the challenges to theorists is to find the reasons why there are global spiral structures, especially since the bright spiral arms are actually marked out by the newly formed young stars. Another challenge is the considerable permanence of the preceding classifications, since the subdivisions are, on the one hand, correlated with a number of physical characteristics (e.g., the bulge size) that cannot change rapidly in time and, on the other hand, with the pitch angle, which appears to be more readily changeable because of differential rotation in the system. This is known as the *winding dilemma*.

Observations, especially of the frequent existence of a regular *grand design* of spiral structure in differentially rotating systems,* have naturally led to the suggestion that spirals are governed by a wave phenomenon. In this view the spiral distribution of stars is a collective mode in the stellar system, which can be studied according to the principles of stellar dynamics just like the collective modes in electromagnetic plasmas. Our chapter is indeed concerned with such collective modes. After the formulation of the *density wave theory* and the analysis of its dynamic implications, many important observations have been made that strengthen the wave interpretation of spiral structure. According to this picture, a density wave *pattern,* often of relatively low amplitude, is present in the stellar component of the galactic disk, and this, in turn, induces shocks in the interstellar medium, which is much less massive than the stellar component. The shocks, in turn, trigger star formation and produce the more striking optical phenomena. Even if it is hard to decide whether a given spiral is a transient or a quasi-stationary structure, on a *statistical* basis the theory of quasi-stationary spiral structure is on firm ground (cf. Fig. 3 and Section II,D). We may add that some transient spiral features (cf. Section IV,B,4) are characterized by a very small amount of relative motion between the pattern and the material. Thus their ability to trigger star formation consistent with observations (cf. Section II,A) is still to be proved.

More recently, the density wave picture has been brought to a relatively complete form, supported by a fairly coherent dynamic basis so that we may attempt to formulate a *dynamic approach* to the classification of spiral galaxies.

We deal with the classification of galaxies into SA, SAB, and SB types, not with the subclassifications such as SAa, SAb, and SAc and their relationship to the size of the nuclear bulge, and to the gaseous content. These have often been discussed in the existing literature. We do not attempt to add to those discussions in the present article, but reserve these elaborations for a later presentation.

Among the numerous dynamic issues that we do not cover are many numerical experiments on the stability of flat stellar systems, the possibility of galactic evolution due to fast internal instabilities, and the detailed behavior of shocks in the interstellar medium in the presence of bar and/or spiral fields. On this last point, a proper discussion of the physics of the interstellar medium would be necessary.

* For a recent study of the frequency of occurrence of grand design spirals, see Elmegreen and Elmegreen (1982). A report on this work was presented at IAU Symposium 106 [see Lin and Bertin (1984)].

In addition to the problem of spiral structure, there are, of course, many other dynamic issues that belong to the domain of galactic dynamics. For instance, the structure of the ellipticals, with their (so far) mysterious three-dimensional structure, provides an example of the richness of phenomena in collisionless stellar systems and may teach us something about the structure of the invisible halos. The dynamic role of halos around galactic disks is being studied actively, especially after the discovery of generally flat rotation curves and the suggestion by Ostriker and Peebles [(1973), see Bertin (1980)] that a global-order parameter ($t$) should govern the overall stability of axisymmetric galactic systems. One important issue is whether global parameters like $t$ can be sufficient to characterize the dynamic properties of self-gravitating systems. The answer may be positive for those systems that possess only one length scale and is likely to be negative for those systems with many scales (cf. the discussion on dual systems in Section VI). Another large-scale phenomenon under active investigation is the warping of the outer parts of galactic disks. Warped disks have been found to be the rule rather than the exception in spiral galaxies. They appear to be related to the presence of a sizable spheroidal halo component, and their collective dynamics may prove to be an interesting probe of the outer structure of galaxies.

It has been suggested that the environment in groups or in clusters of galaxies plays an important role in galactic evolution (see also Section III,E). This would occur via galaxy–galaxy and galaxy–intergalactic medium interactions. Thus there is now revived interest for a deeper understanding of dynamic friction and related phenomena. In this regard also, one important open question is the role of large halos around galaxies in such "environmental" dynamics. All these issues indicate how powerful and complex the collective effects are in galactic systems. They would all deserve a separate and extended discussion. In the present article we have chosen to focus on the problem of spiral structure, where great progress has been made and many detailed cooperative phenomena have been investigated at length.

## II. Observations

Rather than list all the relevant observational material [which has been often reviewed, e.g., see Lin (1971) for an early review and Strom and Strom (1978) for a more recent review in a broader context], we would like here to emphasize only a few cases of great importance in the overall perspective of the theory of spiral structure. These offer strong encouragement for the construction of a dynamic classification of galaxies. The reader is referred to the original literature for other details.

### A. Radio Continuum Observations of M51

Radio observations by Mathewson *et al.* [(1972), see Bertin (1980)] of M51 at 1415 MHz have revealed a clear delineation of a two-armed grand design spiral structure with the emission maxima lying at the inside edges of the bright optical arms in coincidence with the dust lanes. The linear polarization of the observed emission implies that it is due to synchrotron radiation. The inner radio arms are estimated to be less than 250 pc wide. These data strongly support the view that the interstellar medium undergoes a shock in the way predicted by the density wave picture. The clear correlation of the radio continuum data with many other observed properties is indicative of the collective behavior as the basis of the phenomenon of spiral structure.

### B. Gas Motions in M81

Visser [(1978), see Bertin (1980)] studied the dynamics of gaseous motions in M81 (see Fig. 1). After a choice of an optimum model for the optical spiral pattern based on density wave theory, Visser calculated the gaseous response to the spiral field by using the value of the amplitude for the density wave suggested by the photometric data of Schweizer [(1975), see Bertin (1980); cf. Section II,C]. Comparison with the observed gas motions is in remarkable qualitative and quantitative agreement. Visser and Haass (1981) have been able to provide dynamic support to the whole analysis by calculating a self-excited spiral mode for a model of M81 (see Fig. 2) in agreement with the observed pattern and the theoretical pattern used by Visser.

### C. Detection of Smooth, Broad Spiral Arms

By means of photometric techniques, it has been possible to study the structure of the underlying old disk population in a number of spiral galaxies with a grand design [Schweizer (1975), see Bertin (1980); Elmegreen (1981), see Lin and Bertin (1984)]. The consequent detection of smooth, broad spiral arms in this massive component (M51 and M81 are within the sample studied) indicates that an underlying density wave of

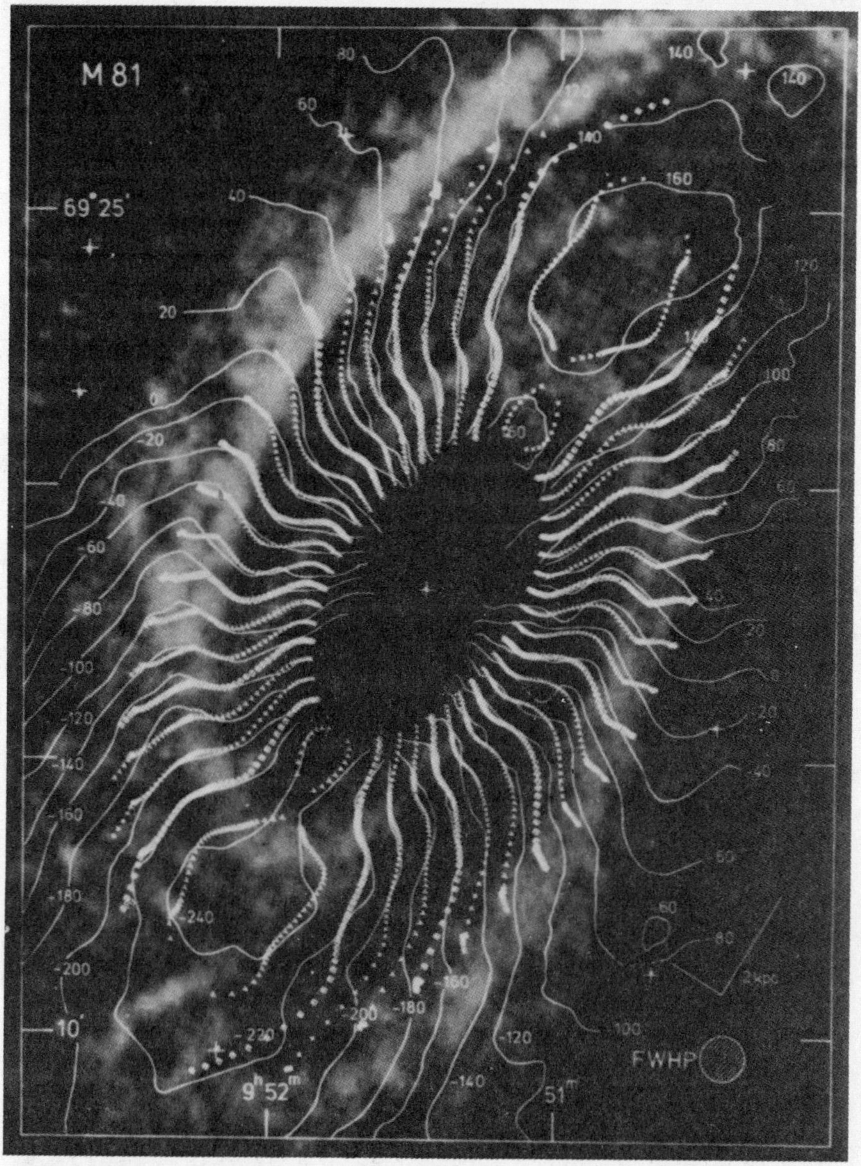


FIG. 1. Radial velocity field associated with density wave model (symbols) compared with observed velocity field superimposed on a radiograph for M81 [from Visser, H. C. D. (1978), Ph.D. Thesis, University of Groningen, Holland].

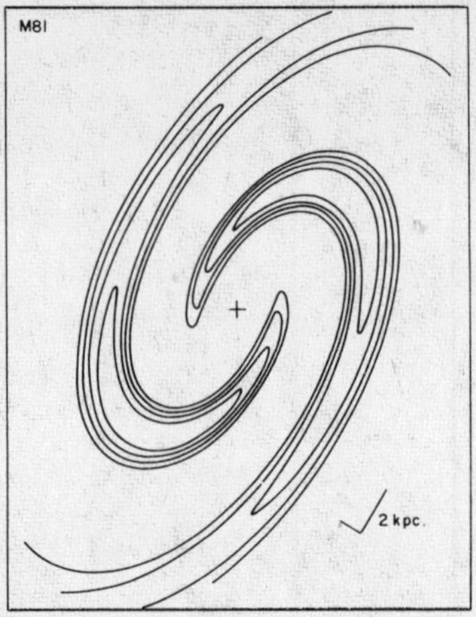


FIG. 2. Density contours for a theoretical spiral mode by Visser and Haass (1981) that provides the dynamic basis for the model of M81 by Visser.

moderate amplitude is behind the scenes and governs the other (less massive and less hot) components of galactic systems to form the observed spiral structure. The smoothness and regularity of these data indicate that a linear analysis for the dynamics of density waves is probably adequate to cover most observed situations.

## D. THE SURVEY BY ROBERTS, ROBERTS, AND SHU

Roberts *et al.* [(1975), see Bertin (1980)] have been able to find a statistical correlation between dynamic (theoretical) parameters and observational classifications: the *morphological* classification mentioned earlier and the *luminosity* classification of van den Bergh (see Fig. 3). This categorization of disk-shaped galaxies on the basis of the theory of density waves (which predicts a general range for pitch angles and shock strength, depending on the axisymmetric basic model of the galaxy) suggests that, at least in a *statistical* sense, the density wave theory of spiral structure is on solid ground. Furthermore, it adds much support to the belief that the current morphological classifications are of a persistent nature and, therefore, gives much confidence to theorists who seek a relatively simple dynamic approach to the classification of galaxies (see Section VI).

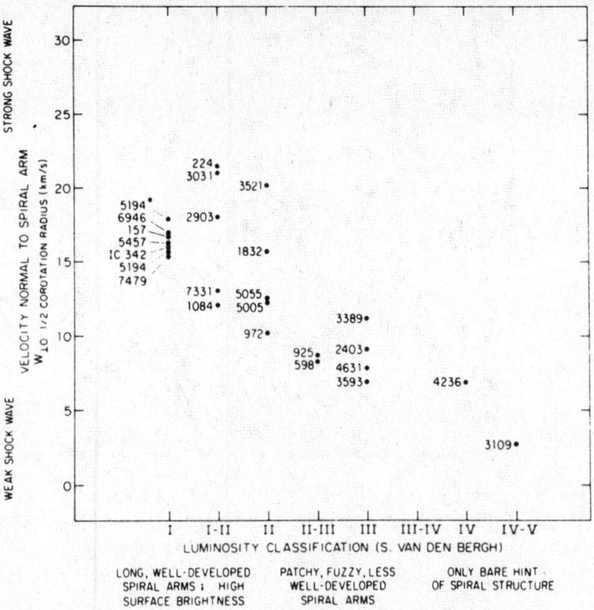


FIG. 3. Correlation between certain dynamic parameters and the luminosity classification of galaxies [after Roberts, W. W., Roberts, M. S., and Shu, F. H. (1975), *Astrophys. J.* **196**, 381].

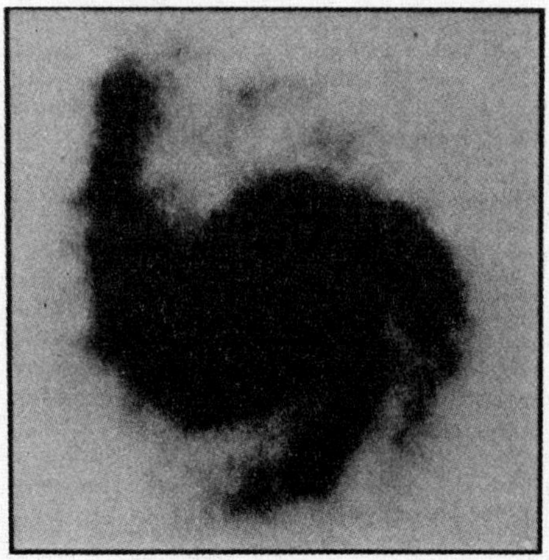

FIG. 4. A galaxy studied by V. Rubin (D100).

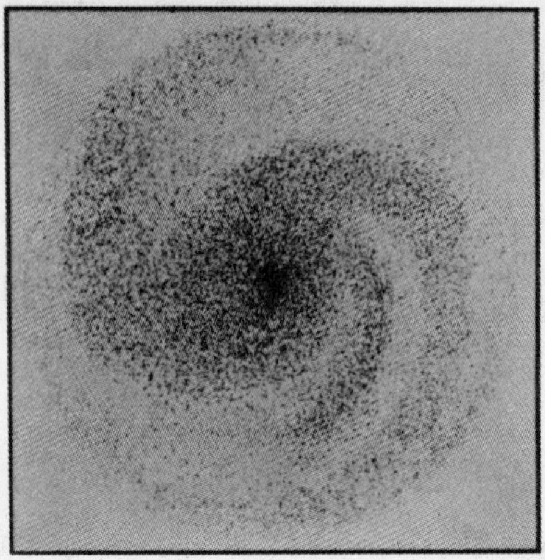

FIG. 5. Effect of superposition of three spiral modes on a galaxy model [from Haass, J., Bertin, G., and Lin, C. C. (1982), *Proc. Natl. Acad. Sci. USA* **79,** 3908]. Spurs, such as the one shown in Fig. 4, could be due to the coexistence of a few modes.

### F. Single-Mode Galaxies and Coexistence

There is no doubt that not all spiral galaxies present a neat grand design. An interpretation is that the more regular the observed spiral structure, the smaller the number of spiral modes involved in its dynamics. Certain observed galaxies seem to demand a single dominant mode (in a sense that will be explained later; cf. Section V), and their nonbisymmetric spiral grand design rules out any tidal or nondynamic origin. However, in general, the observed structure is very often of a global two-armed appearance with some less regular features, such as spurs (see Fig. 4). Haass's example (Fig. 5) of the linear superposition of three global spiral modes (of the kind described in Section V) readily shows that the effects of *coexistence* are likely to play a major role in the comparison of the theory with observations. On the other hand, it would be wrong to conclude that the presence of a number of modes greater than three necessarily implies the fuzzy structures of the NGC 2841 type. In fact, a linear superposition of three regular, two-armed normal modes leads to a coherent (though, in general, evolving in time) two-armed grand design. A further discussion of coexistence is given in Section VII.

## III. Density Wave Theory of Spiral Structure

### A. Some Basic Dynamic Considerations: The Collective Nature of Stellar Systems

Similarities between self-gravitating systems and ordinary plasmas arise from the common long-range nature of the basic forces, whereas differences arise from the opposite signs of these forces, and from the existence, in the gravitational case, of an equivalence principle that identifies the gravitational charge with the inertial mass. Classical and statistical mechanics are common fundaments of stellar dynamics and plasma physics.

Systems dominated by long-range interactions are not "additive" in the sense that, in general, their behavior cannot be described in terms of the properties of their subsystems, when studied separately. Thus, if we are interested in stability properties, it is crucial to distinguish between local and global stability criteria. Furthermore, a critical role in the behavior of such systems is played by their geometry.

It is appropriate to describe the kinematics of large stellar systems (in particular, galaxies) by means of a distribution function indicating the density of stars in a suitable phase space. In general, galaxies are found to be collisionless, so that the evolution of the distribution function is governed by the Vlasov equation of plasma theory. This common description, together with the common nature of the dynamic interactions, is the basis of a considerable number of analogies between stellar dynamics and the physics of high-temperature plasmas.

Gravitation can be naturally balanced not only by pressure but also by rotation. In flat stellar systems rotation is the primary balancing entity. It happens that the Coriolis force in the presence of rotation has the same form as the Lorentz force for a charged particle in the presence of a magnetic field. This is another important source of analogies between stellar dynamics and plasma physics and is the reason for the close similarity of the orbital descriptions in the two cases. In this connection we also mention that rotation, in order to fit in with the gravitational field in flat systems, has to be differential and that such *shear* has important kinematic and dynamic consequences.

Finally we notice that self-gravitating systems, like ordinary plasmas, display the interesting feature of allowing negative energy modes, that is, modes that are amplified by dissipative processes. This feature (cf. following sections) has been recognized as playing a crucial role in the problem of the origin and maintenance of spiral structure.

In recent years, besides the kinetic or "stellar dynamic" approach [Chandrasekhar (1942), see Bertin (1980)], the fluid approach has had a revival of interest. The relevant equations are the standard Euler and continuity equations supplemented by a polytropic equation of state and Poisson's equation [see, e.g., Lin and Lau (1979)]. This approach is relatively simple and, far from the Lindblad resonances (see Section III,C for a definition), results are found to be in good agreement with the more rigorous kinetic analysis.

### B. Disk Galaxies

Disk galaxies have a flat component that is an inhomogeneous system in differential rotation. In the present context the bulge–halo components of a disk galaxy are treated as fixed systems in equilibrium. In general, the disk is basically axisymmetric. In such a system most orbits are quasi-circular and can be described as the superposition of a smooth circular motion with an angular velocity $\Omega(r)$ at radius $r$ on an epicyclic oscillation with frequency $\kappa$, defined by $\kappa^2 = 4\Omega^2(1 - \frac{1}{2}s)$, where $s = -d \ln \Omega / d \ln r$ is the shear parameter. For flat galaxies it is appropriate to introduce a (projected) surface density $\sigma$. A characteristic length for such systems is provided by $\delta = \pi G \sigma / \kappa^2$. In the gaseous approach the Jeans instability against axisymmetric waves occurs when $Q_g \equiv c_s/(\kappa\delta) < 1$, where $c_s$ is the speed of sound. In the kinetic approach the criterion is $Q < 1$, with $Q \approx 0.9354c/(\kappa\delta)$, where $c$ is the radial velocity dispersion of the stars. If the parameter $\varepsilon_0 = (\delta/r)$ is small, an asymptotic treatment of spiral density waves is allowed. Recently, after an improved potential theory had been developed, the asymptotic approach was successfully applied to relatively open structures (with pitch angle less than 30°). The stability of spiral waves in this new regime of open patterns has been found to be dependent on a new dimensionless parameter $J$, in addition to $Q$ (cf. Fig. 9). The parameter $J$ is proportional to $\varepsilon_0$ and to the square root of the shear parameter $s$. Its net effect (which is the combined influence of self-gravity, shear, and tangential forces) can be approximately described in terms of an equivalent stability parameter $Q_e$, which tends to be close to unity even when the actual value of $Q$ departs from such value. More recently, on the basis of the dispersion relation illustrated in Fig. 9, it has been shown that very open waves are supported when both $J$ and $Q$ are relatively high. The reader should consult Lin and Bertin (1984) for a detailed discussion of the regime of open waves.

### C. The Lin–Shu Dispersion Relation

The interpretation of the spiral structure as a density wave phenomenon was originally proposed by Lindblad [see Lindblad (1963); see also

Toomre (1981)]. However, the *collective* behavior of the stars was first put into a satisfactory formulation and calculated by Lin and Shu. A salient point of their "density wave theory" was the hypothesis of a quasi-stationary spiral structure to meet the fundamental need for an explanation of the existence of grand design. There was also, in the beginning, the immediate necessity to provide a consistent description of the spiral phenomenon in sufficiently good agreement with the observational data. This resulted only in the derivation of a dispersion relationship that enables a spiral pattern to be calculated once the angular velocity of the pattern is specified. The problem of the excitation of spiral wave patterns was postponed, and the theory in its primitive form could not provide a method for predicting a well-defined value of the angular velocity $\Omega_p$ of the spiral pattern from given properties of the basic axisymmetric model galaxy.

To be more precise about the dispersion relationship, let $\lambda$ be the spacing between two successive spiral arms and $\nu = m(\Omega_p - \Omega)/\kappa$ be the frequency parameter, where $m$ is the number of spiral arms. Clearly, $|\nu|$ is the dimensionless frequency at which the material rotating at angular velocity $\Omega$ encounters the gravitational field rotating at angular speed $\Omega_p$. The dispersion relationship calculated by Lin and Shu is then of the form $|k\delta| = F(|\nu|, Q)$, where $k \equiv 2\pi/\lambda$. The specific formula defining the dispersion relationship is given in the literature cited; see, e.g., Bertin (1980).

The numerical plot of the dispersion relationship is shown in Fig. 6. When $|\nu|$ assumes integer values, there is a resonance condition. We note

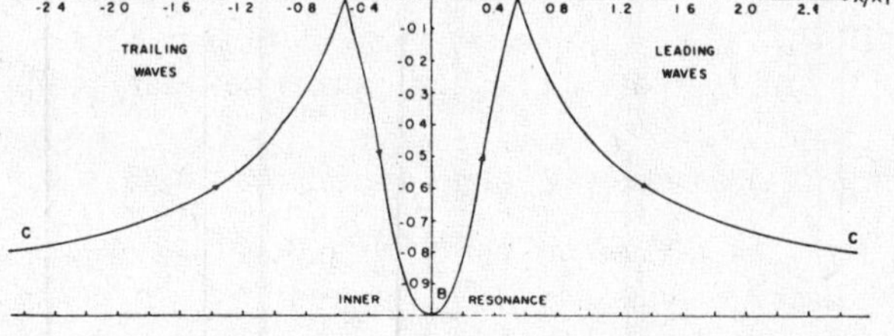


FIG. 6. The Lin–Shu dispersion relation for spiral waves in the case $Q = 1$. Group propagation is indicated by arrows. Different kinds of waves and resonances can be recognized in the diagram.

that in a galaxy $\nu = \nu(r)$. At $\nu = 0$, the wave corotates with the particles; we call this the *corotation resonance*. The *inner* and *outer Lindblad resonances* are defined, respectively, by

$$\nu = -1 \qquad \text{at inner Lindblad resonance,}$$

$$\nu = +1 \qquad \text{at outer Lindblad resonance.}$$

These correspond to locations where the driving frequency is equal to the epicyclic frequency. The conditions at these resonances are more complicated than indicated by the preceding dispersion relationship. The reader is referred to the original literature for the details. Some of the behavior of these waves at these resonances is described below.

In the regions between resonances, wave trains propagate in a way that is defined by the dispersion relation (cf. Fig. 6), and their amplitude is governed by the conservation of wave action. These points are discussed in Section IV. Here we just mention that a proper combination of propagating waves can give rise to stationary rotating structures, that is, to *modes*.

## D. Dynamic Theory of Spiral Patterns

### 1. *Equivalence of Various Mathematical Approaches*

Now we turn our attention to the evolution of *global wave patterns* over a galactic disk. This means that we must heed the boundary conditions at the origin and at the outer edge of the disk, which is often taken to be at infinity.

The analytic treatment of the evolution of a linear dynamic system or of a field can be carried out via three approaches, which are superficially different but substantially the same. These are the solution of the initial value problem by

(1) the superposition of normal modes, including possibly a set with continuous spectrum;
(2) Laplace transformation or other approaches that emphasize evolution from a given state;
(3) the influence function (Green's function).

To initiate the calculation, one may either set an appropriate initial condition or force the system with an external agent that has, for example, a Gaussian variation in time [cf. Toomre (1981)].

The mode approach is the simplest when the situation can be essentially represented in terms of a few (two or three) discrete modes. In such case the behavior of the system is usually not too sensitive to the initial conditions provided that they are "reasonable." The dominant mode soon takes over. However, one has first to make sure that the system does *not* contain a set of modes with a *continuous spectrum,* because their role may not be clearly revealed by simple superposition. For example, a superposition of neutrally stable oscillations (part of a continuous spectrum) often leads to a solution that depends very much on the initial conditions and decays algebraically in time.

### 2. *Absence of Continuous Spectrum*

According to the general spectral theory, a linear eigenvalue problem has no continuous spectrum if the operator is regular and the domain is finite. In the case of an infinite domain, there may be a continuous spectrum, but an eigenfunction of this set corresponds to a distribution with its dominant behavior at infinity. A classical example is the Schrödinger description of the ionized state of the hydrogen atom. On the other hand, when the basic state of a system is such that the amplitude of the small perturbations dies away sufficiently rapidly at infinity, the system is effectively finite in extent.

In one study of the dynamics of galaxies [Hunter (1969), see Toomre (1977)], a set of modes with continuous spectrum has been found to exist not because of the infinite extent of the disk, but because of its infinitesimal thickness. These modes should have a discrete spectrum when the disk thickness is realistically taken to be finite. The amplitude distribution of these modes also suggests that they are not of great physical significance because they are restricted to the edge regions.

We must turn to the other source of continuous spectrum, that is, the possible occurrence of singularities in the interior of the galactic disk, at neither the center nor infinity. In the gas dynamic models (with or without modification of the gravitational field), one has to pay specific attention to the singularities at *corotation resonance* $\nu = 0$ and at Lindblad resonances $\nu = \pm 1$. These singularities are not present in the stellar system. They are, so to speak, smeared out by the dispersive motions of stars. Thus, in using a gaseous model, one must be aware of the pitfalls in the improper handling of the resonances. This issue is further discussed in Section V [cf. Hunter (1969, p. 72), see Toomre (1977)].

Even in the stellar approach, one needs to be aware of another issue suggested by our experience in hydrodynamic stability [as discussed by Hunter and in the following references he cited: Case (1960, 1961), Lin (1961)]. One has to consider the self-adjustment of the system to an initial disturbance through the propagation of signals across the direction of flow. In the problem of hydrodynamic stability of parallel flows, there is a singularity in the comoving layer if the Reynolds number is infinite. Consequently, there is indeed an important set of eigenfunctions with a continuous spectrum. These functions have a mild singularity in the interior of the domain in question. When the Reynolds number $R$ is large but finite, this singularity may not exist, and the corresponding eigenfunctions are wiped out. But this does not mean that the representation of the solution by the discrete modes would be convenient, because the smearing out of the singularity by the viscous forces is done by a diffusive mechanism that acts over a time scale on the order of $R^{1/2}$, and many higher modes are therefore required if the initial value resembles one of the singular modes of the continuous spectrum. In the present case, however, the signal propagation is governed by a wave process. The group propagation of waves across the direction of mean motion is fast. For galaxies with exponential density distribution, as indicated by observations, the time required for signal propagation across the main part of the disk is estimated to be on the order of magnitude of one period of revolution. Hence, there is no need for higher modes (with their fine structures) to be included in abundance to give an approximate description of an evolutionary process for given initial conditions that are not singular in the extreme. In any case, such singular initial conditions are unlikely to be produced naturally.

### 3. *Modes and Their Excitation*

We therefore turn to the modal approach. A consistent dynamic theory for the excitation and maintenance of spiral structure is described in Section V. It provides the basis for our attempting to formulate a dynamic approach to the classification of spiral galaxies. Generally speaking, this implies that definite morphological types are associated with appropriate wave patterns, which are in turn determined by basic dynamic parameters in the equilibrium model. To be sure, nearly stationary wave patterns cannot be expected when several modes coexist, but they may still be in the same general range of pitch angles so that the morphological classification is not changed, even though there may be vacillations in the pitch angle. On the other hand, it is possible for a galactic system to have no unstable spiral modes. In that case an evolving spiral pattern can still be excited by external influences such as a galactic encounter (cf. Section III,E).

Theoretical investigations show that some realistic axisymmetric galaxy models with a disk population can support unstable spiral modes. These modes are *normal spiral modes, barred spiral modes,* or modes in transition between them, depending on the properties of the equilibrium model. All kinds of modes are found to be trailing, consistent with the observations.

## E. Interacting Galaxies

We wish to emphasize that there is no contradiction between the two approaches of internal instability and external excitation. Once the role of internal dynamics is understood, external forcing is just part of the *selection mechanism* of spiral modes. Unless strong collisions are involved (such as those that are supposed to originate *peculiar galaxies*), a tidal encounter can be visualized as a mode selector. This is particularly true if the conditions of the galactic disk under consideration do not have strong spiral instabilities but only damped or weakly growing spiral modes. If the galaxy does not have unstable spiral modes, the spiral pattern excited by the close encounter would eventually die away. If there is a dominant decaying mode, the form of the spiral pattern will settle down essentially to a form that will remain unchanged for quite some time. Otherwise, the form of the pattern will evolve, but the pitch angle may still change over the same general range as determined by the dynamic parameters; so the overall form of the pattern does not change appreciably, and such cases need not disturb the statistical correlation between the morphological and dynamic classifications. One simple general comment should be made: A tidal disturbance is likely to favor bisymmetric structures (i.e., $m = 2$ modes). On the other hand, three-armed and multiple-armed galaxies, spurs, or asymmetric disks (of the $m = 1$ type) are naturally expected in the context of internal dynamics. Indeed, three-armed regular spirals are observed, and these are certainly not of tidal origin.* Finally, we would like to stress that, even in the context of two-armed spirals, a good modeling for a tidal encounter, which could show directly the mechanisms of mode selection and exhibit the amplitude of the induced structures, is not yet available and, indeed, offers an attractive and challenging theoretical problem. For example, a proper inclusion of three-dimensional collective phenomena is still lacking and would be especially necessary in a discussion of a tidal encounter in the presence of sizable halos surrounding galaxy disks.

* Another example in favor of this argument is provided by the observation of NGC 4254 by Iye *et al.* [(1982), see Lin and Bertin (1984)].

Moreover, orbiting satellites need not provide a highly effective mechanism for exciting spiral patterns because there is usually a mismatch of time scales. The inner part of a galaxy has a much faster "clock" than objects traveling at the outskirts of the same galaxy. In controlled numerical experiments, one could adjust the time scale of the exciting force (e.g., that in the Gaussian time variation in Toomre's experiments). It would be interesting to examine the effect of a change in the time scale in a systematic manner to determine the range for most effective excitation and to compare such a time scale with that for the orbiting satellites and galactic encounters. Perhaps Oort [(1970), see Bertin (1980)] was correct, after all, when he made the following remarks on spiral structure at the Basel Symposium in 1969: "But in most cases we cannot invoke interaction. Moreover, if there had been interaction in the past, it might well have contributed to *disturb* the regular wave pattern, but less likely to rebuild it."

Interacting galaxies have been shown [Toomre and Toomre (1972), see Toomre (1977)] to exhibit very interesting behavior responsible for the occurrence of peculiar galaxies. We do not pursue this line in the present article.

## IV. Dynamic Mechanisms

In this section we comment on some dynamic mechanisms, such as wave propagation and wave behavior at turning points and resonances. Only the key qualitative features are discussed. For details and more complete analysis, the reader is referred to the literature cited in the references.

### A. Propagation of a Group of Density Waves; Conservation of Wave Action

The general theory of dispersive waves can be applied to the problems at hand. Toomre considered the propagation of a *group* of density waves and found that short trailing waves propagate away from the corotation circle, whereas long trailing waves propagate toward the corotation circle. The direction of propagation is reversed in the case of leading waves.

The amplitude distribution of such waves is found to conform to the usual principle of conservation of wave action. This was first shown for the case of axisymmetric waves by Toomre, who used Shu's calculation of the amplitude distribution of spiral density waves in a stationary state;

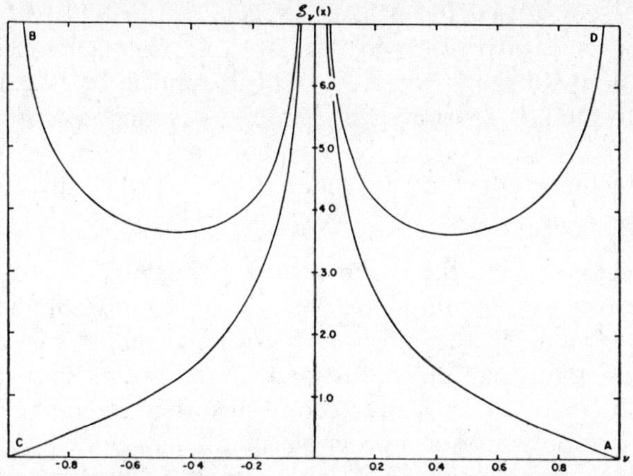


FIG. 7. The wave amplitude [in a dimensionless form defined by Shu (1970)] for different branches of the dispersion relation as calculated by him on the basis of the law of conservation of wave action. Note the amplitude growth as the long trailing wave (C) propagates from near resonance toward corotation.

the demonstration was completed by Shu for the case of spiral waves. The amplitude distribution predicted by this principle is shown in Fig. 7.

When tangential forces are considered, these predictions are slightly altered, but the trends indicated continue to hold. Specifically, we might note that there is a continual rise in amplitude during the outward propagation of the long trailing waves. This rise in amplitude is accompanied by a swing process; that is, the pitch angle decreases as the long trailing wave group progresses (branch C in Figs. 6 and 7). Naturally, the waves would conform to the local dispersion relationship, and it is not difficult to verify that this latter condition requires the wave *packet* to be sheared with the same trend (but at a lower rate) as a material concentration during the process of propagation. It has also been shown in early studies [Goldreich and Lynden-Bell (1965); Julian and Toomre (1966); see Toomre (1981)] that the leading waves have a strong tendency to swing over to the trailing form, with an accompanying transient growth, similar to the vorticity patterns in shear flows.

## B. Propagation of Wave Packets

A closer study of the propagation of density waves can be made in terms of either wave packets or single-frequency sustained wave trains as is reemphasized by Lin [see Athanassoula (1983), p. 117]. Toomre made his pioneering study by using wave packets. He made a number of numerical calculations for his local model (whose application is, however, extrapolated to the neighborhoods of Lindblad resonances $\nu = \pm 1$ without proper justification). The frequency distribution chosen for his group is Gaussian (and hence the time dependence is also Gaussian). He arrived at the following conclusions:

(1) In regions where the group velocity vanishes, the wave propagation changes direction, with a corresponding change in the nature of the wave (e.g., from a short trailing wave to a long trailing wave).

(2) Near regions of Lindblad resonance the waves show an attenuation in amplitude and eventually become absorbed there (similar to what is known as *resonant* absorption in plasma dynamics).

(3) The system appears to be "soft" in the neighborhood of corotation where it allows a free passage of waves. From such a neighborhood, short trailing waves appear to be excited by a barlike forcing function with a Gaussian time variation. The amplitude rises considerably (almost by threefold) during the period when the driving signal is decaying as it passes through the neighborhood of corotation.

(4) More recently, Toomre (1981) carried out similar calculations in the "constant velocity model" used by him and by Zang. In this case, the forcing is done by a leading wave packet propagating outward, away from inner Lindblad resonance. There is found to be a considerable piling up of energy in the nominal turning region (which happens to coincide approximately with the inner boundary of the corotation region in that example) as the wave packet is slowed down, swings to the trailing form, and grows in amplitude because of local instability. There is also considerable "leakage" of trailing waves allowed, so that (the amplified) disturbances are seen to propagate toward the outer Linblad resonance, with an appreciable rise in amplitude partly associated with the principle of conservation of wave action. Toomre refers to this as "swing amplification." The inward propagating trailing waves are eventually absorbed at the inner Lindblad resonance and show only a moderate rise in amplitude.

## C. Propagation of Sustained Wave Trains

The propagation of single-frequency sustained wave trains was studied by Mark, who used an asymptotic approach suitable for galactic disks. As one would expect, the results are more quantitative, not restricted to specific models, and, therefore, more suitable for use as building blocks

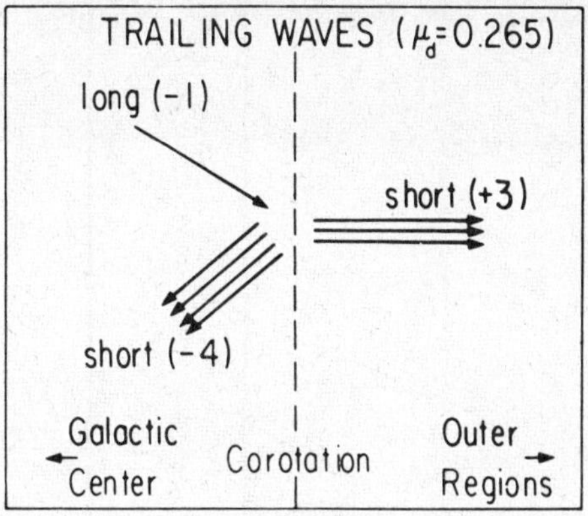


FIG. 8. The WASER mechanism for amplification (excitation) of spiral structure at corotation [from Mark, J. W.-K. (1976), *Astrophys. J.* **205**, 363].

for a theory of spiral modes in a galactic disk (see Section V). Qualitatively, they are of course in agreement with Toomre's conclusions.

(1) The waves are turned back at the turning points, with exponentially decaying solutions in the forbidden zone. Specifically, as first found by Mark, a short incoming trailing wave train is refracted into a long outgoing trailing wave train. (Even though a short leading wave would propagate in the same direction, it is apparently not excited.)

(2) The waves are absorbed at Lindblad resonances. In addition, one can actually calculate the rate of attenuation as the waves approach them.

(3) At the corotation resonance there is the WASER mechanism (Fig. 8), which provides the basis for the excitation of spiral modes. This is similar to Toomre's comment of "softness," but it specifically provides a quantitative calculation of the amount of short trailing waves excited by a long trailing wave train approaching the corotation region from the inside. Specifically, it is found that the incoming wave is not only refracted back as in the case of an ordinary turning point, but there is also stimulated the emission of short waves from the corotation circle in opposite directions. The amplitude of this stimulated "radiation" depends on the parameter $Q$. If $Q$ is high, there is essentially no stimulated emission. If $Q$ is sufficiently low so that there is local Jeans instability, the stimulated wave can have a very high amplitude. It is also found that the presence of a gaseous component and the inclusion of the tangential forces ($J$) enhance the amplitude of the emitted wave so that the amplification mechanism operates even when the system is axisymmetrically Jeans stable (see Figs. 8 and 9).

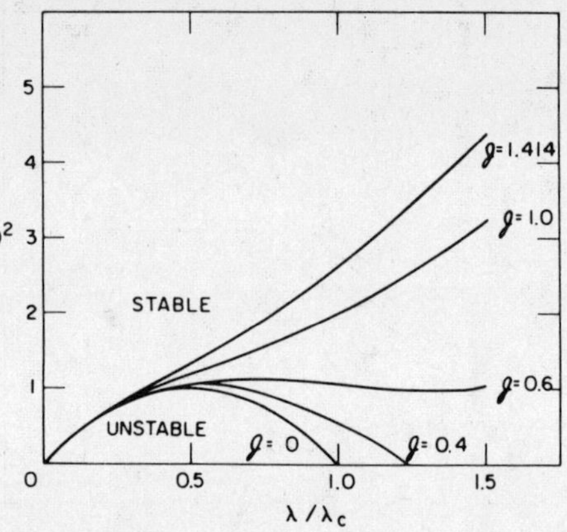


FIG. 9. Local stability diagram shows the role of the parameters $Q$ and $J$ [from Lau, Y. Y., and Bertin, G. (1978), *Astrophys. J.* **226,** 508].

The process described above operates at relatively low values of $J$ (typically up to $J \approx 0.6$). At larger values of $J$ a new kind of overreflection occurs when an open leading wave approaches the corotation circle from inside. In the context of steady wave trains, this process has been discussed by Bertin in Athanassoula (1983, p. 119) and is illustrated more extensively in Lin and Bertin (1984).

These mechanisms of refraction and stimulated emission play important roles in the maintenance of spiral wave patterns and modes, as we shall see in Section V.

## V. Theory of Discrete Spiral Modes

In recent years, galactic disks have been shown to be able to support *discrete unstable spiral modes*. These are regular spiral structures, with a grand design, that rotate rigidly around the galactic center and do not propagate in the radial direction (in contrast to what was found to be the case for density wave packets). In the linear analysis the amplitude of the mode grows exponentially in time. In many cases the growth rate is found to be compatible with the belief that these modes can reach observable amplitudes over a period of time much shorter than the Hubble time scale. Thus they are ideal as bases for the explanation of a *grand design* in spiral structure and for the resolution of the winding dilemma. In addition, it has been noted that the coexistence of more than one mode can be used to explain the existence of less regular galaxies and might even lead (when many modes are present) to situations relevant to the so-called flocculent spirals (like NGC 2841).

The modes owe their excitation to the existence of differential rotation and their capacity to redistribute angular momentum in the galaxy. Mechanistically speaking, their global structure is maintained by proper superposition of different kinds of density waves: leading and trailing waves and, in each case, short, long, and possibly open waves. The transition form normal to barred spiral modes is associated with the different weights of leading and trailing components in the mode composition. Although most of the current developments deal with linear waves, the general conclusions are expected to hold for the nonlinear theory as well, at least for the weakly nonlinear cases.

Even if it is quite obvious from the beginning, it should perhaps be stressed that the growth (or decay) rates of the modes depend quite sensitively on the distributions of the parameters chosen in the basic model. Local instability studies suggest that larger values of the parameters $Q$ and $\varepsilon$ tend to favor more stability, whereas larger values of the parameter $J$ tend to favor more instability. When results obtained by different methods are compared, the ranges of these parameters should be noted, whether we are talking about self-excited modes, externally induced decaying modes, or other processes. Otherwise, misleading conclusions could be improperly deduced.

For example, in the original model of Zang [(1976), see Toomre (1981)], the value of $\varepsilon_0 = \varepsilon/Q$ is 1/4, but in the modified model with "half-mass," its value becomes 1/8. With $Q = 1.5$, this still gives a relatively high value of $\varepsilon = 3/16$, since the observed value of $\varepsilon$ in the solar vicinity is about 1/10, which is about twice as small. The following comments apply to galactic models with values of $\varepsilon$ close to those in the range of the observed value and need not apply, unmodified, to models with larger values of $\varepsilon$ and $Q$. A synthetic view of the different relevant regimes in a $(J,Q)$ diagram is given by Lin and Bertin (1984) where we include a discussion of open waves, as is required when $JQ^2 > 4\sqrt{3}/9$.

## A. Spiral Modes

We present in the following a brief summary of our conclusions. A better appreciation of the theoretical aspects is provided in Sections V,B and V,C.

### 1. *Normal Spiral Modes*

Normal spiral modes are most likely to occur when the typical surface density distribution is low at the center and the basic mass distribution includes a sizable nuclear bulge. In the simplest case, they are formed from the composition of two trailing spiral waves, a short trailing spiral wave train propagating inward and a long trailing spiral wave train propagating outward, in the region inside of the corotation circle. The inward propagating short trailing wave is refracted, at a "$Q$-barrier" in the general neighborhood where the bulge appears, into an outward propagating long trailing wave. At the corotation circle there is a WASER mechanism that results in the amplification (over-reflection) of an inward propagating trailing wave and the transmission of a short trailing wave propagating outward. This outward propagating wave is then absorbed at the outer Lindblad resonance or by the turbulent gaseous medium. The WASER mechanism provides for the growth of these modes, which is found to be aided by the presence of sizable tangential forces (as is the case for open structures).

The eigenvalues of the discrete modes are easily determined by means of a "quantum condition" (reminiscent of many problems in classical and quantum mechanics).

Some examples of the detailed properties of these modes calculated may be found in the references. In some cases, the mode may even have a noticeable "open" component.

### 2. *Barred Spiral Modes and Spiral Modes in Transition*

Barred spiral modes are those that have a barlike structure near the center when $m = 2$. They are expected in those models in which the bulge is not present or not sizable enough to prevent waves from reaching the galactic center. Unlike the normal spiral modes, they include *leading* waves in addition to the trailing waves. In the simplest case, the incoming trailing wave is reflected near the center into a short *leading* wave. The combined structure has a barlike character, rotating at an angular velocity *lower* than that of the material near the center. This is associated with a *trailing* spiral disturbance. Such wave patterns have been obtained by a number of authors using different methods. (The term "barlike modes" has also been extended to include similar modes for $m = 1$ and 3, with reflection of leading waves from the center. Indeed, for $m = 1$, reflection from the center is generally possible.)

To be sure, the incoming wave just discussed can reach the center only if the mass distribution does not have a "hole" in the middle. A study of basic mass distributions with different degrees of "missing mass" in the center shows that there is a continuous transition of spiral modes from the barlike type to the normal type. In Figure 10, we present a sequence of modes given by Haass [(1982), Ph.D. thesis, MIT]. These are the same modes for a sequence of basic states where the neighboring ones are quite close to each other. The transition from bar mode to normal mode is apparent. In the transition range, where the $Q$ barrier is "imperfect," the trailing wave stimulates both a long trailing wave at the $Q$ barrier and a leading wave at the center.

The amplification of the barred spiral modes is still due to an over-reflection mechanism at the corotation circle helped by local Jeans instability associated with a large $J$ parameter, that is, a large influence of tangential forces (cf. subsequent discussion of transfer of angular momentum near corotation).

Not shown in the sequence of Figure 10 is a kind of mode that tends to occur in the *regime of high Q and high J*. Such open bar modes have a

FIG. 10. The same spiral mode is followed in a one-parameter sequence of equilibrium models, which are characterized by different properties of the nuclear bulge. Correspondingly, the mode is found to undergo a transition from a barred (sB) to a normal (sA) spiral mode [from Haass (1982), Ph.D. thesis, MIT].

two-lump appearance and are maintained by a superposition of very open leading waves and very open trailing waves. In the regime just described the long wave branch of the dispersion relation tends to be suppressed. The excitation mechanism can be seen as a new type of overreflection [cf. Athanassoula (1983, p. 119) and Lin and Bertin (1984)] operating at corotation.

To summarize, a spiral mode consists, in general, of a short trailing wave train, a long trailing wave train, and a leading wave component near the center at different levels of prominence. The spectrum of a Fourier analysis of such a spiral mode should tell its precise nature most clearly.

## B. Methods of Analysis

The preceding results were obtained by using the method of asymptotic analysis, combined with a more general fluid dynamic theory. A number of other methods of approach have also been adopted. Among these we should mention the exact numerical integration of stellar models with a high degree of symmetry [e.g., Zang (1976); see Toomre (1981)] and numerical integration of fluid models (e.g., Bardeen, Erickson, Aoki, Iye, and others). Some of these calculations, though important, are somewhat restricted: For example, the distribution of surface density in Zang's elegant model varies inversely with the radial distance, which implies a much larger scale than that of the exponential disk. On the other hand, asymptotic methods have enabled us to develop a flexible and simple theory for calculating spiral modes in a wide class of realistic galaxy models. This has been done both in the simple fluid model [cf. Lin and Lau (1979) and the references therein] and in the more appropriate stellar dynamic theory [cf. Bertin (1980) and the references therein]. In addition, a more recent and general fluid theory, which includes the potential theory in its exact integral form, has enabled us to reach very open configurations and even bar modes (cf. Section V,A). This approach has clearly identified a few parameters (such as $\varepsilon_0$, $J$, in addition to $Q$) that contain much of the physics of the phenomenon and are bound to play important roles in the problem of the dynamic classification of galaxies. It has also proved to be of great help to theorists because of the convenience in including a correct modeling of the complex structures of a galaxy; for example, the mass in the bulge–halo component and the distribution of mass and velocity dispersion in the disk component. In particular, it has been found to be quite easy to construct specific models for comparison with observations (e.g., Milky Way, M81). The physics of important features or effects, such as the thickness of the galactic disk, the presence of a linearly responsive gas component, or the existence of an active halo, can be incorporated into an asymptotic theory. All these go far beyond the infinitesimally thin disk model. In other cases the physics of certain kinetic effects, such as resonance, is included, even in the fluid model, either by modifying the relevant terms according to the results of the stellar dynamic treatment (at corotation) or by a proper choice of boundary conditions (for the Lindblad resonances). These objectives cannot be so easily accomplished with some other approaches.

## C. Resonances: Their Roles in Maintenance of Spiral Modes

Transfer of angular momentum and energy between the wave system and the stars can take place where the stellar response is *resonant,* that is, where the wave resonates either with the bulk motion of the stars (*corotation resonance,* $\nu = 0$) or with the epicyclic motions (*Lindblad resonances,* $\nu = \pm 1$). The inner Lindblad resonance (ILR) ($\nu = -1$) can be absent, depending on the properties of the rotation curve and the value of the pattern frequency of the spiral wave. In general, one-armed waves do not have ILR. In addition, a $Q$ profile rising in the central region (to simulate the presence of a spheroidal bulge) can prevent the wave from reaching the ILR and limit its influence on the dynamics of spiral structure. As a result of the "finite temperature" of the galactic disk, the resonant regions in the galaxy are not circles but annuli with a width of about two epicyclic radii. In analogy with the Landau (1946) discussion of wave–particle interaction in ordinary plasma physics, one finds that resonances are not singular regions but act as sources (or sinks) for wave action (i.e., for wave energy and angular momentum). In general, stars are found to absorb incoming waves at the Lindblad resonances. The presence of the outer Lindblad resonance helps excite the mode because it is consistent with an outer boundary condition of angular momentum radiation. The coupling of the region inside corotation ($\nu < 0$), where the wave energy density is negative, with the region outside corotation ($\nu > 0$), where the wave energy density is positive, through angular momentum transfer is found to be a powerful mechanism (WASER) for mode excitation (cf. previous discussions in Section IV). On the other hand, the absorption of waves at an ILR can interrupt the relevant wave cycle by preventing waves from reflecting back from the central to the outer regions. This is found to damp the spiral mode [cf. also numerical results by Zang (1976); see Toomre (1981)]. Thus the welcomed role of ILR is that of considerably reducing the number of unstable modes that can be supported by a stellar disk (cf. Section V,D). Corotating stars, in general, help excite the mode, but their effect is less dramatic than previously expected.

In general terms, one can summarize the role of resonances by saying that they have no singular properties but act like dissipation. One can thus expect neutral modes to be usually discouraged by their presence.

When the dynamics of galactic disks is studied in the fluid model, singularities appear at the corotation resonance ($\nu = 0$) and at the Lindblad resonances ($\nu = \pm 1$). Thus there is tremendous peculiarity in the behavior of the fluid model at these resonances, especially near the Lindblad resonances, where the amplitude of the disturbance can become very large. Indeed, in the case of neutral oscillations, it becomes infinite. In the case of growing modes, this tendency is somewhat tempered by the growth rate (a detuning process). If the fluid model has no pressure, the tendency toward large or infinite amplitudes is even further enhanced.

A specific example of this disparity is the neutral "*C* modes" calculated by Erickson [(1974), see Toomre (1981)]. In his model pressure is absent, but the gravitational potential is softened. In the axially symmetric case, discrete rings are used. From his results the existence of a continuous spectrum has been suggested. The discrepancy with the stellar system is most easily noticed through an examination of the amplitude distribution of his modes. [See Erickson (1974), Fig. 17, p. 89]. The essential part of his *C* mode is located in the vicinity of the Lindblad resonance, with large amplitudes limited to a small range. But this is a reflection of the behavior of the fluid model and does not properly describe that of a stellar system. In the latter system the Landau damping mechanism should be active, and we would expect discrete damped modes, with very small amplitudes at the circle of Lindblad resonance and outward from there.

In other situations, especially when there are strong gradients of the various parameters in the basic distribution, it may be necessary to use the stellar model to avoid the exaggerated strength of the instabilities that could arise in the fluid model. This issue is especially important in connection with the "edge modes" discussed by Toomre [cf. Toomre (1981, Fig. 12, the *D* mode)].

In conclusion, in studying the large-scale dynamics of galaxies it is of crucial importance to have a proper inclusion of stellar dynamic effects at resonances, because a kinetic treatment (as is appropriate for collisionless systems) incorporates "dissipative" phenomena of the Landau type, which *remove the singularities* (*and hence the related continuous spectrum*) that are present in the fluid model. This is a necessary step to eliminate spurious properties of the singular fluid model, which might be mistakenly attributed to the stability of stellar systems, and to reduce the exaggerated strength of instabilities that could arise in the fluid model, especially when there are strong gradients of the various parameters involved.

### D. Modes with High Quantum Numbers

In the study of a fluid model of a galaxy disk, one usually finds for a given $m$ a large number of low-frequency or "high-$n$," modes. In general, these tend to have lower growth rates with respect to the higher frequency modes (such as those displayed in Fig. 10). Nonetheless, they may give one the impression that too many modes are bound to coexist in a galactic disk, a fact that is hard to reconcile with the appearance of spirals with grand design. Similarly, the investigation of fluid models shows that "high-$m$" modes, being usually quite open, have in general larger growth rates than $m = 2$ modes. This again leaves one with the impression that the theory is not correct, because it would predict many multiple armed spirals ($m \geq 3$), which are not so common in the sky. The answer to both paradoxes lies in the stellar dynamic effects at ILR. The presence of a modest concentration of nuclear mass in the center can easily damp, via ILR, the higher $m$ modes and, at the same time, the low-frequency higher $n$ modes. Indeed, the very fact that bisymmetric spirals are quite common indicates that ILR plays an important role in mode selection. Conclusions of this kind were reached in our earlier experience with pure trailing spirals [Bertin *et al.* (1977), see Bertin (1980)].

## VI. Dynamic Approach to Classification of Galaxies

Our theoretical results, such as those already cited, indicate a simple, promising, working program—partially completed—aimed at the dynamic classification of spiral galaxies. In broad outline, our dynamic approach is as follows. Normal spiral modes and barred spiral modes correspond to normal spiral galaxies (SA) and some barred spirals (SB). In the present perspective, the prominent types of modes that occur depend primarily on the mass distribution in the galaxy, but also on other properties such as velocity dispersion. Barred galaxies, characterized by massive bars, should be treated as a separate category, since the basic mass distribution is not axisymmetrical. *Modes with $m = 2$ are the most commonly observed, because of the nature of the curve $\Omega - \kappa/2$, as already noted in the early phase of the development of theory* (cf. Section V,D).

Those with $m = 1$ should also occur often. One might even speculate that they always coexist with modes with $m = 2$. Certainly in M31, the recent data on the asymmetry of the gaseous distributions and motions suggest the coexistence of such a mode.

We now describe these speculative conclusions in greater detail in the following.

## A. Bar Structure of Type sBB

If the basic mass distribution has a large rotating ellipsoidal (nonaxisymmetric) component and a surrounding disk component with differential rotation, there is naturally to be expected a spiral structure in the disk driven by the "bar." Indeed, it has been shown (by Sanders, Huntley, Roberts, Liebovitch, and others) that even a rotating broad oval distribution of matter with rather small eccentricity can drive a disk of gas into barlike features with rather narrow bars. In such cases *the optical appearance does not coincide with the distribution of mass.*

The spiral feature is caused by the tendency toward epicyclic motion of the material in a rotating system when it is driven by a rotating bar. Stars with low dispersive speeds would respond to such driving just like the gas. This redistribution of matter would produce a spiral gravitational field, which enhances the spiral feature further. Stars with higher dispersive speeds would respond more weakly. But because of their higher total mass, they are also expected to contribute to the spiral gravitational field.

In such cases, however, the spiral waves do not contribute very much to the barlike feature. The oval distortion is perhaps governed by processes similar to those of rotating ellipsoids [cf. Chandrasekhar (1969)], see Bertin (1980)].

## B. Bar Structure of Type sB

When a galaxy is entirely composed of a differentially rotating galactic disk (e.g., like that in the Kuzmin–Toomre sequence) without a nuclear bulge or halo, there is a strong tendency for a barlike instability to develop near the center. This may be expected by analogy with the classical solutions for rotating ellipsoids, since the system has been flattened by rotation. It is expected that the Ostriker–Peebles criterion would apply.

In contrast, there may be bar modes where the spiral density wave pattern is composed essentially of a leading wave and a stronger trailing wave.

Recent calculations made by Toomre (1981) and Haass [(1982), see Lin and Bertin (1984)] have shown that, near the center, the mode may be described as composed of an outgoing leading wave and an incoming trailing wave, if there is a free passage of waves near the center. (Still, keep in mind that the origin is a node of the wave pattern.) There is then a barlike structure near the center with a trailing spiral pattern outside the corotation circle, if the wave pattern can propagate through that neighborhood. We refer to such wave patterns by the designation sB and assume that it essentially falls within the designation SB used by Hubble, Sandage, and de Vaucouleurs on the basis of observational data. A plausible example of such a galaxy could even be NGC 1300. This galaxy also has a small disklike structure near the center. It may be associated with a very small spherical nucleus, which does not effectively interfere with the primary mechanism of the formation of the bar mode. However, many of the galaxies classified as SB according to Sandage and de Vaucouleurs may belong to the sBB category in our dynamic classification.

Galaxies with a mildly ellipsoidal, basic Population II mass distribution may be expected to "force" a prominent barlike sB mode if there is already an instability of this type. The phase angle of the modes is then expected to be correlated with the directions of the principal axes of the ellipsoidal distribution. Thus, the observed spiral features in a galaxy can be used to infer the details of the basic Population II mass distribution.

### C. Bar Structure of Type sAB

If the mass distribution is that of a pure disk, but the velocity dispersion is gradually increased in the central regions from the situation just considered (so that the $Q$ parameter becomes enhanced there), the incoming short waves will not be returned as a pure leading wave from the center. There is also refracted a long trailing wave component propagating outward. As this component becomes more important, the bar structure becomes less prominent, and the spiral pattern gradually takes on the features of a normal spiral. This is then a spiral pattern of a transition type similar to the concept of the S/SB classification or the SAB classification introduced by de Vaucouleurs. We therefore refer to such spiral patterns by the designation sAB.

### D. Normal Structure of Type sA

These are galaxies where the waves propagating from the center outward are essentially long waves of the trailing type, as fully discussed in

Section V,A,1. They are designated as spiral patterns of the type sA to correspond to the classification SA of de Vaucouleurs.

Galaxy models with a substantial nuclear bulge or halo tend to have spiral patterns of this type. In such galaxies, very little mobile mass can be assigned to the central part of the disk component. Thus, the $Q$ parameter can be high even when the dispersion velocity is moderate. Therefore the tendency toward distortions typical of rotating ellipsoids is not so prominent. Indeed, the bulge–halo population would also tend to discourage barlike formation [see Berman and Mark (1979)].

### E. Dual Spiral Structures

A number of galaxies (such as NGC 5364, NGC 210, and NGC 4314) show dual spiral structures. Another such spiral (NGC 1512) has been studied by Lindblad and Jorsater. These structures can be explained in terms of the density wave concept as follows.

Consider, for example, the galaxy NGC 5364. In "The Hubble Atlas of Galaxies" the following description is given:

> NGC 5364 is one of the most regular galaxies in the sky. The thin, bright, internal ring appears to be completely closed. There is spiral structure internal to this ring, but the inner arms starting from the nucleus are not well defined. The regular outer arms begin tangent to the bright ring. Two external arms are present. Their junction on the ring is difficult to find. [Sandage (1961), see Lin and Lau (1979).]

In our view, the outer spiral structure is a regular normal spiral pattern. There is a nuclear bulge and a halo in this galaxy, and hence the ILR ring. This nuclear bulge of the older stars (Population II) is on the scale of the size of the ring (or larger), much larger than the size of the tiny nucleus shown in the photograph. The disk component directly under the nuclear bulge has sufficient mass to develop into an entirely separate spiral structure.

## VII. Concluding Remarks

In applying our theory, we must keep an open mind with regard to other physical processes that produce spiral features. Although the spiral features in neat spirals of the M81 type (which has a regular grand design) and fuzzy spirals such as NGC 2841 (which has no apparent grand design) could both be due to density waves, they may well have been produced by different physical processes. In galaxies of the NGC 2841 type, there appears to be almost complete rotational symmetry. Either there is a superposition of many modes, or other mechanisms (e.g., induced star formation) may be present. An examination of the physical conditions in the interstellar medium is strongly urged for such galaxies in order to decide whether the process of induced star formation can actually continue over long distances. (Induced star formation is probably present in irregular galaxies, but this is outside the scope of our present discussion.)

Observations of large spiral galaxies (with coherent spiral structure up to 100 kpc from the galactic center, as suggested for UGC 2885) offer another dynamic puzzle. In such galaxies the "clock" is running very slowly in the outer parts and there the winding dilemma is not serious; still the connection of the spiral structure with the central regions seems to demand once again a wave interpretation of the observed structures.

In addressing these and other questions arising from observations it is very important to keep in mind the motivation for the original hypothesis of quasi-stationary spiral structure. The hypothesis is specifically directed at the central issue posed by Oort: A pattern with a grand design exists in many galaxies, even when the spiral structure is "hopelessly irregular and broken up." It was also stated in the early papers that the superposition of a few modes would be suitable for the representation of patterns that are not permanent but comparatively transitory [cf. Lin (1965, 1970) cited in Lin (1971) for emphasis on *coexistence*, including the possible presence of material clumps, waves, and wave patterns].

From the original *hypothesis of quasi-stationary spiral structure* the theory has recently developed into a more complete dynamic form as a theory of superposition of spiral *modes*. In the present article we have stressed the point that the modal approach is best justified in the absence of a continuous spectrum of oscillations in stellar disks. Indeed there is as yet no evidence of a continuous spectrum. In support of this view, we have discussed the role of dissipative mechanisms in providing discrete modes. In galactic systems, dissipation can be induced either by gas turbulence or by the collisionless mechanism of Landau damping.

In the present article, on the basis of a number of observational and theoretical results that have accrued during recent years, we have arrived at the following proposal. Starting from the existence of discrete spiral modes, stable or unstable, we have attempted to formulate a dynamic classification scheme for spiral galaxies into categories sA, sAB, sB, and sBB. This proposal is an appealingly simple program, which may require improvements and modifications after more complete observational and dynamic studies are made on the properties of spiral galaxies. An optimistic expectation for the success of such a program will provide the guiding spirit for future investigations.

## ACKNOWLEDGMENTS

This work was partly supported by Consiglio Nazionale delle Ricerche, under the United States–Italy Cooperative Science Program, and partly by the National Science Foundation.

We wish to thank Jon Haass for his help in the preparation of the paper and for his permission to quote some of his results before they are published.

## REFERENCES

Because of space limitations and the wide range of the literature involved, we are unable to list here all the relevant references. This will be done in a more detailed forthcoming publication. The interested reader may find most of the references needed in the following publications.

Athanassoula, E., ed. (1983). "Internal Kinematics and Dynamics of Galaxies," *Proc. IAU Symp.* 100, Besançon, France, August, 1982. Reidel, Dordrecht, The Netherlands.

Bertin, G. (1980). On the density wave theory for normal spiral galaxies. *Phys. Rep.* **61** (1), 1–69 (217 references).

Lin, C. C. (1971). Theory of spiral structure. *Highlights Astron.* **2**, 88–121 (125 references).

Lin, C. C., and Bertin, G. (1984). Formation and maintenance of spiral structures in galaxies. *In* "The Milky Way Galaxy" (H. van Woerden, ed.), *Proc. IAU Symp.* 106, Groningen, The Netherlands, May, 1983. Reidel, Dordrecht, The Netherlands. In press (39 references).

Lin, C. C., and Lau, Y. Y. (1979). Density wave theory of spiral structure of galaxies. *Studies in Appl. Math.* **60**, 97–163 (110 references).

Strom, S. E., and Strom, K. M. (1978). The evolution of disk galaxies. *In* "Structure and Properties of Nearby Galaxies" (Berkhuisen and Wielebinski, eds.), *Proc. IAU Symp.* 77, pp. 69–95. Reidel, Dordrecht, The Netherlands (52 references).

Toomre, A. (1977). Theories of spiral structure. *Ann. Rev. Astron. Astrophys.* **15**, 437–478 (145 references).

Toomre, A. (1981). What amplifies the spirals? *In* "The Structure and Evolution of Normal Galaxies" (S. M. Fall and D. Lynden-Bell, eds.), pp. 111–136. Cambridge Univ. Press, London and New York (18 references).

*Proc. Natl. Acad. Sci. USA*
Vol. 79, pp. 3908–3912, June 1982
Astronomy

# On the dynamical basis of the classification of normal galaxies

(density waves/galactic dynamics/spiral arms/spiral galaxies)

J. HAASS, G. BERTIN*, AND C. C. LIN

Department of Mathematics, Massachusetts Institute of Technology, Cambridge, Massachusetts 02139

*Contributed by C. C. Lin, March 8, 1982*

**ABSTRACT** **Some realistic galaxy models have been found to support discrete unstable spiral modes. Here, through the study of the relevant physical mechanisms and an extensive numerical investigation of the properties of the dominant modes in a wide class of galactic equilibria, we show how spiral structures are excited with different morphological features, depending on the properties of the equilibrium model. We identify the basic dynamical parameters and mechanisms and compare the resulting morphology of spiral modes with the actual classification of galaxies. The present study suggests a dynamical basis for the transition among various types and subclasses of normal and barred spiral galaxies.**

The starting point and some of the important final objectives of any theory of spiral structure in galaxies can be summarized by the questions posed by the morphological classification of Hubble, Sandage, and de Vaucouleurs (1) and by the luminosity classification of van den Bergh (2). The former scheme emphasizes the geometry of spiral structure, the properties of which correlate with a number of apparently independent physical parameters (such as gas content and bulge size). The frequent existence of a grand design poses the winding dilemma—i.e., the problem of how material objects (such as the young blue stars) can maintain their spiral arrangement in the presence of sizable differential rotation. The relevant questions are essentially related to the overall dynamics of self-gravitating disks. The latter classification scheme focuses on the distinction between broad and filamentary arms in different galaxies and on the related correlation with galaxy luminosity (thin-armed galaxies are brighter). The issues raised clearly bear on the mechanisms of shock generation and star formation in the interstellar medium in galaxy disks.

In the last two decades the density wave theory of spiral structure has provided a unifying working program for a number of theoretical and observational investigations on the subject. Specifically, the explanation of the existence of a grand design may be based on the idea that spiral arms are quasi-stationary wave patterns. The crucial basic assumption is that the long-range gravitational interaction among stars is behind the scenes and generates coherent collective phenomena. Even before a full dynamical understanding of the processes that lead to the excitation and maintenance of spiral structure was reached, the density wave theory had been supported by many successful predictions and comparisons with the observations (3, 4). In order to carry out the formidable task of investigating the dynamics of collisionless self-gravitating disks, three principal approaches have been followed: (*i*) *N*-body simulations (see refs. 5–7), (*ii*) stellar dynamical calculations (see refs. 8–14), and (*iii*) fluid models (see refs. 15–20). From the very beginning (5–7), it has been clear that axisymmetric disk models can be subject to instability with respect to regular global modes and thus tend to reach an evolving nonaxisymmetric spiral-like structure.

The hypothesis of quasi-stationary spiral structure stressed from the outset the importance of coexistence and superposition of different spiral features when interpreting the presence of a grand design of spiral structure in spite of some degree of irregularity. But now that unstable discrete spiral modes have been found to be supported in some realistic galaxy models (8, 15) and an evaluation of their implications has been made (21), the coexistence of spiral modes in galactic disks has assumed more importance in comparing theory with observations. In the present paper, we discuss and illustrate an attempt at a dynamical approach to the classification of spiral galaxies via the modal approach, which has yielded a particularly rich set of interesting results.

### Modes, coexistence, and grand design of spiral structure

Not all spirals present a neat grand design. An explanation is that the more regular the observed spiral structure is, the smaller the number of spiral modes that should be involved in its dynamics. Certain observed galaxies should be essentially represented by a single mode, and in some cases a nonbisymmetric spiral grand design (such as the galaxy D32)[†] rules out any tidal interpretation. One remarkable case in which a single wave pattern study has produced a detailed qualitative and quantitative test (23) is that of M81. Recently, H. Visser and J. Haass[‡] have been able to provide dynamical support to the successful observational test by calculating a self-excited spiral mode for a model of M81 in agreement with the observed pattern and with the theoretical pattern used by Visser (23).

However, the observed structure is very often less regular, even if retaining a global two-armed appearance (as is the case of the galaxy D100[†] that is characterized by a spur coexisting with an otherwise fairly regular two-armed grand design). Haass's example (Fig. 1) on the linear superposition of three global spiral modes readily shows that the effects of coexistence are likely to play a major role in the comparison of the theory with observations. On the other hand, it would be wrong to conclude that the presence of a number of modes greater than three already implies the fuzzy structures of the NGC2841 type. In fact, a linear superposition of three regular two-armed modes always leads to a coherent (though in general evolving in time) two-armed grand design (see ref. 21 for additional discussion).

### Discrete unstable spiral modes: Comparison with morphology of galaxies

In this section, we describe the properties of the dominant two-armed modes in different galaxy models and discuss the possibility of a dynamical approach to the morphological classifi-

The publication costs of this article were defrayed in part by page charge payment. This article must therefore be hereby marked "*advertisement*" in accordance with 18 U. S. C. §1734 solely to indicate this fact.

* Permanent address: Scuola Normale Superiore, 56100 Pisa, Italy.

[†] We are grateful to V. Rubin for bringing these galaxies to our attention. Indeed, she pointed out the resemblance between D100 and Fig. 1. Another interesting example of a nonbisymmetric grand design spiral is NGC309 (see ref. 22).

[‡] The spiral pattern was published in ref. 21.

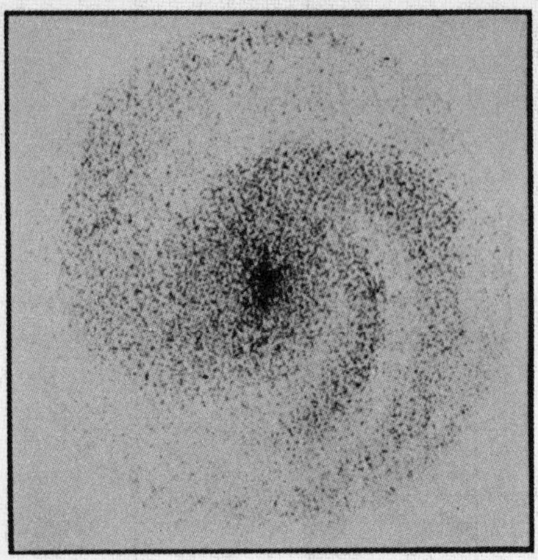

FIG. 1. Effect of superposition of the three fastest growing spiral modes on a galaxy model. Spurs, such as the one observed in the galaxy D100, could be due to coexistence of a few modes.

cation of spiral galaxies. The results presented here represent only a small fraction of an extensive stability analysis, but they serve the purpose of summarizing and illustrating our dynamical conclusions. The modes were obtained by a numerical code that was devised by Pannatoni (19) and modified by Haass.

**Equilibrium Models.** The basic properties of a reference galaxy model ($S^o$) are described in the *Appendix*. The model includes an "active" disk population and an immobile or "inactive" spheroidal bulge-halo component. The model is not chosen to mimic any specific galaxy. It is just a simple realistic case in which we can conveniently vary the (*i*) distribution of the (active) surface density $\sigma(r)$, (*ii*) the distribution of $Q(r)$, the usual $Q$-parameter (defined precisely in the *Appendix*), and (*iii*) the rotation curve $\Theta(r)$.

In line with previous discussions of spiral structure, we focus our attention on the structure of the inner regions of galaxies and the mass density of the active disk population. This explains the general trend adopted for the functions $\sigma(r)$, $Q(r)$, and $\Theta(r)$ as we vary them systematically.

**Morphology of the Modes.** In Fig. 2, we show the dominant two-armed modes for three equilibria characterized by the same $Q(r)$ and $\Theta(r)$; only the surface density $\sigma(r)$ is varied. The *modes* are of the normal type (sA) because the bulge is large enough to effectively refract waves out of the innermost regions of the galaxy. The sequence represents a transition from tightly wound modes (sAa) to modes that are more open (sAb, sAc). They correspond to a gradual increase of the active surface density $\sigma(r)$.

In Fig. 3, we illustrate a similar study for variations on the function $Q(r)$. The first frame is essentially a continuation of the sequence illustrated in Fig. 2. Indeed, it is a quite open normal spiral mode (sAc). The following two frames correspond to variations in the $Q(r)$ distribution so that the role of the bulge is gradually reduced. This sequence of dominant modes suggests a transition to modes of the barred type (sAB). Note that the "bar" is small compared to the corotation circle. The term "bar" mode is justified once the total density distribution is plotted. The resulting shock structure in the interstellar medium would make the identification even more apparent (24).

The modes discussed above do *not* exhibit inner Lindblad resonance. For models in which the modes may include such resonance, a proper treatment of the dynamics of stars should be included. Preliminary results for a sequence of models with variation of $\Theta(r)$ exhibit a gradual approach to the resonance condition. As this happens, the dominant modes show a substantial variation in pitch angle within the disk, the inclination being smaller in the central regions where the resonance condition is approached.

**Comparison with the Morphological Classification of Galaxies.** From the above results (and other material that is not presented in this article) we can draw many interesting conclusions and organize a dynamical approach to the classification of spiral galaxies.

The examples of Fig. 2 and Fig. 3 are indicative of two morphological transitions that are easily recognized in the Hubble diagram, from tightly wound (Sa) to more open (Sc) spirals and from normal (SA) to barred (SB) spiral galaxies through the intermediate SAB class. However, there are complications when we attempt to compare our dynamical framework to actual observed objects. As an example, we may recall that the two well-known galaxies M33 and NGC5364 are both classified as Sc, whereas their contrasting appearances imply that different dynamical phenomena are involved.

Even in our simple systematic scheme to model galaxies, it is easy to realize that, in order to adjust to the proper observational criteria that are commonly used in the morphological classification of galaxies, it would have been necessary to vary *all* three of the controlling parameters simultaneously, rather than change one parameter at a time. In fact, to reproduce the Hubble sequence, the shape of the rotation curve and the related bulge-halo size and structure should define the optimal paths to be followed in the three-dimensional parameter space. With this perspective we have chosen to take the view of defining a (somewhat independent) dynamical approach to the classification of galaxies and then comparing it with the existing observational classification schemes.

To be sure, galaxies with massive bars are not covered in the present linear modal approach. Still some barred (SB) and especially transition (SAB) galaxies may well turn out to be within our reach, as is partly hinted at by recent photometry of the older disk population (22).

In addition, our approach provides some other attractive suggestions, which touch on issues more subtle than those at the basis of the original Hubble program. For instance, we do see a distinction between normal spiral modes that do not reach the center (such as the one applicable to M81) and normal spiral modes that almost reach the center with patterns that resemble that of M51. Another interesting case has been found by exploring the role of the rotation curve. The results obtained tend to support the idea that rapidly rising rotation curves may be compatible with a situation of dual spiral structure (21), even though much more work has to be done in this direction. Finally, we may ask whether the transition from spiral galaxies with a grand design (like M81) to fuzzy spirals (like NGC2841) is within our dynamical program. This may be the case, but other explanations of fuzzy spirals should not be ruled out.

## Dynamical mechanisms and further discussions

Much theoretical work (see refs. 5–21 and 24–30) has been directed towards the understanding of the mechanisms for the excitation and maintenance of large-scale spiral structures. This effort has benefited substantially from the flexible asymptotic approach (see refs. 8 and 15), which has identified the properties of elementary spiral waves and their role in certain global modes (see ref. 21). To extend our ability to calculate discrete global

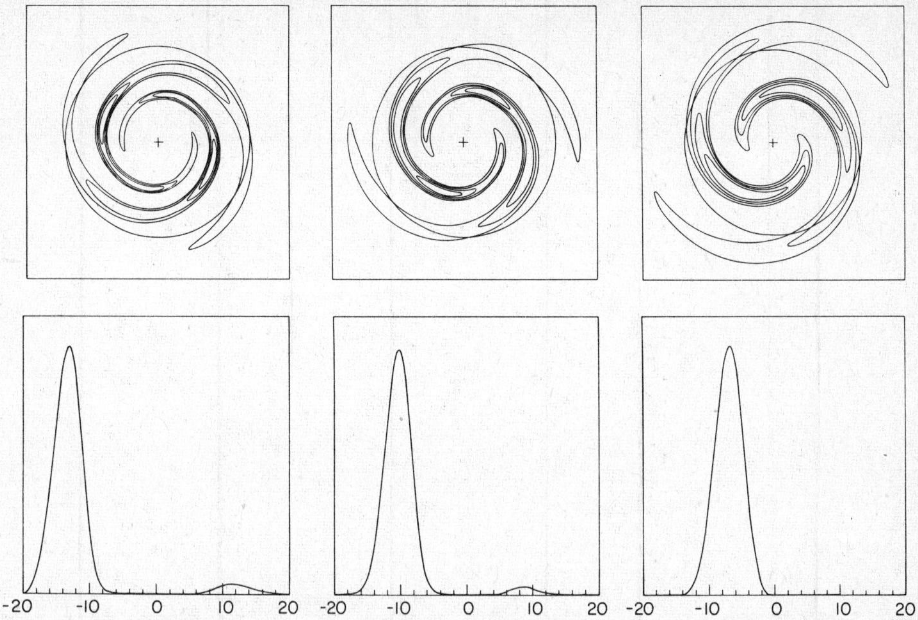


FIG. 2. Transition from tightly wound to open normal spiral modes. The same spiral mode is followed in a one-parameter sequence of equilibrium models, characterized by different values of the active disk mass. The top row provides density contours (corotation circles are also indicated), the bottom row, the corresponding "power" spectra. The density distribution $\sigma(r;\mu)$ in the disk is given by Eq. **A2** in the *Appendix*, with $\mu$ = 7.5, 6.25, and 3.75, respectively, for the three figures from left to right. The parameters in Eq. **A1** and **A3** are: $v$ = 1.75, $q$ = 1. With the choice of units indicated in the *Appendix*, the top frames are 16 × 16 kiloparsecs. The scale for power spectra (the ordinate) may be chosen arbitrarily because we are dealing with a linear theory. The abscissa is a dimensionless radial wave number.

spiral modes beyond the asymptotic theory, a numerical code (19) has been devised. To understand the dynamical mechanisms that sustain these modes, we adopt a spectral analysis (see ref. 21), with results such as those shown in Fig. 2 and Fig. 3. Such an analysis supports the view that the global structure of a mode is maintained by a proper superposition of four different kinds of density waves: leading and trailing waves, and, in each case, short and long waves.

Our present results support the following conclusions (additional discussion is given in ref. 21): (*i*) Normal spiral modes are essentially composed of two trailing spiral waves with opposite propagation properties inside of the corotation circle. (The inward-propagating short trailing wave is refracted back into an outward-propagating long trailing wave from the central regions in the general neighborhood where the nuclear bulge first appears.) The growth of these wave patterns is provided by a wave amplification by stimulated emission of "radiation" (WASER) mechanism (11), which is found to be aided by the presence of sizable tangential forces (14, 17). (*ii*) Unlike normal spiral modes, barred spiral modes include short leading waves in addition to trailing waves of both types. The incoming short trailing wave can reach the center and be reflected into a short leading wave most effectively if the mass distribution does not have a deep "hole" in the middle. The amplification of barred modes can still be interpreted in terms of a WASER mechanism—i.e., super-reflection induced by "radiation" of energy towards the outer regions of the disk. We note that, in the bar structures obtained, there are present not only a superposition of trailing and leading waves (in a way reminiscent of the "anti-spiral" theorem), but also a sizable component of very long "spoke-like" waves ($k \approx 0$). (*iii*) The transition from normal to barred spiral modes is associated with different weights of short leading and long trailing components in the modal composition; both types contain short trailing waves that give galaxies the overall impression of trailing spirals.

We would like to reiterate that the transient approach (9, 16, 18, 25) does not address physical phenomena different from those discussed in the modal approach (8, 15). Only the emphasis is different and the regimes actually investigated do not always overlap in the parameter space. The relationship between these two approaches has been discussed in some detail in ref. 21. It is pointed out that eventually both approaches, as studied by local or asymptotic methods, need some extrapolation when used to describe truly open two-armed spiral features. The main reason for this limitation is the inherent long-range nature of the gravitational forces that dictate a global behavior of inhomogeneous collisionless stellar systems. But decisive progress has been made towards the understanding of open structures despite such restrictions.

Even though great progress has been made in developing a theory of spiral structure, many are the intrinsic limitations of the simplified models that are generally used in the calculation of spiral modes. However, qualitative understanding of specific issues is already available. Among these, we should recall the effects of finite thickness (see refs. 26–28) and the role of stellar dynamical effects on global modes. These factors, besides mod-

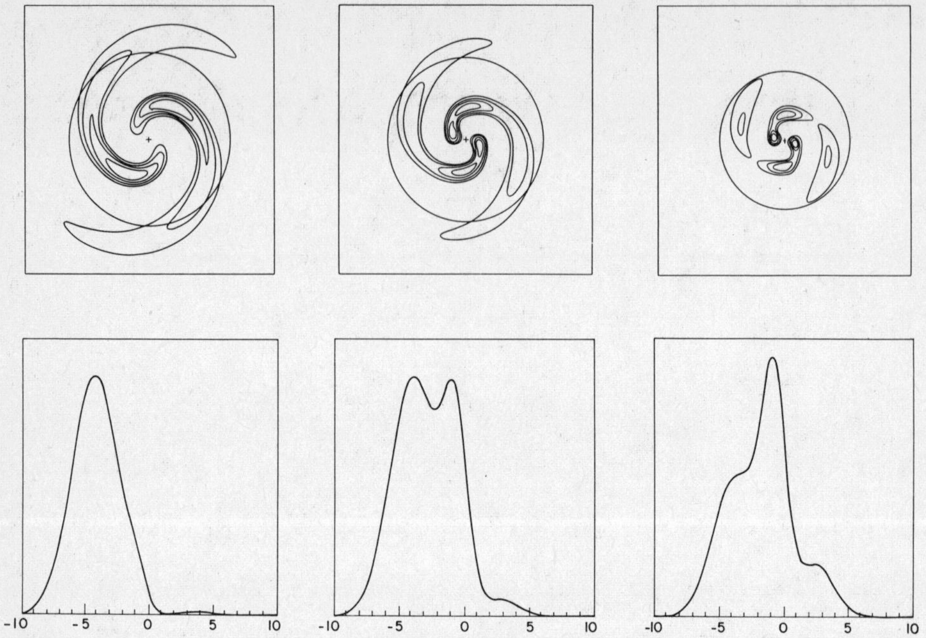


FIG. 3. Transition from normal to barred spiral modes. The same spiral mode is followed in a one-parameter sequence of equilibrium models, characterized by different properties of the nuclear bulge. The top frames illustrate density contours, the bottom frames the corresponding power spectra. The distribution of the stability parameter $Q(r;q)$ is given by Eq. **A3** in the *Appendix*, with $q = 0.35$, 0.15, and 0.05, respectively, for the three figures from left to right. The parameters in Eq. **A1** and **A2** are $v = 1.75$, $\mu = 0$.

ifying quantitative predictions on spiral modes, may play the crucial role of removing the singularities (and hence the related continuous spectra) that are present in more idealized models (21, 29).

One limitation of our approach is that we are relying on results based on a linear theory. The role of nonlinearities in spiral structure has long been discussed (e.g., see ref. 30 and comments and references in refs. 8 and 15). Some support for the relevance of linear theories derives from the fact that observed spiral features often have relatively low amplitude. Still, nonlinear effects at resonances have been claimed to play an important role (e.g., see ref. 30). Here we may remark that the phenomenon of trapping and the stability of trapped orbits (8) are not so obvious. In fact, the very coexistence of linear modes with small or large differences in pattern speeds may eliminate nonlinear trapping altogether, thus inhibiting the so-called nonlinear saturation at resonances.

## Appendix

The family of galaxy models used in the present study essentially originates from the model C3 of ref. 13 and is characterized by three parameters: $\mu$, $q$, $v$. These appear in the definition of the model in the following way. The angular rotation speed $\Omega(r;v)$ in the disk required to balance the gravitational pull at radius $r$ is given by

$$\Omega^2(r;v) = \frac{16}{105}\left[\frac{M_1 G}{a_1^{\,3}} H(x_1)\right] + \frac{M_s G}{v^3}\left[1 + \left(\frac{r}{v}\right)^2\right]^{-3/2}, \quad \textbf{[A1]}$$

in which

$$H(x_1) = \frac{59.0625}{x_1^{\,11}} + \frac{26.25}{x_1^{\,9}} + \frac{16.875}{x_1^{\,7}} + \frac{11.25}{x_1^{\,5}} + \frac{6.5625}{x_1^{\,3}}$$

and $x_1 = [1 + (r/a_1)^2]^{1/2}$; $a_1 = 12$, $M_1 = 2{\cdot}10^{10}$, $M_s = 5.4{\cdot}10^9$. The active disk mass density $\sigma(r;\mu)$ is:

$$\sigma(r;\mu) = \left[\frac{1}{a_1^{\,2}} x_1^{\,-11} - \frac{\mu}{20a_2^2}\left(1 + \left(\frac{r}{a_2}\right)^2\right)^{-11/2}\right]\frac{4.5}{\pi} M_1 \quad \textbf{[A2]}$$

with $a_2 = 8$. Finally, the stability parameter distribution $Q(r;q)$ in the disk is specified by:

$$Q(r;q) = 1 + q\exp[-r^2/a_3^2] \quad \textbf{[A3]}$$

with $a_3 = 1.45$. The reference model $S^\circ$ is characterized by $v = 1.75$, $\mu = 0$, $q = 1$. In the above definitions $G$ is the gravitational constant. In order to have numbers applicable to some realistic galaxies one may consider the (dimensional) lengths $a_1$, $a_2$, $a_3$, and $v$ as measured in kiloparsecs, the masses $M_1$ and $M_s$ as measured in solar masses. We recall that the stability of a galactic disk is essentially determined by three functions: the rotation curve $\Theta(r) = r\Omega(r)$, the disk mass density $\sigma(r)$, and the dispersion speed profile $c(r)$. The rotation curve measures the total mass of the galaxy and provides the local values of the epicyclic frequency $\kappa(r)$ and the shear rate $d\ln\Omega/d\ln r$. The disk density $\sigma(r)$, for a given $\Theta(r)$, determines the global disk to halo mass ratio and the local typical scalelength of the disk—i.e., the dimensionless quantity $\varepsilon_o(r) = \pi G\sigma/r\kappa^2$. For a spiral disturbance with $m$ arms, the above functions completely determine the profile of the parameter $j = m\varepsilon_0(4\Omega/\kappa)|d \ln \Omega/d \ln r|^{1/2}$, which plays a crucial role in the local stability and propagation properties of nonaxisymmetric waves (17). For a given $\Theta(r)$ and $\sigma(r)$ the "pressure" profile $c(r)$ fixes the behavior of the local stability parameter $Q = c\kappa/\pi G\sigma$. In fact, on dynamical grounds, we chose to specify directly $Q$, rather than $c$. Note that the parameters $\mu$, $q$, and $v$ can be varied independently, changing one dynamical function at a time.

This work was partly supported by the National Science Foundation and partly by the Consiglio Nazionale delle Ricerche, under the United States–Italy Cooperative Science Program.

1. Sandage, A. (1975) in *Stars and Stellar Systems*, Galaxies and the Universe, eds. Sandage, A., Sandage, M. & Kristian, J. (Univ. of Chicago Press, Chicago), Vol. 9, pp. 1–35.
2. van den Bergh, S. (1960) *Astrophys. J.* **131**, 215–223.
3. Lin, C. C. (1971) in *Highlights of Astronomy*, ed. de Jager, C. (International Astron. Union, Reidel, Dordrecht, Netherlands), Vol. 2, pp. 88–121.
4. Roberts, W. W., Roberts, M. S. & Shu, F. H. (1975) *Astrophys. J.* **196**, 381–405.
5. Lindblad, P. O. (1960) *Stockholm Obs. Ann.* **21**, No. 4.
6. Miller, R. H., Prendergast, K. H. & Quirk, W. J. (1970) *Astrophys. J.* **161**, 903–916.
7. Hohl, F. (1971) *Astrophys. J.* **168**, 343–359.
8. Bertin, G. (1980) *Phys. Rep.* **61**, **1**, 1–69.
9. Julian, W. H. & Toomre, A. (1966) *Astrophys. J.* **146**, 810–830.
10. Lynden-Bell, D. & Kalnajs, A. J. (1972) *Mon. Not. R. Astron. Soc.* **157**, 1–30.
11. Mark, J. W.-K. (1976) *Astrophys. J.* **205**, 363–378.
12. Zang, T. A. (1976) Dissertation (Massachusetts Institute of Technology, Cambridge, MA).
13. Mark, J. W.-K. (1977) *Astrophys. J.* **212**, 645–658.
14. Bertin, G. & Mark, J. W.-K. (1978) *Astron. Astrophys.* **64**, 389–397.
15. Lin, C. C. & Lau, Y. Y. (1979) *Stud. Appl. Math.* **60**, 97–163.
16. Goldreich, P. & Lynden-Bell, D. (1965) *Mon. Not. R. Astron. Soc.* **130**, 125–158.
17. Lau, Y. Y. & Bertin, G. (1978) *Astrophys. J.* **226**, 508–520.
18. Goldreich, P. & Tremaine, S. (1978) *Astrophys. J.* **222**, 850–858.
19. Pannatoni, R. F. (1979) Dissertation (Massachusetts Institute of Technology, Cambridge, MA).
20. Pannatoni, R. F. & Lau, Y. Y. (1979) *Proc. Natl. Acad. Sci. USA* **76**, 4–7.
21. Lin, C. C. & Bertin, G. (1981) in *Plasma Astrophysics*, eds. Guyenne, T. D. & Levy, G. (European Space Agency, Noordwijk, Netherlands), SP-161, pp. 191–205.
22. Elmegreen, D. M. (1981) *Astrophys. J. Suppl. Ser.* **47**, 229–233.
23. Visser, H. C. D. (1978) Dissertation (Univ. of Groningen, Groningen, Netherlands).
24. Roberts, W. W. (1979) in *Photometry, Kinematics, and Dynamics of Galaxies*, ed., Evans, D. S. (Univ. of Texas at Austin, Austin, TX), pp. 461–471.
25. Toomre, A. (1981) in *The Structure and Evolution of Normal Galaxies*, eds. Fall, S. M. & Lynden-Bell, D. (Cambridge Univ. Press, London), pp. 111–136.
26. Yue, Z. Y. (1982) *Geophys. Astrophys. Fluid Dyn.* **20**, 1–46.
27. Yue, Z. Y. (1982) *Geophys. Astrophys. Fluid Dyn.* **20**, 47–68.
28. Yue, Z. Y. (1982) *Geophys. Astrophys. Fluid Dyn.* **20**, 69–110.
29. Bertin, G. & Haass, J. (1982) *Astron. Astrophys.* **108**, 265–273.
30. Contopoulos, G. (1973) *Astrophys. J.* **181**, 657–684.

# QUASI-STATIONARY SPIRAL STRUCTURE IN GALAXIES

C.C. Lin
Massachusetts Institute of Technology
Cambridge, Massachusetts 02139 USA

The hypothesis of *quasi-stationary spiral structure* in galaxies was explicitly formulated in the early 1960's in papers of Bertil Lindblad and of Lin and Shu. It asserts that the grand design observed in spiral galaxies may be described by the superposition (and interaction) of a *small* number of spiral *modes*. (See Lin and Bertin, 1981 for a fairly extensive review of the theory.) We wish to re-affirm the correctness of this hypothesis in the present contribution. Early numerical experiments by P.O. Lindblad and by Miller, Prendergast and Quirk demonstrated that spiral structures occur naturally in certain models of stellar systems, although it was difficult to control the morphological types of galaxies simulated. We are now able to simulate galaxies of various morphological types in a controllable manner. Numerical fluid-dynamical codes developed by Pannatoni (1979) and improved by Haass (1982) have been used to calculate normal modes of various spiral types (Haass, Bertin, and Lin 1982) in the morphological classification of Hubble, Sandage, and de Vaucoulers. Furthermore, the processes that govern the maintenance and the excitation of these modes simulating both normal spirals and barred spirals, can be understood by using analytical theories which are closely related to the local dispersion relationship, as Bertin will describe in his paper at this conference. Understanding these mechanisms enables us to choose the parameters and the distribution functions in our models more properly in order to exhibit the desired characteristics in the computed modes. Such an approach also has important implications on observational studies. Much of the previous work on comparison between theory and observations in normal spirals used only the short trailing waves. A mode must consist of at least two waves propagating in opposite directions. It has been found that, at least in normal spiral modes, the long wave branch provides essentially only a *modulation* of the amplitude along the short wave branch, which accurately describes the phase. Previous calculations are thereby justified. [These points were not adequately covered in the previous paper reviewing theory of spiral modes.]

I wish now to turn to the discussion of another point of great significance to observers. The powerful "swing" amplification process suggests that the spiral structure might be rapidly varying, on the time scale of an epicyclic period and *not* quasi-stationary. Actually, this perception of swing amplification only holds for wave *packets*. In the context of steady wave *trains* or *modes*, it can be shown that an exactly equivalent amplification mechanism can be described in terms of a process of wave amplification via stimulation of emission of radiation (WASER, see figure). The same analytical and numerical calculations are involved; only the physical interpretation of the symbols are different. Indeed, the amplification factor at the corotation circle of a mode has been systematically calculated and published by Drury (1980) for two standard rotation curves. We have now further clarified the comparison of the two approaches by deriving a second order differential equation for the Fourier transform of the perturbation of the gravitational potential, with a time-*independent* wave number $\xi$ as the independent variable. The parameter $\xi$ then takes the place of the time-*dependent* wave number in the swing formalism. Otherwise, the equations to be solved are identical in form to Eq. (35a) [homogeneous part] and Eq. (35b) in the 1978 paper by Goldreich and Tremaine and to Eqs. (18) and (20) in the 1981 paper of Toomre. The numerical calculations required are also the same. The WASER is, however, more *appropriate* for the description of modes or steady wave trains. Indeed, Goldreich and Tremaine (1978) already demonstrated the equivalence of the swing process with the WASER processes in the special case studied earlier by Mark, who incorporated the amplification mechanism into the asymptotic theory of normal modes (WASER I). Now the equivalence is demonstrated for all spiral modes. Incorporation of the WASER process into the theory of bar modes (WASER II) will now be explained by Bertin.

Figure 1. *WASERS OF TYPES I AND II:* As the incident wave is refracted, there is also stimulated an emission of radiation. This strengthens the returned wave, thereby providing amplification in the feedback cycle.

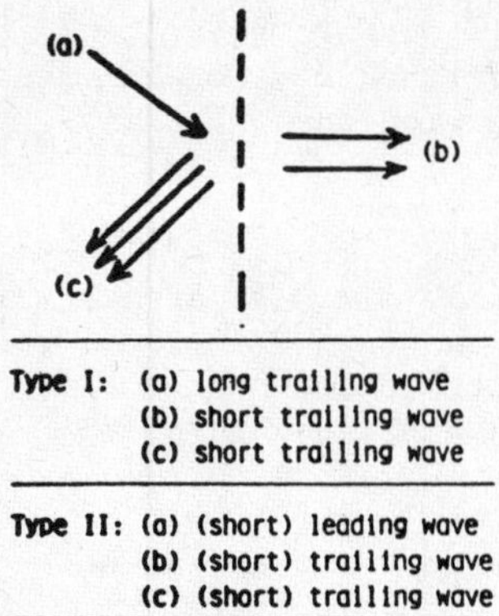


REFERENCES


Drury, L.O'C.: (1980), M.N. 193, 337-343.

Goldreich, P. and Tremaine, S.: (1978), Ap. J. 222, 850-858.

Haass, J., Bertin, G. and Lin, C.C.: (1982), Proc. Nat. Acad. Sci. USA 79, 3908-3912.

Lin, C.C. and Bertin, G.: (1981), in *Plasma Astrophysics*, eds. Guyenne, T.D. and Levy, G. (European Space Agency, Noordwijk, Netherlands), SP-161, 191-205.

Pannatoni, R.F.: (1979), Ph.D. Thesis MIT, Cambridge, Massachusetts, USA.

Toomre, A.: (1981), in *The Structure and Evolution of Normal Galaxies*, eds. Fall, S.M. and Lynden-Bell, D. (Cambridge University Press, London), 111-136.

# FORMATION AND MAINTENANCE OF SPIRAL STRUCTURE IN GALAXIES

C. C. Lin and G. Bertin*
Massachusetts Institute of Technology
Cambridge, MA. USA
*Permanently at Scuola Normale Superiore, Pisa, Italia

ABSTRACT. The formation and maintenance of spiral structure in galaxies is discussed in terms of spiral modes with a general perception consistent with the hypothesis of quasi-stationary spiral structure. The latter is explained in some detail to give it the original proper perspective and to contrast it with recent studies of non-stationary spiral structures. We again emphasize the possible coexistence of fast-evolving spiral features and slow-evolving spiral grand designs. Spiral modes are described in terms of three categories of propagating waves. The reason is given why isolated non-barred galaxies with spiral grand design are found to be mostly two-armed. Some suggestions are made for future research.

## I. INTRODUCTION

Research work on the formation and maintenance of large-scale spiral structure has been pursued along several lines of approach. In this paper, we shall attempt to clarify the relationship among the various approaches by presenting a coherent description of one of them and comparing the conclusions for this approach with those obtained from other alternative approaches on a few crucial issues.

The basic mathematical formalism adopted is the theory of linear spiral modes; these modes are in turn described in terms of steady wave trains. To connect our theoretical analysis with observations, we continue to adopt the hypothesis of quasi-stationary spiral structure (QSSS hypothesis), whose precise nature will be explained in some detail.

To give a proper perspective to our discussions, we again quote from Oort (1962) for his statement of the problem:

> "In systems with strong differential rotation, such as is found in all nonbarred spirals, spiral features are quite natural. Every structural irregularity is likely to be



*H. van Woerden et al. (eds.), The Milky Way Galaxy, 513–532.*

drawn out into a part of a spiral. But this is not the phenomenon we must consider. We must consider a spiral structure extending over the whole galaxy, from the nucleus to its outermost part, and consisting of two arms starting from diametrically opposite points. Although this structure is often hopelessly irregular and broken up, the general form of the large-scale phenomenon can be recognized in many nebulae."

A few comments will now be made to clarify the above statements.

Galaxies may or may not exhibit a grand design. But it is the grand design that challenges the theorists and the observers alike. The theorists should be able to explain why it is possible for grand designs to exist, in spite of the winding dilemma. The observers should find out how the various features in the grand design are related to each other. The theorists should again explain why. Indeed, such efforts to try to explain the grand-design structure lead us to find out more about the astrophysical processes that prevail in the interstellar medium and in the process of star formation. The key step that made such investigations possible is the adoption of the hypothesis of quasi-stationary spiral structure (QSSS) as a working hypothesis in a semi-empirical context. (See Lin, 1971, pp. 89, 91.)

Some unnecessary controversy has arisen around this hypothesis in recent years. There are at least the following two reasons for this controversy: (1) The hypothesis has sometimes been misunderstood and given a role beyond that of a working hypothesis. This confusion can be easily cleared away by referring to the original literature, as is done in this paper. (2) There have been studies of dynamical mechanisms which suggest the possibility of spiral patterns in rather rapid evolution, not quasi-stationary structures. At the meeting in Besançon last year, only one aspect of the latter issue was addressed. In this paper, we shall attempt to deal with the issues more fully. Before we go on with further details, we should perhaps make a few remarks on the general perception.

From laboratory experiments on hydrodynamic stability and turbulence and from the study of weather patterns, -- both of which can be observed to evolve in time, -- we know that some flow patterns are quasi-stationary while others are transient. There are advantages in focussing our attention on regular quasi-stationary structures. In the study of galaxies, the adoption of the QSSS hypothesis has enabled us to pursue the study of a number of astrophysical phenomena and astrophysical processes (see Section V for a short list of examples). We therefore first focus our attention on the study of quasi-stationary phenomena, and leave the study of transient processes to separate investigations. Clearly, the studies of quasi-stationary and transient phenomena are complementary rather than competitive, unless it can be shown that quasi-stationary structures cannot possibly exist or that they can exist only under exceptional circumstances. Observational evidence gives us assurance that this is not so (cf. paper by D.M. Elmegreen at this conference). At the same time, we shall devote a part of this paper (Section II.3) to the discussion of studies which apparently yield transient impressions and point out why they do not, in fact, conflict with the possible existence of quasi-stationary spiral features.

Actually, this issue was already quite carefully examined by Bertil Lindblad (1963) twenty years ago, when he discussed the implications of some pioneering computational studies by Per Olof Lindblad. Specifically, his paper was entitled "On the possibility of quasi-stationary spiral structure in galaxies", and this possibility was recognized despite indications from the computational studies that the spiral structure may be more likely to be "quasi-periodic" (as described in Lindblad's paper), or regenerative (as described in the paper by Goldreich and Lynden-Bell, 1965).

Besides these general issues, two further points should be noted in the above quotation from Oort. (1) Attention is directed towards non-barred spirals as being the more challenging. (2) Even these galaxies preferentially have two-armed structures. In Section IV, we shall show why this is the case even though we know that there are isolated galaxies with other types of symmetry (cf. Iye et al., 1982 for a specific example). As it turned out, Landau damping of waves at inner Lindblad resonance holds the key; multiple-armed spiral structures are more likely to encounter inner Lindblad resonance as already noted in the early nineteen-sixties. Without inner Lindblad resonance, multiple-armed spiral features may indeed overwhelm two-armed spirals (see Section IV.4).

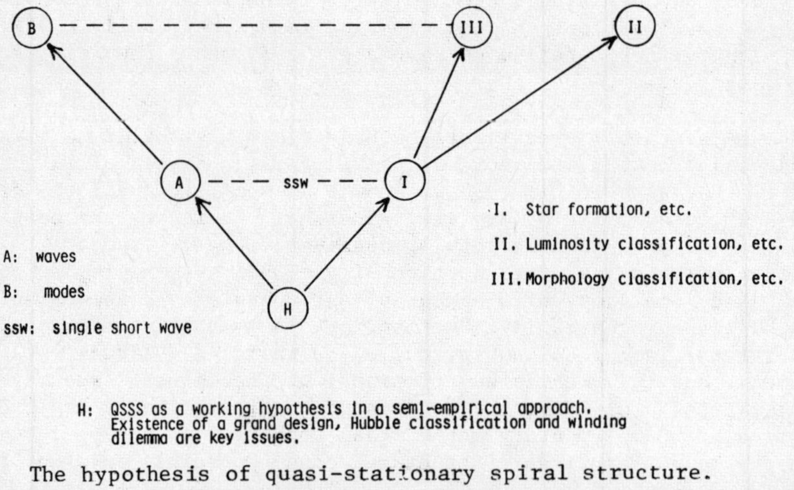


Fig. 1. The hypothesis of quasi-stationary spiral structure.

## II. THE HYPOTHESIS OF QUASI-STATIONARY SPIRAL STRUCTURE

1. The Semi-Empirical Approach. As a working hypothesis in a semi-empirical approach, its correctness is primarily to be judged by the inferences made from it. We require consistency both with observations and with internal dynamical mechanisms. As mentioned above, the QSSS hypothesis has led to a number of successful applications to observed phenomena, a partial list of which will be given in Section V. We call special attention to the application to the process of star formation and to the possibility of providing the basis for the luminosity and the morphology classifications of galaxies. On the other hand, we also derive support from dynamical studies, which will be given in some detail in this paper.

The research efforts may, therefore, be represented in terms of a picture of "mountain climbing", cf. Toomre (1977) and Tremaine (1983). Both these authors appear to emphasize a deductive approach from first principles. From our experience with the much simpler system of homogeneous fluids, we know that it would be quite difficult to "deduce" the existence of regular patterns in the full non-linear theory. Rather, we may attempt to describe such patterns by first accepting their plausible existence on the basis of experiments (see examples of regular and irregular patterns given by Figures 128, 131, 137, 158, 177 in van Dyke's Album, 1982). We then attempt to reach a better understanding of the underlying mechanisms that distinguish irregular flow patterns from regular grand designs. These mechanisms are, however, not always easy to describe.

The QSSS hypothesis receives support from the theory of linear spiral modes which has been quite successfully developed in recent years, since the coexistence of a small number of such modes, slowly growing from a certain basic state, provides a natural basis for the eventual existence of a quasi-stationary spiral structure. However, as is well known in the studies of hydrodynamic stability and turbulence, the calculation of linear modes "merely marks the beginning of the dynamical studies", and the same comment applies to the study of galactic spirals (Lin and Lau, 1979, p. 107). The existence of linear modes does not imply that the spiral structures are always necessarily permanent in shape. "The complicated spiral structure of galaxies indicates the coexistence of material arms and density waves -- and, indeed, of the possible coexistence of several wave patterns. These features [in general] influence but do not destroy one another." (Lin, 1971, p. 91.)

The basic time rate in all dynamical studies is of course the shear rate or the epicyclic frequency. However, linear modes in general evolve at a much slower rate. [Indeed, if the time scale for a feedback cycle in a linear mode were comparable to that of its evolution, the "mode" would hardly be a useful concept from a physical point of view.]

Experience in hydrodynamical studies shows that the observed nonlinear states which are qualitatively similar to slowly evolving linear modes usually also evolve slowly. Thus, the study of linear modes gives a powerful method, perhaps the most convenient, to approach the problems of formation and maintenance of spiral structure on a global scale, since they do represent the intrinsic characteristics of the dynamical system. Thus, even the effect of external excitation may be described in terms of these modes. As an analogy, we may recall that essentially a slowly decaying quasi-stationary mode is excited through the ringing of a church bell.

2. The Single Short Spiral Wave. Here we wish to call attention to one fact which has not always been put in proper perspective. The phenomena listed in Section V have mostly been described by the use of a single short spiral wave. Such a short wave requires a feedback process (must be "replenished", see Toomre, 1969) for its long-term existence as a part of a modal formulation (Shu, 1968). The question is whether this feedback system can be observed. This will be discussed in the section on linear modes (Sections III and IV) where the feedback mechanism will be examined in some detail.

3. Nonstationary Impressions. The plausible existence of quasi-stationary spiral structures can be questioned if all studies suggest that the dynamical processes always imply rapid evolution of spiral structures. Thus, the following three issues deserve our special attention:

a. Rapid dynamical evolution is often found in N-body simulations;
b. strong amplification is associated with a "swinging" process;
c. strong "instability" seems to be present even when the disk is stabilized against axisymmetrical disturbances ($Q \geq 1$).

The first issue will be the main topic of this subsection. The third issue will be taken up in Section III. The second issue, discussed at Besançon (Lin, 1983b), will be briefly reviewed here. As an alternative to the time-dependent "swinging" approach (Toomre, 1981), Lin referred to Drury's work (1980), which does not involve a rapid evolution. It is therefore compatible with growing modes, even though both approaches may still be labelled "swing amplification", in the sense that they both describe the transformation of a leading spiral into an amplified trailing spiral. Bertin (1983b) presented still another analysis of leading-trailing wave amplification in terms of steady wave trains satisfying the local dispersion relationship. Again no transient behavior is implied, i.e., time evolution is not required. Rapid time evolution does occur in certain contexts; for example, as a result of certain special initial conditions (cf. Toomre, 1981). We shall take up this topic again in Sections III and IV in the context of modes.

Let us now turn to the first item. It is well known that there are apparent conflicts between results obtained through the use of numerical simulation and those obtained through analytical methods. There are perhaps three reasons for this. (1) The requirement on the number of particles is highly demanding when simulation of phase-mixing is needed. Often there are simply not enough particles in the phase-space. The simulation of $N_0 \sim 10^6 - 10^8$ particles by a single super-particle severely distorts the distribution function in velocity space. (2) The small number and the large mass of each superparticle magnifies the possible fluctuations of surface density and gravitational field by a factor of the order of $N_0^{1/2}$, which could be as large as $10^3 - 10^4$, thus greatly accelerating the evolutionary processes. These difficulties are not easy to remove. (3) Sometimes artificial physical processes are introduced to control effects such as the rapid rise of velocity dispersion. This could lead us to even worse situations, where it is no longer clear what physical system is being simulated. (The computation may become like a computer game which can be intellectually stimulating and even suggestive, but whose results cannot be accepted at face value. Proper interpretation is needed.)

All these ideas and discussions are not new. Computer experiments in plasma physics have been performed for at least two decades (see an excellent extensive review by Dawson, 1983). Millions of particles have been used, but experts are still demanding more in order to get a proper simulation of processes involving resonances, where phase mixing and Landau damping are important.

4. Some Additional Remarks. In addition to these studies giving nonstationary impressions, there is also the subtle mathematical point of the existence of an infinite number of modes with a continuous spectrum. This issue was examined by Lin and Bertin (1981), and found not to be serious.

In the spirit of the QSSS hypothesis, we therefore adopt the modal approach, because it is mathematically sound and it is convenient for the description of comparatively regular grand designs, including their evolution. The complete description of the evolutionary process must, of course, be based on the use of a nonlinear theory which is yet to be developed. The general nature of such developments can be partly visualized by examining existing theories for some simpler systems. For homogeneous fluids, the mathematical theory can be sketched in the following manner. Any dynamical variable $\psi(\underline{\mathbf{x}},t)$ may be represented in the form

$$\psi(\underline{x},t) = \sum A_n(t)\phi_n(x)$$

in terms of a complete set of eigenfunctions $\phi_n(\underline{x})$. [Integration is implied by $\Sigma$ when a continuous spectrum exists.] In the linear theory, each $A_n(t)$ is exponential in time $t$. In the full non-linear theory, the set of variables $\{A_n(t)\}$ satisfies nonlinear differential equations which describe the process of evolution. Extensive investigations along this line have been carried out, especially in meteorological studies. Galaxies are, of course, complicated systems with many components. Non-linear effects in the gaseous component have been shown to be likely to play an important role.

## III. DYNAMICAL MECHANISMS: WAVES AND MODES

For the study of dynamical mechanisms, we shall focus our attention on linear processes, with one important exception. This is the issue of angular momentum transfer associated with the spiral patterns. Bertin (1983a) has shown that the angular momentum transfer associated with normal spiral modes is not going to produce a significant impact on the basic state over a period of several billion years even when the amplitude of the disturbance is about one-quarter of the basic distribution. Thus, we need not be unduly concerned with any fast evolution from this mechanism.

Linear modes can be described in terms of waves propagating in opposite directions, together with mechanisms of feedback from the central regions and over-reflection of these waves near the corotation circle. In earlier studies, we recognized two categories of waves: long waves and short waves. With leading and trailing configurations in each case, there are altogether four types of waves. Detailed studies based on the Lau-Bertin (1978) dispersion relationship now lead us to the recognition of a third category of waves: the open waves, again with possibly leading and trailing configurations. Open waves are more open than long waves, but propagate like short waves.

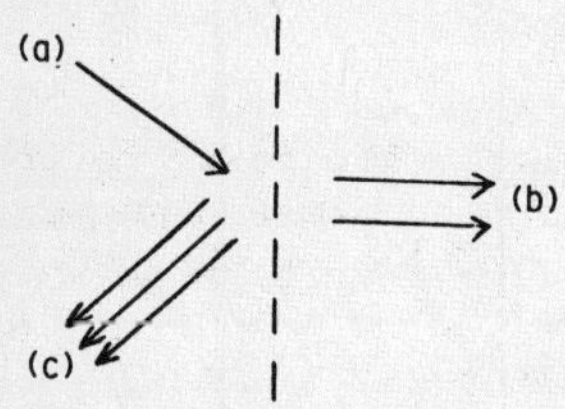


OVER-REFLECTION

| | |
|---|---|
| Type I: | (a) long trailing wave |
| | (b) short trailing wave |
| | (c) short trailing wave |
| Type II: | (a) open leading wave |
| | (b) open trailing wave |
| | (c) open trailing wave |

Fig. 2. Two types of over-reflection

In this new perspective, the two principal types of processes of over-reflection near corotation are shown in Fig. 2, which is amended from Fig. 1 of Lin's Besançon paper. The term over-reflection is used to describe a process in which the reflected wave contains more energy than the incoming wave. Over-reflection of Type I is the WASER (Mark, 1976) where only trailing waves are involved; that of Type II corresponds to the "swing amplification" (Toomre, 1981), in the sense that leading waves are converted into amplified trailing waves. Open waves are especially important for ranges of dynamical parameters for which there is only one real solution. For those cases, the distinction between open waves and short waves becomes less apparent.

The Lau-Bertin dispersion relationship is a cubic, with two parameters $J$ and $Q$ (see their paper for definitions of symbols). For the treatment of all three waves, it is therefore convenient to introduce the $(J,Q)$ plane. The same format can be used for the study of the factor of over-reflection after the value of the shear parameter $s$ is specified. In the case of constant linear velocity of circular motion, the shear parameter $s$ equals unity.

Let us now briefly outline some of the principal results which differ from previous conclusions.

1. Waves in the $(J,Q)$ Plane. First, the cubic dispersion relationship leads to the following overall outlook. The two real solutions representing the long/short waves can exist only if

$$JQ^2 \leq 4\sqrt{3}/9.$$

The third real solution represents an open wave, which is often too long to fit into the disk geometry, and should be excluded. However, this is not always the case. We may have to deal with all three categories of waves under certain circumstances (usually for $J$ not too small). On the other hand, for

$$JQ^2 > 4\sqrt{3}/9 \quad ,$$

there is only one real solution, i.e., the open wave. The solutions which represent long/short waves have become complex conjugate pairs. It is in this range of parameters that the "swing" process is usually discussed.

Thus, in general, neither the long/short wave cycle nor the swing cycle represents the complete picture. We shall present examples of calculated spiral patterns in which the feedback processes include all three categories of waves (cf. Fig. 6 in Section IV).

2. Over-Reflection in Moderate Amounts. The second point is that one can identify a range of parameters in the $(J,Q)$ plane such that the amplification factor (over-reflection) is moderate, whether the over-reflection is of Type I or of Type II (see Fig. 2). Moderate over-reflection is typified by an energy ratio of two (and an amplitude ratio $\alpha = \sqrt{2}$). The curve for this condition has been calculated by two methods. The system of curves shown in Fig. 3 is obtained from the approach used by Bertin in his Besançon paper, for different values of the shear parameter $s = - d\ln\Omega/d\ln r$. We also

checked Drury's calculations of the amplification factor for the case $s = 1$ by using an equation equivalent to Eq. (35) in Goldreich and Tremaine (1978). (Robert Thurstans did the computer work. Drury used a different, though equivalent, procedure.) The results are shown in Fig. 4 as curves of constant over-reflection (logarithmic scale). The two figures may be compared for $s = 1$, $\ln \alpha = 0.35$ [curve (b) in Fig. 4 versus curve marked $\varepsilon = 1$ in Fig. 3]. The numerical values are not identical because in Bertin's formulation the Weber equation is used, just as in Mark's original WASER formalism. The two approaches are indeed two different approximations, and there is no good reason to decide that one approximation is better than the other. Fortunately, they support each other within the proper limits of these approximations.

3. The $(J,Q)$ Parameters for Moderate Over-Reflection. In the new perspective just discussed, one is led to emphasize the role of the parameter $J$ (or $J_1$, its value for $m = 1$) in the determination of the stability characteristics of the spiral modes. As shown before (cf. Lin and Lau, 1979, Bertin, 1980), tightly wound spirals with moderate rates of growth are associated with small or moderate values of the $J$-parameter, and with a value of the $Q$-parameter close to unity. Such modes are maintained by a specific feedback mechanism.

Let us leave the feedback mechanism aside for the moment and focus our attention on the over-reflection factor at corotation. If one follows the curve of "moderate" over-reflection factor, $\alpha = \sqrt{2}$, in Fig. 4, one finds that $Q \simeq 1$ only if $J \lesssim 0.5$ (Axisymmetry implies $J = 0$). In contrast, for $J = 1$, very high over-reflection is reached at $(J, Q) = (1,1)$; the factor becomes moderate ($\alpha = \sqrt{2}$) only when $Q$ is raised to about $\sqrt{3}$. Thus, the condition $Q = 1$ does not always have special significance in all galactic disks. It is essentially the condition to be expected in galaxies with relatively low disk mass (low $J$), typified by normal spirals of types Sa and Sb.

There is another reason for expecting to observe higher values of $Q$ associated with stars. This is the presence of a small amount of gas, which may cause a significant reduction of the effective $Q$-parameter (see Lin and Shu, 1966). For the specific combination of stellar and gaseous disks considered, marginal stability requires that

$$Q_* \simeq 1 + 2(\sigma_g/\sigma_*) \quad .$$

Thus, for 30% gas-to-star mass ratio (which is not large when disk thickness is taken into account), we find that $Q_* = 1.6$. Combining this with the factor $\sqrt{3}$ discussed above, we see that $Q_* = 2.7$ would not be, in some cases, an unreasonable value. Of course, this calculation gives only an indication of the type of situation to be expected and the result should not yet be directly compared with observations (cf. Kormendy's paper at this meeting).

With respect to observational studies, one should note that the usual definition of $Q$ is quite easily influenced by the presence of a

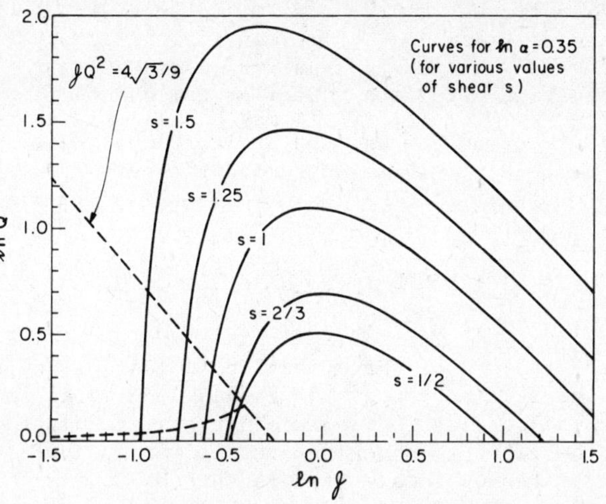


Fig. 3. Curves for a given coefficient of over-reflection ($\alpha = \sqrt{2}$) in the $(J,Q)$ plane for various values of the shear parameter s.

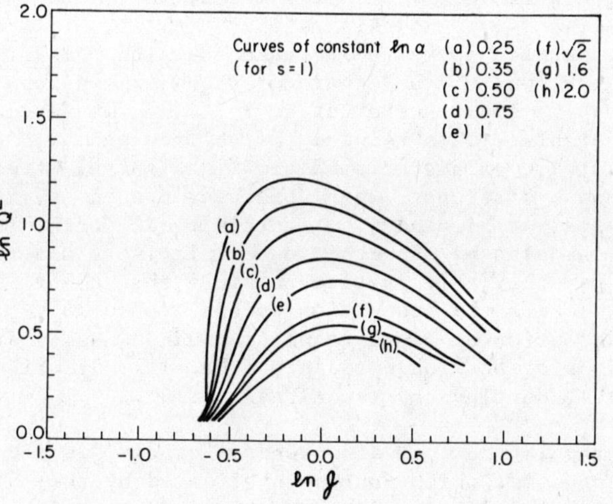


Fig. 4. Curves for various values of the coefficient of over-reflection $\alpha$ in the $(J,Q)$ plane for a given value of the shear parameter (s=1).

small amount of stars with high velocity dispersion. This is the "long-tail" problem common to many statistical distributions. After all, the classical Cauchy distribution implies an infinite value for Q, if one applies the same definition as that for quasi-Maxwellian distributions. "Contamination" from other stellar components (e.g., halo stars) has to be removed. Basically, it may be wise to refine the definition of the stability parameter or even to introduce a new parameter that emphasizes the role of the responsive stars with low dispersion velocities.

## IV. DYNAMICAL MECHANISMS: SPIRAL MODES AND MORPHOLOGICAL CLASSIFICATION

1. A Process of Self-Regulation. Many modal calculations have been done in the fluid model. The process of Landau damping at inner Lindblad resonance is therefore not incorporated in the calculations. An additional step must be taken to check the result for possible resonance effect. Such a step will make it apparent that the number of unstable spiral modes is expected to be reduced, because the feedback mechanism may be seriously impaired by the Landau damping process. (See subsection 4 below for further details.)

If we combine these considerations with the above discussions of the parameters $J$ and Q, we are led to the following scenario for the self-regulation of the instability of spiral modes. Suppose that the galactic system, to begin with, has $J \simeq 1$ (for m = 3, say), and Q = 1. The excitation of spiral waves and modes, possibly violent, will tend to cause the stellar dispersion speed to increase until the disk is "heated up" to such high values of the Q parameter and such a form for its distribution that only a small number of spiral modes may eventually remain moderately unstable. This reduced instability also restricts the increase in the Q-parameter, and the final spiral structure may be expected to be quasi-stationary when nonlinear processes, involving both stars and gas, are included. [In this final state, the over-reflection factor may also be expected to be relatively moderate, but not necessarily very small.] It is also possible that the galaxy will evolve into a state where the dispersion speeds are so large that only a gentle oval distribution of stars results. Such a mass distribution would drive the Population I objects into a bar-like distribution, as shown by Roberts, Huntley and van Albada (1979).

The process of self-regulation just discussed is crucial in the context of the QSSS hypothesis, which can be invalidated by the presence of too many unstable spiral modes. We shall discuss it further in connection with some examples below. Having developed a scenario for the realization of quasi-stationary spiral structures, we are ready to venture to relate our theoretical investigations to one of the major objectives: the search for a dynamical basis for morphological classification. For this purpose, we must first be able to produce modes of various morphological types in the context of the linear theory. Based on experience with hydrodynamical experiments, especially those with rotating fluids, we may then infer that the grand designs in the final state are approximately simulated by the superposition of modes obtained from the linear theory.

2. Modes of Various Morphological Types. In previous publications, (see Lin and Bertin, 1981; Haass, Bertin and Lin, 1982) we have shown how one may use linear modes to simulate the various morphological types of the galaxies by judiciously prescribing three distribution functions: (a) the rotation curve, (b) the mobile component of the disk mass (including the effect of disk thickness) and (c) the distribution of dispersion speed of stars (or the acoustic velocity in the gaseous model). These different morphological types are often associated with different feedback mechanisms. For example, it is clear that the (r), (s) types relate to different degrees of propagation of waves through the center of the galaxies. The time is now ripe for relating the morphological type to dynamical mechanisms and modal calculations. Some examples were published by Haass, Bertin and Lin; some additional examples will now be cited.

Bertin (1983b)contrasted an open mode with a tightly wound spiral mode obtained from the original asymptotic theory. Specifically, their propagation diagrams were compared. The open mode shown by Bertin has an inner turning point for wave propagation. This implies a "reverse swing" of the wave and a rather long feedback cycle. It thus has only a moderate growth rate, smaller than that of the normal mode shown, even though the over-reflection factor is quite high, similar to that of the normal mode. Indeed, this open mode is not represented in Fig. 2 in Norman's Besançon paper (1983). The avoidance of violently unstable modes can be achieved through an increase of the Q-parameter.

For each of the two types of modes discussed by Bertin, we may assign a point in the (J, Q) plane according to the values of these parameters at the corotation circle. In Fig. 5, which shows a portion of Fig. 3, these points are located in the general areas marked I and II, corresponding respectively to the two types of over-reflection. We now note that there is a triangular region in the (J, Q) plane, marked III, corresponding to a gradual transition from one type of feedback to another.

To demonstrate this transition, a sequence of modes has been calculated for which the conditions at the corotation circle vary from Region I to Region II, traversing through Region III. The propagation diagrams show clearly that, in the cases shown in Fig. 6, it is the most open of the three waves that carries the primary signal across and around the corotation circle. The dashed lines show the real part of the complex conjugate pair of comparatively shorter waves. Their roles are, however, not insignificant. Thus, all three categories of waves are present, albeit some in modified forms. The over-reflection process and the feedback process are thus quite complicated; both types are involved. But all of these modes show moderate growth rates. Thus, the presence of over-reflection of Type II does not necessarily imply violent instability. For details, we refer to future publications.

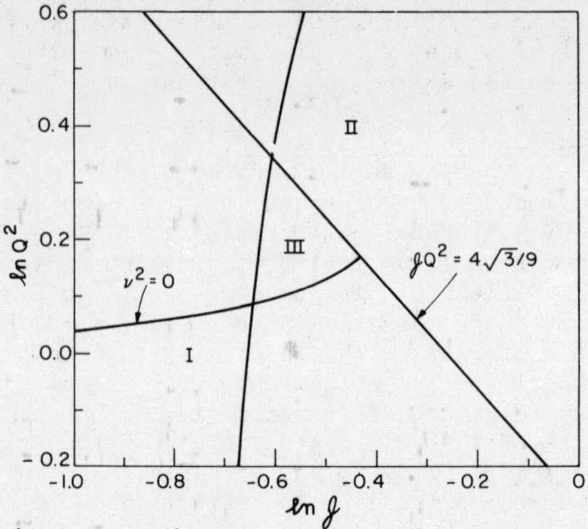


Fig. 5. Typical areas in the (J,Q) plane where over-reflections of Types I and II are important. These are respectively labelled I and II. In domain III, all six waves may be active. The nearly vertical curve is that for $\alpha = \sqrt{2}$ for the open waves. Unstable modes with moderate growth rates have been calculated for all three domains.

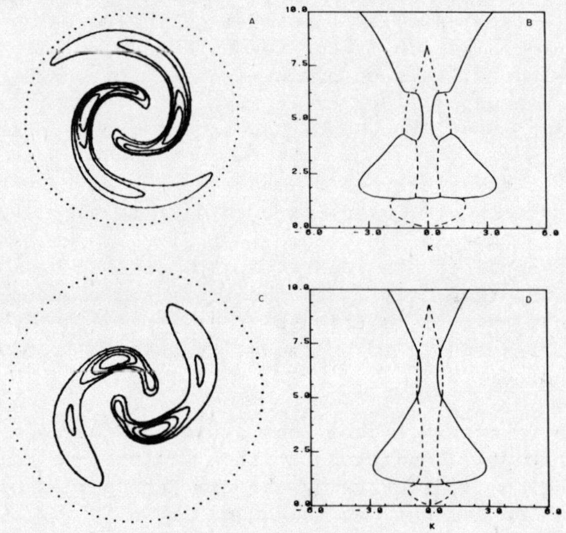


Fig. 6. Spiral patterns and propagation diagrams for two cases, A and C, where both types of over-reflection are active. Note that near co-rotation the open waves, which are real solutions represented by the solid lines, are the most important. For conditions far away from the co-rotation circle, the short waves are the important ones represented by solid lines. In Case A, ($\ell n$ J, $\ell n$ $Q^2$) = (-.511,+.191) near Region I; in Case C, ($\ell n$ J, $\ell n$ $Q^2$) = (-.371,+.365) in Region II.

3. Are Feedback Processes Observable? The short wave is assigned a predominant role in much of the application theory. What can be expected from the feedback process from an observational point of view? There is no general answer to such a question, but the following comments may be helpful.

In the case of normal spirals, the long trailing wave, which is responsible for the feedback process, produces a modulation on the short trailing wave. (See Bertin et al., 1977, p. 4728, and reference to Mark's earlier work.) For empirical purposes, it simply appears that there is a change of intensity of the spiral field with radial distance from the center. The simplicity of the situation results in part from the fact that in the amplified modes, the over-reflection process gives the short trailing spiral a considerable advantage. For a real galaxy, one can construct a scenario compatible with an over-reflection process of the same type by including the damping of such short waves by the shock wave in the gaseous component.

In the case of open modes such as that discussed by Bertin, the situation is somewhat different. The open waves lead to something like an oval distortion in the central regions, which exhibits itself through the gaseous response to a bar-like forcing. Again, the general impression may be that of only one trailing spiral extending out of a bar.

4. Why Two-Armed Modes? We now turn to the following important issue: why do we not see many three-armed grand designs in galaxies even though the example shown by Haass at the Besançon meeting exhibits higher growth for multiple-armed modes? The key to this puzzle lies with Landau damping at inner Lindblad resonance, which is not included in the fluid code used by Haass (cf. III.3, IV.1 above). In the example shown by Haass, the rotation curve has a nearly constant angular velocity at the center ($\Omega \simeq \Omega_0$, somewhat like that of M33). Thus, the curve for $\Omega - \kappa/2$ has very low values at the center, but the curve for $\Omega - \kappa/3$ has a peak value of $\Omega_0/3$ at the center. Thus, if a three-armed mode has a relatively large corotation circle (and hence a low $\Omega_p$), it would tend to suffer from inner Lindblad resonance. Note that the three-armed modes shown by Haass (1983b) have a rather small corotation circle, since he has deliberately chosen the Q-distribution to have a very small scale, smaller than the scale for the velocity distribution, $(V_{max}/\Omega_0)$. In real galaxies, the Q-distribution may have a much larger scale. In such cases, one may regard the corotation circles as being pushed outwards as the scale for the Q-distribution expands. Lindblad resonance follows, and the mode is expected to be damped in the stellar model (even though this is not apparent in the gaseous model calculated). Two-armed modes, being relatively free from Lindblad resonance, thus become predominant. Notice, however, that low frequency two-armed modes would also suffer from inner Lindblad resonance, and consequently the number of unstable two-armed modes can become quite small.

It is obvious that the above line of reasoning applies to four-armed spirals with stronger justification. One-armed spirals do not suffer from inner Lindblad resonance, but they usually have lower growth rates than two-armed spirals. Observationally, their presence together with two-armed spirals is often not as easily detectable. But the well-known asymmetrical motion of the dust lanes in M31 might indeed be due to the presence of a one-armed structure. However, it is not yet entirely clear why galaxies like NGC4254 (Iye et al., 1982) apparently show only a spiral structure with odd harmonics.

## V. CONCLUDING REMARKS

We have surveyed the study of dynamical mechanisms from a semi-empirical perspective. The following is a short list of the successful applications of the theory.

A Short List of Some Physical Phenomena Studied

1. Possible dynamical basis for the Hubble classification
2. Plausible dynamical basis for the luminosity classification (Roberts, Roberts and Shu 1975)
3. Process of star formation
4. Dust lanes: distribution
5. Synchrotron emission (e.g., in M51): distribution
6. Distribution and motion of atomic hydrogen (e.g., in the Galaxy and in M81; including the modal theory of Visser and Haass)

On the theoretical side, the following points appear to be worthy of our attention.

1. We should recognize the possible coexistence of (a) spiral grand designs which evolve rather slowly, and (b) less regular spiral features which evolve more rapidly, generally on the dynamical time scale of $\kappa^{-1}$, $\Omega^{-1}$ or $(s\Omega)^{-1}$. The hypothesis of quasi-stationary spiral structure is thus re-affirmed to provide a basis for the dynamical approach to the morphological classification of galaxies (cf. Fig. 1).

2. The mathematical model adopted and the mathematical formalism used are crucial to the plausible demonstration of the above statements. While the use of wave packets tends to give impressions that favor the less regular spiral features, the equivalent method of using steady wave trains and modes emphasizes the possible existence of long lasting grand designs. Numerical experiments can be misleading unless Landau damping is properly simulated.

3. It is important to recognize different behaviors of the spiral waves and patterns for different physical models and for different ranges of the dynamical parameters. Dynamical processes tend to re-adjust the distribution of the three

basic parameters listed above (Section IV.2), especially the dispersion velocity of stellar motions. This process of self-regulation, aided by the presence of the gaseous component, may be crucial to the quasi-stationary existence of grand designs of various types (cf. Sections IV.1 and IV.2).

Looking into the future, we see that the study of the morphology of galaxies continues to require the collaboration of observers and theorists. In the dynamical approach to the morphological classification of galaxies, we adopt a logical framework based on our understanding of the dynamical mechanisms and on the spiral appearance to be expected under various dynamical situations. We then compare these conclusions with the morphological classification based on observations. Preliminary work (e.g., by Haass, Bertin and Lin, 1982) shows that this approach is highly promising.

Other promising lines of research include the analysis of observed spiral structures into Fourier components and logarithmic spirals, the study of bars and rings, especially in SBa galaxies, and the study of the dispersion of stellar velocities in external galaxies. Some of these studies are being carried out by various researchers. Because of limitation of space, comments on these topics have to be omitted from this paper.

It is perhaps unfortunate that inadequate appreciation of the theoretical points discussed above has led to considerable confusion in the literature. It is hoped that the present paper helps to clear away much of the confusion and to focus our attention on the substantive issues of attempting to explain the observed phenomena such as the different morphology of the galaxies.

## VI. THE MILKY WAY

At this conference, various aspects of the structure of the Milky Way have been discussed, including the successful modelling of the outgoing motion of the 3-kpc arm by Yuan as a flow driven by a rotating bar. The resultant picture is quite similar to that appearing in the inner parts of NGC5364. For this paper, we shall restrict ourselves to some short comments and a few references.

We recall (see Lin and Yuan 1978, Lin 1983a) that there have been rather extensive theoretical examinations of the spiral structure in the solar neighborhood from both the observational and the theoretical points of view. Consistency with the two-armed pattern originally suggested by Lin and Shu (1967) has been found. But there is coexistence of other spiral features together with the two-armed pattern, which is found to be compatible with spiral modes calculated by Haass (1983 a). The Carina arm may be an extra feature associated with the condition $|\nu| = 1/2$ (Shu, Milione and Roberts, 1973). It is not surprising that multiple-armed features can exist in the outer parts of the Galaxy in addition to a dominant two-armed structure in the more massive inner part. It is also not surprising that an additional one-armed spiral structure can lead to asymmetries of the kind that appears to be observed in the fourth quadrant. However, the long stretches of spiral arms observed present an overall picture that shows a predominantly two-armed pattern (see Burton, 1976). Finally, it should be emphasized that the structure of individual spiral features such as the high-velocity stream outside of the Sagittarius arm and other streaming motions (see Burton and Shane, 1970) played an important role in the development of the density-wave theory by giving observational support to the basic concept, -- a support that remains valid today.

## ACKNOWLEDGEMENT

We wish to thank Jon Haass for many helpful discussions and for permitting us to quote his unpublished results on the sequence of models that are calculated for the demonstration of the various types of over-reflection.

## REFERENCES

Bertin, G.: 1980, Phys. Reports 61 (1), pp. 1-69 (217 references)
Bertin, G.: 1983a, Astron. Astrophys. 127, pp. 145-148
Bertin, G.: 1983b, IAU Symposium No. 100, ed. E. Athanassoula (Reidel), pp. 119-120
Bertin, G. et al.: 1977, Proc. Nat. Acad. Sci. 74, pp. 4726-4729
Burton, W.B.: 1976, Ann. Rev. Astron. Astrophys. 14, pp. 275-306
Burton, W.B., and Shane, W.W.: 1971, IAU Symposium 38, eds. W. Becker and G. Contopoulos (Reidel), pp. 397-414
Dawson, J.M.: 1983, Rev. Modern Phys. 55, pp. 403-447
Drury, L. O'C.: 1980, Monthly Notices Roy. Astron. Soc. 193, pp. 337-343
Goldreich, P., and Lynden-Bell, D.: 1965, Monthly Notices Roy. Astron. Soc. 130, 97, pp. 125
Goldreich, P., and Tremaine, S.: 1978, Ap. J. 222, pp. 850-858
Haass, J.: 1983a, in 'Kinematics, Dynamics and Structure of the Milky Way', ed. W.L.H. Shuter (Reidel), pp. 283-287
Haass, J.: 1983b, IAU Symposium No. 100, ed. E. Athanassoula (Reidel), pp. 121-124
Haass, J., Bertin, G., and Lin, C.C.: 1982, Proc. Nat. Acad. Sci. USA 79, pp. 3908-3912
Iye, M., Okamura, S., Hamabe, M., and Watanabe, M.: 1982, Astrophys. J. 256, pp. 103-111
Lau, Y.Y., and Bertin, G.: 1978, Ap. J. 226, pp. 508-520
Lin, C.C.: 1971, in 'Highlights of Astronomy', ed. C. de Jager (International Astron. Union, Reidel, Dordrecht, Netherlands), Vol. 2, pp. 88-121
Lin, C.C.: 1983a, in 'Kinematics, Dynamics and Structure of the Milky Way', ed. W.L.H. Shuter (Reidel), pp. 277-281

Lin, C.C.: 1983b, IAU Symposium No. 100, ed. E. Athanassoula (Reidel), pp. 117-118
Lin, C.C., and Bertin, G.: 1981, in 'Plasma Astrophysics', eds. T.D. Guyenne and G. Levy (European Space Agency, Noordwijk, Netherlands), SP-161, pp. 191-205
Lin, C.C., and Lau, Y.Y.: 1979, Studies in Appl. Math. 60, pp. 97-163
Lin, C.C., and Shu, F.H.: 1966, Proc. Nat. Acad. Sci. 55, pp. 229-234
Lin, C.C., and Shu, F.H.: 1967, IAU Symposium No. 31, ed. H. van Woerden (Academic Press, London), pp. 313-317
Lin, C.C., and Yuan, C.: 1978, 'Astronomical Papers dedicated to Bengt Strömgren', eds. A. Reiz and T. Andersen, published by Copenhagen University Observatory, pp. 369-386
Lindblad, B.: 1963, Stockholm Observ. Ann. 22, No. 5
Mark, J.W.-K.: 1976, Astrophys. J. 205, pp. 363-378
Norman, C.: 1983, IAU Symposium No. 100, ed. E. Athanassoula (Reidel), pp. 163-174
Oort, J.: 1962, in 'Interstellar Matter in Galaxies', ed. L. Woltjer (Benjamin, New York), pp. 234-244
Roberts, W.W., Huntley, J.M., and van Albada, G.D.: 1979, Astrophys. J. 233, pp. 67-84
Roberts, W.W., Roberts, M.S., and Shu, F.H.: 1975, Astrophys. J. 196, pp. 381-405
Shu, F.H.: 1968, Ph. D. Dissertation, Harvard University
Shu, F.H., Milione, V., and Roberts, W.W.: 1973, Astrophys. J. 183, pp. 819-841
Toomre, A.: 1969, Astrophys. J. 158, pp. 899-913
Toomre, A.: 1977, Ann. Rev. Astron. Astrophys. 15, pp. 437-478
Toomre, A.: 1981, in 'Structure and Evolution of Normal Galaxies', eds. S.M. Fall and D. Lynden-Bell (Cambridge University Press), pp. 111-136
Tremaine, S.: 1083, IAU Symposium No. 100, ed. E. Athanassoula (Reidel), pp. 411-416
Visser, H.C.D.: 1978, The dynamics of the spiral galaxy M81. Ph. D. thesis, University of Groningen, Netherlands
Van Dyke, M.: 1982, An Album of Fluid Motion (The Parabolic Press)

## DISCUSSION

F.H. Shu: In this conference many people have commented on the destabilizing contribution of interstellar gas. I would like to remind people that interstellar gas can also have an opposite effect, namely to saturate growing normal modes by dissipative effects. A. Kalnajs pointed out in 1972 that galactic shocks introduce a phase shift between the response of the interstellar gas and the background stellar density wave that drives the gas. This tends to damp the wave; and W.W. Roberts and I suggested that this might turn out to be a good thing, if an instability mechanism lies behind spiral structure. The various instability mechanisms subsequently worked out and described in Lin's talk all have the following features in common: an amplifier (at corotation) plus feedback (via long, leading, or open waves). Thus, consider

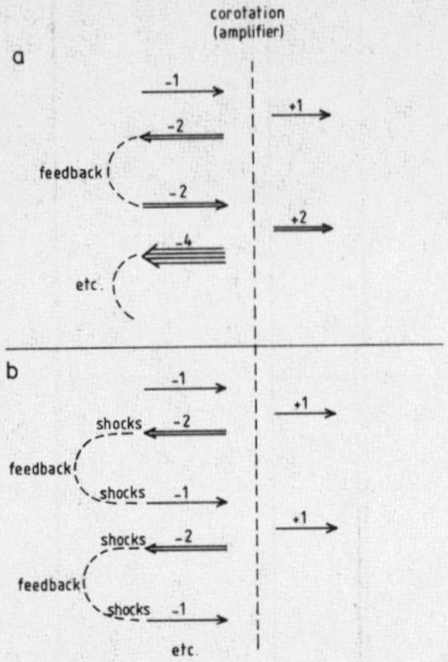


Figure D1 - a) Rapid growth of wave amplitude by amplification at corotation, followed by dissipationless feedback, leading to instability.

b) Quasi-stationary situation: zero growth of wave amplitude; amplification at corotation is followed by dissipative feedback, caused by galactic shocks.

(Figure D1a) the simplest generic example: propagation to corotation of a wave carrying -1 unit of angular momentum, transmission of +1, over-reflection of -2 (to conserve angular momentum), feedback of -2 (if there are no losses), transmission of +2, over-reflection of -4, etc. Thus, the square of the wave amplitude would grow by a factor of 2 per cycle, and may be in danger of growing to catastrophically high values.

Consider, however, how dissipation of the waves by galactic shocks changes this picture (Figure D1b). Start again with a wave approaching corotation and carrying -1 unit of angular momentum. There is transmission of +1, over-reflection of -2, followed now by shock dissipation to cut down the reflected wave as it propagates. By the time the wave is fed back, it might again have only -1 when it comes back to corotation, starting the cycle over again in nearly the same configuration as before. The resulting pattern would be quasi-stationary with evolution only on a slow timescale.

Why should such a quasi-stationary situation obtain? I think it is automatic. Because density waves are dispersive, they do not steepen into shocks until they reach a finite critical amplitude. Until the gaseous density wave reaches this amplitude, the overall wave (stars plus gas) would grow in time. Once the critical amplitude is exceeded significantly, strong shocks develop that halt the growth of the wave, giving a unique solution of finite amplitude.

This consideration may be important in a complementary context. Suppose that in some galaxy there is not enough gas to hold the mode amplitudes to fairly small values. The mode(s) may then grow catastrophically and heat the stellar disk enough to suppress all instabilities. Could something like this have happened to the "flocculent spirals"?

W. Renz: In this connection, I wish to refer to my non-linear calculation of the amplitude to which waves will be limited by damping due to shocks (W. Renz 1982, in "Applications of modern dynamics ...", ed. Szebehely, Reidel, Dordrecht, p. 356). Assuming, with Kalnajs (1972, Astrophys. Letters 11, p. 41), a relaxation time of about $10^9$ years and taking amplification at corotation into account, I found the amplitude of a quasi-stationary wave to be about 5%.

P. Pismis: You stated at the beginning of your talk, Dr. Lin, that there would be little problem in having a driving mechanism for spiral arms in a barred galaxy, as the existence of the bar would give rise to such a mechanism. However, a good many barred spirals have a small, well-defined and tight spiral within the nucleus in the middle of the bar. Good examples are NGC 1097 and 4314. How would the existence of such features agree with the scheme of the density-wave theory?

Lin: I must not give the impression that everything is deductive. In the text of my paper I emphasize the semi-empirical approach of the density-wave theory. I would assume that the structures you describe may perhaps also be quasi-stationary. One has to look for the dynamical process that is responsible.

G.D. van Albada: At least in some cases the nuclear spirals observed in barred galaxies may be an artifact produced by the projection of a curved dust lane on the luminous nuclear region. This conjecture is based on the comparison of recent numerical results with e.g. the pictures shown in the Hubble Atlas.

J.V. Feitzinger: A crucial point is the time scale of growth rates.

Lin: Indeed, that is a very important point. We believe that the growth rates are moderate. Wielen's study of the increase of stellar velocity dispersions in the Milky Way points to the same conclusion. The time scale for these changes may indeed be too short in the numerical experiments (for reasons given in Section II.3 of the text).

D. Lynden-Bell: Sellwood has made thorough comparisons of the modes determined analytically by Toomre and by Kalnajs with his own detailed numerical calculations. He went as far as $10^5$ bodies, without getting accurate growth rates, when starting with Poisson noise in phase space. However when he started with a special smooth procedure, he managed to get almost exact agreement. This may serve as a warning to those in the audience who make numerical simulations: it is hard work to get exact agreement with analytical results.

Lin: Was Lindblad resonance involved in that case?

D. Lynden-Bell: I must ask Sellwood about that.

J.A. Sellwood (answer supplied after Symposium, on request by Editor): No inner Lindblad resonances were involved in these models. The apparent growth rates were too high, because noise interfered with the measurements at early times. The amplitudes also soon became non-linear.

Lin: I would say that both approaches have their own limitations. The analytical theory certainly is now restricted to the linear case; it is a difficult step to go on to non-linear analysis. As to numerical simulations the plasma physicists, with their decades of experience, are very cautious about involving resonances.

# ON THE MORPHOLOGY OF SPIRAL MODES

G. Bertin

*Scuola Normale Superiore, Pisa, Italy*

C.C. Lin & S.A. Lowe

*Massachusetts Institute of Technology*
*Cambridge, Massachusetts, USA*

ABSTRACT

A dynamical basis for Hubble's morphology classification of galaxies is given by combining the existing theory for normal spirals composed of two trailing waves with a similar theory for open spirals composed of one leading wave and one trailing wave. Thus we identify the two classes of spiral modes which form the prototypes for the two prongs of the tuning fork in the Hubble classification. In addition, we recognize the existence of transition spirals and bar-driven spirals as explained in a previous paper (Ref. 7). The relevant physical regimes of the dynamical parameters are those for which modes are moderately growing, consistent with the hypothesis of Quasi-Stationary Spiral Structure. The perceptions presented at IAU Symposium 106 (Ref. 8) are reaffirmed.

Keywords: Galaxies, Spiral Structure, Morphology, Spiral Modes, Over-reflection, WASER Mechanism

## 1. INTRODUCTION

Hubble's morphology classification of galaxies is based on a number of observational criteria. One of the principal challenges in the study of galaxies is to try to understand the physical basis of these criteria and their interrelationships. An essential step in this effort is the study of the morphology of spiral modes and the discussion of the possible existence of quasi-stationary spiral structure in galaxy models which possess a small number of moderately unstable linear modes.

In this abbreviated article on disk galaxies, we sketch a fluid modal theory for those spiral modes that are composed of one leading wave and one trailing wave, each satisfying a cubic local dispersion relationship. Preliminary results and similar conclusions were announced by Bertin (Ref. 2) without derivation. Combining this theory with the existing theory for normal spirals composed of two trailing waves, we have the essential dynamical basis for spiral modes that are representative of the two spiral branches in the Hubble morphology classification. Dynamical models with bar-driven spirals are outside the scope of the present study; they represent galaxies with strong bars in the central regions.

Generally speaking, the characteristics of modal instability of a galactic disk are largely determined by the profiles of the two parameters J and Q, defined respectively by Eqs. (A20) and (A10) in the Appendix. Normal spiral modes are associated with lower values of the J-parameter; barred spiral modes, with higher values. The Q-parameter may be expected to be determined by a process of self-regulation. (See also discussion in Section 4.)

The modal theory is self-contained. The maintenance of spiral modes may be visualized in terms of oppositely propagating wave trains that satisfy a certain local dispersion relationship. The excitation mechanism for such modes can be seen as a process of over-reflection near the corotation circle, in the form of the WASER (See Refs. 9-10). Another reflection mechanism must be present near the central region of the galaxy, so that the two waves are linked in a closed cycle or "feedback loop". In the following paper, which is largely devoted to a study of the excitation mechanism in the corotation zone based on a spectral method, we shall compare the results obtained in this paper, including the over-reflection factor, with those obtained from the "swing" formalism (Ref. 12) and otherwise.

Modes have been calculated for a few selected examples. The numerical results are compared with those obtained from global modal calculations in which the integral form of the solution of the Poisson equation is used. The satisfactory agreement found gives further support to the general perception of the dynamical processes in galaxies described in an earlier paper (Ref. 8) and adopted in the present article.

Stellar dynamic effects should not be ignored. Specifically, we wish to remind the reader of the role of the inner Lindblad resonance. A proper inclusion of kinetic effects would usually show that the number of growing modes in a given galaxy model is limited even if the fluid model may indicate an opposite conclusion. For example, global modes with number of arms $m \geq 3$ are usually damped. With this cautionary note, we shall continue to use a fluid model in the present paper. A similar theoretical development should be carried out on the basis of stellar dynamics. Indeed, a large part of such a theory is already available (See Ref. 1).

*Proceedings of a Course & Workshop on Plasma Astrophysics, Varenna, Italy, 28 Aug – 7 Sept 1984 (ESA SP-207, Nov 1984)*

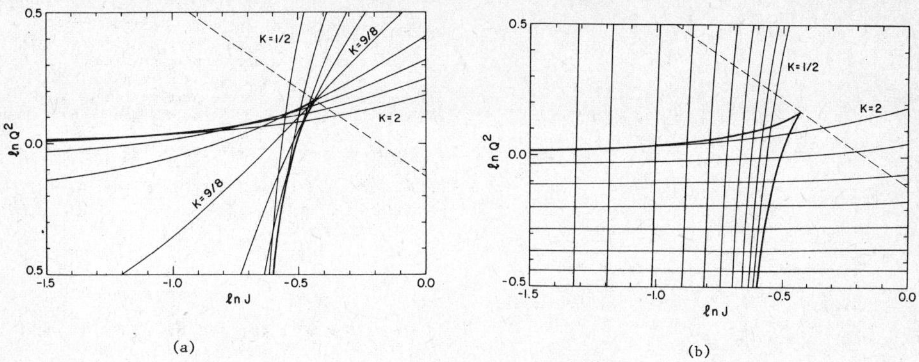


Figure 1. Contour curves of constant dimensionless wave number K in the (J,Q) plane. (a): Curves for $1/2 \leq K \leq 2$. These contours have an envelope which is shown as a thick line. Each contour changes sheets at its point of tangency with the envelope. (b): Curves for $K \leq 1/2$ and $K \geq 2$. The envelope from 1 (a) is reproduced here for reference. These contours do not change sheets.

Finally, the classification of spiral modes only provides one important basis for the classification of observed galaxies. Beyond the present linear theory of modes, we must consider, among other things, (i) the nonlinear mechanisms, (ii) the presence of the interstellar medium, (iii) turbulent motion in the gas, (iv) the dissipative effects associated with the compression of the interstellar medium on a large scale (galactic shocks), (v) the formation of stars, and (vi) the heating of the interstellar medium by these newly formed stars. These are all interesting topics for future research.

## 2. THE LOCAL DISPERSION RELATION

A generalized form of the Lau-Bertin dispersion relation (Ref.5) provides the basis of the present theory. Some of the properties and the implications of this relationship will now be described. Their derivation will be presented elsewhere. This generalized version of the dispersion relation may be written as:

$$\frac{Q^2}{4} = \frac{1}{K} - \frac{1-\nu^2}{K^2 + J^2/(1-\nu^2)} , \qquad (1)$$

where K is the dimensionless wave number, $\nu$ is the dimensionless frequency, Q is a dimensionless measure of the velocity dispersion, and J is a parameter proportional to the surface density and to the number of spiral arms, independent of the velocity dispersion in the basic state, but dependent on the shear parameter s. The precise definitions of all these parameters are given in the Appendix.

For this dispersion relationship, there is a simple transformation [Eq. (4) - (6) ] that reduces the general case to the special case where $\nu^2 = 0$. For the latter problem, which involves only the three parameters K, Q, and J, we find it convenient to discuss the essential characteristics of the dispersion relation by examining, in the (J,Q) plane, (a) the K-contours, and (b) the $\chi$-contours, where $\chi = J^2/K^2$. The former are curves of constant wave number K, and the latter are curves of constant pitch angle, after the shear parameter s is specified. [Cf. Eq. (3) below]. These contour curves, together with their significance, will now be briefly described.

### (a) Characteristics of the K-contours

Contour curves of constant K are shown in Fig. 1. The family of K-contours may be divided into three groups:

(i) short waves, with $2 < K$
(ii) long waves, with $1/2 < K < 2$
(iii) extra-long waves, with $K < 1/2$.

The long waves can be further divided into two subgroups: those with $K > 9/8$ and those with $K < 9/8$.

The heavy contours in Fig. 1(a) are the envelopes of the long wave contours. They delineate a lower left portion of the (J,Q) plane, where the solution surface is three-sheeted. Within this sector, the dispersion relation possesses three real solution for K; outside, it has only one.

For the purposes of this discussion, the three sheets are denoted "red", "yellow", and "blue" in the order of increasing K. Any particular contour of constant K is confined to a single sheet unless that contour is tangent to one of the envelopes. In that case, the curve changes sheets at the point of tangency. Thus, the short waves are always on the "blue" sheet and the extra-long waves are always on the "red" sheet. The long wave contours change sheets from "blue" to "yellow" along the upper envelope (moving from right to left) and from "red" to "yellow" along the lower envelope (moving from top to bottom). Within the sector in the lower left bounded by the curves K = 2 and K = 1/2, the short, long and extra-long contours fall on the "blue", "yellow", and "red" sheets, respectively.

Outside the three-sheeted region, only one real solution exists. For the corresponding wave, the direction of group propagation is the same as that for the short wave. It is important to note that the three-sheeted region extends only as far as the cuspidal point* $P_c$ where J and Q have the values:

$$J_c = \sqrt{27/64} = 0.6495, \; Q_c = \sqrt{32/27} = 1.0887. \quad (2)$$

This means that moderate to large values of J and/or Q will place us squarely in the one-sheeted part of the plane, where the feedback cycle must consist of one leading and one trailing wave.

(b) Characteristics of the $\chi$-contours

These contour curves were presented at IAU Symposium 106 (Ref. 8) as contours of constant shear parameter s (at $\nu^2 = 0$), with the radial wave number parameter $\mu$ set equal to zero. Since (see Appendix)

$$\chi = \chi_o(s)(1 + \mu^2)^{-1} , \quad (3)$$

they are really contours of constant $\chi$ (at $\nu^2 = 0$). These contours are again shown in Fig. 2. For given s, they are contours of constant pitch angle ($\mu$ = constant). Clearly, the condition $\mu^2 \geq 0$ is physically important. It is usually regarded as a geometrical constraint on the waves and defines physical boundaries for both $\chi$ ($\chi \leq \chi_o(s)$) and K ($K \geq 2m\varepsilon_o$). As we shall see in Section 4, the general resemblance of these $\chi$-contours to those for the over-reflection factor has important consequences on the morphology of quasi-stationary spiral modes.

(c) Density waves for arbitrary $\nu^2 > 0$

Referring to Eq. (1), we may introduce the transformation

$$J^2 = J_o{}^2(1 - \nu^2)^3 \quad (4)$$

$$Q^2 = Q_o{}^2(1 - \nu^2)^{-1} \quad (5)$$

$$K = K_o(1 - \nu^2) \quad (6)$$

to bring the dispersion relation for the general case to that for the special case $\nu^2 = 0$, for which the symbols carry the subscript zero.

One specific application of this transformation is the division of the ($\ln J$, $\ln Q^2$) plane into two parts by the transition line

$$\ln J + \frac{3}{2} \ln Q^2 = \ln(16\sqrt{2}/27). \quad (7)$$

Above this line, there is only one solution for all values of $\nu^2$, provided that the condition $\mu^2 \geq 0$ is met. It is therefore possible to characterize a model by plotting its trajectory

$$J = J(r), \qquad Q^2 = Q^2(r) \quad (8)$$

in the ($\ln J$, $\ln Q^2$) plane. We shall find that, if the trajectory lies entirely on the upper side of the dividing line, QSSS can be maintained only for open waves (See Section 4). Quasi-stationary spirals which are tightly wound can only be found below the dividing line, when the conditions specified in the earlier theory for such modes are fulfilled.

## 3. MODAL THEORY

We shall now formulate a theory of open spiral modes in the form of an ordinary differential equation (ODE) of the second order together with the appropriate boundary conditions (ABC). We begin with the cubic dispersion relation and limit ourselves to those conditions where there is only one real root for K, as discussed above.

The parameter $\mu^2$ for the radial wave number is related to the frequency parameter $\nu^2$ and to the basic model parameters $J^2$, $Q^2$ and s through the dispersion relation (1), which has the analytic structure

$$D(K, \nu^2; J^2, Q^2) = 0 \quad (9)$$

Since there is only one real solution for K, there is only one real solution for $\mu^2$. (See Appendix for definition of $\mu^2$). Thus, we have

$$\mu^2 = \hat{\mu}^2(\nu^2; J, Q, s) , \quad (10)$$

where we use $\hat{\mu}^2$ to denote the analytic relationship (which is actually algebraic). We may now write the ODE for open spiral modes as follows:

$$\frac{d^2Y}{dx^2} + (\frac{m}{R})^2 \hat{\mu}^2(\nu^2; J, Q, s)Y = 0 , \quad (11)$$

where

$$x = R \ln(r/R). \quad (12)$$

This is an associated ODE for the dispersion relation (10) in that it leads to the same dispersion relationship. The physical meaning of Y is left open for the moment. (See Section 4).

Equation (11) should be integrated with an appropriate boundary condition (ABC) to yield the eigenmodes. A radiation boundary condition (Ref. 6) should be applied to correspond to the propagation of waves from the corotation zone to infinity without reflection. The zero boundary condition (ZBC) is sometimes applied but there is yet no proof that it is always equivalent to the radiation condition.

(a) Some general conclusions based on numerical examples

By using the method just described, modal calculations have been carried out for a few galaxy models, and compared to existing global calculations. There is general agreement in the value of the pattern speed $\Omega_p$. The real part of the appropriate quantum integral is found to agree well with odd multiples of $\pi$. However, there is considerable variation in the growth rate $\gamma$. We shall therefore examine the influence of the boundary condition (at $r = \infty$) on the growth rates further (Section 4).

(b) Some additional remarks

We wish now to add a few comments concerning some specific issues. Specifically, we shall address (1) the physical meaning of the variable Y, and

---

*In the Lau-Bertin plot (Ref.5) of J-contours for $\nu^2 = 0$, this point $P_c$ lies on the curve $J = J_c$ and corresponds to the location where the point of inflection has a horizontal tangent.

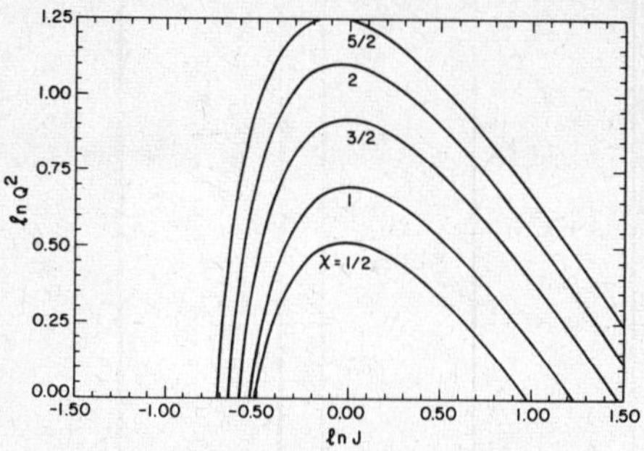


Figure 2. Contour curves of constant $\chi$ (constant pitch angle or shear parameter) in the (J, Q) plane. Note the general resemblance to the curves of constant $I_r$ in the upper part of Fig. 3.

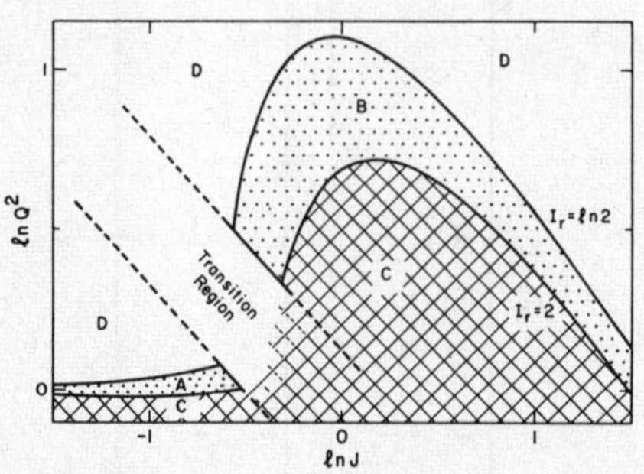


Figure 3. Regions of various amounts of over-reflection in the J-Q plane. In regions A and B, the over-reflection factors are compatible with the QSSS hypothesis (A: normal spirals; B: open spirals). In region C, the strong over-reflection is expected to imply rapid growth of the modes. In contrast, the growth in region D may be too slow to be effective for the excitation of modes. The simple two-wave analysis does not apply in the Transition Region, but it is expected that the trend of strong over-reflection and rapid growth at low values of Q will hold there also.

(2) the possible role that might be played by a complex dispersion relationship. Hunter reminded us (Ref. 4) that the out-of-phase components may have been improperly omitted in the derivation of the Lau-Bertin dispersion relationship.

In Eq. (11), we left the physical meaning of the variable Y open. Comparison with global calculations suggests that, for the present case of open modes, it should differ from the enthalpy perturbation only by a slowly varying factor, as we can confirm from another reasoning to be briefly described. At the same time, it is clear that, for the calculation of the over-reflection factor and for the subsequent discussions on growth rates, the precise meaning of Y is not important.

The answers to the above issues become quite clear when one examines the special case of $\mu^2$ small ($\mu^2 < 1$). In that case, we may expect a good approximation by expanding (9) in powers of $\mu^2$ and neglecting terms of the order $O(\mu^4)$, as was effectively done earlier by Bertin (Ref.2). The case of small $\mu^2$ can be examined by direct application of the Euler-Poisson equations, with the out-of-phase terms included, and this was carried out by one of us (S.A.L.). The resultant differential equation differs from Eq. (11) by the addition of a term in the derivative of the first order. Numerically, however, it has been found that this term being small, has only a minor effect on $\gamma$ and practically no effect on $\Omega_p$, when the ABC is applied.

## 4. THE OVER-REFLECTION FACTOR AND THE GROWTH RATE ESTIMATOR

In order to understand the role of the boundary conditions on the growth rates, one may first relate them to the mechanism of energy transfer and to the WASER process. Here we outline the steps briefly in the context of the present paper.

(1) We consider a feedback cycle of leading and trailing waves. The leading wave approaching the corotation zone is reflected as a departing trailing wave with higher energy density (or flux of angular momentum), accompanied by another departing trailing wave travelling towards large distances, where it is absorbed (radiation boundary condition).

(2) To calculate this over-reflection factor, we use the WASER formalism. In the neighborhood of the corotation circle, we may expand Eq. (10) in powers of $\nu^2$:

$$\mu^2 = \mu_{co}^2 + \mu_1^2 \nu^2 + O(\nu^4), \tag{13}$$

where $\mu_{co}$ is the value of $\mu$ at the corotation circle. If we neglect $O(\nu^4)$, then Eq. (11) leads to the following formulae for the over-reflection factor according to the WASER formalism (Ref. 10):

$$\Gamma_r = \Gamma_f + 1, \tag{14}$$

$$\ln \Gamma_f = \pi \mu_{co}^2 / (\frac{2A}{\kappa} \mu_1). \tag{15}$$

In the above equations, $\Gamma_r$ is the factor of over-reflection, i.e., the ratio of the flux of angular momentum in the reflected (trailing) wave to that in the leading wave approaching the corotation circle from the central regions of the galaxy. In Fig. 3, contours of constant index of over-reflection $I_r = \ln \Gamma_r$ are shown in the ($\ln J$, $\ln Q^2$) plane, for the special case s = 1. Contours above the transition region are evaluated from Eq. (15), those below are calculated from the WASER formulae for normal spirals.

(3) With the over-reflection factor obtained, it is then easy to establish the approximate relationship:

$$2\gamma\tau = \ln \Gamma_r = I_r \tag{16}$$

where $\tau$ is the time of the feedback cycle, properly modified to take into account the difference between the value of $\Gamma_r$ for $\gamma = 0$ and that for $\gamma \neq 0$. We note that (16) provides an <u>estimator</u> for $\gamma$ in the following sense. For each assumed value of $\Omega_p$, we may calculate a value for $\gamma$, even when $\Omega_p$ is <u>not</u> an eigenvalue. We obtain $\gamma(\Omega_p)$ for all <u>potential</u> values of $\Omega_p$, without solving the eigenvalue problem. The relationship (16) is of the same general character as that previously obtained for tightly wound spirals [cf. Ref. 6], but there are also subtle differences.

(4) Clearly, then, the value of $\gamma$ depends on the <u>net</u> flux of angular momentum carried to infinity. Thus, if there is zero flux, we would expect to get $\Gamma_r = 1$ and $\gamma = 0$. In general, we may expect the value of $\gamma$ to be <u>reduced</u> by the imposition of ZBC.

(5) From Eq. (15), it can be seen that, for a given value of the shear parameter s, the over-reflection factor depends strongly on $\mu^2_{co}$, but not otherwise on the parameters J and Q in a sensitive manner. This is empirically confirmed, as shown in Fig. 4 for the case s = 1.

This simple result has interesting implications. It means that for J ≃ 1, modes with <u>moderate</u> growth rates must be <u>open</u> spirals, while spirals with small pitch angles are strongly unstable. If we combine this conclusion with the process of self-regulation discussed elsewhere (see Section 4 of Ref. 8), we see that, for J ≃ 1, the QSSS hypothesis holds only for open spirals with $Q^2 \simeq 2$. For $Q^2 \simeq 1$, there is a plethora of strongly unstable modes: the basic state may be expected to evolve until a state of modest modal instability is restored. Thus, along a vertical line J = 1 in the (J,Q) plane, the lower part corresponds to cases with <u>transient</u> sA spirals; the upper part, to <u>quasi-stationary</u> sB spirals.

(6) To summarize, quasi-stationary spirals are obtained when the parameters $(J,Q^2)$ lie within a definite strip in the $(J,Q^2)$ plane: low values of $(J,Q^2)$, e.g., (0.5,1), for sA spirals, high values of $(J,Q^2)$, e.g., (1,2), for sB spirals, and intermediate values for spirals of the transitional type. Spiral patterns of the last category have also been computed by Jon Haass and presented by Lin and Bertin (Ref 8). [Here we are focussing in our discussion on the specific case m = 2. The reader should be reminded of the importance of Lindblad resonance in limiting wave patterns with $m \geq 3$.]

## 5. CONCLUDING REMARKS

We have developed a quite satisfactory simple theory for calculating and understanding quasi-stationary "bar-like" spiral modes. Together with the earlier theory for normal spiral modes, we are able to characterize the parameters required of the basic state for disk galaxies of various morphologies. The J-parameter holds a key role. As mentioned above, bar-driven spirals require a separate treatment; but that problem is dynamically far simpler than the search for normal and open spiral modes. The original goal of this paper is essentially accomplished.

In the following paper, we shall pursue the mechanism of excitation further by adopting another approximation, which holds in a "corotation zone". Studies of mechanisms of excitation in this context have been carried out in a large class of problems ranging from plasma physics to meteorology. In particular, we shall be dealing with the contrast of behavior between steady wave trains and swinging wave packets, and show how such diverse behavior can be unified in a spectral approach.

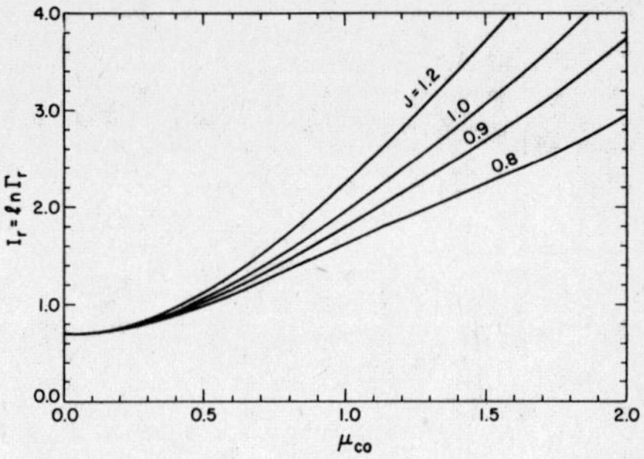


Figure 4. Relationship between the index of over-reflection $I_r$ and $\mu_{co}$, the cotangent of the pitch angle of the wave at the corotation circle. The curves correspond to different values of J with s = 1.


## ACKNOWLEDGEMENT

We would like to thank Jon Haass and Robert Thurstans for many useful discussions. This work was partly supported by Consiglio Nazionale delle Ricerche, under the United States-Italy Cooperative Science Program, and partly by the National Science Foundation.


## APPENDIX

### Definition of Symbols

A.1 <u>Coordinate system</u>

The $(r,\theta)$ plane in the cylindrical coordinate system $(r,\theta,z)$ is often described in the logarithmic system

$$x = R\,\ell n(r/R), \quad y = R\theta. \tag{A1}$$

A.2 <u>Basic state</u>

The distribution functions $\Omega(r)$, $\sigma(r)$, $a(r)$ are the radial distributions of angular velocity, active surface density, and acoustic velocity (simulating velocity dispersion of the stars).

(a) Parameters related to $\Omega(r)$:

(i) shear parameters: $$s = -\frac{d\ell n\,\Omega}{d\ell n\,r} \tag{A2}$$

$$2A = s\Omega \tag{A3}$$

(ii) Epicyclic frequency $\kappa$.

$$\kappa^2 = (2\Omega)^2\,(1-s/2) \tag{A4}$$

(iii) $$\chi_o(s) = s(2\Omega/\kappa)^2 = s/(1-s/2) \tag{A5}$$

(b) Parameters related to $\sigma(r)$:

(i) length scales: $\delta$ and $\lambda_{cr} = 4\pi\delta$, where

$$\delta = \frac{\pi G\sigma}{\kappa^2} \tag{A6}$$

is a typical scale length determined by self-gravity, and $\lambda_{cr}$ is the critical wave length for instability according to Toomre (Ref. 11). From

$\delta$ , we define

$$\varepsilon_0 = \frac{\delta}{r}, \quad \text{(A7)}$$

which is directly related to the typical eccentricity of the stellar orbits, and

$$a_0 = \kappa\delta = \pi G\sigma/\kappa , \quad \text{(A8)}$$

the "critical" velocity dispersion of the stars. We also define the composite parameter

$$J_1 = 2\varepsilon_0[\chi_0(s)]^{1/2} \quad \text{(A9)}$$

which appears in the dispersion relationship through $J = mJ_1$ , $m$ being the number of spiral arms.

(c) Parameters related to $a(r)$:

$$Q = \frac{a}{a_0} = \frac{\varepsilon}{\varepsilon_0} . \quad \text{(A10)}$$

A.3 Perturbations

For each steady wave train or wave pattern of $m$ arms,

$$\psi' = A(r)e^{i\Phi(r)}\, e^{i(\omega t-m\theta)} , \quad \text{(A11)}$$

we define the following symbols.

(a) The frequency parameters, including the complex frequency parameter:

$$\nu = (\omega-m\Omega)/\kappa$$

$$= (\omega_R-m\Omega)/\kappa - i\gamma/\kappa \quad \text{(A12)}$$

where $\omega = \omega_R + i\omega_I = \omega_R - i\gamma$. We introduce also the pattern frequency

$$\Omega_p = \omega_R/m \quad \text{(A13)}$$

(b) The wave number parameters

$$k_r = \frac{d\Phi}{dr}(r), \quad k_\theta = m/r \quad \text{(A14)}$$

for radial and angular directions, and the total wave number $k$ in dimensional and dimensionless forms:

$$k^2 = k_r^2 + k_\theta^2 \quad \text{(A15)}$$

$$K = k(2\delta). \quad \text{(A16)}$$

Note the additional relations for convenience:

$$k_\theta\delta = m\varepsilon_0 \quad \text{(A17)}$$

$$\mu = k_r/k_\theta = r\, k_r/m \quad \text{(A18)}$$

$$K^2 = (2m\varepsilon_0)^2\, (\mu^2+1) \quad \text{(A19)}$$

where the parameter $\mu$ defined by (A18) is the cotangent of the pitch angle.

(c) The additional parameters $J$ and $\chi$ are defined as follows:

$$J = m\, J_1 = 2m\varepsilon_0[\chi_0(s)]^{1/2} \quad \text{(A20)}$$

$$\chi = J^2/K^2 = \frac{\chi_0(s)}{1+\mu^2} . \quad \text{(A21)}$$

REFERENCES

1. Bertin G 1980, On the density wave theory for normal spiral galaxies, *Physics Reports*, vol 61, 1-69.

2. Bertin G 1983, Dynamical mechanisms for discrete unstable spiral modes in galaxies, *Proc IAU Symposium 100*,(Reidel, Athanassoula E ed ), 119-120 (Besancon, 1982).

3. Haass J, Bertin G & Lin C C 1982, On the dynamical basis of the classification of normal galaxies, *Proc Natl Acad Sci USA*, vol 79, 3908-3912.

4. Hunter C 1983, Galactic dynamics, in *Fluid Dynamics in Astrophysics and Geophysics*, Lebovitz N R ed, pp 179-203.

5. Lau Y Y & Bertin G 1978, Discrete spiral modes, spiral waves, and the local dispersion relationship, *Astrophys J*, vol 226, 508-520.

6. Lau Y Y, Lin C C & Mark J W-K 1976, Unstable spiral modes in disk-shaped galaxies, *Proc Natl Acad Sci USA*, vol 73, 1379-1381.

7. Lin C C & Bertin G 1981, Galactic dynamics and gravitational plasmas, in *Plasma Astrophysics*, ESA SP-161, pp 191-205 (Varenna, 1981)

8. Lin C C & Bertin G 1984, Formation and maintenance of spiral structure in galaxies, *Proc IAU Symposium 106*, (Reidel, van Woerden H ed) in press (Groningen, 1983).

9. Lin C C & Lau Y Y 1975, On spiral waves in galaxies - A gas dynamic approach, *SIAM J Applied Math*, vol 29, 352-370.

10. Mark J W-K 1976, On density waves in galaxies. III. Wave amplification by stimulated emission, *Astrophys J*, vol 205, 363-378.

11. Toomre A 1964, On the gravitational stability of a disk of stars, *Astrophys J*, vol 139, 1217-1238.

12. Toomre A 1981, What amplifies the spirals?, in *The Structure and Evolution of Normal Galaxies*, (Cambridge UP, Fall S M & Lynden-Bell D eds), pp 111-136 (Cambridge, 1980).

# ON A SPECTRAL THEORY OF THE EXCITATION MECHANISM OF GALACTIC SPIRALS

**C.C. Lin & R.P. Thurstans**

*Massachusetts Institute of Technology*
*Cambridge, MA, USA*

ABSTRACT

The spectral theory is developed in full to bring out the relationship between swinging wave packets and exponentially growing wave trains in the conventional infinite sheet model with uniform differential rotation. The theoretical results obtained are compared with those presented in Paper I, which is based on the local dispersion relationship for wave trains. General agreement is found. It is pointed out that there may be coexistence of quasi-stationary grand designs with transient spiral features that evolve at a rate comparable to that of shear. The quasi-stationary state may be expected only for certain regimes of dynamical parameters.

Keywords: Galaxies, Spiral Structures, Density waves, Over-reflection, WASER.

## 1. INTRODUCTION

In the preceding paper by Bertin, Lin, and Lowe (hereinafter referred to as Paper I), a coherent theory has been presented on the interrelationships between the morphology of spiral galaxies and the dynamics of the formation and maintenance of spiral patterns with grand design. The theory is based on the local dispersion relationship and now covers both normal spirals and open spirals. In both cases, an appropriate differential equation of the second order has been obtained to describe the spiral patterns.

In the present paper, we shall examine the dynamical mechanisms of the excitation of spiral patterns from a different perspective. We focus our attention on the corotation zone, i.e., an annular region near the corotation circle, in which the mechanism of "over-reflection" has been extensively studied. (See Appendix and the following text for detailed references and definition of terms.) In particular, in the study of "swinging wave packets", Toomre (Ref. 18) has described the process of transient amplification and calculated the factors of over-reflection to show that there are situations where these wave packets undergo very strong transient amplification, in apparent contradiction to the hypothesis of quasi-stationary spiral structure (QSSS hypothesis). It is therefore important to demonstrate and to emphasize that there is indeed no such conflict.

At least two issues should be clarified in this line of reasoning. First, one should demonstrate that the amplification mechanism in question has a counterpart in the context of steady wave trains; that is, the mechanism of over-reflection involving both leading waves and trailing waves is not necessarily transient. Secondly, one should identify the regimes of dynamical parameters which are compatible with the QSSS hypothesis; that is, one should show that the amplification mechanism in question is not necessarily powerful.

The first issue was addressed by Drury (Ref. 2), who showed that the process of over-reflection can indeed be described in the context of steady wave trains. Drury used a spectral approach in which the wave number is time-independent but the spectrum may be time dependent, although he limited most of his analysis to the case of steady wave trains. We shall adopt the same formalism but differ from him in giving time-dependence an even more prominent role. We obtain the general solution for the time-dependent spectral functions in terms of arbitrary functions, including both the convective part and the wave part. We can then specialize these arbitrary functions to obtain two parallel classes of wave solutions: (a) the steady (or nearly steady) wave trains, and (b) the swinging wave packets. The over-reflection factor is however calculated through the use of a differential equation [Eq. (7) below] in the spectral formalism without reference to the arbitrary functions specified. Thus, in two clearly distinct contexts, the same factor of over-reflection is obtained. The first issue is thus fully resolved.

The second issue is, strictly speaking, outside the scope of a paper dealing with the corotation zone alone, since global modes must be considered. However, we may address it by making use of the conclusions of Paper I. In that paper, this issue was addressed in terms of the process of over-reflection of steady wave trains and the amplification of spiral patterns. The over-reflection factor, calculated by using the "WASER formalism" in that paper, was found to show tremendous dif-

*Proceedings of a Course & Workshop on Plasma Astrophysics, Varenna, Italy, 28 Aug – 7 Sept 1984 (ESA SP-207, Nov 1984)*

ferences in the various regimes of dynamical parameters. It was shown that the over-reflection factor can be used as an indicator to examine the compatibility of the QSSS hypothesis with a given regime of dynamical parameters. We shall therefore also use the existing theory for "swing amplification" to make an extensive systematic calculation of the over-reflection factor for various regimes of the dynamical parameters, in order to compare the results with those obtained in Paper I. Preliminary results of these calculations were reported by Lin and Bertin (Ref. 13). The more complete results now obtained are found to be in general agreement with those obtained in Paper I, although different approximations have been used. The results of the present paper, may thus be used to lend further support to the perspective presented in Paper I.

The importance of a corotation zone, or the zone where waves and material objects move approximately together, has been recognized in other dynamical systems. In the classical problem of hydrodynamic stability, the *critical layer* has played a central role. It has also been clearly demonstrated in that context, that the transient amplification of periodic disturbances in an infinite field of flow with *uniform* shear, as shown by Lord Kelvin in 1877 and by Orr in 1907 (See Ref. 15 for the detailed references), leads to impressions and conclusions that *appear* to stand in sharp contrast to those obtained through studies of *modal* instability of flow fields in which the gradient of vorticity is found to play an essential role (Rayleigh, Ref. 17). There is always the need to reconcile these two approaches. In the present context, the issue was specifically raised by Hunter (Ref. 6) in his review of galactic dynamics: "The question of how fully to reconcile this transient kind of instability with a normal mode analysis remains. The same question arose long ago in hydrodynamic stability theory when Orr discovered transient instabilities..." (p. 193 of Hunter's paper). We shall refer to this issue as that of "joining" of solutions near the corotation zone.

In a general perspective, the dynamical processes under consideration are essentially mechanisms concerning linear and nonlinear *dispersive waves in an inhomogeneous medium*, (cf. Ref. 20), which is the basic equilibrium state of the galactic disk. As a general topic in applied mathematics, this problem is still incompletely solved. In the context of density waves, we have made considerable progress in Paper I from the point of view of the modal approach, where inhomogeneity is recognized right from the beginning. The issue of "joining" is essentially resolved in that context. In the present approach using a homogeneous model, this issue remains as yet unresolved.

In a number of physical situations as diverse as plasma physics (Ref. 10) and meteorology (Ref. 19), the spectral approach has been adopted in *homogeneous* models to provide a unifying approach, in order to clarify certain aspects of the relevant dynamical mechanisms, and to display the contrast in behavior between steady wave trains and swinging wave packets. It is unsatisfactory that the mathematical advantage of the homogeneous model (cf. Ref. 20, pp. 369, 380) should be off-set by the difficulty in "joining" of the solutions in the corotation zone with those in the general field. This issue is a general problem that deserves a great deal of future attention.

## 2. SPECTRAL THEORY OUTLINED

We shall now present the spectral theory in order to treat steady wave trains and swinging wave packets on a parallel footing. This usual method of Fourier transform has often been adopted for solving partial differential equations whose coefficients are linear in the independent variable. For this mathematical reason, it is convenient for the infinite sheet model with uniform shear. The detailed steps of this analysis are given in the Appendix. In this section, we shall only briefly outline some of the important conclusions.

### (A) *Decomposition into logarithmic spirals*

In the context of the whole galactic disk, the method of Fourier analysis amounts to the representation of the spiral pattern in terms of logarithmic spirals (cf. Ref. 5, 7, 12, for example). Instead of the cyclindrical coordinate system $(r,\theta,z)$, we introduce the logarithmic coordinate system $(x,y,z)$ with

$$x = R\ell n(r/R), \qquad y = R\theta, \tag{1}$$

where $R$ is a scale parameter at our disposal, and $z = 0$ in the galactic disk. Clearly, near the point $(r,\theta) = (R,0)$, the local Cartesian system is $(x,y)$. Any physical quantity, e.g., the disturbance in the gravitational potential $\Psi'(x,y,t)$, may be written in the Fourier representation

$$\Psi'(x,y,t) = \sum_{m=-\infty}^{\infty} \Psi_m(x,y,t), \tag{2}$$

with

$$\Psi_m(x,y,t) = \int_{-\infty}^{\infty} \tilde{\Psi}(\xi,t)e^{i(\xi x-\eta y)}d\xi, \tag{3}$$

where $\eta = m/R$. We find it convenient also to introduce the notation

$$\Psi(x,t) = \int_{-\infty}^{\infty} \tilde{\Psi}(\xi,t)e^{i\xi x}d\xi \tag{4}$$

so that $\Psi_m = \Psi(x,t)\exp(-i\eta y)$. Eq. (3) is the decomposition of $\Psi_m$ into logarithmic spirals.

### (B) *The infinite homogeneous sheet with uniform shear*

This model has been extensively studied [Goldreich and Lynden-Bell (Ref. 3), Julian and Toomre (Ref. 9), Goldreich and Tremaine (Ref. 4), Toomre (Ref. 18), Drury (Ref. 2)]. It is a restricted analysis, since it is based on the following linear approximation of the angular velocity $\Omega(r)$:

$$\Omega(r) = \Omega'(R)\,(r-R) + \Omega(R)\ . \tag{5}$$

It is also assumed that the other basic physical parameters, such as the surface density and the Oort parameters, may be regarded as constants. Solutions are then obtained by the method of separation of variables in a comoving coordinate system (as was done by Lord Kelvin in 1877 and quoted in Ref. 15). This leads to swinging wavelets and wave packets. In contrast, by using the Eulerian description, the method of separation of variables leads to steady wave trains. The usefulness of this model for the discussion of waves or disturbances over the galactic disk is clearly restricted to the corotation zone, or other limited annular regions. Thus, when Fourier analysis is applied, the spectrum obtained will be very different from that found for the modes even though the same formal procedure is followed. We shall therefore use a different symbol $\psi$ in place of $\Psi$. Otherwise, the equations (1) - (4) retain the same form. There is also an important change of character of the parameter $\eta$. It is now a continuous variable, whereas a discrete parameter m appears in the disk model. We shall solve the same system of basic equations used by the above-named authors with the spectral approach described above, and show that the general solutions obtained can lead, in a simple manner, to both the swinging wave packets and the quasi-steady wave trains. Thus the essential mechanism of over-reflection is the same in both contexts. The results will now be briefly described; the detailed derivation may be found in the Appendix.

The general solution for arbitrary disturbances in the sheet model may be decomposed into one vorticity component and two wave components. The vorticity component has a simple explicit solution involving one arbitrary function as given by Eq. (A17) or (A21) in the Appendix. This solution is indeed a simple extension of Orr's solution. For the wave components, the general solution is given by:

$$\psi(x,t) = \sum_{i=1}^{2} \int_{-\infty}^{\infty} f_i(\xi + \bar{c}t)\, \hat{\psi}_i(\xi) e^{i\xi x} d\xi , \quad (6)$$

where $\bar{c} = 2A\eta$, and $\hat{\psi}_i(\xi)$ $(i = 1,2)$ are two linearly independent solutions of the second order ordinary differential equation

$$\frac{d^2\hat{\psi}}{d\mu^2} + q^2(\mu)\hat{\psi} = 0. \quad (7)$$

In (7), $\mu = \xi/\eta$, and $q^2$ is defined as follows:

$$q^2 = q_1^2 + q_2^2 , \quad (8)$$

$$(2A/\kappa)^2 q_1^2 = 1 - K + \frac{1}{4} Q^2K^2 , \quad (9)$$

$$(2A/\kappa)^2 q_2^2 = -\chi + \chi^2 \cdot \frac{3}{4}(1 - s/2). \quad (10)$$

In the above equations, A is the first Oort constant, s is the shear parameter, $\kappa$ is the epicyclic frequency, K is a dimensionless form of the total wave number, Q is a measure of the dispersion speed, and $\chi = J^2/K^2$. The parameters J and Q are essential for the studies of the local dispersion relationship and of modal instability. All the formulae for defining these symbols may be found in the Appendix to Paper I.

Note that the solution, Eq. (6) requires two arbitrary functions; it is not enough to have a linear combination of solutions to (7). Eq. (7) has essentially the same form as the "oscillator equation" [Eq. (35) in Ref. 4 or Eq. (18) in Ref. 18] in the theory of swinging wave packets, but the independent variable $\mu$ is now time independent, whereas the independent variable in the oscillator equation is linear in time.

The discretion in the choice of arbitrary functions gives us great freedom. As shown in Section (C) of the Appendix, we may get either exponentially growing wave trains or swinging wave packets by simple choices of the arbitrary functions. The resultant amplification in both cases can thus be discussed in a single treatment based on Eq. (7) in spectral space, as will be done in Section 4.

All the above discussions refer to the zone of corotation. As mentioned above, one central issue is to show how the solutions so obtained can be joined with studies of wave propagation at locations outside of the corotation zone. In the wave train approach adopted in Paper I, such a process is quite easy to carry out in the case of open spiral patterns composed of only two travelling wave trains. We shall therefore attempt to accomplish the same goal by first comparing the characteristics of the waves in the present theory with those discussed in Paper I. We shall begin with the development of the dispersion relationship, and then proceed to study the basic mechanism of transfer of energy and angular momentum in the corotation zone based on a wave train approach in which Eq. (7) is also used.

## 3. QUASI-STATIONARY WAVE TRAINS IN THE COROTATION ZONE

In order to fit the uniform sheet model into the context of the galactic disk, it is convenient to study quasi-stationary wave trains, as just explained. For the observer co-rotating with the wave pattern, the steady state solution [cf. Eq. (4)] is given by

$$\psi(x) = \int_{-\infty}^{\infty} \hat{\psi}(\xi) e^{i\xi x} d\xi , \quad (11)$$

where $\hat{\psi}(\xi)$ is defined by Eq. (7). The asymptotic solution of (7) is

$$\hat{\psi}(\xi) = q^{-1/2} \{\alpha\psi_+(\mu) + \beta\psi_-(\mu)\}, \quad \mu = \xi/\eta \quad (12)$$

where $\alpha$ and $\beta$ are constants, and

$$\psi_\pm(\mu) = \exp\{\pm i \int_0^\mu q(\mu)d\mu\}. \quad (13)$$

By applying the method of stationary phase, it can be easily verified that the location x and the dominant wave number $\xi$ at that location are related by

$$x = -q(\xi/\eta)/\eta < 0, \quad \text{for } \psi_+\text{-term} \quad (14)$$

$$x = +q(\xi/\eta)/\eta > 0, \quad \text{for } \psi_-\text{-term} \quad (15)$$

This relates the wave number $\xi$ with the frequency parameter $\nu = (2A/\kappa)\cdot\eta\cdot x$. Indeed, we obtain

$$\nu^2 = (2A/\kappa)^2 q^2(\xi/\eta) \quad (16)$$

as the dispersion relationship where $q^2$ is given by (8). We shall refer to (16) as the S-dispersion relationship, since $q^2$ is proportional to Toomre's "spring constant". This dispersion relationship will now be compared with that obtained by Lau and Bertin (Ref. 11). Both may be written in the form

$$\frac{1}{4} Q^2K^2 = K - (1-\nu^2) + \chi - \chi^2 F \quad (17)$$

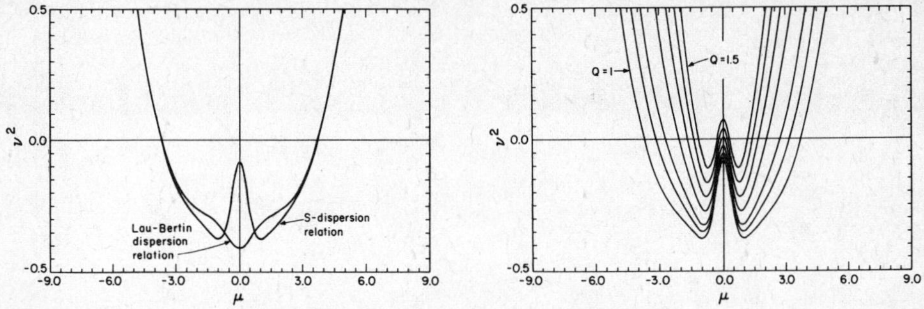


Figure 1. The dispersion relationships

(a) Case of high J ($J = 1.0$; $s = 1$, $Q = 1$). Comparison of the S-dispersion relation with the Lau-Bertin dispersion relation. Close agreement is seen in the relevant range of high $|\mu|$.

(b) Case of high J, additional data for the S-dispersion relation ($J = 1$; $s = 1$, $Q = 1, 1.05, 1.1, 1.2, 1.3, 1.4, 1.5$). There are two roots for $|\mu|$ when $Q > 1.3$, one root for $|\mu| \leq 1.2$,

The symbol F is given by

$$F = \frac{1}{(1-\nu^2)+\chi} \tag{18}$$

in the case of the Lau-Bertin relationship (see Eq. (1) of Paper I) and by

$$F = \frac{3}{4}\left(1 - \frac{s}{2}\right) \tag{19}$$

in the present approximation (16). There are distinct differences. First, in the S-dispersion relationship, the shear parameter s appears explicitly. Secondly, there may be four solutions for $|\mu|$ at $\nu^2 = 0$ in the S-dispersion relationship (17) with F given by (19), but only three in the case of (18). We shall designate these four waves as (i) short (ii) long, (iii) extra-long, and (iv) ultra-long. When all four are present, the directions of their propagation alternate, just as before. The ultra-long wave is absent from the Lau-Bertin dispersion relationship. In the case where the J-parameter is low, we are interested in the short and long waves; there is good agreement between the two approximations. For the case of high J-parameter, we note the appearance of the waves (iii) and (iv). In Fig. 1(a), we show the comparison of the two dispersion relationships in the case $s = 1$, $J = 1$, $Q = 1$. In Fig. 1(b), we show more data for other values of Q. There may now be two real solutions for $|\mu|$ whereas the Lau-Bertin dispersion relation shows only one solution. There is agreement for higher values of $|\mu|$ between the two dispersion relations, but differences exist for low values of $|\mu|$. As one may expect from the restricted nature of these differences, they are not important in the overall picture of wave propagation outside of the corotation zone. Indeed, since the term $\chi^2 F$ is relatively unimportant except when K is small, the general topological nature of the solution surface

$$K = K(\nu^2; J, Q, s) \tag{20}$$

cannot be very different except for very long waves. General agreement is therefore expected for the conclusions obtained from the two different approximations.

We may thus carry out the same type of analysis and present the same diagrams - constant wave-number contours (K-contours) and constant pitch angle contours ($\chi$-contours) - as in Paper I. The results obtained are also quite similar. In particular, we find the cuspidal locus in the ($\ell n\, J$, $\ell n\, Q^2$) plane to be approximately given by the straight lines

$$\ell n\, J + \frac{3}{2}\, \ell n\, Q^2 = \ell n\, C(s), \tag{21}$$

for different values of the shear parameter s, where $C(s) < 16\sqrt{2}/27$. These lines run parallel to the single cuspidal locus in Paper I, but lie to its left. Furthermore, for different values of s, the lines are closely bunched together in a narrow band. [See Fig. 2]. The width of this band is even narrower than its separation from the corresponding single line in Paper I. Thus, the division of the (J,Q) plane is only slightly altered.

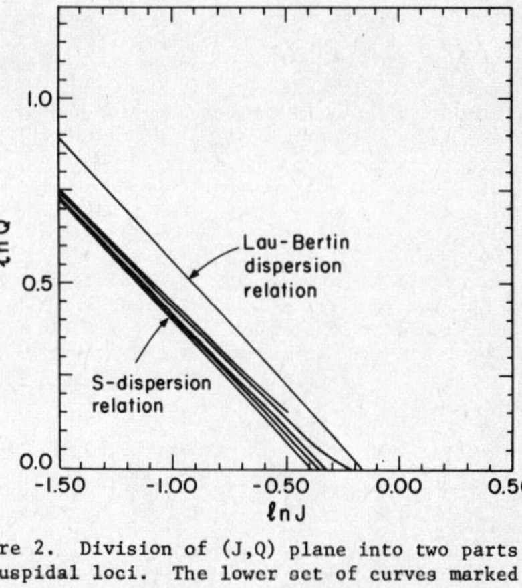


Figure 2. Division of (J,Q) plane into two parts by cuspidal loci. The lower set of curves marked S-dispersion relation are calculated for a few values of the shear parameter ($s = 1/2, 2/3, 1, 5/4, 3/2$). The upper curve is for the cubic dispersion relationship of Lau and Bertin.

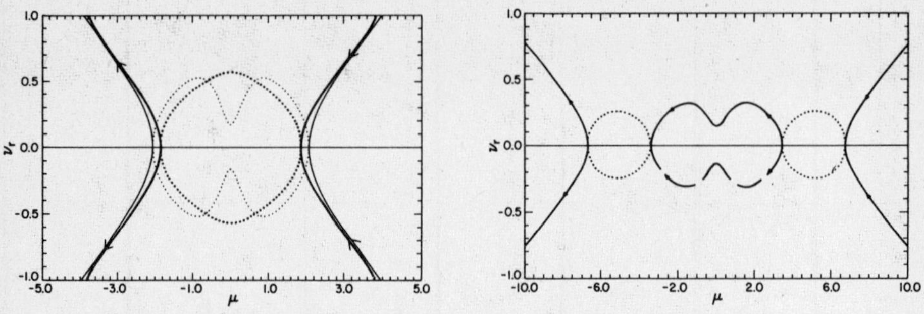


Figure 3. Local propagation diagrams in the corotation zone.

Arrows indicate the direction of propagation of the group velocity. The area under the dotted curve determines the factor of over-reflection.

(a) Case of high J (J = 1; s = 1, Q = 1.2). Comparison of the local propagation diagrams according to the Lau-Bertin dispersion relation $(\mu,\nu)$ and the S-dispersion relation $(\mu,\ 2Aq/\kappa)$. The dotted curves are for $|\mu|$ and $(2A/\kappa)\cdot|q|$, where $\nu, q$ are purely imaginary.

(b) Case of low J (J = 0.5; s = 1, Q = 1). The local propagation diagram $(\mu,\nu)$ according to the Lau-Bertin dispersion relation. The dotted curve is for $|\nu|$.

In view of the closeness between those two approximations, we shall only record some isolated examples of the results obtained.

(i) In Fig. 3(a), we show a comparison of the local propagation diagrams ($\nu$ versus $\mu$), limited to small values of $|\nu|$, for the two dispersion relationships in the case s = 1, J = 1, and Q = 1.2. There are the leading waves approaching the corotation circle ($\nu$ = 0), and the trailing waves departing it. The dotted curve in Fig. 3(a) for the Lau-Bertin dispersion relation is oval shaped; there are dents in the case of the S-dispersion relation. Since the areas under the two curves are nearly the same, the over-reflection factor has been found to be also approximately the same. The propagating part of the diagram (solid lines) is also essentially the same in these two approximations. Thus, in this case, the present homogeneous model can be used approximately for the corotation zone in the context of a global mode over the galactic disk.

(ii) To bring out the contrast between cases of high J and low J, we show, in Fig. 3(b), the local propagation diagram in the case s = 1, J = 0.5, and Q = 1. In this case, only the trailing waves are relevant in the modal context, as shown in the earlier theory for normal spirals. [The feedback cycle does not involve the leading waves in any significant manner. Furthermore, for small values of $\mu$, the lower half of the propagation diagram is usually found to be substantially modified by the inhomogeneity of the basic state, (cf. upper left diagram in the Figure on p. 120 of Ref. 1). Inhomogeneity is an item neglected in the approximation adopted in this paper.]

On the basis of the dispersion relation, and adopting the approach of Paper I, we may also obtain an ordinary differential equation which connects present results in the corotation zone with the modal approach. However, there are complications when ultra-long waves are found. We shall not give the details here. Instead, we shall turn our attention to the magnitude of the over-reflection factor in the present context of the uniform sheet model.

## 4. ENERGY EXCHANGE - OVER-REFLECTION

We now turn our attention to the essential mechanism of exchange of energy between the various wave trains in the zone of corotation.

Before presenting the results obtained from the computations, we first describe some general relationships on energy exchange in the corotation zone. In particular, we shall examine the numerical values of the over-reflection factor as the basic dynamical parameters are varied. We shall present the discussion of the neutral case here. The modification required for the exponentially growing modes becomes clear from the discussions in the Appendix.

### (A) A conservation principle

From Eq. (7), it can be easily shown that

$$\frac{d}{d\mu}\left(\hat{\psi}_* \frac{d\psi}{d\mu} - \hat{\psi}\frac{d\hat{\psi}_*}{d\mu}\right) = 0, \tag{22}$$

where $\hat{\psi}_*$ is the complex conjugate of $\hat{\psi}$. This is a conservation relation. If we now adopt the asymptotic solution (12), we get

$$|\alpha|^2 - |\beta|^2 = \text{const.}, \tag{23}$$

where $|\alpha|^2$ and $|\beta|$ are respectively proportional to the flux of angular momentum inside and outside of the corotation circle [cf. Eqs. (14) and (15)]. Let us denote the leading waves with a subscript $(\ )_1$ and the trailing waves with a subscript $(\ )_2$, and write (23) as

$$|\alpha_2|^2 - |\alpha_1|^2 = |\beta_2|^2 - |\beta_1|^2. \tag{24}$$

Then we may draw the following conclusions.

Case 1. If $|\beta_1|^2 = 0$, we have, in general,

$$|\alpha_2|^2 > |\alpha_1|^2. \tag{25}$$

Thus, there is over-reflection of a leading wave approaching the corotation circle into a departing trailing wave at a higher amplitude, plus an outgoing trailing wave ($\beta_2$). There results a larger flux of angular momentum carried away from the zone of corotation by the wave $\alpha_2$ than that carried into this zone by the wave $\alpha_1$. Balance is achieved by the wave $\beta_2$ which carries away angular momentum of an opposite sign. This is a typical process of wave amplification through the stimulation of emission of radiation (WASER; cf. Mark, Ref. 16, where the possibility of annihilation of the inward propagating wave by two approaching waves is also discussed).

Case 2. In the special case where $|\beta_1|^2 = |\beta_2|^2$, we get

$$|\alpha_2|^2 = |\alpha_1|^2 , \tag{26}$$

the reflection factor is unity. This case could happen if a modal calculation is carried out with zero boundary condition (ZBC) at infinity (which is usually taken to be a large but finite distance from the galactic center).

Case 3. However, it is still possible to have a growing mode with ZBC, because the amplitude $|\beta_1|$ of the reflected wave from infinity could be lower than $|\beta_2|$ when it reaches the corotation zone, since the amplitude of a growing wave train declines in the direction of its group propagation. Thus, we have

$$|\beta_1|^2 = |\beta_2|^2 \exp(-4\gamma\tau_0), \tag{27}$$

where $\gamma$ is the growth rate of the mode and $\tau_0$ is the time of propagation of the wave from the corotation circle to the outer point where the ZBC is applied.

The above discussions show that it is extremely important to impose the appropriate boundary conditions (ABC) in order to get the correct growth rate for the mode.

(B) Influence of basic dynamical parameters in over-reflection

We shall now proceed to the principal goal of determining the dependence of the over-reflection factor on the values of the dynamical parameters at the corotation circle. There are three methods by which this can be achieved: (1) direct numerical integration of Eq. (7), (2) WKBJ approach in spectral space, and (3) derivation of an ordinary differential equation of the second order for the description of wave trains and modes, as mentioned in Section 3. The first method is exact within the context of this model; the two other methods are approximate but have the advantage of simplicity and the consequent convenience for future use. They also lead to better understanding of the basic mechanisms and qualitative features.

Some of the results obtained by the first method were reported at IAU 106 (Ref. 13) in the case where the shear parameter $s = 1$. We now have extensive data for other values of the shear parameter. These data show a systematic decrease

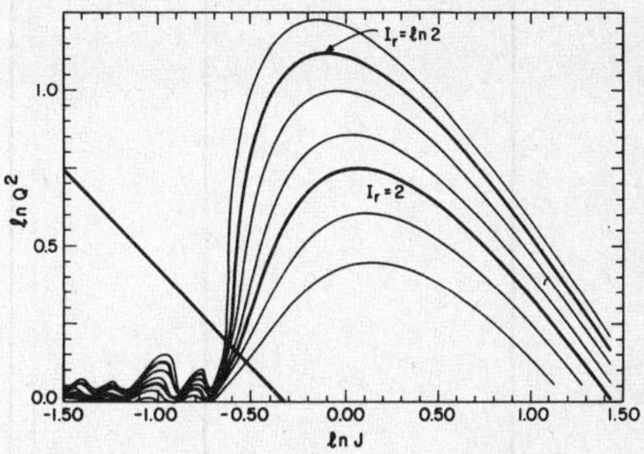


Figure 4. The over-reflection factor calculated according to Eq. (7) in the "swing formalism". The dividing line is taken from Fig. 2, for the case $s = 1$. To the left of this line, the asymptotic theory for normal spirals should be applied.

of the over-reflection factor with decreasing $s$. In Fig. 4, we report the results for the case $s = 1$ in greater detail. There are now also included the data for lower values of the J-parameter. We note the remarkable oscillatory behavior of the contour curves, which can be explained by using the WKBJ method. But these curves are not really relevant, because the validity of the present model becomes questionable in this range of parameters. Instead, the asymptotic theory of normal spirals holds in this case.

In the formulation of these data, we marked out the contour curves $\Gamma_r = 2$ and $\Gamma_r = e^2$ with heavier lines. The strip of parameter space in between is regarded as the domain of moderate over-reflection. This domain moves downwards in the (J,Q) plane as the shear parameter $s$ decreases. In the following discussions, all the numerical results refer to the case $s = 1$. The qualitative trends are however expected to be similar unless $s$ is quite small.

C. Self-regulation

We shall now record here in some detail the arguments of self-regulation which were briefly described in Ref. 13 and in Paper I.

We would like to examine the physical implications of the magnitude of over-reflection on the galactic disk as a whole. This is a difficult issue. We can only make some plausible conjectures at the present time. Also, it is an issue that goes beyond the original restricted scope of the present paper, since the instability of global modes must be considered. The dynamical processes in the corotation zone should not be studied in isolation.

In Paper I, we have seen that the growth rate of the modes is generally correlated with the magnitude of the over-reflection factor $\Gamma_r$. Thus, we may infer that the QSSS hypothesis is likely to hold only when the over-reflection factor has a moderate value. For high values of $\Gamma_r$, it is expected that there is strong instability in the

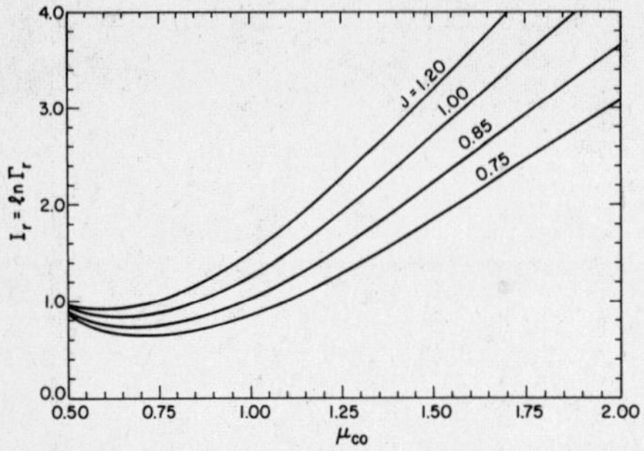


Figure 5. The index of over-reflection $I_r = \ell n \Gamma_r$ as a function of the tightness parameter $\mu_{co}$ for various values of the J-parameter when the shear parameter $s = 1$. The curves for $J = 0.75 - 1.2$ show the same general trend. In all cases, moderate growth rates can be achieved only for low values of the μ-parameter.

modal sense; i.e., there is a plethora of modes with high rates of growth $\gamma$. Let us examine the possible developments under such conditions for different ranges of values of the J-parameter.

We begin our discussion with low values of the J-parameter where the well-developed asymptotic theory of normal spiral modes holds. According to that theory, for low values of the J-parameter ($J < 0.5$), we expect to have a QSSS composed of a small number of normal spiral modes of moderate growth rates only when Q is close to unity. Indeed, the moderate value of the growth rate is related to that of the over-reflection factor $\Gamma_r$, and this happens when Q is restricted to a fairly narrow range close to unity. In contrast, $\Gamma_r$ can become of the order of 10-20 for $Q \simeq 0.9 - 0.8$. In such cases, there is expected a plethora of strongly unstable modes, including the axi-symmetric modes and some quite tightly wound spirals. Such instabilities may lead to an increase of the dispersion speeds of the stars in a stellar disk, and a process of self-regulation sets in; i.e., the disk tends to evolve into such a state that the dynamical parameters again lie in the regime of moderate over-reflection, appropriate to quasi-stationary sA spiral structures.

In the case of high J, the situation is slightly more complicated, but the general outlook is similar. In the Groningen paper (Ref. 13), we already noted a general resemblance between the contour curves of constant over-reflection and those of constant pitch angle. Thus, one may wish to examine whether there is a close relationship between the index of over-reflection $I_r = \ell n \Gamma_r$ and the tightness parameter $\mu_{co}$ (the value of μ at $\nu^2 = 0$), for each given value of the shear parameter s. If this is true, then the experience of Paper I would justify the general correlation of the pitch angle of the waves with the growth rate of the modes.

By calculating the numerical values of the relevant parameters, we now find that an approximate empirical correlation can indeed be established between the parameters $\Gamma_r$ and $\mu_{co}$ for values of the J-parameter lying in the range $0.75 \leq J \leq 1.2$ (Fig. 5). From these data, we may draw the conclusion that, for the parameter J lying in this range, QSSS may be expected to hold only for open modes. The reasoning is similar to that for the case of low values of the J-parameter. Take the case $J = 1$. When Q is slightly above unity, the over-reflection factor is very high. There is again a plethora of unstable modes which are essentially dominated by short trailing waves (μ large). These rapidly growing transient sA spirals cannot be expected to lead to galaxies of the types SAa and SAb. Rather, the conditions of high $J (J \cong 1)$ and low $Q^2 (Q^2 \cong 1)$ do not provide the conditions for QSSS. On the other hand, for a basic state with $Q^2 \simeq 2$, for example, there may only be a few moderately growing modes. The process of self-regulation is expected to bring about the increase of stellar dispersion speeds to change the state $Q^2 \simeq 1$ to the state $Q^2 \simeq 2$.

We have so far covered the two extreme cases, in each of which only two waves play the essential role. In between, there is a transition domain that lies close to the dividing lines Eq.(21). Here all the waves may be important, and the spiral structure of the modes of moderate growth may be of the transitional type sAB. (See examples calculated by Jon Haass in Ref. 13). Thus, we have a fairly good picture of the domain of parameters that correspond to the QSSS that may be expected to be observed in real galaxies.

## 5. CONCLUDING REMARKS

Although we attempted to isolate the study of the mechanism of excitation in the corotation zone from other issues, we cannot fully avoid references to global modes. We shall now make a few remarks on the implications of the above discussions on these modes and on the validity of the QSSS hypothesis.

The main theme of this paper was to examine both the fast evolving spiral features and the slowly evolving wave trains which form the basis of grand designs over a galactic disk. Differences in (i) emphasis on these two types of processes, (ii) mathematical methods adopted, and (iii) the regime of dynamical parameters considered are three of the issues which often lie at the root of recent controversies regarding the validity of the QSSS hypothesis. We shall now briefly comment on these issues, beginning with the last item.

(a) The examination of the mechanism of excitation of spiral modes in the corotation zone brings out the importance of the numerical value of the J-parameter in that zone in the determination of the rate of growth. For low values of J ($J \leq 0.5$), we expect moderate rates of growth if $Q \simeq 1$. On the other hand, for $J \simeq 1$, the corresponding state is $Q^2 \simeq 2$. The values refer to theoretical models with an infinitesimally thin disk composed of a single fluid component. For comparison with real galaxies, the reader is referred to relevant discussions in Ref. 13 and in Paper I. Note that we have taken the shear parameter $s = 1$.

If the Q-parameter is too low, e.g., in the domain near $(J, Q^2) = (1, 1)$, there is very strong over-reflection and hence there is expected to be a large number of strongly unstable modes. Thus, the galaxy may be forced to evolve into a new dynamical state with a substantial rise in the dispersion velocity of the stars. This increase in $Q^2$ will then tend to reduce the instability in the dynamical system. As mentioned above, this process of self-regulation deserves further study. For the present, it is important to note that the QSSS-hypothesis cannot possibly hold for the regime of parameters in question, but the dynamical systems tends to evolve into states in favor of that hypothesis, as a natural consequence of the dynamical processes.

(b) One should have a clear appreciation of the possible coexistence of global modes and local spiral features. In general, they evolve on different time scales and extend to different length scales over the galactic disk. One time scale, namely that associated with the epicyclic frequency $\kappa$ of the motion of the individual stars, is basic. It is essentially impossible for collective behavior of stars to evolve at a faster rate than $\kappa$. Localized spiral features are likely to be material objects and wave packets which evolve at the rate of shear. This rate is comparable or somewhat lower than the epicyclic frequency. On the other hand, there can be unstable spiral modes which grow at much slower rates. There may be a difference of almost one order of magnitude in the time scales (e.g. 40 million years versus 400 million years). The existence of slowly evolving regular grand designs need not be obscured by the coexistence of fast evolving local features.

(c) In the context of the homogeneous model in the corotation zone, the method of spectral analysis adopted in this paper provides a convenient mathematical formalism in which we can place the rapidly evolving wave packet and the slowly evolving wave train in parallel positions in the logical structure. In the context of disturbances over a whole galactic disk, the importance of the choice of the proper mathematical representation is even more obvious. The fast evolving processes can be easily described when a time dependent problem is solved by the use of Laplace transformation or in a step-by-step process of numerical integration. However, in the same approach, it is difficult to bring out the long lasting features. On the other hand, if one chooses the modal approach, it is inconvenient, though still possible, to follow fast time evolution; but this representation is obviously most convenient for bringing out the long lasting features. This contrast in the two approaches is already well illustrated in classical applications of Fourier analysis. [See for example, Ref. 8, pp. 563-565.] The modal approach is clearly to be preferred in the study of grand designs in galaxies. [This issue has been discussed in some detail by Lin and Bertin (Ref. 13), with some comments on nonlinear processes.]

To summarize, the above discussions all suggest that the QSSS hypothesis remains valid despite some of the issues frequently raised against it. In the context of the present paper, we have emphasized the point that the mechanism of over-reflection involving both leading and trailing waves is not necessarily powerful nor is it necessarily transient. In addition, one should keep in mind that the process of over-reflection in the corotation zone forms only one part of the feedback cycle in the modal context. Indeed, further work is still needed to fully reconcile it with the uniform sheet model. At present, we may cautiously claim that the physical mechanisms have been essentially understood and quantitatively determined with reasonable accuracy. The results obtained in this paper are found to be in general agreement with those obtained in Paper I, although differences in detail remain. What is more important, we found further general support for the perception described in Paper I on the dynamical interpretation of the spiral modes. Fairly extensive numerical calculations of modes should now be carried out to confirm these conclusions or to suggest improvements and modifications.

ACKNOWLEDGEMENT

We wish to thank Professor A. Toomre for bringing to our attention the important role of the leading waves in the mechanism of excitation. To a considerable extent, this paper is a follow-up to his earlier work. We would like to thank G. Bertin and S. A. Lowe for many helpful discussions. The work was partly supported by a grant from the National Science Foundation.

## APPENDIX

### The infinite sheet model with uniform differential rotation.

In this Appendix, we give a detailed mathematical analysis for the time dependent solutions of the infinite sheet model. Specifically, we wish to bring out the relationship between two classes of solutions: those that are quasi-stationary and those that vary at rates comparable to the basic dynamic time rate of the epicyclic frequency.

### (A) The basic equations

The basic equations of the model will now be given in the logarithmic coordinate system introduced in Section 2. We shall then introduce the local approximation Eq. (5) for the angular velocity (or the equivalent formula in the x-coordinate). All other basic parameters are taken to be equal to their values at the corotation circle. (The corotation circle is well-defined for any single wave packet, since the basic state is time-independent, cf. Ref. 20, p. 383). Such local approximations can be justified when we consider perturbations of large wave number.

The equations of continuity and of motion are given by

$$\left(\frac{\partial}{\partial t} - 2Ax\frac{\partial}{\partial y}\right)\sigma' = -\sigma\left(\frac{\partial u'}{\partial x} + \frac{\partial v'}{\partial y}\right) \quad (A1)$$

$$\left(\frac{\partial}{\partial t} - 2Ax\frac{\partial}{\partial y}\right)u' = 2\Omega v' - \frac{\partial}{\partial x}(\psi' + h') \quad (A2)$$

$$\left(\frac{\partial}{\partial t} - 2Ax\frac{\partial}{\partial y}\right)v' = -2Bu' - \frac{\partial}{\partial y}(\psi' + h') \quad (A3)$$

where A and B are the absolute values of the Oort constants (respectively for differential rotation and for self-rotation), $\Omega$ is the local value of circular rotation, and $\sigma$ is the local value of the surface density in the basic state. The perturbation quantities $\sigma'$, $u'$, $v'$, $h'$ and $\psi'$ are respectively those of surface density, the components of velocity, the enthalpy and the gravitational potential.

In spectral representation, the above equations become

$$\frac{\partial\tilde{\theta}}{\partial t} - \bar{c}\frac{\partial\tilde{\theta}}{\partial\xi} = -i(\xi\tilde{u} - \eta\tilde{v}) \quad (A4)$$

$$\frac{\partial\tilde{u}}{\partial t} - \bar{c}\frac{\partial\tilde{u}}{\partial\xi} = 2\Omega\tilde{v} - i\xi(\tilde{\psi} + \tilde{h}) \quad (A5)$$

$$\frac{\partial\tilde{v}}{\partial t} - \bar{c}\frac{\partial\tilde{v}}{\partial\xi} = -2B\,\tilde{u} + i\eta(\tilde{\psi} + \tilde{h}) \quad (A6)$$

where $\tilde{\theta} = \tilde{\sigma}/\sigma$. We have a simple approximate solution for the Poisson equation in the spectral representation; i.e. we have

$$\tilde{\psi} + \tilde{h} = a^2\,\tilde{\theta}\{1 - 2k_o/(\xi^2 + \eta^2)^{1/2}\} \quad (A7)$$

where $a$ is the acoustic speed of the basic state and we have adopted the symbols

$$\bar{c} = 2A\eta, \quad k_o = \pi G\sigma/a^2 \quad (A8)$$

The wave numbers $\xi$ and $\eta$ are introduced in the text [Eq. (3)].

(B) The general solution in spectral space.

The system of differential equations (A4) to (A6) is expected to have a general solution that contains three arbitrary functions. By introducing the characteristic variables, one can easily show that the general solution is given by

$$\tilde{\theta}(\xi,t) = \sum_{i=1}^{3} \hat{\theta}_i(\xi)\, C_i(\xi + \bar{c}t) \quad (A9)$$

$$\tilde{u}(\xi,t) = \sum_{i=1}^{3} \hat{u}_i(\xi)\, C_i(\xi + \bar{c}t) \quad (A10)$$

$$\tilde{v}(\xi,t) = \sum_{i=1}^{3} \hat{v}_i(\xi)\, C_i(\xi + \bar{c}t) \quad (A11)$$

where $(\hat{\theta}, \hat{u}, \hat{v})_i$ satisfy the system of equations (where $L$ is a linear operator) of the form

$$\bar{c}\frac{d\hat{\theta}}{d\xi} = L_1(\hat{u}, \hat{v}) \quad (A12)$$

$$\bar{c}\frac{d\hat{u}}{d\xi} = L_2(\hat{v}, \hat{\theta}) \quad (A13)$$

$$\bar{c}\frac{d\hat{v}}{d\xi} = L_3(\hat{\theta}, \hat{u}) \quad (A14)$$

and $C_i(\xi + \bar{c}t)$ are functions at our disposal. They may be chosen to satisfy arbitrary initial conditions. The system of equations (A12) to (A14) is the time-independent form of (A4) to (A6).

The general solution (A9) to (A11) can be decomposed into two parts, one part representing the vorticity disturbance, another part representing the waves. The vorticity disturbance can be isolated by noting that a conservation relation can be obtained by forming a suitable linear combination of (A4) to (A6). We find:

$$\left(\frac{\partial}{\partial t} - \bar{c}\frac{\partial}{\partial\xi}\right)(\eta w) = 0, \quad (A15)$$

where

$$\eta\tilde{w} = \eta\tilde{u} + \xi\tilde{v} + 2iB\tilde{\theta} \quad (A16)$$

Clearly, the general solution of (A15) is

$$\tilde{w} = F(\xi + \bar{c}t) \quad (A17)$$

Let us now consider a fundamental system of solutions satisfying the following initial conditions

$$(\hat{\theta}_1, \hat{v}_1, \hat{w}_1) = (1, 0, 0) \quad (A18)$$

$$(\hat{\theta}_2, \hat{v}_2, \hat{w}_2) = (0, 1, 0) \quad (A19)$$

$$(\hat{\theta}_3, \hat{v}_3, \hat{w}_3) = (0, 0, 1) \quad (A20)$$

Then the third solution is given by (A17) which describes the vorticity disturbance. The waves are then given by linear combinations of solutions specified by (A18) and (A19); for this set of solutions, we have $\tilde{w} \equiv 0$. By introducing this last equation into the system (A4) to (A6), we can simplify that system to yield the results for the waves given by Eq. (6) in the text. The system of equations (A12) to (A14) has been reduced to a single equation of the second order. The general solution is still given in the format (A9) to (A11), but there are now only two arbitrary functions.

(C) Time-dependent solutions in physical space

The solutions in physical space may now be obtained as Fourier integrals. We consider only solutions of the type defined by Eq. (3). For the vorticity disturbance, Eq. (A17) leads to the general solution

$$w(x,t) = e^{-i\bar{c}tx}w(x,0) \quad (A21)$$

For the waves, we have (with i = 1,2),

$$\psi_i(x,t) = \int_{-\infty}^{\infty} f_i(\xi + \bar{c}t)\,\hat{\psi}_i(\xi)e^{i\xi x}d\xi \quad (A22)$$

We shall now discuss a few special cases of (A22) for application to exponentially growing wave trains and modes, and for comparison with solutions obtained from the study of swinging wave packets. Note that the typical time rate in (A22) is the shear rate 2A.

(a) The exponentially growing wave trains

Although the general time rate in the solution (A22) is the shear rate 2A, we can get solutions which vary at much slower rates. For example, if $f(\xi + \bar{c}t) = \exp\{(\gamma/\bar{c})(\xi + \bar{c}t)\}$, then we get the solution

$$\psi(x,t) = e^{\gamma t}\psi(x,0), \quad (A23)$$

where

$$\psi(x,0) = \int_{-\infty}^{\infty} e^{(\gamma/\bar{c})\xi}\hat{\psi}(\xi)e^{i\xi x}d\xi \quad (A24)$$

A truncation of $\psi(\xi)$ at high values of $|\xi|$ is necessary for convergence.

Two remarks should be made here. First, the solution (A23) varies with time at a rate of $\gamma$ which is at our disposal. This can be much slower than the typical rate of 2A or $\kappa$. Indeed, the solution (A23) is in a natural form to be fitted into a growing wave pattern, (or a neutral wave pattern in the case $\gamma = 0$), in contrast to the swinging wave packets to be discussed below.

Another important point to be noted is more technical. It concerns the over-reflection factor discussed in Section 4. The integrand in Eq. (A24) shows that the over-reflection factor for the case $\gamma \neq 0$ may be obtained from that for the neutral case ($\gamma = 0$) by multiplication with the additional factor $\exp\{(\gamma/\bar{c})(\xi_2 - \xi_1)\}$, which may be interpreted in terms of a "swing time" $\tau_s = -(\mu_2 - \mu_1)/2A$; i.e., the additional factor is $\exp(-\tau_s)$. In practice, $\mu_1 > 0$, $\mu_2 = -\mu_1 < 0$; $\tau_s \gtrless 0$. In the context of exponentially growing wave trains, this is similar to the decline of wave amplitude in the direction of group propagation. Its effect on the discussion of the feedback cycle in the modal context can be interpreted as an increase in the length of the feedback cycle, if we insist on using the over-reflection factor calculated for the neutral case.

(b) Swinging wave packets

If we introduce the variable

$$\xi' = \xi + \bar{c}t \qquad (A25)$$

into Eq. (A22) we get the representation of $\psi(x,t)$ in terms of swinging wavelets:

$$\psi(x,t) = \int_{-\infty}^{\infty} f(\xi')\, \hat{\psi}(\xi'-\bar{c}t) e^{i(\xi'-\bar{c}t)x} d\xi' . \qquad (A26)$$

Here, we have omitted the index i. We note that $f(\xi')$ is the amplitude function in the process of superposition, whereas the basic solution is the swinging wavelet:

$$\psi_s = \hat{\psi}(\xi'-\bar{c}t) e^{i(\xi'-\bar{c}t)x} \qquad (A27)$$

The amplitude varies in time; it is obtained by "scanning" over the spectral function, with the wave number given by

$$\xi/\eta = \xi'/\eta - 2At \qquad (A28)$$

This is precisely the type of solution obtained and used by Goldreich and Tremaine (Ref. 4) and by Toomre (Ref. 18) in their treatment of the infinite sheet. To recover (A27) from (A26), we need only take $f(\xi')$ to be $\delta(\xi'-\xi_1)$. The shearing wavelet is

$$\psi_s(x,t)e^{-i\eta y} = \hat{\psi}(\xi_1 - \bar{c}t)\exp\{i(\xi_1 x_L - \eta y_L)\} \qquad (A29)$$

where

$$x_L = x, \quad y_L = y + 2Axt \qquad (A30)$$

are the Lagrangian coordinates for the basic state. We note that (A30) implies a very rapid swinging. Any slow growth implied by a small value of $\gamma$ in (A23) will be "swamped" by this swinging behavior. We also note that (A29) is a solution of the form used by Lord Kelvin in dealing with disturbances in an incompressible viscous fluid in uniform shear.

Gaussian wave packets that have essentially the same kind of swinging behavior may be obtained by setting $f(\xi') = A\exp\{-(\xi' - \xi_1)^2/\ell^2\}$ with $A\ell\sqrt{\pi} = 1$. The solutions obtained can be compared directly with those that appear prominently in the papers cited above. We shall not go into the details because of lack of space.

REFERENCES


1. Bertin, G. (1983) Proc IAU Symposium 100, pp. 119-120 (Besancon, 1982).

2. Drury, L. O'C. (1980). M.N.R.A.S. 193, pp. 337-343.

3. Goldreich, P. and D. Lynden-Bell (1965). M.N.R.A.S. 130, pp. 125-158.

4. Goldreich, P. and S. Tremaine (1978). Astrophys. Journ. 222, 850-858.

5. Haass, J., G. Bertin and C. C. Lin (1982). Proc. Nat. Acad. Sci. USA 79, pp. 3908-3912.

6. Hunter, C. (1983). Fluid Dynamics in Astrophysics and Geophysics, N. R. Lebovitz, editor; being Vol. 20 of Lectures in Applied Mathematics, Am. Math. Soc., pp. 179-203.

7. Iye, M. et. al. (1982) Astrophys. J. 256, pp. 103-111.

8. Jeffreys, H. and B. S. (1956). Methods of Mathematical Physics, Cambridge University Press.

9. Julian, W. H. and A. Toomre (1966). Astrophys. Journ. 146, 810-830.

10. Lau, Y. Y. (1979), Phys. Rev. Letters, 42, 779-781; Lau et. al. (1980), Phys. Fluids 23, 939-941, 1827-1831.

11. Lau, Y. Y. and G. Bertin (1978). Astrophys. Journ. 226, pp. 508-520.

12. Lin, C. C. and G. Bertin (1981); with J. Haass, in Plasma Astrophysics, eds. T.D. Guyenne and G. Levy (European Space Agency, Noordwijk, the Netherlands), SP-161, pp. 191-205.

13. Lin, C. C. and G. Bertin (1984). Formation and Maintenance of Spiral Structures in Galaxies, to appear in Proc. IAU Symposium 106 (Groningen 1983); D. Reidel Publishing Company.

14. Lin, C. C. and Y. Y. Lau (1975). SIAM J. of Appl. Math. Vol. 29, pp. 352-370.

15. Marcus, P. and W. H. Press (1977), J. Fluid Mech. 70, pp. 525-534.

16. Mark, J. W-K (1976) Astrophys. Journ. Vol. 25, pp. 363-378.

17. Rayleigh, Lord (1880). Scientific Papers 1, 474-87; Cambridge University Press.

18. Toomre, A. (1981). In "The Structure and Evolution of Normal Galaxies", ed. S. M. Fall and D. Lynden-Bell, Cambridge University Press (1980), pp. 111-136.

19. Tung, K. K. (1983). J. Fluid Mech., 133, 443-469.

20. Whitham, G. B. (1974). Linear and Nonlinear Waves. (Wiley-Interscience, New York).

# BIBLIOGRAPHY OF THE PUBLICATIONS OF C. C. LIN

[1] *On the dependence of interaction energy upon atomic arrangements in superlattices of binary alloys*
Chinese Journal of Physics, Vol. 4, pp. 182–197 (1939)

[2] *Note on the normalization of Dirac functions*
Physical Review, Vol. 59, No. 10, pp. 841–842 (1941)

[3] *On the motion of vortices in two dimensions*
Proceedings of the National Academy of Sciences, Vol. 27, No. 12, Part I, pp. 570–575, Part II, pp. 575–577 (1941)

[4] *On a statistical model of isotropic turbulence*
(with J. L. Synge)
Transactions of the Royal Society of Canada, Third Series, Section III, Vol. 37, pp. 1–35 (1943)

[5] *On the motion of vortices in two dimensions*
A pamphlet in the University of Toronto Studies, Applied Mathematics Series, No. 5, published by the University of Toronto Press (1943)

[6] *On the motion of a pendulum in a turbulent fluid*
Quarterly of Applied Mathematics, Vol. 1, pp. 43–48 (1943)

[7] *On the stability of two-dimensional parallel flows*
Proceedings of the National Academy of Sciences, Vol. 30, No. 10 pp. 316–324 (1944)

[8] *On the stability of two-dimensional parallel flows*
Quarterly of Applied Mathematics, Vol. 3, No. 4,
Part I, General theory, pp. 117–142 (1945)
Part II, Stability in an inviscid fluid, pp. 218–234 (1945)
Part III, Stability in a viscous fluid, pp. 277–301 (1946)

[9] *On an extension of the von Karman-Tsien method to two-dimensional subsonic flows with circulations around closed profiles*
Quarterly of Applied Mathematics, Vol. 4, No. 3, pp. 291–297 (1946)

[10] *On the steady flow of a gas through a tube with heat exchange or chemical reaction*
(with P. Chambre)
Journal of Aeronautical Sciences, Vol. 13, No. 10, pp. 537–542 (1946)

[11] *Investigation of the stability of the laminar boundary layer in a compressible fluid*
(with L. Lees)
National Advisory Committee for Aeronautics, Technical Note 1115 (1946)

[12] *On the mechanical behavior of metals in the strain hardening range*
(with G. Handelman and W. Prager)
Quarterly of Applied Mathematics, Vol. 4, No. 4 (1947)

[13] *Velocity and temperature distributions in turbulent jets*
The Science Reports of National Tsing Hua University, Vol. 4, Nos. 4–6, pp. 419–450 (1947)

[14] *On the nature of the boundary layer near the leading edge of a flat plate*
(with G. F. Carrier)
Quarterly of Applied Mathematics, Vol. 6, No. 1, pp. 63–68 (1948)

[15] *On the flow behind curved shocks*
(with S. I. Rubinov)
Journal of Mathematics and Physics, Vol. 27, No. 2, pp. 105–129 (1948)

[16] *On two-dimensional non-steady motion of a slender body in a compressible fluid*
(with E. Reissner and H. S. Tsien)
Journal of Mathematics and Physics, Vol. 27, No. 3, pp. 220–231 (1948)

[17] *Note on the law of decay of isotropic turbulence*
Proceedings of the National Academy of Sciences, Vol. 34, No. 11, pp. 540–543 (1948)

[18] *On the law of decay and the spectrum of isotropic turbulence*
Proceedings of the Seventh International Congress of Applied Mechanics, London, Sept. 1948

[19] *On the concept of similarity in the theory of isotropic turbulence*
(with Th. von Kármán)
Review of Modern Physics, Vol. 21, No. 3, pp. 516–519, Jul. 1949

[20] *On the subsonic flow through circular and straight lattices of airfoils*
Journal of Mathematics and Physics, Vol. 28, No. 2, pp. 117–130 (1949)

[21] *Note on the characteristics in unsteady one-dimensional flows with heat addition*
Quarterly of Applied Mathematics, Vol. 7, No. 4, pp. 443–445 (1950)

[22] *On the stability of the boundary layer over a cone*
(with R. H. Battin)
Journal of Aeronautical Sciences, Vol. 17, No. 7, pp. 453 (1950)

[23] *Some problems of flow behind shocks of varying strength*
Proceedings of the Symposium on Shock-Wave Phenomena, U.S. Naval Ordnance Laboratory, Jun. 30, 1949; joint meeting with the Fluid Dynamics Division of the American Physical Society. Published Jul. 1, 1950

[24] *A general integral form of the boundary-layer equation for incompressible flow with an application to the calculation of the separation point of turbulent boundary layers*
(with N. Tetervin)
National Advisory Committee for Aeronautics, Technical Note 2158, Aug. 1950

[25] *Some recent investigations in the theory of hydrodynamic stability*
(with J. R. Foote)
Quarterly of Applied Mathematics, Vol. 8, No. 3, pp. 265–280 (1950)

[26] *On the stability of zonal winds over a rotating spherical earth*
Proceedings of the International Congress of Mathematicians, Vol. 1 (1950)

[27] *On the statistical theory of isotropic turbulence*
(with Th. von Kármán)
Advances in Applied Mechanics, Vol. 2, pp. 2–19, Academic Press, N. Y. (1951)

[28] *Propagation of spherical shock waves in stellar interiors*
(with Z. Kopal)
Proceedings of the National Academy of Sciences, Vol. 37, No. 8, pp. 495–506 (1951)

[29] *On the attached curved shock in front of a sharp-nosed axially symmetrical body placed in a uniform stream*
(with S. F. Shen)
National Advisory Committee for Aeronautics, Technical Note 2505, Oct. 1951

[30] *An analytic determination of the flow behind a symmetrical curved shock in a uniform stream*
(with S. F. Shen)
National Advisory Committee for Aeronautics, Technical Note 2506, Oct. 1951

[31] *A critical examination of similarity theory for incompressible flows*
(with S. F. Shen)
National Advisory Committee for Aeronautics, Technical Note 2541 (1951)

[32] *A similarity theory for turbulent boundary layer over a flat plate in a compressible flow*
(with S. F. Shen)
National Advisory Committee for Aeronautics, Technical Note 2542 (1951)

[33] *A new variational principle for isenergetic flows*
Quarterly of Applied Mathematics, Vol. 9, No. 4, pp. 421–423 (1952)

[34] *On the stability of the boundary layer with respect to disturbances of large wave velocity*
Journal of the Aeronautical Sciences, Vol. 19, pp. 138–139 (1952)

[35] *The stability of the laminar boundary layer in a compressible fluid for the case of three-dimensional disturbances*
Journal of the Aeronautical Sciences, Vol. 19, No. 7, p. 491 (1952)

[36] *A critical discussion of similarity concepts in isotropic turbulence*
Proceedings of the 4th Symposium in Applied Mathematics, pp. 19–27, McGraw-Hill Book Co.

[37] *On Taylor's hypothesis and the acceleration terms in the Navier-Stokes equations*
Quarterly of Applied Mathematics, Vol. 10, No. 4, pp. 295–306, Jan. 1953

[38] *On the stability of the laminar mixing region between two parallel streams in a gas*
National Advisory Committee for Aeronautics, Technical Note 2887, Jan. 1953

[39] *A simplified formulation of the similarity concepts in isotropic turbulence*
Journal of the Aeronautical Sciences, Vol. 20, No. 4, Apr. 1953

[40] *On the role of three-dimensional disturbances in the stability of supersonic boundary layers*
(with D. W. Dunn)
Journal of the Aeronautical Sciences, Vol. 20, No. 8, p. 577 (1953)

[41] *Note on the mean square of integrals in the statistical theory of turbulence*
Quarterly of Applied Mathematics, Vol. 11, No. 3, pp. 367–370, Oct. 1953

[42] *Hydrodynamic stability*
Proceedings of the 5th Symposium in Applied Mathematics (AMS), pp. 1–18 (1954)

[43] *On a perturbation theory based on the method of characteristics*
Journal of Mathematics and Physics, Vol. 33, pp. 117–134 (1954); also
Journal of Aeronautical Sciences, Vol. 21, No. 3, p. 202 (1954)

[44] *Some physical aspects of the stability of parallel flows*
Proceedings of the National Academy of Sciences, Vol. 40, No. 8, pp. 741–747 (1954)

[45] *On periodically oscillating wakes in the Oseen approximation*
Studies in Mathematics and Mechanics, presented to Richard von Mises, Academic Press, Inc., N. Y., pp. 170–176 (1954)

[46] *Theoretical comments on the paper of E. N. Fales*
(with Th. von Kármán)
Journal of the Franklin Institute, Vol. 259, No. 6, Jun. 1955

[47] *On the stability of the laminar boundary layer in a compressible fluid*
(with D. W. Dunn)
Journal of Aeronautical Sciences, Vol. 22, No. 7, pp. 455–477 (1955)

[48] *The theory of hydrodynamic stability*
Monograph published by Cambridge University Press (1955)

[49] *Aspects of the problem of turbulent motion*
Journal of Aeronautical Sciences, Vol. 23, No. 5, pp. 453–461 (1956)

[50] *Stability of laminar flows*
Applied Mechanics Reviews, Vol. 10, No. 1, p. 1 (1957)

[51] *On uniformly valid asymptotic solutions of the Orr-Sommerfeld equation*
IX$^{e}$ Congres International de Mechanique Appliquee, Extrait des Actes, Tome 1, pp. 136–148 (1957)

[52] *On the stability of the laminar boundary layer*
Naval Hydrodynamics, Ch. XIV, Publication 515, National Academy of Sciences, National Research Council (1957)

[53] *Note on Garabedian's paper, "Numerical construction of detached shock waves"*
Journal of Mathematics and Physics, Vol. 36, No. 3, pp. 206–209, Oct. 1957

[54] *On the instability of laminar flow and its transition to turbulence*
Symposium Freiburg, Aug. 26–29, 1957, Boundary Layer Research, edited by H. Görtler, pp. 144–160, Springer-Verlag (1958)

[55] *Note on a class of exact solutions in magneto-hydrodynamics*
Archive for Rational Mechanics and Analysis, Vol. 1, No. 5, pp. 391–395 (1958)

[56] *On the asymptotic solutions of a class of ordinary differential equations of the fourth order*
(with A. L. Rabenstein)
Transactions of the American Mathematical Society, Vol. 94, No. 1, pp. 24–57 (1960)

[57] *Motion in the boundary layer with a rapidly oscillating external flow*
Proceedings of 9th International Congress of Applied Mechanics, University of Brussels, Belgium, Sept. 6, 1956, Vol. 4, pp. 155–167 (1959)

[58] *Hydrodynamics of liquid helium II*
Physical Review Letters, Vol. 2, No. 6, pp. 245–246, Mar. 15, 1959

[59] *On the theory of shallow water waves*
(with Alfred Clark, Jr.)
Tsing Hua Journal of Chinese Studies, Special No. 1, Natural Science, Dec. 1959

[60] *On a theory of dispersion by continuous movements*
The Proceedings of the National Academy of Sciences, Vol. 46, No. 4, pp. 566–570, Apr. 1960

[61] *On a theory of dispersion by continuous movements, II. Stationary anisotropic processes*
The Proceedings of the National Academy of Sciences, Vol. 46, No. 8, pp. 1147–1150, Aug. 1960

[62] *On the secondary motion induced by oscillations in a shear flow*
(with D. J. Benney)
The Physics of Fluids, Vol. 3, No. 4, Jul.–Aug. 1960

[63] *Some mathematical problems in the theory of the stability of parallel flows*
Journal of Fluid Mechanics, Vol. 10, Part 3, pp. 430–438 (1961)

[64] *On the existence of an exact solution of the equations of Navier-Stokes*
(with Th. von Kármán)
Communications on Pure and Applied Mathematics, Vol. XIV, No. 3, Aug. 1961

[65] *Statistical theories of turbulence*
No. 10 – Princeton Aeronautical Paperbacks, from Princeton Series on High Speed Aerodynamics and Jet Propulsion, Princeton University Press (1961)

[66] *On the instability of shear flows*
(with D. J. Benney)
Proceedings of Symposia in Applied Mathematics, Vol. 13, Hydrodynamic Instability (1962)

[67] *Hydrodynamics of helium II*
Estratto da Rendiconti della Scuola Internazionale di Fisica – E. Fermi – XXI Corso. Lectures at the Enrico Fermi International School of Physics, Varenna, Italy, Jul. 1961; published in Italy; reprints received M.I.T. Oct. 1963

[68] *Turbulent flow, theoretical aspects*
(with W. H. Reid)
Handbuch der Physik – Encyclopedia of Physics, Vol. VIII/2 (1963)

[69] *Some examples of asymptotic problems in mathematical physics*
Proceedings of Symposium on Asymptotic Solutions of Differential Equations and Their Applications, at the Mathematics Research Center, University of Wisconsin, May 4–6, 1964, pp. 129–143

[70] *On the spiral structure of disk galaxies*
(with F. H. Shu)
The Astrophysical Journal, Vol. 140, pp. 646–655, Aug. 15, 1964

[71] *The gravitational collapse of a uniform spheroid*
(with L. Mestel and F. H. Shu)
The Astrophysical Journal, Vol. 142, pp. 1431–1446, Nov. 1965

[72] *On the spiral structure of disk galaxies, II. Outline of a theory of density waves*
(with F. H. Shu)
Proceedings of the National Academy of Sciences, Vol. 55, pp. 229–234, Feb. 1966

[73] *On the instability of shear flows and their transition to turbulence*
(with D. J. Benney)
Proceedings of the 11th International Congress of Applied Mechanics, Munich, Germany, Aug. 30–Sept. 5, 1964, pp. 797–802 (Reprints 1966)

[74] *On the spiral patterns of disk galaxies*
Proceedings of the 4th U.S. National Congress of Applied Mechanics, University of Minnesota, Jun. 14–17, 1966, pp. 3–14

[75] *On the mathematical theory of a galaxy of stars*
J. SIAM Applied Mathematics, Vol. 14, pp. 876–921, Jul. 1966

[76] *Stellar dynamical theory of normal spirals*
Lectures in Applied Mathematics, Vol. 9, Relativity Theory and Astrophysics; 2. Galactic Structure, pp. 66–97, AMS (1967)

[77] *Density waves in disk galaxies*
(with F. H. Shu)
IAU Symposium No. 31, Noordwijk, The Netherlands, Aug. 25–Sept. 1, 1966, Paper No. 52, pp. 313–317 (Reprints 1967)

[78] *The dynamics of disk-shaped galaxies*
Annual Review of Astronomy & Astrophysics, Vol. 5, pp. 453–463 (1967)

[79] *Spiral structure in galaxies*
Galaxies and the Universe, edited by Lodewijk Woltjer, pp. 33–51, Columbia University Press (1968)

[80] *Density wave theory of spiral structure*
(with F. H. Shu)
Brandeis University Summer Institute in Theoretical Physics, 1968, Astrophysics and General Relativity, Vol. 2, pp. 239–329; Gordon & Breach (1971)

[81] *On the spiral structure of disk galaxies, III. Comparison with observations*
(with C. Yuan and F. H. Shu)
The Astrophysical Journal, Vol. 155, pp. 721–745, Mar. 1969

[82] *Dynamics of self-gravitating systems: Structure of galaxies*
SIAM Review, Vol. 11, pp. 127–151, Apr. 1969

[83] *On the asymptotic theory of a class of ordinary differential equations of fourth order, II. Existence of solutions which are approximated by the formal solutions*
(with A. L. Rabenstein)
Studies in Applied Mathematics, Vol. 48, No. 4, pp. 311–340, Dec. 1969

[84] *Interpretation of large-scale spiral structure*
IAU Symposium No. 38, Basel, Switzerland, Aug. 29–Sept. 4, 1969, The Spiral Structure of Our Galaxy, edited by W. Becker and G. Contopoulos, pp. 377–390, D. Reidel (1970)

[85] *Theory of spiral structure*
IAU 14th General Assembly, Brighton, England, Aug. 1970, Highlights of Astronomy, Vol. 2, edited by C. De Jager, pp. 88–121, D. Reidel (1971)

[86] *A forcing mechanism for spiral density waves in galaxies*
(with S. I. Feldman)
Studies in Applied Mathematics, Vol. 52, No. 1, pp. 1–20, MIT Press (1973)

[87] *Note on stretching of line and surface elements in isotropic turbulence*
J. SIAM Applied Mathematics, Vol. 25, No. 3, pp. 372–377, Nov. 1973

[88] *Mathematics applied to deterministic problems in the natural sciences*
(book with L. A. Segel; material on elasticity by G. H. Handelman)
The Macmillan Publishing Co., Inc. (1974)

[89] *Influence of a spiral gravitational field on the observational determination of galactic structure*
(with C. Yuan and W. W. Roberts)
IAU 15th General Assembly, Sydney, Australia, Aug. 21–30, 1973, Highlights of Astronomy, Vol. 3, edited by G. Contopoulos, pp. 441–450, D. Reidel (1974)

[90] *Theory of spiral structure*
Lectures presented at NATO Advanced Study Institute, International School of Astrophysics, Erice (Sicily) Italy, Jun. 22–Jul. 9, 1974, Structure and Evolution of Galaxies, edited by Giancarlo Setti, pp. 119–142, D. Reidel (1975)

[91] *On the origin and long term maintenance of spiral structure*
Paper presented at CNRS International Colloquium on Dynamics of Spiral Galaxies, Institute for Advanced Scientific Studies (IHES), Bures-sur-Yvette, France, Sept. 16–20, 1974. Report No. 241, pp. 491–505, Centre National De La Recherche Scientifique, Paris (1975)

[92] *On spiral waves in galaxies – A gas dynamic approach*
(with Y. Y. Lau)
J. SIAM Applied Mathematics, Vol. 29, No. 2, pp. 352–370, Sept. 1975

[93] *On the role of applied mathematics*
Advances in Mathematics, 19, No. 3, pp. 267–288, Mar. 1976

[94] *Unstable spiral modes in disk-shaped galaxies*
(with Y. Y. Lau and James W.-K. Mark)
Proceedings of the National Academy of Sciences, Vol. 73, No. 5, pp. 1379–1381, May 1976

[95] *Theory of spiral structure: Galactic dynamics – spiral structure and star formation*
Proceedings of the 14th IUTAM Congress, Delft, The Netherlands, Aug. 30–Sept. 4, 1976, Theoretical and Applied Mechanics, edited by W. T. Koiter, pp. 57–69, North-Holland (1976)

[96] *Density wave amplitude and the mechanism of long-term maintenance in spiral galaxies*
(with Chi-Bin Li and Nien-Guo Han)
Scientia Sinica, Vol. XIX, No. 5, pp. 665–674, Sept.–Oct. 1976

[97] *On certain eigenvalue problems occurring in physical systems* (Levinson Memorial)
(with Y. Y. Lau)
Advances in Mathematics, 22, No. 1, pp. 120–128, Oct. 1976

[98] *Discrete spiral modes in disk galaxies: Some numerical examples based on density wave theory*
(with G. Bertin, Y. Y. Lau, James W.-K. Mark and L. Sugiyama)
Proceedings of the National Academy of Sciences, Vol. 74, pp. 4726–4729, Nov. 1977

[99] *On the stellar streaming motions and the observational determination of the structural constants of the galaxy*
(with C. Yuan and W. W. Roberts, Jr.)
Astronomy and Astrophysics, 69, pp. 181–198 (1978)

[100] *Star migration, solar vicinity and the density wave theory of spiral structure*
(with C. Yuan)
Astronomical Papers dedicated to Bengt Strömgrem, presented at a symposium held in Copenhagen, May 30–Jun. 1, 1978, edited by A. Reiz and T. Andersen, pp. 369–386, Copenhagen University Observatory (1978)

[101] *On density wave theory of spiral galaxies*
The Ta-You Wu Festschrift: Science of Matter, edited by Shigeji Fujita, Gordon & Breach (1978)

[102] *Education of applied mathematicians*
SIAM Review, 20, Vol. 4, pp. 838–845, Oct. 1978

[103] *Density wave theory of spiral structure of galaxies*
(with Y. Y. Lau)
Studies in Applied Mathematics, Vol. 60, pp. 97–163 (1979)

[104] *Density-wave theory of spiral structure of galaxies*
(with Y. Y. Lau)
Mechanics Today, 5, edited by S. Nemat-Nasser, pp. 233–269, Pergamon Press (1980)

[105] *Some fluid-dynamical problems in galaxies*
(with W. W. Roberts, Jr.)
Ann. Rev. Fluid Mech., 13, pp. 33–55 (1981)

[106] *Galactic dynamics and gravitational plasmas*
(with G. Bertin)
Proceedings of the International School of Plasma Physics, Plasma Astrophysics Course and Workshop, Varenna, Italy, Aug. 27–Sept. 7, 1981
European Space Agency, SP-161, pp. 191–205, Paris, Nov. 1981.
Advances in Applied Mechanics, Vol. 24, pp. 155–187 (1984)

[107] *Turbulence, plasma containment, and galaxies*
Journal of Research of the National Bureau of Standards, 86, pp. 557–563, Nov.–Dec. 1981

[108] *On the dynamical basis of the classification of normal galaxies*
(with J. Haass and G. Bertin)
Proceedings of the National Academy of Sciences, Vol. 79, pp. 3908–3912, Jun. 1982

[109] *Spiral modes and the Milky Way*
Paper presented at the Workshop on the Milky Way, University of British Columbia, Vancouver, B.C., May 16, 1982, edited by W. Shuter, D. Reidel (1983)

[110] *Quasi-stationary spiral structure in galaxies*
IAU Symposium No. 100, Besancon, France, Aug. 9–13, 1982, Internal Kinematics and Dynamics of Galaxies, edited by E Athanassoula, D. Reidel (1983)

[111] *Formation and maintenance of spiral structure in galaxies*
(with G. Bertin)
IAU Symposium No. 106, The Milky Way Galaxy, edited by H. Van Woerden et al, pp. 513–532, D. Reidel (1985)

[112] *On the morphology of spiral modes*
(with G. Bertin and S. A. Lowe)
Proceedings of a Course and Workshop on Plasma and Astrophysics, Varenna, Italy, Aug. 28–Sept. 7, 1984, ESA, SP-207, pp. 115–120 (1984)

[113] *On a spectral theory of the excitation mechanism of galactic spirals*
(with R. P. Thurstans)
Proceedings of a Course and Workshop on Plasma and Astrophysics, Varenna, Italy, Aug. 28–Sept. 7, 1984, ESA, SP-207, pp. 121–130 (1984)